July 30–August 1
Redondo Beach, Califor

Association for
Computing Machinery

Advancing Computing as a Science & Profession

ISLPED'12

Proceedings of the

International Symposium on Low Power Electronics and Design

Sponsored by:

ACM SIGDA

Sponsored by:

IEEE, CAS, SSCS, and EDS

Supported by:

Intel, Texas Instruments, IBM, Broadcom, ARM, and Cadence

Association for
Computing Machinery

Advancing Computing as a Science & Profession

The Association for Computing Machinery
2 Penn Plaza, Suite 701
New York, New York 10121-0701

ISBN: 978-1-4503-1249-3 (Digital)

ISBN: 978-1-4503-1738-2 (Print)

Additional copies may be ordered prepaid from:

ACM Order Department
PO Box 30777
New York, NY 10087-0777, USA

Phone: 1-800-342-6626 (USA and Canada)
+1-212-626-0500 (Global)
Fax: +1-212-944-1318
E-mail: acmhelp@acm.org
Hours of Operation: 8:30 am – 4:30 pm ET

Printed in the USA

Message from the General Chairs

The 2012 ACM/IEEE International Symposium on Low Power Electronics and Design (ISLPED 2012) welcomes you to beautiful Redondo Beach, California, USA. Redondo Beach's scenic coast and beaches bring a unique and relaxing ambience to ISLPED 2012. Its proximity to prominent Southern California's universities and high-tech industry will attract a diverse mix of attendees. We are delighted that you are able to join us.

ISLPED remains the premier international conference in low-power design thanks to the outstanding contributions from the low-power design community. The Technical Program Chairs, Professor Pai Chou (University of California, Irvine) and Dr. Ajith Amerasekera (Texas Instruments, Inc.), have worked hard to put together an excellent Technical Program. They were ably assisted by the ISLPED 2012 Technical Program Committee comprising leading researchers in the area of low-power design. The ISLPED 2012 Organizing Committee has been working tirelessly to bring you a world-class conference experience. We are grateful for their efforts in making ISLPED 2012 a great success.

ISLPED 2012 has been fortunate to receive strong support from industry. In particular, we appreciate the generous financial support from Intel, Texas Instruments, IBM, Broadcom, ARM, Cadence, and Mentor Graphics for various ISLPED 2012 activities.

We hope that you will enjoy an excellent ISLPED 2012 program this year, and have a pleasant, enriching and memorable experience in Redondo Beach.

Naresh Shanbhag and Massimo Poncino
ISLPED'12 General Co-Chairs

Message from the Program Chairs

It is our great pleasure to welcome you to the ACM/IEEE International Symposium on Low Power Electronics and Design – ISLPED 2012. This year's symposium continues its tradition of being the premier forum for presentation of research results and experience reports on leading-edge issues of low-power design. The mission of the symposium is to share novel low-power solutions that fulfill the needs of electronics design and identify new directions for future research and development. ISLPED gives researchers and practitioners a unique opportunity to share their perspectives with others interested in the various aspects of low-power design.

The call for papers attracted 213 submissions from Asia, Canada, Europe, Africa, and the United States. The program committee accepted 66 papers, including 34 for full-length presentations, 14 for half-length presentations, and 18 for poster presentations. They cover a variety of topics, including next-generation technology, architecture, analog and MEMS, design tools, methodology, and optimization techniques for low-power design. We thank the authors for providing the content of the program. We are grateful to the program committee members, who volunteered their valuable time and hard work in reviewing papers, attending the in-person review meeting, and providing feedback for authors. We hope that these proceedings will serve as a valuable reference for researchers and developers on these topics.

This year's program features three keynote speakers over three days. On the first day, Dr. Pradip Bose (IBM) introduces energy-secure computing, a new topic on ensuring robust and secure functionality while meeting the energy-related constraints of the emerging green-computing era. On the second day, Dr. Kaushik Roy (Purdue University) presents using spin (as opposed to charge) as state variable for memory and logic computation to achieve high density and very low power. On the third day, Dr. Uming Ko (MediaTek USA) speaks on ultra-low power challenges for the next-generation ASICs. In addition, the program includes a special session on future needs for low-power electronic design from an industry perspective, a panel on green communication and storage systems, and an embedded tutorial on power and thermal management for smart phones.

We hope that you will find this program forward-looking and thought-provoking and that the symposium will provide you with a valuable opportunity to share ideas with other researchers and practitioners from institutions around the world.

Pai H. Chou and Ajith Amerasekera
ISLPED'12 Program Co-Chairs

Table of Contents

Session 11: Innovations in Low Power Analog

Session Chairs: Rajeevan Amirtharajah *(University of California, Davis)* and Atila Alvandpour *(Linköping University)*

Session 12: Software-driven Techniques for Energy Efficiency in Embedded and Multi-core Systems

Session Chairs: Nikil Dutt *(University of California, Irvine)* and Vivek Tiwari *(Intel Corporation)*

Embedded Tutorial

Keynote Address 3

Session Chair: Ajith Amerasekera *(Texas Instruments)*

Session 13: Processor Design and Implementation

Session Chairs: Lawrence Clark *(Arizona State University & SuVolta, Inc.)* and Sujat Jamil *(Marvell)*

ISLPED 2012 Symposium Organization

General Co-Chairs: Naresh Shanbhag *(University of Illinois at Urbana-Champaign, USA)*
Massimo Poncino *(Politecnico di Torino, Italy)*

Program Co-Chairs: Pai H. Chou *(University of California, Irvine / NTHU)*
Ajith Amerasekera *(Texas Instruments, USA)*

Local Arrangements Chair: Puneet Gupta *(University of California, Los Angeles, USA)*

Publicity Co-Chairs: Luca Benini *(University of Bologna, Italy)*
Hamid Mahmoodi *(San Francisco State University, USA)*
John Donovan *(Low-Power Design, USA)*

Treasurer & Registration Chairs: Yuan Xie *(The Pennsylvania State University, USA)*
Yu Wang *(Tsinghua University, China)*

Industry Liaison: David Garrett *(Broadcom, USA)*

Web Chair: Chien-An Cho *(National Tsing Hua University, Taiwan)*

Design Contest Co-Chairs: Chia-Lin Yang *(National Taiwan University, Taiwan)*
Yiran Chen *(University of Pittsburgh, USA)*

Executive Committee:

Brock Barton *(Texas Instruments, USA)*
Robert Brodersen *(University of California, Berkeley, USA)*
Anantha Chandrakasan *(Massachusetts Institute of Technology, USA)*
Kiyoung Choi *(Seoul National University, Korea)*
Naehyuck Chang *(Seoul National University, Korea)*
Jason Cong *(University of California, Los Angeles, USA)*
Vivek De *(Intel Corporation, USA)*
Giovanni DeMicheli *(EPFL, Switzerland)*
Christian Enz *(Centre Suisse d'Electronique et de Microtechnique)*
Joerg Henkel *(Karlsruhe Institute of Technology, Germany)*
Mary Jane Irwin *(The Pennsylvania State University, USA)*
Rajiv Joshi *(IBM T.J. Watson Research Center, USA)*
Ali Keshavarzi *(Global Foundaries, USA)*
Enrico Macii *(Politecnico di Torino, Italy)*
Diana Marculescu *(Carnegie Mellon University, USA)*
Farid Najm *(University of Toronto, Canada)*
Hiroshi Nakamura *(The University of Tokyo, Japan)*
Vijaykrishnan Narayanan *(The Pennsylvania State University, USA)*
Wolfgang Nebel *(University of Oldenburg, Germany)*
Vojin Oklobdzija *(University of Texas at Dallas, USA)*
Barry Pangrle *(Mentor Graphics, USA)*
Massoud Pedram *(University of Southern California, USA)*
Christian Piguet *(Centre Suisse d'Electronique et de Microtechnique)*
Jan M. Rabaey *(University of California, Berkeley, USA)*

Executive Committee (Continued):

Anand Raghunathan *(Purdue University, USA)*
Kaushik Roy *(Purdue University, USA)*
Takayasu Sakurai *(The University of Tokyo, Japan)*
Mircea Stan *(University of Virginia, USA)*
Christer Svensson *(Linköping University, Sweden)*
Vivek Tiwari *(Intel Corporation, USA)*
Ingrid Verbauwhede *(Katholieke Universiteit Leuven, Belgium)*

Program Committees:

Track 1.1: Technology and Digital Circuits Program Subcommittee

Benton Calhoun *(University of Virginia, USA)* (**subcommittee chair**)
Naveen Verma *(Princeton University, USA)* (**subcommittee co-chair**)
Mohamed Allam *(Qualcomm, USA)*
Kaustav Banerjee *(University of California, Santa Barbara, USA)*
Vikas Chandra *(ARM)*
Aseem Gupta *(Freescale, USA)*
Muhammad Khellah *(Intel)*
Jae-Joon Kim *(IBM T.J. Watson Research Center, USA)*
Eren Kursun *(IBM T.J. Watson Research Center, USA)*
Joyce Kwong *(Texas Instruments, USA)*
Hamid Mahmoodi *(San Francisco State University, USA)*
Ali Muhtaroglu *(Middle East Technical University, Turkey)*
Saibal Mukhopadhyay *(Georgia Institute of Technology, USA)*
Nikola Nedovic *(Fujitsu)*
Borivoje Nikolic *(University of California, Berkeley, USA)*
Venu Puvvada *(Qualcomm)*
Srini Sridhara *(Texas Instruments, USA)*
Carlos Tokunaga *(Intel, USA)*
Bart Zeydel *(Plato Networks)*
Matthew Ziegler *(IBM T.J. Watson Research Center, USA)*

Track 1.2: Logic and Microarchitecture Program Subcommittee

Hans Jacobson *(IBM)* (**subcommittee chair**)
Antonia Zhai *(University of Minnesota)* (**subcommittee co-chair**)
Alper Buyuktosunoglu *(IBM)*
Lawrence Clark *(Arizona State University / Intel)*
David Garrett *(Broadcom, USA)*
Houman Homayoun *(University of California, San Diego, USA)*
Koji Inoue *(Kyushu University, Japan)*
Masaaki Kondo *(University of Electro-Communications, Japan)*
Hai (Helen) Li *(NYU Polytech, USA)*
Vasily Moshnyaga *(Fukuoka University, Japan)*
Hiroshi Nakamura *(The University of Tokyo, Japan)*
Dmitry Ponomarev *(SUNY Binghamton, USA)*
Thomas Wenisch *(University of Michigan, USA)*

Track 1.3: Analog, MEMS, and Mixed Signal Program Subcommittee

Yogesh Ramadass *(Texas Instruments, USA)* (**subcommittee chair**)
Atila Alvandpour *(Linköping University, Sweden)*
Rajeevan Amirtharajah *(University of California, Davis, USA)*
Kamran Eshraghian *(E-Labs, Australia)*
Karthik Kadirvel *(Texas Instruments, USA)*
Jie Gu *(Maxlinear)*
Arjang Hassibi *(The University of Texas at Austin, USA)*
Byunghoo Jung *(Purdue University, USA)*
Suhwan Kim *(Seoul National University, Korea)*
David Wentzlof *(University of Michigan, USA)*

Track 2.1: Design Tools Program Subcommittee

David Z. Pan *(The University of Texas at Austin)* (**subcommittee chair**)
Mustafa Ozdal *(Intel)* (**subcommittee co-chair**)
Vishwani Agrawal *(Auburn University, USA)*
Clive Bittlestone *(Texas Instruments, USA)*
Azadeh Davoodi *(University of Wisconsin, USA)*
Victor Grimblatt *(Synopsys)*
Domenik Helms *(OFFIS Research Institute)*
Taehwan Kim *(Seoul National University, Korea)*
Taemin Kim *(Intel)*
Fadi Kurdahi *(University of California, Irvine, USA)*
Enrico Macii *(Politecnico di Torino, Italy)*
Renu Mehra *(Synopsys, USA)*
Michael Orshansky *(The University of Texas at Austin, USA)*
Barry Pangrle *(Mentor Graphics, USA)*
Jason Xue *(City University of Hong Kong)*

Track 2.2: System and Design Methodologies Program Subcommittee

Tajana Simunic *(University of California San Diego, USA)* (**subcommittee chair**)
Qinru Qiu *(Syracuse University, USA)* (**subcommittee co-chair**)
Elaheh Bozorgzadeh *(University of California, Irvine, USA)*
Davide Brunelli *(University of Trento, Italy)*
Chaitali Chakrabarti *(Arizona State University, USA)*
Naehyuck Chang *(Seoul National University, USA)*
Deming Chen *(University of Illinois at Urbana-Champaign, USA)*
Joerg Henkel *(Karlsruhe Institute of Technology, Germany)*
Tohru Ishihara *(Kyoto University, USA)*
Rakesh Kumar *(University of Illinois at Urbana-Champaign, USA)*
Seok-Jun Lee *(Texas Instruments, USA)*
Jinfeng Liu *(MultiDimension Technology)*
Yung-Hsiang Lu *(Purdue University, USA)*
Diana Marculescu *(Carnegie Mellon University, USA)*
Anand Raghunathan *(Purdue University, USA)*
Li Shang *(University of Colorado at Boulder, USA)*
Yu Wang *(Tsinghua University, China)*
Yuan Xie *(The Pennsylvania State University, USA)*
Chia-Lin Yang *(National Taiwan University, Taiwan)*

Track 2.3: Software Design and Optimization Program Subommittee

Vijay Raghunathan *(Purdue University, USA)* (**subcommittee chair**)
Vivek Tiwari *(Intel, USA)* (**subcommittee co-chair**)
Andrea Acquaviva *(Politecnico di Torino, Italy)*
Yiran Chen *(University of Pittsburgh, USA)*
Ann Gordon-Ross *(University of Florida, USA)*
Jihong Kim *(Seoul National University, Korea)*
Zili Shao *(Polytechnic University of Hong Kong)*
Hiroyuki Tomiyama *(Ritsumeikan University, Japan)*
Lin Zhong *(Rice University, USA)*

Additional reviewers:

Ben Abdallah Abderazek
Mohamed Abu Rahma
Manuel E. Acacio
Juan Luis Aragón
Manish Arora
Andrey Ayupov
Abbas BanaiyanMofrad
Francisco Barat
Ramon Bertran
Paul Bogdan
Steven Burns
Ramon Canal
Yen-Jen Chang
Daniel Ángel Chaver Martínez
Doosan Cho
Yuan-Sun Chu
Florent de Dinechin
Xiangyu Dong
Lide Duan
Mostafa Farahani
David Flynn
Mohammad Gasemazar
Meeta Gupta
Koji Hashimoto
Weifeng He
Enric Herrero
Anup Holey
Ilya Issenin
Weixing Ji
Xiaowei Jiang
Akash Kumar
Vineeth Mekkat
Asit K. Mishra
Carlos Molina
Ragavendra Natarajan
Stuart Oberman
Abbas Rahimi
Anitha Ravi
Glenn Reinman
Jose Renau
Joy Reyes-Madamba
Juan Rubio
Jack Sampson
Ron Sass
Amirali Shayan
Kazuyoshi Takagi
James Tuck
Augusto Vega
Lei Wang
Sying-Jyan Wang
Weng-Fai Wong
Polychronis Xekalakis
Chi Xu
Guan Xuan
Jieming Ying
Haikun Zhu
Cheng Zhuo

ISLPED 2012 Sponsors & Supporters

Sponsors:

Technical Supporters:

Gold Supporters:

Silver Supporters:

Bronze Supporter:

Keynote

Energy-Secure Computing

Speaker

Pradip Bose

IBM T.J. Watson Research Center

Yorktown Heights, NY, USA

pbose@us.ibm.com

Abstract

The "power wall" has forced chip and system architects to design with smaller margins between nominal and worst-case operating points. Localized hot spots and temperature gradients exacerbate lifetime reliability problems. Smaller voltage margins make processors more vulnerable to inductive noise on the voltage rails, as well as soft errors induced by high energy particle incidence. The problem of process variation presents another obstacle to sustained performance growth in the late CMOS design era. At the same time, the emerging phase change memory (a promising technology for future low power, dense storage in systems) is vulnerable to malicious attacks that can reduce the lifetime of an already wear-out prone technology. These issues have all led to R&D in "better than worst-case" design principles. Dynamic power and thermal management control loops have already become an integral part of chip and system design. New research in wearout and general reliability management have recently been published. These new generation management protocols have, however, opened up other sources of concern: e.g. potential security holes exposed by the integrated control loops and system safety issues triggered by potential violations of power or thermal limits imposed by the original specification. We coin the term "Energy-Secure System Architectures" to cover the range of research being pursued within industry and academia in order to ensure robust and secure functionality, while meeting the energy-related constraints of the emerging "green computing" era. This keynote speech attempts to provide a summary overview of the problem and solution spaces around the theme of energy-secure computing – with a special focus on servers and extreme-scale systems.

Categories and Subject Descriptors

B.8.0 [**Hardware**]: Performance and Reliability -- *General*

Keywords

energy-secure computing, power wall, system architecture, servers, extreme-scale systems

Biography

Pradip Bose received his B. Tech. degree in Electronics and Electrical Communication Engineering from the Indian Institute of Technology, Kharagpur, India in 1977 and the M.S. and Ph.D degrees in Electrical and Computer Engineering from the University of Illinois, Urbana-Champaign, in 1981 and 1983 respectively. Since May 1983, Dr. Bose has been a Research Staff Member at the IBM T. J. Watson Research Center, Yorktown Heights, NY. During 1992-94, Dr. Bose was on assignment at IBM Austin, where he served as the lead performance engineer during the definition and evaluation of a processor, which eventually evolved into IBM's POWER3 microprocessor. During the 1989-90 academic year, Dr. Bose was a Visiting Associate Professor at Indian Statistical Institute, Calcutta, India as part of a sabbatical leave assignment from IBM. During that year, he worked on the application of artificial intelligence techniques to problems in VLSI CAD. His current research interests include: high performance computer architectures and their performance evaluation, verification and testing.

ISLPED'12, July 30– August 1, 2012, Redondo Beach, California, USA.
ACM 978-1-4503-1249-3/12/07.

Write-Optimized Reliable Design of STT MRAM

Yusung Kim, Sumeet Kumar Gupta, Sang Phill Park, Georgios Panagopoulos, and Kaushik Roy
School of Electrical and Computer Engineering, Purdue University, West Lafayette, IN 47907
{yusung, guptask, sppark, gpanagop, kaushik}@purdue.edu

ABSTRACT

Spin transfer torque magnetic random access memory (STT MRAM) is a promising non-volatile memory due to its outstanding potential for high integration density and excellent scalability. Despite the attractive features, high write current and power is still a major challenge. As a result, the optimization of the memory for write is critical.

In this work, we analyze asymmetric write currents in STT MRAMs considering process variations, and identify a potential for write power reduction. We propose circuit design techniques 1) bit-line voltage clamping using a pass transistor and 2) 2T-1R dual source-line bit-cell design, to balance out the asymmetric write currents and optimize the memory design from both write-power and reliability aspects. Our proposed techniques can be easily incorporated with previously proposed design techniques without affecting the bit-cell write-ability, read stability, and performance. We analyze the impact of our proposed techniques on write power and MTJ current density and show 30-68% average write power savings and 4-41% reduction in MTJ current density in STT MRAM.

Categories and Subject Descriptors

B.3.3 [**Memory Structures**]: Performance Analysis and Design Aids

General Terms

Design, Reliability

Keywords

STT MRAM, MTJ, write power reduction

1. INTRODUCTION

Spin-Transfer Torque Magnetic RAM (STT MRAM) is a promising candidate for non-volatile on-chip memory applications for its zero stand-by leakage, easy integration with the backend CMOS process, and potential for high integration density and excellent scalability [1, 4, 7–10]. Recently, ITRS has highlighted STT MRAM as one of the two attractive non-volatile memory candidates for accelerated developments and further predicted that STT MRAM has a substantial potential to be ready for manufacture within a five to ten year period [1]. Despite its promising outlook and attractive features, one of the main challenges is the high current required to program information in the storage element, a magnetic tunnel junction (MTJ – Fig. 1) [1, 3, 11, 14, 16].

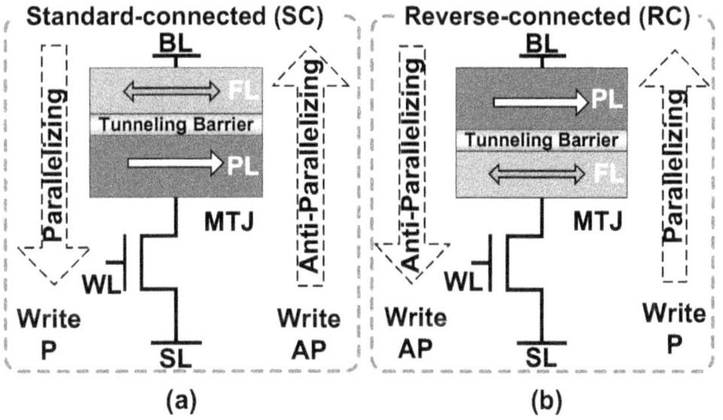

Figure 1: STT MRAM bit-cell (a) standard-connected (b) reverse-connected (FL:free layer, PL:pinned layer)

The high write current (or critical switching current) directly increases the dynamic power dissipation and the overall write energy cost. The high write current can also degrade integration density because the memory cell needs a large access transistor to meet the current requirement. Furthermore, high current injection through the tunneling barrier of the MTJ imposes a severe stress and can lead to a time-dependent dielectric breakdown (TDDB) [12, 15]. Hence, the reduction of write current has been the primary focus for the design of STT MRAM.

The strong need for the reduction of write current has invoked a lot of research efforts at the device-level, to design and engineer the MTJ with lower write current. The use of a Perpendicular Magnetic Anisotropy (PMA) [2] can directly reduce the intrinsic write current by eliminating the conflicting requirements of thermal stability and demagnetization field in in-plane MTJs. Also, a device design to trade off thermal stability or data retention time with the lower write current was proposed by Smullen et al. [14] and it was shown to be effective in reducing the overall write energy. The use of low saturation magnetization material was also shown to be an effective route to achieve low write energy, especially for low-speed STT MRAM designs [11].

While the aforementioned device-level approaches are focused on engineering the MTJ for lower write current, prior circuit and architecture works are focused on the optimization of write operations in a memory array, given a specific MTJ characteristic. At the circuit-level, the Early Write Termination (EWT) technique was proposed to sense and terminate redundant write operations (for example, writing a '0' to a cell storing '0') at early stages [16] to remove unnecessary write operations. As a result, significant write energy savings can be achieved at the cost of extra circuitry and a small increase in the overall memory area. At the architecture level, Nigam et al. [11] makes use of different write energy for writing '0' and '1', and proposed the use of Invert Code scheme to dynamically assign '0' and '1' so that the total write energy can be minimized. However, previous works have not shown how the asymmetry in write currents can be handled at the circuit-level.

In this work, we address the issue of high write current in STT MRAMs by proposing low power circuit techniques. We analyze bit-cell write currents in detail and show quantitatively that the asymmetric bit-cell write currents result in an excessive and wasted write current. This in turn increases the write power and degrades the reliability of the MTJ. To reduce the unnecessary write power we propose two circuit-level design techniques based on (a) clamping the bit-cell voltage using a pass transistor, and (b) using two source-lines of 2T-1R bit-cell. A critical evaluation of each of these techniques and the comparison of write power and current density are carried out, considering process variations. Our proposed techniques do not affect the read stability, cell write-ability and performance. Also, our techniques can be used in conjunction with the previous works.

The contributions of this work are as follows:

- We perform a detailed analysis of the asymmetry in write currents and identify a potential for write power reduction.
- We propose to clamp the voltage across the bit-cell using a pass transistor to reduce the asymmetry in the write currents and achieve power savings as well as an enhancement in MTJ reliability.
- We also propose to use 2T-1R dual-source-line bit cell to optimize the write operation from both power and reliability aspects.

We show that the proposed techniques offer significant reduction in write power and current density (associated with MTJ reliability) in the presence of process variations, without any impact on the bit-cell write-ability, read stability and performance of STT MRAMs.

The rest of the paper is organized as follows. In Section 2, we briefly explain STT MRAM background, process variations, and simulation framework used in our analysis. Section 3 describes the asymmetry in write currents in STT MRAMs. Section 4 describes the proposed design techniques, 1) bit-line voltage clamping and 2) 2T-1R dual source-line bit-cell. In Section 5, we evaluate the proposed techniques, considering process variations and discuss the simulation results. Section 6 concludes this work.

2. STT MRAM BACKGROUND

2.1 Bit-cell operation

The key element of STT MRAM is an MTJ, which is composed of two ferromagnetic electrodes – free layer (FL) and pinned layer (PL) (Fig. 1) separated by a tunneling oxide barrier (AlO_x or MgO). PL has its magnetization fixed while the magnetization of FL can be manipulated by current-induced spin-transfer torque [7]. In most cases, FL is engineered to have two stable magnetization directions – parallel (P) and anti-parallel (AP) to the pinned layer, yielding two stable states of the memory bit-cell. The P configuration is the low-resistance state, while the AP configuration is the high-resistance state. Thus, the binary data is stored as the resistance of the MTJ. To selectively access the stored data in a memory array, a select or access transistor is connected in series with the MTJ leading to the 1T-1R bit-cell structure (Fig. 1). There are two ways in which the MTJ can be connected: (a) FL connected to the bit-line BL (standard-connected (SC) configuration – Fig. 1(a)) and (b) PL connected to BL (reverse-connected (RC) configuration [10] – Fig. 1(b)).

The read and write operations are performed by applying appropriate voltages to the bit-line (BL), source-line (SL), and word-line (WL). Let us define V_{BL}, V_{SL}, and V_{WL} to be the respective voltages. To read out the state of the MTJ, the word-line (WL) is asserted high and the read voltage (V_{READ}) is applied across the bit-cell (i.e. $V_{BL} - V_{SL} = V_{READ}$). The corresponding read current is compared to a reference current to determine the data stored in the MTJ. It is important that the read current be significantly lower than the write current so that the state of the MTJ is not disturbed during the reading operation. While read operation needs only one direction of current flow through the bit-cell, write operation requires bi-directional current flow in order to program the information as the magnetization direction of the MTJ free layer. To write the P state into the MTJ, WL is asserted (i.e. $V_{WL} = V_{DD}$) and a voltage (V_{DD}) is applied across the bit-cell (i.e. $|V_{BL} - V_{SL}| = V_{DD}$), so that current flows from FL to PL (parallelizing current direction – Fig. 1) of the MTJ. To write the AP state into the MTJ, the voltage polarity across the bit-cell is reversed so that the current flows from PL to FL (anti-parallelizing current direction–Fig. 1) of the MTJ. Successful write occurs when the current flowing through the MTJ is greater than the critical switching current, resulting in the switching of the MTJ state within the desired time interval. In 1T-1R bit-cell structure, the optimum access transistor size is chosen to give the lowest failure probability under the given design specifications. In most cases, the write failure probability is the dominant failure and results in up-sizing of the access transistor width [5].

2.2 Bit-cell simulation framework

In order to characterize the STT MRAM bit-cells, we use the simulation framework proposed in [13]. Non-equilibrium Green's Function (NEGF) equations are used to model MTJ resistance over a range of tunneling oxide thickness (t_{MgO}) and the voltage across the MTJ (V_{MTJ}). The MTJ transport characteristics obtained from NEGF are used in the Landau-Lifshitz-Gilbert-Slonczewski (LLGS) SPICE model [13], from which the magnetization switching dynamics of FL are obtained. The MTJ model is calibrated against the experimental results reported in [10]. An MTJ SPICE model [13] is used with 45nm STM models for the access transistor to perform comprehensive transient simulations of the STT MRAM bit-cell, considering process variations in the MTJ and the access transistor. Also, a single voltage plane of V_{DD}=1V is used for the analysis.

3. ASYMMETRY IN WRITE-CURRENTS

Using the simulation framework described above, transient bit-cell simulations are performed to analyze the write currents in STT MRAMs.

When STT MRAM is designed with a single power supply (V_{DD}) write currents are significantly different [3, 8], which can be explained as follows. When current flows from BL to SL, the access transistor has the overdrive voltage of $V_{DD} - V_{Th}$ (i.e., $V_{OV} = V_{DD} - V_{Th}$ where V_{Th} is the transistor threshold voltage) since the transistor source is grounded (source-grounded configuration [8] - Fig. 2 inset (a)). On the other hand, when current flows from SL to BL, the overdrive voltage of the transistor is much smaller

Table 1: Write currents [μA] averaged over T_W = 40ns, for bit-cell design without considering process variations.

	SC (W =200nm)	RC (W =300nm)
AP→P	171.0	83.1
P→P	176.5	132.6
P→AP	105.8	198.7
AP→AP	67.8	189.6

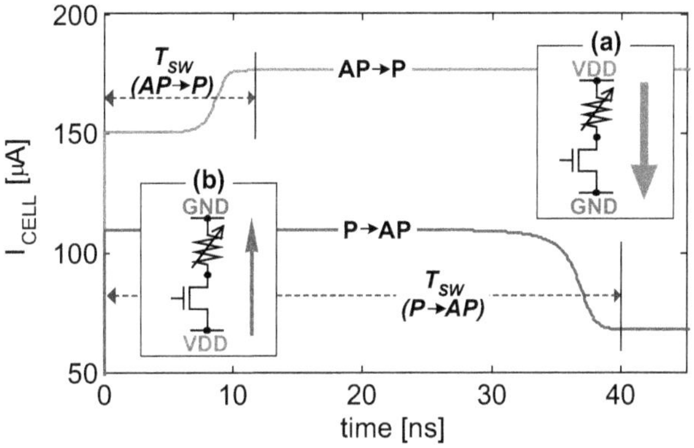

Figure 2: Asymmetry in bit-cell write-currents in magnitude and the switching times for SC configuration. Inset: (a) source-grounded (b) source-follower configuration [8].

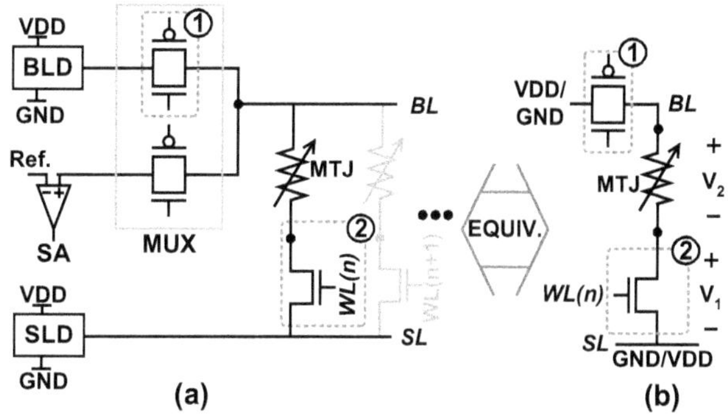

Figure 3: (a) Simplified STT MRAM memory array (b) Equivalent circuit for write operation – baseline, (BLD: Bit-Line write Driver, SLD: Source-Line write Driver, SA: Sense Amplifier)

($V_{OV} = V_{DD} - (V_{Th} + \Delta V_{Th,Body}) - V_S$), because the MTJ functions as a negative feedback resistive element which limits the current flow. Moreover, the body effect ($V_{SB} < 0$) also lowers the overdrive voltage (source-follower configuration [8] – Fig. 2 inset (b)). Since the transistor width has to be designed to supply enough currents in both directions of write operation, the access transistor width for write operation is determined by the current drivability of source-follower configuration.

Before we discuss the impact of transistor sizing on the asymmetry in write currents, we first obtain the transistor width which satisfies the switching time requirement (T_W = 40ns, in this work). For SC configuration, the minimum access transistor width required to complete P→AP is obtained as 200nm and the minimum width required to complete AP→P switching is 82nm. Taking the maximum of the two widths, the access transistor width is chosen to be 200nm. For RC configuration, the minimum widths required for P→AP and AP→P switching are obtained as 110nm and 300nm, respectively. As a result, the access transistor width is selected to be 300nm. Although the write-ability is guaranteed by the above transistor sizes, one can easily note that in one of the write directions, the transistor width is over-designed. For AP→P write in SC configuration, the chosen transistor width is 2.4× larger than what is required. Likewise, for P→AP write in RC configuration, there is a 2.7× over-design in the width. In short, the transistor width is larger than what is required for one direction of write operation. As a consequence, more than necessary write current is supplied, leading to larger write power and aggravated reliability concerns due to time-dependent dielectric breakdown (TDDB) of the MTJ oxide. As shown in Table 1, for SC configuration, the parallelizing write currents (AP→P and P→P) are more than 2× of what is required (critical switching current for AP→P is 82 μA). For RC configuration, excessive write current flows for anti-parallelizing writes. When excessive current is supplied through the cell, higher write speed can be achieved. However, the bit-cell write time or the write-ability is determined by the slower of the two writes. Looking at Fig. 2 which shows the switching characteristics of the SC bit-cell, we see that the P→AP switching is completed at 40ns while AP→P switching is finished at a much earlier time, 12ns. As a result, the extra current supplied for AP→P write operation is wasted and results in a severe stress condition for the MTJ and high write energy consumption.

The discussion above and the results in Table 1 clearly show that in both SC and RC configurations of STT MRAM, write currents for different cases of magnetization transition can be quite different. It is also important to identify the cases when excessive current flows through the MTJ during write, in order to accurately evaluate the power consumption and lifetime of STT MRAMs. For SC configuration, the highest I_W is for P→P case, although no magnetization switching is required in the MTJ. For the RC configuration, the maximum I_W occurs in P→AP write, as shown in Table 1. In the subsequent sections, we propose design techniques which achieve reduction in the write current, thereby, lowering the write power and enhancing MTJ reliability without affecting the cell write-ability, read stability, and performance of the bit-cell.

4. PROPOSED CIRCUIT TECHNIQUES

From the above discussion, it is clear that the cases determining the cell write-ability for both SC and RC configurations is when the transistor is source-degenerated and the current flows from SL to BL (source-follower). On the other hand when the current direction is from BL to SL (source-ground), excessive write current flows through the MTJ. *Hence, the goal of the proposed design techniques is to reduce the excessive write current from BL to SL, with the following constraints: (a) current from SL to BL is not reduced and (b) the reduction in write current from BL to SL is only to the extent that the reduced write current is still greater than the critical current required to meet the switching time constraint.* As a result, write power improvement and enhancement in MTJ reliability are achieved, with no impact on the bit-cell write-ability. Before we discuss the proposed techniques, let us consider Fig. 3, which shows a part of the STT MRAM bit-cell array with BL/SL write drivers (denoted as BLD and SLD, respectively) and a sense amplifier (SA). An analog multiplexer (shared amongst the bit-cells in the same column) is used to connect BL to the write driver (BLD) during write and to the read voltage biasing circuit and sense amplifier during read (Fig. 3(a) denoted as MUX). The equivalent circuit of the STT MRAM bit-cell during write is shown in Fig. 3(b) with a transmission gate in series with the MTJ. We will use the circuit in Fig. 3(b) for the subsequent discussion. Let us now discuss the proposed techniques.

4.1 Bit-line voltage clamping using a pass transistor (PT)

Instead of inserting an additional pass transistor, we propose to modify the existing structure to achieve the goal of reducing the voltage across the bit-cell. When the transmission gate in the analog multiplexer (MUX in Fig. 3(a)) is replaced with a pass transistor (Fig. 4), voltage across

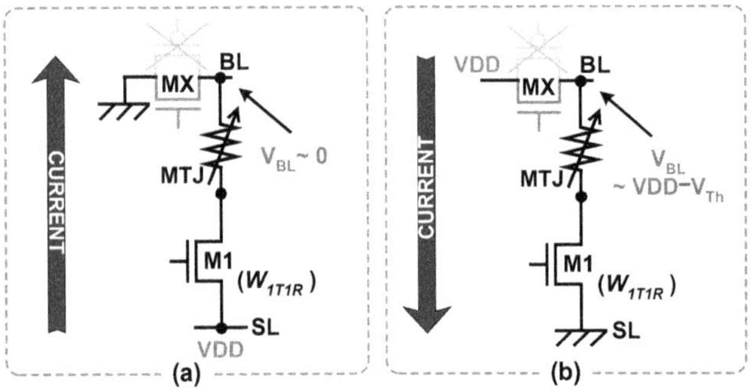

Figure 4: Write oprations in PT bit-cell (a) write AP(SC), P(RC) (b) write P(SC), AP(RC).

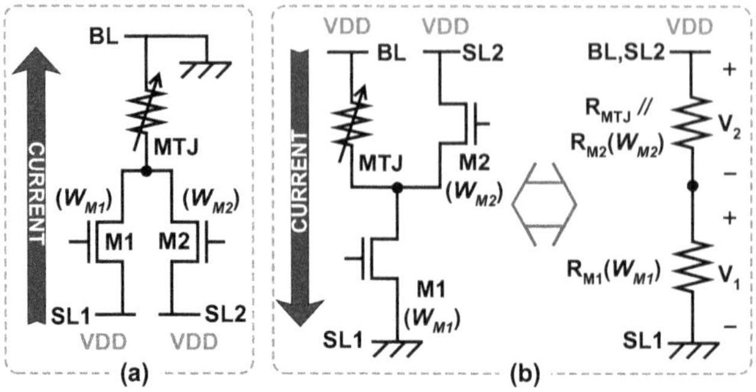

Figure 5: Write operations in 2T-1R dual source-line technique (a) write AP(SC), P(RC) (b) write P(SC), AP(RC).

the bit-cell can be clamped to a lower voltage only when the current flows from BL to SL (the case when excessive current flows). The equivalent circuits for write operations are shown in Fig.4. The proposed modification reduces the current flow from BL to SL only because the pass transistor MX (in Fig.4) is source-degenerated, leading to a reduction in the voltage across the bit-cell ($V_{BL} - V_{SL}$) from V_{DD} to $V_{DD} - V_{Th}$. Hence, the excessive current through the MTJ can be reduced. For the current flowing from SL to BL (case determining the cell write-ability), the pass transistor pulls V_{BL} close to ground as in the case of standard STT MRAM with the transmission gate connecting BL and the write driver. Hence, the proposed technique incurs no penalty on the write-ability of the bit-cell. Moreover, the removal of PMOS from the transmission gate allows increase in the width of MX without any area increase. As a result, the width of MX can be increased to ensure that write-ability is not affected with the proposed technique. The other transmission gate in the multiplexer which connects the sense-amplifier to BL is not changed; hence, there is no impact on the read stability and read performance of the bit-cell. Moreover, as we are modifying the transmission gate to a pass transistor, this technique does not require any extra peripheral circuits and no area penalty is incurred. From hereafter, we will refer to this design technique as PT.

4.2 2T-1R Dual Source-Line (DSL)

As discussed in section 3, the required access transistor width which satisfies the given write-ability for one direction of write is larger than that of the other. To ensure the cell write-ability, the larger width of the two was chosen which led to the problem of excessive write current in one of the directions. To address this issue, we propose to use two access transistors with two source-lines and appropriate bias condition, as discussed later. The proposed methodology achieves reduction in the effective width of the access transistor for write operations that result in the excessive current. The equivalent circuits for write operations are shown in Fig. 5. Let us now look at bias conditions for read and write operations.

During the read operation, SL1 and SL2 are biased at the same voltage, as a result of which, read operation of the 2T-1R bit-cell is the same as that of the standard 1T-1R bit-cell as long as we keep the width of M1 (W_{M1}) and M2 (W_{M2}) equal to that used in the 1T-1R STT MRAM (W_{1T1R}) i.e. $W_{M1} + W_{M2} = W_{1T1R}$. Therefore, the proposed technique incurs no penalty on the read stability and performance.

During the write operation SL2 is tied to V_{DD}, irrespective of the data to be written. SL1, BL and WL are biased in an identical manner as the standard bit-cell (Fig. 3b). To write AP for SC configuration (case determining bit-cell write-ability – Fig 5(a)), $V_{SL1} = V_{SL2} = V_{DD}$ and $V_{BL} = 0$ is used. Then M2 is connected in parallel with M1, and the write operation with current flowing from SL1 and SL2 to BL in the proposed structure is the same as that of the standard STT MRAM. For this reason, the bit-cell write-ability is not affected with the proposed since $W_{M1} + W_{M2} = W_{1T1R}$. To write P in SC configuration (the case when excessive current flows in the 1T-1R bit-cell), $V_{SL1} = 0$ and $V_{BL} = V_{SL2} = V_{DD}$ is used, which places M2 in parallel with the MTJ (Fig. 5b). This results in reduced current through the MTJ due to (a) smaller width of the transistor (M1) draining the current to SL1 and (b) current division between M2 and the MTJ. For this structure, the ideal operating condition is when the voltage across the MTJ (V_2 in Fig. 5(b)) is less than the threshold voltage (V_{Th}) of M2. In this case, M2 is in cut-off and the bit-cell behaves as a 1T-1R bit-cell with smaller transistor width. With the constraint that $W_{M1}+W_{M2} = W_{1T1R}$, the current reduction through the MTJ depends on the widths of M1 and M2.

As the voltage across the MTJ is determined by the resistive divider action between M1, M2, and the MTJ, the magnitude of current reduction through the MTJ can be tuned by appropriately designing the widths of M1 and M2. Looking at the circuit shown in Fig. 5(b), there are two ways to reduce the voltage drop across the MTJ. (1) Increase the resistance of M1 (R_{M1}) by decreasing W_{M1}. (2) Decrease the resistance of M2 or R_{M2} (which in turn decreases $R_{MTJ}||R_{M2}$) by increasing W_{M2}. With the constraint that $W_{M1} + W_{M2} = W_{1T1R}$, both (1) and (2) occur as W_{M2} increases. It follows that the voltage across the MTJ always reduces for the entire range of W_{M2} for both SC and RC bit-cells. As a result, the reduction in the excessive current through the MTJ is achieved. The reduction of the excessive current and voltage across the MTJ mitigate the stress condition for the tunneling oxide of the MTJ, which can be very effective in enhancing the MTJ lifetime [12, 15]. Even though the MTJ voltage and current are always reduced, the reduction in write power occurs when the overall resistance of the bit-cell is increased. If W_{M1} and W_{M2} are designed such that the overall resistance of the bit-cell is decreased, the bit-cell power can increase. Hence, an appropriate value of W_{M2} needs to be chosen so that the reduction of write power can also be achieved using the proposed technique, as we will discuss in Section 5.

Let us now quantify the benefits of the proposed design techniques described above by evaluating the write power and MTJ current density (which determines the MTJ lifetime), considering process variations. We also discuss the impact of the techniques on cell area. This is followed by a detailed comparison of the discussed techniques with the standard 1T-1R STT MRAM.

Table 2: Write currents [μA] averaged over T_W = 40ns, for Baseline design considering process variations.(S:Slow, T:Typical, F:Fast corners)

	SC (W =340nm)			RC (W =900nm)		
	S	T	F	S	T	F
AP→P	**212.7**	**254.6**	**291.7**	83.5	165.6	242.6
P→P	**222.9**	**263.0**	**298.6**	143.7	201.3	272.6
P→AP	100.3	108.4	131.8	**224.6**	**301.7**	**391.9**
AP→AP	58.5	81.1	108.5	**213.4**	**289.6**	**379.0**

5. RESULTS

To compare the effectiveness of our proposed techniques, we first design a baseline bit-cell considering process variations. Since the STT MRAM bit-cell characteristics are primarily affected by the variation in the tunneling oxide thickness (t_{MgO}), cross-sectional area of the MTJ (A_{MTJ}), and variation in the access transistor, we consider ±2.5% deviation from the nominal value for $t_{MgO} = 1.15nm$, ±15% deviation from the nominal value for $A_{MTJ} = 60 \times 60nm^2$ and three different transistor corners. To achieve the desired write-ability under process variations, the access transistor is typically sized considering the slow process corner for write i.e., the corner in which MTJ resistance is high (thicker t_{MgO} and smaller A_{MTJ}) and the drive strength of the access transistor is low (higher transistor threshold voltage (V_{Th})). The discussed design methodology yields larger width of the access transistor compared to that obtained from the bit-cell design without considering process variations (Table 1). We obtained the transistor widths as 340 and 900 nm for SC and RC configurations, respectively. As in the case for the typical corner bit-cell design, P→AP switching for SC and AP→P switching for RC configuration determine the transistor width. Also, note from Table 2 that the magnitude of asymmetry in write current is further aggravated by process variation. It can be observed that the impact of process variations (i.e. the increase in transistor width compared to the typical corner design) is pronounced in RC configuration [5]. This is because the initial MTJ state for AP→P switching in RC configuration is AP (high resistance) and the high resistance provides large negative feedback or source-degeneracy. On the other hand, for P→AP switching in SC configuration the initial MTJ state is P (low resistance) and the negative feedback is weaker compared to that of RC configuration. As a result, the RC configuration results in a larger increase in the transistor width when process variation is considered. Now let us at the baseline design in different process corners.

First, we note that the baseline design yields maximum MTJ current density of 9.1 and 10.2 MA/cm^2 for SC and RC configurations, respectively, in the process corner with lower t_{MgO}, smaller A_{MTJ} and low V_{Th} transistor (see Fig. 6(a)). As discussed in Section 2, the maximum current density is observed in P→P and P→AP case for SC and RC bit-cells, respectively. Also, as a result of unnecessary write power consumption, the average power in the baseline design approaches 207.7 μW and 321.7 μW for SC and RC configurations, respectively, in the process corner with lower t_{MgO}, larger A_{MTJ}, and low V_{Th} transistor.

While the PT bit-cell design requires a minor modification, the sizing of the two transistors in 2T-1R dual source-line bit-cell needs to be properly chosen, as discussed in Section 4.2. The choice of W_{M2} not only determines the write power and the current density but also the layout area. To determine the layout area overhead in 2T-1R DSL designs, we used a general layout design rule (λ-based rules) as discussed in [6]. For the access transistor widths considered in our analysis (W=340nm for SC and W=900nm for RC),

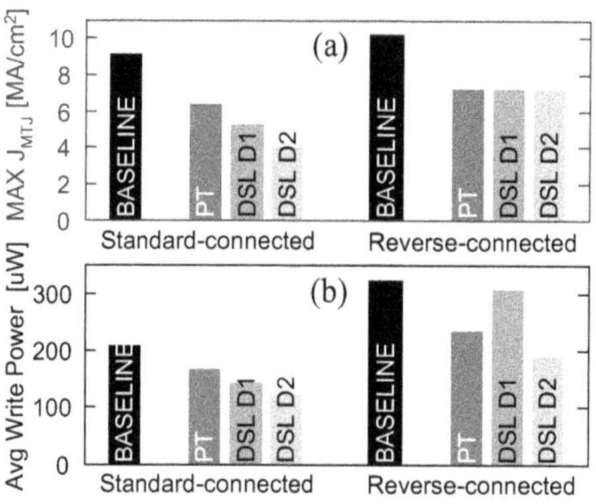

Figure 6: Comparison of worst-case (a) current density and (b) average write power

two-fingered layout yields smaller area compared to that of one-fingered layout based on the analysis presented in [6]. In baseline STT MRAM, the source lines of the two fingers are connected to each other (Fig. 8(a)). In the proposed 2T-1R dual source line bit-cells, separate source lines (SL1 and SL2) for each finger (M1 and M2, respectively) are used as shown in Fig 8(b). One of the source lines (SL1) is shared amongst the bit-cells in the same column, as in the baseline STT MRAMs. The other source line (SL2) is shared amongst the cells in the same row, as a result of which there is no bit-cell area penalty due to the additional source line. As shown in Fig. 8(b-c), there is no area penalty when $W_{M2} = W_{M1}$, but bit-cell area increases when $W_{M2} \neq W_{M1}$. Hence, the lower limit of W_{M2} is chosen to be $W_{1T1R}/2$ at which the bit-cell area is minimum and equal to that of the baseline 1T-1R bit-cell with two-fingered layout. The upper limit of W_{M2} is determined by the switching time. Note that the switching time increases (see T_{SW} in Fig. 7) as we increase W_{M2} in DSL bit-cell, because the current though the MTJ reduced with W_{M2}. Hence, the upper limit of W_{M2} (W_{MAX}) is when the switching time becomes equal to the maximum as per the design specification (40 ns). W_{MAX} for SC and RC bit-cells are 225 and 780nm respectively and the corresponding increase in bit-cell area is 19% and 65%. For the comparison purpose, we consider two extreme cases for W_{M2} ($W_{M2} = W_{MIN}$ and $W_{M2} = W_{MAX}$), denoting them as DSL D1 and DSL D2, respectively. Now let us compare the proposed techniques with the baseline bit-cell to evaluate their effectiveness in improving the write power and the reliability of the MTJ.

To analyze the impact of the proposed techniques on the write power and the reliability of the MTJ due to TDDB, we use the average write power and current density as the metrics to quantify their effectiveness. Although we have used the MTJ current density as a metric to infer the enhancement of the MTJ reliability, the reduction in the voltage across the MTJ shows the same trend when our techniques are applied. For the comparison of the MTJ current density, maximum of AP→P, P→P, P→AP, and AP→AP write current densities is used. Also the write power for each technique is evaluated to an average of AP→P, P→P, P→AP, and AP→AP write power.

When the proposed techniques are applied, P→P write current density is reduced by 30-55% for SC configuration. For RC, P→AP current density is reduced by 39-68%. However, note from Fig.6(a), all the proposed RC bit-cells have the same worst-case current densities, which can be explained as follows. The proposed techniques lead to the reduction in P→AP and AP→AP currents, while P→P and AP→P currents remain the same in RC configuration. As a result,

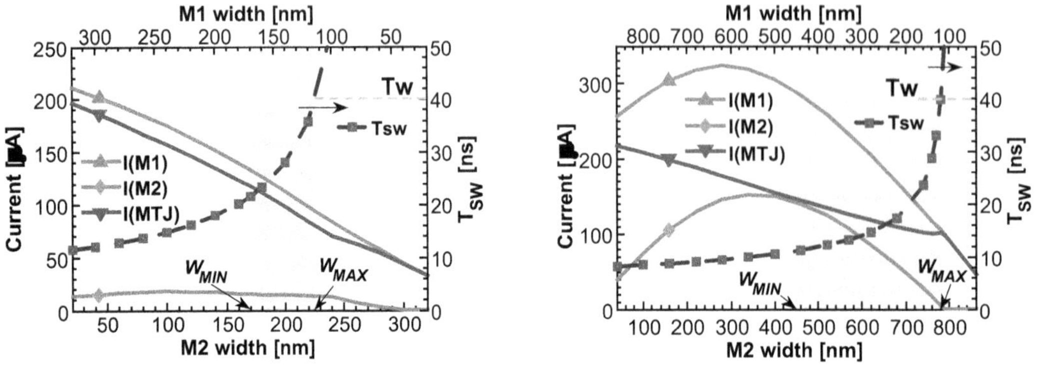

Figure 7: Currents and switching time in 2T-1R dual-source-line bit-cells as a function of W_{M2} (a) SC (b) RC

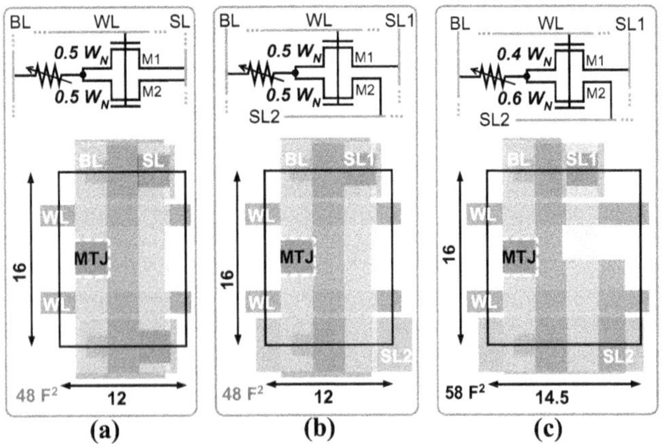

Figure 8: Schematics and bit-cell layouts of (a) Baseline (b) DSL D1 ($W_{M1} = W_{M2}$) and (c) DSL D2 ($W_{M2} > W_{M1}$)

the maximum MTJ current in the proposed bit-cells is observed in P→P case, instead of P→AP, as in the baseline RC bit-cell. Since P→P current is not altered by any of the techniques, the maximum write current density is observed to be the same for all the techniques. The MTJ lifetime enhancement as a result of reduction in excessive current can be very effective from the perspective of reliable design, as shown in [12, 15]. Following this further, with the application of the proposed techniques, the average write power consumption is decreased by 20-41% for SC bit-cells and 4-40% for RC bit-cells as illustrated in Fig. 6. It is also shown that DSL D1 is not very effective for the RC configuration. This is due to non-monotonic trend of current through M1 for small W_{M2} (see Fig. 7(b)). On the other hand, DSL D2 shows larger improvements, albeit with an area penalty. On the other hand, the PT technique requires minimal design change, but still shows considerable improvements over the baseline STT MRAMs.

6. CONCLUSION

In this work, we performed a detailed analysis of the write current in an STT MRAM and showed significant difference in write current for different cases of write operation. We also showed that process variations aggravate the difference in write current. We proposed two design techniques based on using 1) bit-line voltage clamping technique with a pass transistor and 2) 2T-1R dual source line bit-cell, to reduce the excessive and unwanted MTJ current. The strength of the bit-line voltage clamping technique is in its simple implementation with no overhead because the existing transmission gate is replaced with a pass transistor. The bit-line voltage clamping technique led to 20-27% of write power reduction along with decrease in the current density by 31-45%. The 2T-1R dual source-line technique was shown to be very effective in mitigating the stress condition for the tunneling barrier of the MTJ while the reduction in write power was not significant for RC bit-cells, if it is designed to incur no bit-cell area overhead. However, 31% of write power reduction was achieved for SC bit-cells. Both of the proposed techniques were shown to be very effective in reducing the write power without affecting the write-ability, read stability and the performance of the bit-cell. Hence, both low-power and reliability of the MTJ can be achieved by using our proposed techniques.

7. REFERENCES

[1] ITRS report 2011, Available: http://www.itrs.net/.
[2] N. Balke et al. A perpendicular-anisotropy CoFeB-MgO magnetic tunnel junction. *Nature Materials*, 2010.
[3] S. Chatterjee et al. A scalable design methodology for energy minimization of STTRAM: A circuit and architecture perspective. *TVLSI*, 2011.
[4] A. Driskill-Smith. STT-RAM A new spin on universal memory, Future Fab International Report 2008, Available: http://www.future-fab.com.
[5] X. Fong, S. Choday, and K. Roy. Bit-cell level optimization for non-volatile memories using magnetic tunnel junctions and spin-transfer torque switching. *TNANO*, Jan. 2012.
[6] S. K. Gupta et al. Layout-aware Optimization of STT MRAMs. In *DATE*, March 2012.
[7] M. Hosomi et al. A novel nonvolatile memory with spin torque transfer magnetization switching: Spin-RAM. In *IEDM*, 2005.
[8] T. Kawahara, K. Ito, R. Takemura, and H. Ohno. Spin-transfer torque RAM technology: Review and prospect. *Microelectronics Reliability*, 52(4):613–627, 2012.
[9] W. Kim et al. Extended scalability of perpendicular STT-MRAM towards sub-20nm MTJ node. In *IEDM*, 2011.
[10] C. Lin et al. 45nm low power CMOS logic compatible embedded STT MRAM utilizing a reverse-connection 1T/1MTJ cell. In *IEDM*, 2009.
[11] A. Nigam et al. Delivering on the promise of universal memory for spin-transfer torque RAM (STT-RAM). In *ISLPED*, 2011.
[12] Panagopoulos et al. Modeling of dielectric breakdown-induced time-dependent STT-MRAM performance degradation. In *DRC*, 2011.
[13] G. Panagopoulos et al. A framework for simulating hybrid MTJ/CMOS circuits: Atoms to system approach. In *DATE*, 2012.
[14] C. Smullen et al. Relaxing non-volatility for fast and energy-efficient STT-RAM caches. In *HPCA*, 2011.
[15] C. Yoshida et al. A study of dielectric breakdown mechanism in CoFeB/MgO/CoFeB magnetic tunnel junction. In *IRPS*, pages 139–142, 2009.
[16] P. Zhou et al. Energy reduction for STT-RAM using early write termination. In *ICCAD*, 2009.

High-Performance Low-Energy STT MRAM Based on Balanced Write Scheme*

Dongsoo Lee, Sumeet Kumar Gupta, and Kaushik Roy
School of Electrical and Computer Engineering, Purdue University, West Lafayette, IN, USA
{dslee, guptask, kaushik}@purdue.edu

ABSTRACT

It is well known that high write time/energy in STT MRAM are aggravated by the asymmetry in write currents for '0'→'1' and '1'→'0' transitions. This asymmetry is primarily due to the source degeneration of the access transistor during write. In this work, we propose a design methodology which avoids the source degeneration of the access transistor, leading to balanced switching times for '0'→'1' and '1'→'0' transitions. This is achieved by using an additional (negative) bit-line voltage and reduced word-line voltage. The proposed method reduces write time (by ~40%) and write energy (by ~65%), enhances reliability of MTJ, and significantly improves tolerance to process variation. In the proposed scheme, source-line can be directly connected to ground signal leading to a compact cell layout.

Categories and Subject Descriptors

B.3.3 [**Memory Structures**]: Performance Analysis and Design Aids

General Terms

Design, Performance

Keywords

Balanced Write, Low Energy, Process Variation, STT MRAM

1. INTRODUCTION

Spin-transfer torque (STT) MRAM has shown a lot of potential as future on-chip memory owing to promising attributes like non-volatility, zero stand-by leakage, and high density [13]. A typical STT MRAM consists of a magnetic tunnel junction (MTJ) and an NMOS access transistor (NFET) as shown in Figure 1. The MTJ is a storage element that exhibits different resistances depending on the relative magnetization of the free layer (FL) with respect to the pinned layer (PL). Typically, an MTJ has two stable states which correspond to the parallel (P) and the anti-parallel (AP) magnetizations of FL with respect to PL. The data stored in the cell is read using current or voltage sensing techniques to distinguish between the low resistance (P) and high resistance (AP) states [13]. In order to write data (i.e., change the magnetization of FL), a certain amount of spin-transfer torque should be induced by spin-polarized current flowing through the MTJ. P→AP switching is performed by passing the current from PL to FL while for AP→P switching, current is passed from FL to PL. In a standard connected STT MRAM cell (FL connected to bit-line BL - Figure 1), the conventional write technique achieves P→AP switching by applying supply voltage (V_{DD}) to the source-line (SL) and ground to the bit-line (BL). AP→P switching is achieved by applying the opposite voltages, i.e., V_{DD} to BL and ground to SL.

*This research was funded in part by Intel, Qualcomm, Nano Research Initiative, and by the INDEX center.

ISLPED'12, July 30–August 1, 2012, Redondo Beach, CA, USA.

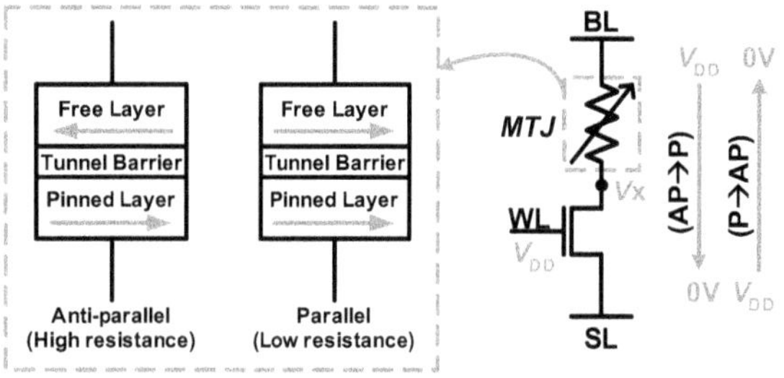

Figure 1: Cell/MTJ structure of STT MRAM.

In spite of the promising attributes of STT MRAMs, designers are faced with some challenging issues. Even though researchers are exploiting various materials for the MTJ and different cell structures, switching the MTJ demands large current density through the MTJ [11, 13]. Therefore, reducing the write time in STT MRAM increases not only the write energy but also the size of access transistors significantly. Moreover, it is important to ensure that the voltage drop across the MTJ (V_{MTJ}) is not higher than a certain limit to reduce MTJ breakdown over time [9]. Also, the cell structure needs to be carefully optimized to mitigate the conflicting design requirements for high read- and write-stabilities [8].

To address these challenges of STT MRAM design, various attempts have been made at the device/circuit/architecture levels of design abstraction. Perpendicular MTJ has been investigated to lower write time of STT MRAM [13]. Redundant writes are early terminated by sampling the resistance of MTJ at early stage of writing to reduce the write energy [15]. Design parameters (such as read current and access transistor size) have been optimized to minimize the read/write failure probability in [8]. Read-preemptive write buffers and SRAM-MRAM hybrid cache have also been introduced to boost the performance [11]. However, in all the previous works, the access transistor is source-degenerated during write, leading to severe degradation in the write speed, write energy, and reliability of the MTJ.

Let us explain in detail the source degeneration of the access transistor during write in STT MRAM. For AP→P switching, the gate-source voltage (V_{GS}) of the NFET is always V_{DD} regardless of the node voltage V_x in Figure 1. However, for P→AP switching, V_{GS} ($=V_{DD}-V_x$) is highly dependent on V_x (source degeneration). If V_{MTJ} ($=V_x$ in P→AP switching) is large, then V_{GS} of NFET is much less than V_{DD}, leading to very weak current driving capability of the NFET. Moreover, body effect of NFET in P→AP switching increases the transistor threshold voltage and further degrades the performance of the NFET. As a result, write time is determined by the switching time of P→AP,

Table 1: Parameters of the MTJ for simulation

Nominal Free Layer Volume (PMA, Ellipse)	$40\times116\times1.5\text{nm}^3$
Energy Barrier	$51k_BT$
Saturation Magnetization	850 emu/cm^3
Damping Factor	0.028
Gyromagnetic Factor	17.6GHz/Oe
Tunneling Oxide Thickness	1.15nm

which is much higher than that of AP→P, as we will show in Section 2. Such unbalance in write also leads to other serious drawbacks such as unnecessarily large V_{MTJ} (leading to aggravated reliability concerns) and power/energy wastage in AP→P switching.

In this paper, we propose a bit-line and word-line biasing technique for write which eliminates the asymmetry in switching times for P→AP and AP→P. In the proposed technique, negative voltage is applied to BL while keeping SL to ground for P→AP switching. As a result the access transistor is not source-degenerated and P→AP switching time is reduced. Note, unlike conventional voltage boosting techniques that can improve writability, we limit the voltage drop across the gate and source/drain of the NFET in a cell to the normal operation range. This is ensured by appropriately reducing the word-line voltage, as discussed later. New biasing points for balanced write can be implemented by a simple level converter and a virtual V_{DD} selection circuit with PMOS headers. Even though we show all results assuming standard connection of the MTJ, reverse-connected STT MRAMs [9] can be designed by following the proposed methodology to achieve lower write energy, faster write speed, improved MTJ reliability, and tolerance to process variations.

The rest of the paper is organized as follows. Section 2 presents the proposed balanced write scheme and a new cell layout that provides reduced lower bound of the cell area. Section 3 shows a process variation tolerant design methodology. Section 4 describes peripheral circuit design for the balanced write scheme and Section 5 concludes the paper.

2. STT MRAM CELL DESIGN WITH BALANCED WRITE SCHEME

2.1 Simulation Setup

To investigate the impact of the balanced write technique on the energy and performance of the cell, we use the simulation framework proposed in [3]. Non-equilibrium Green's Function (NEGF) formulation is used to model MTJ resistance over a range of tunneling oxide thickness and MTJ voltage (V_{MTJ}). MTJ transport characteristics obtained from NEGF are used in the Landau-Lifshitz-Gilbert-Slonczewski (LLGS) SPICE model [7], from which the magnetization switching dynamics of FL are obtained.

The model has been calibrated against experimental data [9]. The MTJ SPICE model is used with commercial 45nm bulk CMOS models for the access transistor to perform comprehensive transient simulations of the STT MRAM bit-cell, considering process variations in the MTJ and access transistor. In this work, we chose MTJ with perpendicular magnetic anisotropy (PMA) because of its low switching current density (and thus, high possibility to be widely accepted for future technologies compared with in-plane magnetic anisotropy (IMA) [10,13]). The design parameters are shown in Table 1. We define switching time of STT MRAM as an interval from an initial state FL magnetization to the time at which magnetization reaches 80% of the final state.

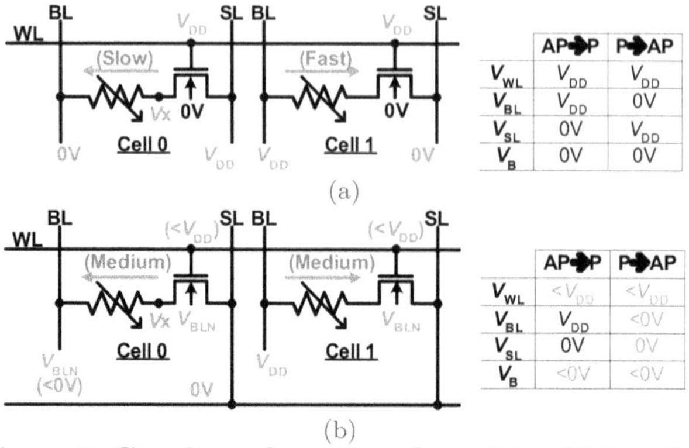

Figure 2: Circuit configurations for writing '1' in cell 0 and '0' in cell 1 using (a) the conventional scheme and (b) the proposed scheme.

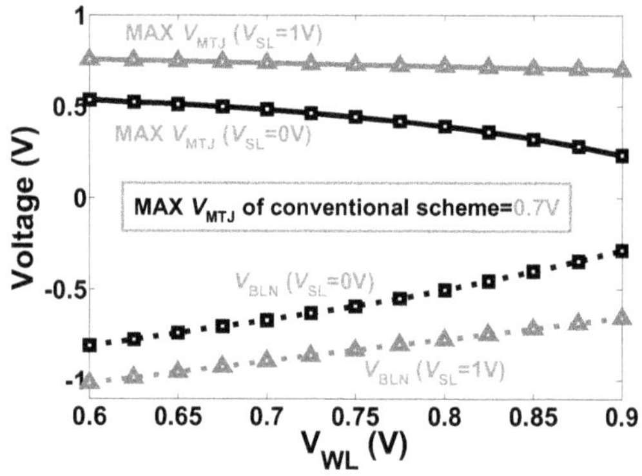

Figure 3: V_{BLN} and the maximum V_{MTJ} when V_{SL} is 0V or 1V (using NFET width=300nm, V_{DD}=1V, and $V_{\text{B}}=V_{\text{BLN}}$).

Keeping WL asserted till the magnetization reaches 80% of the final state significantly reduces the effect of noise components, such as thermal variations that can potentially prevent the FL magnetization to reach 100% of the final state.

2.2 Biasing Technique for Balanced Write

Assuming that P(AP) state represents '0'('1'), Figure 2 shows circuit configurations for writing '10' in one row of STT MRAM. Recall, the major contributor to the asymmetric switching time is the source-degeneration of the access transistor, leading to reduced strength for P→AP switching (cell 0). One potential technique to increase V_{GS} of NFET and mitigate the effect of source degeneration (to lower the switching time) is to boost WL voltage [2]. However, since WL is shared amongst the cells in the same row, V_{GS} of NFET in cell 1 (AP→P) also increases, which leads to higher write current, higher write energy and degraded MTJ reliability in cell 1. On the other hand, our balanced write technique eliminates source degeneration of the access transistor by employing a biasing scheme which results in *lower switching time along with lower write energy and higher MTJ reliability*. The proposed biasing scheme is shown in Figure 2(b) and is explained below:

- **Bit-line voltage (V_{BL}):** In order to lower V_x and increase V_{GS} of NFET for cell 0, we apply a negative voltage (V_{BLN}) to BL of cell 0. This leads to higher strength of NFET of cell 0, thereby lowering the switching time for P→AP transition. For cell 1 (AP→P), BL is biased at V_{DD} as in the conventional technique.
- **Body Voltage (V_{B}):** Body voltage is selected such that the P-N junction (between body and the source)

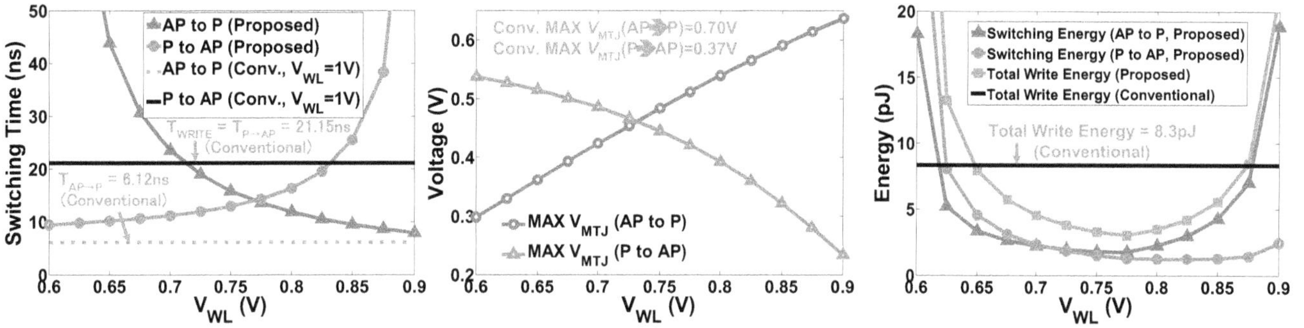

(a) (b) (c)

Figure 4: Simulation results of (a) write time, (b) V_{MTJ}, and (c) write energy using 45nm bulk CMOS, NFET width=300nm, and V_{DD}=1V.

is not turned on in the unselected cells. While this leakage current cannot disturb the MTJ state of the unselected cells during write, the high leakage current can be a roadblock to increasing the number of cells integrated in an array of STT MRAMs. In this work, V_B is equal to V_{BLN}.

- **Word-line voltage (V_{WL}):** Application of negative bit-line to BL can potentially increase V_{GS} of NFET of cell 0 above nominal V_{DD}, which can reduce the oxide reliability of NFET. In order to counter this effect, WL voltage (V_{WL}) is appropriately reduced so that V_{GS} of NFET belonging to cell 0 (P→AP) is at the nominal V_{DD}, i.e., $V_{WL}-V_x=V_{DD}$ (Note, V_x and V_{GS} fluctuate during switching due to the MTJ state change, hence, the maximum value of V_{GS} is considered while choosing V_{WL}). As a result of this constraint on V_{GS}, there is a one-to-one correspondence between V_{BLN} and reduced V_{WL}, for a particular SL voltage. Reducing WL voltage increases the switching time for cell 1 (AP→P). However, by judiciously choosing V_{WL} and V_{BLN}, one can balance the switching time for AP→P and P→AP, leading to overall reduction in write time, as we will show later in this section.
- **Source-line voltage (V_{SL}):** Increasing V_{SL} enhances the switching performance in cell 0. However, V_{MTJ} also increases. Note that the breakdown time of the MTJ exponentially depends on the difference between the breakdown voltage and V_{MTJ} [9]. Hence, reducing V_{MTJ} (while meeting the write time requirements) is crucial to enhance reliability of the MTJ, for which V_{SL} plays a crucial role. Assuming V_{DD}=1V and NFET width=300nm, Figure 3 shows V_{BLN} and V_{MTJ} (determined by maximum V_x during the switching) as a function of V_{WL} for V_{SL}=1V and 0V. Note V_{MTJ} in the conventional scheme is 0.7V (in cell 1 in Figure 2(a)). Hence, the reliability of the MTJ in cell 0 becomes worse than that of cell 1 if V_{SL}=1V is used, as shown in Figure 3. On the other hand, V_{SL}=0V leads to improvement in MTJ reliability. Moreover, if V_{SL}=0V, SL permanently becomes a ground signal and the cell layout becomes simpler (discussed in detail later). Thus, in the proposed scheme, SL is selected to be 0V.
- **$V_{BL}/V_{WL}/V_{SL}/V_B$ during read:** The read scheme including voltage levels in the proposed technique is the same as that of the conventional method. In other words, SL is biased at 0V, BL at read voltage and WL at V_{DD}. Hence, read performance and stability are not affected by the proposed balanced write scheme. Note that V_B can be 0V during read while V_B is changed to be V_{BLN} only during write. Dynamic body bias transition of a sub-array is used in SRAM designs for reducing leakage current and improving process variability tolerance [5]. Assuming 1fF/μm^2 to be P-wll capacitance and 20,000μm^2 to be sub-array size (estimated from a commercial 45nm process technology), capacitance to be charged or discharged is 20pF. For such a case, the body bias transition energy is 7.2pJ if V_{BLN}=−0.6V, which can be ignored compared to the energy required for writing one cache line (e.g., more than 1nJ is required to write 64 bytes as will be shown later). The body bias transition time is much less than 1ns.

In the proposed scheme, V_{SL} is always equal to 0V and V_{BLN} has a one-to-one correspondence with V_{WL}. Therefore, we optimize the voltage biases by exploring the effect of V_{WL} on the write performance, energy consumption, and V_{MTJ} (which determines MTJ reliability). Note that for unselected rows, we apply V_{BLN} to WL in order to turn off access transistors ($V_{GS}=V_{BLN}$, V_{DS}=1V or V_{GS}=0V, $V_{DS}=-V_{BLN}$ for unselected cells). Negative V_{WL} to turn off access transistors during write can raise the drain-gate voltage (V_{DG}) up to ($V_{DD}-V_{BLN}$) which is higher than V_{DD}. However, for the same voltage drop across the oxide, the oxide breakdown time of a transistor in the off-state is several orders of magnitude higher compared to that in the on-state [14].

To optimize and evaluate the proposed technique, we conducted simulations using NFET width=300nm and V_{DD}=1V. Figure 4 shows the switching time, V_{MTJ}, and write energy/power as a function of V_{WL} for the conventional and proposed biasing methods (the corresponding V_{BLN} is shown in Figure 3). In the conventional scheme of Figure 2(a), switching times of '0' and '1' are 6.12ns and 21.15ns, respectively, resulting in highly unbalanced performance. Write time for the conventional cell is equal to the worse of the two switching times (i.e., 21.15ns). In the proposed scheme, we can observe that as we reduce V_{WL} (and decrease V_{BLN}), switching time for P→AP decreases. This is due to the fact as V_{WL} is reduced, the node voltage V_x ($=V_{WL}-V_{DD}$) decreases, which results in increase in V_{DS} ($-V_x$). On the other hand, AP→P switching time increases as V_{WL} decreases due to the decrease in V_{GS} of the NFET. Note for certain value of V_{WL}, the two switching times (AP→P and P→AP) are equal: 13.95ns (34.0% reduction in write time compared to conventional write - Figure 4(a)). Since the overall write time is determined by the worse of the two switching times, write time is minimized when two switching times are balanced. V_{WL} also impacts the reliability of the MTJ. Figure 4(b) illustrates how V_{MTJ} inversely follows the write time (since higher V_{MTJ} implies higher current through the MTJ and hence, lower write time). Using the trends in Figure 4(b), we can determine the voltage V_{WL} that minimizes

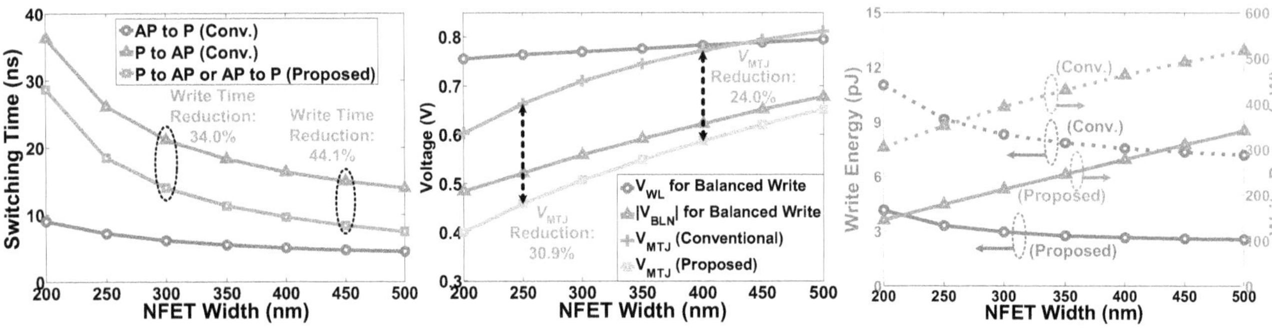

(a) (b) (c)

Figure 5: Simulation results of (a) write time, (b) V_{MTJ}, and (c) write energy/power with various width of NFET.

Table 2: STT MRAM write comparison using 45nm bulk CMOS, NFET width=400nm, and V_{DD}=1V

	Conv. (400nm)		Proposed (400nm)	
	AP→P	P→AP	AP→P	P→AP
V_{WL}	1V	1V	0.783V	0.783V
V_{BL}	1V	0V	1V	-0.622V
V_{SL}	0V	1V	0V	0V
V_B	0V	0V	-0.622V	-0.622V
Switching Time	5.01ns	16.31ns	9.57ns	9.57ns
WL pulse width	16.31ns	16.31ns	9.57ns	9.57ns
Write Energy	4.66pJ	2.89pJ	1.44pJ	1.22pJ
Write Power	285.8μW	176.9μW	150.2μW	127.3μW
Avg. Current	285.8μA	176.9μA	150.2μA	204.7μA
MAX V_{MTJ}	0.772V	0.400V	0.587V	0.498V
V_{GS}	1V	0.600V~ 0.697V	0.783V	0.904V~ 1.000V

Table 3: STT MRAM write comparison using SOI FinFET with 30nm channel length, 15nm fin thickness, 30nm fin height, 5 fingers, and V_{DD}=1V

	Conventional		Proposed	
	AP→P	P→AP	AP→P	P→AP
V_{WL}	1V	1V	0.809V	0.809V
V_{BL}	1V	0V	1V	-0.572V
V_{SL}	0V	1V	0V	0V
Switching Time	5.40ns	25.98ns	10.67ns	10.67ns
WL pulse width	25.98ns	25.98ns	10.67ns	10.67ns
Write Energy	6.73pJ	3.22pJ	1.49pJ	1.17pJ
Write Power	259.0μW	123.6μW	139.7μW	139.7μW
Avg. Current	259.0μA	123.6μA	139.7μA	191.3μA
MAX V_{MTJ}	0.748V	0.342V	0.563V	0.468V
V_{GS}	1V	0.658V~ 0.752V	0.809V	0.913V~ 1.000V

V_{MTJ}. It can be observed that minimum V_{MTJ} is achieved by balancing the V_{MTJ}s corresponding to AP→P and P→AP.

Let us now discuss the effect of the proposed biasing scheme on the write energy. For the conventional biasing scheme (Figure 2(a)), unbalanced switching time leads to significant energy wastage. This is due to the fact that the faster switching cells dissipate power until the switching of the slower cells is complete. On the other hand, in the proposed biasing technique, there is no excessive and unwanted energy consumption due to equal switching times for P→AP and AP→P. Consequently, at V_{WL} corresponding to the balanced switching time (Figure 4(c)), the total write energy (= write energy of AP→P + write energy of P→AP) is minimized.

Table 2 summarizes the comparison of the proposed (Figure 2(b)) and conventional (Figure 2(a)) biasing schemes using NFET width=400nm. For balanced switching time, the average current during write is larger for P→AP switching (by 36.3%) compared to AP→P switching. This is due to different critical current requirements for P→AP and AP→P switching for the same switching time. Despite higher current for P→AP switching, energy/power for P→AP switching is less than that of AP→P because of smaller voltage drop across the cell ($|V_{BL}-V_{SL}|$=0.622V in P→AP). As a results, at the proposed biasing point with NFET width=400nm, the total write energy and write power can be improved by 64.8% and 40.0%, respectively, while V_{MTJ} can be reduced by 24.0%.

To study the effect of width of NFET in the proposed scheme, we compare the switching time, V_{MTJ}, and the write energy/power for different widths of NFET. As shown in Figure 5, write time, write energy, and MTJ reliability are enhanced for the entire range of width of the NFET in our simulations.

Note, although the energy and performance overhead associated with the transition of the body voltage during read and write is small, it can be completely eliminated in silicon-on-insulator (SOI) thin-body FinFET technology. In order to evaluate the proposed technique with SOI FinFET as access transistors, we designed a FinFET with 30nm channel length, 15nm fin thickness, 30nm fin height, and 5 fingers. Table 3 shows the simulation results for FinFET-based STT MRAM using the BSIM-CMG model [1]. It can be observed that the proposed write methodology provides balanced write performance in SOI FinFET technology as well.

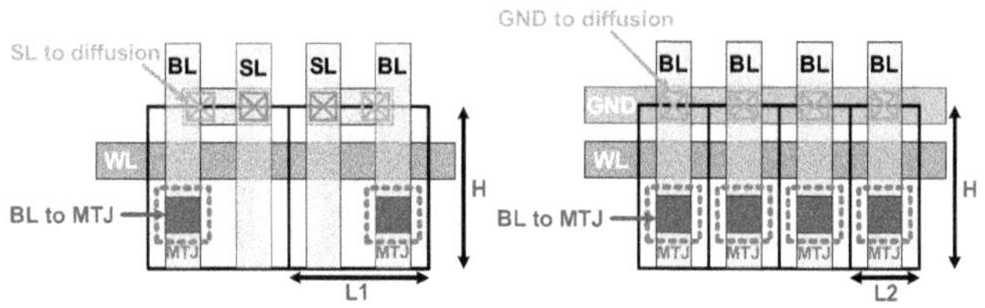

(a) (b)

Figure 6: Cell layout of (a) the conventional scheme and (b) the proposed scheme.

2.3 Cell Layout

Figure 6(a) presents a layout of two cells using the conventional biasing scheme. SL and BL metal layers are routed in the same (vertical) direction to share them amongst the cells in the same column. Note, the minimum cell area is limited by the metal pitch of SL and BL (L1 in Figure 6(a)) rather than NFET width [4]. For the proposed scheme, since SL is connected to the ground rail, the SL metal layer need not be routed in the vertical direction. By routing the SL metal in the horizontal direction (sharing SL amongst cells in the same row), as shown in Figure 6(b), only one metal layer for BL is routed vertically through a cell. As a result, the minimum pitch (L2 in Figure 6(b)) of the cell becomes half of L1. Therefore, the lower bound of cell area determined by layout rules can be reduced by 2X. Note that routing of SL in a power grid form would reduce power delivery problems including di/dt noise.

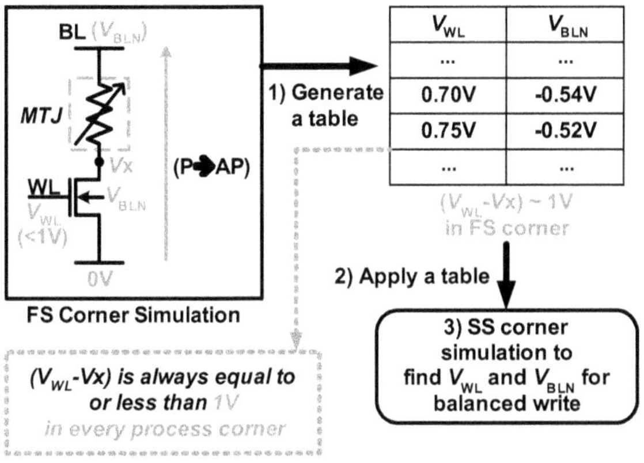

Figure 7: Design procedure for balanced write in SS corner while maintaining reliable voltage operation of NFET in every process corner.

3. PROCESS VARIATION

The random variations in process parameters in nanoscale technologies play a significant role in the design of STT MRAMs like other high-density memories. Design goals, such as write time, energy, and reliability of devices, should be accurately estimated such that STT MRAM can meet the specifications in every process corner. In this paper, we consider variations in the tunnel oxide thickness of the MTJ (t_{MgO}), cross-sectional area of the MTJ (A), and the strength of NFET. Resistance of the MTJ exponentially depends on t_{MgO} [8]. On the other hand, reducing A increases resistance of the MTJ. Therefore, switching in STT MRAM is highly affected by variations in t_{MgO} and A. As mentioned earlier, the proposed method follows the same read scheme as the conventional method. Thus, impact of process variations on the read stability is not analyzed in this section.

To investigate the effects of variations in parameters on write time, energy, and reliability, we simulated the proposed scheme and the conventional scheme (V_{WL}=1V, V_{BL}=1V or 0V, V_{SL}=0V or 1V) in process corners defined as following: In SF (Slow MTJ, Fast NFET) corner, the MTJ has the highest resistance (t_{MgO} is increased by 2.5% and A is reduced by 15%) and NFET has low threshold voltage (high strength). Similarly, FS (Fast MTJ, Slow NFET) corner is defined as the process corner in which the MTJ has the lowest resistance (t_{MgO} is decreased by 2.5% and A is enlarged by 15%) and NFET has high threshold voltage. TT corner is the nominal corner with design parameters shown in Table 1. To design a process variation tolerant STT MRAM, the worst switching time measured in SS (Slow MTJ, Slow NFET) corner is considered as the write time. It is also required to determine the maximum V_{MTJ} and write energy/power (under the write time in SS corner) to find the worst-case MTJ reliability and energy/power consumption. Maximum V_{MTJ} and write energy/power are obtained in the SF corner and the FF corner, respectively.

For the proposed scheme, V_{GS} of NFET is maintained in the "reliable range" (less than or equal to 1V) in every process corner. For AP→P switching, V_{GS} (=V_{WL}) is always less than 1V. For P→AP switching with certain V_{WL} and V_{BLN}, V_x is minimum in the FS corner among all the process corners (Note, node voltage V_x is determined by the voltage division between the MTJ and NFET). Therefore, for the same V_{WL} and V_{BLN}, V_{GS} (=$V_{WL}-V_x$) is maximum in the FS corner among all the process corners. As the first step of process variation tolerant design, we generate a table describing pairs of V_{WL} and corresponding V_{BLN} that provide V_{GS} of 1V in the FS corner as shown in Figure 7. If

Table 4: STT MRAM write comparison using 45nm bulk CMOS, NFET width=400nm, and V_{DD}=1V in the worst case corners

	Conv. (400nm)		Proposed (400nm)	
	AP→P	P→AP	AP→P	P→AP
V_{WL}	1V	1V	0.725V	0.725V
V_{BL}	1V	0V	1V	-0.586V
V_{SL}	0V	1V	0V	0V
V_B	0V	0V	-0.586V	-0.586V
Switching Time (SS)	5.70ns	28.3ns	16.1ns	16.1ns
WL pulse width	28.3ns	28.3ns	16.1ns	16.1ns
Write Energy (FF)	9.69pJ	4.27pJ	2.68pJ	1.84pJ
Write Power (FF)	342.1μW	150.5μW	165.9μW	113.8μW
Avg. Current (FF)	342.1μA	150.5μA	165.9μA	194.2μA
MAX V_{MTJ} (SF)	0.842V	0.456V	0.657V	0.498V
V_{GS} (SS)	1V	0.588V~0.671V	0.725V	0.827V~0.906V

we store and utilize this table for the other process corners, then NFET always operates in a reliable voltage range. Using this table of V_{WL} and V_{BLN} pairs, one pair is chosen for balanced switching time (in SS corner) in order to get the best "worst case" write time.

Figure 8 shows the simulation results of write time, V_{MTJ}, and write energy/power for the worst case process corners (SS for write time, SF for V_{MTJ}, and FF for write energy/power). Note that the write time in the conventional method is very high in the SS corner. This is due to the fact that source degeneration increases the sensitivity of the node voltage V_x (that plays a key role in determining the overall performance) to variations in the resistance of the MTJ and the strength of the NFET. On the other hand, the proposed method shows smaller difference in the write times in the SS corner and the TT corner since source degeneration of the access transistor is avoided in the balanced write scheme. Note that in the TT corner, the proposed method (using bias conditions determined for the FS corner) still offers faster switching performance compared to the conventional method.

Table 4 summarizes the comparison of the proposed (with the design procedure of Figure 7) and conventional biasing schemes in the worst case corners. Note V_{GS} of the proposed scheme in the SS corner is less than 1V since a pair of V_{WL} and V_{BLN} is generated in the FS corner. Despite reduced V_{GS}, the write time (in the SS corner) is reduced by 43.0% in the proposed scheme. The write energy and write power (in the FF corner) are saved by 67.6% and 43.2%, respectively, and V_{MTJ} (in the SF corner) is reduced by 22.0%.

4. PERIPHERAL CIRCUIT DESIGN

The proposed balanced write scheme requires two additional voltage levels: (a) reduced V_{WL} during write ($V_{WL,OPT}$) and (b) negative V_{BLN}, besides V_{DD} and ground. To apply these voltage levels to appropriate signals (as described below), additional peripheral circuits are required. We discuss the peripheral circuits for WL drivers and BL drivers in this section. Recall, SL drivers are not required in the proposed scheme since SL is connected to the ground rail.

In the proposed biasing methodology, BL driver is required to drive BL to V_{DD} and V_{BLN} for writing logic '1' and '0', respectively. This is achieved by employing a level converter that converts the write data 0/V_{DD} to V_{BLN}/V_{DD}. The circuit implementation of the level converter is shown in Figure 9 and is based on the high speed level converter proposed in [6]. However, instead of using a cross-coupled PMOS pair as in [6], we use a cross-coupled NMOS pair which is more suitable to convert 0 to a negative voltage V_{BLN}. Since this level converter is operated in a higher voltage range, thick-oxide transistors may be needed to prevent oxide breakdown [12]. Thick-oxide transistors can be imple-

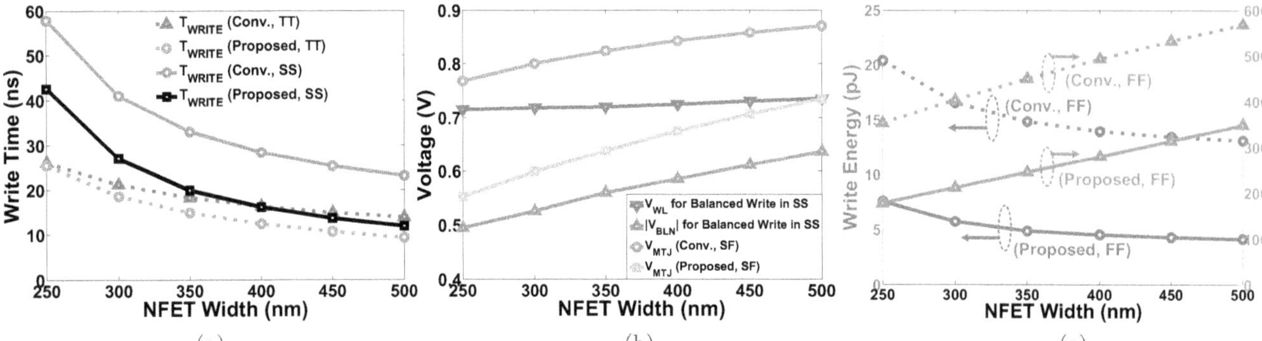

Figure 8: Simulation results of (a) write time, (b) V_{MTJ}, and (c) write energy in the worst case corners.

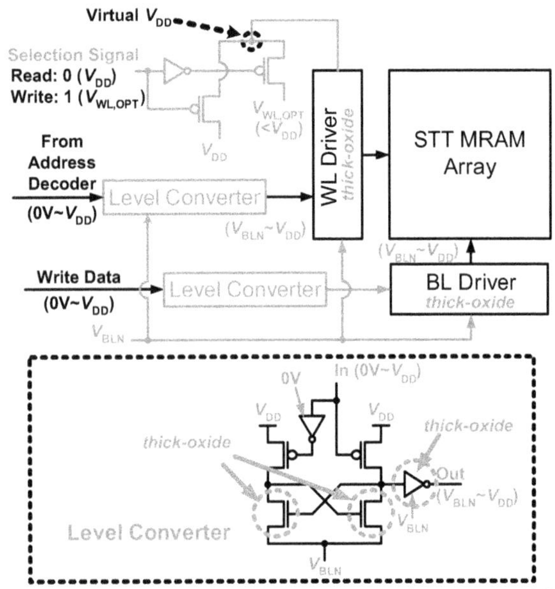

Figure 9: Virtual V_{DD} selection circuit for V_{WL} and level converter for V_{WL} and V_{BLN}.

mented by design rules of I/O circuits (without additional process technology) or by multiple oxide thickness technology (Arsenic implantation into the silicon substrate before thermal oxidation [12]). In order to maintain the current driving strength of BL and WL drivers, thick-oxide transistors may be sized up.

For the WL driver, the output is required to be V_{BLN} if the signal from the address decoder is 0V (This is required to drive the access transistor of the un-accessed cells). If the address decoder signal is V_{DD}, the output of WL driver needs to be V_{DD} during read and reduced WL voltage (optimal for balanced switching times - $V_{WL,OPT}$) during write (This signal drives the access transistor of the accessed cells). This is achieved by employing (a) a multiplexer to select between V_{DD} and $V_{WL,OPT}$ to drive the supply node of the WL driver (labeled as virtual V_{DD} in Figure 9) and (b) a level converter that converts the address decoder signal $0/V_{DD}$ to V_{BLN}/V_{DD} (The implementation of the level converter is identical to that described for the BL driver).

5. CONCLUSION

In this paper, we presented a balanced write scheme and associated peripheral circuits for high-performance and low-energy STT MRAM design. We showed that it is possible to achieve equal switching times for AP→P and P→AP using an additional (negative) BL voltage (V_{BLN}) and appropriately reducing the word-line voltage V_{WL}. At the optimal bias point, we achieve reduction in write time and write energy along with enhancement in the reliability of the MTJ (characterized in terms of V_{MTJ}). We also presented a process variation tolerant design methodology, in which biasing conditions are obtained in the FS corner in order to maintain reliable operation of the NFET. We showed significant improvement in write time, write energy, and MTJ reliability considering the worse case process corners. In addition, the proposed design technique does not require source-line drivers. We also showed that the lower bound of cell area is reduced by 2X compared to conventional design techniques.

6. REFERENCES

[1] Berkeley BSIM Group. *Berkeley Short-Channel IGFET Model.* www-device.eecs.berkeley.edu/bsim/, 2012.

[2] S. Chatterjee et al. A scalable design methodology for energy minimization of STTRAM: a circuit and architecture perspective. *IEEE Trans. VLSI*, 19(5):809–817, May 2011.

[3] X. Fong et al. KNACK: A hybrid spin-charge mixed-mode simulator for evaluating different genres of spin-transfer torque MRAM bit-cells. In *Proc. SISPAD*, pages 51–54, Sep. 2011.

[4] S. Gupta, S. Park, N. Mojumder, and K. Roy. Layout-aware optimization of STT MRAMs. In *DATE*, 2012.

[5] F. Hamzaoglu et al. A 3.8GHz 153Mb SRAM design with dynamic stability enhancement and leakage reduction in 45nm high-k metal gate CMOS technology. *IEEE J. Solid-State Circuits*, 44(1):148–154, Jan. 2009.

[6] F. Ishihara, F. Sheikh, and B. Nikolic. Level conversion for dual-supply systems. *IEEE Trans. VLSI*, 12(2):185–195, Feb. 2004.

[7] K. J. Lee, O. Redon, and B. Dieny. Analytical investigation of spin-transfer dynamics using a perpendicular-to-plane polarizer. *Appl. Phys. Lett.*, 86:022505–1–022505–3, 2005.

[8] J. Li, P. Ndai, A. Goel, S. Salahuddin, and K. Roy. Design paradigm for robust spin-torque transfer magnetic RAM (STT MRAM) from circuit/architecture perspective. *IEEE Trans. VLSI*, 18(12):1710–1723, Dec. 2010.

[9] C. J. Lin et al. 45nm low power CMOS logic compatible embedded STT MRAM utilizing a reverse-connection 1T/1MTJ cell. In *IEEE IEDM Dig. Tech.*, pages 279–282, Dec. 2009.

[10] A. Nigam et al. Delivering on the promise of universal memory for spin-transfer torque RAM (STT-RAM). In *Proc. ISLPED*, pages 121–126, Aug. 2011.

[11] G. Sun, X. Dong, Y. Xie, J. Li, and Y. Chen. A novel architecture of the 3D stacked MRAM L2 cache for CMPs. In *Proc. IEEE HPCA*, pages 239–249, Feb. 2009.

[12] M. Togo, K. Noda, and T. Tanigawa. Multiple-thickness gate oxide and dual-gate technologies for high-performance logic-embedded DRAMs. In *IEEE IEDM Dig. Tech.*, pages 347–350, Dec. 1998.

[13] S. Wolf et al. The promise of nanomagnetics and spintronics for future logic and universal memory. *Proceedings of the IEEE*, 98(12):2155–2168, Dec. 2010.

[14] E. Wu and E. Nowak. Voltage-splitting technique for reliability evaluation of off-state mode of MOSFETs in ultrathin gate oxides. *IEEE Electron Devices Lett.*, 25(6):414–416, Jun. 2004.

[15] P. Zhou, B. Zhao, J. Yang, and Y. Zhang. Energy reduction for STT-RAM using early write termination. In *Proc. IEEE ICCAD*, pages 264–268, Nov. 2009.

Design Benchmarking to 7nm with FinFET Predictive Technology Models

Saurabh Sinha, Brian Cline, Greg Yeric, Vikas Chandra, Yu Cao*
ARM Inc., *Arizona State University, Tempe, AZ
saurabh.sinha@arm.com

ABSTRACT

The coming ten years promise great changes in silicon technology, with the end of planar bulk CMOS and the rise of interconnect parasitics to true significance. With such shifts in the underlying technology, the simple extrapolation of performance metrics may lead to pronounced prediction errors in design pathfinding. In this work, we utilize newly developed Predictive Technology Models for FinFETs aligned to the 2011 ITRS. Together with predictive interconnect models, we project performance and power landscape for the technology nodes from 20nm to 7nm. We present an overview of models, assess the advantage of FinFET over bulk CMOS devices, benchmark the scaling of critical design metrics, and illustrate major design barriers toward the 7nm node.

Categories and Subject Descriptors

B.7.2 [**Hardware**]: Integrated Circuits – *Design Aids, Simulation.*

General Terms

Performance, Design.

Keywords

Predictive Technology Models, FinFET, power, performance, integrated circuits, scaling trends, BEOL.

1. INTRODUCTION

Device-circuit technology forecasts such as [1] in 2009 provide valuable insight into design prediction and help guide design pathfinding. With similar intent, we update to follow the 2011 International Technology Roadmap for Semiconductors (ITRS) [2]. The time period examined, 2012-2020, encompasses the transition from planar bulk CMOS to silicon FinFET, so a key supporting initiative for this update has been the development of Predictive Technology Models for Multi-Gate transistors (PTM-MG). The details of these PTM-MG models are discussed in [3], and the models are available at the PTM website [4]. In this work we employ these models to assess possible scaling of circuit-based power and performance metrics. We overlay simulation results from 32nm and 22nm planar PTM models for continuity.

In spite of the performance gain by moving to a 3D multi-gate structure with FinFETs and continued dimension scaling, interconnect and FEOL (front end of line) RC are expected to have an increasingly degrading impact on the power performance metrics of future circuits [5][6]. This work explores the potential of low power FinFETs by incorporating the effects of intra-cell and interconnect parasitic, allowing early-stage design-technology co-optimization (DTCO).

Section 2 discusses the PTM-MG model development and the methodology used to obtain FEOL and BEOL parameters. Linear scaling of interconnect resistance and capacitance according to the ITRS 2011 roadmap is used in conjunction with the predictive multi-gate models to quantify the behavior of representative design modules such as standard cells, ring oscillators, stacked devices and interconnect repeater design. We discuss the results of our simulation study in Section 3 where the advantages and limits of technology scaling are quantified providing us an insight into process and design choices.

2. PREDICTIVE TECHNOLOGY MODELING

2.1 PTM-MG Device Models

The PTM-MG model parameters have been developed using the BSIM-MG model format [7] based on MOSFET scaling theory, the ITRS roadmap and early stage silicon data available from publications. Predictive models corresponding to the years 2012-2020 on the 2011 version ITRS roadmap have been developed. Details of the model development and verification with published data are presented in [3]. The BSIM-CMG model includes up-to-date FinFET modeling parameters, including the ability to perform geometry-based parasitic resistance and fringing capacitance calculation [7].

Figure 1 shows the trend of I_{DSAT} vs. technology node for PTM-MG HP (high performance) and LSTP (low stand-by power)

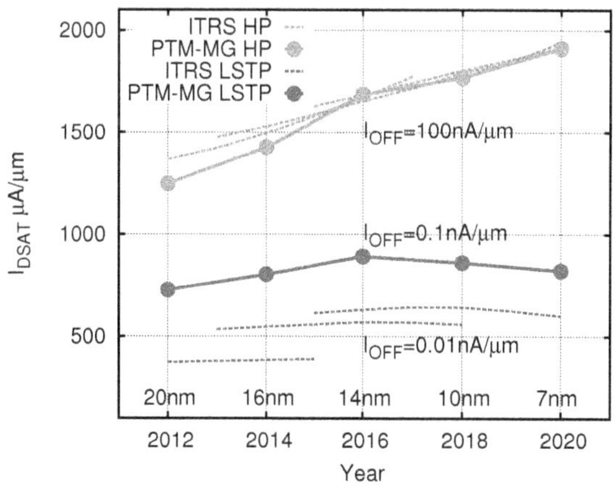

Figure 1. I_{DSAT} vs. technology node/year for PTM-MG HP and LSTP devices. The I_{DSAT} from the ITRS 2011 roadmap are plotted for comparison.

Table 2. FEOL Parameters

Year	2012	2014	2016	2018	2020
Technology	20nm	16nm	14nm	10nm	7nm
R_{GCONT} [12]	150	180	210	250	280
R_{WPT} [2]	1.7	2.6	3.7	5.4	8.2
R_{SDCONT} [6]	100	133	168	213	267
$C_{IN\text{-}OUT}$ (aF) [11]	37.5	28.1	22.3	17.6	14.1
$C_{IN\text{-}RAILS}$(aF) [11]	29.5	22.1	17.5	13.8	11.1
$C_{OUT\text{-}RAILS}$ (aF) [11]	38	28.5	22.6	17.8	14.3

model versions. While the PTM predictions for HP transistors have been set to closely match the ITRS values, we have chosen to target an I_{OFF}=0.1nA/um for PTM LSTP devices while ITRS LSPT targets I_{OFF}=0.01nA/um. Any single device target will be inadequate to estimate SoC power/performance scaling due to increasing use of multiple VT options in design, but in our judgment the higher performance 0.1nA/um target is more useful. For these PTM models, we have tuned the leakage simply by adjusting the metal work function, and at [4] we document the work function adjustments that would more closely align to the ITRS LSPT targets. Note that the ITRS roadmap curves have three separate trend lines associated with planar transistors, FDSOI and FinFETs respectively, while all pathfinding simulations in this work use the LSTP transistor models.

2.2 FEOL / BEOL parameter assumptions

The ITRS 2011 roadmap forms the primary basis for determining the technology node and BEOL parameters. Table 1 lists the important technology parameters such as M1 pitch and supply voltage. The ITRS does not provide guidance for Fin pitch or fin 3D-factor. The effective fin-width is calculated as 2*HFIN+TFIN where HFIN is fin-height and TFIN is fin-thickness. Our choices here have been guided by PTM-MG modeling insights [3] and trends from publications [8][9]. Achieving a fin-pitch smaller than the effective fin-width allows higher FinFET width than the corresponding planar area (3D factor greater than one). Moving from a planar technology to 3D FinFETs impacts standard cell characterization, power and performance as will be seen in the subsequent sections.

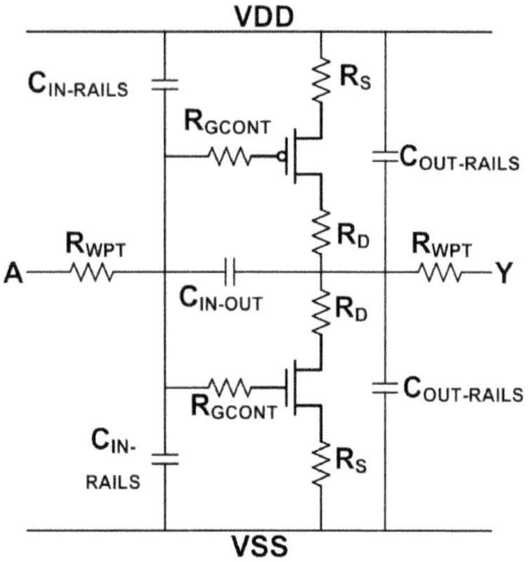

Figure 2. Schematic of sample inverter (1X strength) netlist with parameterized FEOL parasitics.

Table 1. Key Technology Parameters

Year	2012	2014	2016	2018	2020
Technology	20nm	16nm	14nm	10nm	7nm
VDD (V)	0.9	0.85	0.8	0.75	0.7
M1 Pitch (nm)	64	48	38	30	24
Fin Pitch (nm)	60	42	32	28	22
Effective Width (nm)	71	64	56	50	38.5
3D Factor	1.2	1.524	1.75	1.78	1.75

In order to account for FEOL parasitics, a parameterized netlist incorporating the important coupling capacitances and contact resistances is used for simulation. Figure 2 shows a sample inverter schematic with the relevant parasitics. This inverter is referred to as "1X" because it contains one gate finger per device. A standard cell with twice the drive strength would contain 2 gate fingers per device and be referred to as "2X", and the corresponding parameterized netlist would be expanded to reflect the increase in physical complexity. The coupling capacitances are layout dependent and values for the 20nm node are obtained from TCAD based experiments presented in [10] and listed in Table 2, which lists the key FEOL RC parameters as they scale through the ITRS nodes. Also missing from the ITRS is guidance regarding the contact resistance, so we have obtained values from [6], [11] for the 20nm node and have extended using simple extrapolation. Methods to reduce this resistance by replacing the tungsten via with copper, which offers lower resistivity, have been proposed [12], and are generally reflected in the relaxed scaling trend we have assumed for $R_{S/D}$.

To scale the other parameters in Table 2 beyond the 20nm node, we scale the R and C values according to ITRS M1-pitch scaling trend of about 0.75 ~ 0.8 from 2012-2020.

3. RESULTS AND DISCUSSION

3.1 FinFET-based transistor scaling

FinFETs will be the technology of choice for extending CMOS scaling beyond the 20nm node. Improved short channel control through a fully-depleted fin, reduced random dopant fluctuation, improved mobility and area efficiency are some of the primary advantages of FinFETs over planar transistors [13].

Figure 3 shows normalized delay vs. fan-out for Inverter, NAND2 (2-input) gate and NAND3 (3-input) gate comparing 22nm planar PTM and 20nm PTM-MG devices. All delay lines are annotated with their respective logical efforts. FinFET based NAND2 and NAND3 have lower logical effort, closer to their ideal values compared to their planar counterparts. The PTM planar device models have been tuned to have I_{OFF}=0.1nA/um similar to the PTM-MG LSTP models for a fair comparison and both simulations were run with identical effective device width and FEOL parasitics.

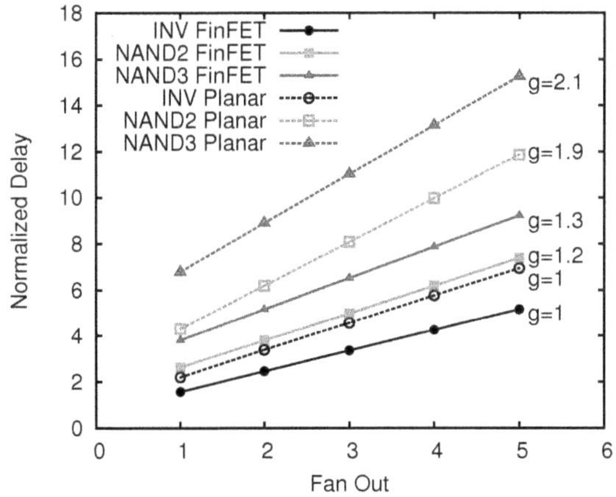

Figure 3. Normalized delay of Inverter, 2-input NAND and 3-input NAND for 22nm planar and 20nm FinFET technologies. FinFET stacked devices are inherently better gates with lower logical effort.

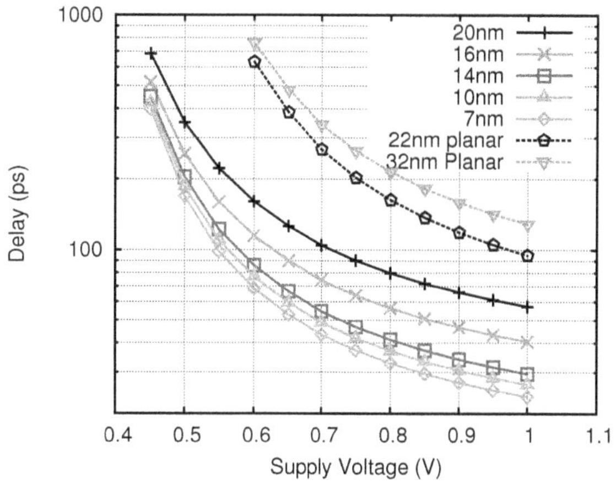

Figure 5. Device Delay trends of RO for 20nm to 7nm nodes with 10 track wire loading between stages. 32nm and 22nm planar devices show inferior performance with V_{DD} scaling.

Near ideal sub-threshold characteristics allowing lower V_T and better gate overdrive, significantly lower DIBL in FinFETs compared to planar transistors, and negligible body effect are the main effects resulting in excellent stacked device performance [14][15]. Comparing FinFET 20nm NAND3 to planar 22nm NAND2, we see that using higher fan-in circuits with FinFETs can be considered, giving area savings with minimal delay penalty.

To understand the impact of fin 3D factor, Figure 4 shows a plot of the planar transistor width (PFET and NFET) for a 9-track standard cell library, where "9-track" refers to the space occupied by 9 M1 pitches. A 75% fill ratio has been assumed, meaning that 75% of the standard cell height is occupied by active transistors and the remaining 25% is isolation between transistors. For reference, we extend the planar trend according to M1 pitch scaling. The effective FinFET width for a 9-track

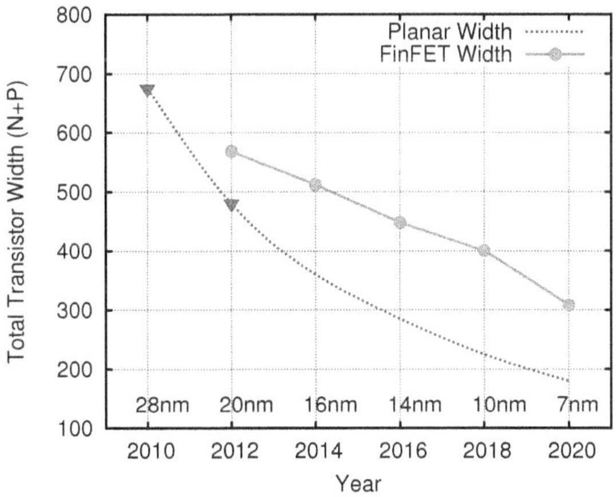

Figure 4. Total Transistor Width (NFET+PFET) for a 9-track standard cell assuming 75% fill-ratio. Owing to FinFET 3-D factor and fin-pitch scaling, the effective transistor width in a 10-track standard cell will be higher than planar scaling assumptions.

cell is also plotted in Figure 4. Based on the effective fin width and fin-pitch assumptions in Table 1, we find that a 9-track inverter can accommodate about 8 active device fins (4 fins per NFET/PFET). Transitioning from planar technology to FinFETs, the total transistor width in a 9-track cell can be much higher owing to the 3D structure of the transistors, e.g., the 16nm node standard cell FinFET transistor width could be equal to or greater than the 20nm node standard cell planar width. This translates to standard cells with significantly higher drive strength as compared to standard process scaling assumptions, although practical issues associated with sub-20nm lithography will also be important. This has implications to the end SoC power. Smaller size standard cells could potentially be deployed to reduce power but yet maintain performance, and the need for larger drive strength cells in loaded critical paths can be reduced. These SoC-level implementation aspects are examples of technology pathfinding which can be considered in future studies, as enabled by these new PTM-MG models.

3.2 Pathfinding Design Analysis

Using PTM-MG model files for FinFET transistors and the FEOL and BEOL assumptions described in Section 2.2, we conducted SPICE simulations of inverter ring oscillators (RO) with different wire-loading to understand the active device and interconnect scaling trends.

Figure 5 shows delay vs.V_{DD} across 10-stages of an inverter based ring oscillator with a light 10-track wire loading for different technology nodes using our BSIM-MG models. For comparison, delay trends with supply voltage scaling using 32nm and 22nm PTM planar models were simulated as well. We have tuned the device planar model parameters for I_{OFF}=0.1nA/um, similar to the PTM-MG models. As demonstrated in previous work [13][14], FinFETs show excellent delay vs. V_{DD} scaling owing to superior electrostatics. Scaling V_{DD} from 1V to 0.6V, FinFET delay degrades by about 3X, while 32nm/22nm planar technologies suffer a corresponding degradation of up to 7X.

In Figure 5 we observe delay down to 0.45V and we find that FinFET delay at 0.45V is similar to that of a 0.6V planar device, demonstrating its potential for very low power near-threshold

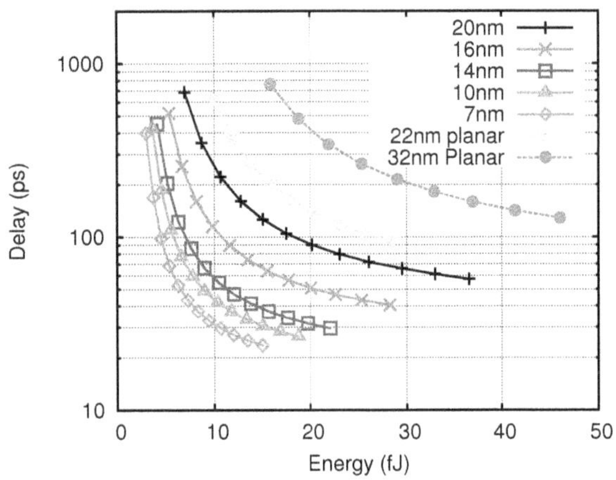

Figure 8. Energy vs. delay trends for RO with V_{DD} scaling. VDD points are the same as in Figure 5, with 1.0V VDD corresponding to the right-most data points.

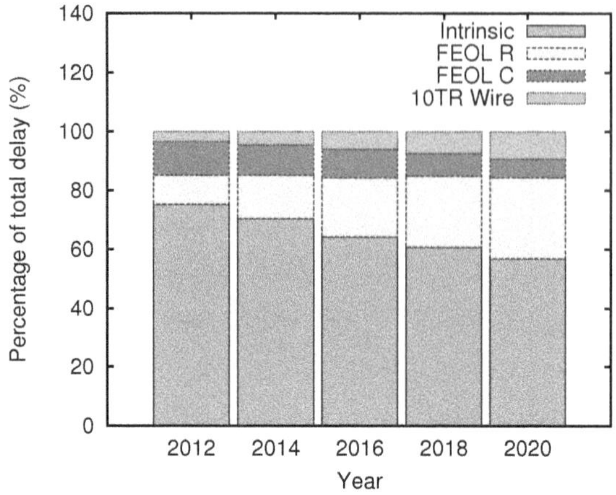

Figure 6. Decomposition of RO Delay as a percentage of the total delay consisting of active devices, FEOL RC and 10-track wire RC.

operation. However, practical V_{DD} scaling studies would require quantification of manufactured device variance, which is an example of potential future work.

Figure 8 shows the energy vs. delay across 10-stages of the inverter based ring-oscillator. In spite of improved delay vs. supply voltage scaling trend, switching energy of 22nm planar circuit is lower compared to 20nm FinFET for the same V_{DD}. This is due to the higher effective transistor width within a standard cell, and a slight increase in parasitic capacitance associated with the FinFET structure. Figure 8 then underscores the requirement that sub-20nm FinFETs need utilize their natural ability to operate at reduced V_{DD} in order to realize an energy savings. Alternatively, smaller cells with reduced fin count can be considered, as could device re-targeting to lower leakage set points to below comparable planar device leakage.

In Figure 7 we take the delay information from Figure 5 for the specific V_{DD} values across technology nodes as listed in Table 1. To understand the impact on delay scaling due to interconnect parasitics, we have investigated different transistor loading

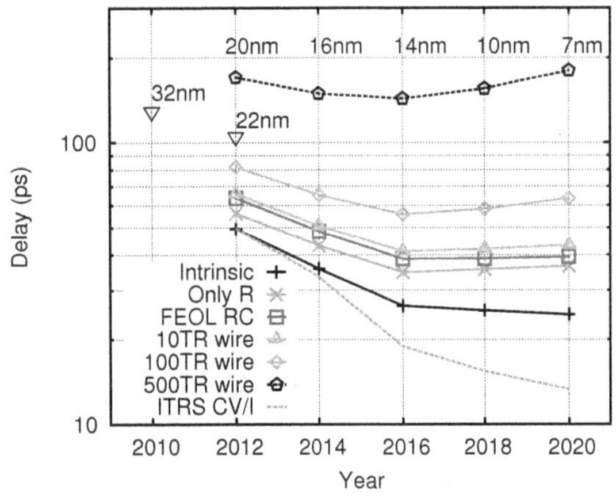

Figure 7. INV RO delay across 10-stages with intrinsic, no-wire load, with 10, 100 and 500 track wire load between stages. 32nm and 22nm planar RO delay with 10 track wire load is plotted for

scenarios. Intrinsic gate delay means we have removed all of the R and C values associated with the parameterized netlist of Figure 2. We then add only the R values, then also the C values to get the "RC" curves, and then we also add wire loads between stages of 10-, 100-, and 500- tracks (M1 pitches). Given that standard cells are typically 7-12 metal tracks in dimension, these wire lengths are representative of the range of "local" wire lengths that would be seen in a logic block. The ITRS CV/I delay trend normalized to 20nm FinFET intrinsic delay is plotted for reference. It is evident from the deviation between the ITRS CV/I trend and the intrinsic RO delay trend that first order CV/I metric does not adequately capture transient device behavior. Changing CV/I from I_{DSAT} to I_{EFF} would not materially change the divergence between static I metric and AC device simulation [17].

Additionally, the ITRS roadmap does not take into account practical circuit FEOL and BEOL parasitics. Adding contact resistance to the intrinsic devices as shown by the red curve in Figure 7 actually reverses the delay scaling trend beyond the 2016 node. With FEOL RC and 10-track wire-load, the delay penalty increases from 33% to 76% from the 20nm to 7nm node, as compared to intrinsic delay. Additionally, a 10-track wire load adds about 15% delay penalty at the 7nm node.

Figure 6 shows the decomposition of the components from Figure 7 as a percentage of total delay of the ring-oscillator for the case of 10-track wire load between stages. The plot highlights the effect of contact resistance on delay scaling and the importance of continued reduction of contact resistance. Potential improvements include full copper contacts, self-aligned contacts [11], and aggressive contact aspect ratio scaling, with the latter two options representing a trade-off in contact R vs. C scaling.

Figure 11 shows the power-performance trends for the ring-oscillator with 10-track load across technology nodes corresponding to the years 2012-2020. The nominal V_{DD} trajectory shown by the dashed line reiterates the Figure 8 conclusion that scaling reverses beyond 14nm at 2016 due to limits from interconnect and FEOL RC while alternate trajectories with higher V_{DD} would lead to improved performance at the cost of reduced power scaling.

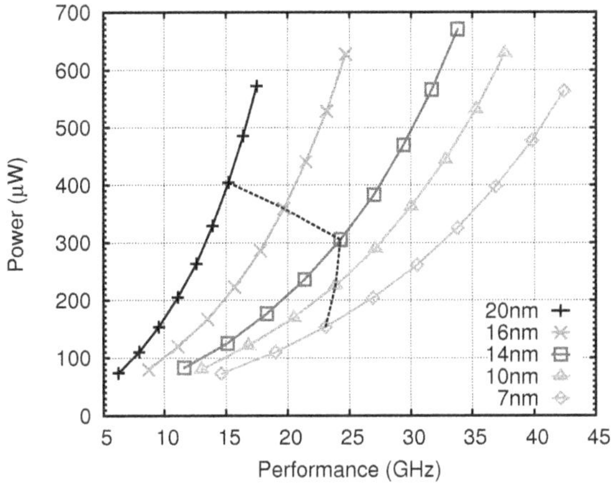

Figure 11. Power vs. Performance of INV RO with 10-track wire load from 2012-2020. The dash-line shows the power performance trajectory with the nominal V_{DD} at that technology node.

When considering conclusions such as in Figure 9, it should be emphasized that this present work considers only silicon-based FinFETs through the 7nm node. Device scaling options such as high mobility channels [16] and/or gate-all-around (GAA) nanowires [18] hold the potential to improve device scaling in this time frame, and are obvious areas for future work as research clarifies these potential device paths.

Several studies have predicted that power scaling will continue to lag area scaling, resulting in increased power density. In many designs, power is already at critical levels, so an increase in power density can only be met by powering down increasing portions of the chip, a trend referred to as "dark silicon" [19]. Figure 12 compares normalized power density scaling to two power density factors derived from the 2011 ITRS. The first ITRS power density factor takes the SoC-level power for Consumer Portable SoC from Figure SYSD6 and assumes a constant die size for the SoC area scaling. The second ITRS power density factor multiplies the PIDS LSPT CV^2 values by a normalized clock frequency that is increasing at 4% CAGR as per the 2011 ORTC chapter to derive power, and then calculates power density by dividing this power number by the logic gate area w/overhead listed in ORTC table 3-C.

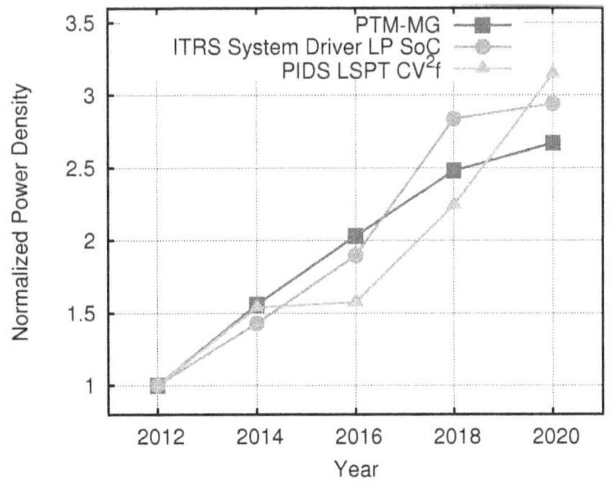

Figure 12. Power density vs. technology node comparing PTM-MG INV RO with 10-track wire load, ITRS system driver projections and ITRS PIDS projections.

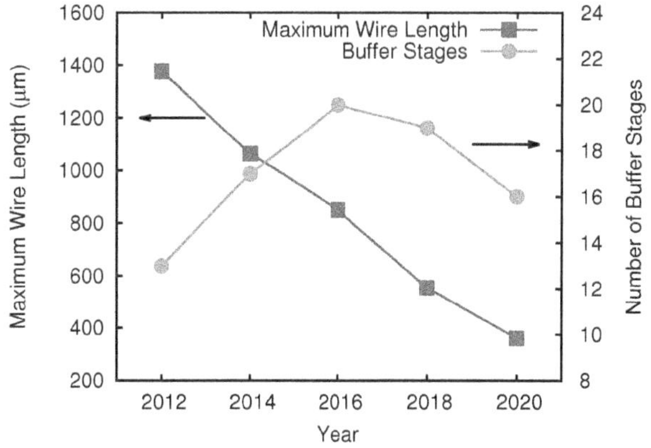

Figure 9. Maximum wire length that can be driven using buffers (left axis) and the number of buffer stages (right axis) at a target frequency vs. technology node. The buffer stages consist of INV (4X minimum strength) cells driving global wires.

Another important design characteristic is the maximum wire length, which is the longest wire that a set of buffers can drive in a given clock period. This metric is important to clock network design and other long distance communications (e.g., core-to-core IO, core-to-cache IO, etc.). Any wire lengths that exceed the maximum value incur additional power overhead, since flip flops have to be introduced to essentially "pipeline" the path. In order to benchmark the scaling of maximum wire length, we again use the 4% CAGR for clock frequency and simulate a number of INV strengths. We found that the rising interconnect parasitics (wire resistance and capacitance) combined with the increased intra-cell parasitics cause the maximum wire length to decrease by ~4.2X on average, whereas the clock frequency only increases by ~1.4X from 2012 to 2020. An example of this trend is shown in Figure 9 for a 4X-minimum-strength INV cell. Normalized to Metal1 track length, the maximum wire length translates to about 21,500 tracks in 2012 down to 15,000 tracks in 2020. The decreased

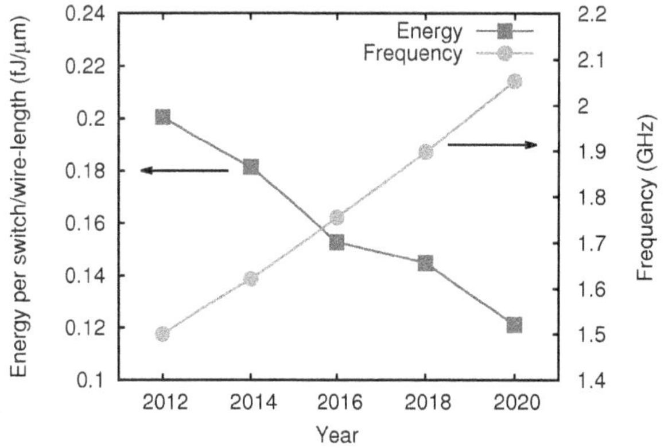

Figure 10. Energy per switch per unit wire length for buffers (left axis) and target frequency (right axis) vs. technology node.

performance scaling seen after the 2016 technology node is also reflected in Figure 9. The ideal number of buffer stages needed to obtain the maximum wire length is shown on the right-hand Y-axis of Figure 9. From 2012 to 2016, it is beneficial to spread the wire resistance and capacitance between a larger number of buffer stages. However, after 2016, inserting more repeaters actually degrades performance because of the increased intrinsic delay of the gates, themselves. In reality, we would optimize buffer strength for each technology node, but in order to do so would require accurate layout parasitic extraction, including stress effects, which is outside the scope of this work.

In the same set of "maximum wire length" experiments, we did observe a ~1.7X average decrease in "Energy-per-wire-length," as shown in Figure 10. Despite the increasing clock frequency, the repeater experiments showed a ~7% year-on-year scaling in energy/μm. This decrease can be attributed to decreasing supply voltage and the ~4.2X total decrease in maximum wire length.

4. CONCLUSIONS

Using new PTM-MG FinFET device models and incorporating the effect of both FEOL parasitic within standard cells and interconnect wire scaling trends from ITRS 2011 roadmap, we have evaluated key performance metrics for scaled CMOS technologies in the 2012-2020 time period. We highlight the difference between ITRS CV/I metric and actual ring oscillator simulations, which include FET geometry-based parasitics. We show the increasing penalty from the contact resistance, ultimately leading an anticipated decrease in local circuit speed after 2016 without further process improvements. Potential FET improvements in beyond 2016 include high mobility channel materials and/or silicon nanowires. Commensurately, contact resistance should be aggressively pursued with R/C tradeoffs from contact aspect ratio. We confirm a trend in power density scaling that highlights increasing dark silicon in future technology nodes, unless (for instance) V_{DD} can be scaled more aggressively than the ITRS roadmap. Investigating the interconnect, we show that the maximum distance serviceable by wire + repeater decreases, even on a per-track basis, although a general trend of reducing energy per switch and increasing clock frequency can be supported through 2020. We have highlighted areas for future work, and look forward to use of these new models by the technology community.

5. REFERENCES

[1] X. Li, Y. Cao, Z. Zhu, J. Song, et al. "Pathfinding for 22nm CMOS Designs using Predictive Technology Models", IEEE Custom Integrated Circuits Conf. (CICC), 2009, pp, 227-230.

[2] International Technology Roadmap for Semiconductors (ITRS), 2011 Edition, www.itrs.net.

[3] S. Sinha, G. Yeric, V. Chandra, B. Cline, Y. Cao; "Exploring Sub-20nm FinFET design with Predictive Technology Models," *in Design Automation Conference (DAC), 2012 49th ACM/EDA/IEEE*, June 2012.

[4] Predictive Technology Models, Arizona State University, Nanoscale Integration and Modeling Group. http://ptm.asu.edu/latest.html.

[5] Lan Wei; Jie Deng; Li-Wen Chang; Keunwoo Kim; Ching-Te Chuang; Wong, H.-S.P.; , "Selective Device Structure Scaling and Parasitics Engineering: A Way to Extend the Technology Roadmap," *Electron Devices, IEEE Transactions on* , vol.56, no.2, pp.312-320, Feb. 2009

[6] Tada, M.; Inoue, N.; Hayashi, Y.; , "Performance Modeling of Low- k /Cu Interconnects for 32-nm-Node and Beyond," *Electron Devices, IEEE Transactions on* , vol.56, no.9, pp.1852-1861, Sept. 2009

[7] M. V. Dunga, C.-H. Lin, D. D. Lu, W. Xiong, C. R. Cleavelin, P. Patruno, J.-R. Hwang, F.-L. Yang, A. M. Niknejad, and C. Hu, "BSIM-MG: A Versatile Multi-Gate FET Model for Mixed-Signal Design," in *2007 Symposium on VLSI Technology Digest of Technical Papers*, 2007, pp. 60–61.

[8] C.-C. Yeh et al., "A low operating power finfet transistor module featuring scaled gate stack and strain engineering for 32/28nm soc technology," *in IEDM*, dec. 2010, pp. 34.1.1 –34.1.4.

[9] C.-Y. Chang et al., "A 25-nm gate-length finfet transistor module for 32nm node," *in IEDM*, dec. 2009, pp. 1 –4.

[10] Scholze, A.; Furkay, S.; Seong-Dong Kim; Jain, S.; , "Exploring MOL design options for a 20nm CMOS technology using TCAD," *Simulation of Semiconductor Processes and Devices (SISPAD), 2011 International Conference on* , vol., no., pp.99-102, 8-10 Sept. 2011

[11] S.-C. Seo, L.F. Edge, S. Kanakasabapathy, M. Frank, et al., "Full Metal Gate with Borderless Contact for 14 nm and Beyond", *IEEE Symp. VLSI Technology (VLSIT)*, 2011, pp. 36-37.

[12] Topol, A., et.al, "Lower Resistance Scaled Metal Contacts to Silicide for Advanced CMOS," *VLSI Technology, 2006. Digest of Technical Papers. 2006 Symposium on*

[13] M. Jurczak et al., "Review of finfet technology," in *SOI Conference*, oct. 2009, pp. 1 –4.

[14] von Arnim, K., et al., "A Low-Power Multi-Gate FET CMOS Technology with 13.9ps Inverter Delay, Large-Scale Integrated High Performance Digital Circuits and SRAM," *VLSI Technology, 2007 IEEE Symposium on*, vol., no., pp.106-107, 12-14 June 2007.

[15] Venugopal, R.; et. al.; , "Design of CMOS Transistors to Maximize Circuit FOM Using a Coupled Process and Mixed-Mode Simulation Methodology," *Electron Device Letters, IEEE* , vol.27, no.10, pp.863-865, Oct. 2006.

[16] K. C. Saraswat, C. O. Chui, D. Kim, T. Krishnamohan and A. Pethe, "High Mobility Materials and Novel Device Structures for High Performance Nanoscale MOSFETs", IEEE Intl. Devices Meeting (IEDM), 2006, pp. 1-4.

[17] G. Yeric, "Technology Roadmaps and Low Power SoC Design", IEEE Intl. Electron Devices Meeting (IEDM), 2011, pp. 15.4.1- 15.4.4.

[18] C.-H. Park, S.-H. Lee, Y.-R. Kim, C.-K. Baek and Y.-H. Jeong, "Gate-All-Around Silicon Nanowire Field Effect Transistors", IEEE Intl. Conf. Nanotechnology, 2011, pp. 255-259.

[19] H. Esmaeilzadeh, E. Blem, R. St. Amant, K. Sankaralingam, and D. Burger, "Dark Silicon and the End of Multicore Scaling", IEEE Intl. Symp. Computer Architecture (ISCA), 2011, pp.365-376.

TSV Array Utilization in Low-Power 3D Clock Network Design*

Xin Zhao and Sung Kyu Lim
School of ECE, Georgia Institute of Technology, Atlanta, GA, USA
{xinzhao, limsk}@ece.gatech.edu

ABSTRACT

This paper focuses on low-power clock network design for 3D ICs, where through-silicon vias (TSVs) form a regular 2-dimensional array. This TSV array style is shown to be more manufacturable and practical than layouts with TSVs located at irregular spots. However, due to limited TSV resources in TSV arrays, TSV utilization in a 3D clock network significantly affects the final clock power. A straightforward extension on existing works for TSV arrays cannot guarantee power efficiency. Therefore, we develop a decision-tree-based clock synthesis (DTCS) method to generate low-power and reliable clock networks by efficiently exploring the entire solution space for the best TSV array utilization. Our DTCS method has been applied for both gate-level chip-scale 3D clock designs and block-level global clock designs. Experimental results show that our algorithm effectively finds close-to-optimal solutions for power efficiency with skew minimization in short runtime. Our DTCS method achieves up to 13.5% average power reduction with more than 50% fewer TSVs compared with the straightforward extension on the existing algorithm.

Categories and Subject Descriptors

B.7.2 [**Hardware, Integrated Circuit**]: Design Aids

Keywords

TSV array, 3D clock network design, low power

1. INTRODUCTION

Three-dimensional ICs (3D ICs) is one of the most promising technologies that enables higher integration and further miniaturization. However, through-silicon vias (TSVs), may cause reliability and cost issues that delay mainstream acceptance [8, 15]. TSVs can squeeze or stretch adjacent transistors and interconnects. This deformation may lead to mobility change and thus performance variation [11, 15]. It also causes mechanical reliability issues, causing open hole, short, or even crack. TSV-to-TSV and TSV-to-device coupling affect timing and signal integrity [9, 14, 16]. All these TSV-related issues require extra design efforts.

TSV array, defined as a group of TSVs placed in regular positions either in 1D or 2D grid fashion, is shown to be more manufacturable and practical to address the TSV-related reliability issues. As shown in Figure 1, multiple TSV arrays can be found in block-level design. It is also possible that a single 2D TSV array covers the entire layout area in gate-level design. Recent studies show that placing TSVs at any desired locations (= irregular positions) during placement [7] or routing [12] leads to shorter wirelength and better timing results compared with regular locations (= TSV arrays). However, this irregular placement may result in TSVs crowded in a certain region and cause problems in coupling [9, 13], timing variations [1, 16], and mechanical reliability [4, 11].

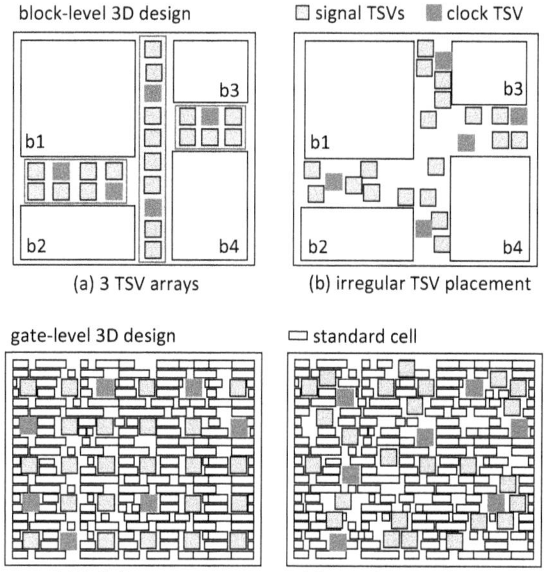

Figure 1: TSVs at regular locations (TSV arrays) vs. irregular locations in block-level and gate-level 3D designs.

Previous works on 3D clock synthesis focus on clock power minimization [5, 18], pre-bond testability [6, 17], and fault tolerance [10]. However, none of them consider TSV arrays, but all insert TSVs at any desired positions during routing. Simply extending existing work for TSV array design style cannot guarantee power efficiency. In the TSV array 3D design, the TSV count and locations have been determined BEFORE clock routing, where any TSV movement or additional TSV insertion is not allowed. Consequently, the clock network is limited to utilize these given TSVs. The final clock power is significantly affected by the TSV array utilization (how many and where). Thus, a practical question is *what is the optimal TSV array utilization for power efficient 3D clock network design with skew minimization and slew constrained?*

This paper addresses the 3D clock routing problem for TSV array utilization to construct low-power and reliable 3D clock networks. The contributions of this paper are summarized as follows.

*This research is funded by SRC ICSS Task 1836.075 and SRC CADTS Task 2239.001.

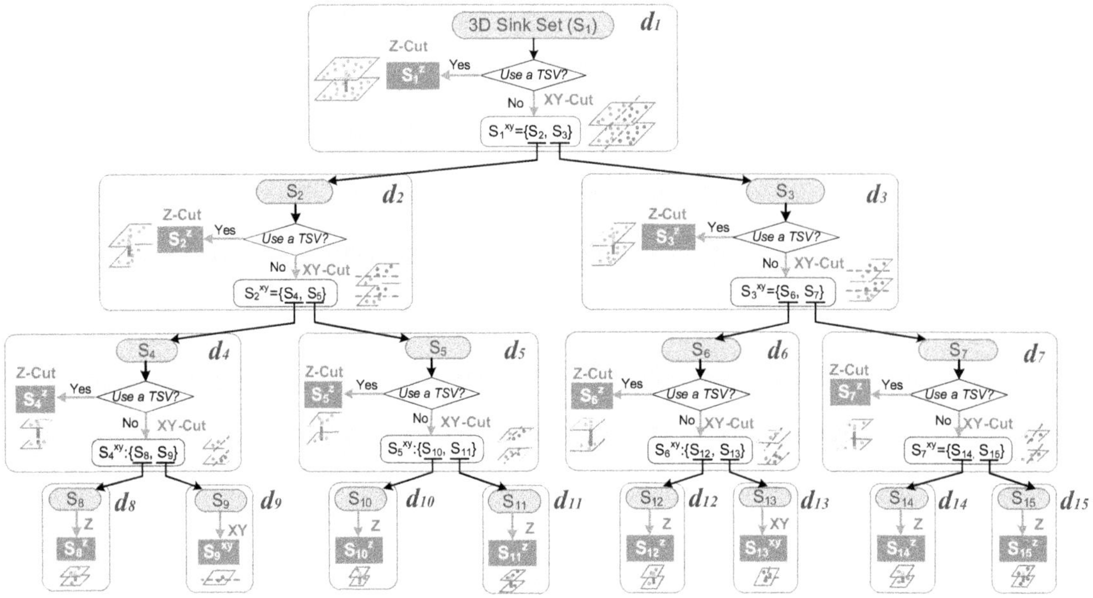

Figure 2: Illustration of our decision tree that shows the entire solution space of TSV array usage for low power. Each node (except leaf nodes) can choose between using one TSV (= Z-cut) or multiple TSVs (= XY-cut) in the array. Once the entire decision tree is built, we obtain different 3D clock trees by visiting all possible sink-to-root paths during our clock tree construction step.

(1) We present a novel method named decision-tree-based clock synthesis (DTCS) that generates small-skew and low-power clock trees by efficiently exploring the entire solution space for the best TSV array utilization. (2) We extend the existing 3D clock synthesis method for TSV array utilization. We observe that it cannot guarantee power efficient clock network designs for all cases. (3) The efficiency of our proposed DTCS method is verified for both gate-level chip-scale 3D clock designs and block-level global clock designs.Experimental results show that our algorithm effectively finds close-to-optimal solutions of power efficiency in short runtime. Our DTCS achieves significant reduction in power, wirelength, buffer, and TSV count.

2. CLOCK DESIGN METHODOLOGY FOR TSV ARRAYS

2.1 Problem Formulation

Given a set of clock sinks and an $M \times N (= K)$ TSV array for each die in a 3D stack, the goal of 3D Clock Routing with TSV Arrays (3D-CRTA) problem is to build a buffered 3D clock tree while using up to K TSVs for each die so that clock skew and power are minimized under clock slew constraint. If an additional constraint that the TSV bound $B < K$ is given on the total TSV count, the clock tree should not use more than B TSVs in each die.[1] In both cases, we do not add additional TSVs into the arrays.

2.2 Overview

We develop a **D**ecision-**T**ree-based **C**lock **S**ynthesis (**DTCS**) method to construct a low-power 3D clock tree with minimum skew while satisfying the slew and TSV count constraints. This method explores the entire solution space of TSV array utilization to find out a power-optimal design based on a decision tree. Meanwhile, the clock routing and buffering method [18] is integrated into DTCS method so that the clock topology is balanced and buffer insertion is performed to satisfy skew and slew constraints.

Our DTCS method consists of three steps. (1) *Decision Tree Construction*: We generate a decision tree that contains all the feasible solutions of TSV array utilization. The decision tree evaluates clock power overhead in various 3D clock trees under consideration. (2) *Clock Tree Construction*: We construct an initial 3D clock tree with lowest power under no TSV bound constraint. If no TSV count constraint is given, this initial tree becomes the final tree, and our algorithm terminates at this point. (3) *Clock Tree Refinement*: If the initial clock tree exceeds a TSV bound, we remove some TSVs by modifying the cut orientation of some decision nodes. Meanwhile, the resulting power and skew are kept minimal.

To verify the efficiency, our DTCS method focuses on finding out the optimal TSV utilization, but tries to keep using existing routing and buffer insertion techniques [18]. We use two-die stack to describe the proposed method and algorithms, which is then extended to handle more-than-two dies in Section 3.4.

2.3 Our Decision Tree

Our decision tree, as shown in Figure 2, is represented as a binary tree which visualizes the entire solution space for TSV array utilization in low-power 3D clock design. A decision node d_i (shown in a gray box) for a sink set S_i contains the following information: (1) cut orientation of S_i (z_i is 0 for XY-cut, 1 for Z-cut); (2) a sink set S_i^{XY} after an XY-cut is applied onto S_i, where $S_i^{XY} = \{S_{2i}, S_{2i+1}\}$ that contains two children nodes; (3) a sink set S_i^Z after a Z-cut is applied onto S_i; (4) power and TSV utilization for S_i. The root node d_1 contains the original 3D sink set, and the leaf nodes represent the subsets that have a unique solution.

A decision tree is built in a top-down recursive fashion, where we explore all feasible partitioning options, both Z-cut and XY-cut, for a given subset of clock sinks. XY-cut and Z-cut are two partitioning steps to define the clock abstract topology. An *XY-cut* partitions the sinks based on their X or Y coordinates, where the 3D sinks are flattened into 2D. The median value of the given sink set is the cutline for the XY-cut. An XY-cut results in multiple TSVs to

[1]A practical purpose of this TSV count constraint B is to reserve $K - B$ TSVs in each array for signal and/or P/G routing later.

connect descendant sink sets. A *Z-cut* separates the sinks die-wise so that the sinks in the same die are assigned to the same subset. A Z-cut immediately requires one TSV for connecting two subsets. This top-down partitioning and decision tree construction continue until the current sink set has a unique TSV utilization.

In our decision analysis, we call a sink set *unique* if it does not require any further partitioning and exploration of TSV utilization. We define a unique set based on the following conditions (shown in red rectangles in Figure 2). *Condition 1*: S_i is a 2D sink set (e.g., S_9 in Figure 2); *Condition 2*: S_i is a 3D sink set, but it requires a Z-cut due to the limited availability of TSVs in the TSV array; *Condition 3*: S_i^Z is obtained by applying a Z-cut on the 3D sink set S_i that does not satisfy Condition 2 (e.g., S_1^Z-S_7^Z in Figure 2). The second condition requires us to look ahead one more partitioning level down: when the bounding box of S_i contains only one TSV in the array or at least one of the two subsets (S_{2i} or S_{2i+1}) is a 3D set, but does not contain any TSV in the bounding box (e.g., S_8 in Figure 2), S_i must select S_i^Z (= Z-cut). Our decision tree exploration is terminated at a unique solution satisfying one of above three conditions. The leaf nodes satisfy Condition 1 or 2. A 3D clock tree can be obtained by traveling from the root node to the unique decision nodes.

The clock trees for all the unique sink sets are generated as follows. First, in abstract tree generation, we apply 2D-MMM algorithm [3] for a 2D sink set, or 3D-MMM algorithm [18] with TSV bound of 1 for a 3D sink set. We then perform clock tree embedding and buffering using classical DME method [2] or 3D slew-aware deferred-merge buffering and embedding (sDMBE) method [18] for zero skew under the Elmore delay model and satisfying the slew constraint.

2.4 Power Minimization with Decision Tree

Our DTCS methodology determines the power-optimal TSV array utilization based on a decision tree, where applying Z-cut vs. XY-cut at different partition levels leads to different power consumption and TSV array usage. Especially, applying an XY-cut on subset S_i ($i = 1, 2, \cdots, n$) results in two subsets S_{2i} and S_{2i+1}. The corresponding power estimation $P(S_i^{XY})$ is represented as follows:

$$P(S_i^{XY}) = P(S_{2i}) + P(S_{2i+1}) + P(S_{2i}, S_{2i+1}) \quad (1)$$

where $P(S_{2i})$ and $P(S_{2i+1})$ are clock power of subsets S_{2i} and S_{2i+1}, respectively. $P(S_{2i}, S_{2i+1})$ is the power of merging S_{2i} and S_{2i+1} that includes clock wires, buffers, and TSVs. Since the partitioning options for S_{2i} and S_{2i+1} are not determined yet, $P(S_i^{XY})$ can not be accurately determined at this point. Therefore, we need to first explore all possible cut orientations (XY-cut or Z-cut) for all descendant 3D subsets, which is done with our decision tree introduced in the previous Section 2.3. In addition, our clock tree construction algorithm to be discussed in Section 3.2 performs bottom-up traversal and accurately computes the power value for all nodes in the decision tree.

3. DECISION-TREE-BASED CLOCK SYNTHESIS ALGORITHMS

3.1 Decision Tree Construction Algorithm

Our decision tree construction algorithm explores all the feasible solutions of TSV array utilization by recursively dividing a given sink set into two subsets in a top-down fashion. Given a sink set S_i, we first add a decision tree node d_i. For a node with Index i, the left and right child node indices are $2i$ and $2i + 1$, respectively. The root node has index 1. Second, we explore two candidate partitioning styles: Z-cut or XY-cut. We apply a Z-cut on S_i, obtain S_i^Z, and estimate the power consumption ($P(S_i^Z)$). Specifically, this is done by performing 3D-MMM algorithm [18] on S_i with the TSV bound of 1, where we select a TSV in the array with the lowest wiring cost in the bounding box of S_i^Z. Then, we embed the tree and insert buffers using sDMBE algorithm [18]. Next, we apply an XY-cut on S_i and obtain two subsets S_{2i} and S_{2i+1}. The $P(S_i^{XY})$ for 3D sink set can not be estimated at this point because the cut orientations for all descendants are not determined yet. If S_i is unique satisfying Condition 1 or 2, the exploration is terminated, and node d_i obtains one solution only (i.e., S_i^Z or S_i^{XY} with estimated $P(S_i^Z)$ or $P(S_i^{XY})$ for 3D or 2D set, respectively); otherwise, we continue our top-down recursion on S_{2i} and S_{2i+1}.

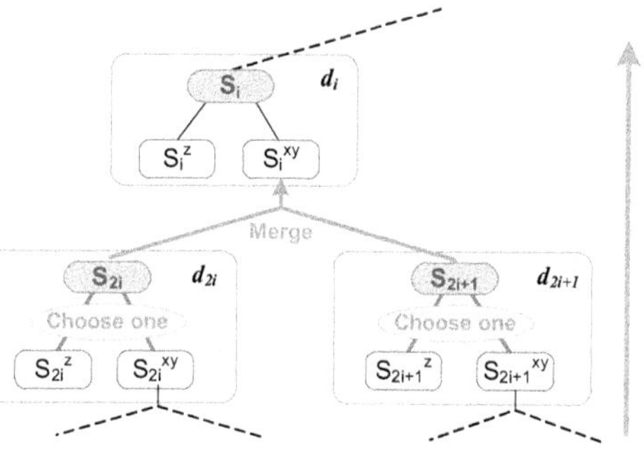

Figure 3: Bottom-up merging for node d_i, where we decide (1) clock tree and its power value for d_i for XY-cut (= S_i^{XY}), and (2) cut orientations for its children d_{2i} and d_{2i+1}.

3.2 Clock Tree Construction Algorithm

Our clock tree construction is to build an initial clock tree for power minimization using TSV arrays. If no TSV count constraint is specified, this initial tree becomes our final result. Otherwise, we perform clock tree refinement discussed in Section 3.3 to reduce the TSV count further down to satisfy the given TSV count constraint.

The input to our clock tree construction is a decision tree, where the cut orientations, clock trees, and their power values are determined for all leaf nodes. For internal nodes, only the clock trees and their power values are determined for Z-cuts. Thus, the goal is to determine the missing information: the clock tree and power value for XY-cut, and the cut orientation for all internal nodes. We accomplish this goal by visiting each internal decision node d_i in a bottom-up fashion and determining the follows: (1) A clock tree and its power value for d_i for XY-cut (note that the clock tree and power are already available for the Z-cut for d_i); (2) cut orientation for its children (not for d_i yet).

To compute the clock power for XY-cut ($P(S_i^{XY})$), we compare following 4 possible trees for d_i and choose the one with the lowest power (see Figure 3): (1) merge S_{2i}^Z and S_{2i+1}^Z; (2) merge S_{2i}^Z and S_{2i+1}^{XY}; (3) merge S_{2i}^{XY} and S_{2i+1}^Z; (4) merge S_{2i}^{XY} and S_{2i+1}^{XY}.

$$P(S_i^{XY}) = \min \left\{ \begin{array}{c} P(S_{2i}^Z) + P(S_{2i+1}^Z) + P(S_{2i}^Z, S_{2i+1}^Z) \\ P(S_{2i}^Z) + P(S_{2i+1}^{XY}) + P(S_{2i}^Z, S_{2i+1}^{XY}) \\ P(S_{2i}^{XY}) + P(S_{2i+1}^Z) + P(S_{2i}^{XY}, S_{2i+1}^Z) \\ P(S_{2i}^{XY}) + P(S_{2i+1}^{XY}) + P(S_{2i}^{XY}, S_{2i+1}^{XY}) \end{array} \right\} \quad (2)$$

We select the merging combination that results in the lowest power for S_i^{XY} and assign the corresponding cut orientation decisions (Z+Z, Z+XY, XY+Z, or XY+XY) to the children d_{2i} and d_{2i+1}. The reason we consider these 4 merging options instead of simply propagating the minimum power bottom-up from the leaf nodes is due to the third term: $P(S_{2i}, S_{2i+1})$. This power overhead is caused by the wires, buffers, and TSVs used to merge the two children, and depends on the first two power terms $P(S_{2i})$ and $P(S_{2i+1})$. Thus, we build all possible merging and then pick the one with minimum power for accurate power evaluation.

Note that if we visit an internal node d_i whose children are leaf nodes, the cut orientation for the leaf nodes has been fixed already. In this case, the clock tree and XY-cut power of d_i are only based on a single combination. For all other internal nodes that do not have leaf nodes as children, the XY-cut power values are determined based on the 4 merging combinations shown in Equation (2). Especially, the root node d_1 selects the cut orientation (z_1) that results in the minimum power from $P(S_1^Z)$ and $P(S_1^{XY})$.

3.3 Clock Tree Refinement Algorithm

Note that our clock tree construction algorithm builds a 3D clock tree that uses no more TSVs than what is available in the TSV arrays in each die. However, if the TSV count is further bounded by a constant B, this constraint may not be satisfied. Therefore, we develop Clock Tree Refinement algorithm to reduce the TSV count below B while maintaining low power. The basic idea is to choose a subset of nodes in the decision tree and convert their cut orientation from XY to Z. Note that this conversion may lead to a TSV count reduction because an XY-cut node uses more-than-1 TSVs while a Z-cut node uses a single TSV. We develop a binary-integer-linear-programming (BILP)-based algorithm for this purpose to choose an optimal set of decision nodes.

The input to our BILP method is a decision tree with the cut orientation for all the nodes fixed and uses more TSVs than allowed. Each decision node d_i contains a sink set S_i, a total number of TSVs (= t_i) used in d_i for either S_i^{XY} or S_i^Z (which is 1 in this case), and the cost (c_i) if we convert the cut orientation from XY-cut to Z-cut. We first present our binary non-linear integer programming (BNLIP) as follows:

Minimize

$$\sum_{i \in D} c_i \times z_i \tag{3}$$

Subject to

$$\sum_{i \in D} (t_i - 1) \times z_i \geq T - B \tag{4}$$

$$z_i \times \sum_{j \in child(i)} z_j = 0, \text{ for } i, j \in D \tag{5}$$

$$z_i = \{0, 1\}, \text{ for } i \in D \tag{6}$$

The binary variable z_i represents the cut orientation for d_i with 0 for XY-cut and 1 for Z-cut. We define D as the set of decision nodes that its XY-to-Z cut orientation conversion leads to a TSV count reduction. Note that for each $d_i \in D$, the $z_i = 0$ initially. The objective function Equation (3) is to minimize the total cost from cut orientation conversion. The cost c_i is the power overhead from the conversion and is expressed as:

$$c_i = P(S_i^Z) - P(S_i^{XY}) \tag{7}$$

where $P(S_i^Z)$ and $P(S_i^{XY})$ are the power of a Z-cut (i.e., using one TSV) or an XY-cut (i.e., using multiple TSVs) for S_i. Thus, our goal is to choose low-cost nodes for TSV count reduction.

Equation (4) ensures that the total TSV count should be no more than the given upper bound B, where T is the initial TSV count that exceeds the bound B. Once d_i is converted from XY-cut to Z-cut, S_i^Z is selected and one TSV is used instead. As a result, the total number of TSVs reduces by $t_i - 1$. Equation (5) shows that if node d_i is converted to Z-cut, the cut orientation of its children nodes z_j will not affect the overall power. This is because S_i^Z is a unique solution, and the decision tree is pruned at this node. Consequently, Equation (5) ensures that if $z_i = 1$, z_j for all descendant nodes are changed to 0.

Note that Equation (5) is not linear. For binary integer variables a and b_i, where $i = \{1, 2, .., n\}$, the quadratic constraint $a \times (b_1 + b_2 + ...b_n) = 0$ can be expressed as n linear constraints as $a + b_i \leq 1$ for $i = \{1, 2, .., n\}$. Thus, our binary integer linear programming (BILP) is formulated as Equations (3), (4), (8), and (6).

$$z_i + z_j \leq 1, \text{ for } i, j \in D, j \in child(i) \tag{8}$$

3.4 Extensions

Our methodology can be easily extended to handle more-than-two dies. A more-than-two-die 3D design can be decomposed into several pairs of two adjacent dies. Then, each two-die pair obtains a low-power 3D clock tree using our DTCS method. Finally, the 3D clock trees in those two-die pairs are connected through a single stacked TSV at the roots. Take a four-die stack (die-1 to die-4) as an example, we apply DTCS on die-1+die-2 and die-3+die-4, separately, and then connect the sources in the two 3D clock trees using a stacked TSV. As a result, the clock tree will use multiple TSVs mostly from two dies. We show 4-die stack clock tree designs in Section 4.3.

We extend the existing method 3D-MMM [18] to utilize TSV arrays. 3D-MMM is a top-down partition procedure, which assigns TSV bound for each subset based on the size, and determines Z-cut primarily based on the given TSV bound in current partitioning. It chooses Z-cut if the current TSV bound is 1. This method inserts TSVs at any desired location. Thus, we modify it to handle TSV arrays in the following way. We first introduce additional constraints for Z-cut decision. Our major goal is to ensure that each 3D sink set will be assigned at least one TSV in the array within its bounding box under small routing overhead. A Z-cut will be applied to the current 3D sink set if any of the following conditions is satisfied: 1) if the bounding box of the 3D sink set contains only one TSV in the array; or 2) look ahead one more partitioning level down and check the availability of TSVs in the array. We apply an XY-cut at current partitioning and check for the bounding box of each subset. If a subset contains 3D sinks but has no TSV in the bounding box, we apply a Z-cut at current iteration. Note that if the available TSV is outside the bounding box of a 3D sink set, a routing detour will be added. Thus, we use bounding box of a 3D sink set to determine the availability of the TSVs in the array.

4. EXPERIMENTAL RESULTS

4.1 Simulation Setting

In our experiments, we compare the following four algorithms to demonstrate the effectiveness of our DTCS method:

- **ALG-D**: our decision-tree-based clock synthesis (DTCS) algorithm that is specific for 3D clock routing with TSV arrays.
- **ALG-M**: extension on the existing method 3D-MMM [18] to utilize TSV arrays, which is described in Section 3.4.
- **ALG-F**: existing method 3D-MMM [18] that freely inserts TSVs at desired positions without the TSV array constraint.
- **ALG-X**: results in the optimal design. It exhaustively enumerates all feasible solutions of TSV array utilization, and selects the tree with minimum power satisfying the TSV bound.

We implemented the above four algorithms in C++/STL, and performed experiments on a 64-bit Linux server with Intel 2.5GHz CPU. 3D clock trees operate at 1GHz frequency and 1.1V supply voltage. We report clock power, slew, and skew from SPICE simulation, where skew and slew are constrained below 30ps and 100ps, respectively. Our DTCS (=ALG-D) solves the BILP problem using MOSEK. We use 45nm PTM model. The TSV parasitic resistance and capacitance are $35m\Omega$ and $15fF$[2], respectively.

[2]Our DTCS takes into account the TSV parasitics in power evaluation, and is efficient for any given TSV parasitics.

Table 1: Benchmark designs. Footprint area is in $mm \times mm$.

gate-level							
		2-Die			4-Die		
ckt	#sinks	footp area	TSV array	TSV bnd	area	TSV array	TSV bnd
ckt1	1089	3.7 × 3.7	32 × 32	120	2.6 × 2.6	20 × 20	130
ckt2	3204	5.3 × 5.4	36 × 36	250	3.8 × 3.7	28 × 28	180
ckt3	12404	12.3× 12.1	52 × 52	271	8.7 × 8.6	46 × 46	544
ckt4	1090	4.0 × 3.7	32 × 32	70	2.7 × 2.6	24 × 24	80
ckt5	12340	6.1 × 6.1	48 × 48	331	4.3 × 4.3	32 × 32	180
ckt6	17616	7.0 × 7.0	52 × 52	320	4.9 × 4.9	42 × 42	400
block-level, 2-Die only							
ckt	#2D blocks	footp area	#TSV blocks	TSV bnd			
ckt7	33	7.9×8.7	16	5	-	-	-
ckt8	49	3.9×4.8	39	5	-	-	-
ckt9	300	3.9×5.0	161	35	-	-	-
ckt10	51	5.7×6.6	30	7	-	-	-

We performed verifications on various types of benchmark circuits (see Table 1): 12 gate-level circuits with 1K to 17.6K clock sinks, where each of the ckts contains both 2-die and 4-die designs; and 4 block-level 2-die circuits (ckts 7-10). These benchmark circuits come from IWLS05, MCNC, GSRC, and ISPD09. For ckts 4-10, we perform 3D placement [7] for gate-level designs and 3D floorplanning for block-level designs. Both are Cadence Encounter-based 3D physical design tool chains. For ckt 1 (from ISPD09) and ckts 2-3 (from GSRC), since they only include the information of clock sinks for 2D ICs, we duplicate the circuits, scale the footprint area by $1/\sqrt{2}$ and $1/\sqrt{4}$ for 2-die and 4-die, respectively, then randomly assign sinks on different dies, and insert TSV arrays accordingly. The TSV bound constraint for each die is set as 5% to 20% of total TSV count in the arrays per die for gate-level and 10% of 2D block count for block-level design.

Two 3D clock trees generated by our ALG-D for the TSV arrays are shown in Figure 4, where Figures 4(a) and 4(b) are for block-level ckt8 using 8 TSVs, and Figures 4(c) and 4(d) are for gate-level circuit with 121 sinks in two-die stack with 11 TSVs. The top die contains a complete 2D tree, and the bottom die contains many subtrees. The utilized clock TSVs are highlighted by red circles.

4.2 Comparison with ALG-X

Table 2 shows comparisons between ALG-X and our ALG-D in power (mW) and runtime (s) for a 2-die circuit with 121 sinks under TSV bound of 7. The TSV array size increase from 4x4 to 7x7. Our ALG-D efficiently finds close-to-optimal solutions in short runtime for both designs with and without TSV bound. First, our ALG-D results in no more than 0.4% power increase compared with the optimal design obtained in ALG-X. This demonstrates the effectiveness of our clock tree construction and refinement in DTC-S. Second, our ALG-D finishes routing within 0.3 seconds. But, ALG-X may cost more than tens of hours depending on the TSV array size. The runtime of ALG-X is unaffordable for two reasons. The solution space tremendously expands in larger TSV array: ALG-X synthesizes 677 clock trees in 20 seconds for 4×4 TSV array, but 2648145 trees in 79813 seconds (>22 hours) for the 7×7 TSV array. Next, a larger circuit with more than 10k sinks requires longer runtime for each run of 3D clock routing. Lastly, as expected, we observe that our clock tree with a TSV bound utilizes fewer TSVs in the TSV array, and consumes more power and longer runtime than the design with no TSV bound constraint. This extra runtime comes from our BILP-based clock tree refinement.

4.3 Comparison with Related Work

Table 3 shows the comparisons of our ALG-D with ALG-F and ALG-M for gate-level and block-level circuits. Results are obtained under three TSV bound constraints, i.e., single, no TSV bound, and with a bound. Detailed clock routing results are also presented for the designs with no TSV bound.

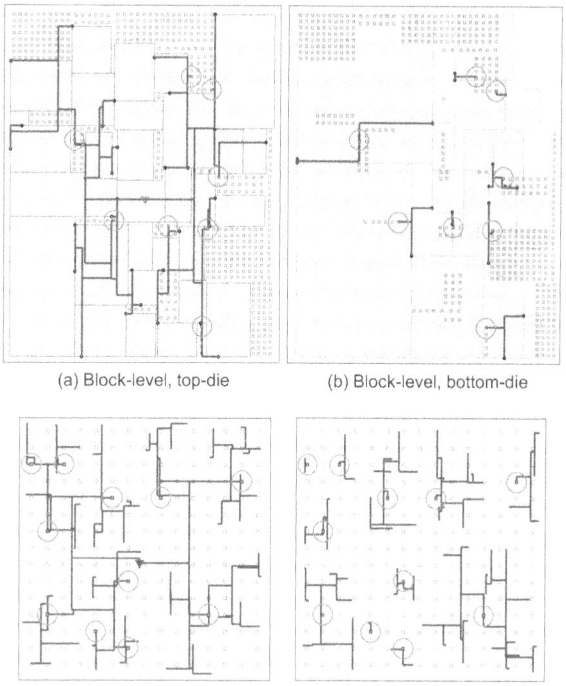

(a) Block-level, top-die (b) Block-level, bottom-die

(c) Gate-level, top-die (d) Gate-level, bottom-die

Figure 4: Clock trees generated by our ALG-D using TSV arrays. We show 3D clock trees for block-level ckt8 ((a) and (b)) and a gate-level ckt ((c) and (d)) in top- and bottom-die, respectively. TSV arrays are denoted as squares. Clock TSVs are shown in red circles.

Table 2: Comparison between ALG-X and our ALG-D in power (mW) and runtime (s).

	No TSV Bound				With TSV Bound			
	Power		Runtime		Power		Runtime	
TSV array	ALG X	**ALG D**	ALG X	**ALG D**	ALG X	**ALG D**	ALG X	**ALG D**
4x4	64.28	**64.34**	20	**0.12**	67.03	**67.27**	20	**0.16**
5x5	64.88	**64.93**	130	**0.11**	65.54	**65.54**	130	**0.17**
6x6	63.82	**63.85**	2357	**0.14**	64.53	**64.53**	2357	**0.18**
7x7	62.31	**62.31**	79813	**0.15**	63.48	**63.57**	79813	**0.21**
Inc.		**0.05%**				**0.13%**		

Limitations in ALG-M: As discussed early, TSV array utilization significantly affects the clock power consumption. The straightforward extension for TSV arrays (ALG-M) cannot efficiently reduce clock power when using many TSVs. When TSVs are irregularly inserted at any desired location (with no TSV array limitation), ALG-F [18] achieves 12% average power reduction compared with using single TSV in a 3D clock network. However, to support the TSV array design style, the ALG-M, a straightforward extension on existing method [18], did not efficiently reduce the power by using multiple TSVs. Instead, it may waste tens to hundreds of TSVs. The major reason is that the existing algorithm does not take into account the TSV array limitations, where using many TSVs may result in detour and extra TSV parasitic capacitance.

Our ALG-D versus ALG-M: Table 3 presents detailed results of wirelength, TSV count, buffer count, and power consumption

Table 3: **Comparison among ALG-F [18], ALG-M, and our ALG-D with no TSV bound and with TSV bound. Detailed results of power (mW), wirelength (μm), TSV count, buffer count, and skew (ps) in no TSV bound designs are shown.**

	Power	With No TSV Bound											With TSV Bound				
	single	Power			#TSVs		WireLength		#BUFs		skew		Power			#TSVs	
ckt_#die	TSV	alg-F	alg-M	**alg-D**	alg-M	**alg-D**	alg-M	**alg-D**	alg-M	**alg-D**	alg-M	**alg-D**	alg-F	alg-M	**alg-D**	alg-M	**alg-D**
ckt1_2die	183.5	160.0	178.9	**155.2**	454	**181**	273769	**246760**	1279	**1105**	26.3	**28.0**	169.0	175.5	**158.8**	120	**120**
ckt2_2die	609.5	553.1	595.3	**523.0**	818	**318**	748353	**651223**	4164	**3791**	28.9	**28.1**	581.6	580.0	**532.0**	250	**250**
ckt3_2die	1445.1	1273.4	1518.6	**1304.8**	1378	**526**	2348205	**1974101**	10027	**9034**	28.9	**26.1**	1372.9	1499.9	**1331.4**	271	**271**
ckt4_2die	189.7	176.5	192.1	**170.8**	220	**88**	388123	**361190**	732	**580**	25.8	**23.1**	188.0	187.6	**171.2**	70	**70**
ckt5_2die	539.4	495.3	554.7	**495.8**	662	**447**	1369811	**1252539**	1970	**1626**	24.9	**21.5**	522.5	552.7	**499.2**	331	**330**
ckt6_2die	970.5	893.3	1087.5	**877.2**	1908	**434**	2111595	**1758220**	3894	**3073**	25.8	**24.1**	924.6	1007.8	**882.7**	320	**320**
ckt1_4die	178.3	162.1	175.2	**151.4**	358	**159**	271870	**241801**	1212	**1041**	26.2	**18.7**	167.6	169.6	**153.7**	129	**130**
ckt2_4die	560.2	488.6	538.8	**466.0**	828	**360**	674576	**583082**	3712	**3298**	26.2	**9.7**	506.4	547.8	**485.1**	180	**180**
ckt3_4die	1413.7	1227.7	1386.6	**1217.6**	1664	**736**	2156004	**1884300**	9552	**8614**	28.6	**25.0**	1295.3	1380.3	**1231.4**	544	**544**
ckt4_4die	194.2	172.9	201.4	**177.6**	327	**105**	414419	**376671**	701	**592**	22.8	**22.1**	183.3	191.7	**179.0**	80	**80**
ckt5_4die	530.7	493.2	596.9	**504.3**	584	**259**	1493940	**1288392**	2085	**1659**	25.1	**12.6**	513.0	552.7	**506.8**	173	**177**
ckt6_4die	987.1	905.6	1100.1	**899.8**	2679	**579**	2070177	**1767473**	3879	**3111**	24.9	**21.2**	940.3	1002.4	**906.3**	392	**400**
ckt7_blk	30.5	24.4	29.4	**27.6**	11	**8**	80410	**76182**	96	**89**	11.8	**18.8**	26.5	30.1	**28.5**	5	**5**
ckt8_blk	19.0	17.1	19.9	**16.2**	11	**8**	53340	**45546**	67	**49**	10.5	**5.4**	18.4	18.8	**17.5**	5	**5**
ckt9_blk	58.2	46.3	55.7	**51.2**	94	**47**	142370	**135997**	189	**168**	9.3	**10.5**	51.8	54.6	**51.4**	35	**35**
ckt10_blk	26.5	21.0	28.1	**23.8**	20	**9**	76091	**67653**	92	**70**	22.1	**22.1**	24.8	27.9	**24.4**	7	**7**
Avg. Ratio	1.00	0.88	1.02	**0.89**	1.00	**0.45**	1.00	**0.89**	1.00	**0.84**			0.94	0.99	**0.90**		
% Improv. of alg-D w.r.t. alg-M:				**13.5%**		**55.1%**		**11.3%**		**15.7%**					**9.1%**		

for the comparisons between ALG-M and our ALG-D. Our observations are as follows. First, in the design with no TSV bound, our ALG-D efficiently minimizes clock power by utilizing TSV arrays. Compared with ALG-M, our ALG-D achieves 13.5%, 11.3%, 15.7% reduction in clock power, wirelength, and buffer count, respectively. In addition, our ALG-D uses 55.1% fewer TSVs than ALG-M. Second, in the design with bounded TSVs, our ALG-D achieves 9.1% power reduction on average. Third, clock skew is well controlled below 30 ps. The ALG-D runtime of the designs is in the range of 5 to 40 seconds based on the circuit size.

Our ALG-D versus ALG-F: Note that ALG-F [18] freely inserts TSVs at any desired position. Our ALG-D results in comparable or even lower power than ALG-F. In the design with no TSV bound, ALG-F and our ALG-D can achieve 12% and 11% average power reduction compared with the single-TSV solution, respectively. With a TSV bound, our ALG-D generates clock designs with 0.90 average power ratio, whereas, ALG-F obtains 0.94 average power ratio.

5. CONCLUSIONS

In this paper, for the first time, we studied low-power 3D clock design with TSV arrays. The TSV array design style is essential for reliable 3D ICs, but it significantly affects the power efficiency due to the constrained TSV locations. We presented a novel methodology, so called decision-tree-based clock synthesis (DTCS), to generate high-quality and low-power 3D clock trees by efficiently exploring the entire solution space for the best TSV array utilization. Our DTCS algorithm consists of decision tree construction, clock tree construction, and clock tree refinement. We demonstrated the effectiveness of our DTCS method for both chip-scale gate-level and block-level 3D IC designs. The following conclusions have been drawn. First, our DTCS algorithm obtains close-to-optimal solutions in short runtime, compared with the method of exhaustive searching TSV utilizations. Second, a straightforward extension on the existing algorithm for TSV arrays can not generate low-power 3D clock network, but waste many TSVs. Third, compared with the extension of existing algorithm, our DTCS algorithm achieves 13.5% and 9.1% power reduction in various given TSV bounds, uses 55.1% fewer TSVs, and obtains 11.3% shorter wirelength and 15.7% fewer buffers on average.

6. REFERENCES

[1] K. Athikulwongse, *et al.* Stress-Driven 3D-IC Placement with TSV Keep-Out Zone and Regularity Study. In *ICCAD*, pages 669–674, 2010.

[2] K. D. Boese and A. B. Kahng. Zero-Skew Clock Routing Trees with Minimum Wirelength. In *Proc. IEEE Intl. Conf. on ASIC*, pages 1.1.1–1.1.5, 1992.

[3] M. Jackson, *et al.* Clock Routing for High-Performance ICs. In *DAC*, pages 573–579, 1990.

[4] M. Jung, *et al.* TSV Stress-aware Full-Chip Mechanical Reliability Analysis and Optimization for 3D IC. In *DAC*, pages 188–193, 2011.

[5] T.-Y. Kim and T. Kim. Clock Tree Embedding for 3D ICs. In *ASPDAC*, pages 486–491, 2010.

[6] T.-Y. Kim and T. Kim. Clock Tree Synthesis with Pre-Bond Testability for 3D Stacked IC Designs. In *DAC*, pages 723–728, 2010.

[7] D. H. Kim, *et al.* A Study of Through-Silicon-Via Impact on the 3D Stacked IC Layout. In *ICCAD*, pages 674 –680, 2009.

[8] J. H. Lau. TSV Manufacturing Yield and Hidden Costs for 3D IC Integration. In *ECTC*, pages 1031–1042, 2010.

[9] C. Liu, *et al.* Signal Integrity Analysis and Optimization for 3D ICs. In *ISQED*, pages 42–49, 2011.

[10] C.-L. Lung, *et al.* Fault-Tolerant 3D Clock Network. In *DAC*, pages 645–651, 2011.

[11] A. Mercha, *et al.* Comprehensive Analysis of the Impact of Single and Arrays of Through Silicon Vias Induced Stress on High-k / Metal Gate CMOS Performance. In *IEDM*, pages 2.2.1–2.2.4, 2010.

[12] M. Pathak, *et al.* Through-Silicon-Via Management during 3D Physical Design: When to Add and How Many? In *ICCAD*, pages 387 –394, 2010.

[13] I. Savidis and E. Friedman. Closed-Form Expressions of 3-D Via Resistance, Inductance, and Capacitance. *IEEE Transactions on Electron Devices*, 56(9):1873–1881, 2009.

[14] T. Song, *et al.* Analysis of TSV-to-TSV Coupling with High-Impedance Termination in 3D ICs. In *ISQED*, pages 122–128, 2011.

[15] G. Van der Plas, *et al.* Design Issues and Considerations for Low-Cost 3-D TSV IC Technology. *JSSC*, 46(1):293–307, 2011.

[16] J.-S. Yang, *et al.* TSV Stress Aware Timing Analysis With Applications To 3D-IC Layout Optimization. In *DAC*, pages 803–806, 2010.

[17] X. Zhao, *et al.* Pre-bond Testable Low-Power Clock Tree Design for 3D Stacked ICs. In *ICCAD*, pages 184–190, 2009.

[18] X. Zhao, *et al.* Low-Power and Reliable Clock Network Design for Through-Silicon Via (TSV) Based 3D ICs. *IEEE Transactions on Components, Packaging and Manufacturing Technology*, 1(2):247–259, 2011.

Practically Scalable Floorplanning with Voltage Island Generation

Song Chen, Xiaolin Zhang, and Takeshi Yoshimura
Graduate School of Information, Production and Systems, Waseda University
Hibikino 2-7, Wakamatsu, Kitakyushu, Fukuoka, 808-0135, Japan
{chensong@aoni., zhang.xiaolin@fuji., t-yoshimura@}waseda.jp

ABSTRACT

In this paper, we propose a method of floorplanning with voltage island generation for system-on-chips (SoC) designs, which is deeply coupled with voltage island partitioning and voltage assignment, and has a good scalability. Floorplans with voltage islands are represented using nested Sequence Pairs, where the cores involved in the same voltage island consecutively appear in the sequences. Starting from a randomly generated initial floorplan, where each non chip-level core occupies an individual voltage island, we iteratively improve the solution by removing a core from the floorplan, then, inserting back the core by trying all the possible $O(n^2)$ *block positions*, which are defined as the combinations of insertion points and voltage islands. An almost linear algorithm is devised to roughly but quickly filter many worse *block positions*, in each iteration, considering fixed-outline constraints, wirelength, power, power/ground routing resources, and level shifters. Compared with the latest work, the proposed method shows 9.74% and 21.04% improvements respectively on the wirelength and power when all the blocks are hard. When soft blocks are involved, the proposed method shows 12.38% wirelength reduction and 3.10% more power saving with a penalty of 7.5% whitespace. Moreover, more than 20X speedups can be obtained for the large test cases.

Categories and Subject Descriptors

B.7 [**Integrated Circuits**]: Design Aids - Layout

General Terms

Algorithms, Design

Keywords

Sequence Pair, Floorplanning, Voltage Islands, Low Power

1. INTRODUCTION

The power issue is one of the important factors that slow down the scaling of the IC technologies. One of the most effective ways is lowering the supply voltages since the leakage power and dynamic power are, respectively, proportional to the supply voltage and its square. In a SoC design, the performance-critical components have to work at standard supply level, whereas the other components can work at lower supply levels. Accordingly, the power consumption of the circuits can be reduced without degrading the performance.

The voltage island technique is widely used in multiple supply voltage designs [8] [7]. A voltage island occupies a contiguous physical region and is powered by a supply voltage, independently from the chip-level one. Voltage islands make the static timing analysis, power routing, floorplanning more complicated. Particularly, the complexity grows significantly with the number of allowed islands. The voltage island generation traditionally involves three sequential sub-problems: Island partitioning, voltage assignment, and floorplanning. However, because the solutions to these sub-problems have strong dependence on one another, they have to be solved simultaneously to get optimal solutions.

The methods of floorplanning with voltage islands in the literature can be classified into two categories. The first category focuses on voltage islands, whose regions could be not contiguous, and uses penalty items in the objective functions to penalize the scattered placement of the cells in the voltage island [9, 11, 18]. Basically, this category pursues transforming the timing slacks of non-critical paths into the power saving, for cell-level design, by assigning low voltages to some blocks (cells). Lee et al. [9] proposed a voltage island aware floorplanning method, which included a dynamic programming-based voltage assignment under performance constraints followed by a power-network aware floorplanning. In this method, the voltage assignment and the floorplanning were done separately with a tight feedback loop governed by timing convergence. Yu et al. [18] integrated the voltage assignment and floorplanning under performance constraints. Ma et al. [11] proposed network-flow based methods to assign voltages under timing constraints and embedded the method into the floorplanning. Generally, voltage islands with non-contiguous region might cause a very complicated power-ground network design.

The second category focuses on generating voltage islands with contiguous regions and core-level floorplanning, in which each block (core) has a lowest supply voltage that ensures the performance requirements [5, 6, 12–14]. Since the maximum power saving could be achieved by operating every block at the lowest supply voltage, this group of works pursue a tradeoff between the power saving and the costs of voltage islands that are defined by the power-ground network resources, number of voltage islands, level shifters, etc [7]. Hu et al. [5] have considered island partitioning, voltage assignment, and floorplanning problem in SoC designs, simultaneously. Hung et al. [6] presented a voltage island assignment method considering thermal issue. In both works, however, the floorplanning included a chip-level floorplanning and a separated

ISLPED'12, July 30–August 1, 2012, Redondo Beach, CA, USA.

island-level floorplanning (only for area), which is neither effective nor efficient for exploring the solution space of voltage island design. Given floorplans, Mak and Chen [13] formulated the voltage assignment and island partitioning problem as a 0-1 integer linear program. In the approach, a few candidate floorplans were generated based on metrics like area and interconnect cost. There was not enough interaction with the floorplanning stage. D. Sengupta et al. [14] presented an application-driven voltage-island partitioning for low-power system-on-chip design, where the voltage assignment, island partitioning, and floorplanning were done sequentially. Given voltage island partitions, Lin et al [10] proposed a voltage island-driven floorplanning considering level shifter placement.

Qiang and Young et al. [12] proposed a core-based voltage island driven floorplanning method, which tightly integrated the island partitioning, voltage assignment and floorplanning. The floorplan configurations are represented by normalized Polish expression (NPE) [17]. However, the NPE can only represent slicing floorplans, which are exponentially less than the general floorplans [15] as the core number increases. On the other hand, the optimal voltage assignment and island partitioning were performed on every encountered floorplan configuration even with a small perturbation, which cause a poor scalability. Moreover, the voltage island number has to be fixed as an input. The method had to be run many times to explore the proper number of voltage islands.

In this paper, we propose a method of floorplanning with voltage-island generation for SoC designs, which is deeply coupled with island partitioning and voltage assignment and has a good scalability. Floorplans with voltage islands are represented using nested Sequence Pairs, where the cores involved in the same voltage island consecutively appear in the sequences. Starting from a randomly generated initial floorplan, where each non chip-level core occupies an individual voltage island, we iteratively improve the solution by removing a core from the floorplan, then, inserting back the core by trying all the possible $O(n^2)$ block positions, which are defined as the combinations of insertion points and voltage islands. In each iteration, the island partition, floorplan, voltage assignment are all changed slightly and an elaborated almost linear algorithm is used to evaluate all the block positions considering fixed-outline constraints, wirelength, power, power/ground resources, and level shifters, which partly cause a better scalability compared with [12]. The voltage island number is determined along with the floorplanning. Compared with the latest work [12], the proposed method shows 9.74% and 21.04% improvements respectively on the wirelength and power when all the blocks are hard, and 12.38% wirelength reduction and 3.10% more power saving with a penalty of 7.5% whitespace when including soft blocks. Moreover, more than 20X speedups can be obtained for some large test cases.

In the remainder of the paper, we will show the problem formulation and some important definitions in Section 2. Section 3 discusses the whole methodology flow and the objective functions used for the evaluations. An almost linear algorithm for evaluating block positions is discussed in Section 4. Experimental results and conclusions are, respectively, shown in Section 5 and 6.

2. PROBLEM FORMULATION

Let $S = \{b_i | 1 \leq i \leq n\}$ be a set of rectangular blocks, among which connections (nets) exist, and each block b_i has specified width w_i and height h_i, and a lowest operation voltage $v_{b_i}^{min}$. The floorplanning with voltage island generation is an assignment of a triple $(x_i, y_i, \mathscr{V}_k)$ to each block b_i such that

a. There is no overlapping between any two blocks;

b. All the blocks are packed into the specified chip region;

c. A set of voltage islands $VI(b_i.\mathscr{V} = \mathscr{V}_k \in VI)$ is generated;

d. The objective function, which is a linear combination of voltage island cost, wirelength, area cost, etc., is minimized.

Table 1 summarizes some notations used in this paper.

Table 1: Some Notations

$P[i]$ or p_i:	$1 \leq i \leq n$, *the i-th block in a block sequence P.*
P_i:	$0 \leq i \leq n$, *the prefix of sequence P ending at* $P[i]$, P_0 *means an empty sequence.*
P^R:	*reverse of sequence P.*
$L^P[b]$:	*position of block b in P, e.g.,* $L^P[P[i]] = i$.
$b_i.\mathscr{V}$:	*the voltage island to which* b_i *is assigned.*
$x_{(i,j)}$:	*x-distance of insertion point* (i, j).
$\mathscr{V}_j$:	*a voltage island.*
$\mathscr{V}_j.v$:	*voltage of* $\mathscr{V}_j$
$\mathscr{V}_j.ps$:	*the starting block of the substring in P for* $\mathscr{V}_j$.
$\mathscr{V}_j.pe$:	*the ending block of the substring in P for* $\mathscr{V}_j$.
$\mathscr{V}_0$:	*the dummy voltage island including all the blocks assigned the chip-level voltage.*
$\mathscr{V}_{\lvert VI \rvert+1}$:	*a new created voltage island.*
S_{CBP}:	*a set of candidate block positions.*
$S_{CBP}.worst$:	*the biggest costs of block positions in* S_{CBP}.

Given the total area A of the blocks and a maximum whitespace fraction γ, we construct a specified chip region, **fixed-outline**, with an **aspect ratio** $\lambda \geq 1$ [1] as follows:

$$W_0 = \sqrt{(1+\gamma)A\lambda}, \quad H_0 = \sqrt{(1+\gamma)A/\lambda}.$$

The aspect ratio λ is defined as $W_0/H_0 (> 1)$ in this paper.

The well-studied Sequence Pair (SP) [4] is used as floorplan representation. An SP is a pair of sequences of n elements representing a list of n blocks and can be transformed into a floorplan in $O(nloglogn)$ time [16]. It imposes the relationship between each pair of blocks as follows.

$(< \cdots b_i \cdots b_j \cdots >, < \cdots b_i \cdots b_j \cdots >) \Rightarrow b_i$ is left to b_j,
$(< \cdots b_i \cdots b_j \cdots >, < \cdots b_j \cdots b_i \cdots >) \Rightarrow b_i$ is above b_j.

Definition 1. *An **insertion point**, in an SP* (P, M) *of n blocks, means a two-tuples* (i, j), $0 \leq i, j \leq n$, *where i is the position immediately after* p_i *and j is the position immediately after* m_j. $i = 0$ *(or* $j = 0$*) means the position before the first block of the sequence.*

Hence, there are totally $(n+1)^2$ *insertion points* to insert a block into a given SP.

A **voltage island** (power domain) [7] [12], $\mathscr{V}_j$, is a rectangular physical region, which involves a set of blocks that can operate at the same voltage level and share the same power supply. The voltage of a voltage island $\mathscr{V}_j \in VI$ is defined as

$$\mathscr{V}_j.v = \max_{b_i \in \mathscr{V}_j} v_{b_i}^{min}. \quad (1)$$

A **floorplan with voltage islands** can be naturally represented by a *nested sequence pair* (nested SP), where the blocks belonging to the same voltage island consecutively appear in both sequences since a voltage island, as a whole, could be regarded as a block. Fig.1 shows an example of voltage islands.

When considering voltage islands, not all the insertion points are feasible. For all insertion points $(i, j), 1 \leq i, j < n$, satisfying the following condition are infeasible since a block cannot belong to two islands.

$$(p_i.\mathscr{V} = p_{i+1}.\mathscr{V}) \wedge (m_j.\mathscr{V} = m_{j+1}.\mathscr{V}) \wedge (p_i.\mathscr{V} \neq m_j.\mathscr{V})$$

For example, in Fig..1, the *insertion point* $(3, 8)$, that is, the position after the 3rd block, 9, in P and the position after the 8-th block,

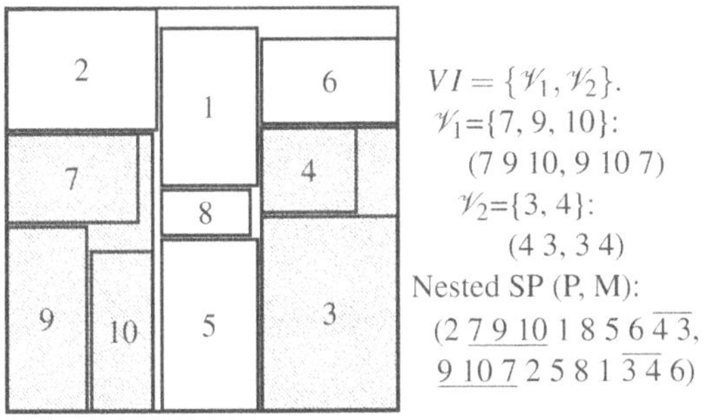

Figure 1: A Floorplan with islands represented by a nested SP

3, in M, is not feasible since $p_3.\mathcal{V} = p_4.\mathcal{V}$ the position in P causes a block involved in $\mathcal{V}_1$, whereas the position in sequence M causes a block involved in $\mathcal{V}_2$.

Given a nested SP, the set of candidate voltage islands for a feasible insertion point $(i,j), 0 \le i,j \le n$, $S_j^{\mathcal{V}}$, is defined as follows.

$$\begin{cases} \{m_j.\mathcal{V}\},\ if (m_j.\mathcal{V} = m_{j+1}.\mathcal{V} \neq \mathcal{V}_0). \\ \{m_j.\mathcal{V}\} \cup \{m_{j+1}.\mathcal{V}\} \cup \{\mathcal{V}_0\} \cup \{\mathcal{V}_{|VI|+1}\},\ otherwise. \end{cases} \quad (2)$$

$m_0.\mathcal{V}$ is defined as $\mathcal{V}_0$. We can see that $\forall j, |S_j^{\mathcal{V}}| \le 4$.

Definition 2. A ***block position*** *is defined as a triple* $(i,j,\mathcal{V}_k)$, *where* (i,j) *is an insertion point,* $\mathcal{V}_k \in S_j^{\mathcal{V}}$, *and the value set of i is defined as follows.*

$$\begin{cases} \{t | L^P[\mathcal{V}_k.ps] - 1 \le t \le L^P[\mathcal{V}_k.pe]\},\ if\ \mathcal{V}_k \neq \mathcal{V}_0 \wedge \mathcal{V}_k \neq \mathcal{V}_{|VI|+1}; \\ \{0,n\} \cup \{t | p_t.\mathcal{V} \neq p_{t+1}.\mathcal{V} \vee p_t.\mathcal{V} = p_{t+1}.\mathcal{V} = \mathcal{V}_0\},\ otherwise \end{cases} \quad (3)$$

In Fig.1, for the insertion point $(i,3)$, $S_3^{\mathcal{V}} = \{\mathcal{V}_1, \mathcal{V}_0, \mathcal{V}_3\}$, $\mathcal{V}_3$ is the probably generated new empty voltage island. And the possible *block positions* are as follows:

$(i,3,\mathcal{V}_1), i \in \{1,2,3,4\}$; $(i,3,\mathcal{V}_0), i \in \{0,1,10\} \cup \{k|4 \le k \le 8\}$; and $(i,3,\mathcal{V}_3), i \in \{0,1,10\} \cup \{k|4 \le k \le 8\}$.

Let n_{bp} be the number of *block positions*. we have $n_{bp} = O(n^2)$. In Section 4, we will show an algorithm to enumerate and evaluate all the *block positions* in almost linear time.

3. OVERVIEW

3.1 Methodology flow

In this work, we modify the perturbation method, Insertion-after-Remove (IAR) in [2], to deal with voltage island generation during the floorplanning. The IAR perturbation method has been successfully used in multi-layer floorplanning [3]. The solution space for the floorplanning with voltage island generation is much larger than the traditional floorplanning. The proposed method is expected to accelerate the searching greatly by skipping many worse solutions.

Figure 2 shows the flow. The initial solution is generated randomly and every block that can work at non-chip-level voltages occupies an individual voltage island. Then, the simulated annealing engine is invoked. The key step is "Select the 'best' *block position* $(j,k,\mathcal{V}_l)$ for b_i", where An empty island could be removed or a new island could be created. The detailed steps are as follows.

a. Compute the floorplan without b_i;

b. Select a set of *block positions* S_{CBP} for b_i by **roughly** (fast) evaluating all the *block positions*;

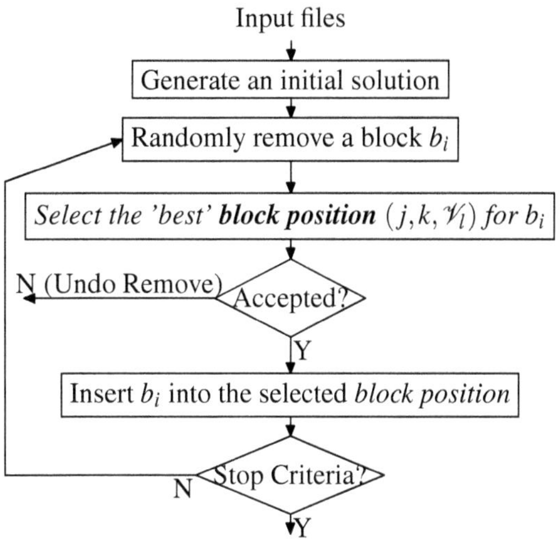

Figure 2: Optimization flow of Floorplanning.

c. Select from S_{CBP} the best *block position* $(j,k,\mathcal{V}_l)$ for b_i by **accurate** (slow) evaluations.

In step *b*, the block positions are evaluated by the linear combination of the area costs, the wirelength of the nets related to the removed block, and the voltage island costs. In this step, the area costs and the voltage island costs are calculated accurately. For acceleration, the wirelength is calculated, roughly, based on the floorplan without b_i. In step *c*, all the *block positions* in S_{CBP} will be evaluated, accurately, based on the whole floorplan and the best one will be chosen as the candidate *block position*.

3.2 Objective functions

Just getting power supply to the different voltage islands can be a challenge for designs which use multiple supplies since every voltage island requires an independent local power supply grid, and a low impedance power connection to supply pads. For flip-chip designs, this problem is mitigated somewhat because power can be delivered locally by a pad located in the power domain. For traditional chips, where the power has to come from the chip periphery, the system designer need to minimize the number of voltage regions to those that provide significant dynamic power and energy savings [7]. In this work, we define the costs of a voltage island $\mathcal{V}_k$ as the linear combination of the power, the half perimeter length (HPL) of the voltage islands, and the level shifter number.

$$VIC = \sum_{\mathcal{V}_k \in VI} (\delta \cdot POWER_{\mathcal{V}_k} + \theta \cdot HPL_{\mathcal{V}_k}) + \lambda \cdot LS\_num \quad (4)$$

With this definition, we can easily make tradeoff between the power and voltage island number. Because merging two voltage island **definitely reduces** the HPL, which could be an evaluation of the power-network resources [9], and **might reduce** the power saving (increase the power consumption) since one of the merged voltage island might get the supply voltage increased.

The power of each block b_i assigned to $\mathcal{V}_k$ is evaluated by

$$\mathcal{V}_k.v^2 \times w_i \times h_i. \quad (5)$$

The counting of level shifter will be discussed in Section 4.3.

The objective function is defined as follows.

$$COST = \alpha \cdot AC + \beta \cdot VIC + \gamma \cdot HPWL. \quad (6)$$

where AC is the area cost for punishing the violations of the fixed-outline constraints and HPWL is the half perimeter wirelength. They are calculated by the methods similar to those in [2].

4. EVALUATION OF BLOCK POSITIONS

In this section, in order to determine S_{CBP}, we discuss an almost linear algorithm to evaluate all the $n_{bp} = O(n^2)$ block positions with consideration of power/ground resources, level shifter, power, wirelength, and fixed-outline constraints.

Given an SP (P, M) of n blocks, let $LCS(P_i, M_j)$ be the block-width-weighted longest common subsequence [1] of P_i and M_j, and $lcs(P_i, M_j)$ be the weight sum of the elements in $LCS(P_i, M_j)$ [16]. We have the following definitions.

Definition 3. ***x-distance*** *[2] of an insertion point(i,j): If a block is inserted into (i,j), in the corresponding floorplan, the minimum distance from the left boundary of the block to that of the chip is defined as the x-distance of (i,j), denoted as $x_{(i,j)} = lcs(P_i, M_j)$.*

We may similarly define *y-distance*, $y_{(i,j)}$, which is the minimum distance to the chip bottom boundary, x^r*-distance*, $x^r_{(i,j)}$, which is the minimum distance to the chip right boundary, and y^r*-distance*, $y^r_{(i,j)}$, which is the minimum distance to the chip top boundary.

Let P^R denote the reverse of a sequence P and $(P^R)_i$ be the prefix of P^R. Similarly, we have

$y_{(i,j)} = lcs((P^R)_{n-i}, M_j)$; $x^r_{(i,j)} = lcs((P^R)_{n-i}, (M^R)_{n-j})$;

$y^r_{(i,j)} = lcs(P_i, (M^R)_{n-j})$.

When computing $y_{(i,j)}$ and $y^r_{(i,j)}$, we use block-height-weighted longest common subsequences.

4.1 Enumerating block positions

Let b_l be a block that can work at non-chip-level voltages. The selection of S_{CBP} for b_l is shown as the following algorithm.

Algorithm 1*: SELECT_S_CBP($P \backslash b_l$, $M \backslash b_l$, x, x^r, y, y^r)*

1. *for* $j \leftarrow 0$ *to* $n-1$ *do*
2.   *Compute* $S^{\mathcal{V}}_j$ *;//equation 2*
3.   *for* $\mathcal{V}_k \in S^{\mathcal{V}}_j$ *do*
4.   *if* $\mathcal{V}_k = \mathcal{V}_0$ *or* $\mathcal{V}_k = \mathcal{V}_{|VI|+1}$ *then*
5.   *for* $i \leftarrow 0$ *to* $n-1$ *do*
6.   *if* $(p_i.\mathcal{V} \neq p_{i+1}.\mathcal{V}$ *or* $p_l.\mathcal{V} = p_{l+1}.\mathcal{V} = \mathcal{V}_0)$ *then*
7.   *Calculate the costs* $(COST)$ *of* $(i, j, \mathcal{V}_k)$*;*
8.   *if* $|S_{CBP}| < k_c$ *then*
9. $S_{CBP} = S_{CBP} \cup \{(i, j, \mathcal{V}_k)\}$.
10.   *else if* $S_{CBP}.worst > COST$ *then*
11.   *replace the worst block position in* S_{CBP} *by* $(i, k, \mathcal{V}_k)$.
12.   *else // if in the step 4*
13.   *for* $j \leftarrow L^P[\mathcal{V}_k.ps]$ *to* $L^P[\mathcal{V}_k.pe]$ *do*
14.   *repeat steps 7 to 11.*

In Algorithm1, steps from 4 to 14 enumerate all the *block positions* defined in equation 3. The size of $|S_{CBP}|$ is a number between 2 and 25 in the experiments. In step 7, we have to calculate the area cost, voltage island cost, and wirelength when inserting b_l into *insertion point*$(i, j, \mathcal{V}_k)$. The chip dimension can be accurately calculated in constant time as follows.

$W = x_{(i,j)} + x^r_{(i,j)} + b_l.width$; $H = y_{(i,j)} + y^r_{(i,j)} + b_l.height$.

[1] For example '1345' is one of the longest common subsequences of two sequences '13245' and '12345'. If the elements of the sequences are weighted, the $LCS(S_1, S_2)$ is defined as the common subsequence with the maximum weight $lcs(S_1, S_2)$.

The area cost can be calculated based on W and H. The evaluation of wirelength can be done in constant time similar to that in [2]. Considering the power model defined Equation 5 and equation 1, the power of voltage island can be calculated in constant time as follows.

$[max(v^{min}_{b_l}, \mathcal{V}_k.v)]^2 \times (\sum_{b_i \in \mathcal{V}_k} w_i \cdot h_i + w_l \cdot h_l)$.

Here we discuss how to calculate HPL and level shifter number.

4.2 Calculating HPL for voltage islands

When the insertion point (i, j) locates inside a voltage island $\mathcal{V}_k \neq \mathcal{V}_0, \mathcal{V}_{|VI|+1}$, whose blocks constitute the substring $M_{[r,s]}$, $r \leq i \leq s$, the half perimeter length of the voltage island can be calculated in constant time as follows.

$$\mathcal{V}_k.width = x_{(i,j)} + x^r_{(i,j)} + b_l.width - (x_{(L^P[m_r]-1, r-1)} + x^r_{(L^P[m_s], s)}),$$

$$\mathcal{V}_k.height = y_{(i,j)} + y^r_{(i,j)} + b_l.width - (y_{(L^P[m_r]-1, r-1)} + y^r_{(L^P[m_s], s)}).$$

4.3 Counting level shifters

A level shifter is needed whenever a wire is connecting two blocks operating at different voltages [7]. We assume the sinks of a net involved in the same voltage island can share a level shifter.

The level shifters can be counted as two disjoint sets: the level shifters for the pins on the moved block b_l, $LS^{\mathcal{V}_k}_{mb}$, and the level shifters for the other pins, LS_{fb}.

4.3.1 Computation of $LS^{\mathcal{V}_k}_{mb}$

Let n_{vi} be the number of voltage islands and n_{pin} be the number of pins involved in the nets that have pins on the block b_l. Given a target voltage island $\mathcal{V}_k$, $LS^{\mathcal{V}_k}_{mb}$ can be calculated in $O(n_{pin})$ time and $O(n_{vi})$ space using a linked list and a flag array. Since there are at most $n_{vi} + 1$ possible voltage islands that b_l could be assigned to, we can pre-compute $LS^{\mathcal{V}_k}_{mb}$ for all the voltage islands in $O(n_{pin} * n_{vi})$ time and $O(n_{vi})$ space.

4.3.2 Computation of LS_{fb}

If b_l is assigned to a voltage island that has a higher voltage level than b_l, LS_{fb} can be pre-computed by scanning all the net pins, which takes approximately $O(n_{nt})$ time, where n_{nt} is the number of nets. However, if b_l is assigned to a voltage island that has a lower voltage level than b_l, the computation is a little complicated since the voltage level of the voltage island has to be raised.

In order to count the level shifters for the pins involved in a voltage island $\mathcal{V}_k$, we pre-compute a voltage distribution vector for $\mathcal{V}_k$. The vector records the voltage distribution of pins outside $\mathcal{V}_k$ but have connections to block pins in $\mathcal{V}_k$.

For example, if a net connects a source in the voltage island $\mathcal{V}_k$ with voltage 0.8 to four sinks, three in the island $\mathcal{V}_i$ with voltage 1.2 and one in the island $\mathcal{V}_j$ with voltage 1.0. Two level shifters will be inserted: one between $\mathcal{V}_k$ and $\mathcal{V}_i$, the other between $\mathcal{V}_k$ and $\mathcal{V}_j$. The voltage distribution vectors of $\mathcal{V}_k$, $\mathcal{V}_i$, $\mathcal{V}_j$, corresponding to 0.8v, 1.2v and 1.0v, will be $(0, \mathbf{1}, \mathbf{1})^k$, $(\mathbf{1}, 0, \mathbf{0})^i$, and $(\mathbf{1}, \mathbf{0}, 0)^j$, respectively. Then, we can compute the required level shifters for the island with any change of the voltages.

If the voltage of $\mathcal{V}_k$ is changed to 1.2. Based on the vector, we can sum all the numbers but the element corresponding to the voltage 1.2 to get the number of the required level shifter, 1. If $\mathcal{V}_i$ has a voltage of 0.8, then,the level shifter number should be 0.

The distribution vectors can be computed by scanning all the nets. The complexity is also approximately $O(n_{nt})$ time. Since the number of voltage levels is very small (in the experiments, just 4 or 5) and can be regarded as a constant, the calculation of level shifter numbers for voltage islands can be done in constant time

Table 2: **Wirelength overhead of voltage islands: $\gamma = 15\%$, $\lambda = 1.0$, and all blocks are hard; all the data are based on 20 runs.**

Circuits	IARFP [2]			Ours (with voltage islands)								
	#succ	wire	time(s)	#succ	wire	power	#p_s(%)	#p_w(%)	#HPL	#HPL_u(%)	#vi	time(s)
ami33	100	65696 (-25.77%)	0.86	60.00	88503	1913102	26.48	99.00	2849	49.36	5.4	1.16
ami49	100	935375 (-28.46%)	1.13	70.00	1307548	55176591	30.81	100	18123	46.4	7.3	1.60
n10	65	39403 (-5.71%)	0.15	30.00	41791	303706	39.11	99.15	1547	79.43	4.5	0.19
n30	90	107377 (-10.96%)	0.32	65.00	120589	513325	25.71	97.34	1156	49.08	4.4	0.36
n50	100	132690 (-16.54%)	1.00	95.00	158989	312654	30.02	95.42	1403	43.66	6.3	1.48
n100	100	206145 (-16.70%)	2.69	95.00	247463	298927	25.99	88.2	1075	26.1	4.9	6.17
n200	100	372099 (-12.95%)	7.88	100.00	427454	298651	24.45	79.73	1100	18.62	5.7	18.36
n300	100	517460 (-13.90%)	17.45	100.00	601010	456163	25.78	81.73	1512	16.41	6.9	39.64
avg.		-16.37%	1		1		28.54%	92.59%		41.13%		+36%

given the voltage distribution vectors. In shortly, with $LS_{mb}^{\mathcal{V}_k}$ and the distribution vectors for voltage islands, the level shifter number can be calculated in constant time for every block position.

4.4 Complexity analysis

In summary, counting level shifter numbers for all the possible block positions, can be done in $O(n_{pin} * n_{vi}) + O(n_{nt}) + O(n_{bp})$ time, where n is the number of blocks. In the experiments, for all the test cases, n_{pin} is less than 50 and could be regarded as a constant, n_{vi} is at most the number of blocks, and n_{nt} is much smaller than $O(n^2)$. As a consequence, the total complexity for counting level shifters is $O(n^2)$.

Because the calculation of the voltage island costs in step 7 can be finished in constant time and $|S_j^{\mathcal{V}}| \leq 4$, the complexity of Algorithm 1 is $O(n^2)$ which is linear with the number of block positions since $n_{bp} = O(n^2)$. The $O(n_{bp})$ distances $x_{(i,j)}$, $y_{(i,j)}$, $x^r_{(i,j)}$, and $y^r_{(i,j)}$ can be computed in $O(n_{bp})$ time [2]. The evaluation of the $O(n_{bp})$ block positions can be done linearly, accordingly.

5. EXPERIMENTAL RESULTS

The proposed method has been implemented in C language and run on a Linux workstation (2.6GHz, 8GB RAM). The MCNC benchmarks: ami33 and ami49, and GSRC benchmarks: n10, n30, n50, n100, n200, and n300, are used in the experiments. The parameters for simulated annealing are as follows: start temperature 200000, end temperature 100, cooling ratio 0.90, and the number of iterations in each temperature is set by the 1.2 times of the block number. We will compare the results with the latest work [12], which also dealt with voltage island with contiguous regions. The comparisons with the other two recent works [11] [9] are not made because their methods focused on cell-level combinational paths and generated voltage islands without contiguous regions.

In Table 3, we compare the proposed method with the method in [12] under the floorplanning with hard blocks. The initial voltage information are from [12]. All the blocks are set hard when running the program of [12]. The program is run with the voltage island number from 1 to 6 and the solutions with the lowest power are shown in the Table 3. In the proposed method, we also extract the voltage information from the test cases and set the white space percentage by 15% for fixed-outline floorplanning. The results show that the proposed method can improve the wirelength and power by 9.74% and 21.04%, respectively, and, at the same time, can get around 20X speedup.

In Table 4, we compare the proposed method with the method in [12] under the floorplanning with soft blocks. The data for [12] were ran on a Sun Blade 2500 with 1.6GHz CPU and 2GB RAM. In small test cases n10 and n30, the method in [12] has better wirelength. Totally, the proposed method increases power saving by 3.10% and reduce wirelength by 12.38% with a penalty of 7.5% whitespace. The average voltage island number is also smaller and the runtime is reduced 60% on average (normalized to the same CPU frequency) without consideration of level shifters.

Because the detailed power/area model of the level shifters used in [12] is not available, we cannot make a direct comparison. Here, we compare the running time of the two methods when the level shifter overhead is considered. Generally, counting level shifters is timing-consuming. In [12], the authors showed 200%-400% increments of the running times for considering level shifters. The running time of n300 shown in [12] is 3321 seconds, whereas the running time is 912.4 seconds (shown in Table 4) without considering level shifters. In the proposed method, we devised a very fast method to count level shifters. Table 4 shows only 32.5% increment of running time on average (The number of level shifters is reduced by 36.92%). For test case n300, the proposed method shows 40X speedup (normalized to the same CPU frequency).

Let P_{max} be the total power with all the blocks operating at the chip-level voltage (the highest one), P_{min} the total power with all the blocks operating at the permitted lowest voltages, and HPL_{max} the sum of half perimeters of all the blocks that can work at non-chip-level voltages. The columns of the tables are defined as follows.

#p_s = $(P_{max} - P)/P_{max}$, where , P is the obtained power;

#p_w = $(P_{max} - P)/(P_{max} - P_{min})$;

#HPL_u = HPL/HPL_{max}, HPL is the calculated HPL;

#vi: the number of voltage islands;

#wire: half perimeter wirelength;

#ws: white space percentage;

#HPL: the obtained HPL sum of the voltage islands.

For fair comparisons, in the above experiments, all the I/O pins of the circuits are removed as done in [12] and the resultant one-pin nets will be ignored. In the following experiments, the I/O pins are included as usual, and we randomly assign low voltages (1.0v, 1.1v, 1.2v, 1.3v, 1.5v) to around 50% blocks as voltage inputs. The chip-level voltage is set at 1.8v.

In Table 2, we also show some results of a traditional floorplanning method IARFP [2] for reference, the data are averages of 20 runs. Because of the constraints from voltage islands, the wirelength overhead is around 16.37% and the success rates in some test cases are lower than those of the traditional floorplanner. The results show that the power saving is 92.59% of the maximum power saving that can be obtained by assigning to the blocks their lowest voltages. The HPL is less than half of HPL_{max} in most cases. By adjusting the balance factors in the equation 4, we can easily make tradeoff between the voltage islands and power saving.

Table 3: **Comparison with [12]: All blocks are hard, $\gamma = 15\%$, $\lambda = 1.0$. (#ws: white space percentage)**

Circuits	[12]					Ours				
	#vi	wire	power	#ws	time	#vi	wire	power	#ws	time
n10	6	13608	285083	10.46	0.61	6	14577	277862	8.41	0.48
n30	6	44279	220076	15.33	7.03	8	41503	219970	13.65	0.53
n50	6	92927	191360	21.48	20.53	5	85756	191360	10.67	1.79
n100	5	167524	234663	19.92	81.09	4	143772	215840	12.90	11.53
n200	6	384693	258434	25.74	355.82	3	272373	235237	11.67	14.82
n300	5	585323	414949	20.78	618.93	5	459382	308072	11.95	28.31
avg.	1	1	1	19%		-8.8%	-9.74%	-21.04%	12%	-94.70%

Table 4: Comparison with [12]: All blocks are soft, $\gamma = 10\%$, $\lambda = 1.0$. (#ws: white space percentage, #ls: number of level shifters)

Circuits	[12]					Ours (w/o LS)						Ours (with LS)					
	#vi	wire	power	#ws	time	#vi	wire	power	#ls	#ws	time	#vi	wire	power	#ls	#ws	time
n10	6	13200	267263	0.46	0.973	4	14315	267263	43	2.91	0.12	4	13915	267263	43	4.73	0.204
n30	6	37540	209293	0.79	10.24	6	40176	206101	80	9.10	0.45	5	38961	206101	80	4.77	0.825
n50	6	88150	191360	0.76	26.87	3	86191	191360	136	7.35	1.23	6	90653	224885	125	8.90	1.37
n100	5	137250	227722	0.89	103.7	5	148690	208065	398	9.1	5.48	5	134167	213855	293	8.35	6.54
n200	6	307060	254794	1.02	487.1	5	236756	258242	597	9.1	20.97	8	269036	241541	546	9.10	26.95
n300	6	507730	342367	3.43	912.4	7	336976	408229	894	9.1	39.64	5	409615	343083	142	9.10	51.80
avg.	1	1	1	1.23%			-12.38%	-3.10%	1	7.49%			-12.34%	0.25%	-36.92%	7.49%	

6. CONCLUSIONS

In this paper, we proposed a method of floorplanning with voltage island generation for SoC designs. Floorplanns with voltage islands are represented by nested Sequence Pairs, where the cores involved in the same voltage island appear consecutively in the sequences. In the iteration of simulated annealing, all the $O(n^2)$ possible block positions can be evaluated in $O(n^2)$ time by the elaborated algorithm with consideration of power, wirelength, level shifters, etc. Compared with the latest work, the proposed method shows 21.04% and 9.74% improvement respectively on the power and wirelength when all the blocks are hard, and 12.38% wirelength reduction and 3.10% more power saving with a penalty of 7.5% whitespace when including soft blocks. In both cases, more than 20X speedup can be obtained for large test cases.

7. REFERENCES

[1] S. N. Adya and I. L. Markov. Fixed-outline floorplanning: Enabling hierarchical design. *Very Large Scale Integration (VLSI) systems, IEEE Trans. On.*, 11(6):1120–1135, 2003.

[2] S. Chen and T. Yoshimura. Fixed-outline floorplanning: Block-position enumeration and a new method for calculating area costs. *IEEE Tran. on CAD of Integrated Circuits and Systems*, 27(5):858 –871, may 2008.

[3] S. Chen and T. Yoshimura. Multi-layer floorplanning for stacked ics: Configuration number and fixed-outline constraints. *Integr., the VLSI journal*, 43(4):378–388, 2010.

[4] H.Murata, K.Fujiyoshi, S.Nakatake, and et al. VLSI module placement based on rectangle-packing by sequence-pair. *IEEE Trans. CAD of Integ. Cir. Sys.*, 15(12):1518–1524, 1996.

[5] J. Hu, Y. Shin, N. Dhanwada, and R. Marculescu. Architecting voltage islands in core-based system-on-a-chip designs. In *Proc. ISLPED*, pages 180–185, 2004.

[6] W.-L. Hung, G. M. Link, Y. Xie, N. Vijaykrishnan, N. Dhanwad, and J. Conner. Temperature-aware voltage islands architecting in system-on-chip design. In *Proc. of ICCD*, pages 689–696, 2005.

[7] M. Keating, D. Flynn, R. Aitken, A. Gibbons, and K. Shi. *Low Power Methodology Manual: For System-on-Chip Design*. Springer, 2007.

[8] D. E. Lackey, P. S. Zuchowski, and et al. Managing power and performance for system-on-chip designs using voltage islands. In *Proc. ICCAD*, pages 195–202, 2002.

[9] W.-P. Lee, H.-Y. Liu, and Y.-W. Chang. Voltage-island partitioning and floorplanning under timing constraints. *IEEE Trans. CAD of Integ. Cir. Sys.*, 28:690–702, 2009.

[10] J.-M. Lin and R. C. J. H. Wei-Yi Cheng, Chung-Lin Lee. Voltage island-driven floorplanning considering level shifter placement. In *Proc. of ASP-DAC*, pages 443–448, 2012.

[11] Q. Ma, Z. Qian, E. Young, and H. Zhou. Msv-driven floorplanning. *Computer-Aided Design of Integrated Circuits and Systems, IEEE Trans. on*, 30(8):1152 –1162, aug. 2011.

[12] Q. Ma and E. F. Y. Young. Multivoltage floorplan design. *IEEE Trans. CAD of Integ. Cir. Sys.*, 29(4):607–617, 2010.

[13] W.-K. Mak and J.-W. Chen. Voltage island generation under performance requirement for soc designs. In *Proc. of IEEE/ACM ASP-DAC*, pages 798–803, 2007.

[14] D. Sengupta and R. A. Saleh. Application-driven voltage-island partitioning for low-power system-on-chip design. *IEEE Trans. CAD Integ. Cir. Sys.*, 28:316–326, March 2009.

[15] Z. C. Shen and C. Chu. Bounds on the number of slicing, mosaic, and general floorplans. *IEEE Trans. on CAD of Integrated Circuits and Systems*, 22(10):1354 – 1361, 2003.

[16] X. Tang, R. Tian, and D. F. Wong. Fast evaluation of sequence pair in block placement by longest common subsequence computation. *IEEE Trans. on CAD of integrated circuits and systems*, 20(12):1406–1413, Dec. 2001.

[17] D. F. Wong and C. L. Liu. A new algorithm for floorplan design. In *Proc. ACM/IEEE DAC*, pages 101–107, 1986.

[18] B. YU, S. Dong, S. Chen, and S. Goto. Voltage and level-shifter assignment driven floorplanning. *IEICE Trans. on Fundamentals*, E92-A(12):2990–2997, 2009.

Thermal-Aware Sampling in Architectural Simulation

Ehsan K.Ardestani, Elnaz Ebrahimi, Gabriel Southern, and Jose Renau
Dept. of Computer Engineering
University of California Santa Cruz
{eka, elnaz, gsouther, renau}@soe.ucsc.edu

ABSTRACT

Thermal behavior of modern processors is a first-order design constraint. However, accurate estimation of thermal behavior is time consuming, and techniques for accelerating performance simulations often yield inaccurate results when directly applied to thermal simulation, or do not reduce the thermal computation at all. This paper is the first to propose thermal sampling techniques. It can be integrated with existing phase-based and statistical-based architectural simulator sampling. The resulting simulator can perform accurate performance, power, and thermal characterization at close to 30 MIPS, on average, instead of 5 MIPS for the fastest sampling technique without thermal-aware sampling.

Categories and Subject Descriptors

C.4 [**Computer Systems Organization**]: Performance of Systems—*Modeling Techniques*

Keywords

Thermal Simulation, Sampling, Architectural Simulation

1. INTRODUCTION

The development of processors relies heavily on architectural simulators which allow researchers and designers to evaluate proposed ideas early in the design phase. However, these simulations are orders of magnitude slower than native execution and it is often necessary to use sampling techniques to reduce simulation time. These sampling techniques, which work well for performance-only simulation, are often slow and/or inaccurate for simulations that model thermal characteristics. The thermal characteristics of a processor affect multiple aspects of processor and system design, including maximum operating frequency, leakage power, performance throttling, and cooling solutions. As a result, without thermal simulation, it is not possible to obtain representative performance or power models.

Sampling solves the problem of slow simulation by reducing the number of instructions that need to be simulated when evaluating a proposed idea. Many studies have been done to evaluate how many instructions need to be simulated in order to accurately estimate performance. Phase-based sampling and statistical-based sampling are the two basic methods of reducing the number of instructions that need to be simulated while still gathering accurate performance statistics [–].

Martínez *et al.* [ ] show that these sampling methods do not work well when directly applied to thermal simulation. Coskun, *et al.* [ ] developed a phase-based simulation methodology to apply sampling to performance and power computations. It provides accurate temperature results, but does not perform thermal sampling, which leaves thermal simulation as the main bottleneck. There is no accurate thermal simulation methodology based on statistical sampling. In this paper, we 1) extend sampling to the thermal domain for both phase-based and statistical sampling, and 2) for the first time develop an accurate statistical sampling based thermal simulation (*TASS*: Thermal-Aware Statistical Sampling). As a result, the thermal computation demand in the simulation is reduced by 25 times.

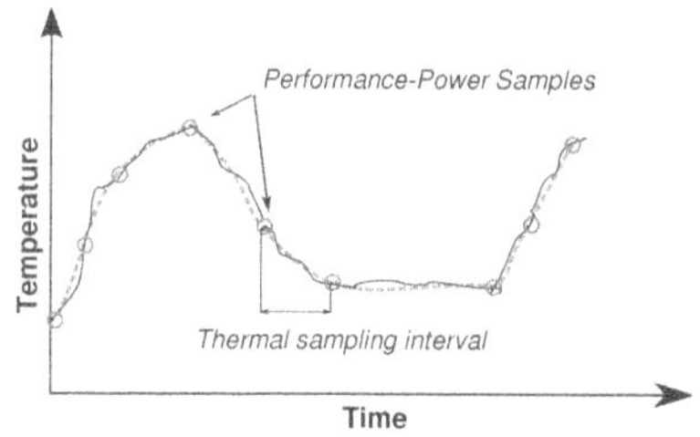

Figure 1: Thermal sampling

The key contribution of this paper is the proposal and development of practical implementations to perform thermal sampling. The proposed methodology leverages the slow state transition in thermal domain to address the sampling unfriendly feature of thermal domain: strong dependency on the previous states. Figure 1 shows a thermal profile for a generic application. The original thermal trace is shown with continuous lines. The performance and power sampling points are shown by circles. Each performance sample is typically thousands of instructions, while the sampling interval is tens of thousands to millions of instructions. To generate the temperature trace, previous works reconstruct the power trace and pass it to the thermal model. Consequently, with or without performance sampling, the number of thermal computations remains the same. While the previous works are oblivious to the sampling at performance stage, our proposed method integrates with it. The dashed line between each two adjacent circle in the figure is one thermal sampling interval. One round of thermal computation is enough in the proposed method to generate the temperature trace for a thermal sampling interval. This benefits the simulation in two connected ways; First, it makes it feasible to extend sampling to thermal by simplifying the integration of performance and thermal sampling. Second, it saves time by significantly reducing the number of thermal computations; We propose a framework to preserve the accuracy while applying this idea.

2. BACKGROUND

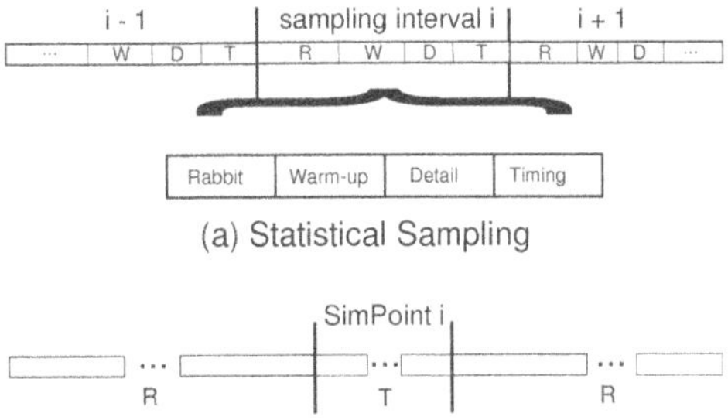

(a) Statistical Sampling

(b) Phase-based Sampling

Figure 2: Execution modes in sampling methods

2.1 Statistical-based Sampling

When sampling is used with performance modeling the goal is to produce an accurate average estimate. Some structures like caches and branch predictors require warm-up when the samples are not long enough [ ]. Figure 2a shows the sequence of execution in statistical sampling process. Each sampling interval is composed of 4 execution modes (or 3 as some methods do not implement *Rabbit* mode, all modes described in Table 1). The *Timing* is where the samples are gathered.

Phase	Description
***R**abbit*	Fast-forward emulation or native co-execution
***W**arm-up*	Memory and branch traces to maintain accurate state
***D**etail*	Cycle-accurate modeling, statistics are discarded
***T**iming*	Cycle-accurate timing modeling

Table 1: Simulation execution modes.

Statistical-based sampling encompasses power sampling as well, but is not commonly used for modeling thermal behavior. However, it has been used by Nookala, et al. for temperature-aware floorplanning [ ]. They did not evaluate the overall accuracy of their sampling methodology, which was only used to help guide a floorplanning optimization algorithm. Our evaluation shows that even an enhanced Nookala methodology that considers CPI variations over time (which we call *SS*) does not yield accurate results as extra thermal-aware adjustments are required for sampling parameters (See Section 3). Dynamic adaptations in the architecture can be captured by statistical sampling at runtime.

2.2 Phase-based Sampling

When the samples are long enough, such as 100M instructions used in SimPoint [ ], it is not necessary to perform microarchitectural warm-up to achieve accurate performance metrics. Hence, phase-based methods go through 2 execution modes, *Rabbit* and *Timing*. The samples, called SimPoints, are gathered during the *Timing* mode. Figure 2b shows the modes in the sampling process. The dotted lines in the figure indicates that the simulation points and the fast-forwarding intervals are much longer in phase-based sampling compared to the statistical sampling technique. Accordingly, the number of samples are much less.

Coskun, et al. [ ] proposed reconstructing the power trace using SimPoints [ ]. This is done in two simulation steps after a round of offline profiling. The first one gathers performance and dynamic power for the SimPoints. The second one reuses the generated dynamic power and performance results, and reconstructs the whole program execution. The program phases that were fast-forwarded are approximated with the most similar SimPoint. The second simulation step reconstructs the power trace and computes the temperature. This technique does not deploy thermal sampling, and requires modeling the temperature for the whole trace. We call this method *PS*.

The phase-based approach of obtaining power traces integrates very well with SimPoint when there is no need for processor adaptation based on temperature or power. However, whenever the processor performs any architectural adaptation based on power or temperature, its performance changes. Coskun, et al. [ ] proposes generating the performance and power SimPoints for each possible processor state. For example, if there are 4 DVFS levels, the SimPoints need to be generated 4 times. The second simulation phase picks the correct trace to build the correct runtime power and thermal profile.

3. THERMAL-AWARE SAMPLING

The existing sampling based thermal simulations perform sampling, and then reconstruct the power trace. Afterwards, a full thermal computation is performed on the reconstructed power to generate a detailed temperature trace. Hence, with or without performance sampling, the thermal stage performs the same amount of computations. The slower state transitions in temperature compared to power and performance suggests the potential for having fewer thermal computations. Previous works have recognized this, and most of them average the power for 10K to 1M cycles. Then the thermal computation is performed once for that timestep [–]. The timestep is set to be around an order of magnitude smaller than the thermal time constant (TC) to provide accurate transients, avoiding thermal computations for every cycle. Nevertheless, while sampling methods drastically reduce the time spent on performance simulation, the execution time for thermal simulation remains a limiting factor to improve the simulation speed.

3.1 TASS Method

We extend sampling to the thermal stage. The main challenge is that unlike performance sampling, temperature sampling requires a very long warmup. In other words, temperature has a strong dependency on previous state, which is at odds with the idea of sampling. However, as mentioned earlier, temperature state changes much more slowly than power and performance. This is leveraged to avoid excessive thermal computations. We start explaining the methodology with the statistical sampling technique. A similar concept is applicable to the phase-based technique which will be discussed later.

In general, precise estimation of temperature at time $T_1 = t$ depends on correct estimation of three parameters:

- temperature at time $T_0 = t - \delta$,
- power consumption for the $(T_0 : T_1)$ interval, and
- the length of the interval, δ.

For sampling, δ is the length of thermal interval. With longer δ, more distribution information will be skipped. Therefore, δ can be adjusted for appropriate trade off between speed and accuracy. We set the length of thermal interval equal to the performance sampling intervals ($TPr = 1$, explained later in this section). Note that thermal intervals shorter than performance intervals do not provide more accurate temperature trace. The reason is that smaller thermal intervals than the sampling interval do not contain more power distribution information. This is something that previous work has neglected, and as a result they use a constant timestep which is independent of the performance sampling interval.

We synchronize the beginning of thermal and performance intervals. The only temperature value computed for the thermal interval is located at the end of the interval. Note that the correct estimation of temperature at T_0 itself is interdependent on the previous power and δ values. Therefore, the main challenge is the correct estimation of *Power* and *Time* for each interval. However, these values are only available for the samples (T) rather than the whole interval. We use a weighted moving average to estimate power and time for a thermal interval, as formulated in Equation 1.

$$\Theta_i = \frac{\sum_{k=0}^{n-1} \alpha_k \times \theta_{i-k}}{\sum_{k=0}^{n-1} \alpha_k}, \alpha_k = \frac{1}{2^k} \tag{1}$$

Θ and θ stand for *estimated* and *measured* value respectively. Note that the estimated value will be used as the representative value for a whole interval (a full *RWDT* sequence). The measured value is the value gathered only during the sampling (*T* mode). n is the *history size*, which determines how many measured samples are used in calculation of the estimated value for the current interval.

3.1.1 Thermal-aware filtering for power

By estimating power values for the fast forwarded points in intervals, the power trace is reconstructed. In our experiments, we set $n = 7$ in Equation 1. We use the *measured* power from intervals i to $i-7$ to obtain an *estimated* power for the whole interval i. Averaging the power values also works as a filtering mechanism to smooth high frequency power spikes. We propose the filtering because power spikes in the samples affect the reconstructed trace, which can increase the error in the reconstructed temperature even more, given that temperature effects tend to stay longer due to the *TC* effect. Instead of applying a filter, we could increase the length of each sample (i.e. simulating more instructions for each sample). However, it would be at the cost of longer simulation time, as detailed timing simulation is slow. The simple filtering process relaxes the demand for longer sampling length, while still providing stable estimations. As the evaluation in Section 5 shows, the recommended performance-driven statistical-sampling parameters (*SS*) yields inaccurate results due to this spike sensitivity during reconstruction.

$$T_i = \sum_{k=1}^{i} \delta_k, \delta_k = \frac{Cyc_k}{ClkFreq}, Cyc_k = CPI_k \times Inst \tag{2}$$

3.1.2 Timeline Reconstruction

The length of sampling intervals are always a fixed number of instructions (Table 4). However, due to the time variant nature of the *CPI*, the actual length of each interval in terms of cycles varies. To estimate the length of each interval, Equation 1 is used with measured *CPI* values from last 7 intervals. Given the estimated *CPI* for interval i, we can estimate the number of cycles (Cyc) for the interval using Equation 2. The timeline of the execution is reconstructed by estimating the length of each interval (δ). Eventually the time at the end of each interval (T) is estimated as formulated in Equation 2.

3.1.3 Sampling Parameters

SMARTS [ ] specifies the length of each sample and the sampling intervals based on the observed variability in the program behavior (IPC). We use the same methodology. In addition, the coefficient of variability of the power samples can also be used to guide the parameter selection. Our observation is that longer samples need to be measured to provide stable temperature results. Table 4 shows the parameters.

Thermal phases are longer than performance phases [ ]. This implies using even longer thermal intervals than performance intervals. We define the natural value $TPr \geq 1$ to be ratio of the thermal to performance sampling interval. For example, $TPr = 2$ means that each thermal interval is as long as 2 performance intervals in terms of number of instructions. We run a set of experiments with different thermal sampling intervals, while performance sampling intervals stay the same. The results are shown in Figure 3. The y-axis shows the Root Mean Square Error (RMSE) of comparing the resultant thermal traces against the shortest thermal interval of 10K. The results confirm that as long as the thermal statistics are gathered at the end of thermal interval, where the temperature is estimated, longer thermal sampling intervals can be tolerated. As a results, $TPr > 1$ can be used to even further accelerate the simulation speed. Practically $TPr = 3$ still generates accurate results. This means that the thermal interval is around 15M cycles, which is longer than thermal time constant of the package we model. This is the first time that a thermal simulation can run with a timestep longer than *TC* and preserve accuracy.

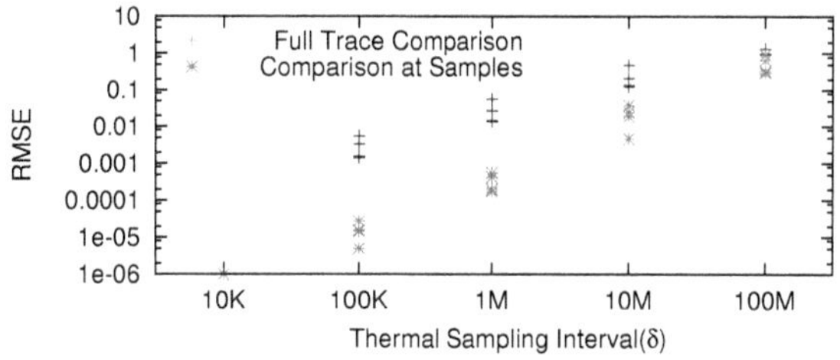

Figure 3: Accuracy of estimated temperature at thermal samples with different thermal sampling intervals

3.2 TAPS Method

As mentioned in Section 2.2, Coskun *et al.* [ ] proposed a power reuse method to reconstruct the power model for phase-based methods. The distance between samples in phase-based sampling is longer and variable (each sample is typically in order of 10M to 100M instructions and the distance between samples could be billions of instructions). Martínez *et al.* [ ] point to the validity of reporting the metrics only at samples as long as the initial temperature for each sample is set to the correct value. So fast-forwarding through the intermediate points and estimating the temperature for the end of the interval between samples (i.e. beginning of next sample) avoids multifold thermal computations, and speeds up the simulation as a result. For fast-forwarding, a faster but less accurate thermal simulation can be achieved by either performing thermal simulation on a coarser grain floorplan, or with a longer time step. We only perform the latter because this is enough to reduce the thermal simulation time to a small percentage of the total simulation time.

4. SIMULATION SETUP

Table 2 lists the methods we evaluate in this work. *SS* uses the same sampling parameters as suggested in [ ]. Both *PS* and *SS* methods perform thermal computation after reconstructing power trace with 100K cycle timestep.

Method	Description
Full	Full simulation, (no sampling)
PS	Phased-base sampling (Perf. + Power sampling)
TAPS	*PS* + (Thermal sampling)
SS	Statistical-based sampling (Perf. + Power sampling)
TASS	Thermal-aware SS, (Thermal sampling)

Table 2: Simulation methods evaluated in this work

For performance simulation, we use a modified version of SESC [ ] that uses QEMU [ ] as the functional emulator executing SPARC instructions. The simulator offers 4 different execution modes, explained in Table 1, to support sampling. We configure

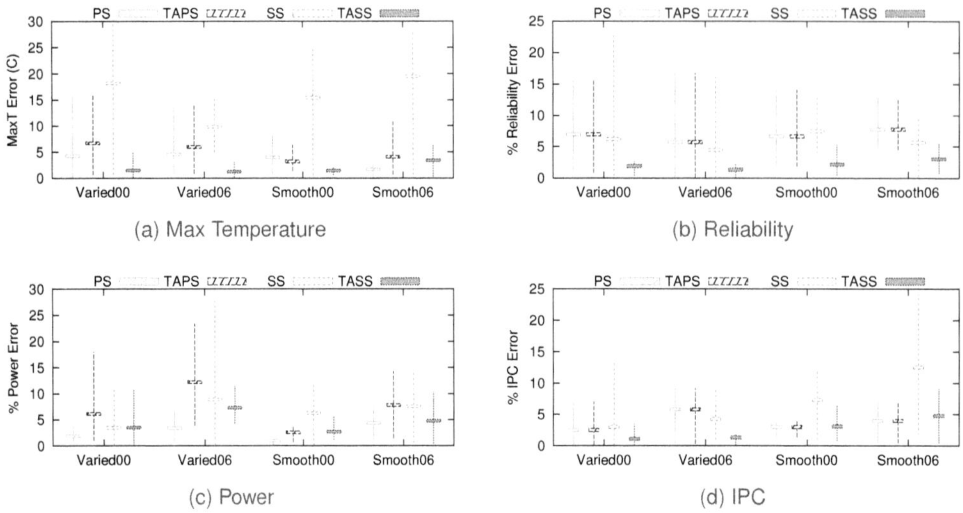

Figure 4: Average, minimum, and max error of different methods. ***TASS*** **outperforms the other methods.** ***SS*** **triggers the DTM method, and as a result has high IPC error, in addition to inaccuracy in estimating thermal metrics.**

SESC to pass activity counters to McPAT [ ] (every 100K instructions max) which we use for calculating power. We use a modified version of SESCTherm [ ] to scale leakage power consumption according to temperature and device properties. We simulate two processors: one is an Intel Nehalem-like high performance (HP) core, and the other is an AMD Bobcat-like low power (LP) mobile processor (Table 3). Table 4 shows the sampling and the thermal simulation parameters respectively. For Dynamic Thermal Management (DTM), we implement thermal throttling. When the chip temperature exceeds a threshold (363K and 348K for HP and LP configurations respectively), the processor gates the clock to prevent physical damage until the temperature drops under the defined threshold.

	Parameter	Value		Parameter	Value
HP	Freq	3.0 GHz	**LP**	Freq	1.6 GHz
	I$	32KB 2w (2c hit)		I$	32KB 2w (2c hit)
	D$	32KB 8w (3c hit)		D$	32KB 2w (2c hit)
	L2	1MB 16w (12c hit)		L2	512KB 4w (12c hit)
	Mem.	180 cyc		Mem.	90 cyc
	BPred.	Hybrid 76Kb mem		BPred.	Hybrid 38Kb mem
	Issue	4		Issue	2
	ROB	256		ROB	56
	IWin.	48		IWin.	20
	Reg(I/F)	128/128		Reg(I/F)	80/64
	Tech.	32 *nm*		Tech.	32 *nm*

Table 3: Architectural parameters

Method	Parameter
PS	BBV = 1e8
TAPS	MaxK = 10 [ ]
SS	R = 0, W = 997e4, D = 2e4, T = 1e4
TASS	R = 257e4, W = 250e4, D = 2e4, T = 7e4, History Size = 7, TPr = 2 unless mentioned otherwise

Table 4: Sampling parameters.

The benchmarks that we selected from CPU2000 and CPU2006 suites are shown in Table 5. We also use the same set of metrics presented in [ ] for the evaluation, namely maximum temperature, reliability (aggregated *EM*, *SM*, *TC*, *TDDB*, *NBTI*), and power. For power, we report the aggregation of leakage and dynamic power.

Suite	Category	benchmark
CPU2000	Varied	swim, gcc, mesa, facerec, lucas, bzip2
	Smooth	gzip, mgrid, applu, vpr, crafty, twolf
CPU2006	Varied	gcc, milc, dealII, mcf
	Smooth	perlbench, soplex, astar, povray, namd, h264ref

Table 5: Selected Benchmarks

5. EVALUATION

For each method we evaluate its accuracy, simulation speed and the impact of thermal triggered adaptations on the overall simulation speed. Thermal throttling and its impact on performance is also evaluated. The results of *LP* and *HP* processor configurations are averaged together, because they are statistically similar.

5.1 Accuracy

We have clustered the benchmarks in categories shown in Table 5. The accuracy of each benchmark is calculated in comparison with *Full* and average result for each category is reported. We also report the minimum and maximum error for each category.

For *SS*, the spikes adversely affect the instantaneous metrics such as *maxT*. *TASS* generates accurate *maxT* results which is indicative of the effectiveness of thermal-aware power filtering and reconstruction. The average metrics like *Power* or *Reliability* are less sensitive to spikes as expected. The *SS* results in very high IPC error. The reason is that temperature spikes trigger the thermal throttling DTM mechanism. Thermal throttling gates the processor clock to allow for it to cool down. This has performance impact. When we disable the thermal throttling, the IPC error of *SS* and *TASS* are similar. On average, *TAPS* reaches the same order of accuracy as *PS*, 6.7% vs. 4.7%. Both *TAPS* and *PS* methods could benefit from extracting shorter SimPoints, *e.g.*, 10M instruction. However, we only run the experiments with the recommended configuration. On average, across all the benchmarks and metrics, *TASS* outperforms other evaluated methods in terms of accuracy with 3.6% average error.

We also compute the breakdown of accuracy of the mean temperature across the chip for two different Confidence Intervals (CI) of 95% and 99.7%. The results show that even choosing a confidence level of 99.7%, the average CI equals ± 0.17 for mean of 40.50 (Max CI = 1.02 for gcc06, mean 38.79). This means that with 99.7% confidence that the result has less than 1C difference even for the worst case application.

5.2 Simulation Speed

The sampling methods are intended to minimize slow *D* and *T* detailed and timing simulation modes, and maximize the use of fast *W* and *R* modes (see Table 1). The thermal-aware methods also try to minimize the number of calls to the thermal model. The last column in Table 6 shows the simulation speed for each method on average across all the benchmarks. *TASS* has a maximum speed of 30 MIPS and an average speed of 18 MIPS, while *TAPS* has a maximum of 50 MIPS and an average of 30 MIPS. *SS* and *PS* are slower as they do not deploy thermal sampling and fast-forwarding. Also the spikes in *SS* trigger DTM, and lowers the simulation speed. *Full* runs at 0.76 *MIPS* on average. These results are generated with $TPR = 2$ (explained in Section 3) for *TASS*, which translates to thermal fast-forwarding of 10M instructions. For *TAPS* there are two different fast-forwarding rates as the samples are much longer and placed farther from each other. Within the samples, the fast-forward is set to 4M instructions, and for the intermediate points between the samples it is set to 40M.

Method	Speed w/o thermal R	W	D	T	Speed w/o thermal	Speed w/ thermal
Full	-	-	-	1.0	1.1	0.8
PS	62	-	-	1.0	38	4.1
TAPS	62	-	-	1.0	38	30
SS	-	49	0.4	0.3	44	2.8
TASS	20	19	0.9	1.0	22.5	18

Table 6: Breakdown of simulation speed in MIPS with and without thermal computations

Extending sampling to thermal stage accelerates simulation speed around 7 times, while the accuracy degrades from 4.7% to 6.7% on average. It also affects the maximum error with the same rate. Table 6 shows the breakdown of execution time for each simulation mode, as well as the simulation speed with and without thermal sampling. The execution time with thermal for thermal-aware methods (*TAPS* and *TASS*) is an order of magnitude less compared to their thermal-unaware counterparts (*PS* and *SS*). *SS* without thermal is faster than *TASS* as the sampling parameter for *TASS* is adjusted for thermal accuracy. The execution time at different stages of simulation is distributed more or less evenly in the same order with thermal time now being around 25% of total execution time for both methods. Longer continuous instruction in each mode provides a better opportunity for QEMU to optimize the functional simulation and chain basic blocks.

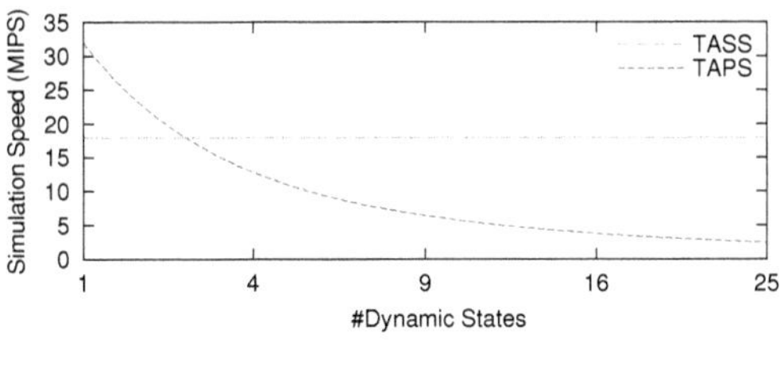

Figure 5: TAPS requires more simulation as processor adapts dynamically, while TASS is insensitive to the dynamic adaptations.

Our experiments show that modeling a system with the number of dynamic architectural states equal or greater than 4, *TASS* achieves faster speed than *TAPS*. In such a case, *TASS* is both faster and more accurate. Figure 5 shows the trend of speed regarding the number of dynamic reconfigurations.

5.3 Full Trace Reconstruction

Thus far, we reported the results with the thermal sampling tuned for speed ($TPr = 2$), which have accurate results at thermal samples. We also tune the parameters to have a fully accurate reconstructed thermal trace, which is important to the $time - varient$ category of studies, e.g. evaluation of Dynamic Thermal Management (DTM) methods. We set $TPr = 1$ for *TASS* to have the same performance and thermal sampling intervals. For TAPS, we apply the same 4M fast-forwarding for all the samples. This results in an average error of 5.7%, while the speed is 14 and 23 *MIPS* for *TASS* and *TAPS* respectively. Figure 6 shows the accuracy of each method, and Figure 7 depicts the thermal trace for four benchmarks.

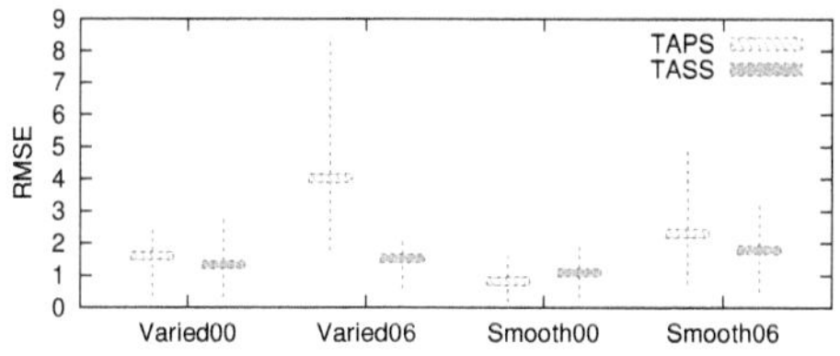

Figure 6: Average, minimum and maximum RMSE across all benchmark categories.

5.4 Thermal Throttling Effect

In addition to thermal behavior, throttling impacts performance as well. Table 7 shows the IPC with and without thermal throttling. It also provides the percentage of time that the processor spends in throttling. For *TASS*, the throttling time is fed back to the performance simulator to calculate the impact on IPC. This is possible because *TASS* does both performance and thermal simulation in one pass.

Benchmark	IPC	Throttled IPC Full	TAPS	TASS	% Time in Throt Full	TAPS	TASS
gzip	0.99	0.99	1.01	0.99	0	0	0
swim	0.45	0.45	0.42	0.45	0.09	0.07	0.06
mgrid	0.75	0.72	0.68	0.75	4.45	7.26	5.9
applu	0.76	0.75	0.72	0.75	1.30	0.82	1.70
vpr	0.87	0.87	0.80	0.85	0	0	0
gcc00	1.29	1.12	1.71	1.21	13.1	17.7	12.4
mesa	1.50	1.35	1.17	1.40	9.87	14.8	11.9
crafty	1.51	1.34	1.21	1.40	11.3	13.8	10.0
facerec	0.72	0.72	0.72	0.69	0	0.98	3.57
lucas	0.65	0.65	0.63	0.64	0.10	0	0.12
bzip2	1.06	1.03	1.09	1.02	2.69	4.26	3.25
twolf	0.88	0.88	0.84	0.81	0	0	0
perlbench	1.29	1.20	1.16	1.20	7.32	17.1	7.57
gcc06	1.00	0.93	0.73	0.94	6.86	5.82	7.61
mcf	0.33	0.33	0.37	0.32	0	0	0
milc	0.50	0.49	0.43	0.50	2.39	4.61	3.01
namd	1.51	1.24	1.09	1.26	17.89	18.9	18.8
dealII	1.12	1.11	0.98	1.11	0.68	1.36	1.08
soplex	0.42	0.42	0.37	0.41	0	0	0
povray	1.88	1.56	1.39	1.48	16.9	21.8	18.1
h264ref	1.34	1.24	1.06	1.25	7.44	9.85	8.76
astar	0.62	0.62	0.43	0.59	0.04	0.14	0.05

Table 7: Impact of thermal throttling on performance.

6. CONCLUSION

This paper shows the necessity of applying sampling to thermal simulation in order to increase productivity by accelerating thermal simulation in architecture level. It proposes a thermal sampling method that is applicable to both statistical and phase-based sampling techniques. The evaluation shows that the newly proposed thermal-aware sampling methods, *TASS* and *TAPS*, outperform existing thermally-unaware methods to estimate performance, power,

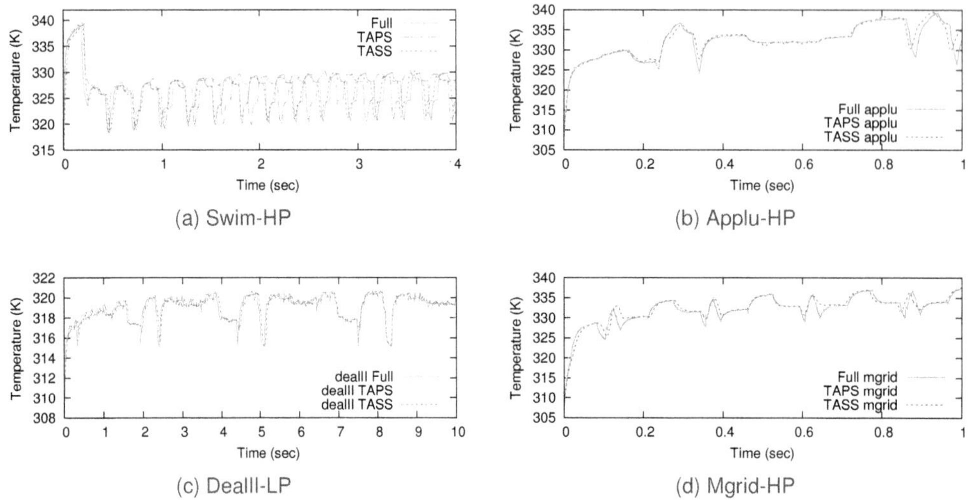

(a) Swim-HP (b) Applu-HP (c) DealII-LP (d) Mgrid-HP

Figure 7: *TAPS* reuses similar simpoints for the intervals with no sample, and does not follow the trend as tightly as *TASS*.

and thermal characteristics, achieving a simulation speeds of 18 and 30 MIPS respectively on average. Without thermal-aware sampling simulation speed is less than 5 MIPS. With thermal-aware sampling the thermal simulation overhead is reduced by 25 times. *TASS* also has advantages in addition to fast simulations. First, it is more accurate than *PS*, *SS*, and *TAPS*. Second, the simulation speed is independent of the number of dynamic states of system (e.g number of DVFS states). We plan to make our simulation infrastructure available for other researchers to use.

7. ACKNOWLEDGMENT

This work was supported in part by the National Science Foundation under grant 1059442; Any opinions, findings, and conclusions or recommendations expressed herein are those of the authors and do not necessarily reflect the views of the NSF.

8. REFERENCES

[1] T. Sherwood, E. Perelman, G. Hamerly, and B. Calder, "Automatically characterizing large scale program behavior," in *Proceedings of the 10th international conference on Architectural Support for Programming Languages and Operating Systems (ASPLOS)*, Oct. 2002, pp. 45–57.

[2] W. Liu and M. Huang, "EXPERT: Expedited Simulation Exploiting Program BehaviorRepetition," in *International Conference on Supercomputing*, St. Malo, France, Jun–Jul 2004, pp. 126–135.

[3] D.G. Perez, H. Berry, and O. Temam, "Budgeted region sampling (beers): do not separate sampling from warm-up, and then spend wisely your simulation budget," in *Signal Processing and Information Technology, 2005. Proceedings of the Fifth IEEE International Symposium on*, Dec 2005, pp. 1 -6.

[4] R. E. Wunderlich, T. F. Wenisch, B. Falsafi, and J. C. Hoe, "Smarts: accelerating microarchitecture simulation via rigorous statistical sampling," in *Proc. 30th Annual Int Computer Architecture Symp*, 2003, pp. 84–95.

[5] Z. Yu, H. Jin, J. Chen, and L.K. John, "Tss: Applying two-stage sampling in micro-architecture simulations," in *International Symposium on Modeling, Analysis Simulation of Computer and Telecommunication Systems (MASCOTS)*, Sep 2009, pp. 1 -9.

[6] F. J. M.-Martinez, E. K.Ardestani, and J. Renau, "Characterizing processor thermal behavior," in *Proceedings of the 15th international conference on Architectural Support for Programming Languages and Operating Systems (ASPLOS)*, 2010, pp. 193–204.

[7] A.K. Coskun, R. Strong, D. Tullsen, and T. Rosing, "Evaluating the impact of job scheduling and power management on processor lifetime for chip multiprocessors," in *Proceeding of the 11th International joint Conference on Measurement and Modeling of Computer Systems (SIGMETRICS)*, Jun. 2009, pp. 169–180.

[8] V. Nookala, D.J. Lilja, and S.S. Sapatnekar, "Temperature-aware floorplanning of microarchitecture blocks with ipc-power dependence modeling and transient analysis," in *Proceedings of the 2006 International Symposium on Low Power Electronics and Design (ISLPED)*, Oct 2006, pp. 298–303.

[9] D. Brooks, V. Tiwari, and M. Martonosi, "Wattch: A Framework for Architectural-Level Power Analysis and Optimizations," in *Proceedings of the Annual International Symposium on Computer Architecture (ISCA)*, June 2000, pp. 83–94.

[10] K. Skadron, M. R. Stan, W. Huang, S. Velusamy, K. Sankaranarayanan, and D. Tarjan, "Temperature-Aware Microarchitecture," in *Proceedings of the 30th Annual International Symposium on Computer Architecture (ISCA)*, Jun 2003, pp. 2–13.

[11] J. Renau, F. Basilio, J. Tuck, W. Liu, M. Prvulovic, L. Ceze, S. Sarangi, P. Sack, K. Strauss, and P. Montesinos, "SESC simulator," January 2005, http://sesc.sourceforge.net.

[12] F. Bellard, "Qemu, a fast and portable dynamic translator," in *Proceedings of the annual conference on USENIX Annual Technical Conference*, Berkeley, CA, USA, 2005, ATEC '05, pp. 41–41, USENIX Association.

[13] S. Li, A. Jung Ho, R. D. Strong, J. B. Brockman, D. M. Tullsen, and N. P. Jouppi, "McPAT: An Integrated Power, Area, and Timing Modeling Framework for Multicore and Manycore Architectures," in *Proceedings of the Annual IEEE/ACM International Symposium on Microarchitecture (MICRO)*, 2009, pp. 469–480.

[14] J. N.-Battilana and J. Renau, "Soi, interconnect, package, and mainboard thermal characterization," in *Proceedings of the 14th ACM/IEEE international symposium on Low power electronics and design (ISLPED)*, 2009, pp. 327–330.

ER: Elastic RESET for Low Power and Long Endurance MLC based Phase Change Memory *

Lei Jiang †, Youtao Zhang ‡, Jun Yang †
† Electrical and Computer Engineering Department
‡ Computer Science Department
University of Pittsburgh, Pittsburgh, PA 15261, USA
†{lej16,juy9}@pitt.edu ‡zhangyt@cs.pitt.edu

ABSTRACT

Phase Change Memory (PCM) has recently emerged as a promising nonvolatile memory technology. To effectively increase memory capacity and reduce per bit fabrication cost, multi-level cell (MLC) PCM stores more than one bit per cell by differentiating multiple intermediate resistance levels. However, MLC PCM suffers from significantly shortened endurance due to its large RESET current that initiates the cell state. In this paper, we propose elastic RESET (ER) to construct non-2^n-state MLC PCM, e.g., 3-state MLC PCM instead of 4-state one for 2-bit MLC. We then adopt data compression and propose fraction encoding to store compressed data using non-2^n-state MLC. By reducing RESET energy, ER significantly reduces write power and prolongs PCM lifetime. On average, we observed 17% RESET power reduction and 32× endurance improvement for 2-bit MLC.

Categories and Subject Descriptors

B.3.4 [**Memory Structures**]: Reliability, Testing, and Fault-Tolerance

General Terms

Design, Reliability

Keywords

Phase Change Memory, Multi-Level Cell

1. INTRODUCTION

Phase Change Memory (PCM) has emerged as a promising nonvolatile memory to alleviate the large leakage and poor scalability problems of traditional DRAM [24]. With advantages such as near zero cell leakage, better scalability and comparable read speed as DRAM, PCM is projected to replace a significant portion of DRAM in main memory system [24]. A PCM cell uses its resistance level to represent logic bits. A single-level cell (SLC) PCM stores one bit (logic '0' or '1'). Given the large resistance contrast between '0' and '1', it is possible to take advantage of intermediate resistance levels such that one cell can store more than one bit, referred to as multi-level cell (MLC) PCM. MLC PCM can effectively increase memory capacity while reducing per bit fabrication cost.

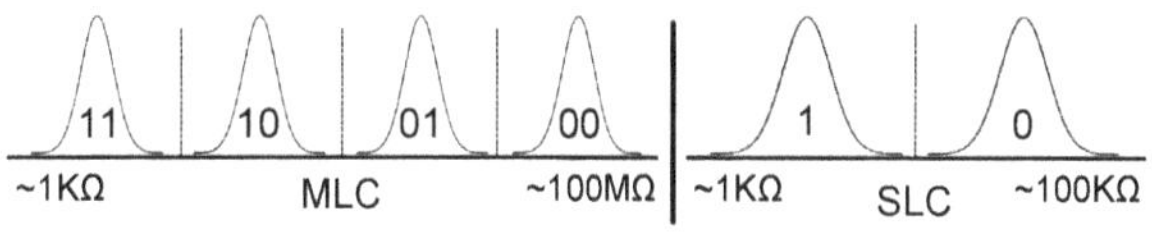

Figure 1: MLC has tighter resistance ranges.

PCM is known for long write latency and short endurance problems. Recently, many architectural techniques have been proposed to attack these two problems. *Differential-write* [24] extends PCM lifetime by removing redundant writes to cells. Wear leveling techniques [17] evenly distribute unbalanced write traffic among the entire chip. Salvaging schemes [20] were proposed to extend chip lifetime when non-negligible number of cells fail. For MLC based PCM, *write pausing* prioritizes read operations to improve system performance [18]. *Write truncation* uses error correction code (ECC) to reduce the number of MLC write iterations [11]. Low power data encodings for MLC PCM were studied in [21]. By transforming MLC to SLC [2, 11], both a longer chip lifetime and a shorter read latency can be obtained on MLC PCM system.

Comparing to SLC, MLC PCM suffers more severely from large RESET power and long read/write latency problems. In order to reliably represent the stored logic value, a PCM cell needs to have its resistance programmed within a tight resistance range. Since MLC has more resistance levels to represent, it often has tighter resistance range per level and/or higher maximum resistance, as shown in Figure 1. Given a PCM write circuit that has certain programming precision, MLC PCM requires a larger RESET current than SLC PCM to initialize its maximum resistance and contain more resistance levels.

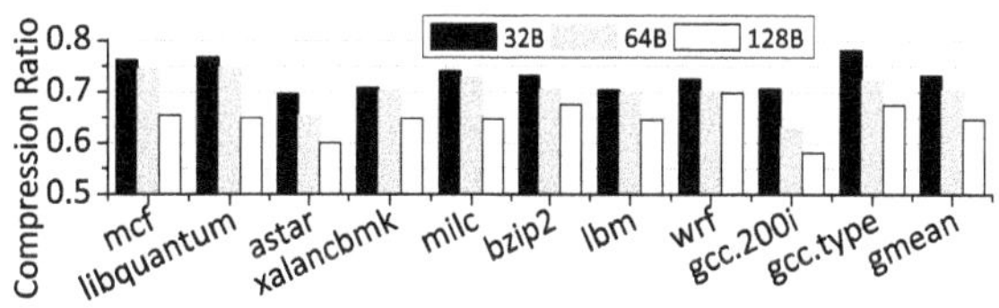

Figure 2: Compression ratio of all line-level writes.

*This work was supported in part by NSF CSR #1012070 and NSF CAREER #0747242.

ISLPED'12, July 30–August 1, 2012, Redondo Beach, CA, USA.

However, larger RESET energy significantly shortens cell endurance. As revealed by device studies [13], 2× RESET energy results in 100× endurance degradation. To address this problem, in this paper, we propose elastic RESET (ER) to construct non-2^n-state MLC PCM. By reducing the RESET energy, we effectively reduce RESET power and prolong PCM lifetime. The existing work that is most close to our design is *MLC-to-SLC transformation* [2, 11] that compresses and stores highly compressible PCM lines in SLC format. *MLC-to-SLC transformation* is only applicable if a line is highly compressible, i.e., the compressed size is ≤ half of the original size. Unfortunately, as shown in Figure 2, many applications have 60%~75% compression ratio and thus cannot benefit much from it.

Our contributions are summarized as follows.

- We propose Elastic RESET (ER) that reduces RESET current to construct non-2^n-state MLC PCM. ER effectively reduces the RESET energy such that it can extend PCM endurance exponentially.
- We propose Fraction Encoding (FE) to store compressed data using non-2^n-state MLC PCM cells. Instead of storing multiple bits in one cell, FE combines multiple cells to store multiple bits, e.g., 2 cells to store 3 bits, and thus can adaptively store compressed data with relaxed compression ratio.
- We evaluate our proposed designs and compare them to related works in the literature. Our experimental results show that ER improves PCM chip lifetime by ×32 and reduces RESET power by 17% on average.

The rest of this paper is organized as follows. Section 2 introduces the models of MLC PCM write. Section 3 elaborates the motivation and the design details. Section 4 presents our experimental methodology. Section 5 analyzes our experimental results. We conclude in Section 6.

2. MODELING MLC PCM WRITE

2.1 MLC PCM Read and Write Operations

Phase Change Memory (PCM) exploits the phase-change behavior of chalcogenide glass (a.k.a GST, $Ge_2Sb_2Te_5$). By injecting electrical pulses, GST can be switched between large resistance state (amorphous state) and small resistance state (crystalline state). MLC PCM exploits intermediate resistance levels between these two states to store multiple logic bits per cell.

Due to process variations and material composition fluctuation, different PCM cells within one memory line respond differently to the programming pulses; and even the same cell responds differently at different times [14]. PCM uses tight resistance ranges to represent the information stored in the cell (and a small unused guardband is often left between consecutive ranges to prevent resistance drift [12]). As we discuss later, all 2-bit MLC cells with resistance between $10^{5.4}\Omega$ and $10^{6.5}\Omega$ are considered to store '01'.

To read a MLC PCM cell, a short current with small amplitude is injected to compare the cell resistance with reference values. It often takes n iterations to read n-bit MLC cells [7] (as shown in Figure 3(a)). To write a MLC PCM cell, a programming and verify (P&V) strategy [4] is often adopted. We first inject a large-amplitude RESET current to initialize cells to amporhous state with large enough resistance, and then inject a sequence of lower-amplitude SET

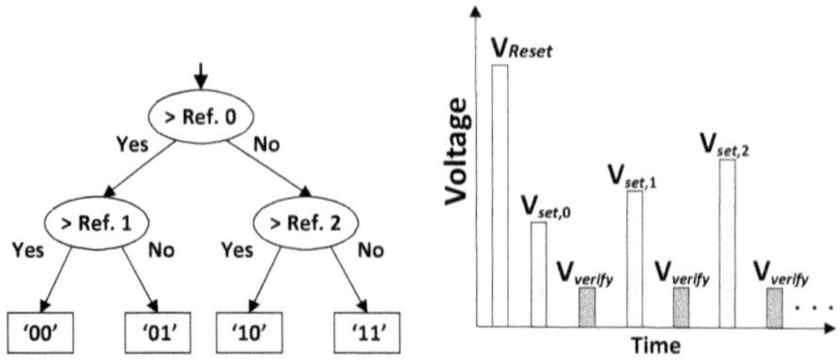

(a) MLC PCM read (b) MLC PCM write

Figure 3: PCM read and write operations.

currents to program each cell to the target resistance range. After each SET iteration, a verify (i.e., read) operation is employed to check if the write operation can be terminated (see Figure 3(b)).

Recently an alternative P&V programming strategy was proposed [6]. The write starts with a SET pulse and then a sequence of RESET pulses (one SET and multi-RESET). While this strategy tolerates better to process variations, it requires larger write energy and longer time to finish. Cell endurance is also reduced, since more RESETs are performed. In this paper, we adopt the traditional P&V programming strategy (one RESET and multi-SET).

2.2 MLC PCM Resistance Ranges

The width of resistance range is often determined by the programming circuit. MLC PCM tends to have tighter ranges and thus requires more precise control. On the other hand, with a given programming precision, MLC PCM has a larger maximum resistance than that of SLC PCM.

Device studies [19] built an accurate RESET current model, whose inputs include the radius of heater (A), the heights of top and bottom electrodes (H_T and H_E), the height of GST (H) and the thickness of amorphous GST fraction (X_0), as shown in Figure 4(a). The resistance of PCM cell is decided by the thickness of amorphous GST fraction. The resistance model [16] described the relation between thickness of amorphous GST and cell resistance, as shown in Figure 4(b).

In this paper, we adopt both two former models with cell parameters in [19]. When X_0 is set to 50nm, the maximum resistance after RESET operation can be 79.4 $M\Omega$. Table 1 shows the resistance ranges and mean values for 2-bit MLC PCM [15].

2-bit MLC	Resistance Range(Ω)	Mean(Ω)
'11'	$[10^3, 10^{4.1}]$	$10^{3.6}$
'10'	$[10^{4.2}, 10^{5.3}]$	$10^{4.75}$
'01'	$[10^{5.4}, 10^{6.5}]$	$10^{5.9}$
'00'	$[10^{6.6}, 10^{7.9}]$	$10^{7.2}$

Table 1: The resistance range of 2-bit MLC.

2.3 The PCM Endurance Model

In this paper, we focus on the hard faults determining PCM endurance. MLC PCM is also vlunerable to soft errors introduced by resistance drift, which can be prevented by guardbands, ECC [22], and/or refresh [3].

PCM cell lifetime has been tested at the device level. In [13], both SET-cycling and RESET-cycling tests were performed to indentify which operation is more critical to cell endurance. The SET-cycling experiment has no failure up to 10^{12} cycles, which is seen as the upper bound of PCM

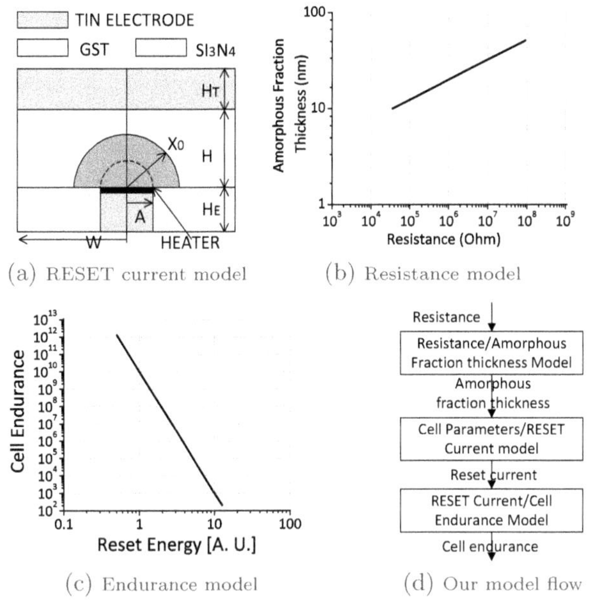

(a) RESET current model (b) Resistance model

(c) Endurance model (d) Our model flow

Figure 4: PCM cell, resistance, endurance models.

cell lifetime. In contrast, hard errors appear after 10^9 cycles in RESET-cycling test. RESET operation energy has been indentified as the most important factor deciding the PCM cell endurance. In this paper, we adopt the RESET-endurance model from [13], with the cell lifetime as a function of RESET pulse energy plotted in Figure 4(c). The power law relationship between the cell lifetime and RESET energy can be summarized in Equation 1, while the conversion between RESET current and RESET energy can be obtained using Equation 2.

$$log_{10}(lifetime)=-7\times log_{10}(\frac{E_{actual\_reset}}{E_{optimal\_reset}})+10 \quad (1)$$

$$E_{reset}=I_{reset}^2\times R\times T \quad (2)$$

where, I_{reset}, E_{reset} denote the RESET current and energy respectively; R represents PCM cell heater resistance; and T is the RESET pulse duration measured in time.

From Figure 4(c), the ideal PCM cell lifetime can be 10^{10} cyclings. But due to process variation and unavoidable over-programming, chip level PCM cell endurance only achieves 10^8 writes for SLC [23, 10] and 10^5 writes for MLC [4]. Through Equation 1, SLC only needs about 40% RESET energy (or power) of MLC.

2.4 Our Work Flow

Our model flow can be found in Figure 4(d). The resistance value is fed into resistance model to get the amorphous fraction thickness. The RESET current model calculates the RESET current based on amorphous fraction thickness. With the RESET current input, the cell endurance model gets the lifetime for each PCM cell. Through our model flow, we can evaluate the impact of the reduced RESET current on PCM lifetime.

3. THE DETAILS OF ELASTIC RESET

In this section, we first discuss the benefits of constructing non-2^n-state MLC cells and then elaborate the design details of ER.

3.1 Constructing Non-2^n-state MLC Cells

An implicit assumption of using n-bit MLC cell is that we need to distinguish 2^n states and thus 2^n resistance ranges, e.g., 4 states for 2-bit MLC PCM. However, this is often not necessary. For example, we need to store a 64B memory line using 2-bit MLC and the line can be compressed to 48B (75% of the original size). There are three alternative approaches to save the compressed line.

- Approach 1: we can store 48B using 4-state 2-bit MLC, which uses 192 cells out of 256 cells in the memory line. By conducting intra-line perfect wear leveling, we can extend the chip lifetime to 133.3%.
- Approach 2: we can store 32B using 4-state 2-bit MLC and 16B using 2-state SLC, as proposed in [2]. Let us optimistically assume SLC lifetime is infinite. By again conducting intra-line perfect wear leveling, we can extend the chip lifetime to 200%.
- Approach 3: we can store 48B using 3-state 2-bit MLC, which uses all 256 cells of the memory line. As one cell has 3 states, two consecutive cells can have 9 combinations, which is enough to record 3 logic bits(i.e., 8 states). Therefore 256 cells can be split into 128 pairs to record 48 bytes (=128 pairs ×3 bits/pair).

Although approach 3 uses all 256 cells, we only need to distinguish 3 states per cell, which requires a smaller RESET current to initialize each cell. According to the write model in Section 2, we reduce memory line level RESET power to 67%, and increase memory line lifetime to ×58 (with an assumption of all bits being changed).

3.2 Elastic RESET

Given the large potential gain from utilizing non-2^n-state MLC cells, we propose elastic RESET (ER) in this paper. ER is a technique initializing a PCM line according to the storage need at runtime. ER is applicable to n-bit MLC. The following illustrates how it works using 2-bit MLC.

To write a PCM line to 2-bit MLC, we first compress the data using *frequent pattern compression (FPC)* [1]. And then (1) if the compressed data (together with FPC meta information) is $\leq$ half of the original size, we save the data in the line and treat each cell as 2-state SLC; (2) if the compressed ratio is $>$ 50% but $\leq$ 75%, we treat each cell as 3-state 2-bit MLC; (3) if the compressed ratio is $>$ 75%, we treat each cell as 4-state 2-bit MLC. To differentiate these cases, a 2-bit SLC cell flag is attached to each line. The flag gets written, when the memory line changes from one format to another.

When reading a memory line written by ER, the data line and the flag are accessed in parallel. Since the flag is SLC cell, its value is known at the end of the first read iteration, which can guide if we should continue the second read iteration for the line.

Recent device study [5] found RESET resistance variation becomes larger when smaller RESET current is applied. The largest variation (σ) can reach $\sim$ 20%. Therefore, we conservatively increase the initial RESET resistance of the non-2^n-state to tolerate variations. To make a n-state cell, we adopt the mean resistance of the original $n+1$ resistance range as the new initial RESET resistance. For example, for 3-state 2-bit MLC cell, we adopt $10^{7.2}\Omega$ (Table 1) as the new initial RESET resistance.

ER, while reducing RESET current, still initializes cells to be written into similar states. It does not require more precise programming circuit and only slightly increases control complexity. ER sets an initial state with lower resitance, which may take less iterations to finish MLC write. However, in this paper, we conservatively assume that writing non-2^n-state MLC takes the same number of write iterations as the original.

3.3 Fraction Encoding

With elastic RESET, we need to store multiple bits in multiple cells. Figure 5 illustrates a 2-bit MLC with three resistance states: '00', '01' and '1X', referred to as 3-state cell. Two 3-state cells have 9 state combinations, which are enough for representing 3 bits (or 8 states). The read latency for 3-state cell is the same as 2-bit MLC cell, since both require two read iterations.

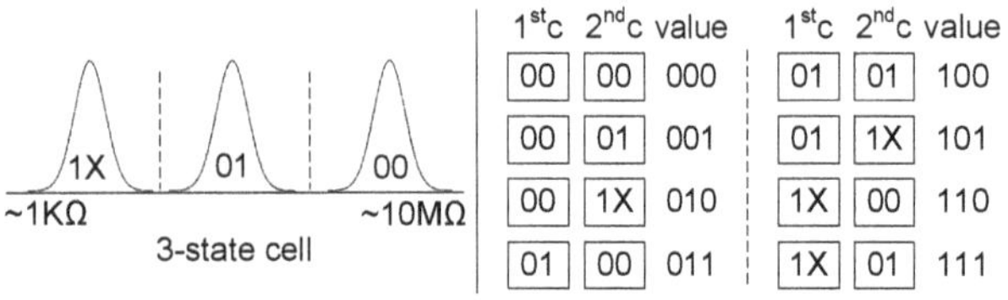

Figure 5: 4/3 fraction encoding.

Figure 5 shows a naive data encoding with two 3-state cells. With naive encoding, one bit flip ('010' to '011') may trigger changes of both two cells. Figure 6 shows the intuitive guess that the rightmost bit is more likely to flip. To minimize cell changes and thus extend chip lifetime, we develop a new encoding that ensures only one cell change if the rightmost bit flips (Figure 7(a)).

We further optimize this encoding by taking write energy into consideration. Since '01' is the intermediate state, it is slow to write and requires more write energy. So, fraction encoding should have as less '01's as possible. Figure 7(a) has six '00's, six '01's and four '1X's. Figure 7(b) shows the improved fraction encoding with only four '01's. Our technique is orthogonal to the energy-efficient data encoding in [21].

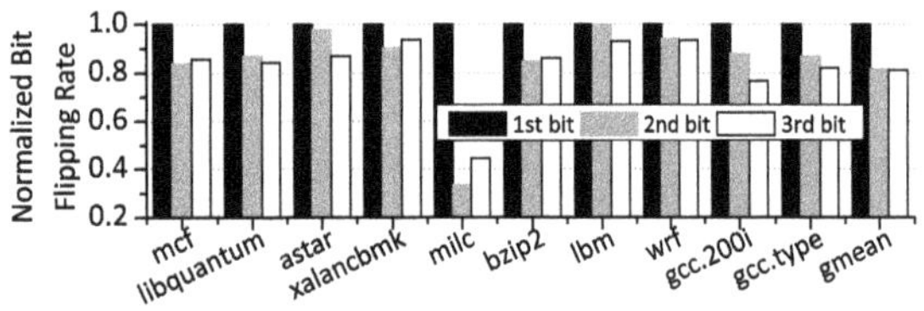

Figure 6: Flipping rate comparison among 3 bits.

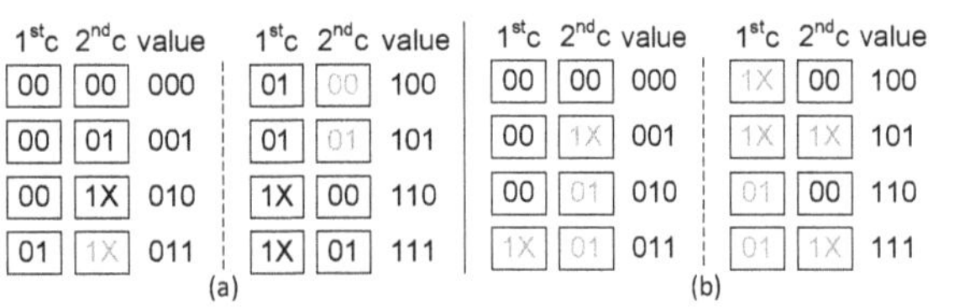

Figure 7: Optimized encoding for cell flipping (a) and energy (b).

3.4 Architectural Designs

Integrating ER in PCM chip faces the following two challenges. First, existing PCM chips usually adopt *differential-write* [24] or flip-n-write [8] to reduce PCM writes at the cell level. Although recent study [11] found that the overall cell changing rate is comparable before and after compression, ER exhibits different pattern of bit changes at the cell level. As Figure 8 shows, ER has larger cell flipping rate than compressed 2-bit MLC. In the worse case, ER may shorten the chip lifetime due to increased cell writes.

Second, when compression ratio changes, a PCM line may switch between different modes. For example, assume the old data in a memory line has a 70% compression ratio such that each cell was treated as 3-state MLC. If the new data has a 78% compression ratio, we need to treat each cell as 4-state MLC, which requires to reset **many cells** due to mode switch. If the compression ratio swings above and below 75%, we may see a large increase of cell changes, resulting in significantly shortened chip lifetime.

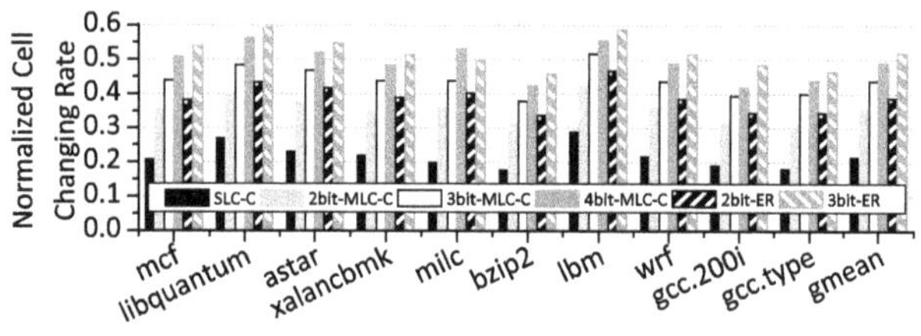

Figure 8: Cell changes for different schemes with compression.

To address these challenges, we propose to dynamically monitor bit flips to prevent ER-introduced cell change increase. The intuitive idea is to compare the number of cell changes with and without ER, and disable ER if the increase is above a preset threshold.

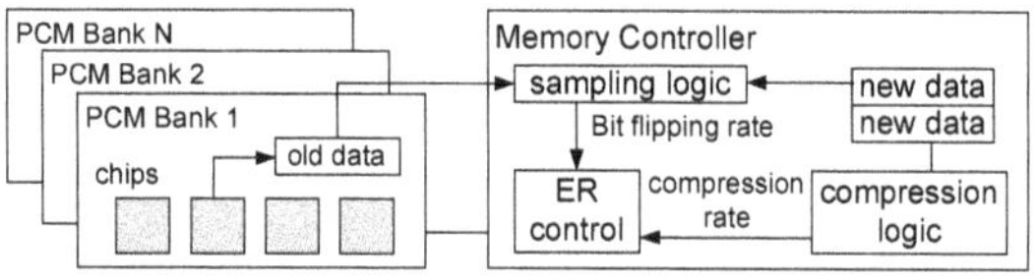

Figure 9: Dynamic sampling for ER.

Figure 9 illustrates our sampling based dynamic monitoring design. A two-entry sample buffer is added to the onchip memory controller while a one-entry buffer is added to each memory bank. We choose to sample one write per 100 offchip writes. We first save the new data in the onchip sample buffer and then read the old data into the PCM chip buffer. The old data is not sent back to the memory controller directly due to the fact that it takes time to switch from write mode to read mode for modern offchip buses. Instead, the old data is buffered in the PCM chip buffer, and is sent back when the bus is in read mode. The memory controller explicitly switches the bus mode to get the old data if the onchip sample buffer is full.

Once the memory controller has both the new and the old data, it computes and compares the number of cell changes (1) if using compressed format; and (2) if using the ER operation. If the compression ratio triggers a mode switch, we pessimistically assume all cells need to be changed. The memory controller disables ER if the increased cell change is above a preset threshold.

4. EXPERIMENTAL METHODOLOGY

To evaluate the effectiveness of our proposed design, we simulated and compared ER with related works using a PIN-based trace-driven simulator. Table 2 lists our baseline configuration. We strictly follow the baseline configuration from [9].The 4GB PCM memory has 32 banks, 8-entry read queue and 16-entry write queue.

For the baseline configuration, we modeled cache and memory bank conflicts, cache and memory bus contention and memory bus scheduling constraints in our simulator. In the case that the write queue is completely full, the memory controller launches a write burst, where only writes are issued and all reads will be held until the write queue is empty. We adopted flip-n-write [8] and power management policy in [9], and MLC-to-SLC form switching [2, 11]. We adopted Frequent Pattern Compression (FPC) [1] and had highly compressible lines saved in SLC in the baseline.

We picked up 5 integer and 5 floating point benchmarks from SPEC2006 with the compression ratio shown in Figure 2. `gcc-200i` and `gcc-type` are `gcc` with two different inputs. The average compression ratio is about 71%. For fair discussion, when using 2-bit MLC PCM, if the compression ratio is $\leq 50\%$, then our scheme is close to FPC-based compression design; if the compression ratio is $> 75\%$, then our scheme is close to the baseline as few opportunities can be exploited. While we did not find a benchmark whose compression ratio fluctuates below and above 75%, we did observe that some memory lines change from compressible to uncompressible at runtime. We evaluated the entire execution for compression/lifetime evaluation. For performance results, we skipped the warmup phase and ran 50 billion instructions to get the IPC.

Processor	4-core, 2GHz, Intel atom-like cores
I/D-L1	private, 32KB/32KB, 4-way, LRU 64B line size, 1-cycle access latency
L2	private, 8MB, 16-way, LRU 64B line size, 10-cycle access latency
PCM Main Memory	4GB, 64B line size, 32 banks Frequent pattern compression frontend 8/16-entry read/write queues Read first, write burst when full 1000-cycle write, 250-cycle read If compressed data size $\leq 50\%$, write in SLC, otherwise write in MLC flip-n-write at line level

Table 2: Baseline Configuration

5. EVALUATION

In this paper, we studied following schemes.

- `n-bit MLC/C` — n-bit MLC with compression. Highly compressible lines are stored as SLC.
- `n-bit MLC/2` — In addition to `n-bit MLC/C`, for compression ratio (50%, 75%], `2-bit (3-bit) MLC/2` stores 25% data line in SLC (2-bit MLC) while the other in 2-bit (3-bit) MLC [2].
- `n-bit ER` — n-bit MLC using ER.

Hardware Overhead. We first evaluated the hardware overhead. FPC overhead is negligible [11]. For sampling-based cell change monitoring, we added 192B onchip SRAM buffer and 96B per bank offchip SRAM buffer. It has negligible timing overhead as it is not on the critical path. The area and energy overhead is also negligible due to its simple control and low activity.

The relatively large hardware overhead comes from 2 SLC bits per PCM line (0.8%) added to indicate the mode of each line. It has no timing overhead due to its parallel access, with the result telling if reading of data cells should continue.

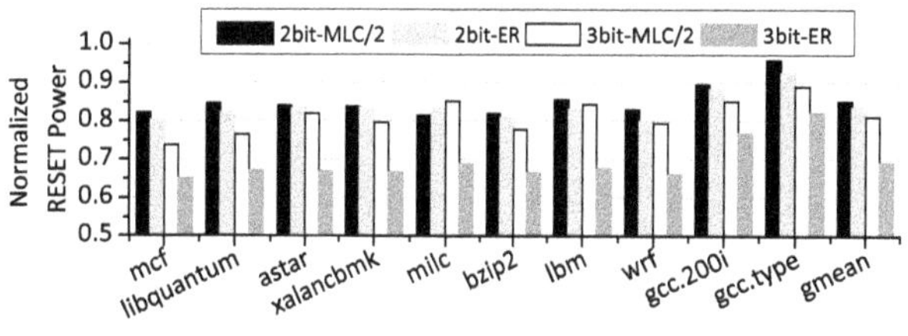

Figure 10: RESET power comparison (normalized to `n-bit MLC/C`)

RESET Power Reduction. Figure 10 compares the RESET power (donimating write power) of different schemes. On average, `ER` reduces 17% and 31% RESET power for 2-bit and 3-bit MLC PCM over `n-bit MLC/C` respectively. Compared to `MLC/2`, `ER` reduces RESET power by 2% and 15% for 2-bit and 3-bit MLC respectively. For 2-bit MLC, `MCL/2` and `ER` have a similar RESET power, as `MCL/2` loses from reseting MLC cells but gains from reseting some cells in SLC mode. `MCL/2` has less lower RESET opportunity to exploit for 3-bit MLC.

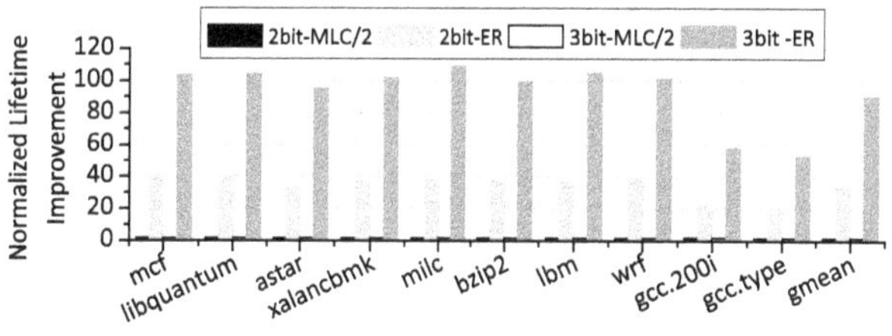

Figure 11: Lifetime improvement (normalized to `n-bit MLC/C`).

Lifetime Improvement. As Figure 11 shows, `ER` significantly improves the PCM chip lifetime by 32× (89×) for 2-bit (3-bit) MLC. Smaller RESET energy per cell produces longer PCM chip endurance. `gcc-200i` and `gcc-type` have less lifetime improvements. We found that they have execution phases in which written PCM lines are less ($> 75\%$) or highly ($\leq 50\%$) compressible (as shown in compression ratio study Figure 12).

Exploiting Low-Power Write for Performance. By reducing write power per PCM write, ER opens the door for either low-power or high-performance designs. For the latter, PCM system usually has tight power budget such that significantly reducing per write power enables more concurrent writes. Figure 13(a) illustrates the potential of improving system performance when using 2-bit MLC PCM. We evaluated with different power budgets. For the power budget of current DDR2 standard, `ER` gets only 5% performance improvement indicating that the current power is sufficient. With lower power budget, `ER` shows significant benefits over `n-bit MLC/C`. With 75% and 50% of the original power budget, `ER` obtains 8% and 11% performance improvements over `n-bit MLC/C`, respectively. However, if we

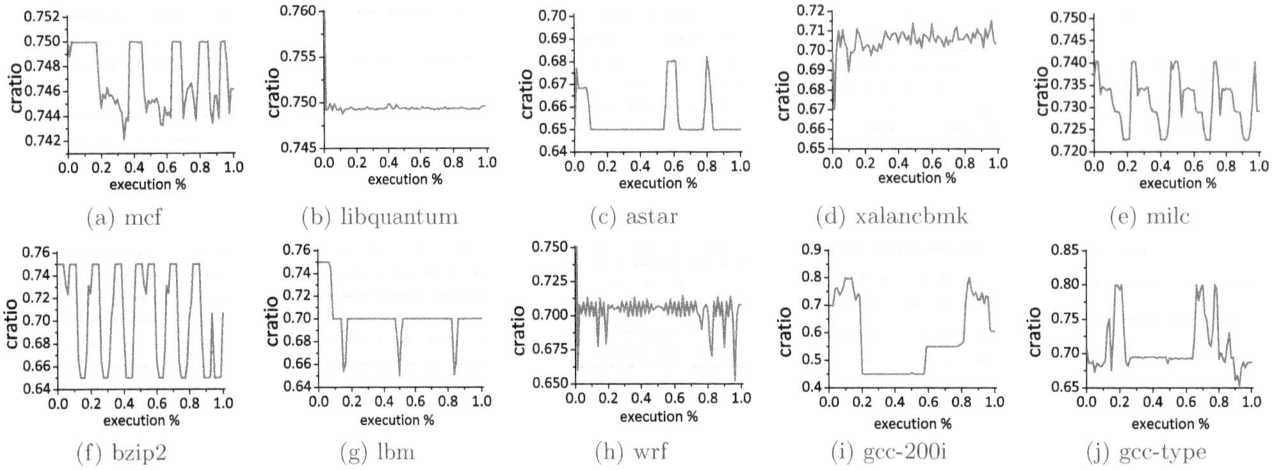

(a) mcf (b) libquantum (c) astar (d) xalancbmk (e) milc
(f) bzip2 (g) lbm (h) wrf (i) gcc-200i (j) gcc-type

Figure 12: Compression ratio of the whole execution (X-axis is normalized execution percentage).

further reduce the power budget, ER achieves less (i.e., 7%) improvement as the extreme low power budget has become a global bottleneck for the whole system. Figure 13(b) compares the performance improvements when using 3-bit MLC PCM. The results showed that ER has the potential to gain better performance improvements for more-bit MLC.

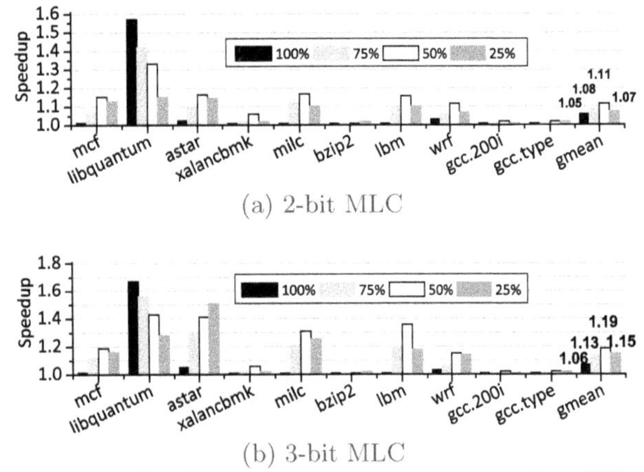

(a) 2-bit MLC

(b) 3-bit MLC

Figure 13: Performance improvement under different power budgets

Compression Study. We also studied the compression ratio during the entire execution. Figure 12 reports the dynamic compression ratio at runtime, i.e., the averaged compression ratio of all write instances for the entire program execution. The results show that the compression ratios are relatively stable such that ER can benefit for the entire execution. Note while the averaged compression ratio may be below 75%, some memory lines are still not compressible.

6. CONCLUSIONS

In this paper, we developed elastic RESET (ER) to construct non-2^n-state MLC cells. Non-2^n-state cells store less information but significantly extend cell lifetime due to using less RESET initialization energy. By adopting compression and fraction encoding, ER can store the compressed line using non-2^n-state cells for PCM lines with different compression ratios. Our experimental results showed that on average, ER gains 17%/31% write power reduction, 32×/89× lifetime improvement, for 2-bit/3-bit MLC PCM respectively.

7. REFERENCES

[1] A. R. Alameldeen and D. A. Wood, "Adaptive Cache Compression for High-Performance Processors," in *ISCA*, 2004.

[2] M. Arjomand, *et al.*, "A Morphable PCM Architecture Considering Frequent Zero Values," in *ICCD*, 2011.

[3] M. Awasthi, *et al.*, "Efficient Scrub Mechanisms for Error-Prone Emerging Memories," in *HPCA*, 2012.

[4] F. Bedeschi, *et al.*, "A Bipolar-Selected Phase Change Memory Featuring Multi-Level Cell Storage," *JSSC*, 2009.

[5] S. Braga, *et al.*, "Modeling of Partial-RESET Dynamics in Phase Change Memories," in *ESSDERC*, 2010.

[6] S. Braga, *et al.*, "Voltage-Driven Partial-RESET Multilevel Programming in Phase-Change Memories," in *ESSDERC*, 2010.

[7] C. Calligaro, *et al.*, "Comparative Analysis of Sensing Schemes for Multilevel Non-volatile Memories," in *ICISS*, 1997.

[8] S. Cho and H. Lee, "Flip-N-Write: A Simple Deterministic Technique to Improve PRAM Write Performance, Energy and Endurance," in *MICRO*, 2009.

[9] A. Hay, *et al.*, "Preventing PCM Banks from Seizing Too Much Power," in *MICRO*, 2011.

[10] L. Jiang, *et al.*, "Enhancing Phase Change Memory Lifetime through Fine-Grained Current Regulation and Voltage Upscaling," in *ISLPED*, 2011.

[11] L. Jiang, *et al.*, "Improving Write Operations in MLC Phase Change Memory," in *HPCA*, 2012.

[12] M. Joshi, *et al.*, "Mercury: A Fast and Energy-Efficient Multi-level Cell based Phase Change Memory System," in *HPCA*, 2011.

[13] K. Kim and S. J. Ahn, "Reliability Investigations for Manufacturable High Density PRAM," in *IRPS*, 2005.

[14] D. Mantegazza, *et al.*, "Statistical Analysis andModeling of Programming and Retention in PCM Arrays," in *IEDM*, 2007.

[15] T. Nirschl, *et al.*, "Write Strategies for 2 and 4-bit Multi-Level Phase-Change Memory," in *IEDM*, 2007.

[16] N. Papandreou, *et al.*, "Estimation of Amorphous Fraction in Multilevel Phase Change Memory Cells," in *ESSDERC*, 2009.

[17] M. K. Qureshi, *et al.*, "Enhancing Lifetime and Security of PCM-based Main Memory with Start-Gap Wear Leveling," in *MICRO*, 2009.

[18] M. K. Qureshi, *et al.*, "Improving Read Performance of Phase Change Memories via Write Cancellation and Write Pausing," in *HPCA*, 2010.

[19] B. Rajendran, *et al.*, "Analytical Model for Reset Operation of Phase Change Memory," in *IEDM*, 2008.

[20] S. Schechter, *et al.*, "Use ECP, not ECC, for Hard Failures in Resistive Memories," in *ISCA*, 2010.

[21] J. Wang, *et al.*, "Energy-efficient Multi-Level Cell PCM System with Data Encoding," in *ICCD*, 2011.

[22] D. H. Yoon, *et al.*, "FREE-p: Protecting Non-volatile Memory against Both Hard and Soft Errors," in *HPCA*, 2011.

[23] W. Zhang and T. Li, "Characterizing and Mitigating the Impact of Process Variations on Phase Change based Memory Systems," in *MICRO*, 2009.

[24] P. Zhou, *et al.*, "A Durable and Energy Efficient Main Memory Using Phase Change Memory Technology," in *ISCA*, 2009.

A Dual-mode Architecture for Fast-Switching STT-RAM

Zhenyu Sun[α], Hai (Helen) Li[α], and Wenqing Wu[β]
[α] ECE Dept., Polytechnic Institute of New York University, Brooklyn, NY, USA
[β] Qualcomm Incorporated, San Diego, CA, 92121, USA

ABSTRACT

In the past, the spin-transfer torque RAM (STT-RAM) suffered from the slow write speed and the high write energy consumption. The latest progress in device engineering has dramatically reduced the write time to a few nanoseconds and hence enabled the fast-switching STT-RAM (FS-STT-RAM). However, the enhancement in write performance results in the degradation of read operations, in terms of both speed and data reliability. Our analysis shows that the read performance becomes critical. Based upon the tradeoff among the read latency, read errors, and system performance, we propose a new FS-STT-RAM architecture, which can switch between two operation modes for either high data accuracy or low power consumption with the support of operation system. In the high accuracy mode, FS-STT-RAM applies the rewrite-after-read scheme to eliminate the data disturbances induced by read current. Even so, with enhancement from *shadow rewrite buffer* and *bit invert scheme*, it gains an average 19% improvement in energy-delay-product (EDP) compared to a conventional STT-RAM. When an application can afford a very low chance of read disturbance, the proposed FS-STT-RAM can operate in the low power mode and further boost the average EDP improvement to 34%.

Categories and Subject Descriptors

B.3.2 [**Memory Structures**]: Design Styles—*Cache memories*

General Terms

Design

Keywords

MTJ, read disturbance, dual-mode architecture

1. INTRODUCTION

As CMOS technology continues scaling down, the leakage power consumption of the conventional SRAM dramatically increases. The latest works have shown that up to 20% to 50% of the total power consumption in CPU systems is spent solely on the *last level cache* (LLC) [1][2]. The high leakage power inspires the adoption of the nonvolatile memory technologies, such as *magnetic RAM* (MRAM) and *phase-change RAM* (PCRAM), as alternatives of the conventional SRAM and eDRAM to break the power wall and to prolong the battery life of portable devices. Among these memory technologies, *spin-transfer torque RAM* (STT-RAM) that is the second generation MRAM becomes the most promising candidate. STT-RAM can be used across multiple layers in memory hierarchies including the upper-level caches requiring high speed accesses, the lower level caches with large capacity, and the stand alone memory for high density [3][4][5].

Compared to the conventional SRAM, STT-RAM has a smaller memory cell size, close-to-zero leakage, and better scalability [6]. However, the write operation is relatively slow and power hungry. For example, a programming time of $10ns$ has been widely adopted in architectural level evaluation for STT-RAM [4][7]. The slow write speed significantly degrades the system performance. Moreover, its higher write dynamic energy, e.g., 10 to 20 times of that of SRAM [5], further tightens the limited on-chip power budget. Many architecture solutions, such as early write termination [8], dual write mode [3], hybrid SRAM/STT-RAM LLC [9] have been proposed to alleviate the impact of the slow write operations of STT-RAM.

Recently, the write performance of STT-RAM has obtained significant improvement, benefiting from the better understanding on magnetic device physics and the enhanced process engineering [10] [11]. Sub-nanosecond switching of in-plane *magnetic tunneling junction* (MTJ), the data storage element of STT-RAM, has been reported [12]. Sacrificing the nonvolatility is another effective way to reduce the MTJ switching time [5][13]. Moreover, the novel perpendicular MTJ technology [14] has demonstrated faster switching and better scalability than the conventional in-plane MTJ. The STT-RAM design based on these *fast-switching* (FS) devices will be the focus of this work.

On the other hand, the importance of read operation, which is not as trivial as what people previously thought of, has been brought out. The stored data of STT-RAM is read out by detecting the MTJ resistance state. The large process variations degrade the resistance differences between the data cell and the reference cell [15]. Consequently, the sensing delay of sense amplifier is enlarged. The situation in *STT-RAM design with fast-switching devices* (called as FS-STT-RAM) is more severe: the improved write performance requires the read current amplitude to decrease accordingly in order to prevent the unintentional data switching, or *read disturbance* (RD). As we shall show in Section 3, the sub-nanosecond read speed under small read current that has been widely used is not true anymore.

We explore the design implication for FS-STT-RAM in this work. Based on the comprehensive cross-layer (device-circuit-architecture) analysis, we first build the relationship amongst the read access latency, read disturbance probability, and the system performance. A novel FS-STT-RAM based memory architecture then is proposed, which switches between the *high accuracy* and the *low power* modes with the support of operating system. Thus, the speed, energy, and data reliability of the overall computing system can be prioritized and balanced according to users' requirement. Moreover, the *shadow rewrite buffer* and the *bit invert scheme* are introduced to enhance the performance in the high accuracy mode and to reduce write energy consumption, respectively.

The rest of our paper is organized as follows. Section 2 intro-

ISLPED'12, July 30–August 1, 2012, Redondo Beach, CA, USA.

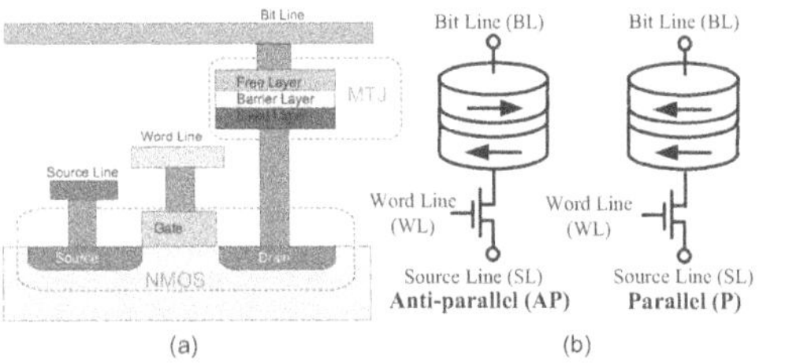

Figure 1: (a) 1T1J STT-RAM cell, and (b) its equivalent circuit.

duces the STT-RAM and FS-STT-RAM; We thoroughly analyzes the relationship between read performance and read disturbance in Section 3, and then describe the need and architecture of the dual-mode FS-STT-RAM design in Section 4. The experimental results are shown in Section 5. At last, Section 6 concludes our work.

2. BACKGROUND

2.1 MTJ and STT-RAM

Fig. 1 shows the scheme of *one-transistor-one-MTJ* (1T1J) STT-RAM cell design and its equivalent circuit. The data storage element of STT-RAM is *magnetic tunneling junction* (MTJ), which is composed of two ferromagnetic layers, namely *reference layer* and *free layer*. The two ferromagnetic layers are isolated by an oxide barrier layer, *e.g.*, MgO. The magnetization direction of the reference layer is fixed while that of the free layer can be changed by passing a current that is polarized by the magnetization of the reference layer. When the magnetization directions of the free layer and the reference layer are *parallel* (P) or *anti-parallel* (AP), the MTJ resistance is low (R_L) or high (R_H), respectively. Accordingly, the data stored in the STT-RAM cell is "0" or "1", respectively. The NMOS select transistor is connected to MTJ in series to control the access to it.

The switching behavior of STT-RAM, such as the critical switching current I_{c0} and the length of the write period t_w, is determined by many factors including MTJ material property, MTJ geometry as well as the size of the select transistor [16]. For the given MTJ property and geometry, increasing the select transistor size can produce a larger switching current through MTJ and reduce t_w. However, the density of STT-RAM is sacrificed. Moreover, as we further enlarge the select transistor, the enhancement on MTJ switching current becomes less effective [5].

2.2 Fast-switching MTJ Technologies

Many efforts on device engineering have been put to make MTJ's switching faster by changing device properties. For instance, decreasing the free layer thickness t_F and/or the magnetization saturation M_s of MTJ can effectively reduce the critical switching current density J_{c0} [5][17]. Adjusting the MTJ device properties such as spin efficiency η and damping factor α helps improve the switching performance. The sub nanosecond switching STT-RAM has been reported with real device fabricated and tested [12]. Recently, the perpendicular technology has gained many attentions. The perpendicular spin transfer torque can cancel out the out-of-plane demagnetizing field so as to reduce the critical switching current density J_{c0}. Meanwhile, the MTJ stability can be maintained by keeping spin efficiency η and damping factor α comparable to those of in-plane MTJ [18].

2.3 Conv-STT-RAM vs. FS-STT-RAM

In this work, we investigate two types of STT-RAM designs, namely, Conv-STT-RAM and FS-STT-RAM. The Conv-STT-RAM uses the regular in-plane MTJs, while the FS-STT-RAM has a better switching performance by properly relaxing its thermal stability and reducing data retention time (*e.g.*, a few seconds) to maintain the trivial performance and energy overhead induced by data refreshing [5][13].

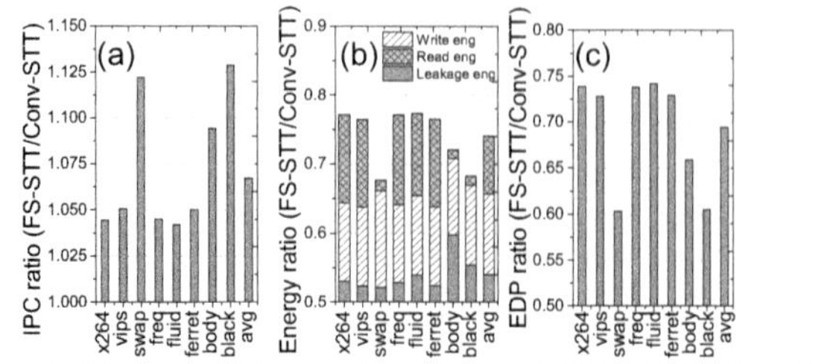

Figure 2: The comparison between Conv-STT-RAM L2 cache and FS-STT-RAM L2 cache.

Fig. 2 demonstrates the comparison between two 4MB L2 cache designs based on Conv-STT-RAM and FS-STT-RAM, in terms of the *instruction-per-cycle* (IPC) performance, the energy consumption, and the *energy-delay-product* (EDP). The simulation results show that the FS-STT-RAM over-performs the Conv-STT-RAM in all the aspects. On average, $\sim$ 34% of EDP improvement can be obtained. The details of the simulation platform and the cache configuration are described in Section 5.

3. READ DISTURBANCE AND READ PERFORMANCE

3.1 Read Disturbance

The magnetization switching of the free layer in a MTJ is a stochastic procedure. Applying a current through an MTJ in a read operation can potentially result in unintentional data flipping, which is called as *read disturbance* (RD). The read disturbance probability P_{rd} of a MTJ at a read current I_R is determined by the critical switching current I_{c0}, the read current pulse width τ_{rd}, and the thermal stability Δ. In theory, P_{rd} can be expressed as [19]:

$$P_{rd} = 1 - exp\{-\frac{\tau_{rd}}{\tau_0} exp[-\Delta(1 - \frac{I_R}{I_{c0}})]\}, \qquad (1)$$

where, τ_0 is the thermally activated reversal time, which is usually assumed to be $1ns$ [13]. The MTJ device parameters such as I_{c0} and Δ are determined by the fabrication process and remain unchanged after a device is made. Therefore, the read disturbance probability P_{rd} indeed is a function of I_R/I_{c0}, under a given τ_{rd}.

In read operations, τ_{rd} should be long enough so that the sense amplifier can complete the data detection. In other words, τ_{rd} is determined by the sense amplifier performance. Prolonging τ_{rd}

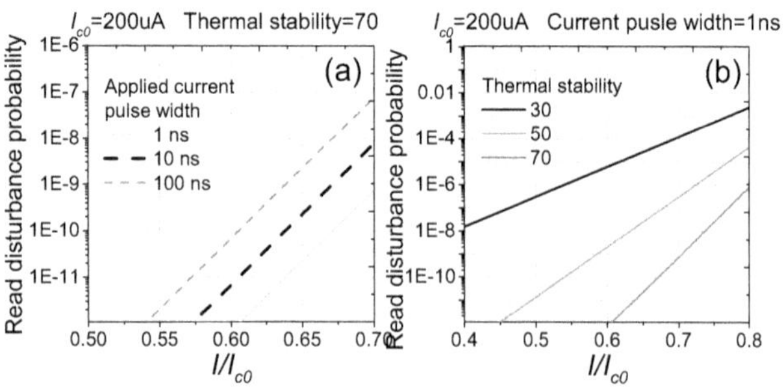

Figure 3: The MTJ read disturbance probability under (a) the different read current pulse width τ_{rd}, and (b) the different thermal stability Δ.

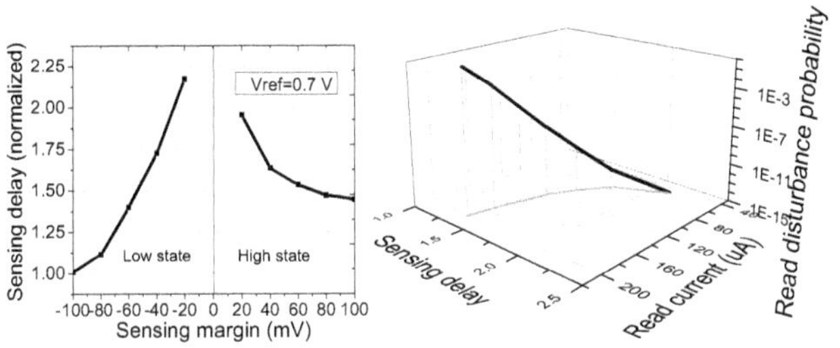

Figure 4: (a) The sensing delay vs. the sensing margin; (b) The read current determines the sensing delay and the read disturbance probability.

significantly increases the possibility of read disturbance as shown in Fig. 3(a).

Under the same operating condition, *i.e.*, the same I_R and τ_{rd}, FS-STT-RAM suffers from a much higher read disturbance probability than Conv-STT-RAM. For example, Smullen and Sun *et al.* proposed to reduce the MTJ thermal stability for better switching performance [5][13]. Correspondingly, P_{rd} increases as Fig. 3(b) shows. Moreover, the smaller I_{c0} of FS-STT-RAM causes a higher I_R/I_{c0} ratio and aggravates the possibility of the unintentional MTJ flipping.

In order to reduce the design complexity of read peripheral circuitry, the read current is always applied along one single direction. As a consequence, the read disturbance is *unidirectional*. For instance, writing "0" into a STT-RAM cell requires a switching current from BL to SL (refer to Fig. 1). The read current flowing in this direction potentially induces the unwanted $1 \rightarrow 0$ flips, but the data "0" remains intact. On the contrary, supplying read current from SL to BL may disturb data "0" only.

3.2 Read Current, Read Speed, and Read Disturbance

In STT-RAM design, a sense amplifier is used to detect the data stored in a target reading cell by comparing the voltage difference between its bit-line and a reference voltage. The voltage difference at the sense amplifier inputs is called as sensing margin. Given a sense amplifier design, the sensing delay is determined by the sensing margin: the smaller the sensing margin is, the longer the sensing delay is required, as shown in Fig. 4(a).

The sensing margin is proportional to the read current I_R. To keep the similar read disturbance probability and hence the induced read errors in Conv-STT-RAM and FS-STT-RAM, I_R/I_{c0} need remain unchanged, as indicated in Eq. (1). Compared to Conv-STT-RAM, the smaller I_{c0} of FS-STT-RAM results in a smaller I_R, which causes the smaller sensing margin and longer read time τ_{rd}. The simulation result in Fig. 4(a) shows that the sensing delay increases dramatically as the sensing margin reduces. Thus, the sub-nanosecond read speed that has been widely used is not true anymore.

On the other hand, we can increase the read current in FS-STT-RAM for better read speed. However, the corresponding higher read disturbance probability results in more unintentional data flipping, or *read errors*. The quantitative relation among the read current, sensing delay, and read disturbance error rate is shown in Fig. 4(b).

4. DUAL-MODE ARCHITECTURE FOR FS-STT-RAM

As we can see, FS-STT-RAM can dramatically improve write performance and reduce the write energy consumption. But it faces severe challenge on data reliability due to the read disturbance. Reducing the read current can alleviate the situation by degrading the overall system performance, especially for the programs with high frequent reads and sensitive to the read speed. In this work, we propose a new architecture with dual operating modes for FS-STT-RAM design.

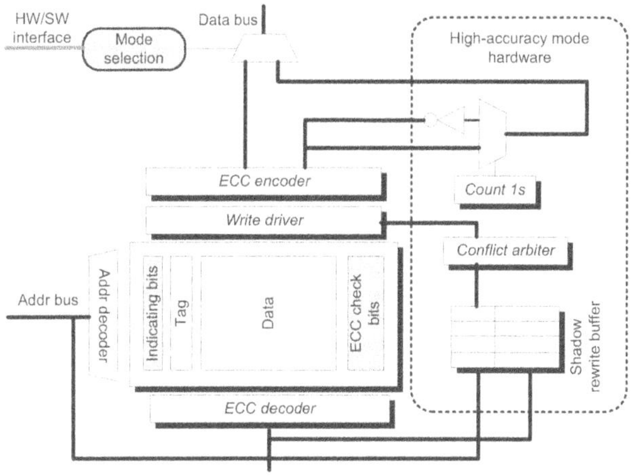

Figure 5: The proposed dual-mode architecture for FS-STT-RAM.

4.1 High Accuracy Mode and Low Power Mode

A general purpose computing system need deal with various tasks with the different data accuracy and throughput demands. For example, the banking and financial transactions require 100% accurate data and relatively low throughput. In contrast, the video stream usually has a high data exchange rate but occasional erroneous pixels usually are not noticeable. Therefore, we propose a FS-STT-RAM architecture that can operate in either the *high accuracy* (HA) mode or the *low power* (LP) mode.

Fig. 5 illustrates the diagram of the proposed dual-mode architecture for FS-STT-RAM. At the beginning of a program, the system determines the STT-RAM operating mode based on the application-oriented data access requirement. Software-level assistance is needed to detect such information and to deliver to hardware as input control.

The low-power mode can be adopted for the applications with relatively low data accuracy request for high throughput and low power consumption. The read operations perform regularly as the conventional STT-RAM design. The regular *error correction code* (ECC), *e.g.*, Hamming code, is utilized that can dramatically lower the raw bit error rate. However the system does not take extra effort to fix the STT-RAM bit flips induced by read current.

For data correctness oriented applications, the FS-STT-RAM design switches to the high accuracy mode, in which every read is followed by a refreshing (rewrite), that is, writing the readout data back to the cache block, to eliminate the impact of the read disturbance. Considering that the read disturbance is unidirectional, we rewrite only those STT-RAM cells with potential to be disturbed (for instance, data 1 when read current flows from BL to SL) and avoid unnecessary energy overhead.

4.2 Hardware and Software Interface

The proposed hardware scheme provides the support and flexility to switch operation between two modes. The mode selection can be controlled at higher system levels. The software interface

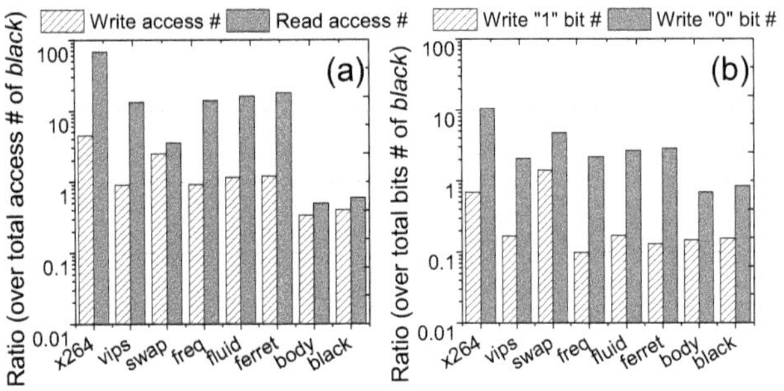

Figure 6: (a) Read access frequency vs. write frequency. (b) Write "1" bits vs. write "0" bits.

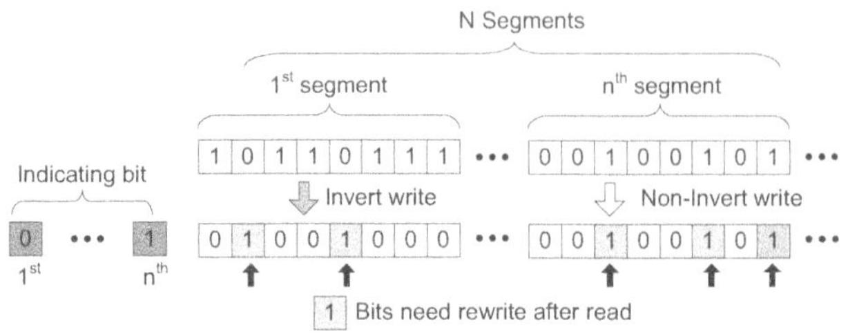

Figure 8: Write bit invert scheme.

shall provide users options to select their flavor of data accuracy. Programmers are able to define the required data accuracy based on the requirement of programs. Extra instructions might be inserted to enable mode switching. Compiler shall also translate and deliver the mode switching information to the hardware. On the other hand, the hardware structure must support the compiler in return. For example, one extra bit data indication operation mode is required in the memory access instructions.

4.3 Enhancement in High Accuracy Mode

We further propose to enhance the high accuracy mode and mitigate the performance and energy overheads induced by the rewrite after each read through two design techniques, namely, the *shadow write buffer* and the *write bit inverting*.

Shadow rewrite buffer: Even in FS-STT-RAM, writing a cell is more difficult and takes longer time than reading out data (refer to simulation data in Section 5). Moreover, the read accesses are more frequent in caches than the write accesses as the cache access statistics in Fig. 6(a) shows. Therefore, the system performance in the high accuracy mode significantly degrades compared to that of the low-power mode.

Fig. 7 illustrates an cache access example in the high accuracy mode. Assume $addr$ 1 and $addr$ 2 belong to the same bank, while $addr$ 3 goes to a different bank. 1) the rewrite at $addr$ 1 in $cycle2$ does not stall the incoming new write $addr$ 3 since they go to the different cache banks; 2) in $cycle3$, the new write and the rewrite goes to the same cache line $addr$ 1. Therefore, the rewrite can be throttled and only the new write need to be executed since the new write will update the value anyway; 3) the conflict happens only when the rewrite operation locates into the same bank but not same cache line, *i.e.*, $addr$ 1 and $addr$ 2 from $cycle7$. At this time, the rewrite stalls the new coming writes and introduce performance overhead.

Instead of sharing the same write queue with regular write requests, we buffer the rewrites after reads in a separated shadow rewrite buffer as shown in Fig. 5(b). Rewrites have a lower priority than the new write requests and are executed when the write driver is available. During a read operation, both cache and its corresponding shadow rewrite buffer are simultaneously accessed, and the data from the latter one, if any, will be taken as the golden copy. The shadow rewrite buffer should be sufficiently large to mitigate performance degradation caused by the conflict between the incoming write and the rewrite. On the other side, it should not be too large after considering the overhead of read latency and energy consumption.

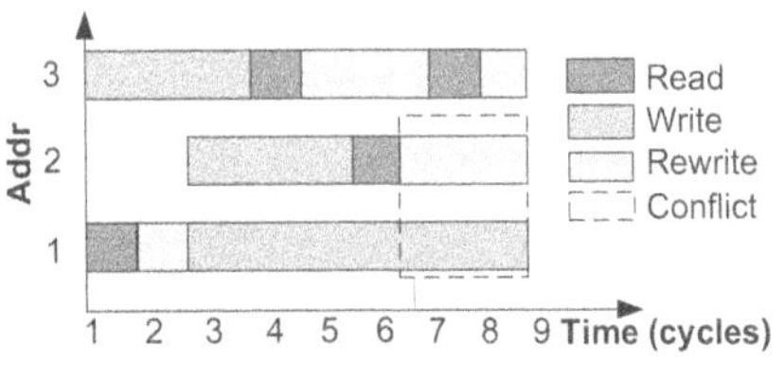

Figure 7: An cache access example in the high accuracy mode.

Write bit inverting: A 1T1J STT-RAM cell provides the different write currents when writing 0 and 1 due to the different biasing condition, which makes the corresponding write energies different [17]. Moreover, the read disturbance affects either data 0 or 1, determined by the read current direction. Therefore, properly selecting the read current direction could help reduce write energy consumption. For example, if writing 1 costs less energy than writing 0, the read current from BL to SL could be more preferable since it causes only $1 \rightarrow 0$ flips.

The write bit inverting scheme is proposed for the same motivation. Previously, the bit inverting technique has been used in STT-RAM [17] and PCM [20] for write energy reduction. Since read accesses are more frequent than the write accesses, the rewrite energy consumption becomes more important FS-STT-RAM operating in high accuracy mode. We utilize the write bit inverting scheme to maintain more 0's and less 1's and hence minimize the rewrite energy.

As illustrated in Fig. 8, when the number of 1's of a cache block is greater than half of the total bit number, we invert all the writing value. An additional bit is used to indicate the inverting status. Dividing a cache block into several smaller segments can help reduce the overall number of 1's. However, it also introduces more indicating bits. Our simulation results in Fig. 6(b) shows that the of write-0 bits are much greater than the number of write-1 bits. Statically the chance of bit inverting is not very high.

5. SIMULATION RESULTS

5.1 Simulation Setup

We performed the evaluations on an 8-core in-order UltraSPARC T1 processor by replacing the SRAM L2 cache with Conv-STT-RAM or FS-STT-RAM. The Simics toolset [21] was used to obtain the cache access trace. More detail of the process configuration is depicted in Table 1. And Table 2 summarizes the cache latency and energy of both Conv-STT-RAM and FS-STT-RAM obtained from SPICE simulation and NVsim [22] calculation based on MTJ in [5]. The multi-threaded benchmarks from Parsec Bench-

Table 1: Processor Configuration

Processors	8 cores, 2GHz, 4 threads/CPU core, 1-way issue (in order)
SRAM L1 Cache	Local, 16KB I/D, 2-way, 64B line, 2-cycle, write-back
STT-RAM L2 Cache	Shared, 4MB, 32 banks, 32-way, 64B line, write-back. 1 read/write port, 4 write buffers.
Main Memory	4GB, 400-cycle latency.

Table 2: Cache Latency and Energy Configurations

Cache type	4MB FS-STT-RAM	4 MB Conv-STT-RAM
Write latency	Peripheral latency: 1.059 ns Switching latency: 5.356 ns	Peripheral latency: 1.121 ns Switching latency: 10.326 ns
Read latency	Peripheral latency: 2.119 ns Sensing latency: 1.5 ns	Peripheral latency: 2.242 ns Sensing latency: 1.5 ns
Write energy	0.932 nJ	1.916 nJ
Read energy	0.083 nJ	0.085 nJ
Leakage power	104 mW	110 mW

mark Suite [23] were adopted in architecture simulations. For each benchmark, we fast-forward to region of interest, warm up the cache for 200 million instructions, and then execute 500 million instructions.

5.2 Read Latency and Read Errors

Due to the large amount of read accesses, the read latency is critical to system's overall performance. In Fig. 9(a), we compare the *instructions-per-cycle* (IPC) performance of FS-STT-RAM in the low power mode by reducing the read current amplitude and hence increasing the read latency cycles. As expected, the system performance degrades significantly as read latency increases. From this perspective, the fast read operation is extremely important. However, as Fig. 9(b) shows that the data error induced by read disturbance, another critical system parameter, is also sensitive to the read current. Even integrating wtih the single-bit ECC, the error rate of the fast read operation (*i.e.*, $1.5ns$ sensing delay) is more than 10^{-12}.

5.3 The Dual-Mode Architecture for FS-STT-RAM

5.3.1 Performance

In the following evaluation, the fast read operation with $1.5ns$ sensing delay is adopted. The LP mode doesn't apply extra data recover scheme. Its data error rate is shown in Fig. 9(b). The hA mode intends to eliminate the read-induced errors but sacrifices performance due to the conflicts between write and rewrite. The performance comparison between LP and HA modes is shown in Fig. 10(a). Without integrating the shadow rewrite buffer, an average 18% of IPC degradation has been observed in HA mode compared to LP mode. By referring Fig. 10(b), we found that the workloads with more conflicts, such as `swap` suffers more on IPC degradation, while there is less impact on the workloads like `body` for its less conflicts.

The introduction of the shadow rewrite buffer can significantly reduce the performance degradation caused by the write and rewrite conflicts. Its effectiveness is related to the entry numbers. Fig. 11 shows three typical workloads with the different sensitivities to the rewrite-induced degradation. In general, a larger shadow rewrite

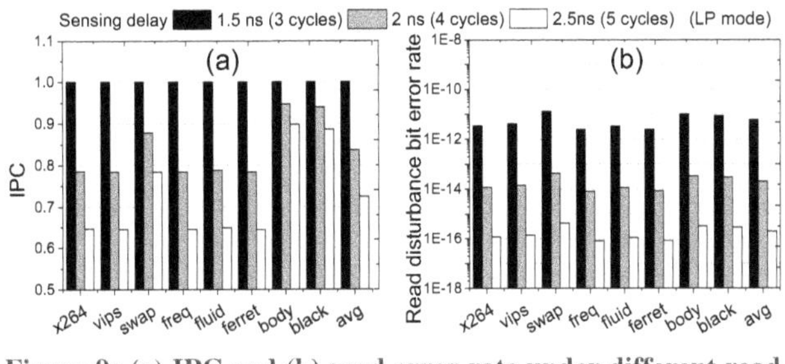

Figure 9: (a) IPC and (b) read error rate under different read latencies.

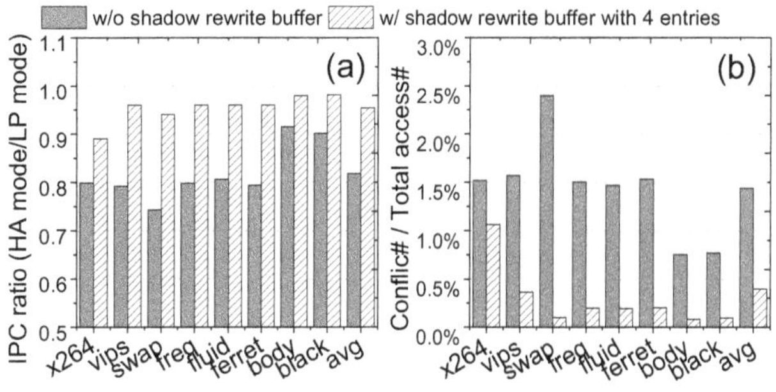

Figure 10: Comparison between LP and HA modes: (a) IPC, and (b) write and rewrite conflicts.

buffer is more effective to reduce the conflicts. When the entry number is above 4, the trend of IPC improvement becomes flat. The only exception is `x264`, which has much more intensive overall accesses as shown in Fig. 6(a). Further increasing the the buffer size can introduce the latency and energy consumption for buffer searching during read operations. After integrating the 4-entry shadow rewrite buffer, the average IPC degradation in HA mode compared to LP mode reduces to 7%.

5.3.2 Energy and EDP

Fig. 12(a) compares the L2 cache overall energy consumptions under the different configurations and different operation modes, with the breakdowns of *shadow rewrite buffer search* (SD search), *rewrite* (Rw), *read* (Rd), *write* (Wr), and *leakage* (leak). In HA mode, the rewrite accounts for about 4% to 12% of overall energy, depending on applications. The leakage energy without the *shadow rewrite buffer* (SD) is much higher than that with the SD because of the prolonged execution time due to performance degradation. The SD increase the rewrite energy slightly and incurs a very small portion of SD searching overhead. Compared to Conv-STT-RAM, the FS-STT-RAM in HA mode with SD can still save $\sim$ 12% total energy on average. The corresponding *energy-delay-product* (EDP) comparison is shown in Fig. 12(b). When the proposed architecture running in LP or HA mode, 34% or 16% of EDP improvement over Conv-STT-RAM can be achieved, respectively.

5.3.3 Reducing Number of Rewrite Bits

Even though only partial of the read bits, *i.e.*, data 1 bits, need to be rewritten after each read operation, the overall rewrite energy overhead is not negligible due to the high volumn of read instructions. The situation is especially severe for those read intensive workloads as we have shown in Fig. 12(a). More explicitly, Fig 13(a) shows the rewrite bits number is close to the total write bits number for some read intensive workloads, such as `x264`. Therefore, the rewrite overhead cannot be ignored.

We proposed the write bit inverting scheme to reduce the rewrite bit number. As shown in Fig. 13(b), a finer granularity can yield a lower rewrite bits number by incurring more bit overhead. When divide a cache block into 16 segments, the number of rewrite bits

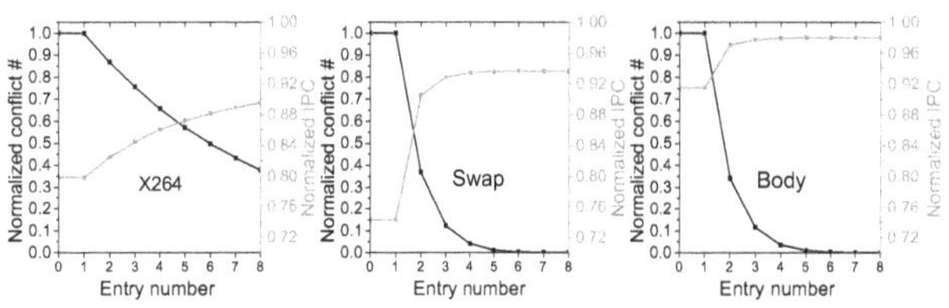

Figure 11: The effectiveness of shadow rewrite buffer at different entry numbers.

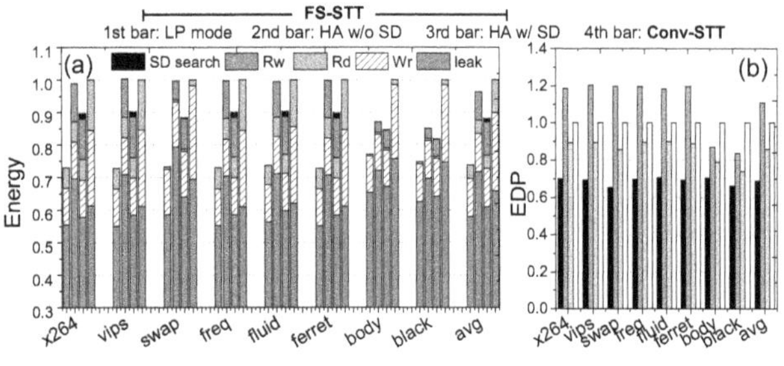

Figure 12: Energy and EDP comparison (normalized to Conv-STT-RAM).

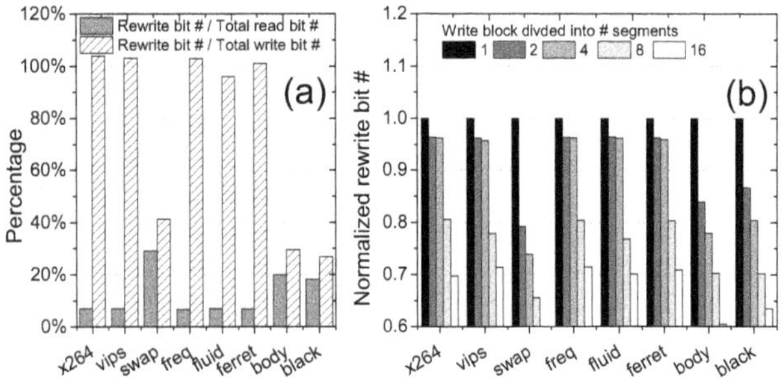

Figure 13: Rewrite bit number statistics and reduction.

can reduce 38% compared to the design with only 1 segments. In other words, we can save $\sim$ 38% of rewrite energy with 16 extra indicating bits per cache block.

6. CONCLUSION

In this paper, we comprehensively study the write and read behavior of FS-STT-RAM at cross layers. Based on the device and circuit exploration, a memory architecture supporting both high accuracy and low power modes is proposed to cope with the read disturbance error for FS-STT-RAM. In the low power mode, data error rate is sacrificed to enable a high performance and low power operation, which on average achieves 6.7% of IPC and 34% of EDP improvements, respectively, over the Conv-STT-RAM. The FS-STT-RAM in the high accuracy mode uses a rewrite-after-read scheme to guarantee the data correctness. Combined with shadow rewrite buffer and write bit inverting scheme, the FS-STT-RAM still demonstrate 19% of EDP improvement over Conv-STT-RAM on average.

7. REFERENCES

[1] R. Ubal, J. Sahuquillo, S. Petit, H. Hassan, and P. Lopez, "Leakage Current Reduction in Data Caches on Embedded Systems," *Proc. of IPC*, pp. 45–50, 2007.

[2] P. Nair, S. Eratne, and E. John, "A quasi-power-gated low-leakage stable sram cell," *Midwest Symp. on Circuits and Systems*, pp. 761–764.

[3] W. Xu and et al., "Design of last-level on-chip cache using Spin-Torque Transfer RAM," *IEEE Trans. on VLSI System*, pp. 483–493, 2011.

[4] G. Sun and et al., "A novel architecture of the 3D stacked MRAM L2 cache for CMPs," *Proc. of 15th HPCA*, pp. 239–249, 2009.

[5] Z. Sun. and et al., "Multi retention level STT-RAM cache designs with a dynamic refresh scheme," *Proc. of Micro*, pp. 329–338, 2011.

[6] X. Dong and et al., "Circuit and microarchitecture evaluation of 3D stacking magnetic RAM (MRAM) as a universal memory replacement," *Proc. of DAC*, pp. 554–559, 2008.

[7] X. Wu and et al., "Hybrid cache architecture with disparate memory technologies," *Proc. of ISCA*, pp. 34–45, 2009.

[8] P. Zhou and et al., "Energy reduction for STT-RAM using early write termination," *Proc. of the ICCAD*, pp. 264–268, 2009.

[9] X. Wu, J. Li, L. Zhang, E. Speight, and Y. Xie, "Power and performance of read-write aware hybrid caches with non-volatile memories," *Proc. of DATE*, pp. 737–742, 2009.

[10] C. Yoshida and et al., "Correlation between microstructure and electromagnetic properties in magnetic tunnel junctions with naturally oxidized MgO barrier," *J. of Applied Physics*, vol. 111, p. 07C716, 2012.

[11] T. Seki and et al., "Switching-probability distribution of spin-torque switching in MgO-based magnetic tunnel junctions," *J. of Applied Physics*, vol. 99, p. 112504, 2011.

[12] H. Zhao and et al., "Low writing energy and sub nanosecond spin torque transfer switching of in-plane magnetic tunnel junction for spin torque transfer RAM," *J. of Applied Physics*, vol. 109, p. 07C720, 2011.

[13] C. Smullen and et al., "Relaxing non-volatility for fast and energy-efficient STT-RAM caches," *Proc. of HPCA*, pp. 50–61, 2011.

[14] A. Driskill-Smith and et al., "Latest advances and roadmap for in-plane and perpendicular STT-RAM," *IEEE InternationalMemory Workshop (IMW)*, pp. 1–3, 2011.

[15] Y. Chen, H. Li, X. Wang, W. Zhu, W. Xu, and T. Zhang, "A nondestructive self-reference scheme for spin-transfer torque random access memory (STT-RAM)," *Proc. of the DATE*, pp. 148–153, 2010.

[16] A. Raychowdhury and et al., "Design space and scalability exploration of 1t-1stt mtj memory arrays in the presence of variability and disturbances," *IEEE Int. Electron Devices Meeting*, pp. 1–4, Dec. 2009.

[17] A. Nigam, C. Smullen, V. Mohan, E. Chen, S. Gurumurthi, and M. Stan, "Delivering on the promise of universal memory for spin-transfer torque RAM (STT-RAM)," *Symp. on ISLPED*, pp. 121–126, 2011.

[18] D. Apalkov, S. Watts, A. Driskill-Smith, E. Chen, Z. Diao, and V. Nikitin, "Comparison of Scaling of In-Plane and Perpendicular Spin Transfer Switching Technologies by Micromagnetic Simulation," *IEEE Trans. on Magnetics,*, vol. 46, no. 6, pp. 2240–2243, 2010.

[19] R. Koch and et al., "Time-resolved reversal of spin-transfer switching in a nanomagnet," *Physical review letters*, vol. 92, no. 8, p. 88302, 2004.

[20] Y. Joo, D. Niu, X. Dong, G. Sun, N. Chang, and Y. Xie, "Energy-and endurance-aware design of phase change memory caches," *Proc. of the DATE*, pp. 136–141, 2010.

[21] "Wind River Simics," *http://www.windriver.com/products/simics/.*

[22] NVsim., "http://www.rioshering.com/nvsimwiki/index.php."

[23] Parsec., "http://parsec.cs.princeton.edu/index.htm."

ASCIB: Adaptive Selection of Cache Indexing Bits for Removing Conflict Misses

Alberto Ros[*], Polychronis Xekalakis[†§], Marcelo Cintra[‡], Manuel E. Acacio[*], and José M. García[*]

[*]Dept. Ingeniería y Tecnología de Computadores, Universidad de Murcia
{a.ros,meacacio,jmgarcia}@ditec.um.es

[†]Intel Labs Barcelona
polychronis.xekalakis@intel.com

[‡]School of Informatics, University of Edinburgh
mc@staffmail.ed.ac.uk

ABSTRACT

The design of cache memories is a crucial part of the design cycle of a modern processor. Unfortunately, caches with low degrees of associativity suffer a large amount of conflict misses, while high-associative caches consume more power per access. We propose ASCIB, a simple technique able to dynamically adjust the bits used for cache indexing so as to minimize conflict misses. By selecting at run time the bits that disperse the working set more evenly across the available sets, ASCIB removes 73% of the conflict misses on average. This results in an improvement in energy efficiency by 17% on average.

Categories and Subject Descriptors

B.3.2 [**Memory Structures**]: Design Styles — Cache memories

Keywords

Cache memories, adaptive indexing, conflict misses, power

1. INTRODUCTION

With the scaling of transistors following Moore's law, the significance of the memory hierarchy to the overall system performance continues growing. The design of the first level cache is an important parameter for achieving high performance systems since it is accessed in the critical path of the processor, which determines the clock frequency of the system. Therefore, first level caches should be as fast as possible, should consume as less power as possible, and should minimize the number of misses.

Cache memories commonly use the least significant bits (LSB) of the block address to form the cache index. For applications that exhibit high spatial locality, i.e., memory requests are mostly consecutive, this indexing function works fairly well because it uniformly distributes requested blocks across the available cache sets. However, for applications that do not exhibit spatial locality, low-associativity caches can suffer lots of conflict misses due to a non-uniform distribution of blocks into cache sets. Although by increasing cache associativity this type of misses can be completely removed, this comes at a high cost in terms of access latency, area required, and power consumption.

Prior work in the field has pointed out that at any given time, 10% of the sets of a first level cache account for 90% of the conflict misses using a typical LSB indexing function [7, 10]. This suggests that a LSB indexing function does not distribute the blocks evenly among the cache sets. As a consequence, several authors have proposed to change the cache indexing function in order to distribute blocks more evenly across cache sets, thus reducing the number of conflict misses [5, 7, 12].These indexing functions are independent of the application that is accessing the cache. However, a careful inspection of the sequence of requested addresses for a particular application can lead to remove a lot of conflicting accesses, for example, by picking the address bits that guarantee a better distribution of blocks among sets to form the index [4].

Finally, it is important to consider that the sets that have the conflicting accesses change not only per application but also within the same application for different program phases [13]. This implies that static indexing policies, such as [4, 5, 7, 12], will not be able to achieve an optimal distribution of blocks. Therefore, we claim that an adaptive cache indexing policy is necessary to minimize the number of conflicting accesses.

In this paper, we propose ASCIB (Adaptive Selection of Cache Indexing Bits), a cache indexing policy that tries to find the address bits that maximize the dispersion of the working set to the available cache sets. This way, most conflict misses can be avoided. Since the working set varies both per application and per phase within the same application, the set of bits that will result in a better dispersion also changes. Therefore, our indexing policy must be able to adapt to such changes at run time.

Experimental results for the SPEC CPU2006 benchmark suite [15] using cycle accurate, full system simulation suggest that our proposal is able to remove 73% of conflict misses and 47% of the total misses for a direct-mapped cache. This is reflected in IPC improvements of 8.5% on average when compared to a conventional LSB indexing policy. Having less accesses to the L2 cache, the proposed scheme also consumes 17% less energy on average. These significant benefits come with less than 2% area overhead.

2. RELATED WORK

There are two main approaches for reducing conflict misses in low-associativity caches. The first approach relies on rehashing conflicting blocks to alternative cache sets [1, 10, 11, 13]. The sec-

[§]This work was done before the author joined Intel, while being at the University of Edinburgh.

ond approach aims at distributing referenced blocks more evenly across the sets by changing the indexing function [4, 5, 7, 12]. Since the mechanism presented in this paper follows the later approach, we start this section by analyzing the later works.

XOR-based mapping policies (e.g., bitwise [5] and polynomial [12]) are used to obtain a pseudo-randomly placement of blocks. These mapping functions require a previous computation of the block address to obtain the cache index, being in both cases fairly simple to implement. However, XOR-based mapping policies are not aware of the particular access patterns of each application and do not perform run-time adaptation.

Kharbutli et al. [7] propose two indexing functions based on operations with prime numbers (prime modulo and prime displacement). Requiring complex logic to calculate the cache index, makes these techniques prohibitive for first level caches, where timing is highly critical. Moreover, similar to the XOR-based mapping, they do not perform run-time adaptation.

The skewed-associative cache [14] is a 2-way cache that uses one hashing function per way. This function is derived by XORing two bit fields from the block address. The most significant limitation for skewed-caches is that they cannot easily use a pseudo LRU (or true LRU) replacement policy, since there is no notion of a cache set. Since we maintain the same indexing function per set, the proposed scheme does not have this issue.

In [4] it was also noticed that the use of some bits for the cache indexing could reduce the miss rate. For detecting which bits to use, they employ offline profiling. This approach assumes that the workloads are known prior to execution, which is generally not the case (for embedded devices this may be a valid assumption).

The B-Cache [17] tries to reduce conflict misses by balancing the accesses to the sets of direct-mapped caches. In order to do this they increase the decoder length and incorporate programmable decoders and a replacement policy to the design. Our proposal is similar in that we also extend the number of bits considered for the cache index. However, the proposed scheme can choose a larger number of bits for the indexing function which is very beneficial. Additionally, the proposed indexing logic is simpler since it only has to select the bits that comprise the cache index.

The column-associative cache [1] uses a direct-mapped cache and an extra bit for dynamically selecting alternate hashing functions. Although this form of semi-associativity is able to remove a large amount of misses it does so at the cost of a large number of second accesses. Column-associative caches can be extended to include multiple alternative locations [3, 18].

The V-Way cache [11] tries to remove many of the conflict misses by allowing more tags than physical sets and using pointers to associate the tags in use with the actual sets. This scheme tries to emulate a global replacement policy for set-associative caches, thus removing conflict misses that are due to poor replacement of blocks.

3. THE ASCIB INDEXING POLICY

3.1 Detecting the Best Address Bits

An appropriate choice of the address bits that comprise the indexing function is essential for achieving a good distribution of blocks among cache sets, and therefore, for avoiding conflict misses. Since the problem of obtaining these bits from all the available ones has been demonstrated to be NP-complete [4], we opt for a simpler heuristic that iteratively improves the indexing function by only swapping one bit upon every indexing function change. After several iterations we will reach a sub-optimal solution for which any bit change will lead to a worse distribution of blocks. Of course, this solution can change dynamically depending on the variations in the working set.

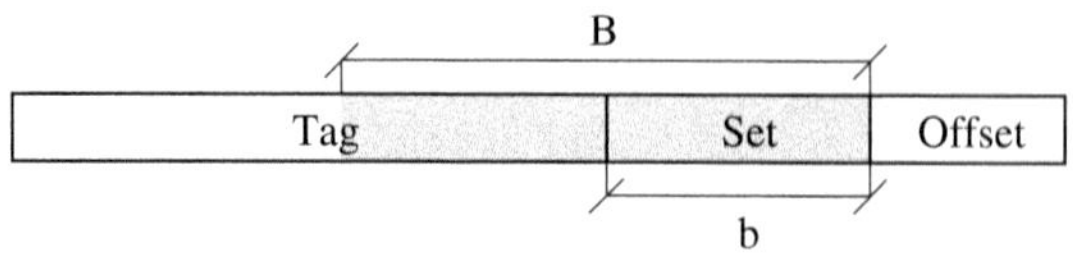

Figure 1: Memory address and bits used to index the cache.

In particular, each iteration of the algorithm is split into three phases: the *bit-victimization phase*, the *bit-selection phase*, and the *idle phase*. During the *bit-victimization phase*, our algorithm selects the bit that will be replaced from the current indexing function. Ideally, this bit should be the one that does not help in dispersing the accessed blocks evenly among the cache sets. When such a bit is found the *bit-selection phase* starts. During this phase, the bit that replaces the victimized one is selected. This bit should be the one that, along with the remaining bits (i.e., all bits forming the index apart from the victimized one), would result in spreading memory references to cache sets as much as possible. Since the bit-selection algorithm also considers the victimized bit, it may finally choose as the new bit the one we had before. This is a nice property of the algorithm, since if the current indexing function can not be improved, it is not changed. Finally, an *idle phase*, where no actions are performed by the algorithm, is introduced in order to save power consumption. Since no measurements are performed, during this phase no extra power is consumed. The underlying assumption is that once the bit-selection phase ends the indexing function will perform properly for some time.

An important decision for designing the proposed scheme is the number of address bits that should be considered as candidates for being part of the indexing function. As expected the more bits are analyzed, the more conflict misses are avoided. However, at the same time if a very large number of bits is examined, our proposal may have a negative impact in both latency, area, and power consumption. Throughout this paper, we will call the number of bits needed to form the cache index as b and the number of address bits considered for selection as B (see Figure 1).

Next subsections describe these three phases in more detail. It is important to point out that the operations performed in each phase are not in the critical path of the cache access. As such the required hardware does not have any impact on the cache access latency.

3.1.1 Bit-Victimization Phase

The goal of this phase is to find which bit should be removed from the current indexing function. We have found that bits that do not help in distributing accesses across the cache have one of the following two characteristics:

- *Low entropy*: Entropy is a metric that measures variability. A bit position that hardly changes its value for a certain set of memory references has low entropy. If a bit position with low entropy is used to form the cache index, most accessed blocks map to half of the sets in the cache, while the other half would remain almost unused. Figure 2a shows an example where bit b_0 presents no entropy at all, so referenced blocks can only map to the shaded sets.
- *High correlation*: Two bit positions that most times have the same value or most times have a different value show high correlation. If the cache index is form from two bits that have high correlation, again, half of the sets in the cache is highly utilized while the remaining half just maps a few blocks. Figure 2b shows an example where bits b_1 and b_2 are totally correlated. Again, referenced blocks can only map to the shaded sets.

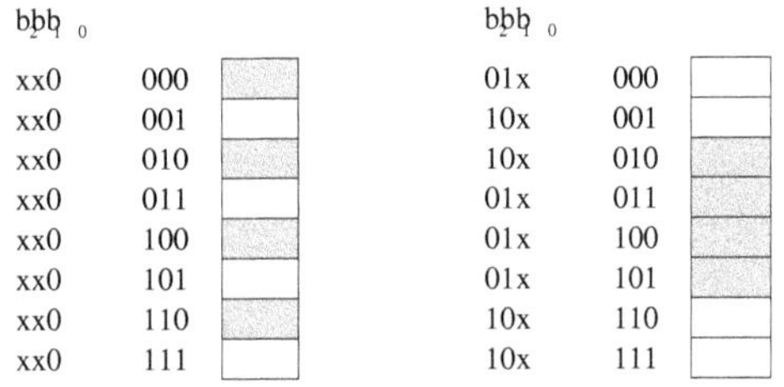

(a) Low entropy example. (b) High correlation example.

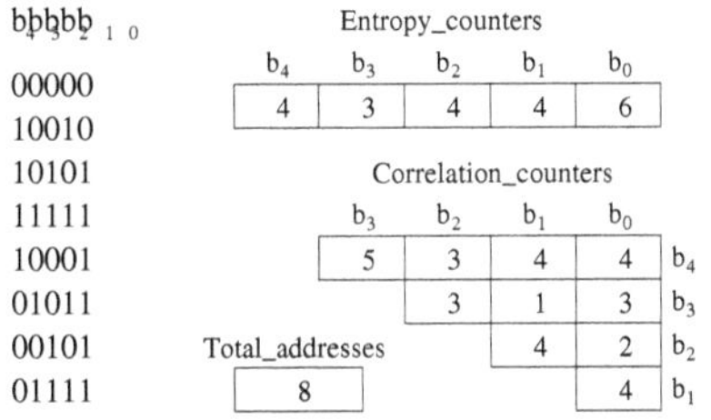

(c) Counters required and example.

Figure 2: Examples and counters for the bit-victimization phase.

During this phase we are going to discard either the bit with low entropy or one of the two bits presenting higher correlation. Both metrics are calculated considering only the memory addresses of the blocks that suffer a cache miss. We need b counters to calculate the entropy of each bit and $\frac{b\,(b-1)}{2}$ counters for the correlation of any pair of bits. This results in a total of $\frac{b\,(b+1)}{2}$ counters. A deep analysis about area requirements can be found in Section 5.3.

The algorithm works as follows. At the beginning every counter is reset. Entropy counters are updated by adding each bit value to the corresponding counter, i.e., $entropy\_counter_i(EC_i)$ is increased when bit_i is 1. Correlation counters are updated by XORing every pair of bits and by adding the result to the corresponding counter, i.e., $correlation\_counter_{i,j}(CC_{i,j})$ is increased when $bit_i \oplus bit_j$ is 1. A final counter is used to store the total number of addresses processed. The victimization phase ends either when any of the counters saturates (each counter is comprised of 14 bits, i.e, it can count up to 16K misses) or after a certain number of cycles (through experimentation we have found that 100000 cycles works properly). At the end of the phase, the selected bit will be the one with lower entropy or higher correlation. In order to compare both metrics we define the usefulness metric (U) for a single bit and a pair of bits, which is calculated as follows:

$$U_i = MIN(EC_i, total\_addresses - EC_i) \quad (1)$$

$$U_{i,j} = MIN(CC_{i,j}, total\_addresses - CC_{i,j}) \quad (2)$$

The bit with lower U is chosen for being evicted from the indexing function. In case the lower U corresponds to a pair of bits (correlation), the one with lower entropy will be evicted. Figure 2c gives an example of the bit-victimization process for a 32-set cache ($b = 5$). The lower U corresponds to $CC_{1,3}$, and from these to bits, the one with lower entropy is bit b_3, so b_3 will be the discarded bit.

3.1.2 Bit-Selection Phase

This phase selects the bit that will be the best replacement for the victimized one. This choice is based on a metric that we name as *mean relative period* (MRP). In order to understand this metric, we will first explain the concept of *mean period* (MP) for an address bit position. By focusing on just one bit position of the address for a certain number of memory accesses, the period of the bit position can be defined as the number of consecutive 0's or 1's. When

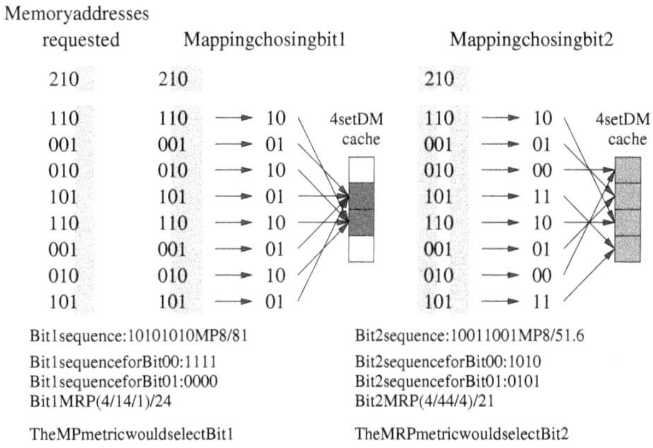

Figure 3: Example of MP and MRP metrics. The cache needs two bits for the indexing. b_o is the one that passed the bit-victimization phase. b_1 and b_2 are candidates to form the index.

there is a bit change (from 0 to 1, and vice versa), a new period is computed. The mean period is then the average of these periods:

$$MP = \frac{\sum_{i=0}^{np} period_i}{np} = \frac{nb}{np} \quad (3)$$

where $\sum_{i=0}^{n} period_i$ corresponds to the number of bits in the sequence (nb), and np is the number of periods, i.e., the number of bit changes (from 0 to 1 or from 1 to 0) plus one. For example, for the sequence 00111011, $nb = 8$ and $np = 4$, so $MP = 2$.

However, although the bit with smaller mean period is the bit that more frequently changes, this metric does not guarantee that this is the best bit to be selected. This is because the new bit could highly correlated with one of the other bits that are going to form the index. Hence, we propose the *mean relative period* (MRP), which is the mean period calculated considering subsequent accesses whose address always has a set of bits kept unchanged. In our case, these set of bits are the $b-1$ bits that passed the bit-victimization phase.

$$MRP = \frac{\sum_{j=0}^{2^{b-1}} mean\ period(j)}{2^{b-1}} = \frac{\sum_{j=0}^{2^{b-1}} \frac{nb_j}{np_j}}{2^{b-1}} \quad (4)$$

Our goal is to find the bit from the remaining $B-(b-1)$ bits with lower MRP, because this will be the bit changing the most while keeping fixed the other ones, i.e., not correlated with any of the used bits. Conceptually, this will also be the bit that will help distribute better the requested blocks among sets. Figure 3 shows an example of the goodness of these two metrics.

The scheme illustrated in Figure 4 is used to compute this metric for each of the candidates. The required structure is a small *tag cache* with the same associativity as the data cache but half number of sets, and it is indexed using the same bits of the data cache after removing the bit that we wish to replace. Therefore, this cache uses a similar indexing logic as the data cache, which will be explained in Section 3.3. Only the address bits that we wish to analyze ($B-(b-1)$ bits) are kept in the tags field of the tag cache along with a valid bit, and as such it is fairly small. This cache is updated on every data cache access. However, due to its small size the power consumption of our proposal does not increase significantly. On a replacement from this structure, the tag of the evicted block and the tag of the new block are bitwise XORed. This provides information about changes of relative periods for each of the tested bits, which are kept in $B-(b-1)$ registers. After a certain number of cycles, the number of periods of each bit are compared. The bit with greater number of periods is selected, since it will be

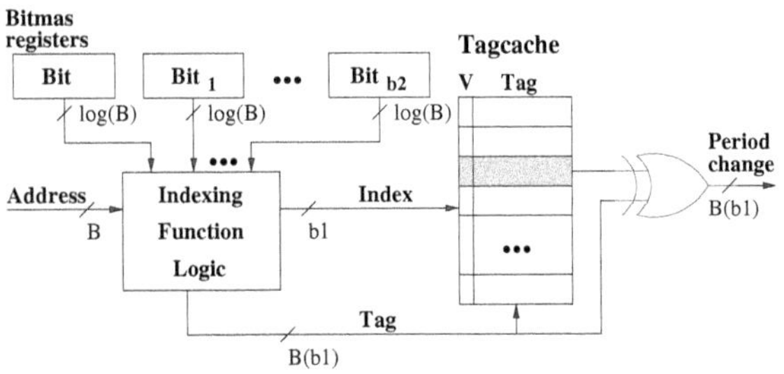

Figure 4: Scheme of the bit-selection process.

the one with smaller MRP. We have found through experimentation that a good length for this phase is 100000 cycles.

Having completed the second phase, the algorithm can proceed to update the indexing function if necessary, first evicting from the cache those blocks that will reside in a wrong set according to the new indexing function, as will be explained in Section 3.2.

3.1.3 Idle Phase

Once the bit-selection phase is completed and the indexing function has been changed (if necessary), we assume that the current function will perform appropriately for some time. Therefore, we decide to introduce an idle phase where no measurements are made, so we can save power. Through experimentation we have found that a length for this phase of 400000 cycles is a good trade-off because the proposed mechanism does not lose effectiveness. When this phase finishes, the algorithm begins a new iteration starting from the bit-victimization phase.

3.2 Updating the Indexing Function

From the point when the indexing function changes, all subsequent memory references will use the new function to access the cache. This can cause consistency issues since memory blocks that resided in the cache prior to the indexing change can be referenced and mapped to a different set.

A simple approach to maintain the consistency of the cache is to evict from cache those blocks that will be mapped to a different set after the index change. Since the proposed algorithm only changes one bit per iteration, some of the cached blocks will be mapped to the same set using the new indexing function.

In order to detect which cache blocks will map to a different set after the index change, it is only necessary to check if the selected bit and the victimized one have the same value for every block stored in cache. Therefore, a cache lookup is required on every index change. Note that the address of a cached block can be obtained by merging its tag and the set where is stored according to the indexing function employed to map the block to the cache.

If there is not a variation in the working set, our indexing function should not change. Therefore, we expect that index changes are triggered on the boundaries of different program phases where the working set also changes and as such many misses would occur in any case. As consequence, the overall impact of the misses caused by index changes should be small. Despite the simplicity of this scheme it works fine, as it will be shown in Section 5.

3.3 Forming the Adaptive Index

A critical component of the proposed mechanism is the one that forms the index out of the available bits. This circuit has to be fast since it will lie in the critical path of the cache access and as such it may impact the access time.

The proposed circuit is depicted in Figure 5. The indexing function is determined by the position of the bits selected for the indexing. This information is stored in b registers (called *bit mask registers*). These registers are updated when the indexing function changes. The size of each register is $log_2(B)$. These B bits are taken from each block address in order to calculate the corresponding cache set for the block. Our logic has b B-to-1 multiplexers, which are lied out in parallel, each of them used for selecting one address bit according to its corresponding bit mask register. The outputs of all the multiplexers form the desired cache index.

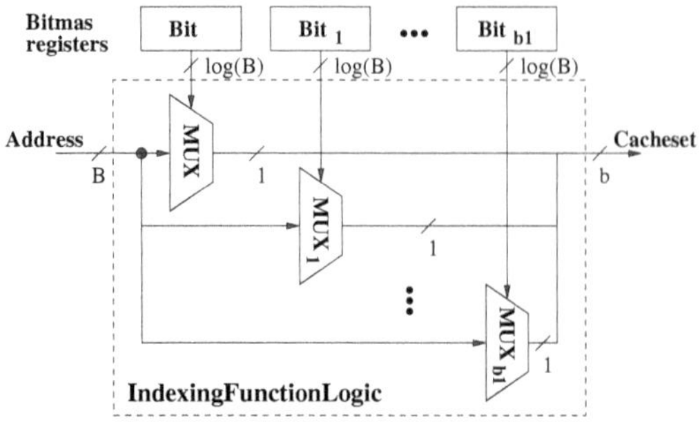

Figure 5: Forming the index using the selected bits

Table 1: Architectural parameters of the system simulated.

Parameter	Value
Core Frequency	2.33GHz
Fetch/Issue/Retire Width	6, 4, 4
I-Window/ROB	64, 100
Ld/St Queue	64, 64
Branch Predictor	YAGS 64Kb
BTB/RAS	1K entries, 2-way, 32 entries
L1 DCache and ICache	16KB, 1-way, 1 cycle
L2 Cache	2MB, 16-way, 7 cycles
Main Memory	150 cycles

The multiplexers shown in this figure can be implemented using transmission gates in order to minimize their delay. Since all multiplexers are processed in parallel, the delay of the indexing logic is represented by the delay of a B-to-1 multiplexer. This delay depends on the number of transistors in the critical path with can be calculated as a function of B. We found that a value for B greater than 16 does not improve the accuracy of the proposal, and therefore we can use 16-to-1 multiplexers, whose delay when implemented using transmission gates is very small (one inverter and four transmission gates in the critical path [2]). But even this small delay can effectively be completely hidden since the time required to perform the tag check is typically more than the one required to access the data component.

4. EVALUATION METHODOLOGY

To evaluate the proposed scheme we employ the Simics full-system simulator [8] extended with the Gems toolset [9] so as to simulate a cycle-accurate out-of-order processor (Opal) and memory subsystem (Ruby). Our area, power and timing estimations are based on Cacti 5.3 [16], assuming a 45nm process technology. We use the SPEC CPU2006 benchmark suite [15] to drive our simulation infrastructure. We fast forward all the benchmarks for the first 4 billion instructions. Throughout this period we only warm up the caches. We then simulate each of the benchmarks for a slice of 500 million instructions for which we keep statistics.

As the base configuration (Table 1) we evaluate a DM cache that uses an LSB indexing scheme (*LSB*). *ASCIB* uses a value $B = 16$. We also evaluate the bitwise Xor function [5] (*Xor*), the prime module function [7] (*PrimeMod*), the near-optimal static indexing [4] (*Static*) and the column-associative cache [1] (*ColumnAssoc*). We also include a section comparing ASCIB to set-associative caches that uses the least significant bits for indexing the cache (*LSB-2ways* and *LSB-4ways*). We pessimistically, for our scheme, assume the same access latency for all schemes.

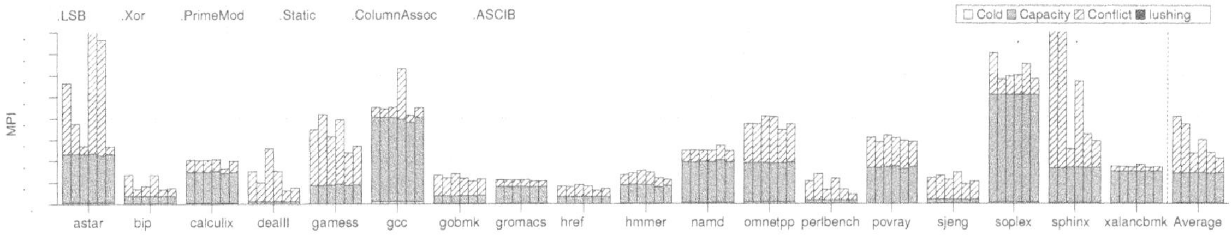

Figure 6: MPKI for the schemes evaluated in this paper.

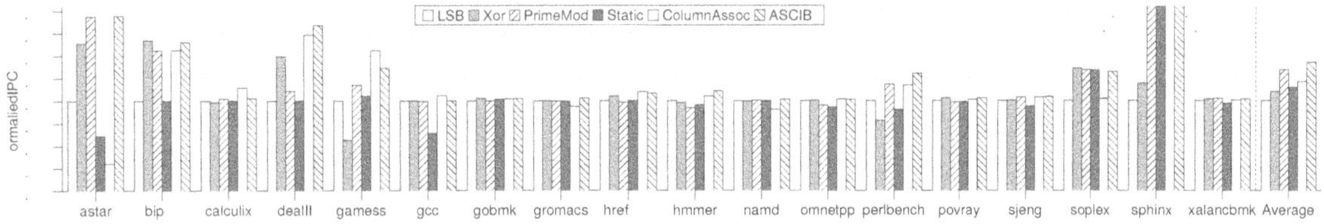

Figure 7: Normalized IPC of the schemes evaluated in this paper.

5. EXPERIMENTAL RESULTS

5.1 Performance Evaluation

Figure 6 presents the MPKI (misses per thousand instructions) for the different schemes evaluated. In this graph, cache misses are split into four categories: *Cold*, *Capacity* and *Conflict* misses corresponds to the typical categories [6], while *Flushing* misses represent the ones caused as consequence of evictions performed upon index changes. The graph shows that for the assumed 16KB DM cache, the amount of capacity misses is quite high.

We can observe that the main benefit of all the schemes over the base case comes in terms of a reduction in the conflict misses. *Xor* reduces the conflict misses by 13% on average. However, this approach increases the number of misses for several applications (e.g., *gamess*, *hmmer*, *perlbench*, and *sjeng*). *PrimeMod* avoids 65% of conflict misses on average, but do not behaves properly for some applications, particularly for *dealII* where the number of MPKI is almost doubled with respect to the base case. The static indexing is able to reduce the number of conflict misses by 41% on average. However, for some applications like *astar*, *gcc*, or *sjeng* it incurs in more misses than the base configuration. The column associative cache is able to reduce conflict misses by 62% on average, although it increases the number of MPKI for *astar* compared to *LSB*. Finally, we can see that our proposal is the one that is able to remove the largest fraction of conflict misses assuming a DM cache (73% on average), reducing the MPKI for all the applications (47% on average). Note that the amount of *Flushing* misses for our proposal is negligible. This supports our assumption that index changes happens on the boundaries of program phases.

In Figure 7, the IPC of the evaluated approaches is presented. As the graph confirms there is a strong correlation between the MPKI and the IPC. The *Xor* scheme is 2% better than the base case. Perhaps the most worrying fact for this scheme is its inconsistent behavior, since it is worse than the *LSB* indexing for some cases (e.g., *gamess* and *perlbench*). *PrimeMod* improves the IPC compared to *Xor*, obviating the fact that the logic for generating the index cannot be efficiently implemented for L1 caches. Particularly, if the hash function could be implemented in less than 1 cycle it would be 6.8% better than the base case. The static indexing is 3% better than the base case. However, it is not behaving better that the base configuration for all the benchmarks (e.g., *astar* and *gcc*). The column-associative improves the IPC by 4.1% on average over the base case. Although it obtains similar reduction in MPKI, sometimes hits require a second cache access –this happens for 22% of the hits on average– which negatively impacts performance. Finally, ASCIB is better than all the previously proposed schemes, improving IPC by 8.5% over the base case.

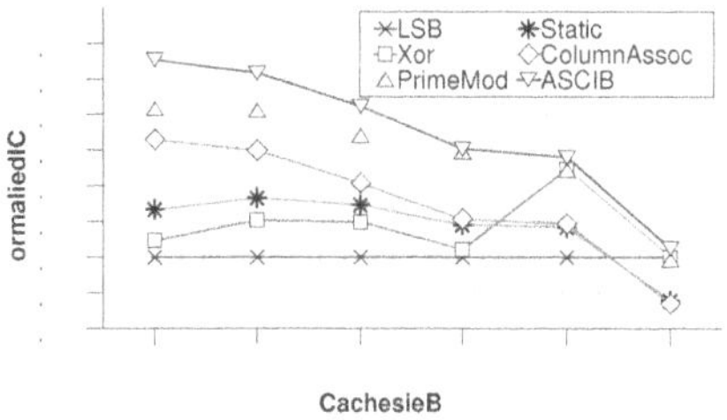

Figure 8: Normalized IPC considering different cache sizes.

To conclude the performance evaluation, Figure 8 shows the IPC of the evaluated proposals for different cache sizes (from 4KB to 128KB). We can see that the adaptive scheme outperforms the other proposals for all cache sizes. Obviously, when the cache size becomes very large, the cache miss rate decreases, and therefore, the improvements in IPC are lower. The schemes that perform closer to our proposal are *Xor* (for large cache sizes), and the unimplementable *PrimeMod*.

5.2 Energy Efficiency

Figure 9 presents an evaluation of the difference in the dynamic energy consumed by the caches for all the schemes. This figure splits the consumption of the three main structures involved in our memory hierarchy: the L1 cache, the L2 cache, and the tag cache and other counters employed by our mechanism. As it is shown, the more the L1 miss ratio is reduced, the more dynamic energy is saved at the L2 cache. The consumption of the L1 cache is similar for all the indexing policies except for the column-associative cache, where some times a cache hit requires two cache accesses. Finally, the energy consumption of the tag cache along with the counters employed by our proposal is negligible. Therefore, our proposal reduces the energy consumption compared to the base case by 17% on average, and it is the more energy efficient scheme from the proposals evaluated.

5.3 Area Overhead

Our algorithm uses the following extra structures to adapt the indexing function at run time: $\frac{b\,(b+1)}{2}$ *counters* used to detect low-entropy and high-correlated bits during the bit-victimization phase, a *tag cache* that computes the MRP for the bits considered in the bit-selection phase, and the other counters for storing the MRP for each bit being analyzed to be selected.

Table 2 shows the extra area requirements of our proposal for several cache sizes. Each of the counters employed in the bit-

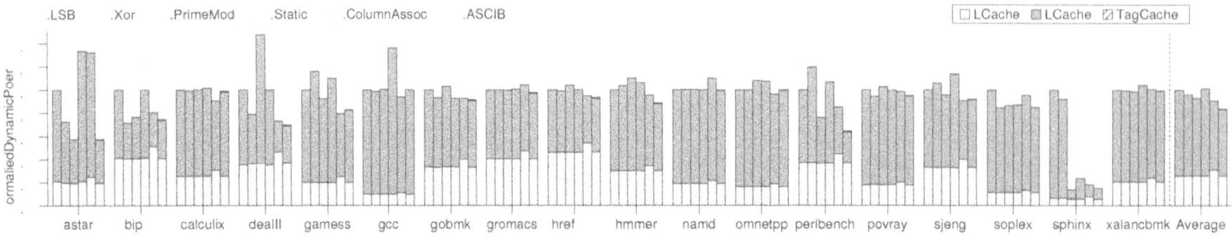

Figure 9: Normalized dynamic energy consumption for the schemes evaluated in this paper.

Table 2: Area overhead of the proposed mechanism.

Cache size (KB)	Cache sets	Index bits (b)	Candidate bits	Bit-victimization bytes (overhead)	Bit-selection bytes (overhead)	ASCIB overhead
4	64	6	11	39.25 (0.90%)	69.00 (1.57%)	2.47%
8	128	7	10	51.50 (0.59%)	107.25 (1.22%)	1.81%
16	256	8	9	65.50 (0.37%)	177.50 (1.02%)	1.39%
32	512	9	8	81.25 (0.23%)	303.75 (0.87%)	1.10%
64	1024	10	7	98.75 (0.14%)	526.00 (0.75%)	0.89%
128	2048	11	6	118.00 (0.08%)	908.25 (0.66%)	0.74%

victimization phase has 14 bits. Since we only consider up to $B = 16$ bits from the address and there are $b - 1$ bits remaining from the bit-victimization phase, only $16 - (b - 1)$ bits plus the valid bit need to be stored in the tag cache, which has half number of sets than the data cache. We can also observe that our proposal scales very well with the size of the cache, i.e., the area overhead decreases as the cache size increases. Overall, the area overhead of our proposal is only 1.39% with respect to a 16KB DM cache.

5.4 Comparison to associative caches

In this section we compare our proposal to 2-way and 4-way associative caches, idealistically considering that the access time of all caches is the same, i.e., favoring associative caches. Figure 10 shows this comparison in terms of CPI, energy, and area. For the energy consumption we consider both the L1 and the L2 caches, since an associative L1 cache incur in more energy consumed per access, but also in less consumption at the L2 cache. Results are normalized with respect to a traditional DM cache (*LSB-DM*). The bars labeled as *Energy_L1* and *Energy_L2* represent the faction of the *Energy_L1+L2* bar consumed by each cache level.

The 2-way associative cache reduces CPI by 11.7%, while the 4-way associative cache reduces it by 13.4%. Although this is a higher reduction than the obtained by our proposal, this numbers consider 1-cycle hit time for the associative cache –for a 2-cycle 2-way cache, the CPI is slightly worse than our proposal–. However, this performance comes at both area and energy costs. The consumption of the L1 cache increases when we use associative caches. Although the number of L2 accesses is reduced, the impact in L1 energy consumption is more important so the overall power increases, up to 52% for a 4-way cache compared to a DM or a 2-way cache. Finally, the area overhead of ASCIB with respect to a 16KB DM cache is less than 2%. A 2-way set-associative cache increases area requirements by 8,7%.

6. CONCLUSIONS

We have proposed an adaptive cache indexing policy (ASCIB) that is able to reduce the conflict misses, by more uniformly spreading the memory references to the available sets. The basic premise of this work is that the non-uniformity in the set usage is caused by a poor selection of the index bits. Instead, by selecting as index bits the bits that disperse the working set more evenly in the available sets, a large fraction of the conflict misses can be removed.

We have shown that ASCIB is able to reduce the percentage of conflict misses by 73%. This is reflected in IPC improvements by 8.5% on average when compared with a conventional cache that used the least significant bits. Finally, when ASCIB is used for DM

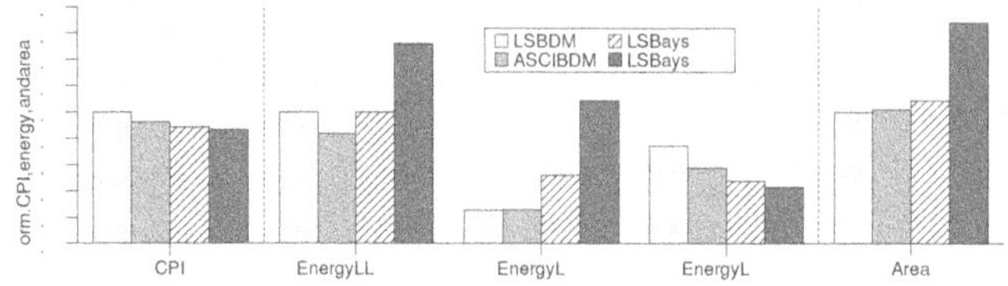

Figure 10: Comparison to associative caches in terms of CPI, dynamic energy consumption, and area. Lower is better.

caches it can obtain a CPI close to associative caches, consuming 17% less energy, and requiring 6% less area than a 2-way associative cache.

7. ACKNOWLEDGMENTS

This research was supported by the Spanish MEC and MICINN under Grant TIN2009-14475-C04.

8. REFERENCES

[1] A. Agarwal and S. D. Pudar. Column-associative caches: A technique for reducing the miss rate of direct-mapped caches. In *20st Int'l Symp. on Computer Architecture (ISCA)*, pages 179–190, May 1993.

[2] K. Chaudhary, P. D. Costello, and V. M. Kondapalli. Programmable circuit optionally configurable as a lookup table or a wide multiplexer. U.S. Patent 7075333, Nov. 2006.

[3] B.-K. Chung and J.-K. Peir. LRU-based column-associative caches. *Computer Architecture News*, 26(2):9–17, May 1998.

[4] T. Givargis. Improved indexing for cache miss reduction in embedded systems. In *40th Design Automation Conference (DAC)*, pages 875–880, June 2003.

[5] A. González, M. Valero, N. Topham, and J. M. Parcerisa. Eliminating cache conflict misses through XOR-based placement functions. In *11th Int'l Conference on Supercomputing (ICS)*, pages 76–83, June 1997.

[6] M. D. Hill and A. J. Smith. Evaluating associativity in CPU caches. *IEEE Transactions on Computers (TC)*, 38(12):1612–1630, Dec. 1989.

[7] M. Kharbutli, K. Irwin, Y. Solihin, and J. Lee. Using prime numbers for cache indexing to eliminate conflict misses. In *11th Int'l Symp. on High-Performance Computer Architecture (HPCA)*, pages 288–299, Feb. 2004.

[8] P. S. Magnusson, M. Christensson, and J. Eskilson, et al. Simics: A full system simulation platform. *IEEE Computer*, 35(2):50–58, Feb. 2002.

[9] M. M. Martin, D. J. Sorin, and B. M. Beckmann, et al. Multifacet's general execution-driven multiprocessor simulator (GEMS) toolset. *Computer Architecture News*, 33(4):92–99, Sept. 2005.

[10] J.-K. Peir, Y. Lee, and W. W. Hsu. Capturing dynamic memory reference behavior with adaptive cache topology. In *8th Int'l Conf. on Architectural Support for Programming Language and Operating Systems (ASPLOS)*, pages 240–250, Oct. 1998.

[11] M. K. Qureshi, D. Thompson, and Y. N. Patt. The v-way cache: Demand based associativity via global replacement. In *32nd Int'l Symp. on Computer Architecture (ISCA)*, pages 544–555, June 2005.

[12] B. R. Rau. Pseudo-randomly interleaved memory. In *18th Int'l Symp. on Computer Architecture (ISCA)*, pages 74–83, June 1991.

[13] D. Rolán, B. B. Fraguela, and R. Doallo. Adaptive line placement with the set balancing cache. In *42nd IEEE/ACM Int'l Symp. on Microarchitecture (MICRO)*, pages 529–540, Dec. 2009.

[14] A. Seznec. A case for two-way skewed-associative caches. In *20st Int'l Symp. on Computer Architecture (ISCA)*, pages 169–178, May 1993.

[15] Standard Performance Evaluation Corporation. SPEC CPU2006. http://www.spec.org/cpu2006.

[16] S. Thoziyoor, N. Muralimanohar, J. H. Ahn, and N. P. Jouppi. Cacti 5.1. Technical Report HPL-2008-20, HP Labs, Apr. 2008.

[17] C. Zhang. Balanced cache: Reducing conflict misses of direct-mapped caches. In *33rd Int'l Symp. on Computer Architecture (ISCA)*, pages 155–166, June 2006.

[18] C. Zhang, X. Zhang, and Y. Yan. Two fast and high-associativity cache schemes. 17(5):40–49, Sept. 1998.

Energy-Efficient Non-Minimal Path On-chip Interconnection Network for Heterogeneous Systems

Jieming Yin[*], Pingqiang Zhou[†], Anup Holey[*], Sachin S. Sapatnekar[†], and Antonia Zhai[*]
[*]Department of Computer Science and Engineering
University of Minnesota, Twin Cities
Minneapolis, Minnesota 55455, USA
{jyin, aholey, zhai}@cs.umn.edu
[†]Department of Electrical and Computer Engineering
University of Minnesota, Twin Cities
Minneapolis, Minnesota 55455, USA
{pingqiang, sachin}@umn.edu

ABSTRACT

Network-on-Chips (NoCs) in heterogeneous systems containing both CPU and GPU cores must be designed to satisfy the performance requirements of both latency-sensitive CPU traffic and throughput-intensive GPU traffic. DVFS and adaptive routing can potentially improve NoC energy and performance efficiency. We further notice that GPU traffic can sometimes tolerate a slack defined as the number of cycles a packet can be delayed without causing performance penalty. In this work, we take advantage of the slack in GPU packets to route packets through non-minimal path, so that routers can operate at a lower frequency without suffering performance penalty.

Categories and Subject Descriptors

C.1.2 [**Multiprocessors**]: Interconnection architectures; C.1.4 [**Parallel Architectures**]: Distributed architectures

General Terms

Design, Performance

Keywords

Heterogeneous Multi-core, Interconnects, NoC, Voltage Scaling, Frequency Scaling, Non-minimal Path Routing.

1. INTRODUCTION

CPU cores nowadays are optimized for single-thread workloads, while GPU cores are optimized for high throughput workloads. Heterogeneous multi-core systems that contain both CPU and GPU cores can potentially achieve the best of both worlds in an energy efficient manner by utilizing different cores as workload characteristics change [20]. In a heterogeneous multi-core system, different components on the die must communicate through a network on-chip (NoC). Since the NoC is the critical resource shared by all the applications running on the processor, it must be carefully designed to satisfy the energy envelop and to meet the requirements for variety of traffic patterns.

In a modern CPU, the processor stalls when requested data are not returned. Therefore, traffic originated from or designated to a CPU core is latency-sensitive. Applications that are computation-bound generate relatively little NoC traffic; while data centric applications can generate significant traffic. The former has a moderate throughput requirement; and the latter has a considerable throughput requirement.

A GPU consists of a collection of data-parallel compute cores referred to as streaming multiprocessors (SMs). A large number of lightweight threads are allocated to each SM and are batch-scheduled on the SIMD pipeline in a fixed size group called a warp. In other words, warps are the basic execution units in a GPU. All the threads within a warp execute the same instruction but operate with different data values. When a thread in a particular warp encounters a load miss, the entire warp is context-switched out, and another available warp is scheduled. Notice that usually there are hundreds of threads executing on a GPU in parallel, so it is highly possible that large numbers of outstanding memory requests are being processed in the interconnection network. Thus GPU traffic requires high network bandwidth. On the other hand, as long as there are enough available warps to schedule in an SM core, the system performance would not be harmed despite the long memory access latency, since the latencies are hidden by scheduling other ready warps. As a result, some GPU traffic is latency-insensitive, and can be delayed in the NoC without performance impact.

Dynamic Voltage and Frequency Scaling (DVFS) is an effective technique for reducing network energy dissipation [12, 17, 21]. The key idea behind network DVFS is to dynamically scale the supply voltage of the routers and links to provide *just-enough* circuit speed to process the traffic workload. Reducing NoC frequency can increase latency and/or reduce throughput, thus can only be applied to workloads with moderate latency and throughput requirements. However, in heterogeneous systems where traffic pattern is mixed, it is difficult to determine the optimal DVFS level.

In this work, we notice that GPU traffic can sometimes tolerate a slack defined as the number of cycles a packet can be delayed without causing performance penalty. Taking advantage of the slack in GPU packets, we propose an NoC design to route packets through non-minimal paths. Such design balances on-chip traffic workloads so that routers can operate at a lower frequency without suffering significant performance penalty.

This paper makes the following contributions:

- We observe that the number of available warps in an SM directly relates to the delay introduced in warp scheduling. Taking advantage of this observation, we propose a mechanism that can dynamically predict the network packet slack. This slack predicting mechanism is applied to GPU messages.
- We propose an energy-efficient network tuning policy that consists of a DVFS mechanism as well as a non-minimal path routing algorithm to reduce the energy consumption of the on-chip interconnection network while mini-

ISLPED'12, July 30–August 1, 2012, Redondo Beach, CA, USA.

mizing the performance degradation caused by network frequency scaling.

2. EXPLOITING SLACK IN GPU

Based on traffic characteristics, different strategies should be applied to different network messages. For CPU applications, a reduction in NoC throughput has relatively little impact on overall performance, but increasing NoC latency can cause significant performance degradation. While for GPU applications, delaying memory requests in the network has little effect on performance, provided a minimum network throughput is guaranteed. However, we cannot increase the NoC latency of a GPU packet limitlessly. Hence, it is necessary to find an upper bound for delaying a network packet for GPU applications.

Since all outstanding memory requests are sent via the on-chip interconnection network in the form of packets, we first define the *slack of a packet* as the number of cycles a packet can be delayed without affecting the overall execution time.

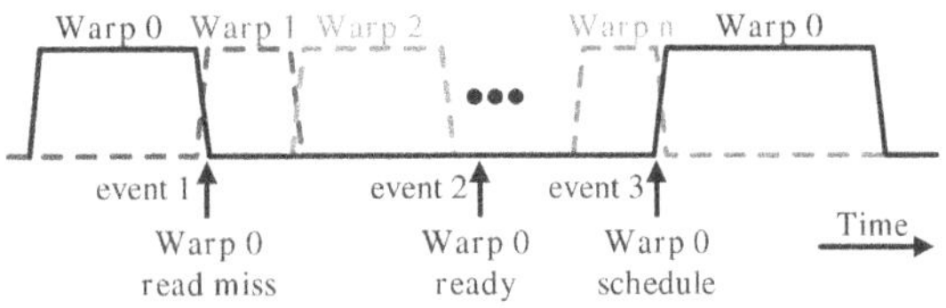

Figure 1: Execution of a GPU warp and packet slack

Figure 1 demonstrates the execution of a GPU warp (Warp 0) and its packet slack. In an SM, a warp is available if all operands are ready. Assume we have n available warps in an SM that can be scheduled at a given time; Warp 0 is scheduled and executed first. After a few cycles of execution, a read miss occurs, which is denoted as Event 1 in the figure. Then a hardware context switch is performed and Warp 1 gets scheduled and executed. Similarly, when Warp 1 encounters a read miss, the next ready warp is scheduled accordingly. Event 2 represents the cycle when the memory request of Warp 0 is sent back, which means Warp 0 is ready to schedule. However, depending on the scheduling algorithm, Warp 0 may not be scheduled instantly since there are other ready warps waiting in the scheduling queue in front of Warp 0. After a read miss of the last available warp, Warp 0 is rescheduled at Event 3. Packet slack for Warp 0 is the time between Event 2 and 3. It is evident that if we delay the network transmission of the packet for a number of cycles smaller than the slack, the overall execution time will remain unaffected.

Intuitively, packet slack is related to the number of available warps ready to schedule in an SM. The more available warps an SM has during a period of time, the longer the slack it can exploit for each packet. However, the number of available warps within each SM varies from application to application. To further understand the relation between packet slack and the number of available warps, we manually delay the scheduling of each warp for a fixed number of cycles after it is ready. For example, in a 10-cycle-delay experiment, if a warp is ready to be scheduled at cycle t_0, we arbitrarily prevent it from scheduling until cycle $t_1 = t_0 + 10$ even if there is no warp available to schedule in between. In the worst case, execution is stalled and system performance is degraded due to the arbitrary delay.

Figure 2 shows the performance degradation due to the arbitrary delay of warp scheduling. We notice that for some of the GPU applications such as BLACKSCHOLES, large schedule delay can be tolerated. By contrast, for applications such as LIB, even a few cycles of delay will result in significant performance degradation. Our hypothesis is that delay tolerable applications have more warps allocated in each SM. To verify our hypothesis, we collect the average number of available warps for each application during the whole execution as shown in Figure 3. Here we set a 95% system speed up as the acceptable performance degradation level. The *tolerable delay cycles* is the maximum number of cycles a particular warp scheduling can be delayed without exceeding the 95% speed up threshold. From the figure, we observe that there is a linear relationship between the average number of available warps and tolerable delay cycles for all the GPU applications.

Based on the observation, we predict the approximate slack of a packet as the number of available warps in the SM in a given cycle. An additional register (*Avail_warp_reg*) which keeps track of the number of available warps for each SM is added in our design. When a packet is sent from/to a particular SM core, the *Avail_warp_reg* is referred and the packet slack is set according to the value register. When possible, packets with slack can be sent through non-minimal paths to less congested routers, balancing the traffic and alleviating congestion in busy routers.

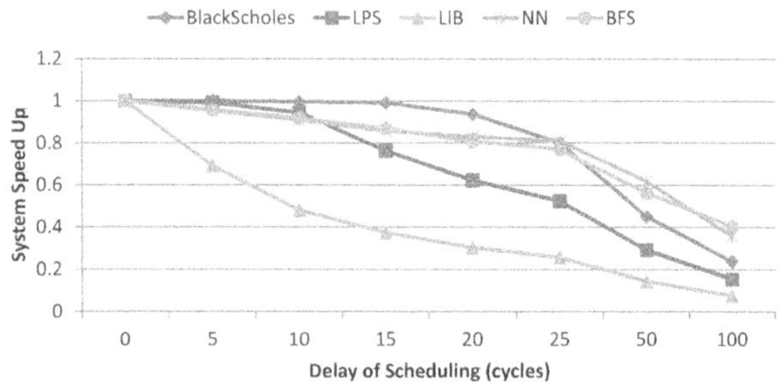

Figure 2: System performance degradation due to delay of warp scheduling

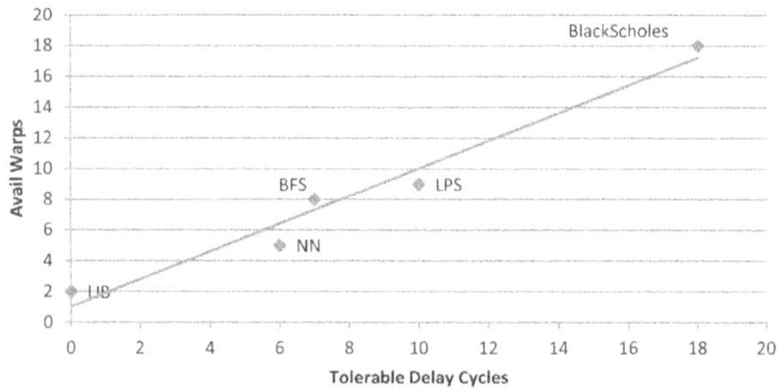

Figure 3: Linear relationship between number of available warps and tolerable delay cycles

3. DYNAMIC NETWORK TUNING POLICY

We enhance *DVFS* and propose *Non-Minimal Path Adaptive Routing* that work interactively to balance network traffic. In particular, the former periodically tunes the router supply voltage as well as the operation frequency based on the NoC workload. The latter takes into consideration the router utilization in conjunction with packet slack allowance, and *selectively* routes packets through non-minimal paths, which can facilitate more aggressive DVFS. By working in tandem, these two mechanisms are able to balance the traffic across the entire NoC and in turn, minimize global network energy consumption.

3.1 Flexible-pipeline Router Architecture

To avoid the latency increase of CPU traffic when passing through routers operating at lower frequency, we adopt the flexible-pipeline router design proposed in [23]. Voltage and frequency scaling (VFS) reduces router energy consumption at the cost of increase in packet transmission time, and degradation in throughput. In flexible-pipeline routers, to maintain a constant latency through routers upon VFS, pipeline stages are reconfigured. However, flexible-pipeline routers cannot avoid the throughput degradation associated with VFS. Thus, traffic designated to cores with high throughput requirement are dynamically rerouted to avoid contention. This is achieved through *DVFS* and *Non-Minimal Path Adaptive Routing* as described in the remainder of this section.

Based on a classic four-stage-pipelined virtual-channel (VC) router with p input/output ports, as used in Garnet [2], providing different supply voltages, flexible-pipeline routers can be operated at three different modes: four-stage-, two-stage-, and single-stage-pipeline modes. We follow the notation of *slow down parameter S* introduced in [23]. A slow down parameter indicates the router operating frequency. For example, S=1 means the router is working at the default frequency without slowing down; S=2 implies that the particular router is operating at half of the default frequency, while S=4 implies that the router is operating at a quarter of the full speed. Correspondingly, a full-speed router works in four-stage-pipeline mode; half-speed router works in two-stage mode; and quarter-speed router works in single-stage mode.

3.2 DVFS Mechanism

The router operation frequency is scaled periodically based on its utilization in comparison to two thresholds, $Threshold_{High}$ and $Threshold_{Low}$. Over a reconfiguration period T, router utilization (μ) is defined as the percentage of active cycles. At the end of each period, if μ exceeds $Threshold_{High}$, the supply voltage and router frequency are scaled up; if μ is below $Threshold_{Low}$, they are scaled down. After a few reconfiguration periods, the network becomes stable without drastic phase changes. And on-chip traffic is as well balanced. While other metrics can also be used to determine the router operation level [17], router utilization is demonstrated effective for our purposes. A thorough evaluation of DVFS metrics is beyond the scope of this work.

Scaling voltage and frequency incurs performance overhead: all existing packets in a router must be cleaned up before VFS; and it takes time to reconfigure circuitry upon VFS, modeled as a 100-cycle delay in our evaluation. The reconfiguration period must be sufficiently long to amortize this overhead. On the other hand, an unreasonably long T limits the opportunities for capturing and responding to real-time traffic patterns. In our experiment, T is set to 20,000 cycles to ensure that the overhead is under 1%.

Between $Threshold_{High}$ and $Threshold_{Low}$ is the stable state where supply voltage and router frequency remain unchanged. If the range is too wide, NoC is unable to fully exploit the potential of DVFS; if it is too narrow, router voltage/frequency toggle frequently, resulting in increasing overhead. In our evaluation, these two thresholds are empirically determined: $Threshold_{High}$=0.6 and $Threshold_{Low}$=0.4.

3.3 Non-Minimal Path Adaptive Routing

This section describes an adaptive routing algorithm allowing packets that can tolerate slack to route through non-minimal paths to reduce NoC energy consumption. Although routing packets through longer paths leads to a longer per-packet delay, targeting only latency-tolerable packets can minimize the performance impact. Meanwhile, non-minimal path routing exploits under-utilized routers and avoids congested ones to balance workload across the NoC, which in turn enables further DVFS. As a result, the NoC is able to meet the desired performance level with less overall energy consumption.

A divergence from the minimal path is referred to as a *Detour*. A detour adds two extra hops to the length of the minimal path. In the absence of contention, each detour adds 10 cycles to the latency of a packet: 8 cycles for traversing through two four-stage-pipeline routers, and 2 cycles for traversing through the two additional links. Thus, packets must be able to tolerate at least a 10-cycle delay to be considered for one detour. Slack level (SL) describes the number of detours a packet is eligible for. For example, packets that can tolerate a longer than 20-cycle delay has a SL of two. Each packet injected into the network is associated with a slack level. The SL estimation of GPU packets is described in Section 2, whereas CPU packets SL is set to zero because CPU messages are latency-sensitive. In the routers, the SL of a packet is decremented once it is detoured.

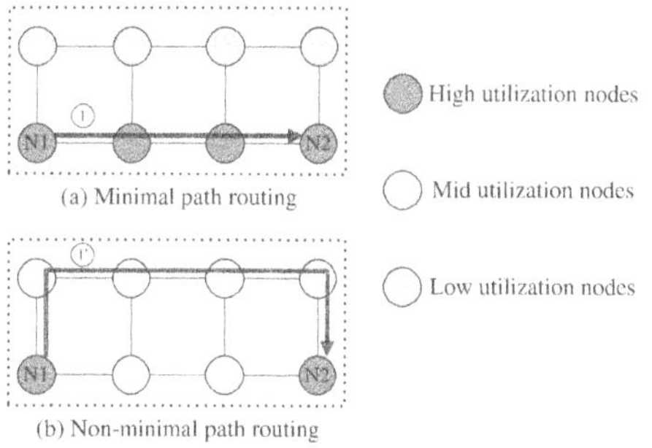

Figure 4: Non-minimal path routing

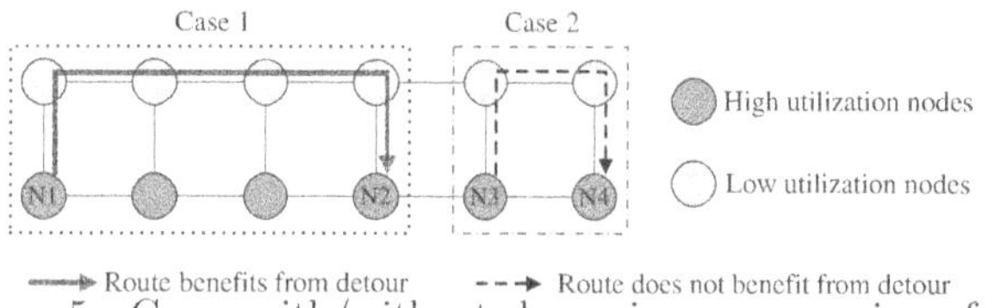

Figure 5: Cases with/without dynamic energy savings from non-minimal path routing

Figure 4 illustrates the benefit of non-minimal path routing. In this example, routers in the top row are under-utilized and are operating at a low frequency; while those in the bottom row operate at a high frequency. This is because a large number of packets are sent from $N1$ to $N2$, while traffic among other nodes is insignificant. In Figure 4(a), packets are sent from $N1$ to $N2$ via the minimal path. They traverse through four highly utilized routers. In Figure 4(b), some packets are sent via the non-minimal path and traverse through a longer route with two highly utilized routers and four under-utilized routers.

Let's consider the impact of DVFS, assuming the NoC can operate at three separate voltage/frequency levels: `high` (voltage = 1.2V; frequency = 1.5GHz), `medium` (voltage = 1.0V; frequency = 0.75GHz) and `low` (voltage = 0.8V; frequency = 0.375GHz). In Figure 4(a), four routers operate at high voltage/frequency level and four operate at the low level; while in Figure 4(b), two routers operate at high voltage/frequency level, and the remaining six operate at medium level. Network energy is categorized into dynamic, static, and clock energy, where static and clock energy correspond to a significant potion of the total energy [10, 23]. Static energy is proportional to supply voltage (V_{dd}), and clock energy is proportional to the product of V_{dd}^2 and number of clock transitions. Compared to routers operating at the high voltage/frequency level, clock and static energy consumption is reduced by 45% when operating at medium level, and by 65% when operating at low level. As a result, in the aforementioned example, allowing non-minimal path routing enables saving in static and clock energy.

Packets routed through non-minimal path can increase dynamic energy consumptions by traversing through additional hops, however they can also reduce dynamic energy consumptions by avoiding routers operating at high frequency. Thus, an important factor that determines whether a detour saves dynamic energy is the distance between the node where detour takes place and the packet's destination. As shown in Figure 5 Case 2, consider a packet sent from $N3$ to $N4$. Routing these packets through a non-minimal path does not provide any benefit in terms of dynamic energy savings. Whereas in case 1, dynamic energy savings is possible since packets traverse through four low frequency nodes and two high frequency nodes through a non-minimal path rather than four high frequency nodes. In our experiment, the minimum displacement towards either x or y dimension between the detour node and packet destination is 3 hops to ensure energy saving. Packets are considered for detouring only when they satisfy such constraint. Furthermore, non-minimal path routing routes packets away from congested routers and can enable further DVFS. In these cases, saving in static and clock energy is often sufficient to offset the slight increase in dynamic energy.

We define the row-wise packet transversal as the x-direction and column-wise transversal as the y-direction. When a packet is injected into the network and processed by a router, the router decodes the packet and obtain its slack information, and adopt different routing options to the packet based on its slack: minimal path routing is applied for packets with a slack of 0; and non-minimal path routing algorithm is performed on packets that satisfy the distance constrains. The algorithm eventually selects a neighbor with the lowest router utilization and forwards the packet. If the neighbor router happens to be a candidate from minimal paths, the packet slack level remains unchanged; otherwise, slack level is deducted by one. Details of the routing algorithm are described in Algorithm 1. Notice that the algorithm itself is not deadlock-free, so methodologies from [7] is used to avoid deadlock. In other words, the base level of VC is designed to be deadlock-free [5]. If the lowest utilized neighbor router is likely to create a deadlock, the packet will be routed over the base level VC.

Algorithm 1 Non-minimal adaptive routing for GPU packets

Require: Current router coordinate, (C_x, C_y)
Destination router coordinate, (D_x, D_y)
In-coming direction of the packet, Dir_{in}
Packet slack level, SL
Ensure: Out-going direction of the packet, Dir_{out}
$Avail\_Dir\_set \leftarrow \{\}$;
$Min\_Path\_set \leftarrow \{\}$;
$e_x \leftarrow |C_x - D_x|$;
$e_y \leftarrow |C_y - D_y|$;
if $(e_x = 0$ and $e_y = 0)$ **then**
  Route the packet to local node and exit;
else
  $Avail\_Dir\_set \leftarrow minimal\_routes()$;
  $Min\_Path\_set \leftarrow minimal\_routes()$;
  $//minimal\_routes()$ returns directions nearer to the packet's destination
  if $(SL > 0)$ **then**
    if $(e_x \geq 3)$ **then**
      $Avail\_Dir\_set = Avail\_Dir\_set \cup \{North, South\}$;
    end if
    if $(e_y \geq 3)$ **then**
      $Avail\_Dir\_set = Avail\_Dir\_set \cup \{East, West\}$;
    end if
  end if
  $Avail\_Dir\_set = \{Avail\_Dir\_set$ - $Dir_{in}\}$;
end if
Select a direction Dir with minimal utilization from $Avail\_Dir\_set$ to forward the packet;
if $(Dir_{out} \notin Min\_Path\_set)$ **then**
  $SL = SL$-1;
end if

4. EVALUATION INFRASTRUCTURE

We build a heterogeneous multi-core simulator that integrates both CPUs and GPUs. CPU and the memory hierarchy is modeled after the Simics-based [14] GEMS [15] simulator. And the GPU cores are modeled after GPGPU-Sim [3]. The interconnection network and its power model is based on Garnet [2] and Orion 2.0 [10], respectively.

In this work, we simulate a 36-tile heterogeneous system with 65nm technology nodes. The architecture parameters can be found in Table 1. Tiles are configured to be comparable in size, and each tile contains approximately 100 million transistors.

Table 1: Baseline Simulation Configuration

Processor Core	Four-way out-of-order, 6 integer FUs, 4 floating FUs, 128-entry ROB
Private L1 Cache	Split private I/D caches, each 64KB, 2-way set associative, 64B block size, 1-cycle access latency
Shared L2 Cache	16M banked, shared distributed, 4-way set associative, 64B block size, 8-cycle access latency
SM Core	32 width SIMD pipeline, 1024 threads, 8CTA, 16KB shared memory, Round Robin warp scheduling
Memory	4GB DRAM, 200 cycle access latency, 4 memory controllers
Router	5 Input/output ports, 4-stage Pipeline, 6 VCs/port, wormhole routing, 64-bit flits, 4-flit buffer depth, 1 flit per control packet, 9 flits per data packet

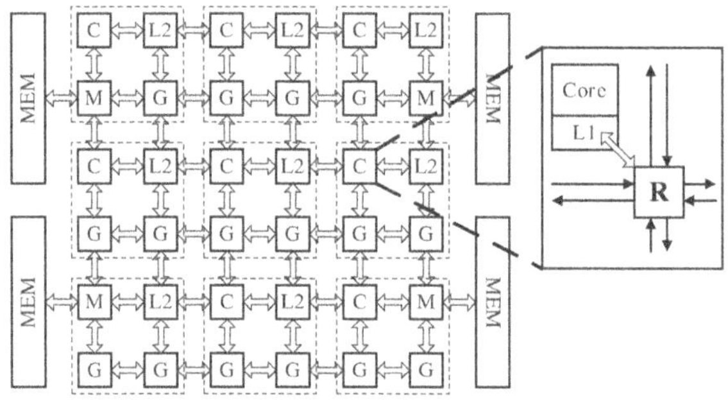

Figure 6: Baseline heterogeneous multi-core system

The optimal topology of heterogeneous multi-core systems remains an open question in the computer architecture community. In this work we choose tile-based architecture since it provides scalability for managing design complexity and can effectively utilize the on-chip resources. Figure 6 shows a 36-tile system connected with routers to a 6x6 mesh network. A tile consisting of a CPU core as well as an L1 cache is denoted by *C*. A tile that consists of a bank of shared L2 cache is denoted by *L2*, and a tile in which a GPU SM resides is denoted by *G*. Off-chip memories are connected via memory controllers labeled with *M*. Components within each tile are connected with a flexible pipeline router *R*, and on-chip traffic is transmitted via routers and links. The rationale for the topology lies in that if we consider four tiles as a cluster (circled with dashed line), except for clusters with memory controllers providing off-chip access, each cluster contains one CPU computation core, one L2 cache and two co-processors, namely SM cores. This structure can be easily extended to a larger system with more tiles, and enables parallelism in finer granularity. Instead of providing CPU and GPU cores with separate networks, sharing the same network guarantees reasonable on-chip resource sharing. An alternative would be all of the components within a cluster sharing a single router, resulting in fewer routers. However, network bandwidth becomes a bottleneck in this design. Although wider links can be provided to guarantee the bi-section bandwidth, energy overhead increases quadratically with the link width.

Restricted by the network size and topology, multiple detours in a single non-minimal path do not provide additional benefits in the evaluated network. Thus, we only consider non-minimal path with a single detour. However, our algorithm can be ex-

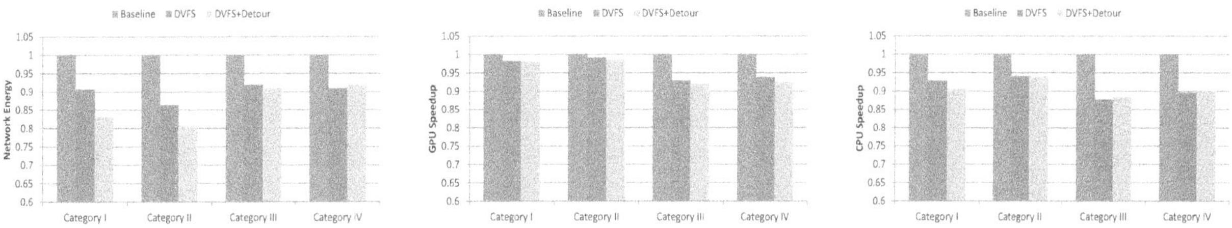

(a) Network energy (b) GPU Application Performance (c) CPU Application Performance

Figure 7: Comparison of energy efficiency and performance impact for all application mix categories. *Baseline* corresponds to network using canonical four-stage-pipeline routers without DFVS, minimal path adaptive routing. *DVFS* corresponds to network using flexible-pipeline routers with DVFS technique deployed, minimal path adaptive routing. *DVFS+Detour* corresponds to network using flexible-pipeline routers with DVFS technique deployed, non-minimal path adaptive routing. *Speedup* is defined as cycles per instruction. All results are normalized to *Baseline*.

tended to multiple detours in a larger system with a different topology.

4.1 Workloads

We use CPU applications from the SPEC OMP 2001 [1] and Nu-MineBench [19] benchmark suites and GPU applications from [3] for our evaluation. We consider 6 CPU benchmarks and 5 GPU benchmarks, where CPU benchmarks include: AFI, AMMP, ART, EQUAKE, KMEANS, and SCALPARC; and GPU benchmarks include: BLACKSCHOLES, LPS, LIB, NN and BFS. The CPU benchmarks are grouped into memory-bound (ART, AMMP, and SCALPARC) and computation-bound (EQUAKE, KMEANS, and AFI) based on their L1 cache miss rate. GPU benchmarks are grouped into latency-tolerant (BLACKSCHOLES and LPS) and latency-intolerant (LIB, NN, and BFS) based on the amount of slack they are able to tolerate. Heterogeneous GPU-CPU workload is created by executing multiple copies of the CPU benchmakrs on multiple CPU cores and one GPU kernel across all SM cores. In each simulation, we sample a period of execution that corresponds to one billion CPU instructions. We enumerate all possible combinations of the CPU and GPU workloads and evaluate 30 workload mixes. The results are presented in four categories: (I) memory-bound CPU and latency-tolerant GPU benchmarks; (II) computation-bound CPU and latency-tolerant GPU benchmarks; (III) memory-bound CPU and latency-intolerant GPU benchmarks; and (IV) computation-bound CPU and latency-intolerant GPU benchmarks.

5. EXPERIMENTAL RESULTS

In this section, we evaluate a heterogeneous multi-core system as described in Section 4, with flexible-pipeline routers [23] as described in Section 3.1. We extend the system with DVFS support (Section 3.2) and non-minimal path routing (Section 3.3), and evaluate their impact on NoC energy consumption as well as overall system performance.

Figure 7(a) shows that the impact of DVFS and non-minimal path routing on NoC energy consumption depends on the workload mix. The impact is most significant for workload mix in Categories I and II: compared to the *Baseline*, DVFS can reduce network energy consumption by up to 14%; incorporating non-minimal path routing further reduces energy consumption by 7%. It is worth pointing out that workload mixes in Categories I and II demonstrate additional energy saving from non-minimal path routing, while workload mixes for Categories III and IV do not. This is because GPU workloads in Categories III and IV cannot tolarate slack, and thus are unable to benefit from non-minimal path routing; whereas GPU workloads in Categories I and II can tolerate slack and are able to benefit. Although non-minimal path routing penalizes GPU packets' transmission latency, its performance impact is minimal, as shown in Figure 7(b). However, DVFS does degrade GPU performance for workload mixes in Categories III and IV. This is because GPU workloads in these mixes cannot tolarate latency, and reducing NoC operating frequency increases congestion and latencies. It is possible to change energy/performance trade-off using different DVFS triggering thresholds.

DVFS degrades CPU application performance, as shown in Figure 7(c), mainly due to the increase in network congestion. Although flexible-pipeline routers avoid packet transmission delay in the absence of network congestion, they cannot eliminate congestion. Non-minimal path routing have minimal impact on CPU performance, with the exception of Category I workload mixes. This performance penalty is introduced by congestion caused by detoured packets. Prioritizing CPU packet can potentially eliminate this impact. However, since this performance impact is relatively small, packet prioritization is not evaluated in this paper.

Among all four categories, the second category, namely, the mix of computation-bound CPU and latency-tolerant GPU applications, suffers from the least performance penalty whereas energy savings is the most significant. This is because of the inherent low utilization in some of the routers for light memory-bound CPU applications. Even if GPU packets are detoured via these routers, there is less likelihood that router utilization exceeds the threshold that requires a boost in frequency. In this sense, low utilized routers remain low in utilization while busy routers are released, leading to more energy reduction. To sum up, DVFS becomes the potential performance bottleneck in heterogeneous systems with significant throughput requirement, while non-minimal path routing can be selectively adopted to achieve better energy efficiency.

Table 2: Percentage of detoured packets

BlackScholes	LPS	BFS	NN	LIB
15.193%	5.287%	0.853%	0.002%	0.000%

Table 2 shows the percentage of the packets that are detoured in different applications averaged across all categories. Both BLACKSCHOLES and LPS, applications we classified into the latency-tolerant category, show higher percentage of detoured packets. This result is consistent with our methodology of exploiting GPU slack suggested in Section 2. The number of packets that are routed with non-minimal paths is proportional to the number of packets that have slack. Allowing more packets to be sent through non-minimal paths guarantees a higher possibility of balancing the network traffic, which potentially enables network frequency scaling.

6. RELATED WORK

DVFS has been widely investigated to allow the network to operate at various frequencies. Previous work in [11,12,21] has specifically focused on link latency/power. Mishra *et al.* [17] enabled DVFS on NoC routers to reduce the network power consumption. Flexible-pipeline [23] and variable-pipeline [8,16] routers are also proposed to save network energy while mitigating the latency penalties brought by network frequency reduction. Our work differs from previous work in that, we apply the DVFS technique in addition to flexible-pipeline routers, allowing aggressive voltage/frequency scaling without causing significant performance degradation.

The concept of CPU packet slack was presented in [6], and they proposed an arbitration mechanism for system performance optimization by predicting packet slack. Our work differs from [6] in that we predict the slack for GPU packets and use the predicted slack information for energy optimization in heterogeneous multi-core systems.

Adaptive and multi-path routing have been well studied. [22] introduced a global adaptive load-balanced routing algorithm, in which the adaptive decisions are based on the length of injection queues. [9] and [13] proposed dynamic routing algorithms that judiciously switch between deterministic and adaptive routing schemes. Both software [4] and hardware [18] multi-path routing approaches have been proposed, enabling full exploitation of on-chip interconnection resources.

7. CONCLUSION

The energy consumption associated with data communication in an NoC can correspond to a significant portion of system energy in future heterogeneous multi-core processors. In particular, with large input/output buffers, static router energy consumption can be significant when operating at a high voltage and frequency. Diverting heavy traffic from busy routers and amortizing it over the entire NoC enables these routers to operate at a lower frequency with DVFS support, and in turn reduce overall energy consumption. In this paper, we have explored this opportunity with traffic generated by cores that tolerate memory latencies through context switching between threads, such as GPU cores. Our results show that in a tile-based heterogeneous multi-core system containing both CPU and GPU cores, it is possible to reduce NoC energy consumption by up to 14% with DVFS and adaptive routing. Routing traffic that is not sensitive to network latency to a longer and slower route can avoid busy routers at the cost of higher network latency. Supporting non-minimal path routing enables more routers to operate at a lower voltage and frequency, and thus further lower NoC energy consumption. In particular, for GPU applications with larger packet slack, we achieve up to 7% additional energy saving. Compared to DVFS, detouring packets in NoCs has a minimal impact on system performance as measured by the speedup of the cores.

ACKNOWLEDGMENT

This work is supported in part by grants from National Science Foundation under CNS-0834599, CSR-0834599, CPS-0931931, and CCF-0903427, a contract from Semiconductor Research Cooperation under SRC-2008-TJ-1819. We would like to thank all anonymous reviewers for their constructive comments that help improve the quality of this paper.

8. REFERENCES

[1] SPEC OMP2001. Available at http://www.spec.org/omp/.

[2] Agarwal, N., Peh, L.-S., and Jha, N. Garnet: A Detailed Interconnection Network Model inside a Full-system Simulation Framework. Tech. Rep. CE-P08-001, Princeton University, 2008.

[3] Bakhoda, A., Yuan, G., Fung, W., Wong, H., and Aamodt, T. Analyzing CUDA Workloads Using a Detailed GPU Simulator. In *2009 IEEE International Symposium on Performance Analysis of Systems and Software(ISPASS)* (2009), pp. 163–174.

[4] Chen, G., Li, F., and Kandemir, M. Compiler-directed Channel Allocation for Saving Power in On-chip Networks. In *Conference record of the 33rd ACM SIGPLAN-SIGACT symposium on Principles of programming languages(POPL)* (2006), pp. 194–205.

[5] Chiu, G.-M. The Odd-Even Turn Model for Adaptive Routing. *IEEE Transactions on Parallel and Distributed Systems 11*, 7 (2000), 729–738.

[6] Das, R., Mutlu, O., Moscibroda, T., and Das, C. R. Aergia: Exploiting Packet Latency Slack in On-chip Networks. In *Proceedings of the 37th annual international symposium on Computer architecture(ISCA)* (2010), pp. 106–116.

[7] Duato, J. A New Theory of Deadlock-free Adaptive Routing in Wormhole Networks. *IEEE Transactions on Parallel and Distributed Systems 4*, 12 (1993), 1320–1331.

[8] Hirata, Y., Matsutani, H., Koibuchi, M., and Amano, H. A Variable-pipeline On-chip Router Optimized to Traffic Pattern. In *Proceedings of the Third International Workshop on Network on Chip Architectures(NoCArc)* (2010), pp. 57–62.

[9] Hu, J., and Marculescu, R. DyAD - Smart Routing for Networks-on-Chip. In *Proceedings of the 41st Design Automation Conference(DAC)* (2004), pp. 260–263.

[10] Kahng, A., Samadi, K., Li, B., and Peh, L.-S. ORION 2.0: A Fast and Accurate NoC Power and Area Model for Early-Stage Design Space Exploration. In *Proceedings of the Conference on Design, Automation and Test in Europe(DATE)* (2009), pp. 423–428.

[11] Kim, E. J., Yum, K. H., Link, G. M., Vijaykrishnan, N., Kandemir, M., Irwin, M. J., Yousif, M., and Das, C. R. Energy Optimization Techniques in Cluster Interconnects. In *Proceedings of the 2003 international symposium on Low power electronics and design(ISLPED)* (2003), pp. 459–464.

[12] Lee, S. E., and Bagherzadeh, N. A Variable Frequency Link for a Power-aware Network-on-Chip (NoC). *Integration, the VLSI Journal 42*, 4 (2009), 479–485.

[13] Li, M., Zeng, Q.-A., and Jone, W.-B. DyXY - A Proximity Congestion-aware Deadlock-free Dynamic Routing Method for Network-on-Chip. In *Proceedings of the 43rd Design Automation Conference(DAC)* (2006), pp. 849–852.

[14] Magnusson, P., Christensson, M., Eskilson, J., Forsgren, D., Hallberg, G., Hogberg, J., Larsson, F., Moestedt, A., and Werner, B. Simics: A Full System Simulation Platform. *Computer 35*, 2 (2002), 50–58.

[15] Martin, M. M. K., Sorin, D. J., Beckmann, B. M., Marty, M. R., Xu, M., Alameldeen, A. R., Moore, K. E., Hill, M. D., and Wood, D. A. Multifacet's General Execution-driven Multiprocessor Simulator (GEMS) Toolset. *SIGARCH Computer Architecture News 33*, 4 (2005), 92–99.

[16] Matsutani, H., Hirata, Y., Koibuchi, M., Usami, K., Nakamura, H., and Amano, H. A Multi-Vdd Dynamic Variable-Pipeline On-Chip Router for CMPs. In *Proceedings of the 17th Asia and South Pacific Design Automation Conference(ASP-DAC)* (2012), pp. 407–412.

[17] Mishra, A. K., Das, R., Eachempati, S., Iyer, R., Vijaykrishnan, N., and Das, C. R. A Case for Dynamic Frequency Tuning in On-chip Networks. In *Proceedings of the 42nd annual IEEE/ACM International Symposium on Microarchitecture(MICRO)* (2009), pp. 292–303.

[18] Murali, S., Atienza, D., Benini, L., and De Micheli, G. A Method for Routing Packets Across Multiple Paths in NoCs with In-Order Delivery and Fault-Tolerance Gaurantees. *VLSI DESIGN 2007* (2007), 1–11.

[19] Narayanan, R., Ozisikyilmaz, B., Zambreno, J., Pisharath, J., Memik, G., and Choudhary, A. MineBench: A Benchmark Suite for Data Mining Workloads. In *2006 IEEE International Symposium on Workload Characterization* (2006), pp. 182–188.

[20] Nickolls, J., and Dally, W. The GPU Computing Era. *Micro, IEEE 30*, 2 (2010), 56–69.

[21] Shang, L., Peh, L.-S., and Jha, N. K. Dynamic Voltage Scaling with Links for Power Optimization of Interconnection Networks. In *Proceedings of the 9th International Symposium on High-Performance Computer Architecture(HPCA)* (2003), pp. 91–102.

[22] Singh, A., Dally, W., Towles, B., and Gupta, A. Globally Adaptive Load-Balanced Routing on Tori. *Computer Architecture Letters 3*, 1 (2004).

[23] Zhou, P., Yin, J., Zhai, A., and Sapatnekar, S. S. NoC Frequency Scaling with Flexible-pipeline Routers. In *Proceedings of the 2011 international symposium on Low power electronics and design(ISLPED)* (2011), pp. 403–408.

Industry Focus Session on Low-Power Design

Session Chair:
Clive Bittlestone
Texas Instruments, USA

Presenters

Paper 1: Jim Kardach
Intel, USA

Paper 2: Renu Mehra
Synopsys Inc., USA

Paper 3: David Flynn
ARM Ltd., UK

Paper 4: Barry Pangrle
Mentor Graphics, USA

Summary

This session consists of three industry papers invited for their viewpoints on the challenges on future needs for low-power electronic design from an industry perspective. The authors are experts on the field of low-power design techniques as well as in CAD tools for implementing these techniques, both areas that are critical to actually achieving the goals of aggressive chip design. In addition, the authors provide a perspective of the lessons learned and the evolution of the techniques that have enabled mobile platforms to become so widely deployed and a big part of our everyday lives.

The first paper is by Jim Kardach from Intel who discusses the technologies and concepts used to dramatically lower platform power for Ultra-book platforms called "Power Optimizer". Renu Mehra from Synopsys then presents a review of commercial low-power EDA tools. This is followed by David Flynn from ARM, UK, who presents lessons learned as an IP provider addressing the transfer of low-power designs and implementation flows and methodologies into energy-efficient System-on-Chip products.

ACM Categories & Descriptors:

B.7.0 [**Hardware**] INTEGRATED CIRCUITS – General ; J.6 [**Computer Applications**] COMPUTER-AIDED ENGINEERING – Computer-aided Design (CAD); C.1.0 [**Computer Systems Organization**] PROCESSOR ARCHITECTURE – General.

Keywords

Low power, circuit design, chip design, SoC, CAD, EDA.

ISLPED'12, July 30– August 1, 2012, Redondo Beach, California, USA.
ACM 978-1-4503-1249-3/12/07.

Advances in Ultrabook™ Platform Power Management

Jim Kardach
Intel Corporation
2200 Mission College Blvd.
Santa Clara, CA 95054, USA
jim.kardach@intel.com

ABSTRACT

In the beginning (at least the early 90s) PC power management was architected such that it traveled with the PC and would work regardless of the Operating System (OS); you loaded DOS™, Windows™ or Unix™ and the power management just worked. This was accomplished through a combination of special CPU modes (System Management Mode/Interrupt [1]), hardware, and BIOS software (e.g. Advanced Power Management, APM, interface) which allowed a special power management software layer to be called to implement the power management features regardless of the operating mode of the processor. The main power management techniques involved turning off or slowing down idle devices, and freezing operation when the computer was not needed for a long period of time (sleeping). As the OS was bypassed, policy was invoked at a very low level and tended to judge activity by what the hardware could see (interrupts, memory accesses) which resulted in many false positives (e.g. screen blanking in the middle of a presentation).

The next step in platform power management came in the guise of OS cooperative power management through the development of the Advanced Configuration and Power Interface, ACPI [2], standard which allowed the operating system to control the hardware, make policy decisions (e.g. doing a presentation, don't turn off the display) and enable finer grain control of the power.

In particular CPU power management could now be controlled through the OS scheduler such that it was only running when there was work to be done. Previously the CPU clock was slowed (divided by "n"), or stopped for short periods of time; but now the OS scheduler worked hand-in-hand with processor clock control by placing the CPU in a C-state when not working. Each CPU would have multiple C-states where the exit latency would increase with higher power savings, allowing the OS to schedule "deeper" c-states while the system was idle and "lighter" c-states with the system was active which allowed the system to adjust the power and performance based on the activity of the system.

This dramatically increased the CPU power savings as instead of scaling the CPU power by the divided clock rate (divide the clock by 2, 4, 8, 16, ...), the power was now related to the C-state duty cycle where an idle system might be active for 1 us for a 15.6 ms preempt period (0.01% ON/C0 duty cycle).

When examining the power of ACPI PC's, the CPU is very well power managed, while we see an offset of platform power referred to as the PC idle power floor.

As the platform consists of many subsystems with their own low power states, this power floor can be seen as the power resulting from the power of these different subsystems, plus, and more importantly, the impact of the behavior of these subsystems on each other (i.e. impact of interrupts, bus cycles and use of computer resources).

As an example, a 1 mW Bluetooth™ radio does not consume much energy itself, but due to its USB [3] interface it can assume exponentially more platform energy. In USB the interface stays connected by periodically reading small descriptors from memory (i.e. every millisecond); this very small amount of activity keeps all of the subsystems, between it and main memory, out of their respective low power states and creates these "platform power effects" that are exponentially higher than the power measured by the Bluetooth subsystem by itself!

This PC platform power floor can be thought of as being caused by the seemingly random and uncoordinated activity of the different platform subsystems on each other.

The goal of the new Intel™ Power Optimizer architecture was to bring this idle power floor down while still being able to follow our legacy ACPI goals: power management travels with the platform, works with any compliant OS, and work seamlessly with other PC standards (USB, PCIe, ...).

To achieve these goals, we looked to power manage the platform similar to how the CPU is power managed in ACPI, where the entire platform could be aggressively scheduled to transition between active and idle states. The main technique was to change the behavior of the system based its activity in order to maximize the opportunities for scheduling these platform power states and thus reducing the idle power floor.

This presentation outlines some of the Power Optimizer techniques used to achieve these goals in Ultrabook platforms.

Categories and Subject Descriptors

C.5.3[**Microcomputers**]: Ultrabook Platform Power Management

Keywords

Ultrabook, platform power management, power optimizer, low power.

Acknowledgements

Wish to thank Barnes Cooper for creating much of the content.

REFERENCES

[1] Kardach et. al, Transparent system interrupt, '92. United States Patent. DOI= http://www.google.com/patents/US5175853

[2] 1996, Advanced Configuration and Power Interface. DOI= http://www.acpi.info.

[3] B. Cooper, P. Diefenbaugh, J. Kardach. 2008. *Making USB a more Energy-Efficient Interconnect, Intel Technology Journal. DOI = http://www.intel.com/technology/itj/2008/v12i1/2-usb/1-abstract.*

ISLPED'12, July 30– August 1, 2012, Redondo Beach, California, USA.
ACM 978-1-4503-1249-3/12/07.

Commercial Low-Power EDA Tools: A Review

Renu Mehra
Synopsys, Inc.
700 E. Middlefield Rd., Mountain View, CA-94043
renu.mehra@synopsys.com

ABSTRACT

Low-power EDA has come a long way in from its infancy almost 20 years ago, when the first low power tools for logic synthesis were introduced. Currently a large number of power-aware tools provide a spectrum of capabilities to address power issues, from analysis of power consumption and IR drops to automated power reduction during chip implementation. The most recent development is the advent of standards to specify power intent along with RTL which resulted in a number of new tools to help users implement and verify their power intent.

This paper takes a look at the different techniques that are commercially available to analyze, optimize, and manage the power consumption of modern day chips. The purpose is not to evaluate the different tools, but to provide an overview of the capabilities available commercially. Since it is impossible to do justice to all the tools that provide low power capabilities in this short space, I focus on a few representative tools. The scope is limited to tools for digital design starting from RTL.

Categories and Subject Descriptors

B.6.3 [**Logic Design**]: Design Aids – *Automatic Synthesis, Optimization, Simulation, Verification;* B.7.2 [**Integrated Circuits**]: Design Aids – *Placement and Routing, Simulation, Verification.*

Keywords

Low power, EDA, Power Analysis, Synthesis, Place and Route, Verification, Multi voltage, Power Management.

1. INTRODUCTION

Commercial tools addressing low power cover several areas including libraries, power analysis, dynamic and leakage power optimization, and support for advanced power management schemes like multi-voltage, state retention, and power shutdown. A large range of tools including simulation, synthesis, physical implementation, static verification, and signoff analysis provide capabilities in these areas.

For the purpose of discussion, we divide the low-power tools into three categories: Section 2 explains power analysis tools, Section 3 covers automated power optimization tools, and Section 4 describes tools that aid global power management in multi-voltage and shutdown designs.

2. POWER ANALYSIS

Power analysis is a critical part of low power design providing crucial information to the user about the power statistics of thedesign and key metrics to optimization tools for power reduction. Current tools cover various aspects of power analysis and apply to various parts of the flow. Figure 1 shows the different power analysis tools and how they are used in the different stages of the ASIC design flow.

2.1 Power-Aware Libraries and IPs

These days, library vendors vie to provide larger variety in their libraries including both low-power and high-speed cells at the same technology node. As a result, a large amount of power related information is embedded in libraries to help tools to use them efficiently.

The basic power information for each library cell includes the state and path dependent dynamic or internal power, state-dependent leakage power, and pin capacitance values. In addition to inputs and output pins, libraries contain description of the supply pins (power and ground or PG pins), bias pins, voltage levels on each pin, relationship of signal pins to PG pins. Multi-corner libraries provide the leakage and power information based on different voltage and temperature corners. Multi-voltage enabled libraries provide voltage values on the PG pins as well as information about the range of voltages over which the cell can operate reliably. Complex macros and IP blocks include information about isolation attributes and the power shutdown modes.

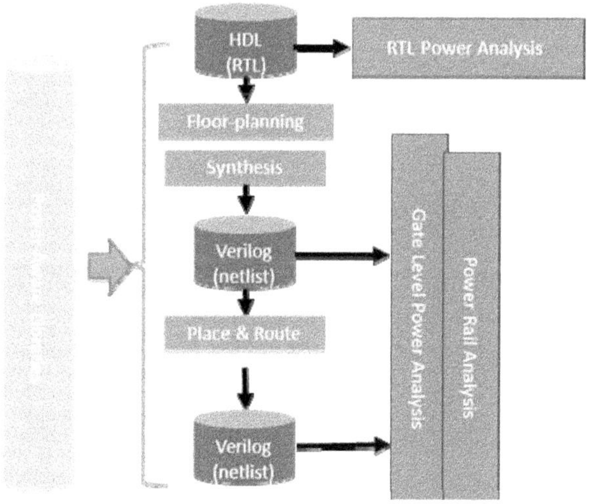

Figure 1: Power analysis tools in the tool flow.

2.2 Power Estimation

The solid foundation of power-characterized libraries led to several tools providing power analysis capabilities at RTL (reduced accuracy) and gate (higher accuracy) levels.

On the RTL power estimation side, some commercial tools provide power estimates [3][14]. However, at such a high level, it is difficult to guarantee a strong correlation between the high level estimates and the final power dissipation. Typically, high-level estimates can be obtained by doing a fast synthesis to gate-level.

At the gate level, power analysis tools are more mature and can provide more accurate power dissipation for a design [13][14].

For accurate average and peak power calculations, tools require switching activity. Switching activity information allows the tools to leverage state-dependent power characterization for improved accuracy. Most tools work with activity collected from RTL simulation and use the SAIF (switching activity interface format) format [1] for average power analysis and VCD (Value Change Dump) format for peak power analysis. More recently, vector-less power analysis is getting more popular where activity information internally derived.

2.3 Power Rail Analysis

As feature sizes shrink, supply voltages are forced to be low causing increased sensitivity to voltage loss from IR drops. Rail analysis is now an integral part of the design flow with both static and dynamic rail analysis capabilities being commercially available [2][14]. Static rail analysis evaluates the IR drop caused by high average currents in the power grid and is often used for generating voltage maps for the chip. At technology nodes above 130nm, static rail analysis was sufficient due to naturally high decoupling capacitance from the supply lines and non-switching logic. At lower geometries (less than 130 nm), it becomes imperative to look at dynamic effects. Dynamic rail analysis evaluates IR drops that happen when large amount of circuitry switches simultaneously causing high peak currents. While traditionally these tools relied on switching activity information, innovative vector-less approaches for dynamic rail-analysis have been introduced more recently.

3. POWER OPTIMIZATION

Power has become a key design metric for all steps in the implementation flow today, optimized together with timing and area. Optimization tools provide reduction techniques targeting both dynamic and static (leakage) components of power dissipation. Figure 2 shows the different stages in the implementation flow where power optimization is done.

3.1 Dynamic Power Optimization

The early commercially available tools for automatic dynamic power reduction leveraged low power synthesis techniques like cell swapping, pin swapping, and rewiring [13]. Today, higher-level activity-based optimizations are available along with logic-level optimizations. These include activity-driven data path gating, operator selection, and operator restructuring. One of the issues that impact all activity-driven approaches is that accurate switching activities may not always be readily available.

By far the most successful technique for dynamic power reduction has been clock gating [5][10][13]. The technique originally was used by designers to reduce activity by shutting off the clock to parts of the chip that are idle. EDA tools took it one step further by clock-gating individual registers and introducing massively clock-gated designs. Figure 3 shows some of the different types of clock gating approaches available in the market today.

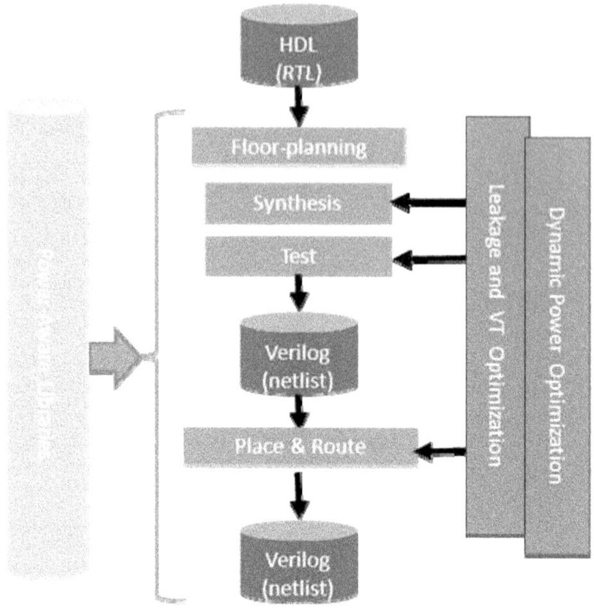

Figure 2: Use of power optimization techniques in the ASIC design flow.

3.2 Dynamic Power Optimization

The early commercially available tools for automatic dynamic power reduction leveraged low power synthesis techniques like cell swapping, pin swapping, and rewiring [13]. Today, higher-level activity-based optimizations are available along with logic-level optimizations. These include activity-driven data path gating, operator selection, and operator restructuring. One of the issues that impact all activity-driven approaches is that accurate switching activities may not always be readily available.

By far the most successful technique for dynamic power reduction has been clock gating [5][10][13]. The technique originally was used by designers to reduce activity by shutting off the clock to parts of the chip that are idle. EDA tools took it one step further by clock-gating individual registers and introducing massively clock-gated designs. Figure 3 shows some of the different types of clock gating approaches available in the market today.

The simplest and most used version is synchronous-enable clock gating. The basic idea is for tools to recognize cycles in which a specific register does not load new data and turn off the clock for those cycles. This technique is mature and works well with the overall tool flow including synthesis, formal verification, test, and clock tree synthesis.

A newer version of clock gating is xor-based self gating which tries to shut off the clock to a register in cases when the data value does not change. This has slightly higher cost than enable based clock-gating due to the extra xor-gate needed for data change detection.

Another approach to clock gating is sequential clock gating. The idea here is to propagate the enable conditions to upstream or downstream sequential elements in order to uncover more opportunities for clock gating. General formal verification tools do not fully support this kind of optimization.

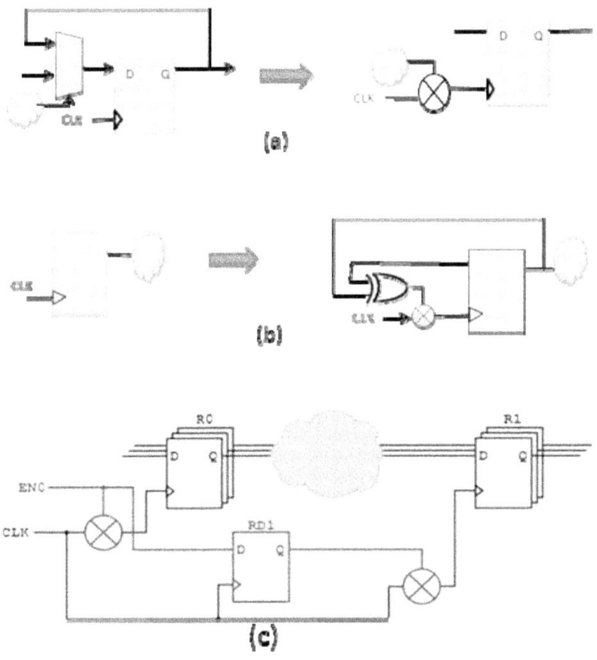

Figure 3: Clock gating techniques: (a) Synchronous-enable clock gating (b) Self gating (c) Sequential clock gating.

3.3 Leakage Power Optimization

At the larger technology nodes (180nm and above), leakage power was mostly ignored by EDA tools since its contribution to the total power was not significant. At technology nodes below 90nm, however, leakage power has become increasingly important. For reliable operation at these nodes, the supply voltages are forced to be low, forcing reduced threshold voltages to maintain speed. This results in huge increase in leakage. In order to address this issue, all foundries now provide multiple threshold-voltage (V_T) devices for the same node – low V_T devices for ultra fast operation and high V_T devices for low power operation. Three or more threshold voltages are common for a single node. In addition library vendors provide variations of channel lengths to provide further control of the speed vs. leakage tradeoff. The onus has increased on design tools to use the large set of devices provided in the most optimum way. Figure 4 shows the different threshold voltages and channel length variants that are available at 40 nm. The power-performance tradeoff between these variants is also shown.

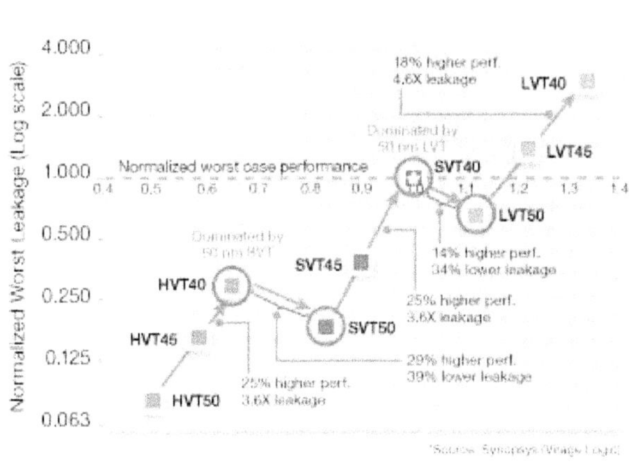

Figure 4: Threshold voltages and channel length variants at 40nm and the associated power-performance tradeoffs.

Implementation tools provide automated approaches to use these devices in an efficient way during synthesis, placement, routing, and post-route phases [10][13]. The tools are based on smart heuristic algorithms for assigning low V_T devices to critical parts of the chip and using high V_T devices where it's possible to do so without hurting the performance. In addition, there are several user controls to dictate the percentage of each device type used in a design.

3.4 Power-Aware Test

On the Test automation front, both Design for Test (DFT) and Automatic Test Patten Generation (ATPG) tools have added power-aware features [9][15]. DFT tools use gating techniques to save power in the compression/decompression logic introduced. Since test vectors are highly compressed, chips often dissipate more power during test mode than during regular operation. As a result, some chips that would work in regular operation may exhibit failures in test mode due to overheating and IR drops. Power-aware ATPG tools distribute vectors in such a way to so that peak currents are minimized. Pseudo-random pattern fills are also done such that switching activity from one pattern to the next is minimized.

4. ADVANCED POWER MANAGEMENT

With the ever increasing demand for functionality in portable and stationary devices, and with growing need for longer battery life, it is clear that it is not possible to power the whole chip all the time. Dark silicon is becoming more and more common with quite complicated shutdown schemes to help keep the power dissipation under control. Most advanced chips designed today use a combination of selective voltage reduction, standby, and shutdown techniques in this context. About 5 years ago, these power management techniques were implemented manually by the designer, with little or no help from automation tools. Power intent standards introduced around 2006-2007 changed this drastically, allowing designers to specify their power intent and tools to automate away much of the design complexity.

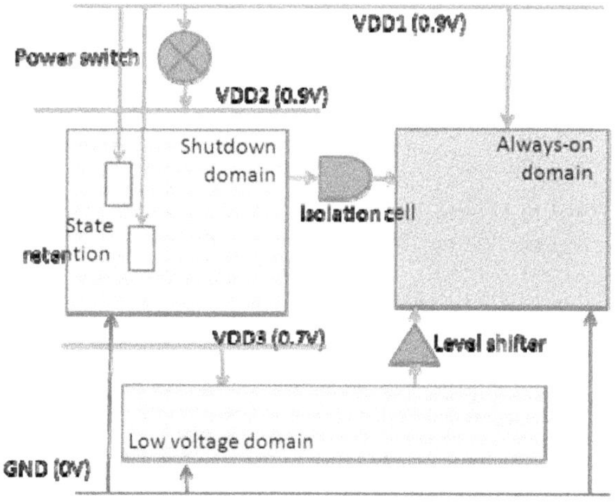

Figure 5: Example of a power managed design using shutdown and multi-voltage.

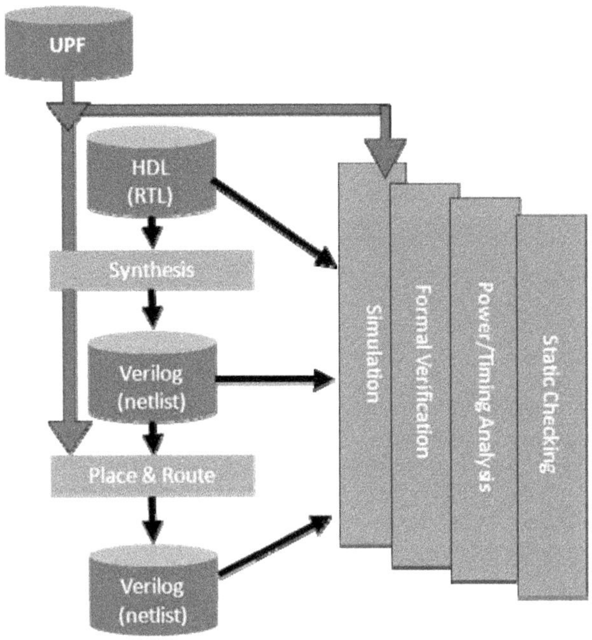

Figure 6: Tool flow support for UPF.

The IEEE-1801 format (Unified Power Format or UPF) [1] is the pre-dominant standard proposed for capturing the power intent of a design. This format allows users to specify a number of power related features including:

- Power domains definitions in the design
- Supply network including the power and ground supplies, power switches, and their connections
- State retention properties for each domain, specifying which state elements should be preserved in the standby and shutdown states of the block
- Voltage levels of power domains and level shifting requirements between domains.
- Isolation between different domain boundaries
- Relationship between power states of different parts of the chip

Figure 5 shows a simple example of a power managed design that can be described in UPF. It shows a supply network consisting of two always-on power sources (VDD1 and VDD3) and a switched supply (VDD2) operating at different voltages. There are three power domains, one of which is a shutdown domain and needs state retention. Signals from this domain to other active domains need isolation cells. Signals from low-voltage domains to high-voltage domain have a level shifter specified.

The power intent specified affects a large number of tools at various places in the flow. Figure 6 shows how the UPF specification is used in verification tools (Power analysis, Timing Analysis, Static checking, Simulation, and Formal verification) and implementation tools (Synthesis, Placement and Routing). The next two sections offer a high level overview of the commercial tools with capabilities in these areas.

4.1 Verifying the Power Intent

The power intent specified in UPF affects the functional behavior of the design in many ways. The supply network determines when a specific domain will power off, which results in corruption of the signals in that domain. Domains may also be put in "corrupt-on-activity" mode where they are able to retain state but cannot sustain new inputs. This can be done using well-biasing. Further, power management cells like state retaining registers and isolation cells change the functionality of the circuit. The sequencing of the different operations is also important. For example, for the circuit to function correctly, it is important that the retention registers go into a retain mode and isolation cells go into isolation mode before the domain is shutdown. Likewise, both retention and isolation elements must restore to original behavior only after wakeup.

Power aware simulators available today simulate the power intent of the design as an integral part of the behavior of the design and account for the effects of power intent described above [8][12]. Further, formal equivalence checkers (RTL-to-gate and gate-to-gate) also understand the power intent as an integral part of the design description [11].

With the new UPF standard, several tools are now available for statically checking the power intent of the design [4][6][12]. These are usually rule-based and focus on ensuring that a set of rules are satisfied. They consider rules of the type: isolation cells must be specified on all signals that go from a shutdown to a powered-up domain, always-on buffers should be powered such that they will not corrupt when its value is being read, etc.

Together these static and dynamic verification tools help ensure correct implementation of the power intent.

4.2 Implementing the Power Intent

Power intent specified in UPF is "implemented" by the synthesis, placement, routing, and clock-tree synthesis tools. This section covers some of the challenges addressed by these tools.

The basic requirement from synthesis is the insertion of power management cells like level shifters, isolation cells, and state retaining elements. In addition, always-on buffering is required on all cross domain paths to ensure that signals that pass through shutdown domains are not unexpectedly corrupted. Synthesis tools must do this while meeting the overall area and performance requirements for the chip [13].

Each power domain has different power grid requirements. One of the common ways to handle this is to specify a physical voltage area for each power domain during floor-planning. Each voltage area has its own power grid with a primary power and a primary ground. The placement tool is required to place all standard cells for a power domain in its corresponding voltage area. The tricky part is to be able to handle the exceptions. While standard cells in a power domain may use only the primary power and ground, the power management cells like level shifters, isolation cells, and retention registers need a second power supply. Other cases that may need a special supply are macro blocks and always-on buffers on feed-through paths. Backup power may be supplied either by using a dual power grid or by using secondary power routing. Secondary power routing requires some designated bounds in each voltage area where the special cells will be placed and appropriate power will be supplied. Various concepts are in use in different tools to solve this including gas stations and power guides [7][10][13].

On the power network synthesis side, switches are implemented in user specified configurations – ring style, fish-bone, etc.; the power grid is created; and the power and ground hookups are provided for standard cells and special cells [10].

In summary, logical and physical implementation tools must handle the power management aspects along with the logic implementation.

5. CONCLUSION

This paper presents a wide range of tools available commercially today to help designers address their low power requirements. In 1995, the first commercial low power tool was introduced, with only power estimation and low-power synthesis technologies. Today's tools cover a large range of design techniques include multi-threshold synthesis, clock gating, and power management. Low power tools are no longer limited to synthesis; power has become an all-pervasive design metric that affects a wide range of tools from implementation and verification to signoff.

6. REFERENCES

[1] 1801-2009 – IEEE Standard for Design and Verification of Low Power Integrated Circuits, March 2009, DOI=10.1109/IEEESTD.2009.4809845, http://ieeexplore.ieee.org/servlet/opac?punumber=4809843

[2] Apache Design Inc, (Ansys) Redhawk http://www.apache-da.com/products/redhawk

[3] Apache Design Inc, (Ansys) PowerArtist http://www.apache-da.com/products/powerartist

[4] Atrenta Inc., Spyglass, http://www.atrenta.com/solutions/spyglass-family/spyglass-power.htm

[5] Calypto Design Systems, Inc. PowerProCG http://www.calypto.com/powerprocg.php

[6] Cadence Design Systems, Inc., Conformal LP, http://www.cadence.com/products/fv/Pages/lp_flow.aspx

[7] Magma Design Automation, (Synopsys, Inc.) http://www.magma-da.com/products-solutions/digitalimplementation/TalusPowerPro.aspx

[8] Mentor Graphics, Questa Power Simulator, http://www.mentor.com/products/fv/questa-power-aware-simulator/

[9] Mentor Graphics, Test Products, http://www.mentor.com/products/silicon-yield/

[10] Synopsys Inc., IC Compiler, http://www.synopsys.com/Tools/Implementation/PhysicalImplementation/Pages/ICCompiler.aspx

[11] Synopsys Inc., Formality http://www.synopsys.com/Tools/Verification/FormalEquivalence/Pages/default.aspx

[12] Synopsys Inc., Low Power Verification, http://www.synopsys.com/Tools/Verification/LowPowerVerification/Pages/default.aspx

[13] Synopsys, Inc., Power Compiler, http://www.synopsys.com/Tools/Implementation/RTLSynthesis/Pages/PowerCompiler.aspx

[14] Synopsys Inc., PrimeRail, PrimeTime http://www.synopsys.com/Tools/Implementation/SignOff/Pages/default.aspx

[15] Synopsys, Inc., Test Automation, http://www.synopsys.com/Tools/Implementation/RTLSynthesis/Test/Pages/default.aspx

An ARM Perspective on Addressing Low-Power Energy-Efficient SoC Designs

David Flynn
Research & Development
ARM Ltd
Cambridge, UK
+44 1223 400438
david.flynn@arm.com

ABSTRACT

In this paper the lessons learned as an IP provider addressing the transfer of low power designs and implementation flows and methodologies into energy-efficient System-on-Chip products are discussed, followed by a preview of the more promising techniques for future deployment.

Categories and Subject Descriptors

B.5.1 [**Register Transfer Level Implementation**] – *design*.
B.6.1 [**Logic Design**]: Design Styles – *sequential circuits*.
B.7.1 [**Hardware**]: Integrated Circuits – *types and design styles*.
C.3 [**Special-purpose and Application-based systems**].

General Terms

Design, Performance, Experimentation.

Keywords

Automatic Test Pattern Generation (ATPG), Central Processor Unit (CPU), Dynamic Voltage and Frequency Scaling (DVFS), eco-system, Electronic Design Automation (EDA), energy-efficiency, Implementation IP (IIP), Intellectual Property (IP), IP-deployment, Logical IP (LIP), low-power, Multi-Threshold CMOS (MTCMOS)Multi-Voltage (MV), Physical IP (PIP), Power-Gating (PG), power intent, standard-cell, State-Retention (SR), System-on-Chip (SoC).

1. INTRODUCTION

As an IP provider, ARM faces interesting challenges in ensuring licensees and customers can address low power product design and implementation in consumer markets where responsive performance must be traded off against active and standby power and battery life[1]. In the market areas where ARM is focused the success metrics for end products are qualified in terms of time-to-market, volume manufacturability, and end-user active and standby experience. This is predicated upon using industry standard EDA tools and design-flows wherever possible for the hardware design, implementation and verification of the SoC component of the product, and the operating system software layers that support the user interface and applications software.

The ARM business model is built on the design and licensing of IP cores and supporting interconnect and sub-systems, typically technology independent and with levels of parameterization and configuration. But this grew out of experience in the 1990's of designing and delivering technology-specific 'hardened' CPU IP cores where full-custom design techniques could be used and the macro-cell was delivered pre-verified and with known (characterized) performance, power and area. The main challenge back then was developing the model abstractions that would allow the physical, functional (simulation), timing, power and test views to work with portability across a wide range of customer design flows and vendor-specific EDA tools. Investment was also put into interconnect standardization to support uniform memory-mapped access to, and hardware debug of, user-developed peripherals and sub-systems.

The lessons learned from hard-macrocell deployment have proved invaluable to building the know-how to deliver synthesizable ('soft') IP cores. A number of years working with EDA companies to develop the implementation and verification support that enables customers to parameterize, build, verify and characterize SoC IP themselves is bearing fruit in allowing multiple customers to develop and optimize designs in parallel[2].

Academic research on mitigating dynamic and static power, such as Dynamic Voltage and Frequency Scaling and Power-Gating respectively is well understood and exploited in full-custom and 'expert' design flows for standard products and integrated circuits. However, enabling such techniques to be deployed successfully by SoC design teams that need to rely on standard-cell design flows and industry standard test and sign-off tools has required significant work with EDA companies and standards committees.

The next level challenge is then at the system design level to ensure optimal control of the hardware added for active and standby power management. This paper explores the "energy management stack" and enabling the eco-system that supports ARM partners and customers to address power and energy efficiency across a wide range of consumer products.

2. LOW-POWER BUILDING BLOCKS

Low-power standard cell libraries and memory compilers are the foundations. These will typically have multiple threshold voltage and/or channel-length families of cells, ideally footprint compatible for optimal cell-swapping EDA tools flows. Additionally a number of 'bolt-on' components are required as described in this section.

2.1 Dynamic Power Reduction

Baseline support is in the obvious areas of clock gating, explicitly at the architectural or micro-architectural level, and inferred at low-level by the EDA tools wherever they can pull out shared clock enable terms across multiple standard-cell registers

ISLPED'12, July 30–August 1, 2012, Redondo Beach, California, USA.
Copyright 2012 ACM 978-1-4503-1249-3/12/07...$10.00.

- "Integrated" Clock Gating cells with a rich set of output drive strengths (with test-override support for clean ATPG test flows)

Additional Physical-IP support for Dynamic Voltage Scaling

- Level shifters, integrated clamps
- Extended characterization of entire base library
- "Current Source" timing/power models (in multiple proprietary EDA formats)

At the system level, the clock management, and especially the interlocked control of voltage and frequency in DVFS, requires explicit control and feedback. Adaptive voltage scaling systems also require Process and Temperature sensing and monitoring. DVFS power domains also need explicit asynchronous clock domain crossing interfaces that have to be handled in the SoC design.

2.2 Static Power Reduction

Power gating is the industry standard technique for cutting static leakage currents when subsystems are un-clocked or deactivated. Power gates are not perfect switches and have non-zero on-resistance and non-infinite off-resistance, so require design and implementation tradeoffs depending on the specific process technology[3].

Additional component support for Power gating includes:

- Power Gates with High-Vth or longer-channel-length devices (PMOS "Header switches, NMOS "Footer" switches) with extended Ion/Ioff /V characterization of the (imperfect) "MTCMOS" switches
- Clamp cells (typically outputs of power-gated regions) for Clamp-Low, Clamp-High, or even Clamp-Hold value functionality
- "Always-on" buffers and cells for power gating and clamp control networks

From an implementation perspective, design for power-gating must also address the real-world issues of decoupling and rush-current management at turn-on, often by logically sequencing weak and strong switch networks, and sequencing power, isolation clamping, resets and clocks to account for the non-instantaneous power gating times. (Currently this is never as transparent as inferred clock gating). Testability for stuck-on and stuck-off power gates or control networks must also be addressed. At the SoC top-level, distributed state-machine controllers are typically required.

Power gating is usually applied to combinational logic and registers, and an explicit reset is required after re-powering due to unknown register state. State-Retention with Power Gating (SRPG) is a valuable extension to basic on-chip power gating in applications where near instantaneous wake-up and response is required, such as low power micro-controllers[4].

The preferred EDA model abstraction for SRPG registers requires explicit state save and restore signaling at each and every register in a state-retained region, plus a form of back-up supply to the retained state element:

- "Split-rail" State Retention registers with secondary power and state save/restore signaling

There is a cost to both area and performance of such SRPG registers, with the additional challenge that hardware state retention schemes must be fully robust. Otherwise they could result in state corruption and potential deadlock on wake-up. The SoC designer is responsible for the careful sequencing of SRPG control and handling of the entry to, and wake-up from retention[5].

RAM compilers can also add the concepts of state retention of the core bit-cell array but with power gating and clamping of the periphery of the memory to reduce static leakage power. The SoC integrator must again manage the sequencing explicitly.

State retention may apply to architectural or selective micro-architectural registers or to the entire state in some, or all, of the logical hierarchy in a design.

2.3 Power Intent and Formats

Bridging the gap between Logical-IP, LIP (configurable RTL typically), and Physical-IP, PIP, is the Implementation-IP, or IIP. This conveys the expertize for the licensee to configure, floor-plan, and constrain timing for the chosen technology, and implement with the appropriate low-power methodology.

EDA industry competitors have come up with more than one power annotation format [6][7] for inferring power intent with IIP, so ARM and industry users have invested time and effort in the evolving IEEE1801 standard[8] that is driving convergence and interoperability[1]. Understanding the IP provider and user IP deployment models has been key to driving successive refinement by design into the power intent standards.

From the IP provider perspective the primary power intent that must be provided can be summarized as:

- The specification of indivisible power domains (which users may merge and combine but not split any further)
- Interface clamping specifications at power domain boundaries (e.g. inactive protocol states)
- State retention levels supported and verified (e.g. no state, full-state or partial state)
- The legal power states and state transitions supported, that may be augmented by control assertions to assist integrators.

The successive refinement model then allows this primary intent to be declared and augmented with:

- the SoC-specific configuration, how the power domains are mapped to regulated supply voltages and power-gated regions
- the technology-specific power components to infer
- the mapping of explicit control signal polarities (etc)
- and hierarchical power-state tables with the system.

This is a considerable improvement over the early UPF and CPF formats where voltages, control signal polarities and power states were effectively bound in the power intent constraints with the RTL.

The layering supported in this successive refinement model enables IP providers to deliver portable reusable and verifiable "constraint" power intent – limits to the multi-voltage mapping and configuration space that are verifiable; the IP-integrator then applies their SoC-specific logical and power configuration that become the "golden" RTL and "configured" power intent for the SoC. The golden configuration can be fully simulated with power-aware simulators, increasing the "verification equity", before The IP implementer then maps the power intent to the technology-specific power management cells – for verification against the golden reference for this SoC.

[1] Chaired from 2011 by John Biggs of ARM Ltd, Cambridge.

3. THE SYSTEM "ENERGY STACK"

The dynamic and static power mitigation techniques described only make a significant energy saving in the end-product when the system level control is in place to understand the application profiles and when to change power states. Figure 1 shows a basic cross section of the different layers of the system that need to be optimized for the foundation technology and hardware, the SoC-specific architecture and power management control and monitoring interfaces up to the operating system policies and management that support the application software layer.

Application software
Power management policies/drivers
System components: memory/display/PMIC
SoC architecture
Power-efficient Logical-IP: RTL
Low-power Physical-IP: standard-cell, RAM
Process technology

Figure 1: Layers of the System Energy Stack

4. BUILDING ON BASIC PG AND SRPG

The multi-voltage tools support and associated power intent now prove to be a foundation for more advanced techniques to improve on the base-line power-gating and state retention support envisaged as the EDA tools were developed.

4.1 Multi-Voltage Power Gating

Industry standard "Multi-Voltage" EDA tools support logic level drive of the gate terminal of the power switches, while more expert approaches have traditionally been required to add Gate Bias to improve the off-current (ratio) of power gates[9].

A Super-Cutoff CMOS "buffered" power gate cell family with integrated level shifting has been developed to work seamlessly with standard EDA MV tool flows (shown in figure 2). Header power gates are of primary interest to facilitate simple generation of gate bias supply voltage (the core voltage rail augmented by small charge pump or regulated from an higher IO supply rail).

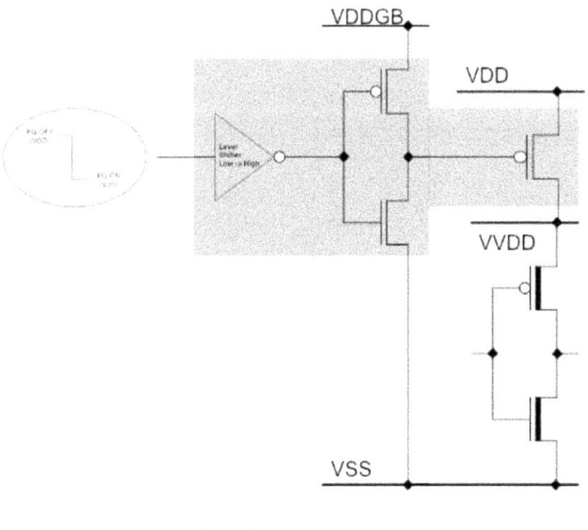

Figure 2: SCCMOS enhanced power gate

The multi-voltage internals of the enhanced switch are hidden from the implementation tools and support lower off-current with High-Vth "MTCMOS" power switches, or lower-IR drop with standard Vth switches.

4.2 State Retention challenges

SRPG is richly supported in UPF and CPF power intent standards but has been of limited use in industry because the area and performance cost of the preferred EDA cell abstraction (where state save and restore control is expected to be added to each and every Flip Flop library element). For ARM® microprocessor-based designs this is typically in the low-power embedded control application area, addressed by Cortex-M® class family CPUs. Low idle power must be balanced with minimal wake-up energy cost and service response latency; hardware supported state retention for small-scale subsystems works well here.

In high-performance CPU designs, expert reuse of the manufacturing scan chains to shift state out and back is possible in hardware, otherwise support has to be deferred to firmware for explicit architectural state saving and restoration after a hardware reset.

4.3 Enhanced State Retention

The experimental approach adopted has been to amortize the cost of state retention across multiple registers by splitting the power rails for high performance flip-flops at near-zero area cost. The retention control cost is amortized by managing the clamping of clocks and resets efficiently in the SOC implementation flow such that the speed and area impacts are minimized over and above the cost of Power Gating that designers well understand.

Figure 3 illustrates how the retention power domain is distributed to manage "live-slave" state retention between clock-gates and registers. For registers with asynchronous set or reset functionality, such controls must also be explicitly clamped similarly.

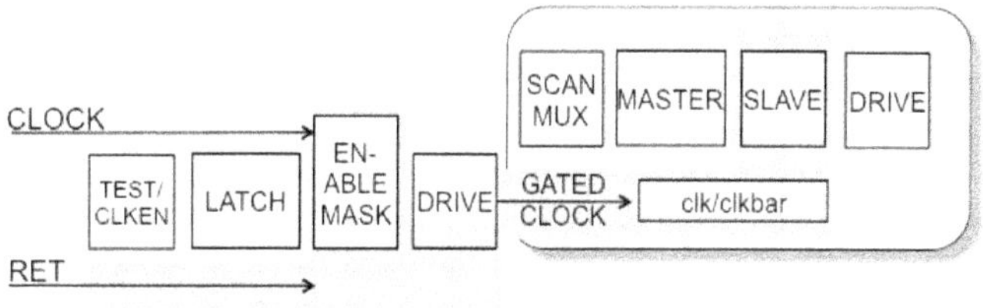

Figure 3: Advanced SRPG distributed retention domain

For short-term SRPG retention the slave latches and associated clamping domains must be kept powered. For deep sleep this domain (shown with gray overlay) is power-gated off as well (state lost PG, typically requiring software state save).

Voltage scaling of the state retention rail is attractive to provide an extended SRPG mode of operation, but simple techniques such as adding a Vth-drop that was safe at higher-voltage process nodes [4] do not provide sufficiently safe state-integrity margin for latch structures on sub 90nm technologies with higher inter-device variation on latch feedback structures. Figure 4 shows the addition of a Boosted-Gate "drowsy" retention to the buffered SCCMOS power-gate of Figure 1 where the raised-voltage gate bias supply provides additional headroom to the scaled retention voltage.

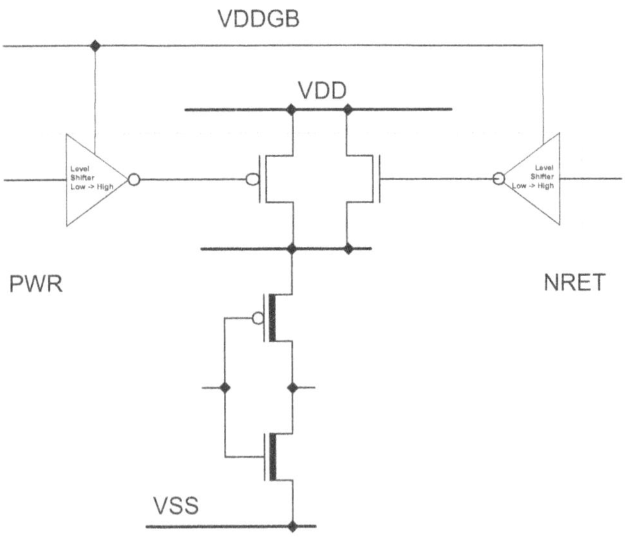

Figure 4: SCCMOS with Boosted-Gate retention

4.4 Sub-clock-cycle SRPG

At the other end of the SoC spectrum are ultra-low-power designs which in some cases are powered only by scavenged energy. Leakage power in such applications consumes power that would ideally be reserved for on-demand processing and communication. Due to low voltage headroom, frequencies are reduced and leakage within the clock cycle can become dominant.

Figure 5 shows the concept of state retention power gating within the clock cycle rather than the conventional power gating between bursts of active clocked operation. In the case of rising-edge clock logic shown, the combinational logic between register or pipeline stages is powered down, and inputs to SRPG elements clamped to avoid crowbar currents. The clock duty cycles may be varied with DVFS, and the standard timing closure requirements of powering back up, evaluating the cones of logic and setup to the next rising edge of the clock are observed.

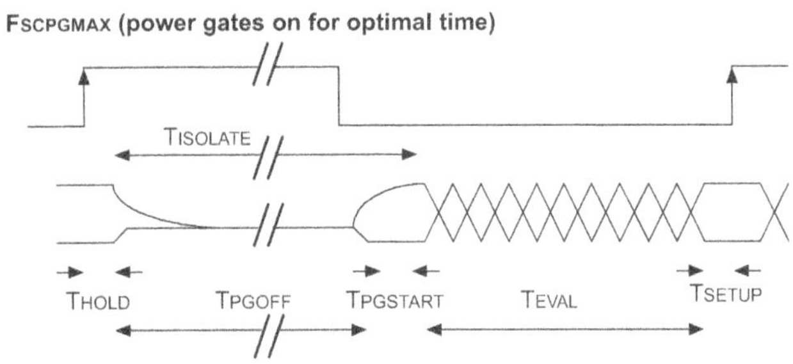

Figure 5: Sub-clock SRPG

The energy savings are obvious for extremely low clock frequencies but hard to model and predict in actual transistor layouts due to the virtual rail collapse behavior and the glitch-energy and cost in re-powering the logic and re-calculating the combinational state logic [10].

Due to the fact that the latest EDA multi-voltage tools support inference of header, footer and clamp cells, it becomes possible to enhance the implementation by adding both header and footer power switches to the combinational state and add in switchable Vth-drop elements to provide additional voltage scaling to the virtual rails across the combinational logic to sub-threshold voltage, reducing the leakage dramatically, while providing symmetric N-well and P-well reverse bias, shown in Figure 6.

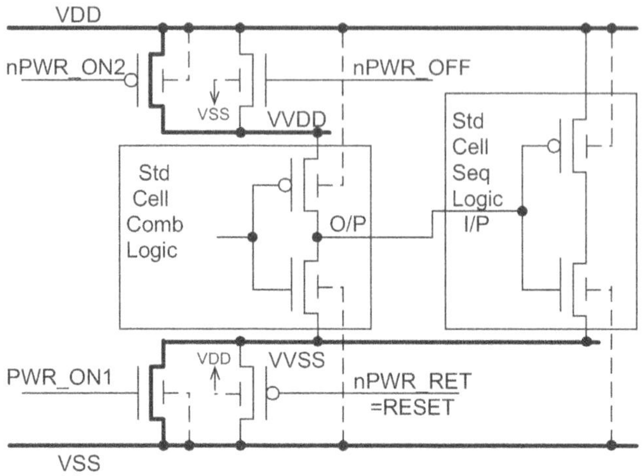

Figure 6: Sub-clock SRPG with drowsy sub-threshold logic

4.5 State Retention Integrity

State retention needs to be 100% robust and reliable in the presence of power-gating transients and noise from neighboring blocks that share a common ground or supply. The underlying retention registers need to be designed to balance retention leakage power with safe retention latch structures. Figure 7 shows an experimental structure to evaluate the integrity of retention registers at reduced voltages.

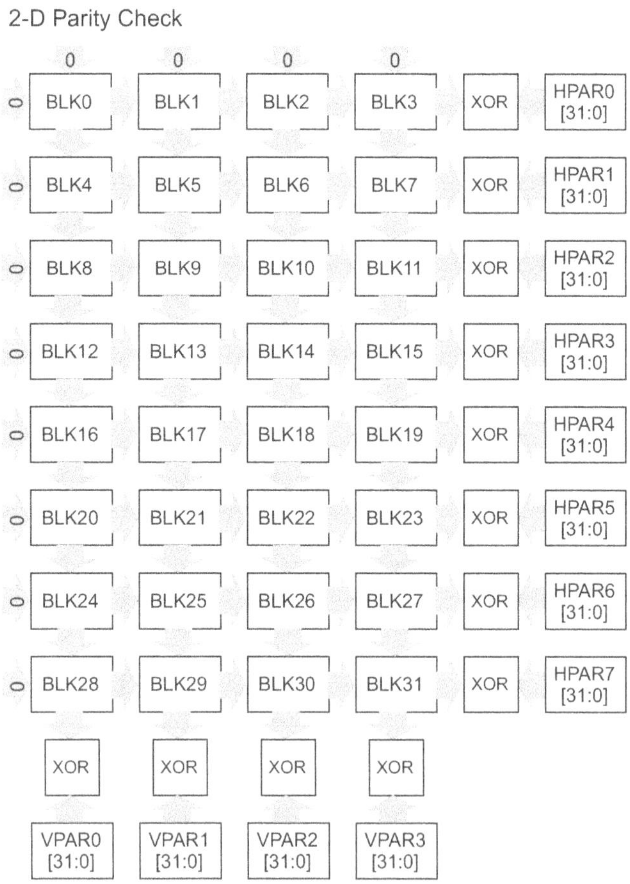

Figure 7: State integrity analysis structures

Such arrays of registers can be configured as read/write synthesized memory regions, parallel switching noise generators or as 2-D parity sense structures, where the Horizontal (HPAR) and Vertical (VPAR) values can capture array parity and then compare this, through level shifters for detecting first failing flop value when held in reduced voltage retention. Power gated arrays that share a common ground, for example, can be used to generate controlled transients and observe arrays for bit corruptions.

5. TECHNOLOGY EVALUATION

Bringing together the experimental IP with standard EDA tools and power intent is important to validate the model abstractions and methodology, and enable evaluation of measured versus predicted data.

Figure 8 shows the layout of test silicon implemented in 2011 to validate the physical IP cell abstractions and EDA flow compatibility. Sponsoring PhD research projects through Europractice[11] and supporting with Processor and Physical IP enabled funding of small (2 x 2mm!) die for rapid prototypes. Due to such chip size constraints, ARM Cortex-M0 CPU macro-cells were chosen. These were constrained for performance to a worst case corner signoff at 330MHz, on the TSMC 65nm Low-Leakage technology. Five matched pairs of CPUs were instantiated with the 4 pairs on the right of the layout to evaluate standard PG and SRPG implementations plus the enhanced live-slave retention as well as SCCMOS gate-bias implementations. A "tracking-pair" design approach allows the implementations to be evaluated at 400MHz+ with each CPU of a pair having critical paths stressed in even or odd clock cycles, while the main SOC runs reliably with zero-wait state memory (top left) at 200MHz.

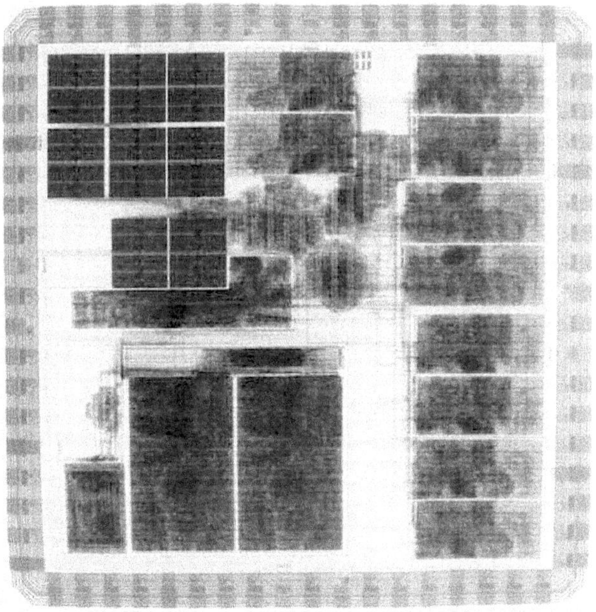

Figure 8: TSMC 65nm SRPG technology demonstrator

The state integrity structures in the lower-left layout analyze state integrity and reliability in the presence of switching noise and power gating inrush, as well as a Cortex-M0 implementation with sub-clock cycle SRPG and sub-threshold drowsy logic.

6. CONCLUSIONS

The multi-voltage extensions to EDA tools and industry collaboration and standardization on power-intent to annotate RTL designs is proving to be a valuable foundation to allow traditionally more "expert" low power design approaches to be applied to SoC designs with standard tool flows with tight time-to-market requirements. This is proving applicable across the range of products from ultra-low-power MCUs to high-performance multi-media systems.

7. ACKNOWLEDGMENTS

Thanks are expressed to Europractice, post-graduate researchers Sheng Yang and Jatin Mistry, at the University of Southampton, UK and to Synopsys Inc for the award of the Charles Babbage prize in 2008 that provided access to the full suite of multi-voltage tools that enabled the implementation and signoff.

8. REFERENCES

[1] Mudge, Trevor, "Power: A First Class Architectural Design Constraint" IEEE Computer, vol. 34, no. 4, April 2001. http://doi.ieeecomputersociety.org/10.1109/2.917539

[2] Keating, M., Flynn, D. et al, "Low Power Methodology Manual - for System-on-Chip Design", Springer 2007 ISBN: 978-0-387-71818-7 http://www.lpmm-book.org/

[3] Shi, K. Flynn, D. "Power Gating Design Tradeoffs and Considerations in Production Low-Power Designs", DesignCon 2009 http://www.designcon.com/infovault/paper.asp?PAPER_ID=474

[4] Mutoh S. et al. "A 1v multi-threshold voltage CMOS DSP with an efficient power management technique for mobile phone applications" ISSCC1996, pages 168–169, 1996.

[5] Flynn, D. Gibbons, A. "Design for State Retention: Strategies and Case Studies" SNUG San Jose 2008, Track TA2

[6] Accellera UPF Standard version 1.0, February 2007, now part of IEEE standard 1801 http://www.accellera.org/activities/p1801_upf

[7] Si2 Common Power Format, CPF, specification http://www.si2.org/?page=811

[8] 1801-2009 IEEE Standard for Design and Verification of Low Power Integrated Circuits http://standards.ieee.org/develop/wg/UPF.html http://standards.ieee.org/findstds/standard/1801-2009.html

[9] Stan, M., "Low-Threshold CMOS Circuits with Low Standby Current," in Proceedings of the International Symposium on Low-Power Electronics and Design. Monterey, CA: IEEE/ACM, 1998, pp. 97–99

[10] Mistry J., et al, "Sub-clock Power-Gating Technique for Minimizing Leakage Power during Active Mode", DATE 2011 http://eprints.ecs.soton.ac.uk/21768/

[11] EUROPRACTICE mini@sic programme: http://www.europractice-ic.com/prototyping_minisic.php

A 55nm 0.55V 6T SRAM with Variation-Tolerant Dual-Tracking Word-Line Under-Drive and Data-Aware Write-Assist

Yi-Wei Lin, Hao-I Yang, Geng-Cing Lin,
Chi-Shin Chang, Ching-Te Chuang, Wei Hwang
Department of Electronics Engineering
National Chiao Tung University, Hsinchu, Taiwan, R.O.C.
ChingTe.Chuang@gmail.com

Chia-Cheng Chen, Willis Shih,
Huan-Shun Huang
Faraday Technology Corporation
Hsinchu, Taiwan, R.O.C.

ABSTRACT

We present a 55nm 128Kb 6T SRAM with a variation-tolerant dual-tracking Word-Line Under-Drive (WLUD) to improve the RSNM and a Data-Aware Write-Assist (DAWA) scheme. Error free full functionality without redundancy is achieved from 1.5V down to 0.55V with area overhead of 4% for WLUD and 14% for DAWA, respectively. The measured power overheads (FF, 25°C) are 1.1% for WLUD and 3.3% for DAWA at 1.0V (3% and 5.3% at 0.6V), respectively. The maximum operating frequency is 940MHz (360MHz) at 1.0V (0.6V) and 25°C. The measured power/performance (FF, 25°C) is 0.117mW/MHz (0.023mW/MHz) at 1.0V (0.6V).

Categories and Subject Descriptors

B.3.1 [**Memory Structure**]: Semiconductor Memories - Static memory (SRAM).

Keywords

Data-Aware Write-Assist (DAWA), Static Random Access Memory (SRAM), Word-Line Under-Drive (WLUD)

1. INTRODUCTION

For conventional 6T SRAM, the cell stability, Static Noise Margin (SNM), and V_{MIN} are limited by leakage, variation and supply voltage in the physical domain; and by the conflicting Read/Write requirements and cell disturbs in the design domain. In the design and optimization of 6T SRAM, the procedure typically starts by reducing Read-disturb to maintain adequate Read Static Noise Margin (RSNM) to make sure the cell won't flip during Read operation. For cost-performance applications and portable/handheld devices where the cost, form factor and weight considerations dictate the use of single supply, Read-assist scheme such as Word-Line Under-Drive (WLUD) (also called Suppressed Word-Line (SWL)) [1-3] can be employed to mitigate Read disturb and Half-Select disturb.

With WLUD, the Read performance and Write-ability degrade. The Write-ability, Write Margin (WM), and Write performance are typically recovered/enhanced by floating power-line Write [4], column based dynamic supply [5], collapsing the cell supply transiently below the SRAM data retension voltage [6], or Negative Bit-Line (NBL) [1-3, 7] techniques. In this work, we present a low power-overhead variation-tolerant dual-tracking Word-Line Under-Drive (WLUD) scheme. The proposed WLUD scheme exhibits excellent process corner tracking capability, and is resistor-less and area-efficient. The Write-ability and Write performance are enhanced by a variation-tolerant Data-Aware Write-Assist (DAWA) [8] scheme with simple virtual cell supply control logic, which requires much smaller area overhead compared with Negative Bit-Line (NBL) [1-3, 7] Write-assist scheme. Compared with the conventional floating power-line Write-assist scheme [4], the proposed DAWA scheme offers faster dynamic supply switching speed and reduces dynamic supply switching power and noise to half. A 128Kb test chip implemented in 55nm Standard Performance CMOS technology achieves error free full functionality without redundancy from 1.5V down to 0.55V.

This paper is organized as follows. Section 2 describes the dual-tracking Word-Line Under-Drive (WLUD) scheme. Section 3 discusses the Data-Aware Write-Assist (DAWA) scheme. Section 4 presents a 128Kb test chip implementation in 55nm Standard Performance CMOS. Section 5 analyzes the measurement results. Section 6 concludes the paper.

2. WORD-LINE UNDER-DRIVE (WLUD)

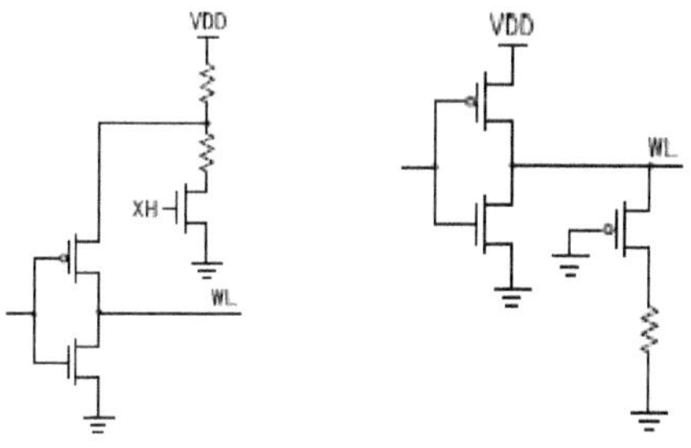

(a) Reference [1] (b) Reference [2]

Figure 1. Previous WLUD scheme in (a) Ref. [1] and (b) Ref. [2].

The commonly used methods to suppress the WL level are: 1) lower/suppress the supply voltage of the WL driver and 2) provide a bleeder to lower the WL voltage. The most straight-forward method is to connect a replica of cell access NMOS between WL and GND [1].

ISLPED'12, July 30– August 1, 2012, Redondo Beach, CA, USA.
Copyright 2012 ACM 978-1-4503-1249-3/12/07...$10.00.

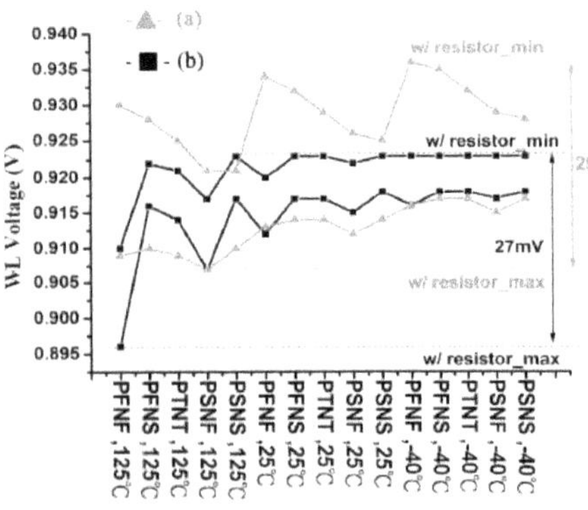

Figure 2. WL voltage level of WLUD schemes in Fig. 1(a) and Fig. 1(b) in 55nm Standard Performance CMOS technology across process and temperature corners at V_{DD} = 1.0V.

This method is quite simple to implement but the WL level suffers severe process corner and temperature variations. A poly-resistor can be added in series with the NMOS to provide feedback to mitigate the process corner and temperature sensitivity. A further improved version utilizes the tracking NMOS and series feedback resistor to control the virtual supply node of the WL driver, as shown in Fig. 1(a) [1]. Notice that in Fig. 1(a), the virtual supply node of the WL driver taps the poly-resistor somewhere in the middle to minimum the sensitivity of WL level to poly sheet-resistance variation, which can be easily ±30% in deep sub-100nm technology. However, due to the presence of resistor in the virtual supply path and the lowered virtual supply for the WL driver, the scheme suffers severe degradation of WL rise time. Another scheme, depicted in Fig. 1(b), uses PMOS and series feedback resistor to lower the WL level [2]. This method could prevent the degradation of WL rise time and provide a stable WL level across process corners, but still suffers from the poly sheet-resistance variation and the large area needed to implement the resistor. Fig. 2 shows the simulated WL levels of the schemes in Fig. 1(a) and 1(b) in 55nm Standard Performance (SP) CMOS technology. The scheme of Fig. 1(a) has smaller temperature dependence, yet exhibits large corner-to-corner variations. The scheme of Fig. 1(b) exhibits larger temperature dependence. At high-T and PFNF corner, the WL voltage drops significantly more than other cases. Finally, The scheme in Fig. 1(a) tracks only the cell access NMOS, and that in Fig. 1(b) tracks only the WL driver pull-up PMOS and the cell holding PMOS. As such, they do not fully track the corner variations, and the WL level is not optimally adjusted across process corners.

Fig. 3(a) shows the basic configuration of the proposed dual-tracking WLUD scheme. The proposed dual-tracking WLUD circuit comprises of one PMOS M2 which tracks the WL driver pull-up PMOS and the cell holding PMOS, and one NMOS M3 which tracks the cell access NMOS. PMOS M2 serves as a variable feedback "resistor" for NMOS M3, and NMOS M3 serves as a variable feedback "resistor" for PMOS M2. This resistor-less implementation eliminates sensitivity to poly sheet-resistance variation [1], and can be easily implemented with minimum device/area overhead. The gate of PMOS M2 can also be connected to the input of the WL driver (Fig. 3(b)). Functionally, it remains the same with the difference being in the layout and possibly area overhead. The discharging NMOS M3 can be shared among WL drivers to further reduce device and area overhead (Fig. 3(c)), and the gate of the shared discharging NMOS M3 can be connected to a "Control Voltage" to offer Die-to-Die programming capability.

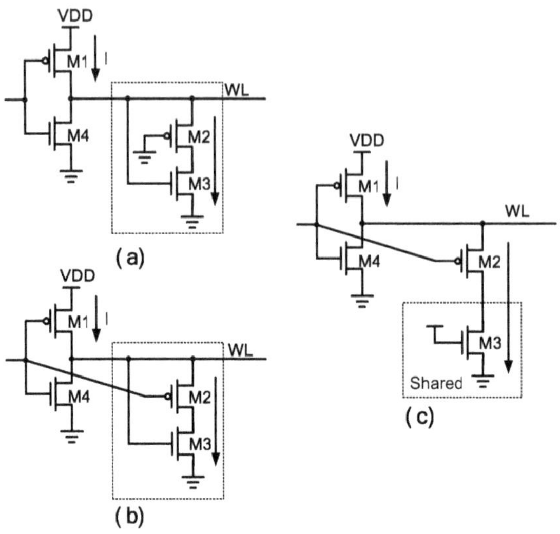

Figure 3. Proposed dual-tracking WLUD schemes: (a) basic configuration, (b) alternative configuration with the gate of PMOS M2 connected to the input of the WL driver and (c) configuration with the gate of shared discharging NMOS M3 connected to a "Control Voltage".

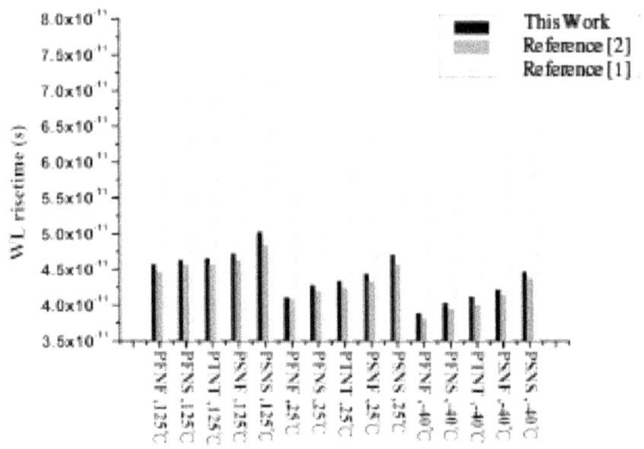

Figure 4. Comparison of WL rise times of the proposed dual-tracking WLUD scheme (Fig. 3(a)) with that in Fig. 1(a) and 1(b) across process and temperature corners at V_{DD} = 1.0V.

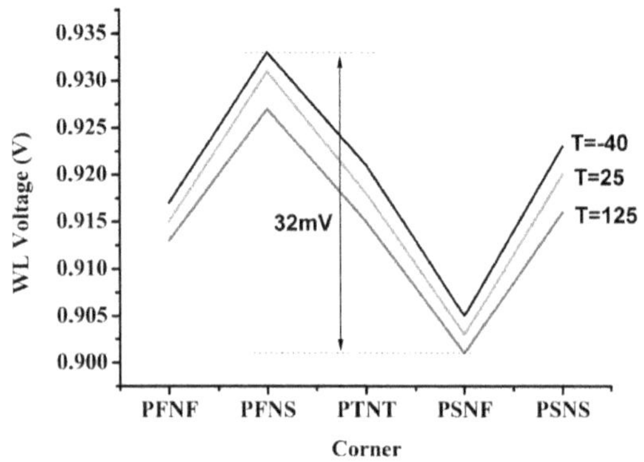

Figure 5. WL voltage level of the proposed dual-tracking WLUD scheme across process corners at V_{DD} = 1.0V.

Fig. 4 compares the WL rise time of the proposed dual-tracking WLUD scheme (Fig. 3(a)) with that of Fig. 1(a) and 1(b) across process and temperature corners at V_{DD} = 1.0V. The proposed scheme and the scheme in Fig. 1(b) [2] can be seen to have significantly better WL rise times than that in Fig. 1(a) [1] as the WL driver is driven by full V_{DD} and there is no resistor in the WL pull-up path. Fig. 5 shows the WL voltage level of the proposed dual-tracking WLUD scheme across process corners at V_{DD} = 1.0V. Notice that as the WL voltage level becomes low, the channel resistance of M3 increases and the current through M2/M3 decreases. This helps to stabilize the WL voltage at a certain level. The tracking of both cell access NMOS and WL driver pull-up PMOS (and cell holding PMOS) offers the capability to optimally place/adjust WL level across process corners and temperature. The proposed scheme provides the lowest WL level at PSNF corner (where the RSNM is the worst), and the highest WL level at PFNS corner (where RSNM is the best while the Write-ability is the worst). At PSNF corner, if there is no M2 to track M1 (i.e. if M2 is replaced by a poly-resistor), the WL level would be pulled-down too low by the strong NMOS M3. With M2 to track M1, the channel resistance of M2 increases and its current decreases, thus limiting the WL pull-down current to maintain a proper WL level. Similarly, at PFNS corner, the slow NMOS M3 allows the WL level to stay higher, thus compensating for the slow (weak) cell access NMOS.

Notice that the WL driver pull-up PMOS M1 works at triode (linear) region, and M2-M3 forms a diode like I-V characteristics (Fig. 6). The diode like I-V curve of M2-M3 intersects the PMOS M1 I-V curve in triode (linear) region where the transistors behave like resistors and have small temperature sensitivity. As a result, the resulting WL level has small sensitivity to temperature as can be seen in Fig. 5.

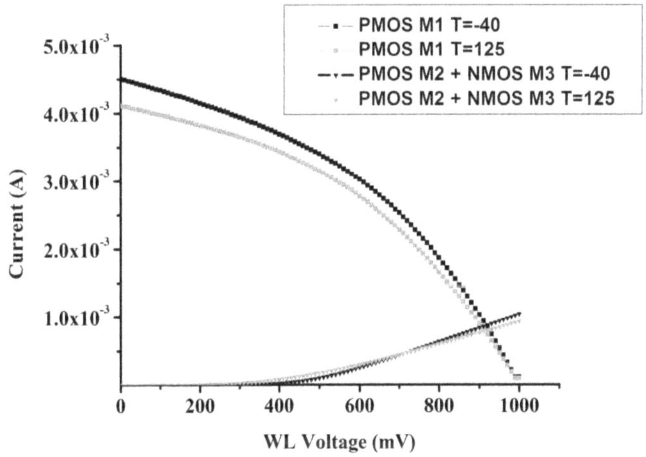

Figure 6. I-V_{WL} curve of the proposed dual-tracking WLUD scheme.

Fig. 7(a) shows that the proposed scheme offers RSNM improvement of 29% at PSNF corner (where RSNM is the worst) at V_{DD} = 0.6V, whereas a constant (90%) suppressed WL level results in RSNM improvement of only 21%. Fig. 7(b) shows RSNM versus V_{DD} at PSNF corner for the proposed dual-trakcing WLUD scheme and that of full V_{DD}. The proposed scheme improves the Read V_{MIN} (defined here as the V_{DD} where RSNM becomes 0) from 0.4V to 0.3V.

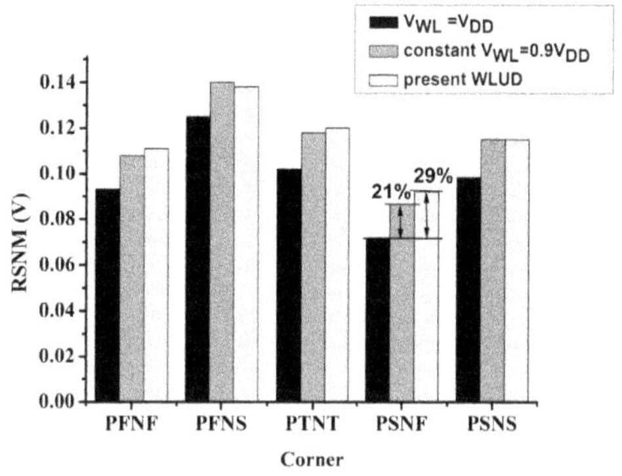

(a)

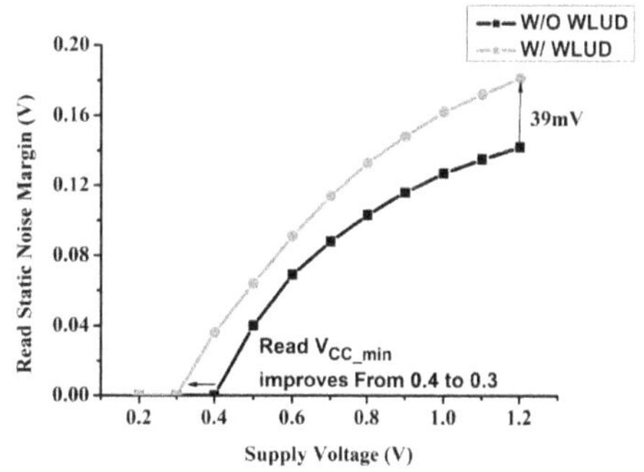

(b)

Figure 7. (a) RSNM of the proposed dual-tracking WLUD scheme compared with that of full V_{DD} and constant (90%) suppressed WL level at V_{DD} = 0.6V, (b) RSNM vs. V_{DD} at PSNF corner for the proposed dual-trakcing WLUD scheme and that of full V_{DD}.

3. DATA-AWARE WRITE-ASSIST (DAWA)

The degraded Write-ability (and WM) due to WLUD is typically recovered/enhanced by Write-assist such as floating power-line Write [4], column based dynamic supply [5], collapsing the cell supply transiently below the SRAM data retension voltage [6], or Negative Bit-Line (NBL) [1-3, 7] techniques. In this work, we propose a low power-overhead variation-tolerant Data-Aware Write-Assist (DAWA) [8] scheme with simple virtual cell supply control logic. The scheme requires much smaller area overhead compared with Negative Bit-Line (NBL) Write-assist which incurs large area overhead due to large boosting capacitor and complicated control logic. Furthermore, NBL Write-assist suffers gate dielectric over-stress and leakage from the unselected transistors in the Column Mux with their source voltages boosted below GND, and extra boost attenuation and control circuits are needed to mitigate the problems [7]. Compared with the conventional floating power-line Write-assist scheme, the proposed DAWA scheme offers faster dynamic supply switching speed and reduces dynamic supply switching power and noise to half.

Fig. 8(a) shows the circuit structure with proposed dual-tracking WLUD and DAWA schemes, while Fig. 8(b) shows an "Alternative DAWA" scheme. In the proposed DAWA scheme, the column-based cell supply is split into 2 virtual supply lines, one for the left-half cells (VDDQB), and the other for the right-half cells (VDDQ). Each of the virtual supply lines is controlled by a large PMOS power-switch (M3/M4) and a small PMOS keeper (M1/M2).

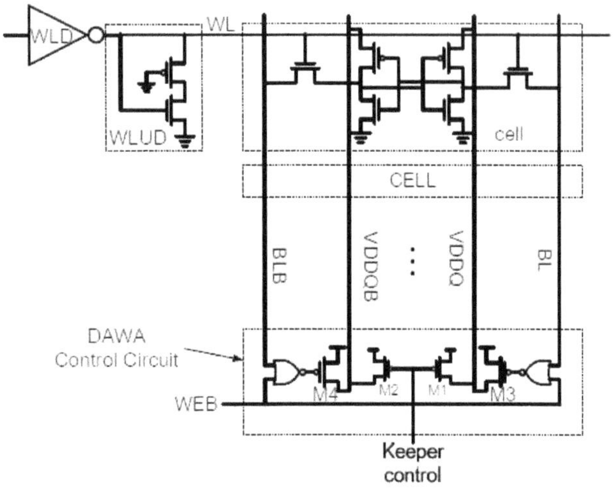

(a)

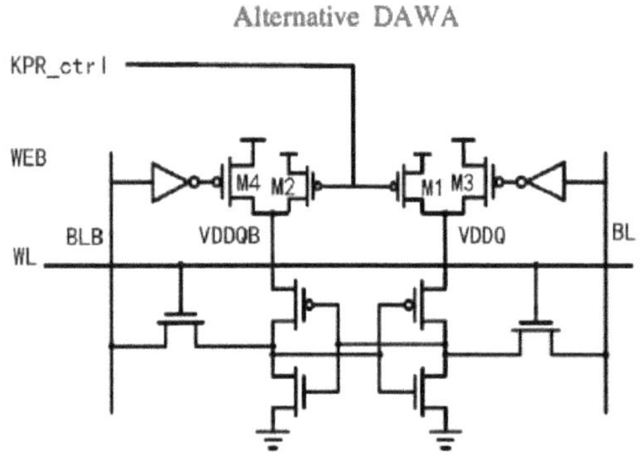

(b)

Figure 8. (a) Circuit structure with proposed dual-tracking WLUD and DAWA and (b) "Alternative DAWA" scheme.

In Standby or Read mode, WEB (Write Enable Bar) = 1, the NOR gates are disabled, and power-switches M3 and M4 are "On" to maintain VDDQ/VDDQB at high level. In Write mode, WEB = 0, the NOR gates are enabled. Depending on Data-in at the bit-line pair, the low-going bit-line turns off either M3 or M4, causing the corresponding virtual supply node (VDDQ or VDDQB) to drop, thus reducing the V_{GS} and contention of the corresponding cell holding PMOS to enhance Write-ability and WM. The opposite half-cell inverter is unaffected and maintains its strength and feedback action. The timing of supply switching is initiated directly from low-going BL, thus tolerant to PVT variations and V_T scatter. Data-aware switching of half-cell supplies reduces dynamic supply switching power and noise to half, and improves supply switching speed. The scheme adds minimum loading to BL (only 1 extra C_{gate}). In 55nm Standard Performance CMOS, splitting the vertical cell supply line into VDDQ and VDDQB does not increase the 6T cell size. The DAWA device/area overhead can be reduced by replacing the NOR gate with inverter as shown by the "Alternative DAWA" scheme in Fig. 8(b). The only difference for "Alternative DAWA" scheme is in Read mode, when either BL or BLB is pulled-down by cell access NMOS and pull-down NMOS, one of the power-switch (M3/M4) may turn off. This does not appear to be a problem since the pull-down of bit-line in Read (by cell transistors) is much weaker/slower than that in Write (by strong Write Driver), and the trip voltage of the control inverter can be adjusted. Furthermore, the gate voltages of cell pull-down NMOS (held at "solid" 1 by unaffected opposite half-cell inverter) and access NMOS remain unchanged, so the Read operation/performance is not affected even if the power-switch for the "0" storage side turns off.

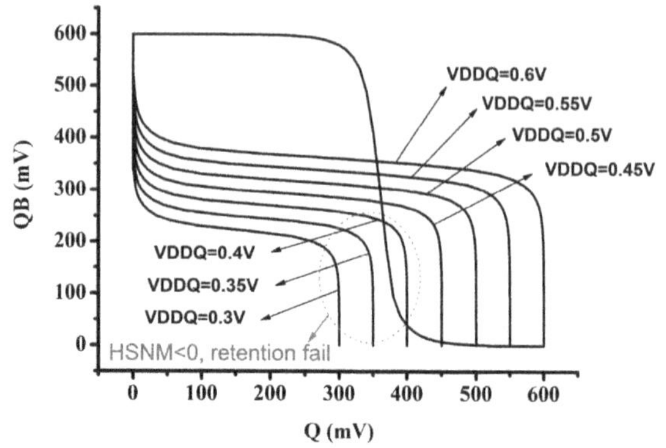

Figure 9. Asymmetrical Hold SNM for half-selected cells on selected BL due to disparity between VDDQ and VDDQB.

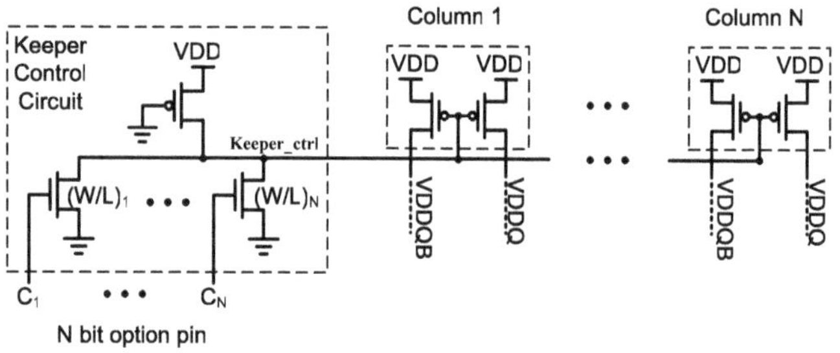

(a)

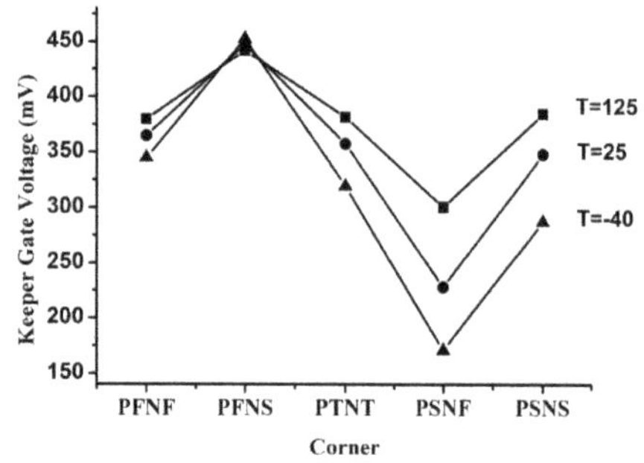

(b)

Figure 10. (a) Dual-tracking level programmable keeper control circuit to control gate voltage of the keepers (M1/M2 in Fig. 8) and (b) keeper gate voltage versus process corners.

Disparity between VDDQ and VDDQB results in asymmetrical Hold SNM for half-selected cells on the selected BL as shown in Fig. 9. To ensure adequate Half-Select stability, a dual-tracking level programmable keeper control circuit (Fig. 10(a), which tracks the keeper PMOS (M1/M2 in Fig. 8) and cell access NMOS, is employed to control the gate voltage of the keepers (M1/M2 in Fig. 8). At PFNS (PSNF) corner where Write-ability is the worst (best), the scheme offers the highest (lowest) gate voltage for the PMOS keepers, hence the weakest (strongest) keeper, as shown in Fig. 10(b). Fig. 11 shows WSNM and WM at PFNS corner (where WSNM and WM are the worst) at V_{DD} = 0.6V. With WLUD, WSNM degrades. With DAWA, WSNM is enhanced/improved to become even better than the original case without WLUD. The WM improves by 74mV at 0.6V.

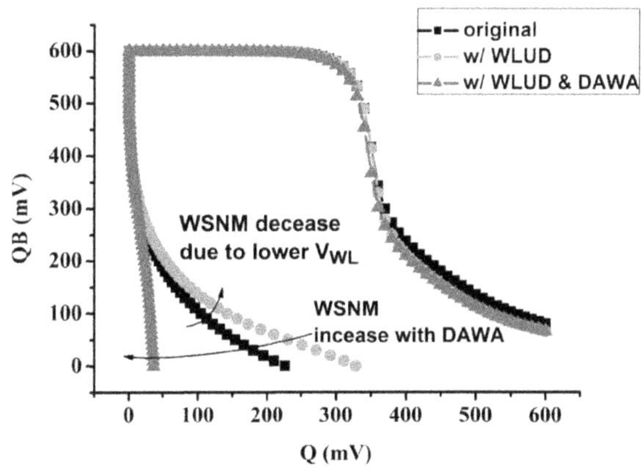

(a)

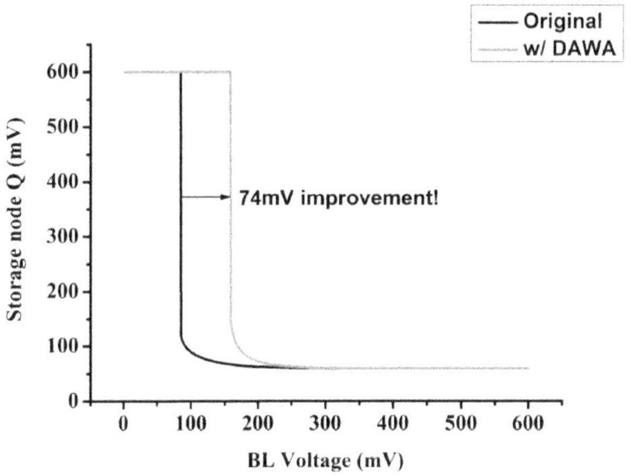

(b)

Figure 11. (a) WSNM and (b) WM at PFNS corner at V_{DD} = 0.6V. DAWA enhances WSNM and improves WM by 74mV.

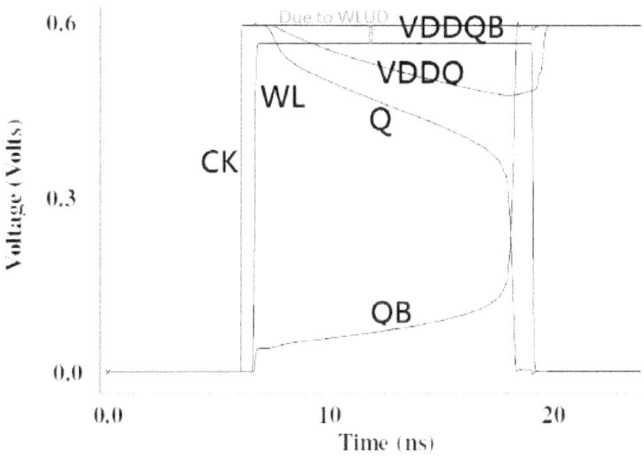

Figure 12. Simulated Write waveforms at V_{DD} = 0.6V.

Fig. 12 shows the simulated waveforms during Write operation at V_{DD} = 0.6V. The under-driven WL voltage and low-going floating VDDQ (which helps the crossing of Q and QB) can be clearly seen.

4. 128Kb TEST CHIP IMPLEMENTATION

The floor plan of the 128Kb test chip implemented in 55nm Standard Performance (SP) CMOS is shown in Fig. 13 with the locations of WLUD, DAWA Control and Keepers marked. The cell size is 0.62 μm^2. The 128Kb array is organized into 1024 Word-Lines and 128 Columns with inter-leaving 8 architecture. The data I/O width is 16 bit. The 1024 rows are divided into 8 local blocks, so the local bit-line length is 128 bit. The conventional small-signal differential sensing is used. We add a set of DAWA control logic and keeper to control the virtual cell supply for each column in local block. The keeper gate voltage control circuit has 4 pin for external (programmable) control. Every word-line is connected with a WLUD circuit.

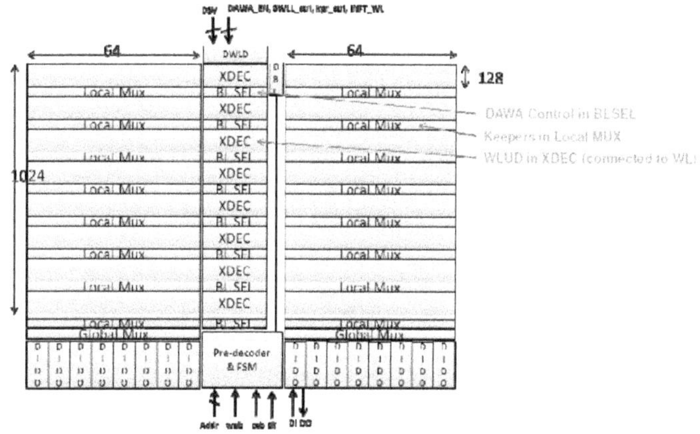

Figure 13. Floor plan of the 128Kb test chip.

5. 128Kb CHIP MEASUREMENT RESULTS

Fig. 14 shows the measured die yield (without redundancy) versus V_{DD} for FF (52 dies), TT (39 dies), and SS (51 dies) corners. Dies are tested with full suits of SRAM compiler qualification patterns including MATS+, MARCH C- , and MARCH C+ test patterns with all High/Low-Read/Write combinations. The design uses a built-in Finite State Machine (FSM) to control the WL pulse width, so the die yield would be the same up to the maximum operating frequency at each V_{DD}. For V_{DD} down to 0.55V, we still have "perfect" die yield of 19.2% (FF), 20.5% (TT) and 7.8% (SS). The failures for V_{DD} down to 0.55V are one or two bit failures per chip, which can be readily repaired using redundancy (not implemented in this test chip). Fig. 15 shows the measure cell Write failure rate versus V_{DD}. The degradation of Write failure rate due to WLUD, and the improvement of Write failure rate by DAWA for V_{DD} < 0.5V can be clearly seen. Fig. 16 shows the micrograph, layout view and features of the 128Kb test chip. The area overheads for WLUD and DAWA (NOR Control) are 4% and 14%, respectively. The DAWA device/area overhead can be reduced by using the "Alternative DAWA" scheme (Inverter Control) in Fig. 8(b) and revoking the programmability of the dual-tracking keeper control circuit in Fig. 10(a). The measured power overheads (FF, 25°C) are 1.1% for WLUD and 3.3% for DAWA at 1.0V (3% and 5.3% at 0.6V), respectively. The maximum operating frequency is 940MHz (360MHz) at 1.0V (0.6V) and 25°C. The measured power/performance (FF, 25°C) is 0.117mW/MHz (0.023mW/MHz) at 1.0V (0.6V).

6. CONCLUSIONS

We presented a 55nm 128Kb 6T SRAM with wide operation voltage range. The design utilized a variation-tolerant dual-tracking Word-Line Under-Drive (WLUD) scheme to improve the Read Static Noise Margin (RSNM). The proposed WLUD scheme exhibited excellent process corner tracking capability, and was resistor-less and area-efficient. The Write-ability and Write performance were enhanced by a variation-tolerant Data-Aware Write-Assist (DAWA) scheme with simple virtual cell supply control logic, which required much smaller area overhead compared with Negative Bit-Line (NBL) Write-assist scheme. Compared with the conventional floating power-line Write-assist scheme, the proposed DAWA scheme offered faster dynamic supply switching speed and reduced dynamic supply switching power and noise to half. Error free full functionality without redundancy was achieved from 1.5V down to 0.55V with area overhead of 4% for WLUD and 14% for DAWA, respectively. The measured power overheads (FF, 25°C) were 1.1% for WLUD and 3.3% for DAWA at 1.0V (3% and 5.3% at 0.6V), respectively. The maximum operating frequency was 940MHz (360MHz) at 1.0V (0.6V) and 25°C. The measured power/performance (FF, 25°C) was 0.117mW/MHz (0.023mW/MHz) at 1.0V (0.6V).

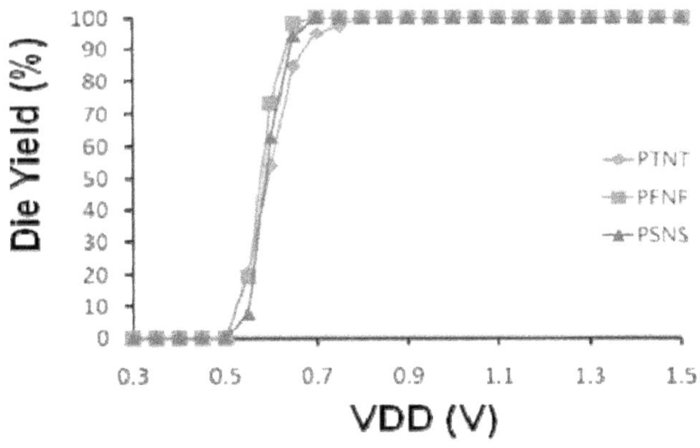

Figure 14. Measured error free full functionality die yield (without redundancy) versus V_{DD} for FF (52 dies), TT (39 dies), and SS (51 dies) corners at room temperature.

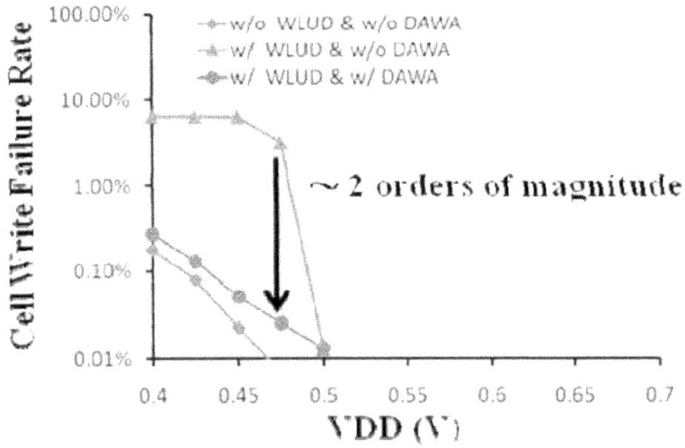

Figure 15. Measured cell Write failure rate versus V_{DD} of a FF die at room temperature.

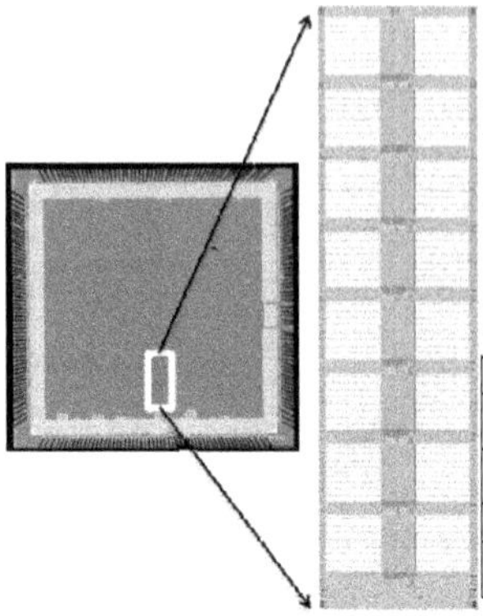

Capacity	1024*128 bits (Total 128 Kbit)
Process	55nm 1P9M SP CMOS
Bit Cell	6T – 0.62 μm²
Max frequency @ 1.0V FF,T=125	940MHz
Average Power @ 1.0V FF,T=25	0.117mW/MHz
Max frequency @ 0.6V FF,T=125	360MHz
Average Power @ 0.6V FF,T=25	0.023mW/MHz
Core Area	714μm * 238μm
V_{MIN}	0.55V

VDD=1.0V	Original	w/ WLUD	w/ DAWA
Power (FF, T=25)	0.112mW/MHz	0.114mW/MHz	0.116mW/MHz
Power (Normalized)	1	1+1.1%	1+3.3%
Area	0.126mm²	0.132mm²	0.146mm²
Area (Normalized)	1	1+4%	1+14%

Figure 16. Die micrograph, layout view and features of the 128Kb test chip.

7. ACKNOWLEDGMENTS

This work was supported by the Ministry of Education in Taiwan under ATU Program and the Ministry of Economic Affairs in Taiwan under Contract 100-EC-17-A-01-S1-124.

8. REFERENCES

[1] K. Nii, et al., "A 45-nm Single-port and Dual-port SRAM family with Robust Read/Write Stabilizing Circuitry under DVFS Environment," Symp. VLSI Circuits, pp. 212-213, June, 2008.

[2] Y. Fujimura, et al., "A Configurable SRAM with Constant-Negative-Level Write Buffer for Low-Voltage Operation with 0.149μm2 Cell in 32nm High-K Metal-Gate CMOS," ISSCC Dig. Tech. Papers, pp. 348-349, Feb., 2010.

[3] D .P. Wang, et al., "A 45nm Dual-Port SRAM with Write and Read Capability Enhancement at Low Voltage," Proc. IEEE Int'l SOC Conf., 211-214, 2007.

[4] M. Yamaoka, et al., "Low-Power Embedded SRAM Modules with Expanded Margins for Writing," ISSCC Digest of Tech. Papers, pp. 480-481, Feb., 2005.

[5] K. Zhang, et al., "A 3-GHz 70Mb SRAM in 65nm CMOS Technology with Integrated Column-Based Dynamic Power Supply," ISSCC Digest of Tech. Papers, pp. 474-475. 2005.

[6] Y. Wang, et al., "Dynamic Behavior of SRAM Data Retention and a Novel Transient Voltage Collapse Technique for 0.6V 32nm LP SRAM," Tech. Digest, IEDM, pp. 741-744, 2011.

[7] H. Pilo, et al., "A 64Mb SRAM in 32nm High-k Metal-Gate SOI Technology with 0.7V Operation Enabled by Stability, Write-Ability and Read-Ability Enhancements," ISSCC Digest of Tech. Papers, pp. 254-255, Feb., 2011.

[8] M. F. Chang, et al., "A Differential Data Aware Power-supplied (D2AP) 8T SRAM Cell with Expanded Write/Read Stabilities for Lower VDDmin Applications," Dig. Tech. Papers, Symp. VLSI Circuits, pp. 156-157, 2009.

A 40-nm 256-Kb Sub-10 pJ/Access 8T SRAM with Read Bitline Amplitude Limiting (RBAL) Scheme

[1]Shusuke Yoshimoto, [1]Masaharu Terada, [1]Youhei Umeki, [1]Shunsuke Okumura,
[2]Atsushi Kawasumi, [2]Toshikazu Suzuki, [2]Shinichi Moriwaki, [2]Shinji Miyano,
[1]Hiroshi Kawaguchi, and [1]Masahiko Yoshimoto
[1]Graduate School of System Informatics, Kobe University, Kobe, Japan
[2]Semiconductor Technology Academic Research Center, Yokohama, Japan
yoshipy@cs28.cs.kobe-u.ac.jp

ABSTRACT
This paper presents a novel read-bitline amplitude limiting (RBAL) scheme which suppresses dynamic energy dissipation caused by random variation. In addition, a discharge acceleration (DA) circuit is proposed to decrease delay overhead of RBAL. The proposed scheme improves the active energy dissipation in a read cycle by 22% at the center–center (CC) corner and 25°C. The maximum delay overhead is 32% at the fast–slow (FS) corner and -40°C. The circuits have been implemented using the 40-nm bulk CMOS process. The implemented 256-Kb 8T SRAM works fine with energy dissipation of sub-10 pJ / access from 0.5–0.7 V.

Categories and Subject Descriptors
B.3.1 [**Memory Structures**]: Semiconductor Memories – *Static Memory (SRAM)*

General Terms
Design

Keywords
8T SRAM, low voltage, low energy, read bitline limiter, discharge accelerator

1. INTRODUCTION
The minimum feature size in transistors continues to decrease with the advance of process technology. Process scaling realizes higher density and lower cost. In the deep sub-micron era, the threshold voltage (Vt) deviation in transistors is, however, increasing to more than 100 mV as $3\sigma_{Vt}$ [1–3]. Consequently, designing a 6T SRAM cell has become increasingly difficult: both read and write margins must be considered [4–7]. The 8T SRAM cell presented in Fig. 1(a) was proposed to eliminate read failures caused by the dedicated read port. Therefore, in the 8T cell, only the write margin must be considered, which can make a layout smaller and less expensive than the 6T cell in future processes [8].

Recent reports have described the effectiveness of low-voltage operation for energy reduction in SRAM degrades at the scaled process [9–10]. Figure 1(b) presents a simulated histogram of the read current in 8T SRAM cell at 0.5 V. The read current of the slowest cell in five thousand Monte-Carlo simulations was 0.016 times slower than the nominal cell. SRAM designers configure a pulse width of the read wordline (RWL) by the slowest cell and the sense-amp offset. Figure 2 depicts simulated waveforms of the read bitline (RBL) including random variation. In most cases, RBLs are fully discharged, although the slowest cell is read out, which increases the dynamic energy by 82% compared to the nominal simulation without random variation [10].

Realizing low-energy operation at low supply voltage demands reduction of the energy increase caused by random variation. An earlier proposed bitline amplitude limiter (BAL) suppresses the energy increase shown in Fig. 3(a) [10], limiting

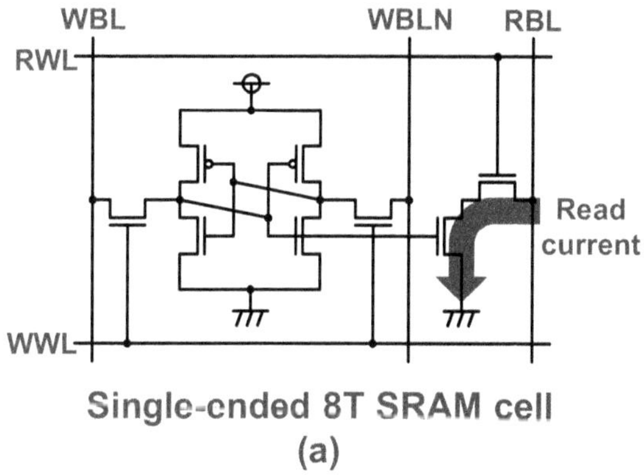

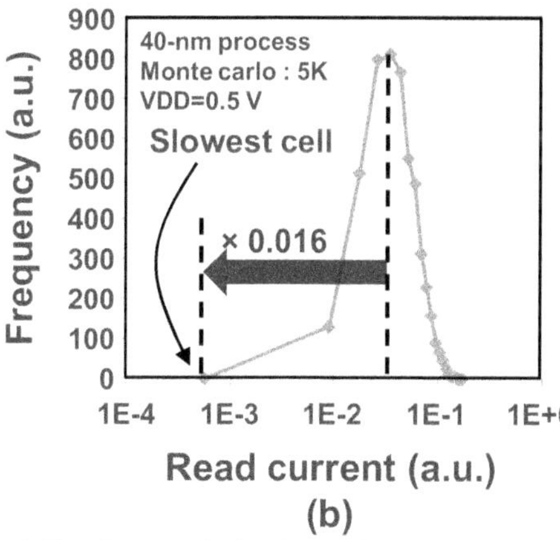

Figure 1. Random variation in read ports of (a) single-ended 8T SRAM cells and (b) histgram of read current.

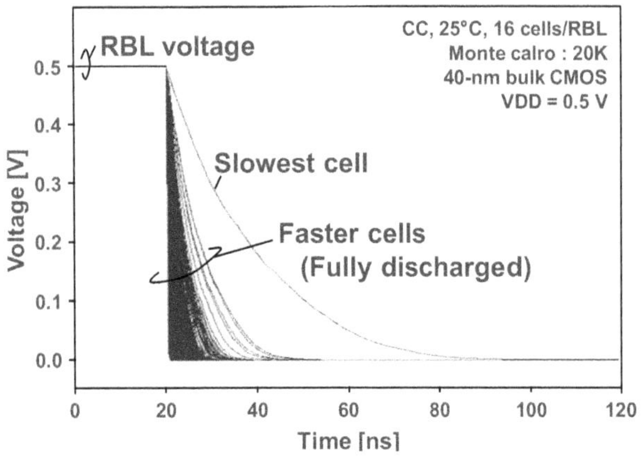

Figure 2. Waveforms of read bitline in single-ended 8T cell at 0.5 V.

the swing of bitlines because half-swing bitlines make the virtual ground of memory cells floating, and reducing the dynamic and leakage power by 26%. However, BAL presents the disadvantage of disturb-margin degradation. The 6T cell uses differential bitlines to read. Consequently, BAL consists of two series of NMOSes, which degrades the cell stability at low voltage.

As described in this paper, we proposed a read bitline amplitude limiter (RBAL) using 8T SRAM cell and a novel discharge acceleration (DA) scheme, presented in Fig. 3(b). RBAL is applied to the disturb-free single-ended read port of the 8T cell. Therefore, disturb-margin degradation and RBAL consists of only one NMOS. The DA scheme is proposed to improve the delay overhead of RBAL.

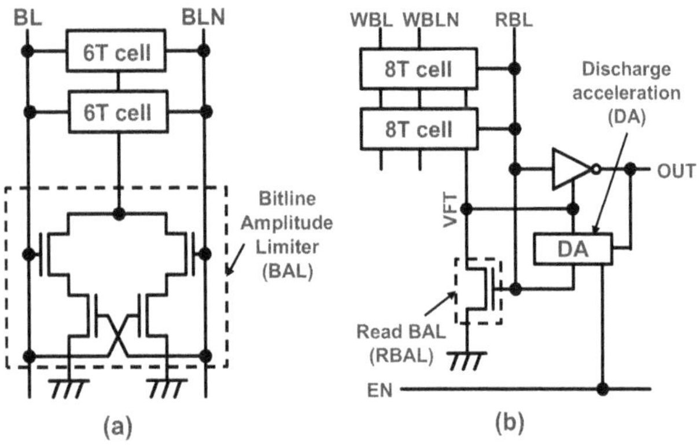

Figure 3. Diagrams of (a) Bitline amplitude limiter (BAL) with 6T cell and (b) read bitline amplitude limiter (RBAL) and distcharge acceleration circuit with 8T cell.

2. READ BITLINE AMPLITUDE LIMITER AND DISCHARGE ACCELERATION SCHEME

Figure 4 portrays an RBAL and novel DA scheme, which consist of only three NMOSes. The gate and drain of RBAL are connected respectively to RBL and virtual footer (VFT). In a 1-read operation, the RBLs are pulled down by the activated read ports. Figure 5 presents the waveforms of RBL with RBAL and without the DA scheme. Although the slowest cell is discharging the RBL, the other RBLs are pulled down faster. However, the discharge is stopped by the threshold voltage of RBAL. The

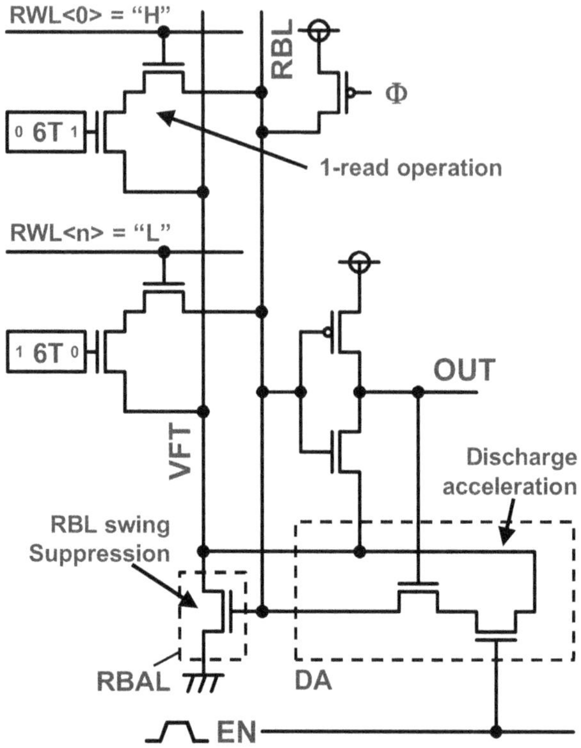

Figure 4. Circuit diagram of RBAL and discharge acceleration (DA) scheme.

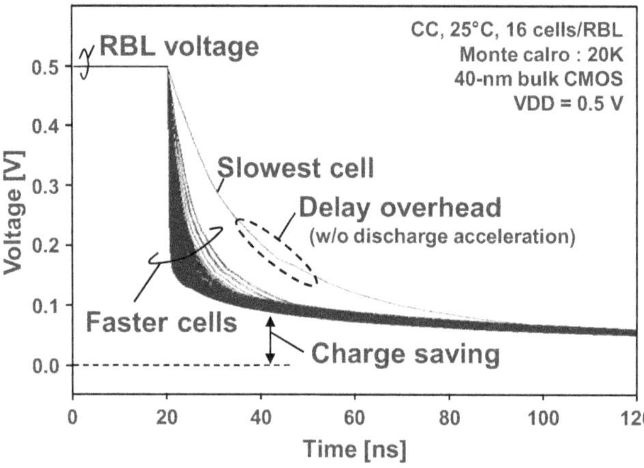

Figure 5. Waveforms of read bitline in proposed 8T cell with read bitline limiter.

decrease of average swing in RBL improves the dynamic energy. The single-ended 8T cell generally uses an inverter as a sense amp. Consequently, there is delay overhead when the RBL ranges under middle voltage. To prevent the delay degradation, the novel DA scheme is proposed. The DA terminals are connected to RBAL, VFT, output of sense-amp (= inverter) and enable signal (EN). When the output voltage is over the threshold voltage of the NMOS and the EN signal is activated, the RBL and the VFT are shorted by the two series of NMOSes. The DA scheme improves the read bitline delay presented in Fig. 6.

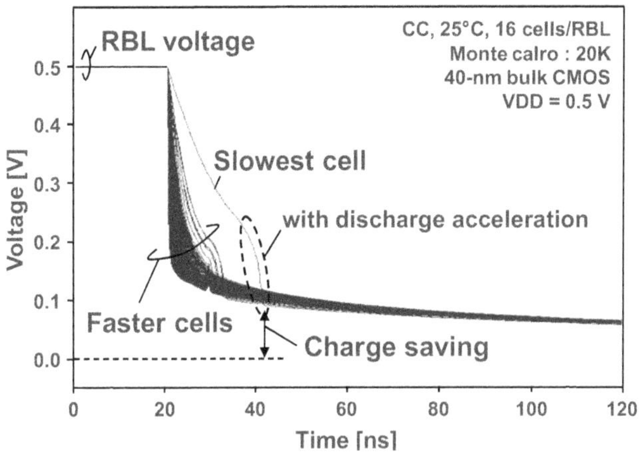

Figure 6. Waveforms of read bitline in proposed 8T cell with read bitline limiter and assist circuit.

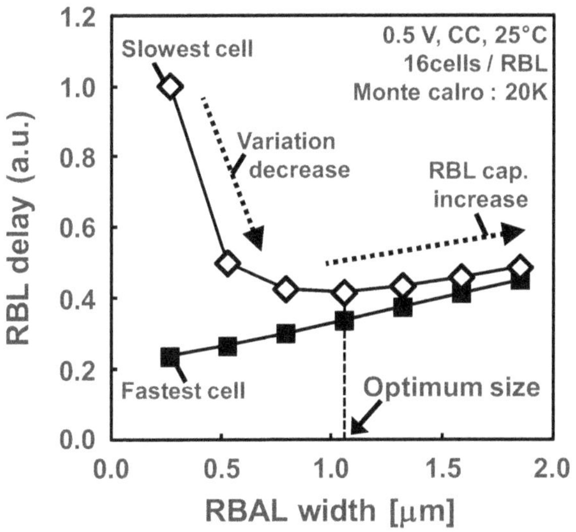

Figure 7. Read bitline (RBL) delay versus the width of the limiter transistor.

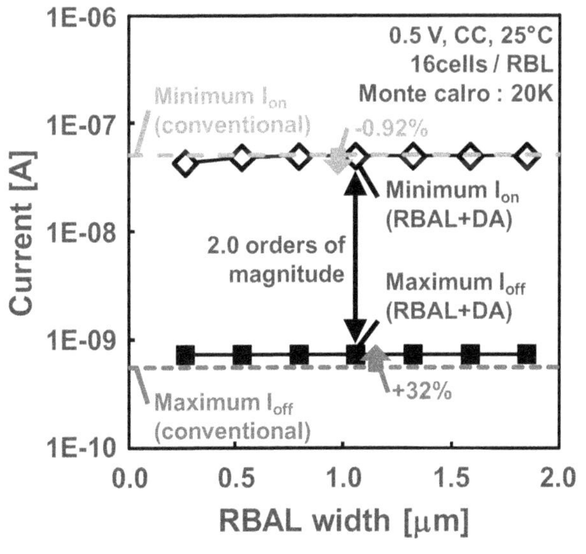

Figure 8. Current margin.

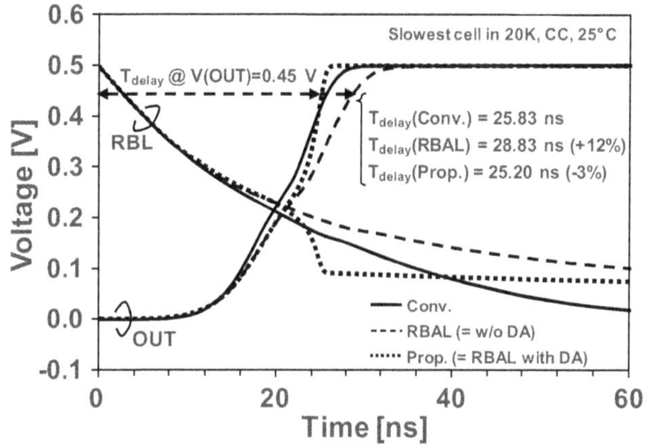

Figure 9. Waveforms of the slowest cell.

Figure 7 portrays RBL delay versus width of the RBAL. Smaller RBAL increases the threshold variation. However, longer RBAL increases the capacitance of RBL. Herein, the width of RBAL is optimized by the RBL delay.

Figure 8 presents the minimum drain current (I_{on}) and maximum leakage current (I_{off}) of the read port in the 8T cell. The number of 8T cells in a RBL is 16. The red line and blue line respectively indicate the minimum I_{on} and I_{off} in a conventional 8T cell. The proposed scheme with RBAL and DA decreases the minimum I_{on} by 0.92% and increases the maximum I_{off} by 32%, however, the I_{on} is two orders of magnitude larger than I_{off}.

Figure 9 shows waveforms of the slowest cell in conventional, RBAL without DA and RBAL with DA when 20-K Monte-Carlo analyses are executed at CC, 25°C. In this paper, T_{delay} is defined by the time to which V(OUT) rises to 0.45 V at supply voltage of 0.5 V. The T_{delay} with only RBAL is increased by 12%. However, the T_{delay} with RBAL and DA is decreased by 3%. Table 1 and 2 respectively show T_{delay} and active energy (E_{active}) improvement at five process corners and temperatures of three kinds. T_{delay} at the FS corner are increased by 14%-32%, however, T_{delay} at SS corner are improved by 2%–5%. E_{active} are improved by 13%–27% at 0.5 V.

Table 1. T_{DELAY}.

		T_{delay} @ V(OUT) = 0.45 V [ns]			T_{delay} ratio
		Conv.	RBAL	Prop.	Prop./Conv.
SS	-40	1761.30	2008.00	1700.30	0.97
	25	145.62	161.31	138.97	0.95
	125	15.12	16.40	14.75	0.98
SF	-40	518.29	547.16	457.84	0.88
	25	55.01	58.33	49.26	0.90
	125	7.28	7.74	7.74	1.06
CC	-40	175.97	205.83	173.40	0.99
	25	25.83	28.83	25.20	0.98
	125	4.52	4.91	4.91	1.09
FS	-40	53.41	86.61	70.62	1.32
	25	11.35	14.54	13.61	1.20
	125	2.69	3.07	3.07	1.14
FF	-40	16.04	18.91	16.78	1.05
	25	4.51	5.02	5.02	1.11
	125	1.36	1.47	1.47	1.08

Table 2. E_{ACTIVE}.

		E_{active} in a RBL [fJ]			E_{active} ratio
		Conv.	RBAL	Prop.	Prop./Conv.
SS	-40	3.1320	2.3595	2.3398	0.75
	25	3.1536	2.4799	2.4507	0.78
	125	3.2174	2.6706	2.6024	0.81
SF	-40	3.1482	2.3205	2.2916	0.73
	25	3.1997	2.4681	2.4204	0.76
	125	3.2968	2.6722	2.6722	0.81
CC	-40	3.1439	2.3748	2.3439	0.75
	25	3.1822	2.5260	2.4716	0.78
	125	3.2884	2.7589	2.7589	0.84
FS	-40	3.1525	2.4406	2.4035	0.76
	25	3.2028	2.6159	2.5188	0.79
	125	3.4060	2.9409	2.9409	0.86
FF	-40	3.1970	2.4570	2.3821	0.75
	25	3.2668	2.6497	2.6497	0.81
	125	3.4509	2.9605	2.9605	0.86

3. CHIP IMPLEMENTATION AND MEASUREMENT RESULTS

We designed and fabricated a 256-Kb SRAM macro in a 40-nm bulk CMOS process, as presented in Fig. 10. Table 3 shows the test chip configuration: The macro size is 0.566 mm × 0.976 mm, the 256-Kb SRAM macro consists of 128 rows × 128 columns × 16 banks, the 8T cell size is 0.706 μm based on logic rule, and the cell density is 463 Kb / mm^2. Figure 11 portrays 16-Kb sub block with the proposed circuit: RBAL and DA. The sub block consists of 16 local cell array (8 columns × 128 rows). Figure 12 portrays a local cell array, which includes the local read circuit, RBAL, DA and low-swing bitline driver (LSBD) which achieves low-power write-back function [11]. At a write cycle, a RWL is activated and an LSBD drives a pair of WBL/WBLN in a half-selected column according to the readout data. Figure 13 portrays the circuit details. The global RBL (GRBL) is not activated during the write cycle to suppress the energy dissipation. The RBAL and DA transfer the readout datum to the LSBD at the write cycle and to GRBL at the read cycle. The LSBDs are activated only in half-selected columns according to the column line enable (CLE) and drive enable bar (DRN). In write target columns, CMOS write drivers activate the write bitlines so the write margin is not degraded. Figure 13 shows the measured energy dissipation per cycle (R:W = 50:50) in the minimum operation voltage (VDD_{min}) of 10 MHz, 20 MHz and 100 MHz at room temperature (RT). The measurement results are investigated using 16 sample chips. Consequently, the energy consumption is less than 10 pJ / access at 0.5–0.7 V.

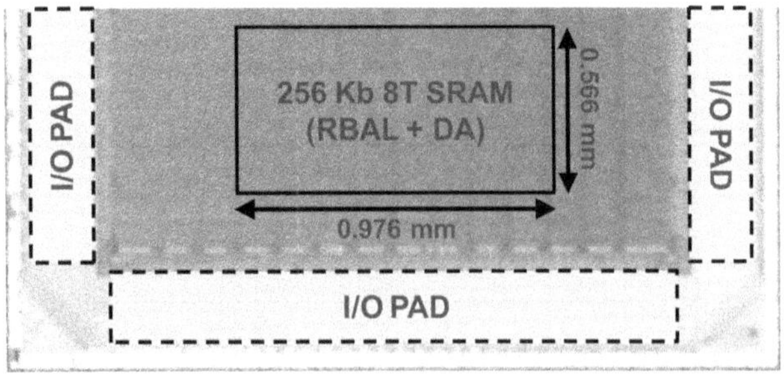

Figure 10. Die photograph of the test chip.

Table 3. Test chip configuration.

Technology	40 nm bulk CMOS
Macro size	0.566 mm x 0.976 mm
Macro configuration	256 Kb (16Kb x 2 x 8), 16 bits/word
Cell size	0.706 μm (logic rule)
# of cells/BL	16 (RBL), 128 (WBL)
Density	463 Kb/mm^2

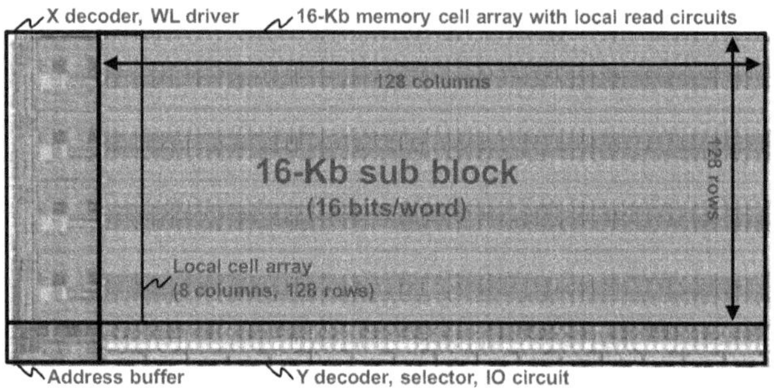

Figure 11. 16-Kb sub block with the proposed circuitry.

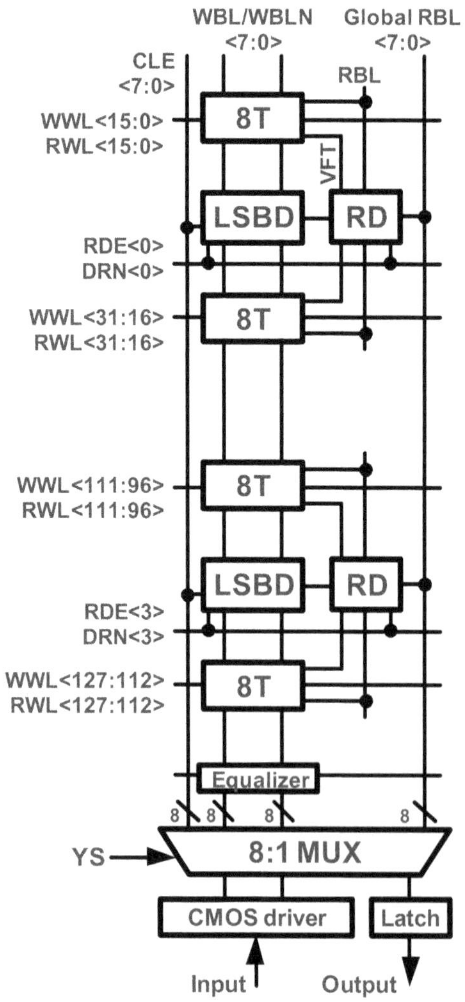

Figure 12. Diagram of local cell array.

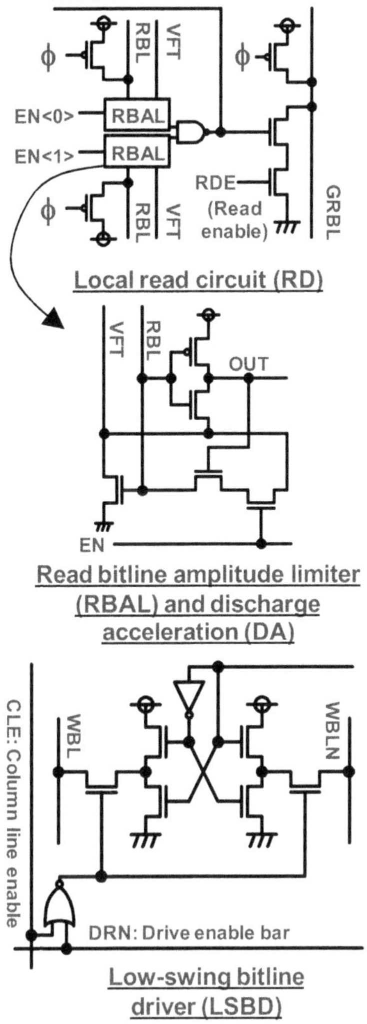

Figure 13. Diagram of the proposed circuits and low-swing bitline driver (LSBD) to prevent the half-select problem [11].

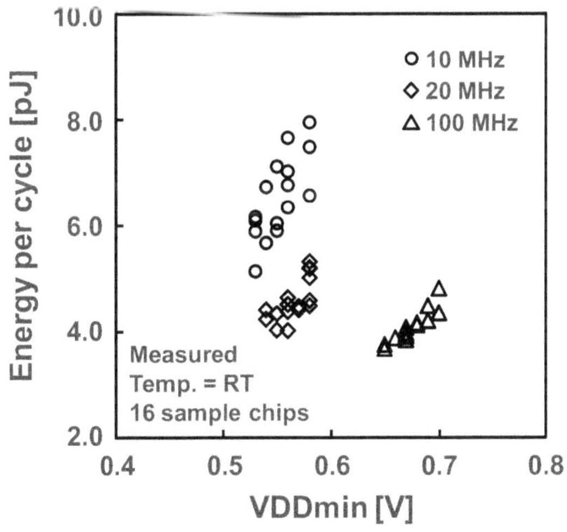

Figure 14. Measured energy dissipation per cycle (R:W = 50:50) at the minimum operation voltage (VDD_{min}) at room temperature (RT).

4. CONCLUSION

As described in this paper, we proposed a read bitline amplitude limiter (RBAL) and discharge acceleration (DA) scheme. The RBAL reduces the active energy dissipation 13%–27% at 0.5 V. The RBAL increases the read delay. However, the DA scheme improves the delay overhead without power penalty. Although the delay overhead with RBAL and DA is 32% at worst case (FS, -40°C), the delay is decreased by 2%–5% at the SS corner. Circuits were implemented to 256-Kb SRAM macros by 40-nm process. The energy dissipation in a cycle is less than 10 pJ / access at 0.5–0.7 V.

5. ACKNOWLEDGMENTS

This work was carried out as a part of the Extremely Low Power (ELP) project supported by the Ministry of Economy, Trade and Industry (METI) and the New Energy and Industrial Technology Development Organization (NEDO).

6. REFERENCES

[1] A. J. Bhavnagarwala, X. Tang, and D. Meindl, "The Impact of Intrinsic Device Fluctuations on CMOS SRAM Cell Stability," *Journal of Solid-State Circuits,* Vol. 36, No. 4, 658–665, 2001.

[2] R. Heald and P. Wang, "Variability in Sub-100nm SRAM Designs," Proc. of *International Conference on Computer Aided Design,* pp. 347–352, 2004.

[3] Y. Morita, K. Nose, K. Noguchi, S. Takami, K. Goto, Y. Aimoto, A. Kimura and M. Mizuno, "Small-defect detection in sub-100nm SRAM cells using a WL-pulse timing-margin measurement scheme," Digest of Technical Papers 2010 *Symposium on VLSI Circuits,* pp. 37-38, Jun. 2010.

[4] H. Pilo, J. Barwin, G. Braceras, C. Browning, S. Burns, J. Gabric, S. Lamphier, and M. Miller, "An SRAM Design in 65 nm and 45 nm Technology Nodes Featuring Read and Write-Assist Circuits to Expand Operating Voltage," IEEE *Symposium on VLSI Circuits,* pp. 15–16, June 2006.

[5] K. Takeda, T. Saito, S. Asayama, Y. Aimoto, H. Kobatake, S. Ito, T. Takahashi, K. Takeuchi, M. Nomura and Y. Hayashi, "Multi-step Word-line Control Technology in Hierarchical Cell Architecture for Scaled-down High-density SRAMs," Digest of Technical Papers 2010 *Symposium on VLSI Circuits,* pp. 101-102, Jun. 2010.

[6] Y. Fujimura, O. Hirabayashi, T. Sasaki, A. Suzuki, A. Kawasumi, Y. Takeyama, K. Kushida, G. Fukano, A. Katayama, Y. Niki, T. Yabe, "A Configurable SRAM with Constant-Negative-Level Write Buffer for Low-Voltage Operation with 0.149μm2 Cell in 32nm High-κ Metal-Gate CMOS," IEEE *International Solid-State Circuits Conference,* pp. 348-349, 2010.

[7] H. Nho, P. Kolar, F. Hamzaoglu, Y. Wang, E. Karl, Y. G. Ng, U. Bhattacharya and K. Zhang, "A 32nm High-κ Metal Gate SRAM with Adaptive Dynamic Stability Enhancement for Low-Voltage Operation," IEEE *International Solid-State Circuits Conference,* pp. 346-347, 2010.

[8] Y. Morita, H. Fujiwara, H. Noguchi, Y. Iguchi, K. Nii, H. Kawaguchi, and M. Yoshimoto, "An Area-Conscious Low-Voltage-Oriented 8T-SRAM Design under DVS Environment," IEEE Symposium on *VLSI Circuits Digest of Technical Papers,* pp. 256–257, Jun. 2007.

[9] Y. Pu, X. Zhang, J. Huang, A. Muramatsu, M. Nomura, K. Hirairi, H. Takata, T. Sakurabayashi, S. Miyano, M. Takamiya, and T. Sakurai, "Misleading Energy and Performance Claims in Sub/Near Threshold Digital Systems," Proc. of *International Conference on Computer Aided Design,* pp. 625–631, 2010.

[10] A. Kawasumi, T. Suzuki, S. Moriwaki and S. Miyano, "Energy Efficiency Degradation Caused by Random Variation in Low-Voltage SRAM and 26% Improvement by Bitline Amplitude Limiting (BAL) Scheme," IEEE *Asian Solid-State Circuits Conference,* pp. 165-168, Sep. 2011.

[11] S. Yoshimoto, M. Terada, S. Okumura, T. Suzuki, S. Miyano, H. Kawaguchi and M. Yoshimoto, "A 40-nm 0.5-V 20.1-uW/MHz 8T SRAM with Low-Energy Disturb Mitigation Scheme," Digest of Technical Papers 2011 *Symposium on VLSI Circuits,* pp. 72-73, Jun. 2011.

An Adaptive Write Word-Line Pulse Width and Voltage Modulation Architecture for Bit-Interleaved 8T SRAMs

Daeyeon Kim[1, 3], Vikas Chandra[2], Robert Aitken[2], David Blaauw[1], Dennis Sylvester[1]
[1] University of Michigan, Ann Arbor, MI
[2] ARM Inc., San Jose, CA
[3] Currently with Intel Corporation, Hillsboro, OR
daeyeonk@umich.edu, {vikas.chandra, rob.aitken}@arm.com, {blaauw, dmcs} @umich.edu

ABSTRACT
We propose an adaptive WWL pulse width and voltage modulation architecture for low voltage bit-interleaved 8T SRAMs. The 8T bitcell offers improved read/write margins but suffers from write and half select concerns when bit-interleaved [1]. Also, low voltage operation leads to a long-tailed write time distribution, requiring large timing margins and limiting V_{min}. To minimize timing margins and reduce V_{min}, both WWL pulse width and voltage level are adaptively modulated by monitoring written values through the read path. In a 65nm CMOS prototype chip, V_{min} is lowered from 700mV to 500mV using this technique, providing 2.55× leakage power reduction and 2.4× active power reduction.

Categories and Subject Descriptors
B.3.1 **[Memory Structures]**: Semiconductor Memories – *Static Memory (SRAM)*; B.7.1 **[Integrated Circuits]**: Types and Design Styles – *Advanced Technologies, Memory Technologies, VLSI (very large scale integration)*; B.8.0 **[Performance and Reliability]**: General

Keywords
SRAM, Low Voltage Design, Noise Margin

1. INTRODUCTION
Low voltage operation is an effective way to reduce power consumption due to the resulting quadratic power savings. Variability at low voltage is a challenge however, and low voltage SRAM in particular is vulnerable to variation and functional failures due to the use of minimum feature sizes in a bitcell and large SRAM array sizes. The 8T bitcell [2] (Figure 1) improves low voltage operation by allowing separate optimization of read and write paths. Bit-interleaving is essential to avoid soft errors, particularly at low voltages, but induces half select disturb [3] as the 6T portion of the 8T bitcell is optimized for write. Since half select and writability create a double-sided constraint on WWL pulse width and WWL voltage, WWL control is key to reducing V_{min} and maximizing yield for bit-interleaved 8T SRAMs.

ISLPED'12, July 30–August 1, 2012, Redondo Beach, California, USA.

To mitigate the trade-off between write and half select, this paper proposes adaptive WWL voltage modulation with regenerative half selected bit-lines. In addition, SRAM bitcells exhibit a wide write time distribution with a long tail at low voltage (Figure 2). Excessive margins are required due to the few bitcells with long write times. To minimize this impact, we propose a multi-cycle

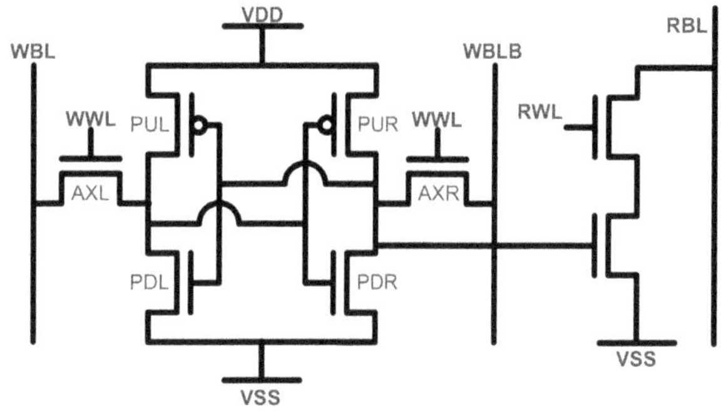

Figure 1. The 8-T bitcell schematic is shown. Write and read are separately optimized.

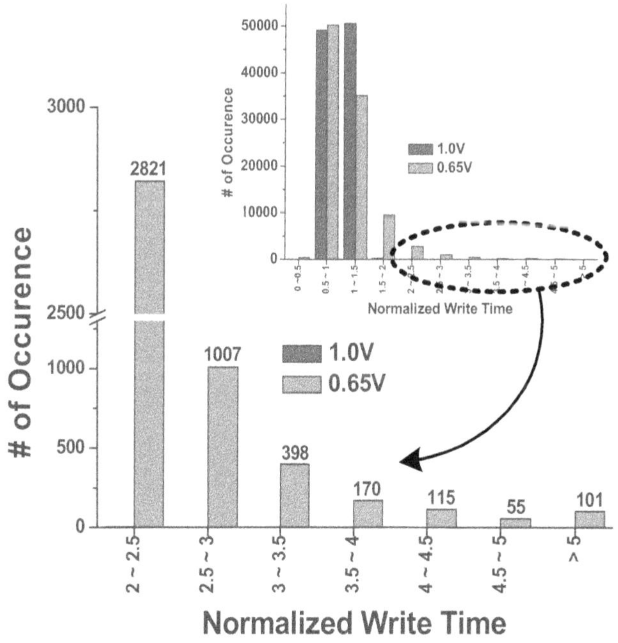

Figure 2. 10K Monte Carlo simulation results of write time distribution at 0.65V and 1.0V. The distribution at 0.65V is wider with a long tail, increasing design margins.

write operation where the number of cycles is adaptively determined via write completion detection, similar to the scheme used in [4]. Overall performance is improved using the multi-cycle write scheme.

The importance of dynamic write stability has been previously discussed [5]. A replica-based adaptive stability enhancement technique [6] was proposed. While [6] is replica-based, this work describes an *in situ* approach (i.e., it uses actual bitcells in the SRAM array, eliminating mismatch issues). Adaptive WWL boosting for 8T SRAMs [7] was also proposed but it does not use a bit-interleaved array and hence is susceptible to soft errors. Half select prevention using a pre-read technique was proposed in [8] but does not address the large write delay margins observed at low voltages in SRAM.

2. 8-T SRAM OPERATIONS AT LOW VOLTAGES

Variation increases as the supply voltage decreases. Figure 3 depicts variation as supply scales. The distribution of FO4 delay is measured using 100K Monte Carlo simulations. At low voltage, performance degrades by larger variation as well as smaller I_{on}. The performance degradation by larger variation limits lowering the supply voltage so variation compensation techniques are required for low voltage operation.

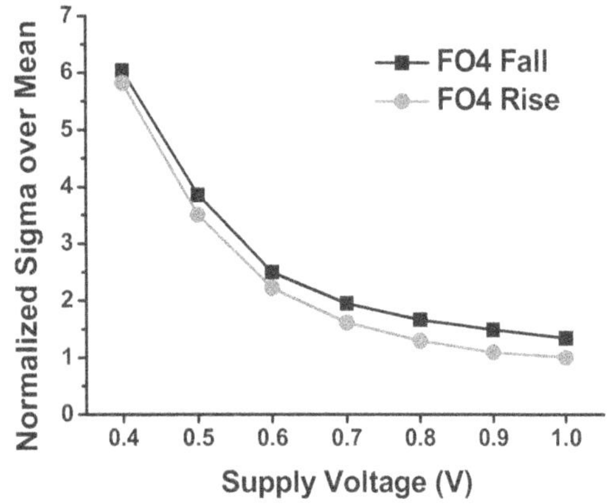

Figure 3. More variation exists at lower voltage.

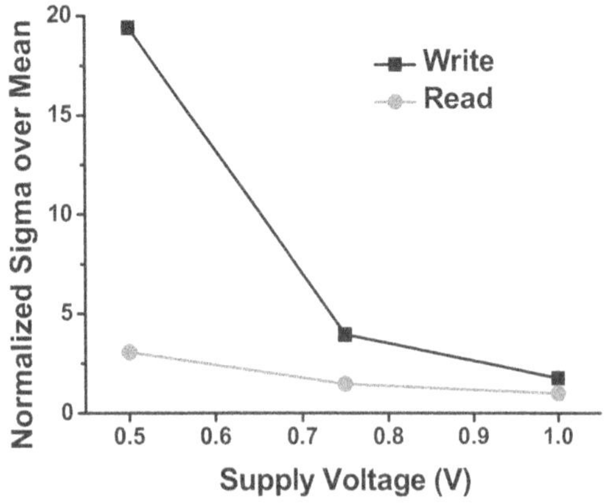

Figure 4. Write operation is critical operation at low voltage because it is more vulnerable to variation.

As already discussed above, the 8T SRAM bitcell is a good candidate as a SRAM bitcell at low voltage since write operation and read operation can be separately optimized. Between read and write, write operation is a critical operation at low voltage because it has more variation. Figure 4 shows 40K Monte Carlo simulation results and it clearly depicts that write operation is more vulnerable to variation. Also, there are five write failures out of 40K at 0.5V while there is no read failure. Because of the five write failure at 0.5V, it is not possible to lower the supply voltage down to 0.5V.

To analyze the write operation of 8T SRAM at low voltage, write time is simulated using Monte Carlo simulation. Figure 5 describes the definition of write time used in this work. WWL is turned on to start write operation. After some time, two internal nodes in an SRAM bitcell are crossed each other. Write time is defined as a time between WWL on and two internal nodes crossing. For successful write operation, WWL pulse width must be larger than this write time. If the write operation fails even though write time is infinitely long, static write failure happens.

The Monte Carlo simulation results with 100K iterations of write time as supply scales are shown in Figure 6. At 1.0V, the worst case write time is ~2.2× larger than typical. If the WWL pulse width is 2.2× larger than typical write time, high yield is expected.

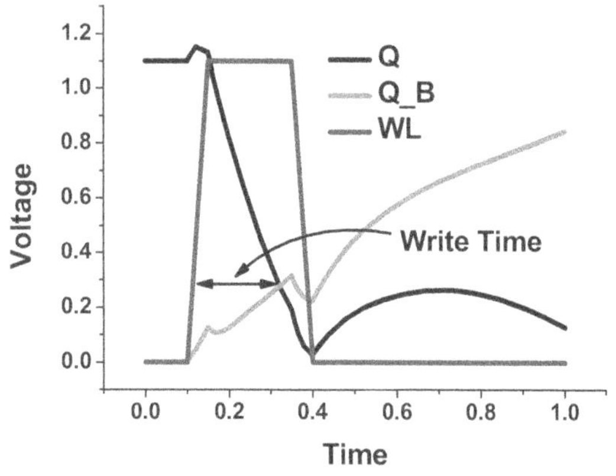

Figure 5. SRAM write time is a time difference from WWL on to two internal nodes crossing.

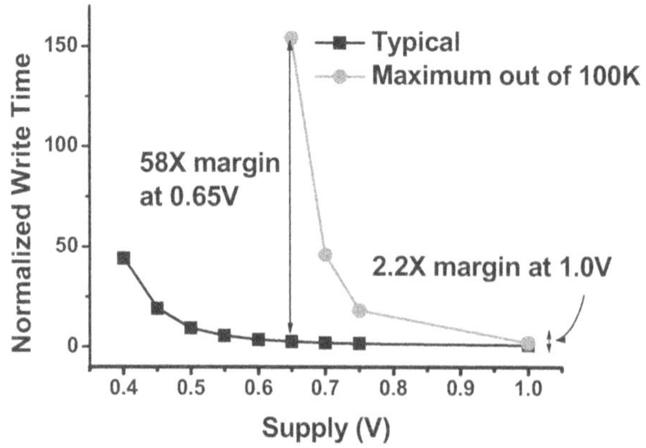

Figure 6. Write time degrades as supply scales. However, the degradation of the worst case write time is much worse than typical cases. Below 0.65V, write failure happens (static write failure).

However, the required margin at low voltage is much larger than nominal voltage. At 0.65V, at least 58× margin is required for the successful write of all 100K iterations. Below 0.65V, the worse case is static write failure: write operation cannot be done even with infinitely long WWL pulse. Based on this simulation, the V_{min} of this SRAM bitcell is 0.65V because of static write failures under 0.65V.

As the long tailed distribution of write time is shown in Figure 2, only a few bitcells need excessive margin. This observance leads the adaptive multi cycle write scheme to increase overall performance while keeping the iso-yield.

3. ADAPTIVE WWL PULSE WIDTH AND VOLTAGE MODULATION

We propose an adaptive WWL pulse width and voltage modulation architecture for bit-interleaved 8-T SRAMs. There are two major WWL parameters: pulse width and voltage level. First, WWL pulse width is modulated by an adaptive multi-cycle write structure. Figure 7 shows the implementation details for adaptive WWL pulse modulation. To write data to worst-case bitcells, frequency is not lowered, but instead additional cycles are used in those cases until write completion is detected. Second, two WWL voltages are applied to fix static failures while avoiding half select disturb. Longer WWL pulses induce more half select disturbs yet still cannot fix static write failures. To address this, a low voltage WWL pulse is initially used to write the majority of bitcells while all half selected bitcells are read. In any subsequent cycles, the WWL voltage is boosted (Figure 8) to write challenging bitcells. Also, half select disturbs are mitigated as half selected bitcells experience only weak read disturb (by a low voltage WWL) during the first cycle while in subsequent cycles half select disturbs are eliminated by driving bit-lines to their appropriate values based on read-out data from the first cycle.

Figure 9 shows a sample timing diagram when a write operation is performed in two cycles. While WWL is on for multiple cycles, RWL toggles every cycle for reading the data repetitively. When read-out data and to-be-written data are the same, WRITE_DONE signal is asserted and the BIST stops the write operation.

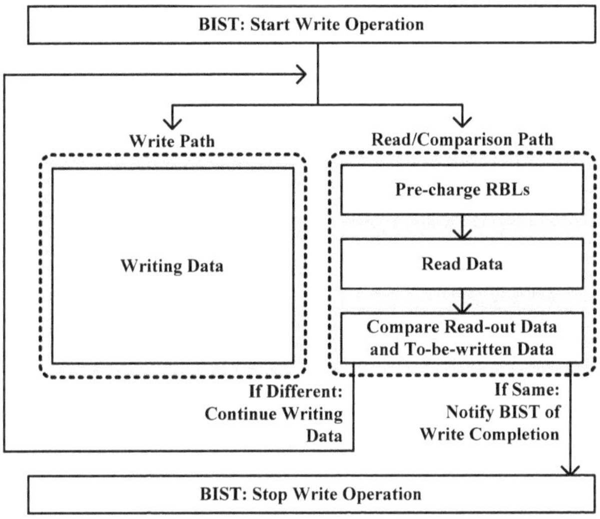

Figure 7. The adaptive WWL pulse width modulation scheme is shown. Write completion is monitored every cycle. When write completion is detected, a BIST stops the write operation.

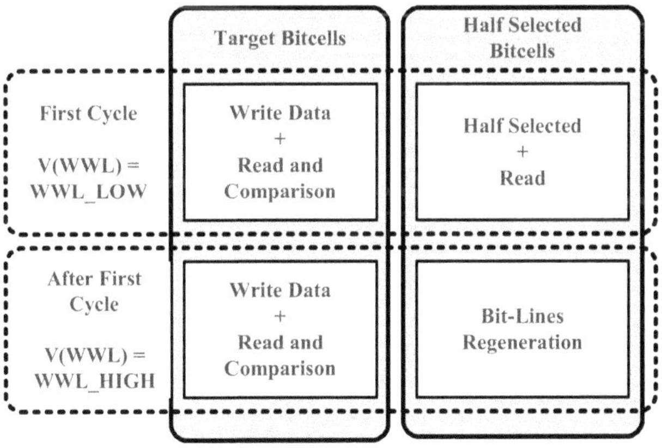

Figure 8. The adaptive WWL voltage level modulation scheme is shown. WWL is boosted after the first cycle but half selected bitcells are not disturbed due to bit-line regeneration.

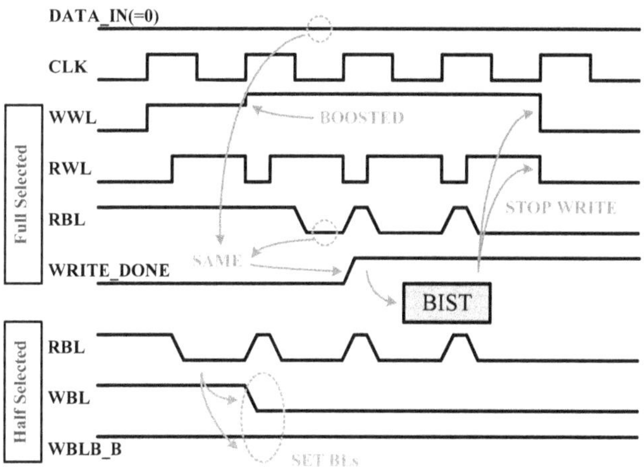

Figure 9. Sample timing diagram when write operation is performed in two cycles.

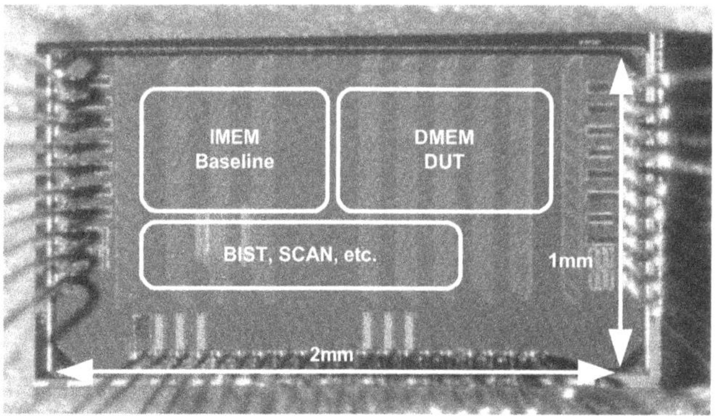

Figure 10. Die photo of the prototype chip fabricated in 65nm CMOS.

4. PROTOTYPE IMPLEMENTATION

Figure 10 shows the die photo of the prototype chip fabricated in 65nm CMOS, which contains a 16kb (128×128) bitcell array (Figure 11). For improved performance and functionality, bit-lines have a cascaded structure with several local blocks. Eight bitcells, a pre-charger, keeper, and tri-state buffer comprise a local block. Sixteen local blocks are connected to form a 128-bit tall global bit-line. The 128×128 bitcell array is bit-interleaved with four 32-bit words. A comparator and a WWL width and voltage controller are also implemented; the controller selects the WWL driver output voltage level from two voltage levels.

5. MEASUREMENT RESULTS

Figure 12 shows a measured shmoo plot of the SRAM array V_{min} when V_{SUPPLY}, V_{WWL_HIGH}, and V_{WWL_LOW} are the same. The results clearly show the double-sided constraint on frequency due to write/read and half select disturb. V_{min} considering half select disturbs is measured to be 0.775V. Below this voltage, half select disturb becomes critical while write and read remain functional. Neglecting half select, the array V_{min} would be 0.5V.

By lowering V_{WWL_HIGH} and V_{WWL_LOW} voltage levels together (i.e., conventional WWL underdrive), half select disturb is partially mitigated (Figure 13). Half select disturb can be eliminated at voltages down to V_{SUPPLY} = 0.7V without multi-cycle write, or 0.6V with multi-cycle write, while write operation remains functional. In the 8T bitcell used in this work, the access transistor and pull-down transistor are sized identically, making write operation strong compared to a traditional 6T bitcell. However, at V_{SUPPLY} < 0.6V, write failures occur when WWL is underdriven sufficiently to fix half select disturbs. In other words, there is no feasible WWL voltage level that avoids both write failure and half select disturbs.

Figure 14 analyzes failures at V_{SUPPLY} = 0.5V as WWL voltage (both V_{WWL_HIGH} and V_{WWL_LOW}) decreases. Initially, half select

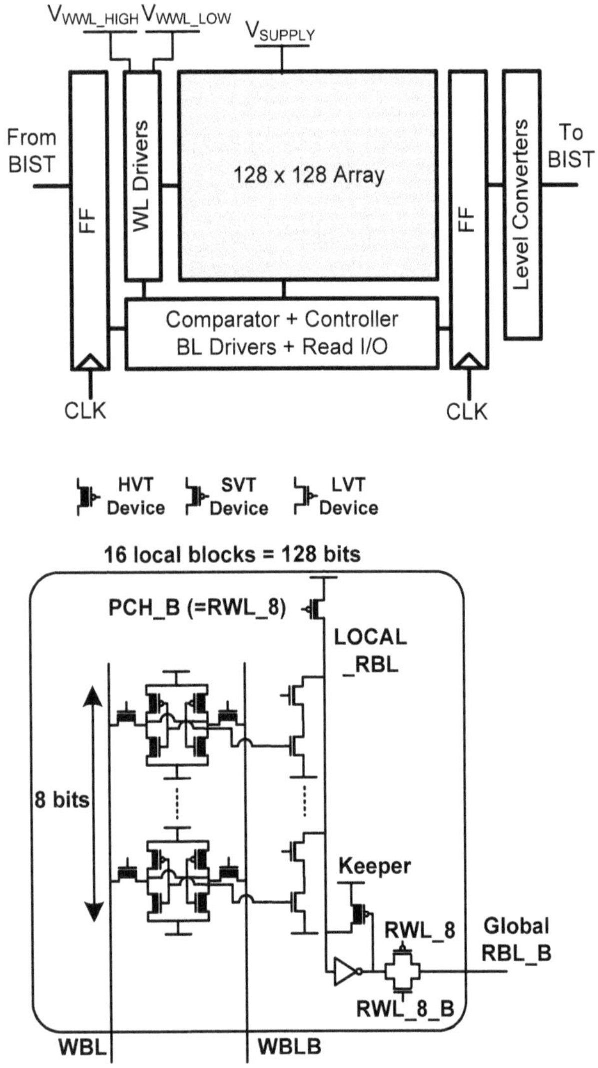

Figure 11. A block diagram of the prototype memory bank is shown at top. A comparator and controller for WWL width and voltage modulation are implemented. A local 8-bit bit-line structure is shown at right with 16 8-bit local blocks. A 128-bit wide row consists of four bit-interleaved 32-bit words.

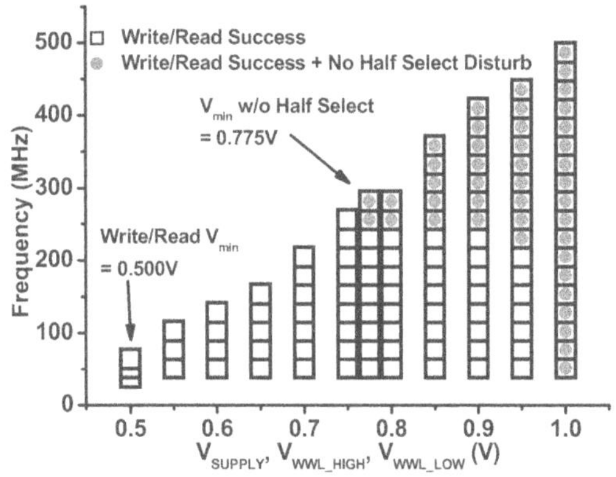

Figure 12. Measured shmoo plot to find V_{min}. V_{min} is 0.775V when all voltage levels are identical.

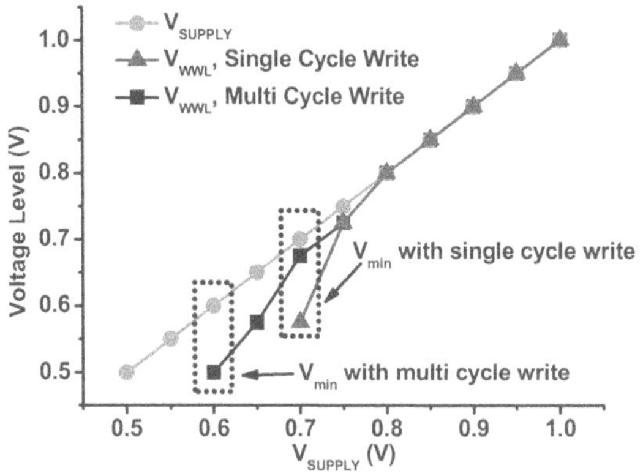

Figure 13. Measured V_{min} is reduced to 0.7V with conventional WWL underdrive, and to 0.6V with WWL underdrive and multi-cycle write. For further V_{min} reduction, V_{WWL_HIGH} and V_{WWL_LOW} must be separately optimized.

disturb dominates. Hence, at higher frequency there are fewer half select events (e.g., at $V_{WWL} = 0.5V$). As WWL voltage decreases, half select disturb is mitigated but write failures become prominent. At $V_{WWL} < 0.425V$, write failures begin dominating, as indicated by a higher number of failures at higher frequencies.

To concurrently fix both half select disturb and read/write failures, the proposed adaptive voltage level modulation scheme is applied at $V_{SUPPLY} = 0.5V$ (Figure 15). If V_{WWL_LOW} is too high, half select may occur during the first cycle (see Figure 8, top right). To avoid this, V_{WWL_LOW} should be ≤ 390mV based on measured results. Also, V_{WWL_HIGH} must be high enough to write data to worst-case bitcells in the second cycle. All failures are fixed at $V_{WWL\_HIGH} = 475mV$. Based on measured results, we can simplify the system by only underdriving V_{WWL_LOW} while setting $V_{WWL\_HIGH} = V_{SUPPLY}$ to maximize writability and half select disturb immunity. With $V_{SUPPLY} = 500mV$, $V_{WWL\_HIGH} = 500mV$, and $V_{WWL\_LOW} = 390mV$, the operating frequency with no failures is 64MHz and the average and maximum number of write cycles for the array is 2.25 and 3, respectively.

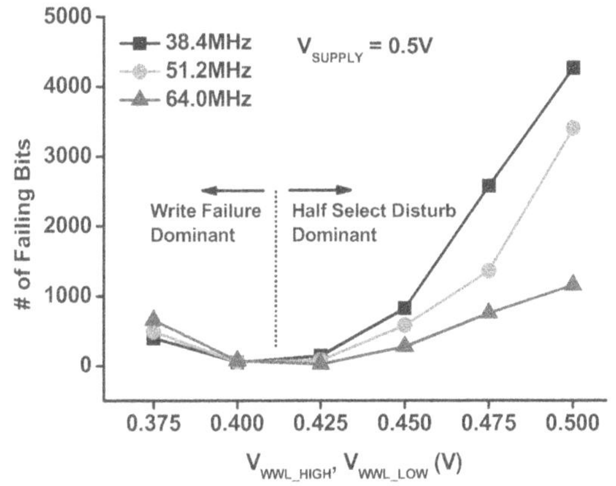

Figure 14. Measured number of failing bits with $V_{SUPPLY} = 0.5V$ and WWL underdrive. As WWL voltage decreases, immunity to half select disturb is improved while writability worsens.

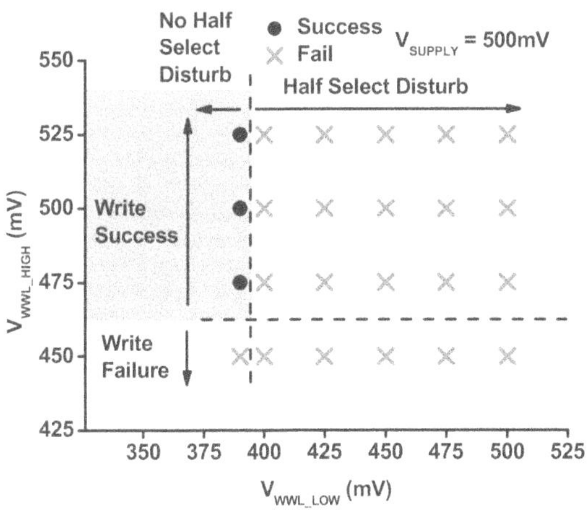

Figure 15. V_{WWL_HIGH} and V_{WWL_LOW} are separately optimized at $V_{SUPPLY} = 0.5V$. Measured results show that all failures are fixed at $V_{WWL\_LOW} = 0.39V$

Overall, V_{min} using the adaptive WWL modulation scheme is measured to be 500mV, offering significant reductions beyond conventional WWL underdrive (700mV) and multi-cycle writes alone (600mV) (Figure 16). Note that this V_{min} (500mV) is equal to the V_{min} when neglecting half select completely. Measured power consumption at 64MHz is summarized in Table 1. The V_{min} reduction from the proposed technique enables 2.55× leakage power reduction and 2.4× active power reduction for the SRAM array, and enables the use of bit-interleaving in low voltage robust 8T SRAMs.

6. CONCLUSIONS

An adaptive write word-line pulse width and voltage modulation architecture is proposed for low voltage bit-interleaved 8-T SRAMs. By minimizing excessive margins and applying regenerative half select bit-lines scheme, V_{min} is lowered from 700mV to 500mV with 2.55× leakage power reduction and 2.4× active power reduction in a 65nm CMOS prototype chip.

7. ACKNOWLEDGEMENTS

The authors at University of Michigan acknowledge the support of NSF, GSRC, and Army Research Laboratory.

8. REFERENCES

[1] D. Kim *et al.*, "Variation-Aware Static and Dynamic Writability Analysis for Voltage-Scaled Bit-Interleaved 8-T SRAMs," *International Symposium on Low-Power Electronics and Design (ISLPED)*, pp. 145-150, Aug. 2011

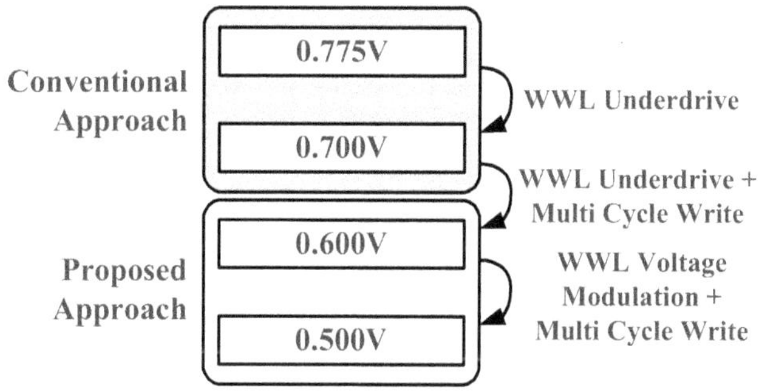

Figure 16. Conventional WWL underdrive can reduce V_{min} to 0.7V. With the proposed methods, V_{min} is further lowered to 0.5V.

Table 1. Measured power/energy improvements (at 64MHz). The conventional method is WWL underdrive. The proposed method includes multi cycle write and WWL voltage modulation. Using the proposed methods, V_{min} is lowered with large gains in leakage power and active power.

Cases	Supply (mV)			Power (μW)	
	V_{SUPPLY}	V_{WWL_HIGH}	V_{WWL_LOW}	Leakage	Active
Conventional	700	575	575	204.7	307.2
Proposed	500	500	390	80.2	127.9

[2] L. Chang *et al.*, "Stable SRAM cell design for the 32nm down and beyond," *IEEE Symposium on VLSI Circuits (VLSI Circuits)*, pp. 128-129, June 2005

[3] R. Joshi *et al.*, "6.6+ GHz Low V_{min}, read and half select disturb-free 1.2 Mb SRAM," *IEEE Symposium on VLSI Circuits (VLSI Circuits)*, pp. 250-251, June 2007

[4] S. Hanson *et al.*, "A Low-Voltage Processor for Sensing Applications with Pico-watt Standby Mode", *Journal of Solid State Circuits (JSSC)*, Vol. 44, pp. 1145-1155, Apr. 2009.

[5] S. O. Toh, Z. Guo, B. Nikolić, "Dynamic SRAM Stability Characterization in 45nm CMOS," *IEEE Symposium on VLSI Circuits (VLSI Circuits)*, pp. 35-36, June 2010

[6] H. Nho *et al.*, "A 32nm High-κ Metal Gate SRAM with Adaptive Dynamic Stability Enhancement for Low-Voltage Operation," *IEEE International Solid-State Circuits Conference (ISSCC)*, pp. 346-347, Feb. 2010

[7] A. Raychowdhury *et al.*, "PVT-and-Aging Adaptive Wordline Boosting for 8T SRAM Power Reduction," *IEEE International Solid-State Circuits Conference (ISSCC)*, pp. 352-353, Feb. 2010

[8] M. Sinangil, H. Mair, A. Chandrakasan, "A 28nm High-Density 6T SRAM wth Optimized Peripheral-Assst Circuits for Operation Down to 0.6V," *IEEE International Solid-State Circuits Conference (ISSCC)*, pp. 260-261, Feb. 2011.

Power-aware Performance Increase via Core/Uncore Reinforcement Control for Chip-Multiprocessors

Da-Cheng Juan
Electrical and Computer Engineering
Carnegie Mellon University
Pittsburgh, PA, USA
dacheng@cmu.edu

Diana Marculescu
Electrical and Computer Engineering
Carnegie Mellon University
Pittsburgh, PA, USA
dianam@cmu.edu

ABSTRACT

Network-on-Chips (NoCs) have emerged as the backbone for the inter-core communication of a chip-multiprocessor (CMP). This paper evaluates and analyzes the advantages of managing the processing cores and the on-chip communication fabric in synergy for the purpose of **performance increase** under power constraints. A semi-supervised reinforcement learning (RL) based approach is proposed for performing dynamic voltage and frequency scaling (DVFS) so as to enable the efficient usage of the available on-chip power budget while maximizing performance. The experimental results show that, on average, overall performance is increased by 11% under iso-power conditions, while a core-only or an uncore-only performance boosting approach can only achieve 7% and 3% improvement in performance, respectively.

Categories and Subject Descriptors

B.8.0 [**Performance and reliability**]: General ; C.1.4 [**Parallel architectures**]: Distributed architectures ; I.2 [**Artificial intelligence**]: Miscellaneous

Keywords

Chip-multiprocessor, On-chip network, Power management, Dynamic voltage and frequency scaling, Reinforcement learning

1. INTRODUCTION

Over the last decade, microprocessor design trends have shifted to chip-multicores from the classic monolithic, single core systems. In a multi-core system, processing elements or cores must communicate with each other under parallel, multithreaded workloads, thereby potentially creating performance bottlenecks in the communication fabric [25]. To reduce this overhead, the network-on-chip (NoC)[1] paradigm [25, 14] has been proposed as a promising solution for on-chip communication for massively-integrated CMPs. However, the enhanced performance and capabilities of such platforms are usually constrained by the on-chip power consumption. While dynamic power management has been extensively studied for multi-core systems in the context of core-only or uncore (on-chip communication fabric) only, the cooperative power management of core and uncore[2] has remained a critical issue not sufficiently explored.

In the context of using dynamic voltage and frequency scaling (DVFS) for minimum power consumption under performance constraints, a possible cooperative power management for both core and uncore resources introduces additional challenges that require maintaining appropriate performance levels for parallel applications executing on the system. For example, in multithreaded applications, spin locks and other synchronization mechanism may amplify small timing differences into very different program execution paths [1], thereby impacting the memory system behaviors substantially. In addition, a significant mismatch between core and uncore frequency may cause unexpected traffic contentions and therefore, may result in significant performance penalty [24]. As a result, how to reliably control cores and uncores in synergy via DVFS while maintaining power constraints remains an open question that needs to be addressed in the context of advanced multi-core systems.

There is a large body of work that has been proposed to address uncore-agnostic power management, especially DVFS, for CMPs [7, 13, 27, 26, 23]. Recently, reinforcement learning (RL) and machine learning (ML) [2] have emerged as popular and robust power management schemes due to their adaptive properties [9, 22, 32, 8, 31]. However, none of the above work has addressed and evaluated the effectiveness of using RL on the power management for NoC-based CMPs, while also including uncore resources. The power management of uncore only or multi-processor systems-on-chip (MPSoC) has also drawn lots of attention from both industry and academia [15, 5, 21, 10, 18]. Although the power management for cores, uncores or MPSoCs has been extensively studied and discussed, none of the previous work has considered synergistic DVFS for NoC-based CMPs to maximize the performance under iso-power conditions.

To the best of our knowledge, this paper brings the following novel contributions:

[1]Network-on-chip (NoC) and on-chip networks will be used interchangeably throughout this paper.

[2]In this paper we refer to uncore resource as representing the communication fabric only, such as routers and links. Caches are included in cores.

ISLPED'12, July 30–August 1, 2012, Redondo Beach, CA, USA.

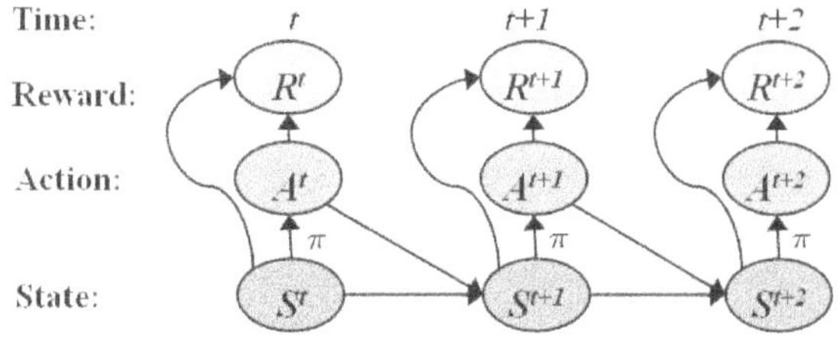

Figure 1: Markov decision process

- We present, for the first time, the evaluation of and comparison among core-only, uncore-only and *cooperative core/uncore DVFS control for performance boosting*. The experimental results confirm that performing DVFS for cores or uncores separately may not be effective for NoC-based CMPs as they are characterized by smaller performance per unit of energy gains.
- Compared to conventional DVFS schemes that address power reduction under performance constraints, we propose a *"reverse"* DVFS: maximize performance while ensuring that power stays within prescribed power constraints.
- To increase the scalability of RL for the adaptive control of advanced CMPs, we proposed a *semi-supervised RL* to reduce the exponential growth of state complexity – the proposed semi-supervised RL maintains a linear complexity in the number of cores. Furthermore, the semi-supervised RL does not compromise the adaptability of the original RL, making it an ideal control scheme for advanced CMPs.
- We evaluate the proposed RL-based, cooperative DVFS control for both cores and uncores with a wide spectrum of parallel, multithreaded applications. The experimental results confirm the effectiveness – on average 10.9% of program execution time is reduced while satisfying given power constraints.

The remainder of this paper is organized as follows. Section 2 introduces the background knowledge. Section 3 provides the problem formulation. Section 4 details the proposed semi-supervised RL. Section 5 presents the implementation flow. Section 6 demonstrates the experimental results, while Section 7 concludes this paper.

2. MARKOV DECISION PROCESS (MDP)

The concept of RL can be described via the interactions between an agent and the environment with uncertainties. The agent attempts to find the best action on the fly for interacting with the various states of the environment, in order to receive the highest reward. Therefore, a MDP-based learning model, such as V learning or Q learning [2], consists of: (1) an agent; (2) a finite state space $S \in \{s_1 \dots s_n\}$; (3) a set of available actions $A \in \{a_1 \dots a_m\}$; and (4) a reward function $r^t = r(S^t, A^t)$ where t represents the time. As a convention in statistics, capital variables such as S represent random variables and lowercase variables such as s_1 stand for the values observed. The goal of the agent is to maximize its expected long-term reward. This can be achieved by learning a policy π which can be viewed as a mapping between the states and the actions. Table 1 lists the parallel between MDP and dynamic power management concepts.

Table 1: MDP and dynamic power management

MDP	Dynamic power management
Agent	Controller
Select an action	Select a Voltage/Frequency
Environment states	Machine states
Rewards	High performance

Figure 1 presents a MDP model. At each time point, the agent chooses an action $a \in A$ based on the state S^t to receive the long-term highest reward. Based on the current state S^t and the action A^t , the state transition probability of S^{t+1} can be calculated as:

$$P(S^{t+1} = s_i | S^t = s_j, A^t = a_k) = \theta_i \tag{1}$$

where $s_i, s_j \in S$ and $a_k \in A$. Here, we define this transition probability as θ_i. More precisely, θ_i represents the probability that state s_i will be reached given that s_j, a_k was reached during the previous time point. Given s_j, a_k, let $\theta = \{\theta_1, \cdots, \theta_{|S|}\}$ represent the probability for each value taken by S^{t+1}, *i.e.*, $\{s_1, \cdots, s_{|S|}\}$. For simplicity of explanation, we assume that the states of MDP in Figure 1 are observable instead of hidden. In other words, on-chip performance or power counters are readily available for all resources; if the states are hidden, expectation maximization (EM) [2] or other algorithms can be used to predict the state. Furthermore, the first-order Markov assumption is used here: the probability of S^{t+1} depends on only S^t and A^t, and therefore $P(S^{t+1}|S^t, A^t, S^{t-1}, \dots) = P(S^{t+1}|S^t, A^t)$. No information before time t is required to calculate the conditional probability.

The new state S^{t+1} provides a reward to the agent, and the agent learns the control policy $\pi : S \to A$ to maximize the long-term expected reward $\sum_{t=0}^{t=\infty} \gamma^t \mathbb{E}[r^t]$, where γ is the discount factor between (0,1). γ is used to discount the future reward so that the long-term reward will converge to a certain value after a sufficiently long time. Since the state transition is not deterministic, the expected value is calculated as the expected reward: $\mathbb{E}_\theta[r^t] = \sum_{s_i \in S} r^t \times \theta_i$. Given $\pi : S \to A$, we can define $V^\pi(s)$ as $\sum_{t=0}^{t=\infty} \gamma^t \mathbb{E}_\theta[r^t]$, where $V^\pi(s)$ represents the long-term expected reward an agent can receive if the agent follows the action sequence chosen according to π, starting at state s. Then, the best policy is $\pi^* = \arg\max_\pi V^\pi(s), \forall s$. Assuming that the state transition probability θ is known, we can calculate the best π^* and the $V^\pi(s)$ in a recursive way to obtain the best control policy $V^{\pi*}(s)$. If θ remains unknown, the learning problem relies on the estimates of the state transition probability. Prior art [32, 18, 22] has shown that Q learning (similar to V learning) can be a good alternative in this context. In this paper, we will demonstrate that, by using *maximum a posteriori* (MAP) estimate with a proper prior distribution of θ, learning the best policy $V^{\pi*}(s)$ can be several times faster than Q learning.

3. PROBLEM FORMULATION

Before elaborating on the proposed semi-supervised[3] RL, we introduce the architecture and the problem formulation used herein. The architectural configuration described here

[3]The definition of semi-supervised learning used in this paper is different from [2].

is to facilitate the explanation of the proposed methodology. Detailed implementation is provided in Section 5.

3.1 Target Architecture

The architecture used throughout this paper is a symmetric, NoC-based CMP that consists of 16 tiles, and each tile contains a Pentium4® core, a private L1 cache, a shared L2 cache and an on-chip router. Table 2 provides the detailed architectural parameters. These 16 tiles are placed in a 4×4 mesh manner. A flit-based mechanism is used for the NoC architecture. The router design follows the standard 5-stage pipeline [17]. Furthermore, each processing core and its corresponding on-chip router are assumed equipped with Intel's Turbo Boost® technology [12]. Therefore, the voltage and frequency pairs, denoted as V/F pairs, can be set to (1) turbo: 1.3V, 3.75GHz, (2) baseline (nominal): 1.0V, 3GHz, (3) low: 0.8V, 2.35GHz and (4) very low: 0.65V, 2.0GHz. These V/F pairs are also listed in Table 2. For conciseness, in the remainder of this paper, when we mention the change of frequency, we actually refer to the change of both voltage and frequency, *i.e.*, V/F pairs. In order to compare with conventional core-only or uncore-only DVFS, we assume that each core and router can be set to different frequencies to best explore the advantages of the cooperative DVFS control.

3.2 Detailed Formulation

The formulation of the cooperative core/uncore DVFS is described as follows. First, the following two inputs are given: (1) a NoC-based CMP, and (2) a parallel, multi-threaded workload that can be executed on the target CMP. The decision variables here are the frequency of each core and router. Under initial conditions, all frequencies are set to the baseline value (3GHz). The objective function to be maximized is overall performance. In this paper, the throughput of the CMP is used as the performance metric. Generally, performance can be expressed as a function of machine states s_i^t and the frequency. Finally, the power consumption of the whole CMP, including both cores and uncores, must be less than or equal to the power constraint (Pow_{const}) at all times.

4. METHODOLOGY

This section elaborates on two main components of the proposed framework: *maximum a posteriori* (MAP) and *semi-supervised* RL.

4.1 Maximum A Posteriori (MAP) Estimate

The key step in MDP is to learn the unknown state transition probability, $\theta_i = P(S^{t+1} = s_i | S^t = s_j, A^t = a_k)$, as mentioned in Section 2. θ heavily depends on the characteristics of the application workload and the underlying processor design. In other words, θ is both machine- and application-dependent. Therefore, θ has to be learned on the fly during the program execution, which is known as the main strength of RL – adaptivity.

Here, we applied *maximum a posteriori* (MAP) to estimate the values of θ. MAP is closely related to maximum likelihood estimate (MLE), but it employs an augmented optimization which incorporates a *prior* distribution over the parameters of interests, θ in this case. Generally, MAP leads to a more accurate and robust estimate than MLE [2]. First, we define d as the observed data of $S^{t+1} = s^{t+1}$ given $(S^t = s_j, A^t = a_k)$, and we want to estimate θ, denoted by $\hat{\theta}$ which equals the maximum value of $P(\theta|d)$ (or called mode in statistics). $P(\theta|d)$ can be interpreted as: after d happened, how should the probability of θ be updated? Based on the Bayes rule [30], $P(\theta|d)$ can be rewritten as:

$$P(\theta|d) \propto P(d|\theta) \times P(\theta) \tag{2}$$

where $P(\theta|d)$ is known as the *posterior distribution*, $P(d|\theta)$ represents the data likelihood and $P(\theta)$ is the *prior* distribution. The normalization term $1/P(d)$ is not shown here since it is a constant (under the piecewise stationary condition) and does not affect estimates. Once we determine the prior distribution $P(\theta)$, the posterior distribution $P(\theta|d)$ can be updated by Eq(2) as the data likelihood $P(d|\theta)$ changes since more state transitions are observed by the agent (or the controller). The likelihood function $P(d|\theta)$ follows the multinomial distribution:

$$\frac{N!}{\prod_{i=1}^{i=|S|} x_i!} \prod_{i=1}^{|S|} \theta_i^{x_i} \tag{3}$$

where $x_i \in \{0, \cdots, N\}$ and $\sum x_i = N$. x_i represents the occurrences of $S^{t+1} = s_i$, given $S^t = s_j, A^t = a_k$. N represents the total number of occurrences of $S^t = s_j, A^t = a_k$. The numerical intuition behind this multinomial likelihood is that, given N, we could calculate how likely S^{t+1} will take on each s_i by using its corresponding occurrence x_i. With Eq(3), the data likelihood can be calculated, but we still need to determine the prior distribution to calculate the posterior distribution in Eq(2).

The prior distribution is usually selected based on one's domain knowledge. By selecting a good prior, the posterior distribution can converge faster, and thus the estimate of state transition probability, $\hat{\theta}$, can be obtained earlier during the program execution. In this work, we propose to use a Dirichlet distribution as the prior distribution:

$$P(\theta) = \frac{\prod_{i=1}^{i=|S|} \Gamma(\alpha_i)}{\Gamma(\sum_{i=1}^{i=|S|} \alpha_i)} \prod_{i=1}^{|S|} \theta_i^{\alpha_i - 1} \tag{4}$$

where α_i represents the "hallucinated[4]" counts of s_i, and $\Gamma(\alpha_i)$ is the Gamma function that equals the factorials of $(\alpha_i - 1)$. α_i is similar to x_i in Eq(3) except that x_i represents the actual counts of s_i , whereas α_i stands for our "belief" or "domain knowledge" of how many times s_i should happen before any data are given. Empirically, the range of α_i is set to $10^1 - 10^2$.

Next, we multiply Eq(3) by Eq(4) to obtain the posterior distribution:

$$\begin{aligned} P(\theta|d = \{x_1 \cdots x_{|S|}\}) &= \frac{\prod_{i=1}^{i=|S|} \Gamma(\alpha_i + x_i)}{\Gamma(N + \sum_{i=1}^{i=|S|} \alpha_i)} \prod_{i=1}^{|S|} \theta_i^{\alpha_i + x_i - 1} \\ &= \frac{\prod_{i=1}^{i=|S|} \Gamma(\beta_i)}{\Gamma(\sum_{i=1}^{i=|S|} \beta_i)} \prod_{i=1}^{|S|} \theta_i^{\beta_i - 1} \end{aligned} \tag{5}$$

where $\beta_i = \alpha_i + x_i$. The numerical intuition of β_i is that, besides the actual observations (x_i), we add the hallucinated counts (α_i) to adjust $P(\theta|d)$. In other words, intuitively the

[4]These hallucinated counts α_i can be any positive integer or 0. If the learning process is sufficiently long, θ will be learned correctly and α_i is irrelevant. This is known as "the prior is forgotten" [2]

Table 2: Architectural parameters

Core Parameters	Values	Uncore Parameters	Values	DVFS V/F Pairs		Values
Number of cores	16	Number of routers	16	Turbo	V_{dd}	1.3V
Core model	Pentium 4®	Nominal frequency	3.0 GHz		Freq	3.75GHz
Nominal frequency	3.0 GHz	Router pipeline stages	5 stages	Baseline	V_{dd}	1.0V
L1-I/D caches	Private 64KB, 8-way SA, LRU	Flit size	16 Bytes	(Nominal)	Freq	3.0GHz
L2 caches	Shared 4MB, 32-way SA, LRU	Number of virtual channels	4 per port	Low	V_{dd}	0.8V
Cache coherence	MOESI protocol [20]	Buffer size	4 Bytes		Freq	2.35GHz
DRAM	2 GB	Network topology	4×4 mesh	Very Low	V_{dd}	0.65V
Technology	45nm node with V_{dd} =1.0V	Routing algorithm	X-Y routing		Freq	2.0GHz

domain knowledge is used to aid the estimate of the state transition probabilities. Furthermore, it can be seen that Eq(5) and Eq(4) are in the same general form of Dirichlet distributions (only the parameters α, β are different), which is known as *"conjugate"* prior and posterior pairs. Each time a new state transition d is observed, the posterior distribution can be updated by the multiplication of new data likelihood and the prior by using Eq(2), and thus, the updated posterior can be fed back into Eq(2) as an *updated prior* ready to estimate the new posterior since they are in the same form. Our experimental results show that by using the MAP with conjugate pairs, the best control policy can be learned several times faster than conventional Q learning.

4.2 Semi-supervised Reinforcement Learning

One critical issue of RL is the lack of scalability due to the exponential growth of machine states [29, 32]. Conventionally, the machine states are defined by three performance counters – instruction per cycle (IPC), misses per kilo instruction (MPKI) and router buffer utilization (BU). These three metrics represent the respective metric of performance from cores, caches and on-chip networks, and thus are sufficiently representative as the machine states of the whole CMP. Therefore, for each tile $|S| = |IPC| \times |MPKI| \times |BU|$ and the total number of states for a CMP is $O(|S|^m)$, where m is the number of tiles. For a dual- or a quad-core system, this issue is not severe. However, for future many-core systems, this exponential growth will require a significant control overhead in terms of memory and computational power.

To increase RL's scalability, we separate the "centralized" agent into several "distributed" agents (one deployed per tile), and arrange a supervisor who can coordinate the actions among agents to best exploit the power budget for the performance increase. Therefore, the agent in each tile has the freedom to learn its control policy dynamically, but at the same time they are "semi-supervised" by the supervisor (implemented as a kernel thread), to work in synergy. By deploying this semi-supervised RL, the complexity of states and actions is reduced to $O(m \times |S|)$ and $O(m \times |A|)$, respectively, which is linear with the number of on-chip resources, instead of exponential.

We illustrate the scheme of the proposed semi-supervised RL in Figure 2. The target NoC-based CMP, shown in Figure 2a, is grouped into four clusters. Each cluster contains one supervisor and four agents (one agent deployed per tile). The power constraint, denoted as Pow_{const}, is treated as the total power budget and distributed equally into four clusters. Therefore, each cluster has the power budget of $\frac{Pow_{const}}{4}$. Figure 2b shows the hierarchy of the supervisor and agents within a cluster. Each agent learns its own best policy by using the RL described in Sections 2 and 4.1. During each control epoch, each agent sends a request to the supervisor,

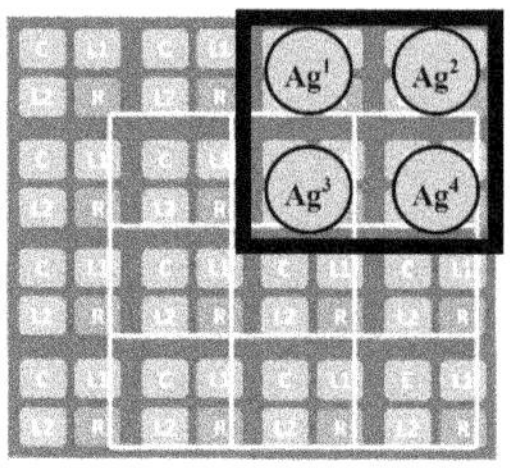

(a) An NoC-based CMP

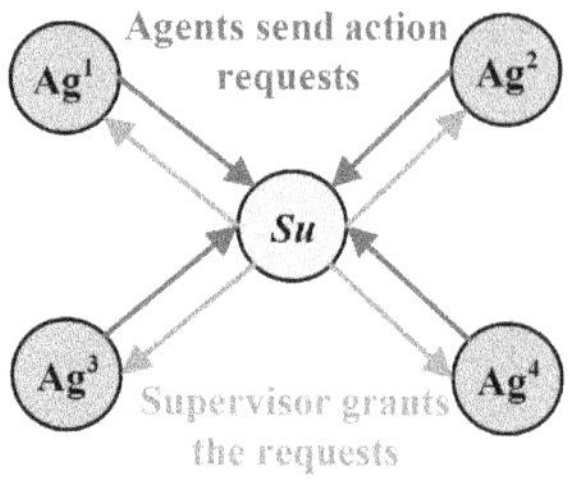

(b) Supervisor and agents

Figure 2: Scheme of semi-supervised RL

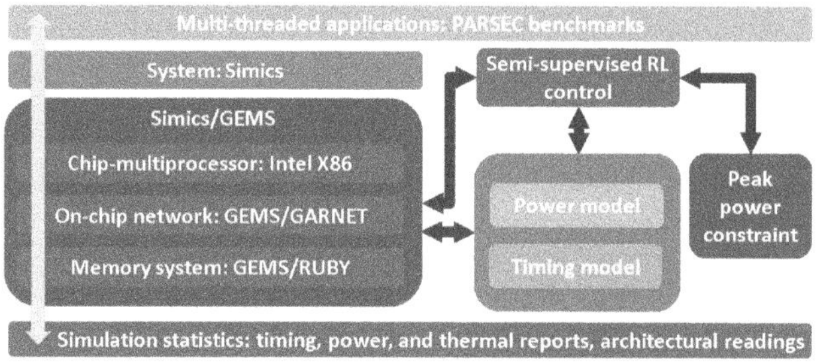

Figure 3: Simulation infrastructure

querying if its own best action (*i.e.*, the frequencies of the core and the router) can be taken. The supervisor will then estimate the total power required. If the power consumption stays within the assigned budget, all requested actions will be granted. Otherwise, the supervisor will constrain each agent from taking its nominal frequency instead of the original best action. The power model used in this paper is described in Section 5.2. With semi-supervised RL, the total state numbers of the CMP is $16 \times |S|$. For the number of actions $|A|$, the core and the router in each tile have four V/F pairs to select from, as described in Section 3.1, which makes $|A| = 4 \times 4$ and the total action numbers is $16 \times |A|$. As a result, the proposed semi-supervised RL reduces the complexity from exponential to linear, and therefore, it is highly scalable.

5. IMPLEMENTATION

In this section, we describe the experimental setup and the simulation infrastructure in detail.

5.1 Performance Modeling Infrastructure

Figure 3 provides an overall view of the simulation infrastructure used in this paper. First, Simics [19] and GEMS [16] are used as a full-system, many-core simulator to evaluate the proposed RL control. The operating system is configured as Linux in Simics. Furthermore, RUBY [16] and GARNET [17] are embedded in GEMS to enable the functional and timing simulations of the cache system and the on-chip communication fabrics, respectively. Detailed parameters are mentioned in Section 3.1. Default values are used as system parameters if not specifically mentioned. For the workloads considered, we use PARSEC [6] as multi-threaded, parallel applications to evaluate the benefits of the proposed control strategy. PARSEC covers a wide spectrum of data sharing and synchronization, which creates both on-chip and off-chip communication close to realistic workloads.

5.2 Power Model

We use the power model proposed by [3] to calculate the power consumptions of processing cores. This power model is calibrated by Intel® Xeon® X7350, and the difference between the actual power consumption and the fitted one from [3] is less than 10%. Therefore, it can accurately calculate the power consumption of NoC-based CMPs emulated by our simulation infrastructure.

For the power model of NoC, we use Orion [11] built in GARNET to provide the power consumption of the on-chip communication fabrics, including routers and links. We further modify and shrink the technology node in Orion to 45nm by using the parameters provided by Orion 2.0 [4]. Finally, CACTI [28] is integrated into RUBY to calculate the power dissipated by L1 and L2 caches in different states (*e.g.*, Read, Write and Standby).

5.3 Controller

The proposed semi-supervised RL-based controller can be implemented as a kernel thread executing on each processing core. Also, the proposed MAP described in Section 4.1 can be implemented very efficiently with a table-lookup technique. We implemented and optimized the controller in C++ and evaluated it on our full-system simulator. The timing overhead is less than 0.03 ms, while each control epoch (or period) is 1.2ms (therefore, controller overhead is around 2%). The power overhead of the proposed control is less than 0.32 Watts per tile, which is less than 1% of the thermal design power (TDP)[5].

For the overhead in memory usage, extra memory space is needed to record all state transition probability. As mentioned in Section 4.2, the number of states per tile is $|S| = |IPC| \times |MPKI| \times |BU|$. Here, we use four states to represent $|IPC|$, three states for $|MPKI|$ and two states for $|BU|$, all based on the utilization rates from [27][5]. Hence, the total number of states per tile is 24 and thus the overhead in memory space is very low.

As a comparison, we re-implement the DVFS controls proposed by [27] and [5]. We select *"Threshold-based"* control of [27] and aggressively upscale the frequency settings for improving the performance. The original baseline frequency in [27] is raised to the Turbo mode (3.75GHz), and the rest of settings are changed in a similar fashion. For uncore-DVFS, the *"freqboost"* of [5] is used to maximize the performance – the frequency is set to 3.75GHz (Turbo) by default. The frequency of a router will be reduced if the occupancy of its downstream routers' buffers is greater than 60% [5], which is considered as potential network congestion. Finally, the power constraint is set to the larger value of the peak power consumption from core-only DVFS or uncore-only DVFS, which is around 350Watts to 470 Watts.

[5]Since the power model [3] is originally for quad-core (Xeon® X7350, TDP=130Watts) and we apply it on the 16-core CMP, the new TDP will be $130 \times 4 = 512$ Watts.

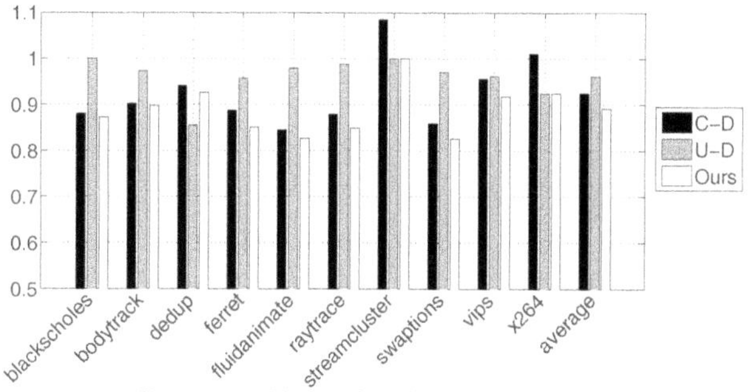

Figure 4: Normalized execution time

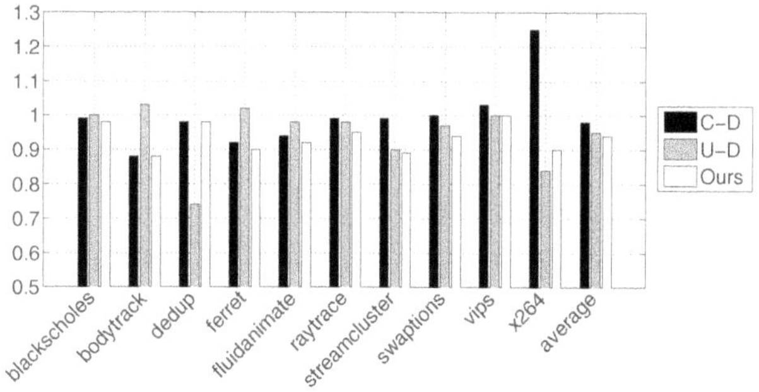

Figure 5: Normalized energy-delay product

6. EXPERIMENTAL RESULTS

This section presents the experimental results, including (1) the overall comparisons, and (2) the network analysis under four control schemes, including baseline (no-DVFS), core-only, uncore-only and cooperative DVFS.

6.1 Overall Analysis

Figure 4 illustrates the analysis of the program execution time for each PARSEC benchmark. "C-D" stands for the Core-only DVFS based on [27], whereas "U-D" is the Uncore-only DVFS based on [5]. All results are normalized to the "Baseline", *i.e.*, the case without any DVFS. On average, the proposed semi-supervised RL achieves 10.9% reduction of the execution time, and outperforms C-D and U-D almost in all cases except *Dedup*. *Dedup* is implemented using a "pipeline programming model" [6] that has two properties: (1) distinct characteristics for each thread, and (2) frequent thread migration: every few thousand cycles. These two facts make the state transition probabilities hard to estimate. However, the proposed control still reduces the execution time by around 7.5%. In most core-bound applications, such as *blackscholes*, both C-D and the proposed control approach outperform U-D. On the other hand, in memory-bound applications, such as *X264*, both U-D and the proposed control have better reduction than C-D. It is also worth mentioning that C-D has longer execution time in *StreamCluster* and *X264*, which will be explained later in Section 6.2. These results clearly demonstrate the effectiveness and robustness of the proposed control, compared to C-D and U-D.

Figure 5 provides the comparisons of energy-delay products. It can be seen that the proposed control achieves the lowest energy-delay product on average – 6% reduction compared to the baseline, whereas C-D and U-D achieve only 2% and 4.7% reduction, respectively. This means the energy efficiency of the proposed control policy is higher than both C-D and U-D. Note that the power constraint is

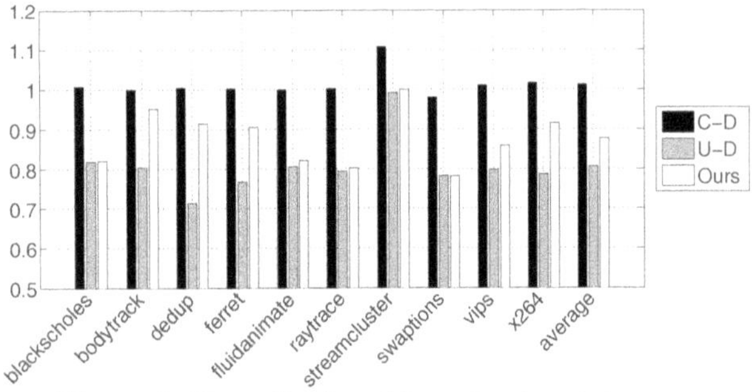

Figure 6: Normalized end-to-end network latency

met at all times and for each benchmark. Furthermore, the peak temperatures (calculated by the thermal model proposed by [3]) of all three control policies stay within 60° C for each benchmark, and thus no thermal emergency occurs due to the "*reverse*" DVFS.

6.2 Network Latency Analysis

Here, we examine the network latency under four different control schemes. Figure 6 presents the average end-to-end network latency for each benchmark. As it can be seen, U-D effectively reduces the network latency by around 19.1%. Considering the reduction of execution time shown in Figure 4, it can be seen that the shorter network latency cannot be directly reflected on the shorter execution time in most cases. This fact reveals that U-D, *i.e.*, uncore-only DVFS can be very effective, but only in certain cases. On the other hand, even if the router frequency is fixed at the nominal value, we can see that C-D has longer network latency in *StreamCluster*, *Vips* and *X264*. This is because the uncore-agnostic DVFS may change the packet injection rate in the network, which results in unexpected congestion and thus longer latency. This also explains why we observe that C-D has longer execution time in *StreamCluster* and *X264* in the previous section. In most cases, the network latency of the proposed control is longer than the latency of U-D (sometimes it is a tie) and shorter than the latency of C-D. This is because the proposed control needs to strike a balance to distribute the power budget to cores or routers for boosting up the frequency, in order to achieve the best performance increase. For example, compared to C-D, the proposed control boosts up the router frequencies (and hence achieves a shorter latency) in *X264*, which in turn reduces the program execution time. All the above results demonstrate that, instead of core-only or uncore-only DVFS, a cooperative control of both cores and uncores is more effective and robust for NoC-based CMPs, which also demonstrates the strength of the proposed semi-supervised RL control.

7. CONCLUSION

In this paper, we evaluate and present the advantages of controlling cores and uncores in synergy for NoC-based CMPs. We also propose a control mechanism based on semi-supervised RL that is highly scalable for advanced CMPs. Experimental results show that the proposed control achieves 11% reduction on the program execution time.

8. ACKNOWLEDGMENTS

The authors would like to thank Dr. Wei-Heng Lo, Dr. Tsung-Yu Tsai, Dr. Andrea Bartolini, Dr. Paul Bogdan and anonymous reviewers for the constructive suggestions. This work was supported in part by an Intel URO grant and NSF Grant CNS-1128624.

9. REFERENCES

[1] A.R. Alameldeen. Ipc considered harmful for multiprocessor workloads. *MICRO*, 2006.

[2] C.M. Bishop. Pattern recognition and machine learning. *Springer*, 2006.

[3] A. Bartolini *et al.* A virtual platform environment for exploring power, thermal and reliability management control strategies in high-performance multicores. *GLSVLSI*, 2010.

[4] A.B. Kahng *et al.* Orion 2.0: A fast and accurate noc power and area model for early-stage design space exploration. *DATE*, 2009.

[5] A.K. Mishra *et al.* A case for dynamic frequency tuning in on-chip networks. *MICRO*, 2009.

[6] C. Bienia *et al.* The parsec benchmark suite: characterization and architectural implications. *PACT*, 2008.

[7] C. Isci *et al.* An analysis of efficient multi-core global power management policies: maximizing performance for a given power budget. *MICRO*, 2006.

[8] G. Dhiman *et al.* Dynamic power management using machine learning. *ICCAD*, 2006.

[9] H. Jung *et al.* Supervised learning based power management for multicore processors. *TCAD*, 2010.

[10] H. Matsutani *et al.* A multi-vdd dynamic variable-pipeline on-chip router for cmps. *ASPDAC*, 2012.

[11] H. Wang *et al.* Orion: a power-performance simulator for interconnection networks. *MICRO*, 2002.

[12] J. Charles *et al.* Evaluation of the intel core i7 turbo boost feature. *IISWC*, 2009.

[13] J. Li *et al.* Dynamic power-performance adaptation of parallel computation on chip multiprocessors. *HPCA*, 2006.

[14] L. Benini *et al.* Networks on chips: a new soc paradigm. *Computer*, 2003.

[15] L. Shang *et al.* Dynamic voltage scaling with links for power optimization of interconnection networks. *HPCA*, 2003.

[16] M. Martin *et al.* Multifacet's general execution-driven multiprocessor simulator (gems) toolset. *SIGARCH Computer Architecture News*, 2005.

[17] N. Agarwal *et al.* Garnet: A detailed on-chip network model inside a full-system simulator. *ISPASS*, 2009.

[18] P. Bogdan *et al.* Optimal power management of multidomain multiprocessor platforms under highly variable workloads. *NOCS*, 2012.

[19] P. Magnusson *et al.* Simics: A full system simulation platform. *Computer*, 2002.

[20] P. Sweazey *et al.* A class of compatible cache consistency protocols and their support by the ieee futurebus. *ISCA*, 1986.

[21] P. Zhou *et al.* Noc frequency scaling with flexible-pipeline routers. *ISLPED*, 2011.

[22] R. Bitirgen *et al.* Coordinated management of multiple interacting resources in chip multiprocessors: A machine learning approach. *MICRO*, 2008.

[23] R. Cochran *et al.* Pack & cap: adaptive dvfs and thread packing under power caps. *MICRO*, 2011.

[24] R. Das *et al.* Performance and power optimization through data compression in network-on-chip architectures. *HPCA*, 2008.

[25] R. Marculescu *et al.* Outstanding research problems in noc design: system, microarchitecture, and circuit perspectives. *TCAD*, 2009.

[26] R. Teodorescu *et al.* Variation-aware application scheduling and power management for chip multiprocessors. *ISCA*, 2008.

[27] S. Herbert *et al.* Analysis of dynamic voltage/frequency scaling in chip-multiprocessors. *ISLPED*, 2007.

[28] S Thoziyoor *et al.* A comprehensive memory modeling tool and its application to the design and analysis of future memory hierarchies. *ISCA*, 2008.

[29] T. Ebi *et al.* Tape: thermal-aware agent-based power rconomy for multi/many-core architectures. *ICCAD*, 2009.

[30] T. Hastie *et al.* The elements of statistical learning: data mining, inference, and prediction. *Springer*, 2009.

[31] Y. Tan *et al.* A framework of stochastic power management using hidden markov model. *DATE*, 2008.

[32] Y. Wang *et al.* Deriving a near-optimal power management policy using model-free reinforcement learning and bayesian classification. *DAC*, 2011.

Power Conversion Efficiency Characterization and Optimization for Smartphones

Woojoo Lee† Yanzhi Wang†, Donghwa Shin‡, Naehyuck Chang‡, and Massoud Pedram†
†University of Southern California, CA, USA, ‡Seoul National University, Korea,
†{woojoole, yanzhiwa, pedram}@usc.edu, ‡{dhshin, naehyuck}@elpl.snu.ac.kr

ABSTRACT

Modern smartphones consume significant power and can hardly provide a full day's use between charging operations even with a 2000 mAh battery. This is in spite of many power management techniques being employed in the smart phones. This paper starts from the observation that modern smartphones waste a significant amount of the battery's stored energy during power conversion from the 3.7V output of a Li-Ion battery cell to different voltage levels needed to power various modules in a smartphone (processors, memory, display, GPS, etc.) Indeed the power conversion efficiency from the battery source to point of use in the smart phone has on average of only 60-75% efficiency. The approach taken to reduce this energy waste in smartphones is to (i) profile the power consumption of each module under different operating scenarios, (ii) build an equivalent DC-DC converter model for each smartphone module and estimate its power conversion efficiency, and (iii) change the parameters of the actual converters in the smartphone to improve the equivalent power conversion efficiencies for all modules. Experimental results demonstrate that we can achieve 6% to 15% power conversion efficiency enhancement, which translates to up to 30% reduction in the power losses incurred during power conversion in smartphones.

Categories and Subject Descriptors

C.5.3 [**Computer system implementation**]: Microcomputers

Keywords

Portable device, Smartphone, DC-DC converter, Power tree

1. INTRODUCTION

Modern smartphones are typically equipped with a multi-core gigahertz processor, gigabytes of high-speed DDR SDRAM, dozens of gigabytes of flash memory, several up to 10 megapixel cameras, 1M+ pixel high-resolution color display, high-power audio, as well as 3G/4G, Wi-Fi and Bluetooth wireless communication devices. Consequently, their power consumption is as high as a small-size notebook computer or a tablet. Although the smartphone battery capacity has been increased from several hundred mAh to over 2000 mAh over the last few years, the actual battery life (i.e., the service time between consecutive charges) has become

ISLPED'12, July 30–August 1, 2012, Redondo Beach, CA, USA.

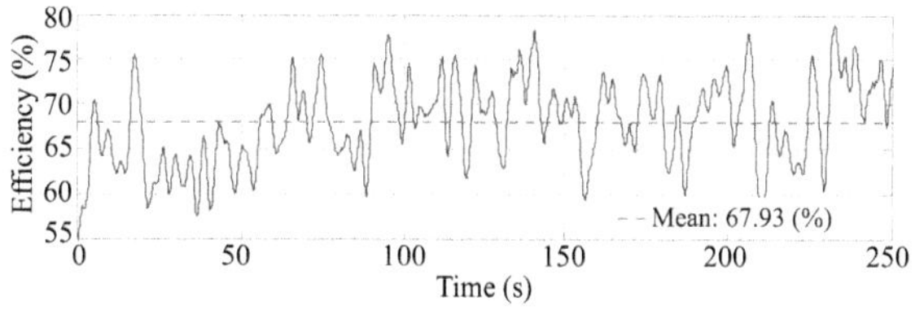

Figure 1: Measured of traces of power conversion efficiency of Qualcomm Snapdragon MDP MSM8660.

shorter due to even faster increase in the power consumption of the smartphones. An examination of the technology trends reveals that smartphone system-on-chips are migrating from dual cores to quad cores, display sizes continue to increase with some high-end smartphones having over 5" display, wireless data communication rates are increasing rapidly with a100MHz aggregated bandwidth, 4G LTE-Advanced providing almost 3.3Gbit peak download rates per sector of the base station, and so on. All this will result in significantly higher power consumption in the smartphones and a smaller battery life (active users of a recent 4G smartphone claim less than 2-3 hours of battery life.) High power consumption of smartphones (and relatively slow rate of increase in the energy storage density of batteries) is thus a serious concern that could derail the smartphone technology development and adoption.

It is not surprising that system-level power optimization and management have been widely investigated and advanced over the last two decades. Some recent works have explicitly focused on smartphones. In particular, references [1, 2, 3, 4] have presented various power dissipation models for the smartphones. Dynamic voltage (and frequency) scaling techniques have also been used to reduce the power consumption of various smartphone modules, including the CPU [5] and the display [6]. LCD backlight scaling has also been widely investigated [7].

Power optimization and management works for smartphones have targeted power savings in embedded processors, memory, display, and so on. None has addressed the issue of power conversion efficiency in the smartphones. This is an important problem that has gone unnoticed. More precisely, modern smartphones are equipped with many modules, each requiring its own supply voltage level which is typically different from those of other modules in the system. The smartphone is powered by a secondary (rechargeable) Li-Ion battery comprised of a single battery cell providing an initial output voltage level of 4.2V for a fully charged cell and as low as 3.0V for a nearly discharged battery (this yields the familiar nominal average output voltage level of 3.6 or 3.7V at 0.5C discharge rate.) This cell output voltage must be converted and regulated to different pre-determined voltage levels and distributed to various modules in the smartphone.

We have performed extensive measurement of power conversion efficiency for Qualcomm's smartphone platform, MDP MSM8660, as shown in Figure 1. Surprisingly, the power conversion efficiency of MDP MSM8660 ranges just from below 60% to slightly over 75%. Improving the power conversion efficiency can achieve significantly longer battery life. This paper thus focuses on power conversion efficiency in smartphones and introduces power conversion efficiency characterization and optimization procedures.

Modern DC-DC converters exhibit very high peak conversion efficiency, but their conversion efficiency can drop dramatically due to the operating conditions, i.e., their output current level [8]. In other words, a low overall conversion efficiency of DC-DC converters is mainly due to mismatch between the power converter characteristics and the load demands.

In this paper we propose a general *equivalent DC-DC converter model* to model different types of converters (such as a buck DC-DC converter, a boost DC-DC converter, low-dropout (LDO) converter) and their series combinations in a power delivery path from the battery cell to the load device. In particular, we use the DC-DC converter model from [8] and identify coefficients of the equivalent power converter models with the aid of the application profiling tool named Trepn™.

We propose device grouping to enhance the accuracy of linear regression used for power efficiency characterization of the equivalent DC-DC converter. We then verify the accuracy of the power conversion efficiency characterization with real measurements. The results point to the fact that power conversion efficiency of the smartphone platform is quite low. Next, we model and verify the current demand distribution for each module in the smartphone platform and derive its expected value. Finally, we adjust the DC-DC converter parameters to ensure that the power converters operate at the most efficient points. Experimental results demonstrate that we can achieve 6% to 15% power conversion efficiency enhancement, which translates to up to 30% reduction in the power losses incurred during power conversion in smartphones.

Notice that the proposed flow can be exploited to do optimization in presently available commercial smartphone platforms, which do not have current sensors to report the component current demands under different applications and usage scenarios. This is possible because we make use of activity profiling-based power estimation [1, 2, 3, 4].

2. POWER CONVERSION LOSS MODELS

Given that the smartphones integrate hardware modules that have their own supply power lines and require their own voltages, it is necessary to generate different voltages for various modules from a single battery that powers the smartphone. A power conversion tree, with the root of the tree being the battery cell, the leaf nodes being the modules, and the internal nodes being DC-DC converters, is designed to achieve this goal. A conceptual example of the power conversion tree for a smartphone is depicted in Figure 2.

Typical DC-DC converters in modern smartphones can be classified into two types, switching-mode DC-DC converters (simply called DC-DC converter) and low-dropout linear regulators (LDO), according to the circuit implementation and operation principles. A DC-DC converter consists of an inductor, capacitors, two MOSFET switches and a pulse-width-modulation (PWM) controller. This type of converter can step-up the output voltage so that it becomes higher than the input voltage (i.e., boost), or step-down the output voltage so that it is lower than the input voltage (i.e., buck). On the other hand, the output voltage of an LDO can only be lower than its input voltage. In general, the LDO has lower power conversion efficiency. Nevertheless, the LDO is an indispensable component in smartphones in that it can provide low-noise output voltage, and therefore, the LDO is the most suitable type of converter to provide power for some noise-sensitive RF or analog modules.

2.1 DC-DC converter power loss model

The power loss model of a PWM DC-DC converter is well-studied in [8]. In general, the major sources of power loss in a DC-DC converter are conduction loss, switching loss in the power switches, and controller power loss, denoted by $P_{conduction}$, $P_{switching}$, and $P_{controller}$, respectively. The power loss of the buck converter,

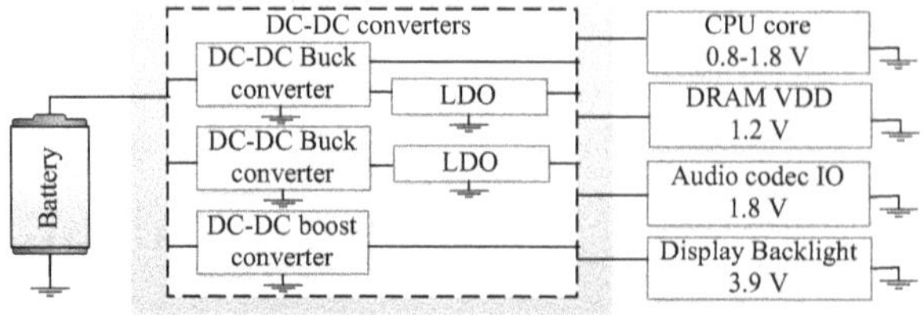

Figure 2: Conceptual diagram of a smartphone power conversion tree.

P_{buck}, can be expressed as:

$$\begin{aligned} P_{buck} &= P_{conduction} + P_{switching} + P_{controller} \quad (1)\\ &= {I_{out}}^2 (R_L + DR_{sw1} + (1-D)R_{sw2}) \quad (2)\\ &\quad + (\Delta I)^2 (R_L + DR_{sw1} + (1-D)R_{sw2} + R_C)/12\\ &\quad + V_{in} f_s (Q_{sw1} + Q_{sw2}) + V_{in} I_{controller}, \end{aligned}$$

where $D = V_{out}/V_{in}$ is the PWM duty ratio of the power switch, and V_{in} and V_{out} denote the input and output voltages, respectively; I_{out} is the output current through the inductor; $\Delta I = (1-D)V_{out}/(L_f f_s)$ is the amplitude of the maximum current ripple at the inductor; f_s is the switching frequency; and $I_{controller}$ denotes the current used in the control logic section of the converter. Series resistances of the inductor L and capacitor C are denoted by R_L and R_C, respectively. Similarly, series resistances of the two MOSFET switches are represented by R_{sw1} and R_{sw2}, respectively, while the amounts of their gate charge are denoted by Q_{sw1} and Q_{sw2}, respectively. The first and second terms of (2) are DC and AC conduction losses, respectively; third term of (2) denotes the switching loss; while the last term of (2) corresponds to the controller power loss.

The power loss of the boost converter, P_{boost}, is modeled similarly except for the switching duty calculation, which is given by:

$$\begin{aligned} P_{boost} &= \left(\frac{I_{out}}{1-D}\right)^2 (R_L + DR_{sw1} + (1-D)R_{sw2} + D(1-D)R_C)\\ &\quad + \frac{(\Delta I)^2}{12}(R_L + DR_{sw1} + (1-D)R_{sw2} + (1-D)R_C)\\ &\quad + V_{in} f_s (Q_{sw1} + Q_{sw2}) + V_{in} I_{controller}, \quad (3) \end{aligned}$$

where $D = 1 - V_{in}/V_{out}$ and $\Delta I = (V_{in}D)/(L_f f_s)$.

As a result, the efficiency of a DC-DC converter, $\eta_{switching}$, can be calculated as:

$$\eta_{switching} = \frac{V_{out} I_{out}}{V_{in} I_{in}} = \frac{V_{in} I_{in} - P_{converter}}{V_{in} I_{in}}, \quad (4)$$

where I_{in} is the input current, and $P_{converter}$ is either P_{buck} or P_{boost} depending on the type of DC-DC converter used.

2.2 LDO power loss model

A typical LDO consists of an error amplifier, a pass transistor, and a feedback resistor network. The power loss of the LDO, denoted by P_{ldo}, is given by:

$$P_{ldo} = I_{out}(V_{in} - V_{ref}k) + I_q V_{in}, \quad (5)$$

where V_{ref} is the reference voltage in the error amplifier; $k = (R_1 + R_2)/R_2$ corresponds to the voltage divider's gain coefficient, and I_q denotes the quiescent current of the LDO. Unlike the switching converter in which the MOSFET switches dominate the total power loss, the pass transistor in the LDO has negligible impact on its total power loss [8]. Therefore, the power loss due to internal resistance of the pass transistor is not accounted for in the model. Thus the conversion efficiency of the LDO, η_{ldo}, may be expressed as:

$$\eta_{ldo} = \frac{V_{out} I_{out}}{V_{in} I_{in}} = \frac{k\, V_{ref} I_{out}}{V_{in}(I_{out} + I_q)} \quad (6)$$

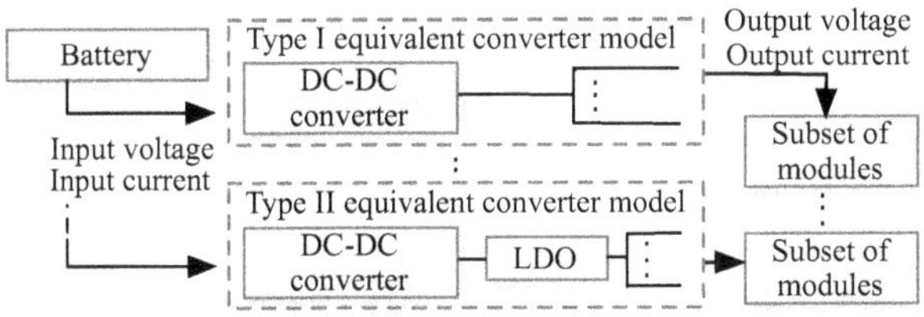

Figure 3: Types I and II equivalent power converter models.

3. CHARACTERIZATION OF THE POWER CONVERSION EFFICIENCY

Power converters including switching-mode DC-DC converters exhibit different conversion efficiency as a function of the load current [8]. A module is often powered through a set of converters from a battery source as shown in Figure 3. The power converter set can be an empty set (direct connection), single DC-DC converter, (more commonly) a cascade (series) connection of a DC-DC converter and an LDO, (rarely) a cascade connection of various DC-DC converters, etc. Characterization of each power converter efficiency is not a trivial work unless the power conversion tree structure and converter specifications, and all the node voltages and branch currents of the conversion tree are available. Such a *white-box* approach is generally not possible for commercial smartphones. Although the Qualcomm MDP provides measurement of the device current values, the power conversion tree structure is not available for examination/measurements.

In this paper, we attempt a *gray-box* approach introducing an *equivalent power converter* concept. The equivalent converter models a set of power converters from the battery source to each module. In other words, the proposed equivalent power converter abstraction treats the set of power converters as a single equivalent converter. The abstraction enables a gray-box approach by which one can group modules in a (smartphone) system by their required supply voltage levels, which can be obtained from datasheets. Power conversion efficiency improvement in the subsequent optimization procedure can be effectively performed once we identify the power conversion efficiency of all the power conversion paths from the battery source to various modules in the system.

3.1 Equivalent converter model

We classify the equivalent converter models either a single DC-DC converter, or a cascaded DC-DC converter and an LDO, named Type I and Type II equivalent converters, respectively. We assume that the battery output current goes through a voltage regulator in order to produce a constant voltage throughout its full discharge cycle. Without loss of generality, Types I and II equivalent converter models can represent most power conversion tree structures [9, 10, 11]. Most digital logic components can be powered by a single DC-DC converter from the battery to the module - this gives rise to Type I converter model. A cascade of two or more DC-DC converters are rare, because increasing the number of cascade DC-DC converters increases the cost and form factor overhead with little (or no) benefit in terms of conversion efficiency. LDOs are often an indispensable component to provide low-ripple output voltage for switching noise-sensitive RF and analog modules. It is uncommon to use a single LDO from the battery to a device due to large dropout voltage. Instead, the conversion of the battery voltage to an initially higher target voltage using a DC-DC converter and subsequent use of LDO for the final power conditioning is a more efficient way of doing this - hence, our emphasis on Type II converter model.

According to (2), (3), and (5), the power loss of the equivalent converter can be expressed as:

$$P_{eqv} = A(\alpha I_q + \sum_{i=1}^{N} I_{mod,i})^2 + \alpha\beta \sum_{i=1}^{N} I_{mod,i} + (B + \alpha\gamma I_q), \quad (7)$$

where N is the number of modules connected to the equivalent converter; $I_{mod,i}$ is the input current of the ith module; A is determined by the type of the DC-DC converter such that $A = R_L + DR_{sw1} + (1-D)R_{sw2}$ for buck converter and $A = (1/(1-D))^2(R_L + DR_{sw1} + (1-D)R_{sw2} + D(1-D)R_C)$ for boost converter; B is the sum of the second, third, and last terms of (2) or (3); $\alpha = 0$ for Type I, and $\alpha = 1$ for Type II; γ is the input voltages of the LDO; and $\beta = (\gamma - V_{ref}k)$. We can further simplify (7) by defining the output current of the equivalent converter, $I_{eqv\_out} = \sum_{i=1}^{N} I_{mod,i}$, and thus, the power loss for both types of equivalent converter models can be expressed as:

$$P_{eqv} = aI_{eqv\_out}{}^2 + bI_{eqv\_out} + c, \quad (8)$$

where the coefficients a, b, and c are derived from (7), and are largely dependent on the power converter design specification such as the power MOSFET gate width, inductor IR loss, controller loss, etc. [8]. Calculating those coefficients is the key step in of the power conversion efficiency characterization.

Table 1: Grouping results for Snapdragon MDP MSM8660.

Group	Modules	Voltage
1 and 2	Group1: CPU core0 and Group2: CPU core1	0.8-1.225 V
3	Internal Memory, Audio DSP, and Digital core (includes GPU and modems)	1.1 V
4	Audio codec Vdd, LPDDR2 Vdd, ISM Vdd, DRAM Vdd2, and Camera digital	1.2 V
5	Audio codec IO , IO PAD3, Display IO, DRAM Vdd1, Camera IO, PLL, and eMMC host interface	1.8 V
6	Audio codec analog, Haptic, SD card, Touch screen, eMMC (Flash), IO PAD2, SD card, and Ambient light sensor	2.85 V
7	Display Memory and Display backlight	3.8 V

3.2 Power converter grouping and regression analysis

Measurement (or estimation) of the output current of all the equivalent power converters enables us to calculate the unknown coefficients of the equivalent power converter model. Once again, the input and output voltage levels of each equivalent power converters can be obtained from the device datasheets. The Qualcomm Snapdragon MDP MSM8660 [12] incorporates embedded power sensors that monitor and report current values of different modules with fine granularity. When the target smartphone does not provide embedded current sensors, we can estimate the module current values by activity profiling [1, 2, 3, 4].

Profiling various applications, which result in diverse usage patterns of the system modules, provides sufficient information and data to perform regression analysis and obtain the unknown coefficients. Linear regression analysis is a widely used method in system identification, requiring (i) a well-designed model and (ii) sufficient experimental data to extract the best-fit model coefficients. In reality, however, independent control of each module is a challenging task due to the lack of direct control knobs. For example, if we run an application that activates a camera module, the CPU, GPU, memory, and other associated component currents also ramp up and down. We must thus apply linear regression analysis to the whole system (including all smartphone modules) simultaneously, while trying to vary the activity level of each module by running different applications. However, this method may not produce sufficient data to cover the full range of activities for all smartphone modules, especially when the number of modules is large (e.g., the Snapdragon MDP MSM8660 has 27 embedded modules.) This is a potential source of inaccuracy for regression analysis due to the *weak training set* issue.

We tackle the problem by doing a module grouping in order to reduce the number of unknown coefficients that must be determined during the characterization process. This grouping procedure reduces the burden in terms of generating sufficient data for performing the linear regression analysis. The idea is that system modules that require the same operating voltage level can be combined into one group, and each group of modules is connected to the battery source via a single equivalent converter, as illustrated in Figure 3.

This method matches well with low power design practices that try to minimize the number of power converters, due to their cost and internal power losses.

Given that the number of different voltage levels required by various modules in a smartphone platform is typically less than 10 [9, 11], the grouping method significantly reduces the number of parameters to be determined in linear regression. For example, the Snapdragon MDP MSM8660 requires only seven groups although the module count is 27.

Finally, the total power loss of the smartphone, P_{loss}, is given by,

$$P_{loss} = \sum_{k=1}^{G} P_{eqv,k} = \sum_{k=1}^{G} (a_k I_{eqv\_out,k}{}^2 + b_k I_{eqv\_out,k} + c_k), \quad (9)$$

where G is the number of groups; $P_{eqv,k}$ is the power loss of the kth equivalent converter corresponding to the kth group of modules; $I_{eqv\_out,k}$ denote the output current of the equivalent converter, which can be measured using embedded sensors in the Snapdragon MDP MSM8660; a_k, b_k, and c_k are the coefficients of the equivalent converter model (to be determined by linear regression.) We treat the battery voltage presented to the power conversion tree as being (nearly) constant, which is valid considering the function of the regulator between the battery cell/pack and the equivalent converter, therefore, we may assume that a_k, b_k, and c_k are constant values.

3.3 Experimental results

3.3.1 Experimental setup

We use the Snapdragon MDP MSM8660 (in short, MDP) as a target platform. It is a cutting-edge smartphone platform equipped with Google Android OS 2.3 on top of Snapdragon 1.5 GHz asynchronous dual-core CPU, a 3D-supporting GPU, 3.61″ WVGA multi-touch screen, 1 GB internal RAM, 16 GB on-board flash, WiFi, Bluetooth, a GPS, dual-side cameras, etc. We perform power measurement of each module using the application profiling tool named Trepn™. Use of Trepn™ ensures higher accuracy of the measurements. Note, however, that our proposed method is independent of the measurement tools, e.g., we may use activity profiling for power measurement provided by Google or based on techniques presented in the literature [1, 2, 3, 4].

3.3.2 Coefficient identification

As shown in Table 1, the MDP modules may be classified into seven groups based on their operating voltage levels. Some modules such as the CPU cores in the MDP use dynamic voltage and frequency scaling (DVFS) techniques that require a range of variable supply voltage levels. Consequently, we keep each CPU core in a separate group but treat the equivalent converters of these groups identical to each other. Group 7 is associated with display, and therefore, the backlight luminance level mostly determines the current demand in this group. Group 7 coefficients are easy to identify because we can independently control the brightness of the display. In other words, we first perform the linear regression to identify coefficients of the equivalent converter model of Group 7, separately from the other groups.

For the remaining six groups, we profile various applications and collect sufficient data for the regression analysis as explained earlier. It is difficult to identify every c_k coefficient of the kth equivalent converters directly from the linear regression process. Rather, we only extract c_{ext} that corresponds to the sum of all the constant terms in (9), i.e., $c_{ext} = \sum_{k=1}^{G} c_k$. We find an approximate value for each c_k as $c_k = c_{ext}(P_{group,k}/P_{group,total})$, where $P_{group,k}$ denotes the power consumption of Group k, and $P_{group,total}$ is the total power consumption of all the groups. The $P_{group,k}$ and $P_{group,total}$ values are available from the embedded sensors in the MDP.

The extracted coefficients of the seven equivalent converters are reported in Table 2. The power conversion efficiency of Group k, derived from $(P_{group,k}/(P_{group,k}+P_{eqv,k}))$, is shown in Figure 4. We verify the characterization results of each equivalent power

Table 2: Extracted coefficients for each group.

k	a_k	b_k	c_k	k	a_k	b_k	c_k
1, 2	0.4427	0.0025	0.0170	5	0.1971	0.5232	0.0128
3	0.4079	0.1742	0.0675	6	0.1814	0.2928	0.0320
4	0.1152	0.1757	0.0077	7	0.4091	0.3871	0.0289

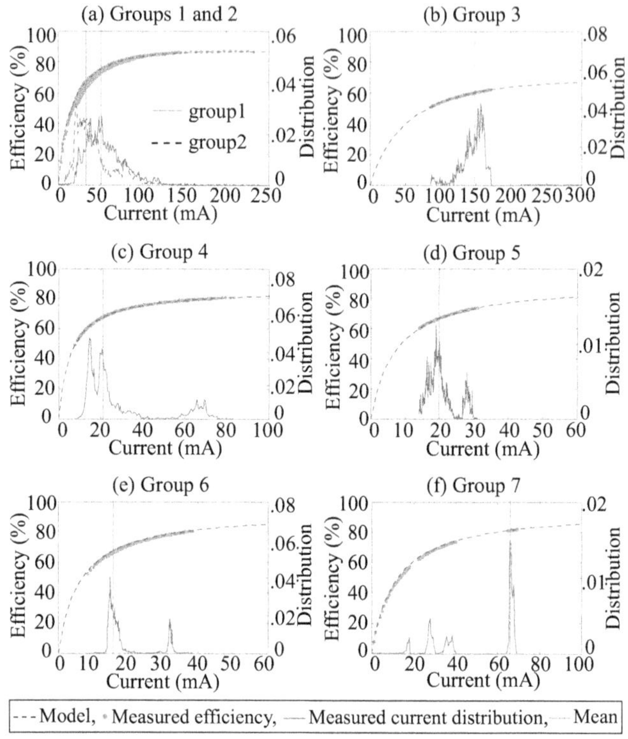

Figure 4: Conversion efficiencies for all groups.

converters. Figure 5 shows the comparison of the system power consumption trace between the real measurement as reported by a built-in battery sensor and the estimation as obtained by our extracted equivalent converter coefficients. We have thus confirmed that the results of the power conversion efficiency characterization process is accurate enough for the subsequent optimization process.

4. POWER CONVERTER TUNING

Power converter tuning reduces power loss without any performance degradation. This is because, unlike typical low-power design techniques that often exploit a tradeoff between performance/service quality and power efficiency, the power converter tuning does not utilize slack time of the system.

Enhancement of the overall efficiency of a DC-DC converter can greatly increase the overall system power efficiency [13, 9]. DC-DC converters show very high overall efficiency under the right operating conditions. However, their efficiency can be low if they are operated outside the recommended range on input and output voltages and load currents [8]. Therefore, ensuring that each DC-DC converter in the system is operating under the right operating conditions is an effective way of improving the system power efficiency. For example, Reference [10] presents a dynamic programing ap-

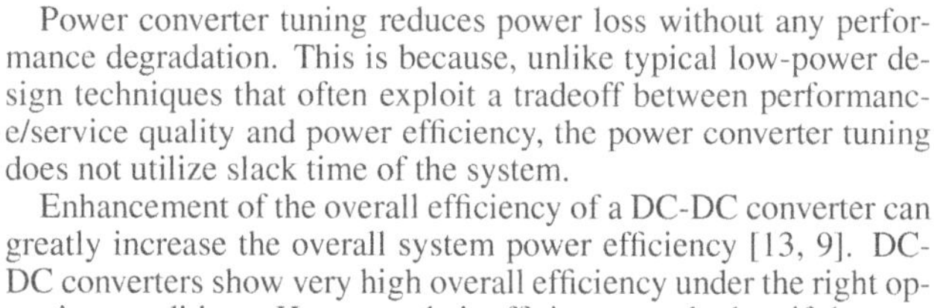

Figure 5: A part of traces of total power consumption: measured data and modeled data.

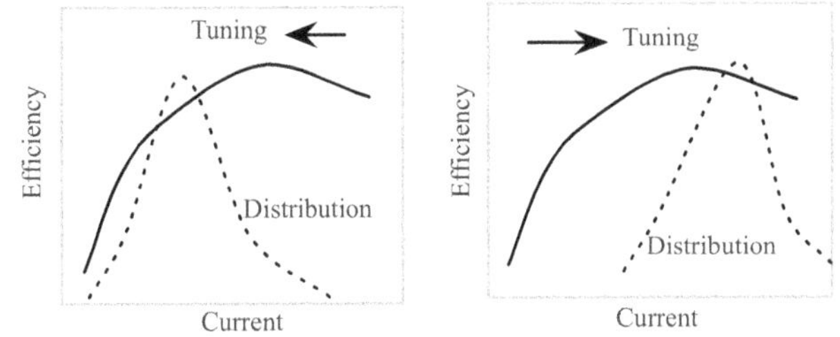

(a) W is over designed. (b) W is under designed.

Figure 6: Concept of the width tuning.

proach to design the structure of the power conversion tree in a system while at the same time selecting the 'optimal' DC-DC converter or LDO for each node of the conversion tree. Reference [14] proposes the concept of parallel connections of high frequency DC-DC converters for distributed energy storage systems. In contrast, the present paper starts with a fixed conversion tree structure, but uses load currents demands and converter characteristics to perform MOSFET switch sizing so as to improve the overall efficiency of the power conversion process in a smartphone system.

As stated above, we focus on the optimal assignment of the gate width, W, of the MOSFET switches in a DC-DC converter. The pass transistor in the LDO has negligible impact on its power loss, therefore, we do not take it into account. (c.f. Section 2.2.) As we previously depicted in Figure 4, DC-DC converters in the MDP are not properly tuned in regard to the actual Android applications running on it. The width tuning, which was introduced in [8], is conceptually explained in Figure 6. The idea is to match the value of W in the equivalent converter such that the desirable operating conditions of the equivalent converter match with the current distribution produced by the actual usage profiles of various Android applications in order to yield the maximum conversion efficiency for the typical daily use of a smartphone.

Sizing down W causes the turn-on resistance of the MOSFET increased and the gate charge decreased. (i.e., $W = (W_o R_o)/R_{sw1,2} = (W_o Q_{sw1,2})/Q_o$; R_o and Q_o correspond to the turn-on resistance and the gate charge of a MOSFET with a gate size of W_o, respectively.) As a function of W, we can express the DC-DC converter power loss models, (2) and (3), as:

$$P_{converter} = \left(\frac{r_1}{W} + r_2\right) I_{out}{}^2 + r_3 W + r_4, \tag{10}$$

where I_{out} denotes the output current of the DC-DC converter; r_1, r_2, r_3, and r_4 are constants.

Given that the two MOSFET switches in a DC-DC converter dominate the power loss of the equivalent converter, and I_q is small, we rewrite (8) as:

$$P_{eqv,k} = \left(\frac{r_{1,k}}{W_k} + r_{2,k}\right) I_{eqv\_out,k}{}^2 + bI_{eqv\_out,k} + r_{3,k} W_k + r_{4,k}, \tag{11}$$

where $P_{eqv,k}$ and $I_{eqv\_out,k}$ are the power loss and the output current of the kth equivalent converter, respectively, corresponding to the kth group of modules; W_k, $r_{1,k}$, $r_{2,k}$, $r_{3,k}$ and $r_{4,k}$ are the coefficients of the equivalent converter model that have been determined

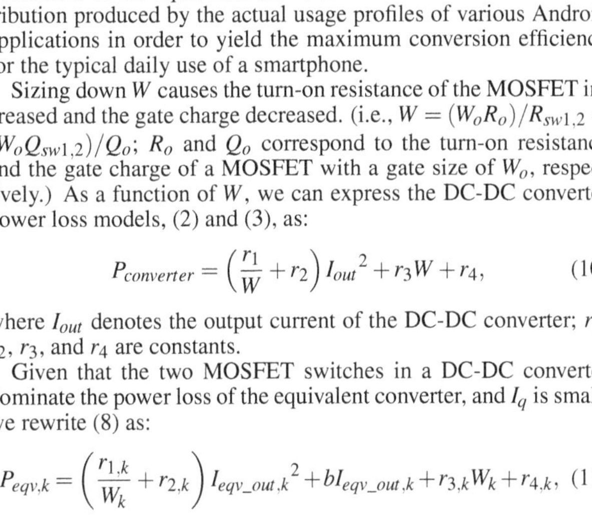

(a) Efficiency : Group 7 (b) Power loss : Group 7

Figure 7: Relation between the power conversion efficiency and W: Group 7.

Table 3: W_{def} of the equivalent converter models.

Group	1 and 2	3	4	5	6	7
W_{def}	1.2401	1.1033	1.3109	1.4033	1.4102	0.7368

based on linear regression. The resultant coefficient W_k is the default value of W (W_{def}) of the kth equivalent converter, which is shown in Table 3.

4.1 Width tuning

Figure 7 (a) shows an example in which W changes the efficiency graph of Group 7. Note that the efficiency of the DC-DC converter increases as W becomes smaller (larger) in the relatively low (high) current demand region. Other groups show similar behaviors. Figure 7 (b) shows the relation between the power loss and W under the condition that the output current of Group 7 is 37 mA. Here, δ is a control variable such that the modified value of W is calculated as $\delta \cdot W_{def}$. The power loss plots have a convex functional form in terms of W, which implies that there is a globally optimal value of W, W_{opt}, that minimizes the power loss of the equivalent converter. Given that I_{eqv_out} is a random variable, our goal of finding W_{opt} for the actual usages of the various applications should be based on the probability distribution of I_{eqv_out} acquired by the usages of the applications. Such probability distribution can be obtained via real measurements, yielding the current profile of each module (thereby, each group of modules) when each type of application is running. We run 10 representative smartphone applications and collect the probability distribution of I_{eqv_out} for each group of modules and for each application.

Throughout the module current profiles collection and (11), the expected power loss of an equivalent converter can be generally expressed as:

$$E[P_{eqv}] = \left(\frac{r_1}{W} + r_2\right) E[I_{eqv\_out}{}^2] + bE[I_{eqv\_out}] + r_3 W + r_4. \tag{12}$$

where $E[I_{eqv\_out}]$ denotes the expected value of I_{eqv_out}, and $E[I_{eqv\_out}{}^2]$ is the expected value of $I^2_{eqv\_out}$.

To find W_{opt}, we limit δ to two decimal places to avoid high computation cost (also in practice, two decimal place precision is quite adequate.) We define the *power conversion efficiency enhancement* by $Gain_\eta = 100\left(\eta_{optimal}/\eta_{original} - 1\right)$ and *power loss reduction* by $Gain_P = 100(1 - P_{optimal}/P_{original})$, where the optimal power conversion efficiency, $\eta_{optimal}$, and power loss, $P_{optimal}$, are derived from (10) with the optimal value of δ, and the original power conversion efficiency, $\eta_{original}$, and power loss, $P_{original}$ are derived from (10) with $\delta = 1$.

4.2 DC-DC converter tuning results

Table 4 shows an example of the width tuning that each W_{opt} is derived from each group and four different applications, 'Clock', 'Call', 'Facebook', and 'Skype-videochat'. As each application causes different current distribution, the values of W_{opt} for the same group are different. Table 4 also shows the tuning results for the four applications. According to the resulted W_{opt} for each group, the $Gain_\eta$ and $Gain_P$ columns show the resultant gains for each group. The upper bound of the power conversion efficiency enhancement ($Gain_{\eta,max}$) and the power loss reduction ($Gain_{p,max}$) are obtained from the distribution of I_{eqv_out} and W_{opt}.

We collect the current distribution data from 10 applications, including the four applications named in Table 4, 'Camera', 'GoogleMap', 'Neocore', 'SMS', 'System setting', and 'Youtube'. All the applications except 'Clock' and 'System setting' are run under the same setup where WiFi is turned on and the backlight level of the display is the highest. 'Clock' is measured under the median level of the backlight and WiFi on, whereas 'System setting' is measured under the lowest backlight and WiFi off. For the case of 'Call', we consider auto turn-off screen during the call. We, then, classify the 10 applications into seven categories, i) *communication* (contains 'SMS', 'Call', and 'Skype-videochat'), ii) *browsing* (contains 'Web browsing'), iii) *media* (contains 'Camera' and 'Youtube'),

Table 4: Example of DC-DC converter tuning results of four types of applications.

Group	Clock					Call				
	W_{opt}	$Gain_{\eta}$	$Gain_{P}$	$Gain_{\eta,max}$	$Gain_{P,max}$	W_{opt}	$Gain_{\eta}$	$Gain_{P}$	$Gain_{\eta,max}$	$Gain_{P,max}$
1	0.2356	11.3508	34.5181	27.2733	43.1362	0.2852	8.7073	30.6121	23.7710	40.8894
2	0.1369	21.1175	42.8718	31.2166	47.0092	0.1991	12.9707	37.3828	27.1296	44.5621
3	0.3751	6.7329	15.7217	13.5717	24.4609	0.3751	6.6543	15.5894	13.5717	24.4609
4	0.1180	15.2996	34.9625	19.4536	38.4971	0.1704	10.2327	28.7502	12.4670	31.7882
5	0.1123	15.5010	30.7496	17.9247	33.0915	0.1263	13.0017	27.9058	15.4971	30.7201
6	0.1128	14.6603	34.4909	15.9501	35.8211	0.1269	14.0235	33.7792	16.0955	35.9118
7	0.0884	6.9630	23.5545	8.1285	25.3117	—	—	—	—	—
Overall	—	9.9381	24.8974	15.0994	30.8775	—	9.3516	23.5594	16.2542	30.9136
Group	Facebook					Skype-videochat				
1	0.4216	4.4201	20.7700	11.9557	32.7724	0.4216	4.3408	20.544	12.8354	33.2795
2	0.2862	8.4227	30.8524	14.4969	38.0183	0.4230	4.1481	20.9525	13.4634	33.9652
3	0.3862	6.0869	14.5694	13.3955	24.1807	0.3751	6.6947	15.6603	13.5717	24.4609
4	0.2097	8.5744	25.9577	14.0064	33.1636	0.5375	1.6203	7.4004	2.5731	10.9563
5	0.1263	12.514	27.2880	15.7297	30.9577	0.1824	8.4489	21.3063	9.7984	23.4784
6	0.1128	14.3447	34.1397	15.9733	35.8437	0.2256	6.0727	24.9642	6.4484	25.8884
7	0.2210	1.9788	10.5936	2.0172	10.7499	0.1179	4.9756	19.7093	6.4110	22.5335
Overall	—	5.7545	19.0159	8.8194	25.0937	—	5.3158	17.9998	9.0870	24.6192

Table 5: DC-DC converter tuning results of Patterns I and II.

Group	$W_{opt,I}$	$Gain_{\eta}$	$Gain_{P}$	$Gain_{\eta,max}$	$Gain_{P,max}$
1	0.2976	7.9717	29.7824	26.2208	40.8462
2	0.1886	14.047	38.7954	43.2898	46.5719
3	0.3751	6.4011	15.1687	13.8601	24.7137
4	0.1704	10.5933	29.3017	22.2939	39.5304
5	0.1263	12.8753	27.7511	17.7792	32.8968
6	0.1269	13.9486	33.7017	19.3969	38.6310
7	0.2137	2.0874	11.0332	6.8824	21.6813
Overall	—	6.2176	20.0263	13.6376	29.4715
Group	$W_{opt,II}$	$Gain_{\eta}$	$Gain_{P}$	$Gain_{\eta,max}$	$Gain_{P,max}$
1	0.3348	6.4977	26.7376	32.3645	39.9298
2	0.2115	6.1590	37.0320	33.1697	44.9638
3	0.3862	9.4790	14.7303	13.6770	24.4264
4	0.1835	12.4529	27.5864	22.0385	39.1593
5	0.1263	13.6757	27.2171	17.7792	32.8968
6	0.1269	2.0615	33.3899	19.3969	38.6310
7	0.2137	2.0874	10.9291	6.8824	21.6813
Overall	—	6.2955	20.0705	15.2870	30.0543

iv) *productivity* (contains 'Clock'), v) *system* (contains 'System setting'), vi) *games* (contains 'Neocore'), and vii) *maps* (contains 'GoogleMap'). We determine their run time according to the two types of usage patterns collected by i) deploying a custom logger on the phones of 33 Android users (Pattern I), ii) 222 Windows Mobile users across different demographics (Pattern II) in [15].

The current distribution from Pattern I is shown in Figure 4. The original power conversion efficiencies of Patterns I and II are 70.7410% and 70.4817%, respectively. Table 5 shows the optimization results for both Patterns I and II, where the values of W_{opt} are $W_{opt,I}$ for Pattern I, and $W_{opt,II}$ for Pattern II. The overall power conversion efficiency enhancements are from 6% to 15%, which correspond to from 20% to 30% power loss reduction during power conversion, respectively.

5. CONCLUSIONS

This paper demonstrated that significant power loss occurs during power conversion from the battery source to various devices in a modern smartphone. To mitigate this problem, the paper introduced a systematic method for system-level power conversion efficiency enhancement in smartphones. First, we proposed the equivalent power converter concept that abstracts the power conversion path from the battery to the modules into a single equivalent power converter. This enabled us to identify the model coefficients from application profiling. We demonstrated the accuracy of power conversion efficiency characterization and how the current power converter setup deviates from the optimal operating conditions. Finally, we proposed a power converter tuning procedure that resulted in 6% to 15% overall power conversion efficiency enhancement, which translates to up to 30% power loss reduction during power conversion. As part of the future work, the optimization procedure will be improved by considering a multitude of different applications and usage scenarios.

6. ACKNOWLEDGEMENTS

This work is sponsored by a grant from the U.S. National Science Foundation, the Brain Korea 21 Project, the Smart IT Convergence System Research Center funded by the Ministry of Education, Science and Technology as Global Frontier Project (SIRC-2011-0031851), and ICT at Seoul National University.

7. REFERENCES

[1] A. Shye, B. Scholbrock, and G. Memik, "Into the wild: Studying real user activity patterns to guide power optimizations for mobile architectures," *MICRO*, 2009.

[2] L. Zhang and et al., "Accurate online power estimation and automatic battery behavior based power model generation for smartphones," *CODES/ISSS*, 2010.

[3] M. Dong and L. Zhong, "Self-constructive high-rate system energy modeling for battery-powered mobile systems," *MobiSys*, 2011.

[4] A. Pathak, Y. C. Hu, M. Zhang, P. Bahl, and Y. Wang, "Fine-grained power modeling for smartphones using system call tracing," *EuroSys*, 2011.

[5] W. Yuan and K. Nahrstedt, "Energy-efficient soft real-time CPU scheduling for mobile multimedia systems," *SOSP*, 2003.

[6] D. Shin, Y. Kim, N. Chang, and M. Pedram, "Dynamic voltage scaling of oled displays," *DAC*, 2011.

[7] I. Choi, H. Shim, and N. Chang, "Low-power color TFT LCD display for hand-held embedded systems," *ISLPED*, 2002.

[8] Y. Choi, N. Chang, and T. Kim, "DC-DC converter-aware power management for low-power embedded systems," *IEEE T. on Computer-Aided Design of Integrated Circuits and Systems*, 2007.

[9] C. Shi, B. C. Walker, E. Zeisel, E. B. Hu, and G. H. McAllister, "A highly integrated power management IC for advanced mobile applications," *CICC*, 2006.

[10] B. Amelifard and M. Pedram, "Optimal design of the power-delivery network for multiple voltage-island system-on-chips," *IEEE T. on Computer-Aided Design of Integrated Circuits and Systems*, 2009.

[11] TexasInstruments, "Handset:smartphone," *http://www.ti.com/solution/handset_smartphone*.

[12] Qualcomm, "Snapdragon MDP MSM8660 datasheet," *https://developer.qualcomm.com/sites/default/files/snapdragon-mdp-8660.pdf*.

[13] J. Xiao, A. Peterchev, J. Zhang, and S. Sanders, "An ultra-low-power digitally-controlled buck converter IC for cellular phone applications," *APEC*, 2004.

[14] Y. Du, M. Wang, R. T. Meitl, S. Lukic, and A. Q. Huang, "High-frequency high-efficiency dc-dc converter for distributed energy storage modularization," *IECON*, 2010.

[15] F. Hossein, M. Ratul, K. Srikanth, L. Dimitrios, G. Ramesh, and E. Deborah, "Diversity in smartphone usage," *MobiSys*, 2010.

Dynamic Reconfiguration of Photovoltaic Energy Harvesting System in Hybrid Electric Vehicles

Yanzhi Wang[1], Xue Lin[1], Naehyuck Chang[2], and Massoud Pedram[1]
[1]Department of Electrical Engineering, University of Southern California, Los Angeles, CA, USA
[2]Seoul National University, Seoul, Korea
[1]{yanzhiwa, xuelin, pedram}@usc.edu, [2]{naehyuck}@elpl.snu.ac.kr

ABSTRACT

Photovoltaic (PV) energy harvesting system is a promising energy source for battery replenishment in hybrid electric vehicles (HEVs.) The PV cell array is installed on different parts of a vehicle body such as the engine hood, door panels, and the roof panel. Non-uniformity of the solar irradiance and temperature on the PV cell array is, however, a serious obstacle to efficient utilization of the PV system in HEVs because such variation, if not managed properly, can result in a significant degradation in the overall output power level of the PV system. This paper presents a dynamic PV array reconfiguration technique with structural support and a dynamic programming-based algorithm with polynomial time complexity to produce the near-optimal reconfiguration of the PV array on the HEV. The goal of this technique is to maximize the PV system output power under any solar irradiance and temperature distribution on the PV array. We demonstrate up to 6X improvement in the output power of a PV system against a conventional fixed configuration PV system.

Categories and Subject Descriptors

B.8.2 [**Performance and Reliability**]: Performance Analysis and Design Aids.

General Terms

Algorithms, Design, Management, Performance.

Keywords

Photovoltaic System, Hybrid Electric Vehicle, Photovoltaic Array Reconfiguration, Dynamic Programming.

1. INTRODUCTION

With growing public consciousness about the greenhouse gas emissions and increase in the price of petroleum, hybrid electric vehicles (HEVs) are perceived as a core segment of future automotive market. HEVs employ both electric motors and internal combustion engines to improve the fuel economy. Batteries powering the electric motor in a HEV must be recharged by a generator in the HEV. On top of a regenerative braking-based battery charging scheme, a photovoltaic (PV) system mounted on the HEV can enable battery charging whenever there is solar irradiance. PV cells can be mounted on the roof panel, engine hood, trunk, and door panels of a HEV.

PV systems have been widely deployed for many applications since solar energy is abundant, renewable, and clean. A PV system consists of a PV array and peripheral charging regulation circuitry (e.g., the charger). A PV array itself comprises multiple PV cells connected in a series-parallel configuration. Ideally, all PV cells in the array experience identical operational conditions and exhibit the same voltage-current (V-I) and voltage-power (V-P) output characteristics. Thus, they can simultaneously operate at their *maximum power points* (MPPs.) We can achieve the maximum output power of the PV array with a *MPP tracking* (MPPT) method [1][2]. Furthermore, the recently proposed *maximum power transfer tracking* (MPTT) technique can guarantee the maximum output power of the PV system under non-ideal behaviors of the charger [3]. In reality, PV cells may receive different solar irradiances due to *partial shading* and variations in the *solar irradiance incidence angle*. In addition, PV cells may experience different temperatures. For instance, PV cells on the door panel have a different solar irradiance incidence angle from those on the roof. PV cells on top of the hood have a higher temperature than PV cells on the roof.

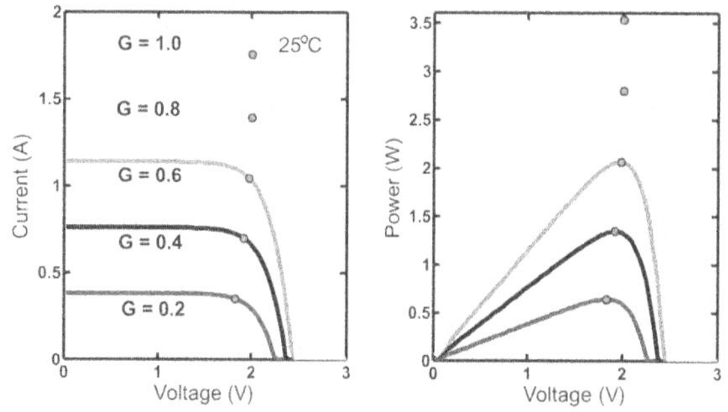

Figure 1. V-I and V-P characteristics of a PV cell under different solar irradiances.

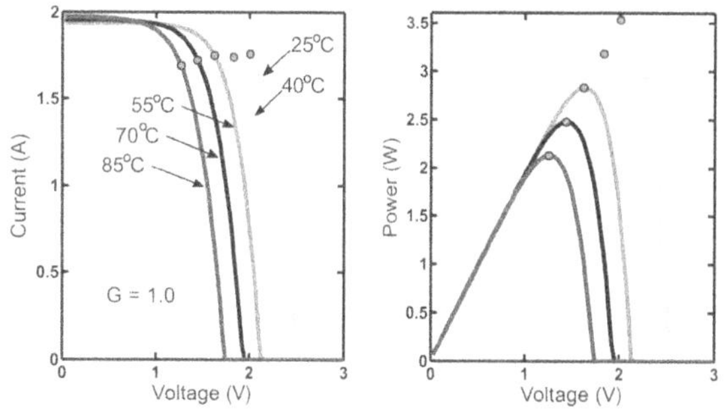

Figure 2. V-I and V-P characteristics of a PV cell under different temperatures.

Figures 1 and 2, where MPPs of a PV cell are marked by red dots, illustrate that a higher solar irradiance or a lower temperature lead to a higher output power at the MPP of the cell. Partial shading not only reduces the maximum output power of the shaded PV cell, but also makes non-shaded PV cells that are connected in series with the shaded one deviate from their MPPs.

Similarly, a hotter PV cell makes PV cells that are connected in parallel with it deviate from their MPPs. Thus PV cells in a PV array with solar irradiance and/or temperature non-uniformity cannot operate at their MPPs simultaneously, and thus, the PV array suffers from significant power inefficiency. In addition, the V-P characteristic curve of such a PV array shows multiple peaks. Therefore, the existing MPPT or MPTT techniques, which are based on the unimodality assumption of V-P characteristics, have to be modified to find the global optimal operating point of the PV array [4][5]. This fact increases the complexity of the PV system controller. Some PV array reconfiguration techniques have been proposed [6][7][8] to solve the PV system power loss problem due to solar irradiance and temperature non-uniformity phenomenon. These techniques suffer from the following shortcomings: i) they require extra PV cells to perform reconfiguration; ii) a systematic and scalable reconfiguration control mechanism is lacking; and iii) they do not consider charger efficiency variation in the PV system.

In this paper, we present a dynamic PV array reconfiguration technique, which updates the PV array configuration according to the change of irradiance and temperature distribution on the PV array during system operation in order to enhance the output power of PV systems. Our reconfiguration technique has the ability to simultaneously meet (to the extent that is possible) the MPPs for all PV cells even under non-uniform solar irradiance and temperature distributions. We provide both a scalable structural support and an effective control algorithm. The control algorithm realizes near-optimal PV array reconfiguration accounting for solar irradiance and temperature distributions and charger efficiency variation. The algorithm is based on dynamic programming with polynomial time complexity so that it can be integrated into PV systems with insignificant computational overhead. Experimental results demonstrate that our proposed dynamic reconfiguration technique results in up to 6X improvement in PV system output power compared with a conventional fixed array structure PV system.

2. PV SYSTEM MODEL AND RECONFIGURATION

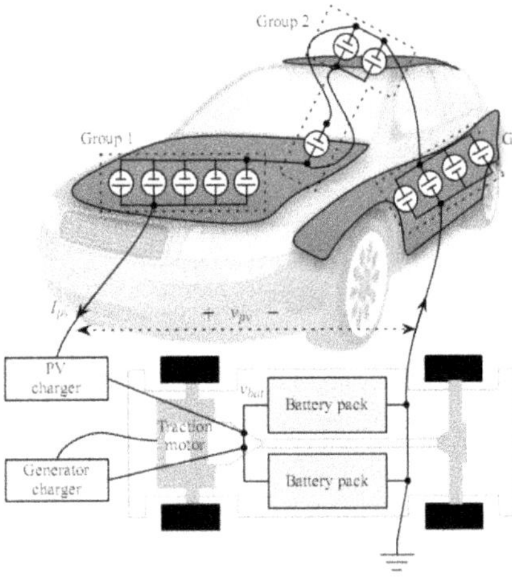

Figure 3. System diagram of a PV-powered HEV.

Figure 3 shows the system diagram of a PV-powered HEV. The PV array consists of a number of PV cells, mounted on the roof panel, engine hood, trunk, and door panels of the HEV. Conventional HEVs are equipped with battery packs as the energy storage system, connected to the traction motor via the generator charger. The PV-powered HEV also has an add-on PV system that consists of a PV array and a PV charger connecting the PV array to battery packs. The generator charger and PV charger can simultaneously charge battery packs. MPPT or MPTT techniques are integrated with the PV charger, and therefore, the PV charger can set the optimal operating point for the PV array. The PV array consists of a number of series-connected PV cell groups where each PV group consists of (potentially a different number of) PV cells connected in parallel. Although mounting positions of PV cells are fixed on the HEV, PV array reconfiguration technique adaptively changes the electric connection of PV cells in response to changes in the environmental condition. Figure 4 provides the equivalent circuit of the PV system shown in Figure 3, where the battery bank in Figure 4 represents the battery packs on HEV.

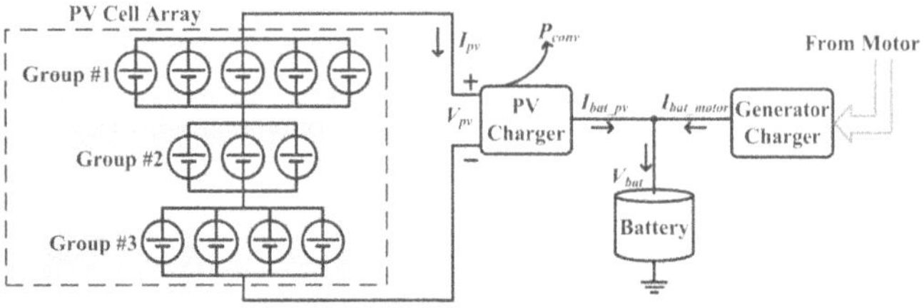

Figure 4. Equivalent circuit of a PV system on a HEV.

2.1 Component Models

It is necessary to have accurate component models, including models for the PV cell, battery and charger, in order to develop an effective PV array reconfiguration algorithm. Given the solar irradiance G and temperature T of a PV cell, the PV cell voltage V_{pv}^{c} and current I_{pv}^{c} satisfy certain V-I characteristics as shown in Figures 1 and 2. The MPP (maximum power point) that maximizes the PV cell output power $V_{pv}^{c} \cdot I_{pv}^{c}$ lies on the V-I curve. In terms of the notation, we have: $V_{pv}^{c} = V_{pv}^{c,MPP}$ and $I_{pv}^{c} = I_{pv}^{c,MPP}$ at the MPP. We obtain the V-I characteristics and MPP of a PV cell under any (G, T) using the PV cell model of reference [10]. We adopt the battery and charger models from [11]. Generally speaking, the power conversion efficiency of a charger is a function of its input and output voltages and currents. The conversion efficiency is maximized when the charger output current is within a small operation range, and its input and output voltages are close to each other.

2.2 Reconfiguration Structure for PV Array

A conventional PV array consists of n series-connected PV groups, whereby each PV group has exactly m parallel-connected PV cells, (all PV groups have the same number of PV cells.) In contrast, the PV array in Figure 4 has three series-connected PV cell groups, and there are different numbers of parallel-connected PV cells within the groups, which is an imbalanced configuration. Reference [9] presented a structural support mechanism for balanced reconfiguration of supercapacitor banks. In this paper, we exploit the structural support in [9] for the PV array. However, we allow more flexible configurations (both balanced and imbalanced) of PV cells in the PV array.

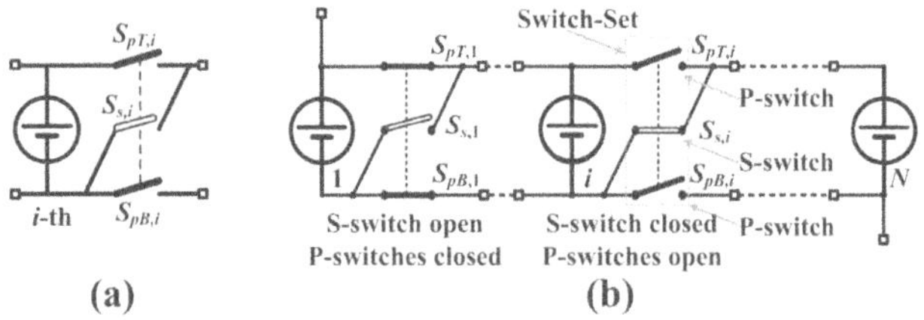

Figure 5. Reconfiguration structure.

We demonstrate the possibility of creating an imbalanced reconfiguration using the reconfiguration structure of [9]. The PV cells are integrated with a series switch S_s and a pair of parallel switches S_{pT} and S_{pB}, as shown in Figure 5 (a). The PV array has a reconfiguration structure as shown in Figure 5 (b), where the PV array consists of N PV cells and all the PV cells except for the last one are integrated with switches. We can realize PV array reconfiguration by controlling the states of switches. For any PV cell, its two parallel switches are always in the same state, so we use $S_{p,i}$ to denote the pair of parallel switches in i-th PV cell of the array. $S_{p,i}$ is used for connecting the PV cells in parallel, and $S_{s,i}$ is used for connecting the PV cell groups in series. If $S_{p,i}$ is on, $S_{s,i}$ must be off; if $S_{s,i}$ is on then $S_{p,i}$ must be off. Figure 6 shows the states of switches to realize the imbalanced configuration shown in Figure 4, where group #1 is formed by turning on $S_{p,1}$ through $S_{p,4}$ to form a group of five parallel connected PV cells whereas groups #1 through #3 are series-connected by turning on $S_{s,5}$ and $S_{s,8}$.

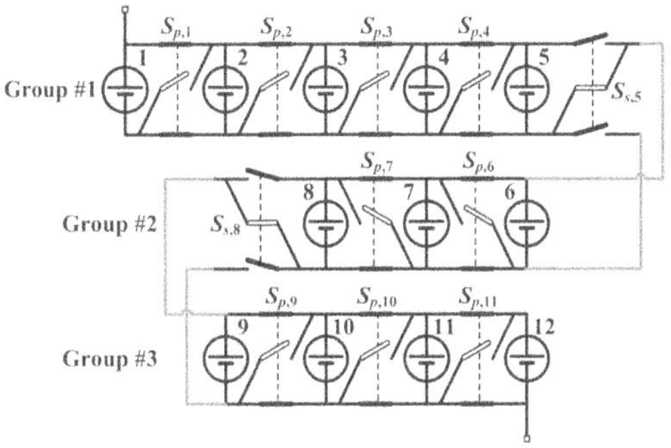

Figure 6. Example of imbalanced configuration.

In the PV array configurations achieved by the reconfiguration structure, the number of groups n in the PV array can be any positive integer less than or equal to N. The number of parallel-connected PV cells $m_j (> 0)$ in each group should satisfy

$$\sum_{j=1}^{n} m_j = N. \quad (1)$$

We denote such a configuration as $\mathcal{C}(n;\ m_1, m_2, \cdots, m_n)$, which could be viewed as a partition of the PV cell index set $\boldsymbol{A} = \{1, 2, \cdots, N\}$, where integers in $\boldsymbol{A}$ denote the indices of the PV cells in the PV array. The partition is denoted by n subsets $\boldsymbol{B}_1$, $\boldsymbol{B}_2$, $\cdots$, and $\boldsymbol{B}_n$, which correspond to the n groups consisting of m_1, m_2, $\cdots$, and m_n PV cells, respectively. The subsets $\boldsymbol{B}_1$, $\boldsymbol{B}_2$, $\cdots$, and $\boldsymbol{B}_n$ satisfy

$$\bigcup_{j=1}^{n} \boldsymbol{B}_j = \boldsymbol{A}, \quad (2)$$

and

$$\boldsymbol{B}_j \cap \boldsymbol{B}_k = \emptyset, \text{ for } \forall\, j, k \in \{1, 2, \cdots, n\} \text{ and } j \neq k. \quad (3)$$

Furthermore, the integer values in subset $\boldsymbol{B}_j$ must be smaller than the integer values in subset $\boldsymbol{B}_k$, for any $1 \le j < k \le n$, due to the structural characteristics of the proposed reconfiguration technique. A partitioning satisfying the above properties is called an *alphabetical partitioning*.

3. PROBLEM FORMULATION

3.1 Problem Statement

The solar irradiance distribution over the PV array on the HEV changes over time, due to driving direction change and shading caused by buildings, clouds, and other moving vehicles. Temperature variation is also severe because of the engine heat dissipation and convection due to airflow. We should perform the PV array reconfiguration algorithm at every decision epoch t to maximize the PV system output power. For notational convenience, we drop the time index t. We denote the solar irradiance and PV cell temperature of the i-th $(1 \le i \le N)$ PV cell by G_i and T_i, respectively. We obtain G_i and T_i values using irradiance and temperature sensors in the system. We denote the output voltage and current of the i-th PV cell by $V_{pv,i}^{c}$ and $I_{pv,i}^{c}$, respectively. The PV array configuration $\mathcal{C}$ is the major control variable of the reconfiguration algorithm.

We develop V-I characteristic of the PV array with a configuration $\mathcal{C}(n;\ m_1, m_2, \cdots, m_n)$ that corresponds to the alphabetical partitioning $\boldsymbol{B}_1$, $\boldsymbol{B}_2$, $\cdots$, and $\boldsymbol{B}_n$ as follows. We use $V_{pv,j}^{g}$ to denote the output voltage of the j-th PV group, and use V_{pv} and I_{pv} to denote the output voltage and current of the PV array, respectively. The V_{pv} and I_{pv} values can be effectively controlled by the PV charger. From the Kirchhoff's laws, we have:

$$I_{pv} = \sum_{i \in \boldsymbol{B}_j} I_{pv,i}^{c}, \ \forall\, j \in \{1, 2, \cdots, n\}, \quad (4)$$

$$V_{pv,j}^{g} = V_{pv,i}^{c}, \text{ for } \forall\, i \in \boldsymbol{B}_j \text{ and } j \in \{1, 2, \cdots, n\}, \quad (5)$$

and

$$V_{pv} = \sum_{j=1}^{n} V_{pv,j}^{g}. \quad (6)$$

We control and maintain the PV array operating point (V_{pv}, I_{pv}) using the PV charger. The power loss of the PV charger, P_{conv}, is calculated from its input voltage, input current, output voltage, and output current, i.e., V_{pv}, I_{pv}, V_{bat}, and I_{bat_pv}, respectively [11]. Due to the energy conservation law, we have:

$$V_{pv} \cdot I_{pv} = P_{conv} + V_{bat} \cdot I_{bat\_pv}, \quad (7)$$

where V_{bat} and I_{bat_pv} are the battery's terminal voltage and charging current from the PV system, respectively. The formal problem statement of the PV array reconfiguration (PAR) problem for HEV is described as follows:

PAR Problem Statement: Given the solar irradiance G_i and the temperature T_i of each i-th $(1 \le i \le N)$ PV cell in the PV array, and the battery terminal voltage V_{bat}, **find** the optimal PV array configuration $\mathcal{C}$ and the PV array operating point (V_{pv}, I_{pv}), **such that** the PV charger output current I_{bat_pv} is maximized.

3.2 Problem Decomposition

We propose a near-optimal solution to the PAR problem using a novel method that consists of a kernel algorithm, an MPTT control algorithm, and an outer loop finding the optimal number of PV groups in the PV array. The kernel algorithm finds the optimal PV array configuration such that all PV cells can work near their MPPs. The kernel algorithm derives the optimal PV array configuration $\mathcal{C}(n;\ m_1, m_2, \cdots, m_n)$ for a given n value to maximize the PV array MPP output power. The MPP output power of the PV array is the maximum achievable $V_{pv} \cdot I_{pv}$ value under a given solar irradiance and temperature distribution on the PV array, and a given PV array configuration. The subsequent MPTT control algorithm aims to find the optimal PV array operating point (V_{pv}, I_{pv}) taking into account the PV charger efficiency variation, such that I_{bat_pv} is maximized. We use the optimal PV array configuration $\mathcal{C}$ obtained via the kernel algorithm for the subsequent MPTT control algorithm. The outer loop enumerates all the possible n values $(1 \le n \le N)$ and finds the optimal one. The outline of this method is given in Algorithm 1. The next section describes the kernel algorithm in detail, which is the key part in Algorithm 1.

Algorithm 1: Near-Optimal Solution of the PAR Problem.

For each possible n value $(1 \le n \le N)$:

> Run the **kernel algorithm** to find the optimal PV array configuration $\mathcal{C}(n;\ m_1, m_2, \cdots, m_n)$, such that the PV array MPP output power is maximized.
>
> Run the **MPTT control algorithm** to find the optimal PV array operating point (V_{pv}, I_{pv}) that maximizes I_{bat_pv}, under the obtained optimal PV array configuration $\mathcal{C}(n;\ m_1, m_2, \cdots, m_n.)$
>
> The maximal PV charger output current is denoted by $I_{bat\_pv,n}^{max}$.

Find the optimal n value that maximizes $I_{bat\_pv,n}^{max}$, and the corresponding optimal PV array configuration $\mathcal{C}(n;\ m_1, m_2, \cdots, m_n)$ and PV array operating point (V_{pv}, I_{pv}).

4. THE KERNEL ALGORITHM

The maximum solar power that can be drawn from a PV array is the sum of MPP output power of all the PV cells in that array. Under the non-uniformity of solar irradiance and temperature, PV cells in a balanced *n*-by-*m* configuration can hardly operate at their MPPs simultaneously. Our reconfiguration technique aims at making all the PV cells operate at, or close to, their MPPs by updating the configuration of the PV array according to current solar irradiance and temperature distributions. In Figure 7, each PV cell is labeled with its MPP output voltage and current. If the PV array employs the $\mathcal{C}(3; 5, 3, 4)$ configuration as shown in Figure 7, all PV cells work at their MPPs simultaneously with the PV array output voltage of 3.6 V (1.5+0.9+1.2=3.6 V) and output current of 4.5 A (1.5+1.5+1.5=4.5 A). The goal of our kernel algorithm is to find a configuration for the PV array such that (i) PV cells within the same PV group have very close MPP output voltages and (ii) the sum of MPP output currents of PV cells within the same group is quite similar among all the PV groups.

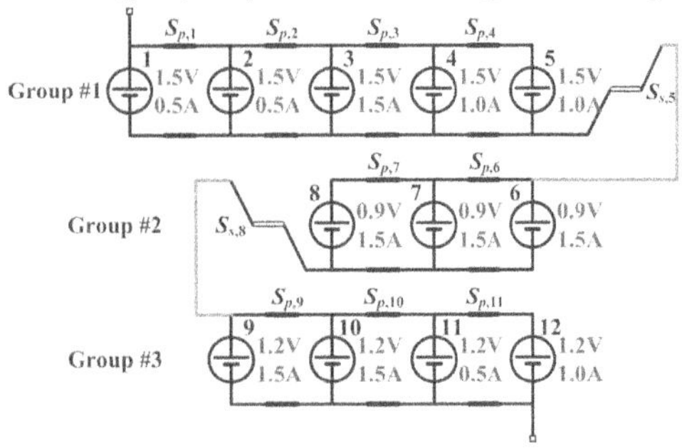

Figure 7. An illustration of the PV array reconfiguration.

4.1 Simplifying the Problem

The kernel algorithm finds the optimal PV array configuration to maximize the PV array MPP output power for the specified number of PV groups in the array. We name the problem *simplified PAR* (SPAR) problem. We further simplify the SPAR problem assuming *ideal PV cells*, and name the new problem *ideal PV Cell-based SPAR* (IC-SPAR) problem. The V-I characteristic of an ideal PV cell is a step function such that

$$I_{pv,i}^{c} = \begin{cases} I_{pv,i}^{c,MPP}, & \text{if } V_{pv,i}^{c} \leq V_{pv,i}^{c,MPP} \\ 0, & \text{otherwise} \end{cases}, \quad (8)$$

where $V_{pv,i}^{c,MPP}$ and $I_{pv,i}^{c,MPP}$ denote the MPP output voltage and current of the *i*-th PV cell in the PV array, which are calculated from the solar irradiance G_i and temperature T_i. The ideal PV cell is an effective approximation of the real PV cell. We use $V_{pv,\text{IC}}^{MPP}(\mathcal{C})$, $I_{pv,\text{IC}}^{MPP}(\mathcal{C})$, and $P_{pv,\text{IC}}^{MPP}(\mathcal{C})$ to denote the MPP output voltage, current, and power, respectively, of the ideal PV cell array with a configuration of $\mathcal{C}(n; m_1, m_2, \cdots, m_n)$. These values are calculated as:

$$V_{pv,\text{IC}}^{MPP}(\mathcal{C}) = \sum_{j=1}^{n} \min_{i\in \boldsymbol{B}_j} V_{pv,i}^{c,MPP}, \quad (9)$$

$$I_{pv,\text{IC}}^{MPP}(\mathcal{C}) = \min_{1\leq j\leq n} \sum_{i\in \boldsymbol{B}_j} I_{pv,i}^{c,MPP}, \quad (10)$$

$$P_{pv,\text{IC}}^{MPP}(\mathcal{C}) = \left(\sum_{j=1}^{n} \min_{i\in \boldsymbol{B}_j} V_{pv,i}^{c,MPP}\right) \cdot \left(\min_{1\leq j\leq n} \sum_{i\in \boldsymbol{B}_j} I_{pv,i}^{c,MPP}\right). \quad (11)$$

Thus the IC-SPAR problem becomes: find the optimal array configuration $\mathcal{C}(n; m_1, m_2, \cdots, m_n)$ (corresponding to the optimal alphabetical partitioning $\boldsymbol{B}_1, \boldsymbol{B}_2, \cdots$, and $\boldsymbol{B}_n$) for a given n value, such that the $P_{pv,\text{IC}}^{MPP}(\mathcal{C})$ value calculated from (11) is maximized.

4.2 The Pareto-Optimal Substructure

We consider a reduced-size PV array consisting of the first L ($L \leq N$) PV cells of the original N-cell PV array. The reduced-size L-cell PV array is composed of l ($l \leq n$) PV groups. We denote a possible configuration of the L-cell PV array by $\mathcal{C}^{L,l}$. Finding the optimal L-cell PV array configuration is equivalent to finding the optimal alphabetical partitioning $\boldsymbol{B}_1^{L,l}, \boldsymbol{B}_2^{L,l}, \cdots, \boldsymbol{B}_l^{L,l}$ of the set $\boldsymbol{A}^L = \{1, 2, \cdots, L\}$. The configuration is optimal when the MPP output power $P_{pv,\text{IC}}^{MPP}(\mathcal{C}^{L,l}) = \left(\sum_{j=1}^{l} \min_{i\in \boldsymbol{B}_j^{L,l}} V_{pv,i}^{c,MPP}\right) \cdot \left(\min_{1\leq j\leq l} \sum_{i\in \boldsymbol{B}_j^{L,l}} I_{pv,i}^{c,MPP}\right)$ is maximized. We use $V_{pv,\text{IC}}^{MPP}(\mathcal{C}^{L,l}) = \sum_{j=1}^{l} \min_{i\in \boldsymbol{B}_j^{L,l}} V_{pv,i}^{c,MPP}$ and $I_{pv,\text{IC}}^{MPP}(\mathcal{C}^{L,l}) = \min_{1\leq j\leq l} \sum_{i\in \boldsymbol{B}_j^{L,l}} I_{pv,i}^{c,MPP}$ to denote the MPP output voltage and current of the PV array with configuration $\mathcal{C}^{L,l}$, respectively. We name the new problem the (L, l) *reconfiguration problem*. The more general (L, l) reconfiguration problem becomes the original IC-SPAR problem when $L = N$ and $l = n$. We find the Pareto-optimal substructure property of this (L, l) reconfiguration problem as follows.

Definition 1 (*Pareto superiority*): Consider two configurations $\mathcal{C}^{L,l}$ and $\hat{\mathcal{C}}^{L,l}$ of the (L, l) reconfiguration problem. Configuration $\mathcal{C}^{L,l}$ is Pareto-superior to $\hat{\mathcal{C}}^{L,l}$ if $V_{pv,\text{IC}}^{MPP}(\mathcal{C}^{L,l}) > V_{pv,\text{IC}}^{MPP}(\hat{\mathcal{C}}^{L,l})$ and $I_{pv,\text{IC}}^{MPP}(\mathcal{C}^{L,l}) \geq I_{pv,\text{IC}}^{MPP}(\hat{\mathcal{C}}^{L,l})$, or $V_{pv,\text{IC}}^{MPP}(\mathcal{C}^{L,l}) \geq V_{pv,\text{IC}}^{MPP}(\hat{\mathcal{C}}^{L,l})$ and $I_{pv,\text{IC}}^{MPP}(\mathcal{C}^{L,l}) > I_{pv,\text{IC}}^{MPP}(\hat{\mathcal{C}}^{L,l})$.

Definition 2 (*Pareto-optimal configuration* and *Pareto-optimal set*): A configuration $\mathcal{C}^{L,l}$ is a *Pareto-optimal configuration* of the (L, l) reconfiguration problem if no other configuration of the (L, l) reconfiguration problem is Pareto superior to $\mathcal{C}^{L,l}$. The set of all Pareto-optimal configurations of the (L, l) reconfiguration problem is the *Pareto-optimal* (*configuration*) *set* of the problem.

Next we present two theorems. Theorem 1 demonstrates that the optimal configuration of each (L, l) reconfiguration problem is an element in the Pareto-optimal set of that problem, whereas Theorem 2 provides the Pareto-optimal substructure property. Proofs are omitted due to limited space.

Theorem 1 (Optimal configuration and Pareto-optimal configurations): The optimal configuration $\mathcal{C}^{L,l}$ (with the maximum MPP output power $P_{pv,\text{IC}}^{MPP}(\mathcal{C}^{L,l})$) of the (L, l) reconfiguration problem is an element in the Pareto-optimal set of the (L, l) reconfiguration problem.

Theorem 2 (The Pareto-optimal substructure): Suppose that a configuration $\mathcal{C}^{L,l}$, corresponding to an alphabetical partitioning $\boldsymbol{B}_1^{L,l}, \boldsymbol{B}_2^{L,l}, \cdots, \boldsymbol{B}_{l-1}^{L,l}, \boldsymbol{B}_l^{L,l}$, is a Pareto-optimal configuration of the (L, l) reconfiguration problem, and the last (the l-th) PV group consists of $m_l^{L,l}$ cells. Then the alphabetical partitioning $\boldsymbol{B}_1^{L,l}, \boldsymbol{B}_2^{L,l}, \cdots, \boldsymbol{B}_{l-1}^{L,l}$ (or the corresponding configuration) must also be Pareto-optimal in the $(L - m_l^{L,l}, l - 1)$ reconfiguration problem.

4.3 The MCHV Function

Based on the above-described Pareto-optimal substructure property, one can find the optimal solution of the original IC-SPAR problem by storing the complete Pareto-optimal set of each (L, l) reconfiguration problem with $L \leq N$ and $l \leq n$, in a way similar to the conventional dynamic programming. However, cardinality of the Pareto-optimal set of the (L, l) reconfiguration problem could be in the order of $L!$, which implies that the Pareto-optimal set cardinality could be exponential. Hence, it is impractical to store the whole Pareto-optimal set of each (L, l) reconfiguration problem to solve the original IC-SPAR problem.

For effectively solving the IC-SPAR problem, we introduce a new function, named *maximal_current_higher_voltage* (MCHV) function for each (L, l) reconfiguration problem, denoted by $f_{MCHV}^{L,l}(V)$. The MCHV function $f_{MCHV}^{L,l}(V)$ derives the maximum $I_{pv,\text{IC}}^{MPP}(\mathcal{C}^{L,l})$ value achieved by any configuration $\mathcal{C}^{L,l}$ of the (L, l) reconfiguration problem satisfying $V_{pv,\text{IC}}^{MPP}(\mathcal{C}^{L,l}) \geq V$, i.e.,:

$$f_{MCHV}^{L,l}(V) = \max_{\mathcal{C}^{L,l}:\, V_{pv,\text{IC}}^{MPP}(\mathcal{C}^{L,l}) \geq V} I_{pv,\text{IC}}^{MPP}(\mathcal{C}^{L,l}). \quad (12)$$

If there exists no configuration $\mathcal{C}^{L,l}$ satisfying $V_{pv,\text{IC}}^{MPP}(\mathcal{C}^{L,l}) \geq V$ for relatively large V values, we have $f_{MCHV}^{L,l}(V) = 0$. The domain of the MCHV function for the (L,l) reconfiguration problem is given by $[\hat{V}_{pv,\text{IC},min}^{MPP,L,l}, \hat{V}_{pv,\text{IC},max}^{MPP,L,l}]$, where $\hat{V}_{pv,\text{IC},min}^{MPP,L,l}$ and $\hat{V}_{pv,\text{IC},max}^{MPP,L,l}$ values are the estimated lower and upper bounds of the $V_{pv,\text{IC}}^{MPP}(\mathcal{C}^{L,l})$ value for any configuration $\mathcal{C}^{L,l}$, respectively. The $\hat{V}_{pv,\text{IC},min}^{MPP,L,l}$ and $\hat{V}_{pv,\text{IC},max}^{MPP,L,l}$ values are calculated by

$$\hat{V}_{pv,\text{IC},min}^{MPP,L,l} = l \cdot \min_{1\leq i\leq L} V_{pv,i}^{c,MPP}, \quad (13)$$

$$\hat{V}_{pv,\text{IC},max}^{MPP,L,l} = l \cdot \max_{1\leq i\leq L} V_{pv,i}^{c,MPP}. \quad (14)$$

Finding tighter upper and lower bounds is an interesting problem, but it is outside the scope of the present paper. An illustration of the MCHV function is shown in Figure 8.

The MCHV function is a non-increasing function with a set of points of discontinuity. Each point of discontinuity in the MCHV function corresponds to a Pareto-optimal configuration of the (L,l) reconfiguration problem. Point $(V_a, f_{MCHV}^{L,l}(V_a))$ shown in Figure 8 corresponds to a Pareto-optimal configuration with $V_{pv,\text{IC}}^{MPP}(\mathcal{C}^{L,l}) = V_a$ and $I_{pv,\text{IC}}^{MPP}(\mathcal{C}^{L,l}) = f_{MCHV}^{L,l}(V_a)$, since we have $I_{pv,\text{IC}}^{MPP}(\mathcal{C}^{L,l}) < f_{MCHV}^{L,l}(V_a)$ for any configuration of the (L,l) reconfiguration problem satisfying $V_{pv,\text{IC}}^{MPP}(\mathcal{C}^{L,l}) > V_a$. Finally, the MCHV function of each (L,l) reconfiguration problem fully represents the Pareto-optimal set of that problem.

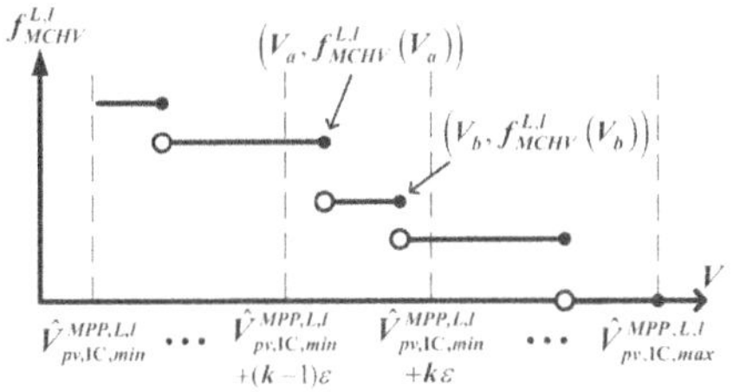

Figure 8. An illustration of the MCHV function.

4.4 Implementation of Kernel Algorithm

In the MCHV function for an (L,l) reconfiguration problem, the total number of points of discontinuity could also be exponential. Hence, we represent the MCHV function in discrete domain. We define a three-dimensional matrix **F_MCHV**, where **F_MCHV**(L,l,k) $(1 \leq k \leq K+1)$ stores the $f_{MCHV}^{L,l}(\hat{V}_{pv,\text{IC},min}^{MPP,L,l} + (k-1)\varepsilon)$ value. In the above expression, $\varepsilon = \frac{\hat{V}_{pv,\text{IC},max}^{MPP,L,l} - \hat{V}_{pv,\text{IC},min}^{MPP,L,l}}{K}$, and K is the predefined precision level in the kernel algorithm. When $K \to \infty$, the array **F_MCHV**(L,l,k) $(1 \leq k \leq K+1)$ fully represents the MCHV function $f_{MCHV}^{L,l}(V)$, which is however not practical since the size of **F_MCHV** is infinite. When K is finite, the Pareto-optimal configurations of the (L,l) reconfiguration problem can be classified into two categories: *observable* and *unobservable* Pareto-optimal configurations. Consider a Pareto-optimal configuration $\mathcal{C}^{L,l}$ of the (L,l) reconfiguration problem satisfying $\hat{V}_{pv,\text{IC},min}^{MPP,L,l} + (k-1)\varepsilon \leq V_{pv,\text{IC}}^{MPP}(\mathcal{C}^{L,l}) < \hat{V}_{pv,\text{IC},min}^{MPP,L,l} + k\varepsilon$. Then $\mathcal{C}^{L,l}$ is *observable* if **F_MCHV**$(L,l,k) = I_{pv,\text{IC}}^{MPP}(\mathcal{C}^{L,l})$, and is *unobservable* otherwise (i.e., **F_MCHV**$(L,l,k) > I_{pv,\text{IC}}^{MPP}(\mathcal{C}^{L,l})$.) An example of observable Pareto-optimal configuration is the one corresponding to the point $(V_a, f_{MCHV}^{L,l}(V_a))$ in Figure 8, and an example of unobservable Pareto-optimal configuration is the one corresponding to the point $(V_b, f_{MCHV}^{L,l}(V_b))$.) The following lemma and theorem prove that we can find an approximate solution of the IC-SPAR problem by considering only the observable Pareto-optimal configurations in the kernel algorithm implementation.

Lemma 1 (Bound on MPP output power of unobservable Pareto-optimal configurations): For each unobservable Pareto-optimal configuration $\mathcal{C}^{L,l}$ of the (L,l) reconfiguration problem satisfying $\hat{V}_{pv,\text{IC},min}^{MPP,L,l} + (k-1)\varepsilon \leq V_{pv,\text{IC}}^{MPP}(\mathcal{C}^{L,l}) < \hat{V}_{pv,\text{IC},min}^{MPP,L,l} + k\varepsilon$, there exists a corresponding observable Pareto-optimal configuration $\tilde{\mathcal{C}}^{L,l}$ with $\hat{V}_{pv,\text{IC},min}^{MPP,L,l} + (k-1)\varepsilon \leq V_{pv,\text{IC}}^{MPP}(\tilde{\mathcal{C}}^{L,l}) < V_{pv,\text{IC}}^{MPP}(\mathcal{C}^{L,l})$ and **F_MCHV**$(L,l,k) = I_{pv,\text{IC}}^{MPP}(\tilde{\mathcal{C}}^{L,l}) > I_{pv,\text{IC}}^{MPP}(\mathcal{C}^{L,l})$. Besides, the MPP output power $P_{pv,\text{IC}}^{MPP}(\mathcal{C}^{L,l})$ of the unobservable Pareto-optimal configuration $\mathcal{C}^{L,l}$ is bounded as follows:

$$P_{pv,\text{IC}}^{MPP}(\mathcal{C}^{L,l}) \leq P_{pv,\text{IC}}^{MPP}(\tilde{\mathcal{C}}^{L,l}) + \varepsilon \cdot I_{pv,\text{IC}}^{MPP}(\tilde{\mathcal{C}}^{L,l}). \quad (15)$$

Theorem 3 (Approximate solution): We use $\mathcal{C}^{near\_opt}$ to denote the near-optimal solution (configuration) of the IC-SPAR problem obtained by considering only the observable Pareto-optimal configurations in the kernel algorithm implementation, with details shown in Algorithm 2. We use $\mathcal{C}^{opt}$ to denote the true optimal solution (configuration) of the IC-SPAR problem. Then the MPP output power $P_{pv,\text{IC}}^{MPP}(\mathcal{C}^{near\_opt})$ of the near-optimal configuration $\mathcal{C}^{near\_opt}$ is bounded as follows:

$$P_{pv,\text{IC}}^{MPP}(\mathcal{C}^{near\_opt}) \geq P_{pv,\text{IC}}^{MPP}(\mathcal{C}^{opt}) \cdot (1 - B(K)), \quad (16)$$

where $B(K)$ is a function of the predefined parameter K and converges to 0 when $K \to \infty$.

Since the number of observable Pareto-optimal configurations for each (L,l) reconfiguration problem is bounded by K, practical implementation of the kernel algorithm is feasible. Details of the proposed dynamic programming based kernel algorithm to solve the IC-SPAR problem are provided in Algorithm 2. The solution of the IC-SPAR problem serves as the near-optimal solution of the SPAR problem, and the original PAR problem can be further solved using Algorithm 1.

Algorithm 2: The Kernel Algorithm.

Maintain three-dimensional matrices **F_MCHV**, **V_Pareto**, **Last_Par_1** and **Last_Par_2**. Each has N rows, n columns and $K+1$ layers.

Initialization:

- **F_MCHV**$(L,1,k) \leftarrow \sum_{1\leq i\leq L} I_{pv,i}^{c,MPP}$ and **V_Pareto**$(L,1,k) \leftarrow \min_{1\leq i\leq L} V_{pv,i}^{c,MPP}$ for $1 \leq L \leq N$ and $1 \leq k \leq K+1$.
- The rest of entries in all four matrices **F_MCHV**, **V_Pareto**, **Last_Par_1** and **Last_Par_2** are initialized to be zero.

For l from 2 to n:
- **For** L from l to N:
  - Calculate $\hat{V}_{pv,\text{IC},min}^{MPP,L,l}$ and $\hat{V}_{pv,\text{IC},max}^{MPP,L,l}$ values using (13) and (14).
  - Calculate ε using $\varepsilon = \frac{\hat{V}_{pv,\text{IC},max}^{MPP,L,l} - \hat{V}_{pv,\text{IC},min}^{MPP,L,l}}{K}$.
  - **For** k from 1 to $K+1$:
    - **F_MCHV**$(L,l,k) \leftarrow \max_{(l-1\leq L_1<L,\ 1\leq k_1\leq K+1)} I_{temp}(L_1,k_1)$, s.t. $V_{temp}(L_1,k_1) \geq \hat{V}_{pv,\text{IC},min}^{MPP,L,l} + (k-1)\varepsilon$.
    - In the above expressions:
      $V_{temp}(L_1,k_1) =$ **V_Pareto**$(L_1,l-1,k_1) + \min_{L_1<i\leq L} V_{pv,i}^{c,MPP}$,
      $I_{temp}(L_1,k_1) = \min\{$**F_MCHV**$(L_1,l-1,k_1), \sum_{L_1<i\leq L} I_{pv,i}^{c,MPP}\}$.
    - **If F_MCHV**$(L,l,k) > 0$:
      - $(L_{opt},k_{opt}) \leftarrow \text{argmax}_{(l-1\leq L_1<L,\ 1\leq k_1\leq K+1)} I_{temp}(L_1,k_1)$, s.t. $V_{temp}(L_1,k_1) \geq \hat{V}_{pv,\text{IC},min}^{MPP,L,l} + (k-1)\varepsilon$.
      - $($**Last_Par_1**(L,l,k), **Last_Par_2**$(L,l,k)) \leftarrow (L_{opt},k_{opt})$.
      - **V_Pareto**$(L,l,k) \leftarrow V_{temp}(L_{opt},k_{opt})$.

The near-optimal MPP output power of the PV array can be calculated by $\max_{1\leq k\leq K+1}\{$**V_Pareto**$(N,n,k) \cdot$ **F_MCHV**$(N,n,k)\}$.

Trace back the three-dimensional matrices **Last_Par_1** and **Last_Par_2** to find the near-optimal configuration $\mathcal{C}^{near\_opt}$ of the PV array.

---

5. Experimental Results

We compare the performances of two PV systems: one is the proposed system with reconfiguration technique, and the other is the baseline system without reconfiguration technique. Both PV systems contain a PV array of 60 PV cells, a charger and a battery bank. The PV array in the proposed system has the reconfiguration structure shown in Figure 5 while the PV array in the baseline system has a fixed 6x10 configuration, where 6 PV cell groups are connected in series and each PV group consists of 10 parallel-connected PV cells. The output power of baseline system is enhanced by the improved MPPT technique [4] and the integration of a bypass diode within each PV cell [6].

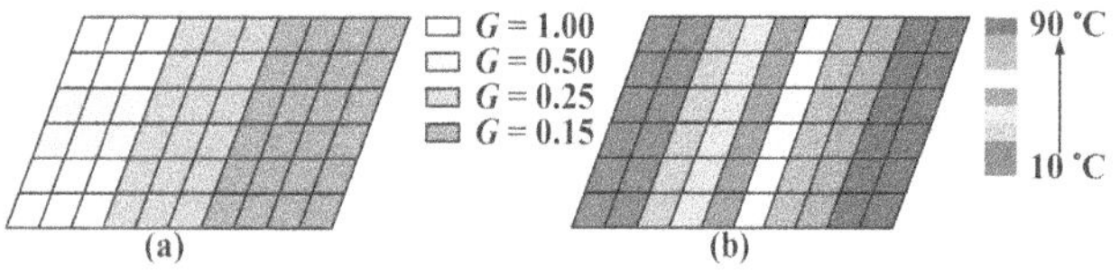

Figure 9. (a) Solar irradiance and (b) temperature distributions on the PV array.

We simulate the output power levels of the two systems under the solar irradiance and temperature distributions as shown in Figure 9. Although the mounting locations of PV cells in the proposed system are the same as those of PV cells in the baseline system, the PV array in the proposed system shows different configurations achieved by the reconfiguration technique, in response to different V_{bat} values. Table 1 summarizes the output power improvement of the proposed system compared to the baseline system with different battery terminal voltage V_{bat} values. The proposed system achieves up to 6X improvement in the output power level with the solar irradiance and temperature distributions as shown in Figure 9 when $V_{bat} = 45$ V. Table 1 also provides near-optimal configurations obtained from the proposed reconfiguration algorithm.

Table 1. Output power improvement of the proposed PV system with different V_{bat} values.

	V_{bat} (V)		
	15	30	45
PV array configuration	$\mathcal{C}$(9; 3, 12, 4, 4, 14, 4, 4, 12, 3)	$\mathcal{C}$(12; 2, 5, 10, 2, 3, 8, 8, 3, 2, 10, 5, 2)	$\mathcal{C}$(12; 2, 5, 10, 2, 3, 8, 8, 3, 2, 10, 5, 2)
Power output improvement	1.99X	2.71X	6.08X

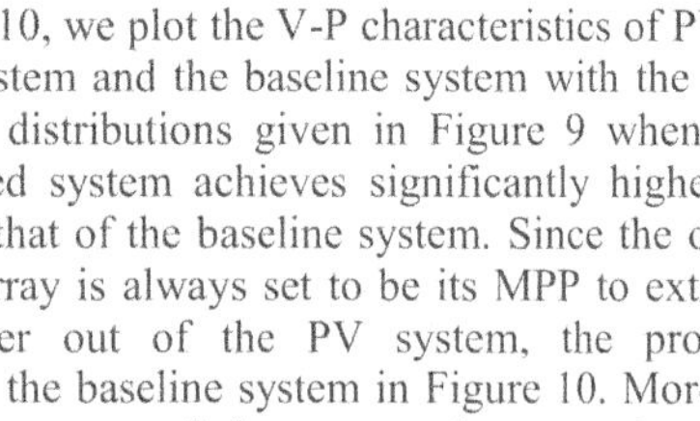

Figure 10. The V-P characteristics of the two PV systems.

In Figure 10, we plot the V-P characteristics of PV arrays in the proposed system and the baseline system with the irradiance and temperature distributions given in Figure 9 when $V_{bat} = 30$ V. The proposed system achieves significantly higher peak output power than that of the baseline system. Since the operating point of the PV array is always set to be its MPP to extract maximum output power out of the PV system, the proposed system outperforms the baseline system in Figure 10. Moreover, the V-P characteristic curve of the proposed system is unimodal even under irradiance and temperature non-uniformity with the proposed reconfiguration technique, while the V-P characteristic curve of the baseline system has multiple peaks. Therefore, the standard MPTT technique could be incorporated into the proposed system without extra modification. However, the MPPT or MPTT technique must be modified to track a global optimal operating point instead of a local optimal point in the baseline system. A typical modification would be searching the whole operating range of the PV array output voltage and current, which results in significantly higher time overhead compared with standard MPPT or MPTT technique based on the perturb and observe approach.

6. Conclusion

Non-uniformity of the solar irradiance and temperature on the PV array is a serious obstacle to efficient utilization of the PV system on HEV. This paper presents a dynamic PV array reconfiguration technique with systematic and scalable structural support and a dynamic programming based reconfiguration control algorithm. Experimental results demonstrate that the proposed reconfiguration method results in up to a factor of 6X output power level improvement compared with a baseline PV system.

Acknowledgements

This work is sponsored by a grant from the U.S. National Science Foundation, the Brain Korea 21 Project, the National Research Foundation of Korea (NRF) grant funded by the Korean Government (MEST) (No. 2011-0016480), and ICT at Seoul National University.

References

[1] N. Femia, G. Petrone, G. Spagnuolo, and M. Vitelli, "Optimization of perturb and observe maximum power point tracking method," *IEEE Trans. on Power Electronics*, 2005.

[2] F. Liu, S. Duan, F. Liu, B. Liu, and Y. Kang, "A variable step size INC MPPT method for PV systems," *IEEE Trans. on Industrial Electronics*, 2008.

[3] Y. Kim, N. Chang, Y. Wang, and M. Pedram, "Maximum power transfer tracking for a photovoltaic-supercapacitor energy system," in *ISLPED*, 2010.

[4] H. Patel, and V. Agarwal, "Maximum power point tracking scheme for PV systems operating under partially shaded conditions," *IEEE Trans. on Industrial Electronics*, 2008.

[5] R. Bruendlinger, B. Bletterie, M. Milde, and H. Oldenkamp, "Maximum power point tracking performance under partially shaded PV array conditions," in *Proc. 21st EUPVSEC*, 2006.

[6] D. Nguyen, and B. Lehman, "An adaptive solar photovoltaic array using model-based reconfiguration algorithm," *IEEE Trans. on Industrial Electronics*, 2008.

[7] G. Velasco-Quesada, F. Guinjoan-Gispert, R. Pique-Lopez, M. Roman-Lumbreras, and A. Conesa-Rosa, "Electrical PV array reconfiguration strategy for energy extraction improvement in grid-connected PV systems," *IEEE Trans. on Industrial Electronics*, 2009.

[8] M. A. Chaaban, M. Alahmad, J. Neal, J. Shi, C. Berryman, Y. Cho, S. Lau, H. Li, A. Schwer, Z. Shen, J. Stansbury, and T. Zhang, "Adaptive photovoltaic system," in *IECON*, 2010.

[9] Y. Kim, S. Park, Y. Wang, Q. Xie, N. Chang, M. Poncino, and M. Pedram, "Balanced reconfiguration of storage banks in a hybrid electrical energy storage system," in *ICCAD*, 2011.

[10] W. Lee, Y. Kim, Y. Wang, N. Chang, M. Pedram, and S. Han, "Versatile high-fidelity photovoltaic module emulation system," in *ISLPED*, 2011.

[11] Y. Wang, Y. Kim, Q. Xie, N. Chang, and M. Pedram, "Charge migration efficiency optimization in hybrid electrical energy storage (HEES) systems," in *ISLPED*, 2011.

Battery Management for Grid-Connected PV Systems with a Battery

Sangyoung Park[1], Yanzhi Wang[2], Younghyun Kim[1], Naehyuck Chang[1] and Massoud Pedram[2]
[1]Seoul National University, Korea, [2]University of Southern California, CA, USA
[1]{sypark, yhkim, naehyuck}@elpl.snu.ac.kr, [2]{yanzhiwa, pedram}@usc.edu

ABSTRACT

Photovoltaic (PV) power generation systems are one of the most promising renewable power sources to reduce the greenhouse gases. Grid-connected PV power systems do not generally have a battery to store the excess charge. However, due to severe imbalance between the peak level of PV power generation and peak load demand, battery-less Grid-connected PV systems are much less effective for the purpose of power generation and demand mismatch mitigation. Grid-connected PV systems equipped with a battery require elaborate management. This is the first paper that introduces a systematic battery management method that accommodates arbitrary electricity billing policies. More precisely, an optimization problem is formulated to determine the battery charging current from the Grid and the PV array taking into account the limited battery capacity, power converter efficiency, battery's internal resistance and rate capacity effect, and the maximum power tracking point of the PV array. Experimental results show that the proposed algorithm effectively reduces the electricity bill by as much as 28% when compared with previous state-of-the-art battery management policies.

Categories and Subject Descriptors

J.6 [**Computer-aided Engineering**]: Computer-aided design (CAD)

General Terms

Algorithms, Design

Keywords

Battery management, Grid-connected systems, Photovoltaic power, Electricity bill

1. INTRODUCTION

The number and capacity of photovoltaic (PV) power system installations are increasing rapidly. Cumulative installations of these systems, which are mainly Grid-connected, have reached 2.15 GW_{DC} in the US alone. Residential PV installations increased at a year-over-year rate of 64% and accounted for 29% of all PV installations in 2010 [1]. However, the growth rate is still slower than desired despite the many advantages of PV systems. This is because of the long break-even time for such systems (the time that is needed for customers to save enough money with lower monthly electricity bills to compensate the initial cost of purchasing and installing the PV system). Maximizing the benefit from the PV system, which is lower electricity bill, is equally important to reducing the installation cost to shorten the break-even time.

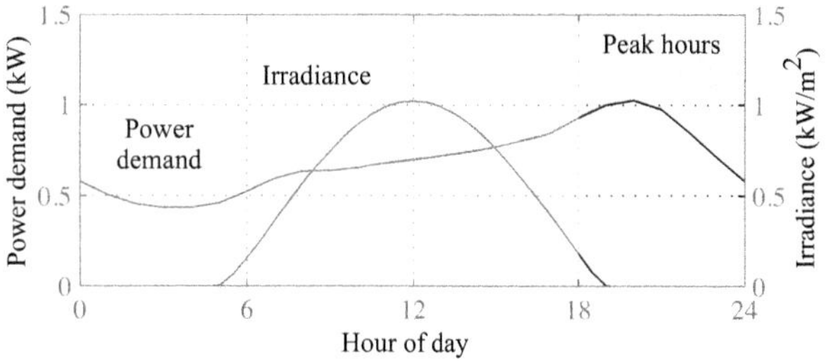

Figure 1: PV array output power and hourly average residential load profile (Southern California Edison territory) [7].

Many previous studies on solar powered systems focused on enhancing the efficiency of the PV system components such as the PV array and PV inverters. The mainstream research is related to maximum power point tracking (MPPT) methods that ensure maximum PV output power in spite of variable solar irradiance [2]. Recent work has presented a maximum power transfer tracking (MPTT) method to maximize the actual energy delivered from a PV array into an energy storage device, considering the power conversion losses [3].

Grid-connected PV systems without a battery do not require elaborate management. Simply performing MPPT or MPTT, and consuming the PV power first and the Grid power second has been sufficient for cases where the PV power is smaller than the load power. Similarly, standalone PV systems equipped with a battery [4, 5] can use a simple policy in which the battery is charged during the day and discharged during the night. In contrast, Grid-connected PV systems should consider complex scenarios such as using the electrical energy stored in the battery when the electricity price is high, i.e., during peak power consumption hours. However, one of the major hurdles in reducing the electricity bill is the mismatch among the PV power generation, load demand, and electricity prices. Figure 1 shows that the peak solar irradiance occurs at noon while the peak residential load demand is at 8pm. Many electricity providers sell the electricity at higher price during the peak hours to control the peak power demand. For example, unit electricity price in the greater Los Angeles are during the peak hours is as much as three times that during the off-peak hours [6]. Grid-connected PV systems without a battery can hardly cope with the peak-hour load demand.

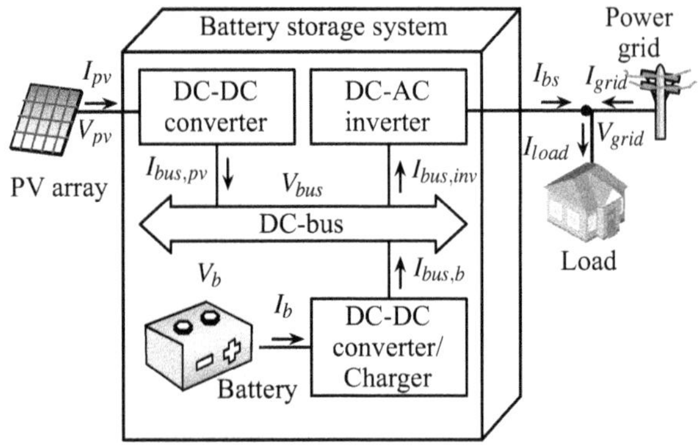

Figure 2: Architecture of the battery storage system for a Grid-connected PV system.

Grid-connected PV systems with a local battery are one way to significantly enhance the usefulness of the solar powered system because it can cope with the peak-hour load demand. Knowing when to charge and when to discharge the battery is the key to success of Grid-connected PV systems with a battery. However, most previous work performs battery management without employing systematic optimization and/or optimality consideration in spite of the significant amount of relevant work. For example, early work simply limits the battery current and considers reselling the excess energy from the PV system to the Grid [8]. *Power leveling*, which controls the power drawn from the Grid [9], and *peak shaving* [10] can mitigate the problem, but they do not provide systematic optimization of the system efficiency or the billing cost. There is a systematic optimization based on Lagrangian relaxation method, but it focuses on issues from the power distribution network such as locational marginal pricing (LMP) and transmission congestion problems [11]. Recent work provides an algorithm that determines when and how to charge and discharge the battery, but the method is ad-hoc without much reasoning about the optimality [12]. An electricity bill minimization algorithm similar to our work has been proposed in [13]. However, the work assumes a different electricity billing policy based on the peak power usage, which makes the proposed algorithm essentially a peak shaving algorithm. Moreover, this work does not consider the rate of PV power generation. In contrast, our work focuses on minimizing the mismatch between the PV power generation and the load demand in a grid-connected PV powered home to minimize the electricity bill.

This paper introduces a holistic optimization framework for residential Grid-connected PV power systems equipped with a battery. Unlike previous work, we develop a systematic optimization method for the battery management, which can effectively mitigate the electricity demand and supply mismatch. We devise an algorithm that determines when and how to store and retrieve energy from the battery to minimize the electricity billing cost. The proposed framework take into account the PV module impedance, converter loss, battery rate-capacity effect, and storage capacity limit for given solar irradiance, load profile, and billing policy. Experimental results show that our technique is capable of reducing 28% electricity bill when compared with previous battery management policies.

2. GRID-CONNECTED PV SYSTEM WITH A BATTERY

2.1 System Architecture

Figure 2 illustrates the overall system architecture considered in this paper. We consider household-scale power consumption as the load. The load is powered by the PV array, power grid, or both. It has an energy storage means to eliminate/reduce the mismatch between power generation and power demand. The energy storage means comprises a battery module to be charged from the Grid when the electricity price is low and from the PV array when the load power is lower than the PV generated power. Together with the Grid, the battery module supplies power to the load device during the peak hours.

The battery module consists of a DC-bus that delivers power to and from the power source, a battery bank, and the load devices. Charging and discharging processes are controlled by a DC–DC converter, a DC–AC inverter, and a charger. This architecture allows use of multiple energy storage banks, even heterogeneous banks, in order to enhance the performance metrics such as power and energy capacity, cycle efficiency, and lifetime [14]. In this paper, a single battery bank is installed.

2.2 Component Models

2.2.1 *Photovoltaic array*

Ideal power source can provide unlimited power capacity and constant voltage or current generation regardless of the environmental and load conditions. The PV array power capacity can be lower than its maximum value due to the environmental conditions such as solar irradiance and temperature. The output voltage of the PV array changes significantly as a function of the load current. We use a single-diode equivalent circuit model [15] and consider such characteristics to maximize the energy efficiency of the system. We keep the PV system efficiency at the maximum, by performing the maximum power transfer tracking introduced in [3].

2.2.2 *Converters and inverters*

The target system consists of two switching power converters that connect the battery to a DC-bus: one for charging the bank and the other for discharging the bank. The PV array also connects to a DC–DC converter, which converts the DC voltage from the PV array to the DC-bus voltage. There also exists a DC–AC inverter and a rectifier for power delivery between the DC-bus and the Grid, respectively. *Efficiency* of a converter or inverter is defined as

$$\eta = \frac{P_{out}}{P_{in}} = \frac{P_{in} - P_{loss}}{P_{in}} = \frac{V_{in} \cdot I_{in} - P_{loss}}{V_{in} \cdot I_{in}}, \tag{1}$$

where P_{in} and P_{out} denote input and output power levels of the converter, respectively, and P_{loss} is the power loss in the converter/inverter. The power loss of a switching power converter comprises three components: conduction loss P_{cdct}, switching loss P_{sw} and controller loss P_{ctrl} [16] such that

$$P_{loss} = P_{cdct} + P_{sw} + P_{ctrl}. \tag{2}$$

The conversion efficiencies of the DC-AC inverter and rectifier can be modeled in a similar way as (2) except that there are additional components P_{trans}, and P_{filter}, which are the power losses from transformer and filters, respectively. The power loss for inverter is given as

$$P_{loss} = P_{cdct} + P_{sw} + P_{ctrl} + P_{trans} + P_{filter}. \tag{3}$$

The power loss components are strongly dependent on the input voltage V_{in}, output voltage V_{out}, output current I_{out}, and the circuit component properties. We derive a closed form expression for η as a function of V_{in}, V_{out}, and I_{out}. The details of power loss equations for each term in (2) can be obtained from [16], and that of inverters are from [17, 18]. We do curve fitting for the converter efficiency as a quadratic function, which can still accurately represent the power converter efficiency in terms of the input voltage V_{in}, output voltage

Table 1: Parameter values for converter/charger and inverter models.

	Converter / charger	Inverter
a_1	2.40E-1	1.78E-1
a_2	-6.63E-5	-3.96E-5
a_3	-5.45E-3	-5.97E-5
a_4	7.13E-4	7.00E-6
a_5	1.16E-2	6.07E-3
a_6	6.89E-3	1.03E-3
a_7	-3.6967E-2	-5.85E-3

V_{out}, output current I_{out} as shown in (4). We denote the equations for calculating efficiencies of the PV converter, battery converter, and inverter as η_{pv}^{out}, η_{b}^{out}, η_{inv}^{out}, respectively. Also, we can calculate the efficiency using the term of I_{in} instead of I_{out}. We denote the equations as η_{pv}^{in}, η_{b}^{in}, η_{inv}^{in}. The regression coefficient values for the quadratic equation are reported in Table 1.

$$\eta^{out} = a_1 V_{in}^2 + a_2 V_{out}^2 + a_3 I_{out}^2 + a_4 V_{in} + a_5 V_{out} + a_6 I_{out} + a_7. \quad (4)$$

2.2.3 *Battery*

Modeling the behavior of battery itself is a challenging task, which has been studied during the past few decades. Battery models from [19, 20] are mainly based on electrochemical process modeling and analysis. Despite the accuracy of the models, they are too complex for use during the system-level design of electronics. Instead we rely on a circuit based model, which captures the most important battery behavior, i.e., the rate capacity effect. The rate capacity effect of batteries is described by the Peukert's formula,

$$\eta_{rate}(I) = \frac{k}{I^{\alpha}}, \quad (5)$$

where k and α are constants.

3. ELECTRICITY BILL REDUCTION

3.1 Power Generation and Usage Models

Residential electrical load demand is fairly periodic although it fluctuates greatly according to time of day. The periodicity of the usage patterns is closely related to consumers' livening patterns, including space heating, cooling, hot water usage, cooking times, television watching hours, and so on. The hourly averaged residential load profile measured in Southern California shows that the peak value occurs at 8pm when residents come home from work [7]. The peak is 2.5 times higher than the minimum value. Many electricity providers sell electricity at different rates at different times of day to control the usage of electricity during peak hours. For example, unit electricity price in Los Angeles is 16.061 ¢/kWh during peak hours, and 4.655 ¢/kWh during off-peak hours [6].

3.2 Battery Management for Electricity Bill Reduction

We minimize the electricity bill, which is the summation of Grid power usage multiplied by the unit cost. We consider two sources of power generation to demonstrate the proposed idea: a PV array and Grid. The optimal policy for the Grid-connected PV system without battery storage is to do maximum energy harvesting. For such systems, simply performing MPPT or MPTT will suffice [3]. On the other hand, achieving the goal of electricity bill reduction in a Grid-connected PV-powered homes with battery storage involves optimization from many aspects. Depending on the ratio between the peak hour and off-peak hour electricity prices, the in-home PV system may either supply power to the load or charge the battery during off-peak hours. Also, we charge the battery when the electricity is cheap and the PV power is not enough to fully charge the battery, preparing to use the battery to perform load shaving during the peak hours. However, charging the battery is less efficient than supplying power to the load due to the battery's internal resistance and charger power losses. Furthermore, high-current charging and discharging result in severe charging and discharging efficiency degradation due to the rate-capacity loss in the battery. We prohibit the case where the battery becomes fully charged and the PV power generation is higher than the load power consumption because we do not consider selling the PV power to the utility company not to lose generality. Not all the electricity companies buy PV power from individual residences.

3.3 Problem Formulation

The electricity bill minimization problem for Grid-connected residential PV power systems is formulated as follows.

Objective:
Minimize electricity bill for a day, that is,

$$cost_{day} = \sum_{n=1}^{N} C[n] p_{Grid}[n], \quad (6)$$

where $C[n]$ is the unit price of the Grid electricity (\$/kWh), $p_{Grid}[n]$ is the power drawn from the Grid (W) at time slot n, and N is the number of time slots per day.

Given:

- Solar irradiance profile for a day $G[n]$.
- Residential load demand profile for a day $I_{load}[n]$.
- Unit price of the Grid electricity $C[n]$.

Control variables:

- DC-bus voltage $v_{bus}[n]$.
- PV operating point $V_{pv}[n]$, $I_{pv}[n]$.
- Battery current $I_b[n]$, where $I_b > 0$ if discharging, and $I_b < 0$ if charging.

The control variables not only determine when to charge or discharge the battery storage, but also set the optimal operating conditions for charging and discharging.

4. ELECTRICITY BILL OPTIMIZATION ALGORITHM

The following observations are the bases of the proposed offline optimization algorithm.

Observation 1: It is beneficial to charge the battery when the electricity is the cheapest, that is, time slot $n_{min} = \arg\min_n(C[n])$, and discharge the battery when the electricity is the most expensive, that is, $n_{max} = \arg\max_n(C[n])$.

Observation 2: Increasing the charging and discharging current of battery storage reduces the charging and discharging efficiency, respectively, due to the battery IR loss and the rate capacity effect.

From the aforesaid observations, it is beneficial to discharge the battery storage at time slot $n_{max} = \arg\max_n(C[n])$, but too much discharging would lower the benefit. We define the variable $C_{com}[n]$, that is the *compensated cost* for effectively determining the time and magnitude of charge/discharge current. Due to the non-ideal characteristics of the power converters and battery storage, charging and discharging efficiency is less than 100%. We reflect the non-ideal characteristics in the variable, $C_{com}[n]$, as follows. The value of $C_{com}[n]$ is initialized to $C[n]$ at the beginning of the algorithm, and it is updated according to the following equation in each iteration.

$$C_{com}[n] = \begin{cases} \dfrac{E_{supplied}}{E_{charge} - E_{loss}} \cdot C[n], & \text{when charging,} \\ \dfrac{E_{extracted}}{E_{extracted} + E_{loss}} \cdot C[n], & \text{when discharging,} \end{cases} \quad (7)$$

where $E_{supplied}$ and $E_{extracted}$ denote the amounts of electrical energy that is supplied to and extracted from the terminal of the battery. E_{loss} is the loss due to the battery IR loss and the rate capacity effect. Higher charging or discharging current makes E_{loss} higher. For example, suppose charging the battery from the Grid at time slot $n_{min} = \arg\max_n(C_{com}[n])$. As our iterative algorithm increases the charging current gradually, $C_{com}[n_{min}]$ increases. As the charging current becomes too high, the compensated cost C_{com} will also become high, and thus further increasing the charging current is avoided. The same approach applies to discharging vice versa when the electricity price is high.

The solution to electricity bill minimization is composed of two parts. The first part of the solution involves solving the electricity bill minimization problem without the maximum battery capacity constraint as shown in Algorithm 1. It assumes that the battery capacity is unlimited. For practical cases, the size of battery is limited, often much smaller than the total load energy consumption throughout a day. Thus, the second part of the solution involves solving the problem with the battery capacity limit constraint based on the solution of the first part as shown in Algorithm 2.

- Operating condition τ : $(v_{bus}, P_{bus,pv}, P_{bus,b}, P_{bus,inv})$ pair that describes the operating condition of the system. The terms correspond to the voltage of the DC-bus, PV converter output power, input/output power to/from the battery charger/converter, and input/output powers of the inverter/rectifier, respectively.
- `VBUS_LUT_1`: lookup table of optimal v_{bus} for given G and I_{load} values constructed at offline.
- `VBUS_LUT_2`: lookup table of optimal v_{bus} for given discharging $P_{bus,b} > 0$ and I_{load} values constructed at offline.
- `VBUS_LUT_3`: lookup table of optimal v_{bus} for given charging $P_{bus,b} < 0$ and I_{load} values constructed at offline.
- `MPTT`: optimal PV operating point (V_{pv}, I_{pv}) for given G and v_{bus}.
- $P^{max}_{bus,pv}(G)$: the PV converter output power for given solar irradiance G and the optimal v_{bus}.
- $P^{min}_{bus,inv}(I_{load})$: the inverter input power when the I_{load} is supplied wholly by the inverter output for the optimal v_{bus}.
- n_s, n_e: the beginning of *active management period* defined by the time when $P^{max}_{bus,pv}$ becomes greater than $P^{min}_{bus,inv}(I_{load})$, and the end of *active management period* defined by 24 hours plus n_s.

Algorithm 1 determines the operating condition τ for every time slot. The power values $P_{bus,pv}$, $P_{bus,b}$, $P_{bus,inv}$ are defined on the DC-bus side of the converter. Three lookup tables `VBUS_LUT_1`, `VBUS_LUT_2`, and `VBUS_LUT_3` are built offline. They contain optimal values of v_{bus} that maximizes the actual power delivered from the source to destination considering the conversion losses. We obtain converter losses from the converter model, and thus, $V_{pv}[n]$, $I_{pv}[n]$, $I_b[n]$, etc. The key idea of Algorithm 1 is to use two threshold values, $c_{th,low}$ and $c_{th,high}$. We initialize the low threshold value to $min(C_{com}[n])$. If we pick all the time slots n with $C_{com}[n] = c_{th,low}$, these are the slots with minimum electricity cost, and thus, are suitable candidates for battery charging. Throughout the algorithm, $c_{th,low}$ gradually increases, as C_{com} increases. We initialize $c_{th,high}$ to $max(C_{com}[n])$, and find all the time slots n with $C_{com}[n] = c_{th,high}$, which are suitable for discharging the battery. We store the cheapest electricity in the battery and use it when the electricity price is the highest. Power from the PV array is free, and always cheaper than the Grid electricity price. The algorithm considers charging the battery with PV power first, and the electricity from the Grid next. Algorithm 1 is optimal if the battery capacity is unlimited and the initial state of charge (SOC) of battery is sufficient.

Algorithm 1: Electricity bill minimzation algorithm without battery capacity limit.

Input: G: irradiance, I_{load}: load current
Output: Operating condition $\tau = (v_{bus}, P_{bus,pv}, P_{bus,b}, P_{bus,inv})$

for $\forall n,\ P^{max}_{bus,pv}(G[n]) > P^{min}_{bus,inv}(I_{load}[n])$ **do**
  $v_{bus}[n] \leftarrow \texttt{VBUS\_LUT\_1}(G[n], I_{load}[n])$
  $(V_{pv}[n], I_{pv}[n]) \leftarrow \texttt{MPTT}(G[n], v_{bus}[n])$
  $P_{bus,pv}[n] \leftarrow V_{pv} \cdot I_{pv} \cdot \eta^{in}_{pv}(V_{pv}[n], v_{bus}[n], I_{pv}[n])$
  $P_{bus,inv}[n] \leftarrow \dfrac{V_{load} \cdot I_{load}[n]}{\eta^{out}_{inv}(v_{bus}[n], V_{load}, I_{load}[n])}$
  $P_{bus,b}[n] \leftarrow P_{bus,inv}[n] - P_{bus,pv}[n]$
  $P_b[n] \leftarrow \eta^{in}_b(V_{bus}[n], V_b, -P_{bus,b}[n]/V_{bus}[n])$
  $SOC_b \leftarrow charge(P_b, SOC_b)$
while $c_{th,low} < c_{th,high}$ && $!endCond$ **do**
  for $\forall n,\ C_{com}[n] = c_{th,high}$ **do**
    if $n = \phi$ **then**
      $c_{th,high} \leftarrow max(C_{com})$
      continue
    $P_{bus,b}[n] = min(P_{bus,b}[n] + P_{inc,d}, \dfrac{P_{load}[n]}{\eta^{out}_{inv}})$
    $v_{bus}[n] \leftarrow \texttt{VBUS\_LUT\_2}(P_{bus,b}[n], I_{load}[n])$
    $(V_{pv}[n], I_{pv}[n]) \leftarrow \texttt{MPTT}(G[n], v_{bus}[n])$
    $P_{bus,pv}[n] \leftarrow V_{pv} \cdot I_{pv} \cdot \eta^{in}_{pv}(V_{pv}[n], v_{bus}[n], I_{pv}[n])$
    $P_{bus,inv}[n] \leftarrow P_{bus,pv}[n] + P_{bus,b}$
    $P_b[n] \leftarrow \eta^{out}_b(V_b, V_{bus}[n], P_{bus,b}[n]/V_{bus}[n])$
    $SOC_b \leftarrow discharge(P_b, SOC_b)$
    $C_{com}[n] \leftarrow decreaseCost(P_{bus,b}$
    if $SOC_b = 0$ **then**
      break
  for $\forall n,\ C_{com}[n] = c_{th,low}$ **do**
    if $n = \phi$ **then**
      $c_{th,low} \leftarrow min(C_{com})$
      continue
    $P_{bus,b}[n] \leftarrow P_{bus,b}[n] - P_{inc,c}$
    $v_{bus}[n] \leftarrow \texttt{VBUS\_LUT\_3}(P_{bus,b}[n], I_{load}[n])$
    $(V_{pv}[n], I_{pv}[n]) \leftarrow \texttt{MPTT}(G[n], v_{bus}[n])$
    $P_{bus,pv}[n] \leftarrow V_{pv} \cdot I_{pv} \cdot \eta^{in}_{pv}(V_{pv}[n], v_{bus}[n], I_{pv}[n])$
    $P_{bus,inv}[n] \leftarrow P_{bus,pv}[n] + P_{bus,b}$
    $P_b[n] \leftarrow \eta^{in}_b(V_{bus}[n], V_b, -P_{bus,b}[n]/V_{bus}[n])$
    $SOC_b \leftarrow charge(P_b, SOC_b)$
    $C_{com} \leftarrow increaseCost(P_{bus,b})$
return $\tau \leftarrow (v_{bus}, P_{bus,pv}, P_{bus,b}, P_{bus,inv})$

Algorithm 2: Electricity bill minimzation algorithm with battery capacity limit $SOC_{b,max}$.

Input: G: irradiance, I_{load}: load current, $SOC_{b,init}$: initial SOC at n_s, $\tau = (v_{bus}, P_{bus,pv}, P_{bus,b}, P_{bus,inv})$: result of Algorithm 1

Output: Operating condition $(v_{bus}, P_{bus,pv}, P_{bus,b}, P_{bus,inv})$

$n_{mark,s} \leftarrow n_s$
$n_{mark,e} \leftarrow n_s$
$SOC_{mark} \leftarrow SOC_{b,init}$
for $n = n_s$ *to* n_e **do**
  $SOC_b \leftarrow updateSOC(P_{bus,b}[n], SOC_b)$
  if $SOC_b > SOC_{b,max}$ **then**
    $n_{mark,e} \leftarrow nextDischarge(P_{bus,b}, n)$
    $\tau[n] \leftarrow reschedule1(SOC_{mark}, SOC_{b,max}, n_{mark,s}, n_{mark,e})$
    $n_{mark,s} \leftarrow n_{mark,e}$
    $SOC_{mark} \leftarrow SOC_{b,max}$
    continue
  else if $SOC_b < 0$ **then**
    $n_{mark,e} \leftarrow nextCharge(P_{bus,b}, n)$
    $\tau[n] \leftarrow reschedule2(SOC_{mark}, 0, n_{mark,s}, n_{mark,e})$
    $n_{mark,s} \leftarrow n_{mark,e}$
    $SOC_{mark} \leftarrow 0$
    continue
return $\tau \leftarrow (v_{bus}, P_{bus,pv}, P_{bus,b}, P_{bus,inv})$

Algorithm 2 reschedules the battery charging and discharging operations so that the battery SOC does not exceed the maximum value or becomes below zero. The charging and discharging schedule from Algorithm 1 might violate both the maximum and minimum battery capacity constraint. Algorithm 2 starts at the beginning of a *active management period*, n_s. Algorithm 2 charges and discharges the battery according to the scheduling result of Algorithm 1 as time passes until the battery capacity constraint is violated. Functions $nextDischarge(P_{bus,b}, n)$ and $nextCharge(P_{bus,b}, n)$ find the timeslot next to n where the closest discharge or charge begins. The functions $reschedule1()$ and $reschedule2()$ derive an operating condition τ schedule that makes SOC_b from SOC_{mark} to $SOC_{b,max}$ and from SOC_{mark} to 0 during time interval $[n_{mark,s}\ n_{mark,e}]$, respectively, while meeting the battery capacity constraints. This rescheduling problem is much simpler than the original problem since the start and end SOC values are given, and the capacity limit is met for the interval $[n_{mark,s}\ n_{mark,e}]$. Function $reschedule1()$ works as follows. It fixes the discharging schedule as derived in Algorithm 1. The next step is to perform initial charging to avoid depletion by the discharging schedule fixed in the previous step. This initial charging is always feasible from the definition of the interval $[n_{mark,s}\ n_{mark,e}]$. Finally, $reschedule1()$ determines the rest of the charging schedule using $c_{th,low}$ and C_{com} to minimize the charging cost until the total accumulated charge at $n_{mark,e}$ becomes $SOC_{b,max}$. Function $reschedule2()$ is defined in a complementary manner. Algorithm 2 is based on Algorithm 1 and ensures the solution quality of the capacity limited problem because Algorithm 1 gives the optimal results for the unconstrained case.

5. EXPERIMENTAL RESULTS

We compare the efficacy of the proposed algorithm with two baseline algorithms. Both baseline algorithms charge the battery from the PV array and Grid during off-peak hours and discharge the battery during the peak hours. For the first baseline, charging and

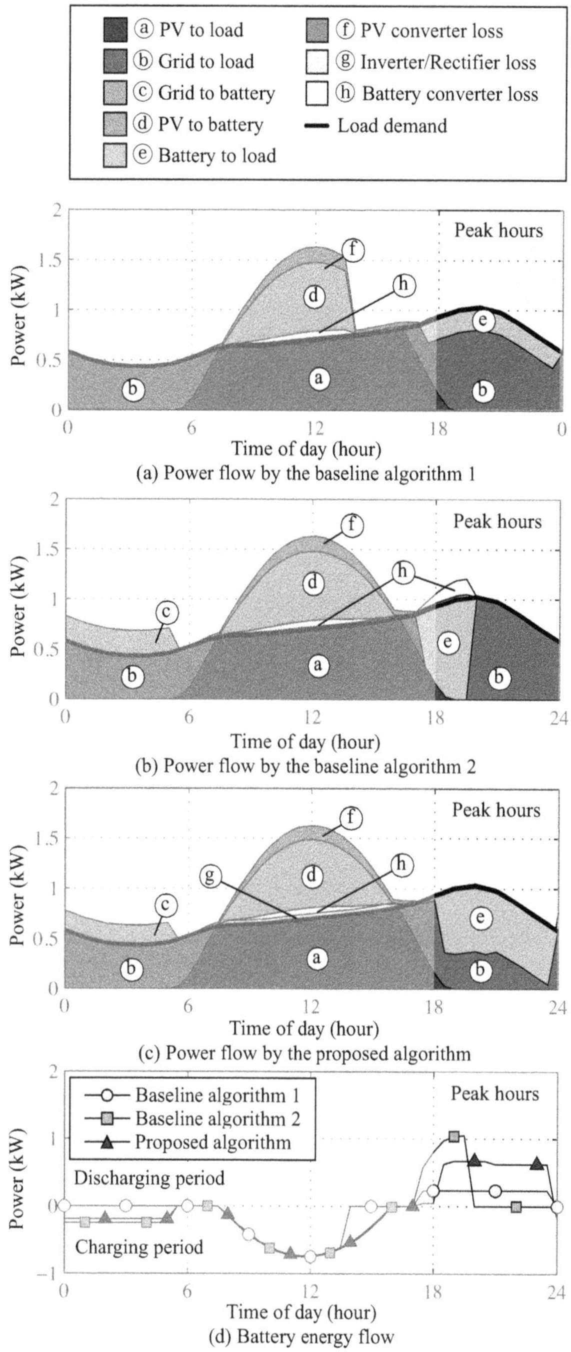

Figure 3: Power input and output variation with time.

discharging current from the Grid is fixed to 1C to maximize cycle efficiency of the battery, and it performs MPTT to determine the operating point of the PV array and DC-bus voltage. For the second baseline, the charging current from the Grid is fixed to 1C and discharging current is determined greedily during the peak hours, and it also performs MPTT.

We use an actual profile of residential load demand and Grid electricity price from Southern California [6, 7]. The peak load demand is 1.025 kW at 8pm as shown in Figure 1. The unit price of the Grid electricity is 4.679 ¢/kWh during peak hours and 1.879 ¢/kWh during off-peak hours. The PV array used for our experiment outputs 1,000 W/m^2 and the maximum power point voltage and current is 67.8 V and 23.6 A. The capacity of battery bank is 0.864 kWh rated at 5 A output current with the output voltage of 48 V. The result of the proposed algorithm is shown in Figure 3.

The resulting battery management policy shows different charging and discharging patterns. The first baseline has fixed charge and discharge current and thus it loses the potential for further gain. Figure 3(a) shows that the first baseline does not fully utilize the PV power even though the battery is not full because the total amount of energy discharged during the peak hours is limited. If the PV power keeps charging the battery during daytime, the net energy throughout the day become positive. Figure 3(d) shows the same results. Baseline 1 does not charge from the Grid during off-peak hours, since it cannot utilize it. Battery current is kept at 1C during the peak hours. This minimizes the rate capacity loss, but it also poses limits on the gain. The second baseline greedily supplies the load during the peak hours, which results in severe inefficiency due to the rate capacity effect. Baseline 2 fully utilizes the battery by charging the battery both from the PV array and Grid as shown in Figure 3(b). Charging from Grid is temporarily disabled at around 6 o'clock in the morning, and it prevents the battery becoming fully charged while the PV power is still available. By the end of daytime, the battery becomes fully charged and is ready for discharging during peak hours. Baseline 2 greedily discharges the battery during peak hours as shown in Figure 3(d), and thus it soon becomes fully depleted. There is no other choice but to use expensive Grid electricity beyond that point, which increases the electricity bill. The proposed algorithm determines the charging and discharging schedule while considering both the loss due to large charge/discharge current and utilization of the Grid price fluctuation. The proposed algorithm fully charges the battery by the end of daytime, just like Baseline 2. To fully utilize the price differences between the peak hours and off-peak hours, the proposed algorithm sets the discharging current to a value between Baselines 1 and 2 as shown in Figure 3(d). Discharging the battery at higher current than Baseline 1 increases the rate capacity loss, but it is beneficial due to the price difference between the peak and off-peak hours. The proposed algorithm reduces the electricity bill during a day to 18.73 ¢ while two baselines give 26.14 ¢ and 25.38 ¢ , respectively.

6. CONCLUSION

PV power generation is promising but not very effective to mitigate demand and supply mismatch of electricity. Grid-connected PV systems with a battery has great potential to resolve the mismatch as long as elaborated battery management ensures optimal charge and discharge policies. This paper is the first work to address holistic optimization of battery management for Grid-connected PV systems. We devise an offline algorithm that schedules battery charge and discharge for solar given solar irradiance and load profile. Our framework allows for arbitrary Grid electricity price function, and all the lossy components, such as converter loss, and rate capacity loss of batteries, in the Grid-connected PV powered system with electrical energy storage. Experimental results show that the electricity price is reduced by up to 28% when compared with baseline policies.

Acknowledgements

This work is sponsored by a grant from the National Science Foundation, the Brain Korea 21 Project, and the National Research Foundation of Korea (NRF) grant funded by the Korean Government (MEST) (No. 2011-0016480). The ICT at Seoul National University provides research facilities for this study.

7. REFERENCES

[1] "U.S. solar market trends, 2010," Interstate Renewable Energy Council, Tech. Rep., 2011.

[2] T. Esram, J. Kimball, P. Krein, P. Chapman, and P. Midya, "Dynamic maximum power point tracking of photovoltaic arrays using ripple correlation control," *Power Electronics, IEEE Transactions on*, vol. 21, no. 5, pp. 1282–1291, 2006.

[3] Y. Kim, N. Chang, Y. Wang, and M. Pedram, "Maximum power transfer tracking for a photovoltaic-supercapacitor energy system," in *ISLPED*, 2010, pp. 307–312.

[4] P. Arun, R. Banerjee, and S. Bandyopadhyay, "Optimum sizing of photovoltaic battery systems incorporating uncertainty through design space approach," *Solar Energy*, vol. 83, no. 7, pp. 1013–1025, 2009.

[5] S. Shaahid and I. El-Amin, "Techno-economic evaluation of off-grid hybrid photovoltaic-diesel-battery power systems for rural electrification in saudi arabia—a way forward for sustainable development," *Renewable and Sustainable Energy Reviews*, vol. 13, no. 3, pp. 625–633, 2009.

[6] Los Angeles Department of Water & Power, *Electric Rates, http://www.ladwp.com/ladwp/cms/ladwp001752.jsp*, 2009.

[7] "Review of residential electrical energy use data," NAHB Research Center, Inc., Tech. Rep., 2001.

[8] G. Palomino, J. Wiles, J. Stevens, and F. Goodman, "Performance of a grid connected residential photovoltaic system with energy storage," in *Photovoltaic Specialists Conference*, 1997, pp. 1377–1380.

[9] S. Chiang, K. Chang, and C. Yen, "Residential photovoltaic energy storage system," *Industrial Electronics, IEEE Transactions on*, vol. 45, no. 3, pp. 385–394, 1998.

[10] F. Giraud and Z. Salameh, "Steady-state performance of a grid-connected rooftop hybrid wind-photovoltaic power system with battery storage," *Energy Conversion, IEEE Transactions on*, vol. 16, no. 1, pp. 1–7, 2001.

[11] B. Lu and M. Shahidehpour, "Short-term scheduling of battery in a grid-connected pv/battery system," *Power Systems, IEEE Transactions on*, vol. 20, no. 2, pp. 1053–1061, 2005.

[12] Y. Gurkaynak and A. Khaligh, "Control and power management of a grid connected residential photovoltaic system with plug-in hybrid electric vehicle (PHEV) load," in *Applied Power Electronics Conference and Exposition*, 2009, pp. 2086–2091.

[13] S. Govindan, A. Sivasubramaniam, and B. Urgaonkar, "Benefits and limitations of tapping into stored energy for datacenters," in *Proceedings of the 38th annual international symposium on Computer architecture*, ser. ISCA '11. New York, NY, USA: ACM, 2011, pp. 341–352. [Online]. Available: http://doi.acm.org/10.1145/2000064.2000105

[14] M. Pedram, N. Chang, Y. Kim, and Y. Wang, "Hybrid electrical energy storage systems," in *ISLPED*, 2010, pp. 363–368.

[15] M. Villalva, J. Gazoli, and E. Filho, "Comprehensive approach to modeling and simulation of photovoltaic arrays," *Power Electronics, IEEE Transactions on*, vol. 24, no. 5, pp. 1198–1208, 2009.

[16] Y. Choi, N. Chang, and T. Kim, "DC–DC converter-aware power management for low-power embedded systems," *Computer-Aided Design of Integrated Circuits and Systems, IEEE Transactions on*, vol. 26, no. 8, pp. 1367–1381, 2007.

[17] J. Kolar, H. Ertl, and F. Zach, "Influence of the modulation method on the conduction and switching losses of a pwm converter system," *Industry Applications, IEEE Transactions on*, vol. 27, no. 6, pp. 1063–1075, 1991.

[18] "MOSFET power losses calculation using the data-sheet parameters," Infineon, Tech. Rep., 2006.

[19] P. Rong and M. Pedram, "An analytical model for predicting the remaining battery capacity of lithium-ion batteries," *Very Large Scale Integration (VLSI) Systems, IEEE Transactions on*, vol. 14, no. 5, pp. 441–451, 2006.

[20] D. Rakhmatov, "Battery voltage modeling for portable systems," *Design Automation of Electronic Systems, ACM Transactions on*, 2009.

Panel

Going Green across Communications and Storage Systems: Control of Power in Non-Mobile Devices

Organizer & Moderator

Kenneth D. Wagner
PMC-Sierra, Inc., Canada

Panelists

Martin St. Laurent
Qualcomm, Inc., USA

Robert Aitken
ARM, USA

Hugh Barrass
Cisco Systems, Inc., USA

Randall Robinson
Rackforce Networks, Inc., Canada

TBD
HP, USA

Panel Topic

There is a growing need to reduce device power to meet environmental, thermal, packaging, operating cost and other constraints in enterprise and carrier class applications. These pressures affect all levels of design targeted towards non-mobile applications: in the data center and its cloud, in the corporate enterprise – storage and WAN, in servers and embedded processing platforms, and in telecom wired transport and wireless backhaul networks.

How will overall energy consumption be reduced with our fast-increasing need for compute power, capacity and performance? How can these trends be tamed with tomorrow's technologies and architectures? What can be leveraged from current sophisticated power reduction technologies in the mobile and multi-core processor domains? And who will look at the big picture of consumption?

Power reduction and hence energy savings have become primary objectives for device suppliers responding to end-customer requirements. Standards such as Greentouch, Open Compute Project and several others have been created for disruption of *business-as-usual*. This Panel will focus on power management and control of plugged-in (non-mobile) systems and devices for process technologies at 40nm and below. It will also discuss the use of USB, POE and other alternative power sources for low power devices in challenging environments.

Categories and Subject Descriptors

B.0 Hardware - GENERAL

Keywords

Green communications, low power, energy consumption, Greentouch, Green Grid, Open Compute Project, Energy Efficient Ethernet

ISLPED'12, July 30–August 1, 2012, Redondo Beach, California, USA.
ACM 978-1-4503-1249-3/12/07.

PANELISTS VIEWPOINTS

Panelist 1, Qualcomm – Martin St. Laurent

There are similarities as well as major differences between the power management and control techniques used for mobile and plugged-in applications. The main low-power design and analysis techniques used on a digital signal processor (DSP) targeted at smartphones will be presented. How these techniques may or may not be applicable for plugged-in applications will also be discussed.

Panelist 2, Cisco- Hugh Barrass

Data center and telecom energy consumption requirements have evolved: equipment was originally constrained by power provisioning limits; then peak power density growth ran into limits for cooling; and now the latest wave of metrics and recommendations constrain the average energy use for equipment over cycles of peak and off-peak usage. These changes have mirrored the evolution of silicon energy consumption constraints that drive the limits of functionality in a device and the packaging costs based on peak power usage but are now also focusing on average energy usage. Metrics published by bodies such as ATIS (telecommunications); SPEC (computing); The Green Grid (data center); and others, are rewarding designs that can scale energy use with workload and are stimulating a new wave of standards such as IEEE 802.3 Energy Efficient Ethernet.

Panelist 3, ARM – Robert Aitken

In electronic systems, "low power" is often thought of as being synonymous with "battery powered" or "mobile", and certainly mobile systems are a major focus of low power efforts. However, concerns about environmentally-friendly products and reduced energy consumption, along with more prosaic issues such as reduced thermal output and cost savings, have created an opportunity for low power design in non-mobile systems as well. Key questions facing designers of such systems include: To what extent are solutions developed in mobile space reusable, and where are new solutions required?

A low power system, whether mobile or not, needs to be designed to be more than the sum of its parts. The more the component pieces work together, from transistors through logic through microarchitecture and architecture, then on into operating systems, workloads, and the software stack, and even into the network for non-mobile systems, the better the overall results. For this reason, standardized approaches from IEEE 1801 at the SoC level up to GreenTouch at the network level are key to effective development and delivery of low power solutions. Many existing low power solutions in mobile space can be reused in non-mobile, but two key differences need to be considered. First, the energy delivery profile of a wired supply is different from a battery in terms of essentially infinite capacity, better flexibility when going from low to high current modes, and lack of degradation between charges. Second, the device needs to be thought of as existing as part of a larger system (its local power network and even the surrounding grid), so thermal effects, for example, cannot be considered in isolation.

For complex systems, such as communication and storage systems, low power needs to be considered as part of the overall architecture requirements, and then applied to the component parts through standards and specifications.

Panelist 4, Rackforce Networks – Randall Robinson

In this Panel, we discuss a methodology for the quantification of Greenhouse Gas emissions (GHG) attributable to ICT where shared compute resources are a factor, such as in the case of cloud computing. With the proliferation of virtualization and cloud computing, the measurement of the power consumption from fractional usage of CPU, RAM, storage and network is at the same time increasingly important and challenging.

For the purpose of this Panel, I will focus on *Project Type 2* and use case 3 which covers: *projects involving improvements to ICT services – Low or Zero Carbon Clouds and Real-Time Workload shifting* in an ISO protocol developed for this purpose.

Some fundamental challenges were encountered around the measurement of power consumption for Virtual Machines (VMs) running in a multi-tenant environment with resources shared amongst other VMs. In this model fractional usage and related power consumption must be calculated for processing, memory, data storage and networking components.

Five variables need to be determined in order to estimate the emissions associated with the project – project ICT power usage, the PUE of the project facility, the emission factor of the source energy for the ICT facility, the energy used by the network involved with transferring data, and the emission intensity of the electricity used for the network.

Further to this, the concept of "work performed" became a key consideration, especially in determining baseline power consumption. For example if the project involves moving a workload from one environment to a new more efficient environment then how can a project proponent compare the power consumption in the new (Project) environment to that of the original (Baseline) environment when you consider that the workload in the Project environment is unlikely to be consistent with the Baseline, realizing that typical workloads are often in flux.

To simplify this it was concluded that the workload could be measured at the project and extrapolated to the baseline. Essentially the workload would have been the same on the original environment as it is in the project environment. To remain conservative we consider the baseline cloud computing environment to consume an equivalent amount of power as the project for the same workload. Therefore we need only substitute project emissions factors and PUE with those from the baseline. If the baseline cloud is not considered to be on par with the project cloud environment regarding its power consumption per unit of workload then the alternate method of using benchmarks can be employed.

The Protocol development team was indeed challenged to find solutions that would be accepted by the industry in general as being both accurate and at the same time not so onerous as to make application of the methods unlikely for the average project proponent to follow. It is expected that additional tools will continue to be developed to aid in the measurement of specific resource utilization metrics and it is important to encourage the link of these metrics to the related power consumption of the ICT workload.

Panelist 5, TBD

TBD

Modeling, Design and Cross-Layer Optimization of Polysilicon Solar Cell Based Micro-scale Energy Harvesting Systems

Elif S. Mungan, Chao Lu, Vijay Raghunathan and Kaushik Roy
School of ECE, Purdue University, West Lafayette, IN, USA
{ebaytok, lu43, vr, kaushik} @purdue.edu

ABSTRACT

This paper presents modeling, design and cross-layer optimization of polysilicon solar cell based micro-scale energy harvesting systems. The proposed design methodology is suitable for energy harvesting systems employed in low-cost applications, such as wireless sensor networks. Our approach is unique in achieving maximum output power by cross-layer optimization of poly-silicon solar cells (thickness, grain boundaries and cell configuration) and power converter circuits. Simulation results indicate that optimizing the solar cell along with the power converter improves the system output power by 16% compared to a baseline approach of optimizing the two components separately.

Categories and Subject Descriptors

B.7.2 [Integrated Circuits]: Design Aids – *Simulation.*

Keywords

Energy scavenging, Power optimization and management.

1. INTRODUCTION

Micro-scale energy harvesting has emerged as a promising and viable option for powering embedded systems in a variety of emerging application domains such as wearable electronics and wireless sensor networks [1][16]. These application domains often have unique operational constraints (*e.g.*, the need to operate unattended for several months to years, extremely small form factor), which mandates an extreme emphasis on power, sizeand cost.

Of the various energy harvesting modalities available, solar energy harvesting through photovoltaic conversion is a popular choice for both indoor and outdoor applications due to the ubiquitous nature of solar energy and its relatively high power density [5][8]. In the block diagram of a micro-scale solar energy harvesting system shown in Fig. 1, a solar cell is used as a power source whose output passes through a power converter (that acts as a power conditioning unit). The power converter is controlled by a maximum power point tracking (MPPT) circuit. The MPPT circuit ensures that the power converter is always configured such that the maximum amount of electrical power is extracted from the solar cells despite the constant change in light intensity

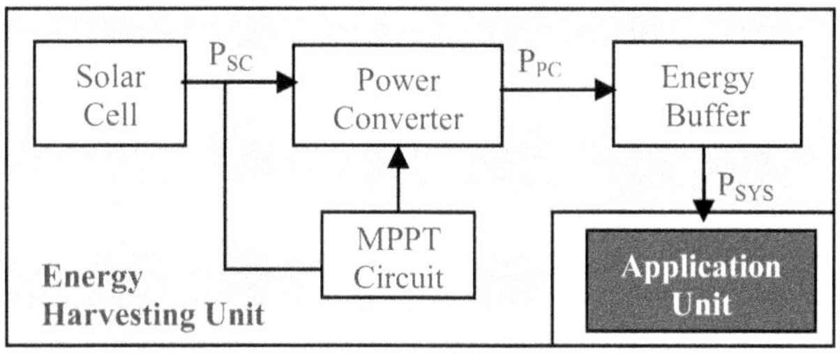

Figure 1. Block diagram of a micro-scale solar energy harvesting system.

throughout the day. The output voltage of this solar cell is converted to the required voltage of an energy storage buffer (*e.g.*, a rechargeable battery) by the power converter. The stored energy is used by an application unit (such as a wireless sensor node or other embedded system) to implement sensing, computation and communication functionalities.

To minimize the overall cost of the system, it is possible to use ultra-thin film polysilicon solar cells fabricated in a low-cost low temperature polysilicon (LTPS) process technology. With this fabrication process it is also possible to use flexible or glass substrates to further reduce the overall cost. Note that, using LTPS technology, it is also possible to fabricate thin film transistors (TFTs) on the same substrate. This leads to the possibility of integrating the solar cell, power converter and application units on the same substrate.

In order to maximize the energy conversion efficiency of such systems, there has been a considerable effort to improve the design of the power-conditioning unit (*i.e.*, MPPT circuit and power converter). For MPPT, the voltage-controlled oscillator (VCO) based [17] and hill climbing [3] methods are preferred for micro-scale energy conversion due to their high tracking accuracy and low power overhead. For power converters, charge pump based designs [14] are more suitable for the applications under consideration because, unlike inductive power converters, they do not require inductors that use up valuable substrate area. One common feature (and limitation) of previous micro-scale energy harvesting designs is that the solar cell is treated as a black box while trying to optimize the power-conditioning unit [2]. The only parameters that designers usually change are the number of solar cells used and the specific array configuration that they are connected in [7]. This design strategy often leads to a larger-than-required solar cell area. In order to achieve a more cost-effective design, one should have control over the physical parameters of the solar cell and should be able to model the effect of those parameters on the cell's behavioral performance. At the same time, one should not design the solar cell in isolation. An improvement in the solar cell may degrade the performance of the power converter if the power converter is not adjusted

ISLPED'12, July 30–August 1, 2012, Redondo Beach, California, USA.
Copyright 2012 ACM 978-1-4503-1249-3/12/07....$10.00.

accordingly [12]. As a result, it is desirable to design and optimize the solar cell and the power converter in a coordinated manner.

To address this goal, we propose a cross-layer framework for the joint design and optimization of the solar cell and its interface circuitry (*i.e.*, the power converter). As part of this optimization framework, we developed an analytical model for polysilicon solar cells and used the analytical model for the power converter circuit that we developed in [10,11,13]. Using these models, the cost-effective solution satisfying the design requirements (power and area) can be determined for different fabrication process parameters for polysilicon solar cells. To the best of our knowledge, this is the first time such a systematic cross-layer optimization has been proposed from the device and circuit design perspective for micro-scale energy harvesting systems.

In the following section the device model for the proposed solar cell is explained in detail while the power converter circuit model and key optimization parameters are discussed in section 3. The proposed cross-layer optimization framework is described in section 4. Finally, an illustrative case study is demonstrated in section 5. Section 6 draws the conclusions.

2. MODELING POLYSILICON SOLAR CELLS

2.1 Modeling Framework

As mentioned earlier, we focus on polysilicon solar cells, which are attractive for low cost applications due to the lower temperatures used during fabrication. Although low temperature reduces fabrication cost, it also results in multi-crystalline structures, as shown in Fig. 2. These structures are composed of grain-shaped crystals with recombination regions called grain boundaries (GBs) between them. These grain boundaries result in polysilicon solar cells having a lower efficiency compared to single crystalline solar cells.

The low cost polysilicon solar cell (Fig. 2) that we first proposed in [13] is designed to have very thin cell thickness (t_{Si}), from 0.1µm to 0.8µm, and narrow device width (200nm). There are two reasons for these design choices. First, it is desirable that the thickness of the solar cell is almost the same as the thickness of the thin-film transistors (TFTs) that are used in the application units. If that is the case, the entire system can be manufactured with the same fabrication steps, reducing the cost. Second, if the device is narrower than the average grain size for a given process technology, the probability of having multiple grain boundaries (GBs) within a single cell is low, leading to enhanced efficiency.

The GB was modeled as a 4nm wide recombination region in a single crystalline silicon layer. The recombination in the GB was modeled using Shockley-Read-Hall (SRH) recombination [9]. The trap energy levels were set to 0.05eV above and below the mid-gap and 0.02eV away from the band edges. Mid-gap states had trap densities of 10^{12} cm^{-2} while the band-tail states had trap densities of 10^{14} cm^{-2} with capture cross-sections of 10^{-15} cm^{2} [9]. The doping concentrations were chosen to be 10^{19}, 10^{15} and 10^{19} cm^{-3} for the n+, p and p+ regions, respectively.

While calculating the output current density (J) of the proposed device, the photocurrent density under short circuit condition (J_{LIGHT}) and the J-V curve calculated for dark condition (J_{DARK})

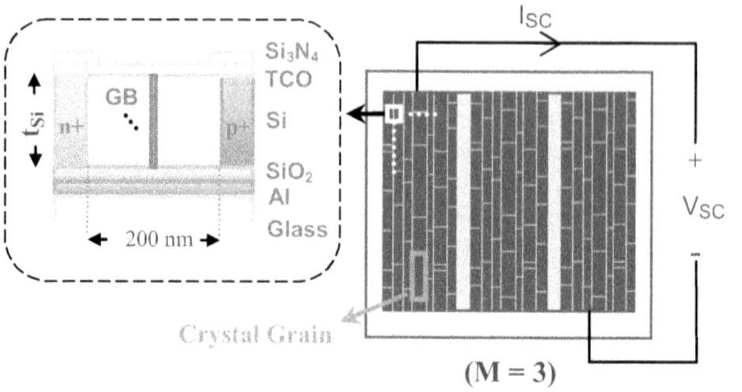

Figure 2. Cross-section of the proposed device structure (left) and top-view of the polysilicon solar cell module divided into M sub-cells connected in series (right).

were superposed as in Eq. 1.

$$J(V) = J_{LIGHT} - J_{DARK}(V) \quad (1)$$

2.2 Photocurrent

In order to optimize the device structure, the relationship between the device parameters and device performance should be known. To obtain this relationship, we first modeled the generation rate (G) as a function of the depth (y) and the cell thickness (t_{Si}) as shown below in Eq. 2.

$$G(y) = A(t_{Si}) \cdot e^{-\alpha(t_{Si})y} + B(t_{Si}) \cdot e^{-\beta(t_{Si})y} + C(t_{Si}) \quad (2)$$

$$X(t_{Si}) = X_1 + X_2 t_{Si} \quad (3)$$

$$C(t_{Si}) = C_1 + e^{C_2 + C_3 t_{Si}} \quad (4)$$

The optical generation profile in a solar cell is obtained from numerical simulations conducted using Sentaurus [18] under AM 1.5 spectrum. The generation profiles were fitted into Eq. 2. In this equation, functions A, B, α and β are of the form shown in Eq. 3 and function C is of the form shown in Eq. 4 where X_1, X_2, C_1, C_2 *and* C_3 are constants. Although it is an empirical fit, Eq. 2 has a physical significance. The first component models the absorption of photons with small wavelengths whereas the second component models the absorption of photons for higher wavelengths. The third component accounts for the reflection from the back surface.

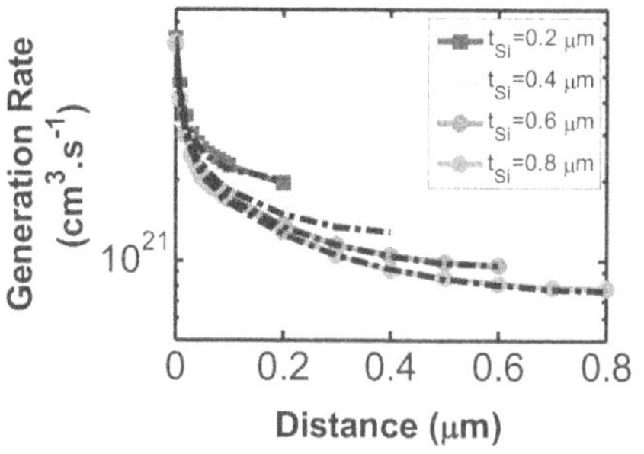

Figure 3. The modeled generation rates within the proposed device for various cell thicknesses (t_{Si}). The black dashed lines are results from Sentaurus.

The generation rates obtained using the above model and plotted in Fig. 3. Using the G values plotted in Fig. 3, J_{LIGHT} can be derived using the following equation:

$$J_{LIGHT} = q\int_0^{t_{Si}} G(y)dy \quad (5)$$

2.3 Dark Current

Due to the narrow width of the device there are two conditions that should be considered while calculating the diode current measured in darkness (J_{DARK}). Since the width is narrow, there is a possibility that a GB might not occur within the device (no GB case, 0GB). In that case, the dark current of the device can be modeled as in a narrow-base diode (Eq. 6).

$$J_{DARK} = q(\frac{D_n}{W_P.N_{A+}} + \frac{D_p}{W_N.N_{D+}})n_i^2.(e^{(qV/kT)} - 1) \quad (6)$$

From the calculated J_{LIGHT} and J_{DARK} values, the I-V curve for 0GB case can be determined using Eq. 1. The second condition to be considered is the probability of having one GB in the device (1GB case). In this case, the dark current can be calculated by solving Eq. 7 and Eq. 8 consistently.

$$|E_{GB-}| = |E_{GB+}| - \int_0^{W_{GB}} \frac{Q_{GB}(x)}{\in} dx \quad (7)$$

$$Q_{GB}(x) = \sum_{i=1}^{i=n_D} q(1-f_i)N_{tD} - \sum_{i=1}^{i=n_A} f_i N_{tA} \quad (8)$$

E_{GB+} and E_{GB-} represent the electric field towards the left and right side of the GB respectively, while f_i is the electron occupation probability of trap states. N_{tA} and N_{tD} indicate the trap density of acceptor-like and donor-like traps. By assuming that energy-band bending between the GB position and the edges of the device is linear, one can account for the change in the electric field within the device due to the trapped charges in the GB (Q_{GB}). As Fig. 4 indicates, the modeled energy conversion efficiencies closely follow the results obtained through numerical simulations.

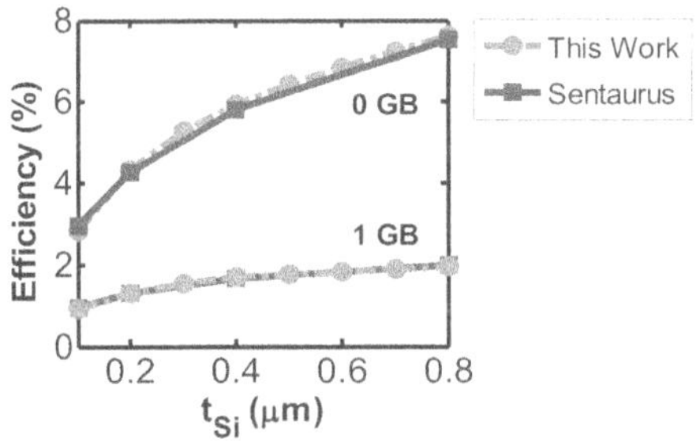

Figure 4. Efficiency as a function of device thickness, computed using our proposed model and using Sentaurus.

2.4 Unified SPICE Model

For the technology under consideration the overall efficiency of the solar cell is between the ones reported for the 0GB and 1GB cases. To determine this efficiency, one can build the SPICE model of the overall cell as a parallel combination of unit cells, some of which have 0GB and some of which have 1GB. In order to do that, the I-V curves of the solar cells with 0GB and 1GB cases are fitted to single-diode SPICE models. Then, given a total area constraint for the solar cell module, we divide this area into $0.2\mu m^2$ sized unit cells. The probability of having a GB within a unit cell is calculated using the formalism in [4].

The assumption here is that all unit cells are connected in parallel. Yet, to increase the output voltage of the proposed solar cell module, its area can be divided into M regions that are connected in series, as shown in Fig. 2. For example, the module shown in Fig. 2 has M=3, which results in an output voltage 3 times higher than a module that is not divided (*i.e.*, M=1).

3. POWER CONDITIONING UNIT DESIGN

As shown in Fig. 1, the solar module is connected to a power-conditioning unit that consists of a power converter and an MPPT circuit. In this section, we address the circuit-level design and modeling of this power-conditioning unit. The output current of the solar cell flows into the power converter and is delivered to an energy buffer for storage. The MPPT circuit acts as a real-time control unit, which dynamically adjusts the operation of the power converter to ensure maximum output power from the solar cell. The power converter can be a switched-capacitor type (*i.e.*, charge pump) or an inductive type (*i.e.*, boost/buck DC-DC converter). Capacitive power converters are the preferred option for low-cost applications since they avoid the area and cost overhead associated with integrating an inductor on chip. A traditional N-stage linear charge pump [14] is shown in Fig. 5.

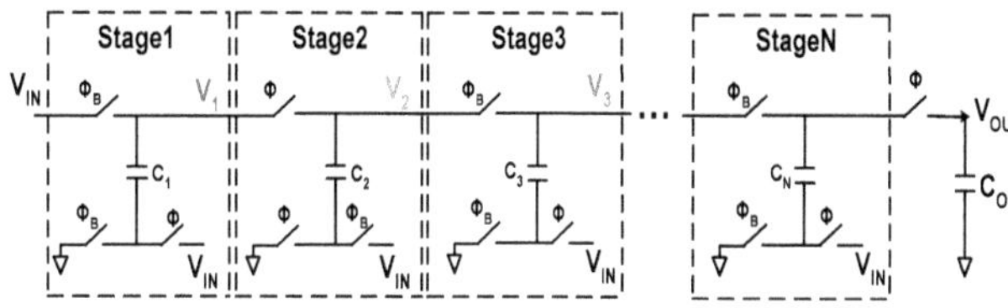

Figure 5. N-stage linear charge pump with gate control

The MPPT circuit provides non-overlapped clock signals (Φ and Φ_B) required to drive the charge pump stages. The MPPT circuit generates the required clock using either a VCO-based [17] or hill-climbing based control mechanism [3]. Here, we use the VCO-based control method due to its simple circuit structure and low-cost implementation.

The design parameters of a power conditioning unit need to be carefully determined and matched with the solar cell and energy buffer used in the system. The charge pump acts as a variable load whose impedance is mainly affected by the number of stages in the charge pump (N) and the applied switching frequency (f_{SW}), making these parameters crucial to the design. In this work, we used the analytical models described below for the power-conditioning unit. These models are also used in our proposed cross-layer optimization framework in Section 4.

The average output current of an N-stage linear charge pump can be expressed as [14]:

$$I_{out} = \frac{f_{SW}C}{N}[(N+1)V_{in} - V_{out}] \quad (9)$$

where C is the capacitor value used at each stage and V_{IN} and V_{OUT} correspond to the input and the output voltage of the power converter, respectively. If N is optimized with the goal of maximum output power per unit cost, the optimum number of stages (N_{op}) is given by [11]:

$$N_{op} = 2(\frac{V_{out}}{V_{in}} - 1) \quad (10)$$

During the day, the intensity of the incident light on the solar cell varies, which affects the output current of the solar cell. To extract the maximum available power from the solar cell at all times, the MPPT circuit changes the clock frequency (f_{SW}) and, thus, the input impedance of the power converter. It also determines the power loss of the converter, which is defined as ($f_{SW} \times \beta$) [17]. Assuming that the maximum available power at the output of the solar cell is provided at an output voltage (V_{MPP}) given by $\alpha \times V_{OC}$, where α is a technology specific constant and V_{OC} is open circuit voltage, the required f_{SW} is determined as [10]:

$$f_{sw} = \frac{I_{sat}(e^{qV_{MPP}/\alpha nKT} - e^{qV_{MPP}/nKT})}{NC(NV_{MPP} - V_{out})/(N-1) + \beta} \quad (11)$$

4. OPTIMIZATION FRAMEWORK

Once the analytical models for the solar cell and the power-conditioning unit are in place, we can optimize the energy harvesting system to satisfy the power requirements of the application unit under various design constraints. In a conventional design approach, a designer would pick a power converter and then select a solar cell array made out of commercially available solar cells. Here, we take a different approach (shown in Fig. 6) of joint optimization of the solar cells and the corresponding power converter unit. Since one of the goals of this optimization is to use a low cost fabrication process, the initial solar cell structure is set to have the minimum available cell thickness for the technology. It is assumed that the cell covers the entire area defined by the area constraint. Thus, the number of sub-cells connected in series within the given area (M) is 1. To start with, the I-V curve of the initial solar cell structure is calculated using the device model defined in Section 2 and the optimum power converter for this specific device is designed using the circuit model given in Section 3. The output power of the power converter for this configuration is stored in a buffer and the inner loop is repeated for a higher value of M. This procedure is repeated until M is equal to the number of sub-cells that requires only a single stage power converter (M_{MAX}). When this occurs, the maximum output power in the buffer (P_{OUT}) is compared with the system power requirement (P_{SYS}). If P_{OUT} satisfies P_{SYS}, the framework exits and reports this as the preferred solution. If not, the cell thickness is increased and the inner loop is repeated for different values of M (1 to M_{MAX}). The outer loop is iterated until P_{OUT} can satisfy the P_{SYS} requirement. Note that there might not always be a feasible solution for any given P_{OUT} and area constraint. In such a case, the algorithm terminates and the only viable option is for designers to relax their design constraints.

The design space output for this procedure contains design solutions with 4 parameters. For each different solar cell thickness (t_{Si}), there exist an optimal number of sub-cells that are connected in series (M), an optimal number of stages (N), and the switching frequency (f_{SW}) for the power converter. The output set varies depending on the area and power constraints as well as the grain size distribution of the process technology available to the user. Our cross-layer optimization approach takes advantage of having more control over the device parameters. Based on the device model, we can determine the exact t_{Si} and M values that will provide P_{SYS}. On the contrary, the conventional approach is limited by the output characteristics of commercially available solar cells. Since off-the-shelf components might not provide the optimized value exactly, this approach might require a larger area to satisfy P_{SYS}. In addition to this, we also optimize the power converter for each cell to make sure that the maximum available power is extracted.

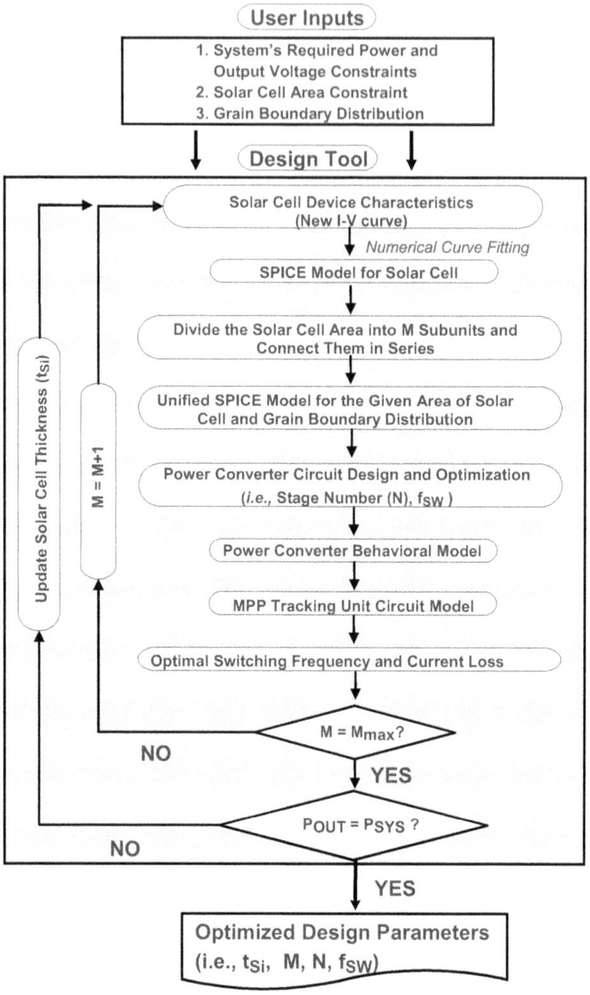

Figure 6. Cross-layer optimization procedure

5. RESULTS AND DISCUSSIONS

5.1 Design Space Exploration

To demonstrate the interdependence between the solar cell and power converter designs, the power converter design procedures for two different solar cell structures have been shown in Figs. 7 and 8. To obtain the results in Fig. 7, the solar cell thickness is fixed to be 0.1μm and its 0.2cm^2 area is divided into M regions to be connected in series. It can be seen that for each M, there is an optimal value of N for the power converter given by Eq. 10. Also, the optimal power converter design has 3 stages for a solar cell with M of 3. For M smaller than 3, the number of sub-cells connected in series is limited, resulting in a low voltage at the output of the solar cell ($V_{OUT\text{-}SC}$). In order to convert this voltage to the voltage requirement of the energy buffer (V_{BUF}), in this case 2.4V as the terminal voltage of two standard AA batteries, higher number of stages is required, that increases the power loss in the converter. On the other hand, for M higher than 3, $V_{OUT\text{-}SC}$ increases as the output current of the solar cell ($I_{OUT\text{-}SC}$) decreases. Thus, a reduction in the output current of the power converter ($I_{OUT\text{-}PC}$) is observed.

For a thicker solar cell (0.8μm), the light induced generation increases. Therefore, the output current of the device increases. Yet, it leads to an increased recombination due to the higher number of carriers generated via absorption and reduces the

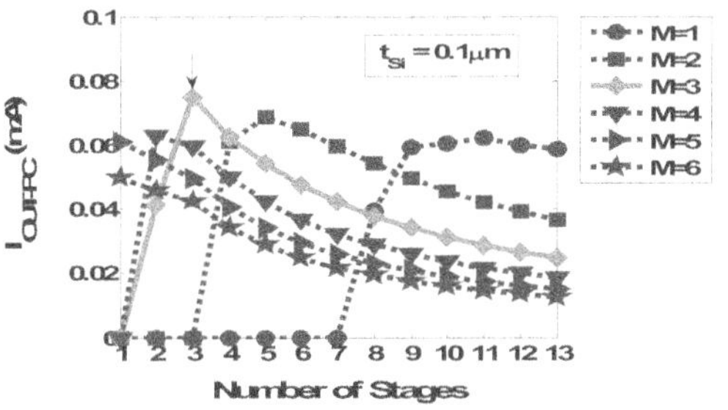

Figure 7. The current at the output of the power converter (I_{OUT-PC}) with respect to the number of stages (N) for a 0.1μm thick solar cell with different M values.

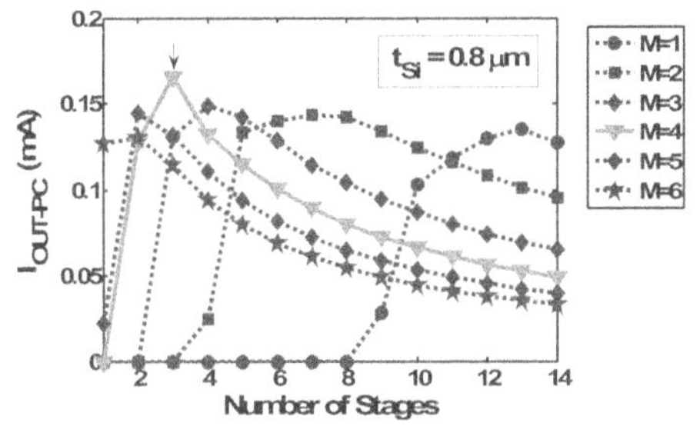

Figure 8. The current at the output of the power converter (I_{OUT-PC}) with respect to the number of stages (N) for a 0.8μm thick solar cell with different M values.

output voltage of a sub-cell. Therefore, when the same design procedure for the power converter is applied to this device, there is a need for more sub-cells to be connected in series to satisfy V_{BUF}. As can be seen in Fig. 8, the optimal power converter design is changed to a 3-stage power converter for a solar cell that has $M=4$. The change in optimal M with respect to the cell thickness (t_{Si}) is shown in Fig. 9. To obtain this figure, N is also optimized to get the maximum I_{OUT-PC} for each data point. Note that, if M is fixed to 3, the design will be optimal only for cell thicknesses smaller than 0.5μm.

Given Figs. 7-9, an optimal design can be obtained by first selecting t_{Si} and then determining the M, providing the maximum I_{OUT-PC}. For this M, N_{OP} and f_{SW} can be calculated using Eqs. 10 and 11.

In case I_{OUT-PC} is not enough, one can increase the area of the solar cell to satisfy the power constraint. From the results depicted in Fig. 10 it can be observed that I_{OUT-PC} scales up with an increase in the device area. It can also be seen that the optimal M and N do not change with the increase in area. When the area is increased for a fixed M, V_{OUT-SC} remains the same and I_{OUT-SC} increases. Since V_{OUT-SC} is fixed, from Eq. 10, N_{OP} is also expected to be fixed. Based on Eq. 9, the increased I_{OUT-SC} can be extracted by the same power converter using a higher f_{SW}. This is a feasible option in the context of micro-scale energy harvesting systems, since the required f_{SW} is usually in the region of a few MHz, which can be easily generated in practice [3][10] [17].

A designer who has control over the device parameters would like to know the optimum design parameters for the specific fabrication process. For LTPS process technologies, a larger average grain size is obtained for higher process temperatures [6]. For the proposed solar cell structure, larger grain size translates to lesser number of GBs in the cell. This leads to a higher V_{OUT-SC}. Hence, less number of sub-cells connected in series can satisfy

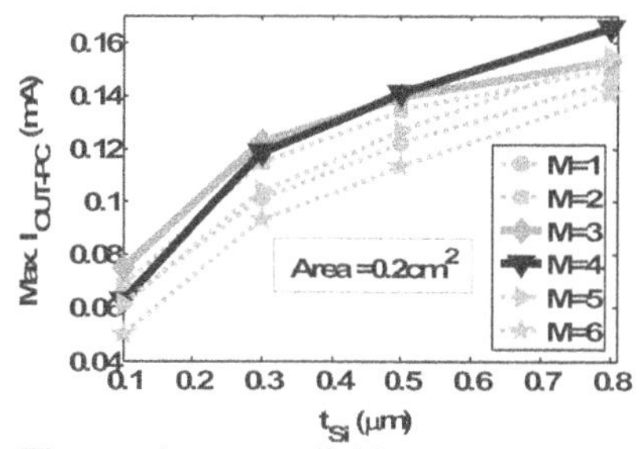

Figure 9. The maximum available current at the output of the power converter (I_{OUT-PC}) with respect to cell thickness (t_{Si}) for different M values.

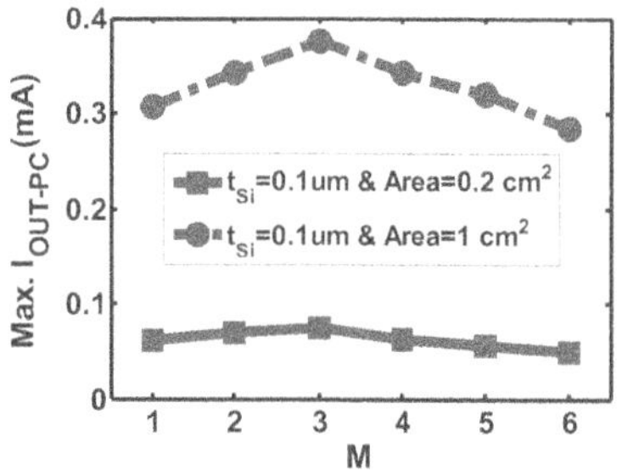

Figure 10. The maximum available current at the output of the power converter (I_{OUT-PC}) for different area constraints.

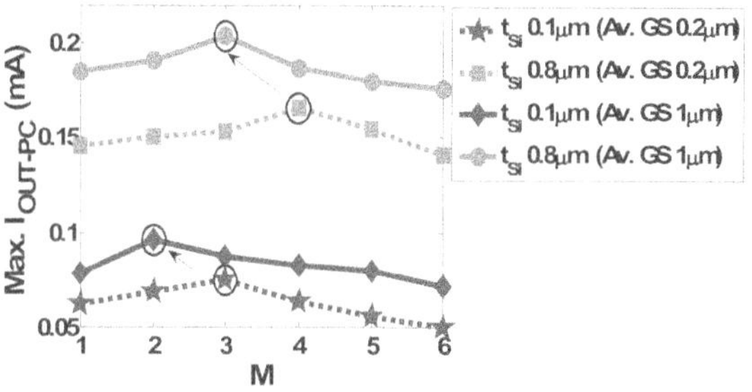

Figure 11. The change in optimum M for different t_{Si} and average grain sizes (Av. GS).

the V_{BUF} requirement. As it can be observed in Fig. 11, once the average grain size is increased from 0.2 to 1μm, the optimal M decreases for both t_{Si} values. The improvement in I_{OUT-PC} is 27% for the 0.1μm thick cell, while it is 31% for the 0.8μm thick cell.

5.2 Comparative Case Study

Next, we present a case study that illustrates the improvement that can be achieved with our proposed cross-layer optimization approach, compared to a baseline design that optimizes the power converter and solar cell separately. For this purpose, we assume that a TELOS wireless sensor node is the application unit that needs to be powered as shown in Fig. 1 [15]. The TELOS node consumes up to 21.8 mA of current in the active mode. If the system was operated with a 5% duty cycle (this is a typical duty cycle for wireless sensor nodes), the average current requirement of the node would be close to 1mA. The module requires a voltage source between 2.1V and 3.6V, so we chose to use two AA batteries (nominal voltage 2.4V) as the energy buffer for this design example.

The following design approach (indicative of how these systems are designed today) is used as a baseline for comparison. The goal is to minimize the total area of the solar panel. Therefore, maximum thickness (0.8μm) is used for the solar cell due to its high efficiency. It is assumed that the area of a sub-module is $0.2cm^2$ and that the designer will build up an array from these sub-modules using as many sub-modules as required to satisfy the power requirement. The power converter is selected to have a single stage since, for a given P_{SYS}, it requires the smallest area when compared to power converters with higher *N*. For such a 1-stage power converter, since the output voltage requirement is 2.4V, the input voltage (which is also the output voltage of the solar cell, V_{OUT-SC}) has to be at least 1.2V. To obtain this V_{OUT-SC}, either 6 or 7 cells should be connected in series. If 7 cells are connected in series, then an array of 1x7 would be enough to satisfy the P_{SYS} requirement. In summary, our baseline is also an optimized design. Note, however, that no cross-layer optimization was used.

If the system is designed with our approach (with the same area as in the conventional design), one obtains the design space shown in Fig. 12. For 1.4 cm^2 of cell area, fabricated with a process technology that has an average grain size of 0.2μm, we can satisfy the same power constraint with t_{Si} of 0.5μm. We can achieve the same P_{SYS} with $7\text{x}10^7$ μm^3 material while the conventional design requires $11.2\ \text{x}10^7$ μm^3. Under this scenario, our approach reduces the material cost of the solar cell by 37.5%. On the other hand, if we are allowed to use a 0.8μm thick solar cell, within the same area, we can achieve an I_{OUT-PC} of 1.16mA, which corresponds to a 16% improvement in the system output power. Finally, if the design is implemented in a process technology with 1μm average grain size, P_{SYS} can be satisfied within a smaller area and the optimized design parameters are significantly different from that of a technology with 0.2μm grain size.

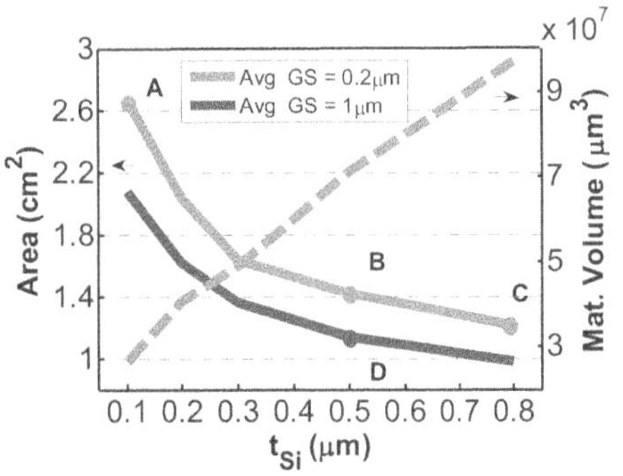

	t_{Si} *(μm)*	*M*	*N*	f_{SW} *(MHz)*
A	0.1	3	3	1.843
B	0.5	4	3	1.62
C	0.8	4	3	1.832
D	0.5	3	2	2.005

Figure 12. The resulting design space to obtain 1mA I_{OUT-PC} and 2.4V V_{OUT-PC} for different average grain sizes (Avg. GS).

6. CONCLUSION

In this work, we demonstrated the need for cross layer optimization of the solar cell and the power-conditioning unit in a micro-scale solar energy harvester. In order to do this optimization, the required analytical models for the polysilicon solar cell have been developed and an existing model has been used for the power-conditioning unit. Our results indicate that the power conditioning circuit design changes with solar cell parameters such as cell thickness, cell configuration and average grain size of the polysilicon material. When compared to a baseline design approach, our approach improves the output power of the energy harvesting system by 16%. Alternatively, the same power constraint can be achieved using a smaller area. Under iso-power conditions, our optimization technique can result in 37.5% decrease in the amount of material used.

7. ACKNOWLEDGMENTS

This research was funded in part by the National Science Foundation, under grant CCF-1018358.

8. REFERENCES

[1] Akyildiz, I. F., Su W., Sankarasubramaniam, Y., Cayirci, E. 2002. Wireless sensor networks: a survey. *Computer Networks*. 38(4), 393-422.

[2] Brunelli, D., Benini, L., Moser, C., Thiele, L. 2008. An efficient solar energy harvester for wireless sensor nodes. In *Proc. DATE* (March 10-14, 2008). 104–109.

[3] Carvalho, C., Lavareda, G., Lameiro, J., Paulino, N. 2011. A step-up μ-power converter for solar energy harvesting applications using hill climbing maximum power point tracking. In *Proc. ISCAS* (May 15-18, 2011). 1924-1927.

[4] Cheng, C. F., Jagar, S., Poon, M. C., Kok, C. W., Chan, M. 2004. A statistical model to predict the performance variation of polysilicon TFTs formed by grain-enhancement technology. *IEEE T. Electron Dev.*, 51(12), 2061-2068.

[5] Hande, A., Polk, T., Walker, W., Bhatia, D. 2006. Self-powered wireless sensor networks for remote patient monitoring in hospitals. *Sensors*, *6*(9), 1102-1117.

[6] Jeon, J. H., Lee, M. C., Park, K. C., Jung, S. H., Han, M. K. 2000. A new poly-Si TFT with selectively doped channel fabricated by novel excimer laser annealing. In *Proc. IEDM* (Dec 10-13, 2000). 213-216.

[7] Kim, Y., Chang, N., Wang, Y., Pedram, M. 2010. Maximum power transfer tracking for a photovoltaic-supercapacitor energy system. In *Proc. ISLPED* (Aug 18-20, 2010). 307-312.

[8] Kimball, J. W., Kuhn, T., Balog, R. S. 2009. A system design approach for unattended solar energy harvesting supply. *IEEE T. Power Electr.*, 24(4), 952-962.

[9] Li, J., Bansal, A., Roy, K. 2007. Poly-Si thin-film transistors: an efficient and low-cost option for digital operation. *IEEE T. Electron Dev.*, 54(11), 2918-2929.

[10] Lu, C., Park, S. P., Raghunathan, V., Roy, K. 2012. Low-overhead maximum power point tracking for micro-scale solar energy harvesting systems. In *Proc. VLSI Design Conference* (Jan 7-11, 2012). 215-220.

[11] Lu, C., Raghunathan, V., Roy, K. 2011. Efficient design of micro-scale energy harvesting systems. *IEEE Journal on Emerging and Selected Topics in Circuits and Systems,* 1(3), 254-266.

[12] Lu, C., Raghunathan, V., Roy, K. 2010. Maximum power point considerations in micro-scale solar energy harvesting systems. In *Proc. ISCAS* (May 30 –June 2, 2010). 273 – 276.

[13] Mungan, E. S. and Roy, K. 2011. 2D Modeling and optimization of excimer laser annealed thin film polysilicon solar cells. In *Proc. PVSC* (Jun. 19-24, 2011). 3059-3064.

[14] Pan, F. and Samaddar, T. 2006. Charge Pump Circuit Design, McGraw-Hill Professional.

[15] Polastre, J., Szewczyk, R., Culler, D. 2005. Telos: enabling ultra-low power wireless research. In *Proc. IPSN* (Apr. 25-27, 2005). IEEE Press, Piscataway, NJ, USA, Article 48.

[16] Qi, Y. and McAlpine, C. 2010. Nanotechnology-enabled flexible and biocompatible energy harvesting. *Energ. Environ. Sci.*, 3(9), 1275-1285.

[17] Shao, H., Tsui, C. Y., Ki, W. H. 2007. A micro power management system and maximum output power control for solar energy harvesting applications. In *Proc. ISLPED* (Aug. 27-29, 2007). 298-303.

[18] Synopsys Inc. SenTaurus Device Simulator. Mountain View, CA.

Static Low Power Verification at Transistor Level for SoC Design

Jérôme Lescot
STEricsson
12, rue Jules Horowitz, BP 217
38019 Grenoble, France
+334-7658-4620
jerome.lescot@stericsson.com

Vincent Bligny
STEricsson
12, rue Jules Horowitz, BP 217
38019 Grenoble, France
+334-7658-5768
vincent.bligny@stericsson.com

Dina Medhat
Mentor Graphics
78 EL Nozha st, Elsbaa Emarat
11361 Cairo, Egypt
+202-2413-5545
dina_medhat@mentor.com

ABSTRACT

This paper presents a transistor-level verification flow to detect electrical overstress, static leakage and ESD-CDM issues in large low power SoC circuits. With innovative features like Spice patterns recognition and static voltage propagation by Calibre® PERC™, this approach brings significant added value to standard digital checkers that cannot capture some parts of the design in essence, such as analog IPs or third-party IPs designed outside power-intent driven flows. The results obtained on a 32nm CMOS circuit using multiple separate supplies with body biasing strategy, demonstrate the ability of this solution to cope with complex design architectures. As a matter of fact, some severe issues like hundreds of missing level shifters and weak input stages inside isolation cells were detected in few hours.

Categories and Subject Descriptors

B.7.1 [**Integrated Circuits**]: Types and Design Styles – *VLSI.*
B.7.2 [**Integrated Circuits**]: Design Aids – *verification.*

General Terms

Design, Reliability,Verification.

Keywords

Reliability, ESD-CDM, low power, verification, pattern matching.

1. INTRODUCTION

Shrinking process geometries have led designers to use various power optimization strategies to minimize static and dynamic power. In fact, advanced low power design architectures implement tens of separate power domains with power gating and body biasing [1], and make power aware verification an increasingly challenging task in standard design flows. Besides, currents leakage is a painful defect to debug on silicon and generally requires complete silicon re-spin. As a consequence, there is an increasing demand from semiconductor industry for robust verification tools to validate power intent and report static leakage issues at system-on-chip (SoC) level.

ISLPED'12, July 30-August 1, 2012, Redondo Beach, California, USA.

Unlike most recent commercial solutions [2,3] used for static checks in low power digital design and based on CPF (Common Power Format) or UPF (Unified Power Format) description [4], the solution presented in this paper can explore all parts of the design, including both digital blocks, analog IPs and the interfaces in between them.

This innovative verification approach works on a transistor-level netlist description of the circuit, such as the one used for final layout versus source (LVS) comparison and does not make any abstraction at block level. Instead, a powerful pattern recognition technique is used to recognize standard logic gates or structures designed for low power purpose, to generate reliable reports. Structural errors such as floating gates, missing isolation cells and level shifters or functional errors such as sleep signal hard-tied to power/ground nets can be captured during the analysis. A user-friendly graphical interface that sets up the verification environment and drives the user from the declaration of the supplies to the analysis of the reports, offers the capability to highlight violations in either a virtuoso schematic view if available, or an automatically generated one otherwise.

The hierarchical extraction of the circuit power setup is described in Section 2. The concept of static voltage propagation is explained in Section 3. Electrical rules implemented for low power checks and Spice pattern recognition of specific topologies are described respectively in Sections 4 and 5. In Section 6, experimental results obtained on a 32nm CMOS SoC design including more than 450millions on-chip transistors, are presented.

2. POWER SUPPLIES DETECTION

The very first step of the verification flow is to define the power scenario of the design. UPF (Unified Power Format) and CPF (Common Power Format) are similar languages designed for the definition of power management in digital design flow. However, the design of some IPs such as analog blocks does not use any of these standards. Therefore, a specific program in Python object-oriented language was developed to parse the transistor-level netlist of the circuit and extract power/ground nets.

2.1 Primary Supplies Detection

As shown in Fig. 1, the implemented algorithm traverses the design hierarchy from the top cell and inspects each cell by summing the number of connections to bulk terminals of MOSFETs for each node. After processing all devices and instances within the cell, only nodes whose number of bulk connections exceeds a user-defined threshold (default: 10), are marked in the cell object and resulting connections number will be preserved for next instances of the same cell. As shown in Fig.1, the instance X0 is the first instance of cell A that is analyzed and 2 bulk connections are marked on port A. Then, all other instances of cell A will be considered as black boxes for the rest of the extraction. For instance, node VBAT connected to pin A of instance X1 will be quickly incremented by two more bulk connections without descending in cell A anymore. Note that both the number of connections and the type of transistors (n, p) are made available to the parent cell by adding these properties on the appropriate ports of the sub cell. Nodes that have connections to bulk terminals of n-type transistors only are automatically declared as grounds and associated with a null voltage value. Other extracted nodes are declared as power supplies and a valid voltage value must be defined by the user.

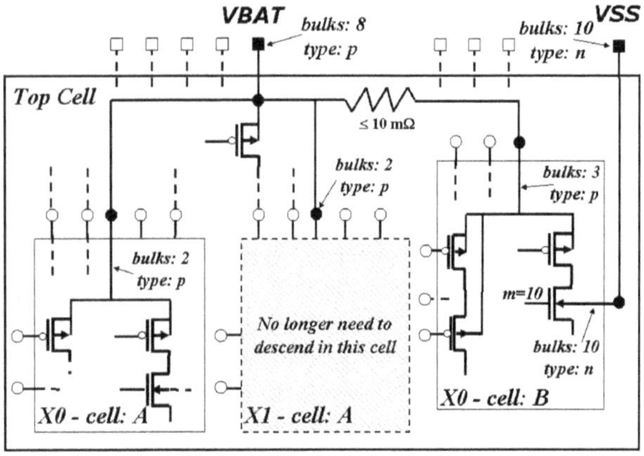

Figure 1. Hierarchical detection of power and ground nets by inspecting every cell only once.

An additional option enables the Python program to consider the source (s) terminal too, which is a mandatory feature in most advanced architectures to catch separate supplies used for reverse or forward body biasing [5,6]. Note that the supplies detection process is not sensitive to short resistors widely used in analog IPs for star routing. As a matter of fact, all resistors whose value is smaller than a user-defined threshold (default: 10mΩ) are simply considered as short connections that do not break power or ground paths.

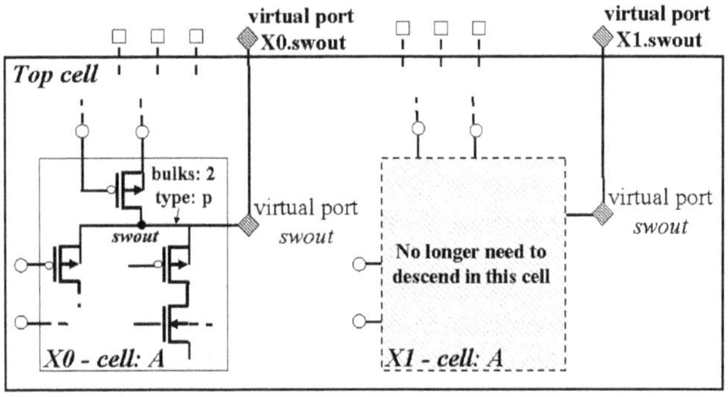

Figure 2. Hierarchical detection of internal supplies with promotion of virtual nodes to circuit top level.

2.2 Internal Supplies Detection

Some internal nodes may serve as supplies for parts of the circuit controlled by MOSFETs switches or specific circuitry such as DC-DC converters or voltage regulators. As these power islands cannot be ignored during static low-power verification, the detection of all internal supplies of the circuit is needed and achieved during the same hierarchical walk as one described in Fig. 1. Whenever a node internal to the cell under analysis is found with a sufficient number of bulk connections (default: 10), then a virtual node is created in the parent cell with the flat name of the original node, as depicted in Fig. 2. At the end of the bottom-up promotion of such virtual nodes to upper levels of hierarchy, flat names of all internal supplies are available at top level and can be captured in power setup. The run time performance overhead of the internal supplies detection is always less than 10% of overall run time for all in-house multimillion transistor ICs analyzed so far.

3. STATIC VOLTAGE PROPAGATION

Once voltages are properly defined on top level pads or internal nets in the design, some basic conduction rules are applied to propagate these voltages across both n-type or p-type channels as shown in Fig. 3. Other components like inductors or resistors are treated as short connections while diodes and capacitors are considered as open circuits. This voltages propagation mechanism is achieved in a hierarchical manner on very large designs with a commercial solution called Calibre-PERC [7,8,9]. In fact, for memory and runtime efficiency Calibre-PERC wherever possible maintains the circuit hierarchy and makes an intensive usage of this optimized circuit structure to quickly assign net voltages.

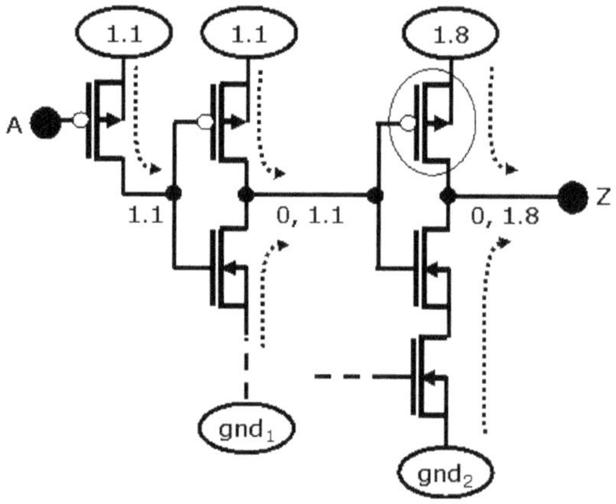

Figure 3. Voltages propagation mechanism across MOSFETs.

Because Calibre-PERC is cell-based and because a cell's voltages are context-dependent, Calibre-PERC constructs a cell signature that represents each unique combination of cell-in context. In this way each cell's voltages only need be computed once per signature. Calibre-PERC allows the user to create a voltage-dependent path through a design. Depending on various conditions, including the voltages between, for example, the gate and source pins of a MOSFET, paths through primitive devices are established. Calibre-PERC then iteratively propagates these voltages up and down the circuit hierarchy until voltages on all nets are known and stable. Propagation of a signal halts when a top level circuit pad, other user-defined port or specifically designated net (e.g. power or ground rail) is reached. As a result of this propagation process, each electrical node of the design network exhibits a range of voltage values. Compared to standard DC simulation, voltage propagation process is obviously much faster and thus can work on complex and large-scaled SoC circuits. Besides, no accurate model of the devices present in the design is required. Only the type of the transistors is needed, which makes this verification flow technology independent.

4. ELECTRICAL RULES CHECKS

Once voltages have been propagated across the design, it is quite straightforward to apply static checks on any type of device by inspecting voltages value on every terminal of the device. The flexibility of Calibre-PERC API in Tool Command Language (TCL) on top of fast built-in procedures, enables the quick and robust implementation of various checks, such as ones described in following sections.

4.1 Electrical Floating MOS Gates Detection

One simple check consists in reporting gate (g) terminals of MOSFETs that could not be reached by voltage propagation. These are electrical floating gates issues that generate unpredictable leakage on silicon. In this vector-less approach, a careful attention must be paid to pins at top level such as the pin A depicted in Fig. 3, that are not part of the supply setup and are left topologically floating during the analysis, which may result in many false electrical floating gates. This issue is fairly simple to solve by removing from the report, all MOS gates whose flat electrical node is a port of the top cell.

4.2 Over-voltage on Thin Oxide Transistors

Another rule is applied to all thin oxide transistors of the circuit to catch any occurrence of a disruptive voltage across the device terminals. Note that the list of thin oxide transistors is selected by the user from the complete list of MOS models automatically detected during the hierarchical walk described in Fig 1. A reliability fault is raised whenever following condition applied on the voltage values of the *d,s,g,b* terminals, is true:

$$max(v_G) > min(v_B, v_S, v_D) + V_{breakdown}$$
$$or$$
$$min(v_G) < max(v_B, v_S, v_D) - V_{breakdown}$$

4.3 Detection of Missing Level Shifters

To detect missing low-to-high (resp. high-to-low) level shifters on the boundary between separate power domains, the following condition is applied:

$max(v_G) < max(v_B)$ *for rule: missing low-to-high level shifter*

$max(v_G) > max(v_B)$ *for rule: missing high-to-low level shifter*

An example of missing low-to-high level shifter is given with the transistor circled in Fig. 3 where the maximum voltage on gate terminal is 1.1V while voltage on (d) and (b) terminals is 1.8V.

4.4 Detection of Missing Isolation Cells

To detect missing isolation cells on the boundary between switch-OFF and power-ON domains, the same rule as one used to detect missing level shifters is applied but a reserved voltage value is defined in the power setup for all OFF power domains.

5. LOCAL STRUCTURES RECOGNITION

5.1 Purpose of Spice Pattern Matching

Electrical rules checks applied at transistor level suffer from the lack of knowledge regarding the intention or functionality of the structure where the violation is found. For instance, a high number of false errors are reported on the boundaries between separate power domains in level shifter structures that include, by construction, some transistors exposed to both low and high voltages, such as devices *M2* and *M3* in Fig. 4-a. Another example of systematic false errors is the one found in isolation cells used between OFF and Always-ON domains where transistors present in *NAND* or *NOR* logic gates will be raised as violations. Therefore, the ability to remove such recurrent false errors by using a Spice pattern recognition technique to identify particular topologies, was a key requirement for wide adoption of this verification tool by design community.

5.2 Spice Pattern Recognition Algorithm

To achieve the reliable recognition of some particular structures, like level shifters, enable level shifters and logic gates, a graph-comparison algorithm was implemented in Python language to solve the theoretical problem of sub-graph isomorphism [10,11]. All topologies are declared in Spice syntax as sub-circuits including a set of standard devices with M, Q, C, L or R prefix, some internal nodes and external ports. The resulting graph contains two types of vertices: devices and nets. Device vertices only connect to net vertices, and net vertices only connect to device vertices. A labeling process is applied during graph generation in order to speed up the circuit comparison. The labels on device nodes recall the type of the device like p- or n-channel. The link labels recall the device terminal, such as drain (d), source (s), bulk (b) and gate (g). The swappable nature of source and drain terminals is made possible by using the same label for (s) and (d) terminals.

A level shifter serving as the sub circuit (i.e. pattern template) is shown in Fig. 4-a. The graph corresponding to this structure is depicted in Fig. 4-b where a device is represented as a square while a net is represented as a circle. The links are the connections between devices and nets through device terminals. The bulk terminal of the transistors is not shown for simplification and can be also ignored in the algorithm. The reduction of parallel MOS devices into a unique device is also achieved during graph generation as matching process must remain insensitive to the number of parallel MOS devices.

(a) (b)

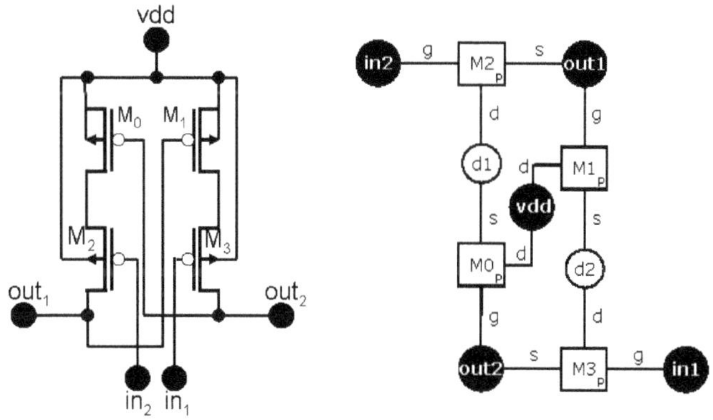

(c)
```
.SUBCKT level_shifter vdd in1 in2 out1 out2
  M0 d1 out2 vdd vdd p
  M1 d2 out1 vdd vdd p
  M2 d1 in2 out1 vdd p
  M3 d2 in1 out2 vdd p
.ENDS level_shifter
```

Figure 4. (a) Core structure of a standard level shifter. (b) Graph representation and (c) Spice netlist of this topology.

In Fig. 5, a schematic is shown as the main circuit, in which we will find the instances of the pattern template. Our goal is to verify that the topology composed of *Mx* and *My* in the main circuit is isomorphic to the pattern structure. The starting point of the pseudo-code of the general algorithm [12] described in Table 1, is the violated transistor to be examined, such as *Mx* and *My* devices that will be compared to *M2* and *M3* in the pattern. The algorithm of Table 1 implemented in Python interpreted language, searches exhaustively regardless of the variable order or choices of neighbors, but both will impact performance and may affect which of potentially multiple matches will be found first. To further speed-up matching process, some additional properties are attached to nets, such as the number of p-type (resp. n-type) MOS devices connected to it along with the Spice index of the connected device terminal.

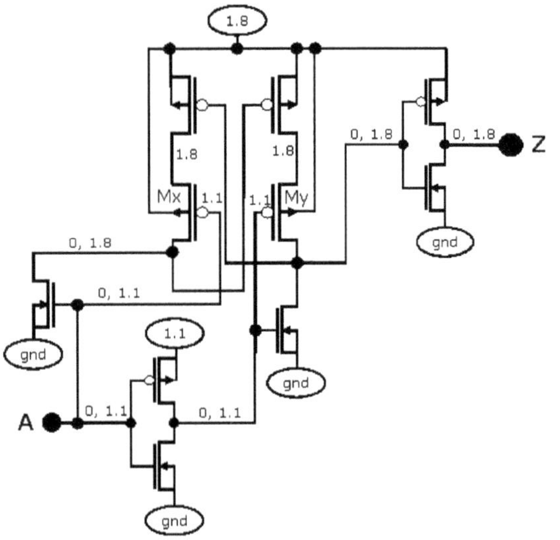

Figure 5. Level shifter circuit where *Mx* and *My* devices are reported as systematic violations in missing level shifter check.

Table 1. General core algorithm of the Spice recognition technique implemented in Python object-oriented language.

```
bound the violation in circuit to the device in pattern that can
be waived. Example: Mx in the circuit bound to M2 or M3 in
the pattern
while (not all variables are bound) {
 select an unbound variable, u, adjacent in the pattern to a
bound variable, b
 for (all objects, o, adjacent in netlist to the object bound to b){
  if (o is already bound to a variable) skip to next o
   for (all bound variables, a, adjacent to u) {
    if (object bound to a not adjacent to the object bound to u)
      {
      skip to next o
     }
   }
   bind o to u
   proceed to next u
 }
 return failure
}
return success
```

5.3 Application to ESD-CDM Checks

A very attractive application of pattern recognition is the implementation of ESD-CDM checks, where the pattern to recognize is the ESD protection cell and the starting point of matching process is the gate terminal of the device to protect. However, conventional ESD protections such as ground-gate CMOS (GGMOS) or double-diode structures [13], can be found anywhere in the design and are usually outside the cell of the device to protect. Therefore, a graph representation of the entire circuit is needed, which may become very memory and run time consuming at SoC level with interpreted language code. Fortunately, Calibre-PERC provides an accurate way of finding desired pattern instances without limitation on the scope of search, by using similar algorithm as one presented in Section 5-2. A pattern template is more than just a Spice sub-circuit with Calibre-PERC, it also has attributes such as power/ground domains and IO pads, that the matching process is sensitive to. As a result, the found pattern instances not only are topologically equivalent to the template, they also meet the condition set by the attributes. For example, if some ports of the ESD protection pattern are declared as power/ground pins, then only instances of this pattern that are also connected to power/ground nets in the design will be reported. Note that these attributes stand also for very useful criteria to selectively prune the scope of search and accelerate the pattern recognition process at IC level. As a result, by using this robust Spice matching process in combination with the condition used for the detection of missing level shifters described in section 4-3, an additional ESD-CDM rule was implemented to report all transistors on the boundary between separate domains without proper ESD protection [13].

6. EXPERIMENTAL RESULTS

A graphical user interface was developed in-house to enable the user to follow all the steps of the entire verification flow shown in Fig. 6, in a user-friendly manner. Results analysis operates under Calibre-RVE environment that is able to create a schematic view (if not available in virtuoso framework) of the cell that contains the violation.

In this section, the interest of the verification tool is demonstrated on a 32nm CMOS SoC including 470 million on-chip transistors. This digital-based band circuit is the main component of a multimedia platform targeting mid to high-end smarphone and tablet applications. The system is divided in three separate voltage domains each supplied by its own SMPS from energy management circuit that can be controlled either in OFF, HighPower and LowPower scenario. Two additional IO domains exist in always-ON mode, or low power when no activity. All power domains can be switched ON/OFF by internal switches. Most of the voltage domains contain additional power islands that can be also switched OFF by use of internal switches.

As previously explained, the flow has the capability of generating standalone power setup files, provided that a transistor-level netlist description of the entire circuit is available. In order to declare the different scenarios of the circuit, power supplies are first extracted with the hierarchical process described in Section 2 on the same circuit netlist as one used for layout versus source (LVS) comparison at SoC level. As reminded in the statistics of Table 2, all power and ground nets, including internal supplies were extracted in less than one hour on a quite standard machine available in-house. At the end of this extraction process, an editor form appears on screen to let the user quickly define the voltage value of each net, with the help of widgets to perform multiple nets selection, immediate probing in schematic views, etc.

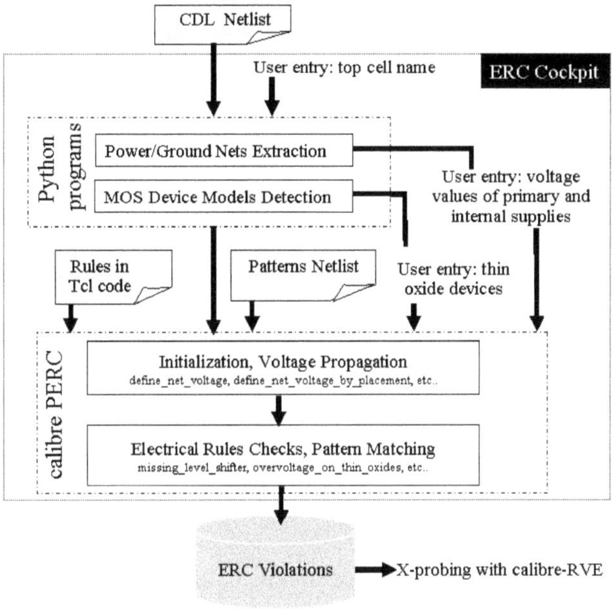

Figure 6. Low Power Verification Flow at Transistor Level.

6.1 Over-voltage on Thin Oxide Transistors

The power of this solution to explore analog IPs in digital context was successfully demonstrated on this circuit by detecting destructive voltages on thin oxide transistors inside a thermal sensor. Such errors were not detected at IP level with Spice simulations because test benches did not capture the actual platform context.

6.2 Missing Level Shifters

About 850 missing level shifters were reported in some portions of the circuit due to a mistaken connection of the low voltage supply, as shown in Fig. 7. Estimated impact on silicon is a current leakage of 15mA WC. Note that more than twenty thousand violations (see Table 2) in level shifter topologies were removed from the final report by using the local pattern recognition technique described in Section 5.2.

Table 2. Key figures for a 32nm SoC

Number of MOS devices	470 Meg
Number of systematic errors found in level shifters and removed with pattern recognition	2 x 14.000
Run time for supplies detection	40mn*
Run time for electrical floating gates detection	3h*
Run time for missing level shifters detection	4h*
Number of missing level shifters violations Estimated leakage on silicon	844 15 mA WC
Number of corrupted isolation cells	>1k
Number of thin oxide devices exposed to overvoltage in analog IP (thermal sensor)	~40
** Machine: Intel Xeon @ 2.93GHz Memory: 144GBytes*	

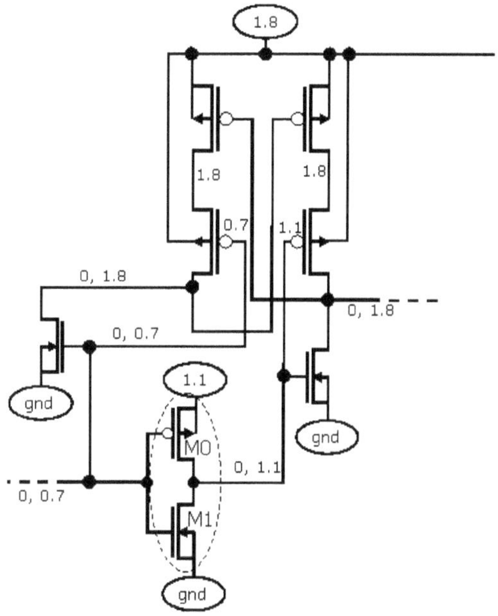

Figure 7. Example of violations found in input stage: {*M0, M1*}

of a level shifter cell due to a wrong connection

of supplies at top level (Vgmax=0.7, Vbmin=1.1).

6.3 Isolation Errors in Unexpected Topologies

This circuit was a very good candidate at demonstrating the interest of the pattern recognition feature, not only to remove systematic violations in level shifters (Fig. 5) but also to identify logic gates in isolation cells between OFF and always-ON power domains. Such isolation cells are expected to be found in *NOR* and *NAND* topologies before removal from the final report. However, since such violations were observed in *INVERTER* cells, a deeper analysis at schematic view level revealed an actual design issue. As shown in Fig. 8, when the input stage of the isolation cell is turn off, then *Mx* and *My* devices are still powered by the always-ON supply and may be subject to a static leakage whose magnitude depends on the undefined voltage of input stage. This design error that digital checkers could not detect by using liberty files of the cell was immediately confirmed and fixed by the standard cells provider.

6.4 Body Biasing Related Errors

Usually, digital designers consider a MOS transistor as a three terminals device where the substrate is tied to the source. However, by making the substrate as a separate terminal and applying a reverse bias, the threshold voltage is increased and the leakage is reduced [5,6]. This reverse body biasing technique is generally used in stand-by mode so as not to impact the performance of the devices. However, the introduction of separate supplies between source and bulk terminals increases the complexity of the power management architecture. More importantly, new connectivity errors may appear and translate in severe leakage issues that cannot be captured by current static digital checkers. In fact, the widely used version 1.0 of UPF does not support the definition of bias supplies. To cope with this limitation that will be overcome with full support of UPF 2.0 [14], the flexibility of our solution made it possible to quickly implement a new rule to compare voltages between source and bulk terminals and raise an error under specific conditions. This new rule check raised about one hundred standards cells as potential sources of leakage, which was confirmed by design team afterwards.

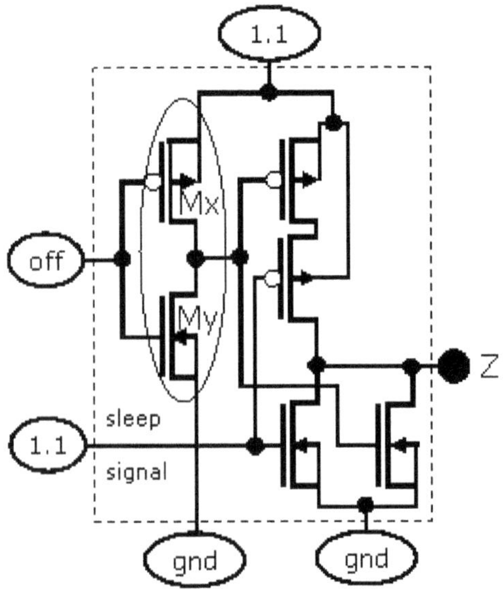

Figure 8. Isolation cell with current leakage issue on inverter structure: {*Mx*, *My*} when input stage is turn off.

7. CONCLUSION AND FUTURE WORK

This paper presented a fast yet robust and accurate solution for the static low power verification of large-scale circuits at transistor level. This vector less and technology independent approach enables a very fast set up of the verification environment inside a graphical cockpit that supports the hierarchical extraction of power/ground nets and the automatic detection of MOS models for a quick user-selection of thin oxide devices. The capacity of Calibre-PERC engine to propagate voltages across multi-million transistors ICs in a few hours enables the verification of multiple power scenarios overnight.

The capacity of this solution to verify some parts of the designs that digital checkers cannot capture due to human mistakes in implementation flow, UPF 1.0 limitations or the nature of the IP itself (i.e. analog IPs), was demonstrated on a 32nm CMOS circuit and was determinant in wide adoption by design community with the full support of the Design Quality Department. Note that the cockpit depicted in Fig. 6 can also be shipped to IP providers so that they can achieve the verification of their IP much sooner than tape out. Main interest is to avoid finding issues in third-party IPs very late in design flow i.e. during final verification at SoC level.

More efforts are currently spent in the development of new features to extract most relevant power scenarios of the circuit from the power state tables (PST) available in UPF files. In parallel, Calibre-RVE used for report analysis is currently enhanced to implement techniques already available in-house like the grouping of violations by topology, by cell name or any type of relevant information extracted during the analysis in order to help the end user quickly sort detected errors by severity.

8. ADDITIONAL AUTHORS

Didier Chollat-Namy (STEricsson, email: didier.chollat-namy@stericsson.com)

Ziyang Lu, Sophie Billy and Mark Hofmann (Mentor Graphics, emails: ziyang_lu,sophie_billy,mark_hofmann@mentor.com).

9. REFERENCES

[1] Keating, M., Flynn, D., Aitken R., Gibsons, A. and Shi K., Low Power Methodology Manual for Systems on Chip Design. *Springer Publications*, (New York, 2007).

[2] Saint-Preux F., MVRC Usage on a Complex Design from the RTL to the Low Power Signoff, *SNUG* (France, 2010).

[3] Dewangan, R., Varadarajan, R., Kumar Gupta, J., Mohindru, N., Soman, S., White paper, Atrenta Inc. SoC Physical Closure Begins at RTL, (2010).

[4] IEEE Standard for Design and Verification of Low Power Integrated Circuits, (March 27, 2009).

[5] Arnim K., Borinski E., Seegebrecht P., Friedler H., Brederlow R., Thewes R., Berthold, Pacha J.,C., Efficiency of body biasing in 90-nm CMOS for Low-Power Digital Circuits, *IEEE Journal of Solid-State Circuits*, (July 2005), 1549-1556.

[6] Meijer M., Pineda de Gyvez J. and Kapoor A., Ultra-Low-Power Digital Design with Body Biasing for Low Area and Performance-Efficient Operation, *Journal of Low Power Electronics*, vol. 6, (2011), 1-12.

[7] G. Hackney, The Future of IC Design Verification, Electronic Products, (Nov. 2010).

[8] Lu Z., Bell D.A., Hierarchical Verification of Chip-Level ESD Design Rules, 32nd Electrical Overstress / Electrostatic Discharge Symposium (EOS/ESD), (Oct. 2010).

[9] Medhat D., Easier Cross-Domain Signal Protection for Mixed-Signal SOCs, EDA Designline/EE Times, (Dec. 2009).

[10] Pelz G. and Roettcher U., Pattern Matching and Refinement Hybrid Approach to Circuit Comparison, *IEEE Transactions on Computer-Aided Design*, vol. 13, no. 2, (1994) 264-275.

[11] Ohlrich M., Ebeling C. and Ginting E., SubGemini: Identifying subcircuits using a fast subgraph isomorphism algorithm, *Proc. IEEE/ACM Design Automation Conference*, (1993), 31-37.

[12] Chanak T.S., Netlist processing for custom VLSI via pattern matching, *Computer Systems Laboratory, Electrical Engineering Department*, Stanford University, Stanford, CA 94305.

[13] Chen S-H., Ker M-D. and Hung H-P., Active ESD Protection Design for Interface Circuits Between Separated Power Domains Against Cross-Power-Domain ESD Stresses, *IEEE Transactions on Device and Materials Reliability*, vol. 8, no. 3 (Sept. 2008), 549-560.

[14] Bailey, S., The Future of Low Power is Here: IEEE P1801, aka, UPF 2.0, *22nd International Conference on VLSI Design*, (2009).

CCP: Common Case Promotion for Improved Timing Error Resilience with Energy Efficiency

Lu Wan Deming Chen
ECE Department, University of Illinois at Urbana Champaign
{luwan2, dchen}@illinois.edu

ABSTRACT

Better-Than-Worst-case (BTW) design has been proposed as an alternative way to operate a circuit by deliberately allowing timing errors for rare cases and rectifying them with error correction mechanisms in order to achieve higher performance, better reliability guarantee, or lower energy consumption. This new design methodology necessitates the analysis and manipulation of signal probabilities during circuit optimization. This paper looks for the solution from the logic synthesis perspective and proposes a new concept, called *Common Case Promotion (CCP)*, to enable effective circuit optimization following the BTW design methodology. CCP consists of: 1) probability-driven re-synthesis that changes a digital circuit's internal structure, 2) a dynamic behavior aware SAT-based redundancy remover that reduces area overhead, and 3) a timed characteristic function (TCF) based circuit dynamic behavior analyzer that provides optimization convergence. The experimental results show that, on average, we can effectively improve circuits' timing error resilience by 24%, which reduces the need of error recovery for BTW circuits considerably, and we can improve the overall energy efficiency by 15%.

Categories and Subject Descriptors

B.6.3 [**Logic Design**]: Design Aids – Automatic synthesis, Optimization

Keywords

Common case, Probability, Dynamic behavior, Timing error resilience, Resynthesis, Voltage overscaling

1. INTRODUCTION

Traditional design methodology exploits a conservative guard band by operating the circuit in a condition that ensures 100% correctness. Such a guarantee, however, may sacrifice the circuit's energy efficiency because it requires either operating the circuit at a higher *Vdd* or at a lower speed. Alternatively, Better-Than-Worst-case (BTW) design [1] was proposed to enable more opportunities for circuit optimizations by removing the guard band and complementing a circuit with error detection and correction mechanisms, i.e. Razor logic [2] or EDS [3] circuitry. The ability of correcting the timing error in-place opens the possibility of operating circuits more energy-friendly by *Vdd* overscaling [1][4][5], in which the supply voltage scales to the extent that timing errors may occur. At detecting of timing errors, the affected operation can be rectified by error correction mechanisms. Scaling down the *Vdd* immediately saves dynamic power. But the overall energy efficiency also depends on how often error correction intervenes.

ISLPED'12, July 30–August 1, 2012, Redondo Beach, California, USA.

To take into account the performance penalty due to error correction, we use throughput (*TR*) to evaluate the BTW circuit performance as a function of operating frequency *f*, the probability (P_s) of getting all *Primary Outputs* (PO) stabilized by the cycle time $1/f$, and the error correction penalty *r*, as described in Equation 1. If razor-logic based error detection is used in a MIPS-like processor design, the error correction penalty may involve flushing the pipeline and replaying the instruction [2]. The overall energy cost as shown in Equation 2 is the product of the expected run time (inverse of *TR*) and the power consumption at *Vdd*. Therefore, overscaling *Vdd* saves energy only if the throughput can keep up.

Traditional way of designing a circuit enforces that the chosen *Vdd* guarantees P_s=100%. With the voltage overscaling, P_s drops as the *Vdd* scales down, thus deteriorating the throughput due to error correction penalty. To keep throughput high, we need to maintain a high stabilization probability P_s even if the circuit slows down. This calls for improving the *timing error resilience* of a circuit. In other words, it is desirable that P_s decreases gracefully as the delay of a circuit deteriorates. Furthermore, this can be translated into the requirement of optimizing a circuit at the design time to have a flat plateau of P_s for the circuit to maintain high throughput over a range of aggressive time points that take longer than the traditionally correctness-guaranteed cycle time.

Equation 1: $$TR = P_s \times f + (1 - P_s) \times \frac{f}{r}$$

Equation 2: $$E = P(Vdd) \times \frac{1}{TR}$$

The relation between P_s and the operating cycle time t_{clk} is called a circuit's *dynamic behavior*. To quantify it, we can plot P_s as a function of t_{clk} as a *dynamic behavior curve* as used in [7]: the *x* axis shows the target operating point t_{clk}, while the *y* axis shows the stabilization probabilities. A point $(t, P_s(t))$ on the behavior curve can tell us the probability $P_s(t)$ of a circuit (or a single PO) being stabilized if operating the circuit at $t_{clk}=t$. As illustrated in Figure 1, behavior curves of two different implementations (A/B) of the same logic function of a PO 'o0' of a MCNC benchmark circuit apex2 are shown: in an ideal environment, both A and B can operate at 450ps with no timing error (P_s=100%). But A has a higher P_s at t_{clk}=300ps while B has a lower P_s at the same t_{clk}. Considering the case that voltage overscaling slows down the signal propagation by 150ps, it would mean that signals cannot reach the POs on time within 450ps. As a result, both behavior curves will shift toward right by 150ps. B will experience a drastic drop of P_s, resulting in a significant throughput drop. In contrast, A still keeps a high P_s because its particular implementation provides a considerable wider plateau of P_s in the range of 300ps-450ps at design time. Therefore, A's better timing error resilience assures much less error correction. Our proposed CCP optimization can manipulate the behavior curve to provide such a wide plateau of P_s by design.

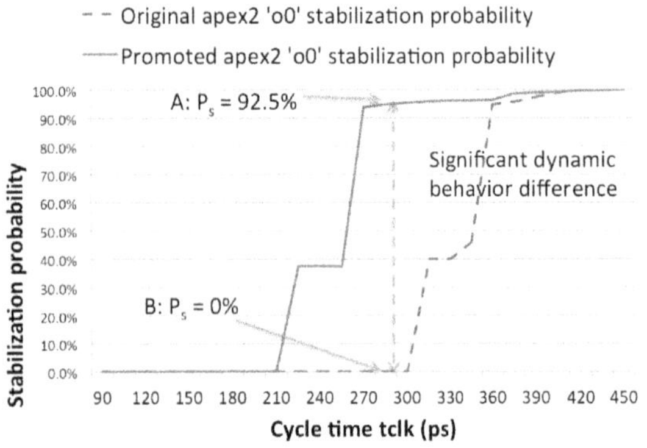

Figure 1: Dynamic behavior curves of two implementations of the same Boolean function 'o0' in apex2

There are existing works targeting to improve a BTW circuit's error resilience from various aspects. Blueshift [6] utilizes a commercial design flow to optimize the dynamically critical nodes for throughput using simulation. DynaTune [7] analyzes a circuit and then improves timing for dynamic critical nodes using low-V_t assignment. Power-aware slack redistribution [5] shifts the slack of frequently exercised and near-critical timing paths in a power efficient manner using gate sizing. The authors of [8] proposed a cost function to penalize frequently switching nodes during logic balance stage. A static-probability-based mux-decomposition method [9] was proposed to change the timing distribution for multi-cycle design.

In this work, we improve a circuit's dynamic behavior from a fresh re-synthesis perspective. We first show the existence of such an optimization opportunity in section 2 (denoted as §2), then we propose a set of logic synthesis techniques (with an umbrella name CCP) in §3, including probability-driven logic re-synthesis, TCF-based circuit dynamic behavior analysis, and a SAT-based redundancy remover. At last, the experimental results are shown in §4.

2. MOTIVATION

In this work, we re-synthesize a circuit in a probabilistic manner that creates shorter logic paths for commonly exercised functions (so-called common cases), while shifts the rarely exercised functions toward longer logic paths. First, we introduce preliminaries:

Definition 1: given an output n and a timing requirement t, we say n is ***stabilized*** no later than t if n has taken either logic 0 or logic 1 after a delay of t since applying the input vector at the rising clock edge, and n will not change its value thereafter within the same clock cycle.

Definition 2: the ***controlling value*** $v \in \{0,1\}$ to a logic cell is the value such that: if taken by an input, the cell output can be stabilized after a delay t_d, the delay between the corresponding input to the cell output, regardless of when or what value the other inputs of this cell are assigned.

As an example, we implement the logic function for an adder carry bit. For simplicity, we omit the subscripts by writing the carry function as follows, where a and b are two bits from two operands, c is the carry-in bit, n is the carry-out bit:

$$n = (a \wedge b) \vee (b \wedge c) \vee (a \wedge c)$$

From the observation made in [1], the higher bits of an adder's operands take 0's more often than 1's for real-life applications. So we assume a and b to have such biased behavior that $Pr(a{=}0){=}80\%$ and $Pr(b{=}0){=}80\%$, where $Pr()$ denotes the probability of a variable being a specified value (0/1). Assume c is unbiased, that is $Pr(c{=}0){=}50\%$.

Next, we can generate a set of common-case input vectors according to these given probabilistic characteristics. Figure 2.C shows five randomly generated input vectors satisfying the given probabilities.

In Figure 2, we have two implementations of this carry function. They both consist of four gates. For simplicity in this example, we assume each gate has a unit delay 1ns, though our experimental results are collected for real delays. We know when an input vector is applied on the PI of a circuit, it needs a certain amount of time t_s to stabilize the PO. Factors that affect t_s include the internal structure of the circuit, the delay of each gate, the sensitization condition of each gate, etc.

In the implementation A shown in Figure 2.A, when {0, 0, 1} is applied at a, b and c, an internal net k will be stabilized after a unit delay by propagating the inputs through the AND gate $g1$. Similarly, OR gate $g2$ will be stabilized after 1ns. At this point the upper input to $g4$ has a non-controlling value 0 from k, so it has to wait until m is evaluated (after 2ns) before it can be stabilized. Therefore, n will be stabilized after 3ns. Now we consider implementation B shown in Figure 2.B. When {0, 0, 1} is applied on a, b and c, q will be stabilized to logic 0 after 1ns through OR gate $g1$. Because $q{=}0$ is the controlling value of the trailing AND gate $g4$, within another 1ns, the PO n will be stabilized to logic 0, resulting in only 2ns to stabilize n.

We can calculate the stabilization time t_s for all five input vectors for circuits A and B, as shown in the right-most two columns in Figure 2.C, respectively. We notice that A needs on average t_s=2.6ns to be stabilized, while B only needs t_s=2.2ns, thus more resilient to timing errors.

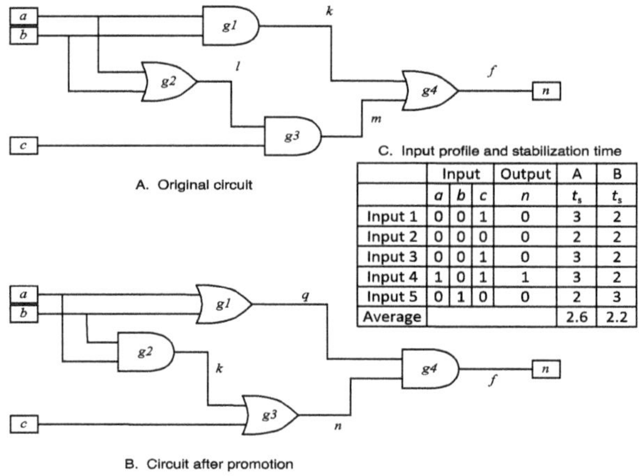

	Input			Output	A	B
	a	b	c	n	t_s	t_s
Input 1	0	0	1	0	3	2
Input 2	0	0	0	0	2	2
Input 3	0	0	1	0	3	2
Input 4	1	0	1	1	3	2
Input 5	0	1	0	0	2	3
Average					2.6	2.2

Figure 2: Different dynamic behaviors of two implementations of $n = (a \wedge b) \vee (b \wedge c) \vee (a \wedge c)$

From this example, we observe that both implementations have the same area and worst-case delay. However, slightly different implementation can improve dynamic behavior in terms of timing error resilience. Motivated by this fact, the problem we are interested to solve can be stated as follows: to find a logic implementation of a circuit that can be stabilized faster for common cases. To achieve this goal, our logic synthesis technique should be able to take common case probabilities as an optimization factor, able to analyze the stabilization conditions for the circuit, as well as to guarantee implementation correctness. In the following sections, we introduce the proposed techniques.

3. COMMON CASE PROMOTION

3.1 Overall flow

To develop an efficient algorithm for optimizing circuit dynamic behavior, we take a divide-and-conquer approach. Sub-circuits are generated as optimization candidates first. Then common case optimization is applied on each sub-circuit. Figure 3 shows the overall flow.

Starting from a netlist, we first generate sub-circuits for timing critical nodes for a given timing distance T_{th} as follows: choose a timing critical node n, then back-trace through its fan-in nodes using a *Depth-First Search* (DFS) algorithm. Further DFS branching stops when it reaches such a node m that its *Required Arrival Time* (RAT) satisfies RAT(n)-RAT(m)>T_{th}. All visited nodes during this DFS traversal are kept in the sub-circuit, thus creating a *timing window* $\boldsymbol{C}_n$ rooted at n, or $\boldsymbol{C}$ for short.

We feed C into "Common cases identification" procedure (§3.2) to identify its probabilistic characteristics. Then it goes through "promote common cases", which changes structure for common cases using logic re-synthesis (§3.3) followed by dynamic behavior analysis using *Timed Characteristic Function* (TCF) to decide if the structure is valid for dynamic behavior improvement (§3.4). If the structure change is accepted we then try to control area overhead using an SAT-based approach (§3.5). Then we iterate for the next timing window. A convergence test (§3.7) will decide whether any further promotion is needed for a logic cone.

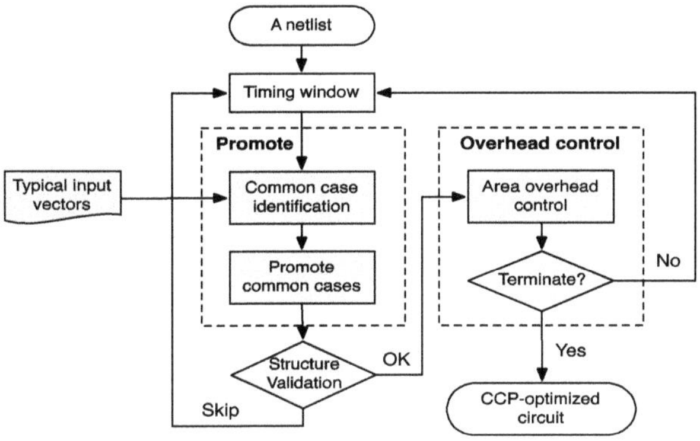

Figure 3: Overall flow of common case promotion

3.2 Identify common-case partial function

In the common-case identification step, we utilize two-level expression forms complemented with a simulation method to generate the sub-circuit's probability profile, which will be used to drive the dynamic behavior optimization.

Each generated timing window C represents a Boolean function f:$B^m \rightarrow \{0,1,*\}$, where m is the number of C's inputs. We use the definitions commonly used for two-level logic optimization [10]: An *implicant* of f is a row vector consisting of the input vector of $\{0,1,*\}^m$ and the associated output $\{0,1\}$ specified by f. The on/off/dc set is the subset of B^m that implies 1/0/* according to f. An on(off)-set *minterm* of f is an implicant whose input vector is $\{0,1\}^m$ and that implies a *true*(*false*) at the output. A *cover* F^{on}(F^{off})of f is a set of implicants that cover all on(off)-set minterms of f. We denote on-set minterm as minterm, F^{on} as F for short when there is no confusion. A *partial cover* F_p of f is a set of implicants that covers some minterms of f. A *prime* is an implicant that is not contained by any other implicant of f. A *prime cover* is a cover consisting only of primes.

To get a probabilistic profile for each timing window, we need firstly prepare a global behavior profile, similar to the table in Figure 2.C, for the whole circuit only once. Then we can reuse this global profile for each individual C. A global behavior profile is prepared using a zero-delay simulation-based method as follows: we generate a set of PI input vectors according to given typical-case characteristics [5][6]. Then we leverage AIG-based simulation engine in *ABC* [13] to propagate these typical input vectors for each cell in the netlist. Note that: 1) Both circuit structure and timing information are decoupled from logic simulation because they will be processed in a separate step called structure validation. Therefore, to generate the profile, the simulation can be performed using a zero-delay model, thus is much faster than timed-simulation used in [5][6]. 2) Rather than applying timed-simulation once for each optimization iteration [6], our behavior profile is prepared only once for the circuit and then is reused across all sub-circuits. 3) We observe from experiments that a fast simulation of a small number of input vectors (e.g., 100) can provide re-synthesis guidance of enough precision. Therefore, it is practical.

From the behavior profile, we carry out *Common Case Identification* as follows: find a partial cover F_p, as a subset of F, such that the implicants in F_p are of high probability being exercised by the common-case input vectors. This F_p, though partially covers the minterms in f, is more often exercised than $F \backslash F_p$, where '\' is set subtraction operator. From a probabilistic point of view, the partial function F_p encodes the common-case portion of f, while $F \backslash F_p$ complement it by encoding the corner cases of f. From F_p, we can later promote this partial function using multi-level logic synthesis techniques to improve circuit dynamic behavior (§3.3). In addition, from logic synthesis point of view, it is preferable to have F_p contain only primes so that unnecessary literals are eliminated to save area and to improve timing. Therefore, we identify F_p within a prime cover F as follows.

Depending on C output's tendency of being logic 1 or 0, we may perform phase-1 or phase-0 identification on C. Here we describe the procedure for phase 1, while phase 0 can be done similarly. For phase 1 optimization, we first generate a cover for the on set by creating a decision diagram of C using a DFS back-tracing. To produce the prime cover F^{on} for the on set, we use Zero-Suppressed Decision Diagram (ZDD) encoding, which suppresses "non-essential" literals along *evaluation paths* (from root node to true terminal in the ZDD) with a special two-phase encoding for each variable [15]. Then we evaluate the probability of each prime in F^{on} being activated using the previously prepared global behavior profile. From the profile, we extract input vectors and the associated outputs for C as row entries. Each such entry is denoted as γ. We say γ can activate a prime p in F^{on} if $\gamma \in p$, that is any literal l_p in p is identical to the associated literal l_γ in γ. We count for each p the number of input vectors in the global behavior profile that can activate it. Then we derive an activation probability.

We iterate common-case identification multiple times for each C. In each iteration, we look for the current most frequently activated prime, denoted as p_c. Note that to identify the next most frequently activated prime in the next round, we need to remove previously identified p_c's effect by starting the next round with $F^{on} \wedge \overline{p_c}$ rather than F^{on}. We collect these common-case primes that will be promoted in a Boolean function f_p and then implement it as the partial implementation of f. Note that we only perform this common case identification within a timing window C, therefore ZDD doesn't run into capacity issue, hence is practical.

In Figure 2.C, for example, the input and output columns can be viewed as a global behavior profile of 5 entries. In this example, it's preferable to identify phase-0 common cases for off-set F^{off} because the output tends to be logic 0. Using the above approach, we can identify that a partial off-set cover $F_p^{off} = \overline{a} \wedge \overline{b}$ is exercised and produces an output 0 frequently. We inverse this off-set partial cover using De-Morgan's law to get a regular on-set function f_p that represents the common cases (in off-set) as: $f_p = a \vee b$. Then we promote it (details in the next section). Note that sometimes, we may promote multiple primes besides the most activated one according to a specified probability threshold.

3.3 Promote common cases

To improve dynamic behavior, we choose common case primes contained in F_p as the sub-function to get promotion. Promotion is a process of implementing a sub-function with a redundant logic cone and then merging it back into the original circuit. In general the sub-function f_p associated with F_p has fewer literals than that of the f, so usually it can be implemented with a shallower logic depth. To control area overhead, a TCF-based structure validation process (§3.4) is applied to avoid unnecessary promotion and a SAT-based redundancy remover is applied to simplify the logic and reduce area (§3.5).

This logic promotion process is mainly performed using multi-level logic synthesis techniques: starting from the partial function f_p, we invoke technology independent multi-level synthesis script "*resyn*" in ABC [13] followed by logic balancing and technology dependent mapping. As a result, a promoted circuit, denoted as $\boldsymbol{C_p}$ is generated to implement the partial function f_p.

Next, we merge C_p with the original implementation C using a merger cell M. If the common-case partial function f_p is generated for phase 1, we correlate it with the original function f with *disjunctive normal form* (DNF) as '$f = f \vee f_p$'. As a result, C_p is merged with C using an OR-type merger cell. Because '1' is the controlling value of an OR gate, as long as the common-case f_p is stabilized to 1, the whole Boolean function will be stabilized regardless of other side inputs. Since this common-case 1 propagates through a shallower logic cone C_p, the output can be stabilized sooner than through the original C.

Similarly, to promote for phase 0, we correlate f_p with f with *conjunctive normal form* (CNF). Then use an AND-type gate as the merger cell to correlate the partial implementation and the original implementation.

Take the circuit of Figure 2.A as an example, after we identify the common-case partial function (in off-set) $f_p = a \vee b$, we implement f_p as C_p with an OR gate p in Figure 4 and merge C_p with the original implementation C using an AND merger cell M.

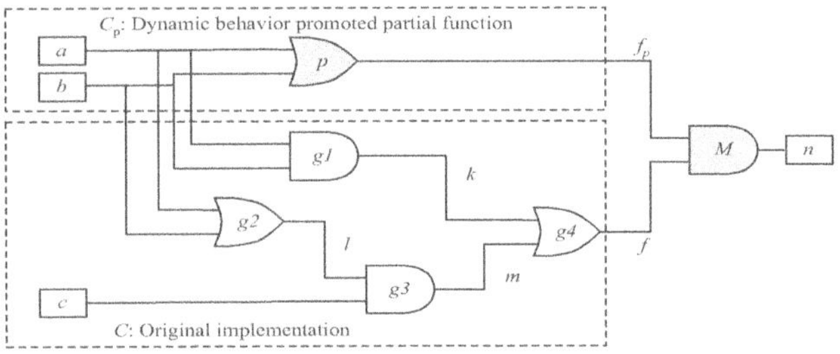

Figure 4: Promote the common cases in $f = (a \wedge b) \vee (b \wedge c) \vee (a \wedge c)$

3.4 Validate structure changes

We should note that, the intentionally introduced redundancy of C_p may evaluate the common cases' input vectors faster than the original design F. It's, however, possible that these identified common-case primes have already been implemented in the original implementation with very short delay to the output. Then, it becomes a waste of resource to re-implement those primes. To avoid introducing unnecessary redundancy, we introduce a TCF-based structure validation technique to determine if the structural change indeed improves the circuit's dynamic behavior.

TCF is commonly used for ATPG [11] and circuit dynamic behavior analysis [7]. Here, we propose to use TCF for structure validation. As with previous work [7][11][12], the TCF is defined in floating mode where input vectors are applied at time t=0, and before time 0, the input vectors are treated as uninitialized.

Given a node n, its stabilization value $val \in \{0,1\}$ and RAT t, we define TCF as:

Definition 3: a ***timed characteristic function*** $T(n=val,t)$ is a Boolean function that evaluates to *true* only for the set of input vectors that can stabilizes the output n to value *val no later than time* t.

TCF of a cell output can be written recursively as TCF of its immediate inputs using sensitization criteria [7]. Take an AND gate "n=AND(a,b)" with cell delay d for example:

Equation 3: $T(n=1,t) = T(a=1,t\text{-}d) \wedge T(b=1,t\text{-}d)$

Equation 4: $T(n=0,t) = T(a=0,t\text{-}d) \vee T(b=0,t\text{-}d)$

TCF for a circuit can be constructed by back-tracing and represented as a BDD [7]. The evaluation paths that lead to true terminal in the BDD encode all the possible input vectors that stabilize n to a required logic value *val* within a specified delay t.

We want a C_p such that it can be stabilized by common-case minterms faster than C is. If C is already capable of being stabilized as quickly as C_p then implementing C_p doesn't improve dynamic behavior. To determine this, our structure validation compares the TCFs of C and the partial implementation C_p. If the common cases F_p is contained in the TCF of C, then it means C is already as capable as C_p regarding stabilizing any minterm in F_p. Hence, we discard the C_p to reduce area overhead. The procedure is carried out as follows:

1. Timing analysis for C_p. In this step, we evaluate the delay of C_p and M as t_c = D(C_p) + D(M), where D() is the delay function from latest AT analysis. It is obvious that C_p guarantees to be stabilized by all minterms in F_p in t_c.
2. Test if the original circuit C can already be stabilize by those minterms of F_p in delay of t_c. We construct a TCF for C as $T(n=val,t=t_c)$, which will contain all minterms that can stabilize C within a delay t_c.
3. Discard C_p if the following test is true:

Equation 5: $Fp \subseteq T(n = val, t = t_c)$

This test being true means C can already be stabilized by minterms in F_p within t_c. Then it's unnecessary to promote F_p into C_p. Next we explain this procedure using the example in Figure 4. To validate the structure of C_p for the common-case off-set partial cover $F_p^{\text{off}} = \overline{a} \wedge \overline{b}$, we first calculate t_c as 2ns=D(p)+D(M). Since we promote for phase 0, we set val=0. Then we construct $T(f=0,t=2)$ for C:

$$\begin{aligned} T(f=0,t=2) &= T(m=0,t=1) \wedge T(k=0,t=1) \\ &= [T(c=0,t=0) \vee T(l=0,t=0)] \wedge [T(a=0,t=0) \vee T(b=0,t=0)] \\ &= T(c=0,t=0) \wedge [T(a=0,t=0) \vee T(b=0,t=0)] \\ &= \overline{c} \wedge (\overline{a} \vee \overline{b}) \end{aligned}$$

After recursively rewriting and expanding about inputs a, b, c, we get the final TCF $T(f=0,t=2) = \overline{c} \wedge (\overline{a} \vee \overline{b})$. Note that $T(l=0,t=0)$ is deprecated (as a *false*) in above derivation because no input vector can stabilize l at t=0 due to the delay of cell $g2$. Now test with Equation 5, we have:

$$\overline{a} \wedge \overline{b} \not\subset \overline{c} \wedge (\overline{a} \vee \overline{b})$$

Now, we know the original implementation C cannot be stabilized by the common-cases in F_p within t_c. Therefore the new structure C_p is valid for dynamic behavior improvement.

3.5 Overhead control

Merging the original implementation C with the promoted partial implementation C_p, we create a new implementation C' that can substitute C. Considering that C_p may introduce redundancy, there exist opportunities to simplify C'. We show how we control the area overhead with redundancy removal.

Traditionally, Automatic Test Pattern Generation (ATPG) techniques were used to identify redundancy. With the success of modern SAT technique, SAT is now widely applied in various areas including redundant wire addition and removal [17]. In this work, we take advantage of the knowledge of the inherent functional connection among C, C_p and C' and then apply SAT to control area overhead in C' without hurting its dynamic behavior.

Given that C_p is a purely redundant partial function that is implemented for better dynamic behavior, applying a general redundancy removal technique [16][17] on it may totally remove this partial implementation, thus defeats our goal of improving dynamic behavior. Therefore, we need to restrict the effort of redundancy removal within the implementation of C rather than C_p. Moreover, the support of C_p is a subset of the support of C. Therefore it's unnecessary to apply SAT testing on all inputs to C.

Our redundancy removal is based on the following observation: if a stuck-at-α fault, where α can be either 0 or 1, on one input is non-observable at the output, we can propagate this constant α thus simplifying the downstream logic. We also observe that both the controlling value and the non-controlling value are potentially useful for redundancy removal. Because tying a gate's input to the controlling value, a gate can be reduced to a constant. While tying an input to the non-controlling value, a gate may be reduced to a simpler gate or even a wire. For example, tying an input to AND2 gate to logic 1 can effectively reduce the AND2 gate to a wire.

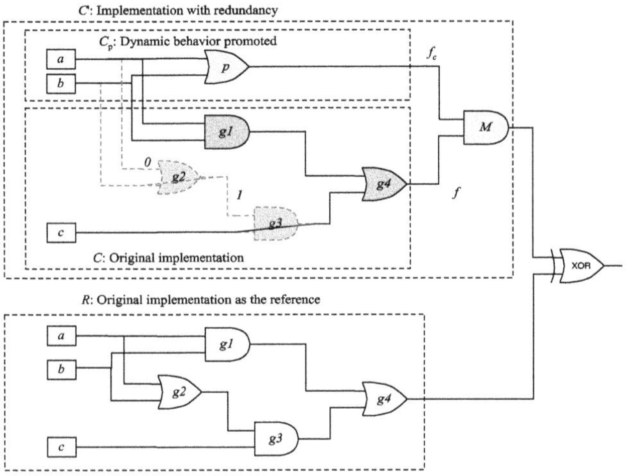

Figure 5: An example of using SAT to remove circuit redundancy during CCP

To control area overhead, we first construct a XOR miter from 1) a clone of the original implementation C as the reference, denoted as R, and 2) the common-case promoted implementation C'. The output of R and the output of C' are fed into two inputs of the XOR gate to form a miter output ϕ. R and C' share the same set of inputs - the inputs of C - denoted as $I=\{i_1, i_2, \dots, i_n\}$. We say R and C' are Boolean equivalent if and only if the miter is non-satisfiable by any Boolean assignment to I. That means no input vector can differentiate R from C' by evaluating them to different values.

After the miter is constructed we solve the associated SAT problem to test the stuck-at-α observability for both α =0 and α =1 on an $i_m \in I$. We pick inputs i_m in the miter's inputs I, such that $i_m \in input(C_p)$. Let $FO(i_m)$ be the inputs of the cells that i_m directly connects to. Moreover, we prevent our redundancy removal from impairing the improved dynamic behavior by preserving the promoted sub-function C_p. So we restrict stuck-at-α tests at non-promoted portion j_m, such that

$$(j_m \in FO(i_m)) \wedge (j_m \in C) \wedge (j_m \notin C_p)$$

We then disconnect j_m from i_m and apply a stuck-at-α value as j_m=α. Then we solve the associated *SAT* problem stated as follows:

SAT problem: For the XOR miter with an output ϕ and an input vector $V=\{i_1, i_2, \dots i_{m-1}, j_m=\alpha, i_{m+1}, \dots i_n\}$ of n variables, is it satisfiable that an Boolean assignment of $\{0,1\}^{n-1}$ to $\{i_1 \dots i_{m-1}, i_{m+1}, \dots, i_n\}$ can evaluate ϕ true.

Obviously, if no input vector assignment to V exists that can satisfy the above SAT instance, then tying j_m to a constant α doesn't differentiate C' and R. Therefore we can simplify j_m to α and propagate this constant to further simplify logic downstream. Because j_m is carefully picked such that it only belongs to C, the simplification applies only on C leaving C_p intact. Hence the promoted dynamic behavior is preserved.

In the example of Figure 5, the upper part is the combined C' consisting of C and C_p, while the lower part is the C's clone R. We perform redundancy removal on nets $\{a, b\}$ for C. The SAT instance can prove that *g2*'s input from a can be tied to zero, which in effect removes the OR gate *g2* shown in dashed line. Furthermore, our SAT instance proves that *g3*'s input from *g2* can be tied to one, which further removes the AND gate.

As a result, we end up with the circuit shown in Figure 2.B, which has a better dynamic behavior but still maintains the same size as the original circuit. This illustrates that our SAT-based area recovery can effectively control area overhead from intentionally introduced redundancy for common cases.

3.6 Correctness

Theory: the proposed common case promotion technique is functionally correct in Boolean domain.

Proof: we prove for phase-1, and phase-0 can be proven similarly. There are two steps that we may change logic structure of the original circuit during CCP. We prove that neither of them alters functionality: (1) promoting common cases doesn't alter functionality. Because the common case primes encoded in F_p belong to set of F, after merging with an OR gate, we have $F_p \vee F == F$. Because F_p, f_p and C_p are different forms of the same Boolean function, and F, f and C are also different forms of the same Boolean function, we can have $f_p \vee f == f$ and $C' = C_p \vee C == C$. That proves C' is Boolean equivalent to C. Since R is a clone of C, C' is also equivalent to R. (2) the SAT redundancy removal doesn't change functionality either. Because we tie j_m to α only if a stuck-at-α fault is non-observable in a XOR miter of C' and R, this will not differentiate C' and R under all possible Boolean input assignments. Indeed, the SAT instance guarantees Boolean equivalence between the original C and the dynamic behavior optimized C'. Therefore, this combined common case promotion keeps Boolean equivalence between CCP optimized version and the original one.

3.7 Convergence

After common-case promotion finishes a round of optimization, we can test if further promotion is needed. This again can be done with the TCF test. For example, given the same common-case prime $F_p^{off} = \bar{a} \wedge \bar{b}$ covering off set, we test whether the redundancy-removed CCP implementation in Figure 2.B needs further promotion by applying the same TCF test (§3.4). Going through the same TCF rewriting procedure, this time we have:

$$\begin{aligned} T(f{=}0,t{=}2) &= T(n{=}0,t{=}1) \vee T(q{=}0,t{=}1) \\ &= [T(c{=}0,t{=}0) \wedge T(k{=}0,t{=}0)] \vee [T(a{=}0,t{=}0) \wedge T(b{=}0,t{=}0)] \\ &= T(a{=}0,t{=}0) \wedge T(b{=}0,t{=}0) \\ &= \bar{a} \wedge \bar{b} \end{aligned}$$

Testing with Equation 5, we have $F_p^{off} \subseteq T(f=0, t=2)$ meaning that the identified common cases can stabilize C' in t=2. Therefore, no further CCP step is needed.

4. Experimental results

The proposed algorithm is implemented in C on top of ABC [13]. MiniSAT [14] is used to perform the area overhead control described in §3.5. Benchmark circuits are compiled into Verilog netlists using Synopsys Design Compiler (DC) ver. 2007-SP3 and TSMC 65nm standard cell library. We set tight delay and power constraints to allow DC to perform high-effort optimization until it gives up. Then we apply the proposed CCP on these netlists to further improve dynamic behavior. For each circuit, a short sequence of 100 common-case vectors are generated according to given PIs' probability profiles to drive the optimization. The runtime of CCP ranges from seconds to several minutes on a

2.4GHz dual-core processor. The relationship between *Vdd* and power is characterized with HSPICE.

To validate the performance gain produced by CCP, we collect sdf-annotated netlist simulation data with Cadence NCSim v5.7 for each circuit by a Verilog Program Interface (VPI) program, which records timestamps of each PO' last switching activity in every cycle. Each circuit is validated by simulating one million cycles. Sometimes, CCP can do an excellent job to improve dynamic behavior without sacrificing timing or area (Figure 2). When penalties are unavoidable, we size up the merger cells for better timing. Furthermore, CCP is only applied for POs whose delays are within 80%-100% of the longest path delay to save area and power.

CCP-optimized circuits have better timing error resilience due to the reshaped dynamic behavior curve. We exploit this effect to achieve a better voltage overscaling solution. Figure 1 shows the effect of applying CCP on a single PO for circuit *apex2*. Clearly, before applying CCP, P_s has decreased to 0 around t=300ps. After CCP, P_s still maintains a high value of 92.5% at the same time point. However, optimizing P_s for the whole circuit is far more difficult than that for an individual PO because of the correlation among POs. Nonetheless, based on the experimental results, we still see a significant boost for a more desirable dynamic behavior as the average P_s increases by 24%.

In Table 1, Col. 5 shows the overscaled *Vdd* for CCP-optimized circuits. Corresponding to these overscaled *Vdds*, the probabilities of producing correct results (P_s) for w/o and w/ CCP are listed in Col. 6-7, respectively. The area cost of applying CCP is shown in Col. 9-11. Thanks to structure validation (§3.4) and SAT-based overhead control (§3.5), the average area cost is merely 3.2%.

Table 1: Circuit profile before and after CCP

			VDD(v)		Ps: Correctness probability			Total area		
	PI	PO	Tra.	CCP	w/o CCP	CCP	Diff	w/o CCP	CCP	Diff
des	256	245	1.0	0.89	81.5%	97.0%	15.5%	14814	15212	2.7%
alu4	14	8	1.0	0.85	89.2%	93.7%	4.5%	2216	2222	0.3%
apex2	39	3	1.0	0.79	76.9%	94.2%	17.3%	598	602	0.7%
apex4	9	18	1.0	0.85	79.3%	94.3%	15.0%	6456	6551	1.5%
dalu	75	16	1.0	0.80	12.9%	74.0%	61.1%	2937	3005	2.3%
ex1010	10	10	1.0	0.84	38.1%	81.2%	43.1%	7083	7396	4.4%
ex5p	8	63	1.0	0.86	65.7%	86.2%	20.5%	923	925	0.2%
misex2	14	14	1.0	0.85	84.4%	97.0%	12.6%	2448	2454	0.2%
pdc	16	40	1.0	0.76	41.6%	93.1%	51.5%	1667	1895	13.7%
seq	41	35	1.0	0.78	70.5%	80.2%	9.7%	5376	5403	0.5%
spla	16	46	1.0	0.81	72.6%	90.4%	17.8%	1576	1720	9.1%
Ave.							24.4%			3.2%

Table 2: CCP optimize performance results

	Power(uw)			Normalized TR			Energy consumption		
	Tra.	w/o CCP	CCP	Tra.	w/o CCP	CCP	Tra.	w/o CCP	CCP
des	3345.1	2918.9	2939.8	1.00	0.85	0.98	1.00	1.02	0.90
alu4	332.8	265.4	265.5	1.00	0.91	0.95	1.00	0.87	0.84
apex2	77.1	53.2	53.5	1.00	0.82	0.95	1.00	0.85	0.73
apex4	930.3	752.4	752.4	1.00	0.83	0.95	1.00	0.97	0.85
dalu	396.1	283.2	288.1	1.00	0.30	0.79	1.00	2.36	0.92
ex1010	827.4	652.1	658.0	1.00	0.50	0.85	1.00	1.56	0.94
ex5p	98.1	79.7	79.9	1.00	0.73	0.89	1.00	1.12	0.92
misex2	321.6	256.7	257.3	1.00	0.88	0.98	1.00	0.91	0.82
pdc	185.7	121.3	132.4	1.00	0.53	0.94	1.00	1.23	0.75
seq	630.6	433.6	433.9	1.00	0.76	0.84	1.00	0.90	0.82
spla	184.7	134.8	140.1	1.00	0.78	0.92	1.00	0.94	0.82
Ave.					0.72	0.91		115.7%	84.5%

To evaluate how much energy can be saved after applying CCP, we use a penalty r=5 to derive *TR* with Equation 1, which has been commonly used in previous studies [2][6][7]. Then we consider the error correction overhead using Equation 2. The energy consumption is the product of 1/*TR* and the power of the circuit. Table 2 listed the data for each circuit. Col. 2 lists the power data for a traditional design. Col. 3 and 4 list the power data for BTW design protected by razor logic with shadow flip-flops' power cost considered. Col. 5-7 list the normalized *TR* for traditional design, BTW design w/o CCP and w/ CCP, respectively. Col. 8-10 list the normalized energy consumption for each circuit to finish the given task.

Table 2 shows that for BTW design w/o CCP, six circuits achieve power savings ranging from 3%-13% through *Vdd* overscaling. The other five circuits consume more energy to finish the same tasks comparing with the traditional designs that do not overscale *Vdd*, because their stabilization probability P_s decreases rapidly as the *Vdd* is overscaled, causing excessive amount of recovery to correct timing errors. For example, *dalu* needs 3.3x (TR=0.3) time to finish the same amount of work, which burns energy unnecessarily. In contrast, CCP optimization creates a flat stabilization probability plateau to improve timing error resilience under *Vdd* overscaling. Therefore, timing errors happen much less frequently; resulting in higher *TR* that directly contributes into energy efficiency. Comparing with traditional design, all CCP-optimized circuits can save energy by 6%-17%. On average, we achieve 15% of energy savings by applying CCP.

5. Conclusion

We proposed a new design technique - Common Case Promotion (CCP) - to improve a circuit's timing error resilience using: 1) probability-driven re-synthesis, 2) dynamic behavior aware SAT-based area recovery, and 3) Time Characteristic Function (TCF) based analysis to provide optimization guidance. Comparing with designs without CCP, CCP-optimized circuits can be operated more energy efficiently under voltage overscaling.

6. REFERENCES

[1] T. Austin, et al., "Opportunities and challenges for better than worst-case design", In *Proceedings of* ASP-DAC, 2005

[2] D. Ernst, et al., "Razor: A Low-Power Pipeline Based on Circuit-Level Timing Speculation", *Annual Intl. Symp. on MICRO,* 2003.

[3] K. Bowman, et al., "Circuit techniques for dynamic variation tolerance", In *Proceedings of* DAC, 2009.

[4] R.A. Abdallah and N.R. Shanbhag, "Minimum-Energy Operation Via Error Resiliency," IEEE Embedded Systems Letters, vol.2, no. 4, pp.115-118, Dec. 2010

[5] A. B. Kahng, et al., "Slack Redistribution for Graceful Degradation Under Voltage Overscaling", In *Proceedings of* ASP-DAC, 2010.

[6] B. Greskamp, et al., "Blueshift: Designing processors for timing speculation from the ground up", In *Proceedings of* HPCA, 2009.

[7] L. Wan and D. Chen, "DynaTune: Circuit-Level Optimization for Timing Speculation Considering Dynamic Path Behavior", In *Proceedings of* ICCAD, 2009.

[8] J. Cong and K. Minkovich,"Logic Synthesis for Better Than Worst-case Designs", In Proceedings of VLSI-DAT, 2009.

[9] S. Ghosh, et al., "A new paradigm for low-power, variation-tolerant circuit synthesis using critical path isolation", In *Proceedings of* ICCAD, 2006.

[10] G. D. Micheli, "Synthesis and optimization of digital circuits", McGraw-Hill. 1994

[11] Y. Kuo, et al., "Efficient Boolean characteristic function for fast timed ATPG", In *Proceedings of* ICCAD, 2006.

[12] G. Silva, et al., "Satisfiability models and algorithms for circuit delay computation", ACM Trans. DAES, 2002.

[13] Berkeley Logic Synthesis and Verification Group, "ABC: A System for Sequential Synthesis and Verification", Release 70930.

[14] Niklas Eén, Niklas Sörensson: "An Extensible SAT-solver", In *Proceedings of* SAT, 2003.

[15] A. Mishchenko, "An Introduction to Zero-Suppressed Binary Decision", In *Proceedings of* Calculemus 2001.

[16] P.R. Menon, et al., "Redundancy Removal and Simplification of Combinational Circuits", In *Proceedings of* VLSI Test Symp., 1991.

[17] C, Wu, et al., "SAT-Controlled Redundancy Addition and Removal – A Novel Circuit Restructuring Technique", In *Proceedings of* DAC 2009

Energy-Optimal Caches with Guaranteed Lifetime

‡Mirko Loghi, †Haroon Mahmood, †Andrea Calimera, †Massimo Poncino, †Enrico Macii
†Politecnico di Torino, Torino, 10129, ITALY
‡University di Udine, Udine, 33100, ITALY

ABSTRACT

This work addresses the aging of the memory sub-system due to NBTI (Negative Bias Temperature Instability) in systems that have to provide a guaranteed level of service, and specifically, a guaranteed lifetime.

Our approach leverages a novel cache architecture in which a smart joint use of redundancy and power management allows us to obtain caches that meet a desired lifetime target with *minimal energy consumption*. This is made possible by exploiting the possibility of putting the cache sub-block used for redundancy into a deep low-power state, thus allowing more energy saving than a regular architecture. Sacrificing a portion of the cache for aging mitigation only marginally affects performance thanks to the non-linear dependency of miss rate versus cache size, which allows to find the best cache size that maximizes the objective.

Simulation results show that it is possible to meet the target lifetime by achieving energy reductions (measured over the lifetime of the system) ranging from 3X to 10X (2X to 8X) for a lifetime target of 15 (25) years, with marginal miss rate overhead.

Categories and Subject Descriptors

B.3.2 [**Memory Structures**]: Design Style

General Terms

Power, Reliability, Performance

Keywords

Aging, NBTI, Leakage optimization, Memory Hierarchy

1. INTRODUCTION

Many mission-critical applications require guaranteed life time of their operations, and therefore of the hardware implementing their functionality. In some cases, a guaranteed level of service must also be granted, for instance in terms of throughput. Such constraints are usually enforced by means of various reliability-enhancing solutions mostly based on redundancy [1], which are typically not energy-friendly, because the replicas consume extra energy (even when power managed) and so does the control logic required for managing the operations.

ISLPED'12, July 30–August 1, 2012, Redondo Beach, CA, USA.

In this work we focus on a particular type of reliability threat, namely the aging of CMOS devices, and in particular the one due to NBTI (Negative Bias Temperature Instability), which is regarded as the most significant source of device aging in nano-scale technologies [2]. More specifically, we target the aging of SRAMs in the memory sub-system, which is the most critical component for warranting reliable and timely operations.

Aging induced by NBTI is a value-dependent effect: when a pMOS device is under negative bias (a logic "0" applied to its gate terminal), this causes the threshold voltage (and propagation delay) to increase over time. Moreover, NBTI is partially reversible: the application of a logic "1" to the gate terminal will cause the device to partially recover its threshold voltage. Therefore, for generic circuits it is essential to evaluate the distribution of "0"'s and "1"'s to determine the actual aging. In SRAMs, conversely, the value dependence of NBTI is weaker than in generic circuits: given the symmetric structure of a SRAM cell there is in practice no recovery effect and the cell ages regardless of the stored value.

Previous works have observed that there exists a correlation between the concept of idleness exploited in power management strategies and aging. In particular, the "hardware" implementation of a low-power state (i.e., voltage scaling for dynamic power and power/ground gating for static power) can be leveraged to reduce NBTI-induced aging [4, 5]. Therefore, several works have properly revisited various traditional power-managed cache and memory architectures under an aging-related metric so to achieve concurrent energy and aging improvements [12, 13].

All these approaches are based on the idea of transforming the idleness resulting from a given workload (which is exploited to reduce energy) into an equivalent benefit for aging as well. Direct use of the idleness is not always possible: while for energy it is the total idleness that matters, for aging (a worst-case metric) it is the *distribution* of the idleness that matters. Such a transformation consists either of a proper arrangement of addressing mechanisms ([12]) or of multi-banked organization of an SRAM ([13]). None of these approaches however relies on widely used reliability-enhancing paradigm, that is, the use of **redundancy**.

In this work we show that, by properly combining the two above approaches and redundancy, it is possible to push the energy reduction beyond the limits of previous works. The rationale of our architecture is to use sub-banking not to reduce the impact of the worst-case idleness ([13]) but rather as an extra memory space over which better distribute idleness. In other terms, we only use a subset of the cache to store values. Energy reduction is achieved because the cache sub-block used for redundancy can be put into a non state-preserving state during standby without compromising performance. As shown in [12], this is not feasible if the whole cache is active, due to the poor exploitability of a non state-preserving state.

The use of a virtually smaller cache, however, affects performance. We show that this impact is marginal, thanks to the non-linear dependence between miss rate and cache size, which saturates for typical cache sizes.
The proposed approach allows, for a given lifetime target, to significantly improve the overall energy consumption of the cache (measured over the target lifetime), with truly marginal degradation of miss rate.

2. BACKGROUND AND RELATED WORK

2.1 Background

For space limits we limit ourselves to a summary of the effects of NBTI-induced aging in SRAM cells. For details about the modeling of NBTI effects and its dependence on technological and environmental parameters, we refer the reader to the classical papers on the subject (e.g.,[2, 3]).
The threshold voltage drift caused by NBTI does not truly affect the *delay* of a SRAM cell but rather it decreases its Static Noise Margin (SNM), i.e., the minimum DC noise voltage required to change the state of the cell. This affects the stability of the cell, because if the SNM of a cell falls below some technology-dependent threshold, the data stored in the cell cannot be safely read or written. Another characteristic of NBTI effects on SRAM cells is the weak dependence on logic values: whatever the values stored, the cell will age anyway. *Distribution* of the values can however impact the cell stability: the best-case occurs when both inverters in the cell exhibit the same amount of degradation, i.e., when a cell stores a 0 and a 1 with equal probability [6]. Based on this consideration and the technology used for our simulations, we define *lifetime* of a SRAM cell as the time after which its SNM has decreased by 20%.

2.2 Related Work

Solutions proposed to mitigate NBTI effects in SRAMs fall in three main categories. The first class includes methods that try to equalize cell value probabilities using various hardware and software strategies [6, 7].
Another type of approach aims at designing customized NBTI-resilient cells [8, 9]. In [8] a new NAND-based cell structure is proposed so that minimum degradation ratio for all PMOS transistors in the cell is obtained. Another solution called *recovery boosting* [9] allows both PMOS devices in the cell to be put into the recovery mode by raising the ground voltage and bit-lines to the nominal voltage through modification of each memory cell.
A third class of solutions is based on the exploitation of the aging benefits provided by low-energy states [13, 12]. Assessment at the architectural level on entire memory blocks of power management solutions (based on both DVS and power gating) were evaluated in [13].
The work of [12] introduces a *dynamic indexing* scheme in which the cache indexing function is modified over time in order to achieve a uniform distribution of idleness over the cache lines; in this way all the leakage saving opportunities can also be used for aging reduction.
One architectural approach based on redundancy is the one proposed in [14], in which a spare cache sub-array is used to replace, on a rotating basis, selected sub-arrays with excessive aging, which are then put into a special wearout-recovery state implemented through a sort of power gating. This approach is more fine-grain than ours, but requires considerable overhead. The "scheduling" of the spare unit must be explicitly managed, and the choice of the unit to be recovered requires per-unit detection of the level of aging. Conversely, our scheme is suitable also for small-scale systems and does not require modification on the internals of the cache.

3. MOTIVATION AND CONCEPT

Consider a system with a cache (for simplicity, direct-mapped with a pre-defined line size) of a given size S, measured in cache lines, determined according to some architectural considerations. This baseline cache will have a given power consumption, lifetime, and performance (e.g., miss rate). Since all these quantities are affected by cache size, we denote them by $P(S), LT(S), MR(S)$. Suppose also that the system constraints include the warranty that the system (in this case, the cache) guarantees operation for a specified target lifetime[1]. Our objective is to *devise an organization the cache sub-system so that the lifetime target is met by using the least possible power and with smallest possible performance penalty.*
One way to approach this problem is to try to extend the default lifetime by using the intrinsic idleness of the application that uses the cache ([12, 13]). For example the "dynamic indexing" proposed in [12] proposes a simple and effective solution that exploit 100% the available idleness by distributing it over the cache lines so that all lines exhibit the same idleness (and lifetime). This solution maximizes lifetime, but the fact that all the cache lines are used limits the possibility of power managing the lines. As a matter of fact, [12] ruled out a low-power state based on power gating (which provided longer lifetime) because it is less exploitable from the energy standpoint due to much longer break-even times caused by miss penalty.
Our method overcomes this limit as follows. We keep only a subset $S' < S$ of the cache lines as "active" cache lines; The remaining $S - S'$ act as "spare" lines that can be used to mitigate the aging of the whole cache. The benefit from the energy standpoint is intuitive: the "inactive" portion of the cache can be put in a non state-preserving low-power state (thus saving more energy). However, this also extends lifetime because aging is virtually removed under footer-based power gating (actually there is a recovery)[12], and the "inactive" part of the cache is less aged when it gets reused.
On the other hand, under this scheme we are virtually using a smaller cache, so performance can be impacted. However, thanks to the dependence of the the three metrics ($E(), LT()$, and $MR()$) versus cache size, it is possible to make this performance overhead negligible.
One final note concerns the concept of "portion". In order to make addressing simple, the above mentioned portions must coincide with power-of-two fractions of the initial cache size. In the rest of the section we illustrate the basic dependencies of the three above metrics versus cache size. In Section 4, we will then discuss architectural issues and the optimization strategy.

3.1 Models

In the following, we consider a power-managed cache in which a cache line is the atomic unit of power management.

[1] We are assuming the regular lifetime of the cache is shorter than the target.

3.1.1 *Lifetime*

By running a given workload with a cache simulator we extract then the *exploitable idleness* (i.e., idle intervals longer than some break-even time) of each line; the latter is a metric of how much energy can be saved. Using the characterization methodology described in [5], it is possible to extract the lifetime (as defined in Section 2) of an SRAM cell as a function of its idleness, that is, the percentage of time in which it is in standby state. The lifetime of a cell determines the lifetime of the line it belongs to, for a line is the atomic unit of access of a cache. The lifetime of the entire cache is the one of the earliest failing line (i.e., the line with smallest idleness). As already mentioned, special architectural arrangement such as the *dynamic indexing* in [12] can distribute the idleness over the cache lines uniformly so that all lines fail at the same time (and worst-case coincides with average case).

Figure 1 shows the result of the characterization for a dynamically indexed 32kB cache. The plot refers to a DVS-based implementation of the standby state ("drowsy"), which is the mechanism used for the "active" portion of the cache in our redundant architecture. Conversely, the "inactive" portion of the cache can be power-gated without incurring the cost of a miss penalty, since it is not used. For the inactive block, idleness is 100% and aging is zero.

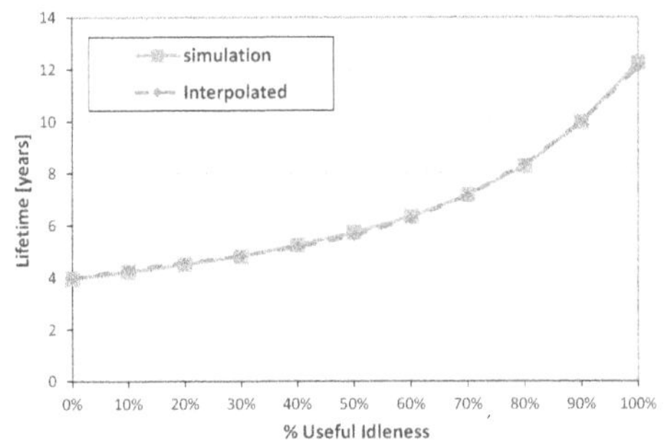

Figure 1: Lifetime of SRAM cell vs. % Idleness.

Lifetime is obviously monotonically increasing with respect to the percentage idleness; the intercept on the Y-axis (3.96 years) denotes the baseline lifetime of the cache (0% idleness). Conversely, 100% idleness (a theoretical value) implies that the all cache lines are always off; the corresponding lifetime value (about 12 years) represents the intrinsic benefit achieved by DVS. The fact that this value is not infinity (as it would be in the case power gating is used [12]) is because DVS does not nullify aging but just mitigates it. The figure also shows the interpolation to be used in our analytical formulation. We found that the cubic function

$$LT(I) = 3.96 + 4 \cdot I - 7.0 \cdot I^2 + 11.1 I^3 \quad (1)$$

well approximates the simulation data (average error = 0.18%, maximum error = +1.97%).

In order to express lifetime in terms of cache size, we need first to establish a relation $I(S)$ between idleness and cache size. Intuitively, a larger cache will have higher **average** idleness (same accesses with more targets available), so $I(S)$ will be a monotonically increasing function. The actual behavior has to be characterized by running a cache simulator under different workloads and for different cache sizes.

Figure 2 shows the result of the characterization.

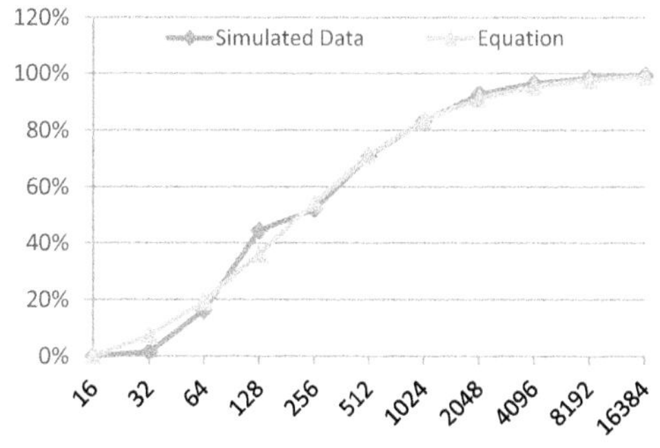

Figure 2: Idleness as a Function of Cache Size.

As one would expect, the idleness is very low for very small cache sizes, then it grows quickly and tends to saturate. Conceptually, idleness becomes 100% for infinite cache sizes. We fitted the curve to a template $f(x) = 1 - \frac{1}{ax+b}$. A good fit is obtained with

$$I(S) = 1 - \frac{1}{0.005 \cdot S + 0.92} \quad (2)$$

For cache sizes larger than 32 bytes, the average (maximum) error is 5.6% (8.3%).

By composing the two functions, we can finally get $LT(S)$ by replacing Equation 2 into Equation 1. Due to the shape of the two curves, the result of $LT(S)$ curve roughly follows the shape of the $I(S)$ curve: lifetime has a steep increase for intermediate cache sizes (between 1K and 4K) and then it quickly saturates.

3.1.2 *Miss Rate*

There are many empirical studies that have tried to correlate cache miss rate versus its size ([15, 16]). A widely accepted model that is relatively accurate over various cache organizations and different workloads states that miss rate roughly goes as $\frac{1}{\sqrt{(x)}}$. We have therefore characterized our embedded applications to extract the actual dependency and have fitted the results to a template $f(x) = \frac{a}{\sqrt{b \cdot x + c}}$, obtaining the following function:

$$MR(S) = \frac{3.6}{\sqrt{0.45 \cdot S + 30}} \quad (3)$$

Figure 3 pictorially shows the simulated data and the fitted curve of Equation 3.

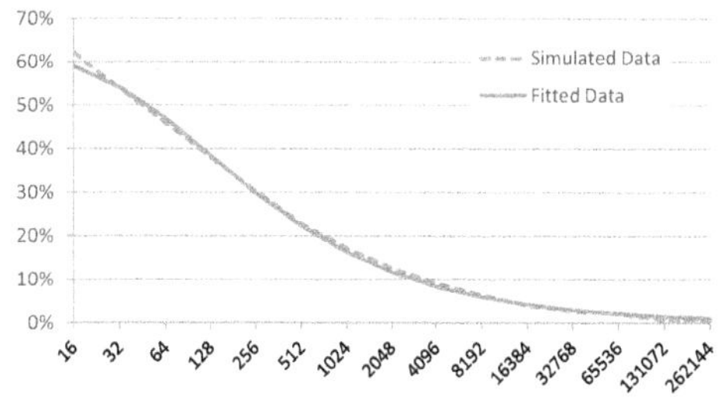

Figure 3: Miss Rate as a Function of Cache Size.

If we consider the complement curve of the hit rate, we could recognize some similitude between the $LT(S)$ and $MR(S)$ curves. This is intuitive since idleness and miss rate are somehow correlated: remember that we are considering *average* idleness. For instance, if idleness is low (as in small caches), very likely the miss rate will also be high. However, the correlation gets weaker as cache size increases: idleness has to do with the *distribution* of accesses, whereas miss rate considers the specific location of accesses. Therefore, we expect our strategy providing more benefits for larger caches.

3.1.3 Power

A line consumes static power only for the fraction of time it is used, that is, $1 - I(S)$. Hence, the static power spent by a cache in "drowsy" mode is:

$$P_{static}(S) = S \cdot (1 - I(S)) \cdot P'(S) + S \cdot P_{idle}(S)$$

where P' is the leakage normalized with respect to line size, and P_{idle} is the (normalized) leakage in the idle state implemented using DVS. In our technology, P_{idle} is not zero and is about one order of magnitude smaller than P'.

Notice that the model explicitly exposes the dependence of both P' and P_{idle} (which is a fraction of P') on S. This dependence accounts for the fact that a given cache size implies a specific aspect ratio (rows and columns) of the tags and data arrays, which affects the capacitance of the interconnects (bit- and word-lines) as well as the structure of the decoder, the number of MUXes and sense amps. It is therefore essential not to consider P' and P_{idle} not as constant values.

Dynamic power, conversely, is not affected by the nature of the idle state: it consists of the sum of the power spent during cache accesses plus the power spent by the background memory when a miss occurs. Since the cache is used at each cycle (at least for the instruction fetch), the dynamic power can be expressed as:

$$P_{dynamic}(S) = P_{cache}(S) + MR(S) \cdot P_{miss}$$

Where P_{cache} is the power spent for accessing the cache, and P_{miss} is the power spent by the lower levels of the memory hierarchy when accessed. Therefore, while exploiting idleness is beneficial for controlling static power, at the same time the miss rate must be kept under control, in order to avoid an increase of dynamic power.

However, when we move to consider energy, and thus include the temporal dimension in the analysis, the choice of the lifetime target significantly impacts the overall energy consumption.

Assume a lifetime target T and that the cache becomes unusable at some time $LT_{cache} < T$. It is clear that after time T, the miss rate jumps at 100% (no cache), and each memory access will imply access to a background memory. Although the cache, once it has become unusable, can now be put into a very low-leakage, the relative benefit is a marginal compensation against the huge increase dynamic power consumption.

The total energy spent by the memory during the system operating time T, can be expressed as:

$$E_{total} = LT_{cache} \cdot (P_{static} + P_{dynamic}) + (T - LT_{cache}) \cdot P_{miss}$$

Since P_{miss} is about two orders of magnitude larger than the power spent by the cache, it is evident that a long lasting cache (i.e., a cache with a larger LT_{cache}) will provide larger energy benefits (or, in other words, will be an energy effective solution for a larger time).

4. ARCHITECTURE AND EXPLORATION STRATEGY

Consider a direct-mapped cache with $S = 2^n$ lines where n is the number of the index bits of the cache address. As discussed in Section 3.1.1, we assume the cache line is the atomic unit of power management. Figure 4-(a) depicts the conventional scenario resulting from the use of Dynamic Indexing (DI) [12]; the original idleness distribution (shown on the left) is transformed by the DI block into a uniform distribution, thus making worst and average cases to coincide. This is achieved by changing the indexing over time; the change is triggered by an Update signal (refer to [12] for more details). Let $I_{avg,1}$ be this average idleness. In this scenario, the entire cache is kept active (blue color), that is, any line is addressable at any time. Clearly, at a given time, some lines will be in a standby state; in this case this state must be implemented through DVS (state-preserving) to avoid the miss penalty in the re-activation.

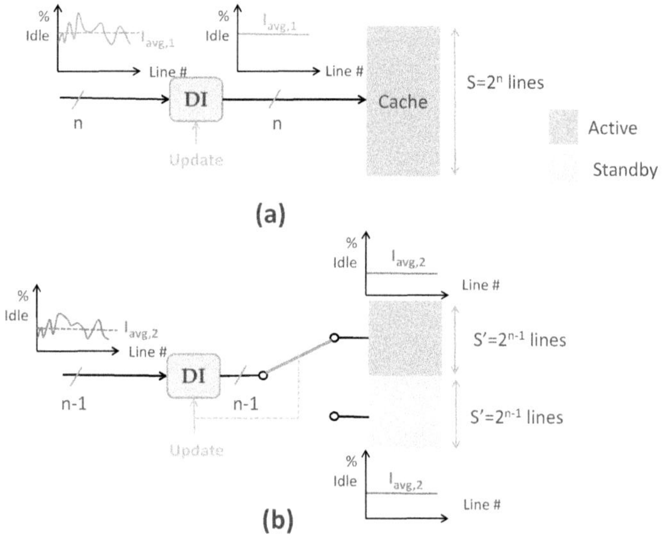

Figure 4: Traditional Dynamic-Indexing Operations (a) and Proposed Architecture (b).

Figure 4-(b) depicts the proposed scenario. The cache is conceptually split into k portions ($S' = \frac{S}{k}$, $k = 2$ in figure), only one of which is active at a given time. This implies that the system effectively uses a smaller cache. The intrinsic idleness (the distribution shown on the left) is now different from the previous case and its average will also be smaller: intuitively, $I_{avg,k} \approx \frac{I_{avg,1}}{k}$, because the same accesses are now distributed over a portion of the address space. Dynamic indexing will again flatten the distribution and make it uniform. Therefore, all lines in the active block will degrade of the same amount.

While a cache portion is accessed, the other ones (in gray) can be put in a non state-preserving state implemented by a footer-based power gating. This state induces an electrical state inside each cell where all nodes drift towards a logic "1", which is the recovery state for NBTI [5]. While this very deep standby state cannot be used for the active block, we can afford it for the inactive ones, which can easily lose their values without impact on performance.

Whenever the Update signal is triggered to change the indexing mechanism, the selector brings up another portion of the

cache (now rejuvenated by having passed some time in the recovery state) and puts the first one into the non-preserving standby state. As in the conventional architecture, updating the indexing needs a flush of the entire cache; this is not a true overhead because updates can be synchronized to architectural events that require cache flushes (e.g., context switches). The overall effect is a sort of *bi-dimensional* and *heterogeneous* indexing over the cache blocks. Each one will age of the same amount, but when not used they save much more energy than in the regular case.
One relevant observation that is worth mentioning is that using one less bit for the index to address a smaller cache ($n-1$ instead of n) implies storing one extra tag bit; the total width of the address is clearly fixed. This is typical in other variants of variable-size caches, like for instance the Dynamically Resizable I-cache [17]. The impact of such a larger tag array, however, will be considered by the accurate explorative analysis below.
Given the above architecture, the objective is to *determine what cache size, and what value of k yields the least energy for a a given target lifetime.* This problem can be viewed as a novel, aging-aware optimization scenario in which lifetime becomes an explicit constraint (as opposed to a traditional performance metrics) and energy as the variable to be optimized. This has also impacts on how we evaluate energy. If the system is forced to operate for a fixed time, energy should be evaluated as the integral of power over the entire lifetime.
Finding an analytical solution to this optimization problem by using the models of Section 3.1 is in theory possible. However, two considerations suggest a much simpler approach. First, the solution would require complex mathematical solutions due to the non strong linearity and the variety of the functions describing the models. Second, and more relevant, the design space is very sparse: not only the main variable S is discretized, but it can assume only a very limited number of values (the mpowers of two). Since we assume S=1KB as the atomic unit of instantiation, we need to evaluate less than 10 solutions.
For this reason, our analysis is based on an explorative approach in which the cache size parameter is swept from the minimum to the maximum value and the values of the other metrics are evaluated.

5. SIMULATION RESULTS

5.1 Experimental Setup

We have assessed the proposed architecture by evaluating a set of applications from the MediaBench suite [18]. We implemented a cache simulator that evaluates the miss rate and, for each cache access, estimates the power consumption, leveraging power models derived from an industrial 45nm design kit provided by STMicroelectronics. The power models also account for the leakage power spent by the cache lines (in the active and in the low-leakage *"drowsy"* state). Furthermore, for each cache miss, the simulator accounts for the energy spent in the background memory to refill the target line. Finally, the simulator keeps track of the useful idleness of each line, in order to evaluate the cache lifetime as depicted in Section 3.1.1.
The following tables report the average power of the whole memory hierarchy (cache and background memory), both static and dynamic, normalized to the power required to access a 8kB cache. Lifetime values (defined as in Section 2 are expressed in years, while miss rates in percentages.
In our analysis, we considered the 1kB cache as the smallest unit that can be instantiated, since the preliminary considerations drawn in Section 3.1 indicate that too small caches incur in too large overhead to benefit from our strategy.

5.2 Power and Lifetime Results

In the first experiment, we compared the non-redundant, power managed caches, with and without dynamic re-indexing, against the proposed redundant architectures where the memory array is split in two or in four blocks. The various configurations are evaluated in terms of power consumption and lifetime.

Cache	*No redundancy*			*2-blocks*		*4-blocks*	
Size	P_{orig}	LT_{orig}	LT_{dyn}	P	LT	P	LT
1 kB	10.12	5.42	8.87	–	–	–	–
2 kB	7.63	5.90	10.53	10.12	17.75	–	–
4 kB	5.76	6.05	11.29	7.63	21.06	10.12	35.50
8 kB	4.34	6.16	11.70	5.76	22.57	7.63	42.12
16 kB	3.21	6.34	11.91	4.34	23.41	5.76	45.14
32 kB	2.66	6.83	12.00	3.21	23.82	4.34	46.81

Table 1: Total Power (Normalized) of the Memory Hierarchy and Cache Lifetime.

Results reported in Table 1 show that while the dynamic re-indexing allows a significant extension of the cache lifetime, still cannot guarantee much more than about 12 years of cache operating time (for the largest cache considered), that is quite near to the upper bound shown in Figure 1. If the required lifetime is longer, the redundant architecture can break this barrier and offer a substantial improvement. A 2-block architecture, can reach about 24 years of lifetime, and extending to a 4-block cache a time horizon longer than 40 years, even for relatively small caches.
Power consumption is clearly larger in redundant architectures; using a smaller "active" set of memory locations implies a larger miss rate. If we measure power over the actual target lifetime of the system, the power balance (initially in favor of the regular architecture) eventually becomes in favor of the redundant one. In other words, there is a power penalty only until the time during which the non-redundant cache is working: after it becomes unusable, the redundant cache becomes also power efficient. This effect can be noticed In Tables 2 and 3, reporting power figures for a $T=15$ and $T=20$, respectively.

	No redundancy			*2-blocks*		*4-blocks*	
Size	P_{orig}	P_{dyn}	MR	P	MR	P	MR
1 kB	37.78	27.81	51.18	–	–	–	–
2 kB	35.41	21.28	38.89	10.12	17.48	–	–
4 kB	34.22	17.57	31.80	7.63	12.96	10.12	17.48
8 kB	33.27	15.13	27.10	5.76	9.36	7.63	12.96
16 kB	32.20	13.56	24.02	4.34	6.57	5.76	9.36
32 kB	30.30	12.82	22.43	3.21	4.29	4.34	6.57

Table 2: Total Power (Normalized) of the Memory Hierarchy and Average Miss Rate for $T=15$.

Results show how the initial (up to regular cache duration) power penalties are compensated by the benefits it provides in the excess operating time. In this second phase, in fact, the non-redundant caches are forcing a 100% miss rate,

	No redundancy			2-blocks		4-blocks	
Size	P_{orig}	P_{dyn}	*MR*	*P*	*MR*	*P*	*MR*
1 kB	44.04	38.06	70.71	–	–	–	–
2 kB	42.62	34.14	63.34	22.69	41.42	–	–
4 kB	41.91	31.91	59.08	14.85	26.67	10.12	17.48
8 kB	41.34	30.45	56.26	10.39	18.17	7.63	12.96
16 kB	40.69	29.51	54.41	7.47	12.52	5.76	9.36
32 kB	39.56	29.07	53.46	5.59	8.83	4.34	6.57

Table 3: Total Power (Normalized) of the Memory Hierarchy and Average Miss Rate for $T = 25$.

which implies a strong increment of the power consumption, because each memory access must be resolved by the background memory. Conversely, the redundant caches are still fully operative and thus they provide a relevant power benefit over the whole time span: if compared with a dynamic re-indexed cache, a 2-bank cache provides a benefit up to 4x for a 15 years operating time and above 5x if such a time is 25 years. Notice, also, that a dynamic re-indexed cache is already the result of a lifetime-enhancing strategy. If we compare against a standard "drowsy" cache, the power benefits increase up to more that 9x ($T = 25$) and 7x ($T = 15$) respectively.

Miss rate values in the tables are **average** miss rates over time, that is, $MR_{orig} \cdot LT_{cache} + 100\% \cdot (T - LT_{cache})$, where MR_{orig} is the miss rate of the regular cache. This explains why the values for the non-redundant caches are quite large and increase as T increases.

Notice also that the values of P and MR for the 4-block architecture are the same in both tables. This is because the corresponding lifetimes (last column of Table 3) are larger than $T = 25$, thus these caches are still active at T.

Finally, Figure 5 illustrates the trend of the power saving as a function of the lifetime target T, for a reference (dynamically indexed) cache with a baseline lifetime of about 12 years.

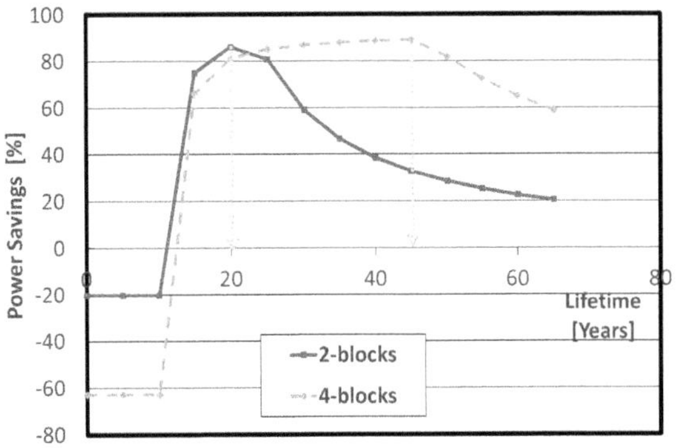

Figure 5: Power Saving vs. Lifetime Target (32 kB Cache).

The plot confirms the results shown in the tables, and also shows that after the redundant cache becomes unusable (the two gray arrows in the plot), the power saving starts dropping and for very long time horizon it asymptotically tends to reach the level of the regular cache.

6. CONCLUSIONS

We have proposed a redundant cache architecture that allows to achieve energy-optimal cache configurations while meeting a given lifetime target. The proposed scheme allows to overcome the limit resulting from non-redundant schemes that are based on redistribution of addresses, and exploits the fact that the drawbacks resulting from effectively using a smaller cache (higher power and higher miss rate) are respectively transformed into a benefit (power) or made negligible (miss rate). For large lifetime targets (15 and 25 years), our redundant architectures can reduce the energy consumed by the memory hierarchy by a factor from 3x to 10x (for 15 years), and from 2x to 8x (for 25 years).

7. REFERENCES

[1] P. Yang, J.-H. Chern, "Design for Reliability: The Major Challenge for VLSI," *Proceedings of the IEEE*, pp. 730-744, Vol. 81, No. 5, May 1993.

[2] M.A.Alam, "Reliability- and process-variation aware design of integrated circuits," *Microelectronics Reliability*, Vol. 48, No. 8, August 2008, pp. 1114-1122.

[3] V. Huard, M. Denais, C. Parthasarathy, "NBTI degradation: From physical mechanisms to modelling," *Microelectronics Reliability*, Vol. 46, No. 1 , Januray 2006, pp. 1-23.

[4] L. Zhang, R. P. Dick, "Scheduled Voltage Scaling for Increasing Lifetime in the Presence of NBTI," *ASPDAC'09*, pp. 492–497, Jan. 2009.

[5] A. Calimera, E. Macii, M. Poncino, "NBTI-Aware Power Gating for Concurrent Leakage and Aging Optimization", *ISLPED '09: International Symposium on Low power Electronics and Design*, pp. 127-132, August 2009.

[6] S.V. Kumar, K.H. Kim, S.S Sapatnekar, "Impact of NBTI on SRAM read stability and design for reliability," *ISQED'06*, March 2006, pp. 213–218.

[7] Y. Kunitake, T. Sato, H. Yasuura, "A case study of Short Term Cell-Flipping technique for mitigating NBTI degradation on cache," *ISQED'10: International Symposium on Quality Electronic Design*, pp. 660–666, March 2010.

[8] J. Abella, X. Vera, O. Unsal and A. González, "NBTI-Resilient Memory Cells with NAND Gates for Highly-Ported Structures", *Workshop on Dependable and Secure Nanocomputing*, June 2007.

[9] T. Siddiqua, S. Gurumurthi, "Recovery Boosting: A Technique to Enhance NBTI Recovery in SRAM Arrays," *ISVLSI'10: IEEE Annual Symposium on VLSI*, July 2010.

[10] M. Powell, et al. "Gated-Vdd: A Circuit Technique to Reduce Leakage in Deep-Submicron Cache Memories," *ISLPED'00: International Symposium on Low power Electronics and Design*, July 2000, pp. 90–95.

[11] K. Flautner, N. Kim, S. Martin, D. Blaauw, T. Mudge, "Drowsy caches: Simple techniques for reducing leakage power," *ISCA'02: International Symposium on Computer Architecture*, May 2002, pp. 148–157.

[12] A. Calimera, M. Loghi, E. Macii, M. Poncino, " Dynamic indexing: Concurrent leakage and aging optimization for caches", *2010 ACM/IEEE International Symposium on Low-Power Electronics and Design (ISLPED)*, pp.343-348, 18-20 Aug. 2010

[13] A. Ricketts, J. Singh., K. Ramakrishnan, N. Vijaykrishnan, D. K. Pradhan. "Investigating the Impact of NBTI on Different Power Saving Cache Strategies," *DATE'10: Design, Automation and Test in Europe*, pp. 592–597, March 2010.

[14] J. Shin, V. Zyuban, P. Bose, T. M. Pinkston "A Proactive Wearout Recovery Approach for Exploiting Microarchitectural Redundancy to Extend Cache SRAM Lifetime," *ISCA'08: International Symposium on Computer Architecture*, June 2008, pp. 353–362.

[15] J. Pal Singh, H. S. Stone , D. F. Thiébaut, "A Model of Workloads and its Use in Miss-Rate Prediction for Fully Associative Caches", *IEEE Transactions on Computers*, vol .41 no. 7, pp. 811-825, July 1992.

[16] A. Hartstein, V. Srinivasan, T. R. Puzak, P. G. Emma, "Cache miss behavior: is it $\sqrt{2}$?" *3rd conference on Computing frontiers*, pp. 313–320, 2006.

[17] M. D. Powell, S.-H. Yang, B. Falsafi, K. Roy, T. N. Vijaykumar, "Reducing Leakage in a High-Performance Deep-Submicron Instruction Cache," *IEEE Transactions on VLSI*, Vol. 9, No. 1, February 2001, pp. 77–89.

[18] M. R. Guthaus et al., "MiBench: A free, commercially representative embedded benchmark suite", *IEEE 4th Annual Workshop on Workload Characterization*, pp. 3–14, Dec. 2001.

Keynote

Spin as State Variable for Computation: Prospects and Perspectives

Speaker

Kaushik Roy
Purdue University
West Lafayette, IN, USA
kaushik@ecn.purdue.edu

Abstract

Recent experiments on spin devices like magnetic tunnel junctions (MTJ's), domain wall magnets (DWM) and spin valves have led to the possibility of having very high density on-chip memories and logic. While the possibility of having on-chip spin transfer torque memories is close to reality, several questions still exist regarding the energy benefits of spin as the state variable for logic computation. Latest experiments on lateral spin valves (LSV) have shown switching of nano-magnets using spin-polarized current injection through a metallic channel such as Cu. Such lateral spin valves with multiple input magnets connected to an output magnet using metal channels can be used to mimic "neurons". The spin-based neurons can be integrated with CMOS and other devices like PCM's to realize ultra low-power data processing hardware based on neural networks (NN), and are suitable for different classes of applications like, cognitive computing, programmable Boolean logic and analog and digital signal processing. In this talk I will first discuss the advantages of using spin (as opposed to charge) as state variable for both memory and logic and then present how a cellular array of magneto-metallic neurons, operating at terminal voltages ~20mV, can do efficient analog computation for applications such as image sensing and processing.

Categories and Subject Descriptors

B.7.1 [**Hardware**]: INTEGRATED Circuits – *Types and Design Styles* (Advanced technologies)

Keywords

Spin device, state variable, low power, ultra low voltage

Biography

Kaushik Roy received B.Tech. degree in electronics and electrical communications engineering from the Indian Institute of Technology, Kharagpur, India, and Ph.D. degree from the electrical and computer engineering department of the University of Illinois at Urbana-Champaign in 1990. He was with the Semiconductor Process and Design Center of Texas Instruments, Dallas, where he worked on FPGA architecture development and low-power circuit design. He joined the electrical and computer engineering faculty at Purdue University, West Lafayette, IN, in 1993, where he is currently a Professor and holds the Roscoe H. George Chair of Electrical & Computer Engineering. His research interests include Spintronics, VLSI design/CAD for nano-scale Silicon and non-Silicon technologies, low-power electronics for portable computing and wireless communications, VLSI testing and verification, and reconfigurable computing. Dr. Roy has published more than 500 papers in refereed journals and conferences, holds 15 patents, graduated 50 PhD students, and is co-author of two books on Low Power CMOS VLSI Design (John Wiley & McGraw Hill).

Dr. Roy received the National Science Foundation Career Development Award in 1995, IBM faculty partnership award, ATT/Lucent Foundation award, 2005 SRC Technical Excellence Award, SRC Inventors Award, Purdue College of Engineering Research Excellence Award, Humboldt Research Award in 2010, and best paper awards at 1997 International Test Conference, IEEE 2000 International Symposium on Quality of IC Design, 2003 IEEE Latin American Test Workshop, 2003 IEEE Nano, 2004 IEEE International Conference on Computer Design, 2006 IEEE/ACM International Symposium on Low Power Electronics & Design, and 2005 IEEE Circuits and system society Outstanding Young Author Award (Chris Kim), 2006 IEEE Transactions on VLSI Systems best paper award. Dr. Roy is Purdue University Faculty Scholar. He was a Research Visionary Board Member of Motorola Labs (2002) and held the M.K. Gandhi Distinguished Visiting faculty at Indian Institute of Technology (Bombay). He has been in the editorial board of IEEE Design and Test, IEEE Transactions on Circuits and Systems, and IEEE Transactions on VLSI Systems. He was Guest Editor for Special Issue on Low-Power VLSI in the IEEE Design and Test (1994) and IEEE Transactions on VLSI Systems (June 2000), IEE Proceedings -- Computers and Digital Techniques (July 2002). Dr. Roy is a fellow of IEEE.

ISLPED'12, July 30– August 1, 2012, Redondo Beach, California, USA.
ACM 978-1-4503-1249-3/12/07.

Register File Write Data Gating Techniques and Break-Even Analysis Model

Eric Donkoh[1], Teck Siong Ong[2], Yan Nee Too[2], Patrick Chiang[3]
[1] Intel Architecture Group, Hillsboro, OR 97124, USA
[2] Intel Architecture Group, Penang, Malaysia
[3]Oregon State University, Corvallis, OR 97331 USA
{eric.donkoh, teck.siong.ong, yan.nee.too}@intel.com, pchiang@ece.oregonstate.edu

ABSTRACT

Register Files account for 30% of 32nm Intel WSM Core dynamic power of which 25% is due to write data distribution. We analyze Register File data gating strategies used to reduce write bitline dynamic power by as much as 96%. We explore the tradeoff of various data gating topologies (Global, Midway, Local), logic implementations (NAND, NOR, Tri-State), and techniques (Stack-Forcing, State-Forcing) to reduce both dynamic and leakage power. We then present a simple and accurate data gating break-even analysis model. The model comprehends "Data" and "Enable" switching activity, signal probability, logic implementation overhead, demonstrating an average error range of ±5%.

Categories and Subject Descriptors

B.7.1 [**Integrated Circuits**]: Types and Design Styles.

General Terms

Performance, Design

Keywords

Data Gating, Low Power, Register File, SRAM

1. INTRODUCTION

As processors continue to scale down to lower operating voltages encompassing new form factors such as Ultrabook and Tablets, power becomes the bottleneck to performance, where performance is redefined to include broader measures such as battery life. Thus, low power design techniques have become major focus in recent CPU designs [1][2], from architecture, through circuit optimization, to power management. Register Files (RFs) are used extensively in modern processors as on-chip memories because of their large storage density and small access latency as well as variability tolerance in low voltage operation [3]. Both high demand on memory and performance has caused the increased number of RF over CPU generations [1][2]. Among the main design blocks in CPU, RFs thus contribute substantially to total CPU power. On the 32nm Westmere (WSM) IA Core [1], RFs accounted for 30% of leakage and dynamic power on a typical benchmark (Figure 1).

With its substantial contribution in total power, a number of power saving ideas, focusing especially on RF read power reduction, have been explored [4]. Within RFs however, the largest single contributor to dynamic power is write data distribution. Write data distribution power has been increasing over process generations as wire scaling challenges and increasing min-vcc constraints have necessitated the use of dual-ended write SRAM bitcell topologies [5].

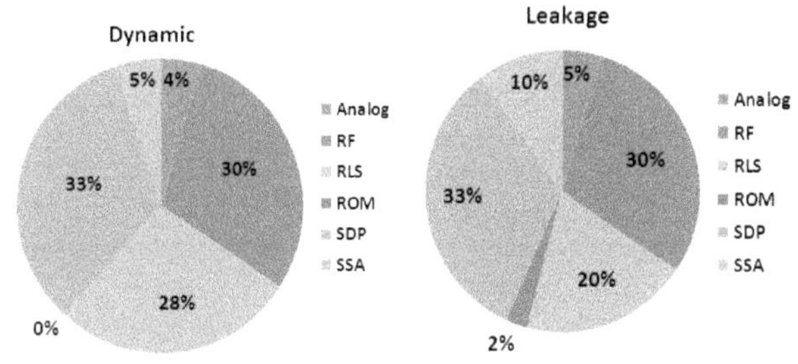

Figure 1: Dynamic and Leakage power distribution in 32nm WSM IA Core by design styles: Register Files (RF), Synthesized Block (RLS), Read Only Memory (ROM), Structural Data Path (SDP), Small Signal Arrays (SSA)

Figure 2 shows power distribution in a 32nm IA core RFs. The write local bitline is the highest dynamic power stage contributing about 18% of total RF dynamic power. The write data distribution (global and local) accounts for 26% and 22% of RF dynamic and leakage power respectively. The SRAM bitcell (excluding read ports identified as "ReadLocaBitline") accounting for 39% of the total RF leakage.

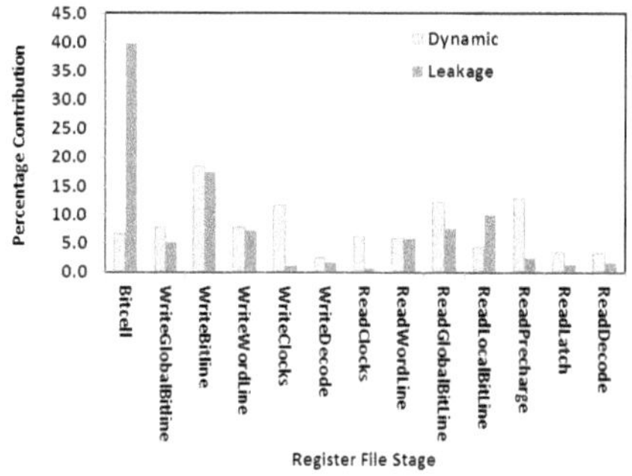

Figure 2: Leakage and Dynamic power distribution in 32nm WSM core Register Files

In this paper, we will focus on write data distribution power. Bitline segmentation to reduce dynamic power has been used on both RF read and write [6]. Other than bitline segmentation, reducing voltage swing on the bitline is also an effective way to decrease power dissipation during writes operation, but it involves additional complexity and challenges [7]. Write data gating with simple logic gate on the other hand is a straightforward static CMOS design, effective yet easy to implement.

In section 2, we further explore different write data gating circuit topologies (NAND, NOR, Tri-State) based on the write data SP as well as the granularity of the gating (Global, Midway, Local) permitted by the area to achieve up to 96% of total write bitline dynamic power reduction. Furthermore, we examine the leakage implications of data gating and present techniques to reduce leakage through device stack-forcing [8][9] and state-forcing based on write data and SRAM bit node probable logic state residency (signal probability). We show that the write data and memory data signal probability (SP) should be a consideration in the selection of data gating strategy. Finally, in section 3, we present a simple and accurate (±5% error) write data gating Break-Even Analysis Model for estimating write data gating implementation gain and evaluation of multiple options, taking into consideration both circuit implementation, architectural, and write data characteristics.

2. WRITE DATA GATING

Write data distribution presents challenges to large RF arrays. Due to performance and write-ability constraints, global and local data buffering are typically needed. The write data is segmented into local (WrBL) and global (WrGBL) bitlines (Figure 3). Each local WrBL drives a sub-set of entries called "bundle". The number of entries per bundle (segment bundle size) is usually influenced by memory write-ability, physical and logical memory size, area, timing, and power constraints. For example, a 128 entry array can be implemented as either 8x16 (8 entries per bundle by 16 bundles), 16x8, 32x4, or 64x2 bundle segmentations. Increasing bundle size reduces the area cost of the data distribution. However, this also poses bitcell write-ability challenges. In a 32nm design, segment sizes of 8 and 16 are commonly used.

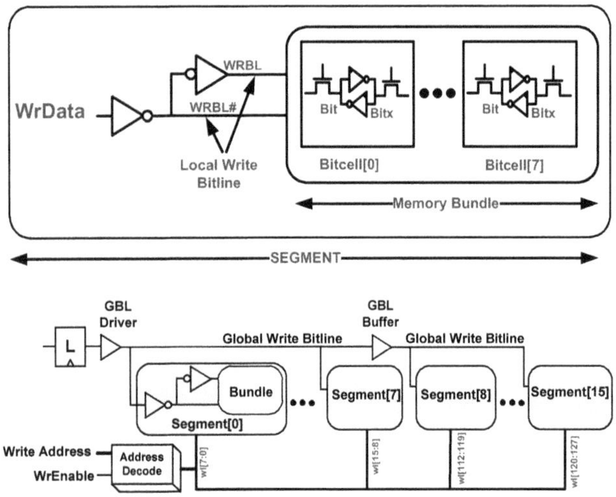

Figure 3: A 128-entry RF write data distribution showing global and local bitline segmentations. Each segment is an 8-entry bundle driven by a local write driver

Each memory bitcell presents a switching capacitance (diffusion load) by way of the memory pass-gate connectivity to the local bitline driver. In addition, the write data buffers (global and local) increases the switching capacitance and leakage associated with data distribution. These devices in addition to the wire capacitance constitute a large capacitance that switch anytime the write data toggles. To control and reduce the switching capacitance, data gating at multiple stages along the data distribution path can be implemented. At each stage area, delay, timing, and power tradeoffs are critical factors that drive the choice of data gating strategy. The three main data gating strategies are: (1) Global Bitline Gating (2) Mid-Way (Partial) Global Bitline Gating and (3) Local Bitline Gating

2.1 Global Bitline Data Gating

Global data gating is needed when the same write data bus is shared by multiple array blocks with independent access. Typically, array data busses have multiple destinations. The data can be toggling any time there is a write to any of the destination blocks.

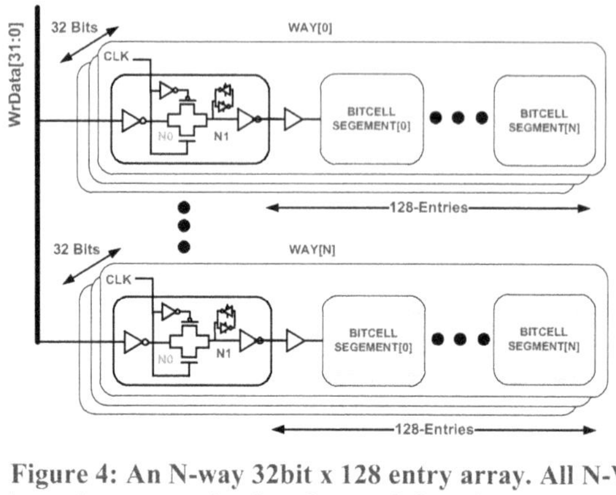

Figure 4: An N-way 32bit x 128 entry array. All N-Ways share the same write data bus and therefore receives every data on the bus regardless of the intended destination

Global data gating prevents the switching of the local distribution in blocks that are not the intended recipient of the data. A common case involves a write data bus shared by multiple ways of an array (Figure 4). Each Way receives the same write data bus but only one way is intended to be written to at any time. The three common gating options on the global write data are (Figure 5): (i) Data Latch Clock Gating (ii) Pre-Latch Data Gating and (iii) Post-Latch Data Gating. Timing is a primary constraint in deciding how early the data can be gated.

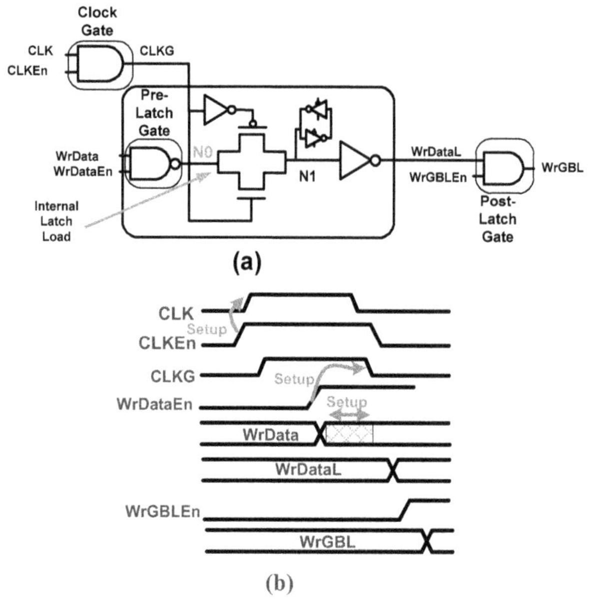

Figure 5: (a) Clock Gate, Pre-Latch, and Post-Latch global write data gate options and (b) timing waveforms

2.1.1 Global Write Data Latch Gating

The global write data latch clock is gated with an "Enable" which can be a global write enable or a more granular enable such as Write Way Enable. This closes the data latch and prevents

WrGBL from toggling whenever WrData toggles unless the block is the intended data destination. The main constraint with latch clock gating is that the enable signal is required a phase before the opening edge of the write latch clock (Figure 5(b)).

2.1.2 Pre-Latch Global Bitline Data Gating

While the clock gating closes the latch and prevents spurious toggling of the global bitline, it does not prevent the toggling of the internal latch node "*N0*" (Figure 5(a)). This internal node toggling can be significant power dissipation in active and highly distributed write data bus. Pre-Latch data gating prevents this internal latch node from toggling. The pre-latch timing is also relatively relaxed compared the clock gating as the Enable can arrive late and still transparently pass through the latch active window without timing failure (Figure 5(b)).

2.1.3 Post-Latch Global Data Gating

The post-latch gating has the significant advantage of relaxed timing constraint as the enable signal does not need to setup to the latch. However, it does not reduce the latch power dissipation.

2.2 Mid-Way Global Bitline Data Gating

Mid-Way gating (Figure 6) is a variant of the global data gating that rely on the address to segment the global bitline into two or more segments. The purpose of this is to prevent a fraction of the global bitline from toggling. The MSB of the address is used to gate the activation of the upper address space. If the entry to be written is in the lower address space, the upper segment is inactive, preventing switching. However, if there is a write to the upper address space, there is no power saving as both the upper and lower bitline segments will toggle. This is especially useful if local data gating is not present due to area constraint. Mid-Way gating can be implemented in addition other global gating options.

2.3 Local Write Bitline Data Gating

Local write data gating is the finest gating granularity (Figure 6). WrBLs are gated per segment using the decoded address. Only the segment that contains the entry to be written is activated for write, decoupling it from the rest of the segments which will remain inactive and hence will not toggle. The timing constraint is relatively more relaxed than global gating.

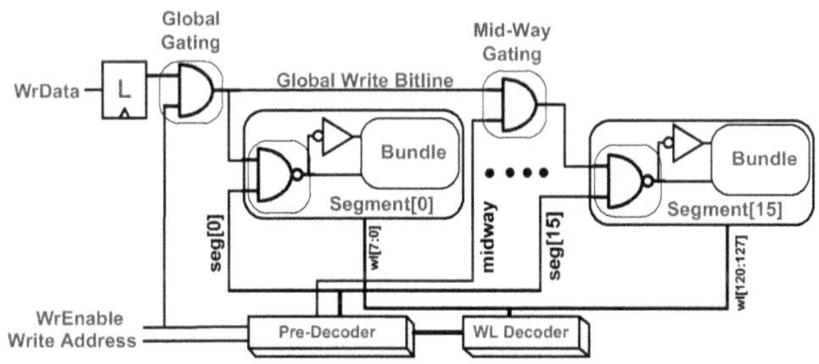

Figure 6:.Global, Mid-Way, and Local Write Data Gating.

Local write data gating is the most effective data gating strategy due to the high local bitline capacitance. The dynamic reduction depends on the granularity of segmentation. More power reduction is achieved with increased segmentation but at an area cost. Figure 7 shows the local bitline dynamic power of a fully extracted 128x32 array with different segment counts and NAND gating. Multiple power benchmarks were run with different write data signal probability (SP). The dynamic power is reduced by 90% when the write bitline is segmented 16 times (8-entry per bundle) with average write data SP=0.5 compared to 38% reduction with 2 segments. The power reduction is also a function of data SP due to reset overhead discussed later. A greater than 2 segments are required to break-even if the SP is close to 1 for a NAND gating logic. The SP therefore becomes important for global gating which is generally a single segment gating.

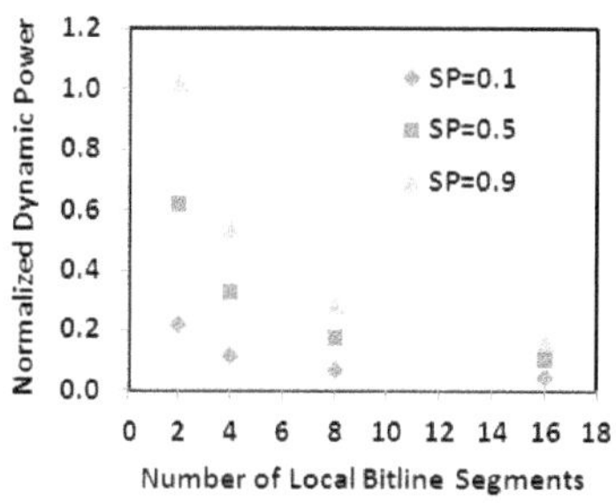

Figure 7: Gated local bitline dynamic power relative to un-gated bitline for 2, 4, 8, and 16-

2.4 Gating Logic Type and Dynamic Power Implications

The effectiveness of data gating is dependent on the gating logic type and the write data characteristics. The characteristic of the write data that impacts data gating are: (i) activity factor (AF) – toggling rate, and (ii) signal probability (SP) – the probable logic state residency of the signal. The data SP relative to the gating logic determines the reset switching overhead. Any time the gating logic is de-activated, the logic output will be reset if the data at the output is opposite polarity of the logic reset state. A NAND gate output will reset to logic state "1", while a NOR will rest to "0". Therefore, the gating logic chosen should take into consideration the SP of the write data to reduce switching cost.

Figure 8 shows the write data SP for different RF blocks in a 32nm IA core across multiple benchmarks. (SP="1" implies WrData="1", SP="0" implies WrData="0"). As shown, average data SP varies across RF blocks. While most RF's average SP are closer to "0" (less that 0.5), some RF block are more likely to write data="1" across multiple benchmarks. This could either be the artifact of the benchmark (random behavior) or systematic occurrence from the RF block designed to store opposite polarity of data. Therefore, the gating logic (NAND/NOR) for each RF block has to take into consideration the most probably data SP. Since the data SP is benchmark dependent and not entirely predictable, an average of a suite of representative application benchmarks are usually used to make the NAND/NOR implementation decision.

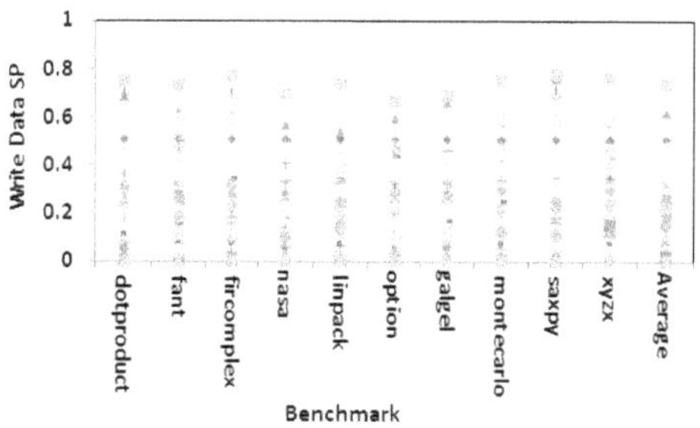

Figure 8: RF Blocks write data signal probability distributions for different benchmarks. Each key on the chart represents a separate RF block

We propose a tri-state gating (Figure 12(h)) that eliminates reset overhead by using an internal keeper on the bitline to hold the written value without resetting. The keeper prevents a floating node input to the memory pass-gate. Figure 9 shows the local data gating dynamic power saving for varying write data SP using a NAND, NOR, and Tri-State logic on a fully extracted 32 bit x 128entry RF block with 16 segments (8-entry per bundle). Dynamic power reduction ranges from 83% to 96%. Maximum reduction using NAND logic is realized with write data average SP < 0.5 while NOR logic gating is preferably at SP > 0.5. The proposed Tri-State is insensitive to SP, showing a consistent 96% reduction. The tri-state however require more area to implement than a NAND/NOR gating.

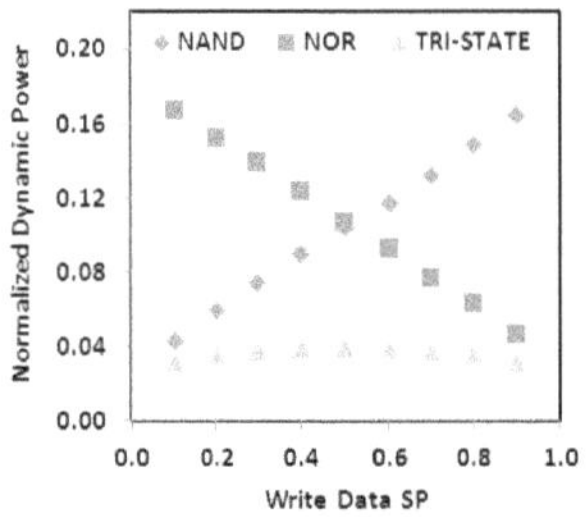

Figure 9: NAND, NOR, and Tri-State logic gating Write Data SP dependency.

2.5 Leakage Impact of Data Gating

Data gating impacts leakage in two ways (1) stack-forcing which reduces the data driver leakage and (2) state-forcing of the memory pass-gate input which impacts memory bitcell leakage.

2.5.1 Gating Logic Leakage and Stack-Forcing

Data gating enforces stacking depending on the write data logic state. Stack-forcing is a common leakage reducing technique [8]. The residency in the stack-forced state depends on the gating logic type used and the write data SP. A WrData="0" enforces stacking on a NAND logic while WrData="1" enforces stacking on a NOR logic (Figure 10). Furthermore, the NAND/NOR gating logic output state is deterministic due to the output reset (NAND="1", NOR="0"). This can be utilized to enforce stacking on the subsequent driver pair to further reduce leakage. Figure 12 (c) (d) shows how this stack-forcing technique is used to further reduce the NAND write data gating leakage (NOR uses same technique).

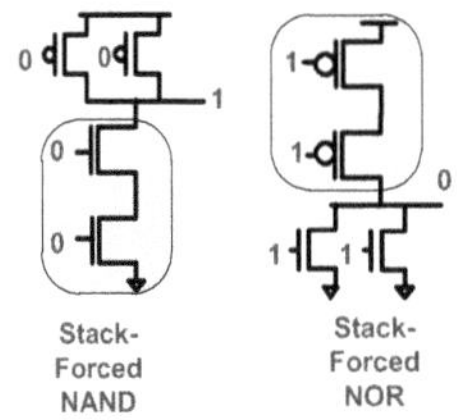

Figure 10: NAND and NOR Stack-Forced conditions

2.5.2 Bitcell Leakage and State-Forcing

The SRAM bitcell leakage sources are shown in Figure 11. The leakage of the bitcell depends on the logic state at the input of the memory pass-gate (WrBL/WrBL#) relative to the internal bitcell node logic state (Bit/Bitx). If the internal node has the same polarity, the channel leakage across the pass-gate is shutoff. However, if the internal nodes have opposite polarity to the pass-gate input, there is channel leakage across both bitcell pass-gates. Using a NAND/NOR logic state-forces the pass-gate input to known states (Figure 12(a)(b)). The leakage impact of gating is dependent on average "Bit" SP. A "Bit" node of SP < 0.5 favors NAND logic gating due to increased probability of same logic state across the pass-gate. On the other hand an SP > 0.5 favor a NOR implementation.

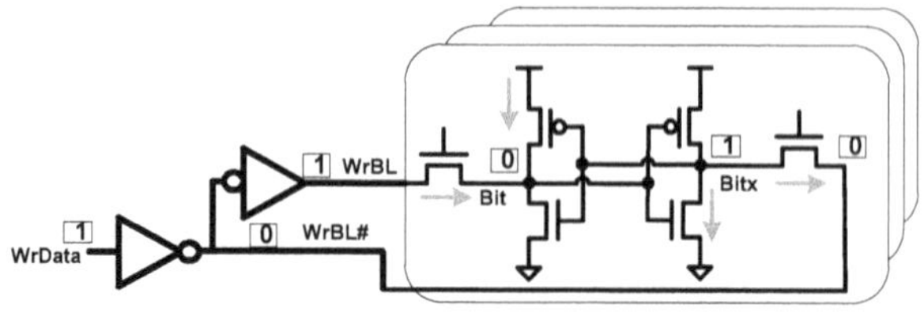

Figure 11: Memory leakage sources and worst-cast condition. Both pass-gates are leaking

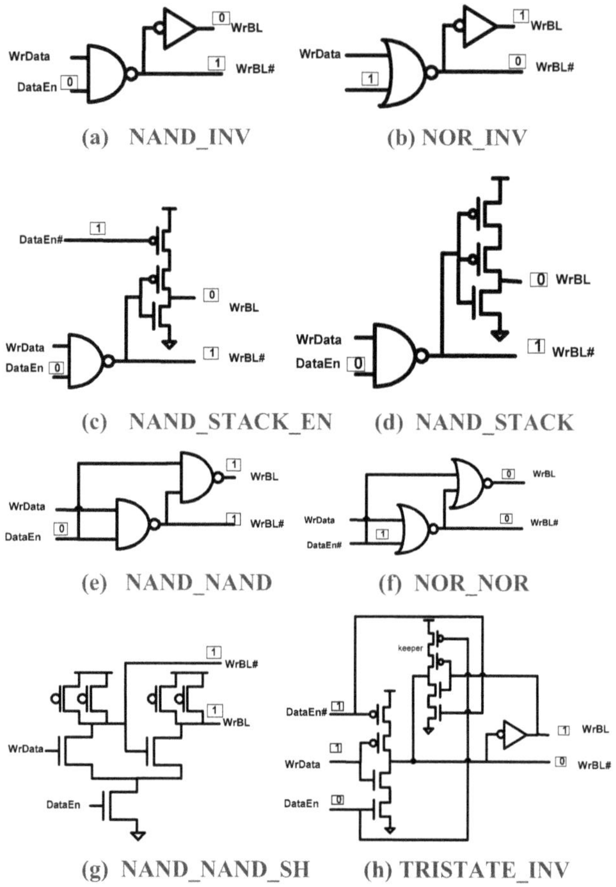

Figure 12: Stack-forcing and state-forcing gating techniques. (a) NAND (b) NOR (c) NAND with stack-forced pair (d) NAND with stacking (e) NAND state-forced pair (f) NOR state-forced pair (g) NAND sharing (h)Tri-State with state retention

While WrBL logic state is deterministic due to state-forcing from enable reset, the state of the memory content ('Bit") is not deterministic across multiple benchmarks. There is a significant bitcell leakage range between the best case leakage condition (WrBL="0", Bit="0", or WrBL="1", Bit="1") and the worst-case leakage condition ("WrBL="0", Bit="1" or WrBL="1", Bit="0") as shown in Figure 13. Therefore, depending on the application benchmark, leakage could be significantly higher than anticipated. Given that the bitcell is the dominant leakage contributor in RFs, this could significantly impact the over array leakage

To force the bitcell leakage to a deterministic value, a back-to-back "NAND_NAND" or "NOR_NOR" logic is used to gate both WrBL and WrBL# (Figure 12(e) (f)). This state-forces logic "1" (NAND) or "0" (NOR) on the input of both pass-gates, guarantee that only one pass-gate will leakage regardless of the bitcell "Bit" state. This enforces a predictable bitcell leakage. Figure 12(g) shows a device sharing technique to reduce the area overhead.

2.5.3 *Simulation Result Comparison*

Figure 13 compares the impact of the various data gating techniques on the SRAM bitcell leakage. The worst-case leakage is as high as 2.7X relative to the best case leakage for conventional NAND/NOR. State-forcing with "NAND_NAND" or "NOR_NOR" results in a deterministic bitcell leakage.

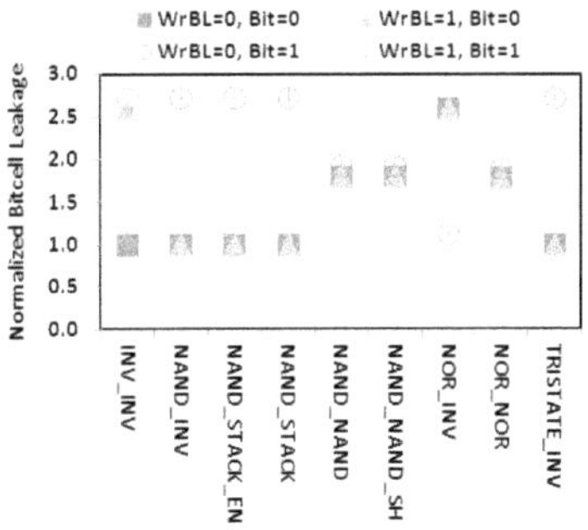

Figure 13: Impact of "Bit" and "WrBL" logic state on bitcell leakage for different gating techniques.

Tri-State leakage exhibits similar characteristic as the NAND. Even though it holds the last written value, since "WrBL" is shared by other entries of the bundle, this value could be opposite the polarity of other bitcells of the same segment. However, it guarantees that at least the last bitcell written to does not leak across the pass-gate, reducing the probability of worst-case leakage condition across all bitcells.

Figure 14(a) shows the comparative driver leakage of the write data gating techniques. The best-case leakage condition for NAND gating with stack-forcing ("NAND_STACK" and "NAND_STACK_EN") is 60% less than the reference "INV_INV". They also show 20% improvement over the conventional "NAND_INV" in both best and worst-case leakage scenarios. The comparative delay and area of the various data gating implementations is shown in Figure 14(b). The same effective driver sizes were used for all gating logic; as such stacked devices incur a higher area overhead for an equivalent INV_INV delay.

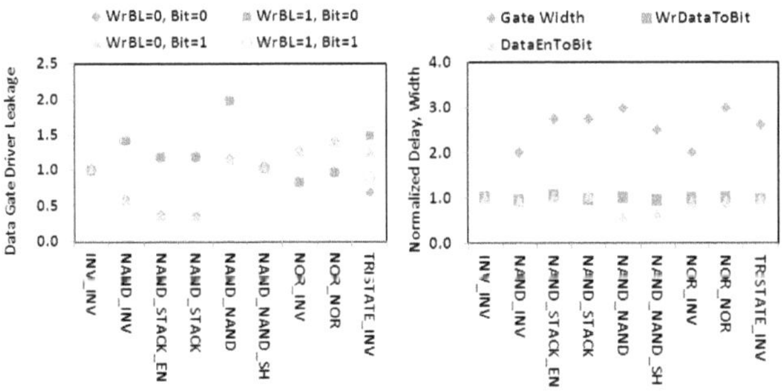

(a) (b)

Fiure 14: Data gating techniques leakage, area, and Delay tradeoff.

3. DATA GATE BREAK-EVEN MODEL

We model data gating break-even by signal transition counts. Each data bit is sub-divided by entry into unit load entities (Figure 15). Each unit load entity is capable of independently switching (transition) and hence can be independently gated. An M bit x N entry array will have *M x N* independent switchable entities. Our model uses the following parameters:

N_{bits}: Number of Data Bits
$N_{entries}$: Number of Entries
$N_{segment}$: Number of Data Gated Segments
AF_{data} : Average Data Activity Factor
SP_{data}: Average Data Signal Probability
AF_{En}: Enable AF,
SP_{En} : Write Enable Effective SP

N_{En}: Number of enable signals, NAND/NOR=1, Tri-State=2
α : This models the increase in bit load cap by the gating logic devices, captured as additional entity load cap.

β: This is loading on the enable from gating. It is normalized to the entity load
$\rho(0,1)$: Data gate logic Enable active (ON) polarity (NAND/Tri-State=1, NOR=0)

$P_r(0,1)$: Reset penalty, indicate whether reset penalty should be applied.
$P_g(0,1)$: Glitch p0enalty indicates whether glitch penalty should be applied.

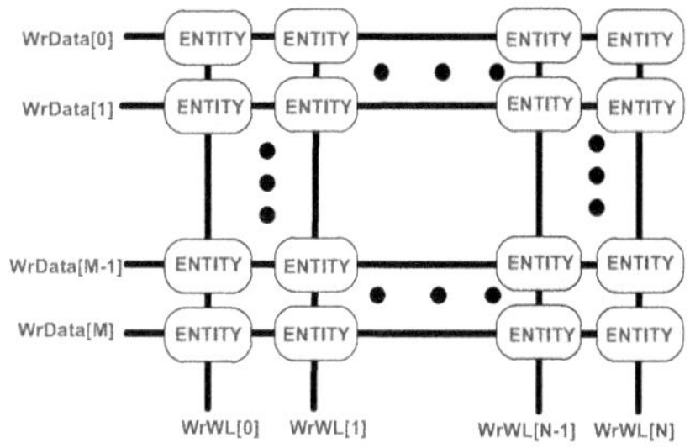

Figure 15: An M bit x N entry arrays. Each associated bitcell load is an entity capable of independently switching and can therefore be independently gated

3.1 Active Data Transition

Each data bit that can transition is considered an "Active Data". The number of transitions is determined by signal activity factor (AF). An AF=1 is considered as 2 transitions. The total number of active transitions of N_{bits} data per entry entity is given by

$$DT_{active} = 2 \times AF_{data} \times N_{bits} \quad (1)$$

3.1.1 *Un-Gated Active Data Transitions*

For an un-gated data, all entry entities are active and can transition during a write. The total number of transitions is:

$$DT_{ungated} = DT_{Active} \times N_{entries} \quad (2)$$

3.1.2 *Gated Active Data Transitions*

When the gating logic is activated by the "Enable" signal, only one segment will be active during access. Segment activation is assumed to be mutually exclusive. Therefore only entities of the activated segment can transition. The total gated active transition:

$$DT_{gated} = \left(\frac{N_{entries}}{N_{segment}} + \alpha\right) \times DT_{active} \quad (3)$$

$$\text{where } \alpha = \frac{\text{Data Gate Device Load per Segment}}{\text{Data Bit Load per Entity}}$$

The logic gate "Enable" signal probability (SP) modulates post-gating transitions depending on its polarity relative to the logic gate. This is captured as:

$$DT_{gatedEn} = DT_{gated} \times [\rho \times SP_{En} + (1-\rho)(1-SP_{En})] \quad (4)$$

3.2 Data Gating Transition Overhead

There are three data gating transition overheads (i) Reset (ii) Glitch and (3) Enable signal transitions.

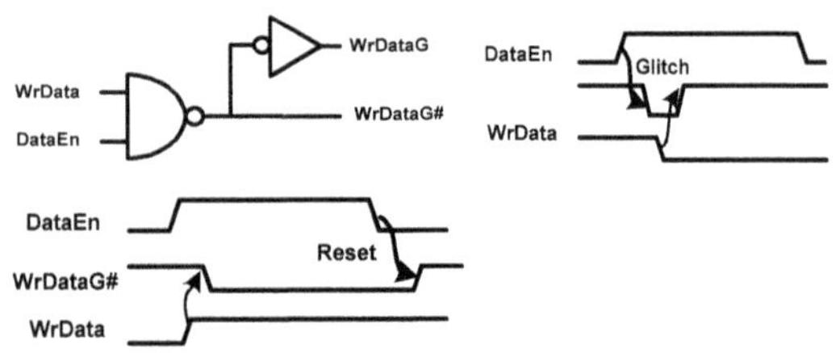

Figure 16: NAND Data gate glitch and Reset overhead

3.2.1 Reset Overhead

Any time the data logic enable signal is turned OFF, all gated output are reset to "1" (NAND) or "0" (NOR). Thus there is a reset cost (Figure 16) associated with the NAND/NOR gating active segment that manifest itself if the logic gate output data polarity is opposite the reset polarity. Tri-State output is not reset. The reset overhead is modeled as:

$$DT_{reset} = P_r \times AF_{En} \times \left(\frac{N_{entries}}{N_{segment}} + \alpha\right) \times N_{bits} \times [\rho \times SP_{data} + (1-\rho)(1-SP_{data})] \quad (5)$$

3.2.2 Glitch Overheard

When the gated logic is enabled, the output will transition if the input and output (from previous reset) have the same polarity (Figure 16). This transition is wasteful if the gating logic "Enable" signal arrives before the new write data. If the data arrives before the "Enable", then no glitch penalty is incurred because the transition that occurs is the actual useful logic transition.

$$DT_{glitch} = P_g \times AF_{En} \times \left(\frac{N_{entries}}{N_{segment}} + \alpha\right) \times N_{bits} \times [\rho \times SP_{data} + (1-\rho)(1-SP_{data})] \quad (6)$$

3.2.3 Enable Signal Transition Overhead

This is the transition overhead of the enable signal. The overhead cost of activating the segment.

$$ET_{overhead} = 2 \times AF_{En} \times N_{bits} \times N_{En} \times \beta \quad (7)$$

$$\beta = \frac{Enable\ Signal\ Load\ per\ Segment\ per\ Data\ Gate\ Logic}{Data\ Bit\ Load\ per\ Entity}$$

3.2.4 Gated To Un-gated Transition Ratio

Total Number of Gated Transition:

$$DT_{totalgated} = DT_{gatedenable} + ET_{overhead} + DT_{reset} + DT_{glitch} \quad (8)$$

The ratio of un-gated to gated transition is:

$$GT_{ratio} = \frac{DT_{totalgated}}{DT_{ungated}} \quad (9)$$

3.3 MODEL VALIDATION

The model was validated with a fully extracted layout of a 128entry x 32 bit RF array. We validated the impact of number of segments on break-even point by implement the array as 2x128, 4x64, 8x32 and 16x8 gated segments. We implemented multiple versions of the gating with NAND, NOR and Tri-state logic. We then generated power benchmark traces with stimulus of different AF and SP using an in-house production power analysis tool. We compared GT_{ratio} from the power benchmarks with our model estimation. The model error was ±5% (Figure 17) showing good prediction accuracy.

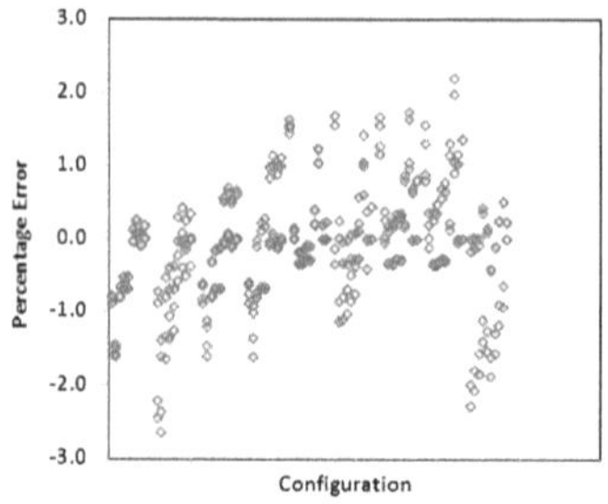

Figure 17: Error distribution of data gate model validation

4. CONCLUSION

Write data distribution is shown to be the highest power consuming stage in a 32nm WSM IA core RFs. We've presented RF write data gating strategies and explored how global and local data gating can be used to reduce write data distribution dynamic power by as much as 96%. The write data SP is shown to be a relevant consideration in the selection of a NAND or NOR data gate implementation due to reset overhead. A Tri-State logic data gating however is independent of data SP and therefore should be the preferred consideration for low dynamic power implementation. This however comes at an area cost. We further showed that using stack-forcing and state-forcing techniques, both write bitline driver and memory leakage can be reduced. In general, area and timing constrain the granularity of data gating. The ideas presented show how to maximize the power benefit in a tradeoff with area and timing.

We also presented a simple and accurate data gating break-even analysis model for determining the best implementation strategy taking into consideration both the Data and Enable characteristics, logic implementation, and overheads. The model can accurately predict data gating power reduction within 5% error margin.

5. REFERENCES

[1] N. A. Kurd et al., "Westmere: A family of 32 nm IA processors," *IEEE Int. Solid-State Circuits Conf. (ISSCC)*, San Francisco,CA, Feb. 2010.

[2] R. Kumar et al., "A family of 45nm IA processors," *ISSCC'09 Digest of Technical Papers*, pp.58-59, Feb. 2009

[3] Chang, L et al., "An 8T-SRAM for Variability Tolerance and Low-Voltage Operation in High-Performance Caches" *IEEE Journal Solid-State Circuits, Apr 2008*

[4] S. Borkar, "Circuit techniques for subthreshold leakage, avoidance, control, and tolerance," *IEDM Tech. Dig., pp. 421-424,Dec, 2004.*

[5] K. Anshumali et al. " Circuit And Process Innovations to Enable High-Performance and Are Efficiency on the Nehalem and Westmere Family of Intel processors" *Intel Technology Journal, vol 14, pp 104-127*

[6] G. Kucuk et al: "Energy–Efficient Instruction Dispatch Buffer Design for Superscalar Processors*" *ISLPED 2001*

[7] K. Kanda et al. "90% Write Power-Saving SRAM Using Sense-Amplifying Memory Cell", *IEEE Journal Solid-State Circuits, Jun 2004*

[8] S. Narendra, et al, "Scaling of Stack Effect and its Application for Leakage Reduction", *ISLPED, 2001*

[9] C.Hua et al. " Distributed Data-Retention Power Gating Techniques for Column and Row Co-Controlled Embedded SRAM" *Memory Technology, Design and testing, 2005*

A Low-Leakage Dynamic Register File with Unclocked Wordline and Sub-Segmentation for Improved Bitline Scalability

Eric Donkoh
Intel Architecture Group
Hillsboro, OR 97124 USA
eric.donkoh@intel.com

Patrick Chiang
Oregon State University
Corvallis, OR 97331 USA
pchiang@ece.oregonstate.edu

ABSTRACT

A register file is presented that uses an unclocked wordline with dynamic bitline sub-segmentation and stack-forcing to reduce both active and standby leakage. The proposed technique improves scalability to a larger number of entries per local bitline segment. For an example 32nm-CMOS 4-write, 6-read port 32 bits x 168 entries register file, the proposed bitline segmentation technique improves local bitline delay by 33%, lowers standby leakage by 70%, and reduces the number of clocked wordlines by 83% when compared with a conventional design. Furthermore, wordline sharing is shown to reduce the number of wordlines by as much as 58%.

Categories and Subject Descriptors

B.3.1 **[Memory Structures]**: Semiconductor Memories.

Keywords

Low Power, Register File, Sub-Segmentation, Unclocked Wordline, Stack-Forcing, Power Gating

1. INTRODUCTION

Register Files (RFs) are ubiquitous in modern microprocessor design, contributing significantly to total chip power and area. Approximately 30% of Intel 32nm Westmere (WSM) IA core [1] power is attributable to RFs [2]. For performance and area efficiency, RF read access is usually implemented with wide-NOR dynamic logic. However, continued device and Vt scaling [3] pose a number of challenges: (1) Increased device leakage from Vt scaling (2) Reduced noise robustness due to increased bitline DC droop from device leakage (3) Increasing wire constrained bitcell size as devices continue to scale better than wires across process generations.

The common approach to reducing leakage and DC droop in the conventional dynamic wide-NOR design are (i) use of high threshold (HVT) devices on the read ports (RPT) (ii) increase in keeper strength and (iii) reduction in the number of entries per local bitline (LBL) segment. Each of these approaches has associated tradeoffs. Increasing the keeper strength increases contention from the RPT during evaluate, increasing LBL fall delay and short circuit power. Using HVT devices also increases LBL fall delay. Reducing the number of entries per LBL segment increase the number of merge logic blocks (RDMRG) and hence array area (Figure 2). The RDMRG block in an 8-entry LBL segmentation configuration can constitutes as much as 30% of the total array bundle. Reducing RDMRG count is therefore a desired design objective. However, due to increasing DC droop from RPTs leakage, the number of entries per LBL segment is typically limited to 8 or 16 entries in a 32nm technology.

Alternative circuit techniques that have been proposed on the other hand generally tradeoff one or more of area, standby leakage, or delay for the other (Figure 1). The leakage-tolerant techniques [4] [5] [6] [7] [8] address LBL noise robustness by reducing LBL active leakage (DC droop) but do not address standby leakage. They rely greatly on keeper downsizing to compensate for the delay degradation resulting from the technique. However, this is limited by process minimum device width and the increasing PMOS/NMOS device strength ratio across process generations [3]. These techniques further compensate for the delay degradation by using all Low-Vt (LVT) devices compared to conventional Dual-Vt (DVT). However, in cases where these techniques do not reduce the standby leakage, HVT devices are still needed.

Power gating [9] is another common leakage reduction technique. The conventional power gating technique however can incur a high delay and area penalty, making its use prohibitive in key logic blocks such as arrays. Moreover, conventional power gating only addresses standby leakage and does not improve active leakage.

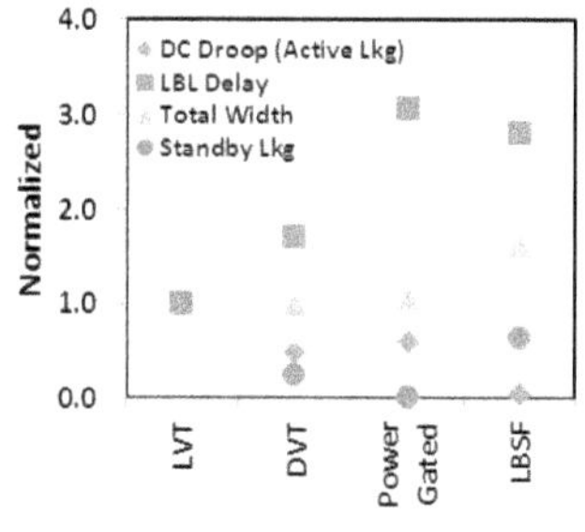

Figure 1: Delay, Leakage, DC droop, and Area tradoff for conventional Low-VT, Dual-VT, Power Gated LBL, and LBSF[4] LBL

Another dynamic LBL design challenge is grid clock loading and wordline clocking distribution. Wordline shielding is typically required to reduce both wordline and bitline noise. As device dimensions continue to scale much better than metal layers in the 32nm process technology nodes and beyond, the physical bitcell width is increasingly constrained by wires. Wordline shielding thus becomes costly, especially in multi-ported arrays.

In this paper, we present an RF bitline technique that (i) uses an unclocked wordline with a dynamic bitline to reduce the bitline delay (ii) uses drain stack-forcing and sub-segmentation to reduce both standby and active leakage for improved noise robustness (iii) reduces clock loading and wire congestion and (iv) provides a scalable LBL sub-segmentation enabling increased number of pull-down devices per bitline segment. We compare the proposed techniques to (i) conventional DVT LBL (ii) conventional LBL

ISLPED'12, July 30–August 1, 2012, Redondo Beach, California, USA.

with modified power gate, and (iii) leakage bypass with stack forcing (LBSF) [4] in a 32nm technology. We show that a fully extracted 4-write, 6-read ports 32 bits x 168 entries RF implemented with 24-entry LBL segment using the proposed techniques improves LBL delay by 33%, reduces standby leakage by 70%, and uses 83% less clocked wordlines relative to a conventional design. We also show that the proposed technique can reduce number of wordlines by greater than 50%.

The organization of the paper is as follow. Section 2 reviews conventional LBL, power gating applied to conventional LBL, and LBSF techniques. Section 3 describes the proposed unclocked wordline stack forced (UWSF) technique. In section 4, we present the comparative simulation results discussing delay, noise robustness, leakage, and scalability. We also present a UWSF implementation results of a 32bit x 168-entries 6-read port array. Summary is presented in section 5.

2. BITLINE TOPOLOGY REVIEW AND RELATED WORK

In this discussion, we define "Active Leakage" as the bitline leakage during read "0" operation (wordline is ON, PCH is OFF). The keeper holds the precharged value and leakage causes bitline DC droop. "Standby Leakage" is defined as the leakage when the bitline is not accessed (all wordlines OFF, PCH is ON).

2.1 Conventional RF Implementation (CONV)

A conventional wide NOR dynamic (domino) RF implementation is shown in Figure 2. RPTs from different memory entries form a segment of LBL in an *Mx1* NOR mux, *M* representing the number of read ports (RPTs). A NAND gate combines two LBL segments which are further merged at the global bitline (GBL), also implemented with wide-NOR dynamic logic. The LBLs are pre-charged to VCC during standby. The precharge, keeper, NAND, and GBL NOR devices typically form the read merge block (RDMRG). The CONV read access decoding and clocking scheme is shown in Figure 3. The read access wordline signals (RWL) are generated by clocking the decoded address. This clocking of all RWL signals creates clock distribution challenges that lead to grid clock loading, and dynamic power dissipation.

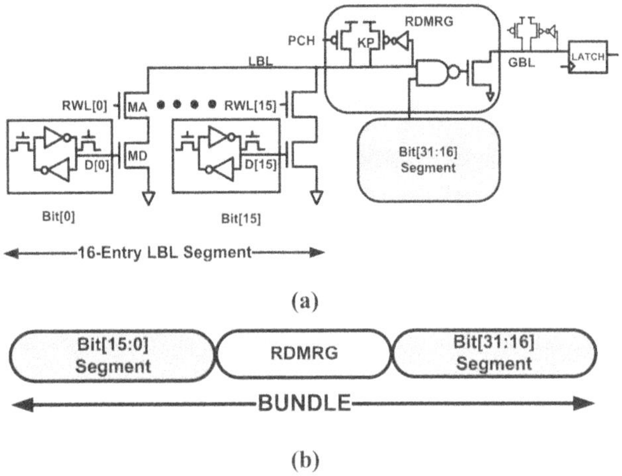

Figure 2: (a) Standard RF dynamic bitline wide-NOR topology with 16-entry per LBL segment and (b) physical organization

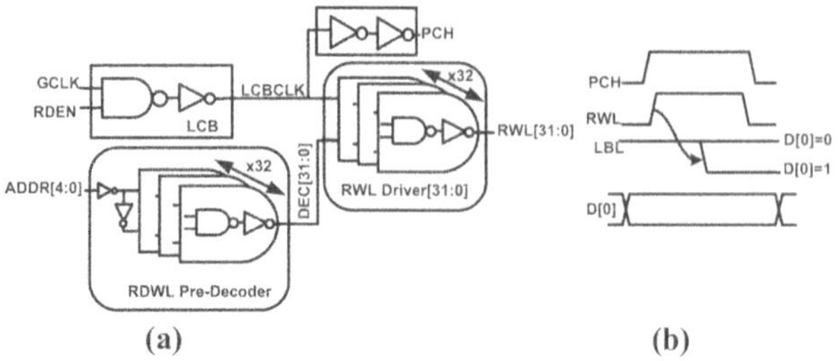

Figure 3: CONV LBL wordline decoding and clocking

The CONV operation is as follows (Figure 3). During read access, the read precharge clock (PCH) goes high and turns OFF the LBL PCH device. The RWL initiate read access to a single entry. If the accessed memory entry is storing data "1", the LBL is discharged to ground, causing a "domino" chain of evaluations along the path through the NAND, GBL, and to the latch downstream. If the stored data is "0", the LBL should retain its precharged value ("1") during the entire read phase with the PCH off. The keeper, *KP*, holds the precharged value to prevent a false evaluation, withstanding any noise impact from charge sharing, LBL interconnect, propagated noise from the RWL signals, and DC droop from RPT device leakage.

2.2 Conventional LBL Footer Power Gating (CFPG) Technique

In a conventional power gating [9] technique, a footer device (MPG) is inserted in series with the RPT devices (Figure 4). When the LBL is inactive, MPG is turned OFF. This enforces stacking on the RPTs. The intermediate node *Vx* is discharged through leakage and converge to a voltage $Vx > 0$, $Vx > Vg$, creating a negative vbias across the top device *MA*. This significantly reduces leakage. During read access, the PGEN signal is asserted high, discharging node *VVS* and creating a virtual ground, before the "RWL" and "PCH" signal are asserted in a similar sequence as the CONV operation.

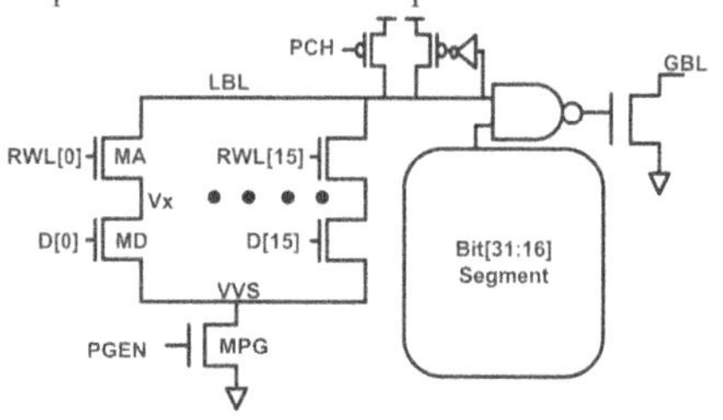

Figure 4: Conventional LBL Power Gating

2.3 Leakage Bypass with Stack Forcing (LBSF) Technique

The LBSF technique proposed in [4] reduces LBL active leakage by precharging the internal nodes *Vx1 and Vx2* of the RPTs (Figure 5). This creates a negative *Vgs* bias across both *MA1* and *MA2* devices, and sets *Vds*='0', cutting off LBL standby leakage through devices *MA1* and *MA2*. However, device *MD* will be leaking due to the internal precharging of node *Vx2*. The timing operation of LBSF is the same as the CONV design.

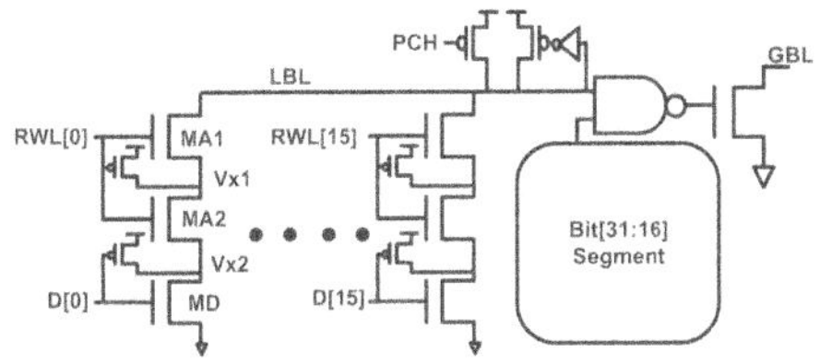

Figure 5: LBSF Technique with internal pre-charging

3. UNCLOCKED WORDLINE STACK-FORCED LBL (UWSF) TECHNIQUE

3.1 UWSF LBL and Decoder

The proposed UWSF LBL technique has two components: (1) removal of wordline clocking and (2) sub-segmentation of the bitline. The RPTs are grouped into sub-segments. A series NMOS device, MSF, is used to mux multiple RPT sub-segments, partitioning the LBL into two: *LBLA* and *LBLB*. The LBL segment is organized as $M \times N$ matrix where M is the number of entries per *LBLB* sub-segment and N is the number *LBLB* sub-segments per *LBLA* segment. A 16-entry LBL segment organized as 4x4 UWSF is shown in Figure 6. The *LBLA* mux select, SFWL, decoded from the address, is clocked to satisfy the LBL domino clocking requirement. Thus, the read wordline (UWL) need not be clocked, reducing the number of clocked signals by a factor of M.

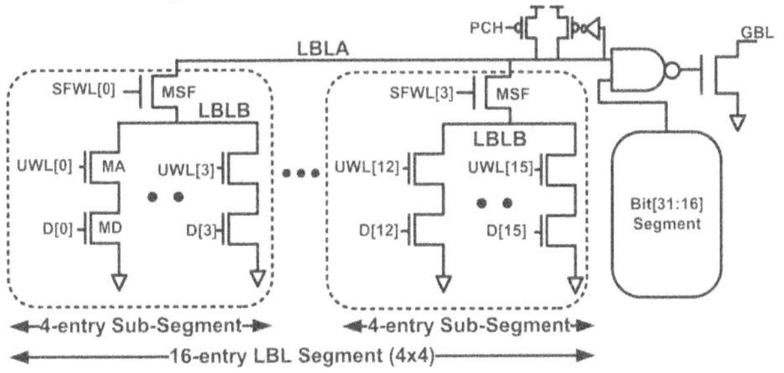

Figure 6: UWSF 4x4 16-entry LBL configuration with stack-forced dynamic bitline and unclocked wordline

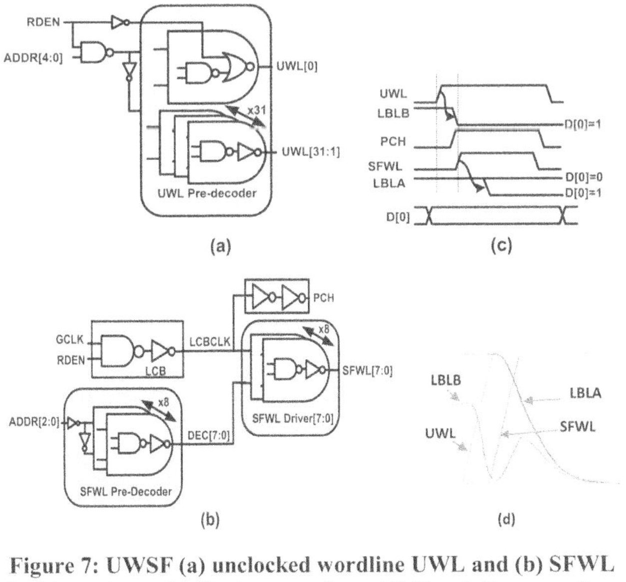

Figure 7: UWSF (a) unclocked wordline UWL and (b) SFWL clocking scheme (c) Timing waveform (d) Simulation waveform

A 32-entry decoding scheme of the proposed UWSF is shown in Figure 7. ADDR[4:2] is decoded and clocked as SFWL[7:0] to select *LBLA* mux while the RPT select wordlines UWL[31:0] from the pre-decoder are unclocked. Thus, only 8 signals are clocked compared to 32 signals in the CONV design (75% reduction). Without the clocking stage, the UWL signal is 2-gate stages faster than the clocked RWL signal of the CONV design and has no required setup to LCBCLK.

The operation of the UWSF is as follows (Figure 7). The faster unlocked wordline (UWL) selects one entry of a sub-segment and starts discharging *LBLB* node. PCH is then turned OFF and SFWL is asserted after 2-gates delay to select an *LBLB* sub-segment, discharging *LBLA*. SFWL timing corresponds to the conventional RWL. In the standby mode, both UWL and SFWL will be reset to "0" to enforce LBL stacking. The SFWL signals are reset by the clock. To reset UWL signals without a clock and at minimal decoder area cost, the read enable signal, RDEN, is used to state-force all addresses to "0". This guarantees a deterministic decoder output, i.e. only the decoder output UWL[0] can be "1". The remaining UWL[n…1] decoder outputs will be "0". UWL[0] is then reset to "0" by RDEN. Alternatively all UWL signals could be reset by RDEN, without state-forcing the address, at a higher area but lower reset switching cost.

3.2 Sub-Segmentation and Active Leakage

The CFPG stack-forcing shares a single power gate footer among all RPTs of a segment. While this reduces standby leakage, it does not improve active leakage since all the RPTs of the active LBL will have their virtual VSS (VVS) enabled during read access. We propose further sub-segmentation of the CFPG LBL to reduce active leakage by only activating one LBL sub-segment during read access while the remaining sub-segments of an LBL segment are kept inactive. An 8-entry 2x4 CFPG LBL sub-segmentation using the UWSF decoding scheme as the power gate control is shown in Figure 8. DC droop is reduced by 58% relative to a single 8x1 LBL segment (Figure 8(b)).

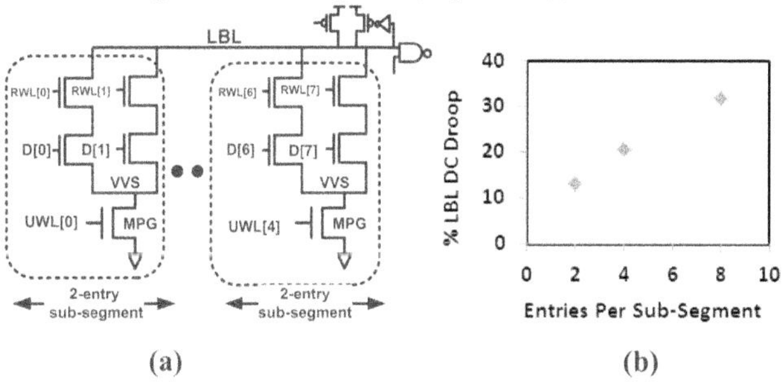

Figure 8: (a) A 2xN sub-segmenation of CFPG. (b) DC droop plot for 2x4, 4x2, and 8x1 CFPG sub-segmentations

3.3 Wordline Sharing

Increasingly, RF bitcell layout sizes are constrained by the number of tracks needed to route wordlines. Wordline clocks have more stringent noise constraints that require extra tracks for shielding purposes. To reduce the number of signals, a scheme that shares wordline among multiple sub-segments is employed. Figure 9 shows a 2-way sharing scheme of a 2xN sub-segmented LBL configuration. Each wordline is shared by two sub-segments. A 16-entry LBL segment will therefore requires a total of 16 signals (SFWL[7:0] and UWL[7:0]) in a 2-way sharing scheme, breaking even with the CONV design.

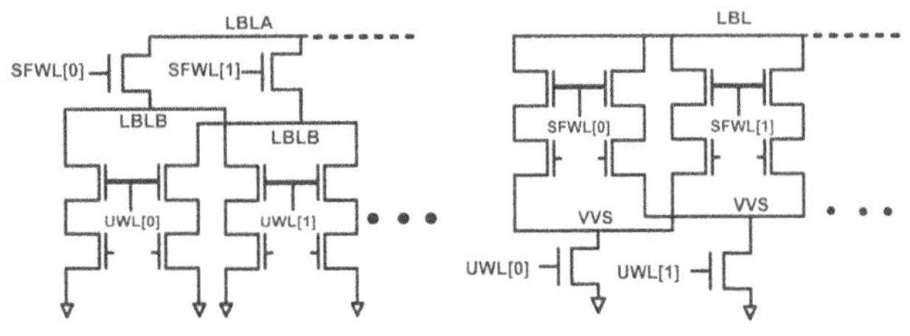

Figure 9: A 2xN sub-segmentation with 2-way sharing to reduce number of wordlines (a) UWSF (b) CFPG

Further sharing and sub-segmentation can reduce the number of wordline signals depending on the decoder scheme used. A 16-entry 4x4 sub-segmented 4-way sharing, for example, can decode address bits ADDR[3:2] as SFWL[3:0] to select one of 4 LBLA muxes, while address bits ADDR[1:0] are decoded as UWL[3:0]. Each UWL signal is shared by four RPTs of alternating LBL sub-segment entries. A total of 8 signals (4 UWL + 4 SFWL) are required to access the 16 entries, a 50% reduction. Table 1 shows the SFWL and UWL mapping to entries.

Table 1: A 4x4 sub-segment with 4-way share mapping

	SFWL[0]	SFWL[1]	SFWL[2]	SFWL[3]
UWL[0]	Entry[0]	Entry[4]	Entry[8]	Entry[12]
UWL[1]	Entry[1]	Entry[5]	Entry[9]	Entry[13]
UWL[0]	Entry[2]	Entry[6]	Entry[10]	Entry[14]
UWL[0]	Entry[3]	Entry[7]	Entry[11]	Entry[15]

4. SIMULATION RESULTS

4.1 Simulation Setup

We compare the different LBL techniques: (1) CONV DVT (2) UWSF (3) CFPG with sub-segmentation and (4) LBSF DVT. Device width is used as a proxy for device area. Devices were sized such that the total device width is equal, accounting for any auxiliary devices required for the topology. Figure 10 shows the device sizes and type used.

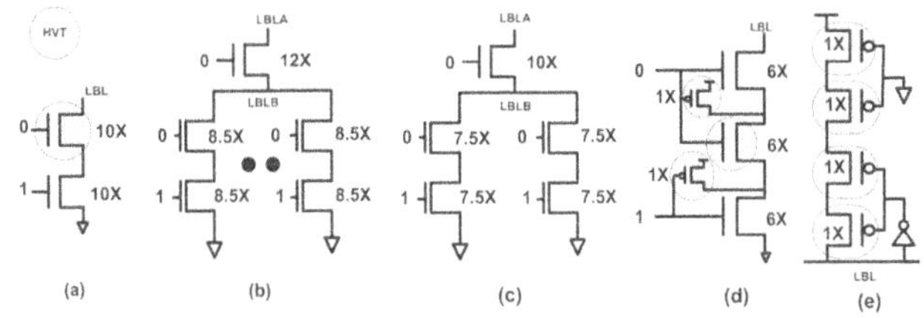

Figure 10: Device size, type, and worst-case leakage conditions for (a) CONV (b) UWSF/CFPG 4xN (c) UWSF 2xN (d) LBSF (e) LBSF 4-stack keeper scheme.

The total device width of UWSF includes the LBLA mux select devices. UWSF and CFPG use the same device sizes. The overhead of the mux select devices are amortized over the *M* RPTs sharing the *LBLB* segment. Thus, the total overhead is less compared to LBSF. Three LBL segment configurations were evaluated for each topology: 8-, 16-, and 32-entry per LBL segment. For a given topology, the same RPT sizes are used for all three configurations of that topology without resizing. We evaluated both *2xN* and *4xN* LBL sub-segmentation configurations of UWSF. The CFPG configuration used was a *4xN* sub-segmented LBL.

4.2 Leakage Consideration

The worst case leakage conditions are as shown in Figure 10. Under CONV leakage condition, the LBL RPTs will be leaking through a single device, *MA* (Figure 2). Both standby and active leakage are reduced by using a HVT device. In the LBSF (Figure 4), LBL leakage through device *MNA* and *MNB* are reduced through the internal precharging of nodes *Vx1* and *Vx2* respectively. However, device *MND* will still be leaking. Hence, both CONV LBL and LBSF require DVT to reduce RPT standby leakage. The worst-case standby leakage condition for UWSF has two series devices, *MSF* and *MA*, turned OFF (Figure 6), enforcing stacking on the RPTs regardless of the memory data content. This and sub-segmentation reduces both standby and active leakage, enabling the use of all LVT devices. CFPG also stack forces *MA* and *MPG* (Figure 4) to reduce standby leakage. Active leakage is also reduced by the proposed sub-segmentation. The standby leakage comparison is presented in Table 2. Leakage is reduced by 91% and 93% in the UWSF and CFPG respectively while the DVT LBSF reduces leakage by 80% relative to the CONV DVT (partly due to smaller sizes).

Table 2: Leakage comparison of the 4 topologies studied

	CONV	UWSF(4x1)	CFPG(4x1)	LBSF
LKG ratio	1	0.089	0.073	0.202
%reduction	0	-91%	-93%	-80%

4.3 Noise Consideration/Keeper Sizing

LBL is subjected to four main sources of noise (i) DC droop from leakage through the RPT, (ii) charge sharing glitch (iii) propagated noise from the RPTs input signals, and (iv) interconnect noise on the LBL wire. Table 3 shows the distribution of noise sources for 4 different arrays from extracted layout in a 32nm design. For a 1-port array with full shielding, the noise is dominated by DC droop. However, for a multi-ported array, interconnect and propagated noise components become significant as full shielding is prohibitive.

Table 3: Noise breakdown of different ported arrays

Noise Sources	1-Port 8-seg LBL	2-Port 8-seg LBL	6-Port 8-seg LBL
% DC Droop	72%	46%	29%
% Interconnect	14%	29%	24%
% Propagated	14%	16%	34%
% Charge Sharing	0%	9%	13%

4.3.1 DC Droop (Active Leakage) and Keeper Sizing

Active leakage manifests as a DC droop during read "0" when the leakage through the RPTs can falsely discharge the LBL in the active window. The keeper should be sufficiently strong to hold the LBL PCH value. Increasing keeper size reduces the droop but also increases the LBL fall delay. Figure 11(a) shows the keeper size impact on DC droop on a 32-entry LBL. The required keeper size for UWSF and CFPG is 2.6X smaller than CONV for a DC droop noise floor of 10% VCC. LBSF has the least DC droop due to the precharging of the internal node.

4.3.2 Propagated Noise

Wordline noise is propagated onto the LBL through the RPTs. Typically propagated noise is reduced by shielding the wordline signals. In a routing constrained array, this could increase the array size. The propagated noise from UWSF wordline signals

(UWL) are attenuated by the mux select device *MSF*. Moreover, inactive sub-segments will have their *LBLB* node at or near "0" voltage. Therefore, propagated noise from UWL does not impact *LBLA* significantly. Propagated noise from SFWL onto *LBLA* is not attenuated. However, since the total number of SFWL is reduced by a factor of *N* relative to CONV, the total noise impact on *LBLA* is reduced. Figure 11(b) shows the LBL sensitivity to propagated noise during read "0". The same keeper size was used for all four techniques. As shown, when noise was only applied to the UWL signal, the propagated noise onto *LBLA* is significantly attenuated. The worst case propagated noise of applying noise simultaneously to both UWL and SFWL is still up to 50X more noise tolerant than the CONV topology at input noise less than 10% VCC. LBSF shows most resilience to propagated noise due to internal precharging.

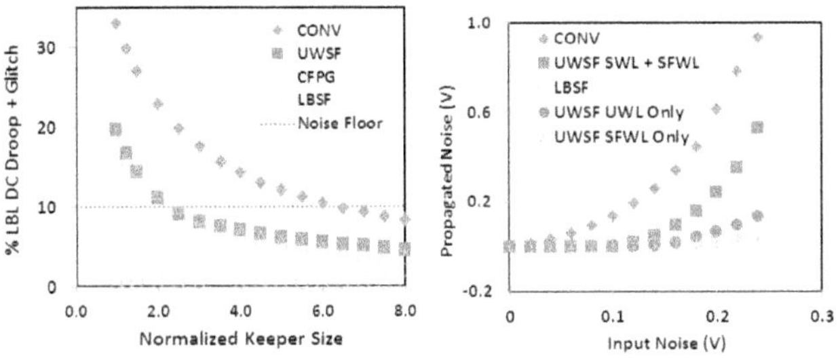

Figure 11: (a) Normalized keeper size vs DC Droop (b) LBL propated noise sensitiviy to input noise.

4.3.3 Glitch Noise

In standby mode, the UWSF *LBLB* node will be discharged to near "0" due to leakage or to "0" from a previous access to adjacent entry sharing the same *LBLB* node. At the onset of evaluation when the SFWL turns ON, a charge sharing glitch occurs between *LBLA* and *LBLB* due to capacitance and voltage differences. Ideally a smaller capacitance on *LBLB* node is required to reduce the glitch; therefore $N > M$ is desirable for glitch reduction (Figure 12(b)). However, for the same total LBL capacitance (*LBLA* + *LBLB*), the best delay is obtained if a significant fraction of the total LBL capacitance is discharged by the early arrival of UWL before SFWL is turned ON. Thus, delay favors $M > N$. There is therefore an optimal $M \times N$ configuration.

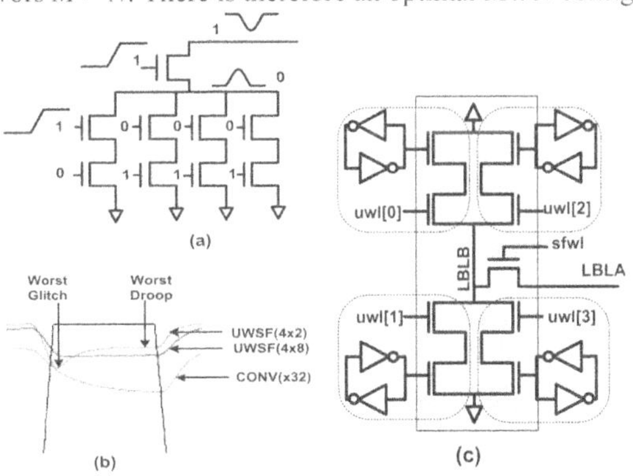

Figure 12: (a) UWSF worst-case glitch condition LBL (b) glitch and DC drop waveform (c) a 4x4 layout organization to minimize *LBLA* routing capacitance

The worst-case charge-sharing glitch and worst-case leakage DC droop occur at different time windows (Figure 12(b)). While the worst charge-sharing occur at the onset of SFWL going active, the DC droop peak occurs at the end of the SFWL active window. The two noise sources are therefore not additive. Figure 12(c) shows a physical organization that groups the shared RPTs into clusters to reduce routing capacitance on the *LBLB* node.

4.4 Delay & Segment Scalability

In comparing the delay using iso-width sizes, the keeper is sized such that LBL DC droop + glitch does not exceed 10% of VCC at the noise corner (ignoring the other noise sources). The UWSF technique has two disadvantages that negatively impact the LBL delay when compared to the CONV topology: (1) a 3-stack pull-down RPTs compared to the 2-stack in the CONV topology and (2) smaller RPT devices for the same total width. However, their impact on LBL fall delay is compensated for by the following advantages: (i) the elimination of UWL clocking to enable an early discharge of *LBLB* node before SFWL is active; this reduces the LBL clock-to-out delay measured as SFWL rise to *LBLA* falls; (ii) the reduced DC droop from reduced leakage, enabling keeper downsizing and less contention; and (iii) the use of all LVT devices due to reduced leakage. While CFPG also shares the same disadvantages over CONV, it does not have advantage (i). LBSF on the other hand does not have advantages (i) and (iii).

4.4.1 Keeper Scalability

Figure 13(a) shows the keeper size required for 10% dc droop + glitch. The 4x2 and 4x4 UWSF configuration keeper size is limited by glitch. Therefore, the required keeper size decreases as *LBLA* capacitance is increased by increasing the number of sub-segments per *LBLA* segment. Comparatively a 2x8 configuration requires a smaller keeper size due to reduced glitch and reduced active leakage from further sub-segmentation. Though the DC droop specification was 10%, LBSF DC droop was only 1% at the minimum possible keeper size. Even though the keeper could theoretically be downsized further, it could not fully capitalize on the increased noise robustness due to minimum device size limit.

4.4.2 Delay Scalability

The delay performance of 8-, 16-, and 32-entry LBL configuration for the 4 techniques is presented in Figure 13(b). The 16-, 32-entry *4xN* UWSF configurations outperform the conventional design by 9% and 34% respectively. The 8-entry *4x2* LBL UWSF shows a 12% delay degradation over CONV. This is primary due to a larger keeper size required for the charge-sharing glitch, a consequence of a relatively lower *LBLA/LBLB* cap ratio. However, an 8-entry *2x4* LBL UWSF configuration only shows 2% delay degradation. The *2xN* 16-entry and 32-entry LBL UWSF also demonstrate 5% and 23% delay improvement respectively over the CONV. As the number of sub-segments per *LBLA* segment is scaled up, the impact of glitch on the *4xN* configuration keeper size is reduced and it outperforms the *2xN* configuration. Increased sharing of *LBLB* improves delay by the early discharge of a relatively larger portion of the total *LBL* capacitance (from the early arrival of the UWL). Increased sharing of *LBLB* also reduces Mux device area overhead, increasing RPTs sizes. The optimal $M \times N$ configurations for the 8-, 16-, and 32-entry LBL segmentations are therefore 2x4, 4x4 and 4x8 LBL sub-segment configurations respectively.

LBSF has the worst iso-width delay, with degradation of 70%, 51% and 13% for 8-, 16-, and 32-entry segments respectively. CFPG shows 11% delay improvement at 32-entry LBL configuration but degrades by 12% and 20% at 16-, 8-enty LBL configuration. The proposed UWSF shows the best LBL segmentation scaling.

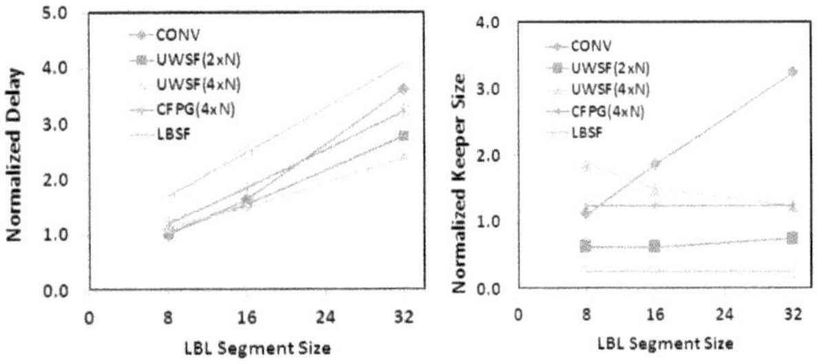

(a) (b)

Figure 13: (a) Keeper sizing and (b) Delay scalability of 8-, 16-, and 32-entry LBL segmentation

4.4.3 Noise Robustness Vs. Delay Tradeoff

Figure 14 shows the noise robustness vs. delay tradeoff for 32-entry LBL configurations with increasing keeper size. The proposed UWSF shows the best tradeoff between noise and delay. LBSF is most noise tolerant but has the worst iso-width delay degradation.

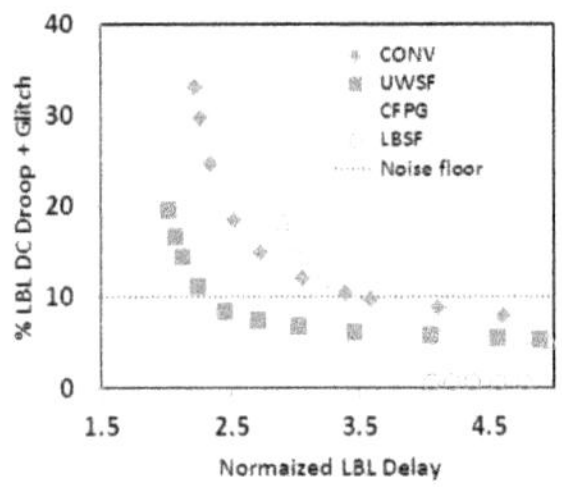

Figure 14: LBL noise robustness (glitch + droop) and delay tradeoff for a 32-entry

4.5 UWSF Multi-Port Implementation Result

The proposed UWSF was implemented on a 4-write, 6-read port 32 bits x 168 entries array. The read ports were clustered [10], making it ideal for sharing. A 6x4 LBL sub-segmentation and a 4-Way UWL sharing scheme was used with each UWL wordline shared by four alternating entries of the 4 sub-segments. Only 10 signals (6 UWL and 4 SFWL signals) were required to access each 24 entry segment. The total number of clocked signals was reduced from 24 to 4.

Table 4: Extracted results of a 4-write, 6-port 32 bits x 168 entries array UWSF LBL compared to CONV LBL

	CONV	UWSF
Delay (Normalized)	1	0.67
Total Noise	1	0.46
Number of RWL per segment	24	10
Number of Clocked Wordlines	24	4
Standby Leakage (Data SP=0.005)	1	0.67
Standby Leakage (Data SP=0.33)	1	0.30

Timing, noise, and power analysis on an extracted layout shows a 54% LBL noise reduction, 33% delay improvement and as much as 70% standby leakage reduction compared to the conventional implementation (Table 4). Leakage reduction is dependent on memory data signal probability (SP) [2] (representing worst-case leakage condition Data=1, RWL="0" residency during the simulation cycle).

5. CONCLUSION

We presented an unclocked wordline technique that in combination with sub-segmentation of the dynamic LBL can implement stack-forcing and improve a 32-entry segmented local bitline delay by 34%, lower worst-case standby leakage by 91%, and reduced number of clocked wordline by over 75% when compared to a conventional design with the same total width. UWSF outperforms LBSF and CFPG in the critical metric of delay, leakage and area. Even though the LBSF was the most noise resilient, this robustness could not be fully converted into reduced keeper sizing to improve delay. Summary of the results are shown in Table 5. The proposed UWSF delay scales with increasing entries per LBL segment. This enables increased LBL segment size, thereby reducing the number of RDMRG blocks and array size. We also showed how sub-segmentation can be applied to conventional stack-forcing to reduce active leakage, improve noise robustness, and hence improve delay. Furthermore, wordline sharing using the proposed technique can reduce the number of wordlines by more than 50%. Finally we implemented the proposed techniques on a production 4-write, 6-read port 32 bits x 168 entries array and extracted result showed 54% LBL noise reduction, 33% delay improvement, 70% standby leakage, 58% reduction in number of wordlines, and 83% reduction in clocked wordlines.

Table 5: Simulation results summary of 32-entry CONV, UWSF, CFPG and LBSF LBL. Keepers are sized for 90% Vcc Droop + Glitch (@ffff corner).

Technique	Keeper Size	LBL Fall Delay	Num of Clock WL	Leakage
DVT CONVx32	3.25	3.63	32	1.00
UWSFx32(2x16)	0.75	2.78	16	0.08
UWSFx32(4x8)	1.25	2.41	8	0.09
CFPG32	1.25	3.23	32	0.07
LBSFx32	0.25	4.09	32	0.20

6. ACKNOWLEDGMENTS

The authors wish to acknowledge Shih-Lien Lu, Bibiche Geuskens for their feedback, Nanda Siddaiah and Kurt Kreitzer for their support.

7. REFERENCES

[1] N. Kurd *et al.*: "Westmere: a family of 32nm IA processors", in *Proc. IEEE Int. Solid-State Circ. Conf.*, 2010, pp. 96-97.

[2] E. Donkoh et al. "Register File Write Data Gating Techniques and Break-Even Analysis Model" *ISLPED 2012*

[3] P. Packan et. al.,"High Performance 32nm Logic Technology Featuring 2nd Generation High-k + Metal Gate Transistors," *IEDM 2009, pp 1-4*

[4] S. Tang et. al., "A leakage tolerant dynamic register file using leakage bypass with stack forcing (LBSF) and source follower", *2002 Symp. VLSI Circuits, pp 320-321*

[5] A. Agarwal et al., "A Leakage-Tolerant Low-Leakage Register File with Conditional Sleep Transistor", *Proc. SOC Conference,2004, pp 241-244*

[6] A. Alvandpour et. al., "A sub-130-nm conditional keeper technique," In *IEEE* Jour. *Of Solid-State Circ.*, May 2002.

[7] R Krishnamurthy et. al., "A 130-nm 6Ghz 256x32bit leakage tolerant register file," In *IEEE JSSC.*, 2002.

[8] S. Borkar, "Circuit techniques for subthreshold leakage, avoidance, control, and tolerance," *IEDM Tech. Dig., pp. 421-424,Dec, 2004.*

[9] S. Narendra, et al, "Scaling of Stack Effect and its Application for Leakage Reduction", *ISLPED,2001*

[10] R. Kumar and G. Hinton, "A family of 45nm IA processors," *ISSCC'09 Digest of Tech. Papers*, pp.58-59.

A Fine-Grained Many V_T Design Methodology for Ultra Low Voltage Operations

Mingoo Seok
Department of Electrical Engineering, Columbia University
mgseok@ee.columbia.edu

ABSTRACT
In this paper, we propose a fine-grained many V_T design methodology for ultra low voltage (ULV) operations of CMOS VLSI circuits. The fine-grained many V_T transistors can be developed through only layout-level technique (e.g. inverse narrow width effects) in a multi-V_T technology without any process modifications. Through SPICE simulations, we confirm that the proposed design methodology can improve performance, energy efficiency and variability of ULV circuits in three important domains, i.e. driving a fixed capacitive load, reducing active leakage energy consumption in non-critical paths, and lengthening short delay paths without energy overhead to aid error detection and correction techniques.

Categories and Subject Descriptors
B.7 [**Integrated Circuits**]: General
Keywords
low power, variability, active leakage energy, many V_T design methodology, near-threshold voltage, sub-threshold voltage, ultra low voltage, VLSI circuits

1. INTRODUCTION
Recently, ultra low voltage (ULV) operations, where the supply voltage of CMOS circuits is scaled down to near or below the threshold voltage (V_T) of transistors, have gained a significant amount of attention due to 5-20× energy savings. Particularly, ULV operations can benefit many energy-constrained applications such as implantable medical devices, remote surveillance systems, wireless sensor networks, and mobile electronics [16].

However, ULV operation has yet been widely adopted due to several challenges to overcome, such as delay variability and low speed. As transistors operate in the weak inversion region, the transistor driving current becomes exponentially smaller and sensitive to process, voltage, and temperature (PVT) variations. Another important challenge of ULV operations is to mitigate leakage energy consumption during active mode. It is well-known that the supply voltage scaling stops improving energy-efficiency below a certain point, referred to as an energy-optimal voltage (V_{opt}) [1][15], due to the rapid increase of active leakage energy consumption. In order to extend energy-saving with further voltage scaling, it is critical to minimize leakage energy consumption during active mode.

Recently, a number of techniques have been proposed to mitigate these challenges. Transistor sizing have been studied by several authors, which include energy-aware sizing [2], variability-aware sizing [3], subthreshold logical efforts [4], length biasing with reverse short channel effects [5], and sizing with inverse narrow width effects [6]. Several publications also propose post-fabrication tuning methods including global body bias [7], multiple clustered body biases [8], and dual-V_{dd} designs [9]. Also, energy-optimal pipelining and cycle borrowing techniques [10] has been proposed to improve energy efficiency, performance and variability.

In this paper, we propose an orthogonal technique, which we refer to as a fine-grained many V_T design methodology. In this methodology, many (e.g. 11 in this study) and fine-grained (ΔV_T=10-20mV) V_T transistors are used in a single design. These fine-grained VT differences, which can be created through only layout techniques, provide a sufficient amount of transistor strength modulations due to its exponential impacts on subthreshold leakage current.

The fine-grained many V_T design methodology can be used in several ways to improve the performance and energy efficiency of ULV circuits. First, it can substitute or complement conventional transistor sizing to modulate transistor driving strength in a more energy-efficient manner. SPICE simulations confirm that the proposed many V_T transistors consume 2.17 × smaller amount of energy at the same performance, compared to conventional sizing. Second, the fine-grained V_T transistors can be selectively assigned in critical paths and non-critical paths to improve performance and also reduce active leakage consumption. In our simulations, 20% improvement in delay and 11-27% improvement in energy efficiency is achieved in several benchmark circuits.

Finally, the proposed methodology can delay short paths in pipelines in energy efficient manners. This enables error-tolerant techniques [11][12] to be used with less risk of short path violations as the amount of error detection window [11] or cycle borrowing [12] is often bounded by short paths. Monte-Carlo SPICE iterations confirm that the proposed methodology can increase the ratio of the minimum short path delay to the standard deviation of critical path delay by ~5× under random process variations, allowing a substantially wider error detection window or cycle borrowing window to be implemented. This delaying is often hard to be achieved only with coarse-grained V_T transistors since a new critical path can be created due to the exponential delay increase. Furthermore this improvement comes even with energy savings while conventional delay padding can cause a considerable amount of energy overhead, particularly in ULV regimes, where a large amount of delay padding is needed.

The remainder of the paper is organized as follows. In Section 2, we develop energy-efficient, fine-grained, many V_T transistors using layout-level techniques in an industrial 65nm CMOS technology. In Section 3, with these new transistors, we propose several domains to use a fine-grained many V_T design methodology, which improves performance, energy efficiency and variability for ULV circuits. Finally, the paper concludes in Section 4.

ISLPED'12, July 30–August 1, 2012, Redondo Beach, California, USA.

2. DEVELOPING FINE-GRAINED MANY V_T TRANSISTORS THROUGH INVERSE NARROW WIDTH EFFECTS

In this section, we investigate developing the fine-grained many V_T transistors using layout-level techniques in an industrial 65nm CMOS technology as currently-available CMOS technologies do not support the fine-grained many V_T transistors due to high fabrication costs. Then, in Section 3, we will discuss the use of these transistors to improve energy efficiency, performance and variability in ULV VLSI circuits.

2.1 Inverse narrow width effects in ULV regimes

Among many techniques [5][6], inverse narrow width effect (INWE), where V_T is reduced with narrower channel width [6], is promising for developing fine-grained many V_T transistors. As shown in Figure 1, the transistors with higher number of fingers per a fixed width, i.e. narrower fingers, can provide higher driving current due to the reduced V_T. The total transistor width is fixed at 1.12μm in this study.

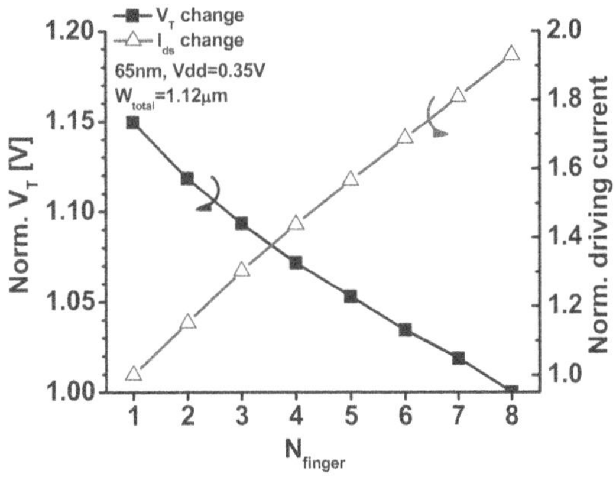

Figure 1. Reduction of V_T and the increase of on-current with inverse narrow width effects

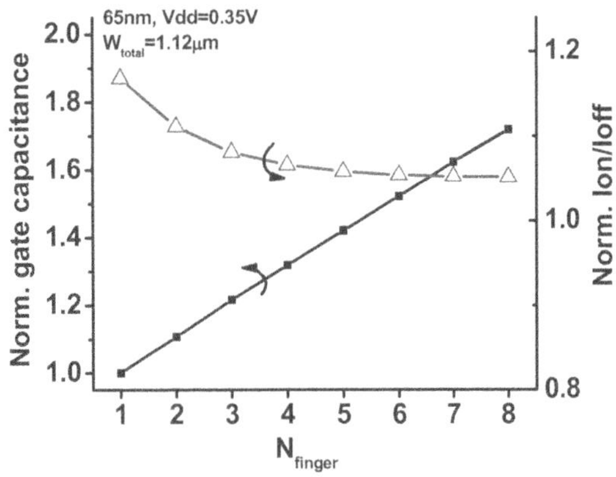

Figure 2. Increase of gate capacitance and degradation of subthreshold slope

However, INWE can cause energy overhead. In ULV regime, differently from super-threshold operations, lowering V_T can lead to larger gate capacitance since it makes channel to be more strongly inverted. As shown in Figure 2, the gate capacitance and thus switching energy consumption increases linearly with the number of fingers (i.e. smaller V_T). In addition to the gate capacitance, the subthreshold slope (on-current to off-current ratio) can be degraded with lower V_T because of non-ideal transistor characteristics of on- and off-current scaling with V_T. Higher gate capacitance and degraded subthreshold slope [7] can collectively suppress energy efficiency when an excessive number of fingers are used. Therefore, a careful investigation is required to develop fine-grained many V_T transistors through INWE in ULV regimes.

2.2 Designing energy-aware, fine-grained V_T transistors

In this section, we investigate the right configurations in terms of the number of fingers and transistor types such that the fine-grained many V_T transistors can provide a smooth delay and leakage scaling at minimal energy overhead. For this investigation, we perform SPICE simulations with FO4 inverter chains, where the number of fingers is swept from 1 to 8 at a fixed total width The total width is fixed at 1.12 μm through all the simulation unless explicitly mentioned. At V_{dd}=0.35V, we perform the finger-sweeping simulations for high-V_T (HVT), medium-V_T (MVT), and low-V_T (LVT) transistors which are provided by the technology. A typical amount of interconnect capacitance are included between inverters in the chains.

As shown in Figure 3, the fingering provides a smooth scaling of delays and energy consumption in all HVT, MVT, and LVT transistors. Without fingering (when N_{finger}=1), the maximum gap between FO4 delays is 4.36×. However the gap is reduced to 1.73× with the fine-grained V_T transistors. We also observe that the energy consumption increases with more fingers due to the change of gate capacitance and subthreshold slope, as discussed in Section 2.1. As shown in Figure 3, 8-finger MVT transistors consume 33% more energy, compared to 1-finger MVT transistors. HVT transistors generally consume less energy, compared to LVT and MVT transistors due to weaker channel inversion.

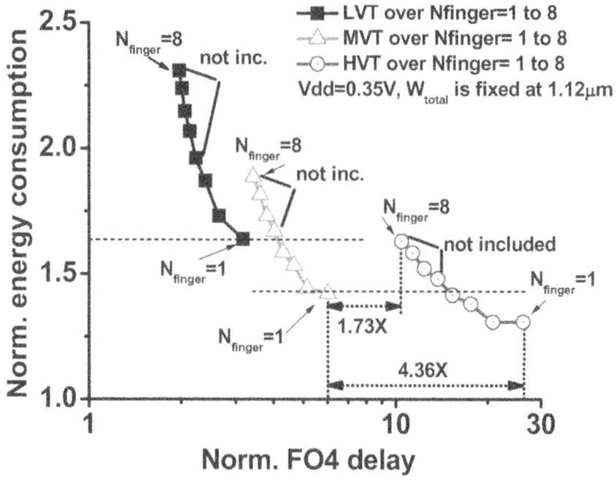

Figure 3. FO4 delays and energy consumptions of fine-grained many V_T transistors

Now we select the finger configurations that provide a good energy and delay tradeoff. For example, as shown in Figure 3, the 5-8 finger MVT transistors consume more energy but are slower than 1 finger LVT transistor. Therefore, it is not beneficial to include the 5-8 finger MVT transistors in our fine-grained, many V_T transistor selection. For the same reason, HVT transistors with 5-8 fingers are not included in the selection. On the other hand, the LVT transistors with 4-8 fingers are preferred to be omitted since they exhibit marginal performance boosts while increasing the energy overhead considerably.

Table 1. 11 Selected Fine-Grained V_T Transistors

	Configuration		Characteristics	
STR	FET	N_{finger}	Norm. FO4	Norm. E
10	LVT	3	1.00	1.43
9.05	LVT	2	1.10	1.32
7.53	LVT	1	1.33	1.25
5.57	SVT	4	1.80	1.21
5.13	SVT	3	1.94	1.17
4.65	SVT	2	2.15	1.10
3.97	SVT	1	2.51	1.09
1.55	HVT	4	6.42	1.08
1.35	HVT	3	7.38	1.05
1.14	HVT	2	8.70	1.00
0.91	HVT	1	11.00	1.00

Based on these rationales, we pick 11 fine-grained V_T transistors, which are summarized in Table 1. The lowest V_T transistors (strength 10 or STR10), using LVT transistors with 3 fingers (W_{finger}=0.37um), shows 11× faster FO4 delay, compared to the STR0.91 transistors in the selection. Although most of the delay gaps between the selected transistors are small, a relatively large gap appears between STR3.97 and STR1.55 in this technology as an energy-efficient finger configuration is not found in this delay range. However, when it is necessary to fill the gap, a less energy efficient configuration like 8-finger HVT transistors could be included in the selection.

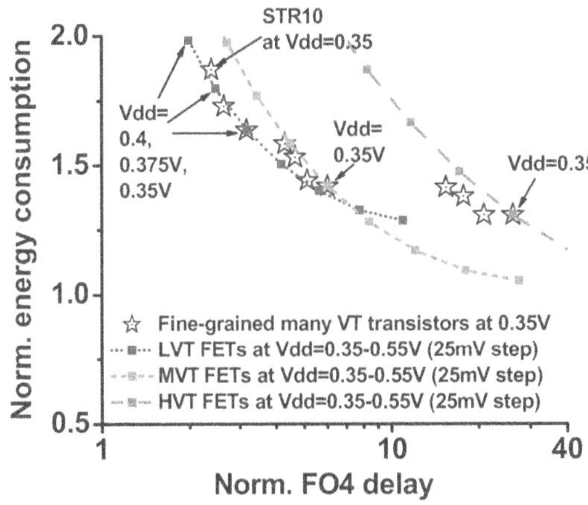

Figure 4. Comparison of the proposed fine-grain V_T assignment and supply voltage scaling

In Figure 4, we compare the delay and energy consumption of the 11 selected transistors at V_{dd}=0.35V to non-fingered HVT, MVT and LVT transistors (provided by the technology) at V_{dd}=0.35-0.55V with a step of 25mV. The STR10, STR9.0 and STR7.5 transistors at 0.35V, which are made out of LVT transistors, exhibits similar delay-energy performance when LVT transistors works at V_{dd}=0.35-0.38V. Similarly, STR1.55 to STR0.91 transistors exhibit slightly better energy-delay tradeoffs than HVT transistors at V_{dd}=0.35-0.375V. *This implies that finely modulating V_T through INWE acts like assigning an individual fine-grained supply voltage for each transistor.* Apparently the fine-grained many V_T design has no power distribution and level conversion overheads, posing an attractive design method, compared to fine-grained multi- V_{dd} design [9] at ULV regimes.

3. FINE-GRAINED MANY V_T DESIGN IN ULV REGIMES

In this section, we propose several design methodologies in ULV regimes with the fine-grained many V_T transistors developed in the previous section. The proposed design methodology can improve performance, energy efficiency, and variability in three primary domains: 1) driving a load in an energy-efficient manner, 2) minimizing active leakage energy consumption, and 3) delaying short paths in pipelines for error-tolerant techniques.

3.1 Energy-efficient load driving, a substitute for transistor width sizing

In super-threshold regimes, transistor width sizing plays an important role in tuning driving strength for its finer granularity and linear dynamic energy and leakage cost. On the other hand, a designer could use multi-V_T transistors provided by CMOS technologies as an option to modulate driving strength, but the usability is limited since only a small number (e.g. ~3) of different V_T's are typically offered. Furthermore, the poor tradeoff between leakage and driving strength, i.e. exponential leakage cost at linear drivability improvements, makes multi-V_T transistors to be less attractive than width sizing for fine-tuning driving strength.

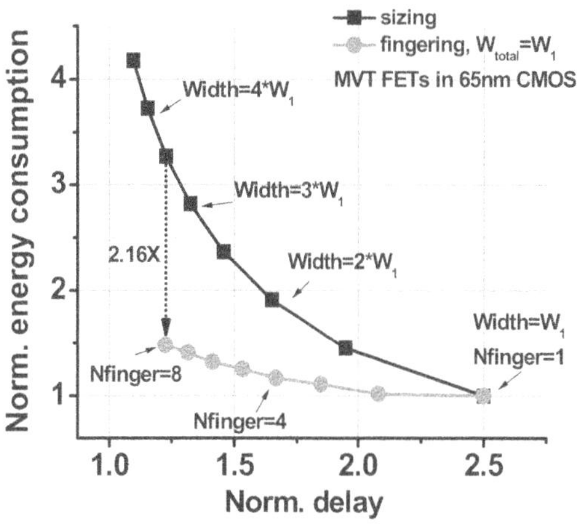

Figure 5. Comparison of narrow width effects and transistor sizing for driving a fixed capacitive load

However, for ULV circuits, the fine-grained many V_T transistors can be a better option than width sizing to modulate driving strength due to the better tradeoff among drivability, leakage and switching energy consumption. In ULV regimes, V_T can exponentially reduce delay at a linear dynamic energy and exponential leakage cost, while sizing only linearly or sub-linearly reduces delay with linearly increasing switch and leakage energy consumption. In addition, in the ULV regimes, *the small (10-20mV) V_T difference out of layout modifications can become sufficient to control driving strength* due to the exponential impact of V_T on transistor drain current.

In particular, the proposed method can benefit driving a large amount of fixed loads such as long interconnects and clock networks. Figure 5 shows the improvement of delay when a buffer drives a 100fF capacitor. In order to reduce delay, in one setup, we modulate strength through the number of fingers at a fixed total width while width up-sizing is used for the other setup. As shown in Figure 5, the V_T modulations reduce delay by 2× at a 50% higher energy cost. However, width sizing needs to consume

3.25× more energy at the same performance gain, confirming the superiority of V_T modulations over sizing in ULV regimes.

3.2 Accelerating critical paths and saving active leakage energy consumption

In super-threshold regimes, multi-V_T design is a well-known strategy, where HVT transistors are used for non-critical paths in order to reduce leakage current. In order to support such multi-V_T designs, modern technologies provide several different V_T transistors. In this technology, 3 V_T transistors are provided with 50-150mV separations among V_T values (Figure 6).

In ULV regimes, however, naively following the conventional multi-V_T design practices could be problematic due to the large delay discrepancy caused by considerable separation of different V_T values. As shown in Figure 6, HVT is 8.27× slower than LVT in ULV regimes, while the difference is only 1.46× in super-threshold regime. These large delay changes limit the use of HVT transistors only in the paths which are substantially slower. Otherwise, they may create new critical paths, slowing pipelines.

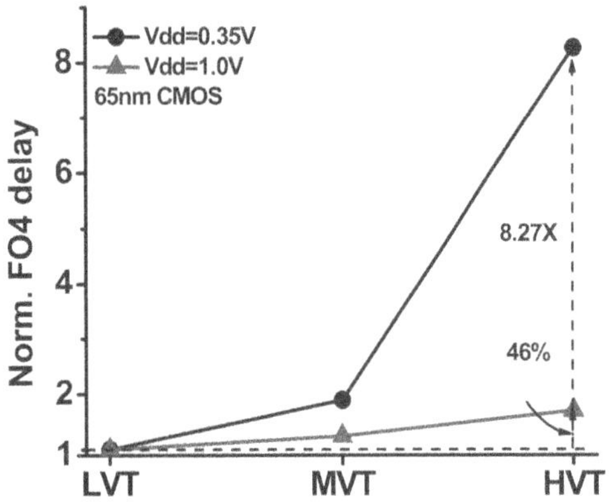

Figure 6. Different impacts of V_T on circuit delay in super-threshold and ULV regimes

However, with the fine-grained many V_T transistors, we can use optimally-chosen V_T transistors for each path in circuits to maximize active leakage energy savings without creating a new critical delay. Additionally, the gates in critical paths can be made out of the lowest V_T transistors to improve performance. The shorter critical delay and the less leaky gates in non-critical paths collectively lead to even higher energy efficiency since the circuits are less dominated by leakage energy consumption and thus more useful voltage scaling, i.e. lower V_{opt}, become available [10].

Table 2. Configurations of a single benchmark circuit

FO4 chains	N_{chain}	Proposed		Baseline1		Baseline2	
		FET	N_{finger}	FET	N_{finger}	FET	N_{finger}
40FO4	2	LVT	3	LVT	1	LVT	8
30FO4	10	LVT	1	LVT	1	LVT	8
20FO4	5	MVT	3	LVT	1	LVT	8
10FO4	3	MVT	1	LVT	1	LVT	8
2FO4	2	HVT	1	LVT	1	LVT	8

In order to confirm the effectiveness of the proposed fine-grained V_T design at ULV regimes, we performed SPICE simulations with a benchmark circuit which consists of FO4 inverter chains. The detailed information on the configuration of the benchmark design, which is a collection of several different length of FO4 chains, is described in Table 2 The number of each FO4 chain (N_{chain}) is chosen such that the benchmark circuits roughly follow the well-known 'wall of slacks' path distribution [13]. However other shapes of path distributions will be considered in Section 3.3.

In the proposed many V_T design methodology, we first use the STR10 transistors, i.e. LVT, N_{finger}=3, in the longest 40 FO4 inverter chains to improve critical path delays. Then we use weaker transistors for the other non-critical paths (30, 20, 10, and 2FO4 chains) in the benchmark circuits in order to reduce active leakage consumption. We choose the proper strength for the non-critical paths such that their delays do not exceed the original critical delay even with 3-sigma random process variations. As comparison points, two baseline circuits are also included: one using only LVT transistors with a single finger, and the other with LVT transistors with 8 fingers, which are referred to as baseline1 and baseline2 designs in Table 2.

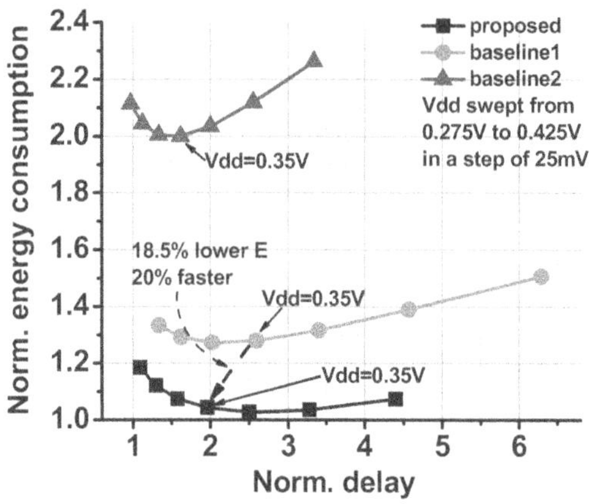

Figure 7. Accelerating critical paths and reducing active leakage energy consumption

Figure 7 shows the results of SPICE simulations over supply voltages, where the proposed design outperforms the baseline1 and baseline2 design, both in energy efficiency and performance. At the same supply voltage of 0.35V, the proposed design shows 20% shorter delay while achieving 18.5% lower energy consumption. Also, note that the proposed design exhibits lower V_{opt} as total energy consumption is less dominated by active leakage. The performance gain of the critical paths is less than the case of driving a fixed load (Section 3.1) since the gate capacitance, which serves as loading, increases with more fingers. The baseline2 design shows that the excessive number of fingers can cause the least favorable energy-delay tradeoff.

So far, we did not consider finer V_T assignment. For example, all the inverters in the same stage of the inverter chain have the same V_T. However, higher V_T transistors can be used for three inverters in the same stage of FO4 chains for further leakage energy reduction since only one inverter in the same stage contributes to the longest path delay. In this fine-grained V_T assignment, ~40% energy savings along with the same performance boost is observed in SPICE simulations.

3.3 Impact of circuit topology and activity

In Section 3, we considered a typical path distribution, known as the wall-of-slacks [13], which is commonly observed in low power, high performance designs. However, it is an important to verify the impact of path distributions on the effectiveness of the proposed methodology. Therefore, in this section, we re-iterate our investigations with new benchmark circuits having different path distributions. Figure 8 shows three circuits with different path distributions: more-short-path (MSP), equally distributed (ED), and more-long-path (MLP) distributions. The MLP distribution is the same one that has been considered in Section 3.

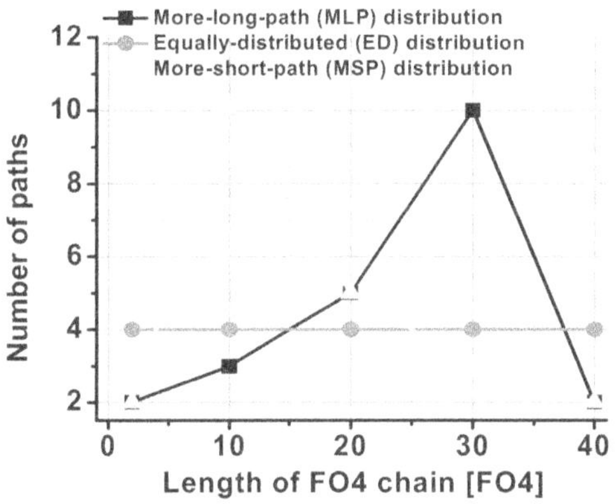

Figure 8. Three different path distributions

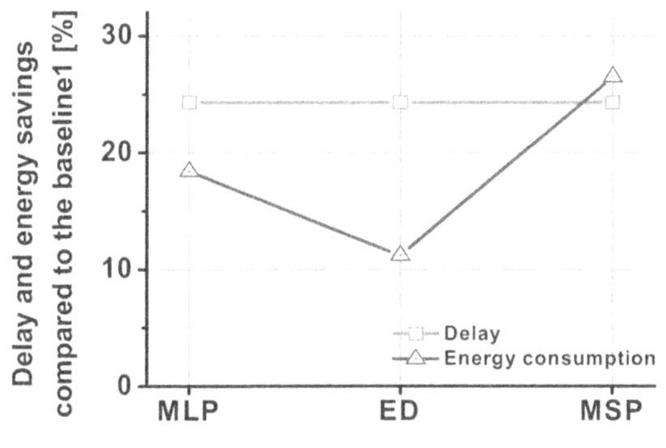

Figure 9. Delay and energy efficiency improvements for different path distributions

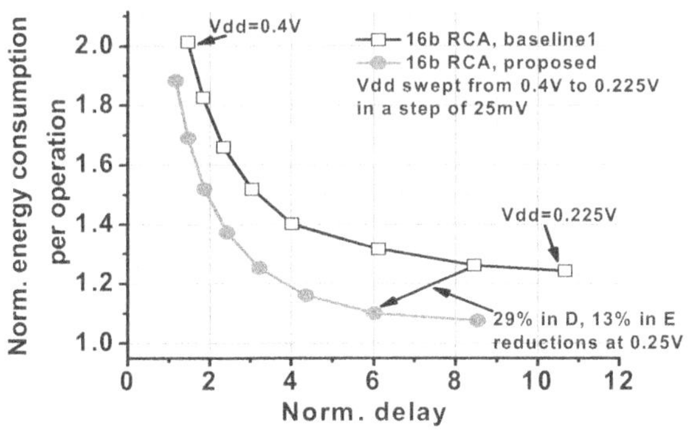

Figure 10. Delay and performance improvement in 16b RCA adders

Among the three domains that we discussed in Section 3, path distributions have little impact on modulating drivability and delaying short paths. However, it can affect the amount of leakage energy savings discussed in Section 3.2. From SPICE simulations, we observe that the energy savings from the many V_T design methodology varies from 27% in the MSP distribution to 11% in the ED distribution (Figure 9). In other words, the energy savings become large when the total number of gates on critical and near-critical paths is a smaller portion of the total number of gates.

As more practical circuits, we also consider two 16b ripple carry adders (RCA), one designed with the proposed fine-grained many V_T transistors, and the other designed with 1-finger LVT transistors, similarly to the baseline1 design in Section 3. Note that RCA has a similar path distribution to ED distribution. The 16-th carry-out bit takes the longest delay while 1-st sum-bit takes the shortest delay. As shown in Figure 10, SPICE simulations show that the proposed methodology improves performance by ~20% and energy efficiency by ~11% at the same supply voltage of 0.3V, which closely matches with the improvement estimations from the FO4 chain studies. At V_{dd}=0.25V, 29% in delay reduction, and 13% in energy savings is observed due to more dominance of active leakage energy consumption.

We also investigate the impact of circuit switching activity on the effectiveness of the proposed methodology. We run SPICE simulations with different switching activities for the proposed and the baseline1 design used in Section 3. The energy savings increase from 18.5% to 23% when switching activity scales from 0.2 to 0.1.

3.4 Delaying non-critical paths

Large delay variability is a critical concern at ULV regimes since it can violate hold time and setup time constraints of pipelines. To improve resiliency to delay variability, several error-tolerant techniques such as Razors [11][17][18] and soft flip-flops [12] have been proposed. Razor has an error detection window just after data-capture clock edge. Once the timing error is detected in this window, it is fixed through architecture replays or supply voltage modulations. On the other hand, soft flip-flops have time borrowing windows, where slightly delayed signals still can be captured during the soft edges.

$$S2C = \frac{\text{min(short path delays in pipelines)}}{\text{std(critical path delays)}} \quad \text{[EQ1]}$$

However, many of these kinds of techniques share an important tradeoff between the amount of error tolerance and the risk of short path violations. For example, the increasing error detection window in [11][17][18] often risks circuits from short path violations or false error detections. Therefore, it is critical to pad sufficient delay on short paths under PVT variations to use those error-tolerant techniques. In this respect, the effectiveness of such techniques can be formulated as the ratio of the minimum of short path delay to the standard deviation of critical path delay under PVT variations, or S2C, defined in EQ1. This illustrates the relative amount of error-tolerance compared to cycle time in circuits under PVT variations. In order to improve S2C, and thus error tolerance, short path delay should be maximized.

In conventional design, adding extra buffers to short paths in pipelines is one of the preferred methods to increase S2C. However, it is often problematic to directly use it in ULV circuits. Due to larger delay variability, more short paths need longer delay padding, which can cause a significant energy overhead due to the divergences of paths in common circuits [14]. Also, the detection window needs to be increased to accommodate widely-varying critical path delay, which again increases the risk of short paths. For limiting the excessive use of delay buffers, it is also proposed to use a low duty-cycle clock for latches, i.e. pulsed-mode latches, [17][18]. However, pulse-mode latches are less preferred in ULV regime due to robustness concerns.

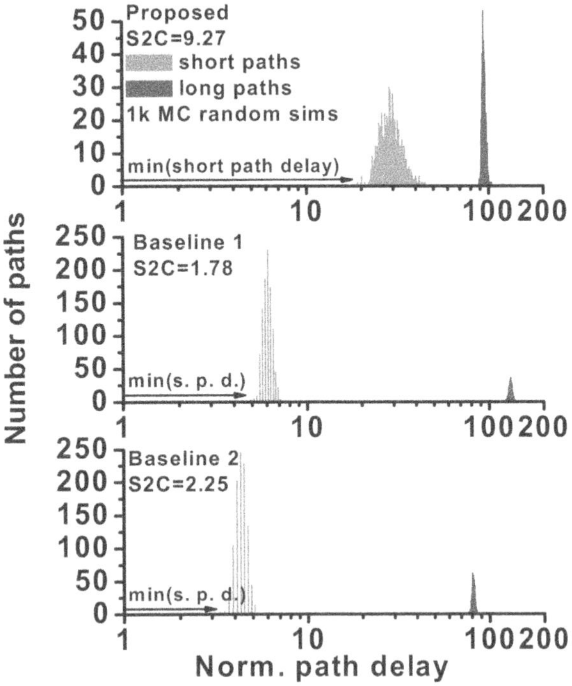

Figure 11. Delay distributions of the shortest path and the longest path of the benchmark circuits over random process variations

The proposed many V_T transistors can be used to delay short paths while saving energy since lowering V_T exponentially slow down circuit delays while reducing both switch and leakage energy consumption. In our benchmark circuits (described in Table 2), we can use STR0.91 transistors for the shortest 2FO4 chains, and other proper (weaker) strength transistors for the 10, 20, and 30FO4 chains such that they do not become either new short paths or longest paths under 3-sigma random process variations.

In order to confirm those short paths are sufficiently delayed, we perform 1k Monte-Carlo SPICE iterations with random process variations. As shown in Figure 11, the short path delay in the proposed design is substantially increased, resulting in the S2C of 9.27. This implies that the design, for example, using [11][17][18] can detect and correct timing errors varying by up to 9.27-sigma. This is 4.1-5.2× larger than two other baseline designs, baseline1 and baseline2 discussed in Section 3.2, which have the S2C of 1.78 and 2.25 respectively.

4. CONCLUSIONS

In this paper, we propose the fine-grained many V_T transistor design methodology in ULV regimes. The energy-efficiency-aware, fine-grained, many V_T transistors are developed exploiting inverse narrow width effects without any modification in fabrication processes. Three domains have been considered (?) where the proposed many V_T design methodology can improve energy-efficiency, performance, and variability. First, the many V_T design methodology achieves 2.17x higher energy efficiency than transistor sizing at the same performance. Second, it improves energy efficiency by 27-11% and performance by 20%, by selectively assigning fine-grained V_T for different delay paths in circuits. Finally, the proposed methodology is able to delay the short path while improving energy efficiency, which potentially enables 4.1-5.2× higher tolerance to delay variability when equipped with the conventional error-tolerant techniques.

Acknowledgements – the author appreciates the technical discussion with Jae-sun Seo at IBM T.J. Watson Research Center

References

[1] B. Zhai, et al, "Theoretical and practical limits on dynamic voltage scaling", *ACM/IEEE Design Automation Conference*, 2004

[2] J. Kwong, et al., "Variation-Driven Device Sizing for Minimum Energy Sub-threshold Circuits," ISLPED, pp.8-13, 2006

[3] Y. Pu, et al., "V_T Balancing and Device Sizing Towards High Yield of Sub-threshold Static Logic Gates," ISLPED, pp.355-358, 2007

[4] J. Keane, et al., "Subthreshold Logical Effort: A Systematic Framework for Optimal Subthreshold Device Sizing," Design Automation Conference, pp425-428, 2006

[5] T.-H. Kim, et al., "Utilizing Reverse Short-Channel Effect for Optimal Subthreshold Circuit Design," IEEE Transaction on Very Large Scale Integration Systems, vol.15, no.7, pp.821-829, Jul, 2007

[6] J. Zhou et al., "A 40nm Inverse-Narrow-Width-Effect-Aware Sub-Threshold Standard Cell Library," Design Automation Conference, pp.441-446, 2011

[7] S. Hanson, et al., "Exploring Variability and Performance in a Sub-200mV Processor," *Journal of Solid-State Circuits*, No. 4, pp. 881-891, 2008.

[8] K. Hamamoto, et al., "Tuning-Friendly Body Bias Clustering for Compensating Random Variability in Subthreshold Circuits," ISLPED, pp.51-56, 2009

[9] M. R. Kakoee, et al., "Automatic Synthesis of Near-Threshold Circuits with Fine-Grained Performance Tunability," ISLPED, pp.401-406, 2010

[10] M. Seok, et al., "A 0.27V 30MHz 17.7nJ/transform 1024-pt Complex FFT Core with Super-Pipelining," ISSCC, pp.342-344, 2011

[11] D. Ernst, et al., "Razor: a low-power pipeline based on circuit-level timing speculation," International Symposium on Microarhitecture, pp.7-18, 2003

[12] V. Joshi, et al., "Soft-edge Flip-flops for improved timing yield: design and optimization," ICCAD, pp. 667-673, 2007

[13] J. Patel, "CMOS Process Variations: A Critical Operating Point Hypothesis," Online Presentation, 2008

[14] V. Srinivasan, et al., "Optimizing Pipelines for Power and Performance," IEEE/ACM International Symposium on Microarchitecture, 2002

[15] B. Calhoun, et al, "Characterizing and modeling minimum energy operation for subthreshold circuits", *IEEE International Symposium on Low Power Electronics and Design*, 2004

[16] M. Seok et al., "The Phoenix Processor: A 30pW Platform for Sensor Applications," IEEE Symposium on VLSI Circuits, pp.188-189, 2008

[17] D. Blaauw et al., "Razor II: In Situ Error Detection and Correction for PVT and SER Tolerance," *IEEE International Solid-State Circuits Conference*, pp.400-622, 2008

[18] K.A. Bowman, et al., "Energy-Efficient and Metastability-Immune Timing-Error Detection and Instruction-Replay-Based Recovery Circuits for Dyanmic-Variation Tolerance," *IEEE International Solid-State Circuits Conference*, 2008

A Programmable Resistive Power Grid for Post-Fabrication Flexibility and Energy Tradeoffs

Kyle Craig[1 2], Yousef Shakhsheer[1], Sudhanshu Khanna[1], Saad Arrabi[1], John Lach[1], Benton H. Calhoun[1], and Stephen Kosonocky[2]
[1] Dept. of Electrical and Computer Engineering, University of Virginia, Charlottesville, VA, USA
[2] Advanced Micro Devices, Fort Collins, CO, USA

ABSTRACT
This paper explores the benefits of splitting a monolithic power gate transistor into parallel, independently controlled, variable weighted power gates to provide programmable post-fabrication power grid resistance. This power gate topology creates energy saving opportunities by providing adjustable localized voltages during active modes and reducing leakage current in idle blocks while retaining data. Measurements show over 30% active energy savings per operation and 90% savings in idle current with retention. A modeling flow for a resistive power grid was also developed that demonstrates the effectiveness of this approach in a Bulldozer processor core.

Categories and Subject Descriptors
I.1 Technologies and Digital Circuits

Keywords
Variable Weighted Headers, Low Power Design, Leakage, Dynamic Voltage Scaling

1. INTRODUCTION
This paper proposes a variable resistance power grid constructed from variable weighted parallel power transistors. This structure provides a dynamic voltage scaling (DVS) approach for active mode at much lower cost than using DC-DC converters and offers a convenient low leakage standby mode with data retention.

Many contemporary and emerging applications such as laptop computers, portable media players, smart phones, and bio-medical devices impose strict constraints on energy consumption due to battery size, yet also demand high performance for short bursts of time to meet timing constraints. Although technology scaling has provided raw performance gains and lower switching capacitances, increasing the battery life by lowering system energy consumption is still an ongoing effort, especially due to increasing energy consumption from leakage current.

Power gating and dynamic voltage scaling are two common solutions to reduce leakage current during standby mode and to tradeoff dynamic energy and delay during active operation, respectively. Power gating uses large transistors in series with the power supply or ground to cut off leakage during idle mode. One disadvantage of power gating is that data stored in registers is lost. A variety of approaches to deal with this problem include putting registers on a separate supply, using high V_T balloon registers in parallel with core registers, or other alternative dual V_T register circuits. All of these incur overhead and added design complexity. DVS during active mode saves power by lowering the frequency and voltage together when timing slack exists. Applying DVS to multiple blocks requires multiple DC-DC converters that adjust the local voltage levels or alternative schemes that allow local V_{DD} selection from among multiple regulated supplies [1]. Again, the overhead of these approaches can be substantial.

This paper explores an approach for providing dynamic system level flexibility by partitioning large, monolithic power gating transistors into parallel, independently controllable power gates with different widths in a scheme called *variable weighted power gates*, as shown in Figure 1. This structure allows a once monolithic power gate to be partially on and/or to vary its effective size dynamically, providing both boot-time programmability and run-time adaptability that enable each block of a system to independently approach an ideal header resistance configuration for all phases and modes of all applications. We can leverage this functionality to implement fine grained DVS at the block level and to enable block level low leakage standby modes with data retention.

The authors of [2] first introduced separating power gate FETs and controlling them individually as a method to reduce noise. This paper expands on this idea and shows how to apply variable weighted power gates in a flexible fashion to provide additional low power operation modes. A variable weighted power gate can provide a controlled power supply resistance to enable a large number of effective voltages at which a block can operate. The block is therefore not constrained by available voltage rails; it can operate at lower voltages than the rest of the system without changing the entire chip V_{DD}, so it avoids the high overhead of

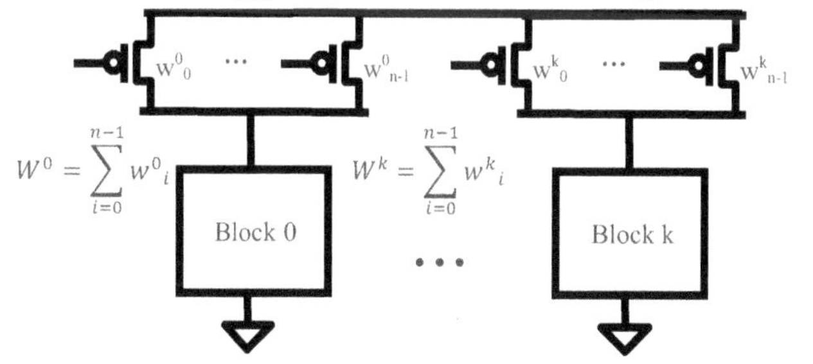

Figure 1. Proposed variable weighted power gate scheme. Multiple header widths are connected in parallel to the same rail with separate gate controls.

extra DC-DC converters. In addition, a low energy standby mode can be provided that reduces leakage current in idle blocks but enables data retention and incurs lower overhead when returning to normal operation than does full power gating. A similar leakage reduction mode was used to give state retention modes in SRAM arrays [3] for reducing idle power.

Extensive simulation results from both a 32nm and 90nm technology demonstrate the effectiveness of variable weighted power gates to create multiple effective voltages through controlled power supply resistance. Measurements from a 90nm bulk test chip and measurements from a 32nm SOI x86 processor show the effectiveness of this technique for dynamic energy savings and leakage reduction respectively. Finally, a modeling methodology was created to enable design and verification of a full commercial processor utilizing a programmable power grid.

2. PROGRAMMABLE POWER GRID RESISTANCE FOR LOCAL DVS

As discussed above, circuits that utilize DC-DC converters for DVS incur significant delay and energy overhead for each voltage transition. Internal regulators require significant area, and external regulators require dedicated V_{DD} pins that increase package and board costs. As a result, DVS schemes usually employ global DVS in which the entire circuit or multiple cores operate at the same voltage.

Variable weighted power gates can be utilized as a controlled resistance to efficiently provide local voltage control without adding new DC-DC regulators, changing the voltage output of the existing regulators, or adding metal routing complexity. These power gates provide the effective voltage by utilizing the voltage drop across the power FET as a controlled resistor. The virtual supply rail (V_{rail}) decreases voltage as the effective power gate width decreases. As seen in Figure 2, there is no feedback loop to adjust the output voltage. Rather, the voltage rail is allowed to settle during operation, thus providing an output voltage on the voltage rail for the circuit that is proportional to the circuit current load and the power gate effective resistance. This virtual rail voltage not only depends on the size of the power gate, but also on the activity factor of the block and extrinsic and intrinsic decoupling capacitance on the virtual rail. This virtual rail voltage sets the delay and energy of the operation.

Assuming frequent operation and no idling, the virtual rail will droop to a voltage below the nominal V_{DD}. Figure 3 shows a 90nm CMOS simulation of the virtual voltage rail of successive operations of a 32b Kogge Stone adder for different fraction of enabled PMOS power gate widths, i.e., the width of enabled power gates enabled divided by the total width of power gates in the

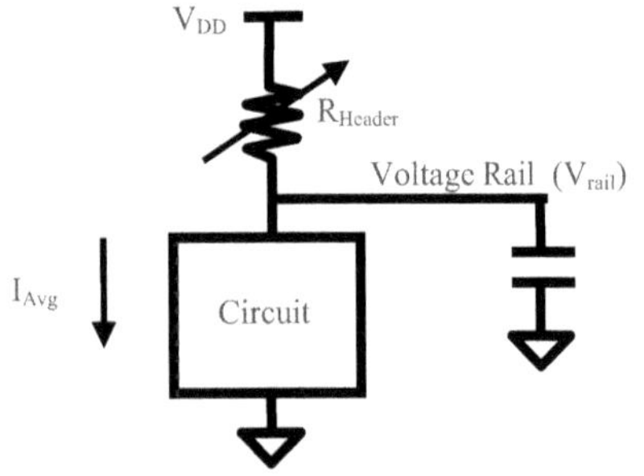

Figure 2. Variable weighted power gate used as a power grid resistance. R_{Header} varies based on header size.

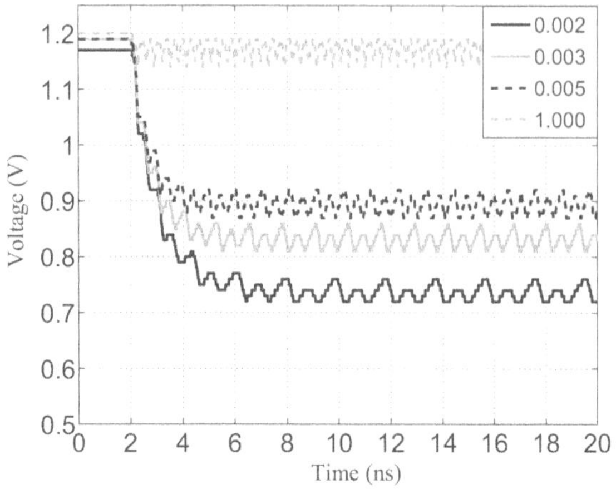

Figure 3. Virtual rail of a 32b Kogge Stone adder during successive adder operations over several header widths.

design, including the extracted virtual rail capacitance and an ideal V_{DD} applied to the circuit. Notice that the rail settles near a certain voltage for each power gate width. In this operation, the energy is reduced by up to 37% in simulation as compared to circuit operation tied directly to V_{DD} with no power gate with a maximum increase of ~2.8x in delay for frequent operations. This effectively implements a light weight DVS mechanism with no additional circuits required (except control), which can be used in lieu of or in addition to conventional DVS methods.

2.1 Activity Factor

Although the effective V_{DD} of the block is set by the unregulated virtual rail voltage, V_{rail}, analysis of the block under maximum current conditions can allow us to select the header width to set the worst case circuit performance. When the circuit consumes lower amounts of current, V_{rail} will not droop as far. This will cut into the active energy savings, but the scheme still saves energy compared to having the full header on. Since the overhead for implementing this scheme is so low, the savings essentially come for free, since the power gates are already utilized in these designs to reduce leakage.

To model the effect of activity factor, a similar methodology as presented in [4] was used. Sixty-four five stage ring oscillators (RO) were simulated in parallel with the ability to disable different ROs with enable signals. With this setup, 64 enabled ROs correspond to the highest activity factor of 1.0, 32 enabled ROs correspond to an activity factor of 0.5 and so on. Each of the different activity factors were simulated with a varied amount of enabled header width. Figure 4 shows the impact of activity factor and header width on V_{rail}. The x-axis is the number of enabled parallel ROs while each bar represents a different amount of total header width enabled. The headers partitions were sized for the 1-RO case, and the same sizes were used for all other RO cases. For the 1-RO case, we are able to see a large potential range of V_{rail} values (~0.6- 1.15V), however as the activity factor increases the V_{rail} range shrinks. For the high RO cases, larger headers widths need to be designed and enabled to regain a large V_{rail} range. In Figure 5, RO performance is shown with varying activity factor and header width. The RO performance is normalized to a RO without any power gates at the nominal V_{DD}. As expected the performance is highly dependent on activity factor and enabled header width and shows similar trends as V_{rail}.

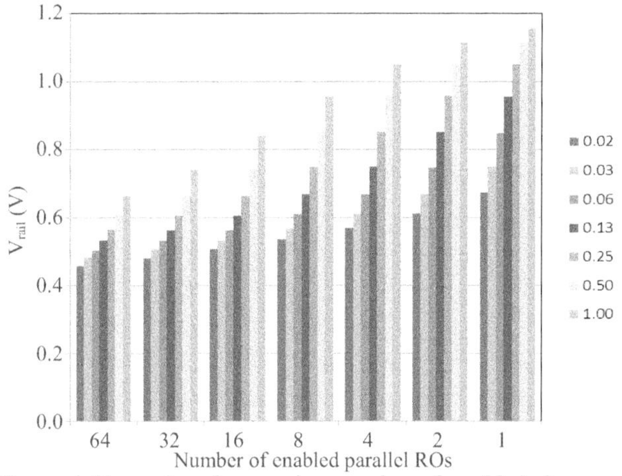

Figure 4. V_{rail} values for varying number of enabled ring oscillators for different normalized header widths

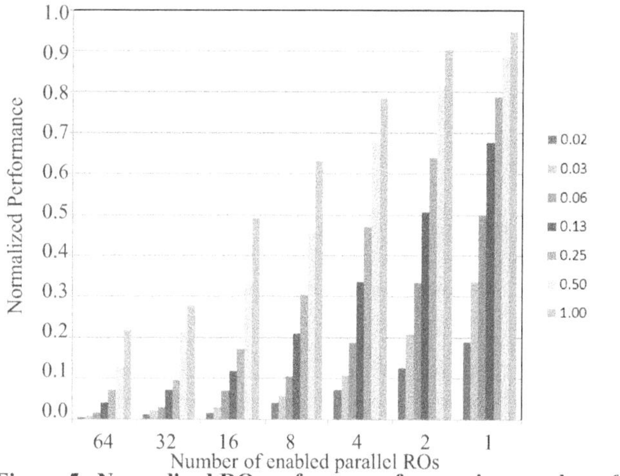

Figure 5. Normalized RO performance for varying number of enabled ROs for different normalized header widths

Traditional header sizing methodology still applies, meaning that the total header size needs to be based off the worst case activity factor and target performance requirement. The header width partitioning is highly dependent on activity factor. In order to achieve a flexible power grid resistance, the header width partitioning should be sized for the range of activity factors expected and the performance range desired based on application. This would involve running simulations varying header widths for different characteristic activity factors to characterize the V_{rail} drop, performance degradation and energy savings. This could be done with low level circuit simulations or high level model techniques as discussed later.

2.2 Energy Savings

This section quantifies the energy savings that the resistive grid scheme can provide when used to implement DVS. The following equation for energy per operation is composed of both the dynamic and leakage energy as a function of V_{DD} (the short circuit component is ignored for simplicity):

$$E_{op}(V_{DD}) = C_{eff}(V_{DD}) * V_{DD}^2 + V_{DD} * I_L(V_{DD}) * t_{op}, \quad (1)$$

where C_{eff} is the effective switching capacitance as a function of V_{DD}, and t_{op} is the delay of the component. I_L is the subthreshold leakage current as a function of V_{DD}. Simple modifications can be made to (1) to include the savings of variable weighted power gates by making the equation a function of both V_{DD} and V_{rail}:

$$E_{op}(V_{DD}, V_{rail}) = C_{eff}(V_{rail}) * V_{DD} * V_{rail} + V_{DD} * I_L(V_{rail}) * t_{op}. \quad (2)$$

As V_{rail} decreases, we expect a greater than linear dynamic energy reduction due to the linear reduction of V_{DD}*V_{rail}, plus a less than linear reduction in C_{eff}, due to device source and drain parasitic junction capacitance being dependent on V_{rail}. Finally, as V_{rail} decreases, I_L will decrease due to the exponential relationship between V_{rail} and the amount of DIBL (drain-induced barrier lowering). To highlight the potential benefits of using this droop to save energy, we connect a ring oscillator to variable weighted power gates and allow the rail to settle for each power gate width.

Figure 6 shows the normalized energy and delay versus the normalized power gate width. As expected, reducing the power gate width decreases the energy and increases the delay. These RO results were confirmed through silicon measurements using a 90 nm commercial bulk technology with four 97 stage ROs in parallel consisting of inverters and delay cells to simulate a high current and activity load. The normalized measured values match the simulated values and show an energy savings of over 30% in silicon.

Additionally, as with any DVS scheme, there is an overhead associated with this variable weighted power gate technique. The virtual rail must eventually recover from the droop during operations when high performance is required. This recovery energy can be defined as:

$$E_{rec} = C_{rail} * V_{DD} * (V_{DD} - V_{rail}). \quad (3)$$

Rail recovery during successive operations should be minimized to save the maximum amount of energy. To minimize this wasteful energy, the clock frequency should be increased, the amount of the

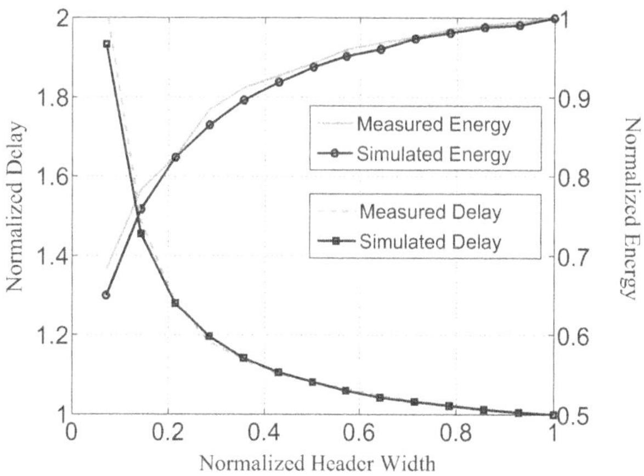

Figure 6. Simulated and measured energy and delay for a ring oscillator with sweeping header size in a 90 nm test chip.

power gate that is enabled should be decreased, or the source voltage should be decreased to reduce the rail's ability to recover. This will allow for most energy efficient operation. We can amortize this recovery energy over multiple cycles by maximizing the number of operations run with this voltage droop, thus reducing energy per operation when the virtual rail is relatively constant.

2.3 Leakage Current Reduction

Many systems and blocks within systems spend large amounts of time idling. Additionally, blocks such as register files or memory may need to retain their data, which is not supported by most power gating schemes. Variable weighted power gates provide a low energy solution that enables data retention with reduced idle leakage current. Leakage current can be reduced through reducing power gate size by dropping the voltage across the active devices which reduces the amount of DIBL, thus reducing device leakage. In Figure 7, the leakage current is measured for a 32nm SOI four-core x86 processor SOC chip which has a variable weighted power gate ring around the core [5] on actual silicon hardware. By changing the power gate width by disabling distributed sections of the footer ring via configuration bits, the idle current can be gradually reduced from 100% to a lowest bound of 10%. Tester characterization can determine the lowest setting to allow core state retention. These measurements confirm that the variable width header scheme supports very low leakage standby modes at low overhead.

3. OPPORTUNITY FOR REGULATION

We investigated the opportunity for dynamic grid voltage control in a commercial x86 four core processor SOC using typical P-state (power state) occupancy data. A P-state defines a voltage/frequency pair independently for each core in the processor, where P0 is the fastest state and P3 is the slowest. To analyze the potential savings of our approach deployed on a larger scale, we applied a model of our variable weighted power gate resistive grid technique (based on equation 2) to our processor simulations. Since all cores share a common V_{DD}, the lowest core P-state sets the operating V_{DD} for all cores, leading to non-optimal V_{DD} for cores with higher P-states. The cores running at a lower P-state than required are only able to use frequency scaling in the absence of a resistive grid, since there is no other mechanism to lower the local V_{DD}. However, Figure 8 shows the opportunity for power savings using variable-weighted power gates. In the figure, the label P1@P0 indicates the total power of core(s) that are running at the P1 frequency while another different core in the SOC is running at a P0 state. During the SysMark trace, the core-wise P-state occupancy is determined by the operating system. By including variable weighted power gates at each core, the cores running at a higher P-state will run at their near optimal V_{DD} during periods of high activity.

The figure shows up to ~15% power savings opportunity by using the variable weighted power gate resistive grid technique. Power/performance results can vary depending on the P-state frequency, voltage settings, and the profile of system activity. By allowing individual core-wise voltage settings, the system has more flexibility to differentiate high performance modes from lower performance modes, which can allow opportunity for additional performance boosting when a single core is running at a low P-state. Definition of the P-state voltage-frequency pair can be determined at SOC characterization time while running a thermal

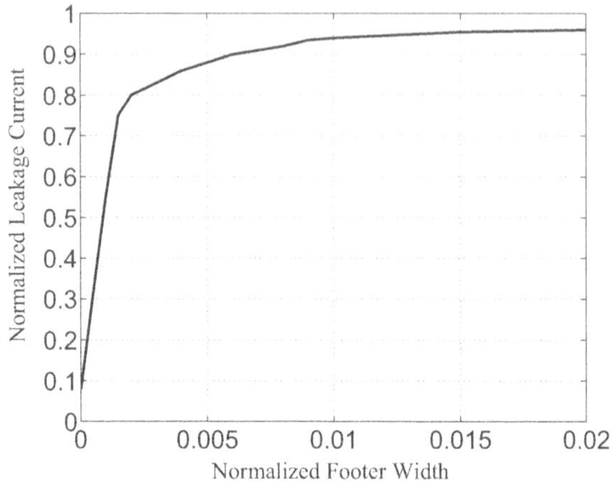

Figure 7. Measured idle current reduction using variable weighted power gates in a 32nm 4-core x86 processor at 1.2V.

design point (TDP) workload in each core to ensure enough voltage margin to accommodate the maximum P-state frequency, at the expense of some power savings.

4. MODELING POWER SUPPLY RESISTANCE

Since the application of this technique requires characterization of the power supply resistance, we developed a design flow, using a commercial power integrity tool, for applying the approach to arbitrary digital designs. We use this design flow to model a full commercial processor using the proposed method for implementing a programmable resistive power grid.

4.1 Route Level Macro Model

A commercial power integrity tool, Apache Redhawk [6], was used to model the effectiveness of variable weighted power gates as a controlled power supply resistance in a large system. In order to demonstrate the feasibility of a controlled power supply resistance to create local effective voltages, a Redhawk set up was created for a single power gated Route Level Macro (RLM) used in a commercial 32nm core. An ideal V_{DD} and V_{SS} were applied to the power grids, which were modeled from metal 11 (M11) to M9 and M11 to M1 respectively, while the virtual V_{DD} grid, modeled from M8 to M2, was observed as the power gate width was varied. The

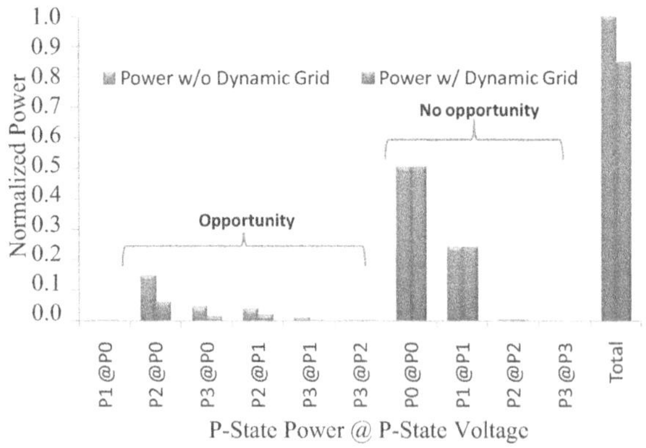

Figure 8. SysMark Trace Segment for a four-core x86 SOC

internal net activity was generated from benchmark simulations of a thermal design point (TDP) benchmark. Figure 9 shows the average virtual V_{DD} and worst case virtual V_{DD} observed for the TDP benchmark. The relatively flat part of the curves is due to the very low impedance sizing required for the maximum header size to enable high frequency operation at the maximum V_{DD}. To achieve variable power gate regulation, fine grain power gate partitioning is needed in the sub 0.2% range of the total power gate width.

4.2 Bulldozer Core Model

Finally, Redhawk was used to model the AMD Bulldozer core [7], which has a similar power gate ring structure seen in [5]. For this simulation, to prevent the simulation time from being prohibitively large, each Route Level Macro (RLM) (roughly 50 in total) in the core was modeled as a time dependent current source and capacitance model, with the exception of the L1 cache and two RLMs without available data. These current profiles were generated from simulations of the Double-precision General Matrix Multiply (DGEMM) benchmark.

A simplified package model was included to capture the real RLC effects seen on hardware. In these simulations, we used power gating footers instead of headers, so the Bulldozer model included real V_{DD} and V_{SS} grids from the C4 bumps down to M10, and virtual V_{SS} (using footer power gates) in M11 and M10. Figure 10 shows a simplified diagram of the setup. This test setup allowed flexibility to vary the power gate width and to observe V_{DD}, V_{SS}, and the virtual V_{SS} during a 25ns window of the DGEMM benchmark. Figure 11 shows the average V_{DD} response to the applied current models in time, showing that our current model profile was functioning correctly.

Figure 12 shows the V_{DD} and virtual V_{SS} voltage profile over time for different normalized power gate widths. The trend in this figure is the same as in Figure 9 the largest change in the virtual V_{SS} happens at very low footer widths. Notice that at 5.07%, the virtual V_{SS} is only slightly above the 100% case, which is expected for a power gate ring system designed for a high performance core.

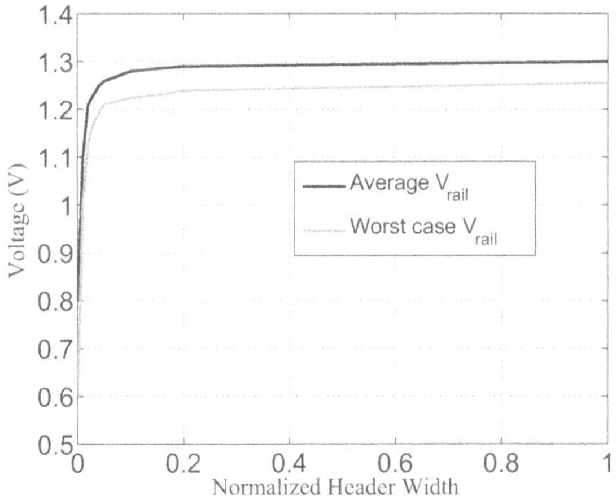

Figure 9. Average and worst case V_{rail} of the single RLM during a TDP benchmark.

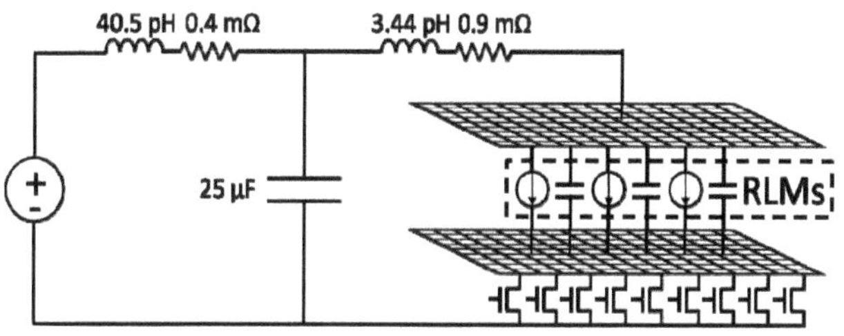

Figure 10. Bulldozer test setup, include RLC package model, RLM current and capacitance models

In this example, dynamic control is required for only a small percentage of the total power gate width. This figure also includes every RLM's V_{DD} and virtual V_{SS} superimposed into a single graph.

The negligible variance in V_{DD} and V_{SS} between RLMs across the chip is due to the robust power grid with low resistance showing that the dominant factor in the virtual V_{SS} droop is caused by the controlled power supply resistance of the power gates. Through variable weighted power gates, we are able to achieve a wide range of virtual V_{SS} supplied to the core.

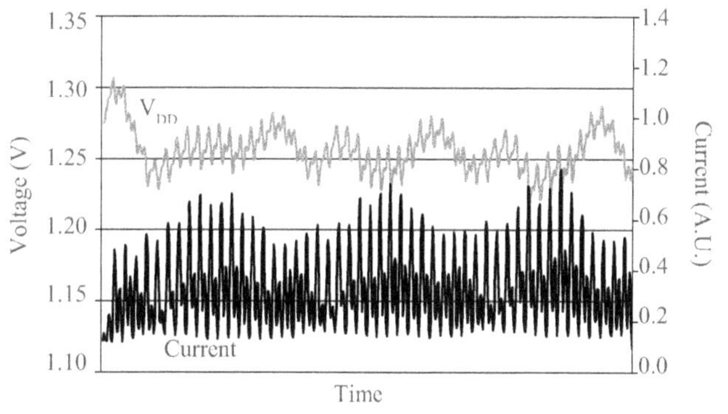

Figure 11. Power grid simulation of V_{DD} and current profiles over time for AMD Bulldozer core.

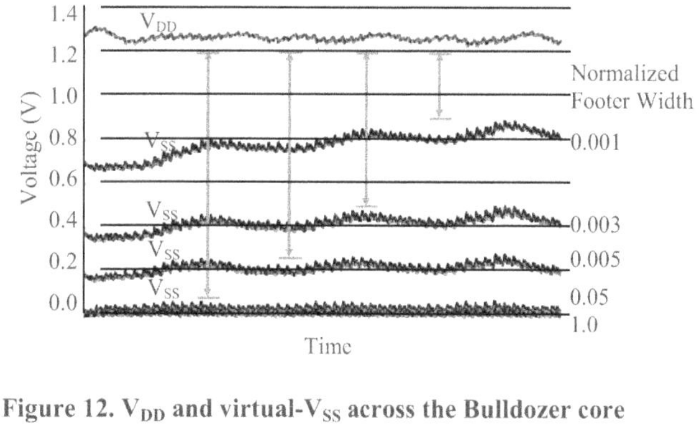

Figure 12. V_{DD} and virtual-V_{SS} across the Bulldozer core during DGEMM benchmark with varying power gate widths.

5. CONCLUSIONS

This work demonstrated the use of variable weighted power gates to provide a controlled power supply resistance for designs using power gating and DVS. We showed how a controlled power grid resistance through variable power gate resistance can be used to provide a large number of voltages to a block during active operation. Through extensive simulation and modeling, we can select the correct amount of width to turn on for various desired operating modes. We discussed the opportunity for using the variable weighted power gate in a commercial x86 four core SOC. Finally, through measurements from a 90nm test chip and a 32nm x86 processor, we showed how variable weighted power gates can be used to reduce energy during active mode and to limit leakage current, respectively. Using variable weighted power gates is a low cost solution for providing both boot-time programmability and run-time adaptability that enables each block of a system to approach an ideal configuration for all phases and modes of all applications.

6. REFERENCES

[1] Shakhsheer, Y., et al., A 90nm Data Flow Processor Demonstrating Fine Grained DVS for Energy Efficient Operation from 0.25V to 1.2V. *Custom Integrated Circuits Conference*, Sept. 2011

[2] Truong, D.N., et al., A 167-processor computational platform in 65 nm CMOS, *IEEE Journal of Solid-State Circuits,* vol. 44, no. 4, pp. 1130-1144, April 2009

[3] Zhang, K., et al., SRAM design on 65-nm CMOS technology with dynamic sleep transistor for leakage reduction, *IEEE J. Solid-State Circuits*, vol. 40, no. 4, pp. 895–901, Apr. 2005.

[4] Wang., A, et al., *Sub-threshold Voltage Circuit Design for Ultra-Low Power Systems*. Springer, New York, NY

[5] Jotwani, R., et al., An x86-64 Core Implemented in 32nm SOI CMOS *International Solid-State Circuit Conference*, pp. 106-107, 2010.

[6] Apache Redhawk, " http://www.apache-da.com/products/redhawk"

[7] Fischer, T., et al., Design Solutions for the Bulldozer 32nm SOI 2-Core Processor Module in an 8-Core CPU, *International Solid-State Circuit Conference*, pp. 78-79, 2011.

Improving Energy Efficiency of Write-asymmetric Memories by Log Style Write

Guangyu Sun[1], Yaojun Zhang[2], Yu Wang[3], Yiran Chen[2]
[1]Center for Energy-efficient Computing and Applications, Peking University
[2]Department of Electrical and Computer Engineering, University of Pittsburgh
[3]Department of Electronic Engineering, Tsinghua University
gsun@pku.edu.cn, yu-wang@mail.tsinghua.edu.cn, {yaz24, yic52}@pitt.edu

ABSTRACT

The significant scaling challenges of conventional memories, i.e., SRAM and DRAM, motivated the research on emerging memory technologies. Many promising memory technology candidates, however, suffer from a common issue in their write operations: the switching processes at different write operations (i.e., $0 \rightarrow 1$ and $1 \rightarrow 0$) are asymmetric. Using a pessimistic design corner to cover the worst case of a write operation incurs large power and performance cost in the existing emerging memory technology designs. In this work, we propose a universal log style write methodology to mitigate this asymmetry issue by operating two switching processes in separate stages. The dedicated design optimizations are allowed on either switching process. The simulation results on the spin-transfer-torque random access memory based last-level cache show that our technique can improve the system performance by 4% while receiving 35% power reduction on average [1].

Categories and Subject Descriptors

B.3.0 [**MEMORY STRUCTURES**]: General

Keywords

Write-asymmetric memory, log write, energy efficiency

1. INTRODUCTION

In order to overcome the memory wall, extensive research has been done to find alternatives of the traditional SRAM and DRAM technologies. Various emerging memory technologies, such as STT-RAM (Spin Transfer Torque RAM), PCM (Phase Change Memory), FBDRAM (Floating Body DRAM), have been proposed to replace SRAM/DRAM in different levels of the memory hierarchy [8, 5, 13, 7, 3]. STT-RAM is normally employed as a competitive replacement of SRAM as on-chip memories because of its advantages of fast read speed, low leakage power, and high density [8, 5]. PCM is widely studied as a potential candidate of the main memory because it has higher density and lower standby power compared to DRAM [13, 7, 3]. As an emerging memory compatible to CMOS technology, FBDRAM is also attracting more attention, recently [4, 1]. With these emerging memory technologies, prior research has shown improvements in performance, power consumption, and reliability.

Table 1: Comparisons of data switching (45nm) .

	STT-RAM		PCM		FBDRAM	
Cell Area	$40F^2$		$16F^2$		$8F^2$	
Switching	$0 \rightarrow 1$	$1 \rightarrow 0$	$0 \rightarrow 1$	$1 \rightarrow 0$	$0 \rightarrow 1$	$1 \rightarrow 0$
Latency	$30ns$	$10ns$	$150ns$	$40ns$	$10ns$	$2ns$
Current	$250\mu A$	$200\mu A$	$300\mu A$	$600\mu A$	$30\mu A$	$30\mu A$

These emerging memories, however, have some common limitations such as long write latency, and high write energy. Although the memory capacity is increased after using these high density memories, the long write latency may offset the benefits and degrade the performance for workloads with intensive write operations. Similarly, although the standby power is reduced by using these emerging memories, the dynamic power can be significantly increased due to high write energy. Some research has been done to hide the long write latency by using hybrid memory [9] or stalling the write and preempting the blocked read [9, 6]. The replacement policy can also be tailored to reduce the write intensity [14]. Recently, Smullen and Li *et al* proposed to scarify the non-volatility to improve the write speed [8, 10].

In most of these techniques, the write operations of these memories are designed for the worst case of the asymmetric data switching processes. The asymmetry means that the switching processes of $0 \rightarrow 1$ (SET) and $1 \rightarrow 0$ (RESET) are quite different from each other, in respect of timing, energy consumption, and even reliability. Table 1 compares the differences between SET and RESET for some typical values of various memory technologies. It is easy to find that a RESET is much faster and consumes much less energy, compared to a SET. In the worst case design, however, the write latency and driving current are designed to satisfy the constraints of the SET. Thus, the RESET is normally overdriven, and much overhead is induced. As the technology scales down, this problem is becoming more severe because the increasing effect of process variations further aggravate the difference between a SET and a RESET. Smullen addressed this issue in a STT-RAM cache by flipping the data to have more RESET in a write operation. This scheme, however, only reduce write energy. Since RESET and SET are still operated simultaneously, the write latency cannot be reduced and the design constraints are not changed.

In this work, we propose a log style write methodology, in which the SET and RESET are processed separately in two stages. This universal methodology is feasible for most write-asymmetric memories designed for various levels of the memory hierarchy. With this methodology, the memory to be written is prepared by a recycle stage, in which only SETs are processed. Then, there are only RESETs when new data are written into the memory. Since SETs and RESETs are separated, they are processed with their own

[1]This work was partially supported by National Natural Science Foundation of China (No.60870001, 61028006).

ISLPED'12, July 30–August 1, 2012, Redondo Beach, CA, USA.

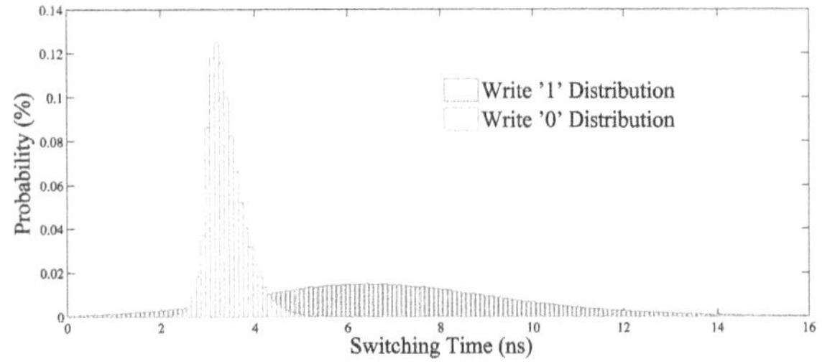

Figure 1: MTJ switching time distributions at the transistor width of 720nm.

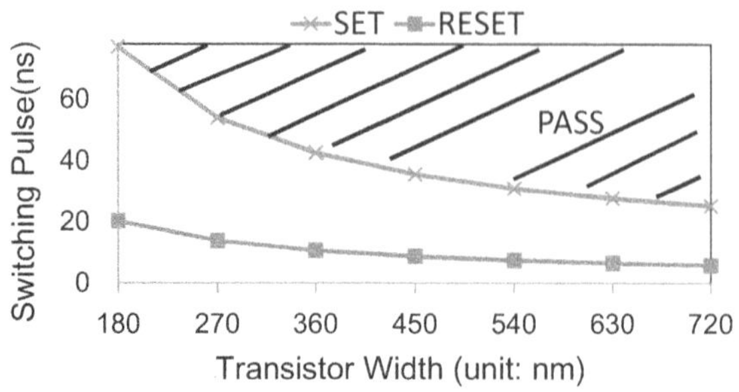

Figure 2: Design space based on transistor width.

constraints. Moreover, more design options and optimization opportunities are provided. We use STT-RAM based cache as a case study of applying this log style write. After using this methodology, we will show that energy efficiency of STT-RAM cache is greatly improved by reducing power consumption and improving performance at the same time. In addition, we also discuss how to extend the methodology for other write-asymmetric memories and compare it with read-before-write scheme.

2. PRELIMINARIES

In this work, we use STT-RAM as an example to study the write asymmetry and its impact. The STT-RAM is chosen because of significant differences between its SET and RESET. In addition, STT-RAM is normally proposed as on-chip memory, which is normally sensitive to long write latency and high write energy.

2.1 Basic of STT-RAM and Its Asymmetric Write

STT-RAM uses magnetic tunneling junction (MTJ) devices to store the information. A MTJ has two ferromagnetic layers (FL) and one oxide barrier layer (BL). The resistance of MTJ depends on the relative magnetization directions (MDs) of the two FLs. When their MDs are parallel or anti-parallel, the MTJ is in its low (bit '0') or high resistance state (bit '1'). R_h and R_l are usually used to denote the high and the low MTJ resistance, respectively. In a MTJ, the MD of one FL (reference layer) is pinned while the one of the other FL (free layer) can be flipped by applying a polarized write current though the MTJ. In the popular "1T1J" (one-transistor-one-MTJ) cell structure, the MTJ write current is supplied by a NMOS transistor. A larger write current can shorten the MTJ switching time by paying the additional memory cell area and increasing the breakdown possibility of the MTJ device.

The asymmetry of STT-RAM's write operations mainly comes from two sources: the asymmetric MTJ switching property and the NMOS transistor driving ability for two switching directions. The asymmetry of MTJ switching property in SET and RESET is mainly due to the different spin-transfer efficiency at the both sides of the oxide barrier. The driving ability of NMOS transistor is also asymmetric due to the different bias conditions at two STT-RAM cell switching directions. In addition, the different resistance in SET and RESET also affect the switching processes[12].

When process variations are considered, the asymmetry becomes more significant between SET and RESET operations because the current through the MTJ is affected by device variations of the MTJ and the NMOS transistor. For instance, the driving ability of the NMOS transistor is affected by the variations of transistor channel length (L)/width (W) and the threshold voltage (V_{th}). The MTJ resistance R_{MTJ}, which determines the voltage drop across the MTJ device, is reversely proportional to the MTJ surface area and exponential to the oxide layer thickness. The ïňĆuctuations of the above device parameters in the chip fabrication introduce bit-to-bit variations of STT-RAM cells [11]. Due to different switching properties and driving abilities, MTJ switching time has a larger mean and a wider distribution in SET than those in RESET under the same switching current[12], as illustrated in Figure 1. Thus, more overhead is induced in performance and power for the worst case design with process variations.

2.2 Impact of Asymmetry in Worst Case Design

In order to study the impact of asymmetry in write operations, we present the design space of SET and RESET operations with process variations and thermal fluctuations in Figure 2. The detailed device parameters of STT-RAM and corresponding process variations are listed in Table 2 and Table 3, respectively. A corner design methodology is usually used to find the boundary of design space with the consideration of device variations. The design corner can be setup as the combinations of device parameters.

Figure 2 shows that, for the same size of access transistor, RESET can be finished much faster than SET. At the same time, the energy consumption of RESET is much less than that of SET. With the worst case design methodology, however, the switching of RESET bits are over-driven for the same latency as the SET bits. Obviously, if there are only RESET in write operations, the performance can be improved with lower latency; the energy consumption can be reduced with lower driving current and less switching time; even the reliability can be improved because the device breakdown possibility is reduced with lower switching current.

3. LOG STYLE WRITE METHODOLOGY

Operating SET and RESET processes separately can improve energy efficiency of write asymmetric memories. Thus, we propose a log style write methodology to separate SET and RESET in two stages. In this section, we introduce this methodology by applying it to a STT-RAM last level cache (LCC). The discussion of applying this method to other memory level and technologies is presented in Section 4.

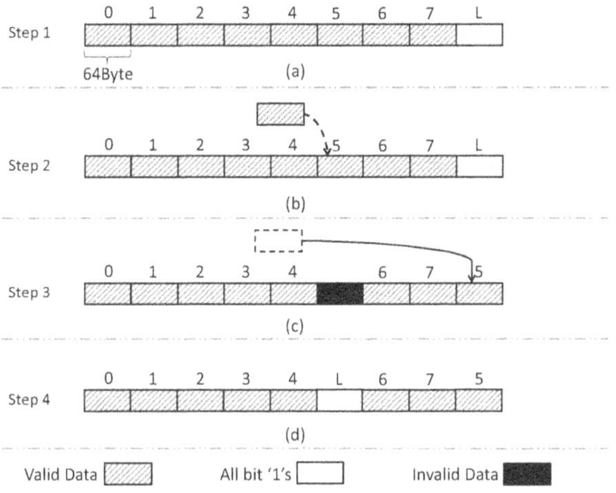

Figure 3: Illustration of the log style write operations.

3.1 Methodology Overview

Figure 3 illustrates the structure view of cache lines in an eight-associativity STT-RAM LLC, which supports log style write operations. The cache line size is set to 64Byte, which is commonly adopted in modern processors. In order to simplify the illustration, we assume that LLC cache line size is equal to that of its upper level cache. With this assumption, the data written back from the upper level cache and the data loaded from the main memory have the same size (64Byte). Note that the log style write also supports variable data lengths. The related issues are also discussed later in Section 4.

Compared to a traditional cache line size, the key difference is that an extra cache line (cache way) is added as the "log way". Thus, the length of one index of cache lines is increased to 576Byte, and there are nine cache lines in total. The eight cache ways storing valid data are labeled with number $0 \sim 7$ in the figure. The log cache way is labeled with L to be differentiated from the others. The cache lines are operated as follows to achieve the log style write:

1. Figure 3 (a) shows initial states of the cache lines. All bits in the log cache way are initialized as bit '1's.
2. In Figure 3 (b), there is an incoming write operation, which will update the $5th$ cache way.
3. In Figure 3 (c), the update data are stored in the log cache way instead of the original position ($5th$ cache way). Since there are only bit '1's in the log cache way, only bit '0's of the incoming data are updated. Thus, there are only RESETs in the write operation. At the same time, the old data in the original $5th$ sector are invalidated.
4. In Figure 3 (d), after the write operation, all bits in the invalidated cache way are programmed to bit '1's in a recycle stage so that the new log cache way is ready for the next write operation.

With the modified cache lines, there are only RESETs in the write operation so that both write latency and energy consumption can be greatly reduced. We call this methodology log style write because the data in each write operation is updated *out of its original place*. This methodology is similar to the log based file system used in NAND flash based SSD. The fundamental difference is that the sector holding invalid data in a SSD cannot be recycled immediately. For the STT-RAM LLC with log style write, the log cache way is recycled as soon as possible after each write operation.

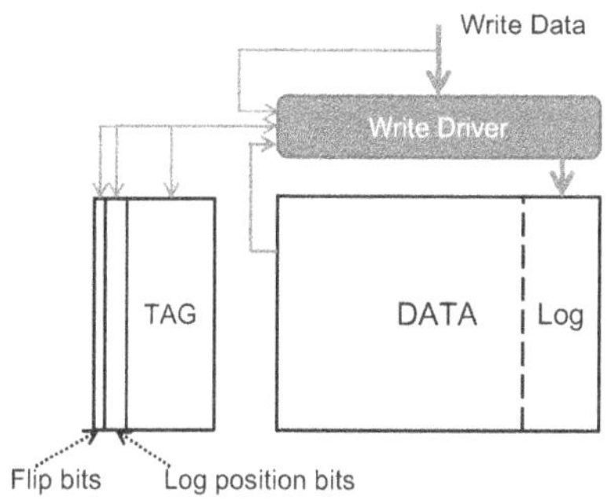

Figure 4: Modified control logic and data path.

A recycle stage is different from a normal write operation. There are no real data written so that the data bus is not occupied and only the local write driver in the sub-array is used. Thus, the access to other sub-arrays will not be blocked by the recycle operation. Theoretically, a read request to other cache lines of the same sub-array can also be operated at the same time in parallel by modifying the control logic. The experiment results (Figure 7) show that the latency of a recycle stage can be well hidden during idle time for most cases. Therefore, the circuitry modification in this work will not support simultaneous read and write requests in the same sub-array. The other design details and corresponding optimization techniques are introduced in the next subsection.

3.2 Design Details and Optimization Techniques

First, we need to decide how to modify cache lines to support the log cache way. As illustrated in Figure 3, we add one extra cache line per set. Thus, the "effective cache associativity", which means the maximum number of cache lines holding valid data in each set, is kept the same as the original cache design. This scheme, however, induces area overhead by adding extra cache lines and increases design complexity of locating cache lines. Alternatively, we can keep the original cache design but sacrifice one cache line per set as the log cache way. It means that the "effective cache associativity" is reduced by one to enable the log style write. In this work, we focus on the former scheme because using log write style can help reduce the cell size so that the area overhead can be offset. This issue will be discussed later in subsection 3.4.

Second, the tag structure of LLC needs to be modified for the log style write. Figure 4 shows corresponding control logic and data path. It should be mentioned that the tag part of the STT-RAM is normally designed with CMOS technology. It is because the STT-RAM based tag has the similar density as the CMOS tag. More important, some status bits of the cache tag (e.g. LRU bits) are updated in read operations. The overhead of long updating latency is too large for read operations. We need to add extra bits for each index of cache to represent the position of the log cache way. For the example of STT-RAM LLC, four bits are needed for each cache set to trace the position of log cache way among nine cache ways shown in Figure 3.

For each write operation, the data are always written to the log cache way, but the cache tag updating and the cache way selected for the following recycle stage are different. For the write hit from the upper cache level, the address bits in the tag of the original cache line is copied to the tag of the log cache way. At the same time, the status bits of the tag are updated in the log cache way, according to the cache management policy. For example, if LRU is used in the LLC of Figure 3, the tag of the last cache line is set to the most recently used one in Figure 3(c). In addition, the fifth cache line is invalidated. For the data write miss from the upper cache level, the tag of the log cache way is updated as in a traditional cache design. At the same time, a cache way (e.g. LRU cache way) is evicted and selected as the log cache way after a recycle stage. Note that this case happens in exclusive and non-inclusive caches. The process is similar for loading data from the main memory. In such a case, a cache line is evicted and erased as the log cache way while LLC is waiting for the data from main memory.

Third, extra control logic is required to operate SET and RESET in two stages efficiently. In order to avoid redundant SETs in the recycle stage, only the bit '0's in the cache line are switched. Thus, the original data need to be first read out and a transistor is needed to control the bit-line driver. Such control logic used for read-before-write has already been proposed in prior research [13]. The drawback of this technique is that the latency of the recycle stage is increased. Fortunately, the recycle stage is not on the critical timing path (shown in Figure 7), and the extra latency for data sensing (read) is trivial compared to the SET latency. For the similar reason, only bit '0's in the incoming data are written to the log sector with RESETs in the write operation. The control logic added in recycle stage can also be used in the write operation to avoid redundant RESETs. Thus, an extra multiplexer is needed to select control signal in different steps.

Forth, optimization technique is needed to reduce energy consumption. With the log style write operation, we can find that the total number of bits switched in two stages are different from that in a traditional write operation. The total write energy consumption in two cases are calculated and compared as follows. Let N_0 and N_1 represent the number of bit '0's and bit '1's in write data, respectively. Let E_0, E_1 and E_0', E_1' represent the RESET and SET energy consumption in a traditional write operation and a log style write operation. The total energy consumption of a traditional write operation is calculated in Equation 1. The total energy consumption in a log style write operation is listed in Equation 2.

$$Energy_{trad} = N_0 \times E_0 + N_1 \times E_1 \tag{1}$$

$$Energy_{log} = N_0 \times E_0' + N_0 \times E_1' \tag{2}$$

The Equation 2 is obtained from the fact that any sector will always be recycled in the future. Although we have $E_1' < E_1$, we cannot promise that $Energy_{log}$ is always less than $Energy_{trad}$. In fact, N_0 in the write data can be much higher than N_1 for many workloads.

The "flip-N-write" [5] technique is employed to solve this problem efficiently. Before the RESET operated in a write operation, the numbers of bit '0's and bit '1's in write data are compared. If there are more bit '0's than bit '1's, all bits in data are flipped before being written into the log sector. An extra "flip-bit" is required for each sector to record whether the data has been flipped before being stored. In fact, we can gain more benefits with this technique in the log style write than in a traditional write operation. With this technique, the total energy consumption of the log style write operation is calculated as in Equation 3. We can find that the total number of switched bits can be much less than the total data length. The energy consumption is always reduced, compared to the traditional write.

$$Energy_{log} = min(N_0, N_1) \times (E_0' + E_1') + 1 \tag{3}$$

3.3 Overhead Analysis

First, there is no extra timing overhead. For each cache access, the position bits for the log cache way are always accessed together with the cache tag. For a write "miss" or a replacement, the extra bits added are updated in parallel with the original bits in a traditional tag . For a write hit, the cache tag is updated simultaneously with the data part. Thus, there is no extra timing overhead for the tag access. For the data access in a read operation, the data may be flipped before being read out. The latency of data flipping is negligible, compared to the total read latency of LLC [5]. Obviously, the write latency is always reduced.

Second, extra bits are accessed in read/write operation, but the total energy consumption is still reduced. In a read operation, an extra flip bit is accessed and the extra tag for the log cache way need to be read/compared. In a write operation, the position bits of the log cache way are updated and the extra flip bit may be updated. Since both the number of total switching bits and the RESET energy consumption are reduced in the write operation, the total energy consumption is reduced. This is proved by experimental results in Section 5.

Third, extra overhead is induced in the area. Most overhead comes from the extra storage space for the log cache way. The extra log cache way induces 12.5% storage overhead for the example of STT-RAM LLC. Besides, extra control logic and bits in the tag also induce space overhead, which is trivial compared to that of the log cache way. Thus, the area overhead is mainly decided by the capacity used for the log. Fortunately, the associativity of modern LLC is normally large (e.g. 16 or 32). It means that the area overhead is moderate. In next subsection, we will show that such overhead can be further compensated.

3.4 Design Trade-off with Log Style Write

With the extra log way, area overhead is induced to reduce write latency and energy consumption. In fact, when SET processes are moved into the recycle stage, we have more opportunities in the new design space for trade-off between capacity and performance/energy. As shown in Figure 2, RESET is much less sensitive to the size of the driving transistor, compared to SET. For example, the RESET latency is only increased a little when the transistor width is decreased from 720nm to 540nm. For memory like STT-RAM, the design area is closely related to the size of the driving transistor. As shown in Figure 5, the area of a 16M STT-RAM LLC can be reduced by about 25%, when the width of the driving transistor is reduced from 720nm to 450nm. However, the SET latency is increased by about 10ns with the smaller transistor. Since the latency of SET is not an critical issue, we can reduce the size of the driving transistor to compensate the log capacity without inducing much degradation in performance.

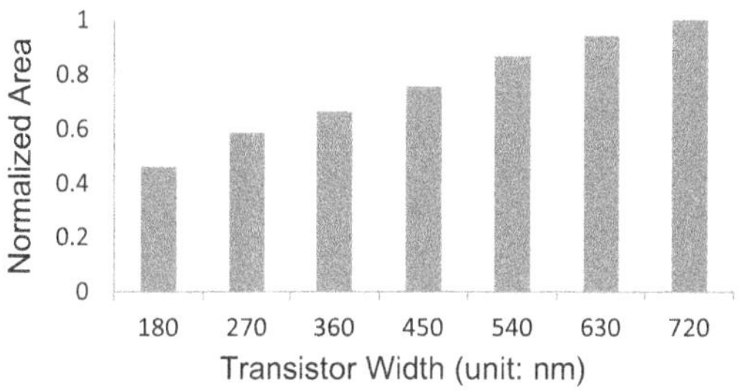

Figure 5: LLC area for different transistor widths ($16MB$ $45nm$).

Another opportunity comes from the error correction mechanisms. The total number of switching bits in each stage is equal to $min(E'_0, E'_1)$, which is less than half of that in the traditional caches. Thus, we can use weaker error correction mechanisms for the same error rates, with a smaller driving current in the RESET operation. The study about this issue is not included due to page limitation.

4. LOG WRITE FOR OTHER MEMORIES

The write style is not just limited to the LLC. For other levels of the cache hierarchy, one important issue is that the write data length may not be fixed. For example, if we want to apply the method to an L2 cache, which works with a write-through L1 cache. The data written through from L1 cache can be less than the size of a cache line, but a whole cache line is loaded from the next memory level. In such scenarios, the log record need to be set to the minimum granularity of the write operation, and the individual log record may be kept for each cache way. Thus, more storage space is needed to trace the log. For a L2 cache used in a 64-bit machine, the cache tag can be increased by about 40%. Fortunately, the cache tag only consumes a small part of the total area (e.g. less than 10%). It is still worth using the methodology because the L2 cache is normally more sensitive to the long write latency of SET.

The methodology is also feasible for other write-asymmetric emerging memories such as FBDRAM. For the capacity based memory such as FBDRAM, the different write voltage is applied the cell in different stages. It should be mentioned that such log write style has an impact on lifetime of the memory like PCM. On one side, the log style help spread the write intensity to different memory space. On the other side, the two-stage write may switch more bits, compared to the read-before-write scheme [13].

Although both log write style and read-before-write schemes can help reduce energy consumption, they cannot work with each other at the same time. However, we believe that these two schemes are used in different scenarios. For PCM, which is normally employed as off-chip memory, a read-before-write scheme is more attractive because the lifetime limitation of PCM makes it more important to reduce write intensity. For STT-RAM and FBDRAM, which are used as fast on-chip memory and have no lifetime issues, it is reasonable to leverage our log write style because read-before-write cannot reduce write latency but increases it instead.

5. EVALUATION

In this section, we evaluate a STT-RAM LLC with log style write in a chip-multiprocessors (CMP) system and compare the results to those of using a traditional one.

Table 2: Device Parameters

Device	Parameters	Mean	Std. Dev
NMOS	Channel Length L	$45nm$	$2.25nm$
	Channel Width W	design dependent	$2.25nm$
	Threshold Voltage V_{th}	$0.466V$	$30mV$
MTJ	MgO Thickness τ	$2.2nm$	2% of mean
	Cross Section A	$45 \times 90nm^2$	5% of mean

Table 3: $3sigma$ corner design parameters.

W	L	V_{th}	τ	A
−15%	+15%	+15%	+6%	±15%

5.1 Configuration Setup

For system level simulation, we use SIMICS simulator to evaluate performance. It is configured to model an eight-core processor. Each core is Ultra-SPARC-like with a 2GHz frequency. There are three levels of caches. The IL1/DL1 and L2 caches are SRAM based and the capacities are set to 64KB and 2MB, respectively. The baseline LLC (L3) is a 16-way 16MB STT-RAM cache. The simulator captures data addresses from all loads, stores, and prefetch operations. We use the information to calculate the memory access intensity, and use that to compute the energy consumption of the cache hierarchy. Our workloads are sets of multi-threaded benchmarks from SpecOMP and PARSEC. We choose benchmarks from the full set and mix them together to help us create a diverse set of intensive cache access patterns. The WPKIs (write per thousand instructions) of these benchmarks range from 1.8 to 9.8.

In order to estimate the access latency and energy consumption of different caches, we extend the widely used tool CACTI [2] to support STT-RAM technology with process variations considered. The parameters used in this work are listed in Table 2 and Table 3. We choose the latency as the value at 3σ of the distribution for both read and write (SET/RESET in STT-RAM) operations. The read/write latency is set to 1ns and 5ns for L1 and L2, respectively. For the traditional STT-RAM LLC, the read latency is 10ns, and the write latency is set to 35ns. As we discussed Section 3.4, we have different design options after using the log style write. In our experiments, we reduce the cell width to compensate the capacity overhead of an extra log cache way, which is about 6.5%. The SECDEC is employed as the error correction mechanisms for both STT-RAM caches, since the ECC optimization is not the focus of this work. Thus, for the STT-RAM with log write style, the read latency is still 10ns and the RESET and SET latency is 15ns and 38ns.

5.2 Evaluation Results

Figure 6 compares the normalized execution time for three different STT-RAM LLCs. The first set of baseline result is for the traditional STT-RAM cache design without using any write optimization techniques. The second result is for the STT-RAM cache using the write-halt technique [6], in order to provide a comprehensive comparison. The third set of result is for the STT-RAM cache using log style write operation. We can find that, after using the log style write, the performance is improved. In addition, using log style achieve even better performance than that of using write-halt technique in prior research. Compared to the baseline case, the performance is improved by about 4% on average after using the log style write. The performance improvement is not significant because the STT-RAM LLC is the

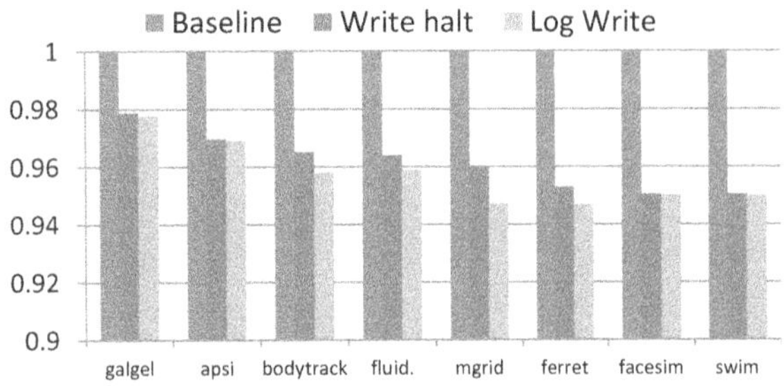

Figure 6: Normalized Execution Time.

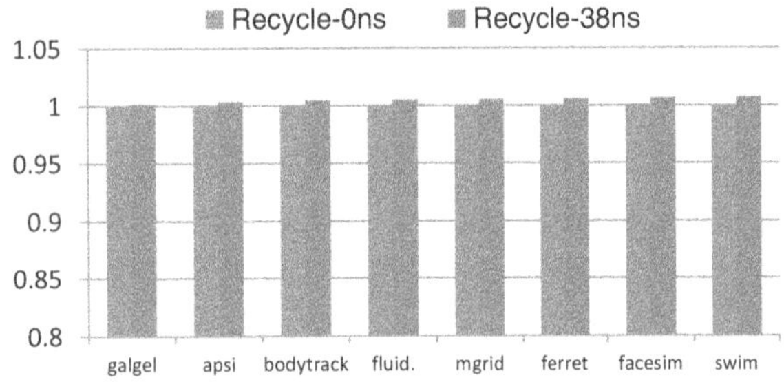

Figure 7: Normalized execution time with different recycle latency.

L3 cache, which is not that sensitive to write latency. We can expect more benefits when applying the log style write methodology in other memory levels with higher write intensity.

In order to analyze the impact of recycle stage on performance, we compare the results of two specific cases shown in Figure 7. The first set of results show normalized IPCs when we set the latency of recycle as zero. It represents a non-real case when recycle causes no overhead. The second set of results show IPCs of the real case in this work with the latency of recycle set to 38ns. The comparison shows that the performance is reduced by less than 1% with the real latency of recycle. It means that the latency of recycle stage can be hidden well and has little impact on performance.

The power consumption results are shown in Figure 8. After using write-halt technique, the total power consumption is increased. It is because the write operation can be canceled and by read operation in such a cache and will be re-written later. Due to the limitation of experimental environment, it is impossible to accurately estimate the energy consumption of an incomplete write. Thus, we assume that an incomplete write operation consumes $0\% \sim 75\%$ energy consumption of a complete write, and we randomly choose a value in this range for each write operation. Figure 8 also shows that the power consumption is greatly reduced after using the log style write operation. First, it is because the RESET energy consumption is greatly reduced. Second, the optimization using flip bit make sure that the total number of write bits are reduced. Thus, the ratios of bit '0' and bit '1' in write data also has an impact on the total energy consumption. For example, the power reduction for workload "facesim" is higher than the others because there are much more bit '0's in the write data. On average, the power consumption can be reduced by about 35% with the log style write.

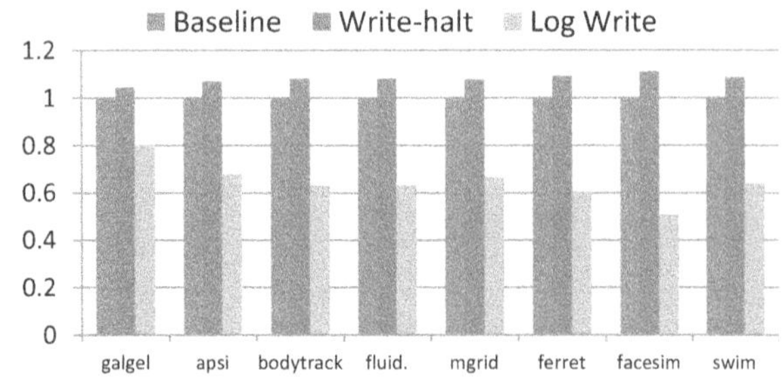

Figure 8: Normalized Power Consumption

6. CONCLUSION

Asymmetric write is a common character for most emerging memory technologies. When both SET and RESET exist in a write operation, the worst case design has to be used. Such a design scheme induces performance, power, and reliability overhead, which is further aggravated by the affect of process variations. In this work, we separate SET from RESET by using a log style write operation. With this methodology, the SET and RESET are optimized and processed individually. Moreover, separating SET and RESET provides more design options and optimization opportunities. After using the log style write, performance is improved and power consumption is reduced for the write-asymmetric memory design.

7. REFERENCES

[1] J. Kim, S. Chung, T. Jang, and *et al.* Vertical double gate Z-RAM technology with remarkable low voltage operation for DRAM application. pages 163–164, 2010.

[2] H. Labs. In *http://www.hpl.hp.com/research/cacti/*, .

[3] B. C. Lee, E. Ipek, O. Mutlu, and D. Burger. Architecting phase change memory as a scalable DRAM alternative. In *Proceedings of ISCA*, pages 2–13, 2009.

[4] Z. Lu, N. Collaert, M. Aoulaiche, B. De Wachter, A. De Keersgieter, W. Schwarzenbach, and *et al.* A novel low-voltage biasing scheme for double gate fbc achieving 5s retention and 10_{16} endurance at 85c. In *IEDM*, pages 12.3.1 –12.3.4, dec. 2010.

[5] A. Nigam, C. Smullen, V. Mohan, E. Chen, S. Gurumurthi, and M. Stan. Delivering on the promise of universal memory for spin-transfer torque ram (stt-ram). In *ISLPED 2011*, pages 121 –126, aug. 2011.

[6] M. Qureshi, M. Franceschini, and L. Lastras-Montano. Improving read performance of phase change memories via write cancellation and write pausing. In *HPCA*, pages 1 –11, jan. 2010.

[7] M. K. Qureshi, V. Srinivasan, and J. A. Rivers. Scalable high performance main memory system using phase-change memory technology. In *Proceedings of ISCA*, pages 24–33, 2009.

[8] C. Smullen, V. Mohan, A. Nigam, S. Gurumurthi, and M. Stan. Relaxing non-volatility for fast and energy-efficient stt-ram caches. In *High Performance Computer Architecture (HPCA), 2011 IEEE 17th International Symposium on*, pages 50 –61, feb. 2011.

[9] G. Sun, X. Dong, Y. Xie, J. Li, and Y. Chen. A novel architecture of the 3d stacked mram l2 cache for cmps. pages 239 –249, feb. 2009.

[10] Z. Sun, X. Bi, H. H. Li, W.-F. Wong, Z.-L. Ong, X. Zhu, and W. Wu. Multi retention level stt-ram cache designs with a dynamic refresh scheme. In *Proceedings of Micro*, pages 329–338, 2011.

[11] X. Wang, Y. Zheng, H. Xi, and D. Dimitrov. Thermal fluctuation effects on spin torque induced switching: Mean and variations. *Journal of Applied Physics*, 103(3):034507 –034507–4, feb 2008.

[12] Y. Zhang, Y. Li, A. Jones, and Y. Chen. Asymmetry of MTJ Switching and Its Implication to STT-RAM Designs. In *Proceedings of the DATE*, 2012.

[13] P. Zhou, B. Zhao, J. Yang, and Y. Zhang. A durable and energy efficient main memory using phase change memory technology. In *Proceedings of ISCA*, pages 14–23, 2009.

[14] P. Zhou, B. Zhao, J. Yang, and Y. Zhang. Energy reduction for stt-ram using early write termination. In *ICCAD*, pages 264 –268, nov. 2009.

Process Variation Aware Data Management for STT-RAM Cache Design

Zhenyu Sun, Xiuyuan Bi, and Hai (Helen) Li
ECE Dept., Polytechnic Institute of New York University, Brooklyn, NY, 11201, USA

ABSTRACT

The *spin-transfer torque random access memory* (STT-RAM) has gained increasing attentions for its high density, fast read access, zero standby power, and good scalability. The recently proposed retention-relax design further improves STT-RAM write access performance and makes it even more promising as an on-chip memory technology. Nevertheless, the process variations could affect the writability of STT-RAM cells. The situation for retention-relax design is even more severe. In this paper, we comprehensively study the impact of process variations, including those from both CMOS and magnetic technologies, on key STT-RAM design parameters. Furthermore, we propose *process variation aware non-uniform cache access* (PVA-NUCA) technique for large STT-RAM cache design. Besides the varying interconnect latencies determined by memory locations, PVA-NUCA compensates write time variations of STT-RAM cells resulted by process variations. Two algorithms, namely, conservative promotion and aggressive prediction, have been introduced and evaluated. A conflict-reduction mechanism is utilized to degrade the data access miss rate caused by conflicts of access-intensive data blocks. Compared to the traditional STT-RAM dynamic nonuniform cache access (DNUCA), our proposed dynamic PVA-NUCA can improve 25.29% of IPC performance and reduce 26.4% of STT-RAM cache energy consumption, with $< 1\%$ of area overhead.

Categories and Subject Descriptors

B.3.2 [**Memory Structures**]: Design Styles—*Cache memories*

General Terms

Design

Keywords

MTJ, process-variation aware

1. INTRODUCTION

The standby power of SRAM on-chip caches has dominated the overall power consumption in the latest computing and embedded systems [1, 2]. Circuit design and architectural solutions, such as V_{DD} scaling [3] and body-biasing [4], have been utilized to reduce the standby power of caches. However, as technology continues scaling down, the exponential increasing of leakage makes these techniques less efficient. As an alternative of future on-chip memory technology, the emerging *spin-transfer torque RAM* (STT-RAM) has been receiving significant attention.

ISLPED'12, July 30–August 1, 2012, Redondo Beach, CA, USA.

Although STT-RAM has the similar read latency as SRAM, its write latency is much longer: a 10ns nominal programming time has been adopted in mainstream STT-RAM research and development [5, 6]. Many techniques have been investigated to mitigate the performance degradation caused by the costly write operation, such as the early write termination [7] and dual write speed [8]. Recently, Smullen *et al.* [9] proposed to relax the retention of STT-RAM cells (*i.e.*, the data retention time) to improve the writability and fasten the programming time. Following it, Sun *et al.* [10] came up a full STT-RAM cache hierarchy with multi-retention levels. The enhancement of write access speed makes STT-RAM more attractive as on-chip memory technology.

Previously, the *non-uniform cache access* (NUCA) [11] has been used to compensate the difference in interconnect latencies determined by cache physical locations across the last-level on-chip cache. Moreover, as technology scales down to 45nm or below, both CMOS and magnetic devices become more subject to process variations. In a STT-RAM cell, the current to switch the resistance state of its data storage element, *magnetic tunnel junction* (MTJ), is determined by MTJ resistance and the NMOS selection transistor. When programming a STT-RAM cell in the sub-10ns region, the required MTJ switching time rises exponentially as the switching current reduces [12]. As we shall show in Section 3, the process variations result in a large distribution of the MTJ switching time that is in the same order of that of interconnect latencies and hence cannot be ignored.

In this paper, we thoroughly analyze the impact of process variations on STT-RAM key design parameters and demonstrate its spatial distribution by using multi-level qual-tree approach [13]. Then two *process variation aware NUCA* (PVA-NUCA) techniques for large STT-RAM cache design are proposed, which include the non-uniformity of ***interconnect latencies*** and adaptively change ***programming duration***. In the *static PVA-NUCA*, a data block stays at a fixed location during its whole life time; while the *dynamic PVA-NUCA* can migrate a data block according to its access frequency. By considering the difference in terms of interconnect latency and programming time at the different physical locations, we propose two date migration algorithms: the first one gradually promotes a data block to a new location with faster access time, while the second algorithm aggressively places an access-intensive data directly into the fastest cache block.

The rest of our paper is organized as follows. Section 2 introduces STT-RAM and the tradeoffs involved in MTJ retention relaxation. Section 3 describes process variation analysis for retention relaxed STT-RAM. We explained the proposed process-variation-aware non-uniform cache access schemes for STT-RAM cache design in Section 4 and the simulation results are shown in Section 5. At the end, Section 6 concludes our work.

2. BACKGROUND

2.1 STT-RAM basics

The popular STT-RAM cell structure is one-transistor-one-MTJ (or 1T1J), in which an MTJ is connected to an NMOS transistor in series. The MTJ is used as data storage and the NMOS transistor is used to control the access of the MTJ. *Magnetic tunnel junction* (MTJ), which is composed of two ferromagnetic layers, namely *reference layer* and *free layer*, and an oxide barrier layer, *e.g.*, MgO. The magnetization direction of the reference layer is fixed while that of the free layer can be changed by passing a current that is polarized by the magnetization of the reference layer. When the magnetization directions of the two ferromagnetic layers are parallel (anti-parallel), the MTJ is in low (high) resistance state with R_L (R_H).

2.2 MTJ writability and data retention

The current amplitude required to reverse the resistance state of an MTJ is determined by many factors including material properties, device geometry, the write switching time, etc. Generally, the larger switching current (I_c) is applied, the shorter the write pulse (τ) is needed to switch the MTJ state. According to the operating range of τ, the relationship between I_c and τ can be approximated as [14]:

$$I_{c,TA}(\tau) = A \cdot J_{c0}(1 - \frac{1}{\Delta}\ln(\frac{\tau}{\tau_0})) \quad (\tau > 10ns); \tag{1a}$$

$$I_{c,PS}(\tau) = A \cdot (J_{c0} + \frac{C}{\tau_{sw}^{\gamma}}) \quad (\tau < 3ns); \tag{1b}$$

$$I_{c,DR}(\tau) = \frac{I_{c,TA}(\tau) + I_{c,PS}(\tau)e^{-k(\tau-\tau_c)}}{1 + e^{-k(\tau-\tau_c)}} \quad (3ns < \tau < 10ns). \tag{1c}$$

Here, C, γ, k, and τ_c are the fitting constants. A is the cross-sectional area of the MTJ free layer. J_{c0}, the *critical current density* at zero temperature, primarily depends on the vertical structure and magnetic properties of the MTJ, such as the MTJ free layer thickness t_f, the saturation magnetization M_s, and the in-plane anisotropy field H_k [14, 15]. Δ is the *magnetization stability energy height*, which determines the data retention time for a MTJ magnetic element. Δ can be calculated by:

$$\Delta \propto \frac{V \cdot H_k \cdot M_s}{T}, \tag{2}$$

where, V is the effective volume considering non-uniform magnetization reversal. T is the ambient temperature.

Some previous research works showed that relaxing the MTJ retention can dramatically improve the write performance of STT-RAM. For example, shrinking the MTJ surface area A can reduce the thermal stability Δ, and consequently, decreases I_c and/or τ [9]. Or, tuning MTJ magnetic properties M_s, H_k, and t_f can lower both Δ and J_{c0}, offering performance improvement over the entire MTJ working range [10].

Figure 1(a) shows the relationship between the nominal (the mean of) τ and I_c of two MTJ designs [10]. "Baseline" is a regular MTJ design with a retention time of 4.27 years at $350K$. Under the same environment condition, "Relaxed" design reduces data retention time to $3.24s$, which is still sufficiently long and induces trivial refresh overhead [9, 10]. However, when the applied switching current is fixed at $200\mu A$, the required write period τ reduces from the baseline's $8ns$ to $4ns$ of the relaxed design. Further decreasing I_c below $200\mu A$ can result in a significant increase of τ.

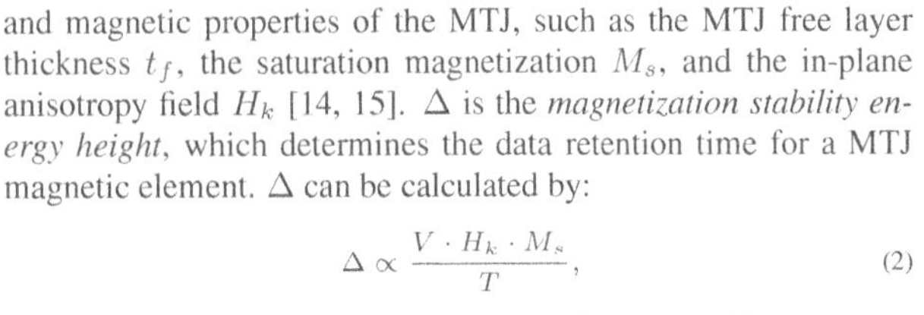

Figure 1: (a) MTJ switching performance comparison of two different MTJ designs at 350K. (b) The STT-RAM cell size vs. the maximal switching current.

Table 1: Process variations in STT-RAM design.

Parameters	Mean (μ)	σ/μ.
NMOS transistor width (W_{tx})	$180nm$	10% (1)
NMOS transistor length (L_{tx})	$45nm$	10% (1)
NMOS transistor threshold voltage (V_{th})	$0.3V$	10% (1)
MTJ resistance area product (RA)	$20\Omega \cdot \mu m^2$	8% (2)
Cross section area of MTJ free layer (A)	$90nm \times 180nm$	5% (2)
MTJ magnetization stability energy height (Δ)	22	27% (3)

Data sources: (1)[16], (2)[17], (3)estimation based on Matlab simulation.

The STT-RAM cell area usually is dominated by the size of the NMOS transistor, which should be sufficiently large to provide enough switching current for MTJ switching. The calculated STT-RAM cell size and the maximal switching current it can provide is shown in Figure 1(b). For instance, a STT-RAM cell with $24F^2$ can supply up to $200\mu A$ switching current. F represents the technology feature size.

3. PROCESS VARIATIONS

3.1 STT-RAM write performance variation

The CMOS process variations, such as the geometry variations of transistor channel length and width, the threshold voltage deviations, *etc.*, results in the driving strength non-uniformity of the NMOS transistor in a STT-RAM cell. The major MTJ device variations include: (1) the deviation of MTJ shape; (2) the variation in MgO thickness; and (3) the normally distributed localized fluctuation of magnetic anisotropy $K = M_s \cdot H_k$. The first two factors cause the variations of the MTJ resistances (R_L and R_H) and the MTJ switching current I_c induced by the discrepancy of the NMOS transistor bias condition. The third factor is an intrinsic feature of magnetic material that affects the MTJ switching threshold current I_{c0} and the magnetization stability barrier height Δ [14]. Table 1 summarizes the key STT-RAM design parameters and the ratio of standard deviation (σ) and mean value (μ) used in this work. We assume each parameter follows a Gaussian distribution.

To analyze the impact of process variations on the STT-RAM design, we conducted 5,000 Monte-Carlo simulations under Cadence Spectre environment. The STT-RAM cell with $24F^2$ was used. The nominal value of the highest possible switching current I_c of this design is $200\mu A$. Figure 2(a) is the distribution of I_c, which varies from $110\mu A$ to $320\mu A$ after including the variations of both NMOS transistor and MTJ. Correspondingly, the required MTJ switching time τ in Figure 2(b) demonstrates a very large distribution from $2.3ns$ to $12.4ns$. The traditional worst-corner design methodology using a uniform write period, *i.e.*, $\tau_{max} = 12.4ns$, will result in a pessimistic design and severe degradation on STT-RAM cache performance. Furthermore, ***the variation of MTJ switching time τ has exceeded the difference in the interconnect latencies***, which is less than $8ns$ for a 32MB STT-RAM based on CACTI [18].

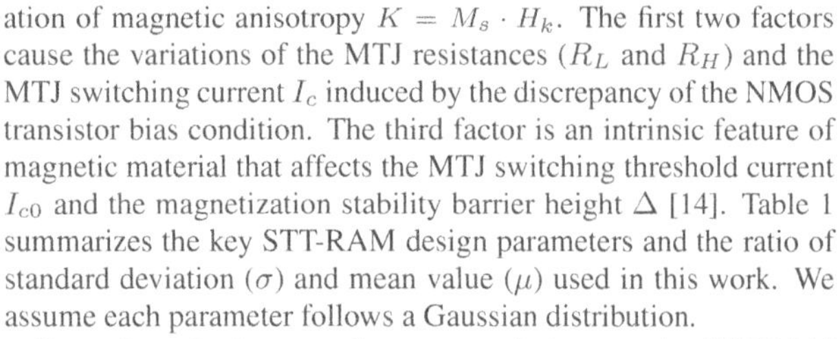

Figure 2: The distribution of (a) MTJ switching current I_c, (b) writing pulse width τ, and (c) MTJ data retention time.

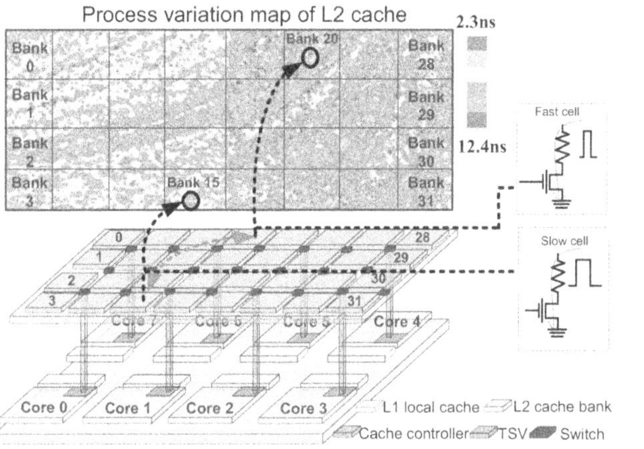

Figure 3: The write pulse period distribution map of a shared STT-RAM L2 cache in 3D stacking structure.

The results in Figure 2(c) shows the minimum retention time is $0.01s$, which corresponds to 20 million clock cycles in a system with $2GHz$ clock frequency. In the following paper, we assume a DRAM-like refreshing scheme [9] with a refresh period of $0.01s$. The overhead on system performance and energy consumption has been considered in evaluations.

3.2 Write performance distribution map

Figure 3 illustrates the study object of this work – an 8MB shared STT-RAM L2 cache stacked on top of an 8-core CPU layer by using 3D technology. The shared L2 cache consists of 32 banks. The signal transportation between two layers is realized by *through-silicon-vias* (TSV).

We generated a distribution map of the required write pulse periods of the 8MB LLC by using the multi-level quad-tree approach [13] and combining the statistical results from Monte-Carlo simulations. The distribution map is also shown in Figure 3. The process variations present strong spatial correlations, that is, the STT-RAM cells close to each other have smaller difference in terms of the required write pulse periods than those far apart.

4. PROCESS VARIATION AWARE NUCA

The *last level caches* (LLC) occupies a big potion of on-chip area. The different core to bank distances and the induced discrepancy of the interconnect lengths make significant variations of cache access latency. Therefore, the *dynamic non-uniform cache access* (DNUCA) has been used to migrate data block from distant bank to nearby one and reduce transmission overhead [11]. Previously, Sun *et al.* extended NUCA to STT-RAM design [5] by separating and adopting the worst-case read and write latencies, *i.e.*, $lat_{c,r,max}$ and $lat_{c,w,max}$, respectively.

Although the STT-RAM read operation is small and not significantly affected by process variations, the distribution of write latencies are large and could be more severe than the discrepancy induced by the different core to bank distances. For example, Section 3 shows that the required STT-RAM write latency $lat_{c,w}$ changes from $2.3ns$ to $12.4ns$ across an 8MB STT-RAM cache, while the latency of peripheral circuitry and interconnect lat_{wire} of the same design varies between $0.5ns$ and $8ns$, based on CACTI [18].

Hence, a *process variation aware NUCA* (PVA-NUCA) with adaptive $lat_{c,w}$ is necessary for STT-RAM cache design to accelerate cache performance and reduce write energy consumption. In this paper, we will present and discuss one static and two dynamic versions of PVA-NUCA, namely, SPVA-NUCA, DPVA-NUCA-1, and DPVA-NUCA-2, respectively.

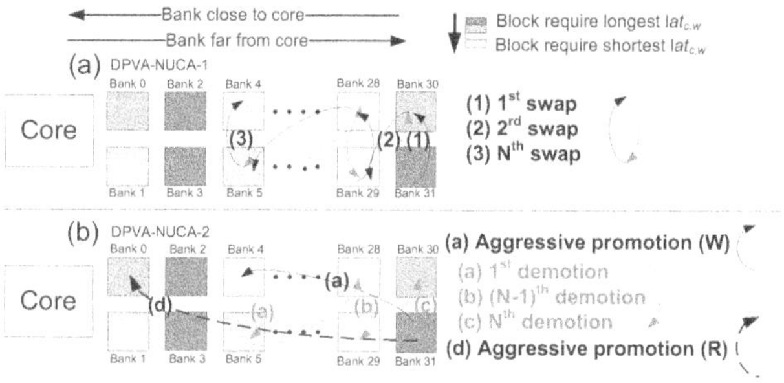

Figure 4: The illustrations of data migration in (a) DPVA-NUCA-1 (b) DPVA-NUCA-2.

4.1 SPVA-NUCA

SPVA-NUCA is a static process variation aware NUCA for STT-RAM design without data migration. Here, a read operation includes the different core to bank distances but uses a fixed, maximum cell read latency $lat_{c,r,max}$. On the contrary, a write access latency is determined by both the physical location and the required write time of the cache block. The overall latency of a cache access then can be expressed as:

$$\text{Read access}: \quad Lat_R = lat_{c,r,max} + lat_{wire}; \tag{3a}$$

$$\text{Write access}: \quad Lat_W = lat_{c,w} + lat_{wire}. \tag{3b}$$

As illustrated in Figure 3, the required write time can be obtained in post-fabrication testing and stored within the LLC data storage array. For the 8MB STT-RAM cache with a $2GHz$ clock frequency in Section 3, a cache block requires 5 extra bits to represent the corresponding $lat_{c,w}$. Considering the spatial correlation, a few cache blocks in close locations (*e.g.*, 64 blocks) can utilize the maximum $lat_{c,w}$ among them to reduce design overhead.

4.2 DPVA-NUCA-1

The dynamic process variation aware NUCA, DPVA-NUCA-1, uses the *conservative promotion* algorithm (see Algorithm 1) for cache block migration. The data promotion/demotion is determined by the frequencies of write hits and write misses: a write-intensive cache block will be promoted to a physical location with the minimum Lat_W.

On each write hit, the Lat_W of 32 ways within the accessing set will be sorted. Based on the sorting results, the data block will be swapped with the data block who is located in the cache block with next smaller Lat_W as the pairs of black and red arrows shown in Figure 4(a). Naturally, the non-write-intensive data blocks will be gradually demoted to cache block with larger Lat_W as shown by the red arrows. During a write miss, the *least recent used* (LRU) data block will be evicted and replaced by incoming request.

Without loss of generality, the example in Figure 4(a) demonstrates how a data block is gradually promoted. Assuming the data block is placed at *bank 31* which is the farthest cache block with longest programming time at very beginning. In this example, it takes *N* swaps for the the data block to promote from *bank 31* with maximum Lat_W to *bank 4* with minimal Lat_W. Cache block in *bank 28* and *bank 29* also have shortest programming time within this accessing set, but their interconnection latencies is too large. Minimum $lat_{c,w}$ doesn't mean Lat_W, so the data won't be moved to *bank 28* or *bank 29* finally. Moreover, the data block won't be swapped to the cache block in *bank 0*, which has smallest lat_{wire} but a larger $lat_{c,w}$. The sorting of Lat_W of 32 ways within the accessing set every time is too costly to be affordable. We use a **sorting recorder** to record sorting results of recent accessed set. The more hits in the **sorting recorder**, the least performance degra-

Definitions:
$CB(Lat_{W,min})$: the cache block with the minimal write latency.
$CB(Lat_{R,min})$: the cache block with the minimal read latency.
$CB(Lat_{W,next})$: the cache block with the second smallest write latency.

Algorithm 1: **Conservative Promotion** used in DPVA-NUCA-1.

if *write(hit)*
  if *data is not in* $CB(Lat_{W,min})$ **then**
    swap data in $CB(Lat_W)$ with *the one in* $CB(Lat_{W,next})$
  end if
end if
if *write(miss)* **then**
  replace the LRU data block
end if

Algorithm 2: **Aggressive Prediction** for DPVA-NUCA-2.

if *hit*
  if *data is read intensive block* **then**
    move data into $CB(Lat_{R,min})$
  else if *data is write intensive block* **then**
    move data into $CB(Lat_{W,min})$
  end if
if *write(hit)*
  if *access address is in the write prediction queue* **then**
    corresponding queue counter ++
  else then
    if *queue is full* **then**
      kick LRU entry and fill the entry with incoming address
    else then
      add new entry with incoming address
    end if
  end if
if *read(hit)* **then**
  do exact same operation in read prediction queue
  end if
end if
if *miss* **then**
  replace the LRU data block
end if

dation caused by the sorting of Lat_W will be. The efficiency of **sorting recorder** will be evaluated in the results part.

4.3 DPVA-NUCA-2

The date migration in DPVA-NUCA-1 is determined by only write accesses. Note that many programs are read access dominant. Moreover, the read access latency variation with its physical location is mainly determined by the difference of lat_{wire} since $lat_{c,r}$ is much smaller and uniform in different banks. Thus, DPVA-NUCA-1 solely based on write accesses might not obtain the best system performance and the least energy consumption. We proposed another dynamic process variation aware NUCA (DPVA-NUCA-2) to balance the read and write requests. Moreover, data blocks are aggressively move to fastest cache block according to the access situation to further improve the performance.

Algorithm 2 is an *aggresive promotion* algorithm used for DPVA-NUCA-2. The data blocks are classified into three types: *read intensive*, *write intensive*, and *not access intensive*. We use two queues, namely, *read prediction queue* and *write prediction queue* to record the read and write access histories, respectively. An incoming read or write hit triggers the address searching in both read queue and write queue. (a) If the block address is in the read queue and its corresponding counter is above threshold, the block is read intensive. We will move it to the cache block with the minimal read access latency Lat_R, as the dash black arrow illustrated in Figure 4(b). (b) Otherwise, if the data is not read intensive, but we can find it in write queue with a counter value beyond threshold, the data is write intensive. We will move it toward the cache block with the minimal write access latency Lat_W, as the solid black arrow illustrated in Figure 4(b). (c) If the data is neither read intensive nor write intensive, no data migration is needed. The proposed DPVA-NUCA-2 grants higher priority to read accesses than that of write ones. This is because usually read is more frequent and more

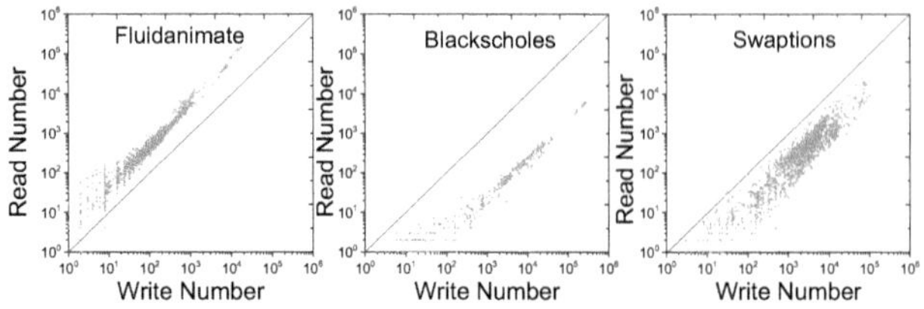

Figure 5: Read and write number of all cache blocks.

critical to system performance. So, if a data block is both read and write intensive, the data migration will be determined by Lat_R. LRU block will be replaced during write or read miss. **Sorting recorder** is needed by DPVA-NUCA-2.

Write and read prediction queues: The two prediction queues have the same hardware structure – have a repository to store recent accessed block address and associated with corresponding counter. An incoming access can trigger the increment of the corresponding counter if the accessing block is already in the queue. Otherwise, a new entry will be added if the queue is not full. If the queue is full, the LRU entry will be evicted. As we shall show in Section 5, a small number of entries, *i.e.*, 32, is efficient enough for read and write queues. The queue is CAM structure, so the searching of the queue is quick and within 1 clock cycle.

Conflict reduction: For DPVA-NUCA-2, it is possible that two or more cores compete for one cache block location with minimum Lat_W. In such a situation, the write-intensive data blocks held by these two cores may take each other replaced alternatively, resulting in unexpected cache misses. To reduce those conflicts, we can simply attach a counter to record its cache miss numbers. If the miss numbers of several cores increase simultaneously, we check if there are conflict of write intensive blocks in their write queues. If it happens to be the scenario, some data will be moved to block with the second minimum Lat_W in a round robin manner. If the miss rate keep on high, the data will be further demoted to cache block with larger Lat_W.

5. SIMULATION RESULTS & DISCUSSION

5.1 Experimental Setup

Our basic architecture configuration is an 8-core in-order Ultra-SPARC T1 processor where the SRAM L2 cache is replaced by STT-RAM. We use the Simics toolset [19] for performance simulations. The details on microarchitecture configurations are depicted in Table 2. UCA, SUNCA, DUNCA and our proposed PVA-NUCA are implemented by modifying the cache model in Simics. We adopted the multi-threaded benchmarks from Parsec Benchmark Suite [20] in our architectural simulations. For each benchmark, we fast-forward to ROI (Region of Interest), warm up the cache for 200 million instructions, and then run 500 million instructions.

The system architecture is illustrated in Figure 3. We use an 8-core in-order Ultra-SPARC T1 processor where the shared SRAM L2 cache is replaced by STT-RAM cache. The shared L2 cache can be stacked on the CPU core layer, benefiting from 3D integration technology. The inter-layer signals are connected by *through-silicon-vias* (TSVs). Each CPU core is connected with the on-chip switch which has four neighboring L2 cache banks. Cache con-

Table 2: Processor configurations.

Processors:	8 cores, 2GHz, 4 threads/CPU core, 1-way issue (in order)
SRAM L1 Cache:	local, 16KB I/D, 2-way, 64B line, 2-cycle, write-through
STT-RAM L2 Cache:	shared, 8MB, 32 banks, 32-way, 64B line, write-back, 1 read/write port, 4 write buffers.
Main Memory:	4GB, 400-cycle latency.

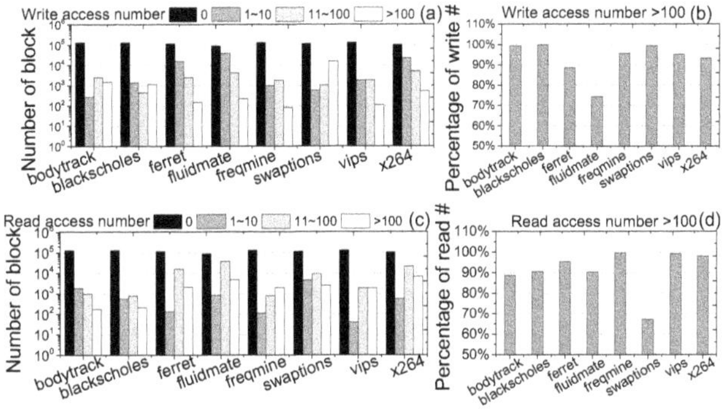

Figure 6: Number of blocks hold different (a) write or (c) read numbers; Percentage of (b) write or (d) read held by access intensive blocks

troller is used to deal with the communication and coherency between L1 and L2 cache, as well as cores.

5.2 Simulation Results

5.2.1 Cache access statistics

Figure 5 compares the overall read and write numbers of three selected benchmarks representing different read and write access distributions. Data was collected without applying any data migration algorithm. Comparably, `Blackscholes` is more write intensive and `Fluidanimate` is more read intensive. Write and read numbers of `Swaptions` are more decentralized. The $x-$axis and $y-$axis are in logarithmic scale. The read and write number distributions of data blocks will affect the hardware optimization efficiencies in the proposed schemes such as conflict reduction, prediction queues and sorting recorder.

Figure 6 demonstrates a more detail write and read access analysis. According to the write access numbers in 500 million instructions, we divide cache blocks into four groups. For example, the first bar in each benchmark of Figure 6(a) represents the number of cache blocks with zero write access, while the last bar shows the number of cache blocks with more than 100 write accesses. Figure 6(b) shows the percentage of writes that fall into those blocks with more than 100 write access. We conducted the similar statistical analysis for read accesses and show in Figure 6(c) and (d). Figure 6 explicitly indicates that a small number of frequently accessed cache blocks hold most of the write and read operations. The goal of the proposed schemes is to allocate these data blocks into fast cache blocks to improve system performance.

5.2.2 Conflict reduction

As discussed in Section 4.3, the conflicts of access-intensive data blocks from different CPU cores could severely degrade overall system performance of DPVA-NUCA-2. Figure 7(a) demonstrates the simulated conflict rate for various benchmarks. Here, the conflict rate is defined as the ratio of the conflict number over the sum up of read hits and write hits. Compared to ideal case without calculating the impact of conflict, the corresponding IPC degradation caused by those conflicts are shown in Figure 7(b). Some benchmarks, such as `Blackscholes` and `Bodytrack`, have strong data locality. The chance to have data confliction is low, which leads to close to zero IPC degradation. However, some benchmarks such as `Fluidanimate` demonstrate high conflictions from different

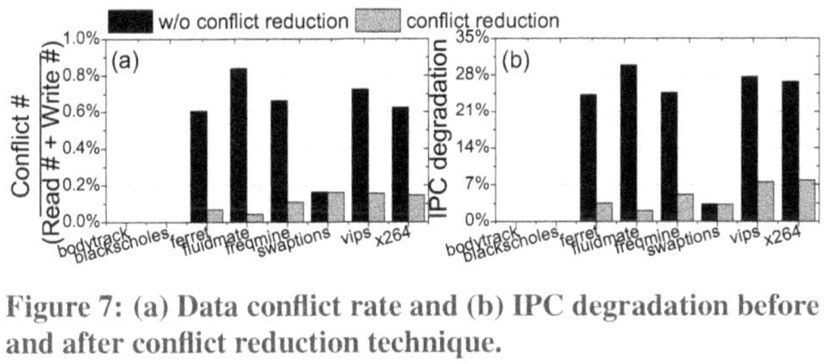

Figure 7: (a) Data conflict rate and (b) IPC degradation before and after conflict reduction technique.

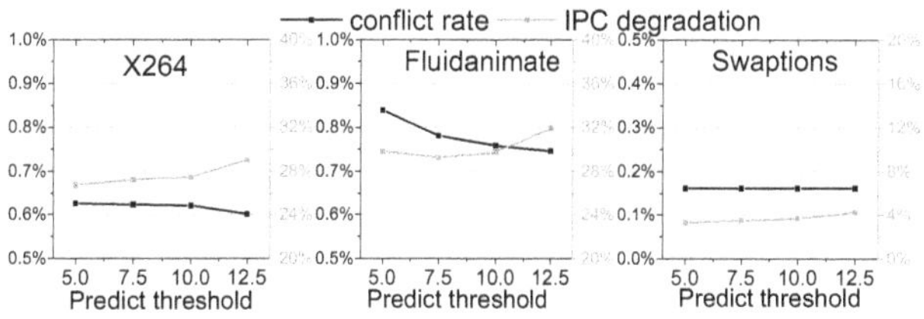

Figure 8: (a) Data conflict rate and (b) IPC degradation with different prediction threshold.

CPU cores and unaffordable IPC degradation. Both conflict rates and IPC degradation have been drastically reduced after utilizing the conflict-reduction mechanism in DPVA-NUCA-2.

The conflict rate and the corresponding IPC degradation could be affected by the prediction threshold. The prediction threshold is the preset threshold of the counter of the prediction queue. Generally, increasing prediction threshold can reduce the number of cache blocks swapped into the fast cache blocks, and hence, decrease the conflict rate. However, the results of `Fluidanimate`, `x264` and `Swaptions` in Figure 8 indicate that merely increasing the prediction threshold will not help improving the IPC degradation, even though the conflict rate decreases. That is because less data blocks are allocated into the fast cache blocks with the leveling up of the prediction threshold. On the contrary, our proposed conflict-reduction mechanism can overcome this difficulty because it dynamically and gradually move write intensive data block away from the fastest block based on the cache miss rate.

5.2.3 Hardware exploration

In DPVA-NUCA-2, we use read and write predction queues to determine if a data block is a read or write intensive block. Intuitively, the larger the queues are, the more access-intensive data block can be allocated in the queue. Figure 9 shows the ratio of read and write hits captured by the queues. Though increasing queue entries can help allocate more accesses in the fast cache blocks, it also generates more conflicts. So keep continuously enlarging the prediction queues does not always help benefit system performance eventually. Our simulation results show that when the entry number of the prediction queues is 32, IPC stops growing even the data access ratio in the fast cache blocks continuously increases.

Similarly we explore the optimal size of sorting recorder. Figure 10 shows the sorting recorder hit ratio and IPC degradation when varying sorting recorder entry number. When the sorting recorder has 32 entries, the hit ratio is less than 60% for all the benchmarks. The IPC degradation is significant due to the respectively reading out sorting information and sorting Lat_W. As entry number increases to 64, the sorting recorder can obtain > 99% of hit ratio for most of benchmarks. The only exception is `Swatpions`, because the recent accessed addresses are frequently replaced by new requests due to its disperse access pattern. Further increasing sorting recorder size does not much benefit on IPC improvement.

Figure 11 shows the sorting recorder hit ratio with time. A small sorting recorder demonstrates a random hit ratio pattern as shown in Figure 11(a). A more stable pattern can be observed as the size of sorting recorder increases. Hit ratio is low at beginning of the

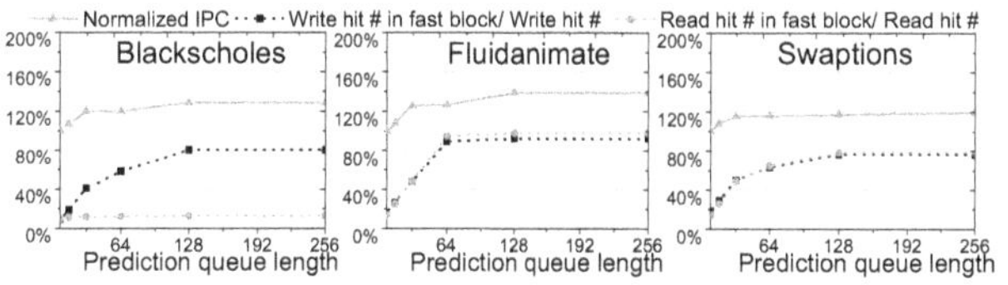

Figure 9: Normalized IPC and write/read ratio in fast cache blocks when varying prediction queue length.

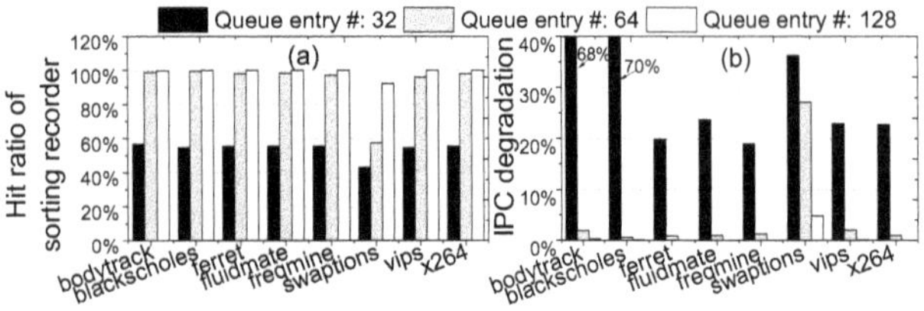

Figure 10: Sorting recorder hit ratio and IPC degradation when varying entry number.

simulation when all the entries are empty. As the incoming request gradually fill the sorting recorder, the hit ratio climbing up to approach 100% except for `Swatpions` who has a decentralized read/write number distribution as shown in Figure 5.

5.2.4 *Performance, energy and hardware*

According to hardware exploration results in Section 5.2.3,we adopt the prediction queue with 32 entries and sorting recorder with 128 entries as the optimal configuration. In Figure 12, the IPC performance of different schemes have been evaluated. All the results are normalized against UCA with consideration of process variation. Because data migration is not introduced in SPVA-NUCA which maximize the effect of the dynamic cell programming latency, so `Bodytrack`, `Blackshoules` and `Swaptions` are more write intensive so that the gaps between SPVA-NUCA and DNUCA as well as SNUCA are enlarged. `Ferret` and `Fluidanimate` have more read intensive data blocks but less write intensive data blocks. Thereby, the their IPCs of SPVA-NUCA don't have too much improvement over DNUCA and SNUCA. The data migration overhead has already been considered in both DPVA-NUCA-1 and DPVA-NUCA-2. The DPVA-NUCA-1 and DPVA-NUCA-2 achieve 21.70% and 29.29% IPC improvement over STT-RAM based DNUCA [5], respectively.

Figure 13 compares the dynamic write energy and the overall energy of three proposed schemes. Since UCA and SNUCA use the worst-case switching time, both dynamic energy and leakage energy of them are obviously higher than the proposed schemes. So we don't show them here. All proposed schemes are compared with DNUCA in terms of energy. Among the three proposed schemes, DPVA-NUCA-1 consumes the highest dynamic write energy because each write hit involve one swap with neighboring cache block. The dynamic write energy of DPVA-NUCA-2 is much lower since it avoids most of the data swaps as DPVA-NUCA-1 does. Compared to DPVA-NUCA-1, DPVA-NUCA-2 saves on average 65% of dynamic write energy and 12.4% of overall STT-RAM cache energy. Since DPVA-NUCA-1 is better than SPVA-NUCA in terms of performance, its leakage is lower than that of SPVA-NUCA. Although dynamic write energy of DPVA-NUCA-1 is much higher than that of SPVA-NUCA, the overall energies of DPVA-NUCA-1 and SPVA-NUCA are very closing on average.

Compared to 8MB ($8 \times 1024 \times 1024 \times 8$ bits) LLC, the hardware overhead including latency indicator for each block (5 bits/block), sorting recorder (maximum 128 entries $\times$ 32 $\times$ 32 bits), read/write prediction queue (2 $\times$ 32 entries $\times$ 40 bits) will be less than 1%.

6. CONCLUSION

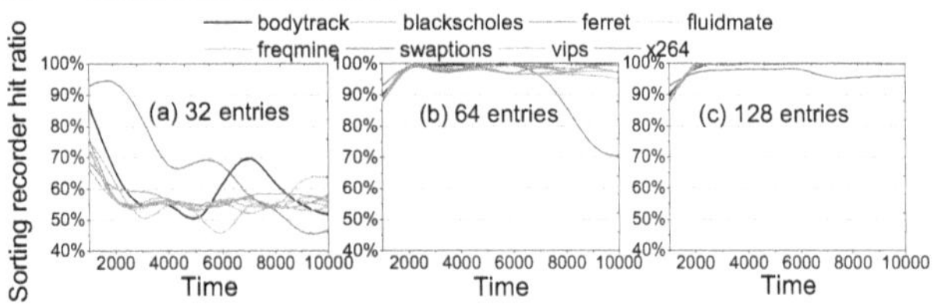

Figure 11: Sorting recorder hit ratio with time of different entry number.

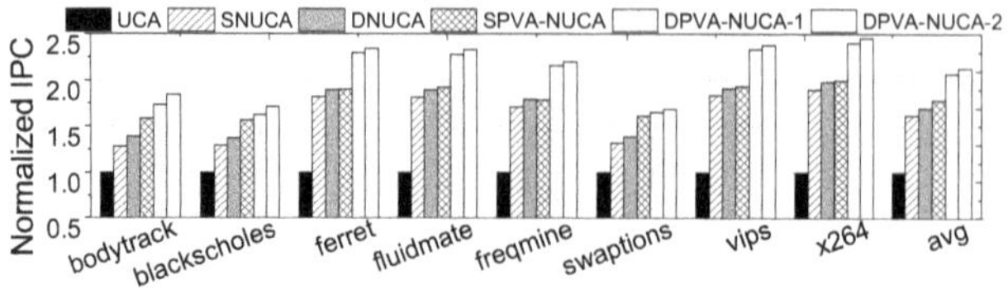

Figure 12: Normalized IPC of different schemes.

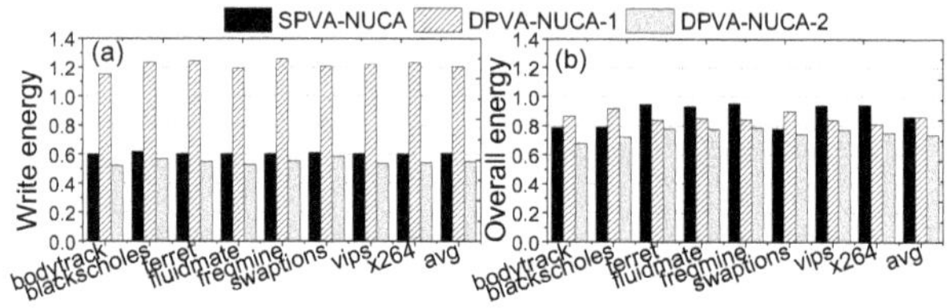

Figure 13: Energy comparison. (a) Dynamic write energy; (b) Overall cache energy.

In this paper, process variation aware non-uniform cache access schemes have been proposed for the retention-relaxed STT-RAM last level cache. By considering non-uniformity of both interconnection latency and cell programming time, all the proposed schemes can achieve better performance and lower energy consumption than previous DNUCA and SNUCA. Among the proposed designs, DPVA-NUCA-2 with conflict reduction techniques can achieve 25.29% IPC improvement and 26.40% overall energy saving when compared to DNUCA.

7. REFERENCES

[1] R. Ubal, J. Sahuquillo, S. Petit, H. Hassan, and P. Lopez, "Leakage current reduction in data caches on embedded systems," in *Proc. of IPC*, 2007, pp. 45–50.

[2] L. Yan, J. Luo, and N. Jha, "Joint dynamic voltage scaling and adaptive body biasing for heterogeneous distributed real-time embedded systems," *IEEE Trans. on CAD*, vol. 24, no. 7, pp. 1030–1041, 2005.

[3] S. Kirolos and Y. Massoud, "Adaptive sram design for dynamic voltage scaling vlsi systems," in *Midwest Symp. on Circuits and Systems*, 2007, pp. 1297–1300.

[4] C. Kim, J. Kim, S. Mukhopadhyay, and K. Roy, "A forward body-biased low-leakage sram cache: device, circuit and architecture considerations," *IEEE Trans. on VLSI Systems*, vol. 13, no. 3, pp. 349–357, 2005.

[5] G. Sun and et al., "A novel architecture of the 3D stacked MRAM L2 cache for CMPs," in *Proc. of 15th HPCA*, 2009, pp. 239–249.

[6] X. Dong and et al., "Circuit and microarchitecture evaluation of 3D stacking magnetic RAM (MRAM) as a universal memory replacement," in *Proc. of DAC*, 2008, pp. 554–559.

[7] P. Zhou, B. Zhao, J. Yang, and Y. Zhang, "Energy reduction for STT-RAM using early write termination," in *Proc. of the ICCAD*, 2009, pp. 264–268.

[8] W. Xu and et al., "Design of Last-Level On-Chip Cache Using Spin-Torque Transfer RAM (STT RAM)," pp. 483–493, 2011.

[9] C. Smullen and et al., "Relaxing Non-Volatility for Fast and Energy-Efficient STT-RAM Caches," 2011, pp. 50–61.

[10] Z. Sun. and et al., "Multi Retention Level STT-RAM Cache Designs with a Dynamic Refresh Scheme," In press.

[11] C. Kim, D. Burger, and S. Keckler, "An adaptive, non-uniform cache structure for wire-delay dominated on-chip caches," in *Proc. of ASPLOS*, 2002, p. 222.

[12] Y. Zhang and et al., "Stt-ram cell design optimization for persistent and non-persistent error rate reduction: A statistical design view," in *Proc. of ICCAD*, 2011, pp. 471–477.

[13] A. Agarwal, D. Blaauw, V. Zolotov, S. Sundareswaran, M. Zhao, K. Gala, and R. Panda, "Path-based statistical timing analysis considering inter-and intra-die correlations," in *Proc. TAU*, 2002, pp. 16–21.

[14] A. Raychowdhury and et al., "Design space and scalability exploration of 1t-1stt mtj memory arrays in the presence of variability and disturbances," in *IEEE Int. Electron Devices Meeting*, Dec. 2009, pp. 1–4.

[15] J. Z. Sun, "Spin-current interaction with a monodomain magnetic body: A model study," *Phys. Rev. B*, vol. 62, pp. 570–578, 2000.

[16] R. Heald and P. Wang, "Variability in sub-100nm sram designs," in *Proc. of ICCAD*, 2004, pp. 347–352.

[17] Z. Sun and et al., "Variation tolerant sensing scheme of spin-transfer torque memory for yield improvement," in *Proc. of the ICCAD*, 2010, pp. 432–437.

[18] CACTI., "http://www.hpl.hp.com/research/cacti/."

[19] "Wind River Simics," *http://www.windriver.com/products/simics/.*

[20] Parsec., "http://parsec.cs.princeton.edu/index.htm."

TapeCache: A High Density, Energy Efficient Cache Based on Domain Wall Memory*

Rangharajan Venkatesan†, Vivek Kozhikkottu†, Charles Augustine§, Arijit Raychowdhury§, Kaushik Roy† and Anand Raghunathan†
† Purdue University, § Intel Corporation
rvenkate@purdue.edu, vkozhikk@purdue.edu, charles.augustine@intel.com, arijit.raychowdhury@intel.com, kaushik@purdue.edu, raghunathan@purdue.edu

ABSTRACT

Domain Wall Memory (DWM) is a recently developed spin-based memory technology in which several bits of data are densely packed into the domains of a ferromagnetic wire. DWM has shown great promise in enabling non-volatile memory with unprecedented density and high energy efficiency. In this work, we propose TapeCache, a first attempt to employ DWMs as last-level caches in general purpose computing platforms. DWMs enable much higher density compared to SRAM, DRAM, and other spin-based memory technologies such as STT-MRAM. However, they also pose unique challenges such as serial access to the bits stored in a DWM cell, leading to variable access latencies. We propose a novel circuit-architecture co-design for TapeCache, consisting of (i) a multi-port DWM macro-cell optimized for read operations considering the asymmetry in applications' read/write characteristics, and (ii) a new cache organization and suitable management policies that mitigate the performance penalty arising from serial access to bits in a macro-cell. Over a wide range of SPEC 2006 benchmarks, TapeCache achieves 7.8X improvement in area, an average energy improvement of 7.3X, and an average performance improvement of 1.2% compared to an iso-capacity SRAM cache. Compared to an iso-capacity STT-MRAM cache, TapeCache obtains 2.3X improvement in area and 1.4X average energy savings with virtually identical performance.

Categories and Subject Descriptors

B.7.1 [**Integrated Circuits**]: Types and Design Styles—*Memory Technologies*; B.3.2 [**Memory Structures**]: Design Styles—*Cache memories*

General Terms

Algorithms, Design, Performance, Theory

Keywords

Domain wall memory, spin memory, cache

1. INTRODUCTION

In modern processors, a majority of the chip transistor count and area is occupied by cache memories. The growing processor-memory gap along with increasingly complex applications and data sets, fuel an ever-increasing demand for larger caches. Caches also account for a significant portion of chip power consumption due to their significant leakage power. Consequently, emerging memory technologies that have very high density and low leakage power are of great interest.

Several new memory technologies – Ferroelectric RAM (FeRAM), Phase Change Memory (PCM), Spin-Transfer Torque Magnetic RAM (STT-MRAM), and Domain Wall Memory (DWM) have been proposed as potential replacements for SRAM and DRAM. Figure 1 compares various key metrics for different memory technologies. DWM, a spin-based non-volatile memory technology, can achieve much higher densities (4X over STT-MRAM) with similar access time and idle power compared to other emerging memory technologies. For these reasons, it is considered to be highly promising and several research efforts have focussed on understanding its device-level characteristics [1,2] and fabricating a functional prototype [3,4]. Considering the potential of DWM, it is critical to understand the design implications and tradeoffs associated with the use of DWMs in future computing platforms.

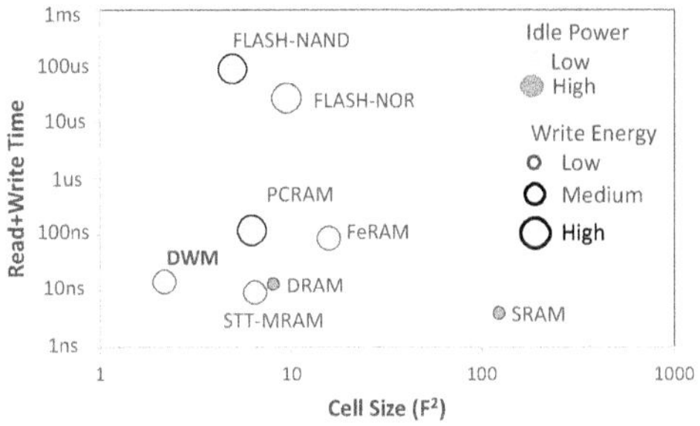

Figure 1: Comparison of different memory technologies (data from [3])

In this work, we present the first attempt to explore the use of DWM as on-chip caches in general purpose processors. Despite possessing a number of favourable features such as very high density, non-volatility, and low leakage, DWM has unique characteristics that pose significantly different challenges from all other memory technologies. From an architectural perspective, a DWM macro-cell looks like a tape, which can store multiple (upto hundreds of) bits. The bits stored in a DWM tape/macro-cell can be shifted in either direction. This enables sharing of read/write ports across the bits stored in a tape, resulting in very high density. However, the time taken to access a bit stored in the tape depends on its location relative to the read/write port, leading to variable access latencies. The overall performance of a DWM cache is determined by the average number of shift operations required per access. Hence, realizing a DWM based cache requires development of suitable circuit/architecture design techniques that exploit its strengths while reducing the performance penalties associated with shift operations. While previous research efforts [5–10] have addressed cache design using other emerging technologies such as STT-MRAM and PCM, DWM poses fundamentally different challenges described above, which must be addressed.

The contributions of this work are as follows:

- We perform the first exploration of the design of on-

*This material is based upon work supported in part by Intel Corporation.

ISLPED'12, July 30–August 1, 2012, Redondo Beach, CA, USA.

chip caches using DWMs. We propose a novel architecture, TapeCache, that consists of (i) a DWM based data array and an SRAM based tag array, and (ii) a bit-interleaved cache organization that reduces the impact of shift latencies.

- Considering the read-write asymmetry and spatial locality of memory accesses found in most applications, we propose a multi-port DWM macro-cell and suitable cache management policies to maximally harness the performance potential of TapeCache.
- We perform an in-depth analysis of the power-performance tradeoffs of TapeCache. Our results show that an iso-capacity replacement of SRAM based cache with TapeCache can result in large benefits in area and energy without any significant penalty in performance. We also demonstrate that in the context of last-level caches, DWM outperforms STT-MRAM, which is one of the most promising emerging memory technologies that has been previously explored.

The rest of this paper is organized as follows. Section 2 outlines the fundamental concepts of DWM. Section 3 presents the proposed multi-port DWM macro-cell design and discusses the corresponding the architectural implications. Section 4 describes the proposed TapeCache architecture. Section 5 presents our methodology for modeling and evaluating TapeCache. Section 6 presents experimental results. In Section 7, we present a brief survey of the related work and Section 8 concludes the paper.

2. BACKGROUND

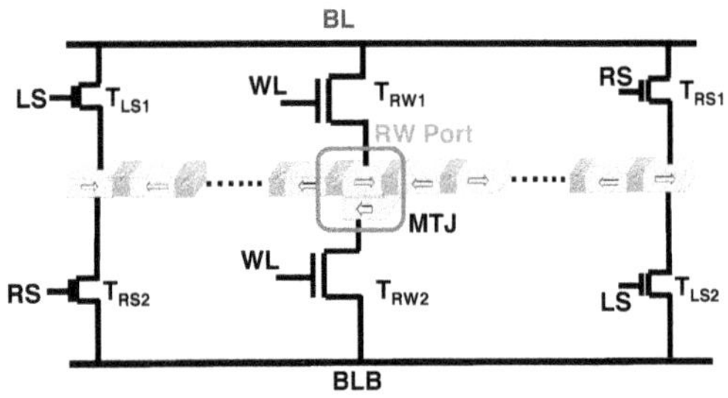

Figure 2: DWM Macro-cell Schematic

Figure 2 shows the schematic of a DWM macro-cell consisting of a ferromagnetic wire, a magnetic tunneling junction (MTJ) and access transistors. The ferromagnetic wire can have multiple domains separated by domain walls. Each domain can be separately programmed to a certain magnetization direction, and can therefore store a single bit. Hence, a DWM macro-cell is capable of storing multiple bits of data. The read and write operations are performed using an MTJ by turning on the corresponding access transistors (T_{RW1} and T_{RW2}) and precharging the bitlines to appropriate voltages. In order to shift the bits right, transistors T_{RS1} and T_{RS2} are turned ON and bitlines BL is connected to V_{DD} and BLB to GND. Left shift operation is performed by turning on transistors T_{LS1} and T_{LS2}. This 'to and fro' movement of bits enables sharing of the read and write ports across all bits in a macro-cell, achieving very high density.

Logically, a DWM can be considered to be a tape that represents the ferromagnetic wire, along with a tape head representing a read/write port. Accessing a bit from a DWM tape involves shifting the tape head to the required location and then performing the required read/write operation. In order to avoid losing data during shift operations, the number of domains in the ferromagnetic wire needs to be twice the number of bits stored in the macro-cell. Note that this does not incur any area penalty as the macro-cell area is determined by the area occupied the access transistors.

3. MULTI-PORT DWM MACRO-CELL

In this section, we present a multi-port DWM macro-cell design enhanced with additional read-only ports that significantly reduces read access latencies, while retaining most of the density benefits associated with DWMs. It also helps expand the design space by introducing additional design parameters that can be utilized to perform more fine-grained energy-performance tradeoffs.

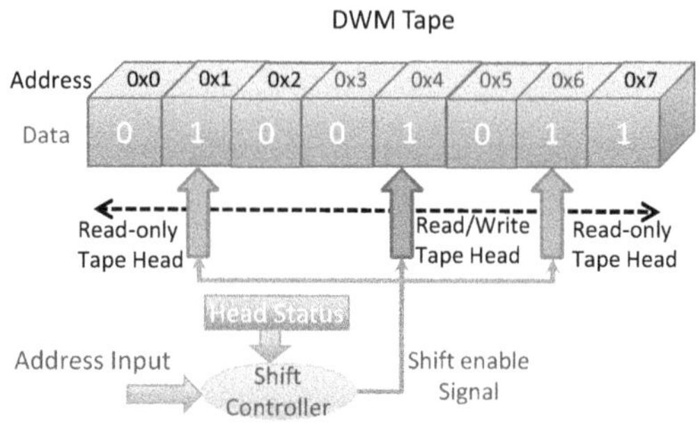

Figure 3: Logical view of a multi-port DWM macro-cell

The logical view of a multi-port DWM macro-cell is shown in Figure 3. It has multiple read-only tape heads distributed across the DWM tape. In order access a bit from the tape, the shift controller determines the appropriate tape head and computes the number of shift operations required by comparing the address bits with the current location of tape heads, referred to as the head status. It is important to note that there can be no relative movement between tape heads in a multi-port DWM macro-cell as it is the bits stored in the tape that physically shift and not the tape heads. The multi-port configuration helps reduce the average number of shift operations per read access, thereby reducing the average read access time. The motivation for having additional read-only ports is two-fold: (i) from an architectural perspective, reads are more performance critical than writes and reads also outnumber writes (across a wide range of SPEC benchmarks store operations account for less than 25% of the total number of memory instructions). (2) Read-only ports can be realized using minimum sized transistors, preserving the density benefits of DWM. For instance, adding a read-only port to a single-port macro-cell achieves 2X reduction in worst case read access latency with only 13% increase in area. On the other hand, halving the number of bits per tape achieves the same reduction in worst case read access latency with a 2X increase in area.

The proposed multi-port macro-cell design expands the design space and offers the following design parameters:

1. Number of bits per tape: Varying the number of bits per tape offers a tradeoff between density *vs.* worst case access latency for both reads and writes. Note that this tradeoff can be performed only at a coarse granularity as the number of bits per tape can be varied only in powers of 2 (to keep the addressing scheme simple).

2. Number of read-only tape heads: Varying the number of read-only tape heads alters only the read access latency (in contrast to bits per tape, which impacts read and write latencies), albeit in a more cost-effective manner. Furthermore, the number of read-only tape heads can be tuned in a more fine-grained manner as the number of read-only ports need not be in powers of 2.

3. Head Selection: The multi-port DWM macro-cell provides the flexibility of selecting the best tape head for a given read operation, depending on the relative location of the tape heads and the bit to be accessed. Architectural policies for head selection are proposed in Section 4.3.

In the next section, we discuss the architecture of a cache designed using the proposed multi-port DWM macro-cell.

4. DWM CACHE ARCHITECTURE

Cache design with DWM differs from traditional cache design in the following respects:

- The basic memory element in a DWM based cache is a macro-cell that is capable of storing multiple bits. As a result, the bits stored in a macro-cell can be organized as belonging to either a single cache block or multiple cache blocks.
- Sharing of read/write ports across multiple bits in a macro-cell introduces the need for shifting bits before performing read/write operations. The addressing logic should not only select the macro-cells but also determine the number of shift operations required to access the required cache block.
- The additional latency, caused by the varying number of shift operations for different bits stored in a macro-cell, necessitates the use of suitable cache management policies to reduce the performance penalty.

In the rest of this section, we provide an overview of the TapeCache architecture and then describe its key features.

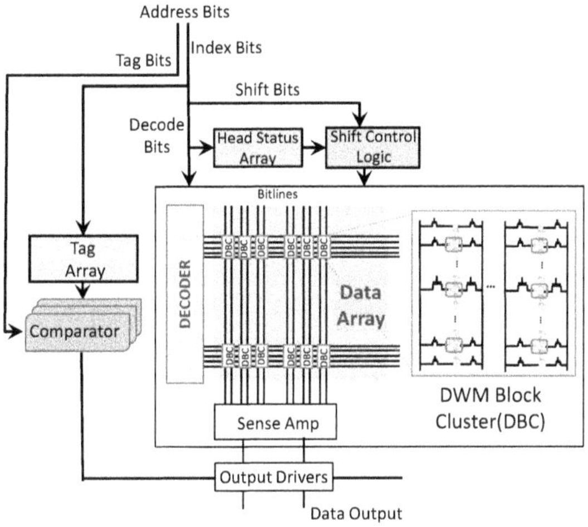

Figure 4: TapeCache organization

Figure 4 demonstrates the overall organization of the proposed TapeCache architecture. The cache consists of a DWM data array, an SRAM tag array and an additional head status array, along with control logic that is used to manage the shift operations required to access data in each DWM macro-cell. The data array is composed of DWM Block Clusters (DBCs), which are groups of DWM macro-cells capable of storing multiple cache blocks in a bit-interleaved fashion as explained in Section 4.2. In order to access a cache block, we need to select the correct DBC and determine the position of the block within that DBC. The traditional index bits are hence subdivided into decode bits and shift bits as shown in Figure 4. The decode bits are used to select the correct DBC and the shift bits are used along with the head status of the DBC to determine the number of shift operations required for accessing the cache block. In the following subsections, we present a detailed description of the cache organization and management policies used in TapeCache.

4.1 Hybrid (DWM+SRAM) Arrays

Let us consider a simple cache organization in which both the data array and the tag array are designed using DWM. This design would result in maximum benefits in terms of both area and leakage power due to the high density and non-volatile nature of DWM. However, the above configuration would require two variable latency operations per access, one for determining the block status from the tag array and the other for fetching the block from the data array [1]. This would considerably degrade the performance of the cache. Further, the area and energy benefits attainable from a DWM based tag array are relatively small, as the tag array represents a small fraction of the total area and power consumption of a cache. For instance, in a 1MB cache, the tag array contributes only 4.8% to the total cache area.

Hence, we propose a hybrid cache organization in which we realize the data array using DWM and the tag array using SRAM. This configuration enables us to leverage the benefits in area and power offered by DWM, while retaining the efficiency of SRAM caches in determining cache hits and misses.

4.2 Data Organization

Let us consider how the multiple bits in a cache block can be mapped to the bits in a DWM tape (macro-cell). One possible scheme involves mapping all the bits in a cache block to the same DWM tape (if cache block is larger than a tape, then it can be stored in multiple tapes). In this scenario, a cache access would involve N serial read/write operations, where N is the number of bits stored in a DWM tape. This mapping would incur a very high access latency compared to a traditional SRAM based cache.

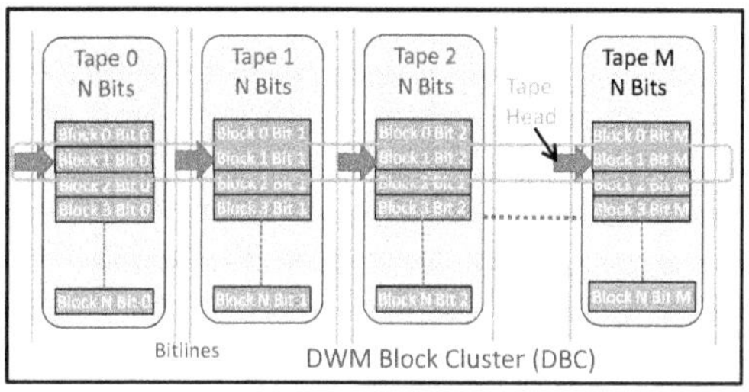

Figure 5: Bit-interleaved data array organization

In order to reduce the high latency overhead, we propose a bit-interleaved data array organization in which each cache block is spread across several DWM macro-cells. Figure 5 demonstrates how N 64-bit blocks are stored in 64 DWM tapes each containing N bits of data. The ith bit of each block is placed in tape number i. Within a tape, the jth bit location stores a bit that belongs to block j. In this scheme, all the tape heads move in a lock step fashion such that at a given time instance, all bits of some block are aligned to their respective tape's heads. Hence, in order to read a cache block, we perform the required number of shift operations and read all the bits belonging to the block in parallel. In this scenario, the number of shift operations required to access a block could vary from 0 to $N-1$, in contrast to the constant access latency of N observed in the previous scheme. The average access latency can be kept small in practice through the use of suitable head management policies, as discussed in the next sub-section. The overhead introduced by this scheme is the need for additional decoding logic to determine the number of shifts required to access a given block. However, our experimental evaluations indicate that this overhead is negligible.

4.3 Head Management Policies

As described earlier, while accessing a cache block stored in a DBC, we need to perform an appropriate number of shift operations. The number of shift operations depends on the location of the block to be accessed relative to the position of the tape heads. Therefore, one needs an effective head management policy in order to reduce the performance overhead due to shift operations. In a multi-port DWM

[1]In lower level caches, the tag and data array access are serialized so as to avoid energy overheads associated with reading all ways of a cache simultaneously.

cache design, head management policies involve (i) selection of the appropriate tape head for accessing the required data (tape head selection), and (ii) positioning of the tape head after the cache access (tape head update).

Tape head selection: In this work, we consider two different tape head selection policies- *Static* and *Dynamic*. In the *Static* tape head selection policy, each cache block is assigned a tape head statically depending on its initial location within the DBC. This policy has the advantage that the tape head can be determined solely based on the address of the cache block. On the other hand, in the *Dynamic* policy, we select the tape head that is nearest to the required cache block depending on the current DBC head status. For this purpose, we use the head status array to store the locations of tape heads and activate the appropriate tape head to access the data.

Tape head update: We explore two different tape head update policies: *Eager* and *Lazy*. In the *Eager* policy, the tape heads are restored to their original default locations after each access. This policy simplifies the tape head selection, since we can assign the nearest tape head for every block statically. Also, this results in simplified shift control logic as the number of shift operations required to access a given cache block can be determined from the block address alone, and the head status does not need to be stored. In the *Lazy* policy, we do not restore the tape head to its default initial position after performing a read/write operation. Instead, we have status bits for each tape head to keep track of its location. When we perform a read/write operation, we calculate the difference between the block location within the tape and the current location of the tape head and then perform the required number of shifts. The *Lazy* policy is motivated by the spatial locality of memory accesses, which implies that the current tape head location tends to be closer to the next block to be accessed.

The tape head selection and tape head update policies described above result in three distinct head management policies: *Static-Eager* (SE), *Dynamic-Lazy* (DL) and *Static-Lazy* (SL). Note that *Dynamic-Eager* would be same as *Static-Eager* as the best port can be assigned statically for each location. Figure 6 illustrates these three cache management

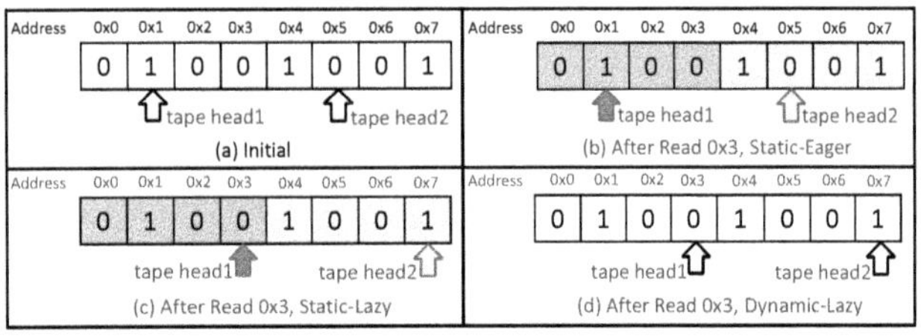

Figure 6: Comparison between different cache management policies

policies using a DWM tape with 2 tape heads. Figure 6(a) shows the initial status. Figures 6(b-d) show the status of the DWM tape after performing a read operation at address 0x3. In Figures 6(b,c), the bits and tape head are shaded to indicate the static head assignment. Let us now consider the tape head selection and number of shift operations required for the next access to address 0x4. The SE policy would use head2 and would require 1 left shift operation. The SL policy will also use head2 but requires 3 left shift operations. The DL policy would choose head1 and require 1 right shift operation. This shows that in the SL policy, a series of accesses to consecutive blocks would activate the worst case access latencies. On the other hand, the DL policy exploits spatial locality, thereby reducing the average case shift latency overhead. However, this policy can skew the position of DWM tape head, thereby increasing the worst case latency for subsequent accesses. This is overcome by restoring the DWM tape configuration during cache idle periods when it gets skewed beyond a certain threshold. For instance, access to address 0x0 after 0x4 would incur a large access latency with the DL policy. However, performing a restore operation after access to address 0x4 would shift tape head2 to address 0x4. This would enable the DL policy to exploit the spatial locality while eliminating higher access latencies due to skewed tape head position.

Note that head management policies are orthogonal to traditional cache management policies. For example, block replacement strategies such as LRU, FIFO, *etc.*, can be used unchanged in TapeCache.

5. EXPERIMENTAL METHODOLOGY

In this section, we present a brief description of the modeling framework and then present the experimental setup used to evaluate TapeCache.

5.1 Modeling Framework

TapeCache differs significantly from traditional memories in terms of both the device structure as well as cache architecture. In order to accurately evaluate the characteristics of the proposed cache design, we have developed a tool, DWM-CACTI, that is based on the CACTI framework [11]. First, we used a self-consistent device simulation framework [2] to accurately capture the domain wall motion using spin-torque and performed device level simulations of the DWM macro-cell. The macro-cell parameters thus obtained were then used as technology parameters in DWM-CACTI to model the characteristics of TapeCache. The DWM-CACTI tool takes the number of bits per DWM tape, the number of read-/write ports, and the number of read-only ports, along with the usual inputs to CACTI, to compute the area, energy and access latencies of TapeCache. DWM-CACTI takes into account the overheads due to the head-status array, additional wordlines, and shift control logic while modeling TapeCache.

5.2 Experimental Setup

In our experiments, we use TapeCache as an iso-capacity L2 cache replacement and compare the area, energy and performance of the proposed design with that of CMOS SRAM and STT-MRAM. All memory technologies considered are based on a 32nm technology node. The processor configuration used in our analysis is provided in Table 1. We evaluate SRAM memories using CACTI [11], STT-MRAM using a modified CACTI tool [12], and TapeCache using our DWM-CACTI tool. We perform architectural simulations over a wide range of benchmarks from the SPEC 2006 suite using SimpleScalar [13] for 1 billion instructions. In all our simulation runs, we warm up the cache by fast forwarding for 1 billion instructions.

Table 1: System Configuration

Processor Core	Alpha 21264 pipeline, Issue Width - 4
Processor Frequency	3 GHz
Functional Units	Integer - 8 ALUs, 4 Multipliers Floating Point - 2 ALUs, 2 Multipliers
L1 D/I-Cache	32KB, direct mapped, 32 byte line size, 1 cycle hit latency
L2 Unified Cache	1MB, 4-way associative, 64 byte line size, hit latency depends on technology

6. EXPERIMENTAL RESULTS

In this section, we present results comparing the benefits of TapeCache with iso-capacity SRAM and STT-MRAM based caches. Note that DWM based caches can also be used in an iso-area scenario wherein the density benefits can be translated into increased cache sizes for achieving higher performance. However, in this work, we focus on improving the energy and area efficiency of last level caches.

Figure 7 summarizes the average benefits of TapeCache compared to iso-capacity SRAM and STT-MRAM caches across a set of 13 SPEC 2006 benchmarks. We consider two different configurations of DWM: DWM32-DL, and DWM16-SE. The DWM32-DL configuration is made up of 32 bit DWM tapes and uses the *Dynamic Lazy* management policy. DWM16-SE has 16 bit DWM tapes with the *Static Eager* policy. Both these configurations have 4 read-only ports and 1 read/write port. These configurations were chosen so as to enable iso-performance (DWM32-DL) and iso-energy (DWM16-SE) comparisons with STT-MRAM. Compared to STT-MRAM, the DWM32-DL configuration achieves 2.3X improvement in area and 1.4X average energy savings without any significant performance penalty. The DWM16-SE configuration achieves 1.5X and 1.6% improvement in area and performance (IPC) respectively, with similar energy, compared to STT-MRAM. Both TapeCache configurations greatly outperform an iso-capacity SRAM cache in all three aspects. The area, energy and performance improvements of DWM32-DL over SRAM are 7.7X, 7.3X, and 1.2% respectively. DWM16-SE achieves 5.5X improvement in area, 5X in energy, and 3.1% in performance compared to SRAM.

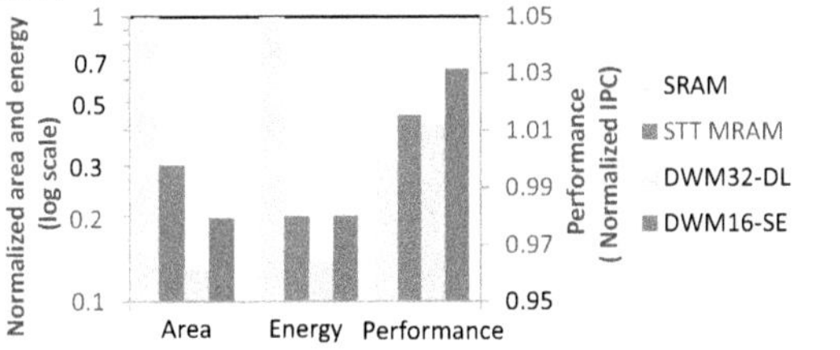

Figure 7: Comparison of area, energy and performance across different memory technologies

Energy consumption of TapeCache: Figure 8 compares the total energy consumed by different Tapecache configurations with SRAM and STT-MRAM caches. Functionally, the different sources of energy consumption are leakage, read and write operations. Due to its non-volatile nature, TapeCache achieves drastic reductions in leakage energy with respect to SRAM. The read energy in DWM is also found to be less than SRAM as (i) the bitlines in DWM are precharged to a value much smaller than V_{DD} and (ii) the bitline capacitance is reduced due to increased density. However, the write energy in a DWM cache is higher due to the high write current requirement similar to STT-MRAM. Among the three components, the major contributor to the total cache energy is leakage. Also, in an L2 cache, reads outnumber writes. Therefore, the reduction in leakage and read energy outweighs the increase in write energy. Overall, TapeCache achieves considerable reduction in the total cache energy consumption compared to the SRAM cache.

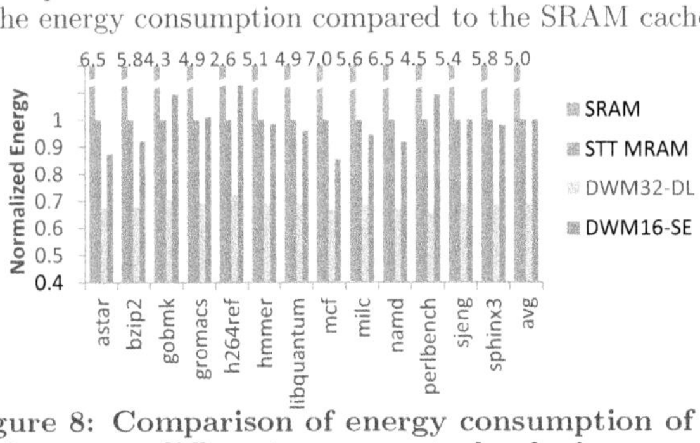

Figure 8: Comparison of energy consumption of L2 cache across different memory technologies

When comparing TapeCache to STT-MRAM, the leakage energy is mainly from the tag array and the decoding circuitry. TapeCache requires smaller decoding circuitry as the number of DBCs is considerably smaller than the number of bit-cells in STT-MRAM, thereby reducing the overall leakage energy. The read/write energies are mostly determined by the charging/discharging the bitlines. In the case of TapeCache, the increased density results in reduced bitline capacitance. However, it requires an additional charging/discharging operation for performing shifts. For the DWM16-SE configuration, the overall read/write energy is higher than that of STT-MRAM due to the shift energy overhead. However, for the DWM32-DL configuration, further reduction in bitline capacitance leads to an overall reduction in read/write energy. As we can see from Figure 8, the benefits of TapeCache far outweigh its overheads, leading to 7.3X and 1.4X energy benefits for DWM32-DL over SRAM and STT-MRAM, respectively.

TapeCache performance evaluation: One of the main challenges in designing caches using DWM is the variable access latency due to shift operations. The performance of TapeCache depends on (i) the access latency to different bits in a DWM tape, and (ii) the access pattern, which determines the number of shift operations required. The shorter bit lines resulting from higher density result in reduced access latency to the bits stored at the tape head compared to SRAM and STT-MRAM. Also, the introduction of multiple ports reduces the number of shifts required to access the bits in a DWM tape, thereby reducing the average access latency. For instance, a 1MB TapeCache in DWM16-SE configuration has a best-case (no shifts required) read latency of 0.92ns and requires a maximum of 2 shift operations to read any block, resulting in the worst-case read latency of 1.58ns. Due to smaller bitline delays, even the worst-case

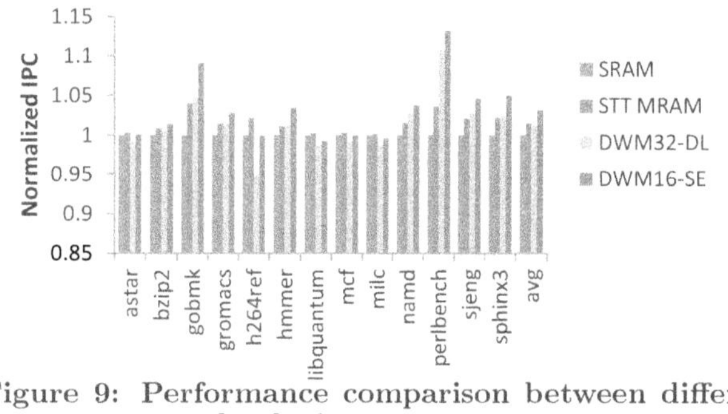

Figure 9: Performance comparison between different memory technologies

read latency for this configuration is less than the read latency of an iso-capacity SRAM cache (1.85ns). As a result, the DWM16-SE configuration achieves 3.1% improvement in performance as shown in Figure 9.

When we consider a denser TapeCache configuration, DWM32-DL, the worst case latency is found to be much higher than an SRAM cache. However, the average number of shift operations is reduced by using the DL policy. Note that in Figure 9, the DWM32-DL case results in performance improvement for benchmarks having high spatial locality between cache accesses (gobmk, namd, perlbench, sjeng, sphinx3). On the other hand, for benchmarks that exhibit low degrees of locality (h264ref, mcf, milc), there is a reduction in performance due to increased shift penalty.

Design space exploration: TapeCache offers a unique parameter –number of read-only ports –to tradeoff energy with performance. We consider a TapeCache configuration comprising 32 bits per DWM tape and 1 shared read/write port, and vary the number of read-only ports and management policy to obtain an energy vs performance Pareto-optimal frontier. As shown in Figure 10, increasing the number of read-only ports reduces the average read latency while increasing the energy consumption of TapeCache. It is interesting to note the relationship between the average number of bits per read port and the best tape head management policy. For configurations with smaller numbers of bits per tape head, the SE policy which reduces the worst-case access latency performs better than the DL policy.

On the other hand, for configurations with higher numbers of bits per tape head (fewer read ports) the DL policy outperforms SE since it exploits the spatial locality in access patterns. The figure also shows the corresponding iso-capacity design points for SRAM and STT-MRAM caches, which are clearly sub-optimal to TapeCache.

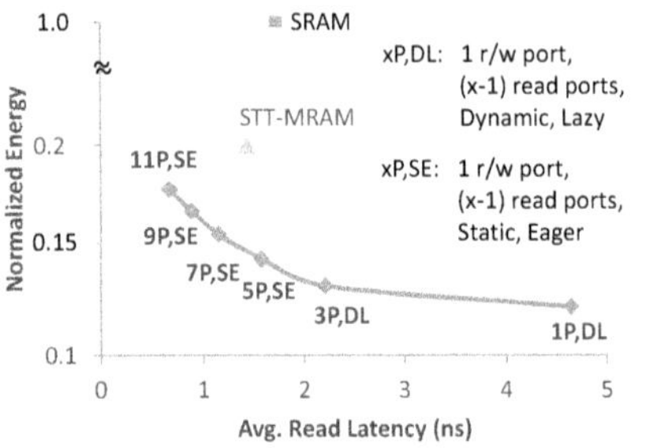

Figure 10: Design space exploration for TapeCache

Table 2: Cache characteristics of different memory technologies at 32nm node

	SRAM	STT MRAM	DWM 16-SE	DWM 32-DL
Area (mm^2)	2.5	0.75	0.49	0.32
Leakage power (mW)	50.45	5.56	3.81	3.53
Read Energy (nJ)	0.13	0.06	0.082	0.044
Write Energy (nJ)	0.05	0.13	0.14	0.09
Read Latency (Best) (ns)	1.85	1.42	0.92	0.89
Read Latency (Worst) (ns)	1.85	1.42	1.58	2.89
Write Latency (Best) (ns)	0.88	2.76	2.8	2.63
Write Latency (Worst) (ns)	0.88	2.76	3.45	3.62

TapeCache Characteristics: We evaluate TapeCache using DWM-CACTI and compare its characteristics with those of SRAM and STT-MRAM based caches. As shown in Table 2, TapeCache shows improvements in area, read energy, leakage power and best case read access latency compared to the SRAM cache. When we compare DWM and STT-MRAM caches, the DWM cache performs better in all categories except for the worst case read/write latency. Our multi-port macro-cell design along with the proposed management policies mitigate the performance penalty from worst case access latencies, resulting in favourable application performance as discussed earlier in this section.

System-level Comparison: We evaluated the effect of TapeCache on the area and energy consumption of the overall system (processor + core + main memory), using the architecture level power analysis tool, Wattch, and CACTI for cache/memory models. We found that TapeCache in the DWM32-DL configuration achieves 7 to 21% energy savings (average of 15.3%) compared to a baseline system with SRAM cache. Note that TapeCache also simultaneously reduces area (processor+cache) by 27%.

7. RELATED WORK

Recent years have seen a considerable increase in the number of research efforts that have investigated the use of emerging memory technologies like STT-MRAM, PCM, *etc.* as potential lower-level cache and main memory replacements [5,8–10,14]. While STT-MRAM and PCM have desirable properties like high-density and non-volatility, they have high write energy, high write latency and limited endurance that needs to be addressed. Earlier research efforts have proposed architectural modifications like hybrid cache design [14],write biasing [15], wear-leveling [9], adaptive line replacement [16] and set-remapping [17] to address the key issues with these memories. While DWM shares some characteristics of STT-MRAM such as low leakage, high write energy *etc.*, it also possesses other unique characteristics. For example, DWM offers an interesting trade-off between density and variable access latency as the number of bits per DWM macro-cell is varied. Our work explores suitable circuit/architecture co-design techniques to address these unique characteristics.

Several research efforts have addressed the issues related to DWM device fabrication and modeling [1–3]. Recently, an application of DWM as FIFOs in a systolic array architecture was demonstrated in [18]. The use of DWM as in reconfigurable logic was proposed in [19]. The tape-like characteristics of DWM make it an ideal match for these application scenarios where random access is not required. Our work is the first to address the challenge of using DWM as random access memory for general purpose computing.

8. CONCLUSION

Domain Wall Memory is an emerging spin-based memory technology that has a much higher density and good energy efficiency compared to current memory technologies (SRAM, DRAM), as well as other candidates for future memories such as STT-MRAM, PCM, *etc.* We explored the use of DWM for the last-level on-chip cache in computing platforms. The proposed cache organization and management policies reduce the performance penalty due to shift latency of DWMs. We performed architectural simulations to evaluate the benefits of a DWM-based cache. Our results demonstrate that DWM based caches offer great potential in improving the energy-performance profile of a wide range of applications, while significantly reducing cache area.

9. REFERENCES

[1] L. Thomas, R. Moriya, C. Rettner, and S. Parkin. Dynamics of Magnetic Domain Walls Under Their Own Inertia. *Science*, 330(6012):1810–1813, December 2010.

[2] C. Augustine et al. Numerical analysis of domain wall propagation for dense memory arrays. In *Proc. IEDM*, pages 17.6.1 –17.6.4, December 2011.

[3] S. Parkin, M. Hayashi, and L. Thomas. Magnetic domain-wall racetrack memory. *Science*, 320(5873):190–194, April 2008.

[4] E. R. Lewis et al. Fast domain wall motion in magnetic comb structures. *Nature*, 9(12):980–983, December 2010.

[5] P. Zhou, B. Zhao, J. Yang, and Y. Zhang. Energy reduction for STT-RAM using early write termination. In *Proc. ICCAD*, pages 264 –268, November 2009.

[6] R. Desikan et al. On-chip MRAM as a High-Bandwidth, Low-Latency Replacement for DRAM Physical Memories. Technical report, In IBM Austin CASC, 2002.

[7] Y. Xie. Modeling, architecture, and applications for emerging memory technologies. *IEEE Design and Test of Computers*, 28(1):44 –51, Jan-Feb 2011.

[8] K. Lee and S.H. Kang. Development of Embedded STT-MRAM for Mobile System-on-Chips. *Magnetics, IEEE Trans. on*, 47(1):131 –136, January 2011.

[9] M. K. Qureshi, V. Srinivasan, and J. A. Rivers. Scalable high performance main memory system using phase-change memory technology. In *Proc. ISCA*, pages 24–33, June 2009.

[10] G. Dhiman, R. Ayoub, and T. Rosing. PDRAM: A hybrid PRAM and DRAM main memory system. In *Proc. DAC*, pages 664 –669, July 2009.

[11] CACTI. http://www.hpl.hp.com/research/cacti/.

[12] X. Dong et al. Circuit and microarchitecture evaluation of 3D stacking magnetic RAM (MRAM) as a universal memory replacement. In *Proc. DAC*, pages 554 –559, June 2008.

[13] T. Austin, E. Larson, and D. Ernst. Simplescalar: An infrastructure for computer system modeling. *Computer*, 35:59–67, February 2002.

[14] X. Wu et al. Design exploration of hybrid caches with disparate memory technologies. *ACM Trans. Archit. Code Optim.*, 7(3):15:1–15:34, December 2010.

[15] M. Rasquinha et al. An energy efficient cache design using Spin Torque Transfer (STT) RAM. In *Proc. ISLPED*, pages 389 –394, August 2010.

[16] A. Jadidi, M. Arjomand, and H. S. Azad. High-endurance and performance-efficient design of hybrid cache architectures through adaptive line replacement. In *Proc. ISLPED*, pages 79–84, August 2011.

[17] Y. Chen, W. F. Wong, H. Li, and C. K. Koh. Processor caches with multi-level spin-transfer torque RAM cells. In *Proc. ISLPED*, pages 73–78, August 2011.

[18] R. Venkatesan et al. Energy efficient many-core processor for recognition and mining using spin-based memory. In *Proc. NANOARCH*, pages 122 –128, June 2011.

[19] W. Zhao, D. Ravelosona, J. Klein, and C. Chappert. Domain wall shift register-based reconfigurable logic. *Magnetics, IEEE Trans.*, 47(10):2966 –2969, October 2011.

A Software Approach for Combating Asymmetries of Non-Volatile Memories

Yong Li
yol26@pitt.edu

Yiran Chen
yic52@pitt.edu

Alex K. Jones
akjones@pitt.edu

Department of ECE, University of Pittsburgh
Pittsburgh, PA, 15261

ABSTRACT

The recent advances in non-volatile memory technologies promise the delivery of future high performance and low power computing systems. While these technologies provide attractive features, they exhibit different degrees of asymmetric read/write behavior, resulting in under utilization of the technology benefits in many scenarios. Future systems should be aware of this asymmetry to fully utilize these memory technologies. In this paper, we propose *software dispatch*, a cross-layer approach to distribute data to appropriate memory resources based on an application's data access characteristics. We demonstrate the application of the proposed technique through a case study system with hybrid memory caches, on which we achieve over 5% performance improvement and 9.8% power savings compared to the state-of-the-art technique.

Categories and Subject Descriptors

B.3 [**Hardware**]: Memory Structures; D.3.4 [**Programming Languages**]: Processors—*Compilers, Optimization*

General Terms

Design, Experimentation, Performance

Keywords

Hybrid Memories, Compilers, Dispatch

1. INTRODUCTION

Non-volatile memory (NVM) technologies are recently undergoing fast development and several of them have already been successfully applied to many consumer electronic products. For example, NAND-flash (Figure 1) based solid-state drives (SSDs) have been widely adopted in embedded and mobile devices due to their power efficiency and high shock/vibration endurance. Figure 2 displays magnetic random access memory (MRAM) of which the spin-torque transfer (STT) variety is receiving considerable attention [4]. STT-RAM offers various attractive features including lower leakage, fast read access and high density. It has been demonstrated that incorporating STT-RAM into mobile applications can dramatically reduce standby power and replacing DRAM with STT-RAM in data centers can reduce power by up to 75% [3]. Other examples of future alternative memories include resistive RAM (RRAM) [10] and phase change RAM (PCRAM) [11], as illustrated in Figure 3 and Figure 4.

Despite the attractive features of the aforementioned memory technologies, several challenges prevent them from immediately and entirely replacing conventional memory/storage such as SRAM, DRAM and HDDs (hard disk drives). One major issue is the read/write asymmetries of these NVMs in terms of power consumption, access latency and endurance, as summarized in Table 1. For example, NAND flash features significant asymmetry in its read versus write operation. While it provides flexible and high performance read operation, its write operation suffers from the limitation of erase-before-write [5], which makes it undesirable for applications with high temporal write reuse patterns. PCRAM, MRAM and RRAM also exhibit different degrees of asymmetric read/write characteristics. Generally, expensive write operations of these memories cause performance and power bottlenecks for write intensive applications.

Table 1: Memories with Asymmetric R/W Characteristics

Memory Type	NAND Flash	MRAM	PCRAM	RRAM
Cell Structure	1T	1T1MTJ	1T1R	1T1R
Cell Area	$4 \sim 6F^2$	$6 \sim 20F^2$	$5 \sim 12F^2$	$4 \sim 10F^2$
Read Latency	$20 \sim 60ns$	$2 \sim 10ns$	$10 \sim 40ns$	$5 \sim 10ns$
Write Latency	$100 \sim 1000\mu s$	$10 \sim 20ns$	$50 \sim 120ns$	$10 \sim 20ns$
Read Energy	low	low	low	low
Write Energy	high	high	high	low
Write Endurance	10^5	10^{16}	10^8	10^8

Prior research efforts have been made to mitigate the overhead incurred by asymmetrically long write penalties for specific memory technologies. In particular, Lee *et al.* proposed in-page logging (IPL) [6] to alleviate the erase-before-write limitation of NAND flash. IPL reduces the overhead of erase operations by introducing a dedicated log space to record the data changes. Once the log space is full, a merge operation is triggered to write data pages to a clean erase block with all of the updates applied. For frequently updated data however, this approach leads to intensive merge operations, causing significant overhead. Other research attempts [16, 9] have been made to mitigate the asymmetry problem by adding conventional memories (CMs) (e.g., SRAM and DRAM) to different levels of the memory hierarchy constructed by NVMs such as MRAM and PCRAM. In these proposed memory systems, data is distributed by hard-

ISLPED'12, July 30–August 1, 2012, Redondo Beach, CA, USA.

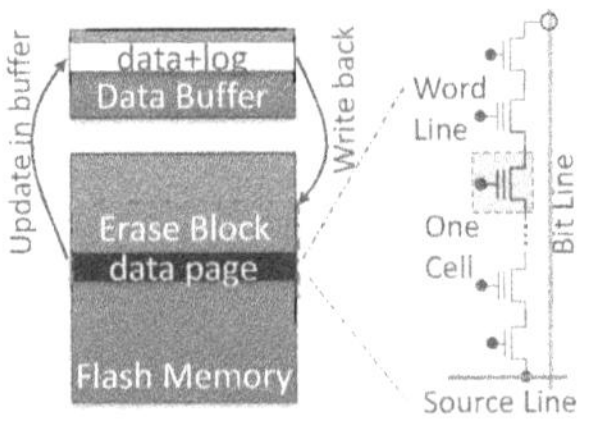

Figure 1: NAND flash

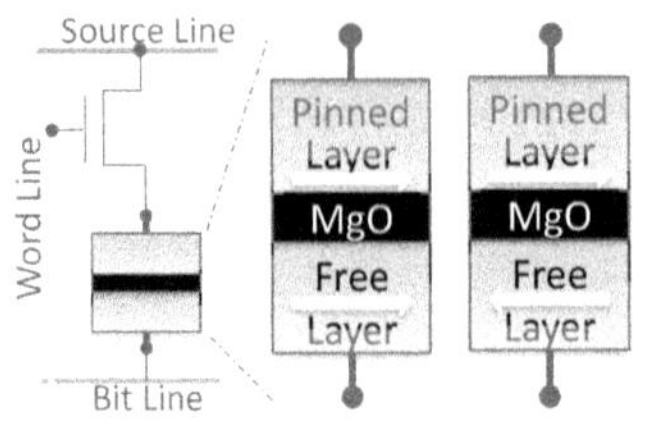

Figure 2: MRAM

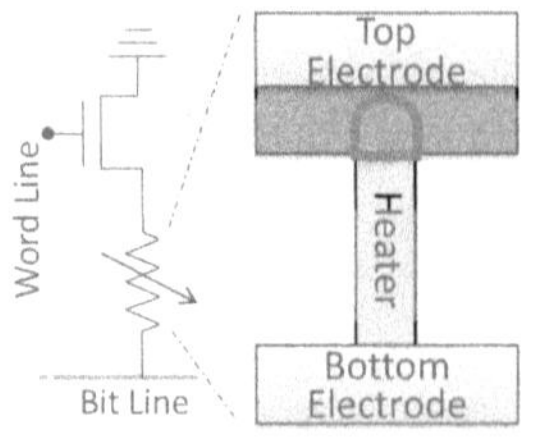

Figure 3: PCRAM

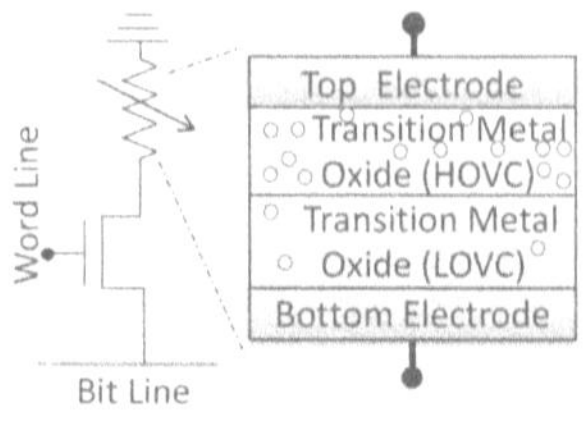

Figure 4: RRAM

ware between NVM and CM to achieve a balanced read/write performance. However, these approaches are typically designed for specific memory levels (e.g., cache and main memory) and do not effectively leverage applications' data access characteristics.

We study the read/write asymmetry as a generic problem of various NVMs used in different levels of the memory hierarchy including scratchpad memories, caches and main memory. In this paper, we propose *SPD* (*software dispatch*), a cross-layer solution including the compiler and OS (operating system) that captures applications' data access characteristics to guide the hardware and distribute data among hybrid memories[1] for optimized performance and power consumption. We evaluate our approach on a case study hybrid memory system and demonstrate that software dispatch ensures that 88.5% of writes are handled in CM, which only comprises 3% of the total cache capacity. Through software dispatch we achieve over 5% performance improvement and nearly 10% power saving compared to the state-of-the-art runtime technique.

2. UTILIZING HYBRID MEMORIES

The availability of various memory technologies brings the potential for building a rich set of hybrid memory systems that trade off among performance, power and density. Due to the relatively high penalty of write accesses, systems utilizing NVMs typically need to incorporate certain amount of CM for data that is write intensive. The methodology for including CM to form hybrid memory systems could be different, depending on scenarios.

One typical organization of hybrid memories is using CM as a buffer for NVM. For example, the entire secondary storage can be constructed using one type of the NVMs with a small block of CM serving as a write buffer. The SSDs discussed in Section 1 (Figure 1) is one type of such hybrid memory systems. Another example is using PCRAM as main memory with a DRAM buffer [11].

Hybrid memories can be also utilized for building scalable, high density and low power on-chip caches with write through policy [16, 14], which is supported by various commercial CPU products including Intel Pentium, Sandy/Ivy Bridge, UltraSparc, etc. For future CMPs with large number of cores, write-through caches are becoming more attractive as they greatly simplify the cache coherence protocol, which could be extremely complex with numerous transient states. Unfortunately, a write-through policy defeats the CM buffer for NVM, exposing write intensive behavior to lower levels in the memory hierarchy that are candidates to utilize NVMs such as last level cache. Thus, a scalable and efficient cache architecture can be built by employing hybrid memories in which each cache set contains a mixture of NVM blocks and CM blocks, as depicted in Figure 5(a). Since the NVM is typically *write-hostile* and the CM is *write-friendly*, it is desirable that write intensive data be dynamically migrated or swapped from the NVM to the CM. A common hardware approach is to keep track of a short history of write accesses using a counter and migrate/swap a data block from the write-hostile memory to the write-friendly memory if the counter indicates the data block is write intensive. This approach however, is easily misled by unpredictable runtime data access behavior. Significant mispredictions can incur an expensive penalty of serving accesses in the wrong type of memory (e.g., write frequent accesses occurring in NVM) as well as large migration/swap overhead. In contrast, a software mechanism has the advantages of taking actions preemptively to hide the migration/swap latency from the critical path and detecting data access patterns more accurately compared to simple run-time approaches.

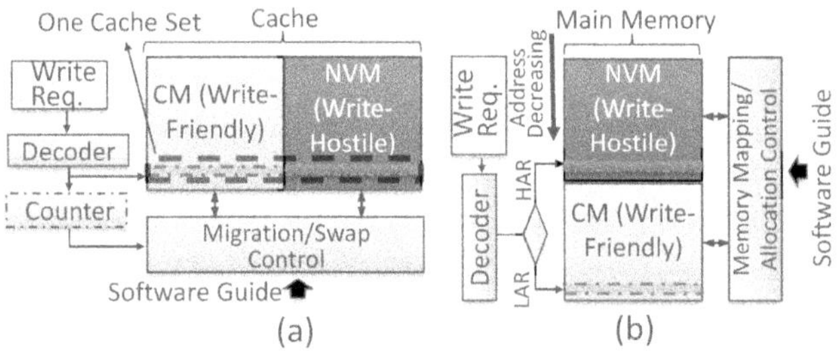

Figure 5: Utilizing hybrid memories at different levels (a): hybrid memories for caches (b): hybrid memories for main memory

Building main memory using NVM and CM has also been studied [9]. One typical organization for hybrid main memory separates NVM and CM into independently addressed memory regions (e.g., high address region (HAR) versus low address region (LAR)), as shown in Figure 5(b). Data with a certain read/write characteristic is allocated or mapped onto a specific address region so that it can be served by an appropriate type of memory. For example, the OS can map a read-only virtual page to a physical page served by the write-hostile NVM to leverage its non-volatility, low leakage and large capacity benefits without bringing negative performance or power impact. To achieve this, the OS needs to be aware of the physical address regions comprised by hybrid memories as well as the application's data access characteristics at the page granularity.

As can be seen from the above discussions, hybrid memory systems can be effectively assisted and optimized by software mechanisms, which will be detailed in the next section.

[1]Different NVMs can comprise hybrid memories. In this paper we restrict the term *hybrid memories* to the systems where both NVMs and conventional memories are used.

3. SOFTWARE DISPATCH

In this section we discuss the proposed software dispatch technique that leverages an application's data access characteristics to optimize the utilization of hybrid memory systems introduced in the previous section.

3.1 Compiler Dispatch for Migration/Swap

In hybrid NVM-CM systems with migration or swap support (e.g., Figure 5(a)), the migration or swap decisions should be made based on various factors including the application's write access patterns and the migration/swap overhead. In order to minimize the overhead, it is important that only the actively write intensive data is migrated or swapped from the NVM to the CM. To achieve this goal, we propose compiler-based dispatch, which identifies the write reuse patterns of heap data objects including arrays and linked structures and inserts instructions to guide the hardware to perform the migration/swap dynamically.

3.1.1 Write Reuse Identification for Arrays

To make migration/swap decisions for a memory block containing continuous bytes, we detect the temporal and spatial write reuses from array accesses in the source program.

Consider Figure 6 as an example. Given the array accesses $A[i+2][j]$ and $B[2][i][2*i+1]$ in the nested loop shown in Figure 6(a), we first convert the subscript functions to the matrix expressions, as illustrated in Figure 6(b). The array access now can be represented as $C*k+O$, where C is the coefficient matrix, k is the index vector and O denotes the offset vector. Determining whether the array access has temporal write reuse now is equivalent to deriving the condition under which the equation $C*k'+O=C*k''+O$ has solutions [15] (k' and k'' represent two different index vectors in the iteration space). The necessary and sufficient condition under which the above equation has solutions is that C is not full ranked. In our example, the coefficient matrix of $A[i+2][j]$ has a rank of 2, indicating no temporal reuse. $B[2][i][2*i+1]$ has temporal reuse since the rank of its coefficient matrix is 1, which is smaller than its dimension.

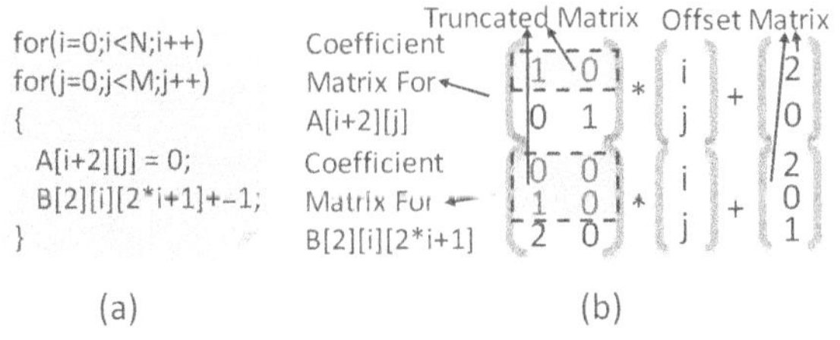

Figure 6: Array accesses and the corresponding matrix representations (a): array accesses (b): matrix representation

An array write access exhibits spatial write reuse when the innermost enclosing loop index varies only the last coordinate of that array. To discover spatial write reuse, we use a truncated coefficient matrix by dropping the last row of the original coefficient matrix, as illustrated in Figure 6(b). If the rightmost column in the truncated coefficient matrix (the coefficients that correspond to the innermost loop index) is a null vector and the rightmost element in the dropped row is nonzero, it is assured that the innermost loop only varies the last coordinate of the corresponding array.

In the above example, $A[i+2][j]$ exhibits spatial reuse since the rightmost column in the truncated matrix (the coefficient corresponding to the innermost loop index j) is a null vector and the rightmost element in the dropped row is nonzero. Using the same rule we can determine that $B[2][i][2*i+1]$ does not have spatial reuse since the innermost loop index j does not vary in the last coordinate of array B.

3.1.2 Write Reuse Identification for Linked Data Structures

To analyze the write reuse pattern for linked data structures such as linked lists and trees, a CFG (control flow graph) of the program is constructed. A CFG $G=(V,E,r)$ is a directed graph, with nodes V, edges E, and an entry node r. Each node v in V is a basic block, which consists of a sequence of statements that have exact one entry point and exit point. To simplify the code structure, a series of traditional compiler optimizations such as expression folding and branch elimination are applied on the CFG. Then the CFG is traversed while the following rules are examined to determine whether a sequence of memory writes (right-hand side of assignments) exhibit memory reuse:

- The analyzed memory writes are common pointer based dereferences. That is, these memory writes only differ in their offsets from a common base pointer.
- There are at least three memory writes whose offsets fall into a specified address range[2]. This is to guarantee the memory writes are consecutive or within small scope in the address space.
- There are no function calls between the analyzed memory writes.
- The memory writes are either in the same basic block or in a set of direct successor basic blocks that meet the above three criteria. If there are conditionals, the second criterion must be satisfied in all branches.

These rules guarantee the analyzed writes are mapped to the same memory block and result in intensive/frequent write behavior at runtime. The pseudocode for identifying write reuse in basic blocks and their successors is presented in Algorithm 1. It iterates over each basic block and collects relevant information on memory writes (i.e., base pointers and offsets). It organizes the collected information from different phases into a table, where each function call initiates a new phase. At the exit of each basic block, the table is traversed and the corresponding entries are marked to indicate the identified write reuses. For unmarked entries in the table, the first phases of all direct successors of the current basic block are further analyzed for potential write reuses across basic blocks in the CFG.

Figure 7 provides various cases to explain the write reuse identification for linked structures. As defined in Figure 7(a), the pointer *nd* is declared to point to a data structure with the type *node_t*. Since the data members *x*, *next* and *prev* have integer/pointer type and are adjacent fields in the same data structure, they are consecutive in the address space and thus, would typically reside in the same memory block such as a cache line. In the case of Figure 7(b), the three memory writes in the same basic block (i.e., *nd->x=5*, *nd->next=A* and *nd->prev=B*) have the common base pointer *nd* and

[2]This range depends on the size of the migration/swap block. For example, in a cache with 64-byte block size, this range is 64-byte.

Algorithm 1: Pseudocode for write reuse identification of linked data structures in CFG $G(V, E, r)$

```
for each basic block b_i ∈ V do
  create table H;
  phase = 0;
  for each statement s_j ∈ b_i do
    if s_j is function call then
      phase + +;
    else
      get the base pointer bp of LHS(s_j);
      get the offset o of LHS(s_j);
      if ∃ entry Y ∈ H such that bp = Y.bp &&
      phase = Y.phase then
        append o to Y;
      else
        push o, phase and bp into H;
      end if
    end if
  end for
  traverse H and mark entries with more than 2 offsets
  in a single phase that are within the specified range;
  for each unmarked entry Y' ∈ H do
    for each block b_j ∈ ⋃ SUCC(b_i) do
      search Y'.bp in the first phase in b_j and compute
      the total number of offsets n that are within the
      specified range;
      if n < 3 then
        continue to process next unmarked entry;
      end if
    end for
    mark entry Y';
  end for
end for
```

there is no interleaved function calls. Thus, Figure 7(b) exhibits write reuse. The program in Figure 7(c) also has write reuse since both successors of the basic block *nd->x=5* lead to write reuse. Figure 7(d) does not exhibit write reuse due to the presence of the function call *foo()*. In Figure 7(e), one of the direct successors *nd->next=A* only has one common pointer based memory write and thus, will not be marked as having write reuse.

3.1.3 Pre-dispatch Instrumentation

Once the write reuse has been determined, the compiler inserts a *pre-dispatch* instruction into the code prior to the memory access to notify the CPU to perform the migration or swap operation. The pre-dispatch instruction can be implemented using the extra bits in the instruction opcode of a particular ISA. For example, in SPARC, the *prefetch* instruction provides dedicated field *fcn* to implement variants of prefetch instruction. The *fcn* value from 16 to 31 is currently reserved and can be used to implement our pre-dispatch instruction. In the ARM architecture, there are similar reserved bits that can be used to implement the migration/swap operation.

3.2 OS/Allocator Dispatch

When data cannot be migrated/swapped on demand due to memory organization limitations (e.g., system illustrated in Figure 5(b)), the OS or the memory allocator (e.g., `malloc()`, `new`) can be used to dispatch[3] the data at the memory

[3]In this scenario, dispatch refers to the software procedure that maps or allocates data to a particular type of memory.

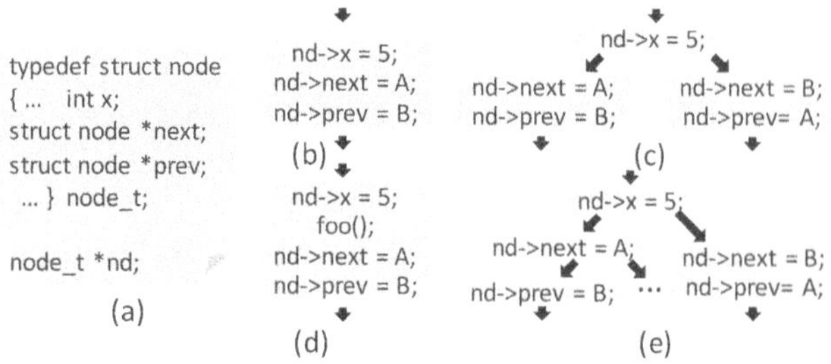

Figure 7: Code and control flow graph examples for write reuse identification. (a): type definition code (b): write reuse in the same basic block (c): write reuse across one basic block and all its successors (d): write reuse broken by function call (e): write reuse broken by one successor

allocation time. This requires a load-time understanding of access characteristics of the memory blocks so that they can be allocated onto appropriate types of memories that best serve them. A heuristic observation is that pages in different virtual segments (e.g., text/code, heap, stack) exhibit distinct access patterns. The pages in the code segment contain only instructions and are predominantly read-only. The pages in the stack typically exhibit a high write reuse pattern due to the way the stack operates. Thus, the OS can dispatch all the instruction pages to the write-hostile memory and stack pages to the write-friendly memory through its virtual-to-physical memory mapping process. In contrast, heap pages consist of dynamically allocated objects exhibiting diverse and application-dependent data access patterns. To analyze write access patterns for heap objects, the compiler techniques presented in Section 3.1 can be adopted. The analyzed results for a particular heap object can be propagated back to the memory allocation point (e.g., the call site of `malloc()`) using data flow analysis [7] to help making the dispatch decision.

4. CASE STUDY AND EVALUATION

To demonstrate the application and impact of software dispatch on hybrid memory systems, we design and evaluate a case study system with hybrid SRAM-STT caches on which we applied the proposed techniques.

Table 2: SRAM and STT-RAM Parameters (45nm)

Size	$128K$ SRAM	$512K$ STT-RAM
Area	$1.339mm^2$	$1.362mm^2$
Read Latency	$1.778ns$	$1.823ns$
Write Latency	$1.868ns$	$8.725ns$
Read Energy	$0.574nJ$	$0.550nJ$
Write Energy	$0.643nJ$	$3.243nJ$
Leakage Power	$0.185W$	$0.013W$

4.1 Case Study and Experimental Settings

Our experimental system consists of 16 cores laid out as a 4×4 mesh. The L1 cache is purely SRAM while the L2 utilizes both SRAM and STT-RAM. The SRAM and STT-RAM layers are stacked using 3D through silicon vias (TSVs) [12] technology, as illustrated in Figure 8. As a software-guided data distribution scheme, we compare with the mechanism that migrates data from STT-RAM to SRAM upon two successive writes (MSW), a relevant hardware approach presented in [14]. We do not compare with other techniques proposed in [14] since they are either orthogonal to data distribution and thus can equally benefit a variety of systems including ours or applicable in different scenarios from ours.

We use HSPICE simulation and modified CACTI [13] to model the latency, power and area of the simulated caches. We observe that a 128K SRAM bank takes up comparable chip area to a 512K STT-RAM bank, as shown in Table 2. The detailed architectural parameters simulated are shown in Table 3.

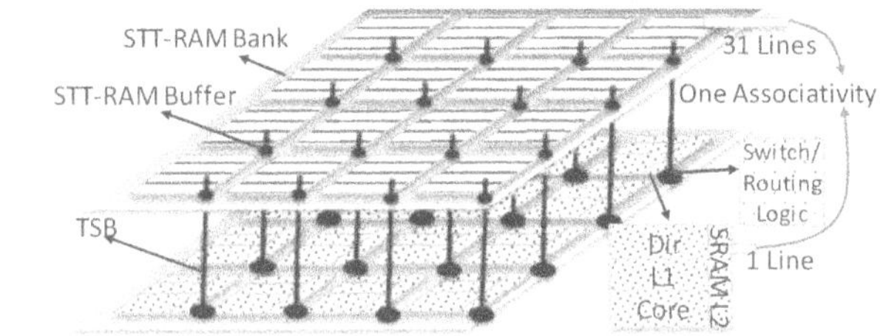

Figure 8: 3-D Architecture with hybrid SRAM-STT caches.

Table 3: Architecture configurations

	SRAM L2	STT-RAM L2
Processor	16 SPARC cores, 2G Hz, 4W/core	
Operating System	64-bit Solaris 10	
L1 Cache	16KB/core, 4-way associative, 64B block size, 1-cycle hit latency, write-through	
L1 Coherence	MESI protocol, in cache directory	
L2 Cache Size	128KB/bank	512KB/bank
L2 Block Size	64B	64B
L2 Associativity	32	32
L2 Read Latency	4 cycles	4 cycles
L2 Write Latency	4 cycles	18 cycles
L2 Energy	refer to Table 2	
Network	4×4 Mesh, packet switching, 3 cycles per hop	
Main Memory	4GB, 150-cycle latency	

In our experiment we apply the proposed technique on a write-back L2 with write-through L1 (see Section 2). Our approach also applies to L1 caches (either write-back or write-through) and scratchpad memory. We utilize multi-threaded workloads from the SPLASH-2 [1] and PARSEC [2] benchmark suites and Wind River Simics [8] as our simulation environment.

4.2 Performance and Power Evaluation

Figure 9 reports the ratio of write operations served by SRAM versus STT-RAM for MSW and software dispatch (SPD). In MSW, two successive writes on the same data block result in a migration and the subsequent writes on that block will be served by SRAM. This allows the SRAM, which comprises just over 3% of the cache capacity to serve 39.6% of the writes. However, MSW does not perform an accurate prediction for write reuse when applications exhibit certain interleaved read/write access patterns. In contrast, SPD dispatches more effectively and results in an average of 88.5% of all the write requests being handled in SRAM.

Another important metric in our evaluation is the number of writes on SRAM per dispatch operation. The larger the number, the lower the dispatching overhead relative to the gain. Figure 10 presents the number of SRAM writes per

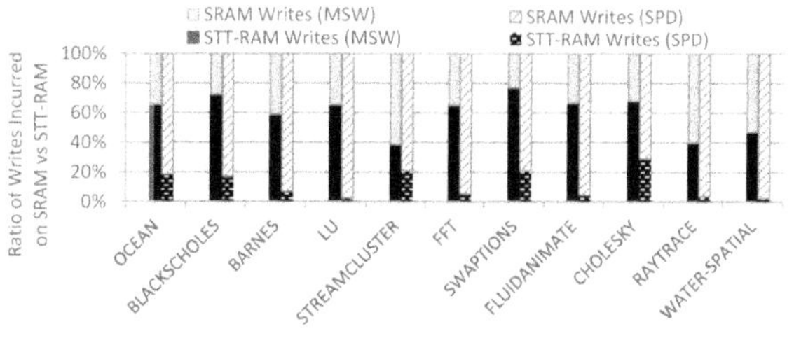

Figure 9: Ratio of writes on SRAM vs STT-RAM for MSW and SPD

dispatch for MSW and SPD. SPD has a much larger number of writes on SRAM per dispatch/migration. On average, 32 writes are served by SRAM per one migration operation in SPD while there are only 8 writes occur in SRAM per one migration in MSW.

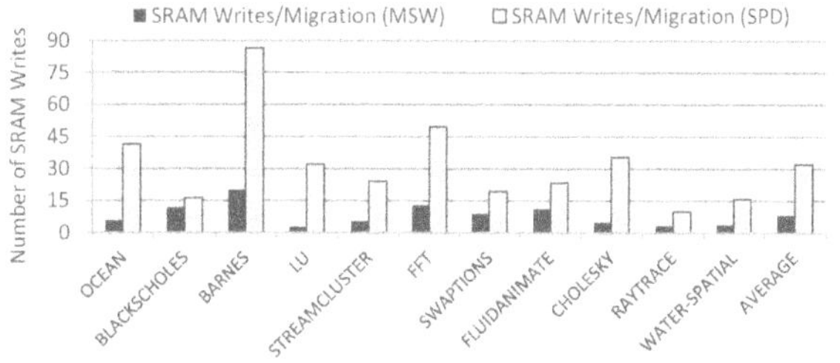

Figure 10: Number of SRAM writes per dispatch (migration)

The off-chip miss rate shown in Figure 11 demonstrates the advantage of reduced off-chip miss rate achieved by employing STT-RAM for on-chip storage. With 4 times more capacity than SRAM-only L2 caches, MSW and SPD reduce the off-chip misses by 38.9% and 40.0%, respectively. The reduction in expensive off-chip misses results in an average of 5% faster memory accesses for SPD, as shown in Figure 12. MSW exhibits a negligible improvement in memory access delay despite the reduction in off-chip misses as this benefit is offset by the large number of writes on STT-RAM (see Figure 9).

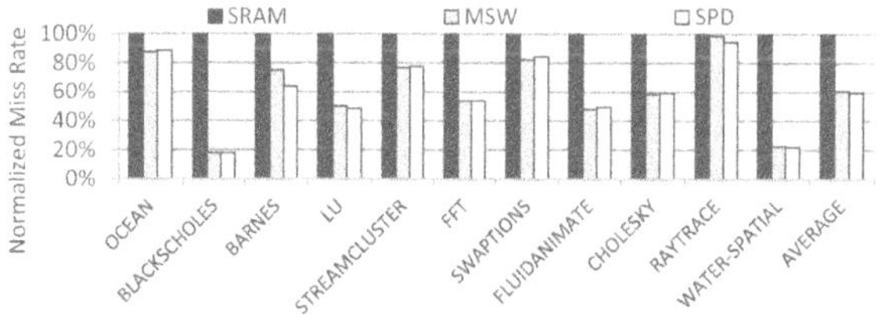

Figure 11: Normalized off-chip miss rate

Figure 13 shows the static and dynamic power breakdown normalized to the total power consumption of the SRAM cache. Due to the leakage problem inherited by CMOS devices, the SRAM-only cache consumes a non-negligible amount of static power while MSW and SPD dramatically reduce it. However, MSW consumes an average of 18.4% more dynamic power than the SRAM-only design due to the high energy overhead incurred by writes on STT-RAM. In contrast, the dynamic power consumed by SPD is close to that consumed by the SRAM-only cache since most of the writes in SPD have been dispatched efficiently onto its

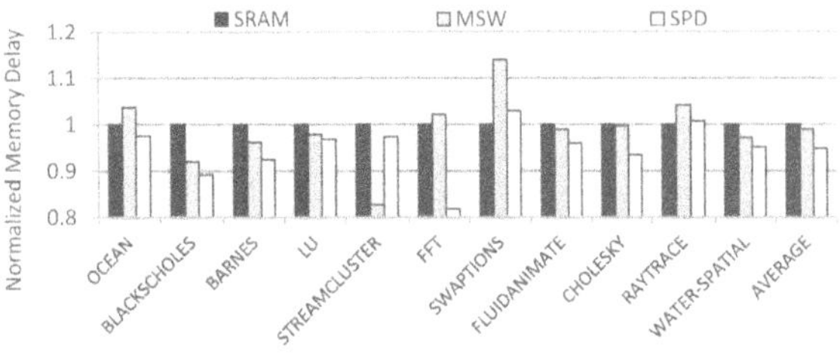

Figure 12: Normalized memory access delay

SRAM fraction. MSW results in an 0.3% overall higher power consumption than SRAM-only in spite of its 86.6% static power reduction. SPD has a similar static power savings of 86.8% as well as a total power savings of 9.8%.

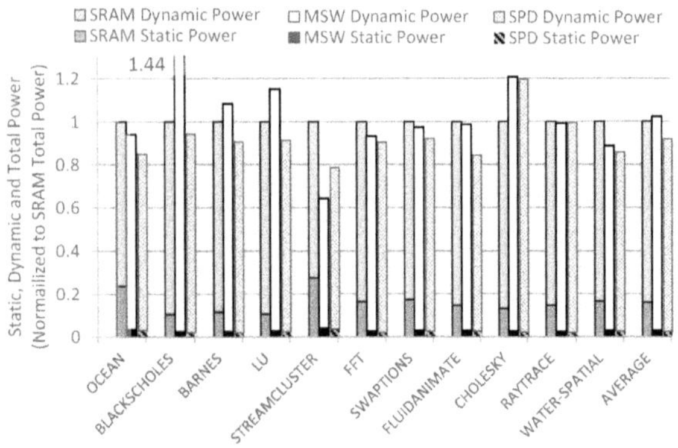

Figure 13: Normalized power consumption

5. CONCLUSION

In this paper, we presented software dispatch, an approach to detect write reuse patterns within the application and use this information to guide the hardware and distribute data accesses to appropriate memory resources in hybrid memory systems. Our evaluation through a case study system with hybrid memory caches demonstrates using software dispatch the number of writes serviced by conventional memory is increased by 81%, memory access performance is improved by 5% and memory power consumption is reduced by 10% over the state-of-the-art runtime system.

6. ACKNOWLEDGEMENT

This work is supported, in part, by NSF award CCF-1064976.

7. REFERENCES

[1] J. M. Arnold, D. A. Buell, and E. G. Davis. Splash 2. In *SPAA '92: Proceedings of the fourth annual ACM symposium on Parallel algorithms and architectures*, pages 316–322, New York, NY, USA, 1992. ACM.

[2] C. Bienia, S. Kumar, J. P. Singh, and K. Li. The parsec benchmark suite: Characterization and architectural implications. Technical Report TR-811-08, Princeton University, January 2008.

[3] A. Driskill-Smith. Latest advances and future prospects of stt-ram. *Non-Volatile Memories Workshop*, 2010.

[4] M. Hosomi, H. Yamagishi, T. Yamamoto, and K. B. et al. A novel nonvolatile memory with spin torque transfer magnetization switching: Spin-ram. *IEDM Technical Digest*, 2(25):459–462, 2005.

[5] J.-U. Kang, H. Jo, J.-S. Kim, and J. Lee. A superblock-based flash translation layer for nand flash memory. In *Proceedings of the 6th ACM & IEEE International conference on Embedded software*, EMSOFT '06, pages 161–170, New York, NY, USA, 2006. ACM.

[6] S.-W. Lee and B. Moon. Design of flash-based dbms: an in-page logging approach. In *Proceedings of the 2007 ACM SIGMOD international conference on Management of data*, SIGMOD '07, pages 55–66, New York, NY, USA, 2007. ACM.

[7] Y. Li, A. Abousamra, R. Melhem, and A. K. Jones. Compiler-assisted data distribution for chip multiprocessors. In *PACT '10: Proceedings of the 19th international conference on Parallel architectures and compilation techniques*, pages 501–512, New York, NY, USA, 2010. ACM.

[8] P. S. Magnusson, M. Christensson, J. Eskilson, D. Forsgren, G. Hallberg, J. Hogberg, F. Larsson, A. Moestedt, and B. Werner. Simics: A full system simulation platform. *IEEE Computer*, 35(2):50–58, February 2002.

[9] J. C. Mogul, E. Argollo, M. Shah, and P. Faraboschi. Operating system support for nvm+dram hybrid main memory. In *Proceedings of the 12th conference on Hot topics in operating systems*, HotOS'09, pages 14–14, Berkeley, CA, USA, 2009. USENIX Association.

[10] C. Muller, D. Deleruyelle, O. Ginez, J.-M. Portal, and M. Bocquet. Design challenges for prototypical and emerging memory concepts relying on resistance switching. In *Custom Integrated Circuits Conference (CICC), 2011 IEEE*, pages 1–7, September 2011.

[11] M. Qureshi, M. Franceschini, and L. Lastras-Montano. Improving read performance of phase change memories via write cancellation and write pausing. In *High Performance Computer Architecture (HPCA), 2010 IEEE 16th International Symposium on*, pages 1 –11, jan. 2010.

[12] K. Sakuma, P. S. Andry, C. K. Tsang, S. L. Wright, B. Dang, C. S. Patel, B. C. Webb, J. Maria, E. J. Sprogis, S. K. Kang, R. J. Polastre, R. R. Horton, and J. U. Knickerbocker. 3d chip-stacking technology with through-silicon vias and low-volume lead-free interconnections. *IBM J. Res. Dev.*, 52:611–622, November 2008.

[13] P. Shivakumar and N. P. Jouppi. Cacti 3.0: An integrated cache timing, power, and area model. Technical report, hp, August 2001.

[14] G. Sun, X. Dong, Y. Xie, J. Li, and Y. Chen. A novel architecture of the 3d stacked mram l2 cache for cmps. In *Proceedings of the High Performance Computer Architecture*, pages 239–249, 2009.

[15] M. E. Wolf. *Improving locality and parallelism in nested loops*. PhD thesis, Stanford, CA, USA, 1992. UMI Order No. GAX93-02340.

[16] X. Wu, J. Li, L. Zhang, E. Speight, R. Rajamony, and Y. Xie. Hybrid cache architecture with disparate memory technologies. In *Proceedings of the 36th annual international symposium on Computer architecture*, ISCA '09, pages 34–45, New York, NY, USA, 2009. ACM.

Design of Low Power 3D Hybrid Memory by Non-volatile CBRAM-Crossbar with Block-level Data-retention

Yuhao Wang[1], Chun Zhang[1], Hao Yu[1], and Wei Zhang[2]
[1]School of Electrical and Electronics Engineering, Nanyang Technological University, Singapore
[2]School of Computer Engineering, Nanyang Technological University, Singapore
ywang29@e.ntu.edu.sg, {chunzhang,haoyu,zhangwei}@ntu.edu.sg

ABSTRACT

As one of the newly introduced resistive random access memory (ReRAM) devices, this paper has shown an in-depth study of conductive-bridging random access memory (CBRAM) for non-volatile memory (NVM) computing. Firstly, a CBRAM-crossbar based memory is evaluated with accurate physical-level model and circuit-level characterization. It is then deployed as NVM component with a 3D hybrid integration of SRAM/DRAM, where one layer of CBRAM-crossbar is designed for data-retention under power gating to reduce leakage power from SRAM/DRAM at other layers. Moreover, a block-level data-retention scheme is designed to only write back dirty data from SRAM/DRAM to CBRAM-crossbar. When compared to the hybrid memory using phase-change random access memory (PCRAM) as data-retention, our CBRAM-based hybrid memory achieves 16x faster migration time and 4x less migration power for hibernating transition. When compared to the FeRAM-based bit-wise data-retention, our approach also achieves 17x smaller area and 8x smaller power under the same data migration speed.

Categories and Subject Descriptors: B.3 [Hardware]: MEMORY STRUCTURES

Keywords: CBRAM, crossbar, data-retention, non-volatile memory, 3D stacking

1. INTRODUCTION

Memory technologies have undergone a rapid development from the traditional CMOS SRAM/DRAM to non-volatile memory (NVM) devices. However, all well-established memory technologies have certain limitations. For example, SRAM/DRAM are fast but volatile and hence large leakage power is experienced to maintain the stored data. FLASH or PCRAM [1] are non-volatile but exhibit low write-speed and limited write-endurance. Moreover, though STT-MTJ device is relatively fast, large foot-print is required which reduces its integration density [2, 3]. As such, many recent design explorations have been focusing on the hybrid memory structure [4] with design constraints of fast accessing time, long endurance time, high integration density and low leakage power.

From the perspective of device technology, one needs to identify a device with fast accessing time and long endurance time. Recently, the conductive-bridging random access memory (CBRAM) has been introduced as another emerging non-volatile resistive random access memory (ReRAM) device [5, 6, 7, 8, 9]. Exploiting the fast ion-migration process, the conductive filament of CBRAM is able to grow and shrink extreme quickly under low voltage supply. As a result, compared to other types of NVMs such as PCRAM, CBRAM has the advantages in speed and power. More importantly, CBRAM has compatible working-voltage and fabrication process with CMOS technologies.

From the perspective of system architecture, one needs to build a system with high integration density and low leakage power. Firstly, one can apply the recent 3D die-stacking [10, 11] to integrate hybrid components together to improve density. Specifically, a hybrid memory system can be built upon layers of different memory technologies stacked by through-silicon-vias (TSVs) [10]. Moreover, as leakage power is the primary concern of a hybrid memory system, one needs to have well-designed data-retention scheme such that power gating can be effectively deployed to reduce leakage power yet without degrading performance. Traditionally, the common approach for data-retention of SRAM/DRAM [12] during sleep-mode deploys a small retention voltage for all memory cells, which still consumes non-negligible leakage power. Recently, the work in [13] applies the non-volatile PCRAM for *system-level* data-retention. However, this work aims at retaining only a small amount of key system status data called check-points, instead of the whole SRAM/DRAM content. Moreover, the relatively slow write-latency of PCRAM makes it infeasible in applications where the large-amount and frequent memory data-retention is required. Another work in [14] has *bit-level* data-retention by embedding each SRAM cell with one FeRAM device. Although concurrent bit-level data migration achieves fast speed, the overhead is overwhelming with additional bitwise data-transfer controllers, which degrades the SRAM performance during the normal active mode.

In this paper, we introduce a novel 3D hybrid memory architecture, in which CBRAM-crossbar is stacked as one layer with SRAM/DRAM for data-retention during power gating. The primary contributions are as follows. Firstly, CBRAM-crossbar based memory is evaluated with accurate device-level modeling and circuit-level characterization for area, accessing time and power. Secondly, the CBRAM-crossbar is deployed for *block-level* data-retention of SRAM/DRAM under power gating, in which one incremental dirty data write-back scheme is deployed for fast data migration. Experiment results show that our 3D hybrid memory with CBRAM-crossbar achieves 16x faster migration time and 4x less migration power for hibernating transition compared to the system-level PCRAM-based data-retention in [13]. Compared to the bit-level FeRAM-based data-retention in [14], our 3D hybrid memory system achieves 17x area reduction along with 2x SRAM performance improvement. Furthermore, 8x transition power reduction and 1.1x transition time reduction are obtained under the incremental dirty data write-back scheme, respectively.

The rest of this paper is organized as follows. In section 2, CBRAM-crossbar memory is evaluated from both device and circuit levels. Section 3 presents the architecture of the proposed 3D hybrid memory with CBRAM-crossbar, and also the according data-retention mechanism for incremental dirty data write-back. The design space exploration of the proposed 3D hybrid memory

ISLPED'12, July 30–August 1, 2012, Redondo Beach, CA, USA.

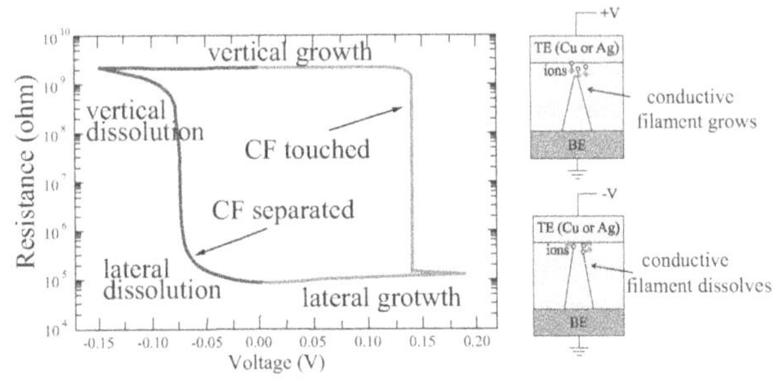

Figure 1: CBRAM device: resistive RAM based on conductive bridge

system is studied in section 4, and experiment results are shown in section 5 with conclusion in section 6.

2. CBRAM-CROSSBAR MEMORY

2.1 CBRAM Device

CBRAM is one emerging two-terminal ReRAM device, whose physical working mechanism is illustrated in Fig. 1. When a positive bias-voltage is applied, the conductive filament grows vertically until the two electrodes are bridged together, which sets the CBRAM into low-resistance or *on* state. Similarly, when a negative bias-voltage is applied, the conductive filament dissolves and disconnects the two electrodes, which sets the CBRAM in high-resistance or *off* state. Specifically, the movement of conductive filament height h is modeled by [5]

$$\frac{dh}{dt} = v_h \cdot exp(\frac{-E_a}{kT}) \cdot sinh(\frac{q \cdot \mathbf{E} \cdot a}{2kT}) \quad (1)$$

where v_h is the vertical growth velocity of conductive filament, kT is thermal energy, E_a is the activation energy, and a is the effective hopping distance. Note that $\mathbf{E}$ is the electric field in the solid-electrolyte region, which can be further obtained from

$$\mathbf{E} = \frac{\mathbf{v}}{L + (\frac{\rho_{on}}{\rho_{off}} - 1) \cdot h} \quad (2)$$

in which $\mathbf{v}$ is the applied voltage on CBRAM device, L is the thickness of CBRAM, and ρ_{on} and ρ_{off} are the conductive filament resistivity and non-conductive solid-electrolyte resistivity, respectively. The equivalent resistance of CBRAM can be calculated as resistance of conductive filament and dielectric in series by

$$R = \frac{\rho_{on}h + \rho_{off}(L-h)}{A}. \quad (3)$$

Two observations from equations (1)-(3) can be drawn for the fast writing-speed of CBRAM. Firstly, there exists an exponential relationship between the state changing-speed dh/dt and the applied voltage $\mathbf{v}$. In the literature, CBRAMs with less than 50ns write-latency have been demonstrated under 1.5V bias voltage [6, 8]. Secondly, the dh/dt increases exponentially when h approaches L, which means the growth speed is accelerated exponentially. As such, by properly setting the thresholds for *on* and off states and operating devices in corresponding regions, the write-latency of CBRAM can be further reduced.

2.2 Stacked CBRAM-Crossbar

Fig. 2 further illustrates the design of the stacked CBRAM-crossbar memory block. Compared to the commonly used 1T-1R structure for PCRAM [15], the crossbar structure demonstrates two major advantages. Firstly, extreme integration density can be realized due to the small pitch size of nano-bars [16]. Secondly, the crossbar structure can be stacked on top of the active transistor layer by 3D technology, which further reduces the area overhead and improves communication bandwidth.

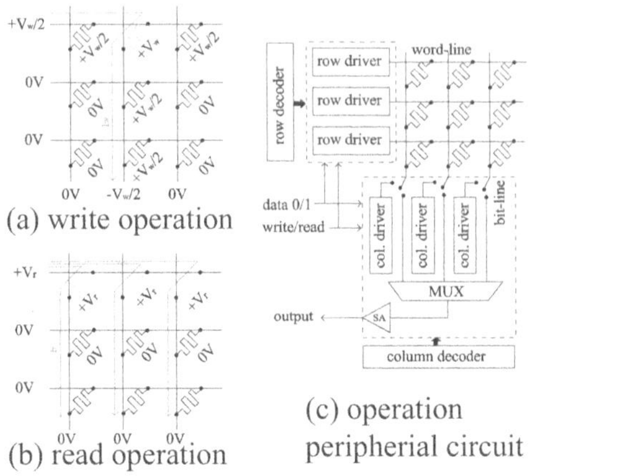

Figure 2: Read/write operations and peripheral circuits for CBRAM-crossbar memory block

The *write* and *read* operations for CBRAM-crossbar are shown in Fig. 2(a) and 2(b), respectively. To write the CBRAM-crossbar, the bias-voltage v_w needs to be applied to the designated cell through multiplexed voltage selector. As such, the corresponding word-line and bit-line are applied with $v_w/2$ and $-v_w/2$ voltages, respectively. By controlling the voltage level and polarity, one can determine the write-speed and change the CBRAM on/off states. To read the CBRAM-crossbar, the bias-voltage v_r (normally $v_r < v_w$) is applied to the corresponding word-line. By measuring the current at the cell bit-line, the corresponding on/off states can be decided.

Fig. 2(c) further illustrates the peripheral circuits design for one CBRAM-crossbar. During write-operation, the corresponding word-line and bit-line are applied with writing-voltage through address decoding. The voltage selection is done by the row drivers and column drivers, which switch among different voltage supplies. During read-operation, word-lines still need voltage selection while bit-lines are switched to current sensing circuit. To sense the current in crossbar structure, transimpedance circuit with current-to-voltage convection is applied [17], where the output voltage is further detected by sense-amplifier (SA). Note the SA can be shared with multiple bit-lines by adjusting the multiplexing level.

3. 3D HYBRID MEMORY WITH CBRAM-CROSSBAR FOR DATA-RETENTION

In this section, we introduce the detailed architecture for the 3D hybrid memory system based on CBRAM-crossbar with an incremental dirty data write-back for data-retention during power gating.

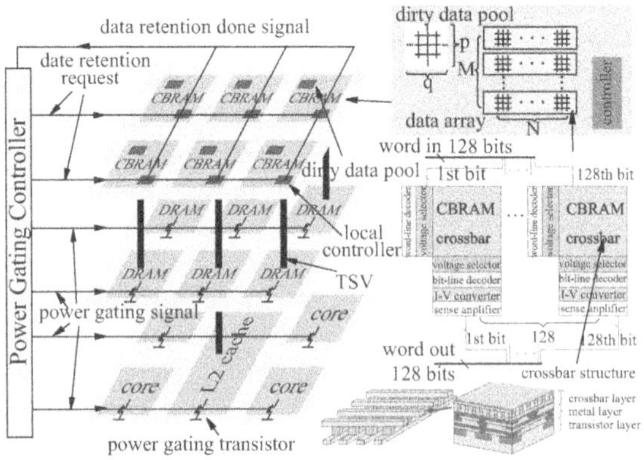

Figure 3: 3D hybrid memory system with CBRAM-crossbar based data-retention

Table 1: Notations

Notation	Descriptions
T_{Hi}/P_{Hi}	Hibernating transition time/power for bank i
T_{Wi}/P_{Wi}	Wakeup transition time/power for bank i
P_b/P_m	Bank/mat power
E_b/E_m	Bank/mat access energy
C_b/C_m	Memory bank/mat capacity
N_b/N_m	Number of memory banks/mats
W_d/W_a	Data/address bus width of each bank
W_D/W_A	Total data/address bus width
M/N	Number of row/column mats of each bank
B_C/B_D	Cache line/DRAM page size

3.1 Block-level Data-retention

Fig. 3 illustrates the overall system architecture of the proposed 3D hybrid memory, which is composed of embedded DRAM, SRAM and CBRAM-crossbar located at three tiers, respectively. They are connected by through-silicon-vias (TSVs). Similar to the common memory organization [18], the entire CBRAM-crossbar memory is broken into *banks*, where each bank can be accessed independently with dedicated data and address buses. Moreover, each row of the memory bank is called a *sub-bank*; and each bank is further broken into a $M \times N$ *mat* array, where each mat consists of one $m \times n$ CBRAM-crossbar. In addition, the terms of *sleep mode* and *active mode* are used to denote if the system is power-gated or not. The active to sleep mode transition is called *hibernating transition*, and term *wakeup transition* is used vice versa. For better presentation, Table 1 lists the notations used throughout this paper.

To reduce the sneak-path power in crossbar structure [16], we propose to distribute the multi-bit data into mats of the same row (i.e., sub-bank), where each mat (i.e., one CBRAM-crossbar) accepts one bit of the written data at each cycle. To achieve this, each bit of the arriving data is fed into voltage selector of each mat, and the address is fed into decoder of every mat. As such, each bit of the data is distributedly stored in same position of each mat decided by the address. During hibernating transition, each SRAM/DRAM bank is associated with one dedicated CBRAM-crossbar bank of the same capacity for data-retention. As such, power gating is employed at bank-level (i.e., block-level) as well. Once the hibernating transition begins, all the data in the specified SRAM/DRAM bank must be copied or migrated to the corresponding CBRAM-crossbar memory bank through dedicated data/address buses implemented by TSVs in the vertical direction. On the other hand, all data must be migrated from CBRAM-crossbar back to the SRAM/DRAM during the wakeup transition.

In this paper, we denote the proposed scheme as *block-level* data-retention because data is migrated between memory blocks (i.e., banks) sequentially through dedicated 3D TSV buses. Compared to the *bit-level* data-retention scheme [14], where each SRAM memory cell is associated with one neighboring FeRAM cell with cell-wise controllers, our block-level approach can achieve much smaller area overhead since the data migration controller is shared by all memory cells of the same block. In addition, our block-level approach will not degrade the SRAM performance since no change is made inside the SRAM memory cell. Furthermore, the *system-level* data-retention deploys PCRAM [13] to keep only small amount of key system status data due to the limited PCRAM write-latency. In contrast, our block-level approach with CBRAM-crossbar memory is capable of fast data migration for the entire SRAM/DRAM content such that even SRAM/DRAM could be totally shut down during the power gating.

To further speed up the data migration process, three techniques are introduced in this paper. Firstly, there is an incremental dirty data write-back deployed to reduce the total amount of data migrated in section 3.2. Secondly, system-level design space exploration is discussed in section 4 to achieve optimal design of memory structure. Thirdly, the 3D integration by TSVs results in high bus bandwidth for data migration as discussed in section 3.1.

3.2 Dirty Data Write-back

The target for data-retention is to synchronize the data of CBRAM-crossbar with corresponding SRAM/DRAM contents at block-level. For any SRAM/DRAM bank i, the time needed to copy the data to CBRAM-crossbar is decided by

$$T_{Hi} = \frac{M_b}{f_b \cdot W_d} \tag{4}$$

where f_b is the write-frequency of CBRAM-crossbar memory limited by its latency, W_d is the bank-level data bandwidth, and M_b is the amount of data to be migrated, which equals to the bank capacity C_b in the brute-force approach. Clearly, reducing M_b directly reduces the T_{Hi}. Considering the common law of locality [19], the system tends to access relatively local-memory regions during a given period of time. As such, between two successive power gating stages, only part of the content in CBRAM-crossbar and SRAM/DRAM becomes unsynchronized, which is denoted as *dirty data* throughout this paper. By only incrementally writing dirty data to CBRAM-crossbar, significant amount of migration data and power can be saved.

As shown in Fig. 3, to keep the dirty data status information, an extra CBRAM-crossbar called *dirty data pool* is embedded to each CBRAM-crossbar bank, where each bit in the pool, referred as *dirty bit*, indicates the dirty status of a few continuous bytes of data in SRAM/DRAM, referred as *dirty data group*. Empirically, we design the group granularity G_d as the cache line size B_C for SRAM or the page size B_D for DRAM.

3.2.1 Dirty Bit Set-up

The dirty bit set-up occurs simultaneously each time when the content of SRAM/DRAM is changed during active mode. As shown in Fig. 4, each CBRAM-crossbar bank *listens* to all memory write-operations issued to its corresponding SRAM/DRAM bank to update the dirty pool during active mode. Once the DRAM write-action is detected, the corresponding data group becomes dirty. As such, the corresponding bit in the dirty pool needs to be SET. The dirty pool size C_p is decided by

$$C_p = \frac{C_b}{G_d} = p \cdot q \tag{5}$$

where p and q is the CBRAM-crossbar dimension of dirty pool. Therefore, the designated dirty bit position can be located by decoding the first $log(p)$ and the following $log(q)$ bits of the physical memory write-address, respectively.

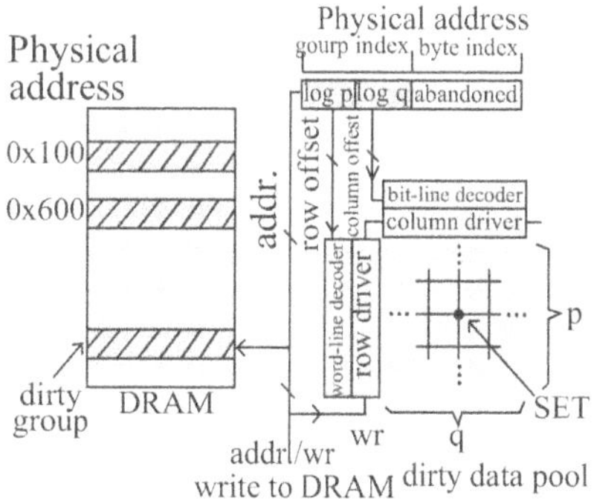

Figure 4: Circuit diagram for dirty bit set-up at active mode

3.2.2 Incremental Write-back

The flow to write back dirty SRAM/DRAM data to CBRAM-crossbar during hibernating transition is illustrated in Fig. 5. Specifically, one address counter is used to check the status of all dirty bits in the dirty data pool. Once the dirty bit in SET state is detected, the corresponding data group needs to be copied to the CBRAM-crossbar. Due to the limited data-bus bandwidth W_d, the group data of size G_d needs to be written back to CBRAM-crossbar in several cycles. As such, the address counter generates

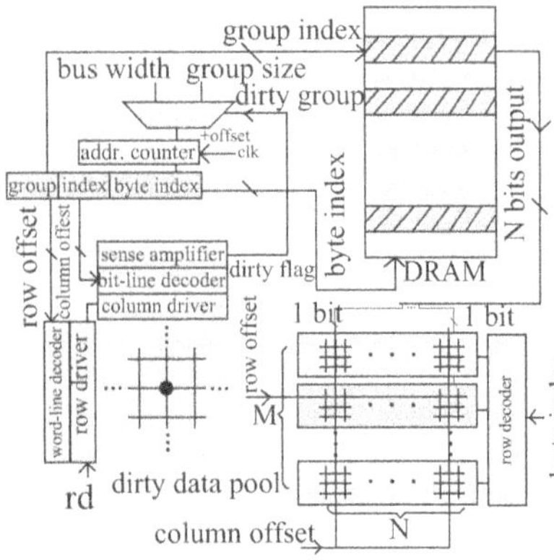
Figure 5: Circuit diagram for dirty data write-back at hibernating transition

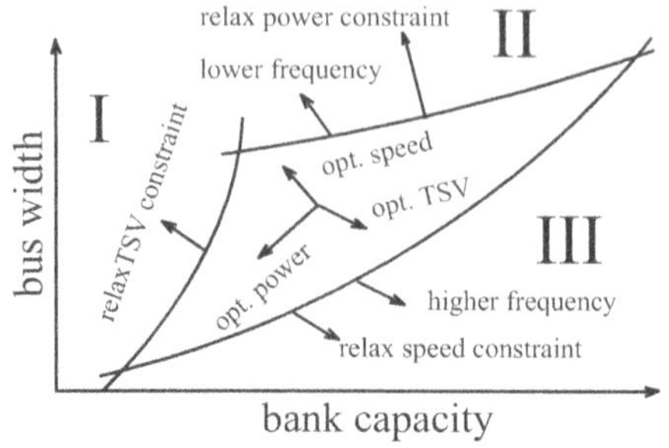

Figure 6: Design space exploration of 3D hybrid memory with CBRAM-crossbar

the memory address of the next piece of data to be copied from SRAM/DRAM by adding W_d-offset each cycle, and the read-signal to DRAM and write-signal to CBRAM-crossbar are issued for data migration. Once finished, the corresponding dirty bit is RESET. During wakeup transition, all data will be copied from CBRAM back to SRAM/DRAM with similar hardware supports.

4. DESIGN SPACE EXPLORATION

In this section, we further show how to optimize the 3D hybrid memory system by performing design space exploration under different optimization objectives.

Given design freedoms or parameters shown in Table 1, the total capacity of SRAM, DRAM or CBRAM is calculated by

$$C_M = N_b \cdot C_b = N_b \cdot N_m \cdot C_m = N_b \cdot M \cdot N \cdot m \cdot n \qquad (6)$$

where $M \cdot N$ is the dimension of mat array for each bank, and $m \cdot n$ is the dimension of crossbar array for each mat. As indicated by equations (4) and (6), different combinations of bank number, mat array dimension and crossbar array dimension need to be explored for optimal performance. For example, based on equation (4) it is desired to maximize the data-bus width W_d and CBRAM working frequency f_b. However, there are several design constraints to be considered as illustrated below one by one.

Data bandwidth constraint: as mentioned in section 3, since each CBRAM-crossbar supports only one-bit-write for each cycle, we need to design the multi-bit data to be distributed into different crossbars at the same row. Consequently, the data-bus width W_d must equal to N, which is the number of crossbars (i.e., mats) that each row contains.

Transition power constraint: during the hibernating and wakeup transition, the concurrent data migration will induce high transition power

$$\sum P_{bi} = \sum f_b W_{di} E_{mi} \leq P \qquad (7)$$

where W_{di} and E_{mi} is the bus-width and energy consumed of bank i. The transition power needs to be limited by reliability concerns for hot-spots of thermal and supply-current.

TSV density constraint: since the TSVs for data transmission occupy certain area, diameter in 5um for example [20], they lead to larger footprint as well as difficulties in placement and routing. As a result, to reduce the TSV overhead, we need to limit the average TSV density

$$\frac{W_D + W_A}{A} \leq D \qquad (8)$$

where A is the chip area.

Moreover, for our 3D hybrid memory system, one can perform design space exploration based on design parameters such as bank capacity C_b, data-migration bus bandwidth W_d and CBRAM-crossbar write-frequency f_b. Fig. 6 shows the performance-objective with different tradeoffs bounded by design constraints, which can be summarized as follows:

- For speed optimization (i.e., minimizing hibernating/wakeup transition time), large bandwidth is desirable. As such, small C_b, large W_d and high f_b are preferred. However, it is limited by TSV density and power constraints.
- For power optimization (i.e., minimizing transition power), small C_b and W_d architecture working in low f_b is favorable, which is mainly limited by TSV density and speed constraints.
- For memory performance optimization (i.e., minimizing memory performance degradation due to high TSV density), small W_d and large C_b helps reducing TSV density, which in turn alleviates memory performance degradation. This is mainly limited by speed constraint.

Regions I, II, III are corresponding to the limiting design constraints of TSV density, power and speed, respectively.

5. EXPERIMENT RESULTS

This section shows the experiment evaluation of the 3D hybrid memory from twofold. Firstly, the evaluation and optimization of the stacked CBRAM-crossbar memory is discussed in section 5.2. Based on the optimized memory design, the comparison between block-level data-retention with dirty data write-back strategy and the other memory systems is discussed in section 5.3.

5.1 Experiment Settings

To evaluate the CBRAM-crossbar memory performance, one CMOS-CBRAM SPICE-like simulator is developed with CBRAM models. The model parameters v_h, E_a and a are based on [5] with $1.8v$ voltage supply. As discussed in section 2, $50nm$ and $10nm$ heights of filament are assumed as thresholds for on and off states. A pitch size of $100nm$ is assumed for the crossbar structure, with $50nm \times 50nm$ cross-section area of nanowire made by platinum. The technology node for CMOS is assumed at 65nm. The simulation of CBRAM-crossbar and its peripheral CMOS circuits is performed to extract the accurate speed and power results. Note that the sneak-path influence can be also taken into consideration due to the circuit level simulation accuarcy. Based on simulation results, lookup-tables are built for different structures and are parameterized by crossbar size, multiplexing level, etc. The constructed lookup-tables are further integrated with the memory design tool CACTI [18] to perform the system-level analysis for the proposed 3D hybrid memory system. Furthermore, a behavioral model of data-retention in section 4 has been built to evaluate the performance of both hibernation transition and wakeup transition. The data-retention behavioral model requires calling the extended CACTI module with specific parameters to obtain the memory block performance data.

The system under investigation is then composed of one 2MB level-2 SRAM cache, one 64MB embedded DRAM and one 66MB CBRAM-crossbar, which are vertically connected by TSVs. Here we adjust the design parameters C_b, W_d and f_b under constraints of the maximal 10W transition power, maximal $60/mm^2$ TSV density and maximal 3ms transition speed. The overall CBRAM-crossbar platform is implemented by C++ on Linux. All presented results were computed on an Intel Core i5 with 2.6GHz clock frequency and 4GB of RAM.

5.2 Memory Design and Optimization

We first discuss the advantage of CBRAM-crossbar memory, and then show the design space exploration to obtain the optimal architecture-level parameters.

5.2.1 Stacked CBRAM-crossbar Memory

Table 2 evaluates and compares the performance of the stacked CBRAM-crossbar memory with the other memory technologies for the same capacity 16MB. The data for SRAM and DRAM are generated by CACTI with default settings, and PCRAM data is from PCRAMsim [15]. As shown in Table 2, the CBRAM-crossbar performance, especially the accessing latency for both read and write, is already close to DRAM, which shows its potential for the future application as main memory. This mainly comes from the 3D die-stacking of CBRAM-crossbar, which leads to a much smaller delay in interconnects. Moreover, compared to PCRAM, the CBRAM-crossbar shows 9x faster write-latency, 1.6x smaller area, 4.5x less write-energy per access, and 1.5x slower read-latency on average. The slightly slower read-latency is due to the crossbar topology for block-level operation of CBRAM.

Table 2: Performance comparison of 16MB SRAM, DRAM, PCRAM, and CBRAM memories

Feature	SRAM	DRAM	PCRAM	CBRAM
area (mm^2)	27.4	4.19	3.77	2.33
read latency (ns)	3.43	2.25	2.54	3.78
write latency (ns)	3.43	2.55	RESET 42.54 SET 102.54	7.99
read energy (nJ)	0.83	0.61	0.73	1.8
write energy (nJ)	0.75	0.61	RESET 6.66 SET 2.28	2.0

5.2.2 Memory Performance Evaluation

As discussed in section 4, one needs to decide the optimal architecture-level parameters for performance objectives under certain constraints. As one example, we show the procedure of transition time (i.e., speed) optimization. Fig. 7 shows the trend of TSV density and P_H/P_W with C_b/W_d. Among the available combinations, the 256KB one is chosen since it has largest W_D and also larger power margin when increasing frequency. When the one works at its allowed maximum frequency, 0.5ms T_W and 1.5ms T_h can be achieved.

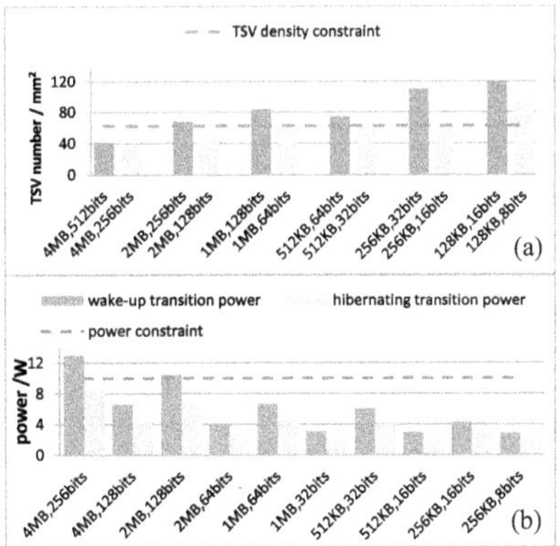

Figure 7: (a) TSV density and (b) mode transition power under different architecture-level parameters

Table 3 demonstrates the optimal results for different performance objectives. An optimal 0.5ms hibernating transition time and 1.5ms wakeup transition time is achieved for the speed optimization. For transition power optimization, 5x less wakeup transition power and 1.6x less hibernating transition power are

Table 3: Optimized performance for different design objectives

Objective	Speed	Power	MP
C_b	256KB	256KB	4MB
W_d (bits)	16	8	128
P_W/P_H (W)	9.0/2.3	***1.8/1.8***	7.1/4.3
T_W/T_H (ms)	***0.5/1.5***	3/3	1.9/3
TSV density ($/mm^2$)	60	48	***26***

achieved with 6x wakeup transition time and 2x hibernating transition time penalties. For memory performance (MP in Table 3) optimization, the lowest TSV density is obtained as the minimized memory performance degradation in normal active mode. The design exploration leaves designer the freedom to choose different design parameters based on system requirements.

5.3 Data-retention Evaluation

In this part, we evaluate the data-retention with the use of incremental dirty data write-back and block-level data migration for our proposed 3D hybrid memory system with CBRAM-crossbar.

5.3.1 Dirty Data Write-back Strategy

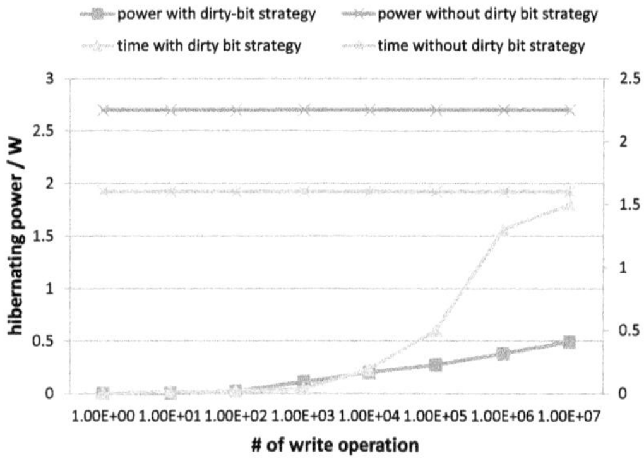

Figure 8: Hibernating power and time reduction by incremental dirty data write-back

Firstly, the cache-access is assumed to follow a Gaussian distribution to demonstrate the locality of memory access pattern. Fig. 8 compares the hibernating transition power and time with and without using dirty data write-back strategy. For the same number of 10^4 memory accesses, system with dirty data write-back strategy achieves 14x reduction in power and 8x reduction in time during hibernating transition. However, the transition time improvement degrades with increased number of memory accesses, since more data has become dirty. For a typical number of 10^6 memory accesses, system with dirty data write-back strategy achieves 8x reduction in power and 1.1x reduction in time during the hibernating transition.

5.3.2 Data-retention Performance Comparison

Next, we compare the performance of data-retention and leakage power in sleep mode for different memory systems. There are following data-retention schemes used for the comparison.

- *STD:* standard SRAM/DRAM without power gating.
- *DPG:* data-retentive power gating (DPG) of both SRAM [21] and DRAM [22] with reduced supply voltage.
- *PCRAM:* system-level data-retention by PCRAM [13]
- *FeRAM:* bit-level data-retention by FeRAM [14].
- *CBRAM:* our proposed block-level data-retention with incremental dirty data write-back by CBRAM.

Table 4 shows the full comparison results in terms of sleep-mode leakage power for SRAM and DRAM (P_s^L/P_d^L), hibernating and wakeup transition time (T_H/T_W), and hibernating

Table 4: Data-retention performance comparison for different leakage reduction schemes

Scheme	P_s^L/P_d^L (mW)	T_H/T_W (ms)	P_H/P_W (W/MB)
STD	209/220	NA	NA
DPG	21/22	1e-4	0
PCRAM	0/0	8/1	0.5/0.003
FeRAM	0/22	1	0.75
CBRAM	0/0	0.49/1.3	0.14/0.005

and wakeup transition power (P_H/P_W). The leakage power for SRAM/DRAM under standard scheme is generated by CACTI at 65nm technology node. Leakage power under DPG scheme is calculated by the leakage reduction factors reported in [21, 22] for both SRAM and DRAM. Block-level CBRAM based data-retention performance is derived from our platform. For fair comparison, bit-level FeRAM based data-retention and system-level PCRAM based data-retention are simulated within our platform as well with models based on [14, 13].

Due to the use of NVM for data-retention, PCRAM, FeRAM and CBRAM based memory systems all outperform the STD and DPG schemes in terms of leakage power reduction. Moreover, our proposed 3D hybrid CBRAM-crossbar memory system allows a shut-down of both SRAM and DRAM during the power gating. As a result, compared to the system-level PCRAM based data-retention, we achieve 16x faster hibernating transition time and 4x smaller hibernating transition power. The hibernating transition performance improvement comes from the advantageous CBRAM-crossbar memory performance as shown in Table 2, and also the block-level incremental dirty data write-back strategy. However, the wakeup transition time is slightly slower and the wakeup transition power is 2x larger for CBRAM. This is mainly because the incremental dirty data write-back strategy does not apply to the wakeup transition. In addition, the proposed CBRAM system also outperforms the FeRAM by 11x transition power saving, and 2x faster hibernating transition time.

Table 5: Cache performance comparison between block-level data-retention (CBRAM) and bit-level data-retention (FeRAM) in active mode

Cache capacity	Feature	Block-level CBRAM	Bit-level FeRAM
128KB	access time (ns)	1	1.8 (-80%)
	access energy (nJ)	0.13	0.21 (-62%)
	area (mm^2)	0.82	7.6 (-827%)
512KB	access time (ns)	1.5	2.8 (-87%)
	access energy (nJ)	0.33	0.6 (-82%)
	area (mm^2)	3.8	28.6 (-642%)
2MB	access time (ns)	2.7	4.3 (-59%)
	access energy (nJ)	0.6	1.3 (-117%)
	area (mm^2)	14.2	113 (-696%)

Compared with the block-level CBRAM based data-retention, the bit-level FeRAM based data-retention induces overwhelming cache performance degradation in normal active mode due to bit-wise controllers, which is further illustrated in Table 5. For bit-level FeRAM based data-retention scheme, the area overhead for retaining one bit data is $6.1um^2$, which is only $0.36um^2$ in our CBRAM system, estimated by the our developed CBRAM-crossbar based CACTI. Due to such a large area overhead, the FeRAM based system shows an 84% SRAM access time degradation, 74% SRAM access energy overhead on average. The performance of bit-level FeRAM based cache is derived from CACTI by replacing SRAM cell with FeRAM-SRAM cell pair. Therefore, our CBRAM-crossbar based data-retention outperforms not only the system-level PCRAM based data-retention but also the bit-level FeRAM based data-retention.

6. CONCLUSION

In this paper, the 3D hybrid memory system with block-level data-retention ability is demonstrated based on the use of stacked CBRAM-crossbar memory. The stacked CBRAM-crossbar memory design platform is developed for device-level characterization of CBRAM device, and also for the performance analysis of the CBRAM-crossbar memory. Moreover, the block-level data-retention scheme is developed with the incremental dirty data write-back ability. Compared to the system-level data-retention based on PCRAM, the proposed system achieves 16x faster time and 4x less power for hibernating transition performance. Compared to the bit-level FeRAM based data-retention, 8x power reduction and 17x smaller area are obtained by the proposed system under the same migration time.

7. ACKNOWLEDGMENTS

This work is sponsored by Singapore MOE TIER-2 fund MOE2010-T2-2-037 (ARC 5/11) and A*STAR SERC-PSF fund 11201202015.

8. REFERENCES

[1] B. Lee and et.al., "Phase-Change Technology and the Future of Main Memory," *IEEE Micro*, 2010.

[2] K. Tsuchida and et.al., "A 64Mb MRAM with clamped-reference and adequate-reference schemes," in *Proc. ISSCC*, 2010.

[3] D. Halupka and et.al., "Negative-resistance read and write schemes for STT-MRAM in 0.13 um CMOS," in *Proc. ISSCC*, 2010.

[4] X. Wu and et.al., "Hybrid cache architecture with disparate memory technologies," in *Proc. ISCA*, 2009.

[5] S. Yu and H. Wong, "Compact Modeling of Conducting-Bridge Random-Access Memory (CBRAM)," *IEEE Electron Devices*, 2011.

[6] S. Dietrich and et.al., "A nonvolatile 2-Mbit CBRAM memory core featuring advanced read and program control," *IEEE JSSC*, 2007.

[7] S. Sheu and et.al., "A 4Mb embedded SLC resistive-RAM macro with 7.2ns read-write random-access time and 160ns MLC-access capability," in *Proc. ISSCC*, 2011.

[8] C. Gopalan and et.al., "Demonstration of Conductive Bridging Random Access Memory (CBRAM) in Logic CMOS Process," in *Proc. IMW*, 2010.

[9] J. Jameson and et.al., "Quantized Conductance in $Ag/GeS_2/W$ Conductive-Bridge Memory Cells," *IEEE Electron Device Lett.*, 2012.

[10] G. Loh, "3D-Stacked Memory Architectures for Multi-core Processors," in *Proc. ISCA*, 2008.

[11] N. Minas and et.al., "3D integration: Circuit design, test, and reliability challenges," in *Proc. IOLTS*, 2010.

[12] C. Hua and et.al., "Distributed data-retention power gating techniques for column and row co-controlled embedded SRAM," in *Proc. MTDT*, 2005.

[13] J. Xie and et.al., "3D memory stacking for fast checkpointing/restore applications," in *Proc. 3DIC*, 2010.

[14] M. Koga and et.al., "First Prototype of a Genuine Power-Gatable Reconfigurable Logic Chip with FeRAM Cells," in *Proc. ICFPL*, 2010.

[15] X. Dong and et.al., "PCRAMsim: System-level performance, energy, and area modeling for phase-change RAM," in *Proc. ICCAD*, 2009.

[16] W. Lu and et.al., "A Functional Hybrid Memristor Crossbar-Array/CMOS System for Data Storage and Neuromorphic Applications," *Nano Lett.*, 2011.

[17] E. Seevinck and et.al., "Current-mode techniques for high-speed VLSI circuits with application to current sense amplifier for CMOS SRAM's," *IEEE JSSC*, 1991.

[18] CACTI: An integrated cache and memory access time, cycle time, area, leakage, and dynamic power model. [Online]. Available: http://www.hpl.hp.com/research/cacti/

[19] J. Hennessy and D. Patterson, *Computer architecture: a quantitative approach*, 2011.

[20] International technology roadmap of semiconductor. [Online]. Available: http://www.itrs.net

[21] H. Qin and et.al., "SRAM leakage suppression by minimizing standby supply voltage," in *Proc. ISQED*, 2004.

[22] T. Nagai and et.al., "A 65nm low-power embedded DRAM with extended data-retention sleep mode," in *Proc. ISSCC*, 2006.

TAP – Token-Based Adaptive Power Gating

Andrew B. Kahng[†‡], Seokhyeong Kang[‡], Tajana Rosing[†‡] and Richard Strong[†] [*]
[†]CSE and [‡]ECE Departments, University of California at San Diego
La Jolla, CA 92093-0404 USA
abk@ucsd.edu, shkang@vlsicad.ucsd.edu, tajana@ucsd.edu, rstrong@eng.ucsd.edu

ABSTRACT

We propose a low-overhead technique, Token-Based Adaptive Power Gating (TAP), to power gate an actively executing out-of-order core during memory accesses. TAP tracks every system memory request, providing a lower-bound estimate for the response time. TAP also tracks the state of every power-gateable core in the system, to provide minimal latency wake-up modes to cores such that voltage noise safety margins are not violated. A power-gating switch that utilizes TAP can deterministically power gate its core with energy savings up to 22.39% and no performance hit.

Categories and Subject Descriptors

B.7.1 [**Integrated Circuits**]: Types and Design Styles—*Microprocessors and microcomputers*; C.0 [**Computer Systems Organization**]: General

Keywords

power gating, core, memory access, voltage noise

1. INTRODUCTION

During every cycle that a core is on, even when stalled, leakage power is consumed via gate leakage, gate-induced drain leakage, junction leakage, and subthreshold leakage. A core may stall quite often if it is intensely accessing the memory subsystem, as every time a thread makes a memory request that misses in the L1 cache, the core is subjected to a variable access latency. This variable latency often translates into a core stall during which no forward thread progress occurs and energy is wasted. For a 32nm out-of-order EV6 core, stall energy can be up to 39.1% of total energy consumption for the Spec2006 benchmarks [12].

Power gating is a technique that drastically reduces leakage power by cutting off the current path from supply to ground through introduction of a transistor switch between them. At one end of the spectrum, functional unit power gating reduces power consumption of unused core functional units [22] with wake-up latencies of several nanoseconds. At the other end, entire cores may be power gated and woken up with latencies of several hundreds of milliseconds to account for saving and restoring all core state from memory [11]. The authors of [9] propose to power gate in-order cores stalled on memory misses. Their programmable power gating switch (PPGS) enables core wake-up latencies as small as 8.06ns in *EV4* cores (with 6.25*M* transistors and 32nm technology). Core state is restored by saving critical sequential cells in retention flip-flops, and retaining SRAM cells with source biasing at a cost of less than 3.5% area overhead. The PPGS is directed by a counter-based controller that power gates the core after a stall is detected for longer than a last-level cache hit. The switch resumes core execution at the predicted memory response time, which dynamically adapts to memory latency. However, without detailed knowledge of all core memory requests, out-of-order execution and hard-to-predict stalls impede use of this power-gating mechanism.

This paper proposes a new system, *Token-based Adaptive Power Gating* (TAP), that deterministically applies power gating during core stalls which are caused by the variable latency of requests to the memory subsystem. TAP achieves this by providing two capabilities. First, TAP informs a core's PPGS about the lower-bound latency of memory requests that miss in a given cache level. The expected latency is sent to each PPGS by enabling the cache controllers to send tokens on a miss that include a lower-bound estimate of the access latency of a next-level memory hit. Second, TAP ensures that each core's PPGS uses a wake-up mode that does not violate supply voltage noise constraints of the system when waking up a core. This is ensured by having each core register its state with a centralized *wake-up controller* (WUC), detailing whether the core is idle or active; the WUC responds with a safe wake-up mode to all registered cores. We determine safe wake-up modes by analyzing every possible combination of core states and locations during core wake-up, and creating an equation that models our findings. Any core's PPGS controller, interfaced with the TAP system, has information regarding a lower bound on the time needed to complete all memory requests and a corresponding safe wake-up mode. Combining these two pieces of information, such a controller may determine when to power gate so as to avoid any performance hit, while still ensuring energy savings.

Our work makes the following contributions:

- A token-based system (TAP) to deterministically and safely manage power-gating constraints with energy savings as high as 22.39% for a wake-up latency of 10.2ns.
- Compared to a previous state-of-the-art technique [9], TAP has 2.58× the average energy savings for an out-of-order core. TAP is also shown to have better energy savings than a practical DVFS policy at 32nm.
- An analytical model that accurately estimates core wake-up modes for an arbitrary multi-core wake-up scenario.
- A new technique that introduces *wake-up stagger* between cores to reduce the minimum feasible core wake-up latency by up to 40.3%.

[*]Authors are listed alphabetically by last name. Principal contributors, to whom correspondence should be addressed: Richard Strong (rstrong@eng.ucsd.edu) and Seokhyeong Kang (shkang@vlsicad.ucsd.edu).

2. RELATED WORK

Power gating has been widely studied at both architectural and circuit levels. The following briefly reviews representative works in these two areas. Hu et al. [8] propose power gating as a technique to reduce functional unit leakage power, when applications underutilize their functional units. Specifically, they power gate the floating-point and fixed-point units according to three different predictors, which are respectively ideal, time-based, and branch-misprediction-guided. The best technique, branch-misprediction-guided, is able to put functional units to sleep for up to 40% of total cycles with only 2% performance loss. The authors of [8] also develop equations to estimate the break-even points for power gating an out-of-order superscalar processor. Although they build a power consumption model with precise analysis of virtual supply voltage during power gating, they do not consider the wake-up energy to restore circuit nodes, which will impact energy savings.

Lungu et al. [13] show that in many cases, the predictor of [8] can lead to increased energy consumption. A monitor that controls the use of power gating is introduced to bound the performance and energy penalty for misbehaved applications. Madan et al. [14] extend the idea of Lungu et al. to the core level, and propose a "guard mechanism" that reduces harmful use of power gating.

Power gating technology is also readily visible in leading commercial products. The recent Nehalem architecture employs power gating at the core level to reduce leakage power on idle cores, but 100ms is required to wake-up a core [10, 11]. In today's systems, the OS only power gates cores in the idle loop which ignores potential for core energy savings during long memory accesses.

At the circuit level, the pioneering work of Horiguchi et al. [7] has been followed by many publications on fundamental circuit design issues related to power gating, including switch cell sizing, data retention methods, physical implementation methodologies, and mode-transition noise analysis and reduction. The recent survey of Shin et al. [17] gives an excellent summary of the history and highlights of power gating technique.

Configurable power gating has been introduced in the past to mitigate process variation, reduce ground bounce noise, and minimize wake-up time. Agarwal et al. [3] and Singh et al. [18] examine multiple sleep modes that feature different wake-up overheads and leakage power savings. Use of multiple sleep modes achieves an extra 17% reduction in leakage power compared to a single power gating mode. Also, one of the sleep modes can reduce leakage power by 19% while preserving circuit state. However, these energy savings are based on static traces of bus activity and do not address the runtime problem of predicting when to power gate. In addition, the reported results are likely optimistic since wake-up noise is neglected, and the overhead of implementing low-voltage sleep control signal distribution is not considered.

The closest work to our own is MAPG [9], which proposes a technique to power gate in-order cores during long memory accesses. However, the MAPG work considers neither out-of-order execution nor the benefits of exploiting core location and state information for determining safe wake-up modes. Our present work addresses these issues, shows the importance of stagger to reduce core wake-up latency, and provides a comparison between TAP, MAPG-Counter, and DVFS for out-of-order cores.

3. POWER GATING AND PDN ANALYSIS

This section provides an analysis of our power-gating methodology, and its impact on the power distribution network. Power gating cuts off leakage current paths between supply and ground by using switch transistors (such as high-Vth or long-channel devices). To power gate a circuit, header switches turn off, and leakage current is reduced. While in the power-gated state, all logic gates connected to the virtual supply lose their logical states. Resuming circuit operation incurs a delay that corresponds to charging circuit capacitance, resetting memory elements, and restoring state.

To maximize power gating opportunities, we use a *programmable power gating switch* (PPGS) [9]. The PPGS offers multiple wake-up modes with different wake-up latencies; this enables greater leakage savings when lower-latency wake-ups are feasible. The PPGS works as a two-stage wake-up controller [6]. In the first stage, the signal *enable_few* turns on a subset of header switches to allow I_{limit} charging current until the circuit nodes are nearly charged. In the second stage, the signal *enable_rest* turns on the remaining header switches, resulting in a triangular charging current profile. The PPGS's wake-up time and inrush current are determined by the number of header switches that are turned on by the first-stage *enable_few* wake-up signal. The two key constraints on the choice of wake-up modes are that the voltage noise must be less than 5% on neighboring active cores, and less than 40% on neighboring idle cores [9].

To study wake-up latency and inrush current, we estimate the total charge for core logic and interconnect capacitance using the methodology of [9]. We determine the core area, power, and inrush current limits with McPAT [12]. Core transistor count, capacitance, and Vdd_core are determined from the 2009-2010 *International Technology Roadmap for Semiconductors* (ITRS) [1]. We summarize these values in Table 1.

Estimated Data	32*nm* HP	32*nm* LOP	22*nm* HP	22*nm* LOP
Design Data				
Vdd_core (V)	1.00	0.77	0.93	0.72
core area (mm^2)	15.894	16.185	8.616	8.608
logic area (mm^2)	11.513	11.568	6.531	6.527
C_{core} (F)	30.7E-9	30.8E-9	18.7E-9	18.7E-9
total charge (C)	30.7E-9	23.7E-9	17.4E-9	13.4E-9
core leakage (W)	0.916	0.143	0.572	0.076
I_{active} (A)	2.665	1.690	1.721	1.152
I_{limit} (A)	13.544	10.712	7.834	6.871
Power Gating and Wake-up				
$T_{min-charge}$ (ns)	7.80	5.40	4.40	3.40
wake-up energy (pJ)	15.40E3	9.16E3	8.11E3	4.86E3
# head switches	126,435	105,295	72,188	67,655
leakage in PG state (W)	21.51E-3	2.01E-3	14.62E-3	1.25E-3
leakage reduction in PG	97.65%	98.60%	97.49%	98.37%

Table 1: Estimated data for 32*nm* and 22*nm* EV6 cores.

We use the detailed PDN model from [9] that includes all package parasitics, to enable realistic noise analysis under various wake-up scenarios. Power is delivered from an external voltage regulator module (VRM) through a printed circuit board (PCB), a package ball, package interconnect, microbumps, on-die redistribution layers, the on-chip PDN, and power gating switches. We model the entire power delivery network, including power gating switches, as a simplified RLC circuit as designed in [9]. Package inductance and series resistance from VRM to bumps for a core are lumped as in-series inductance, $L_{pkg-core}$, and resistance, $R_{pkg-core}$. The PDN in the package, which is shared by multiple cores, is represented as a resistance mesh with a branch resistance of R_{shared}. There are three variant models depending on the state of the core — *core in active mode*, *core being woken up*, and *core in sleep mode*. We use the PDN parameter values from Table I of [9].

Exploiting Spatial Information. Previous PPGS work [9] selected a wake-up mode based on the worst-case wake-up time for each *number* of idle cores. The worst-case wake-up time assumption limits the benefit of power gating. A core's minimum wake-up time is constrained by the voltage noise seen by neighboring active cores – in particular, some *critical* active neighbor core, where the voltage noise constraint is first violated. The voltage noise of an active core is mainly affected by adjacent woken-up cores and the latencies (i.e., associated inrush currents) with which they wake up. In other words, we may exploit knowledge of a core's location to reduce pessimism. We have developed a model that determines the minimum wake-up time based on the number and location of active and woken-up cores. To simplify the model, we assume that all woken-up cores have the same (uniform) wake-up latency.

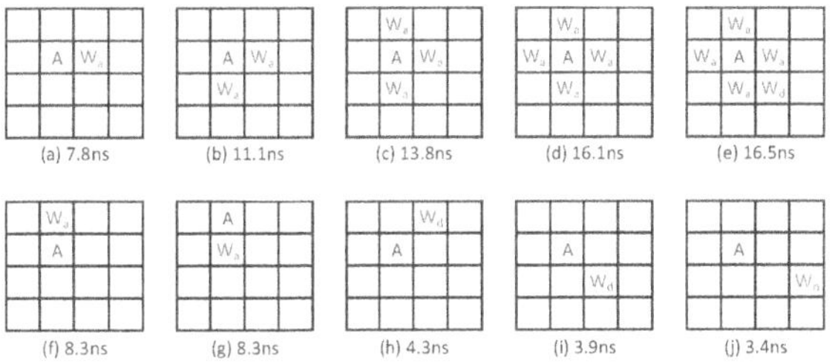

Figure 1: SPICE-calculated minimum wake-up latency for an EV6 16-core CMP with various wake-up scenarios.

Figure 1 shows the minimum wake-up time, according to the location and status of cores, for an example case of an EV6 16-core CMP where A denotes the critical active core, W_a are adjacent woken-up cores, W_d are diagonally adjacent woken-up cores, W_n are non-adjacent woken-up cores, and blank squares are idle or non-critical active cores. The PDN model of [9] is used The wake-up latency increases according to the number of adjacent woken-up cores (Figure 1 (a) - (e)), and is approximately proportional to the square root of the number of adjacent woken-up cores. Woken-up cores in the diagonal (W_d) or non-adjacent positions impact wake-up latency less than adjacent woken-up cores (Figure 1 (f) and (g)). If woken-up cores are located at an edge position (Figure 1 (h)), the minimum wake-up time increases.

From such observations, we have modeled the minimum wake-up latency based on the core status at each location as:

$$T = T_0(w + \beta \cdot x + \gamma \cdot y + \delta \cdot z)^{\alpha} \quad (1)$$

where T_0, α, β, γ and δ are fitting coefficients, w is the number of adjacent woken-up cores, x is the number of diagonal woken-up cores, y is the number of other (non-adjacent) woken-up cores, and z is the number of cores (active core itself or adjacent woken-up cores) which are at edge locations.

We have verified our model with SPICE, and modeled the wake-up time for 4-, 6-, 8-, and 16-core CMPs for all location permutations. Our SPICE simulations measure the voltage drop on active cores to find the minimum wake-up time.

The results in Table 2 show that our model has average error of 2.64%, 1.93%, 2.31% and 1.57% for 4-, 6-, 8-, and 16-core CMP cases, respectively. The parameter values can be adjusted in real IC designs to adapt to process, voltage and temperature variations.

core	coefficient					error	
	T_0	α	β	γ	δ	average (%)	maximum (ns)
4-core	7.9	0.50	0.35	0.15	0.15	2.64	0.37
6-core	7.9	0.50	0.35	0.15	0.13	1.93	1.10
8-core	7.9	0.50	0.30	0.15	0.13	2.31	1.65
16-core	7.9	0.50	0.20	0.10	0.10	1.57	1.40

Table 2: Average and maximum error on the modeled wake-up time for 4, 6, 8, and 16-core cases (EV6, 32*nm* HP).

Sensitivity of Results to PDN Parameter Values. The results of the SPICE simulation have varying degrees of sensitivity to power distribution network (PDN) parameters - the number of bumps, package inductance (L_pkg), package resistance (R_pkg), PDN mesh resistance (R_shared), supply voltage and core capacitance. We have assessed the minimum wake-up time sensitivity to variations in the PDN model. Figure 2 shows the change in the T_0 coefficient when each PDN parameter is scaled from 0.1× to 2× (a 20× range!) with respect to our default values, which are obtained from personal communication with industry experts. Since the actual wake-up latency depends on PDN variations, the T_0 coefficient will be determined by testing the actual packaged chip. Our conclusions regarding energy savings and overheads remain qualitatively the same across the range of PDN parameter values.

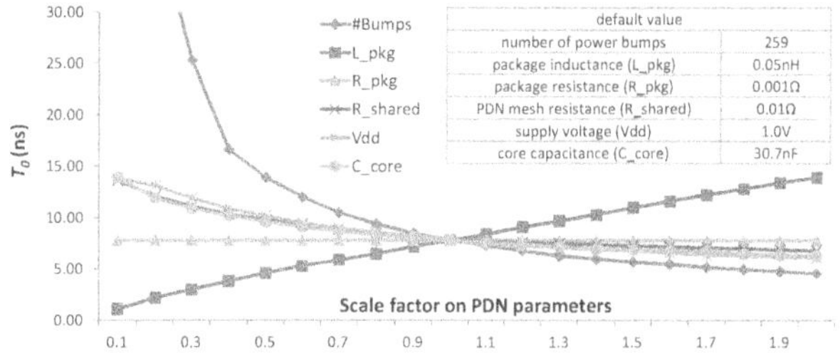

Figure 2: T_0 as a function of PDN parameters.

Benefits of Staggered Wake-up. The above wake-up analysis assumes that all cores wake up simultaneously, which is the worst case. However, wake-up latency is significantly reduced when we *stagger* the wake-up sequence so that two cores wake up at slightly different times (e.g., offset by 1ns). (Staggered wake-up is analogous to multi-stage wake-up control within a core.) We design the wake-up controller to insert stagger between waking cores to reduce wake-up latency. Figure 3 shows minimum wake-up latency for an EV6 16-core CMP when we add stagger between woken-up cores. The minimum wake-up time (y-axis) is reported for the worst case for each number of woken-up cores (x-axis). When stagger is zero, wake-up time increases according to the number of woken-up cores. However, if we avoid simultaneous wake-up, minimum wake-up time reduces greatly. When two, three, and four cores are waking up within an interval of three cycles (0.9ns), we obtain 18.8%, 31.9% and 40.3% wake-up latency reductions, respectively, over simultaneous wake-up. From SPICE results in Figure 3, we can see that the minimum wake-up time does not increase with staggered wake-up when the number of woken-up cores is larger than four. We have modeled the minimum wake-up time with Equation (1) for up to three woken-up cores by changing the parameter α from Table 2. The dotted lines in Figure 3 show the modeled wake-up latency from Equation (1) and its error with respect to SPICE simulation. Our measurements of model accuracy show an average (maximum) error of 2.66% (7.62%), 1.89% (6.61%), 0.93% (3.59%) and 2.51% (3.08%) for the 4-, 6-, 8-, and 16-core CMP cases, respectively.

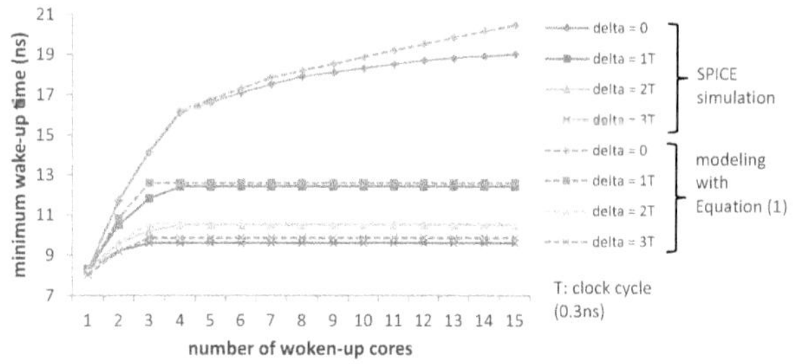

Figure 3: Minimum wake-up latency versus wake-up stagger.

4. SYSTEM DESIGN

We now present our architectural modifications used to control power gating for each core. As noted above, TAP is a system with two purposes, (1) to inform a core's PPGS about the expected latency of memory requests that miss in a given cache level, and (2) to ensure that each core's PPGS uses a wake-up mode that does not violate reliability constraints of the system when waking up a core. Together, these two pieces of information allow each PPGS to power gate its core as soon as a stall on a memory operation is detected, and the expected latency of the memory operation is greater than the break-even point for power gating.

Determining Lower Bound on Stall Operation. We achieve the first goal of informing each PPGS about expected memory latency by modifying the cache controllers to send tokens on cache misses. The tokens include an estimate of the lower-bound access latency for a next-level memory hit, derived from Table 3, and a time stamp of creation. The controllers send the tokens to the PPGS of the core that requested the memory access. Once the PPGS receives the token, it looks at the lower-bound latency to satisfy the request and power gates the core if the core is both stalled and idle long enough to save energy. Should the core receive more than one token for simultaneous memory requests, it will track each expected response separately and schedule the resumption of core execution to satisfy the earliest response. If a token is delayed in the memory subsystem by a controller or queue, the PPGS can compare its arrival time with its generation time stamp and previous tokens, to determine whether the token should be ignored.

Whenever a memory request misses all the way to the memory controller, the response latency experiences a significant amount of variability. This variability is caused by the complexity of DRAM memory [19], which includes bank queues, availability of the data in the row buffer, writing wrong address row buffers, accessing the column in the row buffer, and channel contention between banks. TAP adapts to memory variability by adding a special token. As soon as the last-level cache experiences a miss, a token is sent to the requesting core's PPGS with an estimated completion time of *UNKNOWN*. This is a directive to the PPGS to start power gating its core immediately and to expect one additional token with the ETA of the memory response. Once the memory controller submits the memory access to one of the banks and determines whether the access is a row buffer hit or miss, it sends the second *ETA* token to the core's PPGS with the ETA of the response assuming that there is no memory channel contention. The PPGS schedules the core to wake-up after it receives the second token.

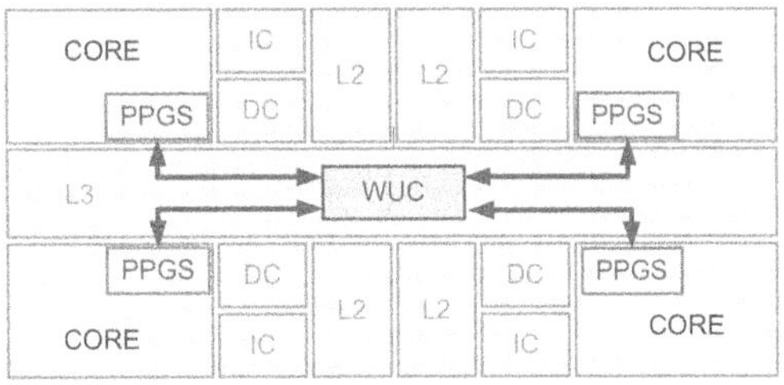

Figure 4: WUC and PPGS integration into a 4-core CMP.

Distributing Safe Wake-Up Modes. To achieve safe and minimal wake-up latencies for a power-gated core, TAP requires each core to register with the wake-up controller (WUC) (see Figure 4), and set whether it is *active* or *idle*. If the PPGS registers with the WUC as *active*, the WUC returns the worst-case lowest latency wake-up mode defined by Equation (1). If the PPGS registers with the WUC as *idle*, the WUC will send the new wake-up mode to the remaining active cores. As soon as the system boots, firmware causes each core's PPGS to send a command packet to the WUC. A core is neither allowed to power gate during a memory stall, nor allowed to wake-up from a power-gated state, until it has registered with the WUC and been assigned a wake-up mode. This prevents any violation of voltage noise constraints.

When a core stalls and power gates, the PPGS may use the worst case wake-up latency it initially received from the WUC and save energy. However, this results in unnecessarily long wake-up latency and reduced energy savings. To enable stagger and more aggressive wake-up modes as a function of whether neighboring cores are active, waking-up, or idle, the PPGS may query the WUC for a lower latency wake-up. If the WUC replies in time, the PPGS may delay its wake-up procedure for the difference between the worst-case and actual wake-up delay defined by Equation (1). Because only adjacent cores have a significant effect on a core's wake-up latency, with appropriate wake-up latency guardband, only adjacent cores must communicate with the WUC, which can be done before the PPGS must start the wake-up sequence. Often, the result is lower core wake-up latency and greater energy savings.

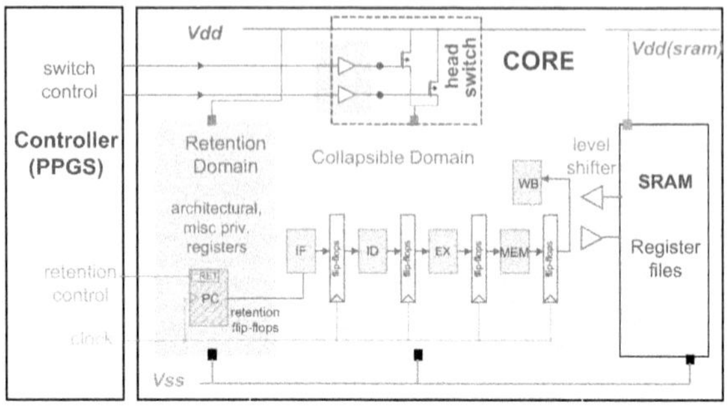

Figure 5: Interface for power gating and data retention.

Retention of Essential Core State. To avoid losing core state that is required for correct and efficient execution, essential sequential and SRAM cells must be retained. We use the technique from [9] which replaces a subset of sequential cells with live-slave retention flip-flops [5] which can be triggered to retain their logical values before a power gating action at a cost of 20% increase in area and power versus a normal flip-flop. Only those sequential cells comprising the architectural registers necessary to refill the pipeline are selected, which results in 3.4% area overhead for the processor. SRAM cells are retained through source biasing [16] in which the supply voltage is reduced to 50% of nominal supply voltage so that SRAM leakage is reduced, but logical state is maintained. This technique allows for saving the contents of L1 caches, TLBs, branch predictor state, physical registers, etc. To provide supply power during power gating, a separate non-collapsible voltage domain provides power to the retention flip-flops and SRAM cells. Thus, as the power is gated from combinational logic and non-essential sequential cells, the separate voltage rail provides power to maintain core state. The overheads from multiple power domains and separate voltage rails already exist for power-gated cores today. Figure 5 shows core design modifications to support power gating and state restoration.

To retain internal data during power gating, additional cycles are required for the power gating and wake-up sequence. These additional cycles account for time to disable/enable the clock, trigger data retention, refill the pipeline, and de-assert/assert the clamps. We model the entire power down and wake-up sequence as in [9].

Figure 6 shows a timing-accurate diagram of a PPGS power gating the core in response to messages from the TAP system. At time 0ns, a memory request occurs that will miss in the cache hierarchy and cause a memory access. The PPGS then receives tokens for the L1, L2 and L3 misses. Just after receiving the L2 Token, the

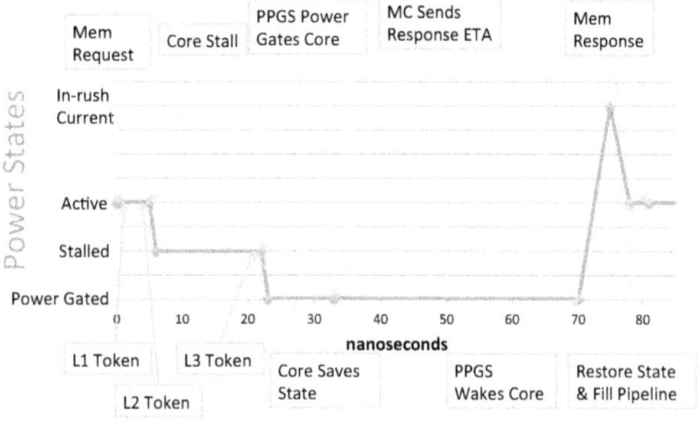

Figure 6: Timing diagram of TAP power gating a core.

core stalls due to a dependency. After the L3 Token is received, the PPGS decides to power gate the core and saves all core state. The core is then power gated and the memory controller (MC) sends an updated ETA for the memory response. At 70ns, the PPGS begins waking up the core. At 78ns, the core state is restored and the pipeline is restarted. The memory response comes back at 81ns and the core resumes execution as if nothing happened.

5. RESULTS

Table 3 summarizes all system values used in our experiments. The system has 4 cores, each with its own private L1 and L2 caches, and a large shared L3 cache. The L3 cache forwards requests to the memory controller through a shared memory bus. The L1 and L2 cache configurations are 32KB-8way and 256KB-8way. The L3 cache is a relatively large, 8MB-16way, which we expect to minimize pressure on the memory subsystem and hence minimize gains we see from our power gating technique. For the core, we model an out-of-order DEC-Alpha EV6 with an issue width of six and clock frequency of 3.3GHz.

We simulate the system with the GEM5 simulator [4]. GEM5 is a full system simulator that can boot an unmodified OS. It features cycle-level models of an out-of-order core, the cache hierarchy, and the interconnect. We integrate GEM5 with DRAMSim2 [2] to provide cycle-level modeling of the memory subsystem including the memory controller, DRAM modules, and shared channels used for communication. We modify GEM5 to support our power gating token methodology described in Section 4. We simulate our system with 21 of the Spec2006 benchmarks using the simpoint methodology [15] in which 100M-instruction representative regions of execution are determined for each benchmark. To simulate each region, we fast-forward to 100M instructions before the region, warm-up the memory and caches, and then perform the detailed simulation.

Once simulation is complete, we feed the system configuration and performance counters to McPAT [12] to model power consumption. McPAT is comprised of a power, area, and timing framework that provides off-line power and area estimates for full systems designed in technology nodes between 90nm and 16nm. McPAT generates values for dynamic power, leakage power, peak power, thermal design power, and area. We update McPAT's *technology.cc* file to accurately reflect the ITRS 2010 update report [1].

Comparison to DVFS. We compare TAP to dynamic voltage and frequency scaling (DVFS). We calibrate our DVFS settings to

parameter	value	notes
Core Model	DEC-Alpha EV6	
Core clock	3.3GHz-1.9GHz	
Execution	6-way out-of-order	
Functional Units	6ALU,2IMULT 2FPALU	
ICache/Dcache	32KB-8way 1cyc	
L2 Cache	256KB-8way 4ns	Private per core
L3 Cache	8MB-16way 13ns	Shared
Memory	DDR3 2GB 50ns	
Core-to-L1 token latency	0.5ns	controller delays
Core-to-L2 token latency	4.5ns	controller delays
Core-to-L3 token latency	17.5ns	controller delays
Core-to-WUC latency	5ns	controller delays
PPGS wake-up modes	4.5ns-16.9ns	SPICE
EV6 pipeline refill latency	2.12ns	7 pipeline stages
EV6 core wake-up energy (EWE)	15,358pJ	Charge cells
EV6 leakage power (ELP)	0.916 Watts	McPAT [12]
EV6 PG leakage reduction (EPLR)	97.65%	[5]
EV6 PG break even point	17.17ns	$EWE/(EPLR*ELP)$
EV6 DFLT core wake-up latency	10.2ns	SPICE
FUPG wake-up energy	9641pJ	McPAT, ITRS [1]
FUPG wake-up latency	6.4ns	SPICE

Table 3: System Configuration Values

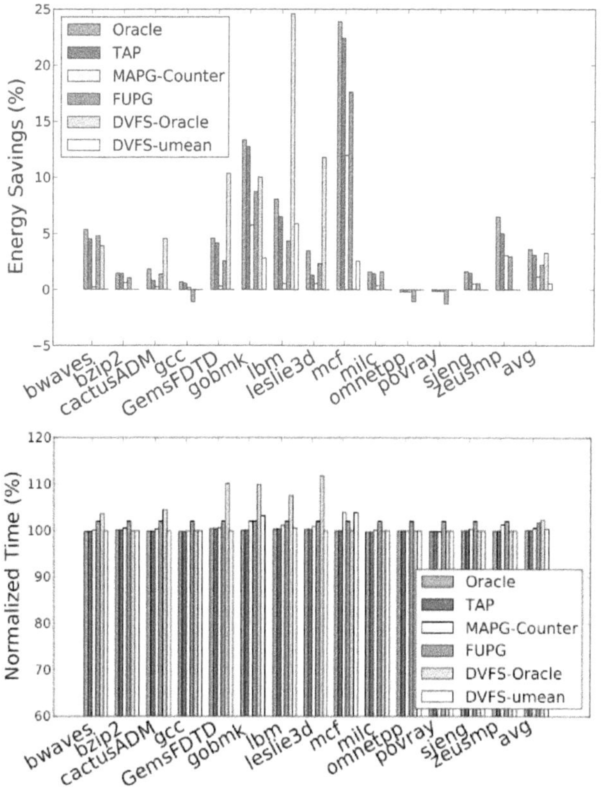

Figure 7: Energy savings and performance overhead of power gating Oracle, TAP, MAPG-Counter, FUPG, DVFS-Oracle and DVFS-μmean. Benchmarks with less than 1% absolute change were filtered out for readability

match those of [21] for the 32nm technology node, in which a 7.5% reduction in voltage follows each 20% reduction in frequency. To direct the DVFS policy, we apply the technique from [20], which uses a cycle-per-instruction based metric, μmean, to detect memory bounded phases of execution. During execution, we sample the application's μmean to determine the most aggressive DVFS setting that may be used to save energy while sustaining at most a 5% performance hit. In addition, we also consider an *oracle* DVFS technique that chooses the DVFS point that results in the lowest energy delay product (EDP). This technique takes an arbitrary performance hit as long as more energy is saved. For both policies, we model the availability of five DVFS modes which include 100%, 95%, 90%, 80%, and 60% frequency.

Comparison to MAPG. We also compare TAP to MAPG-Counter [9]. This scheme works by observing core stall time and predicting its duration. If the core stalls for longer than the average hit latency of the last-level cache, then MAPG-Counter predicts that the stall will last for the latency of a row buffer miss plus δ, where δ is an exponential weighted moving average which adapts to variations in the memory latency. MAPG-Counter uses this predictor to direct its power gating strategy.

For our reported results, we assume a four-core system that is 50% utilized (two cores idle) in a 32nm technology. When determining energy and performance values, we consider core wake-up energy, core wake-up delay, core pipeline refill latency, retention overhead of live-slave retention cells, SRAM leakage during source biasing mode of operation, core-to-WUC communication overhead, staggered wake-ups, and PDN voltage noise safety.

Comparison to Oracle Prediction. Figure 7 compares the energy savings of TAP with the energy savings of an oracle memory predictor (Oracle), MAPG-Counter [9], Functional Unit Power

Gating (FUPG) [8, 13, 14], DVFS-Oracle, and DVFS-μmean. To understand the limit of energy savings from power gating cores during memory stalls, the oracle memory predictor assumes *a priori* knowledge of all memory accesses and latencies and determines the optimal power gating behavior. The EV6 oracle policy achieves a maximum of 23.9% energy savings, and 3.6% savings on average. A few benchmarks show negative energy savings as high as -0.2%. These negative energy savings are caused by the lack of power gating opportunities and the retention cells' power overhead on cpu-bound benchmarks.

In comparison with the Oracle, TAP must determine memory latencies in a running system to ensure that sufficient time is available to power gate a core. TAP EV6 is able to achieve 22.4% (23.9% is Oracle) maximum energy savings, and 3.10% on average. TAP does not achieve the same energy savings as the Oracle because TAP is not able to power gate memory accesses until they miss in the L3 cache, and because lower-bound latencies are used. The result is that TAP avoids any performance hit but misses out on power gating at the beginning of the core stall. TAP also sees a few benchmarks with -0.2% energy savings due to cpu-bound behavior.

MAPG-Counter [9] power gates cores after the core stalls for longer time than the L3 hit latency, and power gates for the predicted stall duration according to its exponential learning algorithm. MAPG-Counter EV6 is able to achieve up to 12.0% energy savings (1.2% savings on average). TAP is able to achieve 2.58× the average energy savings of MAPG-Counter for out-of-order cores. (If method *A* saves 1% energy on average, and method *B* saves 3% energy on average, we say that "method *A* achieves 3.00× the average energy savings of method B".) The reason for this finding is that out-of-order cores stall more randomly than in-order cores, which makes the adaptive counter mechanism unstable and more prone to misprediction.

FUPG EV6 has a maximum and average energy savings of 17.6% and 2.2% with a maximum performance hit of 2% (1.5% average), at which point control logic prevents future power gating actions. The FUPG mechanism does better on average for the out-of-order core than MAPG-Counter, but TAP achieves 1.4× the average energy savings of the FUPG mechanism. FUPG does achieve more energy savings than TAP on a few cpu-bound integer codes in which not all the functional units are being used, but the core does not go idle. Greater energy savings could result from the cooperation of FUPG and TAP.

We also examine DVFS-Oracle using the scaling properties described in Section 5. The maximum and average EV6 core energy savings are 24.6% (lbm) and 3.3%, respectively. DVFS-Oracle on an EV6 core sees less maximum and average energy savings compared to the power gating oracle, but greater savings when compared to TAP. However, these energy savings suffer from two shortcomings. First, DVFS-Oracle has a maximum and average performance hit of 11.8% (GemsFDTD) and 2.4%. Second, these energy savings rely on oracle knowledge.

To understand DVFS under a realistic state-of-the-art policy, we consider DVFS-μmean [20], which predicts the performance hit of DVFS at each interval based on performance counters. DVFS-μmean achieves a maximum and average energy savings of 5.9% and 0.6% respectively. Thus, DVFS-μmean achieves less energy savings than TAP while experiencing a 1.0% performance hit on average (3.9% maximum). This result highlights the challenge of effectively applying DVFS at 32nm to match different application behaviors, and why TAP's determinism can result in higher energy savings.

6. CONCLUSION

With each successive generation of microprocessors, leakage power becomes an increasingly dominant issue. TAP effectively reduces wasted leakage power for out-of-order cores waiting for the memory subsystem. TAP achieves 22.39% maximum energy savings with no performance hit for out-of-order cores. Our method also achieves 2.58× the average energy savings of MAPG-Counter and 5.17× the average energy savings of a practical DVFS policy. Further, we are able to produce an accurate model of safe wake-up modes for an arbitrary arrangement of core states (waking-up, idle, active) in any arrangement of locations; this enables more aggressive wake-up modes for out-of-order cores. Our model reveals the importance of staggered wake-up as a technique to reduce voltage noise fluctuations across the power distribution network, as several cores try to wake up simultaneously. A staggered wake-up of 0.9ns between adjacent cores reduces core wake-up latency by up to 40.3%.

7. ACKNOWLEDGMENTS

This research was supported by Oracle, NSF, the MARCO FCRP (MuSyC and GSRC centers), and Qualcomm.

8. REFERENCES

[1] *International Technology Roadmap for Semiconductors*, 2010, http://www.itrs.net/Links/2010ITRS/2010Update .

[2] *DRAMSim2*, 2011, http://www.ece.umd.edu/dramsim/ .

[3] K. Agarwal, H. Deogun, D. Sylvester and K. Nowka, "Power Gating with Multiple Sleep Modes", *Proc. ISQED*, 2006, pp. 633–637.

[4] N. L. Binkert, R. G. Dreslinski, L. R. Hsu, K. T. Lim, A. G. Saidi and S. K. Reinhardt, "The M5 Simulator: Modeling Networked Systems", *IEEE Micro* 26(4) (2006) pp. 52–60.

[5] D. Flynn, R. Aitken, A. Gibbons and K. Shi, *Low Power Methodology Manual*, Springer, 2007.

[6] K. He, R. Luo and Y. Wang, "A Power Gating Scheme for Ground Bounce Reduction during Mode Transition", *Proc. ICCD*, 2007, pp. 388–394.

[7] M. Horiguchi, T. Sakata and K. Itoh, "Switched-Source-Impedance CMOS Circuit for Low Standby Subthreshold Current Giga-Scale LSI's", *IEEE JSSC* 28(11) (1993) pp. 1131–1135.

[8] Z. Hu, A. Buyuktosunoglu, V. Srinivasan, V. Zyuban, H. Jacobson and P. Bose, "Microarchitectural Techniques for Power Gating of Execution Units", *Proc. ISLPED*, 2004, pp. 32–37.

[9] K. Jeong, A. B. Kahng, S. Kang, T. S. Rosing and R. Strong, "Memory Miss Power Gating", *Proc. DATE*, 2012, pp. 1054–1059.

[10] R. Kumar and G. Hinton, "A Family of 45nm IA Processors", *IEEE ISSCC*, 2009, pp. 58–59.

[11] J. Leverich, M. Monchiero, V. Talwar, P. Ranganathan and C. Kozyrakis, "Power Management of Datacenter Workloads using Per-Core Power Gating", *IEEE Computer Architecture Letters* 8(2) (2009) pp. 48–51.

[12] S. Li, J. H. Ahn, R. D. Strong, J. B. Brockman, D. M. Tullsen and N. P. Jouppi, "McPAT: An Integrated Power, Area, and Timing Modeling Framework for Multicore and Manycore Architectures", *Proc. MICRO*, 2009, pp. 469–480.

[13] A. Lungu, P. Bose, A. Buyuktosunoglu and D. J. Sorin, "Dynamic Power Gating with Quality Guarantees", *Proc. ISLPED*, 2009, pp. 377–382.

[14] N. Madan, A. Buyuktosunoglu, P. Bose and M. Annavaram, "A Guarded Power Gating for Multi-Core Processors", *Proc. HPCA*, 2011, pp. 291–300.

[15] E. Perelman, G. Hamerly, M. Van Biesbrouck, T. Sherwood and B. Calder, "Using SimPoint for Accurate and Efficient Simulation", *Proc. Intl. Conference on Measurement and Modeling of Computer Systems*, 2003, pp. 318–319.

[16] H. Qin, Y. Cao, D. Markovic, A. Vladimirescue and J. Rabaey, "SRAM Leakage Suppression by Minimizing Standby Supply Voltage", *Proc. ISQED*, 2004, pp. 55–60.

[17] Y. Shin, J. Seomun, K.-M. Choi and T. Sakurai, "Power Gating: Circuits, Design Methodologies, and Best Practice for Standard-Cell VLSI Designs", *ACM Trans. on Design Automation of Electronic Systems* 15(4) (2010) pp. 1–37.

[18] H. Singh, K. Agarwal, D. Sylvester and K. Nowka, "Enhanced Leakage Reduction Techniques Using Intermediate Strength Power Gating", *IEEE Trans. on VLSI Systems* 15(11) (2007) pp. 1215–1224.

[19] H. Zheng and Z. Zhu, "Power and Performance Trade-Offs in Contemporary DRAM System Designs for Multicore Processors", *IEEE Trans. on Computers* 59(8) (2010) pp. 1033–1046.

[20] G. Dhiman and T. Rosing, "Dynamic Voltage Frequency Scaling For Multi-Tasking Systems Using Online Learning", *Proc. ISLPED*, 2007, pp. 207–212.

[21] M. Cho, N. Sathe, M. Gupta, S. Kumar, S. Yalamanchilli and S. Mukhopadhyay, "Proactive Power Migration to Reduce Maximum Value and Spatiotemporal Non-uniformity of On-Chip Temperature Distribution in Homogeneous Many-Core Processors", *Proc. Semiconductor Thermal Measurement and Management Symp.*, 2010, pp. 180–186.

[22] O. Wechsler, "Setting New Standards for Energy-Efficient Performance", *Technology@Intel Magazine*, 2006.

Design Trade-Offs for High Density Cross-Point Resistive Memory

Dimin Niu†, Cong Xu†, Naveen Muralimanohar‡, Norman P. Jouppi‡, Yuan Xie†
†The Pennsylvania State University, ‡HP Labs
†{dun118,czx102,yuanxie}@cse.psu.edu
‡ {naveen.muralimanoharnorm.jouppi}@hp.com *

ABSTRACT

With conventional memory technologies approaching their scaling limit, the search for a new technology has gained increased attention in the recent years. Resistive RAM (ReRAM), with its superior write latency and energy, small cell size ($4F^2$ for a single level cell, F is the feature size), and support for 3D stacking, has been a promising candidate among emerging memory technologies. A key advantage of ReRAM comes from its non-linear nature, which enables a cross-point array structure without having a dedicated access transistor for each cell.

While the cross-point structure is effective in improving the memory density, it has inherent disadvantages which introduce extra design challenges. Based on the device characteristics, we perform a comprehensive analysis of issues related to reliability, energy consumption, area overhead, and performance of the cross-point arrays. In addition to the cell-level analysis, we discuss different programming schemes specifically suited for cross-point arrays. We then study the area, energy, and bandwidth of a 256Mbits ReRAM macro in detail for various write schemes. The simulation results enable designers to identify the most performance/energy/area efficient ReRAM organization and cell parameters that meet specific design goals early in the design stage.

Categories and Subject Descriptors

B.3 [**Memory Structure**]: Performance Analysis and Design Aids

General Terms

Design, Performance, Reliability

Keywords

Non-volatile Memory, Resistive Memory, Cross-point Array

1. INTRODUCTION

The scaling of traditional memory technologies, such as DRAM and FLASH, is approaching its physical limit. In the past few years, emerging non-volatile memory technologies (NVM), such as Phase Change RAM (PCRAM), Spin-Transfer-Torque RAM (STT-RAM), and Resistive RAM (ReRAM) have been widely studied as potential candidates for the next generation memory technologies to meet the requirement of higher density, faster access time, and lower power consumption. Among all of these emerging memory technologies, ReRAM has many unique characteristics, including simple structure, nonlinearity, and high resistance ratio, making itself one of the most promising technologies. Researchers have shown that the state-of-the-art single-level-cell ReRAM can achieve $7.2ns$ random access time for both read and write operations with a resistance ratio larger than 100 [1]. Also, HP Labs and Hynix have already announced plans to commercialize memristor-based ReRAM and predicted that ReRAM could eventually replace traditional memory technologies [2].

Unlike other non-volatile memory technologies, ReRAM can be implemented in a cross-point style structure without any access device [3, 4]. Specifically, in a nano cross-point array, each bistable ReRAM cell is sandwiched by two orthogonal nanowires. Thus the area occupied by each cell is $4F^2$ per bit. However, the simplicity of the access-device-free, cross-point structure introduces challenges to the peripheral circuit and memory organization design. While there have been prior studies on cross-point ReRAM arrays [5–9], they do not consider the effect of voltage drivers and programming methods on the array. In addition, detailed area, energy, and performance analysis is also absent. In this work, we address the design challenges of cross-point structure based ReRAM. We use a mathematical model to evaluate memory reliability, energy consumption, and area overhead for different designs and cell parameters. The advantages of nonlinearity K_r and write current I_w scaling are all discussed in detail. In addition, the simulation results of area, energy, and write throughput trade-offs are presented. Our study allows for exploring the most energy/area efficient ReRAM design with different design constraints and cell parameters at the very beginning of the design stage. Moreover, system designers can also leverage the proposed model to provide valuable feedback to device researchers who will in turn adjust ReRAM cell design. We believe that this kind of collaboration will be very helpful to shorten the time to market of ReRAM memory.

2. PRELIMINARIES

This section provides background of ReRAM and cross-point architecture, and discusses the modeling of cross-point ReRAM array.

2.1 Background of ReRAM Technology

As implied by its name, a ReRAM cell uses its resistance to represent the stored information. A ReRAM cell is built on a Metal-Insulator-Metal (MIM) structure and can be switched between a high resistance state (HRS) and a low resistance state (LRS) by applying an external voltage across the cell. The resistance switching behaviors have been observed in many MIM nanodevices with different metal oxide materials. For example, a particular TiO_2 based MIM structure ReRAM, named 'memristor', was developed by HP Labs in 2008 [10]. The proposed memristor-based ReRAM is considered as the first experimental realization and a theoretical model of the fourth fundamental circuit element, which is predicted by Chua [11] about 40 years ago. It has been reported that the memristor-based ReRAM has very small cell size with an access time of less than 50ns [12]. Another H_fO_2-based bipolar ReRAM prototype was fabricated by ITRI with an access time as low as 7.2ns [1].

Although there are several variants of ReRAM cells, all of them can be classified into two broad categories: unipolar ReRAM and bipolar ReRAM. In a unipolar cell, the resistance switching behaviors do not depend on the polarity of the voltage input across the cell and are only

*Dimin Niu and Cong Xu are supported in part by SRC grants, NSF 1147388, 0903432, and 0643902. This material is based upon work supported by the Department of Energy under Award Number DE - SC0005026. The disclaimer can be found at http: www.hpl.hp.com/DoE-Disclaimer.html

related to magnitude and duration of the voltage input. On the other hand, in a bipolar cell, the voltage polarity for ON-to-OFF switching (RESET operation) is different from OFF-to-ON switching (SET operation). The need of different pulse widths for SET and RESET in unipolar ReRAM means that its write latency is determined by the longest pulse. Moreover, the control of SET, RESET, and read operations without any disturbance is another crucial design challenge, especially in high speed ReRAM design. For these reasons, most high performance ReRAM studies are dominated by bipolar ReRAM [1,4,13,14]. In this study, we perform a detailed analysis of design challenges of bipolar ReRAM cross-point arrays.

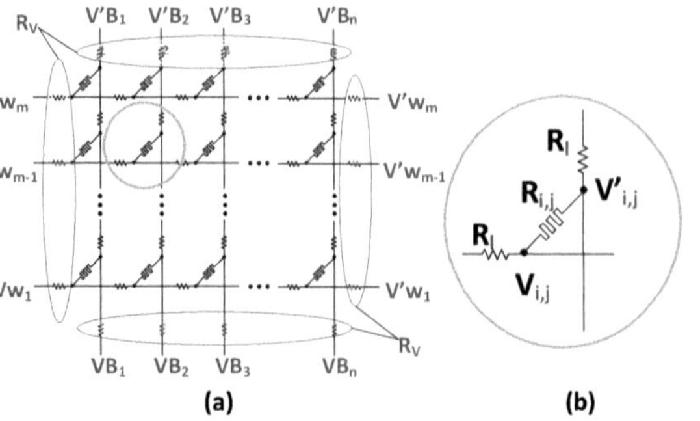

Figure 1: The circuit model of the cross-point array.

2.2 Cross-Point Architecture

There are two possible memory structures for a bipolar ReRAM array: the traditional MOSFET-accessed structure and the cross-point structure. In the former case, a dedicated MOSFET is used as an access device for each memory cell. As the size of a MOSFET access device is typically much larger than the size of a ReRAM cell, the total area of memory array is primarily dominated by MOSFETs rather than ReRAM cells. Also, in order to provide enough drive current, larger than minimum-sized transistor should be used for write operations. Hence, ReRAM's area advantage goes down significantly because of the access devices. Fortunately, we can exploit the non-linear I-V characteristic of some ReRAM devices to eliminate the access device [12, 15]. The I-V characteristic demonstrated in these fabricated devices shows that the resistance of ReRAM significantly increases as the voltage applied on it decreases. Such observation basically indicates effective cut off of the leakage current from the unselected cells in the sneak paths. Therefore, the area-efficient cross-point ReRAM memory array is enabled by the intrinsic property of the device [16]. ReRAM cells are sandwiched between wordlines and bitlines without access devices, which indicates that each ReRAM cell occupies an area of $4F^2$, the theoretical lower limit for a single layer single level memory cell. In addition, this memory density can be further improved by using a multi-layer multi-level cross-point ReRAM array [3] [17].

Although avoiding access transistor is beneficial from cell area standpoint, it introduces other complexities. Following the traditional writing method in which all the bits activated by a wordline are written at once, now needs two steps to prevent unintentional writing [16]. An alternate way is to write one bit at a time but this requires interleaving data across multiple arrays to reduce write latency. Also, while writing to a cross-point array, the unselected wordlines and bitlines can be either left floating or half-biased. In contrast, while reading a cell, the selected wordline should be biased with a read voltage and all the other wordlines and bitlines in the array are shunted to ground. The current in each bitline is then sensed and compared to a reference current to determine the cell content. However, due to the sneak current existing in the cross-point array, the current in bitlines also varies depending upon the data patterns of unselected cells. This read disturbance restricts the size of a cross-point array, since sneak current increases as the number of cells attached to wordlines and bitlines increases, which makes it difficult to sense the current difference of the selected cell at HRS and LRS. Besides, the existence of the voltage drop along the nanowires also limits the length of wordlines and bitlines. Therefore, a cross-point array should be sized carefully to meet the requirements of the read/write reliability. In general, writes are more problematic than reads. The read disturbance problem can be alleviated by adopting a two-level differential sensing scheme, in which the first level reads the background noise followed by a read to data with the background current. Finally, the differential signal is amplified to get the data. In addition to all of these write/read schemes, different cell parameters will also impact the reliability, energy consumption, bandwidth, and area efficiency of the cross-point ReRAM array. In this case, it is not straightforward for a designer to figure out how to design a workable memory array with the minimum energy consumption and area overheads. Thus, the following sections will propose a worst-case oriented methodology to help designers make decisions early in the design flow.

2.3 Modeling of the Cross-Point Memory

The basic circuit model of an M by N cross-point ReRAM array is shown in Figure 1. This model is built upon Kirchhoff's Current Law (KCL) and its validity can be guaranteed by deductions from basic circuit theory. The horizontal lines are wordlines and the vertical lines represent bitlines. The ReRAM cells are located at each wordline and bitline cross-point. A detailed cross-point structure is also shown in Figure 1 (b). The resistance of the ReRAM cell at the cross-point of ith wordline and j^{th} bitline is represented by $R_{i,j}$. We assume the resistance of the wire connecting two cross-points to be R_{line}. The input resistance of each wordline or bitline driver is R_v and the resistance of a sense amplifier is R_s. In order to set up the KCL equations, the voltage at each cross-point is indicated as $V_{i,j}$ for the wordline layer and $V'_{i,j}$ for the bitline layer. In addition, the input voltage for the ith wordline is V_{Wi} and for the ith bitline is V_{Bi}. In the case that a wordline is driven from both sides, the voltage at the other end of the ith wordline is represented as V'_{Wi}.

Based on this model, the current equations for each cross-point can be obtained. All of the cross-points have similar structure with no more than three current branches, and therefore it is very easy to set up the KCL equations for each cross-point. Since the cross-points at the edges of the array have different write/read conditions, the KCL equations of these cross-points should be adjusted according to each write/read scheme. All of the KCL equations can be considered as a system of linear equations, which has the form of $A \cdot V = C$, where A is a $2mn \times 2mn$ coefficient matrix and C is a $2mn \times 1$ vector, containing the constant terms of these equations. Thus, with parameters such as the resistance of ReRAM cells, the resistance of interconnect wires, program voltages, and write/read schemes, voltages at various cross points can be obtained by solving the system of linear equations. With detailed voltage values, $V_{2mn\times1}$, we can analyze the array at a fine granularity. These values are also critical to evaluate the reliability, energy consumption, drive current density, and area overheads of a cross-point array.

To validate the analytical model, we compare the results of our model

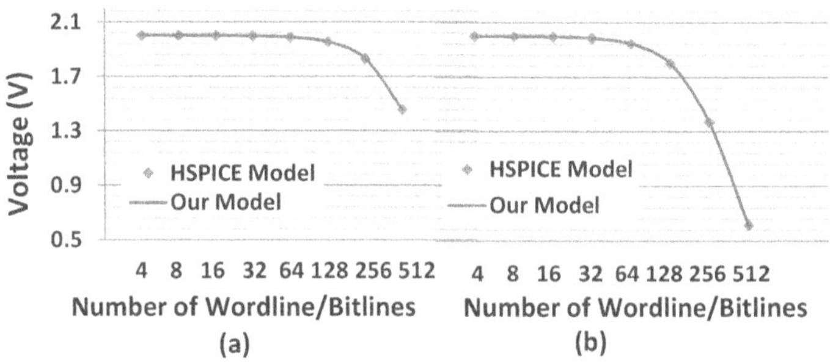

Figure 2: Validation of the analytical model against SPICE simulation. The two figures show the voltage drops obtained from our model and SPICE (a) with a nonlinearity factor of 5 and (b) without nonlinearity.

with HSPICE [18] simulations using a resistor model in cross-point memory arrays. The results of eight cross-point arrays with different array sizes and specific data patterns are shown in Figure 2, which shows that the voltage drop on the selected cell derived from our analytical model are consistent with the HSPICE simulation results.

3. ANALYSIS OF DESIGN CONSTRAINTS

In this section, we study the effect of various schemes on cross-point ReRAM arrays in detail. Specifically, we evaluate the design constraints on array size, energy consumption and area overhead in worst case scenarios. The results of this study will be useful when designing a cross-point array.

3.1 Overview

In order to write or read a cross-point array, proper voltages should be applied across the ReRAM cell. Although the goal of a read operation is different from a write operation, both of them are realized by fully biasing the selected wordlines/bitlines and floating (or half biasing) unselected wordlines/bitlines. Thus, the coefficient matrix A and the constant vector C are very similar for both. In addition, their energy consumption and area overhead will also have a similar trend. Therefore, in this section, we first study the write operation comprehensively. After that, for read operation, we mainly focus on the read margin analysis since it is unique for read operations.

Table 1 shows the circuit parameters of our baseline 50nm design. The data is derived from the recently published studies on ReRAM [16, 19, 20]. The nonlinearity coefficient is defined as

$$K_r(p, V) = p \times R(V/p)/R(V), \quad (1)$$

where $R(V/p)$ and $R(V)$ are the equivalent resistance of the cell biased at V/p and V [16]. Therefore, the resistance of a ReRAM cell with nonlinearity is not constant but varies with the applied voltage. For example, for a ReRAM cell with nonlinearity of 20, the resistance of half biased cell is 10 times larger than resistance of fully biased cell. By using these parameters, we study reliability, energy consumption, and area overheads for four different write schemes, and discuss the sensitivities of these schemes to the data pattern of HRS and LRS ReRAM cells and cell nonlinearity. In this section, the baseline design uses a cell with write current of $40uA$ and nonlinearity $Kr = 20$. A sensitivity study varying the nonlinearity coefficient and the write current is presented in Section 4.

3.2 Write Operation

To write a ReRAM cell, an external voltage is applied across the cell for a certain duration. Intuitively, there are four possible schemes for the write operation: *FWFB* scheme activates the selected wordline and selected bitline, and leaves all of other lines floating; *FWHB* scheme activates the selected wordline and bitline, leaves all the unselected wordlines floating, and half biases the unselected bitlines; *HWFB* scheme activates the selected wordline and bitline, leaves all the unselected bitlines floating, and half biases the unselected wordlines; *HWHB* scheme activates the selected wordline and bitline, and half biases the unselected wordlines and bitlines. However, the FWFB scheme has an inherent problem that may result in severe write disturbance [8]. Therefore, only the FWHB, HWFB and HWHB schemes are workable for programming a cross-point array. We analyzed these three schemes and found that they all have the same worst case voltage drop. Besides, we found that the HWHB scheme is the most energy/area efficient among these schemes. Therefore, in the following discussion, we only show the simulation results of HWHB schemes. The results for the other two schemes have the similar trend as that of HWHB scheme.

Besides, during the write operation, we can write only one bit per access (single-bit write), write several bits on one wordline at the same time (multi-bit write), or even write all of the cell on one wordline (whole-wordline write). We found that, at the array level, the energy consumption and area overhead increase monotonically with the increase of the number of bits per access. Therefore, in this section, we provide simulation results of two extreme instances: single-bit write and whole-wordline write operation. Detailed analysis of multi-bit write operation is discussed in Section 5.

Table 1: Parameters of the baseline Cross-Point Array

Metric	Description	Typical Values (Range)
A_{cell}	Cell Size	$4F^2$
R_l	Interconnection Resistance	0.65Ω
V_{RESET}	Threshold voltage for RESET	$2.0V$
V_{SET}	Threshold voltage for SET	$-2.0V$
V_{READ}	Read Voltage of Cell	$0.5V$
I_{on}	Write Current for LRS Cell	$40uA$ ($40 \sim 200uA$)
$V_W(R)$	Wordline Voltage during Read	$0.5V$
$V_W(W)$	Wordline Voltage during Write	$\pm 2V$
$V_W(H)$	Half Selected wordline Voltage	$1V$
$V_B(R)$	Bitline Voltage during Read	$0V$
$V_B(W)$	Bitline Voltage during Write	$0V$
$V_B(H)$	Half Selected bitline Voltage	$1V$
K_r	Nonlinearity of ReRAM Cell	20 ($2 \sim 40$)
M, N	Number of wordlines/bitlines	512 ($8 \sim 1024$)

Reliable Write Operations. Write reliability is a serious concern in cross-point arrays. In an ideal condition, the resistance of wires and the sneak currents in unselected cells are negligible. In such a scenario, all the write schemes discussed above can make sure that the write voltage $V_W(W) - V_B(W)$ is fully applied across the specified cell. However, in reality, both wire resistance and sneak current are non-trivial. Hence, the voltage applied across a cross-point varies based on the location of the cell as well as the data pattern stored in all of the ReRAM cells in the array. A write is considered reliable if it modifies the content of the selected cells to the new value without disturbing other unselected cells. Correspondingly, there are two potential problems with writes: *write failure*, an unsuccessful write on selected cells, and *write disturbance*, an undesirable write to unselected cells. It is necessary to ensure that a write scheme guarantees reliable operation even in the worst case (w.r.t the location of cells to written and the data pattern stored in the cross-point array).

Write failure typically results from the voltage drop at the interconnect wires along the wordline and bitline. It has been shown that, for single-bit write operation, the worst case voltage drop occurs when writing the cell at the cross point of the Mth wordline and the Nth bitline with all of the other cells in the array are in LRS [7]. In order to avoid write failure and successfully program the selected ReRAM cell, the drive voltage should be boosted to a higher level, making sure that the voltage across the selected cell exceeds the threshold voltage even at the worst case. Figure 3 shows the lower bounds of the drive voltage for different sizes of cross-point array. The minimum wordline/bitline voltage increases from 2.01 V for a 32×32 array to nearly 7 V for a 1024×1024 cross-point array. However, boosting the drive voltage also increases the voltage applied at unselected cells. Therefore, a write disturbance may occur when the voltage applied at an unselected cell exceeds the threshold voltage for SET or RESET operation. According to our analysis, the maximum voltage applied at unselect cells is exactly the same as half of the drive voltage. Thus, only arrays with drive voltage less than 4V are allowable. Otherwise, the array is unreliable because it cannot avoid write failure and write disturbance at the same time. The unreliable array sizes are denoted as red bars in Figure 3. The array size limitation provided by Figure 3 is a hard constraint, and all of the following energy and area trade-offs are bounded by this constraint.

Additionally, the cross-point array can be organized with a different number of wordlines and bitlines. For example, a 256Kbit cross-point array can be implemented either by a 512×512 array or by a 64×4096 array. In the latter case, the voltage drops along the wordline will be much worse than along the bitline. Our analysis shows that from a reliability point of view, a cross-point array with the same number of wordlines and bitlines is the best choice. Thus, in the following discuss, we assume the array has the same number of wordline and bitline.

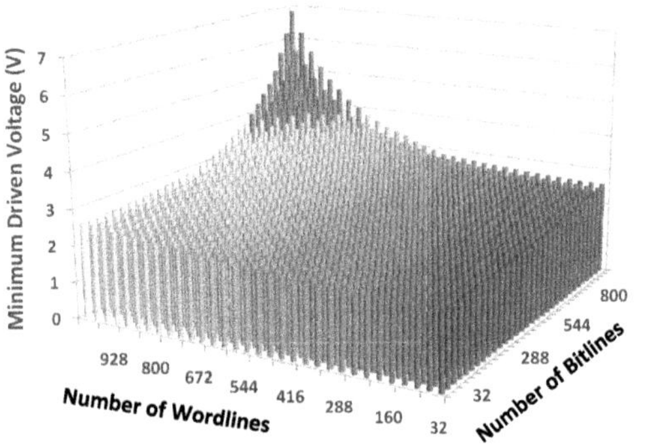

Figure 3: Required write voltages for different cross-point arrays (threshold voltage = 2V.).

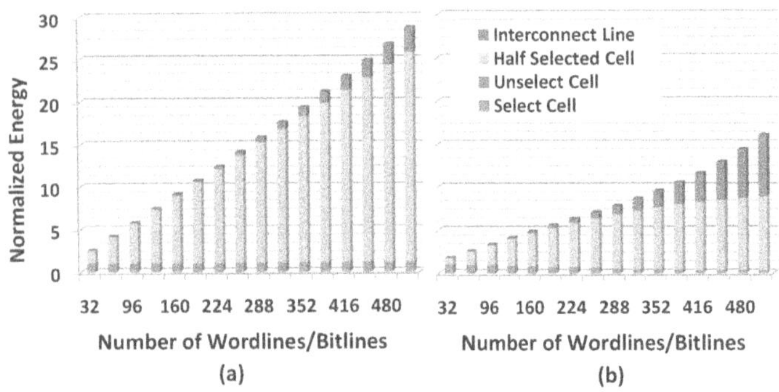

Figure 4: The normalized energy consumption with different array size. (a) Single-bit writing. (b) Whole-wordline writing.

Energy Consumption of Write Operations. The energy consumption of a write operation includes: the energy consumed to change the state of the selected cell, the undesired energy wasted at the half selected cells and unselected cells, and the energy consumed by the interconnect lines. Figure 4 (a) shows the decomposed energy consumption for single-bit write operation. Obviously, the undesired energy consumed by half-selected cells takes a great part of the total energy consumption. Besides, with the increase of array size, the energy dissipated at interconnect lines also becomes significant. Also, this part of the energy wasted during the write operation is a greater part of the total energy for larger array sizes. For example, the undesired energy consumption for writing a 512×512 array is more than 15 times larger than that of a 32×32 array. For whole-wordline write operation, we evaluate the energy consumption of write operations that program the entire wordline at one time. In order to fairly compare the energy consumption, we compare the energy-per-bit instead of the total energy. For example, in order to write a wordline with size of 512 bits, the energy-per-bit can be calculated as: $E_{ave} = E_{total}/512$. Figure 4 (b) shows the energy-per-bit of the whole-wordline write. Compared with the single-bit write operation, we conclude that for large cross-point array sizes, the whole-wordline write operation is much more energy efficient. This is because the energy wasted at the unselected and half-selected cells are amortized by multiple bits and the average energy for one bit is therefore reduced.

Write Current and Area Overhead of Write Operations. The write operation for a $M \times N$ array requires M wordline voltage drivers and N bitline multiplexors. The drivers and multiplexors should be sized such that they can provide the worst-case current of wordline current and bitline current. The transistor sizing of the wordline/bitline circuitry is achieved using HSPICE simulations. We further calculate the area overhead for the drivers and multiplexors by referring to the CACTI area model. Figure 5 (a) shows the maximum write current with different ReRAM array sizes. Not surprisingly, the current requirement increases as the array size increases. Figure 6 (a) illustrates the area overhead for the wordline and bitline circuitry. This show that drivers and multiplexors occupy a smaller area than the cross-point array. Only in this case

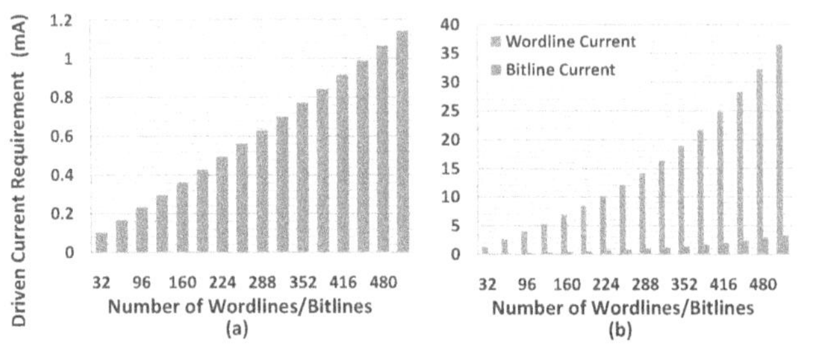

Figure 5: The requirements for wordline and bitline drive currents. (a) One bit per write. (b) One wordline per write.

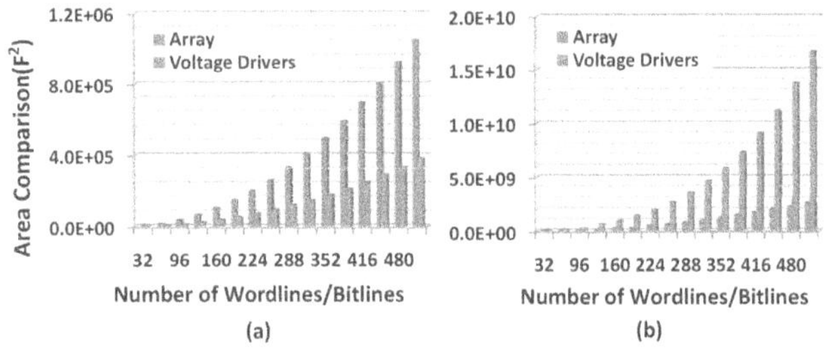

Figure 6: Area overhead comparison. (a) Single-bit write. (b) Whole-wordline write.

can voltage drivers and multiplexors be implemented beneath the array, resulting an ideal cell size of $4F^2$.

Although whole-wordline write operation has the advantage of lower energy consumption, the maximum current requirement for each wordline also increases. As demonstrated in Figure 5 (b), although the maximum drive current for each bitline is almost the same as when writing one bit, the driving current requirement for each wordline in a whole-wordline write scheme is more than 10 times that of a single-bit write scheme. Since the area of the voltage driver increases proportionally with its driving current, the area overhead for whole-wordline writing is much larger than that of single-bit writing. As shown in Figure 6 (b), the peripheral circuitry area is much larger than that of the array. In this case, the total area of the memory array is dominated by the peripheral circuitry rather than the cells. In addition to the extra area overhead, writing multiple bits at one time also worsens the voltage drop along the wordline. Our simulation results show that, in order to program an entire wordline when writing, the maximum reliable array size reduces from 800×800 to 352×352. This is because the current passing through the interconnect wires in the whole-wordline write scheme is much larger than that of the single-bit write scheme, causing more severe voltage drops on the wire resistance.

Therefore, we conclude that although the whole-wordline write operation is more energy efficient, from the standpoint of reliability and area overhead, single-bit write operation is preferred.

3.3 Read Operation

In this section we apply a similar sensing scheme as [6] and [7]. In order to read cell $R_{i,j}$, the ith wordline is biased at V_{READ} and all of the other wordlines and bitlines are grounded. Then the state of the selected cell is read out by measuring the voltage across R_s. The energy consumption for a read operation can be analyzed similarly as a write operation. Since the read voltage is much smaller than write voltage, the read energy is expected to be at least one order of magnitude smaller than for a write operation. Considerable sensing margin is achieved by implementing a current-to-voltage converter and sensing the voltage signal using traditional or more recent sense amplifier designs. The input resistance of the current-to-voltage converter is extracted from HSPICE simulation results. Read sensing margin is defined as $\Delta V = \Delta I \times R_{converter}$ where $R_{converter}$ is the input resistance of the converter. The read reliability is determined by the voltage swing for reading HRS and LRS cells. Detailed results will be shown in Section 4.

4. NONLINEARITY AND WRITE CURRENT SCALING

One of the most distinct features of ReRAM is its nonlinearity. Normally, the K_r value for memristor-based ReRAM is larger than 20, meaning that the resistance of a half-biased cell is at least 10 times larger than a full-biased cell. Clearly, ReRAM cells with larger nonlinearity coefficients result in a better memory cell since the sneak current in half selected cells will be significantly reduced. In addition, the increased resistance at half-selected and unselected cells can also mitigate the voltage drop along the activated wordline and bitline. Also, we find that the cross-point array design can benefit from the scaling of the write current. Figure 7 shows the influence of different nonlinearity coefficients and write currents on the array size requirements for a single-bit HWHB writing scheme. This figure shows that the array size limitation is relaxed as the nonlinearity increases or the write current scales. As we can see from the figure, the maximum array size exceeds 1024×1024 when we have a nonlinearity of 30, together with a write current of $40\mu A$.

Moreover, the increase of nonlinearity or scaling of write current can also reduce the energy consumption and area overhead of the cross-point array. As shown in Figure 8 (a), for a 512×512 array, the energy consumption for the write operation decreases dramatically with the scaling of nonlinearity coefficient K_r. For example, for a ReRAM cell with write current of 50uA, the write energy is reduced by 98.3% when K_r increases from 2 to 40. The area overhead of the voltage drivers is illustrated in Figure 8 (b). As a baseline design ($K_r = 20$ and $I_w = 40\mu A$), the driver area overhead is about 35% of the area of the memory array cells. To design a memory array with an effective cell size close to $4F^2$, we need to make sure that the nonlinearity and write current satisfy certain conditions so that the driver overhead is less than 100% and the wordline drivers can be almost "hidden" underneath the ReRAM cells. As nonlinearity and write current continues to scale, the area overhead can be as low as 10%. In that case, the introduction of 3D stacking of multi-layer cross-point arrays is productive in further reducing the effective cell size to $4/N_l F^2$ where N_l is the number of layers.

Unlike the write operation, the read operation suffers, rather than benefits, from scaling of nonlinearity or write current. This is because the scaling of nonlinearity and write current will reduce read current, degrading the read signal ratio. Figure 9 (a) shows the read noise margin with different array sizes for the baseline design in Section 3. As can be seen, the read noise margin is reduced for large array sizes. The impact of nonlinearity and write current on read noise margin is illustrated in Figure 9 (b). A large K_r value and small write current are harmful to the read noise margin. For example, given a 512×512 array, the read noise margin is less than $10mV$ for $K_r = 40$ and $I_w = 40\mu A$, which makes it very difficult to sense the state of the selected memory cell using traditional sense amplifiers.

Therefore, by knowing the array size and read noise margin constraints, an "optimal cell" with nonlinearity of K_{r_opt} and write current of I_{on_opt} can be determined. For example, when the array size is fixed at 512×512 and the minimum noise margin is $50mV$, a cross-point

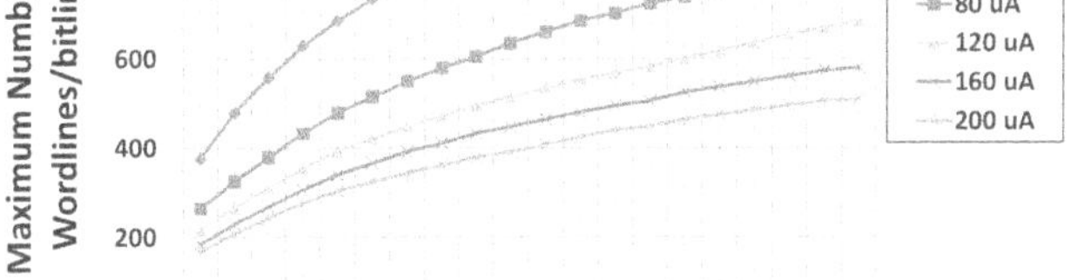

Figure 7: The maximum array size with different nonlinearity coefficients.

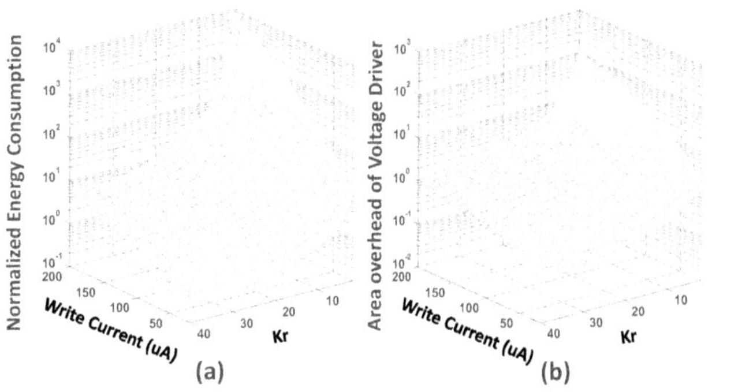

Figure 8: Energy and area overhead comparison. (a) Energy consumption (normalized to baseline). (b) Area overhead of voltage driver (normalized to the area of cross-point array).

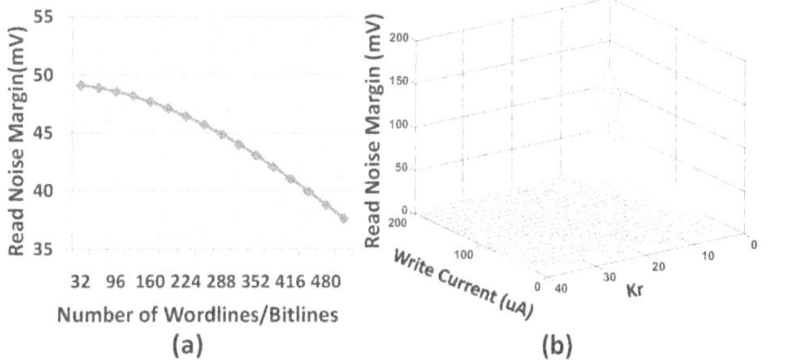

Figure 9: Read noise margin with (a) different array size and (b) scaling of nonlinearity and write current.

array with ReRAM cells which have $K_{r\_opt} = 9$ and $I_{on\_opt} = 40mA$ is the most energy and area efficient design.

5. A CASE STUDY OF CROSS-POINT RERAM MACRO DESIGN

Since the array size of a cross-point ReRAM array is limited by the reliability requirements, the design of a ReRAM macro is different from the traditional DRAM design. In this section, we evaluate the area, energy consumption, and bandwidth of a 256Mbits ReRAM macro. We use an organization similar to Kawahara's design [4], where a 256Mbits ReRAM macro is divided into eight planes. Each 32 Mbit plane has separate wordline decoder, bitline selectors, sense amplifiers, and write circuity. Due to space constraints, we present results for only four typical cell parameters: $(Kr = 20, I_w = 40uA)$, $(Kr = 20, I_w = 200uA)$, $(Kr = 40, I_w = 40uA)$, and $(Kr = 40, I_w = 200uA)$. For each of them, we vary the number of bit per write to investigate the relation among the area, energy consumption, and bandwidth of the ReRAM macro.

Table 2 shows the total area, energy consumption, and bandwidth of the 256Mbits ReRAM macro. Consistent with our earlier discussion, as the device nonlinearity improves, both area and energy goes down. The only downside is the noise margin restriction imposed for reads. Similarly, as the drive current increases, the overhead goes up due to large wordline drivers and bitline multiplexors. Hence, bandwidth improvement comes at the cost of area and energy.

To better understand the ideal design choice for a given device parameter, we investigated three metrics: bandwidth per nanojoule (BW/nJ), bandwidth per square millimeter(BW/mm^2), and bandwidth per nanojoule per square millimeter ($BW/(nJ \cdot mm^2)$). Figure 10 (a) shows how BW/mm^2 scale as we increase the number of bits modified per write operation. From the figure, for a given energy budget, writing one bit at a time provides at least 48% better bandwidth compared to the best performing muti-bit writes. Hence, with the right choice of global interconnect, interleaving writes across multiple sub-arrays is an interesting design point. With multi-bit writes, as the number of bits per write increases, the energy efficiency also increases. However, as the

Table 2: Area, energy, and bandwidth results of 256 Mbits ReRAM macro

Kr	$I_w(uA)$		Number of bit per write at array level							
			1	2	4	8	16	32	64	128
20	40	Area(mm^2)	3.888	3.944	4.056	4.288	4.752	5.688	7.544	11.752
		Energy(nJ)	4.375	12.379	19.860	33.660	59.430	111.404	232.437	576.496
		Bandwidth(Mbit/s)	66.687	72.618	144.747	287.534	567.290	1103.991	2089.847	3649.583
20	200	Area(mm^2)	6.512	6.816	7.424	8.640	11.096	16.144	27.288	N/A
		Energy(nJ)	24.584	67.126	106.848	182.409	338.781	716.141	1845.355	N/A
		Bandwidth(Mbit/s)	90.038	113.028	217.351	401.199	685.218	1018.788	1213.958	N/A
40	40	Area(mm^2)	3.616	3.672	3.776	3.992	4.752	5.688	7.544	11.432
		Energy(nJ)	2.065	5.550	8.711	14.485	25.601	49.322	107.519	280.121
		Bandwidth(Mbit/s)	69.594	74.309	148.111	294.199	580.357	1129.126	2136.164	3777.482
40	200	Area(mm^2)	3.984	4.288	4.888	6.088	8.464	13.328	24.088	N/A
		Energy(nJ)	11.645	29.739	46.939	80.813	155.025	343.658	933.231	N/A
		Bandwidth(Mbit/s)	115.781	131.827	253.686	469.243	800.280	1174.237	1362.144	N/A

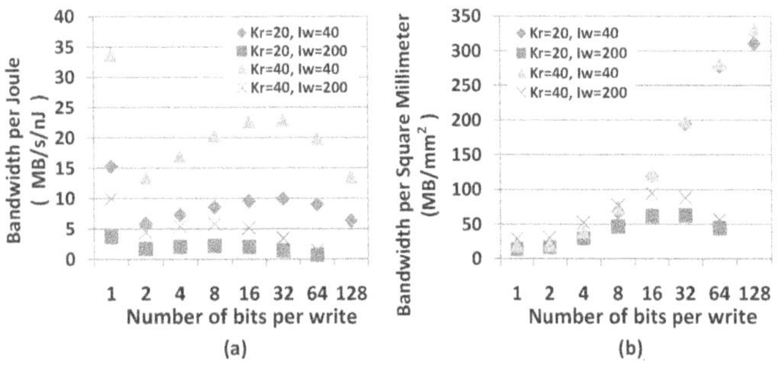

Figure 10: (a) Bandwidth per Joule and (b) bandwidth per square millimeter of 256Mbits ReRAM macro.

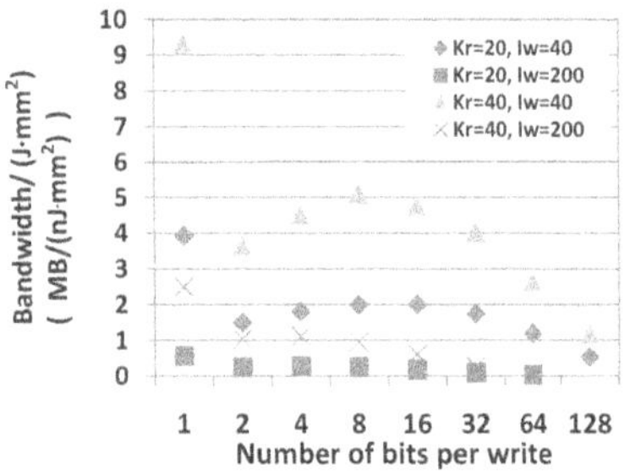

Figure 11: Bandwidth per Joule per square millimeter of 256Mbits ReRAM macro.

word size increases, the voltage drop in the array also increases, which needs to be compensated by increasing the operating voltage of the array (Section 3). Beyond 32 bits, this increase in voltage, outweighing the bandwidth improvement, effectively reducing the energy efficiency. Thus from energy standpoint, multi-bit write is optimal when the word size is 8-32 bits, depending upon the nonlinearity and drive current. Figure 10 (b) shows the effect of multi-bit writes on bandwidth per square millimeter. Unlike energy, as long as the drive current is less, it is beneficial to increase the word size as much as possible to improve bandwidth for a given area. Also, writing one bit at a time is the least attractive option for a design primarily constrained by the area. Figure 11 takes into account both energy and area, and provides a "sweet spot" for multi-bit writes. Thus by understanding the key characteristics of cross-point array, we can identify an optimal configuration that best meets the design constrains.

6. CONCLUSION

In this paper, we use a mathematical model to study in detail how reliability affects the array organization, size, energy consumption, and area overheads of cross-point arrays. The size of a cross-point array is limited by the peripheral circuit overhead as well as the sneak current. Our simulation results show that, with best possible device non-linearity and drive current, the maximum array size cannot exceed 1024x1024 without compromising reliability. We also showed that multi-bit writes is more energy efficient than single-bit write, however, the latter significantly reduces the complexity of peripheral circuits and provides better area efficiency. Both high nonlinearity and low write current are key to reduce energy and area of cross-point arrays. Finally, since memory bandwidth is an important design constraint, we studied various designs that maximizes bandwidth for a given area and energy budget. Through our case study, we show that there is an optimal word size for a given device parameter that has the best energy, area, and bandwidth properties.

7. REFERENCES

[1] S. S. Sheu *et al.*, "A 4Mb embedded SLC resistive-ram macro with 7.2ns read-write random-access time and 160ns MLC-access capability," in *Proc. of ISSCC*, 2011.

[2] "http://www.hpl.hp.com/news/2010/jul-sep/memristorhynix.html."

[3] C. Chevallier *et al.*, "A 0.13 um 64Mb multi-layered conductive metal-oxide memory," in *Proc. of ISSCC*, 2010.

[4] A. Kawahara *et al.*, "An 8Mb Multi-Layered Cross-Point ReRAM Macro with 443MB/s Write Throughput," in *Proc. of ISSCC*, 2012.

[5] M. Ziegler and M. Stan, "Design and analysis of crossbar circuits for molecular nanoelectronics," in *Proc. of IEEE Conf. on Nano*, 2002.

[6] A. Flocke *et al.*, "A fundamental analysis of nano-crossbars with non-linear switching materials and its impact on TiO2 as a resistive layer," in *IEEE Conf. on Nano*, 2008.

[7] J. Liang and H.-S. Wong, "Cross-point memory array without cell selectors -device characteristics and data storage pattern dependencies," *IEEE Trans. on Electron Devices*, vol. 57, no. 10, 2010.

[8] M. Ziegler and M. Stan, "CMOS/nano co-design for crossbar-based molecular electronic systems," *IEEE Trans. on Nanotechnology*, vol. 2, no. 4, 2003.

[9] O. Kavehei *et al.*, "An analytical approach for memristive nanoarchitectures," *IEEE Trans. on Nanotechnology*, vol. 11, no. 2, 2012.

[10] D. B. Strukov *et al.*, "The missing memristor found," *Nature*, 2008.

[11] L. Chua, "Memristor-the missing circuit element," *IEEE Trans. on Circuit Theory*, no. 5, Sep 1971.

[12] J. J. Yang *et al.*, "Memristive switching mechanism for metal/oxide/metal nanodevices," in *Nature Nanotechnology*, vol. 3, Jun 2008.

[13] M. Kim *et al.*, "Low power operating bipolar TMO ReRAM for sub 10 nm era," in *Proc. of IEDM*, 2010.

[14] W. Otsuka *et al.*, "A 4Mb conductive-bridge resistive memory with 2.3GB/s read-throughput and 216MB/s program-throughput," in *Proc. of ISSCC, 2011*.

[15] R. Meyer *et al.*, "Oxide dual-layer memory element for scalable non-volatile cross-point memory technology," in *Proc. of NVMWS*, 2008.

[16] C. Xu *et al.*, "Design implications of memristor-based RRAM cross-point structures," in *Proc. of DATE*, 2011.

[17] M.-J. Lee *et al.*, "Stack friendly all-oxide 3D RRAM using GaInZnO peripheral TFT realized over glass substrates," in *Prof. of IEDM*, 2008.

[18] "http://www.synopsys.com/tools/verification/amsverification/circuitsimulation/hspice/pages/default.aspx."

[19] H. Akinaga and H. Shima, "Resistive random access memory (ReRAM) based on metal oxides," *Proceedings of the IEEE*, vol. 98, no. 12, 2010.

[20] M. Terai, Y. Sakotsubo, Y. Saito, S. Kotsuji, and H. Hada, "Memory-state dependence of random telegraph noise of Ta2O5/TiO2 stack ReRAM," *IEEE Electron Device Letters*, vol. 31, no. 11, 2010.

Performance and Energy-Efficiency Improvement through Modified CPL in Organic Transistor Integrated Circuits

Mingoo Seok
Department of Electrical Engineering, Columbia University
mgseok@ee.columbia.edu

ABSTRACT

We study two logic families, complementary metal oxide semiconductor logic (CMOS) and complementary pass gate logic (CPL) in organic transistor technologies. In nanometer silicon technologies, CMOS logic family generally outperforms CPL in most of the simple gates. However, a significant strength imbalance between p- and n-channel transistors in currently-available organic technologies challenges the use of CMOS logic gates. Direct use of conventional CPL gates is also problematic due to the lack of good inverters for isolations. Therefore, in this paper, we propose a modified CPL logic family with optimized buffers, which can substantially improve both performance and energy efficiency for organic transistor integrated circuits. Benchmark simulations with 8b ripple carry adder also confirm that the modified CPL logic style improves performance and energy efficiency by ~3.2× and ~2.66×, compared to CMOS. We also investigate the impacts of prospective n-channel transistor improvements on the advantages of logic families.

Categories and Subject Descriptors
B.7 [**Integrated Circuits**]: General
General Terms: Design
Keywords: Organic transistor technologies, low power, logic family, CMOS logic, complementary pass gate logic, CPL

1. INTRODUCTION

Organic transistors (OFET), since its invention in 1980's [6], has gained a significant interest for their unique advantages which conventional silicon technologies cannot offer. These benefits include flexibility, biocompatibility, light weight, and low cost [7]. Although the performance and energy efficiency of OFETs are worse by orders of magnitudes than silicon counterparts, those unique advantages make OFET suitable for certain applications such as radio frequency identification (RF-ID) tags, smart packaging materials, wearable electronics, and healthcare systems. One exemplary application can be a wearable health monitor system, which can sense various biological signals such as blood pressure, oxygen saturation (SpO_2) level, and heart rate of patients [8].

Although a significant amount of research is in progress, the current OFETs still face challenges in computational performance, energy-efficiency and area. A recently-demonstrated 8b microcontroller, comprising of ~3400 transistors, operates at a clock frequency of 6Hz [9], which is more than 5-6 orders of magnitude slower than the available 8b-16b microcontrollers in silicon technologies [10]. Although the on-current of the transistors is moderate ($I_{ds,p\text{-}channel}$=120μA at W/L=10, V_{dd}=1.2V) [5], the larger transistor size (L=25μm) and a significant amount of overlap capacitance are some of the root causes for the performance degradation. Power consumption is also significantly higher due to larger capacitance and higher supply voltage (1~20V) (Figure 1). System footprint is also significantly larger; the 8b microcontroller occupies 19.6×17.2mm^2 [9] and 64b DRAM array takes up 2.4×2.5mm^2 [11].

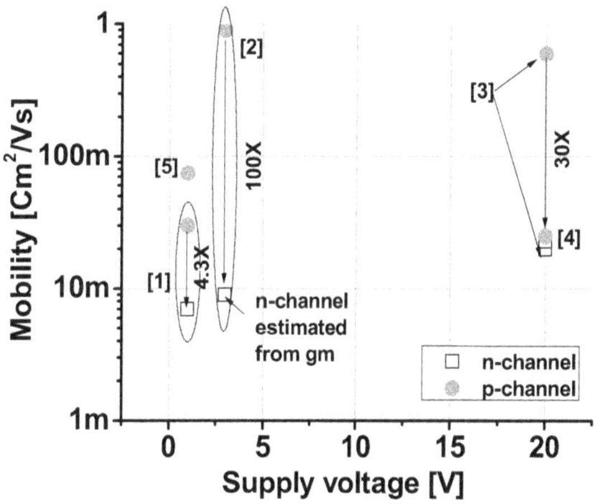

Figure 1. Survey of the supply voltages and mobility's of the recently developed organic transistor technologies

Another technological challenge is that the mobility of n-channel devices are typically ~5-30× lower than p-channel counterparts [1][2][3][4], as shown in Figure 1. Although research on n-channel transistors is underway, introduction of a viable n-channel material has been delayed as potential materials have unstable characteristics and bad contacts. The technology in [4] provides similar n- and p-channel strength, however it has a 26× lower mobility than other technologies with strong p-channel devices.

This imbalance of n-channel and p-channel strength places conventional complementary metal oxide semiconductor (CMOS) style digital gates at a disadvantage in every aspect (delay, power, and area) since n-channel devices should be upsized to roughly balance rise and fall transition delay otherwise excessive short circuit current and long delay can be caused in large-scale integrated circuits. Different logic styles such as resistor-load gates [11] and pseudo-PMOS have been suggested; however they suffer from large static current since transistors are not completely turned off. Also, the output swing is not rail-to-rail, resulting in worse robustness. Due to these challenges digital computations with organic technology are still substantially slower and more power-hungry than desired for many applications.

In order to overcome these challenges, better digital circuit design is paramount. Therefore, in this paper, we investigate two logic families, CMOS and complementary pass gate logic (CPL) style [12][13]. We find that in the currently-available organic technologies with highly-skewed p- to n-channel strength, CPL is promising since its operation is mostly dependent on one type of transistors; the other type of transistors only comes into a switching operation at the last fraction of transitions. CPL gates also provide full-swing outputs.

However, the lack of good inverters and buffers challenges the direct use of CPL in organic technologies. Therefore, we propose a new CPL inverter, where both n-channel and p-channel pull-down transistors are used. The modified CPL gates with CPL inverters, which we refer to as mCPL, outperforms CMOS gates in both delay and energy efficiency in the comparisons of simple gates such as 2-input NAND, NOR, and XOR. We also extend our comparison to larger building blocks such as an 8b ripple carry adder (RCA), confirming mCPL improves performance and energy efficiency by 3.2× and 2.66× over CMOS implementations.

The remainder of the paper is organized as follows. After briefly introducing the basic ideas of logic families in Section 2, in Section 3 we propose circuit modifications and sizing methods for CMOS and CPL logic families in an organic transistor technology. Then, in Section 4, we compare the performance and energy efficiency of 2-input logic gates in the CMOS and mCPL logic styles. In section 5, we extend our comparisons to a larger circuit, an 8b ripple carry adder. Finally, we investigate the implications of supply voltage, temperature and future technology improvements on the performance of each logic style in Section 6. The paper concludes in Section 7.

2. LOGIC FAMILIES

A logic family is defined as a group of transistor-level implementations of logic functions in integrated circuit design. These transistor-level implementations are called logic gates. Logic gates in the same logic family share the common characteristics such as output levels and supply voltages, easing the integration of the gates in the same logic family.

CMOS is one of the most widely-used logic families. Inverters and NAND gates are shown in Figure 2(a) and (c). CMOS logic gates are made of pull-down network (n-channel transistors) and pull-up network (p-channel transistors) to drive outputs. When input values are fixed at either high (supply voltage) or low (ground voltage) values, the gate consumes only leakage current, making CMOS logic an excellent candidate for low power applications. Delay performance is considered to be good, although faster logic families are available. The output is usually a full-swing from ground to supply voltage level. Also, the functional operation is not sensitive to transistor sizing. These characteristics make CMOS to be robust against process variations, supply voltage scaling and noises.

On the other hand, conventional CPL logic family uses two n-channel logic networks and two pull-up p-channel transistors for output swing restoration. N-channel logic networks are preferred due to higher mobility of n-channel transistors in silicon technologies. Figure 2 (d) shows a 2-input CPL NAND gate (p-channel network CPL gates are shown). Different from CMOS logic family, CPL has dual inputs and outputs. The dual rail causes more wiring overhead, but it also provides a free logic inversion. CPL MUX and XOR gates exhibit better performance and energy-efficiency than CMOS counterparts since CMOS implementation of these logic functions are usually inefficient. In general, CPL logic family provide low static power, full output swing, but worse performance, compared to CMOS counterparts.

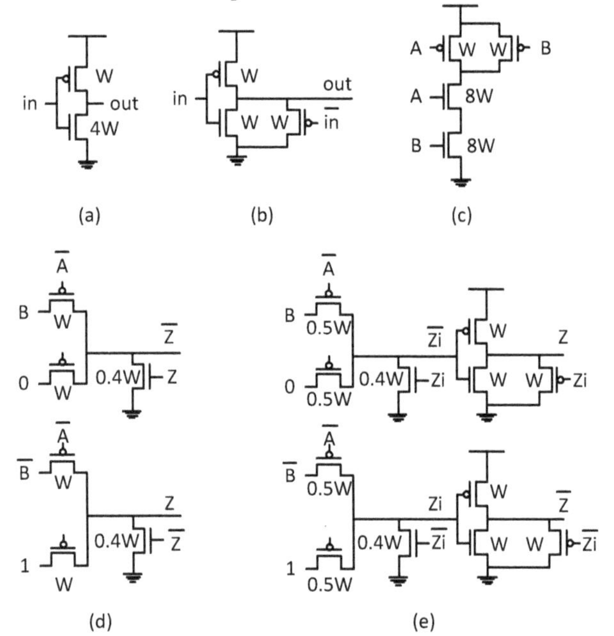

Figure 2. Schematics and sizing information of (a) a CMOS inverter, (b) a proposed CPL inverter, (c) a CMOS NAND, (d) a CPL NAND, (e) a mCPL NAND

3. LOGIC FAMILY DESIGNS IN ORGANIC TRANSISTOR TECHNOLOGIES

In this section, we investigate logic gate design and improvements of CMOS and CPL families in organic transistors. Other logic families including domino logic, although worthwhile to investigate, cannot be included due to space limitation.

As a SPICE simulation setup, we use the design kit available in [5] with the supply voltage of 1.2V. The design kit is used for designing and fabricating memory arrays [11]. However, it has no SPICE model for n-channel transistors; therefore we estimate n-channel model whose mobility is reduced by 10× compared to p-channels while all the other parameters are intact. Although 10× scaling represent typical organic technologies, we investigate the impact of different n-channel strength on logic family design in Section 6.2.

3.1 CMOS Gate Design in Organic Transistors

Currently-available organic technologies have 5-30× higher mobility than the n-channel [3], which should be considered to implement logic gates. In CMOS logic gates, the weaker p-channel transistors are usually upsized for balancing rise and fall delay. The beta ratio (β) of 1.5-2, which is defined as the ratio of n-channel to p-channel strength, is used to upsize weaker transistors in silicon technologies. However, in many organic technologies, the β lies typically between 0.03-0.2, which leads to much larger n-channel pull-down networks in CMOS logic implementations.

In order to find the optimal size of n-channel transistor, we run SPICE simulations for the circuit setup shown in Figure 3(a). As shown in Figure 4, in NAND2 gates, upsizing improves the sum of rise and fall delay up to when ~8× larger n-channel transistors are used, then the gain becomes saturated. Ideally, ~18× larger n-channel transistors are needed to completely balance rise and fall delay, which significantly increase energy consumption. Therefore, we decide to use the β=0.25, i.e. 8× larger n-channel transistors in the NAND2 case, to balance delay and energy tradeoff. The final sizing information is shown in Figure 2(c).

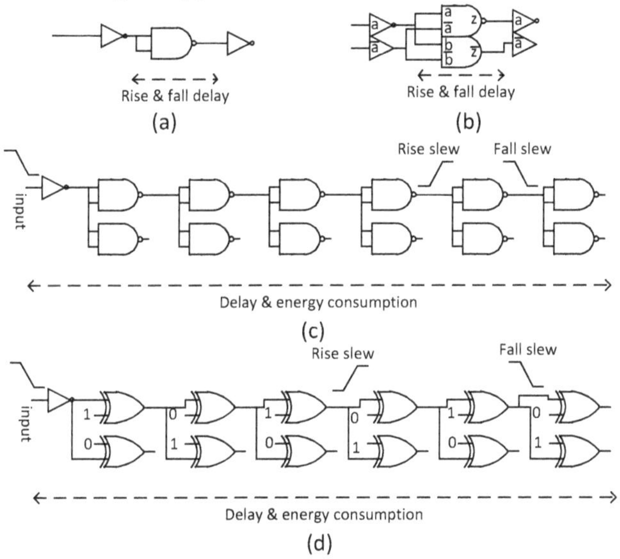

Figure 3. Comparison frameworks (a) for sizing CMOS gates, (b) for sizing CPL/mCPL gates, (c) for measuring delay and energy consumption of NAND2, NOR2 (d) and XOR gates over CMOS, CPL, and mCPL styles

3.2 Modified CPL Gates in Organic Transistors

Although CPL gates can be as simple as Figure 2(d), a good isolation between input and output of a single logic gate is important for a large-scale integration. Usually, CMOS inverters can be added at the outputs. However, the lack of good inverters makes it challenging to directly use conventional CPL gates in organic transistor.

Therefore, we first develop a CPL style inverter, as shown in Figure 2(b). It has an extra pull-down p-channel transistor with a complementary input, which significantly improves fall delay without upsizing n-channel transistors. In SPICE simulations, the proposed CPL inverter improves rise and fall delay for driving a fixed load (1pF) by 3.06× and 3.25×, compared to the conventional inverter (Figure 2(a)).

Then, the CPL inverters are used to isolate each CPL gate. Figure 2(e) shows a NAND2 gates with CPL inverters, which we refer to as modified CPL or mCPL. The mCPL logic gates can provide a better isolation of each gate, easing cell-based design. [13].

The output inverters enable the use of smaller p-channel transistors without delay degradation. About 50% smaller p-channel transistors are used in evaluation networks, compared to CPL gates without inverters (Figure 2(d)). We also investigate sizing level restoring (n-channel) transistors in mCPL NAND2 gates (Figure 2(e)) using the simulation setup of Figure 3(b). Since the main purpose of n-channel transistors is to restore output swing, over-sizing can increase contention and degrade rise delay and slew. Figure 5 show that the fall slew, which is defined as 90% to 10% of output transition time, is improved up to when n-channel transistors are sized similarly with p-channel transistors in evaluation networks. The final sizing information for mCPL NAND2 gates is shown in Figure 2(e).

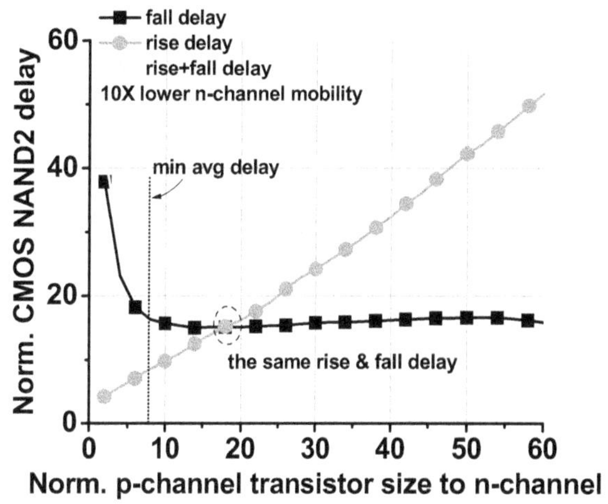

Figure 4. Sizing of CMOS NAND2 gates for the similar rise and fall delay

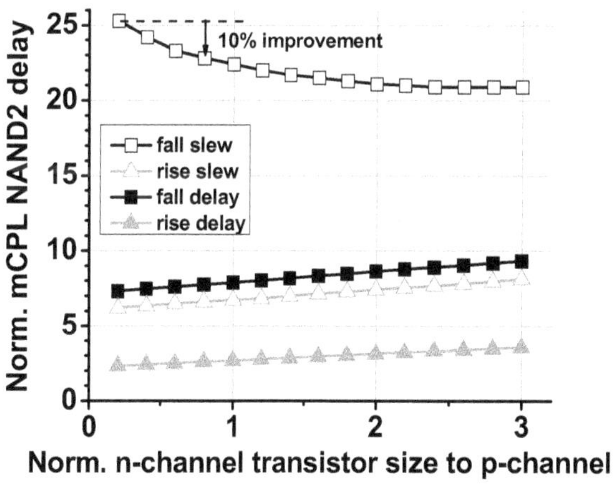

Figure 5. Sizing of mCPL NAND2 gates for the similar rise and fall delay and slew

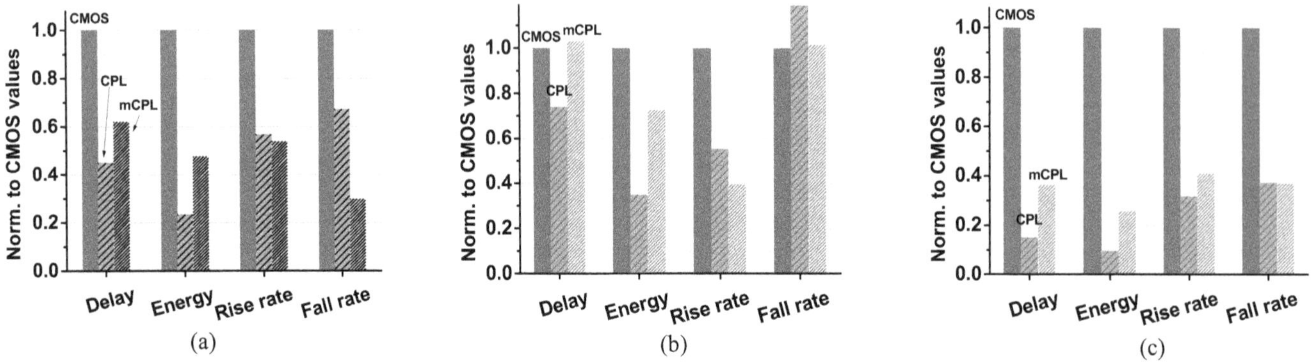

Figure 6. Comparisons of 2-input (a) NAND (b) NOR (c) XOR gates in CMOS, CPL, and mCPL styles

4. COMPARISONS OF SIMPLE GATES

4.1 Comparison Frameworks

In order to compare each logic families in more practical settings, we configure a chain of gates for NAND2 and NOR2, as shown in Figure 3(c). We measure the 50% to 50% delay from the input of the first inverter to the output of the last gates. The energy consumption includes both the input inverter and all the logic gates in the chain. For simplicity, the configuration for CPL and mCPL is not shown. However, the only difference is to use two CPL inverters (Figure 2(b)) for driving the chain of logic gates, instead of one CMOS inverter. For simulating XOR gates, we use a slightly different setup, which is described as in Figure 3(d). This setup includes the non-symmetric effects from the input orders and input levels. The similar configuration is also used in [13] to simulate XOR gates.

4.2 Simple Logic Gate Comparisons

We compare NAND2, NOR2, and XOR2 gates in three different logic families, namely CMOS, CPL, and mCPL. Figure 6 shows the delay, energy consumption, rise slew and fall slew of NAND2 gates. All the values are normalized to CMOS values. The SPICE simulations show that CPL NAND2 gates improve delay by 2.22×, energy efficiency by 4.25×, and average slew rate by 1.6×. mCPL, which should be a better candidate for cell-based design, also reduce delay by 1.6×, and energy consumption by 2.1×. The energy-delay product (EDP) of mCPL NAND2 is improved by 3.36×, compared to the CMOS implementations.

The simulation results for NOR2 are shown in Figure 6(b). Here we find less improvement from CMOS to CPL and mCPL since NOR2 gates have shorter n-channel stacks in pull-down networks. CPL NOR2 provides 27% faster delay and 2.8× better energy efficiency while the slew rate is comparable. mCPL NOR2 gates show 28% better energy efficiency and 2.5× faster rise rates but comparable delay and rise rate, due to the overhead of output inverters.

In the comparisons of XOR2 gates, CPL and mCPL styles significantly outperform CMOS since CMOS style has an inherent inefficiency of implementing XOR2 gates. (i.e. more transistor count) For this reason, transmission gate style XOR gates are often used together with other CMOS logic gates in the standard cell library of silicon technologies. As shown in Figure 6(c), the mCPL XOR2 improves EDP by 10.5×, compared to the CMOS counterpart.

In the conventional silicon technologies, CMOS style usually outperform CPL in delay and energy efficiency for both NAND and NOR gates [13]. However, the simulation results here with organic transistors shows a strong potential to use more CPL and mCPL style gates to improve performance and energy efficiency.

5. COMPARISONS IN 8B RIPPLE CARRY ADDER DESIGNS

5.1 CMOS and CPL 8b Ripple Carry Adders

In this section, we extend the comparisons to an 8b ripple carry adder (RCA), which can be an important building block in low performance microcontrollers in organic technologies [9]. For building a RCA in a CMOS style, we use 28-transistor mirror full adder cells as shown in Figure 7(a). We choose this adder cell for the smallest number of transistors and compact layout. The size of each transistor follows that of the full adders used in [13] except reflecting different ratio of n- to p-channel strength. On the other hand, as shown in Figure 7 (b), mCPL full adder cells are designed. Each transistor is sized based on the investigations in Section 3 and [13].

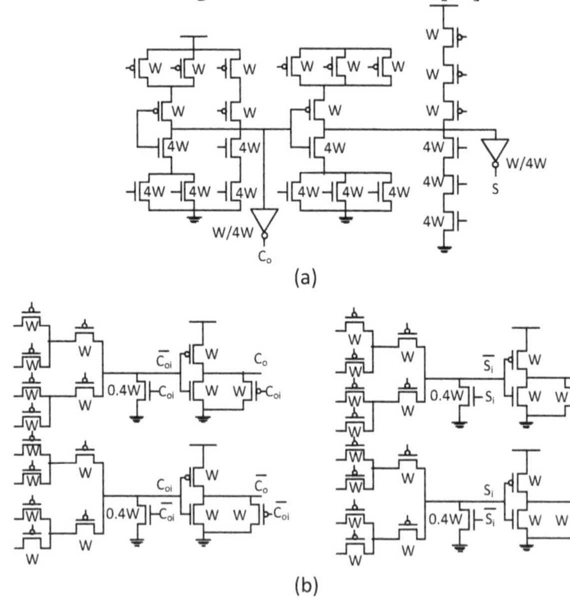

Figure 7. Schematics and sizing information of (a) CMOS mirror full adder and (b) mCPL full adder

Figure 8 shows the SPICE simulation results for CMOS and mCPL style 8b RCAs. The mCPL style improves the critical delay by 2.4× while consuming a 2.1× lower energy. The total transistor width used in 8b adder is also reduced by 1.84× from CMOS to mCPL style, resulting in a smaller footprint. The delay and energy consumption of input buffers are included in the comparisons.

5.2 Impact of Buffering in CPL 8B RCAs

In the previous comparisons, we use mCPL style full adder cells. However, we may be able to remove a part of buffers (i.e. CPL inverters) in the RCA design, leading to better performance and energy efficiency.

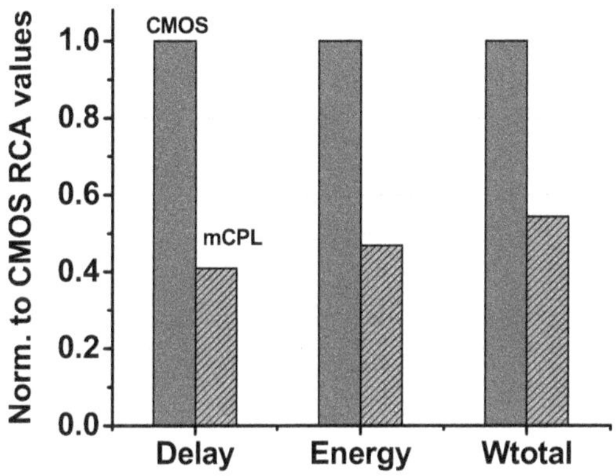

Figure 8. Comparison of CMOS and mCPL styles for the delay, energy consumption, and total transistor widths of 8b RCAs

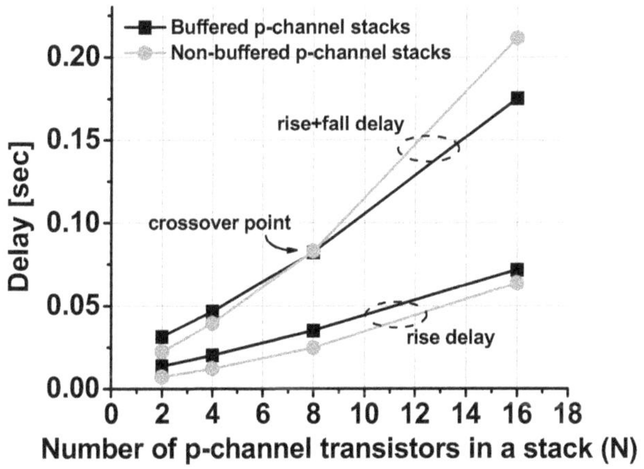

Figure 9. Critical number of stacks for rise+fall and rise-only transitions

For quantitatively investigating the critical number of p-channel transistor stacks which require a buffer, we run a set of simulations. In these simulations, we compare the delays of two stacks of series-connected 'N' p-channel transistors, which model the selected path of the p-channel evaluation networks in CPL style gates. All the transistors in the first stack are driven by a single inverter, while the second stack has another inverter (i.e. a repeater) in the middle of the stack. As shown in Figure 9, as the number of transistors, 'N', in the stack increases more than 8, the stack having a repeater in the middle shows better average delay performance. When N<8, it can cause more delay by adding buffers. When only rise delay is concerned, the crossover point, where the stack with repeaters starts to outperform the stack without repeaters, becomes larger than 16 in the simulations. This extremely high crossover point may be exploited for the circuits with the critical path consisting of only rising transitions. However, a slew rate of signals is expected to be degraded, requiring a careful simulation.

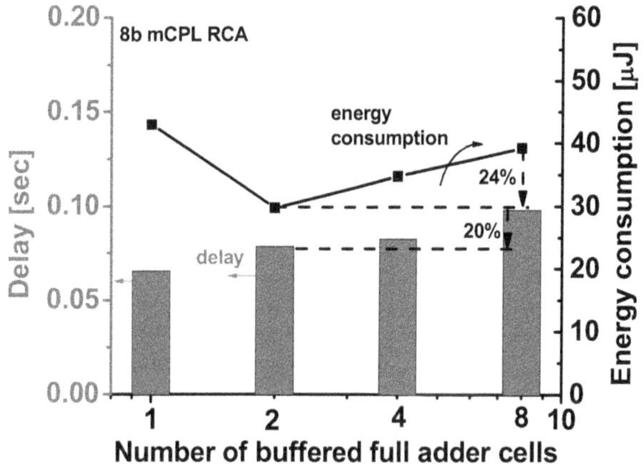

Figure 10. Impact of buffering in an 8b mCPL RCAs

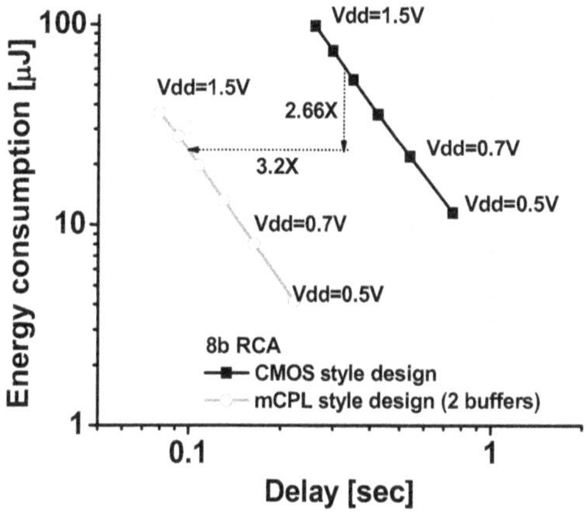

Figure 11. Energy and delay of CMOS and mCPL RCA over supply voltages

Based on the previous analysis, we sweep then number of buffered full adder cells in the mCPL RCA for improving delay and power dissipation. Figure 10 shows the improvements of performance and energy consumption. The simulation shows that it is energy and delay optimal to use only two buffers through the chain of full adders. The delay and energy consumption are reduced by 20% and 24%, respectively, compared to use 8 buffered full adder cells.

6. IMPACT OF SUPPLY VOLTAGE, TEMPERATURE, AND TECHNOLOGY

6.1 Impact of Supply Voltages and Operating Temperatures

As robustness is an important factor for logic families, we investigate the impact of supply voltage scaling and temperature variations on CMOS and mCPL logic family. Figure 11 shows that the mCPL RCA robustly operates over the supply voltages from 0.5V to 1.5V, while maintaining the similar improvements of energy consumption and performance, compared to the CMOS RCA. Figure 12 also confirms that the mCPL logic family is functional over the wide range of temperature changes with consistent delay improvement.

6.2 Impact of Future N-Channel Improvements

Although the arrival of the n-channel organic transistors with a comparable strength to p-channel counterpart is delayed, it is important to investigate the impact of the future improved n-channel transistor on the performance and energy efficiency of logic families. Therefore, in this section, we modulate the mobility of n-channel transistors from $1/20\times\mu_p$ to $1\times\mu_p$. The mobility of p-channel transistor is not changed. Using each modified transistor model, we repeat the comparisons of NAND2 by re-sizing transistors and then re-simulating the energy consumption and delay of the chain of NAND2 gates, as we did in Sections 3 and 4. Figure 13 shows that mCPL style shows better delay performance than CMOS until n-channel transistors are weaker than p-channels by ~5×. However, once n-channels have comparable strength, CMOS logic family shows the best performance among three, which is the same conclusion with silicon technologies.

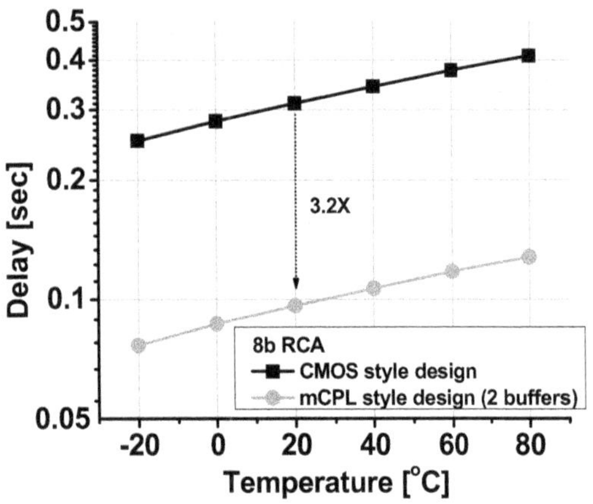

Figure 12. Delay of RCA over temperatures

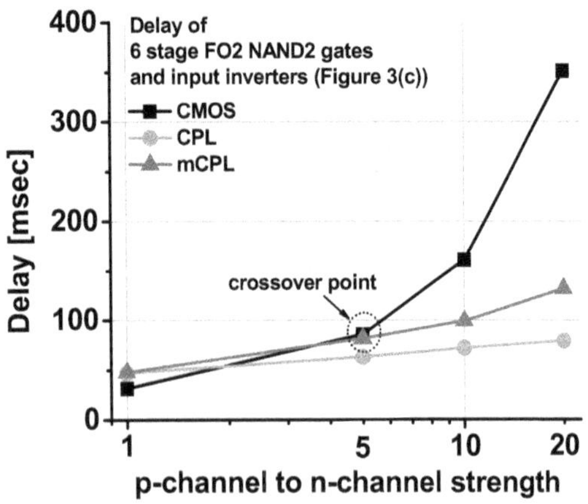

Figure 13. Comparisons of CMOS, CPL and mCPL NAND2 chains over different n-channel mobility

7. CONCLUSIONS

In this paper, we investigate CMOS, CPL, and mCPL logic families in emerging organic transistor technologies. The highly-skewed strength between n-channel and p-channel transistors makes mCPL logic family to outperform conventional CMOS both in performance and energy efficiency. The comparisons are performed for 2-input gates (NAND, NOR, XOR) and a small building block (8b RCA). The optimized mCPL 8b RCA improves performance by 3.2× and reduces energy consumption by 2.66×, compared to CMOS implementations.

Acknowledgements – The author appreciates the technical discussions with Gyouho Kim at University of Michigan and Ioannis Kymissis at Columbia University.

References

[1] L. Herlogsson, et al., “Polyelectrolyte-gated organic complementary circuits operating at low power and voltage,” Advanced Material, 2011

[2] F. Ante, et al., “Submicron low-voltage organic transistors and circuits enabled by high-resolution silicon stencil masks,” International Electron Device Meeting, 2010

[3] B. Murmann, “Design of analog circuits using organic field-effect transistor,” International Conference on Computer-Aided Design, 2010

[4] A. Daami, et al., “Fully printed organic CMOS technology on plastic substrates for digital and analog applications,” International Solid-State Circuit Conference, Feb, 2011

[5] The organic process design kit (OPDK), URL: http://opdk.umn.edu/

[6] H. Koezuka, A. Tsumura, T. Ando, "Field-effect transistor with polythiophene thin film," Synthetic Metals 18: 699–704, 1987

[7] V. Subramanian, et al., “Printed organic transistors for ultra-low-cost RFID applications,” IEEE Transactions on Components and Packaging Technologies, Vol.28, No.4, pp.742-747, Dec, 2005

[8] T. Netter, et al., “Signal processing in a low-power wearable oximeter,” BIOSIGNAL, 2004

[9] K. Myny, et al., “An 8b Organic Microprocessor on Plastic Foil,” International Solid-State Circuit Conference, pp. 322-323, Feb, 2011

[10] MSP430 Microcontroller, Texas Instruments

[11] W. Zhang, et al., “A 1V printed organic DRAM cell based on ion-gel gated transistors with a sub-10nW-per-cell Refresh Power,” International Solid-State Circuit Conference, pp.326-327, Feb, 2011

[12] K. Yano, et al., “A 3.8ns CMOS 16x16b multiplier using complementary pass-transistor logic,” IEEE Journal of Solid-State Circuits, Vol.25, No.2, pp.388-395, Apr, 1990

[13] R. Zimmermann, W. Fichtner, “Low-power logic styles: CMOS versus pass-transistor logic,” IEEE Journal of Solid-State Circuits, Vol.32, No.7, pp.1079-1090, Jul, 1997

[14] D. Harris, M.A. Horowitz, “Skew-tolerant domino circuits,” Journal of Solid-State Circuits, Vol.32, No.11, pp.1702-1711, 1997

Optimal Power Switch Design for Dynamic Voltage Scaling from High Performance to Subthreshold Operation

Kyle Craig, Yousef Shakhsheer, and Benton H. Calhoun
Dept. of Electrical and Computer Engineering
University of Virginia, Charlottesville, VA, USA

ABSTRACT

This work explores optimizing power switch design for dynamic voltage scaling schemes that use headers to connect components to voltage supplies ranging from strong inversion to subthreshold values. We propose using NMOS devices with their gate controlled at the nominal voltage as power switches connected to the subthreshold voltage rail. Measured results show that an NMOS can provide the subthreshold voltage with a power switch size >280X smaller than a PMOS. For architectures targeting operation from subthreshold up to nominal voltage, we show that using an asymmetric transmission gate power switch provides a lower overhead way to enable this flexibility.

Categories and Subject Descriptors

1.1 Technologies and Digital Circuits

Keywords

Power gate, Low Power, Dynamic Voltage Scaling

1. INTRODUCTION

With continued technology scaling and the increase of portability in battery and battery-less electronic devices, energy efficiency has become arguably the most critical metric facing circuit designers. In order to address this metric, there exists extensive research into how to best improve energy efficiency. Some common techniques include dynamic voltage scaling (DVS), power gating for leakage reduction, and subthreshold circuit operation. Since dynamic energy is quadratically proportional to V_{DD} and frequency is proportional to V_{DD}, DVS reduces energy by dynamically reducing voltage when performance requirements allow. Power gating reduces leakage energy by using power switches to cut the circuit off from the supply when idling. Finally, in subthreshold operation the circuit operates below the threshold voltage (V_T). In subthreshold, frequency is exponentially dependent on V_{DD}.

DVS is conventionally deployed at the chip or core level, often using DC-DC converters to adjust the voltage. A different approach uses header switches to connect blocks to one of several power supply voltages, allowing for faster switching and application of DVS to smaller internal blocks. A scheme called Panoptic Dynamic Voltage Scaling (PDVS) uses this approach to achieve

ISLPED'12, July 30–August 1, 2012, Redondo Beach, California, USA.

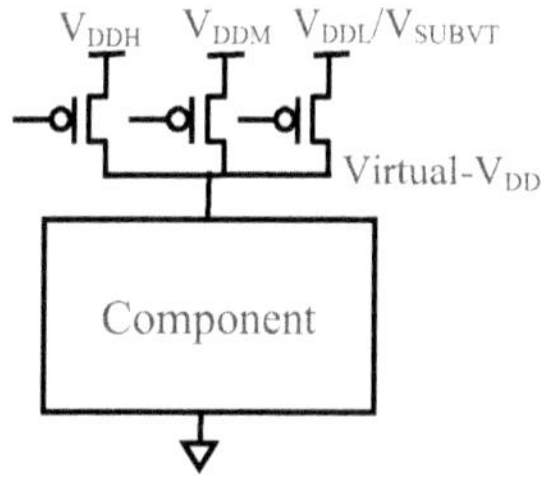

Figure 1. Architecture for using headers to implement DVS across a wide voltage range.

near optimal energy efficiency from high speed to subthreshold operation [1]. The wide range DVS chip in [1] uses multiple voltage supplies at high, middle and low values (V_{DDH},V_{DDM}, V_{DDL}) with PMOS power switches and voltage dithering to select the appropriate V_{DD} for different fine grained blocks/components in the design depending on local workload requirements. Figure 1 shows the generic PDVS architecture. This scheme supports DVS, power gating for leakage reduction, and subthreshold (V_{SUBVT}) operation when appropriate, so we select it as a platform that is representative of header based DVS schemes for our analysis of header switch optimization. In the rest of the paper, we explore power switch design to enable subthreshold operation in the PDVS architecture in the most area and energy efficient manner.

The paper is organized as follows: In Section 2, we will discuss current power switch sizing methodologies. In Section 3, we propose using an NMOS device as a subthreshold power switch and compare it to the conventional PMOS device. In Section 4, we propose a transmission gate based power switch for greater design flexibility. Finally, in Section 5, we discuss some physical design considerations for the proposed power switch.

2. SIZING FOR DELAY

The introduction of a power switch device between V_{DD} and a component creates an IR drop across the header resulting in a reduced virtual-V_{DD} value and performance degradation. Power switch sizing is critical to maintain low power and expected performance. An undersized power switch results in large performance degradation; however, an oversized power switch results in increased leakage and increased area overhead. Power switch sizing methodologies have been examined in depth to support techniques like multi-threshold CMOS (MTCMOS), which used high-V_T power switches to reduce leakage [3]. Optimal power switch design for has been extensively explored, e.g., [4]-[7]. These schemes explore ways to optimize Ion/ Ioff and develop methodologies or tools to implement these schemes. For the purpose of this work, we will build from a commonly used power switch sizing methodology that sets the power switch size

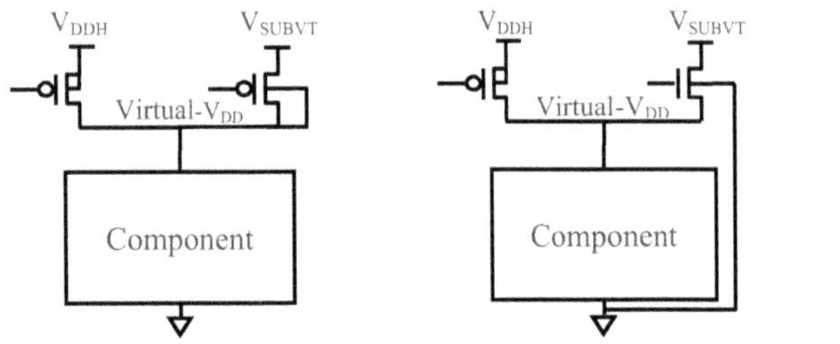

Figure 2. (*left*) Conventional PMOS subthreshold power switch. (*right*) Proposed NMOS subthreshold power switch.

such that the critical path meets an acceptable delay or frequency degradation from the nominal case (i.e., no power switch). The allowable degradation is a design choice chosen by the system designer.

3. NMOS as a Subthreshold Header

In this paper, we propose using an NMOS power switch to connect to a subthreshold voltage rail instead of a PMOS power switch. Figure 2 shows the convention subthreshold power switch architecture as well as the proposed NMOS power switch architecture. The conventional architecture uses a PMOS power switch with the body tied to virtual-V_{DD} to avoid reverse body bias [2]. Since the PDVS architecture has multiple supplies, the power switch control signals are full swing, up to V_{DDH}, providing strong turn off of the off headers. In the proposed alternative, an NMOS device with its body tied to ground is used as the header between the subthreshold rail and the component. During subthreshold operation (i.e., only V_{SUBVT} enabled) the conventional PMOS power switch has a $|V_{GS}| = V_{SUBVT}$; however, the proposed NMOS power switch has a $|V_{GS}|=V_{DDH}-V_{SUBVT}$. The higher $|V_{GS}|$ on the NMOS device provides a much higher current than the PMOS, since the NMOS is in the linear region of operation while the PMOS is in the cutoff/subthreshold region of operation. The higher current from the NMOS device provides a more stable virtual-V_{DD} while potentially using a much smaller device, decreasing the area overhead associated with the power switch.

3.1 Comparison to Conventional

3.1.1 Comparison Circuit Setup

We used a commercial 130nm bulk process to simulate, measure, and compare the conventional and proposed subthreshold header topologies. To provide a flexible, representative load circuit, we used ten 27-stage ring oscillators (ROs) in parallel with each RO capable of being enabled independently. To simplify the comparison, each block of parallel ROs only had two header power switches as shown in Figure 2. In simulation, we swept the widths of the headers to examine the impact of size on header behavior, and in the test chip we describe later, we included programmable sized headers for flexible measurements.

3.1.2 Simulation Results

Figure 3 demonstrates the impact of power switch width on virtual-V_{DD} for two different activity factors. An activity factor of 1.0 corresponds to all 10 ring oscillators enabled in parallel while 0.1 corresponds to only 1 ring oscillator enabled. These two activity factors represent the upper and lower bounds in this design. V_{SUBVT} was set to 0.3V, well below the threshold voltage (V_T) in the technology. Across the wide range of sizes used, the NMOS is

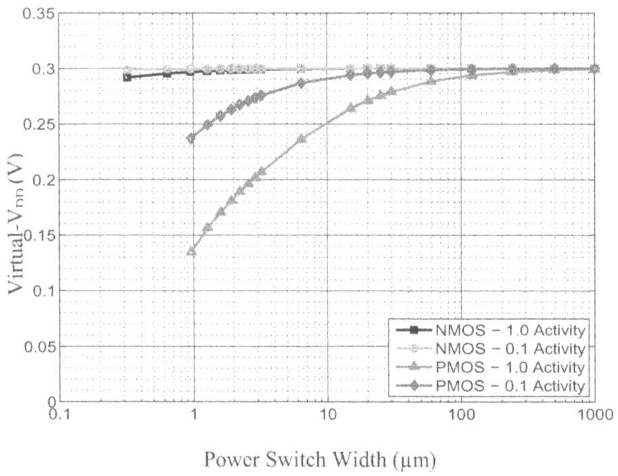

Figure 3. Simulation virtual-V_{DD} at 0.3V for the conventional PMOS and proposed NMOS.

able to keep virtual-V_{DD} at the target 0.3V due to the NMOS being in the linear operating region. The PMOS, however, is unable to keep virtual-V_{DD} at the target 0.3V for small widths since it is in the subthreshold operating region. It is necessary to keep the virtual-V_{DD} near the target V_{DD} because frequency depends exponentially on the virtual-V_{DD} voltage in subthreshold, so voltage droop leads to large slowdowns.

The impact of the power switch width on oscillator frequency is shown in Figure 4. The frequency has been normalized to the frequency at 0.3V without power switches. With near minimum sizing at the lowest activity factor, the NMOS has a worst case frequency degradation of only *3%*, while the minimum PMOS has a worst case frequency degradation of *88%*. At the highest activity factor with near minimum sizing the NMOS has a worst case frequency degradation of *16%*, while the smallest PMOS has a worst case frequency degradation of *93%*. Using the traditional sizing methodology and a target delay degradation of 10%, the required NMOS size is approximately *280X smaller* than a PMOS for the same target degradation at the same worst case activity factor, with sizes of 640nm and 180μm respectively.

The total energy per operation while operating at V_{SUBVT} is defined by the following equation:

$$E_{OP} = E_{DYN_{VSUBVT}} + E_{LEAK_{VSUBVT}} + E_{LEAK_{VDDH}} \quad (1)$$

When we compute this energy equation, we include the overheads of the PDVS architecture associated with having multiple V_{DD}s and power switch devices. Simulated energy per operation versus power switch width for an activity of 1.0 is shown in Figure 5. For both NMOS and PMOS, the energy is normalized to the same value, the energy per operation with no power switches. The shift in energy above the nominal for each of these designs is due to overheads inherent in the PDVS architecture. Specifically, the increase in energy comes from E_{LEAK} through the off V_{DDH} power switch while the V_{SUBVT} power switch is on. This V_{DDH} leakage contributes to an energy overhead of 3% and exists whether the NMOS or PMOS power switch is used. The decrease in energy as power switch size is reduced is due virtual-V_{DD} drop, which lowers the dynamic energy. However for the PMOS design, the energy starts to increase at the lower widths due leakage becoming the

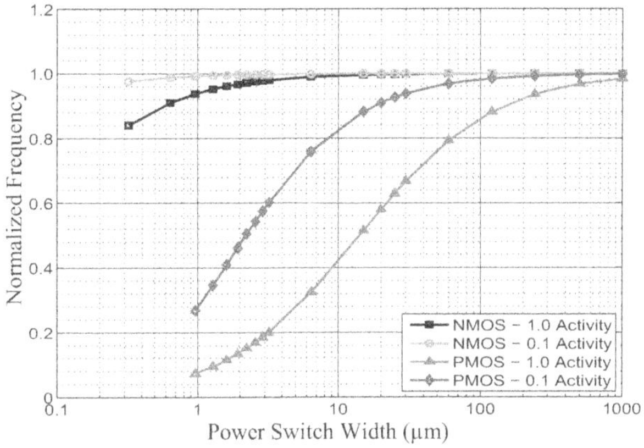

Figure 4. Simulated frequency at 0.3Vfor the conventional PMOS and proposed NMOS .

starts to increase at the lower widths due leakage becoming the dominate factor because of lower operating frequencies. At these lower widths, the virtual-V_{DD} becomes much lower, resulting in initially lower dynamic and total energy as discussed. However, as the width is reduced further, the frequency becomes so slow that E_{LEAK} through the V_{DDH} and V_{SUBVT} power switches becomes dominate, causing an increase in total energy. The use of an NMOS power switch does not adversely increase the overheads associated with the PDVS architecture or have a higher energy per operation than a PMOS power switch. Gate leakage for this technology was not a concern; at the largest power switch size of 1mm, the gate leakage energy was only 0.04% of the total energy.

3.1.3 Measurement Results

A test chip was fabricated in a 130nm bulk commercial process to verify the simulation results. Figure 6 shows the normalized frequency with an activity of 0.1, and Figure 7 shows the normalized frequency at the activity of 1.0. The simulated data was normalized as described earlier, while the measured data were normalized to the largest power switch value for NMOS and PMOS at the largest power switch width possible. Even though the range of width values for the measured data is not as large as the simulated, the same trend is observed for both activity factors. At the lowest activity factor of 0.1, the NMOS has a worst case measured frequency degradation of only *3%*, while the smallest PMOS available in hardware has a worst case frequency measured frequency degradation of only *3%*, while the smallest

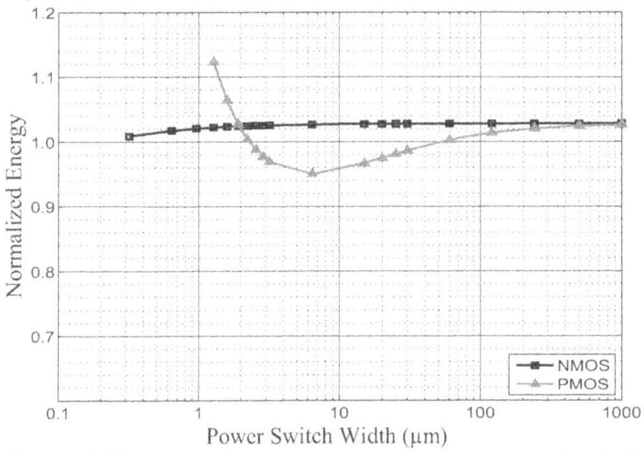

Figure 5. Normalized energy per op vs. power switch width.

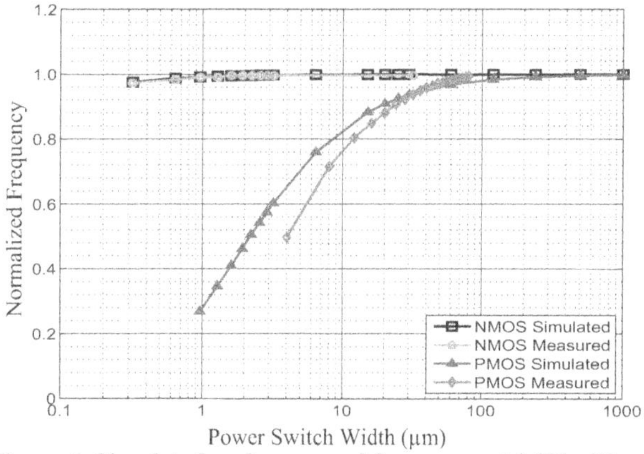

Figure 6. Simulated and measured frequency at 0.3V with an activity factor of 0.1.

PMOS available in hardware has a worst case frequency degradation of *50%*. At the highest activity factor with near minimum sizing, the NMOS has a worst case measured frequency degradation of *13%*, while the minimum PMOS has a worst case frequency degradation of *84%*. The power switch size range in hardware was not large enough to meet the 10% frequency degradation at the highest activity for the PMOS. Using the traditional sizing methodology and a target degradation of 23% (the best achievable by the PMOS in hardware), the required NMOS size is approximately *280X smaller* than a PMOS for the same target degradation at the same worst case activity factor, with values of 320nm and 90μm respectively.

4. Flexible Transmission Gate Design

4.1.1 Motivation

If the V_{SUBVT} rail is always kept at a subthreshold voltage, using an NMOS power switch is optimal since a near minimum sized transistor would provide the target frequency requirement. However, if the rail connecting the header to the component needs to be a flexible and encompass a wide range of V_{DD}s, the NMOS fails as a power switch above V_T due to the V_T drop seen across the NMOS. For designs that require a wide range of voltages on the V_{SUBVT} rail, we propose to use a transmission gate architecture shown in Figure 8. When V_{SUBVT} is near- or subthreshold, the NMOS will be the dominate device. Conversely, when in voltages above V_T, the PMOS will be the dominate device.

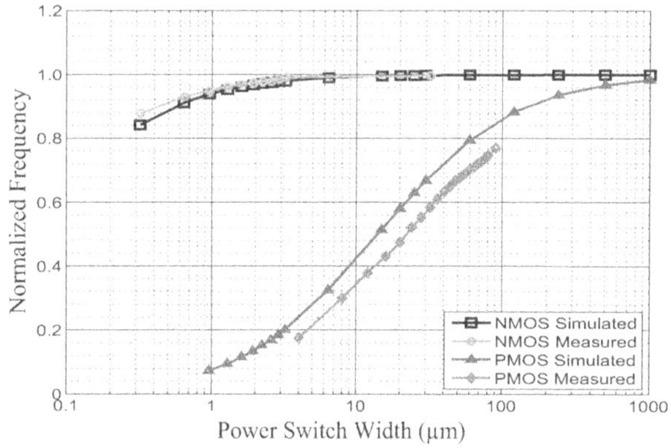

Figure 7. Simulated and measured frequency at 0.3V with an activity factor of 1.0.

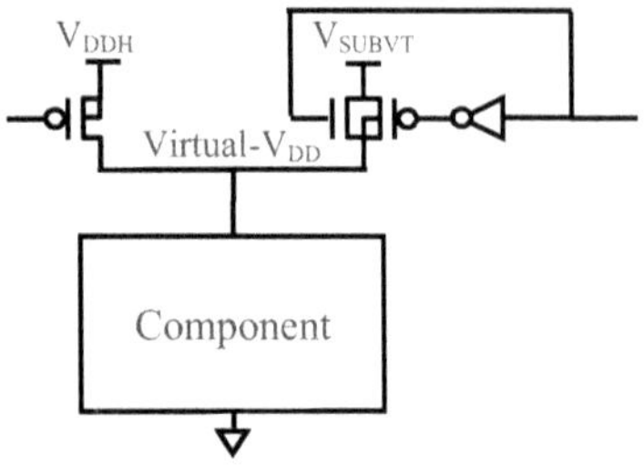

Figure 8. Proposed transmission gate power switch.

4.1.2 Sizing Methodology

Both PMOS and NMOS devices should be sized independently for a given frequency degradation. This will lead to the transmission gate power switch having asymmetric sizing. The PMOS device would be relatively large to meet the frequency target at high voltage; in our design it is in the 10s-100s of µm depending on the frequency target. However, the NMOS device would be in the <1 µm range and still provide the target frequency for the component with the supply at low voltages. For our test chip, the NMOS device provides a frequency degradation target better than designed.

4.1.3 Measured Results

Measured results were taken from the test chip verifying the ability to use a transmission gate for a power switch for a wide range on the supply voltage rail. The V_{SUBVT} rail voltage was varied from 0.3V to 1.2V.The transmission gate PMOS device was limited to a max width of 40µm resulting in a frequency degradation of 30% resulting in a PMOS size of 40µm and a much-smaller NMOS size of 320nm. Figure 9 shows the measured energy-delay (ED) curve for an NMOS-only, PMOS-only, and transmission gate power switch. The NMOS-only ED curve reaches a frequency limitation due to the V_T drop across the NMOS above about 0.6V. The PMOS-only curve becomes slower than the NMOS at about 0.4V due to the virtual-V_{DD} drop causing the increase in delay. Finally, the transmission gate merges the best from the NMOS and PMOS and is able to provide the lowest delay and the lowest energy. In Figure 10, the measured energy delay product (EDP) across the range of V_{SUBVT} is shown for the same three cases described above. Since EDP is a measure of energy efficiency, it is desirable to have a lower EDP. As expected, the NMOS has a lower EDP at lower V_{DD}s, but suffers from a higher EDP at higher V_{DD}s.

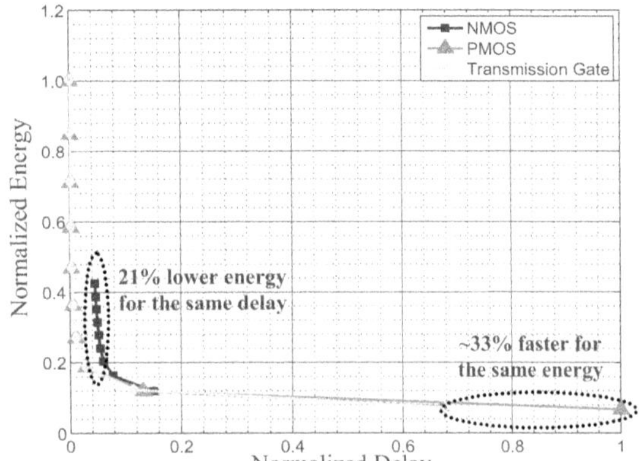

Figure 9. Measured Energy Delay Curves

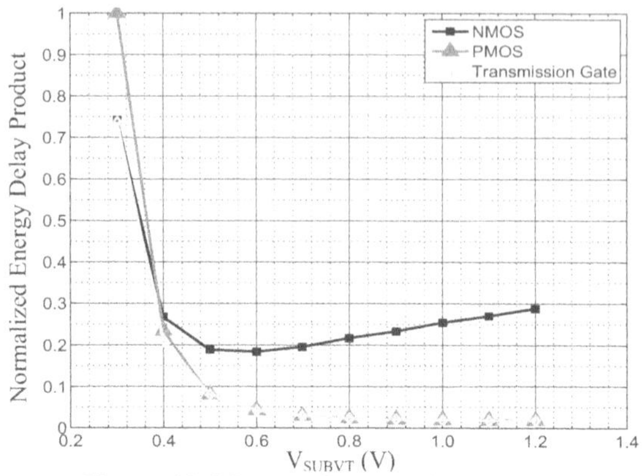

Figure 10. Measured energy Delay Product

Conversely, the PMOS has a lower EDP at higher V_{DD}s, but suffers a higher EDP at lower V_{DD}s. The transmission gate is the pareto optimal curve of the NMOS and PMOS having the lowest EDP across all V_{DD}s.

5. Physical Design Consideration

Traditional methodologies of physical power switch design generally distribute the power switch as standard cell rows, columns, or as rings around design [6] [7]. These help the electrical properties of the power delivery system. Physical implementation of the NMOS device in these methodologies would lead to using above minimum sizing to prevent power delivery system electrical problems, potentially slightly increasing the area associated with the NMOS power switch. This would slightly alter the presented results by decreasing the NMOS area savings prevent compared to the PMOS. This would be especially true for smaller designs where a minimum NMOS could be used, but would need to be increased for the power delivery system. However, for larger designs which require larger than minimum NMOS sizing the impact of potentially needing to increase the physical size would be reduced.

6. Conclusion

We have proposed using an NMOS transistor as a power switch for low rail voltages. Simulations and measurements have shown that an NMOS device with a nominal gate swing (0 – V_{DDH}) is able to provide the target frequency degradation with a size over *280X* smaller than a PMOS. For flexible designs that have a wide range of V_{DD} rail values we propose using an imbalanced transmission gate header with a near minimum sized NMOS device in parallel with a large PMOS. The transmission gate will provide the targeted frequency degradation at the nominal V_{DD} and provide a better than target frequency in subthreshold with minimal additional area.

7. REFERENCES

[1] Shakhsheer, Y., et al., A 90nm data flow processor demonstrating fine grained DVS for energy efficient operation from 0.25V to 1.2V. *CICC*. Sept. 2011

[2] Calhoun, B.H., et al. Flexible Circuits and Architectures for Ultralow Power. *Proceedings of the IEEE* , vol.98, no.2, Feb. 2010

[3] Mutoh, S., et al. 1-V power supply high-speed digital circuit technology with multithreshold-voltage CMOS. *IEEE Journal of Solid-State Circuits*.

[4] Shi, K., Howard, D. Challenges in sleep transistor design and implementation in low-power designs. *DAC*. 2006

[5] Kao, J., Chandrakasan, A., Antoniadis, D. Transistor Sizing Issues And Tool For Multi-threshold Cmos Technology. *DAC*. 1997

[6] Shi, K., Howard, D. Sleep Transistor Design and Implementation - Simple Concepts Yet Challenges To Be Optimum. *VLSI-DAT*. April 2006

[7] Pakbaznia, E., Pedram, M. Coarse-Grain MTCMOS Sleep Transistor Sizing Using Delay Budgeting. *DATE 2008*. 2008

BiN: A Buffer-in-NUCA Scheme for Accelerator-Rich CMPs

Jason Cong, Mohammad Ali Ghodrat, Michael Gill, Chunyue Liu, Glenn Reinman
Computer Science Department, University of California, Los Angeles
{cong, ghodrat, mgill, liucy, reinman}@cs.ucla.edu

ABSTRACT
As the number of on-chip accelerators grows rapidly to improve power-efficiency, the buffer size required by accelerators drastically increases. Existing solutions allow the accelerators to share a common pool of buffers or/and allocate buffers in cache. In this paper we propose a Buffer-in-NUCA (BiN) scheme with the following contributions: (1) a dynamic interval-based global buffer allocation method to assign shared buffer spaces to accelerators that can best utilize the additional buffer space, and (2) a flexible and low-overhead paged buffer allocation method to limit the impact of buffer fragmentation in a shared buffer, especially when allocating buffers in a non-uniform cache architecture (NUCA) with distributed cache banks. Experimental results show that, when compared to two representative schemes from the prior work, BiN improves performance by 32% and 35% and reduces energy by 12% and 29%, respectively.

Categories and Subject Descriptors
B.3.2 [**Memory Structures**]: Design Styles –*Cache memories*.

General Terms: Design, Experimentation.

Keywords: Cache, Memory, Accelerator, CMP.

1. INTRODUCTION
To provide orders of magnitude improvement in power-efficiency over general-purpose processors, recent many-core designs use on-chip accelerators [1][2][3][4] for the majority of computations. According to an ITRS prediction [5], this trend will continue as accelerators become more common and present in greater numbers (close to 1500 by 2022). Under this circumstance, providing a private buffer for each accelerator is very inefficient. Large private buffers occupy a considerable amount of chip area, whereas small private buffers are less effective for reducing off-chip bandwidth. Fortunately, not all accelerators are powered-on at the same time. This provides the opportunity to use a shared buffer [6], or allocate the buffers in the cache on-demand [7][8] so that the space can be used as cache when accelerators are not in use.

As pointed out by [9], increasing the size of the accelerator buffers can reduce more off-chip memory traffic by covering longer reuse distance. This relationship between the buffer size and the off-chip memory bandwidth of an accelerator can be represented by the *buffer size vs. bandwidth curve* introduced in [9] (we call it *BB-Curve* for short in this paper), where each point in the curve shows the required off-chip bandwidth under a given buffer size.

ISLPED'12, July 30–August 1, 2012, Redondo Beach, California, USA.

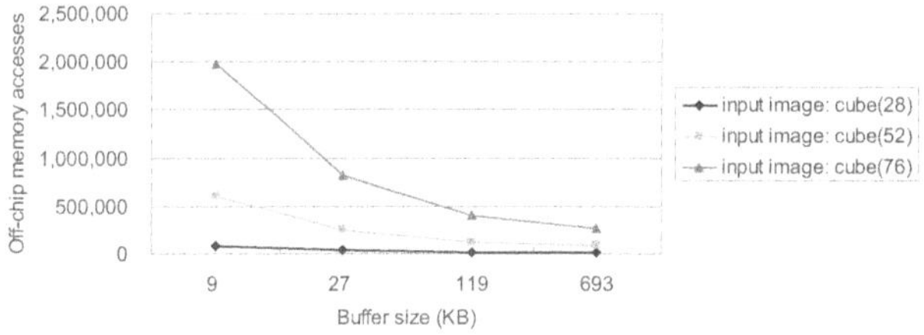

Figure 2. BB-Curves of a *denoise* accelerator.

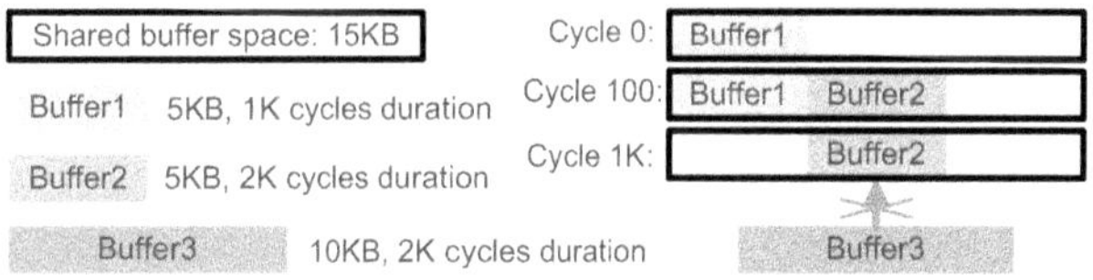

Figure 2. Buffer space fragmentation in a shared buffer.

The BB-Curve not only depends on the accelerator type, but also depends on the input data size. An example of the BB-Curves of a *denoise* [10] accelerator for various input sizes is shown in Figure 2. When we say that an accelerator has higher *buffer utilization efficiency*, we mean that it can reduce more bandwidth per unit increase in buffer size. Given the fact that different accelerators have different buffer utilization efficiencies, no prior work has considered the global buffer demand when allocating buffer space to multiple accelerators. In existing solutions, a fixed-size buffer is allocated to each accelerator statically by the compiler (or designer) by selecting a single, 'best' point in the accelerator's BB-Curve. However the compiler (or designer) has no idea of runtime space utilization behavior, and thus committing to a fixed buffer size a priori often leads to suboptimal solutions.

Moreover, prior work allocates a contiguous space to each buffer to simplify buffer access [6][7], since the address of a buffer block is calculated only as a relative position with respect to the buffer starting address. This may lead to space fragmentation when requested buffers have unpredictable space demand and come in dynamically. As shown in Figure 2, at Cycle 1K, there is 10KB available space in the shared buffer. But since this space is not contiguous, Buffer3 cannot be allocated. In modern CMPs, the last-level cache is typically designed as a NUCA [11][12], which complicates buffer allocations in cache. In addition to the buffer size, the distance of the cache bank in relation to the accelerator also matters. It is possible that the only contiguous space that can satisfy a buffer is quite far from the accelerator and a better choice may be to aggregate several smaller available space segments in the cache banks around the accelerator. One approach to leveraging these fragmented resources is to make use of a paged scheme that adds a level of indirection. This makes physically non-contiguous spaces in NUCA appear to be contiguous, analogous to a typical OS-managed virtual memory. However, accelerators can not afford a large private page table in terms of energy and area; nor can they, for performance reasons, afford a multi-hop scheme to access a centralized shared page table.

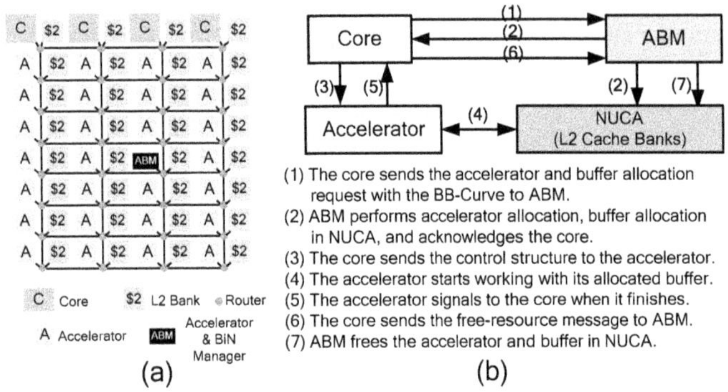

Figure 3. (a) Overall architecture of AXR-CMP with BiN. (b) Communications between core, ABM, accelerator, and NUCA.

In this paper we propose a *Buffer-in-NUCA* (BiN) scheme to dynamically allocate accelerator buffers in NUCA, with the following contributions: (1) a dynamic interval-based global allocation method to assign spaces to accelerators that can best utilize the additional buffer space, and (2) a flexible and low-overhead paged buffer allocation method to limit the impact of buffer fragmentation. To the best of our knowledge, BiN is the first work to offer a cost-effective solution to accelerator buffer allocations in NUCA, with full consideration of both global buffer allocation and buffer space fragmentation.

2. BiN ARCHITECTURE

2.1 Overall Infrastructure

We construct BiN upon the ARC [4]—a hardware-managed accelerator-rich CMP. Figure 3(a) shows the overall architecture of our evaluated ARC, which is composed of cores (with private L1 caches), accelerators, the *accelerator and BiN manager* (ABM), NUCA (shared L2 cache) banks, and NoC routers. The off-chip memory controllers (not shown) are attached to the routers on the four corners. ABM manages accelerator sharing (the same functionality as that of the GAM in [4]) and allocates buffers in NUCA (new contribution of this work). Figure 3(b) shows the communications between a core, ABM, an accelerator and NUCA. The numbered arrows show the order of steps taken during a single accelerator invocation by a core. Buffer allocation (Step 2) is described in the following subsections.

2.2 Dynamic Interval-based Global (DIG) Buffer Allocation

In BiN, the space allocated to a particular buffer is dynamically determined at runtime. To avoid greedily allocating buffer space for each buffer allocation request, we propose a *dynamic interval-based global* (DIG) buffer allocation scheme. The key point is that ABM will collect the buffer allocation requests in a short fixed-time interval and then perform the global allocation for the collected requests to achieve short-time global optimality. By keeping the interval appropriately short, we limit the impact on performance of idle waiting in the interval. In this work, unless otherwise specified, we set the interval to be 10K cycles. Moreover, to avoid having too many accumulated buffer allocation requests, once up to eight buffer requests are collected, the DIG allocation will be immediately triggered.

Given the batch of buffer allocation requests with a set of points in the BB-Curve of request i as $\{(b_{ij}, s_{ij}) \mid 0 \leq j < N_i\}$ (in increasing order of buffer size, where N_i is the number of points in the BB-Curve of request i, b_{ij} and s_{ij} are the bandwidth requirement

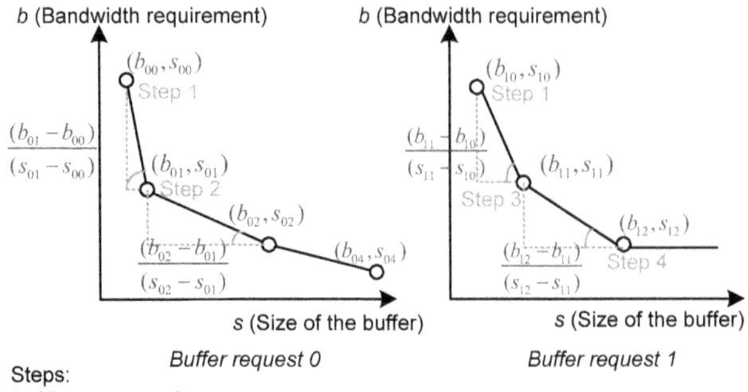

Steps:

1. Allocate s_{00} and s_{10}.
2. Increase buffer 0 to s_{01}, since $\frac{(b_{01}-b_{00})}{(s_{01}-s_{00})} > \frac{(b_{11}-b_{10})}{(s_{11}-s_{10})}$.
3. Increase buffer 1 to s_{11}, since $\frac{(b_{11}-b_{10})}{(s_{11}-s_{10})} > \frac{(b_{02}-b_{01})}{(s_{02}-s_{01})}$.
4. Increase buffer 1 to s_{12}, since $\frac{(b_{12}-b_{11})}{(s_{12}-s_{11})} > \frac{(b_{02}-b_{01})}{(s_{02}-s_{01})}$.
5. Try to increase buffer 0 to s_{02}, but the paged allocation failed, terminated

Figure 4. An example of the DIG buffer allocation scheme.

and buffer size of the jth point of the BB-Curve of request i, respectively), the algorithm first allocates the minimum buffer size for each request and tries the paged allocation (detailed in Section 2.3) to see whether this allocation is valid. If so, it then checks the next point of each curve, and selects the one with the maximum $(b_{ij} - b_{i(j-1)}) / (s_{ij} - s_{i(j-1)})$, i.e., the request that gives the maximum reduction of bandwidth with unit increase of buffer size. Again, the paged allocation is tried to validate this new allocation. This process will go on, until the next point of each BB-Curve makes the resulting allocation not valid. An example of this scheme is shown in Figure 4. To save the computation of the DIG allocation, the $(b_{ij} - b_{i(j-1)}) / (s_{ij} - s_{i(j-1)})$ (buffer utilization efficiency) is pre-computed, so that the curves sent to ABM are actually $\{((b_{ij} - b_{i(j-1)}) / (s_{ij} - s_{i(j-1)}), s_{ij}) \mid 0 \leq j < N_i\}$, where $b_{i(-1)} = 0$, $s_{i(-1)} = 0$. We have developed an optimal pseudo-polynomial algorithm to solve this problem, but the complexity is high [13], thus we resort to the aforementioned linear time greedy algorithm.

Once the allocation decision is successfully made, ABM follows the allocation solution to allocate the buffers in the NUCA banks. As in BiC [7], each L2 cache line has a bit to indicate if it is allocated to a buffer. Allocating a cache line as buffer may result in L1 sharer invalidations and dirty block write-back. If the DIG allocation fails, ABM will leave out the last request and invoke the DIG allocation again on the reduced set of requests. The removed requests are put in an outstanding queue. Once a buffer free event happens, ABM will allocate them. Otherwise, they will be allocated with the requests accumulated in the next interval (as the 'earliest' requests in that interval). If the paged allocation of the minimum buffer size of a buffer allocation request fails at the upper bound of BiN space (discussed in Section 2.4), the core needs to perform the task without calling accelerators. We use the algorithm in [9] to obtain the accelerator BB-Curves for various input sizes. The BB-Curve generation flow is detailed in [13].

2.3 Flexible Paged Buffer Allocation

In BiN, we compose non-contiguous spaces to satisfy the buffer requests. We propose that once a buffer is allocated, the accelerator will use a small local page table to translate buffer addresses into absolute addresses that can be found in NUCA. The key point to achieving this is to set the page granularity for each buffer according to the buffer size; i.e., a larger buffer may have a larger page size so that the total number of pages for this buffer is still a fixed number (32 in our evaluated system). The allowed

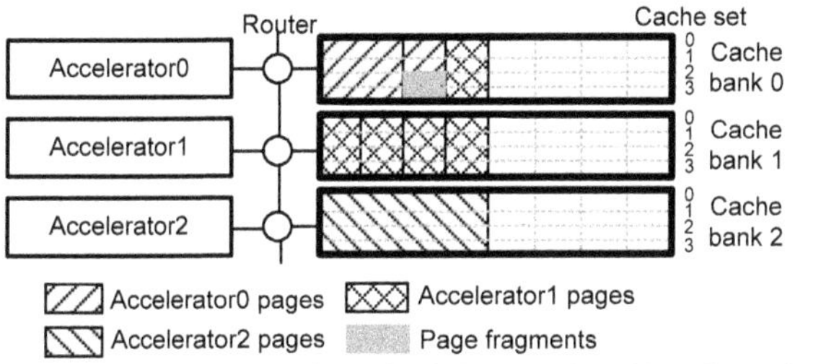

Figure 5. An example of the flexible paged buffer allocation.

page size is always a power of 2 to simplify translation. In our evaluated system it ranges from 4KB to half of the L2 cache bank size (32KB). A page must start at an address that is a multiple of the min-page and should not span cache banks. Since all of the buffer allocation and free operations are performed by ABM, it locally keeps the information about the current contiguous buffer spaces of each cache bank. To allocate a buffer with size *S*, ABM uses the smallest page that is no smaller than *S*/32 to try the allocation, starting from the cache bank nearest to this accelerator, to the farthest cache bank. The amount of cache lines that can be used in each cache bank will be discussed in Section 2.4. To try the allocation of a set of buffers, ABM processes the buffers in a decreasing order of the buffer size, since a larger buffer may be more difficult to fit. If any buffer in this set fails to allocate, the paged allocation for this set fails. An example of the flexible paged buffer allocation is shown in Figure 5, where the min-page is 1 way of an 8-way set-associative cache bank.

To reduce the page fragments, we allow the last page (source of page fragments) of a buffer to be smaller than the other pages of this buffer, since this does not affect the page table lookup. For example, in Figure 5, the last page of Accelerator 0 is only half the size of its other pages. Therefore, the max page fragment for any buffer is smaller than the min-page. Note that the page fragments do not waste capacity since the cache lines in these fragments will remain as cache blocks and be used by the cache. For example, in Figure 5, the shaded blocks denote the page fragments. Then Sets 2 and 3 of Cache bank 0 actually have five cache lines.

2.4 Buffer Allocation in NUCA

In this work we assume a static NUCA design (statically-mapped addresses to banks). Since the buffers are allocated on-demand, the boundary between the cache and accelerator buffers is floating. When BiN allocates buffers in the cache, it could easily consume all or most of the cache space to maximize accelerator gains. To limit the impact on cache performance, we impose an upper bound on the total buffer size that can be allocated. In our implementation where each cache bank is 8-way set-associative, BiN may vary the upper bound from 1/8 to 7/8 of the NUCA size with a 1/8 increment at each step. The upper bound can be controlled by existing cache partitioning schemes (e.g. [14]) where the BiN upper bound is simply one of the partitions competing for space. The accelerators' BB-Curves collected in each partitioning interval can be used to estimate the potential variation of off-chip access counts when making partition decisions. BiN can definitely benefit from dynamic upper bound tuning through smart cache partitioning, but our experimental results show that an upper bound set to half of the NUCA size achieves the best or close to the best performance for most of our evaluated benchmarks. Since the focus of this paper is to show the gain of the DIG buffer allocation and the flexible paged buffer allocation over the prior work, we always set the upper bound to be half of the NUCA size unless otherwise declared, in order to make fair comparisons.

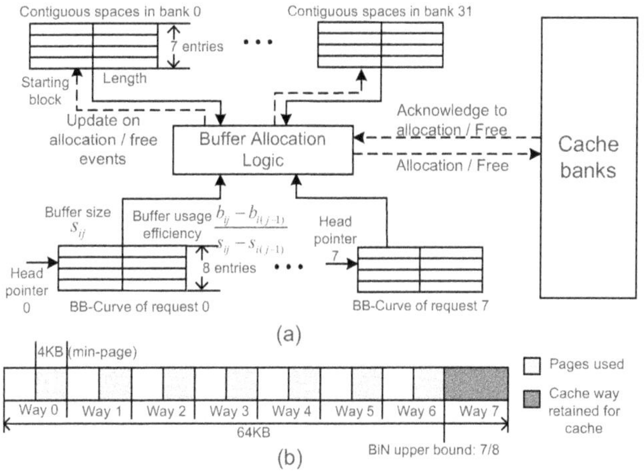

Figure 6. (a) The buffer allocation module in ABM. (b) Worst-case buffer fragmentation in a cache bank.

To avoid creating high contention in a particular cache bank, the upper bound is uniformly distributed to each cache bank; i.e., if the upper bound is half of the NUCA size, then in each cache bank at most half of the cache ways can be allocated as buffers. If high contention still occurs in some cache banks, we use the page re-coloring scheme [12] to remap the OS pages originally mapped to the cache banks to other underutilized banks in order to reduce contention. Other cache bank utilization balancing techniques (such as [11]) can also be used. We would like to emphasize that BiN is orthogonal to and compatible with most state-of-the-art NUCA management schemes since BiN only considers cache bank adjacency while allocating buffer pages in NUCA.

2.5 Hardware Overhead

The hardware overhead introduced by BiN on the evaluated ARC is mainly in ABM (buffer allocation) and the accelerators (look-up of buffer page locations in NUCA).

The block diagram of ABM's buffer allocation module is shown in Figure 6(a). The SRAM table to store the contiguous spaces information for each cache bank is 7-entry. This is because there can be at most 7 contiguous spaces in a 64KB cache bank with a min-page of 4KB, as shown in Figure 6(b). Each entry has 10 bits for the starting block address and 4 bits for the space length in terms of min-page. There are also 8 SRAM tables to store the BB-Curves of the buffer requests (DIG scheme processes at most 8 requests in a batch). We limit BB-Curves to have at most 8 points. Each point uses 2B for the buffer size and 3B for the buffer utilization efficiency. Thus, the total storage overhead is 768B. According to the Synopsys Design Compiler (SAED library, 32nm), the buffer allocation logic has an area of 9,725um^2 under a cycle time of 0.5ns, and the storage area is 3,282um^2 based on Cacti [15] (32nm). Thus, the total area of the buffer allocation module is less than 0.001% for a medium size 1cm^2 chip. An average latency of 0.6us (1.2K clock cycles at 2GHz) is required to perform the buffer allocations. The initial 10K (waiting time in the DIG allocation) + 1.2K cycles of allocation time is only 1% of the typical accelerator runtime (in order of million cycles).

Figure 7 depicts the local page table of the accelerator and the mechanism to generate the block addresses in NUCA. Since we have 32 banks (5-bit bank ID) and each bank has 1024 blocks (10-bit block ID), each table entry is 15-bit and the page table is 64B. This table has an area of 373um^2 based on Cacti [15], which is less than 1% of the area of our evaluated smallest accelerator (*denoise*: 496,908um^2). The table access latency is 0.14ns. Since

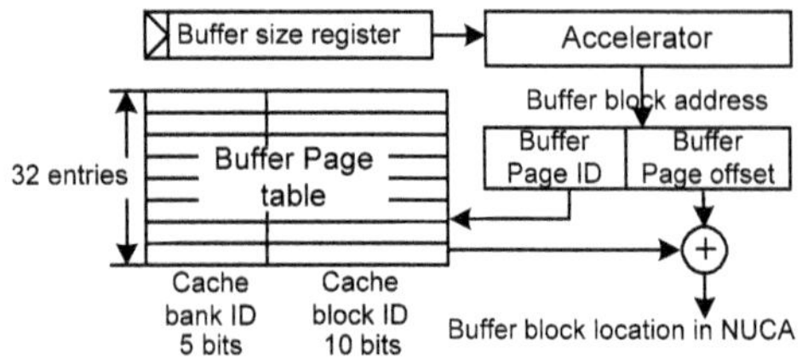

Figure 7. Block addresses generation with the page table.

the clock cycle of the evaluated accelerators is 2ns, this latency can easily fit in the pipeline of block address generation. Note that the 15-bit adder in Figure 7 also exists in a shared buffer design with contiguous space allocation. This is not an overhead of BiN.

BiN does not introduce any more micro-architectural complexity to the L2 cache controller than BiC [7] does. For buffer allocation in a cache bank, BiN uses the same way-based approach as BiC (it uses cache lines uniformly from one way, and then uses the next). Moreover, the overhead of the cache partition scheme and the cache-bank-utilization balancing scheme are not considered as the overhead of BiN, since BiN are orthogonal to these techniques.

3. EVALUATION METHODOLOGY

3.1 Simulation Infrastructure

We extended the full-system cycle-accurate Simics [16] and GEMS [17] simulator to support accelerators and ABM as Simics modules. We also modified Ruby inside GEMS to support the buffer operations in NUCA, such as buffer allocation, free, access, and data transfer from/to global memory space. Table 1 shows our simulated system configuration (also depicted in Figure 3(a)).

Table 1. System configuration of Simics/GEMS simulation

Core	4 Ultra-SPARC III-i cores @ 2GHz
L1 data & instruction cache	32KB for each core, 4-way set-associative, 64B cache block, 3-cycle access latency, pseudo-LRU, MESI directory coherence by L2 cache
L2 cache (NUCA)	2MB, 32 banks, each bank is 64KB, 8-way set-associative, 64B cache block, 6-cycle access latency, pseudo-LRU
Network on chip	4X8 mesh, XY routing, wormhole switching, 3-cycle router latency, 1-cycle link latency
Main memory	4GB, 1000-cycle access latency

3.2 Benchmarks and Accelerators

For this work we chose applications from one domain (medical imaging) to accelerate, since accelerator-rich architectures (such as ARC) are most suitable for domain-specific computing [3][4]. We chose the medical imaging domain because it consists of applications that are both compute- and memory-intensive, and the highly regular computation of the domain makes it an ideal target for hardware acceleration. Furthermore, improved performance in medical imaging has a tremendous potential to transform health care. The chosen applications are: *denoise*, *deblur*, *segmentation*, and *registration*. These applications are explained in detail in [10] and form a *medical imaging pipeline*.

We use the methodology of [4] to extract the computation-intensive tasks of these applications, synthesize these tasks as ASIC accelerators by using a high level synthesis tool AutoESL [18] from Xilinx, and obtain their cycle-accurate modules. We then plug these modules into our simulator. Each of these accelerators has at least 4 copies on the chip to allow threads calling the same type of accelerator to run simultaneously. All of the accelerators work at a frequency of 500MHz. The non-computation-intensive and control tasks of these applications and the Solaris-10 OS are running on the general-purpose cores.

3.3 Reference Designs

We compare BiN to the following representative schemes from prior work. All are evaluated under the same area constraint. To make a fair comparison, we set a fixed upper bound of BiN as half of the NUCA size when comparing to the prior work. We will discuss the impact of dynamic upper bound tuning in Section 4.4.

Accelerator Store (AS) [6]. In AS the shared buffer and NUCA are two separate units. We set the 32-bank NUCA size as 1MB (since we use an upper bound as half of the NUCA size in BiN). Because buffers in cache have some area overhead compared to separate buffers, we set the capacity of the shared buffer in AS as 1.32MB (32% larger than the maximum buffer size in BiN under the same area constraint) based on Cacti [15]. We also partition the shared buffer into 32 banks to increase the buffer access ports and these banks are distributed to the 32 NoC nodes.

BiC [7]. BiC dynamically allocates contiguous cache space to a buffer. Here we set an upper bound for BiC by limiting buffer allocation to at most half of each cache bank. To allow a buffer to span multiple cache banks, the end of the BiC space in one cache bank is considered to be contiguous to the beginning of the BiC space in the next cache bank. The system configurations of BiN and BiC are the same except that the DIG allocation and the flexible paged allocation are not available in BiC. This scheme is used to show that simply allocating buffers in NUCA will result in space fragmentation and underutilization.

We use the off-line compiler analysis in [9] to generate the desired buffer space of a set of potential input sizes for both AS and BiC to achieve their best performance. In addition, to demonstrate the innovations of BiN step by step, we also construct the following schemes. 1) **BiN-Paged**. It only uses the proposed flexible paged allocation scheme (32-entry page table). 2) **BiN-Dyn**. Based on BiN-Paged, it performs dynamic allocation without consideration of near future buffer allocation requests; i.e., it just responds to a buffer allocation request immediately by greedily satisfying the request with the current available spaces. Thus, a new buffer allocation request may not be satisfied since a preceding buffer has consumed most of the space—even if the new requesting accelerator can use the space more efficiently. 3) **BiN-Full**. This is the entire proposed BiN scheme.

4. RESULTS

We conduct experiments for different degrees of buffer pressure by varying the number of medical imaging pipelines that we run in parallel (1, 2, and 4 pipelines). Each pipeline consists of one thread sequentially executing the four medical imaging applications on a different set of imaging data (i.e., no data dependencies between pipelines), where the image sizes also vary. Our benchmark naming convention indicates both the number of concurrent pipelines and the image size for that particular run. For example, benchmark 4P-28 means that there are 4 copies of the pipeline running in parallel and the input to each is a unique image that is 28X28X28 pixels. Benchmarks from 1P-28 to 4P-100 feature pipelines that are running with the same input image sizes. Thus there is no buffer fragmentation problem. These benchmarks are mainly used to demonstrate the gain of the DIG allocation. The "4P-mix" label indicates that the various pipelines in the benchmark have randomly selected input image sizes. Computation over varied image sizes exhibits variation in the

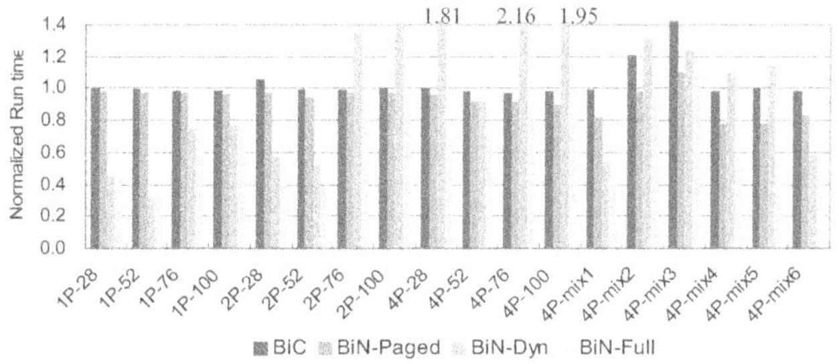

Figure 8. Comparison results of runtime.

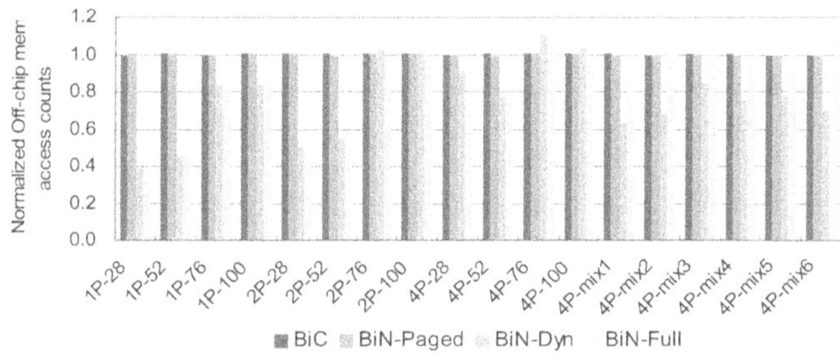

Figure 9. Comparison results of off-chip memory accesses.

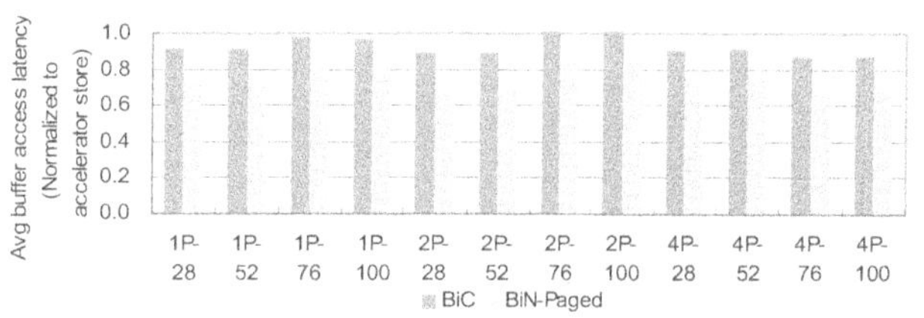

Figure 10. Comparison results of buffer access latency.

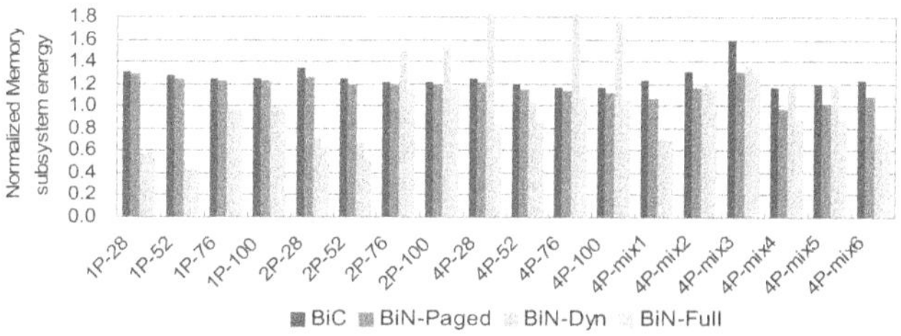

Figure 11. Energy comparison of the memory subsystem.

buffer demand and duration of buffer use, which in turn results in fragmentation; thus they are used to demonstrate the gain from both the flexible paged allocation and the DIG allocation.

4.1 Impact of DIG Buffer Allocation

Figure 8 and Figure 9 show the comparison results of runtime and off-chip memory access counts. All of the results in Section 4.1 to 4.3 are normalized to that of AS (Accelerator Store). In the first 12 benchmarks, since there is no buffer fragmentation, AS, BiC, and BiN-Paged behave similarly (the impact of buffer-to-accelerator distance will be discussed in Section 4.2).

By greedily satisfying the buffer requests with currently available resources, BiN-Dyn outperforms AS, BiC, and BiN-Paged in 1P cases and 2P/4P cases with small inputs, because the shared space can accommodate these small buffers even if they are greedily satisfied. This gain can also be confirmed by the bandwidth reduction in these cases (shown in Figure 9). However, in the cases where greedily satisfying the first buffer requests will severely reduce the available space for subsequent requests that may more efficiently use the buffer space, BiN-Dyn behaves considerably worse compared to the first three schemes, as can be seen in the 2P/4P cases with large input sizes. Interestingly, the off-chip memory access counts do not increase correspondingly. The reason is that, when BiN-Dyn allocates a large buffer for the initial requests, it will delay subsequent requests from accelerators with large input sizes. Eventually these subsequent requests will be assigned to a large buffer (for reducing off-chip bandwidth), but their execution is serialized and thus the performance is impacted. Therefore, a reduction of the total off-chip memory accesses may not necessarily result in an increase in performance.

BiN-Full consistently outperforms the other schemes because of the DIG allocation (and also the flexible paged allocation in the 4P-mix cases). The only exception is in 4P-mix3, where the 1.32X larger capacity of AS is just large enough to accommodate all buffer requests; whereas BiN-Full needs to allocate a smaller buffer size to the accelerator that has the smallest buffer utilization efficiency. But it still outperforms the other three schemes that have the same total capacity. Overall, compared to AS and BiC, BiN-Full reduces the runtime by 32% and 35%, respectively.

4.2 Impact of Paged Buffer Allocation

In the 4P-mix cases where there are buffer fragmentation problems, BiN-Paged and BiN-Full improves the runtime (up to 24% and 56%, respectively) in 4P-mix{1,4,5,6}. However, the bandwidth is not reduced by BiN-Paged and BiN-Full correspondingly, because although AS, BiC and BiN can allocate similar buffer sizes (thus similar bandwidth), BiN can allocate the buffer much earlier than BiC and AS, since BiN can aggregate non-contiguous space to satisfy the parallel buffer requests while BiC and AS can not. In 4P-mix{2,3}, the 1.32X larger capacity of AS can accommodate the buffers even when there is fragmentation. Thus BiN-Paged behaves similar to or even worse than AS. But in both cases, BiN-Paged outperforms BiC, which has the same capacity.

Since the first 12 benchmarks do not suffer buffer fragmentation, BiN-Paged improves their runtime by allocating buffers closer to the accelerator. However, the improvement is small. The reason is as follows. The buffer accesses from our evaluated accelerators are mainly reading in large amounts of input data, performing some calculation on the data and then writing the transformed data out to the buffers. These accesses do not have inter-dependencies. They are pipelined so well that buffer access latency is completely hidden. Therefore, as shown in Figure 10, even though BiN-Paged can improve the average buffer access latency by 19%-32%, the runtime gain is only 2%-9%. We expect that accelerators from other domains which may have dependencies among the buffer accesses can obtain more benefit from adjacent buffer allocation.

4.3 Impact on Energy

We obtain the dynamic and leakage energy data of NUCA and the main memory through Cacti [15] and McPAT [19], and the power data of the ABM module via the Synopsys Design Compiler. We back-annotate these numbers into our simulator to obtain the memory energy results, as shown in Figure 11. By separating cache and buffer units, AS consumes less energy for each cache/buffer access and also consume less unit standby leakage than the other schemes. Therefore, BiC and BiN-Paged consume more energy than AS. BiN-Dyn can save energy in cases where it can reduce the off-chip memory accesses and runtime. However, in the cases where BiN-Dyn significantly increases the runtime, it consumes considerably more energy (more standby energy). By performing DIG and flexible paged buffer allocation, BiN-Full can reduce both the number of off-chip memory accesses and the runtime. When compared to AS, it sees a 12% reduction in energy

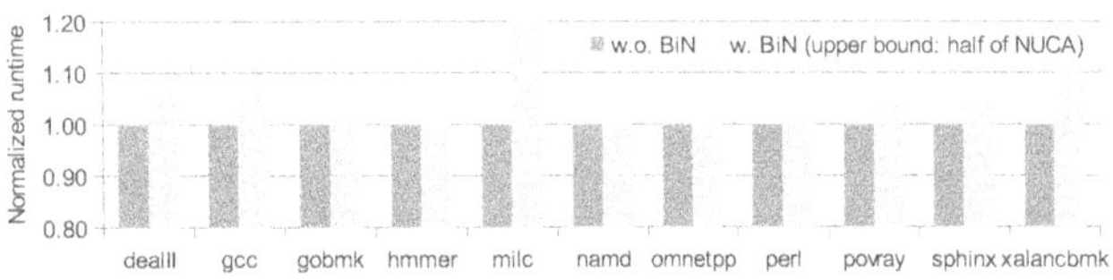

Figure 12. Impact of BiN with a fixed upper bound (half of the NUCA size) on the runtime of SPEC benchmarks.

on average. In cases where the 1.32X capacity of AS can better satisfy buffer requests (4P-mix3), BiN-Full consumes more energy due to more off-chip memory accesses and longer runtime. BiN-Full reduces the energy by 29% on average compared to BiC.

4.4 Impact on Cache

We further quantify the impact of the effective cache capacity reduction due to buffer allocation by running a set of general-purpose applications (SPEC CPU2006 benchmarks) on the cores concurrently. In the previous experiments we use a fixed BiN upper bound which is half of the NUCA size. Thus, we first evaluate the impact on SPEC benchmark runtime when half of the NUCA space is used by BiN, as shown in Figure 12. The results are normalized to full cache capacity performance (additional separate buffers are required for the accelerators). In most cases the runtime increase is within 10%, except *milc* which has a large working set. For *milc*, although a full-capacity NUCA with separate buffers has a remarkably small runtime, it almost doubles the on-chip memory energy and area, and has less flexibility.

The impact of BiN on the cache can be improved via dynamic upper bound tuning. We use the cache partitioning scheme proposed in [14] to dynamically tune the upper bound. We run each SPEC benchmark simultaneously with two medical imaging pipelines for each of the 4 input sizes, respectively. The dynamic tuned BiN upper bounds for each case are shown in Figure 13(a). In most cases an upper bound of half of the NUCA size is selected. In the cases where the SPEC benchmarks can give more space to BiN without considerably impacting the performance, an upper bound of 5/8 NUCA size is selected for BiN. For these cases, the runtime of both SPEC benchmarks and the medical imaging pipelines (normalized to the runtime with an upper bound as half of the NUCA size) and the product of the two are shown in Figure 13(b). All of the runtime variations are within the range of 5%, which suggests that impact of the upper bound tuning is limited.

5. CONCLUDING REMARKS

In this paper a Buffer-in-NUCA (BiN) scheme is proposed to share accelerator buffers in NUCA for accelerator-rich CMPs. It achieves highly efficient on-chip storage utilization through the use of 1) a dynamic interval-based global buffer allocation method to assign spaces to accelerators that can best utilize the additional buffer space, and 2) a flexible paged buffer allocation method to limit the impact of buffer fragmentation. Experimental results show that BiN can significantly improve both performance and energy when compared to representative prior work.

To reduce the complexity of buffer allocation, in this paper, BiN only considers cache bank adjacency when allocating buffers and relies on a uniformly distributed upper bound and existing NUCA bank utilization balancing schemes to mitigate the potential cache bank contentions caused by buffer allocation. An interesting extension of BiN will be to consider both bank adjacency and dynamic bank utilization. Note that it is the flexible paged buffer allocation method proposed in this paper that allows the flexible

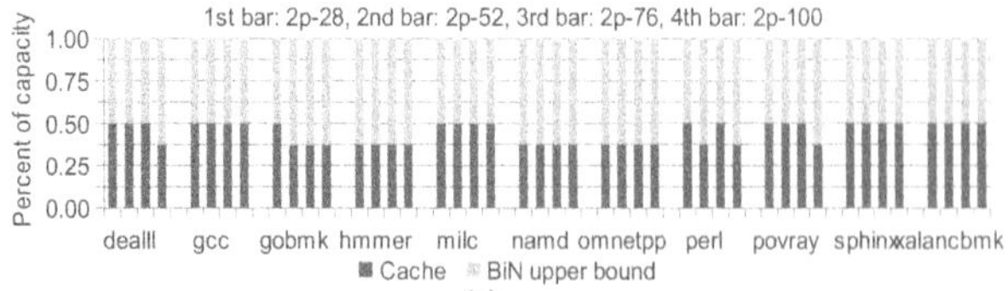

(a)

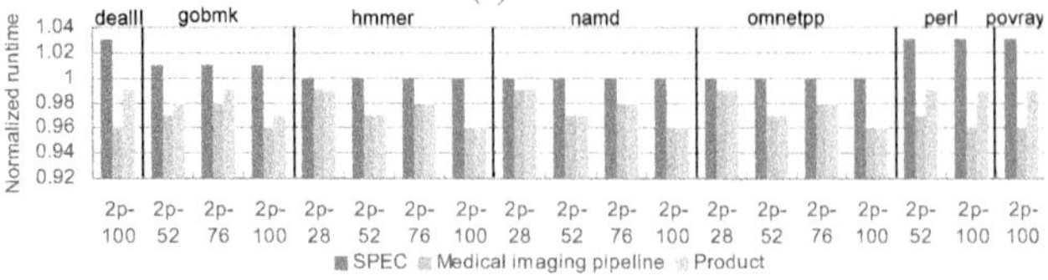

(b)

Figure 13. (a) Partitions via dynamic upper bound tuning. (b) Runtime compared to a fixed upper bound (half of NUCA).

allocation of buffer pages in any cache bank (either more adjacent to the accelerator, or more underutilized, or both).

6. ACKNOWLEDGMENTS

This research is supported by the NSF Expeditions in Computing Award CCF-0926127. The authors thank the anonymous reviewers for their useful comments on this work.

7. REFERENCES

[1] C. Johnson et al. A wire-speed power™ processor: 2.3ghz 45nm soi with 16 cores and 64 threads. *ISSCC* 2010.

[2] L. Seiler et al. Larrabee: a many-core x86 architecture for visual computing. *IEEE Micro*, 29(1):10–21, 2009.

[3] J. Cong et al. AXR-CMP: architecture support in accelerator-rich CMPs. *Workshop on SoC Architecture, Accelerators and Workloads* 2011.

[4] J. Cong et al. Architecture support for accelerator-rich CMPs. *DAC* 2012.

[5] ITRS 2007 system drivers. http://www.itrs.net/.

[6] M. J. Lyonsy et al. The Accelerator Store: a shared memory framework for accelerator-based systems. *ACM Trans. Architecture and Code Optimization*, 8(4):48, 2012.

[7] C. F. Fajardo et al. Buffer-Integrated-Cache: a cost-effective SRAM architecture for handheld and embedded platforms. *DAC* 2011.

[8] J. Cong et al. An energy-efficient adaptive hybrid cache. *ISLPED* 2011.

[9] J. Cong et al. Combined loop transformation and hierarchy allocation for data reuse optimization. *ICCAD* 2011.

[10] A. Bui et al. Platform characterization for domain-specific computing. *ASPDAC* 2012.

[11] B. M. Beckmann et al. ASR: adaptive selective replication for CMP Caches. *MICRO* 2006.

[12] S. Cho and L. Jin. Managing distributed, shared L2 caches through OS-level page allocation. *MICRO* 2006.

[13] J. Cong et al. A shared Buffer-in-NUCA management scheme for accelerator-rich CMPs. University of California, Los Angeles Computer Science Department Technical Report 120012, 2012.

[14] M. Qureshi and Y. Patt. Utility-based cache partitioning: a low-overhead, high-performance, runtime mechanism to partition shared caches. *MICRO* 2006.

[15] HP Cacti, http://quid.hpl.hp.com:9081/cacti/.

[16] P. S. Magnusson et al. Simics: a full system simulation platform. *IEEE Trans. Computer*, 35(2):50-58, 2002.

[17] M. M. K. Martin et al. Multifacet's general execution-driven multiprocessor simulator toolset. *ACM SIGARCH Computer Architecture News*, 33(4):92-99, 2005.

[18] J. Cong et al. High-level synthesis for FPGAs: from prototyping to deployment. *IEEE Trans. Computer-Aided Design of Integrated Circuits and Systems*, 30(4):473-491, 2011.

[19] S. Li et al. McPAT: an integrated power, area, and timing modeling framework for multicore and manycore architectures. *MICRO* 2009.

Adopting TLB Index-based Tagging to Data Caches for Tag Energy Reduction

Jongmin Lee
Department of Computer Science
Korea Advanced Institute of Science and Technology (KAIST)
Daejeon, Korea
square55@kaist.ac.kr

Soontae Kim
Department of Computer Science
Korea Advanced Institute of Science and Technology (KAIST)
Daejeon, Korea
kims@kaist.ac.kr

ABSTRACT

Conventional cache tag matching is based on addresses to identify requested data. However, this address-based tagging scheme is not efficient because unnecessarily many tag bits are used. Previous studies show that TLB index-based tagging (TLBIT) can be used in caches because there are not many different tags at a moment due to spatial locality, and those tags are conventionally captured by TLBs.

In this paper, we show that directly adopting TLBIT is not effective for data caches because TLBIT incurs large overheads in terms of performance and energy consumption due to cache line searches and invalidations. To achieve true potential of TLBIT, we propose three novel techniques: search zone, c-LRU and TLB buffer. Search zone reduces unnecessary cache line searches and c-LRU reduces cache line invalidations. Finally, TLB buffer prevents immediate cache line invalidations on TLB misses. From our experiments, the proposed techniques reduce overall dynamic energy consumption of the data cache by 8.9% on average. Performance impact is small, less than 0.2% on average.

Categories and Subject Descriptors

B.3.2 [**Memory Structures**]: Design Styles—*Cache memories*

Keywords

TLB, Cache Tagging, Energy

1. INTRODUCTION

Cache memories are one of crucial components in modern computer systems because they filter most off-chip memory accesses to improve performance. In conventional address-based cache tagging scheme, upper address bits are used as tags, which is not efficient in terms of area, energy consumption, and latency because unnecessarily many tag bits are used. Previous studies [11, 16] show that required number of tag bits for caches can be effectively reduced by utilizing indexes of Translation Lookaside Buffer (TLB) entries. The indexes are used as tags in place of virtual/physical addresses in caches to identify requested data.

We show in this paper that naively adopting TLB index-based tagging (TLBIT) to data caches can cause large overheads in terms of performance and energy consumption. In TLBIT [11], a TLB miss causes searching all sets in a cache to invalidate cache lines, which use the same index as that of an evicted TLB entry, to prevent conflict use of the index in the cache. This invalidation increases cache access latencies because the cache cannot be accessed during searching the sets. Moreover, the invalidations increase energy consumption and cache miss rates because the cache lines may be not dead (unused) yet. Since TLB misses occur rarely in instruction accesses, performance loss caused by the invalidations is small [11]. However, data accesses typically show lower locality than instruction accesses and, thereby, increase data TLB misses. For these reasons, performance and energy consumption are hurt for the data caches adopting TLBIT.

We propose three novel techniques to overcome those shortcomings of TLBIT for data caches: search zone, c-LRU and TLB buffer. The search zone technique logically divides the data cache into multiple zones and records dispersion of cache lines related to TLB entries across the zones. If a zone has no related cache line, the sets in the zone do not need to be searched on a TLB miss. The c-LRU is a counter-based TLB replacement policy utilizing the block counter technique [16] which stores the number of cache lines related to TLB entries. The c-LRU determines an evicted TLB entry based on the block counters of several candidates from LRU group while conventional LRU selects an evicted entry at the LRU position. The TLB buffer leverages the main concept from victim cache [8] and victim TLB [3] and stores evicted entries to reduce cache line invalidations on TLB misses. Our experiments using a cycle-accurate simulator show that the proposed techniques reduce overall dynamic energy consumption of the data cache 8.9% on average. Performance degradation is small, less than 0.2% on average.

This paper is organized as follows. Section 2 gives necessary background on conventional tagging and TLBIT. In Section 3, we explain the proposed techniques in detail. Section 4 presents evaluation methodology and results. We discuss related work in Section 5. In Section 6, we conclude this paper.

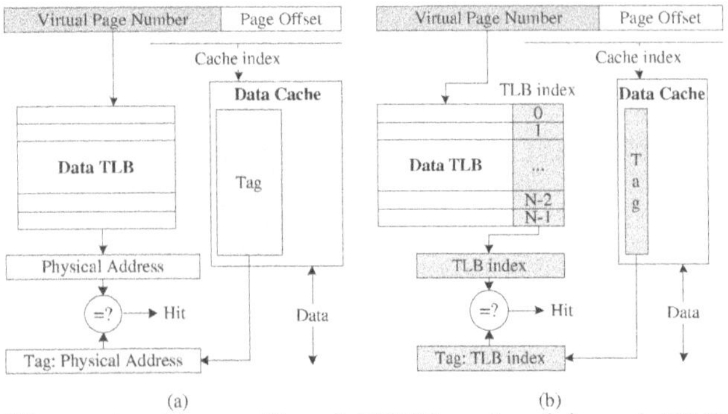

Figure 1: Conventional VIPT cache (a) and TLB index-based tagging scheme for VIPT cache (b)

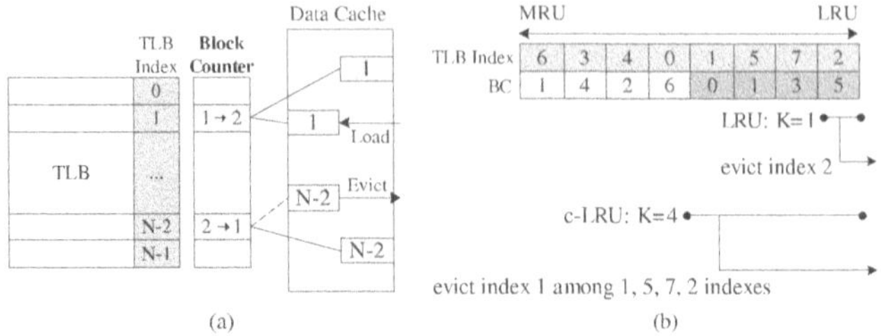

Figure 2: Block counter example (a) and c-LRU replacement policy (b)

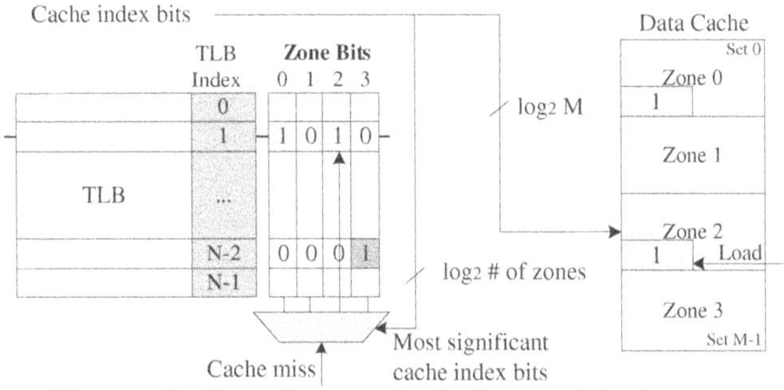

Figure 3: Search zone technique with 4 zones

2. BACKGROUND

2.1 Conventional Tagging

Figure 1 (a) shows conventional Virtually-Indexed and Physically-Tagged (VIPT) cache. The VIPT cache is commonly used in general purpose processors such as Cortex-A series processors for embedded/portable systems [1]. The VIPT cache is indexed by virtual addresses, which are available immediately, and is tagged with physical addresses, which do require translations. To translate virtual addresses, TLBs are commonly used to minimize translation delays. Since a tag comparison occurs after indexing and accessing the cache, the translation using the TLB can be performed in parallel with indexing and accessing the cache. A translated physical address by the TLB is compared with a tag from the cache to determine a hit or a miss.

2.2 TLB Index-based Tagging

Figure 1(b) describes TLB index-based tagging (TLBIT) [11, 16] for the VIPT cache. TLB Indexes are added in each entry of the TLB, which are used as tags in the cache to identify requested data. The VIPT cache supporting TLBIT is indexed using virtual addresses as in the conventional VIPT cache but tagging is performed using the TLB indexes. After address translation, the TLB provides a TLB index corresponding to the translated physical address for a tag matching. On a cache miss, requested data are loaded from the main memory using physical address and the TLB index is stored into the cache for later uses.

When a TLB entry is replaced on a TLB miss, cache lines using the TLB index of the replaced TLB entry become stale because the TLB entry will store a translation for another virtual address. To prevent this situation, all sets in a cache are sequentially checked to invalidate those cache lines. However, these invalidations incur three problems. First, cache access latencies increase because the cache cannot service other accesses during the invalidations. Second, the invalidations increase cache miss rates because cache lines, which are not dead yet, have to be invalidated if they use the same index as that of an evicted TLB entry. Third, if an invalidated cache line is dirty, it must be read to be written back to memory consuming energy.

Since invalidating cache lines occurs on every TLB miss, TLB misses significantly affect performance. In case of the instruction cache, TLB misses occur rarely because of high locality of instruction accesses. Therefore, performance loss due to invalidations is small. However, data accesses typically show lower locality than instruction accesses, which increases TLB misses, and, thereby, increase cache line invalidations.

2.3 Block Counter

A block counter technique is originally proposed in [16]. Its purpose is to reduce cache line invalidation overhead in terms of performance. Searching all sets in a cache for invalidations is not efficient when the actual number of invalidated cache lines is small. A few storage bits are added to each TLB entry as a block counter to count related cache lines as shown in Figure 2(a). The block counter value changes only on cache misses. When a cache line is loaded from the main memory to the cache, the block counter of a corresponding TLB entry increases. The block counter decreases when a related cache line is evicted from the cache. Since the block counter has a few storage bits, it saturates once the counter reaches its max value. Experimentally we found 5- or 6-bit block counter is enough for most cases.

Block counters are utilized for invalidating cache lines on TLB misses. When a cache line is invalidated, the corresponding block counter decreases. If the block counter reaches 0, invalidation is terminated. In the best case, if the block counter is initially 0, invalidating cache lines is not necessary. However, searching all sets in the data cache has to be performed when the block counter is saturated.

3. PROPOSED TECHNIQUES

TLBIT has the potential to reduce energy consumption of the tag array. However, cache line invalidations on TLB misses nullify the potential or even increase energy consumption. To achieve true potential of TLBIT, we propose three novel techniques to minimize cache set searches and cache line invalidations.

3.1 Search Zone

On a TLB entry replacement, related cache lines to the entry must be identified and invalidated, which requires searching all cache sets because we do not know which sets have the related lines. The search consumes cycles and energy by checking each set. To reduce the number of cache set searches, our search zone technique records dispersion of cache lines related to each TLB entry. Figure 3 shows the search zone technique for 4 zones. The data cache is logically divided into multiple zones and search zone bits are added to each TLB entry. When a cache line is loaded into the data cache, the most significant $\log_2$ (number of zones) bits of the cache index bits ($\log_2$ M), where M is the total number of sets, are used as a search zone number currently accessed. Then, the corresponding search zone bit is set to 1 indicating the zone has related cache lines. This search zone bit indicates whether the corresponding zone has to be searched or not. For example, assume that a TLB entry with index N-2 is evicted as shown in Figure 3. Without the search zone technique, all sets in the cache are searched to invalidate cache lines related to the evicted TLB entry. However, only zone 3 is searched with the search zone technique since only that search zone bit for the TLB entry is 1. If a cache line is evicted by the ordinary replacement and no line corresponding to a zone exists, cache sets in the zone will be searched because our search zone technique records only when cache lines are loaded to the cache. When a TLB entry is replaced by a miss, its search zone bits are set as 0 to initialize them for a new page. Since cache lines in a page tend to gather in adjacent sets due to their spacial locality [9], searchning only small portion of sets using zones is effective.

As we divide the data cache into more zones, the number of searched sets decreases because TLB entries generally do not have many related cache lines. However, more search zones require more zone bits increasing area overhead. For evaluations, we vary the number of zones from 8 to 16 because these numbers showed best results.

3.2 c-LRU Replacement Policy

Because evicted TLB entry determines which cache lines are to be invalidated, TLB replacement policy plays an important role in TLBIT caches. We propose a new TLB replacement policy called c-LRU, which is based on the block counter to choose an evicted TLB entry on a TLB miss. A popular replacement policy for fully-associative TLBs is an LRU (Least Recently Used) policy. On a TLB miss, a TLB entry at the LRU position is chosen as an evicted entry. However, this conventional LRU policy does not always work well with TLBIT caches. Since TLB entries have different numbers of related cache lines, TLB replacement policy affects not only TLB miss rates but also cache miss rates by invalidating different numbers of cache lines.

The c-LRU policy increases the candidate size to K from the LRU group. Among K candidates, an entry which has the smallest block counter value is selected as an evicted entry. Figure 2(b) shows an example for the conventional LRU policy and c-LRU policy when K is 4. Selecting the smallest counter value has two advantages. First, it reduces the invalidation latency on TLB misses because the invalidations will finish early if the block counter value is small. Second, it reduces cache miss rate by invalidating small number of cache lines. If an evicted TLB entry has a lot of related cache lines, some cache lines should be invalidated even though they are not expired (dead) yet. Selecting a TLB entry with the smallest block counter value gives cache lines more chances to get expired. For ageing of the TLB entries, we use a 4-bit saturated counter. When a TLB entry is replaced in the c-LRU manner, the counter value increases by 1. Once the counter is saturated, the next evicting entry is selected at the LRU position and the counter is set to 0.

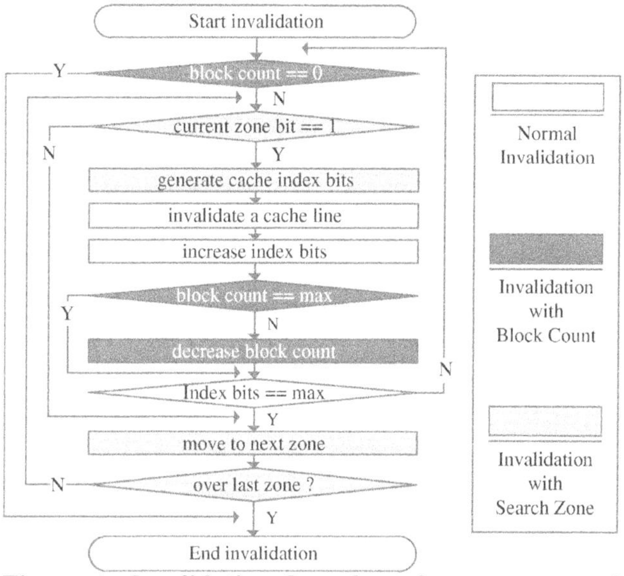

Figure 4: Invalidation flow chart for our proposed techniques

3.3 TLB Buffer

A TLB miss rate is a critical factor in TLBIT. When TLB misses occur frequently with large working sets, cache line invalidation overheads in terms of performance and energy consumption significantly increase. To overcome this problem, we propose a TLB buffer, which shares the main concept with the previously proposed victim cache [8] and victim TLB [3]. The TLB buffer stores evicted TLB entries to prevent immediate cache line invalidations on TLB misses.

An entry size of the TLB buffer is the same with that of TLB (including block counters, zone bits). A 1-bit phase bit is added to each TLB buffer entry, TLB entry, and cache line to distinguish the same index stored in the TLB and the TLB buffer. Because of the 1-bit phase bit, any indexes in the TLB can have the same index in the TLB buffer up to 1. On a TLB miss, the TLB buffer is examined whether the same index of an evicted TLB entry exists or not. If so, the matched TLB buffer entry is evicted and cache line invalidation begins. After copying the entry from the TLB to the TLB buffer, the phase bit of the TLB entry for a new page is flipped. The TLB buffer does not store all evicted TLB entries. If the block counter value of an evicted entry is 0, that entry is not stored in the TLB buffer because it does not induce cache searches for invalidation.

Since the TLB buffer is accessed on TLB misses, accessing the TLB buffer consumes an additional cycle. If the TLB buffer hits, a matched entry provides its index to the cache for a tag matching and that entry is maintained in the TLB buffer. If the TLB buffer misses, an evicted TLB entry is copied to the TLB buffer as explained above. If the TLB buffer is full, one of TLB buffer entries is selected for eviction in the LRU manner.

3.4 Invalidation details

Figure 4 shows a flow chart for invalidations on TLB misses. White-colored shapes indicate the previously proposed invalidation process in [11], which searches all cache sets. Black-colored shapes are for the block counter tech-

Table 1: Processor Parameters

Configuration	Parameter Values
Pipeline stages	7, 8 Stages
Cache configuration	16KB or 32KB, 4-way associative, 32 Byte Block size
Cache latency	2 cycle
Indexing/tagging	Virtually-indexed, Physically-tagged
Instruction TLB	32 Entries Fully Associative
Data TLB	32 Entries Fully Associative
TLB latency	1 cycle
Memory bus	32-bit wide
Memory latency	80 cycles

nique [16] and gray-colored shapes are for the search zone technique. In the block counter technique, the counter value of an evicted TLB entry is examined first and the invalidation process begins if the block counter value is not 0. Otherwise, the invalidation process is terminated because there is no related cache line to the TLB entry. After invalidating a related cache line, the block counter value decreases by 1 if the block counter value is not max. The search zone technique examines current search zone bit first. If the bit is 1, the invalidation process for the current zone starts. While the previously proposed invalidation in [11] sequentially increases the cache index bits from 0 to the number of total sets in the cache, the search zone technique increases the cache index bits up to the number of sets in a zone. Thus, the cache index bits are generated using two counters: a ($\log_2$ (number of zones) + 1)-bit counter, which is to choose a zone, and a (cache index bits $-\log_2$ (number of zones))-bit counter, which is to increase index bits within the zone. After all related cache lines in a zone are invalidated, the next search zone bit is examined. When all zones are invalidated, the invalidation process is terminated.

Note that the control logic involves a few counters and comparators. Therefore, it can be implemented with low cost. The energy overhead of the control logic is low because we can disconnect the power rails to the control logic using a power gating technique [5].

4. EVALUATION

4.1 Evaluation Methodology

Our baseline processor is Intel XScale [7], a high-performance, low-power architecture targeting embedded applications. We evaluate our proposed techniques using XTREM [2], a cycle accurate performance and energy simulator for the Intel XScale. Table 1 shows the parameter values used for evaluation. We used CACTI 6.5 [17] at the 32nm technology to obtain power consumption values. To evaluate benefits of the proposed techniques, we simulated benchmarks from MiBench embedded application suite [4]. For all simulations, benchmarks are fully executed and large input data are used.

4.2 Performance Evaluation

Due to page limitation, we only present the results for 16KB data cache configuration because the results for 32KB configuration show similar patterns to the 16KB configuration. Throughout this paper, block count, search zone, c-LRU and TLB buffer are indicated by BC, Z, L, TB with their sizes, respectively. For example, BC5, Z16, L8, TB8

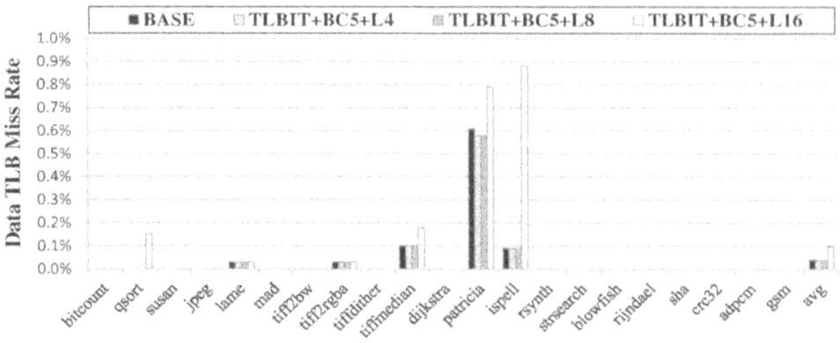

Figure 5: TLB miss rates with the c-LRU replacement policy

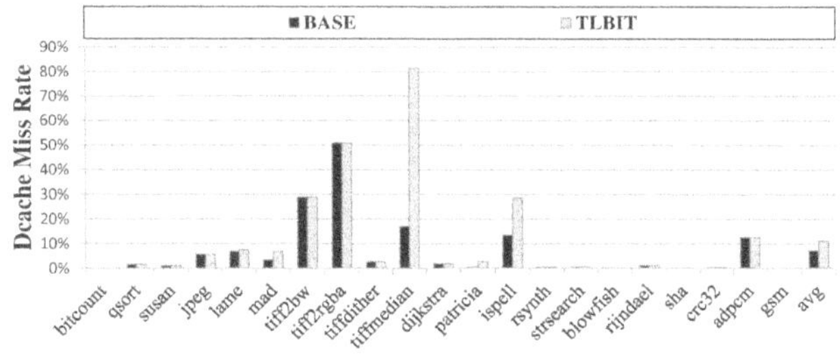

Figure 6: Miss rates of 16KB data cache

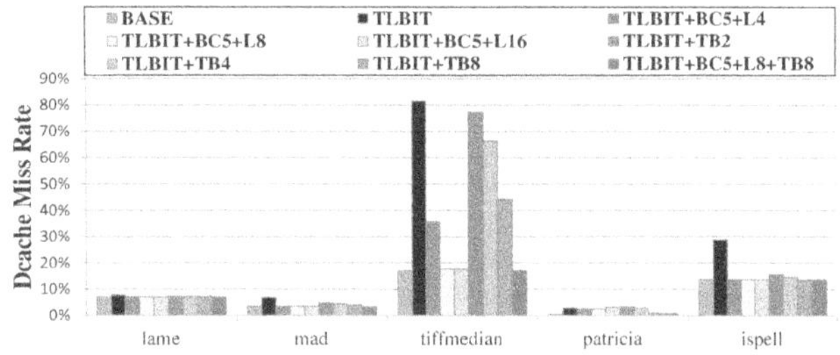

Figure 7: Miss rates of 16KB data cache for TLBIT-sensitive benchmarks

indicate 5-bit block counters, 16 zones in the data cache, 8 candidates for c-LRU and 8-entry TLB buffer, respectively.

Figure 5 shows miss rates of the TLB. While most benchmarks show very low miss rates, some benchmarks show relatively high miss rates, which have large working sets. With the c-LRU technique, miss rates of the TLB do not change much when the size of candidates is under 16. With the L16, the TLB miss rates increase much because it hurts the LRU policy.

Miss rates of the data cache are shown in Figure 6. While most benchmarks show no increase of the miss rates in TLBIT because of very small TLB misses, miss rates of some benchmarks (*lame*, *mad*, *tiffmedian*, *patricia* and *ispell*) increase. Invalidations of related cache lines on TLB replacements negatively affect miss rates of the data cache because live lines are invalidated. Figure 7 shows miss rates of the data cache when candidate size of c-LRU and TLB buffer size vary. Some benchmarks are not shown because they did not show changes in the miss rate of data cache. With the c-LRU technique, the miss rates are successfully reduced when the size of candidates exceeds 4. We chose the size 8 for the c-LRU because it shows good results for both the data cache and the TLB. Though the TLB buffer technique reduces the miss rates by storing evicted entries from the TLB to prevent invalidations, it does not reduce the miss rates much because those entries will be replaced soon by other evicted entries on TLB misses. When using two techniques at the same time shows the best results.

Figure 8 shows how many sets are searched on a TLB miss on average to analyze effectiveness of the proposed tech-

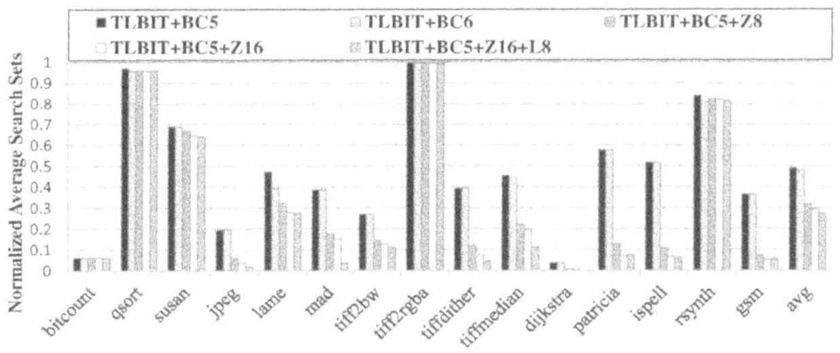

Figure 8: Average searched sets on TLB replacements

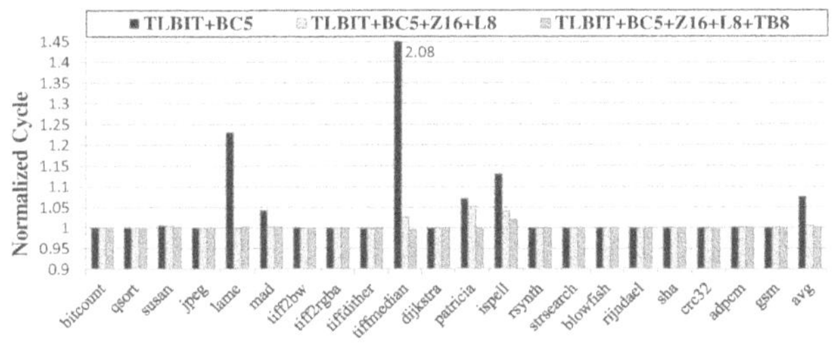

Figure 9: Normalized execution cycles for 16KB data cache

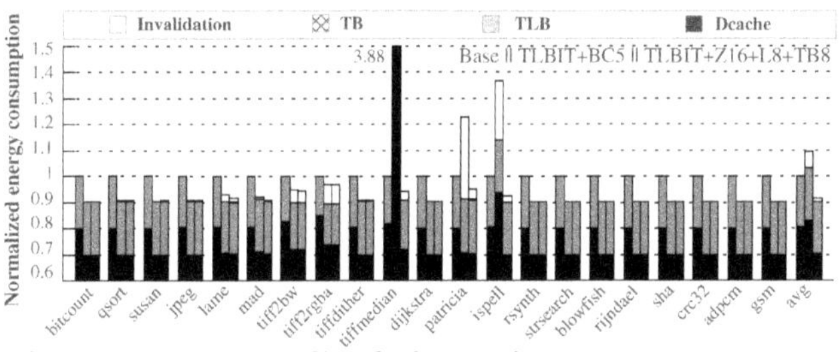

Figure 10: Normalized dynamic energy consumption to the baseline cache. First bar is for the baseline. Second bar and third bar are for TLBIT+BC5 and TLBIT+BC5+Z16+L8+TB8, respectively.

niques. Results are normalized to the number of sets in the data cache. The block counter technique removed almost half of the searched sets on average. However, there is no major difference between the block counter bit sizes. When the search zone technique is applied with the block counter technique, the number of searched sets for invalidations is reduced dramatically because the search zone removes sequential searching of sets. In addition, the c-LRU reduces the number of searched sets further because it selects the smallest block counter values.

Finally, overall execution cycle results for the 16KB data cache are shown in Figure 9. Simulation results are normalized to the baseline 16KB cache. Some benchmarks such as *lame*, *tiffmedian*, *patricia* and *ispell* show increased execution cycles. Especially, *tiffmedian* shows a highly increased execution time because it suffers from the increase in data cache miss rate with TLBIT. In the conventional VIPT cache, the TLB is always accessed in parallel with data cache accesses. Thus, there is no overhead in terms of performance with the TLB index-based tagging scheme on TLB hits. Only when TLB misses occur, TLBIT induces additional cycles for invalidations and/or accessing to TLB buffer. However, the impact of sequential TLB buffer accesses are small (0.07% on average) because of low TLB misses. The proposed techniques effectively reduce execution cycles due to TLBIT. With the TB8, the execution cycles reduce further because of the improved global TLB miss rates. On average, the proposed techniques increase execution cycles by 0.2%.

Table 2: CACTI simulation results (32nm)

Types	Total(Tag) read energy(pJ)	Total(Tag) write energy(pJ)	Total(Tag) area (mm2)	Total(Tag) leakage (mW)
Base 16K $	3.67(0.90)	2.64(0.84)	0.047(0.0037)	3.75(0.44)
TLBIT 16K $	3.21(0.43)	2.22(0.41)	0.039(0.0017)	3.55(0.24)
Base 32K $	5.43(0.99)	3.74(1.13)	0.086(0.0083)	7.20(0.95)
TLBIT 32K $	4.99(0.55)	3.12(0.52)	0.075(0.0032)	6.43(0.38)
TLB32	0.92	0.98	0.0023	0.15
TLBIT TLB32	0.94	1.03	0.0026	0.16
TLBIT TB8	0.51	0.56	0.0014	0.07
BC5	0.04	0.05	0.0001	0.008
Z8	0.05	0.6	0.0001	0.09
Z16	0.09	0.10	0.0002	0.015

4.3 Energy Evaluation

Table 2 shows simulation results using CACTI at 32nm technology. For tag arrays, 52% and 44% of the per-read energy are reduced for 16KB and 32KB caches, respectively. Overall per-access energy consumptions, which include tag arrays and data arrays, are reduced by 12% and 8.1% for 16KB and 32KB caches, respectively. Since the proposed techniques require additional bits in each entry of the TLB, overheads in terms of per-access energy, area and leakage power increase, which are, however, still smaller than tag reduction gains.

$$Cache_{Energy} = Read \times E_{read} + Write \times E_{write} + Replacements \times E_{write} + Invalidations \times E_{read}$$

$$TLB_{Energy} = Read \times E_{read} + Replacements \times E_{write}$$

To calculate dynamic energy consumption of the data cache and the TLB, we use an energy model above. The energy model uses the number of accesses to the cache/TLB from the XTREM simulator and per-access energy on read/write operations, which are extracted from CACTI. *Read*, *Write*, *Replacements* and *Invalidations* are the number of read operations, write operations, replacements and invalidations in the cache, respectively. E indicates per-access energy consumption on read and write operations.

Figure 10 shows energy consumption results normalized to the baseline 16KB cache using the above energy model. There are three bars for each benchmark: the baseline cache, TLBIT cache with the 5-bit block counter and TLBIT cache with all proposed techniques. Each bar is divided into 4 portions: Invalidation, TLB, TB and Dcache. The Invalidation indicates energy consumption in the data cache for searching sets on invalidations. The TLB, the TB and the Dcache indicate energy consumptions in each of them. Some benchmarks such as *tiffmedian*, *patricia* and *ispell* show worse results with TLBIT+BC5 than the baseline cache because energy consumption in the Invalidation and the Dcache dramatically increases due to the increase of misses in the data cache. Thus, the block counter technique is not enough to restrict invalidation overheads on TLB misses. With our proposed techniques, invalidation overheads are effectively reduced. The TLB buffer consumes very small dynamic energy because it is accessed only on TLB misses. Overall dynamic energy consumptions are reduced with our proposed techniques by 8.9% on average.

5. RELATED WORK

Tag Comparison Elimination (TCE) schemes try to reduce the number of accesses to the tag arrays. In [19], a small table for TCE stores recently used address information and minimizes interactions with the instruction cache. A history-based approach expanding the Branch Target Buffer (BTB) is proposed in [6]. The cache utilizes addresses in the BTB to avoid unnecessary tag matchings. Direct addressing scheme is proposed in [18]. This scheme requires ISA extension, an extra hardware structure called Direct Address Register, and compiler support, and does not reduce area for tag arrays.

In [13, 14], application-specific customization of active tag bits is introduced to reduce tag access energy consumption in the data cache for application loops. Tag Overflow Buffering (TOB) scheme [12] exploits the size of tag bits with an external register. The external register captures recently used upper part of tag bits. Only small size of partial tag bits are stored in the cache to reduce energy consumption. However, this scheme suffers when the locality of addresses changes. In [10], a tag energy reduction scheme is introduced with a locality buffer. The locality buffer captures the upper part of tag bits similar to [12]. Remaining tag bits are stored into the cache to save energy. In [15], original physical tag bits are encoded into smaller tag bits. However, this scheme requires a relatively complex encoder and OS supports to maintain the reduced tags.

In contrast to the schemes above, TLBIT proposed in [11, 16], is to reduce tag array size physically and, thereby, to reduce cache energy consumption and area. However, we show that direct extension of the TLB index-based tagging scheme to data caches does not perform well because invalidations on TLB misses incur performance loss and energy consumption increase in data caches.

6. CONCLUSIONS

This paper investigates the use of TLB index-based tagging (TLBIT) scheme for data caches. We show that directly adopting TLBIT to data caches incurs large overheads in terms of performance and energy consumption due to cache line searches and invalidations on TLB misses. To reduce those overheads, we proposed three techniques: search zone, c-LRU and TLB buffer. The search zone technique records dispersion of cache lines related to a TLB entry to reduce unnecessary cache line searches. The c-LRU reduces cache line invalidations by carefully selecting an evicted TLB entry and the TLB buffer stores evicted TLB entries to reduce cache line invalidations. Our evaluations using a cycle-accurate simulator show that our proposed techniques reduce dynamic energy consumption of the data cache by 8.9% on average. Performance degradation is less than 0.2% on average.

7. ACKNOWLEDGMENTS

This research was supported by the National Research Foundation of Korea (NRF) grants funded by the Ministry of Education, Science and Technology (2011-0005378, 2012-0000980).

8. REFERENCES

[1] ARM. Cortex-a processors. http://www.arm.com/products/processors/ cortex-a/.

[2] G. Contreras, M. Martonosi, J. Peng, R. Ju, and G.-Y. Lueh. Xtrem: a power simulator for the intel xscale core. *ACM SIGPLAN Notices*, 39(7):115–125, July 2004.

[3] R. A. Garibay, J. Richardson, M. A. Quattromani, and D. Beard. Address translation unit employing a victim tlb. US Patent 5,752,274, 1998.

[4] M. R. Guthaus, J. S. Ringenberg, D. Ernst, T. M. Austin, T. Mudge, and R. B. Brown. Mibench: A free, commercially representative embedded benchmark suite. In *Proc. of the IEEE 4th Annual Workshop on Workload Characterization*, Dec. 2001.

[5] Z. Hu, A. Buyuktosunoglu, V. Srinivasan, V. V. Zyuban, H. M. Jacobson, and P. Bose. Microarchitectural techniques for power gating of execution units. In *Proc. of the International Symposium on Low Power Electronics and Design*, pages 32–37, 2004.

[6] K. Inoue, V. G. Moshnyaga, and K. Murakami. A history-based i-cache for low-energy multimedia applications. In *Proc. of the International Symposium on Low Power Electronics and Design*, 2002.

[7] Intel. Intel xscale core developer's manual. http://www.intel.com/design/intelxscale/273473.htm.

[8] N. P. Jouppi. Improving direct-mapped cache performance by the addition of a small fully-associative cache and prefetch buffers. In *Proc. of the International Symposium on Computer Architecture*, pages 364–373, 1990.

[9] J. Kim, S. Kim, and Y. Lee. Simtag: Exploiting tag bits similarity to improve the reliability of the data caches. In *Proc. of the Design, Automation & Test in Europe*, pages 941–944, 2010.

[10] J. W. Kwak and Y. T. Jeon. Compressed tag architecture for low-power embedded cache systems. *Journal of Systems Architecture - Embedded Systems Design*, 56(9):419–428, 2010.

[11] J. Lee, S. Hong, and S. Kim. TLB index-based tagging for cache energy reduction. In *Proc. of the International Symposium on Low Power Electronics and Design*, pages 85–90, 2011.

[12] M. Loghi, P. Azzoni, and M. Poncino. Tag overflow buffering: An energy-efficient cache architecture. In *Proc. of the Design, Automation & Test in Europe*, pages 520–525, 2005.

[13] P. Petrov and A. Orailoglu. Data cache energy minimizations through programmable tag size matching to the applications. In *Proc. of the International Symposium on System Synthesis*, 2001.

[14] P. Petrov and A. Orailoglu. Energy frugal tags in reprogrammable I-caches for application-specific embedded processors. In *Proc. of the International Symposium on Hardware/Software Codesign*, pages 181–186, May 6–8 2002.

[15] P. Petrov and A. Orailoglu. Dynamic tag reduction for low-power caches in embedded systems with virtual memory. *International Journal of Parallel Programming*, 35:157–177, April 2007.

[16] A. Seznec. Don't use the page number, but a pointer to it. In *Proc. of the International Symposium on Computer Architecture*, pages 104–113, 1996.

[17] S. Thoziyoor, N. Muralimanohar, J. H. Ahn, and N. P. Jouppi. CACTI 5.1. Technical Report HPL-2008-20, Hewlett Packard Laboratories, 2008.

[18] E. Witchel, S. Larsen, C. S. Ananian, and K. Asanović. Direct addressed caches for reduced power consumption. In *Proc. of the International Symposium on Microarchitecture*, pages 124–133, 2001.

[19] Y. Zhang and J. Yang. Low cost instruction cache designs for tag comparison elimination. In *Proc. of the International Symposium on Low Power Electronics and Design*, pages 266–269, 2003.

Static and Dynamic Co-Optimizations for Blocks Mapping in Hybrid Caches

Yu-Ting Chen, Jason Cong, Hui Huang, Chunyue Liu, Raghu Prabhakar, and Glenn Reinman
Computer Science Department, University of California, Los Angeles
Los Angeles, CA 90095, USA
{ytchen, cong, huihuang, liucy, raghu, reinman}@cs.ucla.edu

ABSTRACT

In this paper, a combined static and dynamic scheme is proposed to optimize the block placement for endurance and energy-efficiency in a hybrid SRAM and STT-RAM cache. With the proposed scheme, STT-RAM endurance is maximized while performance is maintained. We use the compiler to provide static hints to guide initial data placement, and use the hardware to correct the hints based on the run-time cache behavior. Experimental results show that the combined scheme improves the endurance by 23.9x and 5.9x compared to pure static and pure dynamic optimizations respectively. Furthermore, the system energy can be reduced by 17% compared to pure dynamic optimization through minimizing STT-RAM writes.

Categories and Subject Descriptors

B.3.2 [**MEMORY STRUCTURES**]: Design Styles – *Cache memories.*

General Terms

Algorithm, Design.

Keywords

L2 cache, Hybrid cache, STT-RAM, Endurance, Energy.

1. INTRODUCTION

The traditional SRAM caches suffer from huge leakage power, which dominates the total energy consumption in the on-chip memory system. To alleviate this problem, emerging non-volatile memory technologies such as phase-change RAM (PRAM) and spin-torque transfer magnetoresistive RAM (STT-RAM) are used as alternative on-chip memory with the advantages of low leakage and high density. However, non-volatile memories suffer from the challenges of limited endurance, higher write latency and energy. Compared to PRAM, STT-RAM has significantly higher endurance (10^9 versus 10^{12} write cycles) and shorter write latency [1] and is much more promising in the last-level cache design [2] [3][4][5][6][7]. Moreover, due to the intensive writes of caches, hybrid caches consisting of both SRAM and STT-RAM are investigated [2][3][4][7], where the SRAM can accommodate write-intensive data and the STT-RAM can accommodate other data with its dense capacity.

ISLPED'12, July 30–August 1, 2012, Redondo Beach, California, USA.

The STT-RAM endurance is an important issue to be considered in the last-level cache design. Although ITRS predicts the write cycles of STT-RAM will be 10^{15} at 2024 [1], the best available write cycles of STT-RAM are 4 x 10^{12} at present [5]. Supposed we execute *segmentation* [8], a medical imaging application, on a 4GHz CPU with 32 KB L1 cache, 2MB STT-RAM L2 cache continuously, the lifetime of a STT-RAM cache can last only 2.17 years without any optimizations applied. The endurance problem becomes even worse in the multi-level cell (MLC) STT-RAM technology [5]. Block placement optimizations is very important to shrink the large endurance gap between STT-RAM and SRAM in a hybrid cache. Recent work considers either static or dynamic schemes to optimize the block placement to reduce the average write frequency to STT-RAM cells, while maintaining the overall performance by making use of higher density of STT-RAM. Some of the proposed approaches targeted at PRAM, and those ideas can also be applied to STT-RAM with the same objective.

The first category of the prior work uses static schemes. In [9] the authors introduce data migration and re-computation to reduce the write frequency on PRAM main memory. In [10], the partitioning of the application working set into SRAM and PRAM can reduce 79% of the writes to PRAM.

The second category of the prior work uses dynamic schemes. Recent work in [5] uses periodically set-remapping to distribute the writes among sets in a STT-RAM cache. Another set of work migrates the write-intensive cache blocks to other cache lines in the same/different cache set or in the SRAM in order to reduce the average write frequency of the STT-RAM cache lines [4].

However, there exist intrinsic limitations in both approaches, which cannot be resolved independently – The static optimization decisions are made at compile time without run-time information, thus compiler may generate misleading hints to the hardware. On the other hand, pure dynamic optimization use blocked-based counter structures to learn the memory reference patterns on-the-fly. However, the dynamic scheme lacks a global view of the whole program and has no knowledge to future access pattern.

In this paper, we propose a combined approach in which the static and dynamic optimizations can compensate to each other. The compiler tries to provide data placement hints to hardware to reduce STT-RAM write frequency, while the hardware is designed to be able to correct compiler hints based on runtime cache behavior. Experimental results show that the combined scheme improves the endurance by 23.9x and 5.9x compared to pure static and pure dynamic optimizations, respectively, while maintaining similar performance. Furthermore, the system energy can be reduced by 17% compared to pure dynamic optimization since STT-RAM writes are reduced through initial placement from the proposed compiler flow.

2. PROBLEM FORMULATION

In this work, we assume that the L2 cache is a hybrid cache architecture with 4-way SRAM and 12-way STT-RAM, which is similar to the setting described in [4][7]. The block-level initial placement and dynamic migration is allowed to place the data blocks in either SRAM or STT-RAM. The initial placement is given by the compiler hints and the runtime cache pressure while the dynamic migration is designed with hardware mechanisms. Our assumptions in detail are described in Section 4.1 and Section 4.3 for better illustration of our co-optimization strategy.

The objective of this work is to improve the endurance of the hybrid cache and reduce system energy while maintaining performance through the combined scheme. Another meaningful objective is to co-optimize performance and energy while under the endurance constraint. A storage-efficient way to monitoring the endurance of each cache block is required under the second scenario. The discussion of this formulation is not included in this work but may be worthwhile for further investigation.

3. MOTIVATIONAL EXAMPLES

In this section, we use real-life examples to illustrate how pure static optimization and dynamic optimization may produce sub-optimal block placement decisions in a hybrid cache design.

The deficiency of pure static optimization comes from the fact that the runtime write frequencies of L2 cache (which is the hybrid last-level cache in our evaluated system) blocks are input-dependent, which cannot be fully obtained offline. First, the input data may change the control flow of the program, and this will change the write frequency of the data that are affected by the control flow variation. Second, since part of the writes to the L2 cache come from the write-back operations from the L1 caches, the compiler cannot accurately capture the L2 cache write behavior with the existence of the L1 caches. For example, given the LRU replacement policy, the data which is written fewer times in the code may be frequently evicted by the L1 cache and behaves much more write-intensive in the L2 cache than other data which are written more frequently. Note that life time of the STT-RAM mainly depends on the peak write count of all the cells [4][5]. Even static optimizations can reduce the total STT-RAM writes compared to dynamic optimizations via global optimization, the potential mis-predictions can still severely degrade the STT-RAM lifetime, since there is no dynamic scheme to migrate mis-predicted write-intensive blocks into SRAM. This causes a large peak write count to this cell. As an example, Table 1 shows the STT-RAM cell write count distribution of the *segmentation* application [8] for both of the pure static [10] and the pure dynamic [4] scheme. The peak write count of static optimization is significantly larger than that of the dynamic one.

Table 1: STT-RAM write count distribution

#writes	0-100	100-200	200-300	300-400	400-1000	1000-5000	>=5000	max
static	12262	6	5	0	0	10	5	5470
dynamic	12207	38	16	27	0	0	0	395

The deficiency of dynamic migration comes from the fact that it lacks the future memory access information and highly relies on the application to exhibit a bipolar L2 write frequency patterns – the L2 cache blocks are either rarely written or intensively written. Then the write-intensive blocks can swap their places with the rarely written blocks through dynamic migration. However, based on our observation, not all the applications have such characteristics, especially in the three medical imaging applications [8] used in our study. As shown in Figure 1, most of the blocks are uniformly written 2-3 times. Under this circumstance, if the migration threshold is set to be higher than 3, then there will be little migration, both the SRAM and STT-RAM will be evenly written based on the LRU scheme, which may impair the endurance. However, even if the migration threshold is set to be 2 or 3, the migration for most blocks does not save any write, since most blocks behave similarly. A correct approach is to place all these streaming accessed blocks into the SRAM, which can be obtained via static optimization in the compiler. After that, the expensive writes on STT-RAM can be significantly reduced and thus dynamic energy can reduced.

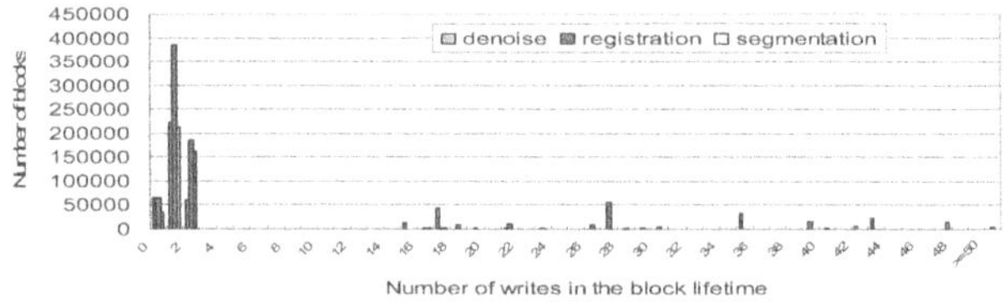

Figure 1: Write frequency distribution of the L2 cache blocks

To overcome these limitations while taking the advantage of both static and dynamic schemes, we use a combined strategy: the compiler tries to guide the hardware in order to rapidly achieve the desired placement, while the hardware corrects the compiler hints based on the run-time cache behavior. To the best of our knowledge, we are the first one taking such a hybrid approach.

4. THE COMBINED APPROACH

4.1 Compiler Support

In this work, we develop an automatic compilation flow to generate data placement hints for each memory reference. Here, we assume LRU replacement policy is used and L2 is an inclusive cache with the same block size of L1, which is widely used in modern processors because of easy coherence implementation.

Similar to [10], our compiler tends to place write-intensive references into SRAM and non-write-intensive data into STT-RAM. Based on our inclusive cache assumption, the write accesses on L2 STT-RAM cells occur at only two situations: (1) L1 dirty evictions due to L1 cache replacement and (2) L2 cache replacement. However, the work in [10] assumes there is no cache in the memory system, thus does not consider the effect of higher-level (L1) cache on the memory access behavior. Figure 2 shows an example code and its corresponding memory access behavior. We can find that both arrays *A* and *B* are written twice. However, since array A is more frequently accessed and can be kept in the L1 cache (we assume LRU replacement policy is used here), neither of the two writes falls into the L2 cache. On the other hand, since array *B* is evicted from the L1 cache before its next access, one write-back operation will be issued into L2.

loop 1: A[i] = …; B[i] = …; (*write array A and B in L1*)
loop 2: … = A[i] …; (*read array A in L1*)
loop 3: … = C[i] …; (*array B is evicted into L2*)
loop 4: A[i] = …; B[i] = B[i]…; (*write array A and B in L1*)

Figure 2: One sample code and its memory access behavior

To capture this effect, we use the concept of *memory reuse distance* (*MRD*) [11], which equals the total size of unique data elements accessed between two references to *X*. A larger memory reuse distance of *X* implies that *X* will not be accessed in the near future, and thus *X* is more likely to be evicted from the L1 cache.

Definition 1: For an write operation w of memory instruction X, assuming the future access sequence of X is w, r_1, r_2, ..., r_n, w', etc (r and w corresponds to read and write operation). w is called an *L1-writeback write* if one of the following conditions is satisfied: (1) there exists $MRD(r_i, r_{i+1}) > dist_{L1}$ (i = 1, ..., n-1) (2) $MRD(r_n, w') > dist_{L1}$ (3) $MRD(w, r_1) > dist_{L1}$. ($dist_{L1}$ is the average reuse distance to keep X in L1 cache)

From Definition 1 we can see, if the memory reuse distance between two accesses into a dirty data X is larger than a threshold value $dist_{L1}$, the compiler will treat the first write (w) to X as a *L1-writeback*, since X will be written back into L2. The other set of L2 write accesses comes from L2 misses and data are written into L2 from main memory. Here we use $dist_{L2}$ to indicate the average reuse distance to keep X in L2 cache. For two adjacent accesses to X, if the memory reuse distance between them is larger than $dist_{L2}$, the compiler will treat the second access as a L2 miss, which will introduce one L2 write operation.

In our flow, we provide a 2-bit compiler hint for each memory instruction to guide its data placement in L2 cache. For each access to reference X, we count the total number of future L2 writes to X including both L1-writeback writes and L2 misses. If there are frequent L2 writes, our compiler will generate hint "01" for X. On the other hand, hint "00" is generated to indicate that X will not be written frequently. For those accesses that the compiler cannot analyze accurately (e.g., due to unknown loop bound), hint "1x" is generated and the data placement is controlled by hardware. Note that a memory instruction in a regular loop is accessed multiple times with repeated access patterns [12], therefore we can apply the same hint to all the accesses to it. It should be noted that the compiler just tries its best to predict the write frequency. The value of $dist_{L1}$ and $dist_{L2}$ can be obtained from profiling on representative input or set to a fixed value by default. However, it is not feasible to profile all input sets. In this work, a conservative approach is used to set $dist_{L1}$ to L1 set associativity. This can ensure that data X will not be evicted out from L1 between two accesses with reuse distance less than $dist_{L1}$. Since the compiler cannot make an optimal decision without knowing the runtime cache behavior, these generated hints may not be followed in the hardware. We will discuss our combined scheme in Section 4.3.1.

4.2 Compiler-Hardware Interface

In our implementation, the compiler passes the hints to the hardware through setting two bits in the 32-bit instruction code. We assume that there are two extra bits in each memory instruction that the compiler can use to assist the run-time cache block replacement. Existing architectures already use these kinds of extra bits in the instruction, such as the *prefetch and evict-next* instruction in the Alpha 21264. We believe that, in most architectures, the increasing speed gap between memory and processor will justify the inclusion of additional bits in the instruction code to facilitate the reduction of this gap. Once a memory reference instruction is executed, if this is a L1 cache hit, these 2 bits will be discarded. If this is a L1 miss, then the bits will be passed to the L2 cache controller, if this is a L2 hit, then these 2 bits will be discarded; if this is a L2 miss, these 2 bits will be used as hints for the initial placement of the new block.

4.3 Hardware Support

In this work, we use a 1MB 16-way hybrid cache including a 4-way SRAM data array and a 12-way STT-RAM data array similar to the configurations used in [2][4][7]. The asymmetric configuration is chosen since smaller SRAM contributes less leakage while the bigger STT-RAM provides the advantage of higher density. We use separate replacement units on SRAM and STT-RAM in order to perform block replacement in these two arrays independently. We also introduce a global replacement unit for them in order to perform a global replacement among them if required. All of these replacement units use LRU policy.

4.3.1 Capacity-Pressure and Compile Hints based Initial Placement

The initial block placement decision is made based on both of the compiler hint and also the SRAM/STT-RAM capacity pressure monitored in the hardware.

Before discussing the decision making process, we first show our capacity pressure assessing hardware. To assess the SRAM/STT-RAM capacity pressure, we introduce two additional hardware structures: *missing tags* (MTs) and *MT counters*. The proposed structures are similar to the missing tags [12] and victim tags [14]. MTs and MT counters are integrated with the tag array design, as shown in Figure 3. In addition to the original 4-way SRAM tag array and the 12-way STT-RAM tag array, a 4-way SRAM MTs and a 4-way STT-RAM MTs are introduced. Moreover, for each cache set, there are a SRAM MT counter and a STT-RAM MT counter, and these counters indicate the capacity pressure of the SRAM and STT-RAM portions in that cache set, respectively. Note that the sizes of MTs are the same for both SRAM and STT-RAM arrays to provide similar pressure monitoring criterion.

We use the SRAM MT to illustrate the MT and MT counter functionality, and the STT-RAM MT and MT counter work in the same way. When a cache miss occurs in the SRAM, the tag of the victim block will overwrite the LRU tag in the same set in SRAM MTs and be marked as most recently reused tag. If there is a miss in the SRAM array and there is a hit in the SRAM MTs, this indicates that a potential hit will occur if the requested block were placed in STT-RAM. Then the SRAM MT counter in the corresponding cache set is incremented by one.

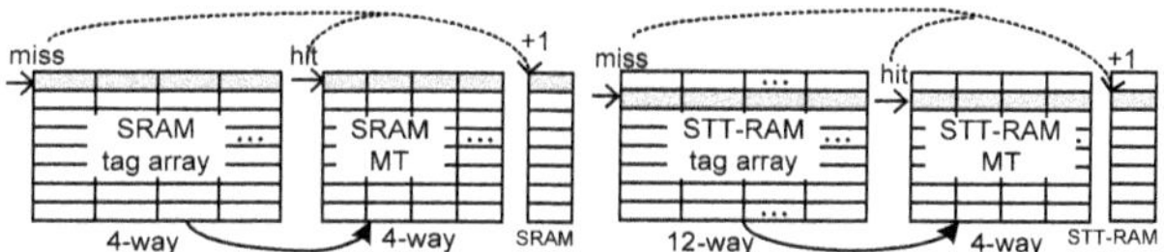

Figure 3: SRAM and STT-RAM missing tag and counter

We use an interval-based assessing approach, i.e., the value of the MT counters in the current interval will be used to guide the initial placement in the next interval. Considering the less-frequent access to the L2 cache, the interval length cannot be too short, but it can also be too long in terms of timely assessment. In this work, we set it to be 10K cycles. At the end of each 10K cycles, the value of all the MT counters will be evaluated to fill the *capacity pressure table* (CPT). The number of entries of CPT equals to the number of sets in the cache and each entry contains two bits: the SRAM capacity pressure and the STT-RAM capacity pressure of that cache set. If the value of the SRAM (STT-RAM) MT counter of a cache set is greater than a threshold, then the SRAM (STT-RAM) bit for that set in the CPT is set to 1 (*high*), otherwise set to 0 (*low*). Then all the MT counters are reset to 0. In the next interval, the CPT is accessed together with the compiler hints to decide the initial block placement.

Given the capacity pressure from the CPT and also the compiler hints, the L2 cache controller makes the initial placement as shown in Table 2. If a block is going to be placed in SRAM (STT-RAM), then the LRU replacement unit of SRAM (STT-RAM) will be triggered to evict the victim in that cache set of SRAM (STT-RAM). If a block is going to be placed globally, the global LRU replacement unit is triggered to evict the LRU block of all the SRAM and STT-RAM cache lines in that cache set.

Table 2: Initial placement decision based on compiler hints and SRAM/STT-RAM capacity pressure

Capacity pressure		Compiler hint		
SRAM	STT-RAM	infreq write	freq write	unknown
High	**Low**	STT-RAM	STT-RAM	STT-RAM
Low	**High**	SRAM	SRAM	SRAM
High	**High**	STT-RAM	SRAM	Global
Low	**Low**	SRAM	SRAM	SRAM

4.3.2 Write-Frequency based Dynamic Migration

As pointed out in Section 4.1, the compiler hints are not absolutely accurate due to the input variation and the L1 cache impact. Moreover, according to Section 4.3.1, when capacity pressure unbalance occurs, blocks may be initially placed in the less-intensive used portion of the hybrid cache, instead of based on the write-frequency of the block itself, as shown in Table 2. Thus it is possible that a block is incorrectly initially placed.

We use dynamic migration to correct the initial placement by migrating the actually write-intensive STT-RAM data blocks to SRAM. We use the dynamic migration scheme similar to [4], which is briefly described as follows. Each L2 cache block is associated with a saturate 2-bit *write counter* to indicate the number of writes during its on-chip lifetime. If the write counter of a STT-RAM block saturates (three writes), the migration unit will check the write counters of the SRAM blocks in the same cache set. If there is any counter that is less than 3, then the corresponding SRAM block is swapped with that STT-RAM block. After that, all the write counters in this cache set are reset to 0. If the counters of all the SRAM blocks in a set are saturated, no migration will be performed. Therefore, the possibility that another write-intensive block could be swapped from the SRAM back to STT-RAM is avoided.

In sum, in our combined approach, if the compiler provides correct hints, the hardware can use them to rapidly achieve correct block placement. If compiler makes mis-predictions, the hardware corrects the compiler hints as shown in Table 3. Note that all the hardware corrections are automatically triggered by our introduced hardware counters.

Compiler mis-predictions: (a) Mis-predicts some write-intensive blocks as non-write-intensive. (b) Generates larger percent of non-write-intensive blocks that it actually is. (c) Generates larger percent of write-intensive blocks that it actually is.

Hardware corrections: (i) Distributes blocks to STT-RAM. (ii) Distributes blocks to SRAM. (iii) Migrates write-intensive blocks from STT-RAM to SRAM.

Table 3: Hardware corrections to the compiler mistakes

Compiler mis-predictions			Hardware corrections		
a	b	c	i	ii	iii
X					X
	X			X	
		X	X		X
X		X	X		X
X	X			X	X

5. EVALUATION METHODOLOGY

5.1 Compilation and Simulation Infrastructure

The compiler support for hint generation is implemented based on LLVM compiler infrastructure [15]. Omega library [16] is used in this flow to perform memory dependency analysis. Given a source program written in C/C++, we parse it into LLVM IR using LLVM's frontend. All standard optimizations in O3 are applied. Our hint-generating flow is invoked as a pass on the optimized LLVM intermediate representation (IR) code and will automatically generate data placement hints for each load/store instruction. We also modify LLVM backend to emit hint-included load/store instructions in the final assembly code. A potential issue of this LLVM frontend analysis is that some load/store instructions cannot be captured in IR level. For example, the loads/stores in pre-compiled library functions cannot be analyzed under this framework. Moreover, the loads/stores from operating system cannot be analyzed during compile time. Therefore, a hardware support mentioned in Section 4.3 is required to provide better optimization.

Table 4: Simics/GEMS simulator configurations

Core	Sun UltraSPARC-III Cu processor core
L1 Instruction/ Data Cache	32KB, 2-way set-associative, 64-byte block, 2-cycle access latency, pseudo-LRU
L2 Cache (Hybrid cache)	1MB, 16-way set-associative (4-way SRAM, 12-way STT-RAM), 64-byte block, access latency: 10-cycle for SRAM, 11-cycle (read) and 30-cycle (write) for STT-RAM
Main Memory	4GB, 320-cycle access latency

We extend the full-system cycle-accurate Simics [17] and GEMS [18] simulation platform to model the proposed hardware support. The system configurations of SIMICS/GEMS are shown in Table 4. We obtain the energy data of the SRAM array and MTs/MT counters through Cacti 6.5 [19] with 32nm process technology at 330K. The energy data of the STT-RAM array are obtained from NVSim [20]. Table 5 shows the energy model we use in our evaluation. Note that the low leakage cells (*itrs-lstp*) are used in SRAM data array and tag array. For peripheral circuitry, we use high performance cells (*itrs-hp*) to optimize performance and area. Note that we also try to implement the peripheral circuitry with low leakage cells for further leakage minimization. However, we observed that considerable area overhead may arise since the width of an *itrs-lstp* transistor should be large enough to provide the enough current for STT-RAM write operation.

Table 5: Energy/power data of the evaluated hybrid cache

	Read energy	**Write energy**	**Leakage power**
SRAM (4-way)	0.0603nJ	0.0603nJ	15.017mW
STT-RAM (12-way)	0.231nJ	1.306nJ	11.173mW
MTs (8-way)	0.0020nJ	0.0020nJ	2.805mW

5.2 Benchmarks

Our testbenchs consist of eight benchmark applications, which have been carefully chosen to represent memory intensive algorithms in the fields of data processing, massive communication, scientific computation and medical applications. The benchmark applications include three memory-intensive applications from SPEC2006 [21] (*bzip2, mcf* and *lbm*) and five applications from the medical imaging domain [8].

5.3 Reference Schemes

To demonstrate the effectiveness of our combined scheme (***combined***), we compare to two representative prior approaches:

Pure static optimization (*static*): The hardware will strictly follow the compiler-generated block placement hint. The compiler hints are generated based on the approach proposed in [10], and we further take the effect of L1 cache into consideration using the techniques discussed in Section 4.1.

Pure dynamic optimization (*dynamic*): We use the dynamic migration scheme proposed in [4]. Our dynamic migration scheme in Section 4.3 uses this scheme with the same migration threshold as 3. There is no compiler hint in this scheme.

Note that the energy overhead of MTs and MT counters is only applied on the *combined* scheme.

6. RESULTS

6.1 Endurance

In this work, we assume that the maximum write cycles of a STT-RAM cell is 4 x 10^{12} [5]. We assume that a workload continuously runs on the system. To model the endurance in a more sophisticated way, one can provide a loading factor, which is the percentage of the overall runtime occupied by the workload. The lifetime is measured from the start of the simulation until the first STT-RAM line becomes defective, which is similar to the estimation methodology proposed in [4] and [5].

Figure 4 demonstrates the lifetime which is normalized to the *static* scheme. The *static* scheme typically performs the worst among the three schemes (up to 1.2x~148x worse than the *combined* scheme). This is because that once a compiler mis-predicts a write-intensive block as a non-write-intensive one and places it into the STT-RAM, this block will be intensively written and there is no dynamic migration to mitigate it. The lifetime of the STT-RAM mainly depends on the peak write count of the cells. The exceptions are *fft*, *lbm* and *denoise* where the program only have negligible input-variation, so that the *static* scheme can have longer lifetime than the other two schemes. Note that the *static* scheme can only reduce the total writes instead of the peak write count among all blocks, as shown in Figure 5. Therefore, the *static* scheme is the worst in terms of endurance but it can save STT-RAM write energy, which is discussed in Section 6.2.

With the dynamic migration to average the writes to STT-RAM blocks, the *dynamic* scheme achieves up to 14x improvement of lifetime compared to the *static* scheme. However, the reduction of the peak write count of STT-RAM is accompanied with the cost of much more total STT-RAM writes, since it lacks global information to reduce the total STT-RAM writes. Figure 5 shows that *dynamic* scheme has 1.6x~36.6x more STT-RAM writes than the *static* scheme. In cases of *fft* and *lbm*, the data blocks are all uniformly written less than 3 times on either SRAM or STTRAM. Therefore, there is little migration in the *dynamic* scheme and it has a life time which is only 4%~13% of that of the *static* scheme.

The *combined* scheme has a 1.6x~14.7x lifetime compared to that of the *dynamic* scheme. By following the correct compiler hints, the *combined* scheme rapidly achieves the optimal block placement without additional migrations, especially in the cases where most of the blocks are uniformly written less than two or three times, as shown in the motivational examples in Section 2. This can save both the peak write count and also the total writes of the STT-RAM. Although the *combined* scheme has 0.8x~4.1x more total STT-RAM writes than *static*, it achieves 1.2x~148x lifetime due to averaging the writes to the STT-RAM cells (except *fft* where *static* has a 1.8x longer lifetime than *combined*).

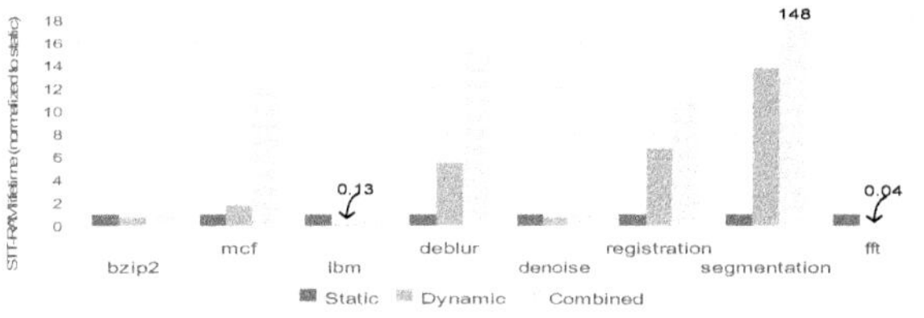

Figure 4: Comparison results of STT-RAM lifetime

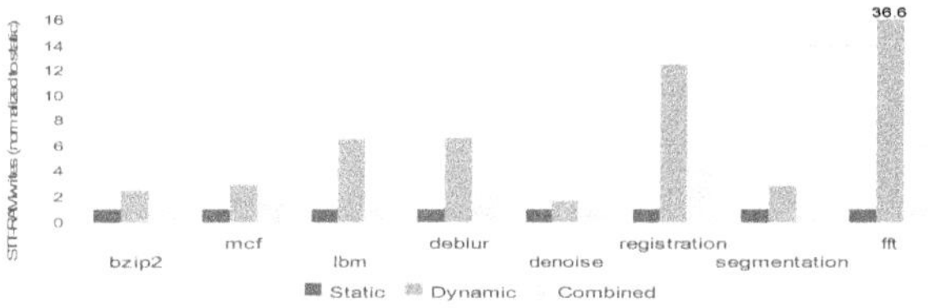

Figure 5: Comparison results of total STT-RAM writes

6.2 Energy

Figure 6 shows the distribution of hybrid cache (L2 cache) energy that is normalized to the *static* scheme. The leakage consumption of three schemes is similar. This is because leakage is proportional to program runtime and the runtime (as shown in Figure 7) of the three schemes is similar. Therefore, the key factor that influences the system energy is the L2 STT-RAM dynamic energy. The *static* scheme has the least energy, because the reduced STT-RAM writes (as shown in Figure 5) bring in considerable dynamic energy savings. Without the hints of initial placement, a large number of writes arises in the *dynamic* scheme, leading to 9%~80% energy overhead (38% overhead on average) compared to the *static* scheme.

The *combined* scheme achieves similar energy consumption to that of the *static* scheme (7%~20% energy overhead, 11% overhead on average) and outperforms the *dynamic* scheme (2%~39% energy reduction, 17% reduction on average). Note that the energy overhead of the *combined* scheme comes from both the leakage of the introduced MTs and the extra dynamic STT-RAM writes energy compared to the *static* scheme.

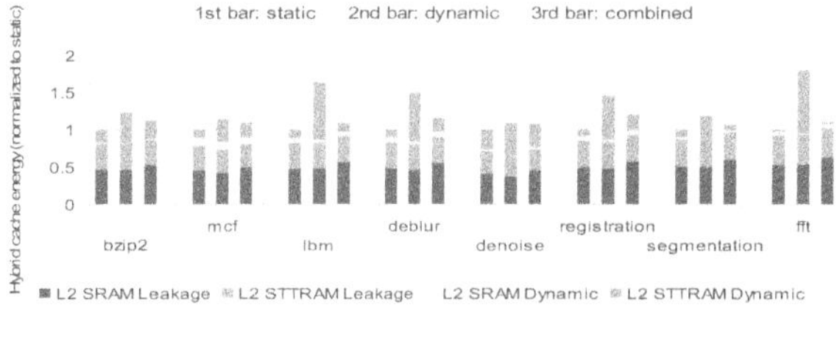

Figure 6: Comparison results of hybrid cache energy

6.3 Performance

Performance is measured by the runtime of a workload (in terms of number of clock cycles obtained from our simulation infrastructure). Figure 7 shows the comparison results of runtime that are normalized to the *static* scheme.

Since the total cache size for the three schemes are the same, the runtime does not varied significantly. The differences among the three schemes come from how efficiently they make use of the aggregate capacity of both SRAM and STT-RAM to reduce the cache misses. The *dynamic* scheme typically performs the best due to the equivalent initial placement to SRAM and STT-RAM, which best utilizes the STT-RAM capacity. As mentioned in Section 4.3.1, compiler may generate larger write-intensive data on SRAM due to the input variation, thus impose high capacity pressure to the SRAM and result in high cache misses (as shown in Figure 8). Therefore, the *static* scheme performs -1%~9% worse than the *dynamic* scheme (with a -1% ~30% increase in the L2 cache misses). The only exception is *fft* where *static* outperforms *dynamic* due to accurate compiler hints.

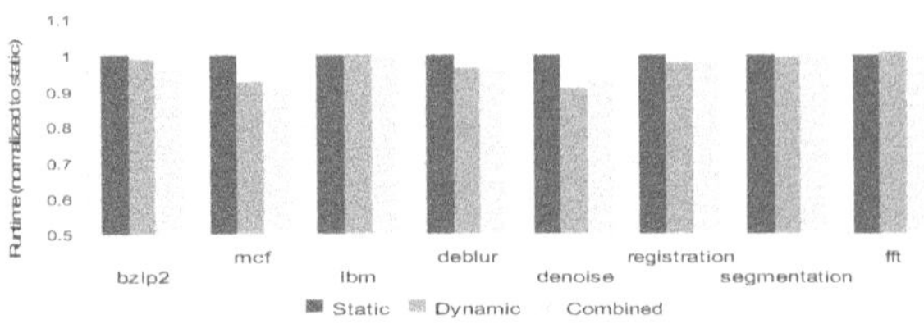

Figure 7: Comparison results of runtime

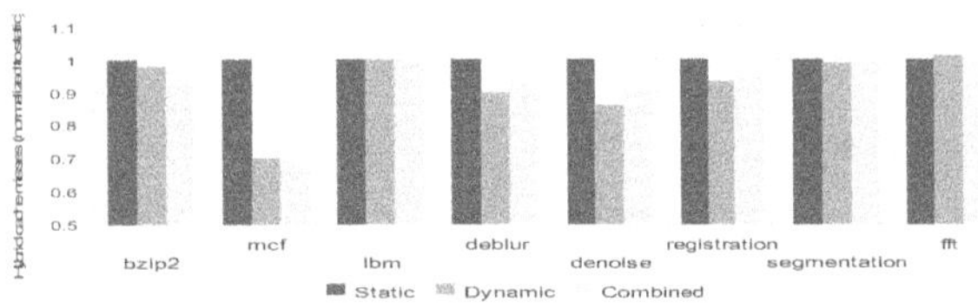

Figure 8: Comparison results of hybrid cache misses

In the *combined* scheme, the hardware can automatically correct the compiler mis-predictions as discussed in Section 4.3. Therefore, it achieves similar runtime to that of the *dynamic* scheme (within a -5% ~ 5% variation). These analyses are summarized in Table 6.

Table 6: Comparison summary of the experimental results

	static	*dynamic*	*combined*
Endurance	worst	fair	best
Performance	fair	best	best
Energy	best	worst	~best

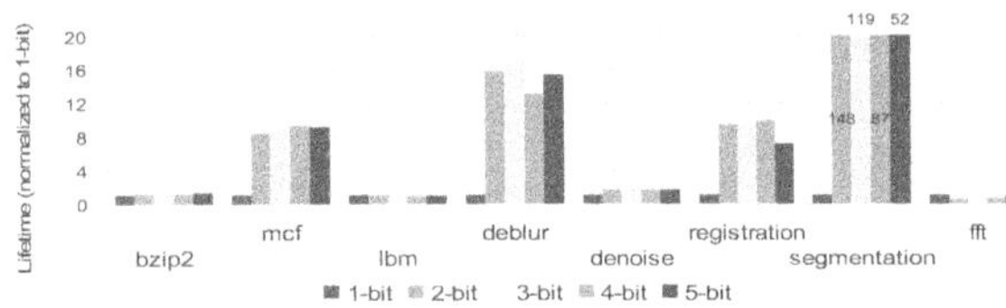

Figure 9: STT-RAM lifetime over different bit widths

6.4 Different Bit Widths of Write Counters

We perform the sensitivity analysis on different bit widths of the write saturation counters in our proposed *combined* scheme. The write counters are used for dynamic migration to improve the endurance. We justify that 2-bit counters are adequate enough for write counters. Figure 9 shows that the lifetime can be significantly enhanced in *mcf*, *deblur*, *registration*, and *segmentation* when 2-bit counters are applied. The 1-bit counters are inefficient since the SRAM blocks in the same set may easily saturate and thus prevent the migration of write-intensive STT-RAM blocks into SRAM ones. For the rest of workloads, the lifetime is insensitive to the bit width. According to our experimental results, the bit widths of write counters are insensitive to both energy and runtime among all workloads (less than 1% difference). In terms of energy, the only exception is *mcf*, where most of write intensive blocks cannot be migrated into SRAM when 1-bit counters are used. Therefore, the STT-RAM energy increases by 15% in the 1-bit counters case compared to the others (2- to 5-bit). For performance, it is insensitive to the widths of write counters since performance is maintained through cache capacity pressure monitoring, as described in Section 4.3.1.

7. CONCLUSIONS

In this paper, a combined static and dynamic scheme is proposed to optimize the block placement in a hybrid SRAM and STT-RAM cache, so that endurance and energy are co-maximized. The compiler tries to guide the hardware to rapidly achieve the desired placement, while the hardware corrects the compiler hints based on the runtime cache behavior. Experimental results show that the combined scheme improves the endurance by 23.9x and 5.9x compared to pure static and pure dynamic schemes, respectively, while maintaining similar performance. Meanwhile, the system energy can be reduced by 17% compared to the pure dynamic scheme.

8. ACKNOWLEDGEMENTS

This work is partially supported by the SRC Contract 2009-TJ-1984, and the Center for Domain Specific Computing (NSF Expedition in Computing Award CCF-0926127).

9. REFERENCES

[1] *International Technology Roadmap for Semiconductors. http://www.itrs.net/*: Semiconductor Industries Association, 2011.

[2] G. Sun, X. Dong, Y. Xie, J. Li, and Y. Chen, "A novel architecture of the 3D stacked MRAM L2 cache for CMPs," In *HPCA*, 2009, pp. 239-249.

[3] X. Wu, J. Li, L. Zhang, E. Speight, R. Rajamony, and Y. Xie, "Hybrid cache architecture with disparate memory technologies," In *ISCA*, 2009, pp. 34-45.

[4] A. Jadidi, M. Arjomand, and H. Sarbazi-Azad, "High-endurance and performance-efficient design of hybrid cache architectures through adaptive line replacement," In *ISPLED*, 2011, pp. 79-84.

[5] Y. Chen, W. Wong, H. Li, and C. Koh, "Processor caches built using multi-level spin-transfer torque RAM cells," In *ISLPED*, 2011, pp. 73-78.

[6] J. Li, C. J. Xue, Y. Xu, "STT-RAM based energy-efficient hybrid cache for CMPs," In *VLSI-SoC*, 2011, pp. 31-36.

[7] Y. Chen, J. Cong, H. Huang, B. Liu, C. Liu, M. Potkonjak, and G. Reinman, "Dynamically Reconfigurable Hybrid Cache: An Energy-Efficient Last-Level Cache Design," In *DATE*, 2012.

[8] A. Bui, K.Cheng, J. Cong, L. Vese, Y. Wang, B. Yuan, and Y. Zou, "Platform Characterization for Domain-Specific Computing," In ASPDAC, 2012.

[9] J. Hu, C. J. Xue, W.-C. Tseng, Y. He, M. Qiu and E. H.-M. Sha, "Reducing write activities on non-volatile memories in embedded CMPs via data migration and recomputation," In *DAC*, 2010, pp.350-355.

[10] T. Liu, Y. Chao, C. J. Xue, and M. Li, "Power-aware partitioning for DSPs with hybrid PRAM and DRAM main memory," In *DAC*, 2011, pp. 405-410.

[11] C. Ding and Y. Zhong, "Predicting whole-program locality through reuse distance analysis," In *PLDI*, 2003, pp. 245-257.

[12] J. Cong, H. Huang, C. Liu and Y. Zou, "A reuse-aware prefetching algorithm for scratchpad memory," In *DAC* 2011, pp. 960-965.

[13] M. Zhang and K. Asanovic, "Fine-grain CAM-tag cache resizing using miss tags," In *ISLPED*, 2002, pp. 130-135.

[14] J. Cong, K. Gururaj, H. Huang, C. Liu, G. Reinman, Y. Zou, "An energy-efficient adaptive hybrid cache," In *ISLPED*, 2011, pp. 67-72.

[15] LLVM compiler. http://llvm.org/

[16] Omega library. http://www.cs.umd.edu/projects/omega/

[17] P. Magnusson, M. Christensson, J. Eskilson, D. Forsgren, G. Hallberg, J. Hogberg, F. Larsson, A. Moestedt, and B. Werner, "Simics: A full system simulation platform," In *IEEE Computer*, vol. 35, pp. 50-58, 2002.

[18] M. Martin, D. Sorin, B. Beckmann, M. Marty, M. Xu, A. Alameldeen, K. Moore, M. Hill, and D. Wood, "Multifacet's general execution-driven multiprocessor simulator (GEMS) toolset," In *Computer Architecture News*, pp. 92-99, 2005.

[19] HP Cacti, http://quid.hpl.hp.com:9081/cacti/

[20] C. Xu, X. Dong, N.P. Jouppi, and Y. Xie, "Design implication of Memristor-Based RRAM Cross-Point Structures," In DATE, 2011

[21] SPEC CPU2006, http://www.spec.org/cpu2006/

Design Space Exploration of Workload-Specific Last-Level Caches

Karthik Swaminathan, Emre Kultursay, Vinay Saripalli, Vijay Narayanan and Mahmut Kandemir
Dept. of Computer Science and Engineering
The Pennsylvania State University
University Park, PA
(kvs120, euk139, vxs924, vijay, kandemir)@cse.psu.edu

ABSTRACT

Leakage power of last-level caches constitute a significant part of overall power consumption. Various circuit-level and technology-based methods have been proposed to reduce cache leakage. However, from a system designer's perspective, for a particular configuration and workload, it is not clear which method will be most effective. In this work, we make a detailed evaluation and comparison of cache energy reduction techniques. Our results show that when energy is very scarce and important, the best results are obtained with highly energy efficient Tunnel-FET caches. When the available energy increases and performance becomes a bigger concern, there is no single winner. While a small number of capacity sensitive workloads benefit from increased capacity of STT-RAM caches, latency sensitive workloads prefer solutions with smaller latency penalties such as drowsy caches.

Categories and Subject Descriptors

Area 1 [**Architecture, Circuits, and Technology**]: Subarea 1.1 [**Technologies and Digital Circuits**]

Keywords

Cache leakage, Power gated/Drowsy caches, HP, LSTP transistors, STT-RAM, Tunnel-FETs

1. INTRODUCTION

In embedded systems, power consumption is a very important metric that directly impacts battery lifetime. When sources of power consumption in processors are investigated, large last-level caches appear to make an important contribution [10, 9]. Further, it has been shown that the majority of power consumption in these caches is due to leakage [18].

Reducing cache leakage has been the subject of significant amount of research in recent years [7, 15, 4, 22]. Circuit-level techniques such as power gating, drowsy caches, and voltage scaling have been proposed to reduce cache leakage. Alternatively, technology replacement oriented studies proposed emerging devices such as Spin-Transfer Torque RAMs (STT-RAMs) [7, 8] and Tunnel-FETs [13] to replace CMOS SRAM with similar goals.

Although various methods to reduce leakage in caches have been proposed, each method has its own strengths and weaknesses. Clearly, from a leakage energy perspective, they are all better than a baseline SRAM cache built using transistors that are optimized not for low power operation, but for high performance. However, it is not apparent which technique is most effective or on what basis should these different techniques be compared. In fact, considering the wide range of possible architectural configurations, workloads, and power-performance constraints, there may not even be a single winner in all cases. While one technique can bring higher energy saving, another can bring less performance penalty, both of which can be considered as the *best* solution from different perspectives. Given an architectural configuration and a target workload, identification of the most suitable energy reduction method requires a detailed evaluation and a fair comparison of *all* alternative cache designs. Although the caches proposed in the literature have been evaluated individually in their own context, a comprehensive evaluation and comparison of these approaches is missing.

In this work, we explore the last-level cache design space for embedded systems and perform a detailed evaluation of various energy efficient last-level caches. In addition to the methods and technologies already proposed by researchers, we also model and evaluate caches built using the emerging TFET technology with the goal of identifying its potential in implementing last-level caches. We model various cache designs and their effects on system performance and energy using full-system simulations. Our system-level evaluation is performed for three configurations, each configuration representing a different usage scenario (ultra-low, low, and medium energy). We present energy, performance, EDP, and ED^2 numbers for *iso-area* last-level cache designs using benchmarks from MiBench [5] and SPEC CPU2006 [6] suites. The main motivation for carrying out an iso-area analysis was that our evaluations are geared towards the embedded systems domain, where chip area is at a premium. We also evaluate private/shared and homogeneous/heterogeneous caches for dual-core processors running multiprogrammed workloads. Our results show that, for ultra-low energy systems where energy is of utmost importance, TFET caches show the highest energy efficiency. As we move to

ISLPED'12, July 30–August 1, 2012, Redondo Beach, CA, USA.

higher energy systems with relatively less concerns on energy, the importance of performance rises. In this case, designs which show a better balance in access latency degradation and energy improvement, such as STT-RAM caches, start to become more effective. When energy becomes even more abundant, cache energy is relatively less of a concern when compared to access latency, in which case, fast drowsy caches give the best results. Our results also indicate that the "best" cache type is strongly dependent on the target configuration and workload. While one cache type can achieve great EDP improvements in one configuration and workload, it can lead to significant EDP degradation in another configuration or workload.

2. ENERGY-EFFICIENT CACHE IMPLEMENTATIONS

In this section, we present various cache designs proposed to reduce energy consumption in last-level caches. These caches either are implemented using new technologies or employ special energy reduction techniques. Although our evaluation has been done on a 2-level cache system, it can be easily extended to deeper cache hierarchies since the underlying considerations remain the same.

2.1 Baseline: HP CMOS SRAM

High performance (HP) CMOS transistors can deliver large drive currents (as high as 1628 uA/um for a 21nm FinFET [2]), which makes them preferred in implementing latency-critical components such as cores. The disadvantage of HP CMOS transistors is that they have a large off-current (in the order of 100nA/um) which contributes to the high leakage power of the device.

Caches built using HP CMOS transistors typically employ the standard 6T SRAM cells. Large drive currents provided by these devices enable low cache access latencies. As L1 cache latency significantly impacts the performance of a processor, these caches are typically designed using HP CMOS devices. If a fast L2 cache is also needed, HP CMOS transistors can be used in L2 caches as well. Such an L2 cache can achieve a latency of 1-2 ns with a capacity of 512KB (per core when private or per bank when shared) which is typical for today's systems. Unfortunately, such a fast L2 cache comes with a significant power penalty. Even when the SRAM cells in the cache are idle, the large leakage current passing continuously through a large number of transistors in a large L2 cache can result in very high leakage power.

2.2 Voltage Scaled HP CMOS SRAM

The supply voltage of HP CMOS caches can be statically set to a lower value than the nominal voltage to reduce power consumption [22]. In order to employ caches with voltage scaling, supply voltage must be delivered to the caches from on-chip voltage regulators. Based on the configuration of on-chip voltage domains, a single voltage level can be assigned to an entire cache or separate voltage levels can be assigned to different banks within a cache. Reducing supply voltage brings both leakage and dynamic power savings, yet these savings come with the cost of reduced drive current which increases cache access latency. Near-threshold caches can lead to significant reduction in power consumption, but the associated latency can be prohibitive.

2.3 LSTP CMOS SRAM

Threshold voltage is a very important factor in determining the off-current of a transistor. Low stand-by power (LSTP) devices [2] are designed to have much higher threshold voltages (0.28V in HP vs. 0.63V in LSTP) which results in about 4 orders of magnitude reduction in off-current (from 100nA/um to 10pA/um). By replacing the leaky HP transistors in the cells of an SRAM cache with low leakage LSTP transistors, the cache leakage can be significantly reduced. However, this increase in threshold voltage also reduces the on-current of the transistor, which leads to significant increase in cache access latency. To prevent larger performance penalties, HP transistors are still preferred for peripherals, the use of LSTP transistors is restricted to SRAM cells, and these LSTP SRAM cells are provided slightly higher supply voltages. Subsequently, this increased supply voltage leads to a quadratic increase in the cache dynamic energy ($P_{dyn} \propto V_{cc}^2$). LSTP devices also require additional process steps to manufacture transistors due to their increased threshold voltage.

2.4 TFET SRAM

When power becomes a significant concern, Inter Band Tunnel-FETs (TFETs) [13] can be considered as a technology replacement. TFET devices show much higher energy efficiencies than CMOS devices below 0.5V [1]. The benefits of TFET devices come from the fact that at low voltages they can provide higher drive currents than CMOS devices. Further, TFETs show extremely low leakage, which means that they can be good candidates to implement caches in low power applications. TFET technology is compatible with the CMOS process flow and requires a small number of additional steps in the flow. On account of their low operating voltages, TFET-based SRAMs are more susceptible to variation-induced errors. There are several existing variability mitigation techniques for 6T CMOS SRAM cells operating at near-threshold and sub-threshold voltages [23][11]. However, these techniques do not work well for the corresponding TFET-based designs. [17] discusses the design of a variability-resilient 10T TFET SRAM cell. However, the increase in the number of transistors in the TFET SRAM cell results in a reduction in capacity. Under the same area, a variation-aware 10T TFET SRAM cache can realize about 67% of the capacity of a CMOS SRAM cache.

2.5 STT-RAM

STT-RAM is a magnetic storage technology where the spin-transfer torque (STT) effect is used to invert the data stored in the magnetic storage element [7, 8]. Typical STT-RAM cells employ the standard 1T-1MTJ structure, where the polarity of the magnetic tunnel junction (MTJ) represents the stored data. The data stored in a cell can be read by applying a small voltage across the terminals of the cell and measuring the cell current. Writing into an STT-RAM cell is achieved by applying a strong and long current pulse that re-aligns the free layer in the MTJ in the opposite direction. This results in STT-RAM caches suffering from high write energy and write latency. Currently, typical values for the magnitude and duration of this write pulse are 50-100 uA and 10-35 ns [2, 3].

Although the underlying 1T-MTJ cell structure in STT-RAMs does not suffer from leakage current, the leakage power of the peripheral circuitry (write amplifiers etc.) is much more than that of HP CMOS caches. This is because, the peripherals must be sized bigger to deliver the large currents needed to write into the STT-RAM cell. Another disadvantage of STT-RAM is that, although it is considered to be compatible with CMOS process flow, it still requires an additional 3-4 masks [1]. This increases the chip manufacturing cost.

There are some parameters to be considered in the design of STT-RAM caches, such as the organization of the subarrays and the sizing of the peripherals. By tuning these parameters, STT-RAM caches can be optimized for access latency or capacity. Capacity-optimized STT-RAM caches can have upto 4x capacity of HP CMOS under the same area. However, optimizing for capacity results in high access latencies which can penalize performance. Alternatively, latency-optimized STT-RAM caches have lower access latencies, but they cannot realize capacities as high as the capacity-optimized STT-RAM cache.

2.6 Power Gated/Drowsy HP CMOS SRAM

Power gating [15] and drowsy modes [4] are two methods used to reduce the leakage power of HP CMOS caches. In power gating, an idle cache line is put to sleep by disconnecting the cells from the power supply. Although the cache line suffers almost zero leakage power, its contents get lost. Thus, the next access to this block becomes a cache miss, requiring an extra memory access.

In the drowsy mode, the supply voltage is reduced to a level that can barely keep the stored data [16]. While this does not eliminate cell leakage entirely, it can reduce it significantly. When a block is put into drowsy mode, the data it contains is not lost, but an access to this cache line requires the supply voltage to be increased to the full level before serving the request. Drowsy caches can be implemented by delivering two supply voltages (a full voltage and a reduced voltage) to all cache blocks. Switching between modes requires disabling one supply and enabling the other, which results in an increase in the cache access latency. The power savings and performance degradation obtained by these approaches depend on the chosen idleness duration to enter drowsy or sleep modes. Power gating and drowsy caching techniques require high threshold power gates to block current flow into the cells, which would increase the area overhead.

3. EXPERIMENTAL SETUP

This section presents our simulation infrastructure, the systems/workloads we simulate, and the metrics we used to compare the caches discussed in Section 2.

3.1 Core and Cache Modeling

For cores, we assumed a simple in-order core model and calculated its energy using the method in [20]. In order to model our caches, we used various simulators. First, we used TCAD Sentaurus [21] to obtain device parameters for 22nm FinFETs (since the state-of-the-art CMOS process uses 22nm FinFET transistors) and TFETs. We modified CACTI-6.5 [14] to use the 22nm FinFET technology parameters in modeling CMOS SRAM caches and to use 22nm TFET technology parameters with a 10-T cell structure in modeling TFET SRAM caches. In this simulator, we also implemented cache voltage scaling to model CMOS SRAM caches operating at reduced voltages (0.7, 0.6, and 0.5V). For our power gated and drowsy caches, we used three sleep/nap intervals (100, 100K, 10M cycles). We assumed that an extra 3 cycles are required to wake up from drowsy mode. We modeled STT-RAM caches using NVSim [3]. *All caches we experimented with are configured to be ISO-AREA.* As a result of these simulations, we obtained the capacity, leakage power, dynamic energy per access, and access latency values for all of our target caches. A list of caches we evaluated in this work is given in Table 2.

3.2 Simulation Infrastructure

In order to investigate the impact of using different caches on overall system performance and energy, we performed full-system simulations. Using the core and cache latency/energy numbers we obtained using the above methods, we carried out simulations using the Simics full-system simulator. We modified this simulator to accumulate the leakage and dynamic energy of the core(s) and cache(s). To represent different usage scenarios, we constructed the following three configurations:

- *Ultra-low Energy Domain:* This domain represents the scenario where the total available energy is extremely scarce and the battery lifetime is very critical. The available power budget is assumed to be less than 100mW. Systems used for wireless sensing and some embedded applications can be considered as examples.
- *Low Energy Domain:* This domain represents the scenario where energy is important, but performance cannot be sacrificed indefinitely for lower energy. This domain is assumed to operate within a power constraint of around 500mW. Most embedded applications (e.g., handheld devices whose battery can be recharged, but improved lifetime is useful) fall into this domain.
- *Medium Energy Domain:* This domain represents the scenario where performance becomes more important than improvements in energy. Consequently the operating power of the core and cache is assumed to be around 1-2W. Some performance-critical embedded applications and low-end laptop computers are in this category.

For each domain, we used a different configuration and a different set of applications to better capture its needs and characteristics. Details of our target configurations are shown in Table 1. We assumed a latency-optimized HP-CMOS L1 cache and a capacity-optimized off-chip DRAM to be constant across all cache configurations and power domains. We also experimented with dual-core systems running multiprogrammed workloads. With dual-core configurations, we expanded our target design space to include *shared/private* and *homogeneous/heterogeneous* caches. We also analyzed the effect of uneven distribution of area across private L2 caches of different cores on overall system performance and energy.

3.3 Evaluation Metrics

The criteria for efficient execution can differ widely across energy domains. While performance may be compromised on platforms with a strict limit on available energy, other platforms may expect the highest performance with little emphasis on the power consumed. In order to make domain-specific comparisons, for the ultra-low energy domain, we concentrated only on energy and assumed that performance can be sacrificed as long as it gives better energy. When we moved to the low energy domain, performance and energy were given equal importance, and hence, we used energy delay product (EDP) as our metric. In the medium energy domain, performance becomes even more important and the metric to use must reflect this. Therefore, we evaluated caches in the medium energy domain using energy delay-square (ED^2).

Table 1: System Parameters.

Parameter	Ultra-low	Low	Medium
Core Frequency	500 MHz	1 GHz	2 GHz
L1 Capacity		32 KB	
L1 Latency		1 cycle	
Base L2 Capacity (HP)	128 KB	256 KB	512 KB
Base L2 Latency (HP)	1.30ns	1.40ns	1.58ns
Memory Latency		55 ns	
Applications	MiBench	MiBench,SPEC	SPEC

Table 2: Evaluated Cache configuration. Capacity is normalized to HP CMOS SRAM.

Cache Type	Symbol	Parameters	Capacity
HP SRAM	HP	-	1.00
Voltage Scaled	HP-VS	0.5-0.7V	1.00
LSTP SRAM	LSTP	-	1.00
TFET SRAM	TFET	0.3-0.6V	0.64
Area Opt. STTRAM	STT-Area	-	4.00
Lat. Opt. STTRAM	STT-Lat	-	2.00
Power Gated	HP-PG	100,100K,10M	1.00
Drowsy	HP-D	100,100K,10M	1.00

4. EXPERIMENTAL RESULTS

We organize our experimental results into two parts. First, we explore the cache design space for a single core system and identify the preferred cache type for each of our target system-application combinations. Next, we evaluate the scenario where workloads consisting of two applications are executed simultaneously on a dual-core system. Finally, we provide a discussion on two emerging technologies, namely, STT-RAM and TFET.

4.1 Single Core Results

Ultra-Low Energy. In the ultra-low energy domain, the core executes at a relatively low frequency (500MHz) and the capacity of the L2 cache is kept small (128KB) to minimize leakage and dynamic energy. The left part of Table 3 shows the energy savings obtained in this domain, which indicates that up to 20% energy savings can be obtained using LSTP, TFET, and HP-PG configurations. Although our comparison metric for the ultra-low energy domain is only energy, we also give the associated performance values in the same table. While the performance degradation with LSTP and TFET-based caches are limited to 1%, HP-PG's performance significantly suffers due to increased number of writebacks and misses. Looking at individual application results, the top-left part of Table 4 lists the cache type that has the lowest energy and the corresponding energy with that cache type. These results show that, TFETs at extreme low voltages consistently provide large energy savings.

Low Energy. When we move to the low-energy domain, the frequency of the core and the capacity of the L2 cache are increased (1GHz and 256KB). Further, a metric that also considers overall performance is used, which is EDP. Average EDP values for this domain are given in Table 3, which show that STT-RAM, LSTP, and TFET caches have the best EDP. Results with individual MiBench and SPEC benchmarks are shown on the top-right and bottom-left parts of Table 4, respectively. While for MiBench applications, LSTP and TFET-0.25V have the best EDP, for SPEC applications that put more pressure on the L2 cache, STT-RAM starts to appear as the best choice in 4 of 10 benchmarks.

Medium Energy. The importance of caches in medium energy configuration is significantly higher than other configurations. This is because (i) using larger caches leads to higher cache access latencies, and (ii) higher core clock frequencies make caches appear relatively even slower. Therefore, higher access latencies can be expected to lead to higher degradation in performance. Table 3 shows average ED^2 values for all cache types and indicates that only HP-D is able to achieve a better score the baseline on average. Energy benefits with other schemes do not appear to be high enough when compared to the performance degradation suffered. However, when we examine individual benchmarks, many of the cache designs appear to be better than the baseline in at least one case. On the other hand, in this domain, we also observed that most designs also have at least one case where it significantly degrades performance, penalizing the average ED^2 scores.

Table 3: Average energy, performance, EDP, and ED^2 results normalized to HP CMOS cache in each domain.

Type	Ultra-Low		Low			Medium		
	E_{tot}	Perf	E_{tot}	Perf	EDP	E_{tot}	Perf	ED^2
HP	1.00	1.00	1.00	1.00	1.00	1.00	1.00	1.00
HP-VS	0.96	1.00	0.98	0.99	0.98	0.98	0.94	1.06
LSTP	0.80	0.99	0.89	0.99	0.90	0.94	0.92	1.04
TFET	0.80	0.99	0.89	0.95	0.93	0.94	0.83	1.15
STT	0.85	0.98	0.91	0.99	0.93	0.97	0.92	1.09
HP-PG	0.81	0.83	0.89	0.80	1.25	0.94	0.67	1.54
HP-D	0.86	0.98	0.93	0.98	0.95	0.96	0.97	0.98

Table 4: The best cache types and the corresponding evaluation normalized metrics for (top) MiBench and (bottom) SPEC benchmarks.

Benchmark	Ultra-Low Energy		Low Energy	
	Best Config.	E_{tot}	Best Config.	EDP
adpcm	TFET	79%	LSTP	94%
basicmath	TFET	79%	LSTP	90%
bitcount	TFET	80%	LSTP	91%
blowfish	TFET	79%	LSTP	91%
fft	TFET	80%	TFET	91%
gsm	TFET	81%	TFET	91%
lame	TFET	80%	TFET	91%
patricia	TFET	80%	TFET	91%
qsort	TFET	80%	TFET	91%
stringsearch	TFET	80%	TFET	91%

Benchmark	Low Energy		Medium Energy	
	Best Config.	EDP	Best Config.	ED^2
cactusADM	LSTP	84%	HP-D	99%
calculix	STT-area	91%	STT-Area	94%
dealII	LSTP	90%	LSTP	97%
gamess	STT-Lat	74%	STT-Lat	99%
lbm	LSTP	85%	HP-D	99%
mcf	STT-Lat	84%	LSTP	93%
milc	LSTP	87%	TFET	95%
namd	LSTP	90%	LSTP	97%
omnetpp	STT-Area	87%	STT-Lat	98%
sjeng	LSTP	91%	LSTP	98%

4.2 Multicore Results

On a multi-core platform, we have an additional parameter which states that the L2 cache can be private or shared. While a private cache provides isolation of applications, a shared cache can achieve better utilization of the total capacity. In case of private L2 caches, different cache types can be implemented on different cores, which can lead to a heterogeneous private L2 cache. Further, under iso-area constraint, the total cache capacity can also be distributed unevenly across the private L2 caches. Such an uneven distribution can enable capacity sensitive applications to receive higher capacity and operate at a lower miss rate, while resulting in a marginal degradation of the less capacity-sensitive co-runner application.

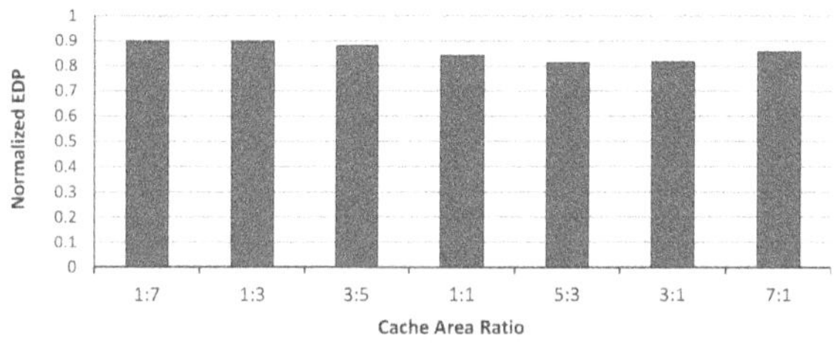

Figure 1: Variation of EDP in *omnetpp-lbm* workload with varying cache area ratios.

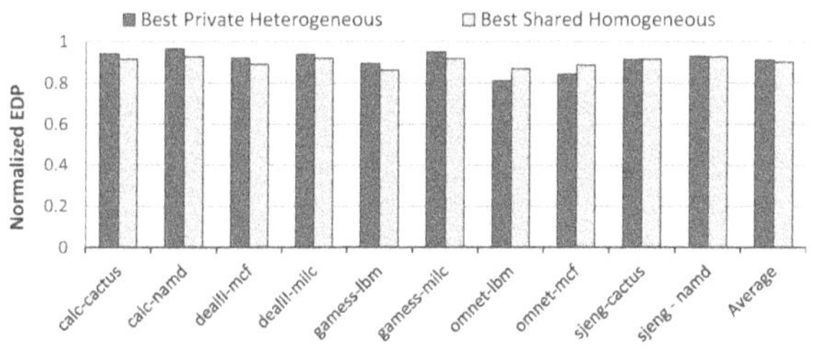

Figure 2: EDP of multiprogrammed workloads in low energy domain with best private and shared caches (normalized to iso-area shared HP cache).

When working with shared caches, we evaluated all possible caches presented in Section 2 in a *homogeneous* setting, i.e., we did not consider bank-level or way-level heterogeneity. However, in case of private caches, the best private cache need not have a symmetric area distribution across the individual caches. We performed dual-core experiments in the low-energy domain using 10 randomly selected workload combinations from the SPEC2006 suite. Figure 1 shows the variation in EDP for one of our workloads (*omnetpp-lbm*) as the area distribution across two private caches is varied. Note that, for this workload, optimal EDP is obtained at the ratio of 5:3. To take this into account, we also analyzed asymmetric distributions of area across the two private caches of our processor. We also considered the possibility of using different technologies for each private cache. Note that all cache configurations are statically assigned and we do not work with dynamically reconfigurable caches as in [19].

Figure 2 shows the EDP values of the best private and the best shared cache configurations for all workloads. Note that, in a multiprogrammed setting, we redefine EDP to be the product of total energy and weighted speedup. In workloads consisting of capacity sensitive and insensitive benchmarks such as *omnetpp-mcf*, we observe that the best private cache has a better EDP than the best shared cache. This is because the best private cache has a heterogeneous structure where each application is free to operate on its own best cache implementation. For instance, the capacity sensitive benchmark (*omnetpp*) uses an STT-RAM cache to improve its hit rate while the capacity insensitive benchmark (*mcf*) uses an LSTP cache to reduce its energy. On the other hand, for some other workloads such as *calculix-namd*, the shared cache configuration is preferred due to better dynamic utilization of the total cache capacity.

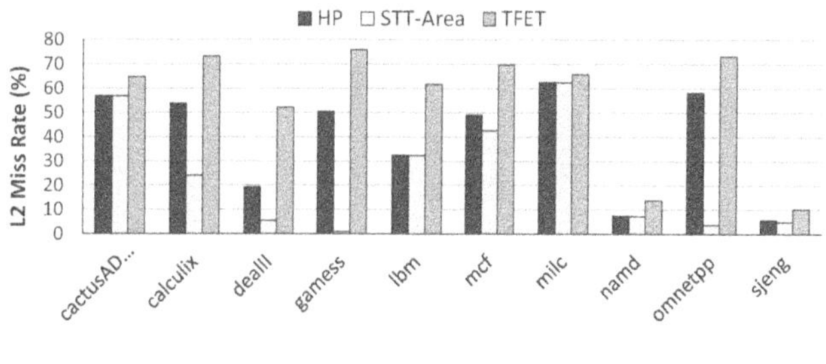

Figure 3: L2 cache miss rates of HP, STT-Area, and TFET (SPEC on low energy domain).

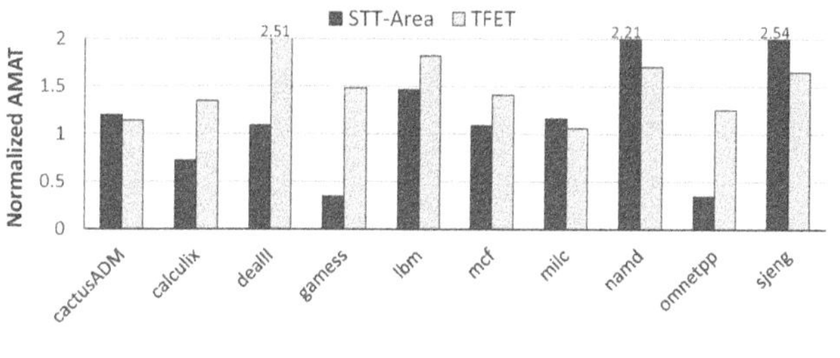

Figure 4: Average memory access time of STT-Area and TFET normalized to HP (SPEC on low energy domain).

4.3 Discussion on Emerging Technologies

STT-RAM Behavior. When the 4x density and zero cell leakage advantages of STT-RAM over SRAM are considered, one can expect STT-RAM caches to consistently achieve very good results. However, looking at Figure 3 which compares the L2 cache miss rate of an HP cache with that of STT-Area and TFET caches, one can see that only 4 out of 10 benchmarks (*calculix*, *dealII*, *gamess*, and *omnetpp*) have a significant reduction in their miss rates. As observed in [12], read-intensive applications benefit more from the capacity increase of STT-RAM, whereas the long write latency offsets this capacity benefit in case of write-intensive applications. This means that not all applications actually benefit from the increased L2 cache capacity of STT-RAM. Thus, it should be noted that a lower cache miss rate need not imply a better memory system performance: one must also consider the increased access latency (due to long STT-RAM write pulse) and analyze average memory access time (AMAT). Figures 4 and 5 show the AMAT and the IPC obtained with STT-RAM normalized to HP baseline. Among the four benchmarks with reduced miss rates, *dealII* has an AMAT value worse than the baseline and only *gamess* shows a significant IPC improvement (18%). Further, from the energy perspective, Figure 6 shows the energy distribution the STT-Area cache normalized to HP. It is clear that the core leakage energy and the cache dynamic energy are the only visible components. Although the cache leakage is reduced significantly, we can still observe a non-zero leakage (around 6% of overall energy) in STT-RAM.

These results indicate that the high write latency associated with STT-RAM restricts the cases where it brings significant performance benefits. Further, due to the leakage of its peripheral circuitry, STT-RAM does not eliminate cache leakage completely.

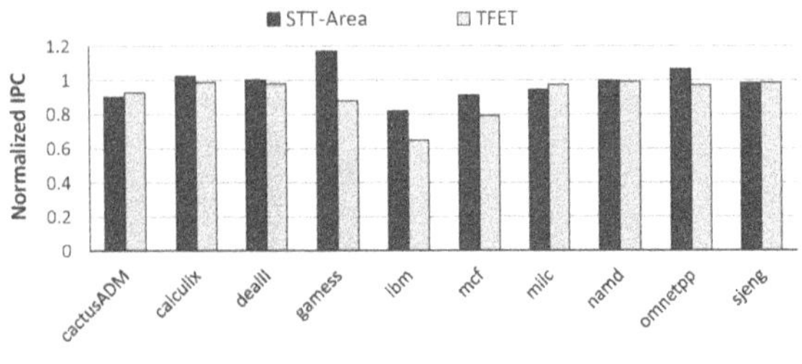

Figure 5: IPC of STT-Area and TFET normalized to HP (SPEC on low energy domain).

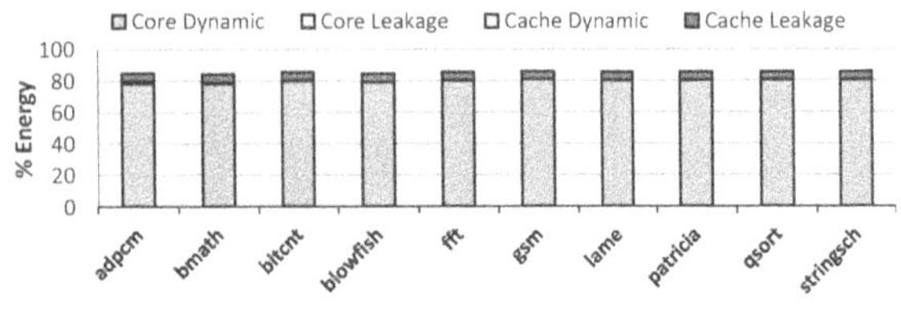

Figure 6: Distribution of energy with STT-RAM cache normalized to HP baseline (SPEC on low energy domain).

TFET Behavior. Due to its 10T cell structure, TFET caches have 33% less capacity than iso-area HP caches. Therefore, the miss rates with TFET caches, given in Figure 3, are higher than those with HP caches. Further, as TFETs are operated at very low voltages (0.25V for TFET vs. 0.8V for HP), they also suffer from higher access latencies. This manifests itself as even higher degradations in AMAT as shown in Figure 4. Figure 5 indicates that, the overall impact on system performance is an average IPC degradation of 7%.

However, while TFET caches suffer high performance penalties, they are also very energy efficient. Due to the orders of magnitude reduction in off-current, leakage energy is almost entirely eliminated, which gives very significant energy improvements. Analyzing the data in Table 3 for TFETs in the low-energy domain, we observe that for a 5% reduction in performance, an energy improvement of 11% is obtained. On average, this leads to a significant 7% EDP improvement. Further, in the ultra-low energy domain, the performance degradation reduces to less than 1% and the energy savings reach as high as 20%. Based on these results, one can conclude that when energy is a critical player, using TFET caches can lead to significant energy savings with small performance penalties.

5. CONCLUSION

Although various circuit-level and technology-based methods to reduce cache leakage energy have been proposed in the literature, from a system designer's perspective, it is not clear which method shall be expected to be the most effective under a target architectural configuration and workload. In this work, for various energy domains and workloads, we evaluated the effectiveness of various leakage reduction techniques. Our results indicate that in a strictly energy constrained domain, TFET caches achieve the lowest energy levels. On the other hand, in a relaxed energy constraint scenario, while capacity sensitive applications prefer larger capacity STT-RAM caches, latency sensitive applications prefer lower latency drowsy caches.

Acknowledgment

This work was supported in part by NSF Awards 1147388, 0903432, 0903432, 1152479, 1017882, 0702617 and 0963839.

6. REFERENCES

[1] ITRS Emerging Research Devices (ERD) Chapter. In *ITRS*, 2011.

[2] Process Integration, Devices, and Structures (PIDS). In *ITRS*, 2011.

[3] X. Dong et al. NVSim: A Circuit-Level Performance, Energy, and Area Model for Emerging Non-Volatile Memory. In *TCAD*, Dec. 2011.

[4] K. Flautner et al. Drowsy caches: simple techniques for reducing leakage power. In *ISCA*, 2002.

[5] M. R. Guthaus et al. MiBench: A free, commercially representative embedded benchmark suite. In *WWC*, 2001.

[6] J. L. Henning. SPEC CPU2006 benchmark descriptions. *SIGARCH Comput. Archit. News*, 34(4), Sep 2006.

[7] M. Hosomi et al. A novel nonvolatile memory with spin torque transfer magnetization switching: Spin-RAM. In *IEDM*, 2005.

[8] T. Kawahara et al. 2 Mb SPRAM (SPin-Transfer Torque RAM) With Bit-by-Bit Bi-Directional Current Write and Parallelizing-Direction Current Read. *JSSC*, 43, Jan. 2008.

[9] N. S. Kim et al. Circuit and microarchitectural techniques for reducing cache leakage power. *TVLSI*, 12(2), Feb. 2004.

[10] J. Kin et al. The filter cache: an energy efficient memory structure. In *MICRO*, 1997.

[11] J. P. Kulkarni et al. A 160 mV, fully differential, robust schmitt trigger based sub-threshold SRAM. In *ISLPED*, 2007.

[12] A. K. Mishra et al. Architecting on-chip interconnects for stacked 3D STT-RAM caches in CMPs. In *ISCA*, 2011.

[13] S. Mookerjea et al. Experimental demonstration of 100nm channel length In0.53Ga0.47As-based vertical inter-band tunnel field effect transistors (TFETs) for ultra low-power logic and SRAM applications. In *IEDM*, 2009.

[14] N. Muralimanohar et al. Architecting Efficient Interconnects for Large Caches with CACTI 6.0. *IEEE Micro*, 28(1), Jan. 2008.

[15] S. Mutoh et al. 1-V power supply high-speed digital circuit technology with multithreshold-voltage CMOS. *JSSC*, 30(8):847–854, Aug 1995.

[16] H. Qin et al. SRAM leakage suppression by minimizing standby supply voltage. In *ISQED*, 2004.

[17] V. Saripalli et al. Variation-tolerant ultra low-power heterojunction tunnel FET SRAM design. *NanoArch*, 2011.

[18] L. Shifren. Leakage power, it's worse than you think. In *EE Times*, 2011.

[19] S. Srikantaiah et al. MorphCache: A reconfigurable adaptive multi-level cache hierarchy. In *HPCA*, 2011.

[20] K. Swaminathan et al. Improving energy efficiency of multi-threaded applications using heterogeneous CMOS-TFET multicores. In *ISLPED*, 2011.

[21] Synopsys. *TCAD Sentaurus Device Manual, Release: C-2009.06*, 2009.

[22] W. Wong et al. VOSCH: Voltage scaled cache hierarchies. In *ICCD*, 2007.

[23] B. Zhai et al. A sub-200mV 6T-SRAM in 0.13um CMOS. In *ISSCC*, 2007.

Low-Power Adaptive RF System Design using Real-time Fuzzy Noise-Distortion Control

Debashis Banerjee, Shreyas Sen*, Aritra Banerjee, Abhijit Chatterjee
Georgia Institute of Technology, Atlanta, GA, USA, *Intel Circuit Research Lab, Hillsboro, Oregon,USA
debashis.banerjee@gatech.edu, chat@ece.gatech.edu

ABSTRACT
Earlier research has demonstrated that the power in a RF front-end can be traded off for performance in real-time to operate at the threshold of acceptable operation using a lookup table driven controller. Such a controller needs careful calibration and suffers from modeling inaccuracies which must be guardbanded during real-time operation resulting in "less than optimal" power consumption. In this work we propose a real time fuzzy noise-distortion control algorithm for low-power adaptation of the RF front-end to continuously changing channel conditions. As opposed to prior techniques, the proposed control algorithm handles adaptation to *channel attenuation and fading as well as in-band and out- of- band interference*, taking advantage of signal and interferer estimation techniques that are already incorporated in modern transceivers. In the proposed technique, the RF front-end controller automatically finds an optimal power-performance trade-off point in real-time across a large gamut of channel conditions. It is seen that about 23% savings in power consumption can be obtained over RF systems that are not capable of real-time adaptation.

Categories and Subject Descriptors
B.7.1 [**Integrated Circuits**]: Types and Design Styles–*Algorithms implemented in hardware.*

General Terms
Performance, Algorithms, Design

Keywords
Low-power, Fuzzy Logic, Radio Frequency, Adaptive, Receiver

1. INTRODUCTION
Wireless sensor networks are proliferating rapidly across applications such as smart camera systems, green buildings, immersive computing, mobile (cloud) computing platforms and cyber-physical systems that support image and video processing. The wireless channel and available power in the transmitter as well as power *minimization* and *battery life improvement* is of vital concern to the design and operation of such wireless sensor networks.

We desire to solve the following problem: *Minimize power consumption* of wireless transceiver systems by tuning the active and passive components of the relevant wireless subcircuits dynamically in the field, taking into account the effects of real-time changing wireless channel conditions and end-to-end bit-error limits. The tuning mechanisms (e.g. bias control) effectively trade off module-level (e.g. LNA, mixer, PA, ADC) signal noise and distortion vs. power consumption across the RF and baseband subsystems. The feasibility of such a system is motivated by the fact that modern wireless systems are designed to adapt to wireless channel quality by adjusting wireless data transmission rates, adjusting the output power of the RF power amplifier in wireless transmitter subsystems and by automatic gain compensation in RF receiver subsystems. However, the intrinsic noise/distortion parameters of *all* the RF modules and the baseband signal processing modules are designed for the *worst case* channel. These worst case conditions include weak received signal power and interference and are particularly demanding on power consumption. This leads to a scenario in which the typical wireless transceiver *consumes more power than necessary* for a specified end-to-end bit error rate (BER) *when the wireless channel is not the worst case channel* corresponding to the date rate at which the transceiver is operating.

Power can be saved as follows. Without loss of generality across different modulation rates, consider 16-QAM operation of a WiMax system for signal SNR values lying between 16 dB and 22 dB. If the signal SNR is just below 22dB (say 21.9 dB for argument's sake and corresponds to the best channel at which the system can be operated at 16 QAM), then the SNR of the signal can be degraded to 16 dB by trading the fidelity of the RF circuit components for power consumption *without compromising system level bit error rate.* Previous work has demonstrated that RF front end systems can be tuned dynamically to minimize power consumption using a lookup table (LUT) based proportional controller driven by real-time wireless channel monitoring algorithms[1][2][3]. Such techniques have also been investigated at higher protocol level[4] and in digital processors [5]. When the wireless channel is "good", power is saved by decreasing the fidelity (noise/distortion) of the RF modules and vice versa, while staying within a specified end-to-end system level bit error rate (BER) limit at all times. The LUT based approach requires extensive simulation and characterization of the front-end to find the optimum "tuning-knob" combinations for each sub-block of the front end across a large range of channel conditions. Moreover, the control system in a LUT based approach exhibits oscillatory behavior around its quiescent operating point for any specific channel condition. This leads to less-than optimal power consumption for any given wireless channel.

In this work we propose the design of an adaptive RF front end that can adapt to changes in channel conditions by modulation of the noise-distortion characteristics of embedded RF modules. As opposed to prior designs, a new fuzzy logic based system level control mechanism is proposed that, for the first time, facilitates RF front end adaptation to *both channel attenuation & fading and in-band and out-of-band interference* in a power-optimal manner. A simple control strategy is proposed that alleviates the need for extensive characterization and guardbanding of the RF system

ISLPED'12, July 30–August 1, 2012, Redondo Beach, California, USA.
Copyright 2012 ACM 978-1-4503-1249-3/12/07...$10.00.

while ensuring good control stability and performance. It must be noted that RF architectures, like super-regenerative receivers, have demonstrated low power operation but they are limited in their scope and may not be suitable for many wireless standards. However this work demonstrates low-power operation for existing architectures without compromising on receiver performance.

The rest of the paper is organized as follows: Section 2 describes the adaptation metric and reconfiguration strategy, Section 3 formulates the fuzzy logic control mechanism, Section 4 describes the RF system of interest and provides system level simulations. Section 5 discusses the design of a fuzzy controller, Section 6 describes the transient performance of the proposed adaptation scheme followed by hardware implementation of the scheme in Section 7.

2. ADAPTATION METRIC AND RECONFIGURATION STRATEGY

2.1 BER based adaptation

The BER of a transmitter-channel-receiver combination is determined by the net SNR present at the input of the demodulator and the transceiver must be operated within stipulated BER limits. For the same BER, the minimum required SNR is dependent on the data rate determined by the modulation used (e.g. BPSK, 16 QAM, 64 QAM). As an example in order to achieve reliable transmission for QPSK we need 9 dB of SNR. However since the probability of occurrence of worst case channel is very small the SNR is usually much higher and this is a case of overdesign. In these situations we can degrade the performance of some of the sub-blocks of our system without exceeding the minimum SNR bound, thereby saving power. Thus if we can calculate the BER of a system in real-time we can adaptively reconfigure the circuit for low-power.

In order to maintain the BER below a stated limit across all channel conditions, it is desirable to monitor a signal quality metric that is the *sum of the signal degradation* due to the *wireless channel* and the (degraded) *performance of the transceiver circuitry*. Accurate estimation of BER is difficult in real-time due to the large volume of data needed to compute BER with high confidence. We overcome this problem by using Error Vector Magnitude (EVM) as a signal fidelity metric. It has been shown[2] that EVM requires much smaller time to compute, varies monotonically with BER (and SNR) and that bound on BER can be translated to a bound on computed EVM. From experiments we can determine that for a BER of 5×10^{-4} the maximum threshold EVM (EVM_{th}) is 35% for QPSK and is 14% for 16-QAM.To overcome the inherent uncertainty in calculation of a metric like EVM we can put a guardband of δ to get EVM_{gbd}. In LUT based approach a much larger guardband over and above this δ is required due to system oscillations. In our case:

$$EVM_{gbd} = EVM_{th} - \delta \quad (1)$$

2.2 Reconfiguration strategy

The two major specifications that are important in a modern radio receiver are its linearity and the input referred noise added by the receiver. These two are measured by the IIP3 and Noise Figure (NF)/Gain of the receiver. Static radio designs optimize the IIP3 and NF for the worst case channel conditions. In this work, both these specifications are changed dynamically based on the fidelity of the received signal.

For receiver adaptation, we are interested in determining the received signal SNR and the strengths of in-band and out-of-band interferers. Interferers cause signal degradation in the receiver due to receiver saturation. This results in significant intermodulation distortion and the resulting intermodulation components raise the noise floor of the received signal. On the other hand if the received signal has low signal power then the NF & gain specifications of the receiver become critical due to the fact that any noise contributed to the received signal by the receiver circuitry can significantly decrease the SNR of the received signal. In modern receivers, both received signal power and the signal strength of interferers are estimated by embedded RF detectors (see Section 4.1). These detectors can be used to aid system level adaptation. We now discuss practical constraints on the linearity & NF of a receiver. These are used later to formulate the control law for dynamic adaptation of the RF front-end.

2.2.1 Linearity (IIP3) constraints

The linearity of a receiver becomes critical when either the signal or the interferer strength becomes high. One of the key goals of the adaptation law is to ensure that the IIP3(≈P1dB+9.6 dB) of the signal should be high enough so that in the presence of a strong signal or strong interferer, gain compression or intermodulation components do not push the EVM (and BER) above the maximum allowable level. Thus we can say that

$$IIP3 = f(Signal\ Strength, Interferer\ Strength) \quad (2)$$

More specifically if the signal or interferer is sufficiently below the IIP3 of the receiver then it has very little effect on the EVM. In dBm scale this means that for IIP3 to have an effect on the EVM:

$$\text{Signal Strength} > IIP3 - \Delta_1 \quad (3)$$

$$\text{Interferer Strength} > IIP3 - \Delta_2 \quad (4)$$

Since a high signal strength causes saturation only and a high interferer strength causes both saturation and intermodulation the margins Δ_1 and Δ_2 are different. They are constants (in dBm) whose values are system specific and can be determined from very simple system simulations.

2.2.2 Noise figure (NF) Constraints

The NF of the system becomes critical when the signal strength is close to the noise floor (SNR is low). In this case the goal is to demodulate the signal without any further significant addition of noise from the receiver. The NF of a system is given by:

$$NF = NF_1 + \frac{NF_2 - 1}{G_1} + \frac{NF_3 - 1}{G_1 G_2} + \cdots \quad (5)$$

Here NF_i and G_i are the Noise Figure and gain of the ith stages in the RF chain. Thus the NF and Gain of the Low Noise Amplifier (LNA) becomes critical in this case. However the noise added by the system has very little effect on the EVM of the system when the signal strength is high. Only when the signal falls below certain strength, will we start seeing an appreciable effect of NF on the system performance. So assuming a fixed channel thermal noise floor we can say that NF of the system becomes critical when

$$Signal\ Strength < \Delta_3 \quad (6)$$

Again Δ_3 is a constant (for a particular system) and can be determined from the system design through simple simulations.

2.2.3 *Control Strategy*

Based on the discussion in Sections 2.2.1 and 2.2.2 we can now formulate certain heuristics for modulating the noise and distortion in the front-end so as to power-optimally operate at the threshold of acceptable operation (i.e. at guard-banded EVM threshold EVM_{gbd}). Let us take a snapshot of the system at a time instant T_0 and formulate the strategy as to how to change the IIP3 and NF of the system. Let $IIP3_0$ and NF_0 be the IIP3 and NF of the system at this instant and let the signal and interferer strengths be S_0 and I_0 respectively. We formulate this strategy in the form of a rule table shown in Table 1. Here an 'X' denotes a Don't-Care condition while a '-' means that decision regarding the variable cannot be taken based on that piece of information. When taking a decision, all the rows of the table are evaluated and a decision is taken regarding as to how to change the IIP3 and the NF of the system.

Another goal of the control system is to make the transitions in IIP3 and NF dependent on the difference between the current EVM and the desired one (EVM_{gbd}). We have assumed that NF and gain of the sub-blocks are correlated (as is usually the case). An increase in NF is linked with a decrease in gain and vice versa.

Table 1: Heuristics for Noise-Distortion Modulation

S_0	I_0	$(EVM_{gbd}$-EVM)	IIP3	NF
$>IIP3_0-\Delta_1$	X	-ve	↑	↑
		+ve	↓	↑
X	$>IIP3_0-\Delta_2$	-ve	↑	-
		+ve	↓	↑
$<\Delta_3$	X	-ve	-	↓
		+ve	↓	↑

3. FUZZY CONTROL FORMULATION

We can visualize the RF receiver system as a control system as shown in Figure 1.

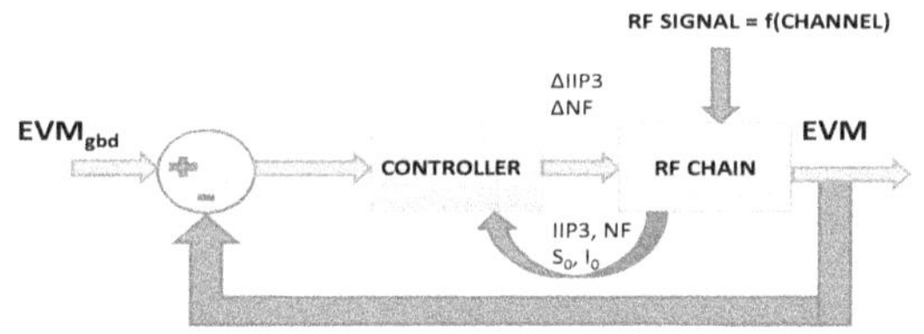

Figure 1: A Control Perspective of the RF chain and feedback controller

One of the disadvantages of the control strategy formulated in section 2.2.3 is that it is essentially based on decisions taken depending on the relationship between the input variables (Signal Strength, Interferer Strength and EVM) and their decision boundaries($IIP3_0$-Δ_1, $IIP3_0$-Δ_2, Δ_3 and EVM_{gbd}). This leads to a "Yes" or "No" decision as to changing the IIP3 and NF and the amount of their change is also fixed. The disadvantage of such a control is that the system keeps on oscillating around the desired value like in a thermostat based room heater(bang-bang control). Moreover the choice of decision boundary becomes very critical in system performance. A more desirable feature of the control algorithm should be that the increment in IIP3 and NF for the next interval (T_1) should be a function of the deviation of the input variables from their decision boundaries.

$$\Delta IIP3 = g(\Omega) \qquad (7)$$

$$\Delta NF = h(\Omega) \qquad (8)$$

Here Ω is the set of deviations of input variables from their boundaries i.e.

$$\Omega=\{IIP3_0\text{-}\Delta_1\text{-}S_0,\ S_0\text{-}\Delta_3,\ IIP3_0\text{-}\Delta_2\text{-}I_0,\ EVM_{margin}\} \qquad (9)$$

$$EVM_{margin} = EVM_{gbd} - EVM \qquad (10)$$

This would ensure that the system rapidly corrects any deviation from the desired steady position (i.e. EVM $\approx EVM_{gbd}$) quickly and settles down to a steady state. This type of control is known as proportional control. In this case the choice of decision boundaries becomes less critical than in "Bang-bang control"

In the field of control theory such problems are dealt with by designing a Proportional controller for optimum performance. However to design such a "controller" one needs to know the transfer function governing the "plant" (which in our case is the RF front-end) and hence the exact input-output relation i.e. we need to know the equation k where:

$$EVM = k(Channel,\ IIP3, NF) \qquad (11)$$

However in this case the relationship is highly non-linear and difficult to formulate. Hence we cannot design a Proportional control using traditional techniques. This problem can be solved by the use of intelligent control algorithms like Fuzzy Controller[11][12]. A Fuzzy controller "attempts to mimic human control logic" and creates a robust control logic which is very easy to implement and is computationally simple and it corrects the problems with the "bang-bang control". A schematic of a Fuzzy controller is shown in Figure 2.

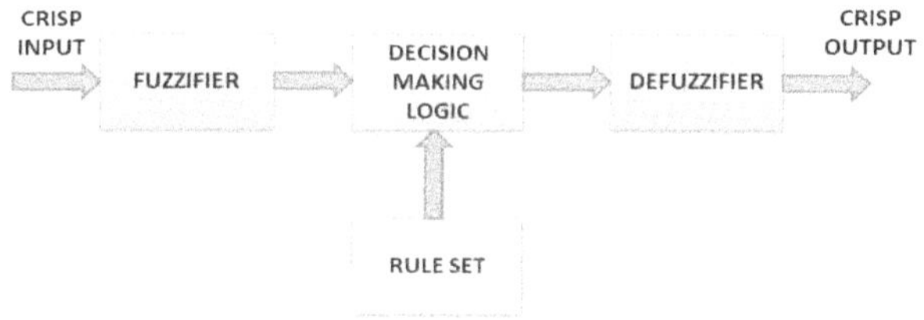

Figure 2: Block Diagram of a Fuzzy Controller

The basic chain of fuzzy control starts with the fuzzification of the input. A group of membership functions is constructed which gives the grade, or degree, of "*membership*" within the set, of any element of the universe of discourse. The *membership function* maps the elements of the universe onto numerical values in the interval [0, 1]. Fuzzification is the process of decomposing a system input into one or more fuzzy sets. For example a decision making question "Is $S_0 < \Delta_3$?" no longer yields a binary value of 0 or 1(as in bang-bang control) but can be anything in the interval [0 1]. Thus they convert the input a "Fuzzy value". The process of fuzzification allows the system inputs to be expressed in linguistic terms so that rules can be applied in a simple manner to express a complex system. The next step is to evaluate the rules of the system using the fuzzified inputs. All the rules that apply are invoked and using the membership functions and truth values obtained from the inputs the result of the rule is determined. This result in turn will be mapped into an output membership function and truth value controlling the output variable. The final step is to combine the fuzzy output of each output rule into a "crisp" value which is the output of the controller.

4. SYSTEM SIMULATION

In order to demonstrate the concepts discussed in previous sections we consider the test vehicle of a 2.4 GHz OFDM WLAN direct conversion receiver. The block diagram of the receiver is shown in Figure 3.

4.1 Signal and Interferer Strength Estimation

In order to adapt to channel conditions the receiver must first estimate the quality of signal it receives. Modern radio front-ends already have sophisticated sensors and baseband algorithms built –in which sense the signal and interferer power.

One such widely-used metric is Received Signal Strength Indicator (RSSI), which gives an estimate of the in-channel power received. [6] discusses a method of detecting the power of in-band interferers. This is done by sensing the total in-band power by using a power detector between the mixer and the channel-select filter. The RSSI can then be subtracted from the in-band power to get the in-band interferer power. [7] discusses the design of a wide-band jammer detector which can be used to sense the total power received at the receiver. Thus combining both the outputs of interferer power detectors one can get the idea about the total interferer power at the input of the LNA.

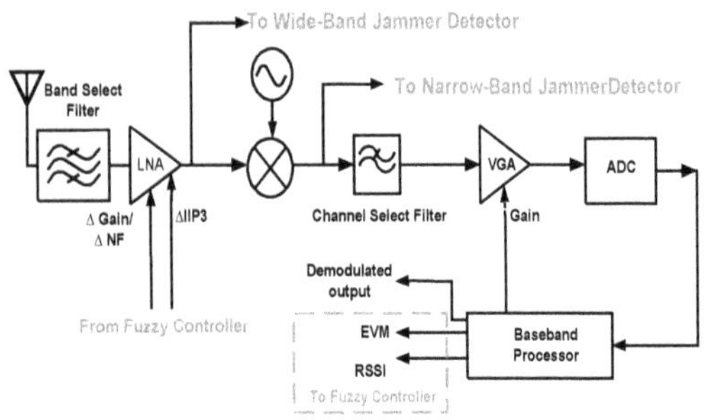

Figure 3: Detailed schematic of the proposed RF chain and baseband circuit showing signal & interferer estimation points.

It must be noted that if interferer power is estimated to be x dBm it could mean that there is one or more interferers causing this effect. However the worst case can occur when there are 2 interferers of (x-3) dBm each causing saturation and 3rd order intermodulation products which can fall in-channel corrupting the desired signal and increasing the noise floor. We shall always adapt the system for this worst case interferer combination for a particular interferer power.

4.2 Orthogonally tunable LNA

One of the most important components of the front-end described is the LNA. The NF of the system is dominated by the NF of the LNA and the linearity of the LNA must be sufficient to allow strong input saturation and intermodulation components from corrupting the signal of interest. [8][9] demonstrate the design of an orthogonally tunable LNA where the Gain/NF and the IIP3 of the LNA can be tuned independent of each other. The NF can tune from 2 dB to 8 dB, while the Gain tunes from 6 dB to 17 dB. At the same time the IIP3 can be varied from -9 dBm to nearly -28 dBm. The schematic and the NF, OIP3 and current consumption variation with tuning knob voltages for the LNA are demonstrated in Figure 4. In the circuit shown in Figure 4, due to the orthogonal nature of tuning knobs, we can say with high confidence that NF is a function of V_{Gain} and IIP3 is a function of V_{IIP3}. So we can approximate these functions by polynomials. Given a desired value of NF and IIP3 we can then solve these equations on the baseband processor to get the required value of tuning knob voltages. Henceforth, in this paper we would discuss about NF and IIP3 tuning directly. However, actually we are changing the tuning knob voltages to produce these changes. In our system we attempt to dynamically vary the NF/Gain and IIP3 of the LNA. As opposed to [2][3] we want to preclude the locus based adaptation approach which requires extensive characterization and a bang-bang control algorithm. [10] discusses adaptation at Variable Gain Amplifier(VGA) and filters keeping the power in RF blocks (LNA, Mixer) unchanged. Our approach fills in this void by tuning the power in the RF blocks and along with [10] provides a complete tuning solution for low-power RF receiver.

4.3 System Characteristics

In order to formulate the control law we need to exercise the circuit so as to get an intuition about its behavior. Using the result of a few simple experiments we can determine the values of the constants discussed in Section 2.

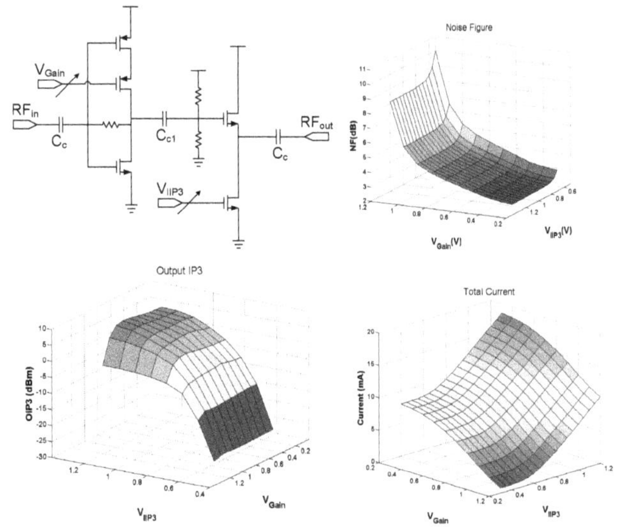

Figure 4: Orthogonally tunable LNA and variation of NF, OIP3 and Current with tuning knobs.[8][9]

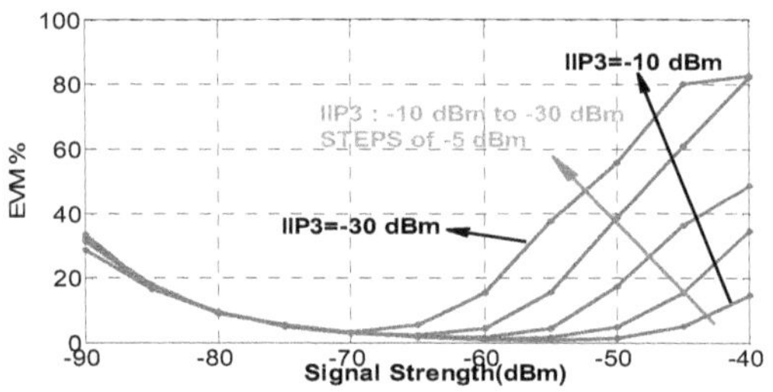

Figure 5: EVM variation with Signal Strength for different IIP3 values

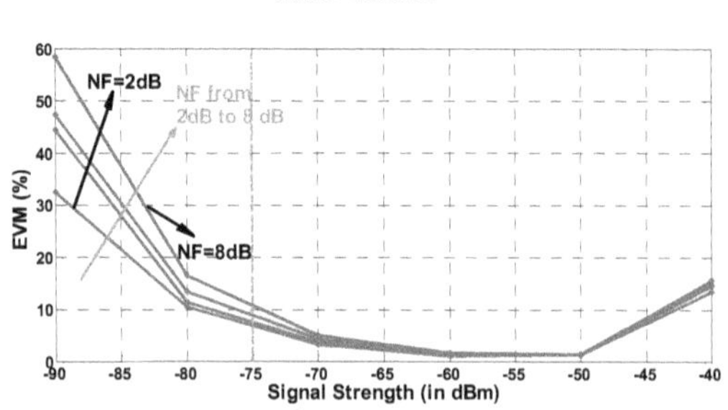

Figure 6: EVM variation with signal strength for varying NF

To estimate value of Δ_l we sweep the value of the signal strength over the entire Spurious Free Dynamic Range (SFDR) of the receiver and observe the system EVM for different values of LNA IIP3. It can be seen that non-linearity effects start to assume significant proportions once the signal crosses the level (IIP3-35) dBm. Hence Δ_l=35 dB. Since interferers behave similarly but also have the added effect of intermodulation components we can keep Δ_2=50 dB. In order to determine Δ_3 we sweep the signal over the entire SFDR for various Noise Figures of the system. It can be seen from Figure 6 that noise figure really starts affecting the signal below Δ_3=-75 dBm. The amount of system simulation required to find these constants is significantly less than gradient based search for optimum knob setting for each channel condition in a LUT based approach.

5. FUZZY CONTROLLER DESIGN

Using the heuristics formulated in Table 1 and estimating the values of the system parameters from Section 4.3 we formulate input and output membership functions for the controller (Figure 7). Each input/output membership function is constructed to have considerable overlap with the adjacent functions in order to prevent abrupt step changes in the output of the Fuzzy controller. Here PVL, PL, PS, NL, NVL and NS stand for Positive Very Large , Positive Large, Positive Small, Negative Large, Negative Very Large and Negative Small. Given a value of S_0, I_0 the Fuzzy controller uses the rule set and membership functions to calculate the desired change in IIP3(i.e. ΔIIP3). Here ΔIIP3 is a function of (IIP3-I_0), (IIP3-S_0) and EVM_{margin}. The surface for ΔIIP3 at EVM_{margin}=3% is shown in Figure 8. The use of a Fuzzy controller has ensured the smooth transition of ΔIIP3 across the range of input variables instead of a step jump at hard decision boundaries (as is for "bang-bang" control). As there are no hard boundaries (and hence discrete switching points), hysteresis need not be implemented in the system. Similarly, input and output membership functions are constructed for NF controller as a function of signal strength and EVM_{margin}.

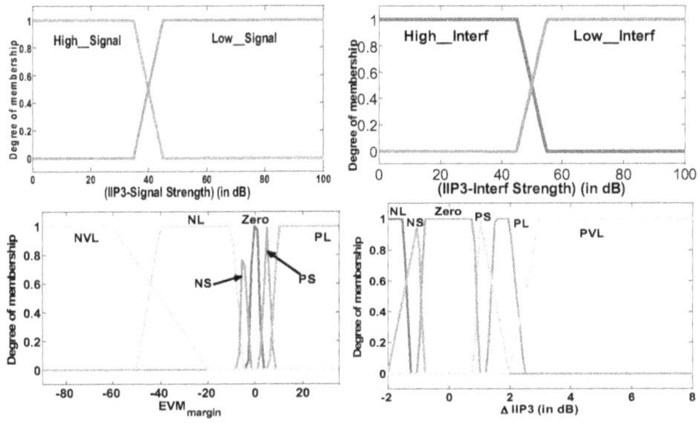

Figure 7: Input and Output Membership Function for IIP3 controller

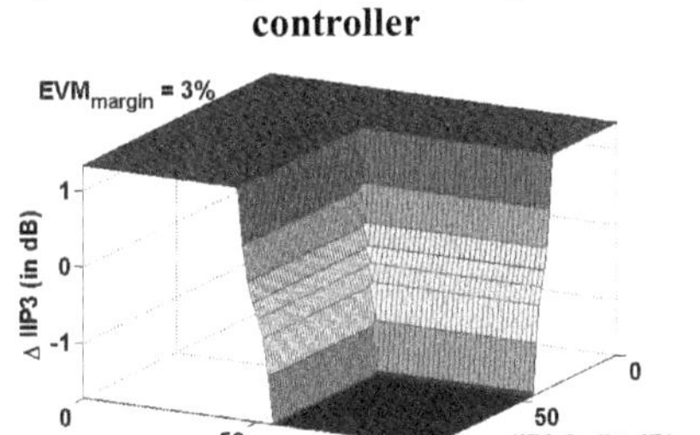

Figure 8: Variation of ΔIIP3 with Interferer and Signal Strength for a fixed (3%) EVM margin

For the sake of brevity, we do not show the membership functions of the NF controller. Figure 9 shows the ΔNF surface as a function of signal strength and EVM. Given an input vector (set of signal strengths, current LNA specifications and EVM), the controller maps the vector to a particular point on the surface. This corresponds to the desired change in IIP3 and NF over the next cycle. One must remember that this fuzzy controller is computationally inexpensive and can run as software on the baseband processor. Thus no additional hardware overhead is required. Moreover the controller needs to update the state variables only once per EVM calculation and hence should consume negligible power as compared to the front-end.

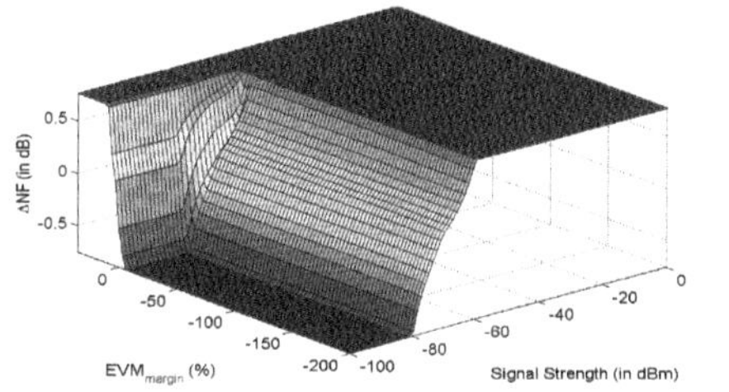

Figure 9: Variation of ΔNF with EVM margin and Signal Strength

6. SIMULATION RESULTS

In order to demonstrate the operation of the Fuzzy Controller based system adaptation for low-power 2 simple experiments were devised. We set up a RF receiver chain model with an orthogonally tunable LNA as the tunable element. An OFDM QPSK signal at 2.4 GHz passes though channel (with variable attenuation, variable strength interferer addition and constant thermal noise floor) before reaching the receiver. We vary the channel conditions to observe the performance and power consumption of the RF-chain-Fuzzy feedback control combination.

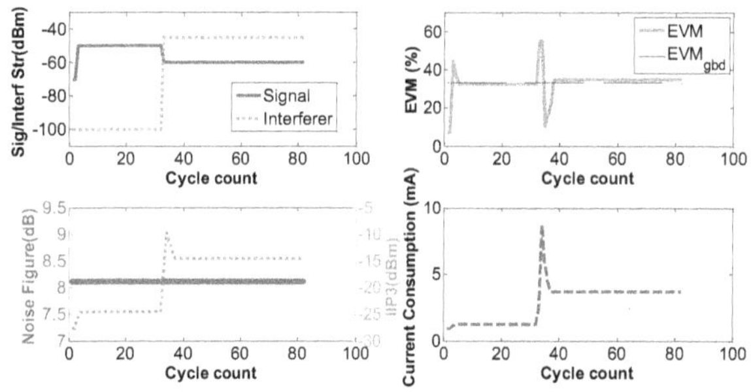

Figure 10: Transient Simulation for Experiment 1 showing Signal and Interferer Power, EVM and LNA IIP3, NF and Power Consumption

(1) Experiment 1: We vary the signal strength and interferer strength of the received signal to show how the system/LNA adapts to the changing conditions. The experiment is demonstrated in Figure 10. Initially the signal becomes strong enough to hit non-linearity and hence EVM overshoots the threshold (by a small amount). In order to correct this, the Fuzzy controller increases the IIP3. It should be noted that in each decision cycle of the controller (denoted in Figure 10 as cycle count), the rate of rise of IIP3 is dependent on the deviation of the EVM from EVM_{gbd}. Therefore IIP3 changes slowly when EVM is close to EVM_{gbd} equilibrium position and fast when the system EVM is too low or too high. Once the EVM reaches EVM_{gbd}, the IIP3 is maintained constant. At cycle count 32, the signal strength decreases but interferer strength increases to a large value. The controller again adapts to this situation by adjusting the IIP3 to a high enough level. It must be noted that all through this experiment the NF remains at the highest possible value due to reasons discussed in Section 2.2.3. The LNA current consumption shows 4x variations (1 mA to 4 mA) in this experiment but remains low as compared to the maximum current consumption

(≈11mA) because NF constraints are relaxed and the circuit is not fully stretched to its maximum capabilities.

(2) Experiment 2: In this experiment the signal strength is made much smaller so that the NF becomes critical. Initially there is no interferer and the NF of the LNA reduces in order to maintain an acceptable BER. Next a high power interferer is introduced and the Fuzzy controller detects the overshoot in EVM and adjusts both the IIP3 and the NF simultaneously so that EVM can be pulled down below EVM_{gbd}. As is evident the highest power consumption occurs when the signal is weak (requiring low NF) and the Interferer is strong (requiring high IIP3) simultaneously.

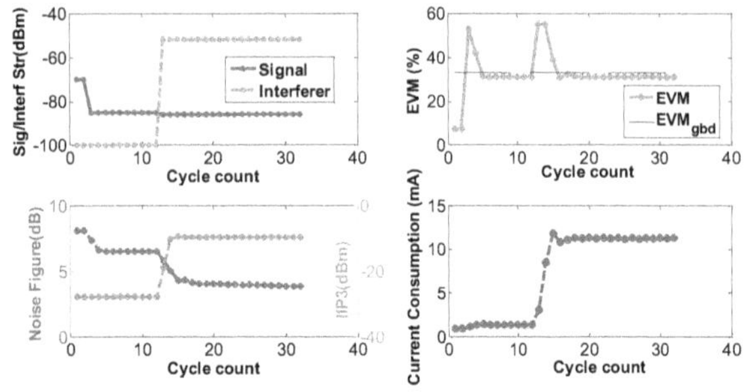

Figure 11: Transient Simulation showing Signal and Interferer Power, EVM, LNA NF and IIP3 and LNA Current for Experiment 2

In such a circumstance both the specifications of the LNA are stretched to their maximum to ensure acceptable BER performance. However this is just a corner case and would occur rarely. Thus most of the time the circuit can operate in a low power mode. As seen from Figure 11 the savings in current can be as large as 10mA as the highest current is ≈11mA and the lowest current being ≈1mA. Assuming the LNA consumes nearly 25% of the total current in a receiver chain we can achieve a maximum power savings of nearly 23% for the receiver.

7. HARDWARE IMPLEMENTATION

An experiment is performed in hardware to show the feasibility of the proposed concept. A 802.11 RF front end (PA / LNA) from Texas Instruments is used in the experiment where supply voltage is varied from 2.6 to 3.9 volts in steps of 0.1 volts and the signal strength is varied from (0.2*fullscale) to the fullscale swing.

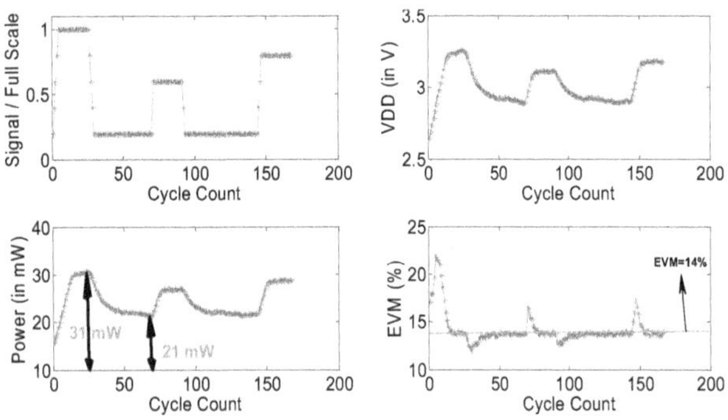

Figure 12: Relative signal strength, V_{DD} control, LNA Power consumption and EVM in the hardware setup.

A mixer AD5382 is used for downconverting the 2.4GHz RF signal. With change in the supply voltage, linearity of the front end also changes which affect the EVM for a given signal strength. On the other hand, as the signal strength increases, the front-end goes into nonlinear region and EVM starts increasing. We set a desired EVM threshold of 14%(pertaining to a BER of $5x10^{-4}$ for 16-QAM) and formulate a Fuzzy control law which tunes the V_{DD} control knob to modulate linearity of the device as a function of the EVM margin and input signal strength(known).

The system is excited with signal of temporally varying strength and the EVM is measured in the baseband. Figure 12 shows the results of the experiment. We observe that the Fuzzy control law allows for a smooth EVM response (without oscillations). Due to the supply voltage modulation the maximum power savings that is obtained is 33%(10 mW out of a total of 31mW). Thus these results corroborate the software simulation of proposed system.

8. CONCLUSION

In this paper the authors have presented a methodology for low-power adaptation of the RF front-end with on-the-fly reconfiguration using Fuzzy Controller. We use the test vehicle of an orthogonally tunable LNA to demonstrate the concept of locus-less receiver adaptation over changing channel conditions. One can implement tuning knobs in different front-end blocks, each of which tunes a circuit performance parameter orthogonally to the other parameters. Thus the concept itself is more generic than the implementation and can be expanded to incorporate multiple blocks in the system. The proposed intelligent control algorithm is a significant improvement over Bang-bang control in terms of steady state response. The system demonstrates significant power savings (≈23%) while doing away with the need for extensive pre-characterization. This concept provides a low-power solution for the power conscious systems for modern wireless communication.

9. ACKNOWLEDGMENTS

This work was funded in part by GSRC/FCRP 2009-DT-2049 and NSF CCR - 0916270.

10. REFERENCES

[1] Debaillie, B. Bougard et. al., "Energy-scalable OFDM transmitter design and control", 43rd IEEE Design Automation Conference, July 24-28, 2006. pp. 536-541.

[2] R. Senguttuvan, S. Sen, A. Chatterjee, "Multi-Dimensional Adaptive Power Management for Low-Power Operation of Wireless Devices" in IEEE Transaction on Circuits and Systems II: Express Briefs (TCAS II), vol.55, no.9, pp.867-871, Sept. 2008.

[3] Sen, S.; Senguttuvan, R.; Chatterjee, A.; , "Environment-Adaptive Concurrent Companding and Bias Control for Efficient Power-Amplifier Operation," Circuits and Systems I: Regular Papers, IEEE Transactions on , vol.58, no.3, pp.607-618, March 2011

[4] Woesner, H.; Ebert, J.-P.; Schlager, M.; Wolisz, A.; , "Power-saving mechanisms in emerging standards for wireless LANs: the MAC level perspective," Personal Communications, IEEE , vol.5, no.3, pp.40-48, Jun 1998.

[5] D. Ernst, S. Das, S. Lee, D. Blaauw, T. Austin, T. Mudge, N. S. Kim, K. Flautner, "RAZOR: Circuit-Level Correction Of Timing Errors For Low-Power Operation," IEEE Micro, Vol. 24, Issue 6, Nov-Dec 2004 pp:10 – 20.

[6] S. C. Ciccarelli, "Dynamically programmable receiver," US Patent No.: 7130602, Oct 2006.

[7] H. Muthali, S. Sen, "Wideband Jammer Detector" through Qualcomm Inc. Provisional Application # 082557P1, US. Pat. No. 12/582,090, Oct 20, 2009.

[8] Sen, S.; Verhelst, M.; Chatterjee, A.; , "Orthogonally tunable inductorless RF LNA for adaptive wireless systems," *Circuits and Systems (ISCAS), 2011 IEEE International Symposium on* , vol., no., pp.285-288, 15-18 May 2011

[9] Sen, S.; Banerjee, D.; Verhelst, M.; Chatterjee, A.; , "A Power-Scalable Channel-Adaptive Wireless Receiver Based on Built-In Orthogonally Tunable LNA," Circuits and Systems I: Regular Papers, IEEE Transactions on , vol.59, no.5, pp.946-957, May 2012

[10] Behbahani, F.et.al. , "Adaptive analog IF signal processor for a wide-band CMOS wireless receiver," Solid-State Circuits, IEEE Journal of , vol.36, no.8, pp.1205-1217, Aug 2001

[11] L. A. Zadeh, "Fuzzy sets," *Information and Control*, vol. 8, no. 3, pp. 338-353, June 1965.

[12] L. A. Zadeh, "Fuzzy logic and the calculus of fuzzy if-then rules," in Proc. 22nd Intl. Symp. on Multiple-Valued Logic, Los Alamitos, CA: IEEE Computer Society Press, 1992, pp. 480-480.

Designing for Dark Silicon: A Methodological Perspective on Energy Efficient Systems

Jason Allred
jason.m.allred@aggiemail.usu.edu

Sanghamitra Roy
sanghamitra.roy@usu.edu

Koushik Chakraborty
koushik.chakraborty@usu.edu

USU BRIDGE LAB
Electrical and Computer Engineering
Utah State University
Logan, Utah 84322

ABSTRACT

The emergence of dark silicon—a fundamental design constraint absent in the past generations—brings intriguing challenges and opportunities in microprocessor design. To gracefully embrace dark silicon, design methodologies must adapt themselves to identify progressive systems that can effectively exploit the growing dark silicon. We demonstrate that relying on traditional design metrics may lead to suboptimal design choices with the rise of the dark silicon area. We provide a new metric to guide a dark silicon aware system design and propose a stochastic optimization algorithm for dark silicon aware multicore system design. Our design approach shows **7-23**% benefit in upcoming technology generations.

Categories and Subject Descriptors

C.1.3 [**Processor Architectures**]: Other Architecture Styles—*adaptable architectures*

Keywords

Dark Silicon, Multi-core, Energy Efficiency

1. INTRODUCTION

Technology scaling has been the major driver in producing faster chips in the past three decades. In recent years, however, transistor scaling has strayed from the ideal as voltage levels have failed to scale proportionally owing to reliability and leakage concerns [2, 9]. This trend will continue to dramatically increase the chip power density going forward. The ITRS scaling projections imply that power density will increase 4X by 8nm [1]. More conservative scaling projections [5] based on Borkar's research [3] indicate a 7X increase in the power density by the same technology node.

This alarming increase in chip power density forces current and future designs to be power limited. Such limitation is the key factor in the creation of *dark silicon*—the fraction of a chip that must be powered down to meet the power budget [5]. The percentage of dark silicon in a chip is projected to increase with every forthcoming technology generation. According to ITRS scaling projections, this percentage may reach 75% by 8nm, assuming power budgets remain constant. Models developed by Esmaeilzadeh et al. are more optimistic, estimating a dark silicon level of 50% by 8nm [5].

The emergence of dark silicon—a fundamental design constraint absent in the past—simultaneously brings intriguing challenges and opportunities in microprocessor design. To gracefully embrace dark silicon, design methodologies must adapt themselves to identify progressive system designs that can effectively exploit the growing dark silicon to improve power-performance and reliability characteristics. Thus, a key technological necessity going forward will be a systematic approach towards designing systems for dark silicon.

In this paper, we present a methodological perspective in designing energy efficient multicore systems geared towards the forthcoming dark silicon. Using a concrete case study, we demonstrate that optimizing for traditional design objectives like energy efficiency can hurt the optimality of a design in the context of on-chip dark silicon. Fundamentally, with the growing dark silicon, design choices that appear superior in one technology generation may become inferior in subsequent technology generations. Thus, to sustain economic viability across multiple technology nodes, it is critical to select progressive designs that can efficiently mitigate the dark silicon challenge.

Our paper makes several contributions for progressive dark silicon aware design methodologies. First, we demonstrate the critical need for a dark silicon aware design with a case study (Section 2). Second, we propose a new Dark Silicon Utilization-Efficiency (DSU) metric for measuring dark silicon exploitation and present a Dark Silicon-Aware (DSA) design methodology for multicore systems based on this metric (Section 3). Third, we present a formalized robust approach for designing a DSA multicore to optimize its DSU (Section 4). Fourth, using a detailed methodology combining physical design flow with architectural simulation, we analyze the energy efficiency benefits of our scheme (Sections 5 and 6). Overall, we observe **7-18**% and **9-23**% energy efficiency improvements in near and long term technology generations, respectively.

2. MOTIVATION

In this section, we motivate the critical need for a systematic approach towards designing multicores for dark silicon. To this effect, we devise a motivational case study com-

ISLPED'12, July 30–August 1, 2012, Redondo Beach, CA, USA.

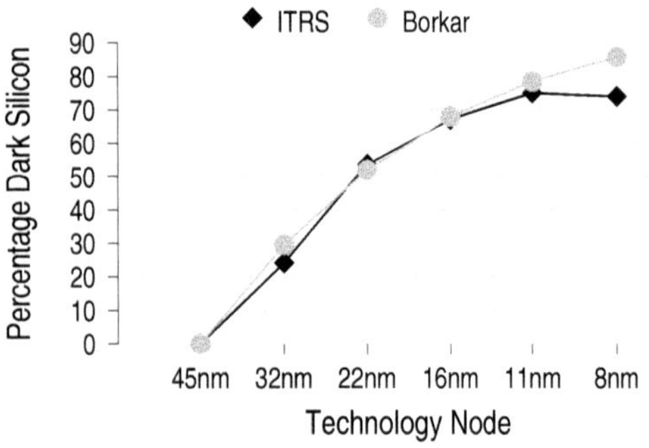

Figure 1: Estimated dark silicon levels for future technology nodes based on current scaling projections.

prising two components. First, we use existing projections to quantitatively estimate the dark silicon challenge in the forthcoming technology generations (Section 2.1). Second, we present results showing how different styles of multicore designs can yield varying energy efficiency benefits under a wide range of dark silicon levels (Section 2.2). Our goal is to demonstrate the challenges in identifying efficient designs in the dark silicon era.

2.1 The Effect of Dark Silicon

Dark silicon—the fraction of a chip that must remain powered off—is an outcome of steady technology scaling coupled with limited progress in cooling. To estimate dark silicon levels for upcoming technology nodes, we use ITRS [1] and Borkar [3, 5] scaling projections to calculate the projected rise in power density. Assuming a constant power budget and chip area across upcoming technology nodes, we can model the active portion of the chip as being inversely proportional to the rise in power density. Figure 1 presents these estimated dark silicon levels.

2.1.1 Significance

The rise of dark silicon presents an intriguing design challenge to forthcoming technology generations. Systems can now be composed of many different active and inactive cores, where the level of dark silicon dictates the aggregate area of inactive cores. Furthermore, the set of active cores can be dynamically altered so as to map software threads on appropriate cores in the pursuit of energy efficiency. With higher levels of dark silicon, multicore systems can accommodate a greater diversity in hardware, thereby improving energy efficiency on a diverse range of workload characteristics. A key question then is how do various design styles compare against each other across different technology generations in the dark silicon era? We investigate this question through an extensive case study next.

2.2 Case Study

For our motivational case study, we compare a conventional multicore design with a recently proposed energy efficient multicore under increasing levels of dark silicon. Our goal is to demonstrate the critical need to adopt a dark silicon aware design for forthcoming technology nodes.

2.2.1 Multicore System

For this study, we use the recently proposed Topologically Homogeneous Power-Performance Heterogeneous (THPH) multicore system [4]. A THPH system is composed of architecturally identical cores designed to be power-performance optimal for different voltage-frequency (VF) domains. Such ground-up design approach has been shown to offer better energy efficiency than the Dynamic Voltage Frequency Scaling (DVFS) technique as lower frequency cores can use less leaky device saving substantial static power, as well as reduce the gate sizes to save dynamic power.

In the THPH system, lower frequency threads are assigned to slower, low power cores and higher frequency threads are assigned to faster, power hungry cores. However, when workload demands are not consistent with the system configuration, the system will at times be forced to schedule high frequency threads on low-performance cores, hurting performance. Likewise, occasionally, low frequency threads will be scheduled on high performance cores, degrading the system energy efficiency. Increasing dark silicon offers an opportunity to a THPH-like system to encounter fewer such scenarios by effectively utilizing the additional hardware resources for greater runtime flexibility. Eventually, it can approach an ideal setup where each thread can be assigned to a core designed optimally for that thread.

2.2.2 Methodology

We use a system with 16 ALUs for this motivational study using 32-bit ALUs from the ISCAS benchmark suite (c7552). Each ALU belongs to a separate core in the multicore system. In Section 6, we forego this simplified system and present results for a high performance industrial microprocessor. The level of dark silicon is varied between 0% to 62.5% representing several forthcoming technology generations. We synthesize the following design styles using the Synopsys Design Compiler and a TSMC 45nm technology library to measure their power and performance:

- **Conventional**: All 16 ALUs are designed for the nominal frequency.
- **Style A**: Eight ALUs are designed for nominal and eight are designed for 50% of nominal frequency.
- **Style B**: Four are designed for nominal, four for 75%, four for 50%, and four for 25% of nominal frequency.
- **Style C**: Six are designed for nominal, six for 75%, two for 50%, and two for 25% of nominal frequency.
- **Style D**: Three are designed for nominal, five for 75%, five for 50%, and three for 25% of nominal frequency.
- **Style E**: Five are designed for nominal, five for 75%, three for 50%, and three for 25% of nominal frequency.

On each design style, we simulate 1000 workloads, each consisting of a set of software threads with frequency demands chosen from a Gaussian distribution with a mean at 75% of the nominal frequency. This distribution captures the application diversity in typical workloads. The number of threads in each set is equal to the total number of active cores (ALUs). Using the technique described in Section 4.2.2, we use a thread-to-core assignment that maximizes the system energy efficiency.

2.2.3 Results

Figure 2 presents data for the improvement in energy efficiency over a conventional multicore measured using the $Energy \times Delay^{\lambda}$ metric (here we chose $\lambda = 4$ due to the strict timing requirements of the ALU). We make two key observations from this figure. First, energy efficiency over

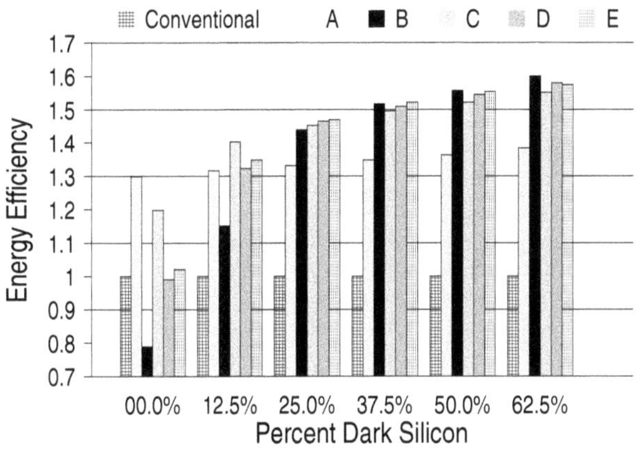

Figure 2: Energy efficiency improvement over the conventional multicore system (higher is better).

a conventional system improves with higher levels of dark silicon. This is because increasing levels of dark silicon provide the system with greater flexibility in choosing a set of cores that better matches the software demands. Second, we observe that a design style that is optimal for 0% dark silicon (style A) can become sub-optimal for 62.5% dark silicon with respect to traditional design objectives like energy efficiency. On the other hand, design style (style B) that shows limited promise in low levels of dark silicon, can outperform with growing levels of dark silicon.

These results demonstrate the emergence of an intriguing design challenge as technology scaling brings forth the possibility of dark silicon. To effectively tackle this challenge, we need a systematic design approach that helps us to identify progressive design styles that optimally exploit their own dark silicon. In the next section, we propose a metric for measuring how well a design takes advantage of the underlying dark silicon in the hardware resources, leading to the design of a *Dark Silicon-Aware (DSA)* system.

3. DSA SYSTEM DESIGN

In this section, we first describe how a core can be specialized for a range of workloads. We then introduce the *Dark Silicon Utilization-Efficiency* (DSU) metric that measures how effectively a design exploits its own existing dark silicon. Finally, we discuss the implementation of this metric in a new DSA design methodology.

3.1 Range of Optimality of a Specialized Core

When a core is designed optimally for a set of thread types, it can operate more efficiently for threads in that set. For example, if a core is designed ground up for a specific target frequency, it can provide energy efficient operation for threads demanding a range of frequencies centered around its design frequency. In this paper, we refer to this range as the *range of optimality* (RoO) of a specific core—the range of software threads for which a given core is specialized. In a system with many cores, the RoO of a specific core is also dictated by the target frequency levels of other cores in the system. Thus, the RoO of a core can be changed by altering its own specialization, or the specialization of other cores. For example, it can be increased or decreased by varying the voltage frequency domain of the core or of the competing cores in the frequency spectrum. Figure 3 shows

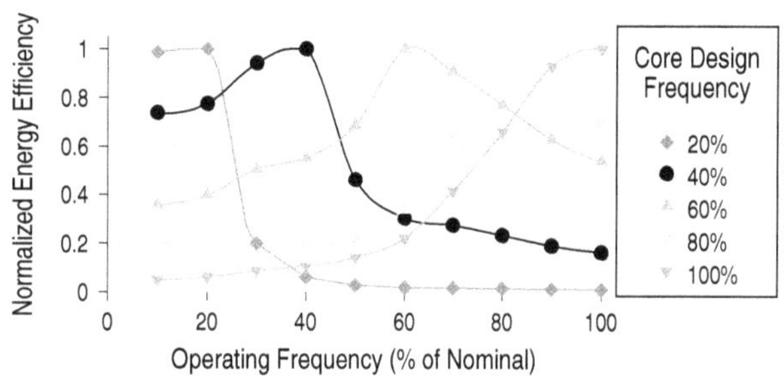

Figure 3: Core RoOs for various target frequency levels.

the energy efficiency distributions of various cores specialized for different target frequencies, using the same modeling and measurement methodology described earlier in Section 2.2. Essentially, each core has a RoO, the range of frequencies where it is more energy efficient than other competing cores.

The RoO of cores can also be altered in ways other than voltage and frequency. For example, if architectural specialization is employed, the RoO can be altered by designing application-specific cores [6] or QsCores [12] for a different range of function types.

3.2 Dark Silicon Utilization-Efficiency (DSU)

When a processing core becomes more specialized, it provides better energy efficiency for a decreasing range of target workloads. This idea is exploited with our new DSU metric. We measure DSU (Equation 1) as the product of two sub-metrics: the core-level *dark silicon utilization* metric (U_{DS}) and the core-level *efficiency* metric (E_{cores}). Each of these metrics scales the other to provide a trade-off between an individual core's level of specialization and the range of threads for which it is optimally designed. We next discuss these two metrics in more detail.

$$DSU = U_{DS} \times E_{cores} \tag{1}$$

3.2.1 *Dark Silicon Utilization* (U_{DS})

Dark silicon provides the current software workload with the choice of cores to activate. In a multicore system with a combination of specialized cores, each software workload will prefer certain cores over others. In such a scenario, a system that more uniformly utilizes all of its cores can better extract the best dynamic match between a software thread and a specialized core.

With a dark silicon level of γ and a multicore with n cores, there will be $\gamma \times n$ inactive cores at any given time. As lower-level caches (e.g., L2 or L3) have substantially low power densities, we assume in this work that dark silicon is applied on the power hungry cores. Overall, the average core utilization will equal the active silicon level $(1-\gamma)$. Using this information, Equation 2 presents our dark silicon utilization metric (U_{DS}):

$$U_{DS} = 1 - \frac{\sum_i^{|O|}(utilization(O_i) - (1-\gamma))}{\gamma \times (n \times (1-\gamma))} \tag{2}$$

where O is the set of over-utilized cores—those with an average utilization rate higher than $(1-\gamma)$. The utilization of a core is the portion of time that the core is active.

The U_{DS} value ranges from 0 to 1. It is 0 when software workloads always prefer the same set of cores, substantially

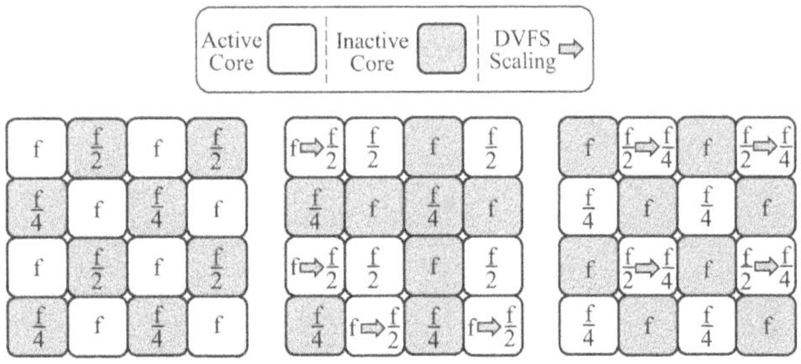

(a) full frequency (b) 1/2 frequency (c) 1/4 frequency

Figure 4: An example DSA system showing 3 of many possible operating configurations.

under-utilizing the inactive cores. U_{DS} is 1 when all cores are utilized equally. Over-utilization of certain cores indicates that these cores can improve their efficiency by increasing their level of specialization and decreasing their RoO, catering to a smaller set of threads. The under-utilized cores, on the other hand, can increase their utilization by increasing their RoO. Overall, there is a trade-off between a core specialization and its utilization, and it is critical to understand this trade-off for designing systems for dark silicon. We now elaborate on our metric for core-level specialization.

3.2.2 Core-level Efficiency (E_{cores})

The second part of the DSU metric is design-dependent, based on the goals of the multicore design. It is a measure of the efficiency of the system as compared to the efficiency of an ideal system, which always allows each thread to execute on a core optimally designed for that thread. In this study, we focus on energy efficiency giving us the following efficiency metric:

$$E_{cores} = \frac{[Energy\ Efficiency]_{actual}}{[Energy\ Efficiency]_{ideal}} \tag{3}$$

We determine energy efficiency by measuring the $Energy \times Delay^{\lambda}$ product. The E_{cores} value ranges from 0 to 1 and is higher for thread-to-core assignments that provide a better software-to-hardware match. The E_{cores} metric can also be used for other design objectives like average core temperature or chip reliability based on design goals. The combined DSU metric allows us to develop a DSA design to optimally meet those design goals.

3.3 DSA Multicore System: An Example

A DSA multicore system is designed to exploit its underlying dark silicon. By choosing various combinations of active cores, a DSA system is able to operate in various configurations, each providing a high level of efficiency for different workloads. Figure 4 explains the operation of such a DSA multicore system. As an example, we show a 16-core multicore system that must maintain a dark silicon level of 50% (i.e. at most 8 active cores at any time). This system improves power efficiency by designing cores for different VF domains. The multicore has 8 cores designed to operate optimally at the nominal frequency, 4 cores designed optimally for 50% of the nominal frequency, and 4 cores designed at 25% of the nominal frequency.

The 8 active threads are allowed to choose any 8 of the 16 cores to optimize energy efficiency. Figure 4 shows three of the many possible scenarios: all threads can be executed at the nominal frequency (Figure 4a), at half frequency (Figure 4b), or at quarter frequency (Figure 4c). If not enough of the lower-frequency cores are available, as in scenarios 4b and 4c, the higher performance cores can be DVFS scaled to the appropriate level. However, the cores that are designed for sub-nominal frequencies provide significant power benefits as compared to DVFS scaling on a high frequency core.

Dark silicon is best exploited by a DSA system that properly designs these cores with appropriate ranges of optimality. The optimal configuration of the DSA cores depends on the expected workload characteristics and the number of cores, as well as the spectrum of dark silicon that the multicore system is expected to encounter. For a general purpose microprocessor, several applications can combine in various phases creating a very large workload space, making the DSA design problem computationally hard. To efficiently choose our DSA cores, we need a formalized algorithm. We present such an algorithm in the next section.

4. OPTIMIZING A DSA SYSTEM DESIGN

In this section, we present a stochastic optimization algorithm for choosing the design parameters of the cores in a DSA multicore system. Our goal is to optimize the DSU for a given range of dark silicon levels and expected workload characteristics.

4.1 Defining the Solution Space

With n cores and k possible core design choices, there are $\binom{n+k-1}{k}$ possible solutions. An exhaustive search is computationally prohibitive. To tackle this computational complexity, we present a stochastic optimization algorithm using simulated annealing in Section 4.2. This algorithm finds the range of optimality for which each core in a DSA system is specialized, allowing us to design the multicore to better exploit the dark silicon.

4.2 A Simulated Annealing Approach

Algorithm 1 shows our algorithm for the DSA design problem using simulated annealing. This algorithm is based on the traditional simulated annealing process, beginning with an anneal temperature T initialized to T_0 and cooled down to δ. Through a series of random moves, accepted or rejected by the annealing schedule, the multicore design is altered to explore the solution space and find an optimal configuration. We define the multicore design as the set of all core configurations $C = \{c_1, c_2, c_3, ..., c_n\}$, where c_i is the configuration of core i.

4.2.1 Annealing Schedule

We choose an initial setup C_0 with each core designed for the nominal frequency. The annealing algorithm maximizes DSU by applying the annealing moves described in Section 4.2.3. $ANNEAL\_CONDITION\_TRUE()$ checks whether DSU_{new} is greater than DSU, or smaller with a probability decreasing with the annealing temperature. This allows occasional uphill moves to better explore the solution space.

4.2.2 Scheduling Threads to Cores

We determine the DSU of a configuration by measuring the utilization and energy efficiency of each core while simulating the execution of 1000 random workloads chosen from a set of expected applications (Section 5.2). Energy efficient execution of applications on a DSA system depends heavily on correct thread-to-core assignment, especially in the presence of dark silicon. In order to correctly measure the

Algorithm 1 DSA Core Configuration Selection

```
Initialize: C ← C_0; T ← T_0
Calculate DSU ← GET_DSU(C)
while T > δ do
    while moves < M do
        C_new = ANNEAL_MOVE(C)
        DSU_new ← GET_DSU(C_new)
        if ANNEAL_CONDITION_TRUE() then
            C ← C_new; DSU ← DSU_new
        end if
    end while
    T ← αT
end while
```

effectiveness of a specific multicore configuration, we must have a method of assigning threads to available cores in a way that minimizes cost. A popular solution to this problem is the *Hungarian Algorithm* which produces a job-to-worker assignment that minimizes cost [8].

We use the Hungarian Algorithm within our simulated annealing framework to activate cores and schedule threads while minimizing the energy efficiency of the system. We then take the harmonic mean of this value across all of the sample workloads to produce $[Energy\ \ Efficiency]_{actual}$, which is used in the formulation of the DSU metric to measure a particular core configuration's effectiveness.

4.2.3 Annealing Moves

$ANNEAL\_MOVE()$ randomly applies a move selected from four moves that combine to explore the entire solution space.

- Decrease the RoO to increase the level of specialization for a set of over-utilized cores.
- Increase the RoO to decrease the level of specialization for a set of under-utilized cores.
- Split a set of cores into two randomly-sized sets of cores and divide its RoO between the two.
- Merge two sets of cores, designing the new set to be optimal for the combined RoOs of the original sets.

In the following sections, we provide the power-performance benefits of an example DSA system employing this annealing algorithm to maximize the DSU.

5. EXPERIMENTAL METHODOLOGY

Our methodology integrates a physical design tool flow capturing the technology mapped power-performance characteristics of synthesized hardware with a cycle accurate full system architectural simulation to precisely model the application level utilization characteristics in a multicore system.

5.1 Physical Design Flow

To precisely model the power-performance characteristics of the hardware, we synthesize the critical datapath and control components of the Alpha 21264 microprocessor [7]. We use the RTL available from the Illinois Verilog Model, and synthesize the following components using the 45nm TSMC library: L1 Caches, Scheduler, Register File and the ALUs. Collectively, these components consume bulk of the power in the microprocessor core. We synthesize these components for several target frequencies ranging from 3.5GHz to 1.5GHz, respectively using the Synopsys Design Compiler.

Several power-performance characteristics like frequency and dynamic and leakage power of the synthesized hardware are measured at each VF operating point. Subsequently, we combine the utilization information from our architectural simulation, described next, to improve the power consumption estimation based on application characteristics.

5.2 Architectural Simulation

Our architectural simulation infrastructure consists of a full-system simulation support built on top of WindRiver SIMICS [10]. SIMICS provides the functional model of several popular ISAs, in sufficient detail to boot an unmodified operating system. Without any loss of generality, we use the SPARC V9 ISA, and use our own detailed timing model to extract hardware utilization characteristics of several SPEC CPU2006 benchmarks on a superscalar out-of-order microarchitecture.

Our on-chip cores loosely represent an Alpha 21264 pipeline micro-architecture. The 12-stage pipeline has 132 instruction window entries, 4-wide fetch-issue-commit engine, with 64KB 2-way instruction and data caches, respectively. The multicore consists of 16 cores, with a 16 MB backing L2. The pertinent hardware utilization characteristics of several SPEC CPU2006 benchmarks are collected through cycle accurate simulation of their representative SimPoint phases [11]. In addition to the nominal frequency, we also simulate these benchmarks at different frequencies within the range of 3.5-1.5 GHz. Since the memory is off-chip, we assume that with processor clock frequency scaling, memory latencies do not scale (thus returning data at the same wall-clock time throughout).

5.2.1 Multicore Workload

To efficiently study a wide range of multicore workloads, we focus on several SimPoint phases of SPEC CPU2006 applications. We find that based on their intrinsic characteristics, our chosen workloads demand a range of target frequencies (1.5GHz-3.5GHz) for respective energy efficiency (ED^{λ}, where λ=2). Actual applications used are: *gcc, astar, bzip2, dealII, GemsFDTD, gobmk, h264, hmmer, libquantum, mcf, milc, omnetpp, perlbench, povray, sjeng, sphinx3, and xalancbmk.*

In the 16-core system, we then randomly combine these applications to create a wide pool of workloads. In our proposed algorithm (Section 4), we use 1000 such workloads in the Monte Carlo simulation, each comprising a random mix of SPEC CPU2006 applications. The outcome of our algorithm is two different DSA systems, one optimized for the near term and the other for the long term technology generations (Table 1). We compare these optimized systems against two other DSA systems with ad-hoc core configuration designs, referred to as *System A* and *System B*.

After selecting the optimal core configurations employing our algorithm, we compare these systems using 4 groups of workload sets, each consisting of 3000 workloads. These workloads within a given set are uniformly spread across multiple technology generations, covering a range of dark silicon levels. The first workload set, referred to as *High*, comprises high IPC benchmarks. The second group consists of medium IPC benchmarks, and is hence referred to as *Medium*. The third group, termed as *Low* consists of low IPC benchmarks. Finally, the fourth group, termed as *Mix*, comprises benchmarks from all IPC ranges. *Overall, we*

System	Core Configuration
Conventional	16 @ 3.5 GHz
System A	8 @ 3.5 GHz, 8 @ 2.4 GHz
System B	4 @ 3.5 GHz, 4 @ 3.0 GHz, 4 @ 2.4 GHz, 2 @ 2.0 GHz, 2 @ 1.5 GHz
DSA (near)	3 @ 3.5GHz, 7 @ 3.0 GHz, 6 @ 2.4 GHz
DSA (long)	4 @ 3.5 GHz, 5 @ 3.0 GHz, 5 @ 2.4 GHz, 1 @ 2.0 GHz, 1 @ 1.5 GHz

Table 1: Core Configurations

evaluate our system using 12000 *multicore workloads, covering a spectrum of benchmark combinations.* Note that none of the workload sets used for evaluation (presented in Section 6) match the set chosen for driving our algorithm in Section 4.

6. EXPERIMENTAL RESULTS

In this section, we show the efficacy of our proposed design methodology for DSA multicore systems through a rigorous experimental analysis. We show results for the near term technology generations where dark silicon varies from 12.5–37.5%. In addition, we show results where dark silicon is substantially larger (50%–75%), as expected in the long term. Using our formalized algorithm (Section 4) and the methodology (Section 5), we design a DSA system for each of these dark silicon spectrum. Table 1 presents the core configurations in the various schemes we evaluate.

Figures 5 and 6 show the energy efficiency improvement in the near term and long term, respectively, normalized to the conventional system. We observe that in the near term, our proposed DSA design consistently out-performs all other schemes across a range of workload combinations (except in *Low* where B performs better). Similar trends are also seen in the long term results. Compared to the conventional design, our proposed schemes offer 7–18% and 9–23% energy efficiency improvements in the near term and long term, respectively. Overall, these results demonstrate the robustness of our design approach in tackling the imminent dark silicon challenge spanning several upcoming technology generations.

7. CONCLUSION

We demonstrate that the emergence of dark silicon–a fundamental design constraint absent in the past generations—ushers a need for a systematic approach in designing systems that can simultaneously mitigate and exploit dark silicon. We propose a new metric, *Dark Silicon Utilization-Efficiency*, that allows us to capture the tradeoff between utilization and specialization in a system with dark silicon. Subsequently, we propose a stochastic optimization algorithm based on simulated annealing to guide the design of a power efficient DSA system, while also considering the power-performance characteristics of the core hardware and application level utilization characteristics. Overall, our technique demonstrates **7-18%** and **9-23%** benefits in near term and long term technology generations, respectively.

Acknowledgment

This work was supported in part by National Science Foundation grant CNS-1117425.

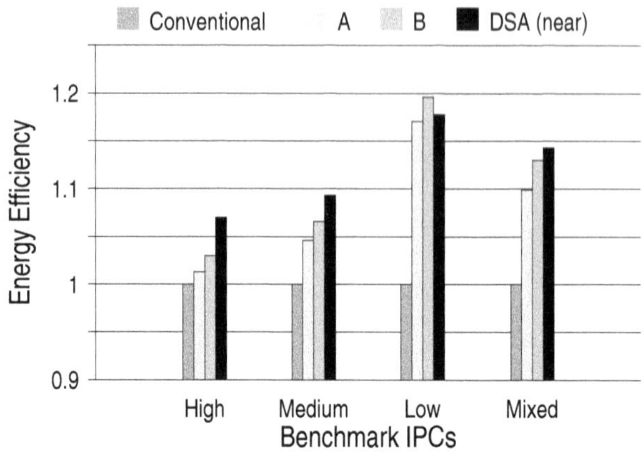

Figure 5: Comparison of DSA systems on four workload sets: near term technology generations. Each workload set consist of 3000 different multicore workloads.

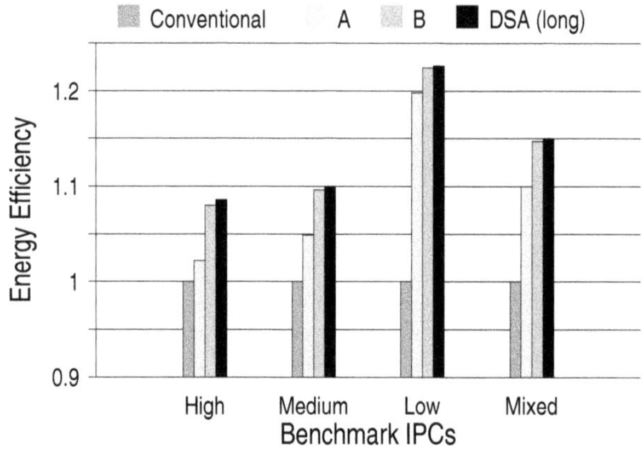

Figure 6: Comparison of DSA systems on four workload sets: long term technology generations. Each workload set consist of 3000 different multicore workloads.

8. REFERENCES

[1] International technology roadmap for semiconductors, 2010 update. Tech. rep., ITRS, 2011.

[2] Borkar, S. Design Perspectives on 22nm CMOS and Beyond. In *Proc. of 46th Proc. of DAC* (2009), pp. 93–94.

[3] Borkar, S. The Exascale challenge. In *VLSI-DAT* (2010).

[4] Chakraborty, K., and Roy, S. Topologically Homogeneous Power-Performance Heterogeneous Multicore Systems. In *Proc. of DATE* (march 2011).

[5] Esmaeilzadeh, H. and others Dark silicon and the end of multicore scaling. In *Proc. of ISCA* (2011), pp. 365–367.

[6] Hardavellas, N. and others Toward Dark Silicon in Servers. *Micro, IEEE 31*, 4 (july-aug. 2011), 6–15.

[7] Kessler, R. The Alpha 21264 microprocessor. *Micro, IEEE 19*, 2 (1999), 24–36.

[8] Kuhn, H. W. The Hungarian method for the assignment problem. *Naval Research Logistics Quarterly 2* (1955), 83–97.

[9] Lee, J., and Kim, N. S. Optimizing total power of many-core processors considering voltage scaling limit and process variations. In *ISLPED* (2009), pp. 201–206.

[10] Magnusson, P. S. and others Simics: A Full System Simulation Platform. *IEEE Computer 35*, 2 (Feb 2002), 50–58.

[11] Sherwood, T. and others Basic Block Distribution Analysis to Find Periodic Behavior and Simulation Points in Applications. In *PACT* (2001), pp. 3–14.

[12] Venkatesh, G. and others QsCores: trading dark silicon for scalable energy efficiency with quasi-specific cores. In *Proc. of MICRO* (2011), pp. 163–174.

HANDS: Heterogeneous Architectures and Networks-on-Chip Design and Simulation

Davide Zoni, Simone Corbetta and William Fornaciari
Politecnico di Milano – Dipartimento di Elettronica e Informazione
Via Ponzio 34/5, 20133 Milano, Italy
{scorbetta, fornacia}@elet.polimi.it

ABSTRACT

In current multi-core scenario, Networks-on-Chip (NoC) represent a suitable choice to face the increasing communication and performance requirements, however introducing additional design challenges to already complex architectures. In this perspective, there is a need for flexible and configurable virtual platforms for early-stage design exploration. We present the *Heterogeneous Architectures and Networks-on-Chip Design and Simulation* framework for large-scale high-performance computer simulation, integrating performance, power, thermal and reliability metrics under a unique methodology. Moreover, NoC exploration is possible from a reliability/performance and thermal/performance trade-offs.

Categories and Subject Descriptors

B.3.3 [**Perfomance Analysis and Design Aids**]: Simulation

Keywords

Multi-core, Network-on-Chip, Simulation, Reliability

1. INTRODUCTION

Continuous technology scaling of recent decade leads to an exponential increase in processor performance, with power consumption going as faster as clock rate. The transition to multi-core architectures introduced an opportunity for performance to grow faster than power consumption, while the need for even more performance and integration of cores in a single chip leads to the definition of novel architectural solutions to cope with unmanageable communication contention. In this scenario NoC truly became the appropriate design paradigm to manage increasing performance and reliability requirements [2]. However such on-chip networks are expected to consume significant part of the total chip power [4]. In particular a few commercial designs show a NoC power consumption up to 28% of the total chip power [5].

ISLPED'12, July 30–August 1, 2012, Redondo Beach, CA, USA.

Meanwhile, increasing operating temperature caused by increasing power consumption density is continuously affecting the reliability of VLSI systems: experimental results show that high temperature is responsible for more than 50% of failures in CMOS integrated circuits [12].

The need to consider such a huge amount of architectural design aspects requires appropriate methodologies for accurate analysis. Such methodologies can be focused on simulation and analytical modeling. Simulation represents the most accurate method to extract valuable information on the architecture, while can be very time consuming. On the other side, analytical models reduce evaluation time even if the output data can be affected by greater errors. Moreover, analytical models provide an intrinsic suitable way for further optimization methodology, while simulation data are almost raw. However analytical models are usually employed for the analysis of specific parts of the architecture after a system-wide analysis, since they are difficult to extract and their characterization requires low level information from real or simulated architecture. In this scenario, this work proposes a novel framework for joint thermal, performance and power analysis to be used both at early design stages, while the extracted information can be used for further localized platform optimizations and trade-off exploration.

1.1 Novel contributions

The proposed work focuses on an accurate, flexible and extensible tool that allows to explore different design space dimensions, i.e. performance metrics, thermal and power profiles as well as reliability issues, during early design stages, covering: *(i)* NoC thermal impact and *(ii)* trade-off analysis and optimization. Indeed traditional approaches of solely caring about performance is nowadays superseded by more critical and multi-dimensional constraints. In this perspective the thermal issues are of paramount importance for both performance and architecture lifetime maximization. Moreover, on-chip interconnect contribution to the chip temperature cannot be neglected, since it represents a consistent part of the total heat flow generated. The proposed simulation framework allows for an accurate thermal chip evaluation accounting for both computational and communication blocks, at different levels of detail.

Furthermore, current multi-core architectures exhibit complex architecture design, since multiple design dimensions should be accounted at the same time. Moreover, the analysis and optimization of orthogonal design dimensions must be conducted both at design-time and dynamically. The proposed framework allows to analyze both compile-time and run-time situations. First, our framework can impose a fine control on detail levels and number of optimization dimensions to be extracted from the simulation to manage design-time trade-offs. Moreover, the complexity of both current multi-core architectures as well as parallel applications, negatively impact the possibility for a complete and accurate design-time optimization, since an amount of metrics can be evaluated at run-time only. In this perspective, the proposed framework allows for run-time policy evaluation allowing for further policy optimization, safety analysis and cutting edge stress tests.

The remainder of the paper is organized as follows. Section 2 provides an overview of the state-of-the-art simulation frameworks, highlighting their major features and limitations, focusing on Network-on-Chip analysis. Section 3 provides an in-depth discussion of the proposed framework, discussing the advantages as compared to state-of-the-art solutions. Experimental results are reported in Section 4, focusing on a detailed analysis of relevant use cases. Conclusions are finally drawn in Section 5.

2. RELATED WORKS

Several proposals can be found in literature for power, performance and thermal estimation of single-core and multi-core processors. Nevertheless, only a few are focused on a comprehensive approach to jointly estimate multiple design dimensions considering Network-on-Chip. This work provides an accurate and flexible design tool that accounts for all the different design aspects for high-performance multi-core architectures, with particular emphasis on NoCs.

Different works that can be found in literature are meant for single-core analysis. Nevertheless, the advent of multi-core architectures is pushing strict requirements on multi-core simulation: power, performance and temperature should be be jointly analyzed at system-level, and considering the on-chip networks due to their increasing relevance in parallel architectures. In this perspective, literature lacks a suitable simulation methodology general enough to deal with all the design aspects described above. The `SESC` simulator [13] provides cycle-accurate simulation of bus-based multi-core processors, based on MIPS architecture. However, it does not support Network-on-Chip architectures. The `Polaris` framework [15] can provide power and area estimates of Network-on-Chip architectures. Since the focus is on the interconnect, it lacks for a detailed power estimation for both processors and memory hierarchy. Although precise, it is not suitable to be employed in system-level computer architecture research. Power, area and thermal modeling are accounted for also in the `SST` framework [6]. The work focuses on large-scale systems, but application traces are emulated, rather than collected from cycle-accurate simulation, with higher simulation rate at the cost of much lower accuracy. Power and thermal models are proposed in [3], based on the `Simics` functional simulator. The interesting achievement in this approach lies on the possibility to develop, analyze and tune different control algorithms for thermal and power management, based on high-level Matlab descriptions. The work is suitable for designing control-theoretic thermal management solutions, although bound to a particular architecture, ISA and floorplan (precisely, the reference architecture is an Intel©Xeon X7350 system). The work presented in [10] is meant to simulate large-scale architectures, and exploits parallel simulation on physical hardware. It can simulate several cores based on the MIPS in-order architecture. However, the output thermal map refers to the only communication infrastructure, without providing a system-wide perspective from a thermal view-point. In addition, each of the above approaches lacks in the reliability aspects, i.e. MTTF projection that is jointly coupled to chip temperature profile. Table 1 finally summarizes the advantages and pitfalls of these works, and reports a comparison with the framework presented in this paper.

3. PROPOSED EVALUATION FLOW

We provided an appropriate set of modifications to third party tools, while a complete set of other tools has been developed from scratch.

The logical snapshot of the proposed virtual platform is given in Figure 1. Steps are executed in pipeline fashion, i.e. each step provides input to the subsequent one, and requires output from the previous, ensuring a flexible and extensible tool. The four steps involved are: cycle-accurate simulation, power consumption estimation, floorplan generation and thermal/reliability models. Cycle-accurate simulation is required to provide relevant access and usage statistics at architecture and microarchitecture level. During simulation, the most relevant information from the modeled architecture are acquired, e.g. accesses to instruction fetch unit, number of committed integer instructions, number of stall cycles in the pipeline and the like. For processors, this means that we can collect sensible metrics about the status of the hardware pipeline while executing the software application; for NoC routers, on the other hand, we collect raw metrics about network interface accesses and traffic patterns. Memory-related statistics are also required, providing a system-wide set of performance and access statistics. Accurate power consumption estimates are required for both processing cores and interconnect primitives, as well as for storage blocks. Last, temperature profile is required to evaluate the impact of hardware or software design choices on the reliability and power consumption of the entire architecture. There are also three different feedback paths: temperature feedback exists from the thermal model to the power model for leakage power analysis; an additional temperature feedback is used to back-annotate the simulator, useful for temperature-aware scheduling policies evaluation; last, power back-annotation is used to allow for power management policies to be developed and estimated along the flow. These feedback paths can be employed in a more general fashion to provide an analysis of power-related and temperature/reliability-related design choices, either at design-time or run-time. In Figure 1, white boxes are those related to third-party software, gray boxes represent in-house developed tools, while striped boxes are those tools for which relevant modifications have been performed. The rest of this section gives an insight of each step, the tools involved in each stage and the modifications that have been applied to third-party tools.

3.1 Cycle-accurate simulation

We employ `GEM5` cycle-accurate performance simulator (`http://gem5.org/`), using the *syscall emulation* approach that models bare-metal execution and can simulate the underlying hardware with precise processor models and collecting statistics at microarchitecture level. `GEM5` can also supports *Full-system* simulation mode, that requires and simulates an OS to support application execution, but introduces a

Table 1: **Features of state-of-the-art multi-core simulation frameworks compared to the proposed flow.**

Framework	Cycle-accurate simulation	NoC support	Power support	Thermal support	Reliability projection	Floorplan exploration	Objectives
Renau et al. (SESC) [13]	✓	✗	✗	✗	✗	✗	multi-core simulation, parallel applications
Soteriou et al. (Polaris) [15]	✗	✓	✓	✗	✗	✗	Network-on-Chip design-space exploration
Hsieh et al. (SST) [6]	✗	✓	✓	✓	✗	✗	microarchitecture, power and thermal
Lis et al. (HORNET) [10]	✗	✓	✓	✓	✗	✗	many-core processors, mainly NoC interconnect
Bartolini et al. [3]	✗	✓	✓	✓	✓	✗	run-time control policies evaluation
Our flow	✓	✓	✓	✓	✓	✓	microarchitecture, NoC, reliability, design-space exploration

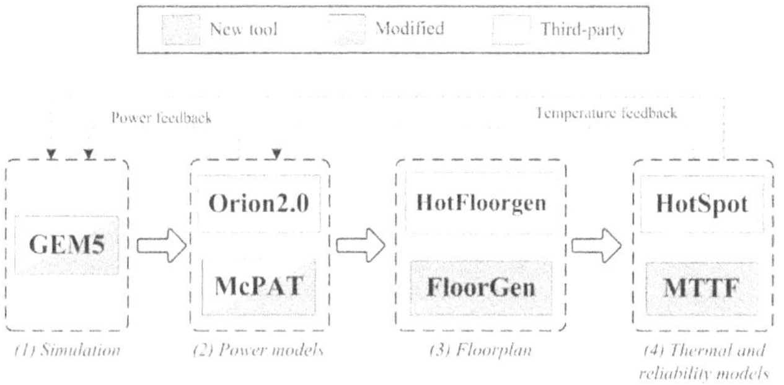

Figure 1: Proposed estimation flow.

great computational overhead. We are mainly interested in bare-metal execution, providing HW-based simulation of synchronization primitives where required; we also introduce the support for clock-toggling processors. Indeed, current and future multi-core architectures have to face with thermal and reliability issues, as well as performance and power ones. Most of the run-time thermal management and power management techniques rely on dynamic voltage and frequency scaling (DVFS) to control both chip temperature and power consumption, or on clock-gating hardware support to cool down processors. We provide a per-CPU clock-toggling configurable implementation, in which the user can synthesize the desired duty-cycle for thermal/performance tradeoff exploration, either at design-time or run-time;

3.2 Power estimation

Power estimation is accounted for using different tools for processor and Network-on-Chip router, as a function of the access statistics provided by the cycle-accurate simulation phase. `McPAT` [9] is used to generate power estimates of the core and memory architectures, as a function of the access statistics provided by the cycle-accurate simulation phase presented in Section 3.1. `Orion2.0` [8], on the other hand, is employed for computing power contribution of NoC routers and links, as a function of traffic and packets traversing the network. Two significant improvements have been made to these tools. The original version of `McPAT` takes as input a single temperature value to compute leakage contribution: the entire chip region is assumed to work at the same operating temperature. This assumption does not fit well with architecture modeling, for two reasons. At first, it is impractical that different regions of the chip experience the same amount of temperature [14], due to the asymmetric load assignment, especially in multi-core architectures. In addition, the chip temperature profile is an aspect of paramount importance for reliable design, while the assumption provides an overlay simplistic scenario. The proposed flow, on the other hand, is able to annotate the correct temperature to each microarchitecture block in the processor, thus providing a way to better estimate leakage power contribution. In addition, `McPAT` provides a discretized amount of leakage levels, ranging from 300K to 400K temperatures at steps of 10K. The available temperature range is thus reduced to 11 values, providing an impractical scenario for aggressive thermal simulations. In our flow, on the other hand, the temperature range is fully covered, and leakage curve within temperature steps at distance 10K are approximated linearly. This aspect, along with the feedback from thermal model (refer to Section 3.3) provides a comprehensive and more accurate estimation.

3.3 Temperature estimation

Temperature estimation is an essential phase in the computer architecture research, due to the increasing relevance of temperature-aware designs. We use `HotSpot` [14] thermal model for multi-core thermal map evaluation. This model requires a chip floorplan, and a set of power measurements to compute steady-state temperature. The main improvement we propose in this work is related to a flexible and customizable floorplan generation tool coupled to `HotFloorplan`, part of the `HotSpot` model release. Since `HotFloorplan` does provide single-core floorplan only, we developed `FloorGen` to generate the floorplan for the desired multi-core architecture. We focused on two main aspects: to provide flexibility to generate any desired floorplan, and to provide the flexibility to generate the floorplan at any desired level of details. Up to now, we target only 2D mesh topologies, based on the Alpha-21364 network architecture [11], but we are able to generate the floorplan of each core according to user-defined requirements: the user can thus specify core floorplan, and let the tool generate a multi-core architecture with core replication. An interesting support in this direction has been made to integrate the output from `HotFloorplan` with `FloorGen`. Notice that this step is entirely decoupled from the cycle-accurate simulation, since core access statistics are not affected by the floorplan since the wires are assumed to be dense in their respective microarchitecture block. The floorplan impact on the power consumed by router links is disscussed in Section 3.2.

Table 2: Processor and router setup.

Processor core	4GHz, in-order Alpha21264 core
Int-ALU	4 integer ALU functional units
Int-Mult/Div	4 integer multiply/divide functional units
FP-Mult/Div	4 floating-point multiply/divide functional units
L1 cache	64kB 2-way set assoc. split I/D, 2 cycles latency
L2 cache	1.75MB per bank, 8-way associative (shared)
Router	3-stage wormhole switched (Garnet network [1])
Topology	2D-mesh, based on Alpha21364 network processor
Technology	65nm at 1.2V, and 45nm at 1.1V

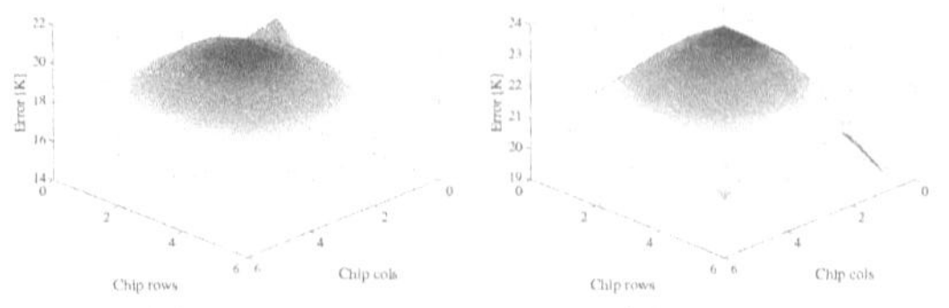

(a) 65nm (b) 45nm

Figure 2: The estimation error obtained without considering the impact of the Network-on-Chip temperature is maximum at the center of the chip, and diminishes towards the perimeter.

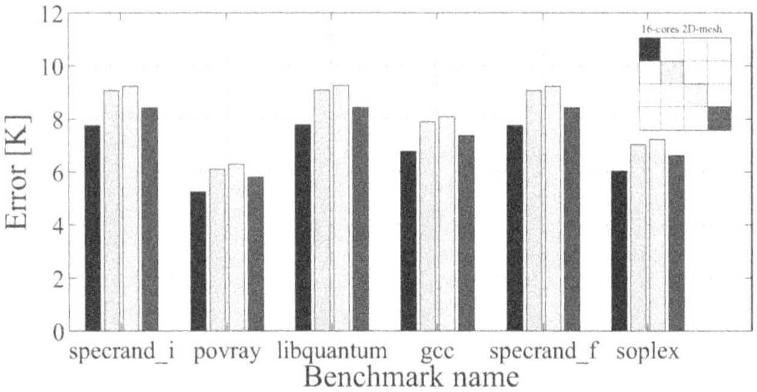

Figure 3: Maximum observed temperature mismatch while considering the NoC impact.

3.4 Reliability analysis

The last logical block from Figure 1 is used to compute reliability projection. Temperature-dependent reliability estimate is done through Mean Time To Failure (MTTF) analysis of different mechanisms: electromigration, stress-migration and thermal cycling. MTTF for these processes is known to be exponentially dependent on temperature, and this library provides an easy way to perform reliability-directed design optimizations with direct input from the simulated architecture. This contribution makes the proposed framework suitable for aggressive reliability projections and hardware/software estimation.

4. EXPERIMENTAL RESULTS

This section highlights the flexibility of the proposed framework when different design dimensions are considered. Three main results are discussed: Section 4.1 shows how the proposed flow can be used to easily analyze the impact of the interconnect on the chip thermal profile. Section 4.2 presents a simple exploration and evaluation of the thermal coupling coefficients accounting for both communication and computational logic. Last, Section 4.3 presents a reliability/performance trade-off analysis exploiting the provided clock toggling implementation.

In the experiments, we consider an Alpha-21364 network processor as reference architecture [11]. This is composed of tiles, organized along a 2D-mesh topology: each tile is composed of an in-order version of the Alpha-21264 processor core, with private L1 cache and shared distributed L2 cache; the router is used to interface the processing core to the local and shared memory.The router is a 3-cycles delay architecture, and local L2 cache bank surrounds both processing core and router as in [11]. The main features and technology parameters used throughout the experiments are summarized in Table 2.

4.1 The role of Network-on-Chip

The NoC has great impact on the temperature distribution within the SoC. This section details how such thermal contribution increases with integration density, and technology node scaling down. Figure 2 shows the thermal impact of the NoC on the architecture. The results are shown for a 36-cores processor, considering 45nm and 65nm technology nodes. The surface reports the absolute estimated temperature mismatch while neglecting the NoC contribution, considering integer-intensive workload from SPEC-CPU 2006 benchmark suite. Each point approximating the surface represents the absolute temperature error in a specific core in the floorplan; thus, points are ordered by means of rows and columns. At first, it is clear how the peak contribution of the NoC routers increases while technology node scales down. As a matter of fact, the maximum temperature error increases from 20K to 24K while going through 65nm and 45nm nodes, respectively. Moreover, the spread is increased, and a higher variation is observed in lower technology nodes. Secondly, the higher error is experienced a the central region of the chip, where coupling is maximum (refer to Section 4.2 for more details on this).

The impact that the NoC has on the processing cores, from a temperature view-point, is also summarized in Figure 3 considering a 45nm 16-cores processor running different applications from the SPEC-CPU 2006 suite. Six different applications are proposed, each core running the same application. The error is reported in temperature degrees, and corresponds to the absolute difference between steady-state temperatures when cores only and cores plus NoC routers are considered. Results are given for the four cores along the diagonal of the architecture, and have been collected after executing 30×10^6 instructions[1]. As it can be seen, the NoC router has a great impact on the temperature of adjacent cores: on average, 6K up to 9K degrees are experienced. This demonstrates the importance of a system-level temperature estimation methodology and thermal design, to ensure proper design choices and increase reliability of the final system. In addition, notice that different applications lead to different impact: in general, the higher the communication required by the applications running in the processor, the higher will be the impact of the NoC.

[1]Notice that for steady-state analysis and single-phase applications, this is almost enough to get realistic temperature values.

4.2 Thermal coupling in multi-core architectures

The increasing availability of multi-core architectures is driven by continuous shrinking technology and achievements in architecture design, e.g. Network-on-Chip supporting hard power/performance constraints. With increased power density, operating temperature increases, and with shrinking technologies the role of mutual influence increases as well. Core-to-core thermal coupling has been shown to get worse with technology scaling [7], however the impact of Network-on-Chip has not been quantified, yet. The proposed estimation flow allows to easy quantify the relative impact of computational cores and communication blocks. To this extent, the multi-core processor is modeled as a set $A = \{a_1, a_2, ...\}$ of $|A|$ blocks being either processor cores, interconnect routers or memory banks. Architecture blocks are connected each other and their physical position is determined by the floorplan. Notice that there might be cases in which the floorplan is either unknown, or its complexity is just unmanageable. For these reasons, we capture the floorplan information indirectly, through the use of observed temperature measurements, and adopt an appropriate thermal coupling model to quantify the mutual influence. The effects of heat transfer between adjacent blocks are seen as a net increase in temperature, induced by proximity to active blocks. The magnitude of this influence is modeled through the thermal coupling coefficient, denoted with $\psi_{ij}(p) = \frac{T^p_{ji} - T^p_i}{T^p_{ji} - T_{amb}}$: a non-linear function of the blocks relative position, the relative temperatures and power levels, and system configuration (package solution and thermal design solution. We denote with T^p_j the *steady-state* temperature of block a_j while consuming (on average) a power of p Watts. This is the temperature value due to self-heating, i.e. while consuming power. We then introduce T^p_{ji} denoting the steady-state temperature of block a_j when block a_i is loaded at power level p, and a_j is assumed to be idle. At quiet-state, each block a_i is assumed to be at T_{amb}.

For each block a_i, we compute the aforementioned coefficient; an extensive procedure is employed for this purpose. The procedure takes as input the chip floorplan and the power traces generated by the cycle-accurate simulation and power/area models, and it is repeated for each tile j in the architecture; notice that we hereby consider the core, the router and the L2 cache as a whole, as done in [7]. We then selectively switching off a portion of the available tiles, and then compute the self-heating contribution T_j. Coupling contribution, on the other hand, is computed by switching off alternatively the remaining tiles k different from j, to get an estimate of T^k_j. Last, coupling coefficient is computed using the aforementioned $\psi_{ij}(p)$ metric.

Figure 4 shows the thermal coupling index considering both cores and routers as the "destination" block of induced temperature. The results are taken considering a single tile for temperature coupling as a whole (Figure 4(c)), focusing on the direct impact that neighbor tiles have on cores (path (a) in Figure 4(c)) and routers (path (b)). The temperature values have been simulated using applications from the SPLASH-2 benchmark suite. The proposed metric is able to capture the spatial correlation of temperature, since coefficients have decreasing value moving far from the destination (tile at position $(0, 0)$ in this case). Also, there is a slightly different absolute value in thermal coupling for the core and router case due to chip geometry, although their magnitude is comparable.

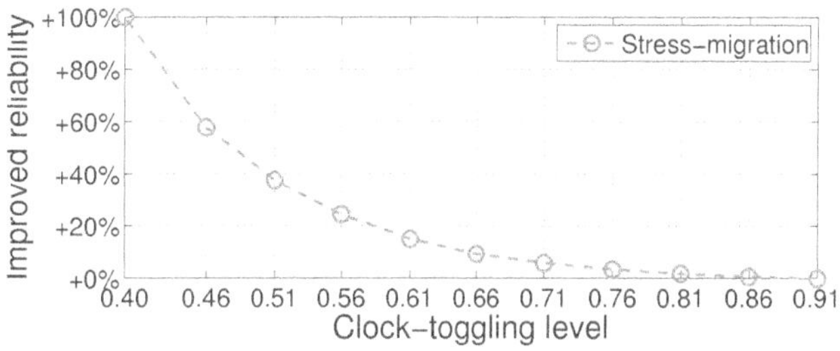

Figure 5: Performance/reliability improvement against base-case scenario (MTTF = 1).

4.3 Reliability/Performance Exploration

Thermal analysis represents a requirement of paramount importance in current multi-core designs. For instance, reliability optimizations focus on minimizing operating temperature, to increase the MTTF and reduce the probability of faults. However, a constrained chip temperature can degrade performance. In this section we discuss the performance/reliability trade-off addressing hard-faults only, while considering MTTF degradation caused by stress-migration. We consider a 45nm 36-cores 2D-mesh architecture and we apply different clock toggling levels to diminish temperature, thus improving MTTF. Even if the current clock toggling implementation allows for per-core clock manipulation, we apply the same clock toggling level to each core to provide the same performance degradation on each core. Clock-toggling in this context is defined as the amount of idleness that is required to synthesize the desired duty-cycle level of the processor: thus, a clock-toggling level of 0.9 allows to synthesize a 10% duty-cycle processor usage. More sophisticated clock toggling designs should consider different clock toggling levels on different cores, based on the running applications, multi-core floorplan and desired temperature profile, however such methodology is out the scope of this work. Starting from the described architecture, we compute, for different clock toggling levels, the MTTF expression $MTTF_{SM} \propto |T_0 - T|^{-n} \cdot exp\left\{\frac{E_{SM}}{k\,T}\right\}$ taken from [16], where E_{SM} is the energy activation for stress-migration, k is the Boltzmann's constant, T the operating temperature, T_0 the reference temperature for (melting temperature), and n is a technology-dependent parameter. We use the values for these parameters as given in [16].

Figure 5 shows the reliability projection of the 36-cores processor, while constraining the clock toggling level. Notice that the hottest core only is reported, having all the other cores at lower temperature, thus with higher MTTF. The dotted line shows the reliability trend (MTTF values) when applying the desired clock-toggling. The horizontal axis reports the clock toggling level required to accommodate the MTTF improvement. The vertical axis reports the target reliability improvement relative to base case when reliability equals 1 and no clock modifications have been applied. For instance to increase by 40% MTTF caused by stress-migration in 36-cores processor, performance level should be diminished to 51%.

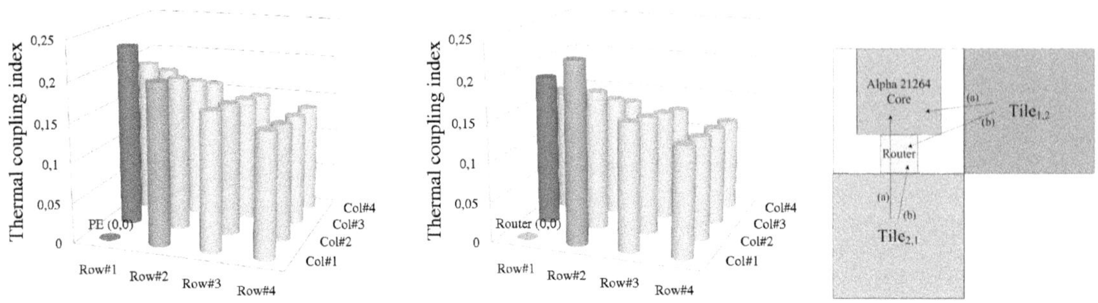

(a) PE coupling (b) Router coupling (c) Configuration

Figure 4: Thermal coupling coefficients trend, considering tiles as source and PEs/routers as destination.

5. CONCLUSIONS

This work presented a cycle-accurate simulation framework for accurate analysis of thermal issues in modern multi-core architectures, focusing on a joint analysis of computational blocks and Network-on-Chip communication fabric. Moreover, the proposed tool allows for exploring different design-time metrics of interest, such that thermal/performance and reliability/performance trade-off optimization. The flexibility of the proposed work allows to perform accurate and detailed joint analysis on power, performance, thermal and reliability metrics. The simulation flow is composed of state-of-the-art tools, as well as tools developed from scratch. We then described and discussed a set of experiments, using SPEC CPU 2006 and SPLASH-2 benchmark suites, showing the flexibility of the framework to support several kind of estimation objectives and methodologies.

6. ACKNOWLEDGMENTS

This research work is supported by European Community Seventh Framework Programme (FP7/2007-2013), under agreements no. 248716 (2PARMA project www.2parma.eu).

7. REFERENCES

[1] N. Agarwal, T. Krishna, L.-S. Peh, and N. Jha. Garnet: A detailed on-chip network model inside a full-system simulator. In *Performance Analysis of Systems and Software IEEE International Symposium on*, 33–42, 2009.

[2] A. Banerjee, R. Mullins, and S. Moore. A Power and Energy Exploration of Network-on-Chip Architectures. In *NOCS '07*, 163–172. IEEE Computer Society, 2007.

[3] A. Bartolini, M. Cacciari, A. Tilli, L. Benini, and M. Gries. A virtual platform environment for exploring power, thermal and reliability management control strategies in high-performance multicores. In *GLSVLSI'10*, 311–316, New York, NY, USA.

[4] S. Borkar. Networks for multi-core chips: a contrarian view. In *Special Session at ISLPED*, 2007.

[5] Y. Hoskote, S. Vangal, A. Singh, N. Borkar, and S. Borkar. A 5-ghz mesh interconnect for a teraflops processor. *Micro, IEEE*, 27(5):51 –61, 2007.

[6] M.-y. Hsieh, A. Rodrigues, R. Riesen, K. Thompson, and W. Song. A framework for architecture-level power, area, and thermal simulation and its application to network-on-chip design exploration. *SIGMETRICS Perform. Eval. Rev.*, 38:63–68, 2011.

[7] M. Janicki, J. H. Collet, A. Louri, and A. Napieralski. Hot spots and core-to-core thermal coupling in future multi-core architectures. In *26th Annual IEEE SEMI-THERM Symposium*, 205–210. 2010.

[8] A. Kahng, B. Li, L.-S. Peh, and K. Samadi. Orion 2.0: A fast and accurate noc power and area model for early-stage design space exploration. In *DATE '09.*, 423 –428, april 2009.

[9] S. Li, J. H. Ahn, R. Strong, J. Brockman, D. Tullsen, and N. Jouppi. Mcpat: An integrated power, area, and timing modeling framework for multicore and manycore architectures. In *Microarchitecture, 42nd Annual IEEE/ACM International Symposium on*, 469 –480, 2009.

[10] M. Lis, P. Ren, M. H. Cho, K. S. Shim, C. Fletcher, O. Khan, and S. Devadas. Scalable, accurate multicore simulation in the 1000-core era. In *Performance Analysis of Systems and Software (ISPASS), IEEE International Symposium on*, 175 –185, 2011.

[11] S. S. Mukherjee, P. Bannon, S. Lang, A. Spink, and D. Webb. The alpha 21364 network architecture. In *Proceedings of the The 9th Symposium on High Performance Interconnects*, Washington, DC, USA, 2001.

[12] M. Pedram and S. Nazarian. Thermal Modeling, Analysis, and Management in VLSI Circuits: Principles and Methods. *Proceedings of the IEEE*, 94(8):1487–1501, August 2006.

[13] J. Renau, B. Fraguela, J. Tuck, W. Liu, M. Prvulovic, L. Ceze, S. Sarangi, P. Sack, K. Strauss, and P. Montesinos. SESC simulator, January 2005.

[14] K. Skadron, M. R. Stan, K. Sankaranarayanan, W. Huang, S. Velusamy, and D. Tarjan. Temperature-aware microarchitecture: Modeling and implementation. *ACM Transactions on Architecture and Code Optimization (TACO)*, 1(1), 2004.

[15] V. Soteriou, N. Eisley, H. Wang, B. Li, and L.-S. Peh. Polaris: A system-level roadmap for on-chip interconnection networks. In *ICCD* , 134 -141, 2006.

[16] J. Srinivasan, S. Adve, P. Bose, and J. Rivers. The case for lifetime reliability-aware microprocessors. In *Computer Architecture, Proceedings. 31st Annual International Symposium on*, 276 - 287, 2004.

XIOSim: Power-Performance Modeling of Mobile x86 Cores

Svilen Kanev Gu-Yeon Wei David Brooks
Harvard University
33 Oxford St., Cambridge, MA
{skanev, guyeon, dbrooks}@eecs.harvard.edu

ABSTRACT

Simulation is one of the main vehicles of computer architecture research. In this paper, we present XIOSim – a highly detailed microarchitectural simulator targeted at mobile x86 microprocessors. The simulator execution model that we propose is a blend between traditional user-level simulation and full-system simulation. Our current implementation features detailed power and performance core models which allow microarchitectural exploration. Using a novel validation methodology, we show that XIOSim's performance models manage to stay well within 10% of real hardware for the whole SPEC CPU2006 suite. Furthermore, we validate power models against measured data to show a deviation of less than 5% in terms of average power consumption.

Categories and Subject Descriptors

C.4 [**Performance of Systems**]: Modeling techniques

Keywords

simulation, x86, power-performance, in-order

1. INTRODUCTION

Architecture simulation has been used extensively both in the early stages of a processor design and in architecture research. During the design process, performance estimates from simulations of various levels of detail are used to motivate architectural decisions. Simulators are also heavily used research tools, because they provide a relatively inexpensive means of exploring new ideas.

After power has established itself as a significant metric in microprocessor design, there has been a tendency to produce simpler cores because of their better performance per watt. Simple, in-order cores have been the norm for mobile applications, but recent industrial designs [10, 14, 15] explore the opportunity of using multiple such cores for parallel workloads found in datacenters, multimedia or networking. The simulator we developed, XIOSim, is intended as an integrated model of such simple in-order cores. Unlike other microarchitectural simulators used in the research community [2, 3], XIOSim targets the IA-32 instruction set due to its widespread adoption. To the best of our knowledge, it is the only publicly available (`http://xiosim.org`) detailed microarchitectural model of in-order IA-32 cores.

Genealogically, XIOSim was derived from the Zesto x86 user-level simulator [8], which in turn originates from SimpleScalar [3]. XIOSim extends the Zesto framework by providing a different, binary instrumentation-based functional execution model, a detailed in-order core model, and by integrating very closely with the McPAT power models [7] for producing power consumption traces.

In relation to other architectural simulators, we believe XIOSim has a niche in the design space. As opposed to other frameworks that support the IA-32 instruction set [2, 11, 16], it is not geared towards full-system simulation, but rather focuses on user-level, CPU-intensive workloads in isolation. Furthermore, its core models, albeit slower, are very detailed and strive for accuracy. While we do not go as far as to claim cycle-accuracy, such a level of detail allows us to validate the simulator against a real machine with Intel Atom cores across a statistically significant sample of the 29 SPEC CPU2006 benchmarks. This allows architectural exploration that is grounded on current industrial designs.

Finally, XIOSim's execution model can support executing workloads that are out of reach for traditional user-level simulators. By using the host hardware as a functional model, it can continue correct simulation in the presence of instructions or system calls that are not explicitly supported by the performance model. Thus, in implementing XIOSim there is no need for heroics such as modelling every instruction in the complex IA-32 instruction set, or explicitly capturing every POSIX system call.

In order to make sure that XIOSim is reasonably accurate, we have developed a validation methodology that enables comparing performance and power characteristics of short benchmark fragments. This allows us to compare power and performance of the exact same instruction sequences both running on real hardware and being simulated with a significant slowdown. Doing such a comparison eliminates the need for sampled fragments to be representative of full program execution and allows for a fair validation procedure.

The contributions of this paper are as follows:

- A simulator execution model that executes instructions on the host hardware and observes their effects through binary instrumentation.

ISLPED'12, July 30–August 1, 2012, Redondo Beach, CA, USA.

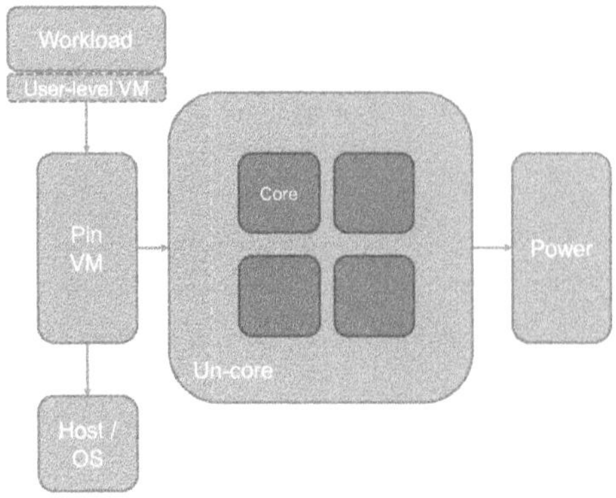

Figure 1: XIOSim execution model.

- A detailed methodology for validating simulators against real hardware based on running full benchmarks, even under a significant speed difference.
- A publicly available very detailed performance model for in-order IA-32 cores, augmented with a power model.

In the beginning of this paper, Section 2 motivates a simulator execution model based on instrumenting instructions running on real hardware. Section 3 describes a mobile IA-32 core model that runs under this execution model. Section 4 describes the extensive validation methodology that we used while developing the power/performance models for XIOSim and finally, Section 5 presents validation results.

2. EXECUTION MODEL

The simulator execution model we propose is a mixture between traditional user-level simulation and full-system simulation. XIOSim is a user simulator in that it does not model instructions past system call boundaries. However, it runs as a pintool, under the Pin virtual machine [9], which translates application instructions to run natively on the simulation host.

Figure 1 sketches out this execution model showing the interactions between functional, performance and power simulation. The user-level workload is run under Pin control, with the Pin VM instrumenting every instruction and capturing the host machine state immediately before it. That instruction context is passed to the performance model, which is expecting it at the fetch stage of the modelled pipeline, effectively achieving execute-at-fetch semantics. At certain cycle intervals, simulated statistics are aggregated and passed to a modified version of McPAT [7] to create a power consumption trace.

The closest analog to this execution model is implemented in PTLSim [16] and MARSS [11]. In these simulators, the performance model is run as an extension to a system-level virtual machine monitor or a full-system emulator. In our approach, keeping the simulator completely in user-level, while limiting the scope of benchmarks whose performance will be accurately modeled, is beneficial from a simplicity and a stability point of view. Full-system simulators need to run a complete OS image, virtual device drivers, etc., all of which affect simulation accuracy and reliability. For microarchitectural studies, it is often beneficial to ignore such behavior and only focus on the characteristics of the benchmark under test in isolation.

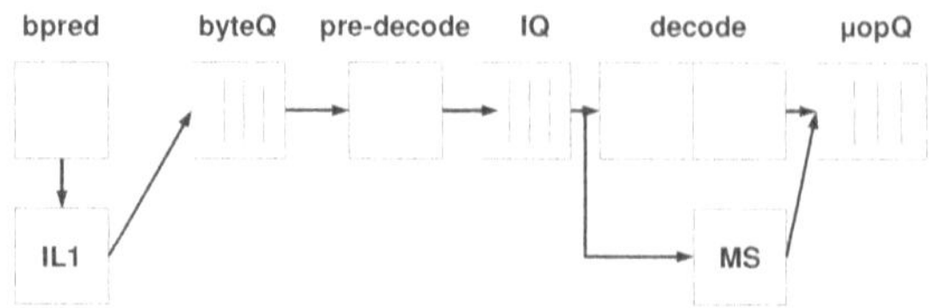

Figure 2: Front-end pipeline model.

Furthermore, this model has benefits over traditional user-level simulation. First, all system code, including system call handling, is directly executed on the host, freeing up the simulator from implementing a pass-through layer and keeping up with relatively fast-changing system call interfaces. This allows simulating a broader range of applications than traditional user-level simulation, where system call support is typically added based on the popularity of the particular call. For example, this has allowed us to simulate workloads under the ILDJIT user-level virtual machine [4] that translates CIL bytecode in IA-32 instructions.

Second, by running under an instrumentation engine, the performance model does not require a complete or correct functional model in order to maintain the correct execution of the workload. Since instructions are executed on native hardware, the performance model can only observe their effects without explicit knowledge of instruction semantics. This allows continuing execution even after encountering esoteric unsupported instructions, which is especially important for simulating a complex and constantly evolving instruction set like IA-32.

Finally, this execution model allows for close-to-native speeds of execution of code outside simulation regions of interest. This enables efficient skipping of non-interesting benchmark phases and integration with statistical sampling tools, such as the SimPoint-based PinPoints [12].

The major disadvantage with such an approach comes from speculative execution. Pin cannot instrument instructions on a speculative path since by definition they are not visible above the architecture level of abstraction. In order to deal with that, XIOSim keeps a legacy functional model that is only used on speculative paths. On simulating a mispredicted branch, this functional model is invoked, and the simulated core continues utilizing it until the branch is resolved in the execute portion of the pipeline. At this point, the functional model is switched back to the instrumentation engine. Note that, to ensure correctness, the legacy functional model does not need to be completely correct or accurate, since the mispredicted branch will get resolved regardless of the instructions on the speculative path.

3. PERFORMANCE MODEL

In this section, we describe the in-order core performance model used in XIOSim. Overall, we model an in-order, multi-issue x86 core, a proxy for Atom-style microarchitectures [5]. Its pipeline is typically deeper than the ones found in mobile cores, partially due to the complexities involved in decoding a variable-length ISA, as well as due to the desire for efficiently handling CISC instructions with memory operands without out-of-order execution.

The front-end model, as seen in Figure 2, is typical for an IA-32 architecture. The pipeline starts with branch prediction, which directs accesses to the instruction cache. It

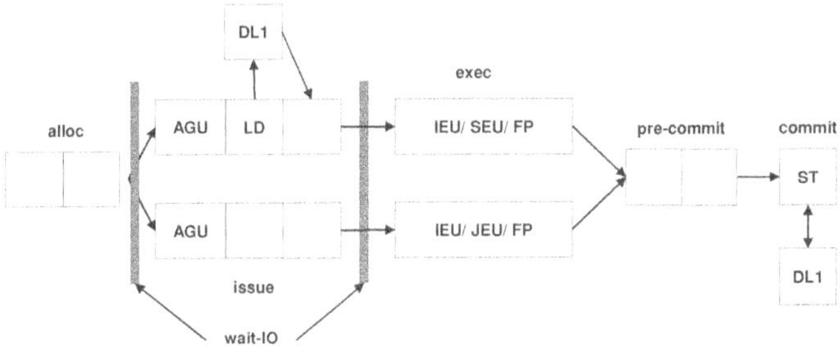

Figure 3: Back-end pipeline model, configured to model an Atom-like architecture.

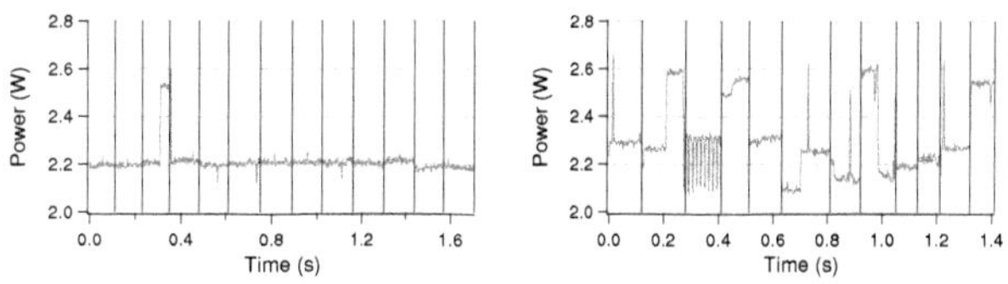

(a) *453.povray*

(b) *447.dealII*

Figure 4: Power traces with qualitatively different behavior, but similar average power.

continues with explicit pre-decode stages, whose aim is to predict instruction boundaries in the raw byte stream that was fetched. Pre-decoded instructions then propagate to the asymmetrical decoders, which break the possibly complex instructions into RISC-like μ-ops. The decoders can process multiple instructions in parallel, in order to keep the rest of the multiple-issue pipeline occupied. Complex instructions are sent to the micro-sequencer for decoding, which incurs an additional penalty of several cycles.

The decoders make heavy use of μ-op fusion – combining multiple μ-ops from a single instruction, such that they progress together through the pipeline, not incurring additional latency and occupying less space in queue structures. Out-of-order x86 microarchitectures have traditionally used load-op fusion, which combines a load with its dependant subsequent operation. In a deep in-order pipeline not incurring latency costs is even more important and fusion can be applied more aggressively – combining loads, their dependant operations and a store from a complex, memory-reference instruction. This results in being able to execute a large fraction of instructions in a CISC-like fashion.

Decoded and possibly fused μ-ops then enter the allocation stages of the pipeline (see Figure 3), where they access the register file, and get assigned to an execution port based on functional unit availability and port loading. Because decoding latencies can vary and different execution ports can stall independently of each other, there is an explicit in-order synchronization point at the end of the allocation pipeline. Its purpose is to enforce program order of μ-ops entering the execution ports, so that data dependencies can be observed without any of the complexities of an out-of-order pipeline.

Once in an execution port, μ-ops go through issue stages, mostly dedicated to memory accesses. These include address generation, load dispatch and return from the data cache. Having dedicated pipeline stages for processing loads is in unison with μ-op fusion, effectively achieving a zero-cycle load-to-use penalty for load-op or load-op-store fusions.

There is another in-order synchronization point at the end of the issue stages. Since all execution latencies after it are deterministic, this point ensures that μ-ops are properly serialized for in-order commit. Once executed and serialized, μ-ops enter a pre-commit pipeline stage which models exception handling. Then they follow to the commit buffer, where stores are sent to the data cache and individual μ-ops are composed back to macro instructions in order to commit atomically, observing instruction boundaries.

4. VALIDATION METHODOLOGY

We will show that the performance model described earlier can be configured to match a real design and is therefore useful as a performance exploration tool, as well as a power model driver. We start by presenting the extensive validation methodology that we used for developing XIOSim.

4.1 Fine-grained comparison

The major difficulty in validating a detailed model against real hardware arises from the discrepancy in speed. While real hardware can often commit instructions at rates of 10s billions instructions per second (BIPS), a detailed performance model typically achieves 10-100s kilo-instructions per second (KIPS). Simulating full real-world benchmarks is therefore prohibitively slow and a sampling approach has to be used. This presents two possible reference points for the sampled benchmark run: (i) either data gathered from the full benchmark execution, or (ii) collected data from a similarly sampled run on a real machine.

The first approach is conveniently simple, but has several issues. First, when doing sampling, the short execution slices are selected with a particular metric in mind. For example, SimPoints [13] targets end performance and the selected slices can be used to estimate the full program's execution time. However, this does not guarantee that SimPoint-derived slices are representative of the full program's branch behavior or power consumption, since these metrics are not always strongly correlated with performance.

Furthermore, even when used with the appropriate metric, the predictions from sampling are within some confidence interval of the full-length values. Patil et al. [12] demonstrate that this error can be as high as 10% for predicting CPI with SimPoints. This makes it impossible to separate simulator errors from sampling errors in a validation scenario.

For these reasons, we choose the second validation approach. Since sampling is inevitable when dealing with the slowdown of detailed simulation, we do sample with end performance in mind for our simulated runs. For the reference runs on a real machine, we use a very similar sampling approach, making sure we gather data only during the same instruction intervals that are simulated. The difficulty in this case lies in gathering data for a short period (measured in millions of instructions) without either modifying the workload source code, or perturbing the experiment with the measurement code.

Without source modification, one can use binary instrumentation to trigger the appropriate measurements once execution reaches the simulated region. However, the instrumentation must be light enough in order not to introduce errors in measurements on the host machine. For example, in the Pin virtual machine even inserting a simple basic

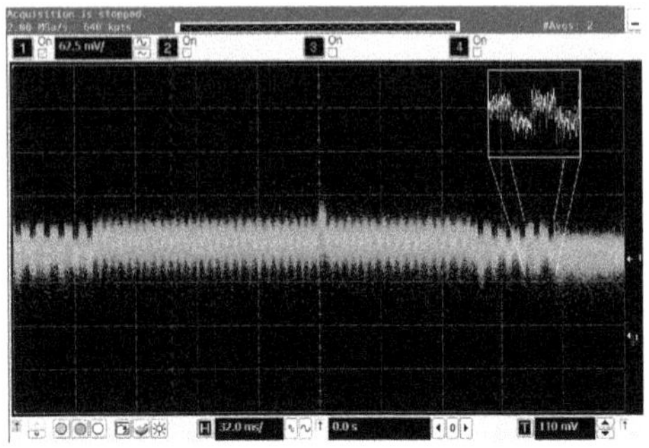

Figure 5: Sample power trace corresponding to a single slice between power markers. Inset shows an expanded view of the pattern used for marking the start and end of slices.

block counting instrumentation routine can have overheads in execution time on average exceeding 100% [9].

In order not to pay this penalty, we use Pin to insert lightweight trampolines that effectively replace routines of interest with versions that have thin instrumentation code inserted. The instrumentation code tracks function call counts and determines when it is appropriate to start or stop data collection. The routines of interest are defined as the closest ones to the beginning and the end of simulation slices, as identified by SimPoints, and the target call counts for them are gathered during an initial profiling run.

The stub code that we insert is a hand-optimized assembly sequence tracking the number of times the routine has been called and toggling data capture when appropriate. The measured average instruction overhead of this scheme over the SPEC CPU2006 suite is 0.7%. For performance validation, the stub code triggers performance counter collection on the host machine through the Linux perfmon2 interface. Its behavior in power validation mode is described in the following section.

4.2 Targeting power

Measurement setup. Our particular power capture system taps in the processor 12V power supply line and measures current consumption through a low-impedance high-precision sense resistor. At a fixed nominal supply voltage this provides enough data to capture processor power consumption. The traces are collected using a high-end Aglient DSA-91304A oscilloscope and a DSA-3100 differential probe. Trace collection is triggered remotely over the network. The high-end collection system allows us to gather power traces for a short amount of time with enough resolution to capture intra-slice power behavior. The sampling rate used in the following experiments is set to 160 kHz.

The benefit of being able to capture single execution slices can be seen in Figure 4, which shows power traces for samples of the SPEC benchmarks *453.povray* and *447.dealII*. If we average power over all execution slices, the difference between the two benchmarks is only 4%. However, in order to perform any meaningful power validation, we need to be able to capture and compare individual slice behavior for benchmarks like *447.dealII*.

Aligning code and measurements. Since executing a single sampled slice on a real machine can take time on the order of 10ms, which is comparable to the network delay to trigger power capture, explicit measures have to be taken to synchronize the power trace collection with the slice execution. We use the low-overhead instrumentation described in Section 4.1 to insert power markers at the beginning and end of a sampled slice. The markers are repeating sequences of a hand-crafted power virus instance (derived from CPUBurn and modified for an in-order pipeline) and suspending the core by executing a `usleep()` system call. By sandwiching the slice execution with long enough markers, we can mitigate the various variable delays and latencies in the measurement system and be also sure that the correct execution slice has been captured. A sample trace captured with this methodology is shown in Figure 5, with two iterations of a power marker in the inset.

In a post-processing step, the power markers are programmatically removed from the collected trace, so that only the slice region-of-interest is accounted for. The marker removal algorithm operates by searching for the specific frequency footprint of the power markers.

5. VALIDATION RESULTS

We applied the validation methodology described in the previous section to the XIOSim implementation. Measurements were performed on a real system with a dual-core Intel Atom 330 processor, with only one core active. Hyper-threading was also disabled during all experiments.

Since XIOSim focuses on detailed modeling of the core microarchitecture, the benchmark suite we use is the CPU-centric SPEC CPU2006 [1]. The sampled slices are collected using PinPoints [12], with a slice length of 100M instructions. Over the whole benchmark suite, this results in $\approx$500 slices, totalling $\approx$50B simulated instructions.

5.1 Performance

The in-order performance model is reasonably accurate. This is demonstrated in Figure 6. It shows measured versus simulated data for overall IPC, as well as for the major microarchitectural events. The dashed unit lines in each subplot represent the ideal case, and each circle marker is measured from a single 100M instruction execution slice. In an ideal, cycle-accurate model, all markers should be clustered along the unit lines.

Figure 6a shows that the performance model tracks end-to-end performance with satisfactory accuracy – the geometric mean of the IPC error over all slices is 11.02%. Furthermore, if we take into account the relative importance of each slice as computed by SimPoints, the mean IPC error decreases to 7.67%. Looking at the distribution of markers in Figure 6a, the simulator is not systematically underestimating or overestimating end performance.

The relative accuracy in predicting individual microarchitectural events is lower, as seen by Figures 6b-6f. For example, the average error in instruction cache miss rates (Figure 6e) is as high as 74%. The explanation of this discrepancy comes from the absolute values of these events – they are often small numbers. In that case, the error between a simulated miss rate of 0.1% and a measured rate of 0.2% is 50%, but the end impact on performance is insignificant because both rates are sufficiently small. This is especially true for miss rates in the instruction cache and the TLB, which are, on average 0.3% and 0.7%.

While most of the event errors also appear unbiased toward under-counting or over-counting, there are two distinct clusters of slices in Figure 6c and Figure 6e, where the

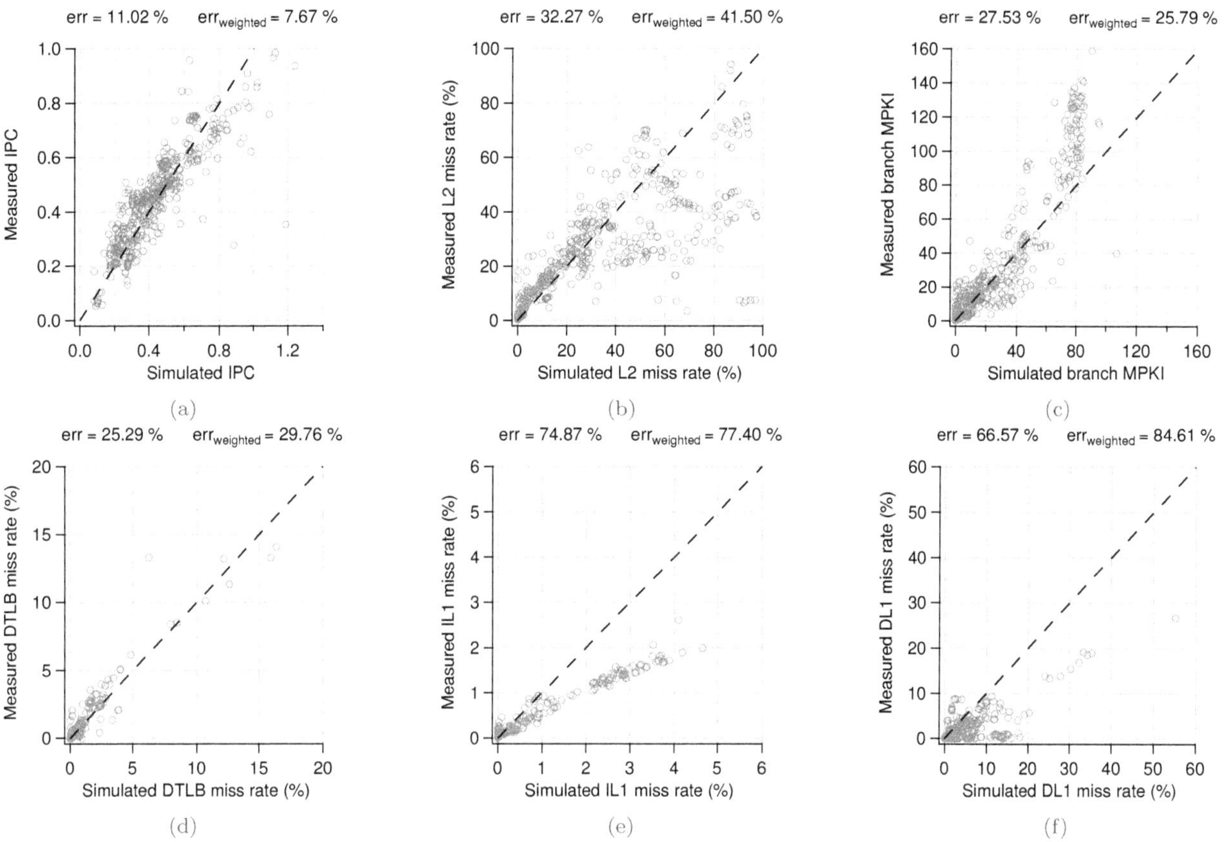

Figure 6: Comparison of microarchitectural statistics between performance counter measurements and simulation for execution slices of the SPEC CPU2006 suite.

simulator underestimates branch mispredictions and overestimates instruction cache misses. Most of those slices can be traced to the benchmark *445.gobmk*, which has been shown to have high branch misprediction rates [6]. On a real machine, the higher number of branch misses can have a prefetcher-like effect explaining the lower miss rates in the instruction cache.

5.2 Power

It is reasonable to expect that the power model is less accurate than the performance model. We first show that it can qualitatively track benchmark behavior on a real machine. We then go on to validate the quantitative predictions for power consumption, starting with leakage estimates, and continuing with the sum of static and dynamic power.

Workload tracking. The methodology described in Section 4.2 allows us to demonstrate that the power model can adequately track workload-related activity. Figure 7 shows that by comparing measured versus simulated power consumption for the benchmarks *400.perlbench* and *401.bzip2*. Vertical lines denote execution slice boundaries. Even though the two benchmarks show qualitatively very different power profiles, XIOSim's power model tracks the shape of both relatively well. As with any model, the synchronization is not perfect – for example, notice how the first simulated slice of *400.perlbench* finishes significantly before its corresponding measured slice. In this particular instance, the misalignment comes from a large IPC error in the performance model, which predicts execution time wrongly.

Leakage. In order to quantify the model accuracy, we start by estimating the leakage of our real machine with a simple experiment. While our test processor is executing an instance of the power virus, we scale the frequency from the maximum supported 1600 MHz to the minimum 200 MHz at 200 MHz increments, measuring average power consumption at each step. Since we are only scaling frequency and not voltage, power consumption scales linearly, and we can estimate the static power as the portion independent of frequency in a linear fit.

Performing this experiment, we estimated that the Atom 330 consumes 1.79 W static power. Our power model leakage estimate is 1.4 W. However, it does not include any overheads associated with simultaneous multi-threading (SMT). Even though we keep SMT disabled through our experiments, it still incurs a leakage cost because it requires additional queues and buffers, or additional access ports to existing structures. Intel data [5] suggests that the power cost of adding SMT to an Atom core is 15%. Adding this overhead to our leakage estimates results in 1.61 W in static power, which is within 10.1% of the measured value. It is logical that the model underestimates static power – a real design includes components that are not explicitly modelled, such as PLLs, bus drivers and memory buses, which could consume a non-ignorable amount of energy.

Total power. After incorporating the leakage correction for SMT, we can look at total processor power across benchmarks. We collapse the dynamic behavior of each execution

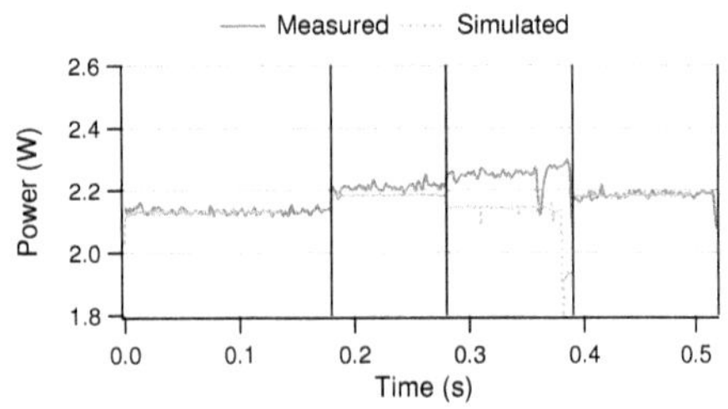

(a) *400.perlbench*

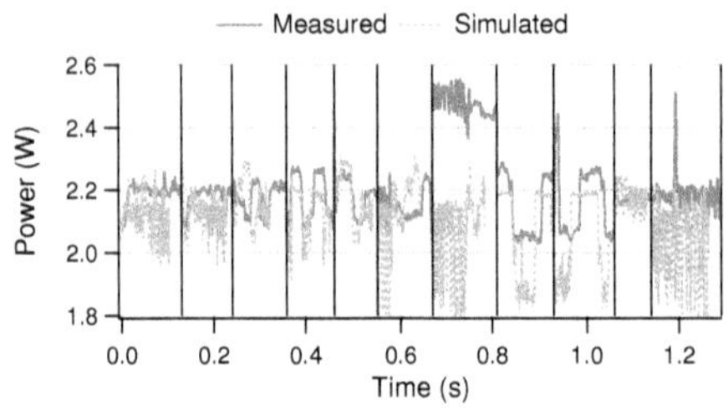

(b) *401.bzip2*

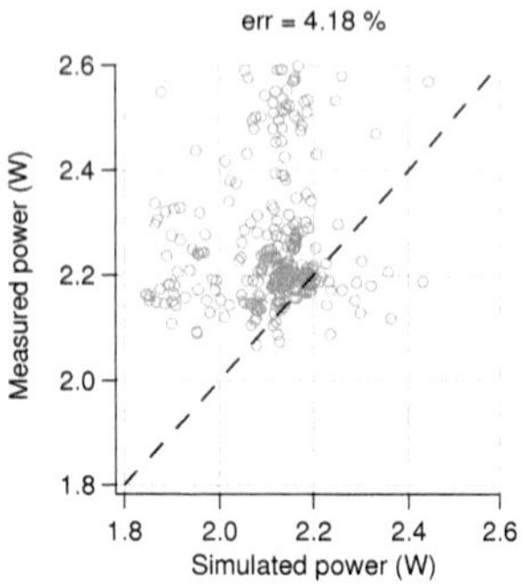

(c) Aggregate for CPU2006

Figure 7: Comparing measured and simulated power consumption for execution slices of SPEC CPU2006.

slice to a single average power value and compare measured and predicted per-slice average power in a manner similar to Section 5.1. Figure 7c shows the results of this validation procedure. The geometric mean power error is only 4.18% (after accounting for SMT). However, as seen in Figure 7c, a large fraction of execution slices cluster around an average power consumption of 2.2 W. We attribute this low variability to the fact that the core under test is in-order and relatively simple.

From Figure 7c, we can clearly notice that the simulation systematically underestimates power consumption. As discussed in the previous paragraph, this is expected behavior, since we do not model all components on a real system. There is a distinctive set of slices with measured power between 2.4 and 2.6 W which correlate very poorly with simulation results. One such example is the 7-th slice in Figure 7b. We have not been able to identify the reason for the discrepancy, but possible candidates are the operating system timer triggering on the real system, or the voltage regulator triggering a feedback loop.

6. CONCLUSION

With the growth of the mobile sector, simple power-efficient cores are becoming increasingly more relevant, even with instruction sets as IA-32 that have been traditionally reserved for high-performance, high-power domains. We have presented XIOSim – a simulation tool that is able to model the power and performance of such cores. In order to validate its correctness, we have developed a rigorous methodology and an experimental setup that allows comparing the same execution slices on a real machine and in simulation.

This paper explicitly focuses on single-core modelling and validation, leaving the multi-core case for future work. In the near future, we are planning to extend our simulation methodology and infrastructure to be able to handle heterogeneous multi-core configurations, keeping true to a rigorous validation approach.

Acknowledgements

This work was partially supported under SRC contracts with task ID 1814 and 1973, as well as partially supported by National Science Foundation grants CCF-0903437, CSR-0720566 and CCF-0702344. Any opinions, findings, conclusions, or recommendations expressed are those of the authors and do not necessarily reflect the views of the SRC or NSF.

7. REFERENCES

[1] SPEC CPU2006, `http://www.spec.org/cpu2006/`.

[2] N. Binkert et al. The GEM5 simulator. *ACM SIGARCH Computer Architecture News*, 2011.

[3] D. Burger and T. Austin. The SimpleScalar tool set, v. 2.0. *ACM SIGARCH Computer Architecture News*, 1997.

[4] S. Campanoni et al. A highly flexible, parallel virtual machine: design and experience of ILDJIT. *Softw., Pract. Exper.*

[5] G. Gerosa et al. A Sub-1W to 2W Low-Power IA Processor for Mobile Internet Devices and Ultra-Mobile PCs in 45nm Hi- κ Metal Gate CMOS. In *2008 IEEE International Solid-State Circuits Conference.*

[6] A. Kejariwal et al. Comparative architectural characterization of SPEC CPU2000 and CPU2006 benchmarks on the intel Core2 Duo processor. *2008 International Conference on Embedded Computer Systems: Architectures, Modeling, and Simulation.*

[7] S. Li et al. McPAT: an integrated power, area, and timing modeling framework for multicore and manycore architectures. In *42nd International Symposium on Microarchitecture (MICRO-42)*, 2010.

[8] G. H. Loh and S. Subramaniam. Zesto: A Cycle-level Simulator for Highly Detailed Microarchitecture Exploration. *2009 IEEE International Symposium on Performance Analysis of Systems and Software.*

[9] C.-K. Luk et al. Pin: Building Customized Program Analysis Tools with Dynamic Instrumentation. In *PLDI '05: Proceedings of the ACM SIGPLAN 2005 Conference on Programming Language Design and Implementation.*

[10] H. McGahn. Niagara 2 Opens the Floodgates. *Microprocessor Report*, 2006.

[11] A. Patel et al. MARSS: A full system simulator for multicore x86 CPUs. In *Design Automation Conference (DAC)*, 2011.

[12] H. Patil et al. Pinpointing Representative Portions of Large Intel Itanium Programs with Dynamic Instrumentation. In *37th International Symposium on Microarchitecture (MICRO-37)*, 2004.

[13] T. Sherwood et al. Automatically characterizing large scale program behavior. In *10th International Conference on Architectural Support for Programming Languages and Operating Systems (ASPLOS)*, 2002.

[14] S. Vangal et al. An 80-tile Sub-100-W Teraflops processor in 65nm CMOS. *IEEE Journal of Solid-State Circuits*, 2008.

[15] D. Wentzlaff et al. On-chip Interconnection Architecture of the Tile Processor. *IEEE Micro*, 2007.

[16] M. Yourst. PTLsim: A Cycle Accurate Full System x86-64 Microarchitectural Simulator. In *International Symposium on Performance Analysis of Systems & Software*, 2007.

LogStore: Toward Energy-Proportional Storage Servers

Wei Zheng Ana Paula Centeno Frederic Chong[†] Ricardo Bianchini

Rutgers University
{wzheng,anapaula,ricardob}@cs.rutgers.edu

[†]University of California at Santa Barbara
chong@cs.ucsb.edu

ABSTRACT

Storage servers consume significant amounts of energy and are highly non-energy-proportional. Fortunately, researchers have proposed disks that adjust their rotation speeds based on utilization. However, these disks either pose significant engineering challenges (by having too many speeds) or adjust behavior only at a coarse grain (by having just two speeds). In this paper, we propose LogStore, a storage system that enables two-speed disks to achieve substantially increased energy proportionality and, consequently, lower energy consumption. Our evaluation with real I/O workloads shows that LogStore can decrease energy consumption significantly, even for very tightly provisioned servers, with a negligible impact on aggregate throughput.

Categories and Subject Descriptors

D.4 [**Operating systems**]: Storage management

Keywords

Energy conservation, multi-speed disks

1. INTRODUCTION

Many enterprises use storage servers, each of which with many high-performance disks and RAID redundancy, behind their database and/or file servers. These servers provide fast and reliable access to data, but also consume significant amounts of energy. For example, small-sized storage servers from multiple vendors provide 16-24 drive bays. When using fast SCSI disks, each of these servers can consume in excess of 400W just from the drives. Larger storage servers can have hundreds or even thousands of disks, which typically represent more than 50% of the total system power. For example, the EMC Symmetrix VMAX [7], an entry-level storage server, is rated at more than 8KW with disks representing 60% of this total power.

High peak power is not a problem in itself, as peak I/O loads occur rarely. The worse problem is that storage servers are highly non-energy-proportional, i.e. they consume a large fraction of their peak power even at low utilization. Recent years have seen advances in processor energy proportionality [11], as well as proposals for improving memory

ISLPED'12, July 30–August 1, 2012, Redondo Beach, CA, USA.

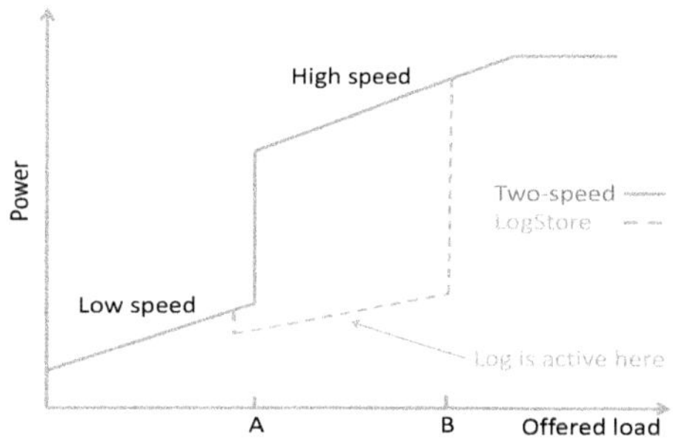

Figure 1: Main idea of LogStore.

energy proportionality [4, 6]. Even if storage servers take full advantage of these innovations, they will still exhibit limited proportionality due to their use of conventional disks.

The reason is that the energy consumption of conventional disks is dominated by the energy required to keep them spinning [9]. In fact, disk energy is a quadratic function of the rotation speed [9]. Based on these observations, researchers have proposed disks that can operate in one of two [2] or more speeds [9]. Under lighter loads, these disks can dynamically change speed and serve requests at lower speed. A few years earlier, Sony had produced a two-speed disk that required a re-start to change speeds [12]. Since these early works, some vendors started producing two-speed disks that can dynamically change speeds, e.g. [10]. However, these disks only perform accesses at the high speed; the low speed is used solely to conserve energy during periods of idleness. Since servers rarely exhibit idle periods that are long enough to justify speed changes, existing two-speed disks are useless for storage servers.

We argue that disks capable of serving accesses at multiple speeds are the key to disk energy proportionality, in the same way that processors capable of Dynamic Voltage and Frequency Scaling (DVFS) are more energy proportional. Our preference is for two-speed disks, since implementing more speeds poses engineering challenges, e.g. adjusting sensing mechanisms to the many distances between the disk head and the magnetic medium, and high manufacturing costs. However, two-speed disks allow only coarse proportionality.

Our goal is then to create *software* to improve the energy proportionality of two-speed disks that can serve accesses in both speeds, and thereby reduce energy consumption. Our storage system, called LogStore, increases energy proportionality by storing data into a (disk-based) log data structure at the utilization boundary between the two speeds. The improved write performance and lower disk utilization resulting from the log enable disks to stay in low speed longer. Figure 1 illustrates the main idea, showing disk power consumption as a function of the disk request rate (utilization). The full curve represents a sample two-speed disk, whereas the dashed curve represents the same two-speed disk when managed by LogStore. As we can see, LogStore allows the disk to serve a higher request rate still in low speed. Under LogStore, the speed transition occurs at offered load B, rather than offered load A. LogStore seeks to delay the transition without affecting throughput. However, the delayed transition means that more requests are served in low speed, which may increase the average response time.

LogStore replays the log against the original disk blocks when utilization is low. Since a disk cannot be accessed during a speed transition, storage systems with two-speed disks require data redundancy. When a transition is necessary, all accesses to the transitioning disk are directed to the disk(s) containing the redundant data. We designed LogStore for RAID 1, since it enables the highest energy savings. Nevertheless, LogStore can be easily adapted to any RAID level that uses block mirroring (e.g., RAID 1+0, RAID 0+1). We leave other RAID levels as future work.

We built a prototype of LogStore in user-space. To mimic two-speed disks that can serve accesses in both speeds, we also built a two-speed disk emulator (via delay injection) into the Linux kernel. We compare LogStore with RAID 1 implemented using: (1) conventional one-speed disks; (2) two-speed disks without LogStore; and (3) two-speed disks plus Diverted Accesses (DIV) [13]. Our results for real write-intensive workloads show that LogStore can decrease energy consumption by roughly 35%, compared to conventional disks, when the storage system is very tightly provisioned for the workload. Compared to two-speed disks without and with DIV, these savings are roughly 20% and 27%, respectively. Under more realistic provisionings, the LogStore energy savings can reach 81%, 57%, and 73%, respectively. Our results for these workloads also show that LogStore has a negligible impact on aggregate throughput. However, as expected, it does degrade response time somewhat (up to 2.4ms). For read-only workloads, LogStore behaves as well as the best of its competitors, still achieving the highest energy savings, having no impact on aggregate throughput, and minimally increasing response times. Thus, LogStore behaves well for throughput-oriented workloads, regardless of their write intensity.

2. BACKGROUND AND RELATED WORK

Two-speed disks. We study two-speed disks that are similar in functionality to those proposed in [2], but with contemporary seek performance. As in that study, the disks can serve requests from both speeds and speed transitions are based on utilization. However, we assume transitions triggered by the kernel, rather than the disk controller. For the kernel to select speeds, it queries the disks about their utilizations. We measure utilization as the disk busy time as a fraction of the wall-clock time. As suggested in [2], we take two measures to improve stability and reduce the impact of extraneous load spikes: filtering and smoothing of utilization observations. We filter the first observation that is more than twice as high/low as the previous one. After this first disregarded spike/valley, the filter takes new high/low observations into account. Using those observations that pass the filter, we smooth them using an exponentially weighted moving average with alpha equal to 0.875 [2].

Disks with two or more speeds continue to draw attention, e.g. [8, 22, 23]. However, a common concern is reliability. We argue that this concern is unfounded for two reasons: (1) some commercial server disks already switch to a lower speed when they become idle (e.g., [10]); and (2) commercial server disks are capable of many hundreds of thousands of stop/start cycles (which impose a heavier burden than speed transitions). For example, many Hitachi disks (e.g., [10]) are rated at 600,000 cycles. Since each disk is typically replaced after 3 years of use, the system would have to change speeds more often than once every 158 seconds on average to exceed the maximum number of cycles during the disk's lifetime.

Although commercial two-speed disks do not serve accesses in both speeds, there is no fundamental reason why disks that can do so cannot be manufactured [9]; the barrier is in achieving a high benefit-to-cost ratio. LogStore demonstrates that system-level techniques can significantly increase the potential benefits of such disks, creating a greater incentive for manufacturers to produce them.

Storage energy conservation. Many works have proposed techniques for reducing disk energy consumption, e.g. [2, 3, 9, 13, 17, 19, 20, 21, 24]. Zhu *et al.* [24] have exploited two-speed disks in enterprise storage servers as we do. However, they changed speeds only at the granularity of hours, relying on data migration for finer-grained adjustments. Our system relies on more frequent speed transitions and avoids data reorganization altogether due to their potentially high energy and performance overheads. Pinheiro *et al.* [13], Weddel *et al.* [20], and Yao and Wang [21] studied how to leverage redundancy to conserve energy using conventional disks. Unfortunately, the use of conventional disks limits the energy proportionality and savings that can be achieved in storage servers. Verma *et al.* [19] used selective data replication to improve proportionality. However, replication can be a significant overhead when workloads have large working sets and/or change frequently.

Still leveraging redundancy, Amur *et al.* [1] and Thereska *et al.* [18] focused on clustered storage systems. They achieve proportionality by intelligently placing data replicas across servers (their conventional disks). The number of disks and the amount of redundancy are typically larger in a clustered system than in a single storage server, enabling finer-grained energy proportionality and greater savings.

3. LOGSTORE

3.1 Design

The key observation behind LogStore is that sequential disk access leads to higher disk throughput, and lower response time and occupancy. Thus, if we could transform many random disk accesses into sequential ones near a transition to high speed, we would not need to effect the transition until a higher load is offered to the disk. A well-known way to achieve this transformation is by using an on-disk log for writes [14]. *LogStore uses one log per disk.* Unfortunately, a log may also make some disk reads more expensive, since sequential read accesses may become random when directed to the log. LogStore mitigates this problem in two ways: the log is only used when the disk is about to transition to high speed; and LogStore sits behind the file system buffer cache, allowing most reads to hit it. In our terminology, each block has a "regular location" selected by the file system and possibly a (temporary) "log location".

In more detail, LogStore implements three modes of operation: regular, log, and log-replay. In the regular and log-replay modes, LogStore writes data to their regular

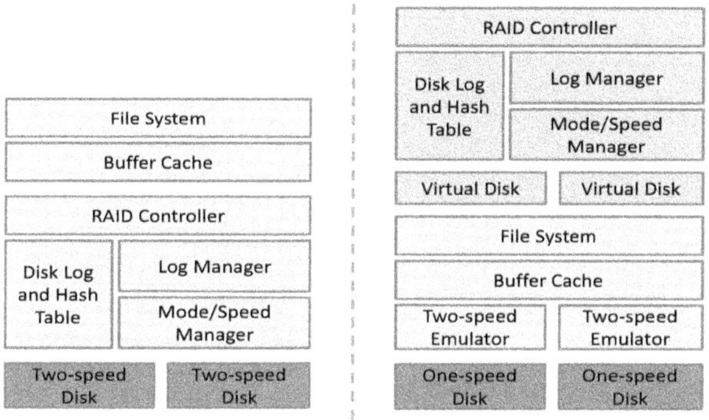

Figure 2: Design (left) and prototype (right). Green = user space; Light blue = kernel; Dark blue = hardware.

locations. In the log mode, it writes data to the end of the log. In the log-replay mode, LogStore replays the log against the blocks' regular locations. We detail these modes and the transitions between them below.

LogStore comprises five components: the on-disk log for each disk, an in-memory hash table for each log, a log manager, a software RAID controller, and software mode/speed manager. Figure 2(left) illustrates these components. As aforementioned, we design LogStore for RAID 1, but it can be easily adapted to handle other RAID levels that use block mirroring. In fact, the RAID controller is the only piece of LogStore that depends on the RAID level.

Data structures. LogStore maintains two data structures for each disk: the on-disk log and an in-memory hash table. The space for the log is statically allocated. When a disk is in the log mode, LogStore appends all block writes to the disk's log. In our experiments, we never need more than 3 GBytes for a log. (LogStore is not a log-structured file system; it does not implement log cleaning or compaction.) Since the log is just an optimization, our implementation limits each log to this size. Each log entry contains the address of the written block and the written data itself. This is enough information to reconstruct the associated hash table, when recovering from a server crash.

Each log has a hash table to track the location of the logged blocks. Each table entry has a field for the block's regular address, one for the log address, and a pointer to any other entry hashing to the same slot. The table is small. For example, each 3-GByte log requires only 15 MBytes (when the log is full) for its table, assuming 4-KByte blocks.

Log manager. It is responsible for accessing the log and replaying it against the blocks' regular locations.

In the regular and log-replay modes, this manager directs a block write to its regular location. If the hash table shows that the log includes this block, the manager removes the table entry and invalidates the log entry. The invalidation is needed for server crash recovery. A block read in these modes accesses the regular location, if the table does not have an entry for the block. If it does, the read is directed to the proper log entry. The modes differ in that this manager replays entries from the log against the regular locations in the log-replay mode, but not in the regular mode.

In the log mode, LogStore appends the disk writes to the log. If an entry for the same block exists in the hash table, the manager updates the table entry to point to the new log entry. If a table entry for this block does not exist, LogStore creates one. The manager handles reads in this mode in the same way mentioned above for the other modes.

In the log-replay mode, the log manager starts replaying the log from its beginning, using a separate thread, as soon as the mode/speed manager changes the mode to log-replay. For each block in the log, the thread checks the hash table. If it finds the block there, it writes the data to its regular location and removes the entry. Otherwise, it skips the log entry. The thread replays requests when disk traffic is low to avoid degrading foreground performance.

Software RAID controller. It forms the top part of LogStore. Our design for RAID 1 with two-speed disks allows the disks of each mirrored pair to spin at different speeds. The controller allows only one speed transition at a time for each pair. Write accesses are directed to both disks in the pair. For disks with non-volatile controller caches, it may be possible to buffer the writes in the caches, regardless of the disks' speeds. Whether this approach would work depends on the write rate and the size of the caches. As we discuss later, our prototype uses a different implementation that is more generally applicable.

The controller is oblivious to the modes of operation, so the mirrored disks can be in different modes as well. *This means that the pair of disks are mirrors in terms of content, but not in terms of the exact location of the latest version of each block.* For example, a write to a block may be written to its regular location on one disk and to the log of the other.

Before deciding on the destination disk for each read, the controller predicts what the utilizations would be in case both mirrored disks were at low speed. The "Balance" threshold defines the utilization starting at which at least one disk must be in high speed. If the sum of the utilizations is lower than this threshold (both disks can be at low speed), the controller evenly distributes the reads across the disks. If not (one disk at least must be in high speed), it concentrates the reads on one disk, only using the other when the first disk is nearly saturated.

LogStore uses the fraction of disk busy time as a measure of utilization. It assumes that the disks provide a standard interface (e.g., SMART [16]) by which the host can query the disk controller about the average busy time, and the average seek, rotation, and data transfer times. Using these data (and assuming the near-future will resemble the near-past), LogStore can predict what the utilization would be at low speed using the following equations: $T_{high} = T_{seek} + T_{rotation} + T_{transfer}$, $T_{low} = T_{seek} + \frac{S_{high}}{S_{low}} \times (T_{rotation} + T_{transfer})$, and $U_{low} = \frac{T_{high}}{T_{low}} \times U_{high}$, where T_{high} and T_{low} are the average access times at high and low speed, S_{high} and S_{low} are the high and low rotational speeds, and U_{high} and U_{low} are the high and low utilizations, respectively. Given U_{high}, we can compute U_{low} and vice-versa.

Mode/speed manager. It is responsible for setting the mode of operation and the rotational speed of each disk. It also divides the execution into epochs.

This manager uses a thread to select each disk's mode based on the policy in Figure 3. It considers 4 utilization thresholds set by the user: MaxRegular, MinRegular, MaxReplay, and MinLog. MaxRegular specifies the maximum utilization before we transition to the log mode. MinRegular specifies the regular-mode utilization up to which we switch to the log-replay mode. MaxReplay specifies the utilization up to which we replay the log. MinLog specifies the minimum utilization before we switch to the regular mode.

```
1. At the beginning of each epoch:
2.    Determine disk util in the last epoch
3.    If curr_speed == low && curr_mode == regular
4.       If util >= MaxRegular
5.          Change mode to log and exit
6.       If util <= MinRegular && log != NULL
7.          Change mode to log-replay and exit
8.    If curr_speed == low && curr_mode == log &&
          predicted util in regular mode <= MinLog
9.       Change mode to regular and exit
10.   If curr_speed == low && curr_mode == log-replay
11.      If log == NULL || util >= MaxReplay
12.         Change mode to regular and exit
13.   If curr_speed == high && curr_mode == log-replay
          && log == NULL
14.      Change mode to regular and exit
```

Figure 3: Policy for mode selection.

```
1. At the beginning of every epoch and more than 3
      epochs since the last speed transition:
2.    Determine disk util in the last epoch
3.    If curr_speed == low && util >= GoUp
4.       Change mode to log-replay
5.       Transition to high speed and exit
6.    If curr_speed == high && predicted util at
          low speed, log mode <= GoDown
7.       Change mode to log
8.       Transition to low speed
```

Figure 4: Policy for speed selection.

This manager may need to predict the utilization of a disk if it were to transition from the log mode to the regular mode (line 8). We predict this utilization by using the formulas above (replacing *high* for *log* subscripts), and the seek and rotational time statistics from before the disk went into the log mode to compute T_{low}.

If the manager does not change the mode, it uses the policy in Figure 4 to select disk speeds. To allow utilization to settle after a speed transition, it does not change speeds again for a few epochs. The GoUp and GoDown thresholds are also pre-defined by the user. They specify the utilizations at which the disk should be transitioned from low to high speed and from high to low speed, respectively.

3.2 Implementation

For simplicity, we implemented the LogStore prototype at the user-level on top of Linux, as depicted in Figure 2(right). Each disk is implemented as a large file, called "Virtual Disk", with 4KByte blocks. The functionality of the prototype components is as described above, except for the software RAID controller.

On a block read access, our prototype RAID controller uses a new system call we added to Linux to find out if the corresponding "mirror block" (i.e., the same block at its regular location on the other disk) is in the buffer cache. If it is, the controller performs the read to the other disk. If it is not in the buffer cache, the controller issues the request for the accessed block. This implementation achieves the same behavior as an in-kernel design (left side of Figure 2) would have. Unfortunately, the same cannot be done for block writes. Writes have to be performed to the two disks, causing two copies of the block to reside in the buffer cache. In addition, writes to the logs also take up space in the cache. Thus, our prototype caches less useful data than an in-kernel design would, i.e. the LogStore performance and energy results would be better for the latter system.

The fact that one of the disks can be transitioning speeds when a write arrives also complicates our RAID controller. Our prototype does not assume that disks have large non-volatile caches. Instead, the controller records each write in

Parameter	Low speed 3K rpm	High speed 10K rpm
P_{idle}	0.5 W	6 W
P_{access}	2.5 W	9 W
T_{change}	7 sec	7 sec
E_{up}	123 J	–
E_{down}	–	17 J

Table 1: Defaults for the two-speed disks. P_{idle} = power when idle; P_{access} = power when accessing; T_{change} = time for speed transition; E_{up} = energy to transition to high speed; E_{down} = energy to transition to low speed.

main memory while the write has not completed at both disks. In addition, the controller also flags a write as complete to the file system as soon as it is performed at either of the disks. If a disk is transitioning speeds when a write arrives, the controller performs the write at the other disk. When the transition completes, it performs the postponed write from the memory record. The prototype does not guarantee that all written data can be recovered after a server crash: blocks with pending writes at the time of the crash may be lost. Except for writes during speed transitions, this is the same semantics that existing software RAID controllers provide [5]. To deal with each transition, the prototype could perform two actions: complete all pending writes at both disks before starting the transition; and set (reset) a persistent flag before (after) the disk starts (ends) a transition. Because transitions are infrequent, these actions would have negligible overhead.

We emulate the disks we model in the kernel using conventional (one-speed) disks. The disks operate at 15K rpm, whereas we model a 10K rpm high speed and a 3K rpm low speed. Our emulation uses careful delay injection between the disk scheduler and the disk driver. Accesses are not allowed to complete until the proper time has elapsed. For accurate timing, we use a dedicated kernel thread that monitors time using performance counters and sleeps using the kernel's precision sleep functionality.

4. EVALUATION

4.1 Methodology

We use a dual-core Xeon server with 2 GBytes of memory, one 10K rpm disk (for the kernel), and two 15K rpm Maxtor Atlas II disks. On this server, we run a customized Linux kernel, version 2.6.18, with the ext2 file system. The kernel implements our system call for checking the buffer cache and our two-speed disk emulator (Section 3.2). We mount the file system with the "noatime" option to eliminate metadata updates that would unnecessarily move the disk arm.

We emulate two mirrored two-speed disks using the Maxtor disks. The speeds are 3K and 10K rpm. We list the disks' default parameters in Table 1. The 10K rpm values were measured by Carrera [2]. From those values, we extrapolated the 3K rpm values based on [9]. The seek time is that of the Maxtor disk itself, whereas the rotational and transfer times are inversely proportional to the speed. Note that the difference in average rotational time (1/2 of a rotation) between the two speeds we emulate is 7ms. We validated that the emulator achieves accurate timing by comparing its response times with those the emulated disks should achieve according to our performance models. In all experiments, the average difference between these times is less than 0.3%. The emulator computes the disks' energy consumption based on their behavior over time.

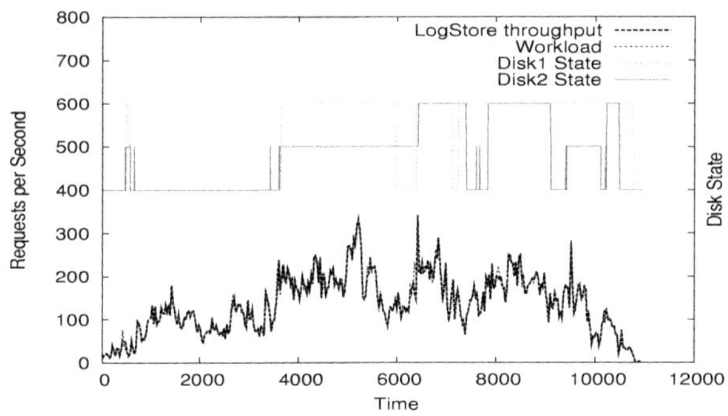

Figure 5: Financial load (4x) and disk state over time.

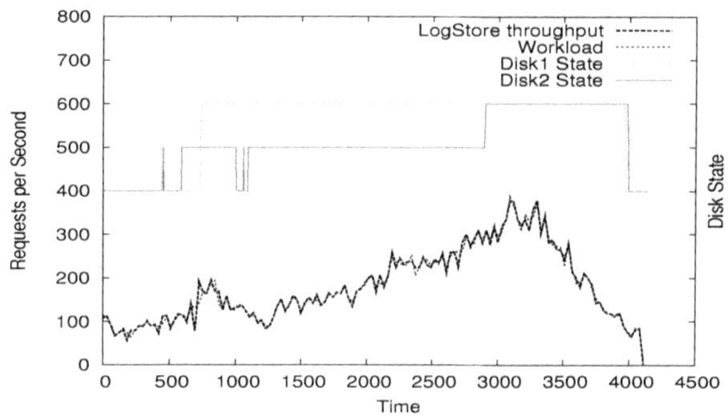

Figure 6: Proxy load (21x) and disk state over time.

Name	Num requests	Avg access size (bytes)	Write %	Write % after cache
Financial	1484205	2914	68%	93%
Proxy	750758	8510	43%	93%
Search	1544375	15218	0%	0%

Table 2: Trace characteristics.

We experiment with the real-world workloads described in Table 2. The first (Financial) is an OLTP trace from a financial institution [15]. The trace is write-intensive and dominated by small requests. The second (Proxy) represents a Web proxy and is also somewhat write-intensive [2]. The vast majority (93%) of the disk accesses in these workloads is writes. The third (Search) is a read-only Web search trace [15]. Since the disk traffic in these traces is light, our trace replayer accelerates (shortens the time between accesses) each of them by a user-defined factor to achieve a desired disk utilization. The maximum factor we use leads to a peak utilization of 90%. We also explore lower factors. Previous papers have also accelerated traces, e.g. [2, 20].

In our experiments, we set the LogStore thresholds to 60% (Balance), 50% (GoUp), 30% (GoDown), 50% (MaxRegular), 30% (MinRegular), 50% (MaxReplay), and 30% (MinLog). The epoch length is 3 seconds. We study the impact of these default values below.

For comparison, we study two energy conservation techniques: Two-speed disks with RAID 1 but not LogStore; and Diverted Accesses with two-speed disks and RAID 1. Diverted Accesses or simply DIV [13] concentrates all read accesses on a single disk and only uses the other if the first disk approaches saturation. The disk that receives most reads typically stays in high speed, whereas the other typically stays in low speed. We compare against DIV because it represents techniques (e.g., [1, 18, 20, 21]) that leverage redundancy to conserve energy. As a baseline, we study a RAID 1 system with conventional disks spinning at 10K rpm. All systems rely on the same code base and have the same reliability and availability characteristics.

4.2 Experimental Results

Figure 5 shows the offered load and LogStore throughput (Y-axis on the left), and disk states (Y-axis on the right) over time for the Financial (4x acceleration) trace. Each point in the graph corresponds to 30 seconds. We represent low speed as the lowest disk state, low speed in the log mode as the middle state, and high speed as the highest state. Figure 6 shows the same data for the Proxy (21x) trace.

We can see that, in both cases, one of the disks stays in the log mode for an extended period. The two-speed disk without LogStore transitions to high speed at those points and consumes substantially more energy. Specifically, in LogStore the disks operate in low speed 58% and 48% of the time in these Financial and Proxy experiments, respectively. In contrast, the same statistics are only 38% and 27%, respectively, when using two-speed disks without LogStore.

We can also see that the offered load and LogStore throughput curves are nearly indistinguishable, showing that LogStore has a negligible impact on throughput at this granularity. At substantially shorter granularities (not shown due to space constraints), the impact of queueing during speed transitions becomes more pronounced in both LogStore and two-speed disks without LogStore.

Table 3 details the disk energy (in KJoules) and average response time (in ms) of the Financial trace under each of the techniques we study. The table shows that LogStore conserves a substantial amount of energy compared to the one-speed results: the highest savings (81%) are for acceleration 2x, whereas the lowest (35%) for acceleration 6x. Regardless of the acceleration, LogStore does not affect the aggregate throughput. The average response time increases by 1.5ms at 2x acceleration and 0.6ms at 6x.

At 2x acceleration, LogStore consumes 57% and 73% less energy than two-speed disks without LogStore and DIV with two-speed disks, respectively. The response time losses are 0.7ms and 1.4ms again compared to these techniques. For the highest acceleration, LogStore consumes 18% and 27% less energy, while degrading response time by only 0.3ms and 0.6ms, respectively.

Interestingly, the intermediate acceleration (4x) leads to higher energy than the other accelerations for LogStore. With the higher load, the disks are more likely to transition to high speed. Whether this increases consumption depends on how long the disks stay in high speed (how much the greater acceleration shortens the execution).

Table 4 shows the same statistics for the Proxy trace. The LogStore energy savings are significant, 75% at 7x acceleration and 36% at 21x acceleration, compared to one-speed disks. Again, regardless of the acceleration, LogStore does not affect the aggregate throughput. The LogStore savings are also significant in comparison to two-speed disks without LogStore and DIV with two-speed disks. At the highest acceleration, LogStore still consumes 21% and 28% less energy than these techniques, respectively.

The results for these write-intensive workloads show that LogStore achieves its main goal: conserving energy while retaining aggregate throughput. LogStore degrades response time for three main reasons: (1) the additional buffer cache space taken by writes to the logs (this effect would not occur in an in-kernel implementation), which causes a higher read

Rate	LogStore		Two-speed		DIV		One-speed	
	E	RT	E	RT	E	RT	E	RT
2x	52.1	2.2	120.0	1.5	195.2	0.8	274.5	0.7
4x	76.3	1.7	99.5	1.3	115.9	1.0	142.9	1.0
6x	64.1	1.9	78.4	1.6	87.8	1.3	98.9	1.3

Table 3: Disk energy (KJoules) and average response time (ms) for the Financial trace.

Rate	LogStore		Two-speed		DIV		One-speed	
	E	RT	E	RT	E	RT	E	RT
7x	38.3	3.7	47.6	2.7	104.0	1.3	155.8	1.3
14x	54.3	2.2	55.9	2.2	66.2	1.7	82.0	1.7
21x	36.0	2.5	45.6	2.0	49.9	1.6	56.0	1.6

Table 4: Disk energy (KJoules) and average response time (ms) for the Proxy trace.

miss ratio; (2) the additional time disks spend in low speed (during this time, each disk access takes 7ms longer); and (3) the improved write latency (and potentially worse read latency) when the logs are active. Reasons (1) and (2) are the most important in our experiments, but (1) is due to our user-level implementation and is not intrinsic to LogStore.

Under read-only workloads, LogStore cannot take advantage of disk logs. In these cases, LogStore behaves the same as the best competing technique. For example, for the Search trace at the highest acceleration, LogStore achieves 39% energy savings and an increase of 0.4ms in average response time, compared to the one-speed system. This slight performance degradation is due to the many requests that we serve in low speed. The two-speed system without LogStore achieves roughly the same energy savings and performance. DIV with two-speed disks consumes 10% more energy but does not degrade performance.

4.3 Sensitivity Analysis

We study the impact of many parameters on the behavior of LogStore on the Financial trace with acceleration 4x.

Lowering MaxRegular and/or MinLog tends to lower energy consumption but increase response times, since the disks stay in log mode longer. For example, for MaxRegular = 30% and MinLog = 15%, energy consumption decreases 28%, while the average response time becomes 2.3ms. Lowering MinRegular and/or MaxReplay has a smaller impact on energy and response time, since these parameters only affect the amount of time the system stays in log-replay mode. Lowering GoUp or GoDown tends to increase energy consumption while lowering response times, since the disks stay in high speed longer. These trends should guide users in selecting values for the LogStore parameters.

As one would expect, shorter transition times lead to lower energy. For a transition time of 3.5s, the LogStore energy savings become 57% (instead of 47%) compared to the one-speed system. Aggregate throughput is not affected. The increase in average response time becomes 0.8ms (instead of 0.7ms), again compared to the baseline.

Finally, the epoch length exhibits a tradeoff between performance and energy. Lengthening the epochs tends to decrease energy consumption while degrading response time. For example, with 5-second epochs, the LogStore energy savings become 64% and the increase in average response time becomes 1.3ms, compared to the one-speed system.

5. CONCLUSIONS

We proposed LogStore, a storage system that improves the energy proportionality of storage servers with two-speed disks. LogStore can conserve up to 81% and 57% energy, compared to one-speed and two-speed disks, respectively.

We conclude that expensive disks with many speeds are not necessary to achieve storage energy proportionality. Good results can be achieved with much simpler two-speed disks and careful software management.

Acknowledgements: This research was partially supported by Google and NSF grant CCF-1017578.

6. REFERENCES

[1] H. Amur et al. Robust and Flexible Power-Proportional Storage. In *SOCC*, 2010.

[2] E. V. Carrera, E. Pinheiro, and R. Bianchini. Conserving Disk Energy in Network Servers. In *ICS*, 2003.

[3] D. Colarelli and D. Grunwald. Massive Arrays of Idle Disks For Storage Archives. In *SC*, 2002.

[4] H. David et al. Memory Power Management via Dynamic Voltage/Frequency Scaling. In *ICAC*, 2011.

[5] T. Denehy, A. Arpaci-Dusseau, and R. Arpaci-Dusseau. Journal-guided Resynchronization for Software RAID. In *FAST*, 2005.

[6] Q. Deng et al. MemScale: Active Low-Power Modes for Main Memory. In *ASPLOS*, 2011.

[7] EMC. Symmetrix VMAX Enterprise Storage. http://www.emc.com/storage/symmetrix/vmax.htm.

[8] R. Garg et al. Markov Model Based Disk Power Management for Data Intensive Workloads. In *CCGrid*, 2009.

[9] S. Gurumurthi et al. DRPM: Dynamic Speed Control for Power Management in Server Class Disks. In *ISCA*, 2003.

[10] Hitachi. Hitachi Ultrastar 7K3000. http://www.hitachigst.com/internal-drives/enterprise/ultrastar/ultrastar-7k3000.

[11] Intel. Intel Core i7 Processor Series, 2011. http://download.intel.com/design/processor/-datashts/320834.pdf.

[12] K. Okada, N. Kojima, and K. Yamashita. A Novel Drive Architecture of HDD: "Multimode Hard Disk Drive". In *ICCE*, 2000.

[13] E. Pinheiro, R. Bianchini, and C. Dubnicki. Exploiting Redundancy to Conserve Energy in Storage Systems. In *SIGMETRICS*, 2006.

[14] M. Rosenblum and J.K. Ousterhout. The Design and Implementation of a Log-Structured File System. *ACM TOCS*, 10(1), 1992.

[15] P. Shenoy et al. UMass Trace Repository, 2011.

[16] SMART. Self-Monitoring Analysis and Reporting Technology, 2011. http://smartlinux.sourceforge.net/smart/attributes.php.

[17] M. W. Storer et al. Pergamum: Replacing Tape With Energy-Efficient, Reliable, Disk-Based Archival Atorage. In *FAST*, 2008.

[18] E. Thereska, A. Donnelly, and D. Narayanan. Sierra: Practical Power-Proportionality for Data Center Storage. In *Eurosys*, 2011.

[19] A. Verma et al. SRCMap: Energy Proportional Storage Using Dynamic Consolidation. In *FAST*, 2010.

[20] C. Weddle et al. PARAID: A Gear-Shifting Power-Aware RAID. *ACM TOS*, 3(3), 2007.

[21] X. Yao and J. Wang. RIMAC: A Redundancy-based, Hierarchical I/O Architecture for Energy-Efficient Storage Systems. In *EuroSys*, 2006.

[22] S. Yin et al. Improving Energy Efficiency and Security for Disk Systems. In *HPCC*, 2010.

[23] Y. Zhang et al. Software-Directed Data Access Scheduling for Reducing Disk Energy Consumption. In *HPDC*, 2011.

[24] Q. Zhu et al. Hibernator: Helping Disk Arrays Sleep Through the Winter. In *SOSP*, 2005.

A Game Theoretic Resource Allocation for Overall Energy Minimization in Mobile Cloud Computing System[1]

Yang Ge, Yukan Zhang and Qinru Qiu
Department of Electrical Engineering and Computer Science, Syracuse University
Syracuse, NY, USA, 13210
{yage, yzhan158, qiqiu}@syr.edu

Yung-Hsiang Lu
Department of Electrical and Computer Engineering
Purdue University
West Lafayette, IN, USA, 47907
yunglu@purdue.edu

ABSTRACT
Cloud computing and virtualization techniques provide mobile devices with battery energy saving opportunities by allowing them to offload computation and execute code remotely. When the cloud infrastructure consists of heterogeneous servers, the mapping between mobile devices and servers plays an important role in determining the energy dissipation on both sides. From an environmental impact perspective, any energy dissipation related to computation should be counted. To achieve energy sustainability, it is important reducing the overall energy consumption of the mobile systems and the cloud infrastructure. Furthermore, reducing cloud energy consumption can potentially reduce the cost of mobile cloud users because the pricing model of cloud services is pay-by-usage. In this paper, we propose a game-theoretic approach to optimize the overall energy in a mobile cloud computing system. We formulate the energy minimization problem as a congestion game, where each mobile device is a player and his strategy is to select one of the servers to offload the computation while minimizing the overall energy consumption. We prove that the Nash equilibrium always exists in this game and propose an efficient algorithm that could achieve the Nash equilibrium in polynomial time. Experimental results show that our approach is able to reduce the total energy of mobile devices and servers compared to a random approach and an approach which only tries to reduce mobile devices alone.

Categories and Subject Descriptors
C.4 [**Performance of Systems**]: Reliability, availability, andserviceability

General Terms
Algorithms, Management, Performance

Keywords
Congestion Game, game theory, mobile cloud computing, power management, virtualization

1. INTRODUCTION
The emerging paradigm of mobile cloud computing (MCC) moves the processing, memory and storage requirements all together from the resource limited mobile devices to the resource unlimited cloud. MCC provides many advantages to the mobile devices [3]. It extends the storage capacity for mobile users [4] and also reduces the risk of data and application lost on mobile device by backing up users data on several computers in the cloud. Security services such as virus scanning and malicious code detection provided by the MCC improves the safety and reliability of the mobile device.

One very important benefit brought by MCC for mobile users is the extended battery life time. The MCC helps the mobile devices to run the computation intensive applications, which normally consume a large amount of battery energy. This is enabled by virtualization technique which allows the cloud infrastructure to run arbitrary mobile applications from the mobile users or service subscribers. We refer to this technique as *computation offloading*. Reference [5] presents a high level analysis on the conditions that computation offloading could save the energy for mobile phones. According to their results, an application with large amount of computation but very limited data communication could benefit most from computation offloading. The authors in [6] propose an architecture called MAUI to dynamically control the computation offloading for .NET applications at runtime. MAUI utilizes some .NET features to partition and profile the applications and formulate the offloading problem as a linear programming (LP) problem. The authors in [7] propose a similar architecture for Android applications.

Although moving the computation energy away from the mobile devices and into the cloud relieves the pressure on the devices' batteries, it will increase the energy consumption of cloud infrastructure such as the servers in the data center. From environmental impact and carbon emission control perspective, energy is task-centric instead of system-centric [1]. Any energy dissipation related to computation should be counted and carefully managed. From cost reduction point of view, reducing energy consumption of cloud can potentially reduce the cost of mobile users, because current pricing model of cloud services is pay-by-usage and energy consumption is a major factor in the operating cost of cloud services.

In this paper, we consider the problem of energy minimization for a mobile cloud computing system under computation offloading. The MCC system consists of a group of mobile devices and a set of servers in the data center. Each mobile device runs an application and tries to upload a portion of its application to one of the servers. The offloading strategy involves two decisions, (1) the amount of computation to be offloaded, and (2) the destination of offloading. A mobile device selects offloading strategy not only to minimize its own energy dissipation, but also to minimize its energy usage in the cloud. The decision of one mobile device changes the status of the cloud infrastructure, for example one particular server will become more congested or wakeup from sleep mode. Since each mobile device chooses its offloading strategy to minimize the overall energy, a status change in the cloud infrastructure will trigger some mobile devices to adjust their offloading strategies. This, consequently, will impose more

[1]This material is based upon work supported by the National Science Foundation under Grant No. CNS-0845947

changes in the cloud and leads other mobile devices to adjust their offloading strategies. From the above analysis an interesting question is raised, if all mobile devices aim at minimizing the end-to-end overall energy and adjust their offloading strategies independently, will the system ever become stable?

We formulate the MCC system energy minimization problem as a class of games called congestion games. In this model, each mobile device is a player and his strategy is to select one of the available servers to offload its computation. All players compete for the same set of resources and their goal is to minimize the combined mobile and server energy dissipated to provide service for its own application. We prove that the Nash equilibrium always exists in this congestion game formulation. Nash equilibrium is the optimum policy in the sense that no player can find better policy if he deviates from current policy unilaterally [2]. We propose an efficient algorithm that achieves the Nash equilibrium in polynomial time.

There have been some works in the literature using game theory to solve general resource allocation problem in cloud computing environment, for example [14]. However, this work only considers the cost in cloud side and ignores the cost on mobile devices. Furthermore, it defines cost as a simple linear function of workload, which is not a suitable model of energy dissipation if we want to capture some nonlinearity and discontinuity introduced by server power management.

The uniqueness of our work is summarized as the following:

- This is the first work that aims to reduce the overall energy of a mobile cloud computing system under computation off-loading context. Although computation offloading techniques have been investigated in some previous work [5][6][7], their goals are only to reduce the energy consumed by the mobile devices and extend their battery lives, without considering the energy consequence of offloading on the cloud computing infrastructure.
- We proposed a game theoretic formulation of the problem. We also proved that the Nash equilibrium of this problem always exists and it could be achieved in polynomial time if each mobile device selects its offloading strategy based on our proposed algorithm. The important implication of Nash equilibrium is that our algorithm will eventually converge to a stable state, where every mobile device finds its current optimal strategy and has no incentive to leave.
- We demonstrate the necessity of joint optimizing the energy of the mobile devices and the cloud infrastructure. Compared to the techniques that only aim to reduce the mobile device energy, our technique is able to reduce the overall energy by 45.42%. Our approach is also able to reduce more than 61.83% energy compared to a random offloading approach.

The rest of the paper is organized as follows: We discuss our system model and introduce the energy minimization problem in Section 2. We introduce the congestion game model and discuss its application in detail in Section 3. We present the algorithm to achieve the equilibrium in Section 4. Experimental results are reported in Section 5. Finally, we conclude the paper in Section 6.

2. SYSTEM MODEL

2.1 MCC System Architecture

On one side of the MCC architecture, mobile devices like smart phones, tablet computers are connected to the cloud through Wi-Fi or 3G networks. On the other side, a group of servers residing in a data center constitute a large distributed computing system which can provide the cloud users with different kinds of services, including infrastructure as a service (IaaS), platform as a service (PaaS) or software as a service (SaaS) [3]. In this paper, we focus on the IaaS, in which the servers in the cloud provide the mobile devices with hardware resources, like CPUs and memories for them to offload computation for battery energy saving.

At a specific execution point of the mobile application, the migration manager decides to move a portion of the application to the cloud (for example, an individual thread). The manager will send the migration request as well as necessary data and program states to the remote server. Upon receiving the migration request, the application data and program states, the server creates a dedicated virtual machine (VM) for the mobile device, loads the application executable and starts execution. In the mean time, the mobile device continues to run other threads or waits for the results return from the remote server. At the end, the migrated portion returns back to the mobile device, and merges back to the original process.

Mobile devices generally can be benefited from code offloading and remote execution because of two reasons. Firstly, the plenty of hardware resources in the cloud can help mobile devices to overcome resource limitation and run some resource intensive applications. Secondly, by executing code remotely, the mobile devices could avoid spending a long time in high power state, and either stay in idle state or go into low power sleep state, thus save the energy and extend the battery life. However, as pointed out in [5], not all kinds of mobile applications could save energy through code offloading. For example, an application with very high data communication volume will spend much more extra energy in transferring the data between local memory and remote servers and not benefit from computation offloading for energy saving. Deciding which portion of the application needs to be offloaded is non-trivial. It usually needs detailed program profiling information and a fast solver for integer linear programming problems. In this paper, we assume that each device has made their own decisions about which part of the mobile applications should be offloaded to the cloud. We focus on how a mobile device selects a server from a group of heterogeneous servers in the data center for computation offloading.

2.2 MCC Energy Minimization Problem

We assume that there are n mobile devices and r cloud servers in the MCC system. The ith mobile device runs an application A_i with computation workload C_i (which can be measured by the number of clock cycles or execution time). We assume that, based on careful application profiling and device characterization, the amount of computation to be offloaded to the dedicated virtual machine VM_i is pre-determined and is denoted as O_i. The amount of computation left for local execution is denoted as L_i, $L_i + O_i = C_i$. We also assume that the mobile device has a performance requirement on VM_i, i.e. the O_i amount computation must be finished within deadline D_i, otherwise offloading computation to server will not bring notable performance benefit. If this performance constraint cannot be satisfied, the mobile device will either reduce the amount of offloading, or move the VM to a faster server or even request to wake up a new server.

We assume the power consumption of the mobile device i during the execution of the application A_i is determined by an application specific factor α_i, which is a high level parameter reflecting its overall power intensity, i.e. $P_{active,i} = \alpha_i P$, where P is the normalized maximum mobile device power consumption.

Therefore, the energy consumed by running the local portion of the application A_i is $E_{local,i} = P_{active,i}L_i$.

We assume that the servers in the data center are also heterogeneous machines. They have different processing speeds and power consumption, so they could accommodate different service requests (delay sensitive or power sensitive) from the mobile clients. Here we assume server j is operating at speed s_j and its power consumption is $f(s_j)$, e.g. $f(s_j) = s_j^3$. Please note that our algorithm does not rely on any properties of f. We assume different offloaded applications have different power even running on the same server. Similar to the factor α_i on the mobile device, we assign each application a parameter β_i. So the power consumption of application A_i running on server j will be $\beta_i f(s_j)$. And the energy consumption for the jth server to run the offloaded code from device i will be calculated as $E^i_{server,j} = \beta_i f(s_j) O_i$.

If only one application A_i running on server j, we assume that the waiting time experienced by the mobile device i is $g(s_j)O_i$, where $g(s_j)$ is a non-increasing function of speed s_j. If there are n_j applications running on the server j, we assume the waiting time experienced by the mobile device i is $g(s_j)O_i h_j(n_j)$, where $h_j(n_j)$ is a non-decreasing function of n_j. For example, if each VM has equal time slice on the server, without considering the overhead of context switch and cache miss, $h_j(n_j)$ can be simplified to n_j. As a server gets more congested, A_i running on it will experience longer delay and might violate the performance constraint D_i. The mobile device has to reduce the computation amount O_i to satisfy D_i or selects another server which is less congested. [12] shows that, with proper configuration, network latency in a data center is independent of network topology and server's location. Thus we ignore the network latency in this problem formulation. Neither do we consider the communication energy on the routers and switches because they are usually small comparing to the energy consumed on mobile devices and servers [13], and they are not affected by the mapping of the computation.

The overall energy related to the execution of application A_i consists of the energy dissipation on both the server and mobile device. It can be calculated as the following:

$$E_{total}(i,j) = E_{local,i} + E^i_{server,j} \quad (1)$$

j is the index of the server to which the application A_i is mapped.

We also assume servers follow time out power management policies, i.e. they will switch to low power mode after a certain period of idle. Let T denote the time out threshold and $P_{static,j}$ denote the static power server j consumes when it is idle. Then the static energy consumption of server j during the idle period is $E_{static,j} = P_{static,j} \times T$. The total energy consumed by the MCC system is $E_{MCC} = \sum_{i=1}^{n} E_{total}(i) + \sum_{j=1}^{r} E_{static,j}$.

The MCC system energy minimization problem can be stated as the following resource allocation problem:

Given n mobile devices and r servers, for each device, find a server for computation offloading, such that the overall MCC system energy E_{MCC} is minimized while the performance constraint is satisfied.

This problem can be formulated as an integer linear programming (ILP). When the number of mobile devices and cloud servers gets large, which is usually the case in reality, solving such ILP becomes exponentially difficult. In the next sections, we introduce a game theoretic formulation which solves this problem using a distributive approach, where each mobile device chooses its own offloading strategy, including the amount of offloading and the destination of offloading in order to minimize its overall energy.

3. GAME THEORETIC FORMULATION

3.1 Congestion Game Model

Congestion games [8] model a group of players sharing a set of resources. In a congestion game, each player chooses one or a subset of resources to maximize his own utility or minimize his own cost. The utility/cost obtained by the player is the sum of the utilities he received from each resource he chooses. The utility/cost received from a resource depends on how many players sharing the same resource and it is generally a non-increasing/non-decreasing function of the number of players sharing it. For example, the more senders in a network share the same link to send packets, the less throughput and longer waiting time they will experience. In this paper, we consider a simple congestion game in which each player only selects one resource, because each mobile device is only allowed to offload its computation to one server.

We introduce the following notation to define congestion game.

- $N = \{1, \dots, n\}$ is a finite set of n players
- $R = \{1, \dots, r\}$ is a finite set of r resources
- Σ_i is the set of strategies for player i, which is a subset of resource set R. We use $\Sigma = \prod_i^n \Sigma_i$ to denote the joint strategy space and $\sigma = (\sigma_1, \sigma_2, \dots, \sigma_n)$ a strategy tuple in which the player i plays strategy σ_i, i.e. it chooses resource $\sigma_i \in \Sigma_i$. Please note σ is the strategy vector, and σ_i with subscript i is the strategy of player i. Sometime we also use J to denote a strategy in our algorithms.
- n_j is the number of players who selects strategy j, and for a strategy tuple σ, $n_j = \sum_{i=1}^{n} I(\sigma_i = j)$, I is the indicator function. We call $(n_1, n_2, \dots, n_r)$ a congestion vector.
- S_{ij} is the cost function when player i selects resource j. S_{ij} is a monotonically non-decreasing function of n_j, i.e. $S_{ij}(n_j) \le S_{ij}(n_j + 1)$ for any positive integer n_j.

For some specific congestion games, there exists Nash equilibrium. Nash equilibrium is a general term in game theory and is defined as a state that no player can benefit by changing its strategy while the other players keep their strategies unchanged [1]. For a congestion game defined as above, a strategy tuple σ leads to Nash equilibrium if and only if

$$S_{i\sigma_i}(n_{\sigma_i}) \le S_{ij}(n_j + 1), \forall 1 \le i \le n, 1 \le j \le r$$

σ_i is also said to be the *best reply* or *best response* for player i against strategy tuple σ. In other words, a strategy vector σ is at Nash equilibrium, if for any user i, moving away from resource σ_i to choose resource j by itself will lead to higher cost.

3.2 Application to MCC Energy Minimization Problem

To apply the congestion game model to our MCC system energy minimization problem, we define each mobile device and the dedicated VM created for it to be a player and each server to be a resource. We use mobile device, VM, player interchangeably in the paper. For each player, the strategy set Σ_i is equal to the resource set R, and if mobile device i selects server j, then $\sigma_i = j$.

One way for the player (i.e. the mobile device) to define its cost function is to use the total energy consumed locally. By reducing his own energy, the player can extend its battery life. However, if every player wants to save his own local energy, he will select faster servers to maximize the offloaded computation. This makes the faster servers more congested than the others and as a result, there might be performance violation and the mobile device have to either reduce the offload amount or move to another server. Furthermore, our objective is not to minimize the mobile devices' energy but the overall system energy. We would like each player be less selfish and take some "social responsibilities". Therefore, we define the first component of cost function as the sum of weighted energy associated with the execution of application A_i at both mobile device side and server side, as show in equation (2).

$$a_i E_{local,i} + b_i E^i_{server,j} \tag{2}$$

By this definition, each player does not only try to minimize its own energy but also the energy that he added on the servers as well. The weight coefficients a_i and b_i in this equation differentiate the *quality* of the energy used by a mobile device and a server. The former, due to the overhead of battery charge and discharge and also because of the limited battery capacity, is usually considered higher quality, therefore we have $a_i \geq b_i$. These weight coefficients do not have to be the same for all mobile devices. For example, when a mobile device's battery is full, it could be more altruistic and reduce the value of a_i to offload less computation; when its battery drops under a critical level, it could increase the weight on $E_{local,i}$ and offload more computation.

Eq. (2) is a non-decreasing function of n_j. Whenn_j increases, server j becomes more congested. As a result, less offloaded computation can be completed within the performance constraint and more may have to be done on the mobile device locally, thus gives an increase in $E_{local,i}$ and decrease in $E^i_{server,j}$. Because servers usually have better energy efficiency than the mobile devices due to their multi-core architecture and powerful co-processors such as GPUs and DSPs, the increase in $E_{local,i}$ is usually greater than the decrease in $E_{local,i}$. Also because coefficient a_i is greater than or equal to coefficient $b_{i.}$ Eq. (2) is a non-decreasing function of n_j.

In order to consider each server's static energy penalty during the time-out period, we add the second component to each player's cost function. It is proportional to the current total static energy, i.e. $w_i(\sum_{j=1}^{n} E_{static,j})$, w_i is a constant weight. Therefore the overall cost function for a player is as following:

$$S_{ij} = a_i E_{local,i} + b_i E^i_{server,j} + w_i(\textstyle\sum_{j=1}^{n} E_{static,j}) \tag{3}$$

Because the static penalty does not depend on the congestion of each server, the overall utility is still a non-decreasing function of n_j. The cost function meets the requirement of a congestion game.

4. NASH EQUILIBRIUM OF THE ENERGY MINIMIZATION CONGESTION GAME

In this section, we present the algorithms which find the Nash equilibrium for the energy minimization problem defined above. The algorithms themselves provide a constructive proof for the existence of the Nash equilibrium. Please note that the actual algorithm to achieve the equilibrium could run by the dedicated VM to relieve the computation burden on mobile devices and reduce communication between mobile devices and the cloud.

Our algorithm adopts an incremental optimization scheme. At each step, only one player is allowed to change its current strategy and choose its best response against other players' current strategies. Consider the initial scenario that the first $n-1$ players have achieved equilibrium and their strategy vector is $\sigma(0) = (\sigma_1(0), \sigma_2(0), \dots, \sigma_{n-1}(0))$. We are interested to find out, after the nth player enters, how will the system regain the equilibrium.

Let R denote the set of all servers and S_{on} denote the set of servers which have already been turned on. Let $J(0) = \sigma_n(0)$ denote the first strategy chosen by player n against the initial strategy vector $\sigma(0)$ of the rest of the system. We consider the following two scenarios.

Scenario 1: $J(0) = \sigma_n(0) \in S_{on}$. Then the only players affected by player n's choice are those players whose current strategy is also $J(0)$ in $\sigma(0)$, because player n makes server $\sigma_n(0)$ more congested. For other players, their current strategies remain to be their best strategies. If every player in server $J(0)$ finds $J(0)$ remaining to be its best choice, then Nash equilibrium is achieved. Otherwise, there will be a player p on $J(0)$ whose best strategy deviates from its current strategy. Assume this player moves from $J(0)$ to $J(1)$ at step 1. We let the process continue as shown in algorithm1. The algorithm returns when all players are in their best strategies or a new server is turned on. Please note that in the algorithm, $J(i)$ represents a strategy that a player chooses at step i. It can be proved that this process will stop in finite step. The claim is given in Lemma1. We skip the proof due to space limit.

Algorithm1: type1_ move **(σ_n(0), σ(0), *R*)**

```
1.  i = 0; J(0) = σ_n(0);
2.  while true
3.      p = null;
4.      for each player p with strategy σ_p(i) = J(i)
5.          if J(i) is not player p's best reply against σ(i) in R
6.              J(i + 1) = player p's best reply against σ(i) in R
7.              σ_q(i + 1) = σ_q(i), ∀q ≠ p,
8.              σ_p(i + 1) = J(i + 1);
9.              break;
10.         end
11.     end
12.     if (∄p) return ∅;
13.     else if (J(i + 1) ∉ S_on)
14.         turn on J(i + 1);
15.         return J(i + 1);
16.     end
17.     i++;
18. end
```

Lemma1: TYPE1_MOVE will stop in at most n steps, where n is the number of players. Furthermore, if type1_move returns at line 12, then Nash equilibrium is achieved.

We define a limited version of type1_move called type1_move_limited. The limited version restricts the move in a given subset of servers denoted as S_{limit}. We set the third input of type1_move algorithm to S_{limit} instead of using R in the limited version.

Algorithm2: type2_ move **(*J*(1), σ(0), *R*)**

```
i = 1;
while true
    p = null;
    for each player p with strategy σ_p(i) ≠ J(i)
        if J(i) is player p's best reply against σ(i) in R
            σ_q(i + 1) = σ_q(i), ∀q ≠ p,
            σ_p(i + 1) = J(i); J(i + 1) = σ_p(i);
            break;
        end
    end
    if (∄p)
        break;
    end
    i++;
end
```

Scenario 2: If $J(0) = \sigma_n(0) \notin S_{on}$. Then all players' strategies will be affected by the newly turned on server because of two reasons. Firstly, all players' utilities are increased by the static energy penalty of turning on $J(0)$. Secondly, $J(0)$ is much less congested than other servers in S_{on}, players could potentially offload more computation if they choose $J(0)$ and achieve energy savings (reduce cost function). For each player, if the current strategy is its best strategy then the system is still in equilibrium; otherwise $J(0)$ will be its best strategy. Assume player p_1 on server $J(1) = \sigma_{p1}(0)$ finds its current best response to be $J(0)$. Then at step 1, we let player p_1 move from $J(1)$ to $J(0)$, i.e. $\sigma_{p1}(1) = \sigma_n(0)$. p_1's move will make server $J(1)$ less congested than before. Then there might be a player p_2 on $J(2) = \sigma_{p2}(1) \neq J(1)$ find its best strategy to be $J(1)$. Then at step 2, we let p_2 moves from $J(2)$ to $J(1)$. We let the process continues and summarize it in algorithm2. We call it type-2 move. We claim type-2 move will stop after finite steps in Lemma2. Again the proof is skipped due to space limit.

Lemma2: TYPE2_MOVE will stop in at most $n \cdot r_{on}$ steps, where r_{on} is the number of servers which has been turned on.

Again, we define a limited version of type2_move called type2_move_limited. The limited version restricts the move in a given subset of servers denoted as S_{limit}. We set the third input of type2_move algorithm to S_{limit} instead of using R in the limited version.

We now present our algorithm for finding Nash equilibrium for the first n player. We again assume the first $n-1$ players have achieved equilibrium. Lemma1 shows that if $J = \emptyset$ (line 8), then the equilibrium is achieved. If $J \neq \emptyset$, from algorithm1 we know that the congestion vector for servers in S_{on} does not change after type1_move (line 5). Because the cost function (3) is an increasing function of the congestion, for each player, either the current strategy is its best strategy or newly turned on server J is its best strategy. Property 2 in Lemma3 shows that the loop 11 ~ 15 will repeat at most n iterations. After this loop, property 1 show that the first $n-1$ players are in their best strategies and player n is either in its best strategy or has its best strategy outside S_{on}. Because there are only r servers in the system, the outer loop will repeat at most r times. After the outer loop terminates, all players are in their best strategy and Nash equilibrium is achieved. We skip the proof of Lemma3 due to space limit.

Algorithm3: find_equilibrium_for_first_n_players(n)

```
1.  while true
2.      J = best strategy of player n; σ = current strategy vector;
3.      assign player n to J;
4.      if (J ∈ S_on)
5.          J = type1_move(J, σ, R);
6.      end
7.      if (J == ∅)
8.          return; // reach the equilibrium
9.      end
10.     S_on = S_on ∪ J; // turned on a new server J
11.     while (∃p, its current strategy is σ_p and its best strategy is J)
12.         move p from σ_p to J;
13.         type1_move_limited(J, σ, S_on);
14.         type2_move_limited(σ_p, σ, S_on − J);
15.     end
16.     if (player n is in its best strategy)
17.         return; // reach the equilibrium;
18.     end
19. end
```

Lemma3: The loop in line 11 ~ 15 have the following properties:

1. After every iteration, for each of first $n-1$ player p, either it is in its best strategy or new server J is its best strategy.
2. The number of players in the server J keeps non-decreasing. If a player p leaves J at an iteration, it never comes back to J.
3. After the loop, for player n, either the current strategy is its best strategy or its best strategy is not in S_{on}.

5. EXPERIMENTAL RESULTS

To demonstrate the effectiveness of our approach, we implement our Nash equilibrium algorithm in Matlab and carry out the experiments on a DELL T3400 workstation. We obtain the power model of mobile applications from [9], which is a linear combination of the activities on different components, including CPU utilization, Wi-Fi module status, the brightness of the LCD etc. Because our purpose is not a detailed power analysis for mobile devices, we just use one factor α to represent the overall activities of the application. We assume the maximum power consumption of a mobile device is 2.4W and the actual power of the application is $P_{app} = \alpha \times 2.4\text{W}$. We set the activity factor α as a random variable with a uniform distribution in the range $[0.3, 1.0]$. We assume the offloaded computation amount is predetermined by careful application profiling as in [6][7].

We assume the servers are heterogeneous and they are operating at different speeds in the range of [2.0GHz, 3.0GHz]. We adopt the server power model from [15], which is based on Intel Xeon processors. According to this model, server power is a linear function against the CPU's frequency at full utilization. It is 200W at 2.0GHz and 240W at 3.0GHz. These data are the full system power including the static power. Based on the data provided in [16], we assume the static power of our server is 100W when it is in idle state. Reference [10] shows that at a given operating frequency, server power is a linear function against the VM system utilization β. We set $\beta = \alpha$ in our experiments for simplicity. Overall, we assume the server power is a linear function against both operating frequency and utilization. This is also confirmed by [17]. At 100% system utilization, the server consumes 100W to 140W more power than in idle state.

For the delay model, we assume all servers provide M times speedup over the mobile devices [5] when running at 2.0GHz. When running at speed s, its speedup ratio is $sM/2$, i.e. $g(s_j) = 2/(s_j M)$. If there are N mobile devices offloading to the server, we assume the slowdown factor is $h(N) = N^{\gamma}, \gamma \geq 1$, γ accounts for the context switching overhead [11]. If a mobile device i offloads an amount of 10 seconds computation on a server with speed s and shared by N mobile devices, device i will wait for $10 \times (sM/2) \times N^{\gamma}$ time to get results. In our experiment, γ is set to 1.05. Please note that our optimization algorithm does not rely on any specific power or delay models.

5.1 Results Analysis

In the first set of experiments, we compare our policy with a policy which randomly assigns the mobile devices to cloud servers. We call this policy *Random*. The proposed policy that minimizes the overall energy is denoted as *Nash-overall*. We set 10 servers and 60 mobile devices in the MCC system. The results are from the average of 10 runs.

Table 1 Comparison between Nash-overall and Random

Mobile weight (a)	Nash overall (J)	Random (J)	Improvement (%)
1	3558.26	9319.07	61.83%
1.5	5695.91	9319.07	38.92%
2	6007.31	9319.07	35.55%
2.5	6637.01	9319.07	28.79%

Table 1 shows the energy comparison between Nash-overall policy and Random policy. We vary the mobile weight, i.e. a in equation (3) from 1 to 2.5. As the a goes bigger, the Nash-overall algorithm tries to complete more offloaded computation on the server side. The random policy is not affected by the weight. As shown in this table, our Nash-overall policy could achieve large energy savings compare to the Random policy. This is because our policy tries to reduce the static energy consumption of the servers by conservatively waking them up and consolidating the offloaded computation to a small number of energy efficient servers, especially when the weight is small. On the other hand, when is big, the first component in equation (3) becomes dominant. Our policy tries to reduce the energy on mobile devices and turn on more servers to accommodate the offloaded computation. Therefore, the static energy increases which causes higher total energy consumption.

This trend can be seen in Figure 1, which shows the energy components of each policy. The blue bar represents the dynamic energy when servers are running the offloaded computation and the red bars represent the static energy when servers. Because the Random policy will blindly turn on servers, it incurs most static energy. However, this behavior is beneficial for mobile devices because mobile devices can be uniformly distributed among all the servers, and each server is less congested. Servers could complete more offloaded computation under mobile devices' performance constraints. Therefore, the Random policy results the minimum mobile devices energy consumption among all five groups. For our policy, when weight is small, mobile device energy is dominant, when is big, the server static energy is dominant. As shown, when , mobile devices' energy is same as the Random policy. Further increasing does not reduce the mobile devices' energy anymore because all offloaded computation has been completed by the servers. In this case, we still achieve 28.79% total energy savings compare to Random policy, because each mobile device is aware of server static energy and will not turn on unnecessary servers. It is the unique feature of our algorithm to get energy tradeoff between mobile devices and cloud servers.

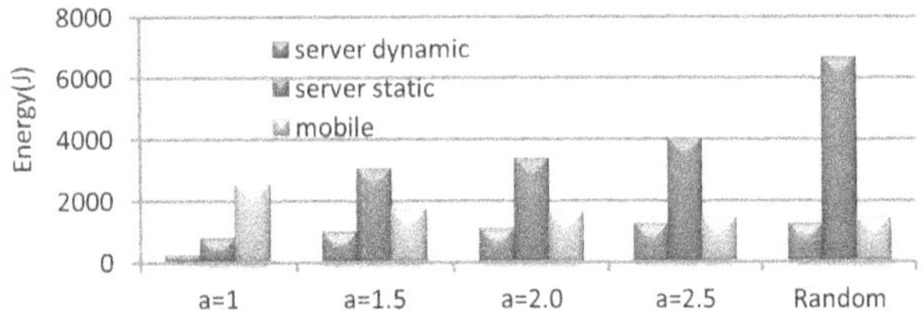

Figure 1 Energy components of Nash-overall and Random policy

Table 2 Comparison between Nash-overall and Greedy

Mobile weight (a)	Nash overall (J)	Greedy (J)	Improvement (%)
1	3659.28	6703.22	45.42%
1.5	5779.22	6703.22	13.80%
2	6153.47	6703.22	8.18%
2.5	6575.44	6703.22	1.90%

In the second set of experiments, we compare Nash-overall policy with a greedy policy which tries to reduce mobile devices energy. For a mobile device, this greedy policy first selects those servers which could mostly satisfy its offload computation within the performance constraint. Among these servers, the greedy policy then selects the server which could minimize its total energy. Table 2 shows the energy comparison between Nash-overall and Greedy policy as the weight gradually increases. This table shows similar trend as in Table 1, because the greedy policy is also relatively aggressive in turning on servers. When is small, the total energy reduction is most significant. When increase to 2.5, the reduction in total energy decrease from 45.42% to 1.9%, because the Nash-overall also become more aggressive in turning on new server and incurs larger server static energy. On the other hand, the energy consumption differences on mobile devices between these two policies decrease from 43.75% to only 1.64%.

6. CONCLUSIONS

In this paper, we propose a game-theoretic approach to optimize the energy consumption of the MCC systems. We formulate the mobile devices and cloud servers energy minimization problem as a congestion game. We prove that the Nash equilibrium always exists in this congestion game, and propose an efficient algorithm that could achieve the Nash equilibrium in polynomial time. Experimental results show that our approach is able to reduce the total energy compare to a random approach and an approach which only tries to reduce mobile devices energy.

7. REFERENCES

[1] Y. H. Lu, Qinru Qiu, A. R. Butt and K. W. Cameron, "End-to-End Energy Management," in *IEEE Computer*, 44 (11), November 2011.

[2] M. Osborne and A. Rubinstein, "A Course in Game Theory" Cambrisdge, MA: MIT, 1994.

[3] H. Dinh, C. Lee, D. Niyato and Ping Wang, "A Survey of Mobile Cloud Computing: Architecture, Applications, and Approaches", in *Wireless Commu. and Mobile Comput.*, 11(11), Oct. 2011.

[4] http://aws.amazon.com/s3/

[5] K. Kumar and Y. Lu, "Cloud Computing for Mobile Users: Can Offloading Computation Save Energy?" in *IEEE Computer*, vol. 43(4), Apr. 2010.

[6] E. Cuervo, A. Balasubramanian, D. Cho, A. Wolman, S. Saroiu, R. Chandra and P. Bahl, "MAUI: Making Smartphones Last Longer with Code Offload," in *Proc. of international conference on Mobile systems, applications, and services,* 2010.

[7] B. G. Chun, S. Ihm, P. Maniatis, M. Naik, and A. Patti, "CloneCloud: elastic execution between mobile device and cloud," in *Proc. of conf. on Computer systems (EuroSys)*, Apr. 2011.

[8] I. Milchtaich, "Congestion games with player-specificpayoff functions," in *Games and Economic Behavior*,13(1):111–124, 1996.

[9] L. Zhang , B. Tiwana, Z. Qian , Z. Wang , R. Dick , Z. Mao and L. Yang, "Accurate online power estimation and automatic battery behavior based power model generation for smartphones," in *Proc. of Inter. Conf. on HW/SW codesign and system synthesis*, Oct. 2010.

[10] M. Pedram and I. Hwang, "Power and Performance Modeling in a Virtualized Server System," in *Proc. of International Conference on Parallel Processing Workshops (ICPPW)*, Sep. 2010.

[11] T. Enokido, A. Aikebaier and M. Takizawa, "A Model for Reducing Power Consumption in Peer-to-Peer Systems," in *IEEE System Journal*, vol. 4, issue 2, pp. 221-229, Jun. 2010.

[12] A. Greenberg, J. R. Hamilton, N. Jain, S. Kandula, C. Kim, P. Lahiri, D. Maltz, P. Patel and S. Sengupta. "VL2: a scalable and flexible data center network," in *Proc. of the ACM SIGCOMM*, 2009.

[13] J. Chabarek, J. Sommers, P. Barford, C. Estan, D. Tsiang,and S. Wright. Power Awareness in Network Design andRouting. In INFOCOM, pages 457 - 465, 2008.

[14] G. Wei, A. Vasilakos, Y. Zheng and N. Xiong, "A game-theoretic method of fair resource allocation for cloud computing services," in *Journal of Supercomput.* 54(2), pp 252-269, Nov. 2010.

[15] A. Gandhi, M. Harchol-Balter, R. Das and C. Lefurgy. "Optimal power allocation in server farms," in *Proc. of intern. joint conf. on measurement and modeling of computer systems*, pp157-168, 2009.

[16] Intel® Xeon® Processor 5600 Series Product Brief: http://www.intel.com/content/www/us/en/processors/xeon/xeon-5600-brief.html

[17] Z. Wang, X. Zhu, C. McCarthy, P. Ranganathan and V. Talwar, "Feedback Control Algorithms for Power Management of Servers," in *Proc. Intern. Workshop on Feedback Control Implementation and Design in Computing Systems and Networks*, Jun. 2008.

A Low-power "Near-threshold" Epileptic Seizure Detection Processor with Multiple Algorithm Programmability

Himanshu Markandeya[1], Shriram Raghunathan[2], Pedro Irazoqui[3,1], Kaushik Roy[1,3]

[1]School of ECE
Purdue University
West Lafayette, IN 47907, USA
{hmarkand, kaushik}@purdue.edu

[2]Cyberonics Inc.
100 Cyberonics Blvd.
Houston, TX 77058, USA
shriram.raghunathan@cyberonics.com

[3]Center for Implantable Devices
Purdue University
West Lafayette, IN 47907, USA
pip@purdue.edu

ABSTRACT

In this paper, we developed and implemented a low-power multiple-algorithm processor for detection of epileptic seizures with high efficacy. Four different algorithms -- The Coastline, Hjorth parameter, Energy and Non-linear energy metric based algorithms -- are used in our analysis. The Boolean logical combinations of these algorithms allow programmability according to the patient specific needs and improve efficacy by almost 10%. The system implemented in 65-nm TSMC technology consumes about 13.1 μW (measured) at scaled supply voltage of 400mV. Programmability using Boolean logic helps in improving the efficacy resulting in almost an error-free detection.

Categories and Subject Descriptors

B.7.1 [**Integrated Circuits**]: Types and Design Styles – *algorithms implemented in hardware, VLSI (very large scale integration), Input/output circuit.*

General Terms

Algorithms, Measurement, Design.

Keywords

Biomedical, ASIC, Epilepsy, Low-power Implants.

1. INTRODUCTION

Epilepsy is the third most common neurological disorder in United States after Alzheimer's disease and stroke. It is characterized by recurrent involuntary physical convulsions viz. epileptic seizures [1] [2]. A third of the 50 million people affected worldwide do not respond well to surgical or pharmacological therapy. Neuromodulation by means of electrical, chemical or even optical stimulation is rapidly emerging as a promising alternative therapy for these patients [3] [4]. At present, the Cyberonics Vagus Nerve Stimulator (VNS) is the only commercially available anti-epileptic prosthesis [5]. VNS is based on continuous neurostimulation which has been argued to be detrimental and ineffective in the long run [6]. Responsive neurostimulation impacts the efficacy of this therapy by critically increasing its temporal and spatial specificity. Neuropace device (Medtronics Inc.) is developed based on this principle and awaits Food and Drug Administration (FDA) approval [4] [7]. However, responsive neurostimulation implies that the epileptic seizure be detected with high efficacy. Implantable epilepsy prosthesis that can deliver neurostimulation at the epileptogenic focus is rapidly emerging as an effective solution for treatment of epilepsy, especially for the population that is medically refractory.

In recent decades, scaling of integrated circuit technology has allowed the development of implantable biomedical applications in order to provide alternative solutions for biological ailments. Scaling has the advantage of implementing complex functionalities on the silicon chip and operating them at lower supply voltages. However, the exponential increase in the leakage implies higher power consumption and poses a significant design challenge. This is undesirable in an implantable application such as epilepsy prosthesis. This is because the implanted battery is a limited source for energy. A wasteful usage of energy by the implant will severely deplete the battery. This may necessitate an expensive and often risk-prone surgery to replace the battery. Apart from this, it is also mandatory to keep the power dissipation of the implant to a minimal to avoid any tissue damage by heat. Thus, the design of any implantable biomedical application is subjected to stringent design constraints. Efficient and low-power seizure detection algorithms are the key to realizing a fully integrated implantable prosthesis which can responsively deliver the therapy to the epileptogenic focus. An efficacious detection will also enable the clinicians and researchers to better understand the disorder.

In this paper, we have implemented a seizure detection processor with multiple algorithms using mathematical metric based algorithm in 65-nm bulk-Si technology. Due to inter-patient variability of neural recordings, it is difficult to maintain high efficacy over a range of patients. We have addressed this by providing programmability in the system to operate according to patient-specific needs. A built-in self-test (BIST) block has been implemented to check the functional nature of each algorithm as well as the peripheral blocks which aide in improved functionality. It will be shown in the results that there is a distinct improvement in efficacy of detection due to provision of low cost programmable logic. Due to low frequency of operation, it is possible to operate the system at a scaled power supply in the vicinity of threshold voltage for the transistor. This greatly reduces the power consumption of the system without affecting the efficacy.

2. BACKGROUND

The literature is abundant with numerous algorithms for detection of the epileptic seizure [8-11]. The characteristic of the seizure is a gradual surge in amplitude of the signal in specific frequency bands. This signature is exploited by number of existing algorithms for seizure detection. Signal processing tools such as

Fast Fourier Transform (FFT), Short-Time Fourier Transform (STFT) and Discrete Wavelet Transform (DWT) have been efficiently utilized to this effect [8] [9]. Some of the FFT and DWT based algorithms utilize the artificial neural networks in order to learn the seizure pattern for detection [8] [9]. This results in a power hungry architecture that is not suitable for implantable application. Furthermore, majority of these algorithms are based on neural recordings from the scalp electrodes i.e. Electroencephalogram (EEG) or Electrocorticogram (ECoG). This implies that the seizure has already spread through the brain before being detected at the surface. In this paper, we use the Local Field Potential (LFP) recordings from a small volume of brain tissue near the epileptogenic focus. This helps in reducing the latency of detection. The event- based algorithm utilizing LFP provides a low power solution for implantable epilepsy prosthesis [10]. However, its efficacy may be debatable over a range of patients [11]. Utilizing the DWT with a quasi-averaging technique helps in providing the programmability and maintain high efficacy as well [11]. However, it has a large number of redundant computations and hence is wasteful in terms of energy consumption. Statistical mathematical parameters have also been devised in order to quantify the amplitude and frequency variations. Some of these parameters are energy, line-length or coastline, hjorth parameter, non-linear autocorrelation etc. [12]. These parameters can also be used to detect the seizures with a fair degree of accuracy [12]. They can be used to demarcate the recorded data into baseline and seizure states. The parameters may be linear or non-linear in nature and may be operated in the time or the frequency domains. Although these mathematical metrics detect seizure with reasonable accuracy, their usage is limited by the fact that, they may not be efficacious over a range of patients due to inter-patient variability of neural recordings. However, by logically combining the detections based upon these parameters, the efficacy can be improved over different patients. In this paper, we have implemented the processor in which the outputs of the implemented algorithms can be combined logically by using simple Boolean AND's and OR's. Hence, the best possible combination can be selected on patient-specific basis to maintain high efficacy of detection.

The paper is organized as follows. Section 3 explains the parameter based algorithms implemented in the processor. It is followed by Section 4 for the explanation about the implemented hardware. Section 5 includes the results from efficacy and power efficiency perspective. Section 6 concludes the paper.

3. SYSTEM OVERVIEW

The implemented system has an algorithm bank consisting of four algorithms. Each of these algorithms is based on a mathematical parameter extracting the seizure feature. These algorithms are selected based on the ability to extract onset feature and proven functionality for seizure detection. These seizure features are used to demarcate the seizure from the baseline. The algorithms operate on real-time input data and produce processed signals. The processed signals are then compared with prefixed threshold values to cause a detection of the onset of seizures. The detection from the four algorithms can be logically combined according to needs of the patient. This maintains high efficacy operation of the processor on a wider range of patients. The thresholds for each of the algorithms and the Boolean logical combination to be used are decided in the training phase. The algorithms, logical combination technique and the training phase are explained below.

3.1 Algorithms

The system consists of Energy, Coastline, Non-Linear Energy and Hjorth parameter based algorithms. These algorithms are chosen on the basis of their onset feature isolation properties. While Energy and Coastline are simple arithmetic parameters which measure the energy content and the line-length of the LFP, Hjorth variance and Non-Linear Energy are more statistical in nature. They capture the non-linear features in the recorded LFP. All these algorithms operate in the time domain on real-time data stream. The recorded LFP data is digitized into 10-bit 2's complement format and streamed into each of the four algorithms. The definition of the four parameters is explained below.

3.1.1 Energy

The "energy" parameter calculates the energy content of the signal in a prefixed window. It is a well known fact that as the onset of seizure approaches, there is gradual surge in the amplitude of the signal in specific frequency bands. This results in an increase in energy content of the signal. This parameter operates on adjacent windows of the streaming data. Hence, a gradual increase in the energy value implies that the signal is fast approaching a seizure onset. The size of the window is taken as 1024 samples in this case. By appropriately thresholding the energy value in the window, a detection of the onset can be made. The mathematical description of the energy metric is given in equation 1. Here E is the energy parameter, x is the input data, N is the size of the window and n is the window index.

$$E_{AVG}[n] = \frac{1}{N}\sum_{i=1}^{N} E(i + (n-1) * N) \qquad (1)$$

where $E(i) = x^2(i)$

3.1.2 Coastline

The coastline parameter is also known as line-length algorithm. Its definition has been adapted from [13]. This feature measures the distance between adjacent data points x in the signal and accumulates them over a window of width N. In the baseline section of the recording, this parameter remains more or less constant. However as we approach the onset of seizure, due to increase in amplitude and frequency of spikes in the signal, the distance between the neighboring data points will increase. The threshold for this parameter can be adjusted accordingly in the training phase. The Neuropace responsive stimulation device by Medtronics Inc. utilizes the coastline or "line-length" as one of the parameters for seizure detection [4] [12]. The mathematical expression for computing the coastline (CL) for k^{th} window is given in equation 2.

$$CL(k) = \sum \left| x[i + (k-1) * N] - x[i - 1 + (k-1) * N] \right| \qquad (2)$$

3.1.3 Non-Linear Energy

The non-linear energy parameter is based on Teager's algorithm [14]. In the conventional energy parameter, all the frequency components in the signal were weighed equally. In comparison, the non-linear energy operator weighs the components at different frequencies non-uniformly. Using square-law weighting, it emphasizes higher frequencies over lower frequencies. Since there is a surge in amplitude at higher frequencies in the seizure phase of the LFP recordings, it expends greater energy during the seizure or ictal phase. This non-uniformity is compensated in the

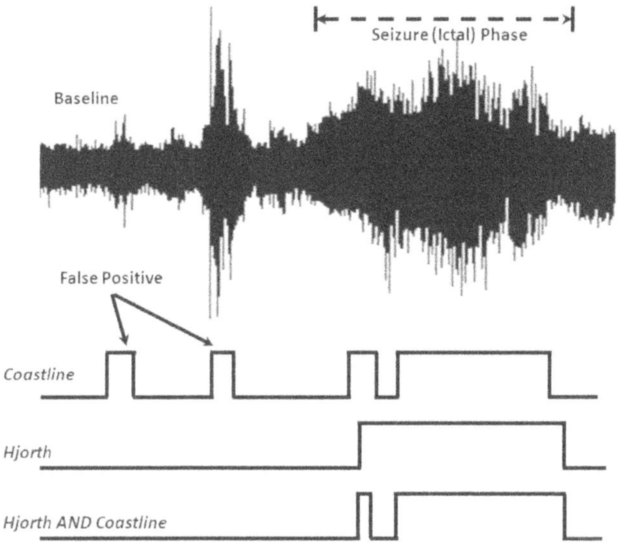

Figure 1 Seizure Detection with independent algorithm and Boolean combination

Teager's algorithm. It has been shown that Teager's algorithm to calculate energy attributes to higher contrast between the energy content in the baseline and the ictal phase [15]. This results in better seizure detection. The non-linear energy Ψ for k^{th} window of size N for data x is given by equation 3.

$$\psi(k) = \frac{1}{N}\sum_{i=1}^{N} x^2[i+(k-1)*N] - x[i+(k-1)*N-1]*x[i+(k-1)*N+1] \quad (3)$$

3.1.4 Hjorth Variance Parameter

Hjorth parameter has been widely used for statistical analysis of EEG [16] [17]. The variance of the signal amplitude also termed "activity" is the first hjorth parameter. It calculates the variance of a window of N samples and averages it to obtain mean variance. The mean of the kth window, μ_k, is also calculated to obtain the first hjorth parameter. The mathematical expression for computing the hjorth parameter is given in equation 4.

$$Var(k) = \frac{1}{N}\sum_{i=1}^{N} x^2[i+(k-1)*N] - \mu \quad (4)$$

$$\text{where } \mu = \frac{1}{N}\sum_{i=1}^{N} x[i+(k-1)*N]$$

3.2 Boolean Logical Combination

The four algorithms described above are used to process the input data and independently detect the onset of the seizure. The system can be programmed to provide a Boolean combination of the detections of any two of these algorithms. By using simple Boolean AND or OR operation, it is shown that the efficacy of detection of seizure is improved with negligible increment in hardware cost. For instance, it can be seen in Figure 1 that the false positives detected in the coastline algorithm are eliminated by using the Boolean ANDing of the coastline and the Hjorth parameter detections. It can also be seen that this might lead to increased delay of detection. Depending upon the nature of training data and neural recording for each patient the best possible combination for the patient can be identified. The detection could be independent using individual algorithm or by the Boolean combinations of multiple algorithms. This ability of programming enables the use of the processor over wide range of patients, thereby increasing the scope of its application.

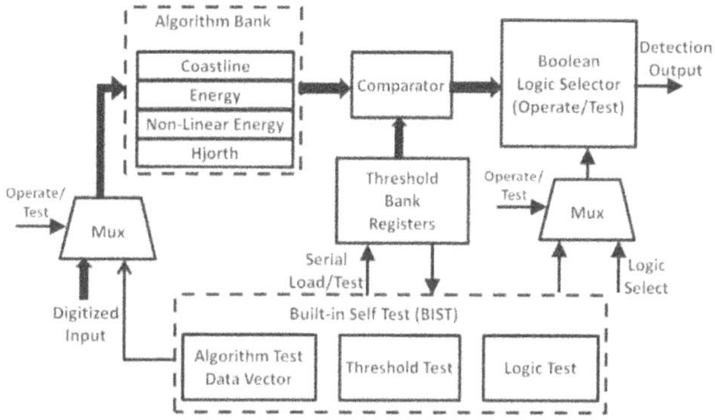

Figure 2 Block Diagram of the Seizure Detection Processor

3.3 Training Phase

The algorithms generate processed values from the input data according to equations 1-4. These values have to be compared with thresholds in order to classify the data as a baseline or seizure. The thresholds are prefixed in the training phase. The training data is randomly selected from the LFP recordings consisting of an equal duration of baseline and seizure phase. This data is then processed using each of the four described algorithm and the threshold is adjusted in order to have the highest efficacy of detection. The definition and method for calculation of efficacy is discussed in Section 5. The training phase is also used to decide the Boolean logic to be used for the combination of multiple algorithms. This is because a patient may not respond positively to individual algorithm but may show a significant improvement if a combination of algorithms is used.

4. HARDWARE IMPLEMENTATION

The multi-algorithm programmable seizure detection processor is implemented as shown in the block diagram in Figure 2. The system basically consists of the Algorithm Bank, Threshold Bank, Boolean logic selector and the Built-in self-test (BIST) blocks. The system has two modes of functions viz. operate mode and the test mode. The operate mode puts the system in the seizure detection mode. The test mode activates the BIST block and performs a self test on the system to verify its integrity. The individual implementation and function of each of these blocks is described in this section.

4.1 Algorithm Bank

The algorithm bank consists of four parameter based algorithms as described in the section 3 viz. Energy, Coastline, Non-linear energy and Hjorth parameter. The hardware implementation is based on the equations 1-4 respectively. The block diagram for each of these algorithms is shown in Figure 3.

The Energy parameter uses the multiplier to square the 10-bit input data and an adder to accumulate it over a window. The window width is 1024 and is controlled by the counter. The counter generates a signal at the end of the window which resets the registers and loads the computed average energy to the output register. Since the window width chosen is a power of two, the average energy is calculated by simple dropping the last 10 bits of the accumulated sum. This avoids a power hungry divider for computing the average. The accumulated sum is passed on to the threshold register bank for comparison and detection.

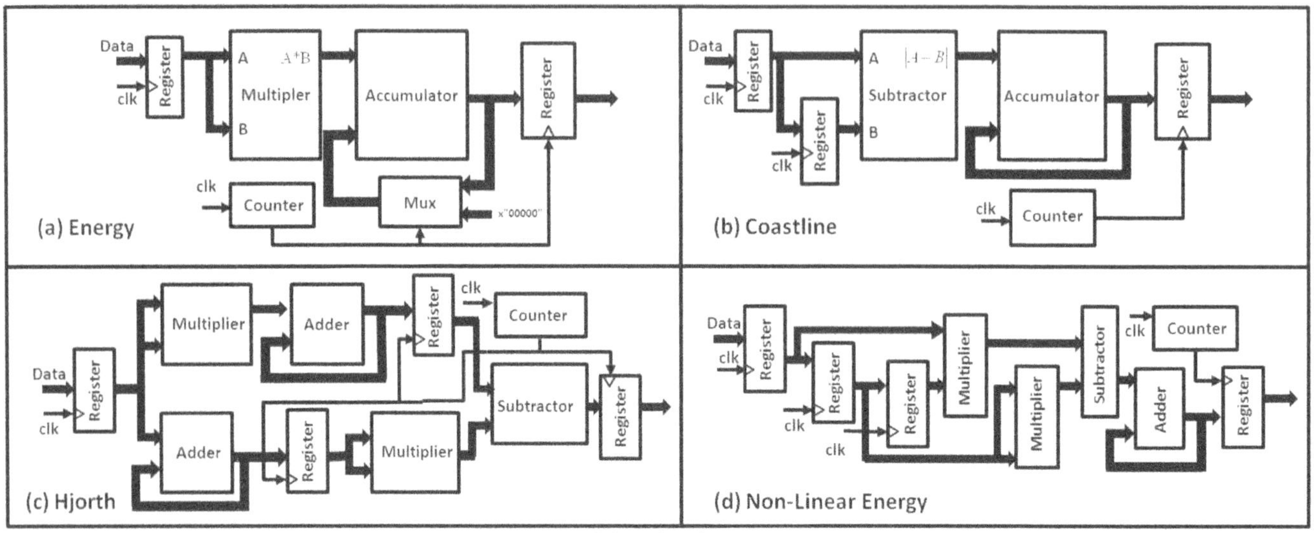

Figure 3 Block Diagram of the Algorithms in the Algorithm Bank

The implementation for coastline parameter is shown in Figure 3(b). The registers at the input store the two neighboring values of the signal. The difference is then accumulated by the adder. The width of the window is controlled using a counter which generates a signal for passing the accumulated difference to the threshold bank.

The Non-linear energy parameter is computed using the values preceding and succeeding each data point as seen in equation 3. This is implemented in hardware by introducing a delay of one clock cycle. The first value is computed only after 3 samples of data are streamed in the system. The counter is used to control the window size and pass the output to the threshold bank.

The Hjorth parameter is calculated by finding the difference between the accumulated squared data and the squared mean in a window of 1024 samples. The counter controls the window size. The Hjorth first parameter is compared to threshold in the threshold bank.

In the operate mode, all the algorithms are simultaneously processing on the 10-bit input data for detecting the seizure. However in the test mode, the algorithms are provided with previously stored data vectors from the BIST block. The intermediate processed values indicate the correctness of computation in each of the algorithms. It should be noted that, the algorithm bank can be generalized to encompass more algorithms. This paper limits the number of algorithm to four for testing the concept.

4.2 Threshold Bank

The processed signals from the algorithm bank are compared to thresholds stored in the threshold bank. The threshold bank consists of four 20-bit registers. These registers hold the threshold values for each of the algorithms. The values in the threshold registers are decided in the training phase as described in the previous section. In the operate mode, the threshold bank is loaded with these threshold values by serially streaming in the bits using a dedicated clock. The processed signals are compared with the threshold values to detect the onset of the seizure. In the test mode, the clock can be used to serially stream out the stored threshold value to verify its integrity. The system has a dedicated clock to operate with threshold registers. This is independent of the system clock and can be turned off after the loading of the threshold registers is complete.

4.3 Boolean Logic Selector

The Boolean logic selector is the programmable block which can be used to select the type of output needed for seizure detection. Depending on the select lines of this block, the output can be either the detection by individual algorithms or their Boolean AND/OR combination. Since the system is designed to be efficacious over a range of patients, the type of logic used for seizure detection is decided in the training phase. In the test mode, by appropriately applying the select values the Boolean logic selector can be tested for its functionality of multiplexing the correct data to the output.

4.4 Built-in Self-Test (BIST)

The seizure detection system implemented on the ASIC includes a built in self testing block. The BIST block is provided to have an on chip testing circuitry to verify the functionality of the system. The BIST block is activated by pulling the test/operate signal low. It provides set of prefixed data vectors to the algorithm bank to test the functionality of the algorithms. The intermediate outputs of each of the algorithms are compared to stored values to verify their correctness. The comparisons generate signals which are given to the Boolean logic selector. By appropriate select signals the output of the system can be used to check the functionality of each of the algorithms. The BIST tests the threshold bank registers by streaming out the stored threshold values through a serial output port by a dedicated threshold clock. The Boolean logic selector block is tested by providing logical inputs to the block and testing the selectivity of the block. The BIST covers all the significant sections of the system to maintain sustained and correct functionality.

5. RESULTS

The seizure detection processor was implemented in 65 nm bulk-Si TSMC technology. Figure 4(a) shows the die photograph of the system. Figure 4(b) shows the test setup. The functionality of the system was tested using the internal BIST block. The efficacy obtained was also verified by comparing it with a MATLAB simulation result. The LFP neural data was recorded in vivo from

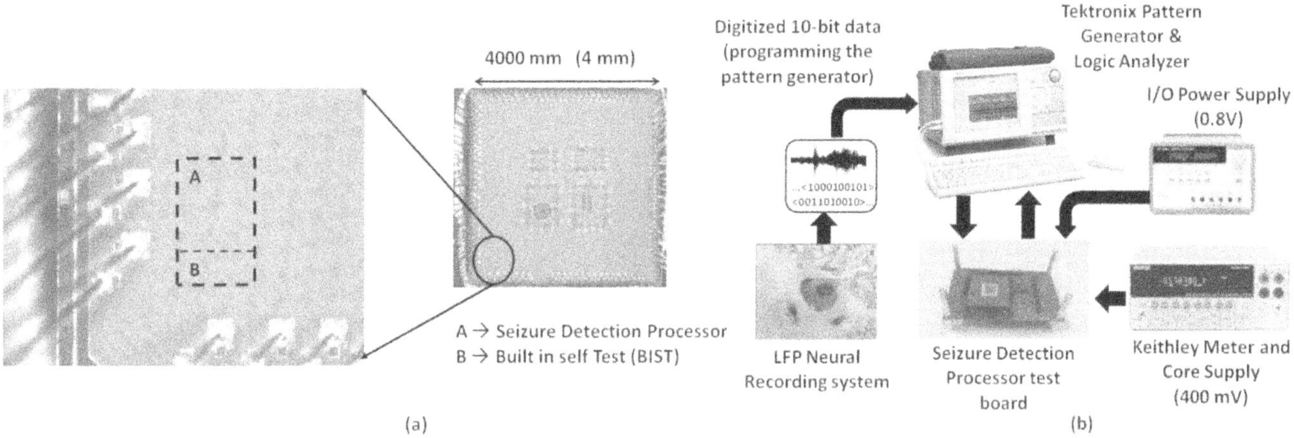

Figure 4 (a) Die Photo of Seizure Detection Processor (b) System Test Setup

rats at a sampling rate of 1.526 KHz along with video. The seizures were induced in the animals using periodic kainic acid injections. The data was classified into the baseline and seizure sections by electrographic and visual evidence. A section of the data was randomly selected as training data set for each animal and used to fix the threshold values. The recorded data was streamed in a digitized form into the 10-bit system input. The system was put in the "operate mode" and the output was observed for detection of the onset of seizure. The Boolean logic selector is programmed according to the training phase to select the best configuration for each set of data.

The efficacy was calculated based on the Specificity, Sensitivity and Average Detection Rate (ADR) as defined by equations 5-7.

$$\text{Sensitivity}=\frac{TP}{TP+FN}\times 100\% \quad (5)$$

$$\text{Specificity}=\frac{TN}{TN+FP}\times 100\% \quad (6)$$

$$\text{ADR}=\frac{\text{Sensitivity}+\text{Specificity}}{2}\times 100\% \quad (7)$$

The Sensitivity is calculated using true positives (TP) and false negatives (FN). It represents the number of false detections caused in the algorithm due to seizure-like events or due to motion artifacts. A high sensitivity indicates that the algorithm would flag any seizure like event and may cause undesirable triggering of neurostimulation. Specificity represents the ability of the

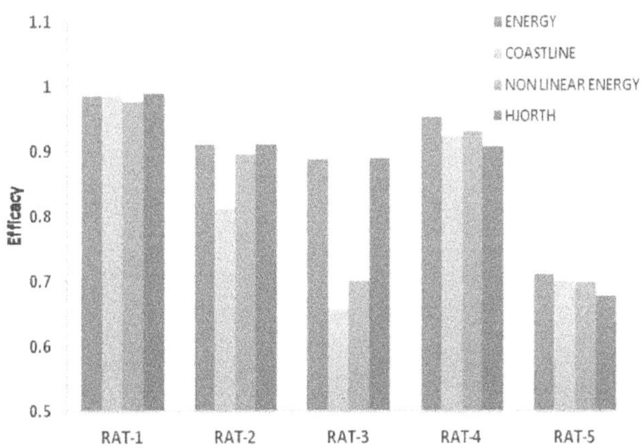

Figure 5 Efficacy of the Algorithm for Large Animal Study

algorithm to reject a FN and is calculated using true negative (TN) and false positive (FP). A higher specificity indicates that the algorithm may reject a non-seizure event with higher confidence and may result in missing a seizure. An increase in the sensitivity indicates a higher percentage of false positive and hence lower specificity and vice versa. Hence, the ADR is used as a fair measure of the efficacy of the algorithm and is the average of the sensitivity and specificity. Neural recordings from five animals were used to compute the efficacy of detection. In Figure 5, the efficacy results are summarized for the four implemented algorithms. It can be seen that the efficacy of the algorithms vary over different animals. On an average, the Hjorth parameter algorithm performs better than the others. However, if we use a Boolean combination of two of these algorithms, the efficacy can be increased for a nominal increase in hardware. This is shown in Figure 6. In this figure, we have plotted the average efficacy of the four algorithms with respect to normalized hardware cost. The hardware cost is calculated by normalizing total power consumption of the algorithm and its area as obtained from simulation using Synopsys and Cadence tools. As seen in Figure 6 we combine the coastline and hjorth parameter using a single OR gate. This configuration increases the average efficacy. By configuring the system for each patient, the best possible efficacy can be achieved.

The seizure detection processor was operated with a clock of 10 KHz. This allowed for voltage scaling in order to reduce the energy consumption of the system. The supply voltage was scaled to 400 mV which is very close to the transistor threshold voltage of 310 mV. This is the most energy efficient region of operation. The power consumption is significantly reduces due to quadratic

Table 1 Measured Results for Power

Technology	65 nm bulk-Si (TSMC)
Dimension	200 μm X 340 μm
I/O supply	0.8 V
Core supply	400 mV
f_{clk}	10 KHz
$P_{leakage}$	12.8 μW
$P_{dynamic}$	360 nW

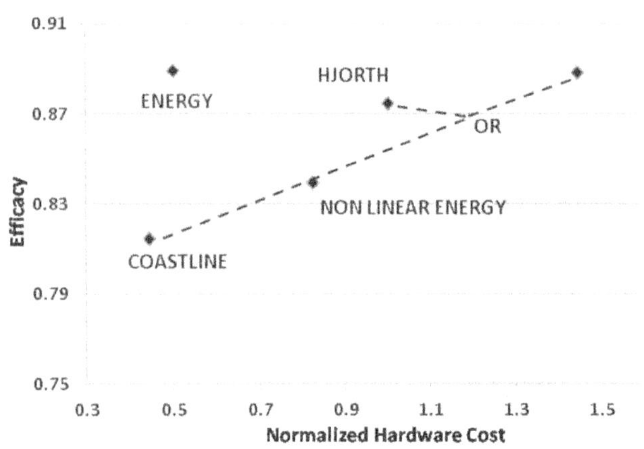

Figure 6 Average Efficacy vs. Normalized Hardware Cost

dependence on the supply voltage. The processor functionality was verified. The power consumption of the system is listed in Table 1. The I/O voltage was scaled to 0.8V. All four algorithms are operational at all times and consume about 12.8μW as leakage power. The dynamic power is 360 nW. This is almost 80% lower than the wavelet based algorithm [11]. Although the parameter based algorithm do not show higher efficacy than the wavelet based algorithm, the programmability can be utilized to improve the efficacy with negligible change in power consumption. The power consumption reported in this paper is without utilization of any leakage control techniques. Programmable logic and near-threshold operation make this seizure detection processor efficient as well as practical for implantable prosthesis.

6. CONCLUSION

In this paper, we have implemented a seizure detection "near threshold" processor which is low in power consumption and has a high efficacy. Furthermore, it is programmable to be adaptable to wider range of patients. The seizure detection processor successfully addresses two major concerns in development of implantable epilepsy prosthesis viz. low power consumption and functionality over inter-patient variation. The processor operates at near threshold voltage supply. This reduces the power consumption significantly and hence increases the longevity of the battery. In future, leakage control techniques such as powering off the unused algorithms can be applied to further reduce the power. Such low-power efficient implementations are the basis for development of practical implantable epilepsy prosthesis

7. ACKNOWLEDGEMENTS

This research was funded in part by Cyberonics Inc. and by National Science Foundation.

8. REFERENCES

[1] http://www.eilepsyfoundation.org

[2] Begley C. E., Famulari M., Annegers J. F., Lairson D. R., Reynolds T. F., Coan S, et al., 2000, The cost of epilepsy in the United States: an estimate from population-based clinical and survey data. *Epilepsia*, 41(3), 342–351.

[3] Ghai R. S., Bikson M., Durand D. M., 2000, Effects of applied electric fields on low-calcium epileptiform activity in the CA1 region of rat hippocampal slices. *J. of Neurophysiology*, 84,274–280.

[4] Sun F. T., Morrell M. J., Wharen R. E., 2008, Responsive cortical stimulation for the treatment of epilepsy. *Neurotherapeutics*, 5, 68–74.

[5] Labar D., Murphy J. and Tecoma E., 1999, Vagus nerve stimulation for medication-resistant generalized epilepsy, *Neurology*, 52, 1510–1512.

[6] Politsky J. M., Estellar R., Murro A. M., Smith J. R., Ray P., Park Y. D., et al., 2005, Effects of electrical stimulation paradigm on seizure frequency in medically intractable partial seizure patients with a cranially implanted responsive cortical neurostimulator. In *Proc. of the Annual Meeting of American Epilepsy Society (AES)*.

[7] D'Allesandro M., Esteller R., Vachtsevanos G., Hinson A., Echauz J., Litt B., 2003, Epileptic seizure prediction using hybrid feature selection over multiple intracranial EEG electrode contacts: a report of four patients. *IEEE Transactions on Biomedical Engineering*; 50, 603–613.

[8] Zarjam P.; Mesbah M. Discrete wavelet transform based seizure detection in newborns EEG signals, 2003, In: *Proc. 7th International Symposium on Signal Processing and Its Applications*, 2, 459- 462.

[9] Webber W. R. S., Ronald P., Lesser, Richardson R. T., Wilson K., An approach to seizure detection using an artificial neural network (ANN), 1996, *Electroencephalography and Clinical Neurophysiology*, 98(4), 250-272.

[10] Raghunathan S., Gupta S. K., Ward M. P., Worth R. M., Roy K., Irazoqui P. P., 2009, The design and hardware implementation of low-power real-time seizure detection algorithm, *J. Neural engineering*, 6(5).

[11] Markandeya H., Karakonstantis G., Raghunathan S., Irazoqui P. P., and Roy K., 2010, Low-power DWT-based quasi-averaging algorithm and architecture for epileptic seizure detection, *Proc. of the 16th ACM/IEEE international symposium on Low power electronics and design (ISLPED '10)*,301-306.

[12] Raghunathan S., Gupta S. K., Markandeya H. S., Roy K., Irazoqui P. P., 2010, A hardware-algorithm co-design approach to optimize seizure detection algorithms for implantable applications, *J. of Neuroscience Methods*, 193(1), 106-117.

[13] White A.M., Williams P.A., Ferraro D.J., Clark S., Kadam S.D., Dudek F.E. et al., 2006, Efficient unsupervised algorithms for the detection of seizures in continuous EEG recordings from rats after brain injury, *J. of Neuroscience Methods*, 152, 255–266.

[14] Kaiser J. F., 1990 "On a simple algorithm to calculate the `energy' of a signal," *International Conference on Acoustics, Speech, and Signal Processing (ICASSP), 1*, 381-384.

[15] Zaveri H. P., W. J. Williams W. J., and Sackellares J. C., Energy Based Detection of Seizures, 1993, *15th Annual International Conference on Engineering and Medicine in Biology*, 363-364.

[16] Hjorth B., An on-line transformation of EEG scalp potentials into orthogonal source derivations, 1975, *Electroencephalography and Clinical Neurophysiology*, 39,526–530.

[17] Pfurtscheller G., Lopes da Silva F.H., 1999, Event-related EEG/MEG synchronization and desynchronization: basic principles, *Clinical Neurophysiology*, 110, 1842–1857.

Understanding the Impact of Laptop Power Saving Options on User Satisfaction Using Physiological Sensors

Matthew Schuchhardt
Northwestern University
matt.schuchhardt@u.northwestern.edu

Ben Scholbrock
Northwestern University
benjamin.scholbrock@eecs.northwestern.edu

Utku Pamuksuz
Northwestern University
utkupamuksuz2011@u.northwestern.edu

Gokhan Memik
Northwestern University
memik@eecs.northwestern.edu

Peter Dinda
Northwestern University
pdinda@northwestern.edu

Robert P. Dick
University of Michigan
dickrp@eecs.umich.edu

ABSTRACT

Several techniques are available to save power consumption in laptop computers. However, their effect on user satisfaction has not been well studied. We analyze how user satisfaction is affected by these techniques and show that, within a fixed power budget, some techniques cause more dissatisfaction than others. Second, we study the use of physiological sensors and show that the sensor readings are stable across times when no technique is applied, whereas they show statistically significant changes when power-saving techniques are employed. Finally, we demonstrate a prediction mechanism using these sensors that predicts user satisfaction with over 80% accuracy.

Categories and Subject Descriptors: C.5.3 [**Computer System Implementation**]: Microcomputers—*Personal computers*; H.1.2 [**Models and Principles**]: User/Machine Systems—*Human Factors*; J.4 [**Computer Applications**]: Social and Behavioral Sciences—*Psychology*;

Keywords: Affective computing, human factors, user-aware computing, power saving techniques, empathic systems and architectures

1. INTRODUCTION

The primary goal of a computer system is to satisfy the end-user. Traditionally, computer designers measure the effect of their optimizations using metrics expected to correlate with user satisfaction (e.g., tasks per time, instructions per cycle, packets per time). However, previous work has shown that a) providing higher performance levels does not necessarily improve the user satisfaction and b) the level of performance necessary to satisfy a particular user varies significantly among users (e.g., [7, 16].) With no knowledge of user satisfaction regarding machine performance, optimizations exercising trade-offs between performance and power consumption will be designed to satisfy the nonexistent "typical" user with a "one-size-fits-all" approach. This will often lead to sub-optimal configuration choices, which could be improved by understanding individual user satisfaction.

Computer optimizations are typically performed from the perspective of a single component or system block. For example, Dynamic Voltage and Frequency Scaling (DVFS) [12], which reduces power at the cost of CPU performance for a period of time, is controlled independently from screen brightness, which may be adjusted to conserve power. User satisfaction is a function of both CPU performance and screen brightness, but current control schemes do not consider combined effects. This paper aims to lay the groundwork for a new method of measuring computer performance that would allow the integration of such effects and their derivation from easy-to-acquire user measurements. This new method of evaluating computer systems describes how well a computer performs, not only in traditional terms of raw computational capabilities or power efficiency, but from the perspective of end-user satisfaction.

In this study, we first construct a linear regression-based power model to estimate power consumption of a computer system using software-visible metrics. We leverage this model to estimate the impact of various power-saving techniques on total system power. We then perform user studies, activating combinations of these power-saving techniques during machine use. We request that users in the study verbally report dissatisfaction with the computer system. We also employ several physiological and behavioral sensors to monitor the user. These sensors measure the galvanic skin response (GSR), pupil dilation and movement, the amount of wrist movement, the pressure applied to the keyboard during key presses, and the frequency of key presses. Finally, we analyze the sensor readings based on the reported dissatisfaction.

We next demonstrate the impact of power-saving techniques on user satisfaction. We find that the frequency of user-reported dissatisfaction varies by power-saving technique, even though the same reduction in power consumption is achieved. We also find that changes in user satisfaction

This work was supported by the United States National Science Foundation through grants CCF-0916746, CCF-0747201, CNS-0720691, and DGE-0948017.

ISLPED'12, July 30–August 1, 2012, Redondo Beach, CA, USA.

are reflected in measurable physiological responses. In situations where saving power does not affect user satisfaction, physiological sensor data remains constant. However, when a power-saving technique affects reported user satisfaction, sensor data changes. These findings motivate satisfaction-oriented power/performance optimizations in light of constrained power requirements on personal computers.

We also create a prediction system that predicts whether a power-saving technique changes the user satisfaction. Our system achieves over 80% accuracy. Finally, we perform a supplementary study which demonstrates that our prediction system performs equally well whether users verbally indicate dissatisfaction or not.

The rest of the paper is organized as follows. In Section 2, we present our experimental setup. In Section 3, we present our analysis of the sensor data, as well as the results of our prediction system. In Section 4, we discuss related work in the field, and we conclude in Section 5.

2. EXPERIMENTAL SETUP

We now describe the environment in which we carried out our studies, and the studies themselves.

2.1 Power Model

Before correlating user satisfaction with power savings techniques, we first develop a model for accurately estimating the power saved by each technique. Techniques may then be parameterized to save roughly equivalent levels of system power.

Our target platform is an IBM Thinkpad T61 with a 2.2 GHz Intel Core 2 Duo T7500 processor and 2 GB DDR2 SDRAM running Microsoft Windows XP. This computer is also the platform used during user studies.

The predictors we target in the power model follow: RAM bus transactions, disk accesses, processor utilization percent, screen brightness, screen on/off state, and WiFi on/off state. We also directly measure the power being used by the system; this is our dependent variable.

One of the most common techniques for saving power is reducing CPU frequency. However, changing the CPU frequency substantially alters the amount of power that the other predictor variables would contribute to the power model. Hence, we generate separate power models for the system running at 1.6 GHz and 2.2 GHz.

To create the power model, we run a variety of benchmarks, each which primarily stress one component on the system. We then create a linear regression model from this data.

We also perform four user-driven verification workloads, each which stress a different aspect of the computer. Using these workloads, we compare the actual power readings with the power levels predicted by our model. Averaged across all readings and verification workloads, our mean absolute error is less than 2%.

2.2 Physiological Sensors

In this work, we utilize a GSR sensor, a 3-axis accelerometer, an eyetracker, and piezoresistive force sensors to measure the physiological and behavioral effects of different power saving techniques. Additionally, we also record keyboard button presses via software hooks.

The GSR sensor and the accelerometer are both mounted on a wrist strap, the signals from which are fed into an Arduino-based microcontroller. The force sensors are attached to the arrow keys on our keyboard, and their signals are also fed into an Arduino device. The eyetracker is made by MobileEye [2], and consists of a head-mounted camera which is controlled by a dedicated laptop.

From those sensors and logging software, we extract seven different metrics:

- AccelMag: sum of squares of X, Y, and Z accelerometer axes
- DeltaGSR: change in GSR value since the last sensor reading
- Keypress: time since the last keyboard button press
- MaxForce: largest current value from the force sensors
- NormalMaxForce: same as MaxForce, but normalized to each key's highest force reading
- PupilMovement: change in position of the pupil since the last pupil reading
- PupilRadius: the radius of the pupil.

We tried to match the metric to what we expected to be the most interesting behavior of the sensor. For instance, GSR sensors tend to exhibit a low-frequency change in sensor values on top of the more interesting high-frequency signal, so we only look at the change in GSR since the last sensor reading to reduce the low-frequency effects.

2.3 User Studies

In this section, we describe the techniques that are used to reduce power consumption. To simplify the discussion, we call these techniques 'annoyance events' because they are applied for short durations during our user studies, and they may cause user dissatisfaction/annoyance.

From the results of the power model, we decided to focus on the CPU utilization (CPUUtil), CPU frequency (CPUFreq), and screen brightness (ScreenDim and ScreenGrad). CPUUtil limits the scheduling time of the target application without changing the frequency. CPUFreq reduces the frequency of the system's processor. ScreenDim reduces the brightness of the screen to save power. ScreenGrad does the same, but does so gradually over 10 seconds. The rest of the predictors either would completely disrupt the experiment (e.g., WiFi on/off), or are not easily throttleable (e.g., RAM accesses).

We select a variety of different annoyance events which total up to 4, 8, or 12 Watts of power savings for our target system (Table 1). If multiple annoyances are in effect at the same time, the power savings is split between them (e.g., if 3 annoyances are active for 4 Watts of total savings, each annoyance saves 1.33 Watts). We also include 'pseudo annoyances', where no annoyance is initiated by our system. We initiate each annoyance event twice during user studies. Every annoyance event is active for 30 seconds after which the computer reverts back to the original state for 25-40 seconds. The order of annoyance events are randomized to eliminate any possibility of biasing.

The workload that each user in this study performs is a racing game called Motocross Madness 2. We use a racing game as our workload because it is relatively CPU-bound, and requires consistent attention to play. Future work will likely explore additional workloads. The laptop typically consumes 40-41 Watts of power while running the game without any active annoyance events.

Table 1: **Total number of times (out of 40 instances) that the annoyance event combination was indicated as being annoying. If multiple events are active, each contributes an equal amount to the total amount of power saved.**

Power Saved	Annoyance Events Active	# Times User Indicated Annoyance
0W	None	1
4W	CPUFreq	1
4W	CPUUtil	15
4W	ScreenDim, CPUFreq, CPUUtil	15
4W	ScreenDim	30
4W	ScreenGrad	31
8W	CPUFreq	0
8W	CPUFreq, CPUUtil	18
8W	ScreenDim, CPUFreq, CPUUtil	27
8W	ScreenDim, CPUUtil	30
8W	ScreenDim, CPUFreq	31
8W	CPUUtil	31
12W	ScreenDim, CPUFreq, CPUUtil	33

As the users are playing the game, we request that they verbally notify us when they become dissatisfied. Other methods of reporting dissatisfaction are possible, such as using a physical button or periodically rating machine performance on a scale of values. Simple verbal reports require minimal user action, which is desirable since gameplay is continuous through annoyance events.

We ran this study on 20 users. The studies are conducted in a well-lit room with overhead fluorescent lighting and drawn shades, to minimize the effect that the time of day and screen brightness have on the user's pupil dilation.

The advertisements for the study were placed around various parts of Northwestern's campus, and the respondents were all university students. Of the 20 users, there were 12 males and 8 females. 10 were undergraduate students, and 10 were graduate students.

3. RESULTS

We now examine the results of our study in order to derive a model for predicting user satisfaction from the physiological sensors.

3.1 Annoyance Event Analysis

In this section, we examine how users react to the annoyance events. As the users are playing the game, we request that they verbally notify us when they are dissatisfied. We then note how many times users indicated annoyance for each annoyance event. In Table 1 we list how frequently the users indicated annoyance.

One interesting piece of information to note from this table is that different users have very different reactions to the CPUFreq and CPUUtil annoyances. Users very rarely express annoyance when only the CPU frequency is reduced, even though the corresponding CPU utilization reduction saves the same amount of power. Reducing the frequency not only reduces the number of cycles the processor can complete in a given amount of time, but also allows the processor to reduce its voltage, which may explain this discrepancy in indicated annoyance events.

It's also interesting how users readily notice screen brightness changes. Users frequently report annoyance with the system when the screen is dimmed. One notable exception is the 4-Watt ScreenDim + CPUFreq + CPUUtil annoyance: even though there is a ScreenDim annoyance active in that set, the screen isn't dimmed for this annoyance as much as other cases. This seems to suggest that there is a particular threshold for when users indicate annoyance with a system component. This result alone shows the importance of understanding user satisfaction with power saving techniques: a naïve optimization would dim the screen (with or without a change in CPU frequency) to save power. However, our results show that for this target application, the best method is to partially reduce the CPU frequency, and partially dim the screen . A system that can understand user satisfaction with different optimizations will choose this option over the rest.

3.2 Statistical Sensor Value Analysis

To analyze the data collected from the biometric sensors, we first note that there are two distinct sets of data that relates to every individual annoyance event: data during the non-annoyance phase preceding the introduction of the annoyance event, and data during the annoyance phase.

Intuitively, we would expect the sensors to either substantially change before and during an annoyance event, or to remain approximately the same. We hypothesize that events which the users either don't notice or don't care about will not substantially change the sensor metrics. Conversely, we expect that annoyance events which the user perceives as annoying would cause a measurable change between the two sets of sensor readings.

Since the data before and during each annoyance event consists of hundreds of sensor readings, we compute the standard deviation, mean, and median of each of the metrics before and during each annoyance event.

We do not know for certain how the users' biometric data will change for each annoyance event. There's a good chance that we would miss some sensor responses if we only looked at one static span of data for each event; some sensors may have a delayed reaction, or the change may happen over a larger or smaller window of time. To improve our likelihood of capturing the timespan that is most indicative of annoyance for each sensor, we consider multiple data windows and offsets to see how the different sensors respond. We look at window sizes of 10, 15, 20, and 25 seconds for multiple offsets from the start of the annoyance event. Please note that the larger windows do not have as much room to move within the 25-second window of non-annoyance, and hence, we do not analyze as many offsets for those windows.

Since we are only concerned with detecting changes in metrics when the user is actually annoyed, we first only consider event instances which are indicated by the user as unsatisfactory. For each event, we have data from seven sensor metrics (Section 2.2). For each metric, we collect data from the various window/offset combinations, as described above. On each of those, we then calculate a standard deviation, mean, and median. Each of these data frames consists of the mean, median and standard deviation from the window before and during one annoyance event instance.

All of those similar data frames are then compared from before and during the annoyance using a paired 2-tailed t-test. 2-tailed t-tests with a relevant p-value ($p<0.1$ in this study) suggest a statistical difference between two sets of samples.

Due to space constraints, we do not present graphs for every sensor, but the results for the selected set of sensor metrics are shown in Figure 2.

One interesting thing to note in these graphs is that the metrics and windows which give statistically significant t-tests differences vary for each sensor. Some of the sensors respond best with a larger window, while others seem to react best using a smaller one. Additionally, it seems that lower offsets perform better than larger offsets, which would suggest that these sensors reflect external stimuli fairly quickly. Many of the sensors show promise for our future studies, as we can statistically show the difference between the data before and during an annoyance event with a high degree of confidence for many of the windows and offsets.

3.3 Equivalence Analysis

A false positive occurs whenever the user annoyance prediction system indicates that the user is annoyed when they actually are not. A theoretical system could be devised which always detected a difference in some sensor metric. This system would always indicate that the user is annoyed, and so would be 100% effective in detecting annoyance, but would never detect non-annoyance. Thus, we have to prove that our sensor metrics are not only good at detecting differences in sensor data due to annoyance events, but we also must show that they perform reasonably well at detecting non-annoyance. In other words, the system needs both a low false negative rate and a low false positive rate.

'TOST' [15], or 'two one-sided tests', is one way of demonstrating equivalence between two sets of data. TOST works by running two paired one-tailed t-tests with some offset ϵ.

When proving dissimilarity (Section 3.2), we only analyze data sets that the user deems annoying. Conversely, for this equivalence analysis, we only consider data sets in which the user did not indicate annoyance. From the previous t-test analysis, we take the sets of data with the 10 lowest p-values, but at least one and no more than two from each sensor. We then determine the smallest possible ϵ value that is required to achieve a confidence of 90% that the two sets of data are equivalent. In Table 3, we present the results of our equivalence analysis. We represent the relative size of the offset by dividing ϵ by the average of the means.

The required ϵ values for all of the metrics we analyze are relatively small, compared to the rest of the data. However, a few of the t-tests (PupilMove, one of the AccelMag) are statistically significant: this would suggest that those particular sensor/window/offset combinations may not be the best choice for an eventual prediction system. The rest of the metrics behave well, however.

3.4 User Annoyance Prediction

In this section, we determine how accurately we can actually predict user annoyance.

To perform this analysis, we employ standard data mining techniques using labeled instances of data. We employ the Weka [1] data mining suite to assist in handling the data and building the models. In this analysis, we label any event where the user expressed annoyance as 'True', and instances where they did not as 'False'.

We then create a large number of attributes on those labeled instances. To generate our attributes, we take data from 0–2 sec, 2–4 sec, 4–6 sec, 6–8 sec, 0–5 sec, 5–10 sec, 10–15 sec, 15–20 sec, 0–10 sec, and 10–20 sec, from before and from after each annoyance event. For each of those time periods, we find both the means and standard deviations. Additionally, we take the difference from before and after. This gives us a total of 60 attributes. We run this attribute generation method for each of our 7 sensor metrics.

If we were to simply use all of those attributes in a data mining algorithm, the most useful attributes would get lost in the noise of the rest of the attributes. Thus, to improve our eventual model's accuracy, we employ CFS [8], a feature selection algorithm, which reduces the number of attributes we end up using in the actual model.

After feature selection, we then run the actual data mining algorithm. To build our model, we use logistic regression [18], which is a type of generalized linear model frequently used on binary labeled data. For all of our analyses, we use 10-fold cross-validation.

We generate a different model for each user. We also include a 'baseline' set of data. This baseline was generated by using ZeroR, which is a naïve algorithm that always chooses the most common label (True or False). The results are presented in Figure 4.

As the graph shows, all of the individual sensors have medians better than the ZeroR baseline case. However, when all of the sensors are combined in the model, the prediction accuracy is significantly better. We would like to note that some users did not indicate any annoyance, which gives ZeroR a slight advantage (since it is always correct for those users), but we did not discard the data; we assume that they could not tell a difference in performance.

We must also know how accurate the system is at individually predicting 'True' and 'False' instances, as a system which has a very high 'True' prediction accuracy, but very low 'False' prediction (or vice-versa) is not as useful. Using the combined sensors model, our system predicts 'True' accurately 212/263=80.6% of the time, and 'False' accurately 203/257=79.0% of the time (false positive=19.4%, false negative=21%).

3.5 Results Verification

In our original experiment, the users verbally notified us when they were annoyed. However, their verbal notification could alter the readings on our biometric sensors. To address this possibility, we run a small supplementary user study on five users to verify that we can detect when the user is annoyed without a verbal indication. The primary difference in this study is that we do not have the users notify us when they are annoyed; we simply have them play the game as the performance changes.

Since we already have determined which annoyances the user perceives as annoying and non-annoying, we only use the top 5 most annoying events, as reported by users in our original study (see Table 1), as well as 5 'pseudo' annoyance events where there is no change in performance. We run each of those annoyance events two times. We increase the time of non-annoyance to 55–65 seconds, with 30 seconds of annoyance.

We then run a similar analysis to our original study, but instead of using the user-reported annoyances to label the annoyance events as 'True' or 'False', we label all of the intentionally annoying events as 'True', and all of the pseudo annoyance events as 'False'. The graphs in Figure 5 are the

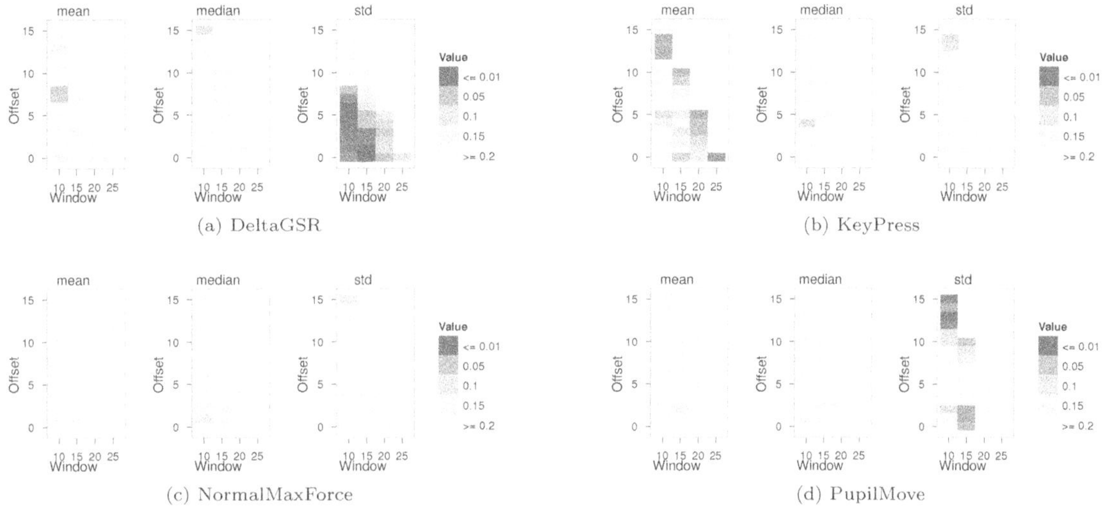

Figure 2: Windowed t-test analysis for selected sensors. Low p-values, represented by dark areas, indicate metric/offset/window combinations where sensor data differs before and during annoyance events. Only user-indicated annoyances are analyzed.

Table 3: Confidence intervals for equivalence test analysis, using data from the non-annoying events. Only includes selected metrics which performed well in the previous t-test analysis.

Sensor	Metric	Window	Offset	Average Before Annoyance	Average During Annoyance	Required ϵ for 90% Confidence	ϵ / avg of means	t-test P-value
AccelMag	std	15	3	3.0123	3.2871	0.7048	0.2238	0.1853
AccelMag	std	20	0	3.0288	3.4610	0.9376	0.2890	0.0949
DeltaGSR	std	10	5	38.9661	39.3882	6.8749	0.1755	0.7906
DeltaGSR	std	10	6	39.0690	39.4917	6.9296	0.1764	0.7889
Keypress	mean	25	0	0.1968	0.1878	0.0249	0.1297	0.2565
MaxForce	median	15	1	142.7626	138.3891	20.1704	0.1435	0.5764
NormalMaxForce	std	10	15	0.1468	0.1431	0.0130	0.0896	0.3681
PupilMove	std	10	13	7.6988	6.7270	1.7515	0.2428	0.0267
PupilMove	std	10	15	7.5278	6.8014	1.4574	0.2034	0.0717
PupilRadius	std	15	2	25.7464	25.2551	2.1452	0.0841	0.5001

result of our prediction analysis. The baseline for this graph is exactly 50%: there were exactly 10 annoying events and 10 non-annoying events for each user, so ZeroR would always be correct 50% of the time.

As the graph shows, all sensors perform much better than the baseline, and again, the combined sensor model performs the best, with an average accuracy of 79%. This shows that our system works, regardless of whether there are any verbal user indications of annoyance or not.

Using the combined sensors model, we are able to predict 'True' instances 38/50=76% of the time, and 'False' instances 41/50=82% of the time (false positive=24%, false negative=18%).

4. RELATED WORK

Mouse and keyboard actions alone are often used to determine user state in human-computer-interaction scenarios. Fox et al. [6] try to determine user satisfaction based on these in the context of web searching. Macaulay [10] uses mouse click frequency as an indicator for user anxiety.

Endo et al. [5] explore using OS event handling latency, rather than computational throughput, as a measure for performance. Video frame rates and application response times are also used as a user-perceivable performance metric for the system, as a proxy for user satisfaction [11]. Gupta et al studied the effects of resource scheduling decisions on directly reported user satisfaction [7]. Even further studies investigate leveraging biometric data to adjust microprocessor power level, to account for user satisfaction [17]. However, none of these studies analyzed the impact of different power saving methods.

Picard et al. [13], proposes a method of using a variety of sensors to reduce the negative impact that interruptive technologies have on users. That same group also proposed a method of predicting user frustration using sensors in a learning environment (Kapoor et al. [9]), but use a different set of sensors, and do not consider power saving methods in this case.

Considerable work has been put into systems which use sensors to acquire affect data from users in order to enable computers to perform certain user-facing tasks. For example, the educational field has looked into using sensors to improve tutoring systems. Cooper et al. [4] describes a tutoring system which uses multiple sensors to improve the relevancy

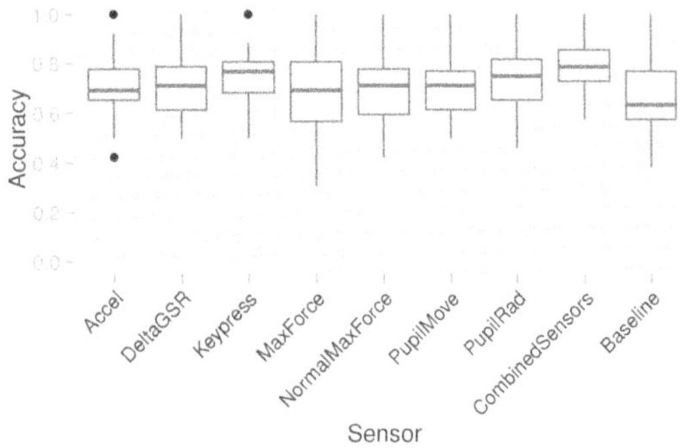

Figure 4: Model prediction accuracy using individualized user models. We include results for 7 individual sensors, a combined model which uses all 7 sensors, and a baseline prediction accuracy. The box's bottom and top are the 25th and 75th percentile, the middle line is the median, and the whiskers go to the highest and lowest points within 1.5 IQR of the higher and lower quartiles. Dots represent outliers beyond 1.5 IQR.

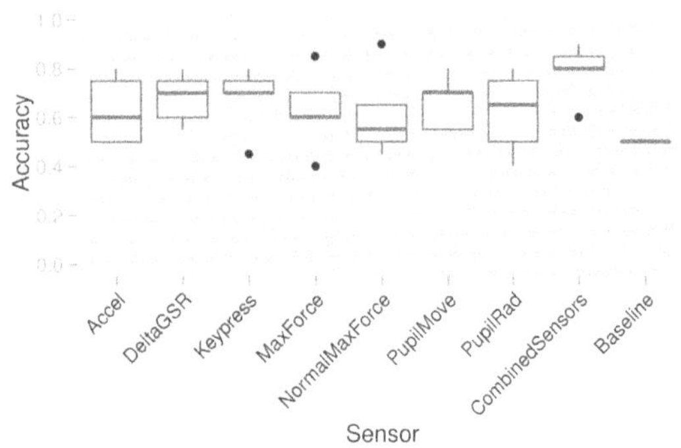

Figure 5: This boxplot is similar to the plot in Figure 4, but the users don't verbally indicate if they are annoyed. Data is labeled based on if the annoyances were typically annoying or not.

of questions on a standardized test studying platform by maximizing user engagement. Burleson describes a system [3] where a virtual agent helps guide the user through the Towers of Hanoi problem; the agent uses sensors to try to empathize with the user's current emotional state. Predinger and Ishizuka describe another agent-based interface that empathizes with the user through the use of physiological sensors [14]. However, these studies do not consider computer system performance or power saving methods.

5. CONCLUSION

We have argued for measuring and understanding end-user satisfaction using physiological sensors, and supported our argument through a user study whose data allow us to correlate voiced satisfaction, biometric information, and intentionally-introduced power-savings events. We found that different power-savings techniques reduce user satisfaction to different degrees, and that saving a given amount of power using a combination of techniques is generally preferable to using a single technique alone.

We have also shown that it is possible to accurately measure changes in user satisfaction via various physiological metrics. When the performance is constant, biometric readings remain constant, while when performance varies, biometric readings also vary. This correlation can be used to predict user satisfaction from the metrics, and we showed how to do so with 80% accuracy through the use of per-user models developed through data-mining.

In a design environment where power and performance trade-offs directly impact user experience, understanding the relationship between user satisfaction and power-saving techniques is critical. This paper highlights some of these effects, and indicates that biometric sensors are a powerful way to acquire feedback about the individual end-user without requiring any effort on his part.

References

[1] Weka 3 - data mining with open source machine learning software in java. http://www.cs.waikato.ac.nz/ml/weka/.

[2] ASL. Applied science laboratories - mobile eye: Lightweight tetherless eye tracking, 2008.

[3] W. Burleson. Advancing a multimodal real-time affective sensing research platform. *New Perspectives on Affect and Learning Technologies*, pages 97–112, 2011.

[4] D. Cooper, I. Arroyo, and B. Woolf. Actionable affective processing for automatic tutor interventions. *New Perspectives on Affect and Learning Technologies*, pages 127–140, 2011.

[5] Y. Endo and M. I. Seltzer. Using latency to evaluate interactive system performance. In *Proc. of the Intl. Conf. on Measurements and Modeling of Computer Systems*, 2000.

[6] S. Fox, K. Karnawat, M. Mydland, S. Dumais, and T. White. Evaluating implicit measures to improve web search. *ACM Trans. Inf. Syst.*, 23:147–168, April 2005.

[7] A. Gupta, B. Lin, and P. A. Dinda. Measuring and understanding user comfort with resource borrowing. In *Proceedings of the 13th IEEE International Symposium on High Performance Distributed Computing (HPDC 2004)*, June 2004.

[8] M. Hall. *Correlation-based feature selection for machine learning*. PhD thesis, The University of Waikato, 1999.

[9] A. Kapoor, W. Burleson, and R. Picard. Automatic prediction of frustration. *International Journal of Human-Computer Studies*, 65(8):724–736, 2007.

[10] M. Macaulay. The speed of mouse-click as a measure of anxiety during human-computer interaction. *Behaviour and information Technology*, 23(6):427–433, 2004.

[11] A. Mallik, J. Cosgrove, R. Dick, G. Memik, and P. Dinda. PICSEL: Measuring user-percieved performance to control dynamic frequency scaling. In *Proceedings of the Intl. Conference on Architectural Support for Programming Languages and Operating Systems*, March 2008.

[12] Microsoft. Windows native processor performance control. In *Windows Platform Design Notes*, November 2002.

[13] R. Picard and K. Liu. Relative subjective count and assessment of interruptive technologies applied to mobile monitoring of stress. *International Journal of Human-Computer Studies*, 65(4):361–375, 2007.

[14] H. Prendinger and M. Ishizuka. The empathic companion: A character-based interface that addresses users'affective states. *Applied Artificial Intelligence*, 19(3-4):267–286, 2005.

[15] D. Schuirmann. A comparison of the two one-sided tests procedure and the power approach for assessing the equivalence of average bioavailability. *Journal of Pharmacokinetics and Pharmacodynamics*, 15(6):657–680, 1987.

[16] A. Shye, B. Ozisikyilmaz, A. Mallik, G. Memik, P. A. Dinda, R. P. Dick, and A. N. Choudhary. Learning and leveraging the relationship between architecture-level measurements and individual user satisfaction. In *Proceedings of the Intl. Symposium on Computer Architecture*, June 2008.

[17] A. Shye, Y. Pan, B. Scholbrock, J. S. Miller, G. Memik, P. A. Dinda, and R. P. Dick. Power to the people: Leveraging human physiological traits to control microprocessor frequency. In *Proceedings of the Intl. Symposium on Microarchitecture*, December 2008.

[18] I. Witten and E. Frank. *Data Mining: Practical machine learning tools and techniques*. Morgan Kaufmann, 2005.

MultiScale: Memory System DVFS with Multiple Memory Controllers

Qingyuan Deng David Meisner[†] Abhishek Bhattacharjee
Thomas F. Wenisch[‡] Ricardo Bianchini

Rutgers University [†]Facebook Inc. [‡]University of Michigan
{qdeng,abhib,ricardob}@cs.rutgers.edu meisner@fb.com twenisch@umich.edu

ABSTRACT

The fraction of server energy consumed by the memory system has been increasing rapidly and is now on par with that consumed by processors. Recent work demonstrates that substantial memory energy can be saved with only a small, tightly-controlled performance degradation using *memory* Dynamic Frequency and Voltage Scaling (DVFS). Prior studies consider only servers with a single memory controller (MC); however, multicore server processors have begun to incorporate multiple MCs. We propose MultiScale, the first technique to coordinate DVFS across multiple MCs, memory channels, and memory devices. Under operating system control, MultiScale monitors application bandwidth requirements across MCs. It then uses a heuristic algorithm to select and apply a frequency combination that will minimize the overall *system* energy within user-specified per-application performance constraints. Our results demonstrate that MultiScale reduces system energy consumption significantly, compared to prior approaches, while respecting the user-specified performance constraints.

Categories and Subject Descriptors

C.5 [**Computer System Implementation**]: Miscellaneous

Keywords

Memory system, energy conservation, dynamic voltage and frequency scaling

1. INTRODUCTION

Processors have historically dominated server power consumption. However, main memory power has been growing substantially as multicore servers require increasing memory capacity and bandwidth [4, 18, 21, 31]. Today, main memory accounts for roughly 40% of server energy [31], which is comparable to or slightly higher than processor energy.

Most prior work on memory energy conservation creates memory idleness through scheduling, batching, and layout transformations, so that *idle* low-power modes can be exploited [9, 11, 12, 14, 17, 26]. Techniques that reduce the number of DRAM chips accessed together [2, 33] or change DRAM chip microarchitecture [7, 30] have also been considered. Unfortunately, these prior techniques either provide limited benefit for DDR* memories or require changes to DIMMs (Dual In-line Memory Modules) and/or DRAM chip microarchitecture.

In contrast, *active* low-power modes for main memory have shown great promise in trading memory performance for energy savings [8, 10]. In these techniques, the hardware and the operating system (OS) collaborate to assess the benefits of scaling the voltage and frequency of the MC, and scaling the frequency of the memory channels and devices. The ultimate goal is to provide maximal energy savings under a user-specified performance degradation constraint.

These past memory DVFS proposals select a single performance setting for the memory system and are thus ideal for systems with a single MC. However, chip multiprocessors (CMPs) are increasingly integrating multiple on-die MCs [1, 3, 32]. Furthermore, recent work has demonstrated the benefit of *deliberately* skewing traffic across multiple MCs to preserve fair performance among applications judged likely to interfere (i.e., by placing data such that memory-intensive and non-memory intensive applications access disjoint MCs/channels) [25]. Such asymmetric traffic patterns will call for correspondingly asymmetric DVFS control.

Recent hardware trends also suggest that traffic skew across MCs will grow. For example, as servers increasingly rely on multi-socket configurations, inter-socket MC bandwidth requirements will vary significantly [31]. In addition, the advent of heterogeneous processors incorporating sophisticated superscalar out-of-order cores with simpler in-order cores (e.g., ARM's big.LITTLE architecture [13]), and graphics processing units (e.g., AMD's Fusion and Intel's Sandybridge architectures [28]), will fundamentally increase traffic skew across MCs. Therefore, it is critical to explore novel multi-MC active low-power mode management techniques. The straight-forward extension of prior work—selecting the same frequency for all MCs based on their average bandwidth requirement—will lead to sub-optimal savings under skewed traffic.

Thus, this paper presents MultiScale, a set of software policies and hardware mechanisms for coordinating DVFS across multiple MCs, channels, and devices. Under OS control, MultiScale monitors per-application traffic across MCs and estimates their varying bandwidth and latency requirements. It then uses a heuristic algorithm to quickly select and apply an optimized MC frequency combination. Like prior work on memory active low-power modes [10], MultiScale's goal is to minimize the overall system energy, without degrading performance beyond a user-specified limit. Unlike past work however, MultiScale is able to do so effectively and consistently for multi-MC systems under a variety of traffic skews.

We evaluate MultiScale using detailed simulation on a diverse set of workload mixes constructed from the SPEC benchmark suite [29]. We quantify MultiScale's benefits across a range of traffic-skew patterns, showcasing its consistently higher energy efficiency versus past single voltage/frequency approaches [10].

This work is the first to study techniques to apply memory system active low-power modes to multiple MCs in a coordinated manner. First, we develop a set of low-overhead, yet effective software policies and hardware mechanisms that monitor per-application, per-MC bandwidth/latency requirements. Our readily-implementable performance counters, allied with low-overhead OS support, can be used to realize MultiScale's performance and energy models. Second, we quantify MultiScale's ability to exploit user-defined per-application performance degradation constraints across a range of traffic skew patterns. Our results show that MultiScale's ability to coordinate per-MC DVFS based on the workloads' dynamic memory bandwidth requirements allows it to achieve up to 4.5× greater energy savings than prior approaches. MultiScale is effective even in situations when performance constraints are tight; for example, when the allowable degradation is capped at just 1%, MultiScale can still achieve energy savings over 9% whereas past work achieves savings of merely 2%.

2. BACKGROUND AND RELATED WORK

We summarize DRAM operation and architecture; more details can be found in [23, 24]. In single-MC CMPs, a single controller services memory requests to all addresses, which are interleaved over one or more memory channels. Each channel has its own data and address bus and can be independently accessed. A channel connects to one or more DIMMs, each of which usually includes 16 or 18 DRAM chips. Each memory access targets a rank, a set of chips that collectively respond to the access. Each rank comprises multiple banks, each of which is a two-dimensional DRAM cell array. A channel's banks can be accessed in parallel, though they share the channel's address and data buses.

Several recent commercial processors integrate MCs on-chip. For example, Nehalem [15] integrates four cores and one MC with three channels to DDR3 memory. However, MC complexity and resource contention limit the number of channels a single controller can effectively manage. As systems accommodate higher core counts, multiple on-chip MCs will become the norm. For example, Power7 uses two MCs connected to eight channels [31], whereas Tile64 uses four MCs to service 64 cores [32].

2.1 Memory Power Management

Increasing memory capacity demands and growth in the fraction of overall system power attributable to memory [4, 6, 21, 31] have motivated studies of memory low-power modes. Numerous studies investigated the use of idle low-power modes (e.g., precharge powerdown, self-refresh) [9, 11, 12, 14, 17, 26]. However, past work has shown that active low-power modes are generally more successful at saving energy under performance constraints, while being readily implementable [8, 10]. These techniques observe that while server workloads are often highly sensitive to memory access latency, only rarely do they demand peak memory bandwidth [22]. In our prior work [10], we used this insight to propose MemScale, an energy management technique for the memory system. MemScale leverages dynamic profiling, performance and power modeling, DVFS of the MC, and DFS of the memory channels and DRAM devices to realize significant energy savings under user-specified performance degradation constraints.

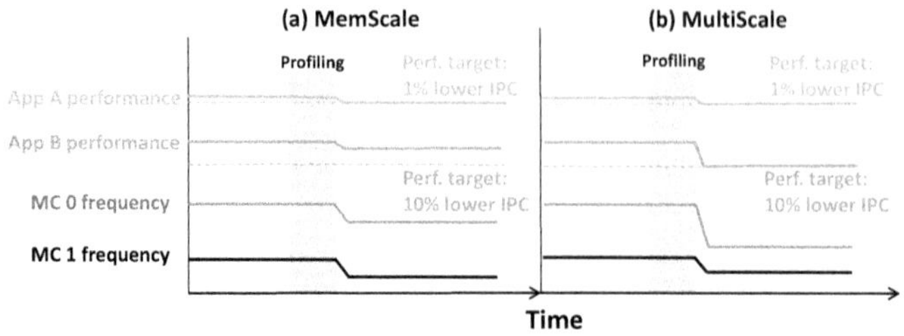

Figure 1: Because it independently manages each MC, MultiScale can select a lower frequency for MC 0 and thus save more energy than MemScale while remaining within the prescribed performance bounds.

2.2 Cross-MC Traffic Skew

MemScale selects a single performance setting for the memory system based on aggregate application requirements. As such, it is unable to exploit time-varying or asymmetric traffic patterns across MCs in multiple-MC systems. Figure 1 illustrates a scenario where two applications, A and B, run on a dual-MC system. Suppose the memory allocation is skewed such that 80% of application A's accesses are directed to MC 1, whereas application B's accesses are predominantly to MC 0. Further, suppose that application A is under a tight performance constraint, and can tolerate only a 1% degradation, whereas application B can tolerate up to 10% slowdown. Application A's tight performance constraint will require MC 1 at a high frequency. However, there might be substantial opportunity to slow MC 2 without violating the performance constraint of either application. MemScale (Figure 1 left) selects only a single frequency for both MCs, based on the tighter performance constraint of application A, limiting energy savings. In contrast, MultiScale (Figure 1 right) independently assesses the bandwidth requirements of each MC and selects an appropriate frequency/voltage to maximize energy savings while respecting each applications' degradation constraint.

While the OS maps virtual to physical addresses, the interleaving of physical addresses has the greatest impact on the traffic skew across MCs. Cache line and page interleaving seek to distribute traffic evenly. While such interleaving balances traffic across MCs, there are many strong arguments and hardware/software trends pointing toward addressing schemes that skew traffic across channels. For example, recent work by Muralidhara and co-authors has demonstrated the benefit of deliberately skewing traffic across multiple MCs to preserve fair performance among applications judged likely to interfere [25]. Under this scheme, data from memory-intensive and non-memory-intensive applications are mapped to disjoint MCs/channels. This isolates non-memory-intensive workloads from interference effects of intense memory traffic, while clever scheduling techniques manage access among the memory-intensive workloads. Similarly, Awasthi and co-authors propose mechanisms to place data in multi-MC systems to improve performance [3]. Their approach, which balances the benefits of allocating application data to the MC closest to the corresponding core against queuing delays, on-chip latencies, and row buffer hit rates, often results in traffic skew. Traffic skew is also common in heterogeneous platforms with accelerators, such as graphics processing units and sophisticated out-of-order cores coupled with simpler in-order cores, which might impose differing bandwidth requirements across MCs.

MultiScale adapts better than MemScale to traffic skew across MCs. So, in characterizing MultiScale, we have two primary goals: (1) We demonstrate its effectiveness across a range of traffic patterns. To this end, we quantify

MultiScale's operation across a spectrum of traffic skews. (2) We aim to showcase MultiScale's effectiveness at realizing energy savings even under the tightest performance degradation constraints. To this end, we study the benefits of MultiScale across a range of degradation constraints, as small as 1% allowable slowdown.

3. MULTISCALE DESIGN

MultiScale seeks to maximize energy savings while adhering to per-application user-specified performance degradation constraints. We now detail the hardware mechanisms and software policies that make this possible.

3.1 Hardware and Software

Hardware mechanisms. For each MC, MemScale adjusts its frequency and voltage and the frequency of its associated memory channels and DIMMs; for expediency, we shall refer to these operations collectively as "adjusting the MC frequency". The DIMM clocks lock to the bus frequency (or a multiple thereof), while the MC frequency is fixed at double the bus frequency. While MultiScale can readily be applied to systems where each MC controls multiple channels at different frequencies, assessing the hardware overheads of such an approach is beyond the scope of this work. We therefore treat each MC, together with all its memory channels, as the unit for frequency selection. A frequency change requires the system to briefly suspend operation and reconfigure to run at the new target [10]. Our experiments model the associated recalibration delays.

Similar to [10], MultiScale frequency scales the MCs, buses, and DIMMs. Voltage scaling is restricted to the MCs, and is set according to the selected frequencies.

Performance counter monitoring. Our management policies require input from a set of performance counters implemented on each core and on-chip MC. Specifically, we require counters tracking the amount of work pending at each MC's memory banks and channels. Counters similar to those we require already exist in most modern architectures, and are accessible through the CPU's performance-monitoring interface. For further details on the exact performance counters used, we refer the reader to [10] as MultiScale uses identical counters.

Energy management policy. Our goal is to minimize overall system energy consumption without degrading performance beyond user-specified bounds. Therefore, as in previous schemes [10, 20, 26], MultiScale exploits the notion of performance *slack*: the difference between a baseline execution and a target slowdown that the each application may incur to save energy. Our control algorithm exploits this allowable slack to reduce memory system performance and save energy. The per-application performance target is defined such that the application incurs no more than a pre-selected maximum slowdown relative to its execution without energy management (i.e., at maximum frequency). Formally, the slack is the difference in time of the program's execution (T_{Actual}) from the target (T_{Target}).

$$\begin{aligned}\text{Slack} &= T_{\text{Target}} - T_{\text{Actual}} \\ &= T_{\text{MaxFreq}} \cdot (1+\gamma) - T_{\text{Actual}}\end{aligned} \tag{1}$$

where γ defines the target maximal execution time increase.

In exploiting this slack, MultiScale's control algorithm divides execution into fixed-sized epochs. We typically associate an epoch with an OS time quantum. MultiScale splits each epoch into four distinct phases. First, applications are profiled by collecting statistics from the performance counters. We find that profiling for 300 μs in an epoch of 5 ms suffices. Second, the OS uses the profiling information to select new MC frequencies (as detailed in the next subsection). Third, each MC, its channels, and DRAM devices are transitioned to their new frequency. Finally, the epoch completes at this new frequency configuration. At the end of the epoch, we again query the counters and estimate the performance that would have been achieved had the memory system operated at maximum frequency. The difference between this estimate and the achieved performance is used to update the slack and is carried forward to calculate the target performance in the next epoch.

3.2 Performance and Energy Models

Performance model. Our control algorithm utilizes a performance model extended from [10] to account for per-MC frequencies and the traffic directed by each application to each MC. The performance model predicts the relationship between CPU *cycles per instruction* (CPI) of an application, and the per-MC frequency. There are three steps in our modeling approach. First, we estimate the memory-boundedness of each running application, and thus estimate the target access latency that satisfies the performance target of each application. Second, we estimate each MC's contribution to this average latency (which requires the MCs to be aware of the hardware thread that issued each access). Finally, based on the first two steps, we calculate per-MC frequencies. Next, we detail each step.

Step 1: Under our performance model, the runtime of a program is defined as: $t_{\text{total}} = t_{\text{CPU}} + t_{\text{Mem}} = I_{\text{CPU}} \cdot E[TPI_{\text{CPU}}] + I_{\text{Mem}} \cdot E[L_{\text{Mem}}]$, where I_{CPU} represents the number of instructions, and I_{mem} is the number of last-level cache (LLC) misses stalling the pipeline. TPI_{CPU} represents the average time that instructions spend on the CPU (including L1 cache hits, L1 cache misses, and L2 cache hits in a two-level cache hierarchy), and L_{Mem} is the average memory latency of each application.

Since runtime is not known a priori, we model the rate of progress of an application in terms of CPI. The average CPI of a program is defined as: $E[\text{CPI}] = (E[TPI_{\text{CPU}}] + \alpha \cdot E[L_{\text{Mem}}]) \cdot F_{\text{CPU}}$, where α is the fraction of instructions that miss in the L2 cache and stall the pipeline, and F_{CPU} is the operating frequency of the core. The value of α can easily be calculated as the ratio of instruction to LLC miss counts, accessible through performance counters. Given a target CPI, we can compute $E[L_{\text{Mem}}]$ assuming other components in the above equation are constant. By substituting α into the equation and dividing by the frequency, we can compute the target average per-application latency needed to compute per-MC latency.

Step 2: Having calculated the target average memory access latency per application, we now focus on the latency breakdown per MC. To understand this, consider a simple scenario where there are two applications, A and B, and two MCs, MC 0 and MC 1. Suppose that for A, $Perc_{A0}$% of total memory accesses go to channels under MC 0, while $Perc_{A1}$% go to channels under MC 1. Further, assume that $Perc_{B0}$% of accesses from B go to channels under MC 0, and $Perc_{B1}$% of accesses go to channels under MC 1. We have:

$$\begin{cases} Perc_{\text{A0}}\% + Perc_{\text{A1}}\% = 100\% \\ Perc_{\text{B0}}\% + Perc_{\text{B1}}\% = 100\% \end{cases} \tag{2}$$

Assume that the average access latency to channels under MC 0 is L_0, and to channels under MC 1 is L_1. Furthermore, denote the average memory access latency of A and B is $E[L_A]$ and $E[L_B]$, respectively. We then have:

$$\begin{cases} E[L_{\text{A}}] = Perc_{\text{A0}} \cdot L_0 + Perc_{\text{A1}} \cdot L_1 \\ E[L_{\text{B}}] = Perc_{\text{B0}} \cdot L_0 + Perc_{\text{B1}} \cdot L_1 \end{cases} \tag{3}$$

We can now use these equalities to calculate per-MC memory access latencies. Using the information from step (1) on each application's $E[L_{\text{Mem}}]$, we can cap each application's latencies so as to ensure the correct performance targets. Specifically, assuming that $L_{\text{Target\_A}}$ and $L_{\text{Target\_B}}$ are the threshold latencies that guarantee the performance targets of A and B, we have the following inequalities.

$$\begin{cases} E[L_{\text{A}}] \leq L_{\text{Target\_A}} \\ E[L_{\text{B}}] \leq L_{\text{Target\_B}} \end{cases} \tag{4}$$

Solving this system provides L_0 and L_1 values which can then be used as input to our next step.

Although this simple example assumes two applications and two MCs, this approach can be generalized to any number of applications and MCs. In general, this entails solving a linear programming (LP) problem where the number of MCs is likely smaller than the number of running applications. (MultiScale only needs to deal with the applications running during the next epoch.) Standard LP solvers can be used to calculate these latencies efficiently. The overhead of this computation is negligible for realistic numbers of MCs, since it only occurs once per epoch. For our setup (4 MCs and 16 cores), the overhead is less than 50 μs on a Xeon 5520 machine.

Step 3: Having solved for L_0 and L_1 and the latencies of all other MCs (which we collectively denote as L_{Mem}), we model the relationship between channel frequency and memory access latency. We take the approach of [10]: $E[L_{\text{Mem}}] = \xi_{\text{bank}} \cdot (S_{\text{Bank}} + \xi_{\text{bus}} \cdot S_{\text{Bus}})$, where ξ_{bus} represents the average number of requests waiting for the bus and is approximated by the counters capturing the queuing impact of waiting for bus transfers; ξ_{bank} represents the average number of requests waiting for the bank and is approximated by the counters capturing per bank queuing; S_{Bank} is the average time, excluding queueing delays, to access a bank (including precharge, row access and column read, etc); and S_{Bus} is the average data transfer (burst) time across the bus. Since the values of S_{Bank}, ξ_{bank}, and ξ_{bus} can be obtained by profiling performance counters, we can calculate S_{Bus}, which is a function of the frequency. Finally, we can calculate the target frequency from S_{Bus}.

Full-system energy model. Simply meeting the CPI loss target for a given workload does not necessarily maximize energy efficiency. In other words, though additional performance degradation may be allowed, it may save more energy to run faster. To determine the best operating point, we construct a model to predict full-system energy usage. For memory frequency $f_{\text{MC\_1}}, f_{\text{MC\_2}}, \ldots, f_{\text{MC\_N}}$, we define the *system energy ratio* (SER) as:

$$\text{SER}(f_{\text{mem}}) = \frac{T_{f_{\text{Mem}}} \cdot (\sum_i P_{f_{\text{MC\_i}}} + P_{NonMem})}{T_{\text{Base}} \cdot P_{\text{Base}}} \tag{5}$$

$T_{f_{\text{Mem}}}$ is the performance estimate for an epoch at frequency $f_{\text{MC\_1}}$ through $f_{\text{MC\_N}}$. Memory power is calculated with the memory power model in [24], and P_{NonMem} accounts for all non-memory system components and is assumed to be fixed. T_{Base} and P_{Base} are corresponding values at a nominal frequency. At the end of each epoch's profiling phase, we calculate SER for all memory frequencies that can meet the performance constraint given by the slack, and select the frequency that minimizes the SER, from the range calculated by the latency model as described earlier in this section.

4. EVALUATION

In this section, we demonstrate the efficacy of MultiScale over a range of traffic skews relative to MemScale.

Table 1: Workloads.

Name	MPKI	WPKI	Applications (x4 each)			
MIX1	2.93	2.56	applu	hmmer	gap	gzip
MIX2	2.34	0.39	milc	gobmk	facerec	perlbmk
MIX3	2.55	0.80	equake	ammp	sjeng	crafty
MIX4	2.41	1.41	lucas	vpr	h264ref	eon
MIX5	2.35	1.38	swim	ammp	twolf	sixtrack
MIX6	2.91	1.57	libquantum	twolf	vpr	sjeng
MIX7	3.12	1.48	mcf	astar	gzip	sixtrack
MIX8	1.83	0.77	mgrid	fma3d	crafty	eon

Table 2: Simulation parameters.

Feature	Value
CPU cores	16 in-order, single thread, 4GHz Single IALU IMul FALU FMulDiv
L1 I/D cache (per core)	64KB, 2-way, 1 CPU cycle hit
L2 cache (shared)	16MB, 16-way, 10 CPU cycle hit
Memory configuration	4 MCs, 1 channel/MC, 8 2GB DIMMs
tRCD/tRP/tCL	15ns, 15ns, 15ns
tFAW/tRTP/tRAS/tRRD	20/5/28/4 cycles
tXP/tXPDLL/Refresh	6ns/24ns/64ms
Row buffer read, write	250mA, 250mA
Activation-precharge	120 mA
Active standby	67 mA
Active powerdown	45 mA
Precharge standby	70 mA
Precharge powerdown	45 mA
Refresh	240 mA
VDD	1.575 V

4.1 Methodology

Simulator and workloads. Our evaluations are based on a two-step simulation methodology. First, we use M5 [5] to collect memory access traces (consisting of L1 cache misses and writebacks), and per-core activity traces. Second, we feed these traces into a detailed simulator modeling a 16-core CMP with a shared L2 cache (LLC), on-chip MCs, memory channels, and DRAM devices. We also feed core activity traces, with the run-time statistics from the L2 module, into McPAT [19] to dynamically estimate the CPU power. Overall, we simulate in detail all aspects of core microarchitecture, caches, memory controller and memory devices relevant to our study, including memory device power and timing, and row buffer management.

Table 1 lists the main characteristics of our 8 workloads. The workloads are formed by combining applications from the SPEC 2000 and SPEC 2006 suites. We analyze the best 100M-instruction simulation point for each application (selected using Simpoints 3.0 [27]). The workload terminates when the slowest application has executed 100M instructions. We report the LLC misses per kilo instruction (MPKI) and LLC writebacks per kilo instruction (WPKI) of each workload in Table 1.

Simulation parameters. Table 2 details our baseline memory system with 4 MCs, each connected to one DDR3 channel. Each channel is populated with two registered, dual-ranked DIMMs with 18 DRAM chips each. We further assume the 16 cores are interconnected using as 4x4 mesh with the MCs on the corners. Every set of 4 cores has a *local* MC, which is at the adjacent corner.

Table 2 also shows the timing and power parameters of the DRAM chips, register, PLL, and MC [24], assuming a frequency of 800 MHz. We also consider frequencies of 733, 667, 600, 533, 467, 400, 333, 267, and 200 MHz. Frequency transitions take 512 memory cycles plus 28ns [16]. MultiScale further assumes that the MC frequency is double the channel frequency, which in turn is double the DRAM device frequency. The voltage range of each MC is the same as that of the cores (0.65V-1.2V). Each MC can be set at a different voltage. Each MC's power ranges from 2W to 4W, depending on the voltage, frequency, and utilization. Based on these values, our memory system (including MCs) accounts for 44% of the total system power on average.

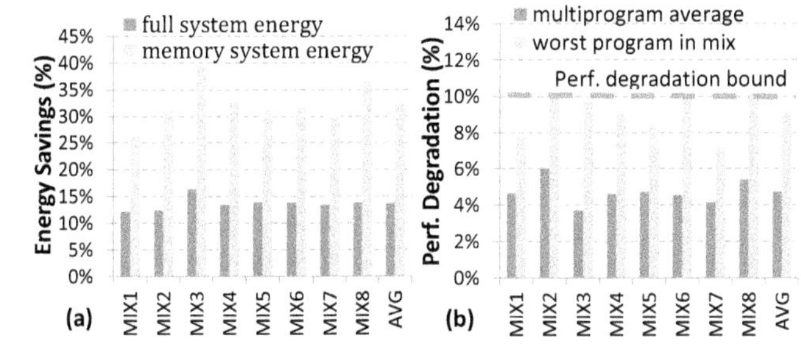

Figure 2: (a) MultiScale's energy savings assuming that 80% of an application's pages are mapped to its local MC and a 10% performance loss bound; (b) MultiScale's actual performance loss in this scenario.

Experiments. We compare MultiScale to MemScale for different traffic distributions and performance degradation bounds. We distribute traffic by controlling how many pages of each application are allocated to its local MC. In our experiments, we vary this from 100% (the extreme form of channel partitioning proposed in [25]) to 80%, 60%, 40%, 25%, and 20%. Non-local MC pages are allocated randomly across the remote MCs. Thus, the 25% case represents the scenario in which each MC is responsible for (roughly) the same number of pages.

For each of these cases, we investigate 1%, 3%, 5%, 7%, and 10% allowable performance slowdowns.

4.2 Results

Figure 2(a) shows MultiScale's memory and full-system energy savings across the workload mixes, assuming maximum allowable performance degradations of 10%, and a distribution where 80% of an application's pages are allocated to its local MC. The figure indicates that MultiScale is successful in saving energy across all workload mixes under skewed traffic. Although the exact savings vary across the workloads, on average, MultiScale saves 13% of the baseline *full-system* energy.

MultiScale's energy savings do not come at the cost of excessive performance degradation. Figure 2(b) shows the average and worst-case performance degradation across all applications in a workload. The results indicate that MultiScale saves energy without exceeding the 10% performance loss constraint across all workload mixes.

Figure 3 depicts the MultiScale and MemScale energy savings across the entire spectrum of performance loss bounds and page allocation schemes. The horizontal axis plots the page allocation scheme, whereas the vertical axis captures the full-system energy savings of each approach. For every allocation scheme and approach, there are five bars, each illustrating the full-system energy savings for a different performance loss bound.

From this figure, we make the following observations. First, MultiScale saves energy across *every* considered allocation and performance bound scenario. The exact energy savings depend upon the amount of traffic skew across MCs and the performance bound. As expected, greater skew and performance bounds allow MultiScale to save more energy. For example, with 100% local page allocation and 10% performance bound, MultiScale saves over 14% of full-system energy. In contrast, with 40% of pages allocated to the local MC and a 5% bound, the energy savings are 8%.

Second, MultiScale conserves at least as much energy as MemScale on *every* considered scenario. MultiScale's advantage increases with greater traffic skew across MCs. This is expected since MultiScale is better able to detect the traffic pressure on each MC and adjust each MC to an appropriate frequency. Interestingly, Figure 3 also shows that MultiScale outperforms MemScale substantially when the performance bounds are low. For example, at a 1% performance bound and 100% local MC allocation, MultiScale can still achieve energy savings of over 9%, whereas MemScale manages merely 2%. The reason is that MultiScale provides finer-grained control of the required memory system performance for a given slack; as such, it is easier for MultiScale to exploit any available slack.

Finally, we consider the performance loss that MultiScale and MemScale incur across the same spectrum of page allocations and loss bounds. Figure 4 depicts these data, showing the performance loss of the *most* degraded application. The figure demonstrates that MultiScale degrades the performance of the worst-hit application slightly more than MemScale. These differences are most pronounced (but still lower than 2%) with lower traffic skew and higher performance bounds. In fact, MultiScale slightly violates the bound in a few cases, but always by less than 0.23%. These slight violations occur because MultiScale pushes *all* applications to the edge of their performance bounds, whereas MemScale pushes only the application that is most sensitive to memory performance to its bound. Thus, MultiScale is more prone to (slight) violations.

5. CONCLUSION

We proposed MultiScale, a set of hardware mechanisms and software policies for using active low-power modes to manage multiple MCs in a coordinated fashion and under performance constraints. MultiScale yields greater energy savings than the best previous approach, MemScale, by gauging the traffic requirements of each MC and setting it to the appropriate DVFS level. As our results demonstrate, MultiScale is particularly effective in scenarios where traffic is skewed across MCs and when the allowable performance degradation is low. As such, MultiScale is an ideal energy management approach for multi-MC systems with scheduling and allocation policies that promote traffic skew.

6. ACKNOWLEDGEMENTS

This research was partially supported by Google and the National Science Foundation under grants #CCF-0916539, #CSR-0834403, and #CCF-0811320.

7. REFERENCES

[1] D. Abts, N. D. Enright Jerger, J. Kim, D. Gibson, and M. Lipasti. Achieving Predictable Performance Through Better Memory Controller Placement in Many-Core CMPs. In *ISCA*, 2009.

[2] J. H. Ahn, J. Leverich, R. Schreiber, and N. P. Jouppi. Multicore DIMM: An Energy Efficient Memory Module with Independently Controlled DRAMs. In *IEEE Computer Architecture Letters*, 2009.

[3] M. Awasthi, D. Nellans, K. Sudan, R. Balasubramonian, and A. Davis. Handling the Problems and Opportunities Posed by Multiple On-Chip Memory Controllers. In *PACT*, 2010.

[4] L. A. Barroso and U. Hölzle. The Datacenter as a Computer: An Introduction to the Design of Warehouse-Scale Machines. *Synthesis Lectures on Computer Architecture*, Jan. 2009.

[5] N. Binkert, R. Dreslinski, L. Hsu, K. Lim, A. Saidi, and S. Reinhardt. The M5 Simulator: Modeling Networked Systems. In *IEEE Micro*, 2006.

[6] S. Blagodurov, S. Zhuravlev, and A. Fedorova. Contention-Aware Scheduling on Multicore Systems. In *ACM Trans. Comput. Syst.*, 2010.

[7] E. Cooper-Balis and B. Jacob. Fine-grained Activation for Power Reduction in DRAM. In *IEEE Micro*, 2010.

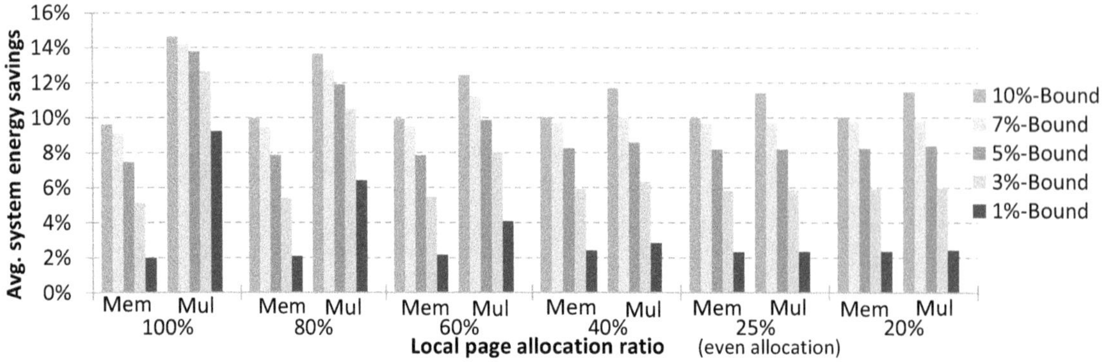

Figure 3: MultiScale's energy savings versus MemScale across a spectrum of traffic skews and performance degradation bounds. MultiScale consistently provides greater energy savings than MemScale.

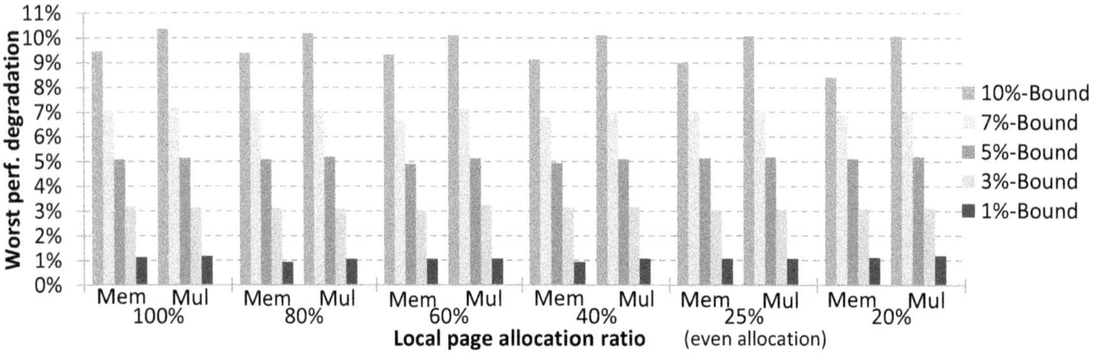

Figure 4: MultiScale's worst performance degradation versus MemScale across a spectrum of traffic skews and performance degradation bounds. MultiScale leads to slightly higher degradations than MemScale.

[8] H. David, C. Fallin, E. Gorbatov, U. R. Hanebutte, and O. Mutlu. Memory Power Management via Dynamic Voltage/Frequency Scaling. In *ICAC*, 2011.

[9] V. Delaluz, M. Kandemir, N. Vijaykrishnan, A. Sivasubramaniam, and M. J. Irwin. Hardware and Software Techniques for Controlling DRAM Power Modes. In *IEEE Transactions on Computers*, 2001.

[10] Q. Deng, D. Meisner, L. Ramos, T. F. Wenisch, and R. Bianchini. MemScale: Active Low-Power Modes for Main Memory. In *ASPLOS*, 2011.

[11] B. Diniz, D. Guedes, W. Meira Jr, and R. Bianchini. Limiting the Power Consumption of Main Memory. In *ISCA*, 2007.

[12] X. Fan, C. Ellis, and A. Lebeck. Memory Controller Policies for DRAM Power Management. In *ISLPED*, 2001.

[13] P. Greenhalgh. big.LITTLE Processing with the Cortex-A15 and Cortex-A7 Processors, 2011.

[14] H. Huang, K. Shin, C. Lefurgy, and T. Keller. Improving Energy Efficiency by Making DRAM Less Randomly Accessed. In *ISLPED*, 2005.

[15] Intel. Intel Xeon processor 5600 Series, 2010.

[16] JEDEC. DDR3 SDRAM Standard, 2009.

[17] A. R. Lebeck, X. Fan, H. Zeng, and C. Ellis. Power Aware Page Allocation. In *ASPLOS*, 2000.

[18] C. Lefurgy, K. Rajamani, F. Rawson, W. Felter, M. Kistler, and T. W. Keller. Energy Management for Commercial Servers. *IEEE Computer*, 36(12), 2003.

[19] S. Li, J. Ahn, R. Strong, J. Brockman, D. Tullsen, and N. Jouppi. McPAT: An Integrated Power, Area, and Timing Modeling Framework for Multicore and Manycore Architectures. In *Micro*, 2009.

[20] X. Li, Z. Li, F. M. David, P. Zhou, Y. Zhou, S. V. Adve, and S. Kumar. Performance-Directed Energy Management for Main Memory and Disks. In *ASPLOS*, 2004.

[21] K. Lim, J. Chang, T. Mudge, P. Ranganathan, S. K. Reinhardt, and T. F. Wenisch. Disaggregated Memory for Expansion and Sharing in Blade Servers. In *ISCA*, 2009.

[22] D. Meisner, C. M. Sadler, L. A. Barroso, W.-D. Webber, and T. F. Wenisch. Power Management of Online Data-Intensive Services. In *ASPLOS*, 2011.

[23] Micron. 1Gb: x4, x8, x16 DDR3 SDRAM, 2006.

[24] Micron. Calculating Memory System Power for DDR3, July 2007.

[25] S. P. Muralidhara, L. Subramanian, O. Mutlu, M. Kandemir, and T. Moscibroda. Reducing Memory Interference in Multicore Systems Via Application-Aware Memory Channel Partitioning. In *MICRO*, 2011.

[26] V. Pandey, W. Jiang, Y. Zhou, and R. Bianchini. DMA-Aware Memory Energy Management. In *HPCA*, 2006.

[27] E. Perelman, G. Hamerly, M. V. Biesbrouck, T. Sherwood, and B. Calder. Using SimPoint for Accurate and Efficient Simulation. In *SIGMETRICS*, 2003.

[28] S. Sawant, U. Desai, G. Shamanna, L. Sharma, M. Ranade, A. Agarwal, S. Dakshinamurthy, and R. Narayanan. A 32nm Westmere-EX Xeon Enterprise Processor. In *ISSCC*, 2011.

[29] Standard Performance Evaluation Corporation. SPEC CPU 2006.

[30] A. N. Udipi, N. Muralimanohar, N. Chatterjee, Rajeev Balasubramonian, A. Davis, and N. P. Jouppi. Rethinking DRAM Design and Organization for Energy-Constrained Multi-Cores. In *ISCA*, 2010.

[31] M. Ware, K. Rajamani, M. Floyd, B. Brock, J. Rubio, F. Rawson, and J. Carter. Architecting for Power Management: The IBM POWER7 Approach. In *HPCA*, 2010.

[32] D. Wentzlaff, P. Griffin, H. Hoffmann, L. Bao, B. Edwards, C. Ramey, M. Mattina, C.-C. Miao, J. Brown, and A. Agarwal. On-Chip Interconnection Architecture of the Tile Processor. *IEEE Micro*, 2007.

[33] H. Zheng, J. Lin, Z. Zhang, E. Gorbatov, H. David, and Z. Zhu. Mini-Rank: Adaptive DRAM Architecture for Improving Memory Power Efficiency. In *Micro*, 2008.

Semantics-driven Sensor Configuration for Energy Reduction in Medical Sensor Networks

James B. Wendt, Saro Meguerdichian, Hyduke Noshadi, and Miodrag Potkonjak
Computer Science Department, University of California, Los Angeles
Los Angeles, CA 90095, USA
{jwendt, saro, hyduke, miodrag}@cs.ucla.edu

ABSTRACT

Traditional optimization methods for large multisensory networks often use sensor array reduction and sampling techniques that attempt to reduce energy while retaining full predictability of the raw sensed data. For systems such as medical sensor networks, raw data prediction is unnecessary; rather, only relevant semantics derived from the raw data are essential. We present a new method for sensor fusion, array reduction, and subsampling that reduces both energy and cost through semantics-driven system configuration. Using our method, we reduce the energy requirements of a medical shoe by a factor of 17.9 over the original system configuration while maintaining semantic relevance.

Categories and Subject Descriptors

C.3 [**Special-Purpose and Application-Based Systems**]: Real-time and Embedded Systems; B.8.2 [**Performance and reliability**]: Performance Analysis and Design Aids; J.3 [**Life and Medical Sciences**]: Medical Information Systems

General Terms

Design, Algorithms, Performance

Keywords

Energy optimization, sensor networks, wireless health

1. INTRODUCTION

The evolution of the wireless sensor network has enabled its practical application for remote monitoring and analysis of a wide variety of environments (e.g. oceanic, urban, wildlife). More recently, these wireless sensor networks have also enabled remote sensing and remote monitoring of the human body. There is great potential for the medical community to benefit from such wearable sensing systems by utilizing their capacities for remote surveillance to observe and diagnose patient ailments and disease. These wireless health systems allow doctors to remove the constraint that they rely solely on in-person patient checkups and interviews in order to diagnose patient illness. By utilizing non-invasive wireless health monitoring, doctors are able to incorporate information gathered from the patient's day to day activities and routine into their professional medical diagnoses.

However, such multisensory systems are often expensive and have rigid energy and power requirements due to their wireless framework and remote deployment. Due to the often complex design, expensive cost, and energy demands that can accompany such wireless sensor networks, wireless medical sensing systems have not yet made headway into widespread use. Medical sensing systems can contain very large sensor arrays; for example, a commercial medical shoe might contain as many as ninety-nine sensors [1]. Not only are these sensors expensive and make up a shoe that costs thousands of dollars, but they also draw power and consume energy. However, a medical shoe is inherently a mobile device, attaching a large battery pack or requiring frequent recharges are a strong deterrent to the adoption of such a medical shoe by the common patient.

In order to minimize the cost and energy demands of these sensor networks, techniques for sensor coverage, selection, and sampling have been proposed. These techniques often focus on maintaining full predictability of the original array while reducing the sensor count and coordinating sensor subsampling.

In a great majority of sensor network applications, readings from a single sensor are not sufficient for extraction of information, knowledge, or decision processes. Furthermore, we are rarely interested in the raw sensed data, but rather more interested in the semantic information it provides. For example, in a sensor network distributed among the trees of a wildlife preserve, we are rarely interested in the temperature of many or even all points in a given area but rather more concerned with the early detection of fire.

Traditional approaches to sensor selection for energy reduction in multisensory systems remove redundant sensors from the array while maintaining full sensor predictability [2]. However sensor predictability (i.e. raw data prediction) is not necessary when the essential information is the semantic information itself (e.g. fire detection). Thus, our key conceptual contribution is the reduction of energy using semantics-driven sensor configuration; we reduce the energy requirements of the sensor network while maintaining system relevance, in particular, for medical diagnosis.

This is achieved through a two step process, using coarse

ISLPED'12, July 30–August 1, 2012, Redondo Beach, CA, USA.

grained energy reduction techniques followed by fine grained techniques. We explain these steps in detail in Section 4.

In coarse grained energy reduction we introduce sensor fusion and bottom-up selection techniques that maintain semantic relevance. In Section 4.1 we describe our sensor placement algorithms, including: sensor selection, which iteratively chooses sensors that most improve semantic prediction accuracy; adjacent sensor fusion, in which sensors are physically or electronically combined; search space pruning for runtime reduction; and a generic method for comparison of orthogonal semantics.

Our fine grained energy reduction technique employs a very similar approach to the coarse grained technique, but applies to sampling instead of sensors. Ultimately, we find a minimal subsampling configuration that reduces energy while maintaining semantic relevance.

Our technical contributions are algorithms and methods for (i) sensor selection with an accompanying method for ranking and sorting of arbitrary semantics, (ii) sensor fusion, and (iii) sampling configuration, all of which employ the key conceptual idea that energy reduction is semantics-driven.

We use as our driving example, a medical shoe fitted with ninety-nine pressure sensors on the sole. The semantics this system is designed to compute are gait characteristics, shown by VanSwearingen et al. to be highly correlated with disease and the risk of falling [3].

2. RELATED WORK

The emergence of embedded sensor networking has introduced new scientific and engineering challenges. Much attention is now focused on energy and power reduction in wireless sensor networks due to the often large networks and their constant power demands [4]. Current energy optimization methods focus on hardware design, signal processing, and sensor selection [5] [6] [7].

Recent attention in wearable sensing systems has fostered a growing interest in medical-based sensing devices [8] [9]. Like research in other wireless sensing systems, current attention in the wireless health domain focuses on the utility and convenience of such systems as well as their cost and energy demands [8] [10] [11] [12] [13].

Investigations have been made into the application of gait analysis in wearable sensing systems such as sensor-equipped shoes [11] [12] [13]. These provide solutions for semantic computation from raw sensor data, but do not leverage gait analysis for energy optimization. Efforts employing gait analysis for energy optimization are preliminary and limited to analysis of same size sensors and prediction of only single semantic measurements [14] [15] [16].

3. PRELIMINARIES

3.1 Semantics-driven Energy Reduction

In general, sensor selection is best applied on large multisensory arrays for maximal energy savings. Traditional approaches remove redundant sensors from the original array while maintaining full sensor predictability [2]. However, in semantics-driven devices, such as medical sensors for diagnosis, sensor predictability (i.e. raw data prediction) is not necessary. The essential information that the system is intended to measure are the diagnostic semantics (e.g. gait characteristics in a medical shoe).

Our semantics-based sensor configuration technique for energy reduction benefits sensor networks that are ultimately designed for measuring semantics from the raw data. For example, if a sensor network is required to measure temperature in a forest using thermometer sensors, traditional techniques for sensor selection are sufficient to reduce energy and cost of the system. If, however, the purpose of the forest sensor network is to detect fire through the analysis of temperature measurements, we claim that using a semantics-driven approach to sensor selection—rather than raw data predictability—yields stronger semantic prediction and energy reduction over traditional approaches.

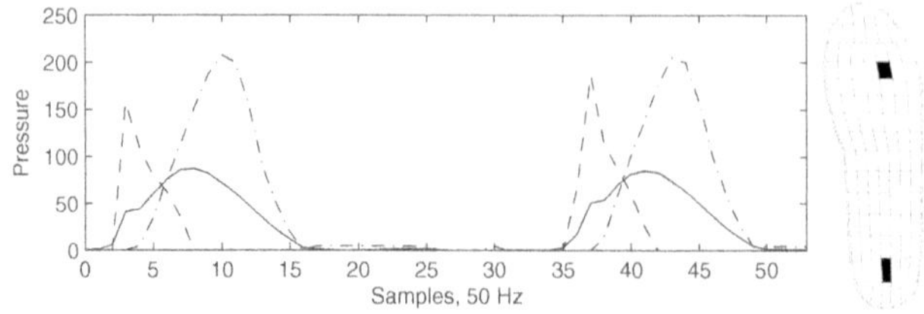

Figure 1: Pressure measurements of two steps of a single foot measured by sensors on the heel (dash), toe (dash-dot), and averaged over all ninety-nine sensors (solid). The heel and toe sensors are shaded on the Pedar mapping [1].

3.2 Medical Shoe

Medical sensor networks are inherently semantics-driven systems. A doctor is very often not concerned with the raw measurements of the sensors (e.g. accelerometers, pressure sensors, electrochemical biosensors); but rather, more concerned with the semantic information derived from those sensors (e.g. distance/speed of an impact, gait characteristics, glucose levels).

We perform our semantics-driven energy reduction methodology on a wireless medical wearable sensing system containing a large multisensory array, known as the Hermes shoe [17]. This platform is designed with the purpose of assessing balance and instability in patients through the measurements of ninety-nine passive resistive pressure sensors distributed on the sole of the foot using the Pedar plantar mapping [1].

The processing unit samples data from these pressure sensors at 50 Hz using a 16-bit analog-to-digital converter. An example of the sensor mapping and typical step profile is depicted in Figure 1. The data consists of hundreds of steps from five subjects.

3.3 Gait Characteristics

Medical shoes similar to the Hermes platform have been used in a variety of applications, from sports science to elderly care. In this paper, we choose gait characteristics, such as step stride, change in step stride, maximum pressure, lateral pressure difference, and guardedness (time between heel and toe landing) as the semantics for our study. VanSwearingen et al. have concluded that these gait characteristics correlate to a number of ailments and diseases in the elderly and directly contribute to the prediction of risk of falling in this population [3]. This strong correlation between gait and risk is a powerful means to help medical professionals diagnose these ailments with the availability of such gait data.

We normalize and extract these gait characteristics measured collectively by all ninety-nine sensors as well as measured by the individual sensors independently. Our sensor selection procedure conducts metric prediction using the metric measurements at each sensor, while our sampling solution determines the best sampling of raw data for a given set of sensors and their metric prediction model. We separate our data into training (80%) and testing (20%) subsets.

4. ENERGY REDUCTION

Energy reduction is achieved by applying two sequential steps to the design space: coarse grained optimization, which includes sensor fusion and selection, followed by fine grained optimization, which includes subsampling configuration.

In describing the coarse grained technique, we introduce our semantics-driven sensor selection algorithm that reduces the original multisensory array to a minimal subset of sensors that accurately predicts the semantics introduced in Sections 3.1 and 3.3. This approach is a bottom-up selection process which begins with an empty set, then iteratively adds sensors until a sufficient minimal set is found that accurately predicts the semantics to a specified threshold.

Our fine grained energy reduction technique applies a similar approach, however is performed on sensor-samples instead of sensors. A sensor-sample is a single measurement of a unique sensor at a discrete time step. In this approach, sensor-samples are iteratively removed from the minimal sensor set (found via coarse grained optimization) while maintaining semantic prediction accuracy.

Theoretically, it is possible to bypass the coarse grained sensor selection process and only apply fine grained subsampling to the original design. However, the exponential search space of this problem, along with the huge number of sensor-samples that accompanies a large multisensory array, renders this problem near impossible to solve in a reasonable amount of time. Given s samples and n sensors there are 2^{sn} possible solutions. By applying coarse grained sensor selection first, we reduce n to a small constant and reduce the problem size.

Additionally, while our primary concern is the reduction of energy usage in these medical devices, another equally important concern is the reduction of their cost, which is proportional to the number of sensors. Even if in a reasonable amount of time we could find an optimal solution using only sensor-sample selection, it is not immediately evident that this solution would ultimately reduce the number of required sensors. For example, a hypothetical solution could be that only a single sample measurement is taken at each of the ninety-nine sensors per step. While this drastically reduces energy usage, the cost of the shoe remains exorbitantly high since no sensors are removed. Coarse grained sensor selection ensures cost reduction in addition to energy reduction.

4.1 Coarse Grained Optimization

Our coarse grained optimization technique reduces both energy consumption and cost of the sensor network by combining and removing sensors from the original array while maintaining prediction accuracy of the aforementioned gait characteristics. A key observation in Figure 2 is that while clear correlations might exist between the measurements of a few individual sensors and the individual semantics, between the four semantics it is not immediately apparent which small subset of sensors can predict all semantics simultaneously well.

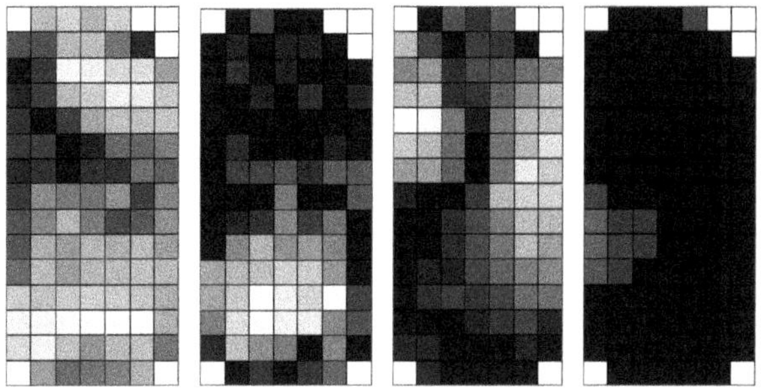

Figure 2: Individual sensor coefficients of determination for (a) average maximum step amplitude, (b) change in step stride, (c) lateral pressure difference, and (d) guardedness. The lighter the sensor, the more correlated it is to the metric; darker shadings denote weaker correlations.

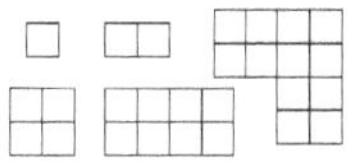

Figure 3: Shapes used in sensor fusion, pre-computed prior to sensor selection.

We explore two degrees of freedom in searching for a minimal set of optimal predictors: sensor *fusion* and sensor *selection*. Combining sensors electronically or physically reduces the total energy spent in sensing since the fused sensor acts as a single sensor and is sampled as one. We find that these fused sensors are often better semantic predictors than individual sensors.

We employ three steps to achieve coarse grained energy and cost reduction. The first step is sensor fusion. The second step is pruning, in which we eliminate combinations of sensors and fused sensors whose semantic measurements are highly correlated with one another. This ensures that pairs of sensors that yield no significant advantage in prediction over each other are not considered for combination, thus reducing the problem size. The final step is sensor selection.

4.1.1 Sensor Fusion

A fused sensor is a set of adjacent sensors physically or electronically combined with one another to create a single, larger sensor whose measurements are averaged over its new area. Including sensor fusion as an additional dimension in sensor selection provides benefits in semantic prediction. A fused sensor effectively becomes a new sensor to consider, it helps reduce noise between its constituent sensors, and simultaneously reduces system energy requirements.

However, for the purposes of sensor selection, coupling the fused sensor solution space (2^n) with that of sensor selection (2^n) yields a hugely exponential search space. In order to reduce this effect we pre-select a number of fused sensor shapes. By choosing a static set of these combinations and treating them as unique sensors, we restrict the search space.

We use application specific knowledge to pre-construct a number of fused sensor shapes. For example, the correlations of determination for the semantics in Figure 2 reveal that potentially good sensor combinations might be squares, rectangles, or L-shapes. Thus, from the existing ninety-nine sensors, we construct five new fused sensors, as depicted in Figure 3, and apply them and their rotations across the

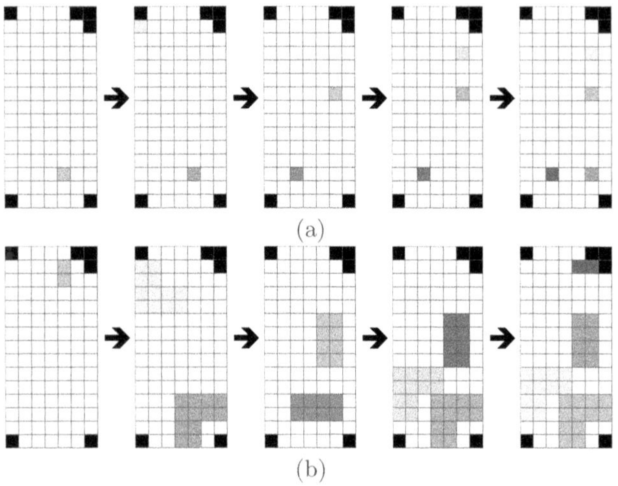

Figure 4: Top sensor configurations at iterations $1 \leq i \leq 5$. Solution (a) limits sensor selection to individual sensors, (b) includes sensor fusion.

Algorithm 1 Sensor Selection

Input: S_1 = set of sensor configurations of size 1
Input: k = # of configurations retained between iterations
Input: G = pruned graph of sensor-semantic correlations as described in Section 4.1.2, $N_G(s)$ is the of set neighboring sensors to s in G
1: $i = 1$
2: **while** $\epsilon > threshold\_error$ **do**
3: $i = i + 1$
4: T = GetGreedyRep(S_{i-1}, k)
5: **for** $1 \leq j \leq k$ **do**
6: **for** all sensors s, $s \notin T[j]$, $s \notin N_G(T[j])$ **do**
7: Append $T[j] \cup s$ to S_i
8: **end for**
9: **end for**
10: S_i = SortByPredictionError(S_i)
11: ϵ = GetPredictionError($S_i[0]$)
12: **end while**
Output: Top sensor configuration, $S_i[0]$

sensor mapping, effectively adding 544 fused sensors to the selection pool.

4.1.2 Pruning

With 643 total sensors to consider, the exponential search space remains large. Even with our approach to sensor selection that reduces the number of potential sensor groupings considered at each iteration to the constant k (described in Section 4.1.3), sensor selection is still a very time consuming task. By pruning the search space and reducing the number of sensors considered at each iteration, we instead use that computation time to increase the value of k and thereby drive our algorithm to find a more optimal solution. The key observation is that if two sensors give similar predictions for each metric, then there is likely no benefit in having them both in the same predictive set. In experimentation, pruning had minimal, if any, impact on sensor selection while still reducing overall computation time.

We use this observation to formulate the pruning problem as a complete weighted graph, where each sensor is a node and each edge has weight equal to the maximum difference between the predicted values for each metric by the two corresponding sensors. We then prune all edges with weight greater than ω, where ω is a specified similarity threshold. Now, adding a sensor s to a predictive set eliminates all sensors with edges to s as future candidates for that set, since those sensors will not add any additional information about the gait metrics.

4.1.3 Sensor Selection

Our semantics-driven sensor selection procedure, described in detail in Algorithm 1, is a bottom-up selection process. It systematically selects the best combinations of sensors until a minimal subset that accurately predicts the required semantics is found.

Figure 2 represents significantly strong linear correlations between a sizeable portion of sensors and three of the semantics. Thus, we use linear regression for semantic prediction; however, our method can easily be implemented using almost any prediction model. For example, it could be postulated that a linear regression model is not best for guardedness prediction given the low correlation values per sensor in Figure 2d. Interestingly, while no single sensor is an adequate predictor for guardedness using a linear model, after a couple iterations of sensor selection, guardedness prediction error drops drastically. This intuitive result derives from the fact that the guardedness metric is inherently derived from the measurements of at least two sensors.

The prediction error of sensors varies from metric to metric. Due to the very application-specific nature of these semantics, some are inherently harder to predict (change in step stride) while others are very well suited to the sensor and system design (amplitude). Because of these discrepancies, it can be very difficult to determine the relative prediction accuracy of a single sensor against two different metrics. We overcome this barrier by mapping the prediction error of a given sensor for a given metric to the cumulative distribution function of the prediction errors of all the sensors for that same metric. This binds the error to a normalized value that is relative to the rest of the system's prediction capabilities. Now, we are able to compare metric predictions and correctly rank our sensors by how well they predict each metric relative to one another. Our ranking function weights each metric equally, since ultimately we are most interested in designing a medical device that can provide the doctor with the most information. This functionality is embodied in the *SortByPredictionError* and *GetPredictionError* functions in lines 10 and 11 of Algorithm 1.

While sensor fusion and pruning help to reduce the search space, they still only do so nominally. The process of combining all previously considered sensor combinations with the set of remaining sensors at each iteration still grows exponentially. Thus, at each iteration, instead of looking at all possible sensor combinations, we only investigate combinations with the best k size subset of the previous iteration's results; for every set in the best k solutions from the previous iteration, we create $O(n)$ new sensor configurations. This reduces the original exponential problem formulation complexity to $O(kn^2)$.

It is important that the k-size best subset retained between two iterations contains both top predicting sensor sets as well as a representative distribution of the current solution space. This ensures that our selection process drives to

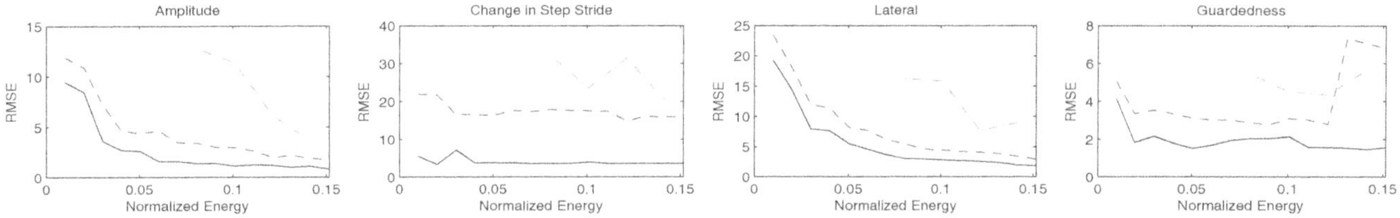

Figure 5: Coarse grained optimization. Root mean squared testing error in prediction using best selected sensors using only single sensors (dash), using sensor fusion (solid), and results from [2] (dash-dot). Units are *pressure* for amplitude and lateral, and *samples* for step stride and guardedness.

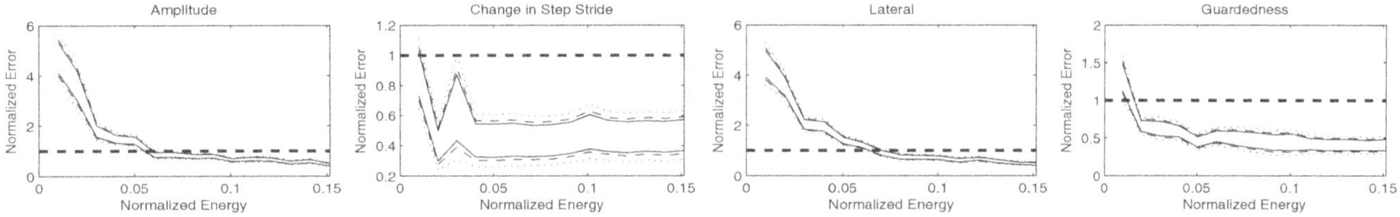

Figure 6: Coarse grained optimization using sensor fusion. Semantic prediction error as a percentage of desired error threshold (thick dashed); 90% confidence interval (solid), 95% (dashed), 99% (dotted). When the prediction error reduces to the threshold for all semantics, sensor selection is complete.

an optimal solution quickly while providing neighboring solutions that could potentially find other local minima. This is done through a greedy selection of the top $k/2$ configurations and an evenly distributed selection of $k/2$ configurations from those remaining. This functionality is embodied in the *GetGreedyRep* function in line 4 of Algorithm 1.

4.2 Fine Grained Optimization

While sensor selection is a crucial step for reducing the cost and complexity of wearable medical sensing systems, energy is ultimately spent mainly in sampling. Therefore, the sampling strategy is of utmost importance to energy optimization of any sensor network. We conduct subsampling post-selection, based on the following key observations: (i) during ambulation, the foot spends a majority of the time in the air and therefore applying no pressure to any sensors; (ii) a single sensor is sufficient to detect the start and end times of a step; and (iii) during a step, applied pressure follows multimodal behavior predictable from semantic information, as described in [2] and [18].

Based on the first two observations, we add a sensor that covers the entire sole and sample it at the full rate (50 Hz) solely to detect the start and end times of steps. Note that this sensor is large and therefore subject to a high signal-to-noise ratio, rendering it ineffective for predicting gait metrics. Without loss of accuracy, we can begin sampling all other selected sensors only when the foot lands, and stop sampling as soon as the foot leaves the ground.

We formulate the sampling reduction problem as a simple variation of our sensor selection algorithm presented in Algorithm 1. Again, this is an iterative process, but now, at iteration i, the k strongest predicting and representative sets of $n-i$ sensor-samples are returned, where a sensor-sample is a single sampling point of a single sensor, and n is the number of coarse grained selected sensors times the number of samples in the step. Therefore, we make the following modifications: S_i now represents a set of sensor-sample configurations at iteration i, s is a sensor-sample in question to be removed, and line 7 now reads, Append $T[j] - \{s\}$ to S_i. Note that by using this approach, each sensor can have a different sampling rate, as only one sample of one sensor is removed at each iteration. Furthermore, the same strategy for pruning can be applied as in sensor selection, with sensor nodes replaced by sensor-sample nodes.

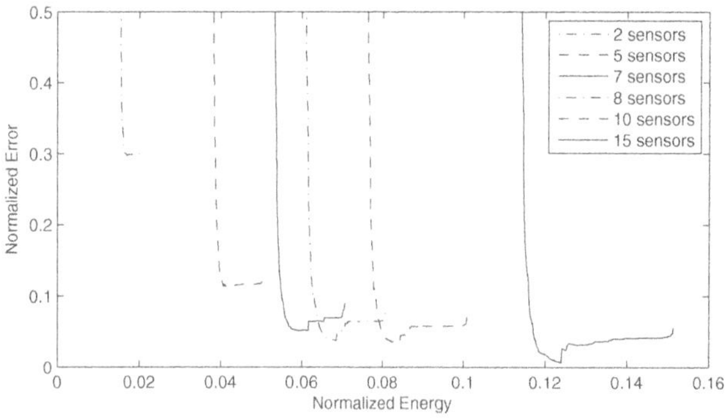

Figure 7: Fine grained optimization applied to the best configurations found in coarse optimization. Curves are constructed from right to left as sensor-samples are removed iteratively. The four semantic errors are normalized for equal comparison between semantics. Normalized energy is the fraction of energy expended on the new configuration over the original array of ninety-nine sensors sampled at 50 Hz.

5. RESULTS

5.1 Coarse Grained Optimization Results

We perform the semantics-driven sensor selection algorithm with pruning on individual sensors only, then incorporating sensor fusion, and compare both results to a traditional sensor selection technique that retains raw data predictability. In both Figures 4a and 4b the selected sensors are well distributed over the inside and outside of the foot, cover both the heel and the toe, and generally attempt to cover the highest correlated areas depicted in Figure 2.

Semantics-driven sensor selection clearly performs better than traditional selection (see Figure 5). For the same level of energy reduction, semantics-driven selection predicts a factor of 5.8 times and 1.6 times better than the traditional approach for the change in step stride and guardedness metrics, respectively. Analysis of the amplitude and lateral semantics shows that semantics-driven sensor selection provides a factor of about 4 times better energy reduction over traditional methods for a constant prediction error rate. Between semantics-driven selection using individual sensors and fused sensors, the latter provides much better prediction capabilities, and thus, greater energy gains.

Figure 6 shows that semantics-driven sensor selection reaches the desired error threshold by the seventh iteration, returning a factor of 14.1 in energy reduction over the original ninety-nine sensor system design. Recall that while the correlations between individual sensors and the guardedness metric are nearly non-existent (Figure 2d) the semantics-driven sensor selection process reduces the prediction error of this metric immediately in the second iteration, a testament to the semantics-driven procedure.

5.2 Fine Grained Optimization Results

We apply fine grained optimization to the best sensor configurations of size 2, 5, 7, 8, 10, and 15 sensors found in the coarse grained step. In all cases, fine grained optimization further reduces energy from the coarse grained solution by the equivalent of one to three sensors sampled at 50 Hz. For example, applying fine grained optimization to the 10 sensor solution reduces the energy consumed by the configuration to the equivalent of employing 8 sensors at 50 Hz (see intersection of 8 and 10 sensors in Figure 7).

As we reduce energy in Figure 7, prediction error dips slightly, but for the most part, remains relatively flat until those sensor-samples that are most crucial for semantic prediction are removed. This result is expected due to the predictable nature of the pressure signals, gait metrics, and physiological events, in general, as observed by [2] and [18].

As we reduce sensor-samples the prediction error initially decreases slightly. This is because in the initial iterations, the sensor-samples that are removed are those that actually impair the prediction model. Though as expected, eventually the iterative removal of sensor-samples begins to have a negative effect on prediction and the error rises sharply. At this point we stop the fine grained optimization process and present the final configuration.

While coarse grained optimization reduces the original array to seven best predicting sensors, applying fine grained optimization further reduces energy to 5.6% of the original sensor array, a factor of 17.9 overall improvement.

6. CONCLUSION

We have presented a novel approach for energy reduction in medical sensor networks using a semantics-driven approaches to sensor selection and subsampling. We leverage the key observations that the raw sensed data is unimportant, that only the metrics relevant to diagnosis are needed, and that the important metrics can be easily derived from the raw data. Consequently, our key contributions are as follows: (i) a bottom-up iterative approach to selection of a minimal set of best predicting sensors; (ii) a novel procedure for physically or electronically combining adjacent sensors to reduce sampling cost while improving prediction strength; and (iii) an extension of our sensor selection algorithm to minimize the sampling rate of individual sensors while maintaining accuracy. Our approach yields a cost reduction of 93%, and furthermore reduces energy by a factor of 17.9 over the original system configuration of ninety-nine sensors sampled at 50 Hz.

7. REFERENCES

[1] Novel.de, Pedar, 2007, http://www.novel.de/

[2] H. Noshadi et al., "Energy optimization in wireless medical systems using physiological behavior," *Wireless Health*, pp. 128-136, 2010.

[3] J. M. VanSwearingen et al., "The modified gait abnormality rating scale for recognizing the risk of recurrent falls in community-dwelling elderly adults," *Physical Therapy*, 1996.

[4] G. Anastasi et al., "Energy conservation in wireless sensor networks: A survey," *Ad Hoc Networks*, vol. 7, no. 3, pp. 537-568, 2009.

[5] V. Leonov, P. Fiorini, S. Sedky, T. Torfs, and C. Van Hoof, "Thermoelectric mems generators as a power supply for a body area network," *Solid-State Sensors, Actuators and Microsystems*, vol. 1, pp. 291-294, 2005.

[6] Y. Liu et al., "Critical-path based low-energy scheduling algorithms for body area network systems," *RTCSA*, pp. 301-308, 2007.

[7] S. Xiao et al., "Transmission power control in body area sensor networks for healthcare monitoring," *IEEE Communications Journal*, pp. 37-48, 2009.

[8] K. Lorincz et al., "Mercury: a wearable sensor network platform for high-fidelity motion analysis," *SenSys*, pp. 183-196, 2009.

[9] A. Krause et al., "Unsupervised, dynamic identification of physiological and activity context in wearable computing," *ISWC*, pp. 88-97, 2003.

[10] M. Rofouei, M. A. Ghodrat, M. Potkonjak, and A. Martinez-Nova, "Optimization Intensive Energy Harvesting," pp. 272-275, *DATE*, March 2012.

[11] S. Popovic et al., "A reliable gyroscope-based gait-phase detection sensor embedded in a shoe insole," *IEEE Sensors Journal*, vol. 4, no. 2, 2004.

[12] K. Oshima et al., "A human probe for measuring walkability," *SenSys*, pp. 353-354, 2009.

[13] V. Erickson, A. U. Kamthe, and A. E. Cerpa, "Measuring foot pronation using RFID sensor networks," *SenSys*, pp. 325-326, 2009.

[14] J. B. Wendt and M. Potkonjak, "Medical diagnostic-based sensor selection," *IEEE Sensors*, pp. 1507-1510, 2011.

[15] J. B. Wendt, S. Meguerdichian, H. Noshadi, and M. Potkonjak, "Energy and Cost Reduction in Localized Multisensory Systems through Application-Driven Compression," *DCC*, pp. 411, 2012.

[16] J. B. Wendt, S. Meguerdichian, and M. Potkonjak, "Small is Beautiful and Smart," to appear in *Tele-Healthcare Computing and Engineering: Principles and Design*, ed. by Fei Hu, Science Publishers, 2012.

[17] H. Noshadi et. al, "HERMES: Mobile balance and instability assessment system," *BioSignals*, 2010.

[18] S. Meguerdichian et al., "Semantic multimodal compression for wearable sensing systems," *IEEE Sensors*, pp. 1449-1453, 2010.

Voltage Droop Reduction for Multiple-Power Domain SoCs with On-Die LDO Using Output Voltage Boost and Adaptive Response Scaling

Tetsutaro Hashimoto, Satoshi Tanabe, Kouichi Nakayama and Hisanori Fujisawa
Fujitsu Laboratories Limited
1-1, Kamikodanaka 4-chome, Nakahara-ku,
Kawasaki 211-8588, Japan
{thasimo, tanabe.s, nakayama.koichi, h.fujisawa}@jp.fujitsu.com

ABSTRACT

Low power techniques such as clock gating and dynamic frequency scaling cause a sudden surge in power supply current. To reduce the voltage droop induced by such a surge in the load current of an LDO regulator, we propose output voltage boost and adaptive response scaling techniques that utilize clock activation detection. Measured results from a test chip fabricated in 65-nm CMOS technology show that a combination of the two techniques reduces the worst-case output voltage droop by 63% compared to operation without them. This results in a voltage offset reduction from 45% to 15%, which leads to 20% power savings.

Categories and Subject Descriptors

B.7.1 [**Integrated Circuits**]: VLSI (very large scale integration)

General Terms

Measurement, Performance, Reliability, Experimentation

Keywords

VLSI, voltage droop reduction, voltage droop cancel, on-die LDO regulator, low dropout linear regulator, power integrity

1. INTRODUCTION

With advances in VLSI technology, more processor cores have been integrated into a single die. However, this scaling has increased the power consumption of SoCs. Therefore, low power design techniques such as clock gating, power gating, adaptive voltage scaling (AVS), and dynamic voltage and frequency scaling (DVFS) have become essential for SoCs. When AVS and DVFS are applied to each processor core integrated in a die, the same number of voltage regulators as the number of processor cores is required. Off-chip components and external pins of a chip are also required for the voltage regulators, as shown in Figure 1(a), which shows a power system diagram for the case when DC-DC converters are used. This power system shown in Figure 1(a) requires such a large area for off-chip components that it is difficult to implement it on a small board for a tiny device. Moreover, many external pins of a chip and many voltage regulators are required, which lead to increased costs. To eliminate external pins and off-chip components and reduce the number of voltage regulators, voltage regulators and capacitors should be fully integrated on a die as shown in Figure 1(b). Low dropout (LDO) linear regulators are slightly inferior to DC-DC converters in terms of power efficiency, but more suitable for integration compared to DC-DC converters in terms of the small area, fast load transient response, and off-chip-component-free feature.

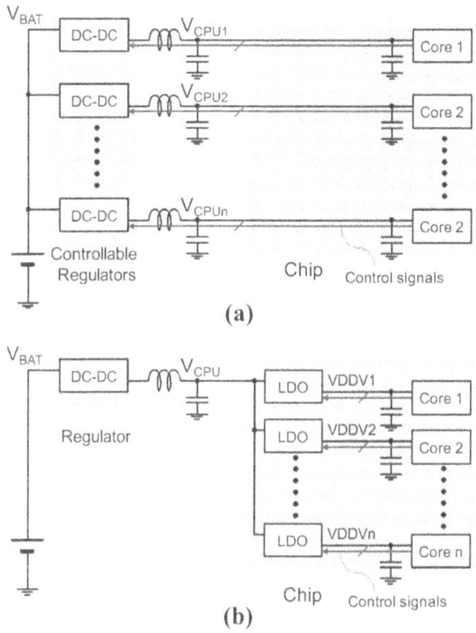

Figure 1: Power system diagram for multiple-core multiple-power domain SoCs (a) composed of off-chip controllable DC-DC converters and (b) composed of one off-chip DC-DC converter and on-die programmable LDOs.

However, it is well known that LDO regulators are negatively affected by a trade-off between feedback stability and fast load transient response. In particular, off-chip-capacitor-free LDO regulators are extremely difficult to stabilize for all load conditions because on-die capacitance is smaller than off-chip capacitance by three orders of magnitude. Moreover, low power design techniques such as clock gating and dynamic frequency scaling generate a sudden surge in power supply current. Since the current surge substantially exceeds the LDO regulator's load transient response, a large voltage droop is induced, which causes low power digital VLSIs to fail.

Therefore, various LDO structures have been developed to overcome the trade-off of off-chip-capacitor-free LDO regulators. These structures are based on the flipped voltage follower (FVF)

[1-3], frequency compensation schemes [4-6], and voltage droop direct feedback schemes [7-9]. The FVF-LDO structures in [1-3] improve feedback stability, but their load transient response is slower than that of a generic LDO structure under the same quiescent current. The frequency compensation schemes in [4-6] also improve the feedback stability of a generic LDO regulator, but do not reduce the output voltage droop induced by the current surge. The schemes in [7-9] boost the output voltage after detecting voltage droop, so they do not sufficiently prevent the output voltage from drooping. In addition, since the circuits for detecting voltage droop in [7, 8] are a kind of amplifier, their feedback is slower than the current surge. The voltage droop detector in [9] is also slower because it is based on a ring oscillator (ROSC). This results in a large voltage droop, and VLSIs are not prevented from failing.

We propose an output voltage boost (OVB) technique that utilizes clock activation detection for cancelling an output voltage droop induced by a clock-gating-caused sudden surge in the load current. This technique boosts the output voltage of an LDO regulator just before the output voltage starts drooping. We also propose an adaptive response scaling (ARS) technique for improving feedback stability and minimizing an output voltage droop induced by a workload-change-caused sudden surge which cannot be detected by utilizing clock activation detection. We experimentally demonstrated that a combination of these two techniques reduces the worst-case output voltage droop. This paper is organized as follows: In Section II, we discuss an off-chip-capacitor-free on-die LDO regulator negatively affected by low power techniques such as clock gating. Section III describes our proposed OVB and ARS techniques. Section IV presents measurement results for verifying our proposed techniques. Finally, Section V concludes this paper.

2. CRITICAL ISSUE OF OFF-CHIP-CAPACITOR-FREE ON-DIE LDO

In this section, we discuss the impact of low power techniques on power integrity, and describe the case in which an off-chip-capacitor-free LDO regulator is negatively affected by these low power techniques.

Clock gating, one of the most common techniques for reducing dynamic power consumption, generates a sudden surge in power supply current when some global clocks become active. If a digital core has micro-farad off-chip-capacitors on the power system as shown in Figure 1(a), the current surge is somewhat covered by the capacitors. However, with on-die decoupling capacitance of only a few nano-farads, the electrical charge of the on-die capacitor is not sufficient to cover the current surge; therefore, the power supply voltage of the digital core droops severely, as shown in Figure 2. VDDV is the output voltage of the LDO regulator and the power supply voltage of a digital core.

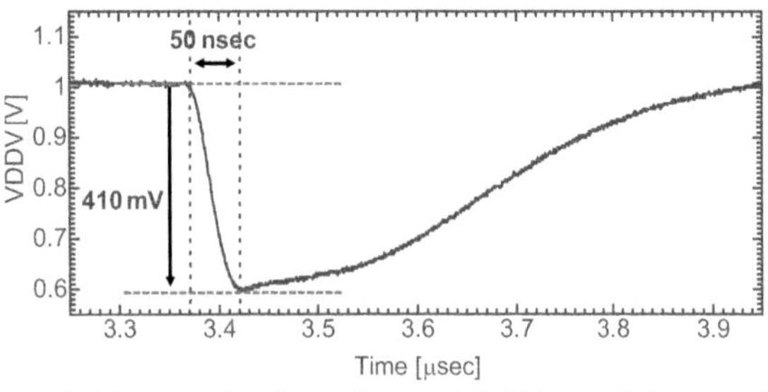

Figure 2: Measured voltage droop of LDO regulator's output voltage (VDDV) caused by clock gating.

On-die voltage droop cancellers that use a voltage droop detector have been proposed for reducing power supply voltage droop [8, 9]. However, these detectors cannot keep up with a voltage droop because such a voltage droop at a rate of 8.2 mV/ns (Figure 2) is too fast for the detectors. If the detectors take 10 ns to detect a voltage droop, the cancellers start to cancel the voltage droop after the power supply voltage droops more than 82 mV. The cancellers in [8, 9] are aimed not at the power system shown in Figure 1(b), but at the power system shown in Figure 1(a); therefore, from the beginning, they are inadequate for reducing the output voltage droop of an off-chip-capacitor-free on-die LDO regulator.

Therefore, the power system shown in Figure 1(b) requires techniques other than voltage droop detection to reduce the voltage droop.

3. VOLTAGE DROOP REDUCTION USING CLOCK ACTIVATION DETECTION

3.1 Output Voltage Boost Technique

We observed that a sudden surge in power supply current occurs when global clocks of a digital core become active. Therefore, we propose an OVB technique that uses clock activation detection. Figure 3 shows the principle of this technique and a comparison between an OVB using voltage droop detection and our proposed OVB using clock activation detection. Point A in Figure 3 is the start of global clock activation. A current surge occurs at point A and the voltage of the VDDV starts to severely droop after point A. The voltage boost technique using voltage droop detection starts to boost the voltage of the VDDV after the voltage droop detector detects a voltage droop at point B. Thus, the voltage boost using voltage droop detection cannot prevent a voltage droop in principle as shown in Figure 3. On the other hand, the voltage boost technique using clock activation detection starts to boost the voltage of the VDDV at point A; therefore, a voltage droop can be cancelled in principle.

Figure 4 shows the circuit implementation of an LDO regulator control scheme with our OVB technique that uses clock activation detection. VDD, VSS, and V_{EAMP} are the input supply voltage of the LDO regulator, the ground voltage, and the output voltage of the error amplifier, respectively. The clock activation detector in Figure 4(a) sends one positive pulse to the BOOST signal when the first clock pulse arrives. The BOOST signal selects V_{EAMP} or VSS as bias voltage to the gate of the power PMOS transistor. While the BOOST signal is high, the power PMOS transistor is turned on and the voltage of the VDDV is boosted to the level of the VDD. This voltage boost cancels the voltage droop of the VDDV caused by clock gating. When the BOOST signal is low, the voltage of the VDDV is regulated by the LDO regulator.

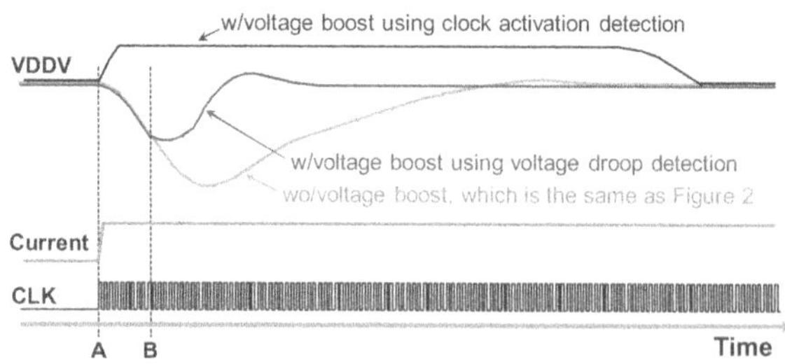

Figure 3: Principle of voltage boost techniques, and comparison between conventional technique using voltage droop detection and our proposed technique using clock activation detection.

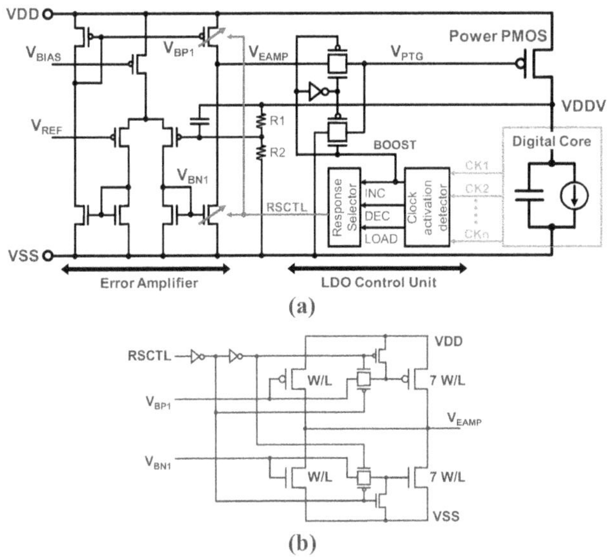

Figure 4: (a) Simplified circuit diagram of LDO regulator control scheme with OVB and ARS techniques using clock activation detection. (b) W/L ratios scalable output PMOS/NMOS transistors of error amplifier.

Of course, the voltage of the VDDV has to be boosted before the voltage starts drooping due to global clock activation. The clock activation detector shown in Figure 5 enables this. The detector sends one pulse to the BOOST signal without any clock latencies after the first clock pulse arrives at its input ports. Therefore, the voltage of the VDDV can be boosted before the voltage droop is induced on the VDDV by global clock activation. Moreover, the clock activation detector features a reference-clock-free-topology so that its switching power consumption is reduced to zero while all clock domains are inactive.

The clock activation detector is composed of clock detectors (CKDET), pulse generators (PULSEGEN), positive edge detectors (PHDET), negative edge detectors (NHDET), and ring oscillators (ROSC), as depicted in Figure 5(a). The CKDET is designed to detect clock pulses without requiring control signals from a clock gating management unit, as shown in Figure 5(b). The CKDET eliminates the need for changing the user logic of a SoC, and makes it easy to implement the OVB technique into a SoC, e.g., an ASIC. The OUT signal of the CKDET changes from low to high without any clock latencies after the first clock pulse arrives. The OUT signal remains high while clock pulses are arriving at the CK port. After no more clock pulses arrive, the OUT signal changes to low. As shown in Figure 5(c), the PULSEGEN is designed to generate one pulse whose width T_{PULSE} is specified by

$$T_{PULSE} = T_{ROSC} \times WD,$$

where T_{ROSC} is the oscillating cycle time of the ROSC shown in Figure 5(f), and WD is the value set in the PULSEGEN. As with the CKDET, the OUT signal of the PULSEGEN changes from low to high without any clock latencies after the IN signal of the PULSEGEN changes from low to high. The PHDET is designed to detect a positive edge and to output one shot pulse, as shown in Figure 5(d). The NHDET is designed to detect a negative edge and to output one shot pulse as well, as shown in Figure 5(e).

As shown in Figure 5(a), the clock activation detector sends one pulse, whose width is specified by T_{ROSC} and WDB, to the INC port without any clock latencies after the first clock pulse arrives at any CK port. WDB is the value that is given to the WD port of the upper PULSEGEN. One pulse whose width is specified by T_{ROSC} and WDD is sent to the DEC port of the clock activation detector after no more clock pulses arrive at the CK ports. WDD is the value that is given to the WD port of the lower PULSEGEN. The LOAD signal is set to high while clock pulses are arriving, and set to low when no clock pulses arrive.

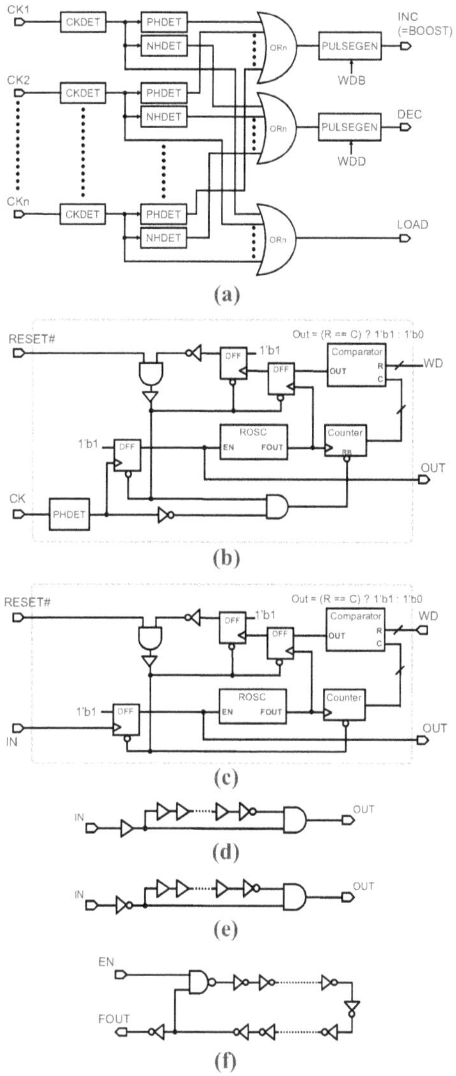

Figure 5: Circuit diagram (a) overview of clock activation detector (b) CKDET (c) PULSEGEN (d) PHDET (e) NHDET (f) ROSC.

3.2 Adaptive Response Scaling

We have explained how an OVB technique using clock activation detection cancels a voltage droop induced by a global-clock-activation-caused sudden surge in power supply current. Nevertheless, voltage droops are not only caused by a current surge induced by global clock activation. Voltage fluctuation induced by a rapid decrease in power supply current caused by global clock deactivation, undershoot that occurs after output voltage boosting, and voltage droop induced by a surge in power supply current caused by a workload change cannot be cancelled using the OVB technique that utilizes clock activation detection.

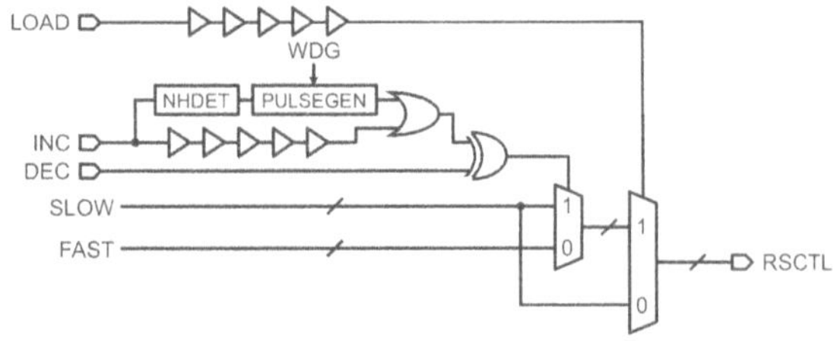

Figure 6: Circuit diagram of response selector.

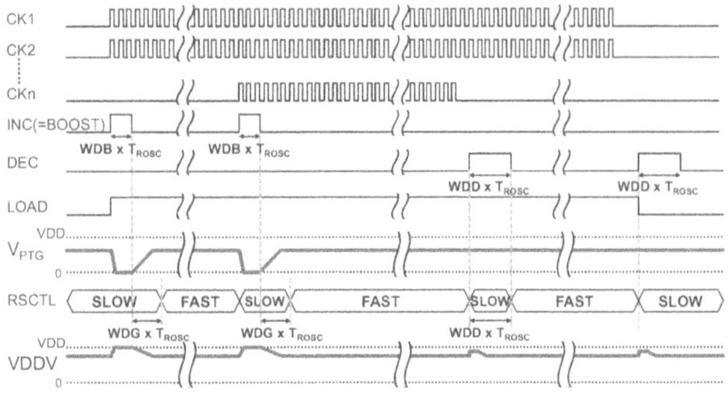

Global clocks	State	INC	DEC	LOAD	RSCTL
All inactive	No-load	0	0	0	SLOW
Some become active	Load increased	1	0	1	SLOW
Active	Loaded	0	0	1	SLOW
* Time duration of WDG x T_{ROSC} after INC changes to 0 from 1					
Active	Loaded	0	0	1	FAST
Some become inactive	Load decrease	0	1	1	SLOW
All become inactive	Load decrease to no-load	0	1	0	SLOW

Figure 7: Timing diagram of clock activation detection, clock deactivation detection, output voltage boost, and adaptive response scaling.

The ARS technique minimizes such voltage fluctuations. The clock activation detector and the response selector shown in Figure 4 enable this. The transient response of the LDO regulator is scalable by scaling the W/L ratios of the output PMOS and NMOS transistors of the error amplifier as shown in Figure 4(b). The ratios of the output PMOS and NMOS transistors are set to W/L according to the SLOW value so that the LDO regulator does not oscillate under no-load operation. The ratios are set to 8 W/L according to the FAST value so that the transient response of the LDO regulator is the fastest. A circuit diagram of the response selector is shown in Figure 6. Figure 7 shows the timing diagram of our OVB and ARS techniques utilizing clock activation detection. The details of adaptive response scaling are described as follows.

When the LOAD signal is low, which means that the LDO regulator is unloaded, the response selector outputs the SLOW value to the LDO regulator. When the LDO regulator is unloaded, its phase margin is the smallest and the LDO regulator is unstable. Then, the load transient response of the LDO regulator is set to SLOW to stabilize the LDO regulator. When the LOAD signal is high, meaning that the LDO regulator is loaded, the response selector outputs the FAST value to the LDO regulator. When the LDO regulator is loaded, changes in workload occur frequently. The load transient response of the LDO regulator is then set to

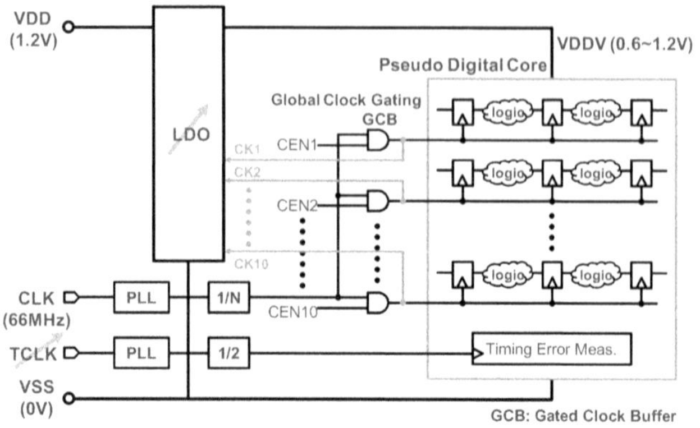

(a)

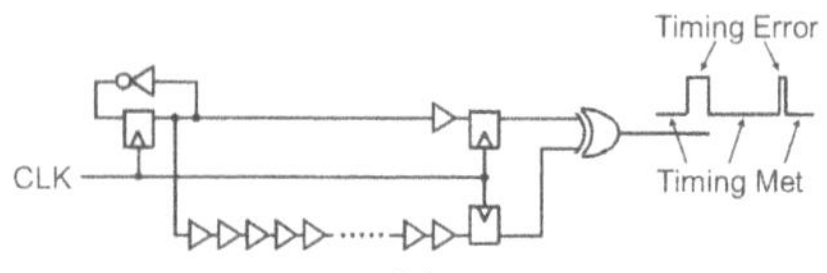

(b)

Figure 8: (a) Overview of a test chip fabricated in 65-nm CMOS technology (b) Timing error measurement circuit.

FAST to be responsive to the current increase induced by workload-change.

When the LOAD signal is high and the INC signal changes from low to high, meaning that global clocks become active and the voltage of the VDDV is boosted, the response selector outputs the SLOW value to the LDO regulator. After the boosting of the VDDV is stopped, the INC signal changes from high to low. After outputting the SLOW value for the time specified by T_{ROSC} and WDG, the response selector outputs the FAST value. WDG is the value that is given to the WD port of the PULSEGEN in Figure 6. This is to minimize undershoots after stopping output voltage boosting.

When the LOAD signal is high and the DEC signal changes from low to high, which means that some global clocks become inactive and the load of the LDO regulator decreases, the response selector outputs the SLOW value. After the DEC signal changes from high to low, the response selector outputs the FAST value. This is to minimize voltage fluctuations after a rapid decrease in power supply current.

4. EXPERIMENTAL RESULTS

A test chip depicted in Figure 8(a) was fabricated in 65-nm CMOS technology to verify our OVB and ARS techniques that utilize clock activation detection. The test chip consists of an LDO regulator with OVB and ARS techniques, a pseudo digital core, and a timing error measurement circuit. The pseudo digital core has 10 clock domains and each clock distribution is implemented with a global clock gating scheme. Figure 8(b) shows the circuit diagram of the timing error measurement circuit. When setup violations occur, the output signal of the circuit turns high. A die photo and specifications of the test chip are shown in Figure 9. The power PMOS transistor of the LDO regulator encloses the pseudo digital core, as pictured. Off-chip capacitors for the VDDV were not on a test board, but on-die decoupling capacitance of 2.5 nF was integrated to reduce simultaneous switching noise of the pseudo digital core. The clock activation detector and the response selector consume 4.1 mW at maximum

Technology	65-nm CMOS
Chip size	4.2 mm x 4.2 mm
Supply voltage	Core 1.2 V, I/O 3.3 V
On-die decoupling	2.5 nF
Package	208-pin LQFP

Figure 9: Die photo and specifications of test chip.

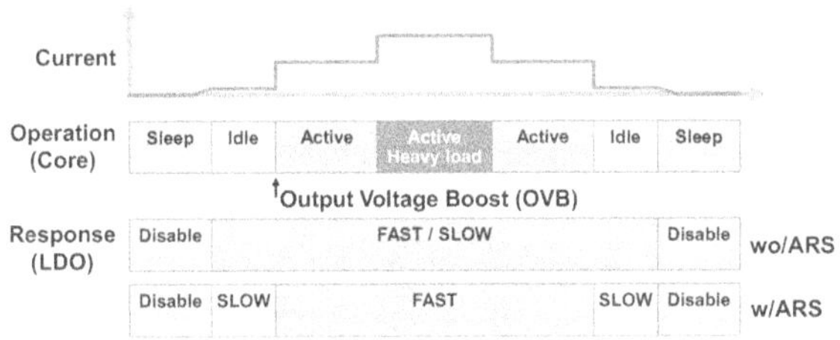

Figure 10: Operation sequence and power supply current.

while 10 clock domains are active. Measurements were conducted under the operation sequence shown in Figure 10.

Figure 11 shows the measured waveforms of the VDDV, where the voltage of the VDDV was set to 0.9 V. Four waveforms in each graph were measured using the testing parameters listed in Table 1.

Figure 11(a) presents the waveforms of the VDDV, which were measured under the condition that power supply current is increased by global clock activation. The measured waveform labeled "Normal+FAST" shows that a global-clock-activation-caused current surge induces a voltage droop of 320 mV without OVB. In contrast, the voltage of the VDDV is boosted with OVB before the voltage starts to droop, and the voltage droop induced by a current surge is cancelled, according to the measured waveforms labeled "OVB+FAST", "OVB+SLOW" and "OVB+ARS". This indicates that our OVB technique that uses clock activation detection prevents the voltage droop induced by a global-clock-activation-caused current surge. However, after stopping voltage boosting with the LDO's load transient response set to FAST, a large undershoot of 150 mV occurs, as indicated by the measured waveform labeled "OVB+FAST" in Figure 11(a). This indicates that the OVB technique interacts adversely with fast transient response. The measured waveform labeled "OVB+ARS" shows that ARS technique reduces the undershoot from 150 mV to 34 mV although the settling time is eight times longer than with the load transient response set to FAST. The ARS technique significantly reduces the adverse interaction between the OVB technique and fast transient response. The time duration of the voltage boost is about 8 μs; therefore, if global clock activation occurs every 1 ms, the power overhead of the OVB technique is 0.27%.

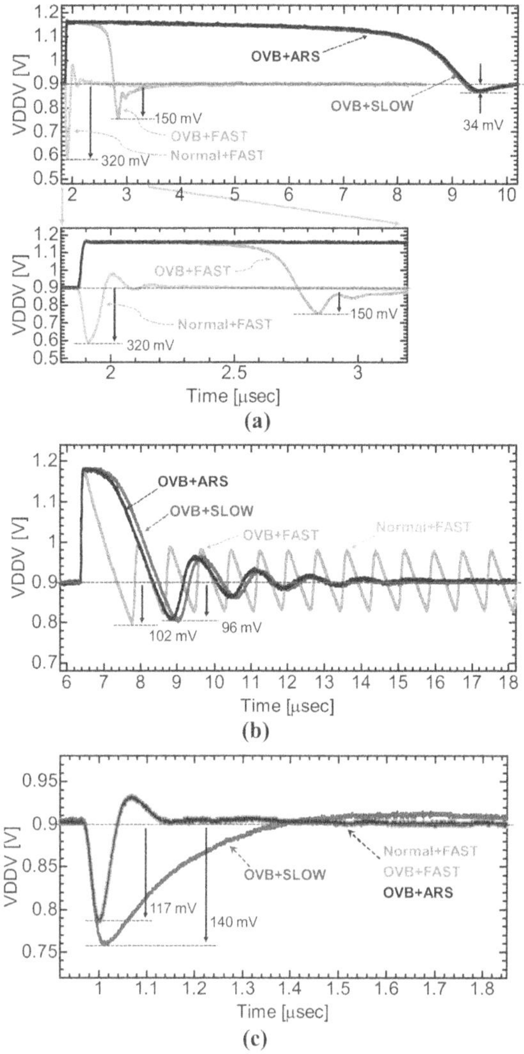

Figure 11: Measured waveforms of VDDV. (a) Current increase from 2 mA to 136 mA caused by clock activation; (b) current decrease from 136 mA to 2 mA caused by deactivating all clocks; (c) current increase from 136 mA to 264 mA caused by workload change.

Table 1: Testing parameters in Figure 11

Label	Output Voltage Boost	Response
Normal+FAST	Don't use	FAST
OVB+FAST	Use	FAST
OVB+SLOW	Use	SLOW
OVB+ARS	Use	Adaptive Scaling

Table 2: Summary of results of worst-case voltage droop for VDDV set to 0.9 V

Normal+FAST	OVB+FAST	OVB+SLOW	OVB+ARS
320 mV	150 mV	140 mV	117 mV

Figure 11(b) presents the waveforms of the VDDV, which were measured under the condition that power supply current is decreased by global clock deactivation. These results indicate that too fast of a transient response causes voltage fluctuations, and ARS technique minimizes the voltage fluctuations after the rapid decrease in power supply current.

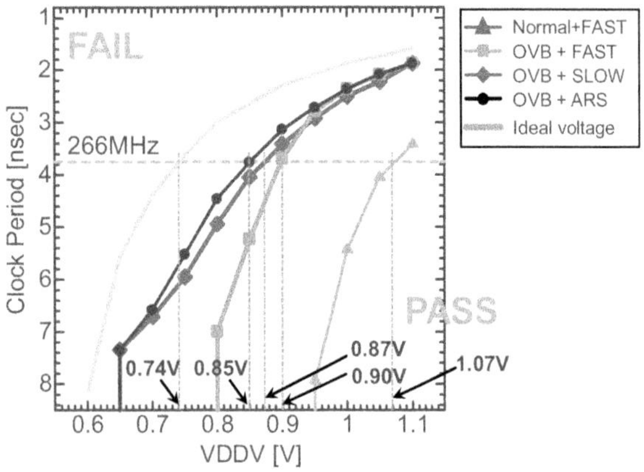

Figure 12: Measured minimum clock period (inverse of maximum frequency) versus voltage of VDDV.

Table 3: Summary of results of voltage offset for 266-MHz operation

Normal+FAST	OVB+FAST	OVB+SLOW	OVB+ARS
+0.33 V (+45%)	+0.16 V (+22%)	+0.13 V (+18%)	+0.11 V (+15%)

Figure 11(c) presents the waveforms of the VDDV, which were measured under the condition that power supply current is increased by workload change. Because the OVB and ARS techniques are triggered by clock gating, three measured waveforms, "Normal+FAST", "OVB+FAST", and "OVB+ARS", overlap. The voltage droop of "OVB+SLOW" is the largest among the measured waveforms.

In conclusion, the measured results listed in Table 2 indicate that a combination of the two techniques reduces the worst-case voltage droop from 320 mV to 117 mV, a 63% reduction.

Figure 12 shows the measured minimum clock period as a function of the voltage of the VDDV. The testing parameters are also listed in Table 1. The measured line labeled "Ideal voltage" represents the minimum clock period versus the voltage of the VDDV under the condition that the voltage of the VDDV stays flat. According to Figure 12, the OVB and ARS techniques do much to broaden the operating range. If the target operating frequency is to be 266 MHz, an operating voltage of 1.07 V is needed without the OVB or ARS techniques. This means that a voltage offset of 0.33 V, in percentage terms; 45%, is needed. On the other hand, with both OVB and ARS techniques, an operating voltage of 0.85 V, i.e., a voltage offset of 0.11 V, is sufficient. This results in a voltage offset reduction from 0.33 V to 0.11 V, in percentage terms, a reduction from 45% to 15%, as listed in Table 3. In addition, this voltage offset reduction leads to 20% power savings.

5. CONCLUSIONS

We proposed OVB and ARS techniques that use clock activation detector for reducing the output voltage droop of on-die LDO regulators. The OVB technique is implemented when global clocks become active in a clock gating technique, and it cancels the output voltage droop induced by a clock-gating-caused sudden surge in the load current of an on-die LDO regulator. However, this technique is not implemented if a current surge cannot be detected by clock activation detection. To suppress a voltage droop caused by factors other than clock gating, the ARS technique is applied. This technique adapts an on-die LDO regulator's transient response to its load current.

We fabricated a test chip with an on-die LDO regulator in 65-nm CMOS technology to verify the effectiveness of our proposed techniques. The power overhead of the clock activation detector and the response selector in the test chip was 4.1 mW at maximum. The measured results show that the OVB technique almost completely cancels the voltage droop induced by a clock-gating-caused sudden surge in the load current. In addition, the results demonstrate that the ARS techniqe reduces voltage fluctuations induced by factors other than global clock activation. Thus, a combination of the two techniques reduces the worst-case output voltage droop by a 63% compared to the operation without them. This results in a reduction from 45% to 15% in voltage offset which is necessary for preventing failures due to voltage droop. This decreasing nominal operating voltage leads to 20% power savings.

6. REFERENCES

[1] P. Hazucha, T. Kamik, B. A. Bloechel, C. Parsons, and S. Borkar, "Area-Efficient Linear Regulator with Ultra-Fast Load Regulation," *IEEE J. Solid-State Circuits*, vol. 40, no. 4, pp. 933-940, Apr. 2005.

[2] T. Y. Man, K. N. Leung, C. Y. Leung, P. K. T. Mok, and M. Chan, "Development of Single-Transistor-Control LDO Based on Flipped Voltage Follower for SoC," *IEEE Trans. on Circuits and Systems* I, Regular Papers, vol. 55, no. 5, Jun. 2008.

[3] M. Lüders, B. Eversmann, J. Gerber, K. Huber, R. Kuhn, D. Schmitt-Landsiedel, and R. Brederlow, "A Fully-Integrated System Power Aware LDO for Energy Harvesting Applications," in *Symposium on VLSI Circuits Digest of Technical Papers*, pp. 244-245, 2011.

[4] K. N. Leung and P. K. T. Mok, "A Capacitor-Free CMOS Low-Dropout Regulator with Damping-Factor-Control Frequency Compensation," *IEEE J. Solid-State Circuits*, vol. 38, no. 11, pp. 1691-1702, Oct. 2003.

[5] R. J. Milliken, J. Silva-Martínez, and E. Sánchez-Sinencio, "Full On-Chip CMOS Low-Dropout Voltage Regulator," *IEEE Trans. Circuits and Systems* I, Reg. Papers, vol. 54, no. 9, Oct. 2007.

[6] H.-C. Lin, H.-H. Wu, and T.-Y. Chang, "An Active-Frequency Compensation Scheme for CMOS Low-Dropout Regulators with Transient-Response Improvement," *IEEE Trans. on Circuits and Systems* II, Express Briefs, Vol. 55, no. 9, Sep. 2008.

[7] W. Oh and B. Bakkaloglu, "A CMOS Low-Dropout Regulator with Current-Mode Feedback Buffer Amplifier," *IEEE Trans. on Circuits and Systems* II, Express Briefs, Vol. 54, no. 10, Oct. 2007.

[8] Y. Nakamura, M. Takamiya, and T. Sakurai, "An On-Chip Noise Canceller with High Voltage Supply Lines for Nanosecond-Range Power Supply Noise," in *Symposium on VLSI Circuits Digest of Technical Papers*, pp. 124-125, 2007.

[9] Y. Kanno, K. Yoshizumi, Y. Yasu, K. Ishibashi, and H. Mizuno, "Dynamic Voltage Boost (DVB) Method for Improving Power Integrity of Low-Power Multi-Processor SoCs," in *Symposium on VLSI Circuits Digest of Technical Papers*, pp. 148-149, 2008.

A 33 μW 42 GOPS/W 64x64 Pixels Vision Sensor with Dynamic Background Subtraction for Scene Interpretation

Nicola Cottini
Fondazione Bruno Kessler
38100 Trento, I
cottini@fbk.eu

Massimo Gottardi
Fondazione Bruno Kessler
38100 Trento, I
gottardi@fbk.eu

Nicola Massari
Fondazione Bruno Kessler
38100 Trento, I
massari@fbk.eu

Roberto Passerone
Universitá di Trento
38100 Trento, I
roberto.passerone@unitn.it

Zeev Smilansky
Emza Visual Sense Ltd.
44422 Kfar Sava, IL
zeev@emza-vs.com

ABSTRACT

A 64×64 pixel vision sensor performs adaptive background subtraction and event detection at very low power consumption. The chip is based on a VLSI-oriented vision algorithm, implemented at pixel-level, mimicking the basic process of pre-attentive visual perception. Anomalous pixel behaviors are detected and coded into a 2-bit/pixel. Each pixel integrates two programmable Switched-Capacitors Low-Pass Filters and two clocked comparators, which are fundamental blocks for the execution of the vision algorithm. The 45T square pixel has a pitch of $26\mu m$ and a fill factor of 12%. The vision sensor consumes $33\mu W$ at $13fps$ and $3.3V$. This turns into a computing performance of 42 $GOPS/W$ and 4 $GOPS/mm^2$, which are values aligned with the most advanced computational vision sensors.

Categories and Subject Descriptors

B.7.1 [**INTEGRATED CIRCUITS**]: Types and Design Styles—*VLSI*

Keywords

Vision sensor, low-power sensor, early image processing.

1. INTRODUCTION

Sensors are becoming increasingly pervasive in our everyday life. With minimum dimensions and less infrastructures, sensory networks need to embed computing resources and wireless data communication at a minimum energy budget, maximizing its operating lifetime and minimizing environmental footprint. In particular, vision is the sensing technology with the largest information density, which is required to be processed in real-time through high-performance computing platforms. Unlike consumer market (cameras, mobile phones), asking for high-quality and high-resolution images, there are many other application areas where visual information is used for control purposes, ranging from surveillance and people and traffic monitoring, energy-saving in buildings, elderly care and many other. Vision computation did not make over the years significant progress in energy-autonomous applications, despite big advances in microelectronic energy-efficiency. In fact, standard imagers continuously deliver sequences of images with large redundancy, although only a small amount of information is used to perform any kind of visual task. This overloads the processor, which is required to undertake the high level part of the visual task.

Custom digital processors have been proposed for early visual processing [1],[2],[3], thanks to their high parallelism. They are mainly based on Single-Instruction Multiple-Data (SIMD) architectures, offering massive parallel processing capabilities. However, their performance are limited because they need to be fed by the video signal, which is slow.

Embedding some intelligence at sensor level, making it able to recognize and deliver only those data related to image features in the scene which are of interest, would drastically increase the energy efficiency of the entire system, without losing performance.

Despite research activity on custom vision sensors is fairly mature, a lot of work has still to be done on the ultra-low power side.

In literature, several recent examples of custom vision chip implementations are reported, targeted to low-power applications [4],[5],[6],[7]. Although most recent works report very aggressive figures of merit in pixel power consumption, ranging from 460pW/frame.pixel [4] up to 84pW/frame.pixel [7], it has to be pointed out that these works only refer to imagers where the vision processing is accomplished outside the chip and typically takes a large power consumption. On the contrary, a vision sensor embeds a custom image processing and can tolerate a larger consumption at chip-level, in exchange for reduced external computational resources. This main characteristic turns into a significant power saving of the entire vision system [8],[9],[10]. For this reason, dealing with vision sensors, it is more appropriate to talk about computing performance over power consumption, P_E ($GOPS/W$), and/or computing density, P_A ($GOPS/mm^2$), rather than absolute power consumption.

In this paper, the architecture of a low-power computa-

ISLPED'12, July 30–August 1, 2012, Redondo Beach, CA, USA.
Copyright 2012 ACM 978-1-4503-1249-3/12/07 ...$10.00.

tional vision sensor is reported, performing pixel-level adaptive background subtraction as a basic image processing for pre-attentive visual perception [11]. Under normal operating conditions, the sensor delivers two binary images, detecting temporal gradient with sign. During an event detection in the scene, the obtained output data compression ratio is about 0.1 - 0.2, while under stationary condition, no data is delivered off-chip.

2. SENSOR OPERATING PRINCIPLE

The proposed sensor extracts temporal contrast at pixel level, embedding an algorithm for dynamic background subtraction. In the presented implementation, the pixel is the main building block, executing in parallel the most power consuming operations of the vision algorithm. Its basic operating principle is shown in Fig.1, plotted over 20 frames. The activity over time of the current signal V_P, acquired during the integration time, is monitored over a certain number of frames by means of two voltage signals, V_{Min} and V_{Max}. These two voltages slowly keep track of the low values and the high values of V_P, respectively. The basic function of the two signals, V_{Min} and V_{Max}, is to define a voltage gap inside which V_P is to be considered normal, *cold-pixel* ($V_{Min_i} < V_{P_i} < V_{Max_i}$). This means that until the pixel behavior is in line with its past history, no anomalous events are detected. Outside the voltage boundary, the signal is recognized as anomalous (*hot-pixel*), detecting a potential alert situation ($V_{P_i} > V_{Max_i}$ or $V_{P_i} < V_{Min_i}$). Depending on the binary status of the pixel, *hot-pixel* or *cold-pixel*, the voltage thresholds V_{Max} and V_{Min} will be updated as follows:

Max Value:

$$Q_{Max} = 1 \rightarrow V_{Max_{i+1}} = \alpha_H \cdot V_{P_i} + (1-\alpha_H) \cdot V_{Max_i}, \quad (1)$$

$$Q_{Max} = 0 \rightarrow V_{Max_{i+1}} = \alpha_C \cdot V_{P_i} + (1-\alpha_C) \cdot V_{Max_i}; \quad (2)$$

Min Value:

$$Q_{Min} = 1 \rightarrow V_{Min_{i+1}} = \alpha_H \cdot V_{P_i} + (1-\alpha_H) \cdot V_{Min_i}, \quad (3)$$

$$Q_{Min} = 0 \rightarrow V_{Min_{i+1}} = \alpha_C \cdot V_{P_i} + (1-\alpha_C) \cdot V_{Min_i}; \quad (4)$$

with $\alpha_H > \alpha_C$. According to the four equations, when the current value of the pixel exceeds one of the two thresholds, ($V_P > V_{Max}$ or $V_P < V_{Min}$), an *hot-pixel* is detected and the corresponding threshold is rapidly updated (α_H), to absorb the unusual signal variation within few frames. In case of a *cold-pixel*, the threshold slowly (α_C) converges toward V_P. The main advantage of this approach, over the classic temporal difference [12], consists of its adaptive capability, which makes it very reliable in detecting true alert situations.

3. PIXEL ARCHITECTURE

The basic pixel schematic is shown in Fig. 2. The photodiode (PD), working in storage mode, is buffered by a source follower (M1,M2,M3), which is turned ON by V_{p_clk} only when capacitors C_{1M} and C_{1m} have to be pre-charged. The two Switched-Capacitors Low-Pass Filters (SC-LPF1/SC-LPF2) are fed by V_P and compute V_{Max} and V_{Min} respectively, clocked at the sensor frame rate. The two filters share the

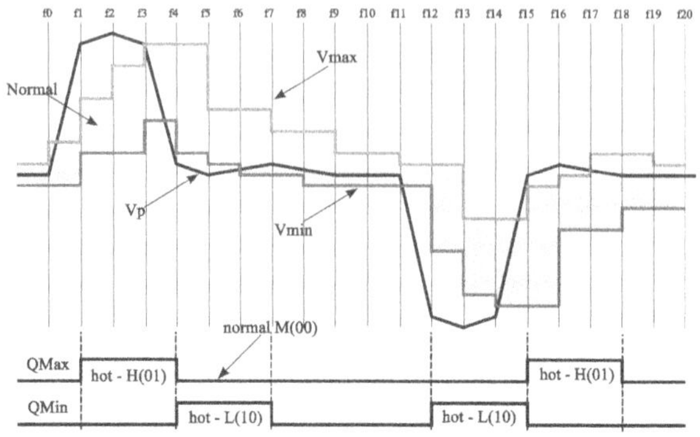

Figure 1: Basic operating principle of the APS with thresholds adapting capabilities.

first clock $SETV_P$, which is used to store the current value V_P on the two capacitors C_{1M} and C_{1m}. The SC-LPF1 (SC-LPF2) operation is managed by a digital control block (UPDATE REGISTER, Fig. 4), which is placed outside the imager and properly activates the phase MM (Mm), according with the high-level algorithm for scene interpretation. The output of the filter is stored onto the PMOS capacitor C_{2M} (C_{2m}). The switches MSH1 (MSH2) and MSW1 (MSW2) are PMOS transistors with the bulk tied to V_{blk}, generated by the structure shown in the blow-up of Fig. 2 and placed in a dummy pixel. Here, the photodiode is shorted to V_{RES} and the output buffer is always on, generating $V_{blk} = V_{RES} - V_{gsM1}$. This makes all the switches of the filters to always turn off into the accumulation region, minimizing the the leakage current through the transistor channel [?].
The SC-LPF transfer function is:

$$H(s) = \frac{1}{1+s\tau_n} = \frac{1}{1+s\cdot\left(\frac{C_{2M}}{C_{1M}}\cdot\frac{n}{f_0}\right)}; \quad (5)$$

where C_{1M} and C_{2M} are two NMOS filter capacitors, with $C_{2M}/C_{1M} = 220fF/150fF \sim 1.5$. The integer n makes the filter to be clocked once over n frames, thus changing the SC-LPF response. With $n = 1$, the filter exhibits the fastest response, being clocked at each frame. Referring to the equations in Section II, the value of n directly affects the values of α_H and α_C, hence the behavior of the filters. The larger the n, the larger is the time constant of the filter (τ_n). The value of n can be digitally changed by properly clocking the MM (Mm) phase of the filter.

After the exposure time, a new value of V_P is acquired and the analog outputs of the two filters (V_{Max} and V_{Min}) are simultaneously compared with respect to V_P, by clocking the comparators CMP1, CMP2 with CLKCMP. The pixel provides two binary outputs (Q_{Max}, Q_{Min}), which are directly available on the two bit-lines of Fig. 2 (B_{Min}, B_{Max}):

- $Q_{Max} \otimes Q_{Min}$ = L $\rightarrow$ cold pixel;
- $Q_{Max} \oplus Q_{Min}$ = H $\rightarrow$ hot pixel.

Depending on the pixel status, the analog memories C_{2M} and C_{2m} are updated through the activation of MM and/or Mm respectively by means of a proper control block (UPDATE REGISTER in Fig. 4), placed outside the pixel array.

In order to prevent false *hot-pixel* detection, the two comparators have a symmetric built-in offset (V_{off}) (Fig. 3).

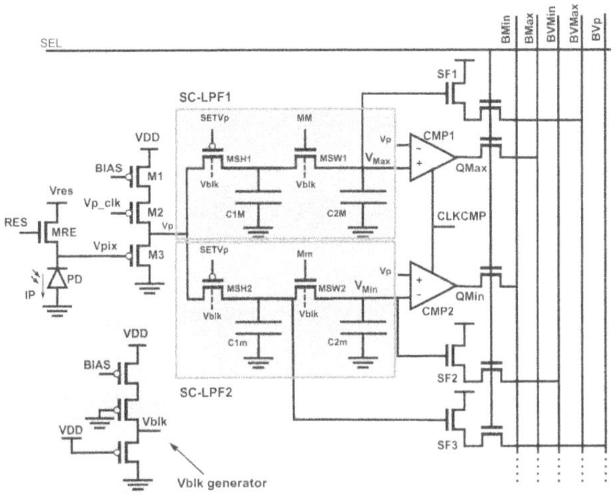

Figure 2: Schematic of the Active Pixel Sensor. In the blow-up, the reference pixel generates the bias voltage for the bulk (Vblk) of the PMOS switches of the filters (MSH1, MSW1, MSH2, MSW2). Here , the photodiode is always reset to Vres and the buffer is always ON (Vp_clk=0).

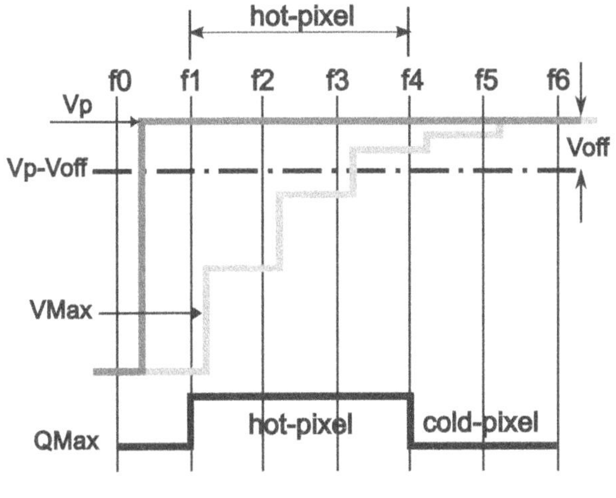

Figure 3: Comparator offset defining a safe margin. For $V_{Max} > V_p - V_{off}$, the pixel is considered a *cold-pixel*. In this example, signal Vp changes abruptly before f1, setting an *hot-pixel* and forcing the pixel to adjust the VMax threshold in order to compensate for the alert status.

The grey-zone represents the range of values inside which the pixel is detected as a *cold-pixel*. Under an *hot-pixel* detection (Q_{Max}), the related memory is updated, changing the threshold voltage (e.g. V_{Max}) toward the current value V_P. The *hot-pixel* status starts from frame $f1$ until $f4$, being $V_{Max} > V_P - V_{off}$.

Moreover, the pixel is equipped with three NMOS source-followers (SF1, SF2, SF3) buffering the three analog voltages (V_{Max}, V_{Min}, V_P). The analog bit-lines, BV_{Max}, BV_{Min}, BV_P, deliver the analog signals outside the imager and feed three operational amplifiers, which are directly connected to the output pads. This last functionality was implemented for debug and calibration purposes, while during the sensor operating condition, it is completely turned off.

4. SENSOR ARCHITECTURE

The proposed sensor consists of a 64×64 pixels array with peripheral circuitry for array scanning, data readout and pixel control for the algorithm implementation. The ROW DECODER progressively selects the rows of the imager. The COLUMN DECODER delivers the binary output of the pixels in asynchronous way. The 64-stages UPDATE REGISTER can deliver both digital and analog signals coming from the bit-lines. Each stage consists of two D-type flip-flops, storing the binary values of the two bit-lines (BMj, Bmj), and three analog switches, connecting V_{Max_j}, V_{Min_j} and V_{P_j} to the three opamps. The temporal filter, implemented into the sensor, can work in two different modes: *learning mode* and *adaptive mode*. The learning mode is typically used during the system setup. The background of a static scene is stored into the two memories and compared with the current signal. It takes several frames to settle V_{Max} and V_{Min} by continuously updating the filters. In the *adaptive mode*, the memories are quickly updated under *hot-pixel* and slowly under *cold-pixel*, mimicking the eye adaptive behavior.

After the image acquisition phase and after comparing V_P with the two analog memories, V_{Max} and V_{Min} need to be updated before the binary data to be delivered off-chip. This operation takes place at each row selection. In this case, the bit-line pairs (B_{Max}, B_{Min}) of each column are used as masks for the memories update phase.

There are two ways to read out the binary informations from the sensor: asynchronous and mixed sync/asynchronous. In the asynchronous mode, after pulsing START, the first row of the array is selected by the ROW DECODER. As soon as the bit-lines are settled and the update phase has been accomplished, the COLUMN DECODER sequentially looks fro bit-line pairs disparities and for each pixel delivers signals signals, MIN, MAX synchronized on WRN, at a data rate of about 100MHz. One row will therefore take about $650ns$ to be read out. If the current pixel is hot, $V_P > V_{Max}$, the corresponding output will be (MIN=0, MAX=1). After 64 WRN pulses, the End Of Row (EOR) is provided and the readout process stops, waiting for the acknowledge (ACK) to be provided off-chip, selecting next row. At the end of the readout process, an End Of Frame (EOF) is provided.

The second operating mode of the sensor is mixed, synchronous/asynchronous. At the START, the first row is selected and the binary bit-lines (BM_j, Bm_j) of the current row are uploaded into the 2×64 Flip-Flops stages UPDATE REGISTER, ready to be read out through DOUT, by providing an external clock (CLK). Next row is selected at the rising edge of ACK. EOR is provided by the chip at the end of each row readout. In this case, two binary images are delivered for each frame: one for the MAX hot-pixels and one for the MIN hot-pixels. An analog multiplexer is embedded into the UPDATE REGISTER, multiplexing the three analog voltage signals (V_{Max}, V_{Min}, V_P) of the pixels of the selected row. This capability is only available under the synchronous/asynchronous operating mode, and needs a relatively slow clock rate ($f_{CLK} < 2MHz$ typ). Although adding 6 NMOS and 3 bit-lines to each pixel will turn into a significant electronic overhead, this additional functionality is worth to be embedded, allowing sensor's self-test and diagnostic capabilities.

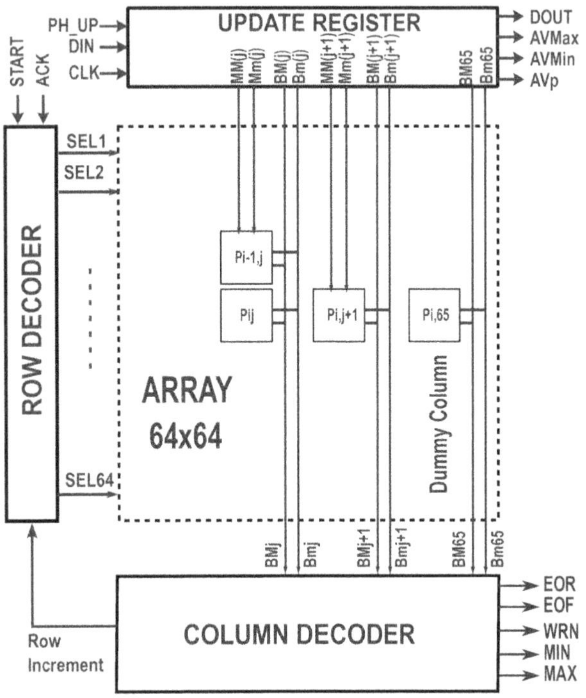

Figure 4: Block diagram of the vision sensor architecture.

5. EXPERIMENTAL RESULTS

Electro-optical tests have been carried out on the sensor, measuring the most important pixel characteristics: filter's transfer function and memory retention time, power consumption and computational performance.

5.1 Memories Update

The filter transfer function has been measured through an electrical test, applying at first a positive voltage step on the photodiode, thus simulating an abrupt light transition, from light-to-dark. Right after the two filters has been settled, a negative voltage step has been applied on the photodiode, mimicking a dark-to-light transition (Fig. 5). In the first case, an *hot-pixel* on V_{Max} ($Q_{Max} = 1$) has been forced. Hence, the memory C2M is updated at every frame ($n = 1$), by clocking MM, until the pixel returns into the cold-pixel status ($Q_{max} = 0$), which takes about 15 frames. During this phase, the memory C2m, storing $V_{Min} < V_P$, is updated once every 4 frames ($n = 4$), slowing down its time response. During the second phase, an *hot-pixel* is detected on V_{Min} ($Q_{Min} = 1$) and the memories update process is complementary, as shown in Fig. 5. It is possible to notice a difference in the time response between the V_{Max} and V_{Min} memories. In the first case, the *hot-pixel* compensation takes about 15 frames, while in the second case, it takes more than 20 frames. This is due to some offset between the two filters. One reason could be that, despite both channels are identical, with similar leakage currents and coupling effects, the signals they have to deal with are complementary. However, this is not a serious problem, because each pixel is intended to work in a closed control loop, adjusting its time response according with the output of the high-level algorithm.

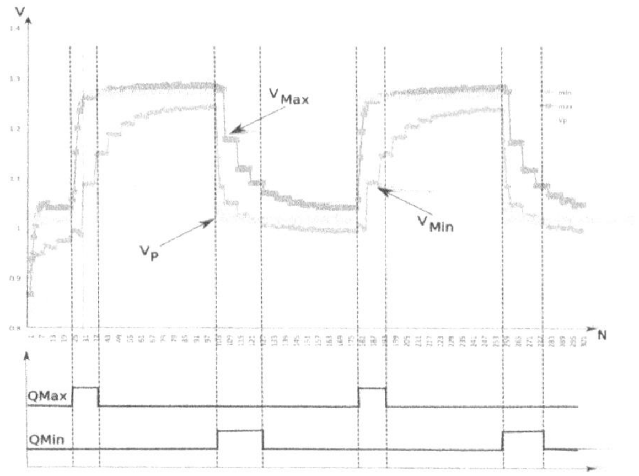

Figure 5: Example of the measured pixel response to a light step stimulus. *Hot-pixels* are updated with $n = 1$; '*cold-pixels* with $n = 4$. In this way, an asymmetric functionality is implemented, mimicking the eye behavior in presence of sudden light stimuli.

5.2 Memory Retention Time

Another important pixel parameter is the analog memory retention time. The two memories C2M and C2m are expected to store the two voltage thresholds, V_{Max} and V_{min}, for number of frames that can range from 1 up to 50, depending on the application. Making the sensor to operate at a frame rate of 10Hz, it means that the memory must retain information, without being updated, for about $5s$. The pixel analog memories have PMOS switches with leakage current which makes the signal to drift toward increasing voltage values. Therefore, the worst case is given by the memory C2M pre-charged at the lowest value $V_{Max}(t_0) = V_{sat}$, with no update applied. In this case, V_{Max} drifts toward larger values. We measured the retention time by forcing V_P at its highest voltage ($V_P = V_{dark}$), making the pixel to be hot ($V_{Max}(t_0) < V_{dark}$). The comparator is clocked at 10Hz, comparing V_P with V_{Max}. $Q_{Max} = 1$ for the time taken by V_{Max} to cover the voltage range of 750mV, from V_{sat} to V_{dark}. This time takes about 37 s (Fig. 6). It is obtained with the sensor placed under a light source making the photodiode to saturate in $100\mu s$, which is very close to the worst case conditions. If we consider $5s$ to be the useful memory retention time, we can see that after $5s$ the memory loses 1LSB on a 4-bit scale.

5.3 Power Consumption

The minimum power consumption of the chip has been measured under mixed synchronous/asynchronous operating mode, with a $2MHz$ readout clock (CLK) and a frame rate of $13fps$. The three operational amplifiers dedicated to the analog imager readout, have been turned off, as well as all the three analog source-followers inside each pixels of the array (SF1, SF2, SF3), by forcing the analog bit-lines to VDD (BVp, BVMax, BVMin). The $1\mu A$ pixel buffer is turned-on for $10\mu s$ only after the integration time, needed to charge V_P on the two capacitors C1M and C1m. In this operating condition, the sensor exhibits a total power consumption of $66\mu W$. However, it has to be considered that the BIAS block, shown on the top-left side of Fig. 8, providing the temperature compensated bias voltage to the pixels (BIAS in Fig. 2),

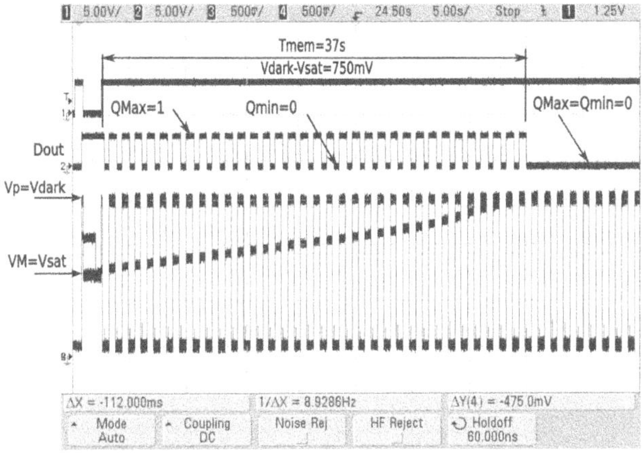

Figure 6: Retention time of the analog memory C2M (C2m) measured under an impinging light forcing the photodiode PD to saturate in $100\mu s$. The memory covers the entire signal range (750mV) in about 37s.

Table 1: Main chip characteristics

Parameter	Value
Technology	$0.35\mu m$ 2P-4M
Sensor resolution	64×64 pixels
Chip Size (pads included)	$6.6\ mm^2$
Pixel Pitch	$26\mu m \times 26\mu m$
Number of transistors/pixel	45
Fill Factor	12%
Memory retention time	0.2 LSB/s 4-bits scale
Supply Voltage	3.3 V
Pow. Cons./pixel	$620pW/frame \cdot pixel$
Computing performance - P_E	$42\ GOPS/W$
Computing density - P_A	$4\ GOPS/mm^2$

consumes $33\mu W$ ($10\mu A$). This means that, the vision sensor alone burns $33\mu W$. Based on this value, the derived power consumption per pixel per frame is $620pW/pixel \cdot frame$. This value is larger than that of other low power vision sensors [6],[7]. However, the absolute value of power consumption is not to be consider as an appropriate figure of merit for this class of sensors. More interestingly is the computing performance related to the power consumption. Considering that each pixel executes 26 operations per frame on a gray-scale, the total estimated computing performance of the sensor is 42 $GOPS/W$ and about 4 $GOPS/mm^2$. The obtained values are competitive with respect to most advanced computational vision sensors [2], [14], [15], [16].

5.4 Operating Functions

The sensor has been tested on moving objects to validate the vision algorithm. In Fig. 7 the sensor looks at an hand moving left and right periodically. At the beginning, this movement activates a large number of hot-pixels in the sensor. After a certain number of frames, the sensor starts to progressively absorb these pixels until the image returns to be black.

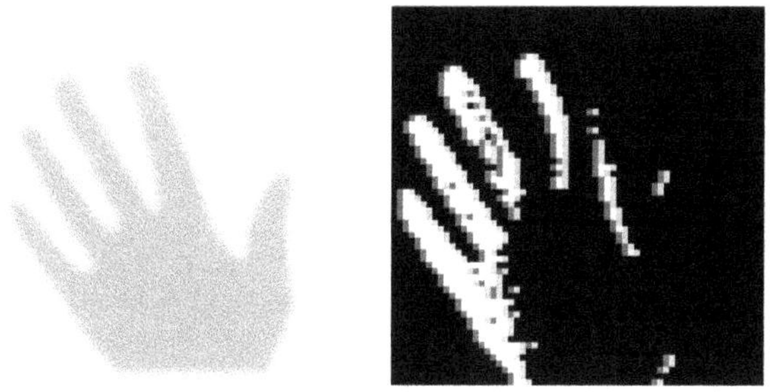

Figure 7: Example of hot-pixel detection. a) periodically left and right moving hand acquired by means of one of the three analog channels of the sensor; b) binary image resulting from the sensor image processing. The hot-pixels are white.

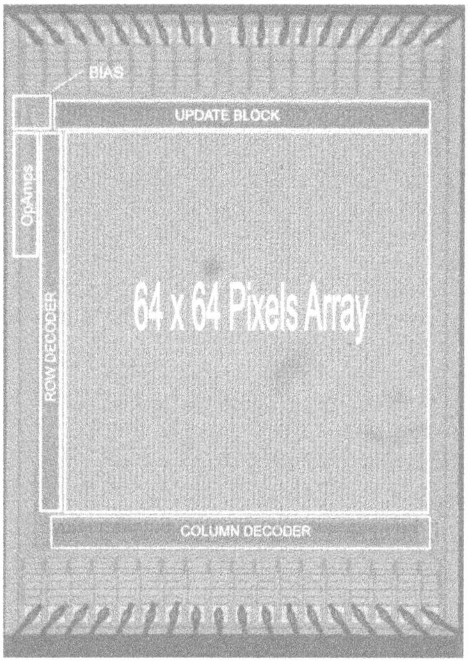

Figure 8: Photograph of the vision sensor chip.

6. CONCLUSION

A 64×64 pixels low-power vision sensor has been presented, performing $26\mu m$ pixel pitch and 12% fill-factor in a $0.35\mu m$ CMOS technology. The sensor executes an algorithm for adaptive background subtraction, aimed at detecting anomalous pixel behaviors, which could be part of a potential event, occurring in the scene. The algorithm is mainly implemented at pixel-level, mimicking the eye adaptation behavior. Each pixel embeds two switched-capacitors low-pass filters (SC-LPF1/2), which can be individually controlled from outside the imager, thanks to a custom control logic (UPDATE REGISTER), which can properly mask each filter's clock signal. The sensor power consumption has been minimized, reducing the operating duty-cycle of the voltage buffer inside each pixel, which is the main source of dc power consumption. Thanks to the pixel-level advanced image processing, the output of the sensor can be directly binarized, allowing the successive image processing to be fully binary and much easier to be implemented at a very low supply voltage and reduced energy budget. The computing performance of the sensor over power consumption is 42 $GOPS/W$.

7. ACKNOWLEDGMENTS

This work was partially supported by the Project BOViS ("A Battery Operated Vision System for Wireless Sensor Network Applications"), within the Italy-Israel Cooperation Program 2009.

8. REFERENCES

[1] F. Gregoretti, R. Passerone, L. M. Reyneri, and C. Sansoe,*"A High Speed VLSI Architecture for Handwriting Recognition,"* Journal of VLSI Signal Processing 28, 259-278, 2001.

[2] T. Komuro, I. Ishii, M. Ishikawa, and A. Yoshida, *"A digital vision chip specialized for high-speed target tracking,"* IEEE Trans. Electron Devices, vol. 50, no. 1, pp. 191-199, 2003.

[3] Komuro, T.; Kagami, S.; Ishikawa, M.; , "A dynamically reconfigurable SIMD processor for a vision chip," IEEE Journal of Solid-State Circuits, vol.39, no.1, pp. 265- 268, Jan. 2004.

[4] K. Kagawa, S. Shishido, M. Nunoshita, and J. Ohta, *"A 3.6 pw/frameá pixel 1.35 v pwm cmos imager with dynamic pixel readout and no static bias current,"* in Solid-State Circuits Conference, 2008. ISSCC 2008. Digest of Technical Papers, IEEE, 2008, pp. 54-55.

[5] K. Cho, D. Lee, J. Lee, G. Han,*"Sub-1-V CMOS image sensor using time-based readout circuit,"* IEEE Transactions on Electron Devices vol. 57, No.1, Jan. 2010, pp. 222-227.

[6] S. Hanson, D. Sylvester, *"A 0.45-0.7V Sub-Microwatt CMOS Image Sensor for Ultra-Low Power Applications"*, Symp. on VLSI Circuits, 2009, pp. 176-177, 2009.

[7] F. Tang, Y. Chao, A. Bermak, *"An Ultra-Low Power Current-Mode CMOS Image Sensor with Energy Harvesting Capability,"* Proc. of European Solid State Circuits Conf., pp. 126-129, Sept. 2010.

[8] Dongsoo Kim, Zhengming Fu,Joon Hyuk Park, and Eugenio Culurciello,*"A 1-mW CMOS Temporal-Difference AER Sensor for Wireless Sensor Networks,"* IEEE Transactions on Electron Devices, vol. 56, no. 11, pp. 2586-2593, November 2009.

[9] M. Gottardi, N. Massari, S.A. Jawed, *"A 100uW 128x64 Pixels Contrast-Based Asynchronous Binary Sensor for Sensor Network Applications,"* IEEE Journal of Solid State Circuits, vol. 44, no. 5, pp. 1582-1592, May 2009.

[10] Suat U. Ay, *"A 1.32pW/frame·pixel 1.2V CMOS Energy-Harvesting and Imaging (EHI) APS Imager,"* Digest of ISSCC 2011, February 2011, San Francisco.

[11] Miniature Autonomous Agents For Scene Interpretation, US Patent 7,489,802 B2, Filing Date Feb. 10, 2009.

[12] V. Gruev, R. Etienne-Cummings, *"A pipelined temporal difference imager,"* IEEE Journal of Solid-State Circuits vol.39, N. 3, pp. 538-543, March 2004).

[13] M. O'Halloran, and R. Sarpeshkar, *"A 10-nW 12-bit Accurate Analog Storage Cell With 10-aA Leakage,"* IEEE Journal of Solid-State Circuits, Vol. 39, No. 11, pp. 1985-1996, Nov. 2004.

[14] A. Lopich, P. Dudek, *"A SIMD Cellular Processor Array Vision Chip With Asynchronous Processing Capabilities,"* IEEE Trans. on Circuits and Systems, vol. 58, no. 10, pp. 2420-2431, Oct 2011.

[15] G. C. Linan, A. Rodriguez-Vazquez, and R. C. Galan et al., *"A 1000 FPS at 128 x 128 vision processor with 8-bit digitized I/O,"* IEEE J. Solid-State Circuits, vol. 39, no. 7, pp. 1044-1055, 2004.

[16] M. Wei, L. Qingyu, Z. Wancheng, and W. Nan-Jian, *"A programmable SIMD vision chip for real-time vision applications,"* IEEE J. Solid-State Circuits, vol. 43, no. 6, pp. 1470-1479, 2008.

Process and Temperature Invariant Bandwidth and Gain, Low-Area, Low-Power and High Swing G_m-C Filter for Multichannel Neuro-potential Signal Conditioning

Anvesha Amaravati
Indian Institute of Technology-Bombay
Powai, Mumbai-400076, India
anvesh@ee.iitb.ac.in

Maryam Shojaei Baghini
Indian Institute of Technology-Bombay
Powai, Mumbai-400076, India
mshojaei@ee.iitb.ac.in

ABSTRACT

A novel first order ultra low power and area efficient bandpass G_m-C filter aimed for varieties of neuro-potential signal conditioning applications, in 180nm CMOS process, is presented. The proposed G_m-C filter takes advantage of an ultra low-power constant G_m/C circuit, which leads to constant gain and bandwidth performance. Total current consumption of the bandpass G_m-C filter is 4.03μA and operates at 1.8V supply voltage. The proposed filter has a tunable bandwidth from 160Hz to 7.2kHz and exhibits input referred noise voltage of 4.17μV for a bandwidth of 10.2kHz. The filter provides output voltage swing of 1.55V and occupies area of $250\mu m \times 250\mu m$.

Categories and Subject Descriptors

B.7.1 [INTEGRATED CIRCUITS]: Types and Design Styles—*VLSI*

Keywords

Neural amplifier, G_m-C filter, bio-potential

1. INTRODUCTION

Demand for technologies that enable neural scientists to observe the activity of large number of neurons in the brain is very high. Neuropotentials can be divided into high frequency signals like action potentials (neural spikes) and low frequency signals like local field potentials (LFPs), electrocarticograms (ECoG) and electroencephalograms (EEG). Neural spikes have frequency components in the range 1Hz-10kHz. LFPs and ECoG signals have frequency content up to 200Hz. Neural potentials have amplitudes ranging from $50 - 500\mu$V[9]. Therefore, neural signals are having wide range of amplitude and frequencies. Multichannel recording of neural signals in the order of 128 channels require precise amplifiers with low current in terms of tens of micro amperes and least possible area. The amplifier should be able to reject DC offset voltages caused by electrode-tissue interface.

Commonly-used architecture of a neural amplifier is shown in Fig. 1 [3]. Gain of this amplifier is set by the ratio of capacitors C1 and C2. The amplifier uses total capacitance value of tens of pF to set the gain. As a result in multichannel neural recording such amplifiers will occupy a large on chip-area and hence increases the cost. Bio-amplifier shown in Fig. 2, reported in [2], occupies area of $0.05mm^2$ but it uses an amplifier in open loop configuration and hence it is susceptible to large gain and bandwidth variations due to interference, PVT. Moreover, it [2] uses single ended architecture which is susceptible for common mode interference. The gain of the bio-amplifier reported in [2] varies by a factor of $\pm$ 4.2% and bandwidth varies by a factor of $\pm$13.2% across 5-chips. To correct gain variations programmable gain amplifiers are used in the subsequent stages [11]. Additional gain correcting stages will cause extra power and area overhead. Although [4] acheives a low input referred noise of 2.3μV, but the bandwidth of the circuit varies by $\pm$22.1% which is very large. In low power, low noise applications the transcondutors are biased in subthreshold region and hence they are susceptible to process, voltage and temperature (PVT) variations. Constant gain and bandwidth are important factors for recording low frequency neural signals like EEG and ECoG. In this paper we propose a technique to achieve a constant bandwidth (constant G_m/C) for G_m-C filter, which is power and area efficient. The proposed G_m-C Filter gives 2x reduction in area, as compared to the reported neural amplifier works[1][3][4]. The proposed G_m-C filter can be used in the input stage of acquisition of neural, EEG, LFPs signals.

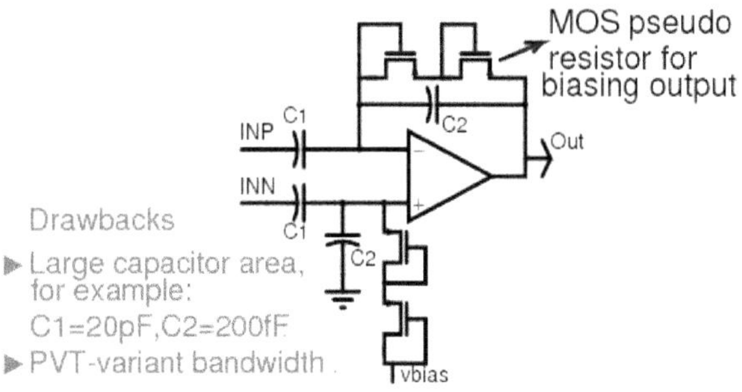

Figure 1: Traditional neural amplifier[3]

ISLPED'12, July 30–August 1, 2012, Redondo Beach, CA, USA.

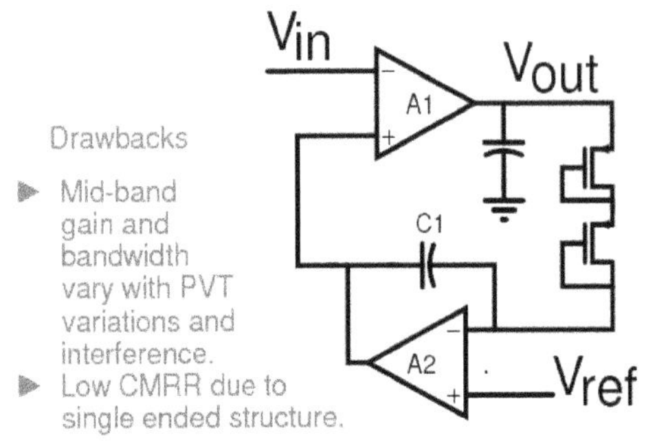

Figure 2: Bio-amplifier[2] used for neural recording and biopotential signal acquisition

2. PROPOSED G_M-C FILTER

We propose a novel bandpass G_m-C filter and its associated architecture, as shown in Fig. 3. Midband gain of the filter is independent of process and temperature, set by the ratio of two capacitively-coupled transconductances (G_{m1}/G_{m2}). Here, input coupling capacitors are of value 1.5pF which is much smaller than 20pF used in neural amplifier [3]. Internal circuitry of G_{m1} and G_{m2} are shown in Fig. 5. Every transconductor uses transistors, biased in sub threshold region, and ultra low power switched-capacitor PTAT current reference generator. Therefore the achieved transconductance will be proportional to the switching capacitor C and independent of temperature. Hence, the gain (G_{m1}/G_{m2}) will be independent of process and temperature. We have designed this architecture in 180nm technology and simulation results show that the gain and bandwidth of the proposed circuit changes less than $\pm 2\%$ across all process corners, temperature range of -10° C to 60^oC. Value of G_{m2} is chosen in such a way that on-chip filter capacitance of 2pF is enough for EEG applications. In Fig. 3 QFG (Quasi Floating Gate)[14] module uses MOS pseudo resistors to bias the capacitively-coupled input transistors of G_{m1} and G_{m2}. QFG input helps to reject DC offset voltages coming from electrodes. It also acts as a high pass filter with cut-off frequency of $f_{HP} = 1/(2 \cdot \pi \cdot R_p \cdot C_1)$ where R_p is the resistance of MOS pseudo resistor and C_1 is the input coupling capacitor.

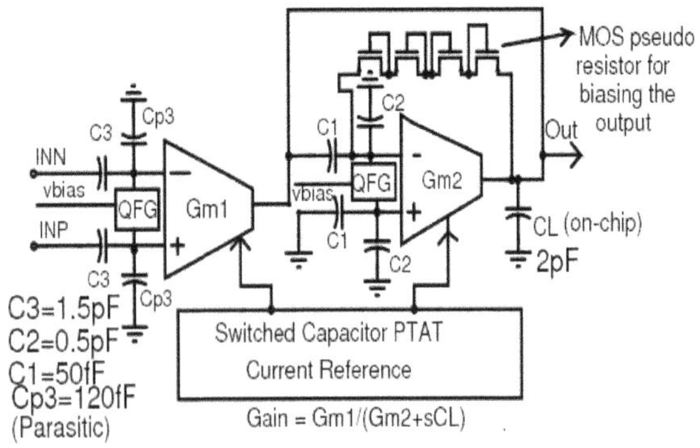

Figure 3: Proposed G_m-C filter for nuero-potential signal conditioning

2.1 Low power constant G_m/C

To achieve a constant bandwidth (i.e. constant G_{m2}/C) in G_{m2} module, shown in Fig. 3, we need a PTAT voltage reference and switched-capacitor PTAT current reference as shown in Fig. 4 to cancel effect of C and V_T [5]. PTAT voltage generator, reported in[5], consumes 7 μA current and also is not area efficient due to on-chip resistors. Other PTAT reference voltage generator, reported in [6] as shown in Fig. 4, is highly process variation-tolerant. It is made use as the reference voltage for switched-capacitor current reference. Hence, an ultra low power switched capacitor PTAT current reference is generated as shown in Fig. 4 which is used as the tail current source of transconductors G_{m1} and G_{m2}.

The PTAT voltage is given by Eq. 1.

$$V_{PTAT} = (N \cdot \eta V_T ln(K)) + (\eta V_T (ln(N) + ln(N-1) + \cdot\cdot + ln(1)) \quad (1)$$

where, η is the sub-threshold slope constant. For N=8 in Eq.1, we have $V_{PTAT} = 620mV$.

3. GM1 AND GM2 TRANSCONDUCTORS

A novel capacitively coupled bulk transconductance boosting technique is used in G_{m1} to achieve low input-referred noise and to reject electrode-offset. Amplifiers A1 and A2 are used to boost the output impedance by local series feedback. Capacitively coupled bulk-transconductance enhances the transconductance of M1 (g_{m1eff}) as explained here. The effective transconductance of the transistor M1 is given by,

$$g_{m1eff} = \frac{C_3 \cdot g_{m1}}{(C_3 + C_{p3})} + \frac{C_4 \cdot g_{mb1}}{(C_4 + C_{p4})} \quad (2)$$

where, g_{m1} is the transconductance of M1, g_{mb1} is the bulk transconductance of M1, C_{p3} and C_{p4} are parasitic capacitances at the gate and bulk of transistor M1, respectively. The values of capacitors are $C_1 = C_2 = 1.5pF$, $C_{p3} = 120fF$ and $C_{p4} = 84fF$ respectively. K is current mirror gain. The effective transconductance of the transconductor G_{m1} is given by Eq. 3.

$$G_{m1} = \frac{C_3 \cdot g_{m1}}{(C_3 + C_{p3})} \cdot K + \frac{C_4 \cdot g_{mb1}}{(C_4 + C_{p4})} \cdot K \quad (3)$$

Relation for the tail current source of $G_{m1}(I_{ref})$, provided by switched-capacitor current reference, is given by Eq. 4.

$$I_{ref} = (2 \cdot f_{clk} \cdot C \cdot V_{PTAT}) \quad (4)$$

Sub threshold g_m of a transistor is given by Eq. 5.

$$g_m = \frac{I_D}{\eta V_T \cdot 2 \cdot (1 + \sqrt{1 + 4 \cdot IC})} (EKV model) \quad (5)$$

where, IC is the inversion coefficient which is less than 0.1 for sub-threshold operation. According to Eq.6 g_{m1} is derived as follows.

$$g_{m1} = \frac{f_{clk} \cdot C \cdot V_{PTAT}}{\eta V_T \cdot 2 \cdot (1 + \sqrt{1 + 4 \cdot IC})} \quad (6)$$

Bulk transconductance of M1 is given by Eqn.7 [7].

$$g_{mb1} = \frac{\lambda_0}{\sqrt{2\Phi_F + V_{SB}}} \cdot g_{m1} \approx 0.25 \cdot g_{m1} \quad (7)$$

Finally G_{m1} is obtained from Eq. 8.

$$G_{m1} = \frac{f_{clk} \cdot C \cdot V_{PTAT}}{(1 + \sqrt{1 + 4 \cdot IC})2 \cdot \eta V_T} \left(\frac{C_3 \cdot K}{C_3 + C_{p3}} + \frac{0.25 \cdot C_4 \cdot K}{C_4 + C_{p4}} \right) \quad (8)$$

We can observe that G_{m1} is independent of temperature and directly proportional to switched-capacitor C.

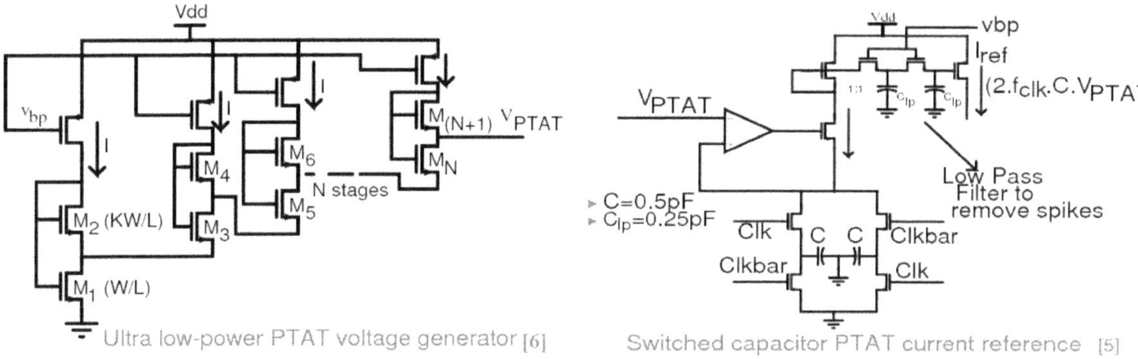

Figure 4: Ultra low power switched capacitor PTAT current reference

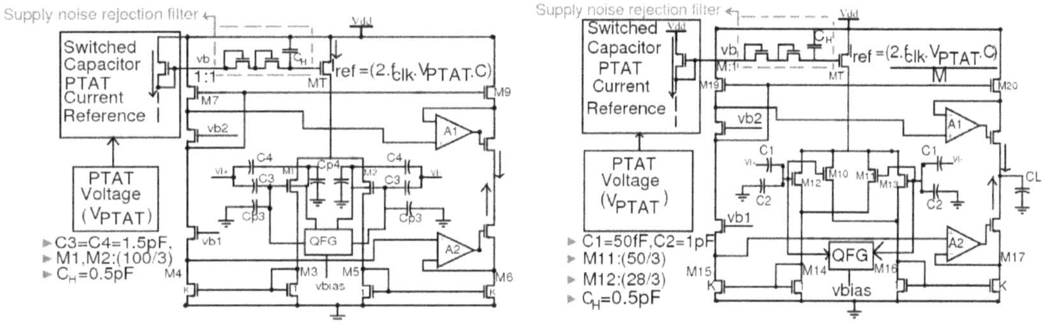

Figure 5: (a)Proposed Architecture of G_{m1} and (b) Proposed architecture of G_{m2}

3.1 Gm2 Transconductor

Bandwidth of the bandpass filter is set by G_{m2} and CL. Spectrum of many neural signals like EEG and ECoG includes low frequencies mainly from 0.2Hz to 200Hz [17]. Tail current source of transconductor G_{m2} carries M times less current as compared to tail current source transconductor G_{m1}. To reduce the required value of CL a very low G_{m2} is realized by deploying two techniques: current cancellation[8] and capacitive attenuation. As a result an ultra-low transconductance of $2.5nA/V$ is achieved. This low value of G_{m2} leads to the value of 2pF for CL, practical for on-chip implementation. Therefore cutoff frequency of the order 200Hz is obtained without compromising the area. The effective G_{m2} of the entire transconductor is given by Eq. 9.

$$G_{m2} = \frac{(C_1 \cdot g_{m11} \cdot K \cdot (N-1))}{(C_1 + C_2) \cdot \eta V_T \cdot 2(1 + \sqrt{1 + 4 \cdot IC})(N+1) \cdot M} \tag{9}$$

where, K is the current mirror gain and N is the ratio of aspect ratio of M11 to aspect ratio of M13. As a result ratio of DC bias current of these two tarsnsitor will be N+1. Since the tail current source of G_{m2} is also a switched capacitor current source, therefore G_{m2} can be given by Eq.10.

$$G_{m2} = \left(\frac{C_1}{C_1 + C_2}\right) \frac{2 \cdot f_{clk} \cdot C \cdot V_{ptat} \cdot (N-1)}{\eta V_T \cdot 2(1 + \sqrt{1 + 4 \cdot IC})(N+1) \cdot M} \tag{10}$$

We can observe that G_{m2} is also independent of temperature and directly proportional to switched-capacitor C. We can also observe that, ratio of transconductances G_{m1} and G_{m2} will be independent of process and temperature.

3.2 Gain boosting Amplifiers A1 and A2

Amplifiers A1 and A2 are used to boost the output impedance and open loop gain of G_{m1} and G_{m2}. Amplifier A1 and A2 constitutes an input level shifter followed by a differential amplifier [16]. Level shifter adjusts input common-mode voltage of the differential amplifier. Hence drain of transistors M6 and M9 can have a low DC voltage, which enhances output voltage swing of G_{m1} and G_{m2}.

3.3 Proposed supply noise rejection technique

Fig. 6 shows the proposed power supply noise rejection technique. v_s resembles power supply noise and V_b is the bias voltage to the transistor M_T which is used as tail current source for transconductors G_{m1} and G_{m2} shown in Fig. 5. Transistors M_{R1} and M_{R2} form MOS pseudo resistors with large resistance in the orders of Terra Ohms. The noise rejection filter forms a high pass filter from v_s to V_g with large time constant and hence very small value of cut-off frequency. As a result gate and source of M_T experience same supply noise level. Similarly, a low pass filter is formed from V_b to V_g with a very small cut-off frequency. As a result no noise is passed from V_b to V_g. C_H of 0.5pF is used to set the low pass filter cut-off frequency to 0.5Hz. As a result cut-off frequency of high pass filter will be the same. The tail current source transistor (M_T) is of long channel length (L=10μ) so as its current will be mainly dependent upon its gate to source voltage. As V_g follows V_s at non-zero frequencies, the current from the tail current source will be independent of power supply noise. PSRR of G_{m1}, G_{m2} and overall G_m-C filter improves by 24dB.

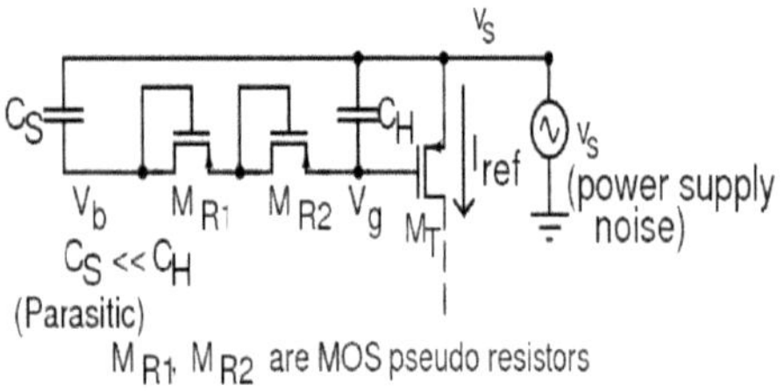

Figure 6: Proposed power supply noise rejection technique

4. INPUT REFERRED NOISE CALCULATIONS

Input referred noise of the G_m-C filter is divided into two parts: Noise contributed by transconductor G_{m1} and Noise contributed by transconductor G_{m2}.

Noise contribution by transconductor G_{m1}: Current noise at the output of G_{m1} is given by Eq. 11.

$$I^2_{out,G_{m1}} = \frac{8kT\cdot 2G_{m1}}{3} + \frac{8kT\cdot 2g_{m5}K^2}{3} + \frac{8kT\cdot 2g_{m6}}{3} + \frac{8kT\cdot 2g_{m9}}{3} + \frac{V^2_{in,A2}}{r_{o6}} + \frac{V^2_{in,A1}}{r_{o9}} \tag{11}$$

Where, $I^2_{tot,G_{m1}}$ is total current noise at drain terminal of transistor M1 and $V^2_{in,A1}$,$V^2_{in,A1}$ are the input-referred voltage noise of Amplifiers A1 and A2. We calculate the input referred voltage noise by dividing Eq.11 through G^2_{m1},

$$V^2_{in,G_{m1}} = \frac{16kT}{3G_{m1}}\left(1 + \frac{g_{m5}K^2}{G_{m1}} + \frac{g_{m6}}{G_{m1}} + \frac{g_{m9}}{G_{m1}} + \frac{V^2_{in,A1}}{G_{m1}\cdot r_{o9}} + \frac{V^2_{in,A2}}{G_{m1}\cdot r_{o6}}\right) \tag{12}$$

Where, K=2 for transconductor G_{m1}. The capacitively bulk transconductance boosting technique helps to improve the noise efficiency by 24% and rejects electrode offset. Noise contributed by Amplifiers A1 and A2 are only 3%, because their noise contribution will get divided by $(G_{m1}\cdot r_{o9})$.

Noise contribution by transconductor Gm2: Current noise at the output of G_{m2} is given by Eq. 13.

$$I^2_{out,G_{m2}} = \frac{8kT\cdot 2g_{m11}\cdot(N-1)}{3(N+1)\cdot M} + \frac{8kT\cdot 2g_{m14}}{3\cdot M} + \frac{8kT\cdot 2g_{m17}}{3\cdot M} + \frac{8kT\cdot 2g_{m18}}{3\cdot M} + \frac{V^2_{in,A1}}{r_{o20}\cdot M} + \frac{V^2_{in,A2}}{r_{o17}\cdot M} \tag{13}$$

We can observe from Eq. 13 due to low transconductance of g_{m11} due to current cancellation and large capacitive attenuation, the current noise at the output of transconductor Gm2 will be reduced. The total input referred noise of the G_m-C filter is the sum of input referred voltage noise of transconductor G_{m1} and output current noise of G_{m2} which is divided by G^2_{m1} and is given by Eq. 14.

$$\begin{aligned} V^2_{inn,G_m-C} &= V^2_{in,G_{m1}} + \frac{I^2_{out,G_{m2}}}{G^2_{m1}} \\ &= \frac{16kT}{3G_{m1}}\Big[1 + \frac{g_{m5}K^2}{G_{m1}} + \frac{g_{m6}}{G_{m1}} + \frac{g_{m9}}{G_{m1}} + \frac{V^2_{in,A1}}{G_{m1}\cdot r_{o9}} + \frac{V^2_{in,A2}}{G_{m1}\cdot r_{o6}}\Big] \\ &+ \Big[\frac{16\cdot kT\cdot g_{m11}\cdot(N-1)}{3(N+1)\cdot G^2_{m1}\cdot M} + \frac{16kT\cdot g_{m14}}{3\cdot G^2_{m1}\cdot M} \\ &+ \frac{16kT\cdot g_{m17}}{3G^2_{m1}\cdot M} + \frac{16kT\cdot g_{m18}}{3G^2_{m1}\cdot M} + \frac{V^2_{in,A1}}{G^2_{m1}r_{o20}\cdot M} + \frac{V^2_{in,A2}}{G^2_{m1}r_{o17}\cdot M}\Big] \end{aligned} \tag{14}$$

5. MISMATCH SIMULATION USING WORST CASE OFFSET

Due to unavailability of Monte Carlo models for UMC 180nm technology, we have simulated worst case CMRR and PSRR by applying manual offset at the gate of input differential pair of transconductor G_{m1}. The random offset at the input of differential pair of G_{m1} (Fig. 5) is given by Eq. 15 [6].

$$V_{os} = \Delta V_{th1} + \frac{g_{m3}}{g_{m1}}\Delta V_{th3} + \frac{\eta kT}{q}\left(\frac{\Delta\beta_1}{\beta_1} + \frac{\Delta\beta_3}{\beta_3}\right) \tag{15}$$

$\Delta V_{th1} = V_{th1} - V_{th2}$ and $\Delta V_{th3} = V_{th3} - V_{th5}$. $\Delta\beta_1$ indicates the mismatch between current factor($\beta = \mu_n C_{ox} W/L$) of input pairs M1 and M2. g_{m1} and g_{m3} are the transconductance of the transistors M1 and M3. Standard deviation of threshold voltage mismatch and current factor mismatch, i.e. σ_{Vth} and σ_β are given by, Eq. 16 and Eq. 17 [13].

$$\sigma_{Vth} = \frac{A_{Vtho}}{\sqrt{WL}} \tag{16}$$

$$\sigma_\beta = \frac{A_\beta}{\sqrt{WL}} \tag{17}$$

$A_{Vtho} = 6mV\mu$m and $A_\beta = 0.88\%.\mu$m for 180nm technology [13]. $W = 100\mu$m and $L = 3\mu$m for input transistors having transconductance of G_{m1}. $W = 2\mu$m and $L = 50\mu$m for load transistors (M3 and M4). The ratio of transconductances of M3 and M1 ($\frac{g_{m3}}{g_{m1}}$) is set to 0.12. By substituting the values in Eq. 16 and Eq. 17 we have: $\sigma_{Vth1} = 0.346mV$, $\sigma_{\beta1} = 0.050\%$, $\sigma_{Vth3} = 0.6mV$ and $\sigma_{\beta3} = 0.088\%$. Accordingly the variance of input referred offset voltage of transconductor G_{m1} is calculated by Eq. 18.

$$\sigma^2_{Vos} = \sigma^2_{Vth1} + \left(\frac{g_{m3}}{g_{m1}}\right)^2\sigma^2_{Vth3} + (\frac{\eta kT}{q})^2\left(\sigma^2_{\beta1} + \sigma^2_{\beta3}\right) \tag{18}$$

After substituting the values of standard deviations in Eq. 18 value of σ_{Vos} is obtained as 0.355mV. Therefore the worst case offset voltage is = $\pm 6\sigma_{Vos}$ =$\pm 2.13mV$.

Layout of the proposed G_m-C filter: Fig.7, shows the layout of proposed G_m-C filter. Entire G_m-C filter circuit occupies an area of $250\mu m \times 250\mu m$.

6. SIMULATION RESULTS

The entire G_m-C filter circuit is simulated in UMC 180nm mixed mode CMOS technology at 1.8V supply. From Fig.8(a) and Fig.8(b), worst case variation of gain w.r.to process corners and temperature (-10^oC to 60^oC) is $\pm 1.2\%$ and worst

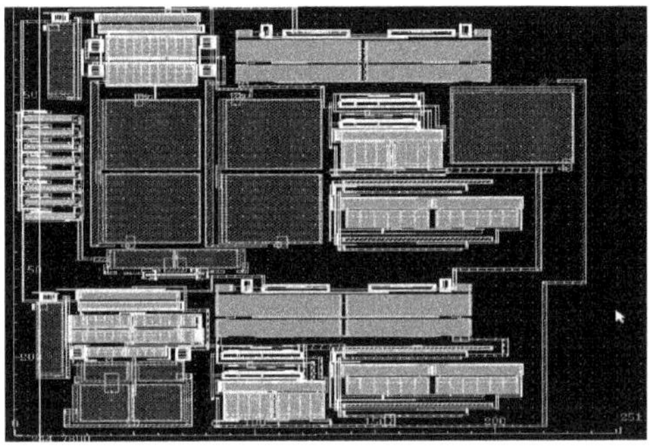

Figure 7: Layout of the proposed G_m-C filter

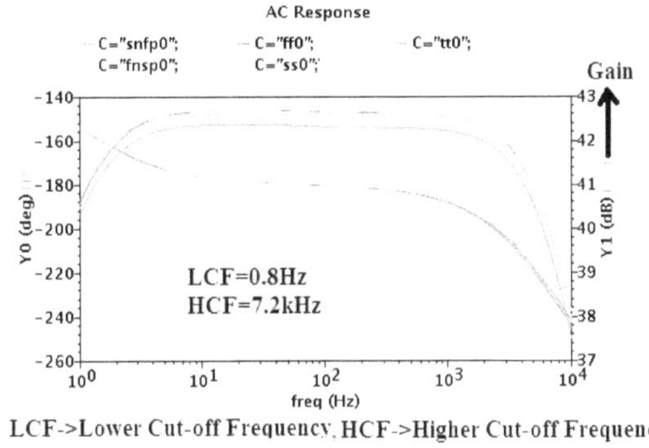

(a)

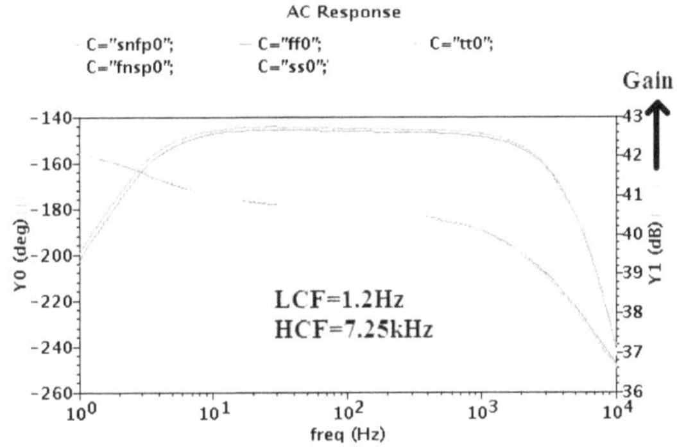

(b)

Figure 8: AC response of the proposed G_m-C filter across all process corners at a) $T = -10^oC$ b) $T = 60^oC$

case variation of the bandwidth under same conditions is ±1.9%.

From Fig. 9, the flicker noise corner frequency for the G_m-C filter is 65Hz and thermal noise floor is $44nV/\sqrt{Hz}$. From Fig. 10(a) CMRR of the G_m-C filter up to 5.2kHz frequency and under worst case input offset voltage of 2.13mV is 95dB. From Fig. 10(b) PSRR of the G_m-C filter up to 5.2kHz frequency and under under worst case input offset voltage of 2.13mV is 70dB. Output swing of the proposed G_m-C filter for T.H.D. (total harmonic distortion) of 1% is equal to 1.55V(p-p). The corresponding maximum output swing for 7.7mV(p-p) input sine wave is shown in Fig.11.

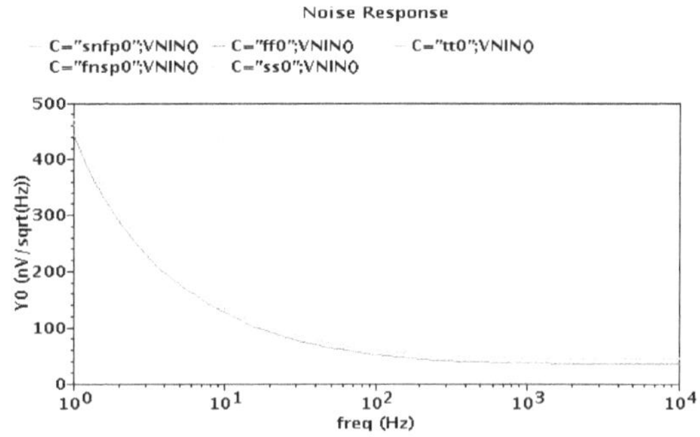

Figure 9: Input referred noise plot of proposed G_m-C filter across process corners

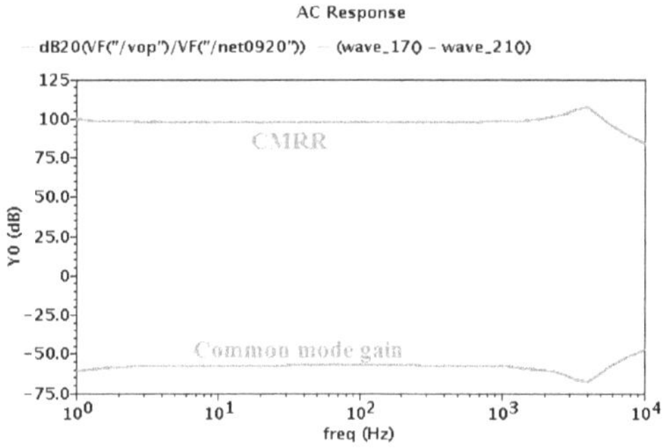

(a)

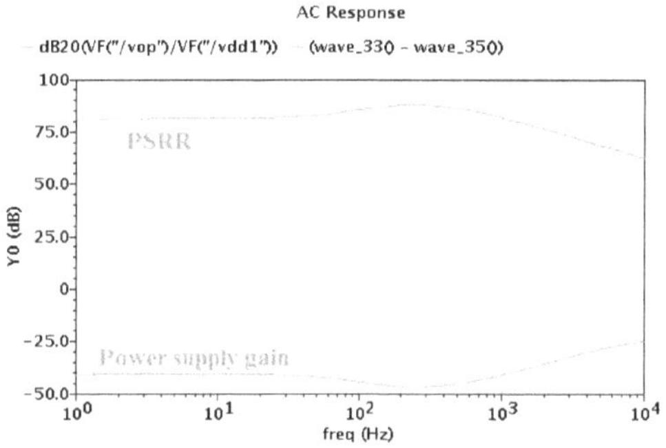

(b)

Figure 10: a)CMRR b)PSRR plot with frequency

7. COMPARISON WITH REPORTED NEURAL AMPLIFIERS

Table.1 compares performance of the state of the art reported neural amplifier circuit with the proposed circuit.

8. CONCLUSION

A novel area efficient G_m-C filter having process and temperature invariant gain and bandwidth with output voltage swing of 1.55V at 1.8V supply voltage, is presented. The presented circuit can be used for neuro-potential signal conditioning and also for other bio-potential measurements such as precise ECG and EMG recording. The proposed G_m-C filter exhibits only ±1.2% variation of gain across all process corners and temperature range of -10^oC to 60^oC. At the same time it shows only ±1.9% variation of bandwidth

Table 1: Comparison with reported neural amplifiers

	Borghi[12] Measurement	Gosselin[2] Measurement	Harrison[3] Measurement	Liu[15] Simulation	Sarpeshkar[4] Measurement	Chengliang[10] Measurement	This work Simulation
Process/Supply	0.35μm/3.5V	0.18μm/1.8V	1.5μm/5V	0.18μm/1.8V	0.5μm/3V	0.6μm/2.7V	0.18μm/1.8V
Current/Power	NA/248μW	4.7μA/8.4μW	16μA/80μW	18μA/7.6μW	2.7μA/8.2μW	0.87μA/2.4μW	4.03μA/7.2μW
C_{total} (On-chip)/Area	22pF NA	8pF 0.05mm^2	42pF 0.160mm^2	40pF NA	37pF 0.160mm^2	40pF 0.130mm^2	11pF 0.0625mm^2
Gain	39.9dB	49.5dB	39.5dB	100dB	40.8dB	39.4dB	39.8dB
Bandwidth	10kHz	9.1kHz	7.2kHz	9.2kHz	5.32kHz	1.3kHz	7.2kHz
Bandwidth Variation	NA	±13.2% across 5-chips	NA	NA	±22.5% across 4-chips	NA	±1.9% across all process corners & temp. range $(-10^oC, 60^oC)$
CL	NA	0.4pF	17pF	15pF	9pF	11pF	2pF
$V_{in,n,rms}$	3μV (20kHz)	5.6μV (9.2kHz)	2.2μV (50kHz)	8.5μV (15kHz)	3.06μV (98kHz)	3.07μV (30kHz)	4.17μV (10.2kHz)
$V_{in,max}$(p-p)/ $V_{out,max}$(p-p) (THD=1%)	NA/ NA	2.4mV/ 0.7V	16.7mV/ 1.6V	NA/ NA	7.3mV/ 0.76V	10mV/ 1V	7.7mV/ 1.55V
CMRR	NA	>50dB	>66dB	>66dB	>80dB	>90dB	>100dB (up to 5KHz and at $V_{os,in}$=2.13mV)
PSRR	NA	>50dB	>85dB	NA	>75dB	>75dB	>70dB (up to 5KHz and at $V_{os,in}$=2.13mV)

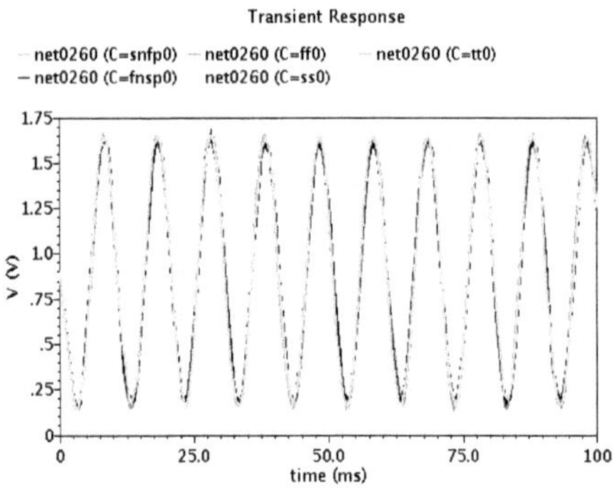

Figure 11: 1.55V output voltage swing with T.H.D. of 1% for input sine wave of 7.7mV(p-p) across all process corners

under same conditions. Area of the proposed G_m-C filter is less than similar circuits due to smaller values of the capacitors at a given technology node.

9. REFERENCES

[1] P. Mosheni et al.,"A fully integrated neural recording amplifier with DC input stabilization," *IEEE Trans. Biomed. Eng.*, May 2004.

[2] B. Gosselin et al, "A low-power integrated bioamplifier with active low-frequency suppression," *IEEE Trans. Biomed. Circ. Syst*, Sep. 2007.

[3] R. Harrison et al., "A low-power low-noise CMOS amplifier for neural recording applications,"*IEEE JSSC*, 2003.

[4] W. Wattanapanitch et al., "An energy-efficient micro power neural recording amplifier," *IEEE Trans. Biomed. Circ. Syst*, 2007.

[5] Q. A. Khan et al., "A tunable $g_m - C$ filter with low variation across Process, Voltage and Temp.," *Int. VLSI Design Conf.*, 2004.

[6] W. M. Sansen et al., "A CMOS temperature-compensated current reference," *IEEE JSSC*, June 1988.

[7] R. Fried et al., "Bulk-driven MOStransconductor with extended linear range," *Electron. Lett*, 1996.

[8] J. Silva-Martinez et al., "IC Voltage to Current Transducers with Very Small Transconductance," *Analog Integ. Circ. and Sig. Proc.*, (1997).

[9] K. D Wise, "A multi-channel microprobe for neural recording," *Ph.D.dissertion*, Stanford Univ., 1969.

[10] K. Qian, "A Micropower Low-Noise Neural Recording Front-End Circuit for Epileptic Seizure Detection," *IEEE JSSC*, June 2011.

[11] V. Chaturvedi et al., "A Low-Noise Low-Power Noise-Adaptive Neural Amplifier in 0.13um CMOS technology," *VLSI Design Conf.*, 2011.

[12] T. Borghi et.al, "Compact Multichannel Sys. for Acquisition and Processing of Neural Signals," *IEEE EMBS*, 2007.

[13] M. P Flynn et.al, "Achieving Analog Accuracy in Nanometer CMOS," *World of Scientific Journals*, 2005.

[14] J. R Angulo et.al, "Very Low-Voltage Analog Signal Processing Based on Quasi-Floating Gate Transistors," *IEEE JSSC*, 2004.

[15] W. Liu et.al, "A Neural Recording for Monitoring Shark Behaviour," *IEEE ISCAS*, 2006.

[16] J.C Baker, "CMOS: Mixed-Signal Circuit Design," *Wiley-IEEE Press*, 2009.

[17] M. Mollazadeh et.al, "Micropower CMOS Integrated Low-Noise Amplification, Filtering, and Digitization of Multimodal Neuropotentials" *IEEE TBCAS*, 2009.

A Charge Pump Based Receiver Circuit for Voltage Scaled Interconnect

Aatmesh Shrivastava
University of Virginia
Charlottesville,
VA, USA
as4xz@virginia.edu

John Lach
University of Virginia
Charlottesville,
VA, USA
jlach@virginia.edu

Benton H. Calhoun
University of Virginia
Charlottesville,
VA, USA
bcalhoun@Virginia.edu

ABSTRACT
This paper presents a charge-pump based low swing interconnect receiver circuit. The interconnect circuit is single ended and supports swings of 300mV or lower. A charge pump front end at the receiver boosts the arriving signal before restoring it to the full logic level, improving the performance of the interconnect. For a 10mm long interconnect wire in a 45nm CMOS process, the proposed scheme provides 3X energy reduction at constant speed and 3.5X delay improvement at constant energy relative to prior art. We deploy the interconnect scheme as the data bus between the L1-L2 caches of a 4-core Alpha processor. Over a set of Splash benchmarks, the proposed architecture reduces total energy consumption by 70% while maintaining the same performance.

1. INTRODUCTION
Studies have shown that 50% of total chip power is dissipated in interconnect wires and circuits in a modern microprocessor [2]. This number is close to 90% for reconfigurable architectures like FPGAs [3]. Interconnect power is going to become even a bigger concern for exascale computing [1] where 10 billion transistors are expected to be present in one square centimeter of chip area. Over the past decade, voltage scaling has been employed to reduce the power of interconnects [4-10]. In a voltage scaled interconnect, the interconnect wire is driven at a much lower voltage than the logic. A receiver circuit converts the low swing signal on the interconnect back to the full swing logic level. Various architectures for the interconnect driver and receiver have been proposed in the literature. These can broadly be categorized as single ended, differential, and capacitive interconnects. Table 1 shows an approximate energy-delay comparison of interconnect circuits reported in the literature. The basic and differential interconnect show the best performance, but they have higher energy consumption. The basic interconnect does not employ voltage scaling, and therefore consumes higher energy, while differential interconnects [8-10] use a differential amplifier and two wires per interconnect signal, which increases the energy consumption. The single ended interconnects [4-5,7] show reduced energy consumption, but their performance is poor because the lower input swing reduces drive current in the receiver. The capacitive interconnect schemes (e.g., [6]) have driver and receiver circuits capacitively coupled to the wire using series capacitors. The charge distribution between the wire and capacitor reduces the swing, which saves power. The receiver circuit in this scheme is either differential or single ended. The best reported work here [6] claims a bandwidth of 250MHz, much lower than the desired on-chip signal rate of a GHz or higher.

Table 1: Energy and Delay of existing interconnects

Schemes	Speed (GHz)	Swing (V)	Normalized Energy
Basic	>1	1	1
Single-ended [4,5,7]	<0.25	0.6	0.6
Differential [8-10]	>1	0.05	0.8
Capacitive [6]	<0.25	0.05	0.2

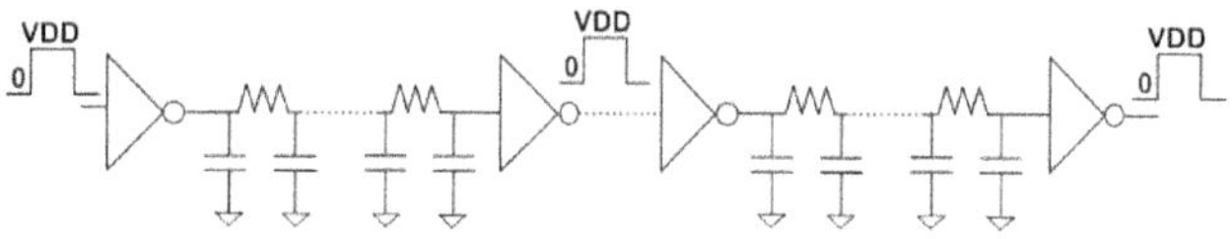

Figure 1. Basic interconnect circuit

One of the primary reasons for the lower performance of single ended interconnects is the lower voltage swing at the receiver input. In this paper, we employ a charge pump to increase the swing at the receiver's input. The receiver sees three times the interconnect swing voltage at its input. This saves power without impacting performance.

2. PREVALANT INTERCONNECTS
Figure 1 shows the most basic type of interconnect architecture also known as CMOS interconnect. It does not employ voltage scaling. The metal interconnect between two points inside a chip can be approximated as a distributed π-RC network as shown Figure 1. Its delay increases quadratically with length due to Elmore delay. Repeaters are inserted at regular intervals to obtain the optimal delay point. We simulated a π-RC model of a 10mm long wire in 45nm CMOS with repeaters. By controlling the number of repeaters in the path, either a minimum delay or a minimum energy point can be achieved. However the overall energy consumption of this interconnect architecture is high. Voltage scaling has been employed to reduce this power in [4-10].

2.1 Voltage-scaled Interconnect Architectures
Figure 2 shows the architecture of a voltage scaled interconnect.

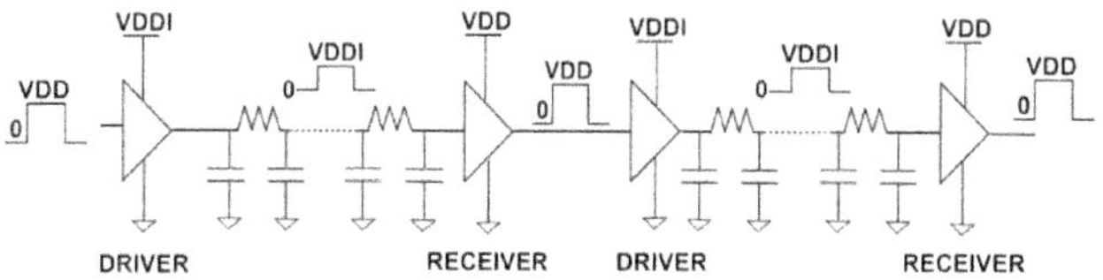

Figure 2. Voltage-scaled interconnect circuit

ISLPED'12, July 30– August 1, 2012, Redondo Beach, CA, USA.

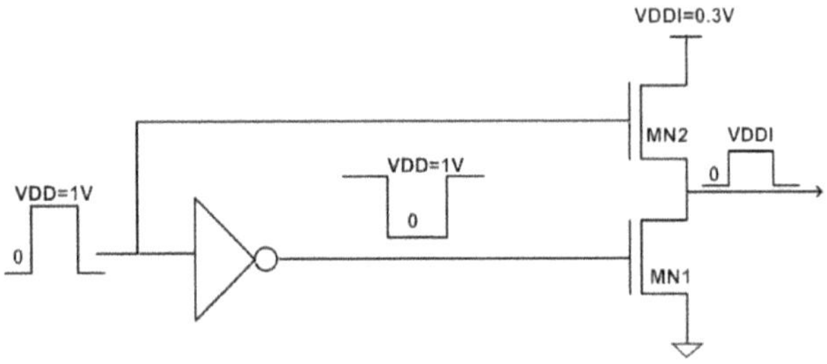

Figure 3. Asymmetric source follower driver [4]

It consists of an interconnect driver that converts the signal at the logic level (V_{DD}) to the lower interconnect voltage (V_{DDI}) and a receiver that converts the signal back to the logic level. The driver-receiver pairs are called repeaters as shown in Figure 2.

Interconnect Driver: Figure 3 shows an asymmetric source-follower driver [4], which is commonly used for low voltage interconnects. It uses NMOS devices to drive both *high* and *low* voltages. When *high* is to be passed, MN1 is off and MN2 acts as a source follower to pass V_{DDI} to the interconnect. The delay of transmitting a *high* through the NMOS is not increased because MN2 is over-driven. When a *low* is to be passed, MN2 is off, and MN1 turns on as in a regular inverter. This architecture is useful if the interconnect voltage is scaled down to very low voltages in the range from about 0.1V to 0.5V.

Interconnect Receiver: Receiver design plays an important role in the overall performance of interconnects. Figure 4 shows the energy-delay points of different interconnects reported in the literature for a 10mm long wire. We scaled the data accordingly to compare with the 45nm CMOS process with nominal operating voltage at 1V. The single ended interconnect implementations use a source follower receiver [4][5][7]. The performance of these receivers is poor because of the source degeneration. Also the interconnect signal should be at least $2*V_T$ for the receiver to operate correctly. This puts a limit on the amount of power that we can save using voltage scaling. These interconnect circuits fall on the right side of Figure 4. Usually, basic interconnect or differential interconnect [8-10] is used for higher performance. These schemes form the left side of Figure 4. They can have higher performance than the basic interconnect, but that costs additional energy. We find that existing solutions do not improve power and performance at the same time. There is a need to bring down the interconnect power while maintaining the performance [1]. The proposed interconnect circuit addresses this issue and gives a high performance yet lower power interconnect.

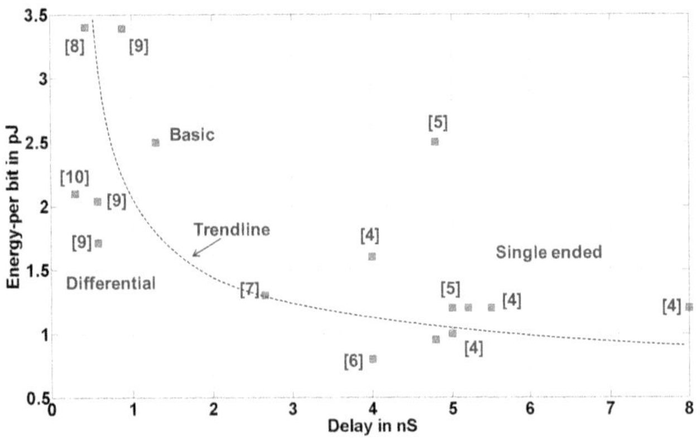

Figure 4. Energy-Delay points of interconnects in literature

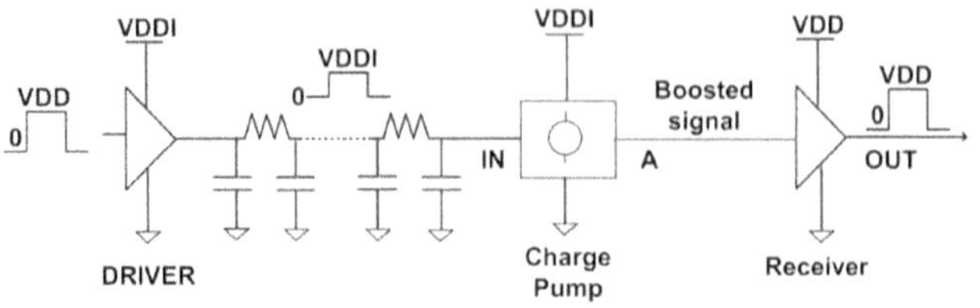

Figure 5. Proposed interconnect architecture

3. PROPOSED INTERCONNECT CIRCUIT

Figure 5 shows the block diagram of the proposed interconnect scheme. It can operate at an interconnect swing of 300mV or lower. We use an asymmetric source follower driver [4], which is well suited for this case. The receiver is novel and uses a charge pump, which boosts the signal and improves its performance. We will discuss each of these components in turn.

3.1 Charge-Pump

Figure 6 shows the charge pump. It has two capacitors (C_{CH} and C_{CL}) connected in series with the interconnect wire, IN. Nets A and C are dynamic nodes. Net A is controlled by NMOS MN4 and MN5. These transistors turn on for a very small duration of time to set the voltage at A to 0.3V or 0V. This voltage is then dynamically held at A. Net A is precharged to 0.3V before IN makes a high transition. As IN goes *high*, A gets charged to 0.6V. Similarly, A is precharged to 0 when IN is at 0.3V. Net A goes to -0.3V as IN makes a transition to zero. Therefore the overall swing at A is boosted to 0.9V. Similarly, C has a PMOS connected to ground and an NMOS to V_{DD}. It can swing from V_T to V_{DD}-V_T. It makes a transition from V_T to V_T+0.3V when IN goes *high* and V_{DD}-V_T to V_{DD}-V_T-0.3V when IN goes *low* through the capacitive coupling of C_{CH}.

The series capacitors are implemented using NMOS in an Nwell. A 1.5µm by 1µm capacitor gives roughly 35fF of series capacitance, which was good enough to get the desired swing at A and C. The control and precharging of A and C is implemented using feedback discussed in Section 3.2. The proposed interconnect receiver circuit uses the charge pump and a pulse generator circuit in a feedback manner to properly set the voltage at A. This assists in receiving the incoming low swing signal and converting it to full swing. In the following section, we will explain the functioning and the circuit diagram of the receiver.

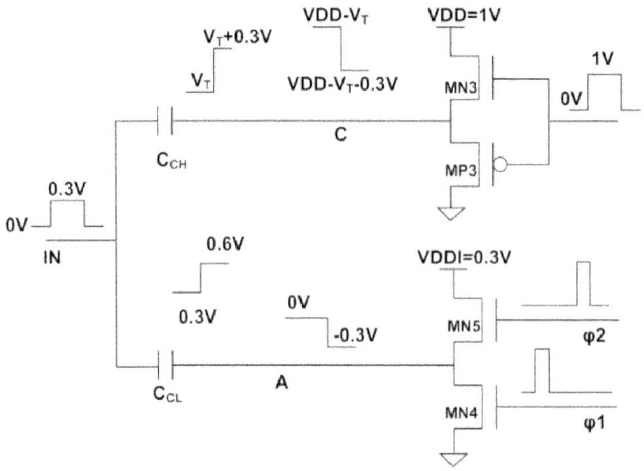

Figure 6. Proposed charge-pump circuit

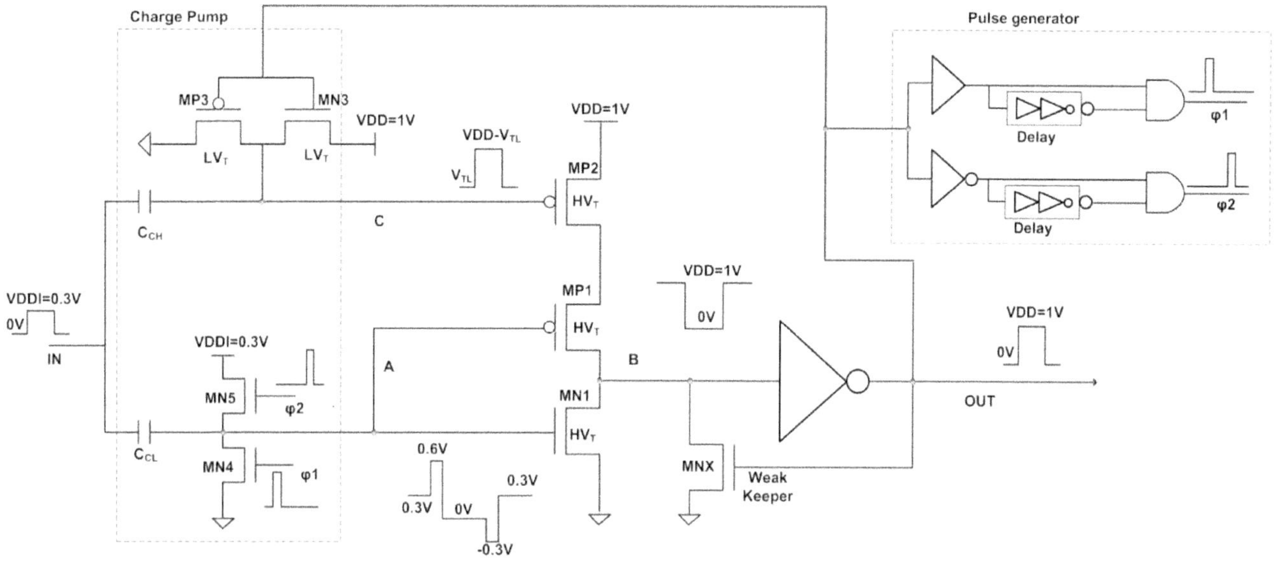

Figure 7. Proposed interconnect receiver circuit with charge-pump based pre-amplification

3.2 Proposed Interconnect Receiver

Figure 7 shows the full circuit diagram of the proposed receiver. It consists of a cascoded PMOS inverter (MP1, MP2, MN1) using HV_T transistors, a regular inverter, and a positive feedback NMOS transistor (MNX) along with the charge pump made of LV_T transistors and pulse generator circuit. Figure 8 shows the associated timing diagram of the receiver. We will explain the functioning of the circuit with the aid of this idealized timing diagram.

Transmitting HIGH*:* Consider the case at time=0, when IN=0.

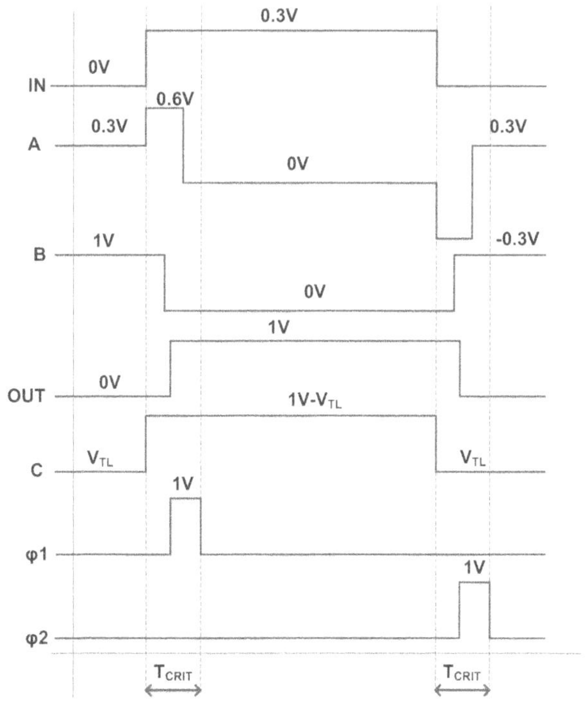

Figure 8 Timing diagram of the receiver

At this point, OUT should equal 0. Assuming OUT=0 and therefore B=1, so Net C should be equal to V_{TL}. Also, we assume that Net A is charged to V_{DDI}=0.3V. Since A=0.3V and C=V_{TL}, MP1 and MP2 are on while MN1 is close to off, which means that our assumption for B=1 is consistent with this circuit. Now consider the case when IN goes high to 0.3V from 0V. Net A gets charged to 0.6V through the capacitive coupling of C_{CL}. Net C gets charged through capacitor C_{CH} to V_{TL}+0.3V. This causes the current drive of transistor MN1 to increase, while the drive of MP1 and MP2 decreases. The change in drive strength causes Net B to discharge to ground. As B goes to ground, OUT gets charged to 1V. Also, as OUT goes to 1V, MNX turns on and keeps B at ground. MNX is a weak keeper transistor and holds the state at net Net B in the absence of any other signal driving it. At this point, we have propagated a low to high transition through the receiver.

As OUT goes high, two more transitions take place. Net C will get charged to V_{DD}-V_{TL} through MN3 in the charge pump, which turns off MP2. Also, the pulse generator circuit produces a small pulse at φ1, which pulls A to ground. Once φ1 goes to ground, Net A remains charged to 0, but it becomes high impedance. The state at B is maintained by MNX through the positive feedback.

The delay from IN going *high* to φ1 going *low* is called the critical delay, T_{CRIT}. It limits the maximum operating frequency of the circuit since the receiver is not ready to receive a low before φ1 goes low.

Transmitting LOW: At this point, we have OUT=1V, C=1V-V_{TL}, A=0, and B=0. Now consider the case when IN goes to 0 from 0.3V. Net A will go to -0.3V through capacitor C_{CL}, and Net C will go to V_{DD}-V_{TL}-0.3V through capacitor C_{CH}. This will turn on MP2 and increase the drive strength of MP1. The increased drive will overcome MNX (MNX is a weak keeper) and put a 1V at B, making OUT to go to 0V as shown in Figure 9. As OUT goes to 0V, C will be pulled to V_{TL} through MP3. At this point, the pulse generator circuit produces a pulse at φ2 that pulls A to 0.3V making the receiver ready to receive a *high*.

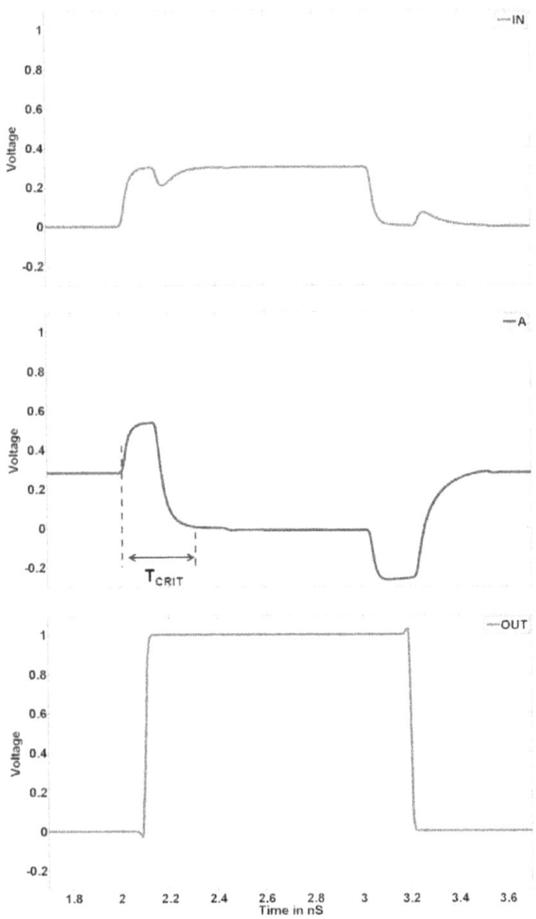

Figure 9. Simulation results of the receiver

Simulation: The interconnect circuit was designed in a 45nm CMOS process. We used HV_T and LV_T transistors and used NMOS in Nwell capacitors. The wire used is a distributed π-RC network that models the wire in our 45nm CMOS process. We selected a 10mm long wire and introduced the proposed interconnect driver and receiver circuits at regular intervals. The driver circuit reduces the swing from the logic level (1V) to the interconnect level (0.3V) and launches it to the long interconnect wire. The receiver circuit receives this signal and converts it back to the logic level. Figure 9 shows a typical simulation result in 45nm CMOS for key signals. Typical delay from A to OUT is close to 80ps. The receiver circuit gave the desired performance at

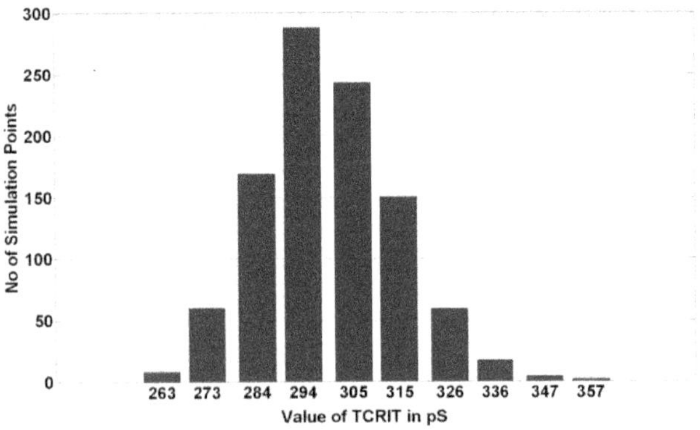

Figure 10. Histogram of T_{CRIT} from Monte-Carlo simulation

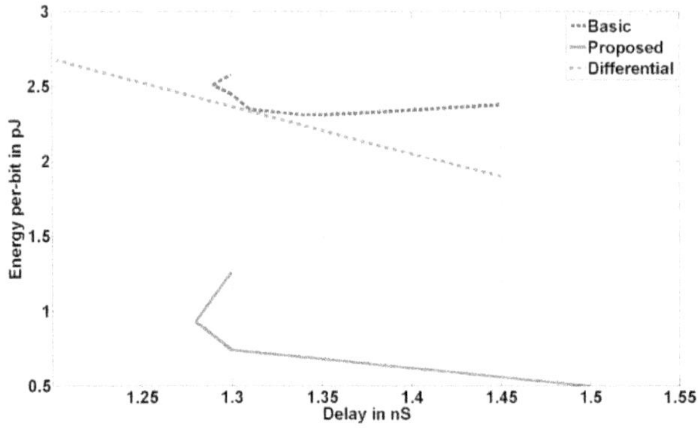

Figure 11. Energy-Delay curve of the proposed interconnect

fast, slow, typical, and skewed corners. Net A does not reach the ideal voltages shown in Figure 9 because of charge sharing, but it is good enough to obtain the desired performance. The small kink in the waveform of IN is caused by charge coupling from Net A. The T_{CRIT} of the circuit is 300ps, setting the maximum operating frequency at ~3GHz. Figure 10 shows the histogram of T_{CRIT} obtained from 1000 local mismatch Monte-Carlo simulations of the circuit. The charge pumping technique can ideally make Net A of Figure 7 swing up to (0.6+0.3V) 0.9V. Increased swing at A comes from the charge-pumps and not by increasing the interconnect swing, as was done in [4-5]. The lower voltage swing on the wire reduces energy dissipation, while the elevated swing at the receiver level ensures good performance.

Figure 11 shows the energy delay curve of the interconnect scheme obtained by varying the number of repeaters. We also used the differential interconnect of [8] and basic interconnect for comparison. Other single ended interconnects in literature cannot be used for this experiment owing to their high delay. We chose the same 10mm long wire discussed earlier. The proposed circuit has much lower energy at the same performance points as that of differential or the basic interconnect.

Initial conditions and leakage: The proposed circuit needs to be initialized properly before we can start transmitting data. This is because of the dynamic nature of the circuit. For example, Net A should be at 0.3V before the receiver can receive a *high*. At the very beginning this cannot be ensured by the feedback through the pulse generator. A RESET signal is used to initialize the voltage of the dynamic nets. Similarly, if the interconnect is left idle for very long, leakage can affect the dynamic nets. These nets are refreshed at a regular interval of time to ensure fidelity. We needed to refresh the circuit if it is idle for more than 100µs using the same RESET signal. Figure 12 shows the initialization

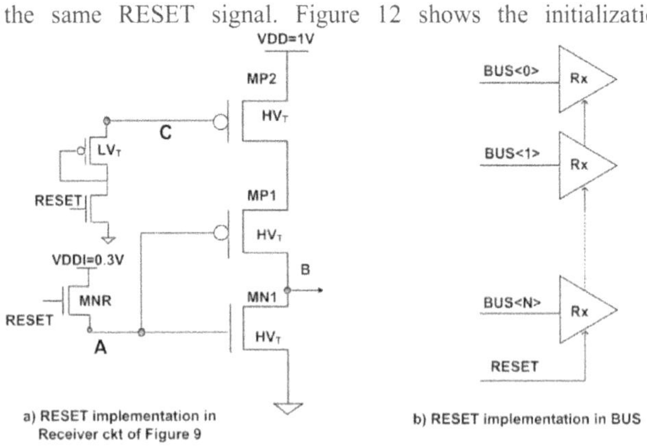

Figure 12. Initialization scheme for the receiver circuit

scheme using RESET. When RESET is held high, Net A is forced to 0.3V while Net C is taken to ~V_{TL}, making B go high, which sets OUT to 0 (cf. Figure 7). This makes the circuit ready to receive a high. The RESET signal is low frequency, and only one signal is needed per bus. Therefore it does not contribute a significant area or power overhead.

Static current: The interconnect circuit consumes a small amount of static current that is present because pull-up and pull-down paths are not fully disabled. Figure 13a shows the first stage of the receiver circuit. When B is high, MP2 is connected to V_{TL} while MN1 is connected to 0.3V. MN1 is not completely cut off, which causes static current. However since the high V_T (V_{TH}) is greater than 0.3V, the static current is very small. Similarly, when B is low, MP2 is at V_{DD}-V_{TL} which will result in static current. However V_{TH}>V_{TL}+(~0.2V) ensures that leakage current is small.

Figure 13b shows the Monte-Carlo simulation result of the leakage in the receiver circuit. The simulation was performed at 30°C. The maximum leakage is less than 1μA, and average leakage is around 100nA. A basic interconnect receiver made of inverters has leakage in the range of ~1nA. The LHOS receiver of [5] will have leakage of ~200nA, while the HOA [5] will have leakage in the range of ~1nA. Static current is also present in differential interconnect receivers [8-10] in the form of bias current, which ranges from a few 100μA to a few mA. The leakage current in our receiver is an overhead that increases power consumption if the interconnect is not switching. However, as switching activity on the interconnect increases, this power will become insignificant. At 1GHz, for a 10mm long wire the switching energy of the proposed interconnect is 0.8pJ/bit, while leakage is 0.1fJ/bit. Energy benefits can be realized at switching activity of 0.03% and above. Later in the paper we show energy benefits in a real system. Also, if the interconnect is idle for a long time, then it can easily be power gated to save this power.

Voltage sensitivity: The interconnect circuit is sensitive to the variation in V_{DDI}. An increase in V_{DDI} increases the leakage in the first stage of the receiver as explained in previous section. We simulated the circuit with V_{DDI}=0.35V and Figure 14a shows the Monte-Carlo simulation results across process. The average leakage increases to 316nA and maximum leakage goes to 1.5μA. However, this is still a small overhead when compared to the overall energy savings. In the other case when V_{DDI} goes lower, the drive to MN1 goes low, causing the receiver to lose performance. We simulated the circuit at V_{DDI}=0.25V and measured the propagation delay from IN to OUT (Figure 7).

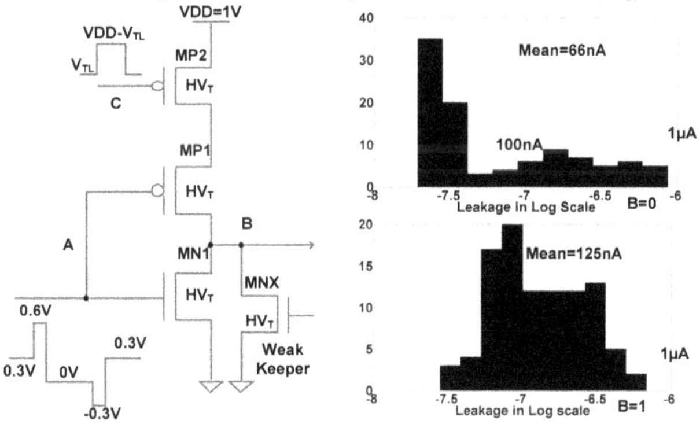

a) First Stage of receiver having high leakage

b) Monte Carlo result of leakage for B=0 and B=1 cases

Figure 13. Static current consumption in the Receiver

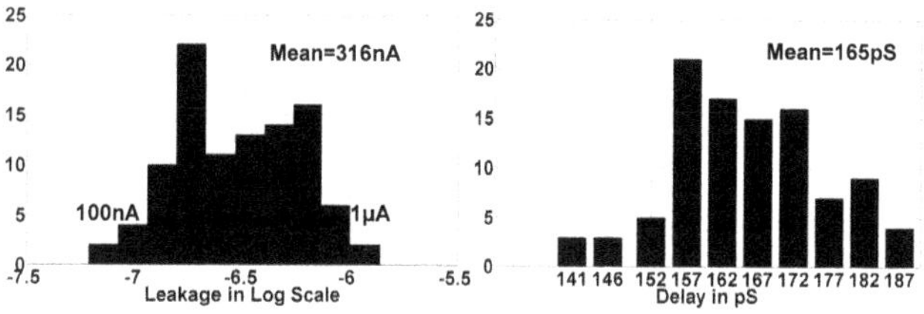

a) leakage at VDDI=0.35V

b) Propagation delay (IN→OUT) at VDDI=0.25V

Figure 14. Leakage and delay with varying V_{DDI}

Figure 14b shows the Monte-Carlo simulation result. The receiver performance drops and average delay goes to 165ps. However, this is not a significant drop in performance because the overall delay of the interconnect will be dominated by wires and is close to 1ns for 10mm long wires. These simulations show that the proposed receiver circuit performs well for V_{DDI} varying from 0.25V to 0.35V (30% variation in V_{DDI}).

Parasitic and negative voltage on net A: The voltage seen at Nets A and C depends on the value of the series capacitance and input capacitance seen at those nets. Parasitics can increase the capacitance, reducing the swing seen at A and C. Increasing the value of series capacitances will make sure that the swing is not attenuated at these nets because of the parasitics. Note that this will only increase the area and will not affect the power. We simulated the circuit with 5fF of additional parasitic cap at both the nets A and C. All the simulation results in this paper include an additional 5fF of parasitic load on A and C.

Another concern can be the negative voltage of -0.3V at Net A, which can turn on the body diode of MN4 in Figure 7. However, the cut-in voltage for the body diode ranges from 0.5V-0.7V, and a voltage -0.3V will not result in a significant reliability issue. The duration of this negative voltage is small too.

Noise Performance: The interconnect circuit has better noise performance than alternative receivers. To understand the noise performance, let us consider the cases when the receiver circuit receives a *high* and a *low*. The receiver is designed to receive a *high* when Net A makes a transition from 0.3V to 0.6V. Therefore V_{IH} of the receiver can be anywhere between 0.3V to 0.6V; suppose for example that it is at 0.45V. Similarly, V_{IL} can be between 0 to -0.3V; suppose for example that it is at -0.15V. Therefore, the total hysteresis of the receiver can be (V_{IH}-V_{IL}) 0.6V. The worst case hysteresis is 0.3V. The receiver can tolerate a noise of 0.3V on A or equivalent 0.1V on IN. The high hysteresis in the receiver is produced because of the feedback path and charge-pumping technique. Most differential and single ended receivers in the literature do not have any hysteresis. Some receivers [4] [5] have hysteresis of 50mV or lower.

The interconnect circuit consumes approximately three times lower energy than the prevalent interconnect circuits at the same performance points. The use of a charge pump circuit in the receiver enables this energy benefit without any significant performance penalty.

4. RESULTS

Figure 15 compares the proposed interconnect with existing architectures. [4-5,7-8] have results based on simulation while [6][9-10] have silicon results. The differential interconnects [6, 8-10] use a single supply, and the interconnect swing is restricted by IR drop in the diffamp. [4-5][7] are single ended interconnects. In

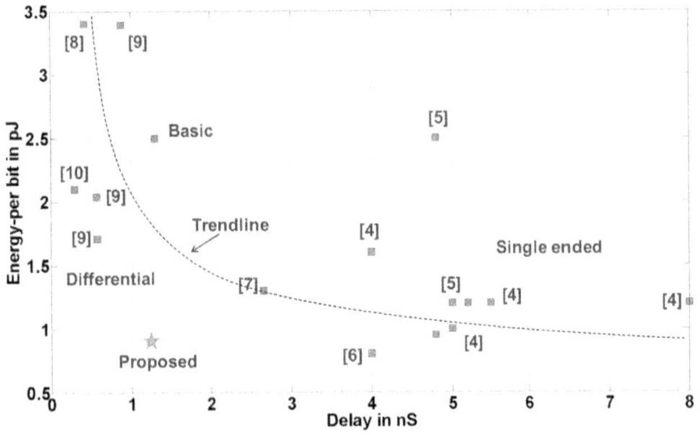

Figure 15. Proposed work in comparison with prior art

[4], authors present multiple circuits with different input swing on the wires. They use one logic supply and one or more interconnect supply voltages for their circuits. [5] and [7] present circuits with only one supply, and the interconnect wire swings from V_{DD} to V_{DD}-V_T, resulting in higher swing and hence higher power.

The proposed interconnect circuit uses a dedicated interconnect supply along with a supply voltage for the logic. The circuit has the best energy number and achieves very high performance. Table 2 compares the proposed circuit with existing architectures

Table 2: Energy, Delay and area of interconnect

Schemes	B/W (GHz)	Swing (V)	Norm. Energy	Area of 1 repeater
Basic	>1	1	1	2X
S-E [4,5,7]	<0.25	0.6	0.6	15-24X
Diff [8-10]	>1	0.05	0.8	100-250X
Cap[6]	<0.25	0.05	0.2	NA
This Work	**>1**	**0.3**	**0.3**	**22X**

We used the proposed interconnect to design the data bus connecting the L1 and L2 caches of a 4 core Alpha processor. Each core has a local L1 cache, while L2 is shared among all cores. The data bus between L1 and L2 will form a long interconnect, which makes the case for our experiment. We simulated the Alpha using m5 [11] and a spice model for the interconnect circuits implementing the data bus inside the Alpha.

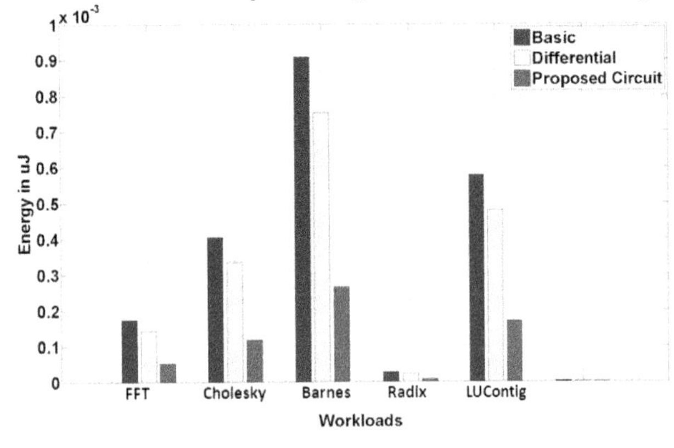

Figure 16. Interconnect energy dissipated in the data bus while simulating Splash workloads on an Alpha processor

We ran different Splash workloads to see the actual energy consumption in the interconnect during operation of a real processor. The energy consumption includes leakage as well as switching energy for the given interconnect circuits. The waveforms on the data bus were fed to the spice model of the interconnect circuit. This was done for differential, basic, and the proposed interconnect schemes set at the same delay constraint. Figure 16 shows that the proposed architecture saves up to 70% energy.

5. CONCLUSIONS

A new low power interconnect circuit was proposed and demonstrated. The proposed interconnect uses a charge-pumping technique to achieve high performance at 3X less energy than alternatives at comparable speeds. Simulations of a four core Alpha processor running Splash workloads show up to 70% energy savings at constant performance over alternative interconnect implementations.

6. REFERENCES

[1] P. Kogge, K. Bergman, S Borka, et. al, "ExaScale Computing Study: Technology Challenges in Achieving Exascale Systems" *DARPA/IPTO*, September 2008

[2] D. Liu and C. Svensson, "Power Consumption Estimation in CMOS VLSI chips" *IEEE Journal of Solid-State Circuits*, Vol-29 No-6, June 1994.

[3] E. Kusse and J.M. Rabaey, "Low-Energy Embedded FPGA Structures" *IEEE International Symposium on Low Power Electronics Design*, August 1998 .

[4] H. Zhang, V. George and J.M. Rabaey, "Low-Swing On-Chip Signalling Techniques: Effectiveness and Robustness" *IEEE Transactions on Very Large Scale Integration (VLSI)*, Vol-8 No-3, June 2000

[5] J.C.G. Montesdeoca, J.A. et. al, "CMOS Driver Receiver Pair for Low Swing Signalling for Low Energy On-chip Interconnects" *IEEE Transactions on Very Large Scale Integration (VLSI)*, Vol-17 No-2, February 2009.

[6] R. Ho, I. Ono, F. Liu, et. al, "High Speed and Low Energy capacitively driven wires" *IEEE International Solid State Circuits Conference*, February 2007.

[7] M. Ferretti and P.A. Beere "Low Swing Signaling Using a Dynamic Diode-Connected Driver" *European Solid-State Circuits Conference*, September 2001.

[8] A. Narshimha, M. Kasotiya and R. Sridhar "A Low-Swing Differential signaling Scheme for on-chip Global Interconnects" *International Conference on VLSI Design*, January 2005.

[9] N. Tzartzanis, W.W. Walker "Differential Current Mode Sensing for Efficient On-Chip global Signaling" *IEEE Journal of Solid State Circuits*, Vol-40 No-11, November 2005.

[10] H. Ito, M. Kimura, K. Miyashita, et. al, "A Bidirectional and Multidrop Transmission Line Interconnect for Multipoint to Multipoint On-Chip Communication" *IEEE Journal of Solid State Circuits*, Vol-43 No-4, April 2008.

[11] Binkert, N.L. , Dreslinski, R.G. , Hsu, L.R. , Lim, K.T. , Saidi, A.G. , Reinhardt, S.K. , "The M5 Simulator: Modeling Networked Systems" *IEEE Micro*, July 2006

0.35V, 4.1μW, 39MHz Crystal Oscillator in 40nm CMOS

Akira Saito[1], Yunfei Zheng[2], Kazunori Watanabe[1], Takayasu Sakurai[2], and Makoto Takamiya[2]

[1] Semiconductor Technology Academic Research Center (STARC), Yokohama, Japan,

[2] University of Tokyo, Tokyo, Japan

[1]{saito.akira, watanabe.kazunori}@starc.or.jp, [2]{yfzheng, tsakurai, mtaka}@iis.u-tokyo.ac.jp

ABSTRACT

A design methodology for sub-0.5V crystal oscillators is shown to realize an all-sub-0.5V ultra low power RF transceiver for wireless sensor networks. To reduce the minimum operating voltage (V_{DDmin}) of the crystal oscillator, both the optimization of the gate width of the CMOS inverter in the crystal oscillator and the reduction in gate length by CMOS technology scaling are required. In accordance with the developed design methodology, a 39MHz crystal oscillator is designed and fabricated in a 40nm CMOS. The measured power consumption is 4.1μW at 0.35V and 39MHz, and the power supply voltage is the lowest among the previously reported crystal oscillators.

Categories and Subject Descriptors

B.7.1 [**Hardware**]: Integrated Circuits, Types and Design Styles –*Advanced technologies*

General Terms

Measurement, Performance, Design

Keywords

Crystal oscillator, Pierce oscillator, Sub-threshold characteristics, Low Power, Low Voltage, Wireless Sensor Networks

1. INTRODUCTION

An ultra low power RF transceiver is required for wireless sensor networks. Reducing the power supply voltage (V_{DD}) is effective for reducing the power consumption of the transceiver. Sub-0.5V RF receiver circuits [1-4] have been reported. V_{DD} is 0.5V in [1-3] and 0.25V in [4]. A sub-0.5V FBAR oscillator using a forward-biasing bulk has been reported [5]. A sub-0.5V crystal oscillator, however, has not yet been reported. Without a sub-0.5V crystal oscillator, all-sub-0.5V low power RF transceiver cannot be realized. The feasibility of the sub-0.5V operation of a crystal oscillator, however, has not yet been clarified. Therefore, the purpose of this study is to clarify a design methodology for the sub-0.5V crystal oscillator and to demonstrate the sub-0.5V ultra low power operation of the crystal oscillator in a 40nm CMOS. The target frequency of the crystal oscillator is 39MHz. This frequency is multiplied by 8 by a 0.6V injection-locked frequency multiplier [6], and the generated frequency of 315MHz is used as the carrier frequency.

The remainder of this paper is organized as follows. Section 2 presents the design of the sub-0.5V crystal oscillator. Section 3 presents the measured results of the crystal oscillator in a 40nm CMOS. Finally, Section 4 presents the conclusion of this paper.

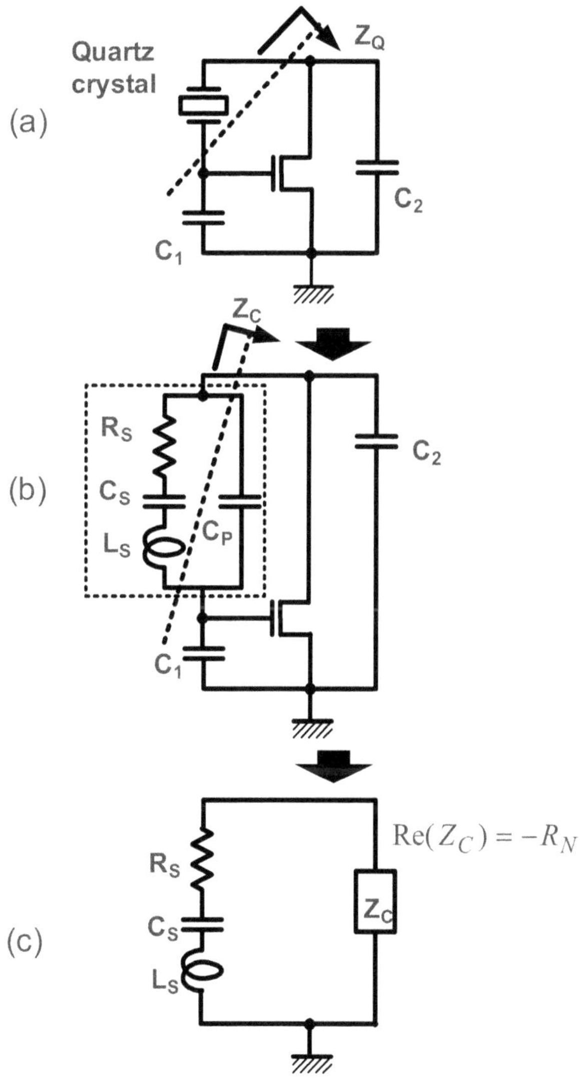

Figure 1: Pierce crystal oscillator. (a) Basic circuit (biasing omitted). (b) Equivalent circuit. (c) Equivalent circuit of (b).

2. DESIGN FOR SUB-0.5V CRYSTAL OSCILLATOR

2.1 Principle of Pierce Oscillator

In the design of a crystal oscillator, the negative resistance is an important design parameter. An oscillator circuit oscillates when the magnitude of the negative resistance is larger than the resistive loss in the quartz crystal. However, an MOS transistor itself does not exhibit I-V characteristics with a negative resistance as observed in a tunnel diode. The Pierce oscillator produces a negative resistance with a transistor and capacitors, and is a typical crystal oscillator.

Figure 1(a) depicts the Pierce crystal oscillator. The quartz crystal and load capacitors (C_1 and C_2) are connected to an nMOSFET. Figure 1(b) shows an equivalent circuit of the Pierce oscillator. The quartz crystal is represented as a serial connection of R_S, C_S, and L_S, and a shunt capacitor (C_P). In the Pierce oscillator, the circuit impedance Z_Q excluding that of the quartz crystal, is expressed as [6]

$$Z_Q = \frac{1}{j\omega C_1} + \frac{1}{j\omega C_2} + \left(\frac{1}{j\omega C_1}\right) \times g_m \times \left(\frac{1}{j\omega C_2}\right), \quad (1)$$

where ω is the angular frequency of oscillation and g_m is the transconductance of the transistor. The real part of Z_Q corresponds to the negative resistance ($-R_{NQ}$, $R_{NQ} > 0$) observed from the quartz crystal. $-R_{NQ}$ is expressed as [7]

$$\mathrm{Re}(Z_Q) = \left(\frac{1}{j\omega C_1}\right) \times g_m \times \left(\frac{1}{j\omega C_2}\right) = -\frac{g_m}{\omega^2 C_1 C_2} \equiv -R_{NQ}. \quad (2)$$

The negative resistance is realized by the multiplication of $1/j\omega C_1$ and $1/j\omega C_2$. To increase R_{NQ}, g_m should be increased and C_1 and C_2 should be decreased.

Figure 1(c) shows a circuit equivalent to that in Figure 1(b). The total impedance, excluding R_S, C_S, and L_S, is denoted as Z_C. Z_C consists of Z_Q and C_P, and is thus expressed as [8]

$$Z_C = \left(\frac{1}{j\omega C_P}\right) // Z_Q. \quad (3)$$

The real part of Z_C is the negative resistance ($-R_N$, $R_N > 0$) observed from the serial connection of R_S, C_S, and L_S. By substituting Equations (1) and (2) into Equation (3), R_N is expressed as [6]

$$\mathrm{Re}(Z_C) = -\frac{g_m C_1 C_2}{(g_m C_P)^2 + \omega^2 (C_1 C_2 + C_P C_2 + C_P C_1)^2} \equiv -R_N. \quad (4)$$

When $C_P = 0$, Equation (4) is equal to Equation (2).

In chip implementation, a Pierce oscillator with a CMOS inverter is often used. Figure 2 shows a Pierce crystal oscillator with a CMOS inverter. The nMOSFET in Figure 1(a) is replaced with the CMOS inverter and a feedback resistor (R_F). R_F induces half V_{DD} biasing to the CMOS inverter. The Pierce oscillator with the CMOS inverter in Figure 2 is equivalent to the Pierce oscillator in Figure 1, assuming that g_m in Equation (4) is determined by the parallel connection of the nMOSFET and pMOSFET.

2.2 Design Strategy for Sub-0.5V Operation

In this section, a design strategy for the sub-0.5V crystal oscillator is discussed. Figure 3 shows the relationships between the condition required for oscillation and the circuit parameters in the Pierce crystal oscillator. The crystal oscillator oscillates when it satisfies the condition

$$R_N \geq R_S, \quad (5)$$

where R_S is the parasitic resistance of the quartz crystal shown in Figure 1(b). Equation (2) shows that R_{NQ} is proportional to g_m. Equation (4) also shows that R_N depends on g_m. In the above-threshold region, where the gate-to-source voltage (V_{GS}) is larger than the threshold voltage (V_{TH}), g_m of the MOSFET is expressed as

$$g_m \propto \frac{W}{L}(V_{GS} - V_{TH}), \quad (6)$$

where L is the gate length and W is the gate width. When V_{DD} and V_{GS} are reduced, g_m and R_N decrease, terminating the oscillation. To keep g_m constant when V_{DD} is reduced, W/L should be linearly

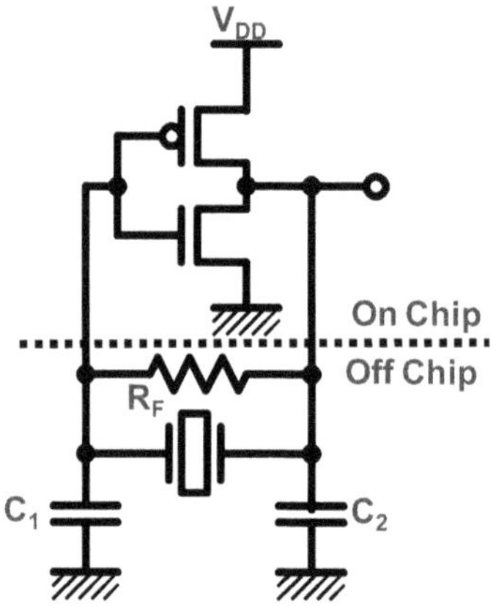

Figure 2: Pierce crystal oscillator with CMOS inverter.

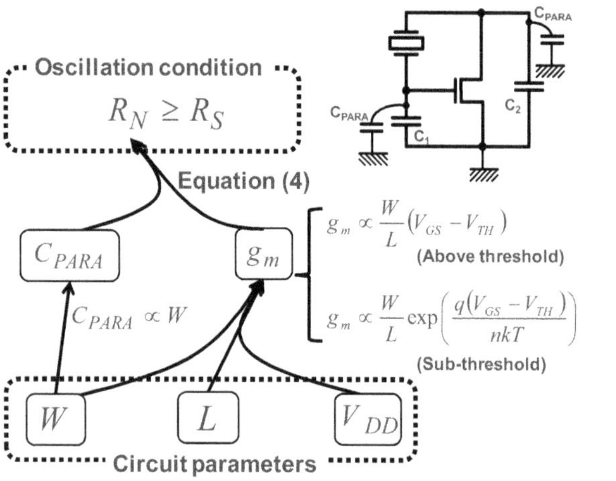

Figure 3: Relationships between condition required for oscillation and circuit parameters in Pierce crystal oscillator.

increased approximately proportionally to $1/V_{DD}$. In contrast, in the sub-threshold region, where V_{GS} is smaller than V_{TH}, g_m of the MOSFET is expressed as

$$g_m \propto \frac{W}{L}\exp\left(\frac{q(V_{GS}-V_{TH})}{nkT}\right), \tag{7}$$

where q is the elementary electric charge, n is constant (1<n<2), k is the Boltzmann coefficient, and T is the absolute temperature. To keep g_m constant when V_{DD} is reduced, W/L should be exponentially increased approximately proportionally to $1/\exp(V_{DD})$. When W is increased, the parasitic capacitance of the transistor (C_{PARA}) also increases because

$$C_{PARA} \propto W\ . \tag{8}$$

According to Equation (4), R_N decreases as C_{PARA} increases, because C_{PARA} is added to C_1 and C_2. Therefore, a large W increases R_N by increasing g_m, while it reduces R_N by increasing C_{PARA}. The trade-off relation suggests that an optimal W exists to minimize V_{DDmin}. The reduction of L increases g_m, and the minimum L is determined by CMOS technology. CMOS technology down scaling also reduces C_{PARA}, thereby increasing R_N.

2.3 Gate Width Optimization

In this section, optimization of the gate width (W) to reduce V_{DDmin} for the crystal oscillator is discussed. Figure 4 shows the simulated dependence of the negative resistance (R_N) normalized by R_S on V_{DD} in the crystal oscillator. In this SPICE simulation, the W values of the nMOSFET and pMOSFET are the same. L is fixed to 40nm. The resonant frequency of the quartz crystal is 39MHz. The other simulation conditions are R_S=60Ω, C_1=C_2=5pF, and R_F=500kΩ. The crystal oscillator can oscillate when R_N/R_S > 1. This figure clearly shows that an increase in W reduces V_{DDmin}. In the case of $W \leq 1$ μm, the crystal oscillator requires V_{DD} above 1.0V to oscillate. The sub-0.5V oscillation requires $W \geq 100$ μm. For example, in the case of $W = 100$ μm, V_{DDmin} (= V_{DD} at R_N/R_S = 1) is 0.49V and the maximum operating V_{DD} (V_{DDmax}) (= V_{DD} when R_N/R_S = 1) is 0.97V. The operating range of V_{DD} is 0.48V (=V_{DDmax}-V_{DDmin}).

In Figure 4, R_N/R_S at W = 1μm *monotonically* increases with V_{DD}, because the transistors operate in the above-threshold region. As shown by Equation (6), g_m *linearly* depends on V_{GS}, and R_N monotonically increases with V_{GS}, as shown by Equation (4). In contrast, R_N/R_S at W > 10μm does not *monotonically* increase with V_{DD}, because the transistors operate in the sub-threshold region. As shown by Equation (7), g_m *exponentially* depends on V_{GS}, and R_N increases in the low-V_{DD} region and decreases in the high-V_{DD} region with increasing V_{GS}, as shown by Equation (4).

Figure 4 also shows that an increase in W reduces the peak value of R_N/R_S. As W increases, both V_{DDmin} and V_{DDmax} decrease and the V_{DD} range (V_{DDmax}-V_{DDmin}) also decreases. At $W \geq 20$ mm, the crystal oscillator cannot oscillate for all values of V_{DD}. This can be explained by the increased C_{PARA} caused by the increases in W, because C_{PARA} is added to C_1 and C_2.

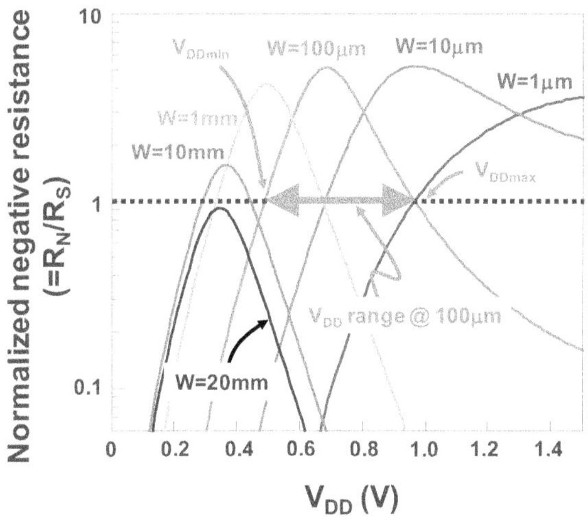

Figure 4: Simulated dependence of negative resistance (R_N) normalized by R_S on V_{DD} in crystal oscillator. The crystal oscillator can oscillate when R_N/R_S > 1.

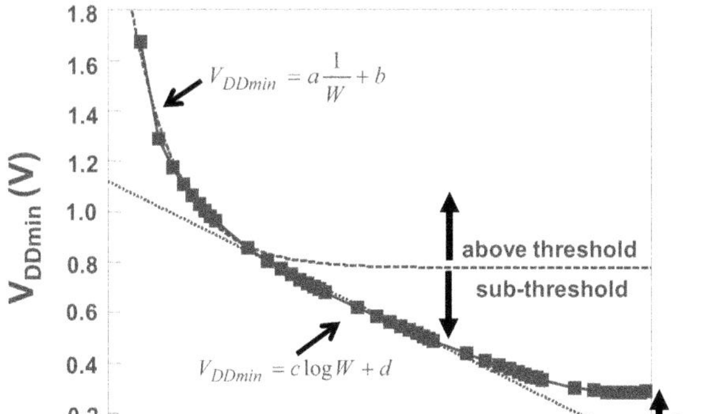

Figure 5: Simulated dependence of V_{DDmin} on W at L=40nm.

Figure 5 shows the simulated dependence of V_{DDmin} on W. The squares at high V_{DDmin} values (in the above-threshold region) can be fitted as

$$V_{DDmin} = a\frac{1}{W} + b\,, \tag{9}$$

at $W \leq 1$ μm, as shown by the dashed line in Figure 5, where *a* and *b* are constants. The squares at low V_{DDmin} values (in the sub-threshold region) can be fitted as

$$V_{DDmin} = c\log W + d\,, \tag{10}$$

at $W > 1$μm, as shown by the dotted line in Figure 5, where *c* and *d* are constants. Equations (9) and (10) can be derived from Equations (6) and (7), respectively, assuming that g_m is constant. This curve fitting means that V_{DDmin} at $W \leq 1$ μm is determined by the above-threshold operation, while V_{DDmin} at $W > 1$μm is determined by the sub-threshold operation. A V_{DD} of 0.8V is twice of the threshold voltage. At the gate width $W \geq 200$ μm, the squares shift away from the dotted line. This shift can be explained by the reduced negative resistance caused by the increase in C_{PARA} with increasing W. This results in the lower V_{DDmin} limit of 0.28V.

2.4 CMOS Technology Scaling

In this section, reduction of the gate length (L) by CMOS technology scaling to reduce V_{DDmin} for the crystal oscillator is discussed. Figure 6 illustrates the SPICE-simulated dependence of V_{DDmin} on W in the case of three different CMOS technologies. The L values are 40nm, 160nm, and 360nm in the 1.1V 40nm, 1.8V 160nm, and 3.3V 360nm CMOSs, respectively. The curve of the 1.1V 40nm CMOS is the same as that in Figure 6. As L decreases, V_{DDmin} decreases at a fixed W, because g_m is proportional to 1/L, as shown in Equations (6) and (7). As discussed for Figure 5, the lower limit of V_{DDmin} is determined by C_{PARA}. The lower limit of V_{DDmin} in the 40nm, 160nm, and 360nm CMOSs are 0.28V, 0.41V, and 0.78V, respectively. In this manner, the lower limits of V_{DDmin} is reduced by CMOS technology scaling, because C_{PARA} is reduced by CMOS technology scaling. Therefore, the CMOS technology scaling is effective for reducing V_{DDmin} for the crystal oscillator. For example, to achieve a sub-0.5V crystal oscillator, a 40nm or 160nm CMOS is required, as shown in Figure 6.

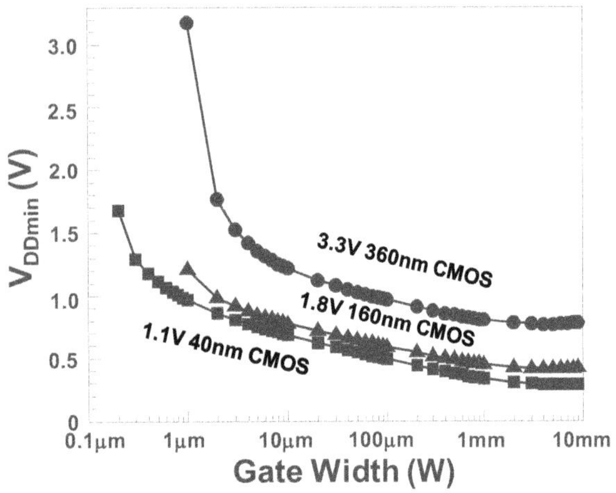

Figure 6: Simulated dependence of V_{DDmin} on W for three different CMOS technologies.

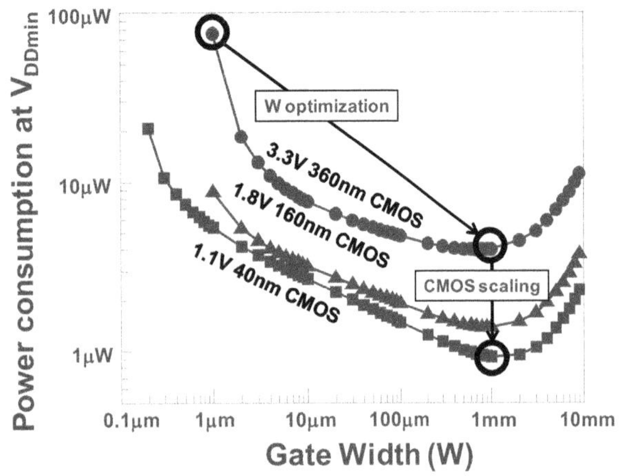

Figure 7: Simulated dependence of power consumption at V_{DDmin} on W for three different CMOS technologies.

2.5 Low Power Design

In this section, the reduction of power by reducing V_{DDmin} for the crystal oscillator is discussed. Figure 7 depicts the SPICE-simulated dependence of the power consumption at V_{DDmin} on W for three different CMOS technologies. The power consumption is minimum at $W = 1$ mm in all three different CMOS technologies. At $W \geq 1$ mm, although V_{DDmin} is constant, as shown in Figure 6, the power consumption increases with increasing W, because C_{PARA} increases and the relevant switching power also increases.

In summary, as shown in Figure 7, the design strategy to reduce V_{DDmin} for the crystal oscillator involves (1) the optimization of W and (2) the reduction of L by CMOS technology scaling. W has an optimum value, because the large W increases R_N by increasing g_m, but it reduces R_N by increasing C_{PARA}. The reduction of L increases g_m and reduces C_{PARA}, thereby increasing R_N and reducing V_{DDmin}, as shown in Figure 3.

3. MEASUREMENT RESULTS

3.1 Chip Implementation

In accordance with the developed design methodology, a 39MHz crystal oscillator is designed and fabricated in a 1.1V 40nm CMOS. Figure 8 shows the chip microphotograph and layout of the crystal oscillator. The core area is 80µm x 130µm. The worst-

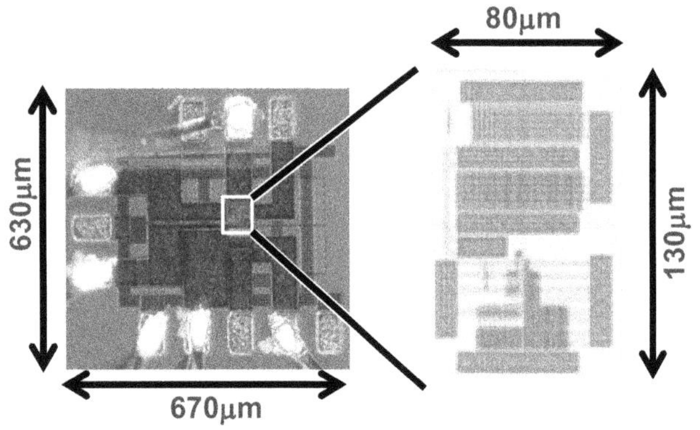

Figure 8: Chip photograph and layout of crystal oscillator.

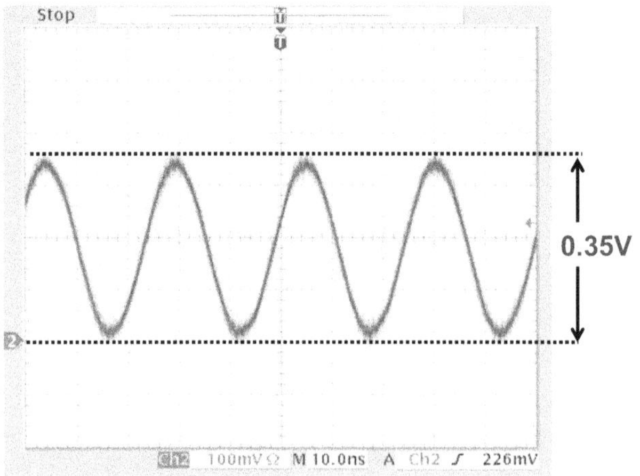

Figure 9: Measured output waveform of 39MHz crystal oscillator at V_{DD}= V_{DDmin} = 0.35V.

case R_S of the quartz crystal shown in the spec sheet is 60Ω. The value of W for the crystal oscillator is optimized to 220μm by considering the worst-case R_S and the oscillator margin for tolerances. This chip is implemented in a PCB with off-chip components such as capacitors (C_1=C_2=5pF).

3.2 Measurements

The measured V_{DDmin} of the 39MHz crystal oscillator is 0.35V. Figure 9 shows the measured output waveform of the 39MHz crystal oscillator at V_{DDmin} = 0.35V. It shows a clear sinusoidal waveform with full swing. Figure 10 shows the measured spectrum of the 39MHz crystal oscillator at V_{DDmin} = 0.35V.

In the measurement, oscillation was observed in the V_{DD} range from 0.35V to 1.1V. The measured V_{DD} range is wider than the simulated V_{DD} range from 0.45V to 0.85V at W=220μm. The mismatch is due to the error in R_S. In the S-parameter measurement of the quartz crystal using a vector network analyzer,

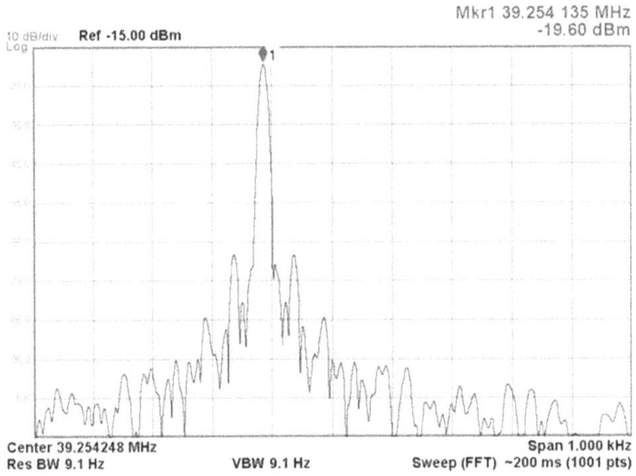

Figure 10: Measured spectrum of 39MHz crystal oscillator.

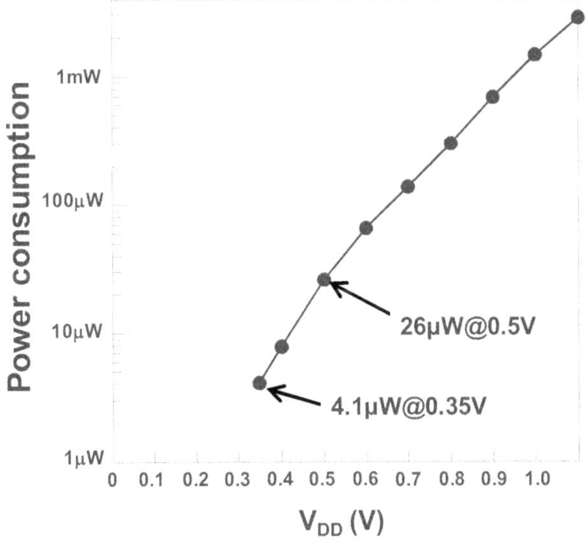

Figure 11: Measured dependence of power consumption on V_{DD} in 39MHz crystal oscillator.

Table I: Performance comparison with previous crystal oscillators.

Reference		[9]	[10]	[11]	[12]	This Work
Frequency	MHz	2.1	2.1	2.1	19	39
CMOS process	μm	3.0	NA	2.0	0.1	0.04
V_{DD}	V	1.5	1.3	1.8	1.2	0.35
Power	μW	1.4	3.0	0.7	21.6	4.1

the measured R_S is 10Ω, which is less than the worst-case R_S of 60 Ω shown in the spec sheet. The smaller than expected R_S results in the wider measured V_{DD} range than that obtained by simulation.

Figure 11 shows the measured dependence of the power consumption on V_{DD} in the 39MHz crystal oscillator. The curve exhibits an exponential dependence on V_{DD} due to the sub-threshold operation. The power consumption is 26μW and 4.1μW at V_{DD} values of 0.5V and 0.35V, respectively.

Table I shows a performance comparison with previously reported crystal oscillators [9-12]. In this work, V_{DD} is the lowest among the previously reported crystal oscillators.

4. CONCLUSION

In this paper, the design methodology for sub-0.5V crystal oscillators is shown to realize an all-sub-0.5V low power RF transceiver for wireless sensor networks. To reduce V_{DDmin} for the crystal oscillator, both the optimization of the gate width (W) and CMOS technology scaling are required. The optimum W is determined by g_m and the parasitic capacitance of the transistors. In accordance with the developed design methodology, a 39MHz crystal oscillator is designed and fabricated in a 40nm CMOS. The measured power consumption is 4.1μW at 0.35V and 39MHz, where both the power supply voltage is the lowest among the previously reported crystal oscillators.

5. ACKNOWLEDGMENTS

This work was carried out as a part of the Extremely Low Power (ELP) project supported by the Ministry of Economy, Trade and Industry (METI) and the New Energy and Industrial Technology Development Organization (NEDO).

6. REFERENCES

[1] N. M. Pletcher, S. Gambini, and J. M. Rabaey, "A 2GHz 52μW Wake-Up Receiver with -72dBm Sensitivity Using Uncertain-IF Architecture", IEEE International Solid-State Circuits Conference, pp. 524-525, Feb. 2008.

[2] X. Huang, S. Rampu, X. Wang, G. Dolmans, and H. de Groot, "A 2.4GHz/915MHz 51μW Wake-Up Receiver with Offset and Noise Suppression", IEEE International Solid-State Circuits Conference, pp. 222-223, Feb. 2010.

[3] A. Saito, K. Honda, Y. Zheng, S. Iguchi, K. Watanabe, T. Sakurai, and M. Takamiya, "An all 0.5V, 1Mbps, 315MHz OOK transceiver with 38-μW carrier-frequency-free intermittent sampling receiver and 52-μW Class-F transmitter in 40-nm CMOS, " IEEE Symposium on VLSI Circuits, June 2012. (To be presented)

[4] A. Heiberg, T. Brown, K. Mayaram, and T. S. Fiez, "A 250 mV, 352μW Low-IF Quadrature GPS Receiver in 130 nm CMOS", IEEE Symposium on VLSI Circuits, pp.135-136, June 2010.

[5] A. Nelson, J. Hu, J. Kaitila, R. Ruby, B. Otis, "A 22μW, 2.0GHz FBAR Oscillator", IEEE Radio Frequency Integrated Circuits Symposium, June 2011.

[6] L. Liu, M. Takamiya and T. Sakurai, "315MHz Energy-Efficient Injection-Locked OOK Transmitter and 8.4μW Power-Gated Receiver Front-End for Wireless Ad Hoc Network in 40nm CMOS", IEEE Symposium on VLSI Circuits, pp.164-165, June 2011.

[7] B. Razavi, Design of Analog CMOS Integrated Circuits, McGraw-Hill, Oct. 2003.

[8] E. Vittoz, Low-Power Crystal and MEMS Oscillators -The Experience of Watch Developments-, Springer-Verlag, Sep. 2010.

[9] E. Vittoz, M. Degrauwe, and S. Bitz, "High-Performance Crystal Oscillator Circuits: Theory and Application", IEEE Journal of Solid-State Circuits, pp.774 -783, June 1988.

[10] D. Lanfranchi, E. Dijkstra, and D. Aebischer, "A Microprocessor-Based Analog Wristwatch Chip with 3 Second/Year Accuracy", IEEE International Solid-State Circuits Conference, pp. 92-93, Feb. 1994.

[11] D. Aebischer, H. J. Oguey, and V. R von Kaenel, "A 2.1-MHz Crystal Oscillator Time Base with A Current Consumption under 500 nA", IEEE Journal of Solid-State Circuits, pp. 999– 1005, July 1997.

[12] R. van Langevelde, M. van Elzakker, D. van Goor, H. Termeer, J. Moss, and A. J. Davie, "An Ultra-Low-Power 868/915 MHz RF Transceiver for Wireless Sensor Network Applications", IEEE Radio Frequency Integrated Circuits Symposium, pp. 113– 116, June 2009.

A Study of the Effectiveness of CPU Consolidation in a Virtualized Multi-Core Server System*

Inkwon Hwang and Massoud Pedram
University of Southern California
Los Angeles CA 90089
{inkwonhw, pedram}@usc.edu

Timothy Kam
Intel Corporation
Hillsboro OR 97124
Timothy.kam@intel.com

ABSTRACT

The focus of this paper is on dynamic power management in virtualized multi-core server systems. The paper starts by analyzing the effect of virtualization and CPU consolidation on power dissipation and performance (latency) of such systems, and concludes by presenting two new CPU consolidation algorithms for multi-core servers. The paper also reports an extensive set of experimental results founded on a realistic multi-core server system setup and well-developed benchmarks, i.e., *SPEC2K* and *SPECWeb2009* and obtained through hardware measurements.

Categories and Subject Descriptors

D.4.1 [**Process Management**]: Scheduling

Keywords

Energy efficiency, virtualization, consolidation, and scheduling

1. INTRODUCTION

Today's servers consume large amounts of energy so there is a growing need for energy-aware resource management in multi-core server systems. Virtualization has emerged as a promising solution for eliminating "computing waste" through physical resource sharing in data centers. In this study, we focus on the CPU (i.e., physical core), which is one of top energy consumers in a virtualized system. A common power saving technique for CPUs is *Dynamic Voltage and Frequency Scaling* (DVFS) [1-3]. However, the additional power savings possible through DVFS is becoming smaller and smaller in part due to lower supply voltage levels and a shared power and clock distribution network for the cores. In addition, it is not easy to apply existing DVFS techniques to cores in a virtualized multi-core server system. This is mainly because the existing DVFS techniques require information about running applications to make decisions about the voltage and clock frequency setting, but a virtual machine manager (hypervisor), which resides in a privileged domain, does not have information about applications running on the guest domains because of abstractions [4].

Another energy saving technique is *Dynamic Power Shutdown* (DPS). In particular, some modern CPUs support *Core-level Power Gating* (CPG), which allows individual core to be put into very low power, but non-functional, state. Such CPUs have their own *Power Control Unit* (PCU), which performs DPS; however, we expect that more power savings are possible if there is software level assistance for DPS. This is because the PCU in current servers does not have enough information about the application running on the system. Hence, there is a pressing need to understand when and how the CPU consolidation works with respect to power savings without violating performance constraints.

A common deployment model for high-end multi-core server systems is in data centers, which forms the computational and storage backbone of the digital information provided to end users via the Internet. As shown in Figure 1 , there is a large variance in workload intensity of commercial data centers [5]. To guarantee a required service level agreement (SLA) under the worst-case conditions, servers are typically designed to handle a peak workload condition even if the servers are under-utilized at times. There are very few hours that all servers in the data centers are running at their peak utilization levels. Motivated by this observation, many studies have suggested the use of *virtual machine migration* (VMM) for energy saving [3, 4, 6-8]. In theory, the VMM promises high energy saving, but it is difficult to apply the technique to servers in a data center because of the high overhead of the VMM, e.g., the large system boot time, network traffic caused by the need to transfer the running application and its local context to a new server, and so on. In general, the VMM does not aggressively address the server under-utilization because of its conservative nature (in order to avoid violating SLA.)

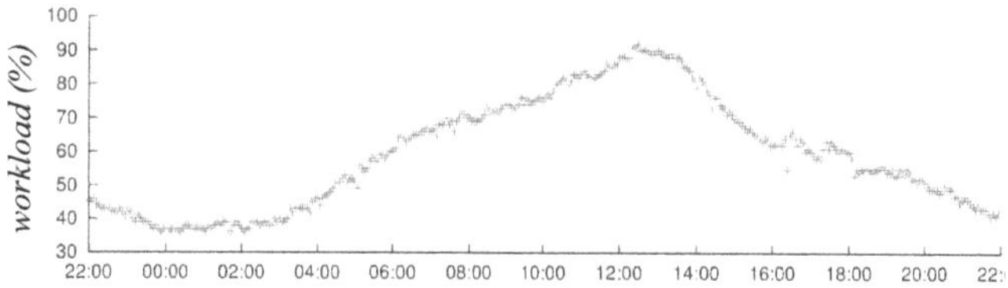

Figure 1 Typical server workload in Facebook data centers [5]

After an investigation of the effect of consolidation on the power dissipation and performance (latency) in virtualized multi-core server systems, this paper presents a consolidation technique for assisting DPS. There are a number of studies that have investigated energy savings due to consolidation. In [9], the authors showed that consolidation across cores in a single four-core-per-processor, two-processor-per-server system offers very small energy savings. However, their server system did not support CPG, so the energy saving potential of core-level consolidation needs to be investigated. In [10], Jacob *et al.* compared CPG and DVFS and showed that CPG can result in 30% energy saving compared to DVFS. However, the results were calculated from a combination of real measurements and estimated leakage power (the adopted leakage power model was quite simple.) In [5], the authors presented a technique called Core Count Management (some variant of the consolidation technique), and reported 35% energy saving. However, they reported both power and performance results based on simulations performed by using simple power and performance models. In our study, we show energy savings and

*This work is sponsored by grants from the National Science Foundation and the Intel Corp.

ISLPED'12, July 30–August 1, 2012, Redondo Beach, California, USA.
Copyright 2012 ACM 978-1-4503-1249-3/12/07...$10.00.

latency impacts of consolidation in a virtualized system based on realistic benchmarks and obtained through hardware.

Contributions of our work are as follows: 1) we perform extensive experiments (using well-developed benchmark sets, *SPEC2K* and *SPECWeb2009*) under various conditions in terms of workload intensity, number of virtual CPUs (vCPUs), set of active CPUs (i.e., CPUs at an operational state), and so on. This information is useful for developing consolidation algorithms. 2) We rely on real data to evaluate the energy savings of consolidation, instead of using simulators. Virtualized systems are more complex because of abstraction, e.g. vCPU, so it is difficult for simulators to model the system very accurately. Hence, real measurement is essential for proving effectiveness of consolidation. 3) We introduce two consolidation algorithms and assess their efficacies on the *SPECWeb* benchmark suite. The results demonstrate that the proposed consolidation algorithms can result in improvement on energy efficiency under very realistic environments.

The remainder of this paper is organized as follows. In section II we show analysis which shows how consolidation affects energy and performance. The experimental system setup is explained in section III. Following section IV shows experimental results. Finally, we summarize the results and give guidelines in section V.

2. POWER AND LATENCY MODELS

In this section we show how consolidation affects the average power dissipation and latency through mathematical analysis. As a result we can derive insight about how consolidation affects the power dissipation and latency of tasks. Our analytical predictions about the power/latency tradeoffs in a multi-core server system have been empirically shown to be valid for realistic cases. Note that consolidation is reasonable only when the system is under-utilized, thus the model does not consider thermal issue, e.g. leakage power variation with respect to temperature.

2.1 Average power dissipation

In this section we show the relationship between consolidation and power dissipation. Assume we have a system that has '*N*' CPUs with only '*n*' of them being active (i.e., all workloads have been consolidated to run on '*n*' active CPUs and that the remaining '*N-n*' CPUs are turned off or put in a deep sleep state.) The average power dissipation of the ith CPU (P_i) is intrinsically related to the average CPU utilization (U_i) and can be modeled as follows [6]:

$$\begin{aligned} P_i &= aU_i + b, \quad && if\ U_i > 0 \\ &= 0, && otherwise\ (turn\ off) \end{aligned} \qquad (1)$$

where U_i denotes $utilization$ of the ith CPU, $0 \le U_i \le 1$.

U_i is a function of workload (k_i), which can be defined as the number of tasks served by the ith CPU (2). '*K*' tasks are sent to the system every second, and a CPU scheduler evenly distributes the tasks to the '*n*' active CPUs, so each active CPU takes '*K/n*' tasks per second. U_i is thus linearly proportional to the total system workload and inversely proportional to the number of active CPUs. However, this statement may not be valid when the workload is very high. As an example, consider a scenario whereby the utilization is '*m*' for *H=K/n* memory-bound tasks per second sent to the target CPU. If more tasks (say *2*H*) are sent to the target CPU per second, the cache miss rate on that CPU will increase (more precisely, the working sets of the *2*H* tasks will not fit on the cache and hence every task will experience a higher cache miss rate.) This means that execution time of the tasks increases, and therefore, the CPU utilization will increase by more than a factor of two. Coefficient '*c*' captures this non-linear effect:

$$U_i = ck_i^2 + dk_i, \quad \text{where } k_i = K/n \qquad (2)$$

Total power dissipation is the sum of average power dissipations of individual core (P_i.) Assuming all cores consume the same amount of power, we have:

$$\begin{aligned} P &= \sum_{i=1}^{N} P_i = \sum_{i=1}^{n} aU_i + b \\ &= acK^2/n + adK + bn \end{aligned} \qquad (3)$$

Figure 2 depicts the relationship between the number of active CPUs and total average power dissipation for different coefficients, '*c*' and '*b*', under the same overall workload. Note that lower average power dissipation for the same workload means higher energy efficiency. If '*c*' is equal to zero, the total power dissipation decreases monotonically as more CPUs become inactive, which implies that, in this case, consolidation always reduces power dissipation. However, for higher '*c*', the total power dissipation reaches a minimum, and subsequently, goes up with fewer active CPUs. Hence, '*c*' should be carefully considered when we determine the number of active CPUs to consolidate workload. In addition, higher '*c* decreases power saving (Figure 2 a.) The slope of the graph is related to '*b*'. Larger '*b*' results in steeper slope to the right of the minimum power point and therefore, higher power saving. Note that '*c*' is application-dependent whereas '*b*' is a hardware-dependent parameter (with CPG, we expect bigger '*b*'.)

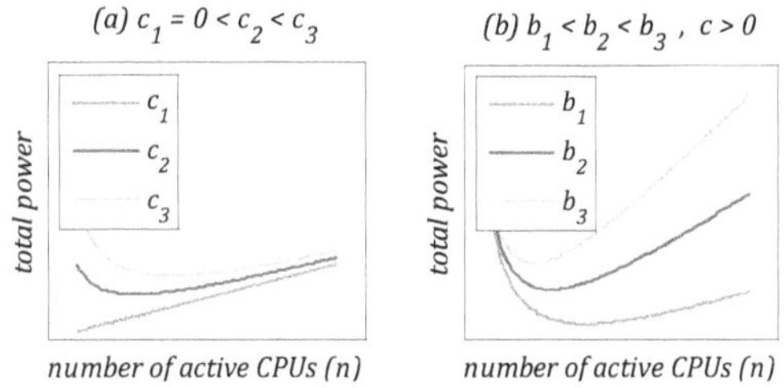

Figure 2 Power dissipation vs. the number of active CPUs under the same overall workload (*K* tasks/s)

2.2 Latency (delay)

As shown in (4), our task latency model has two terms. The first term makes delay larger for higher CPU utilization [6]. This term dramatically increases delay when the system is almost fully utilized. For higher utilization, fewer power state transitions occur, which in turn tends to decrease delay because of lower power state switching overhead. The second term of (4) shows this effect.

$$\begin{aligned} D = D_i &= D_{i,1} - D_{i,2} = \left(\frac{e}{1 - U_i} + f\right) - gU_i \\ &= \left(\frac{e}{1 - (ck_i^2 + dk_i)} + f\right) - g(ck_i^2 + dk_i) \end{aligned} \qquad (4)$$

where U_i denotes the $average\ utilization$

As shown in the power dissipation analysis, higher '*c*' results in a decrease in potential power saving of consolidation. Figure 3 a. shows the relationship between '*c*' value and delay: higher '*c*' results in larger delay. Note that overall workload of results in Figure 3 is the same (*K* tasks/s.) This implies that consolidation may incur potentially high delay penalty if '*c*' is big.

Figure 3 b. depicts the effect of '*g*'. With a positive value of '*g*', delay decreases as the numbers of active CPUs decreases. This suggests that it is possible to decrease both delay and power dissipation by consolidation if '*g*' is large enough. However, as shown in Figure 3 b, delay increases rapidly when the number of active CPUs becomes very small (since then the CPUs are almost fully utilized), and delay penalty of consolidation becomes significant. Thus, one must carefully choose the number of active CPUs. Coefficient '*g*' is also dependent on the application type.

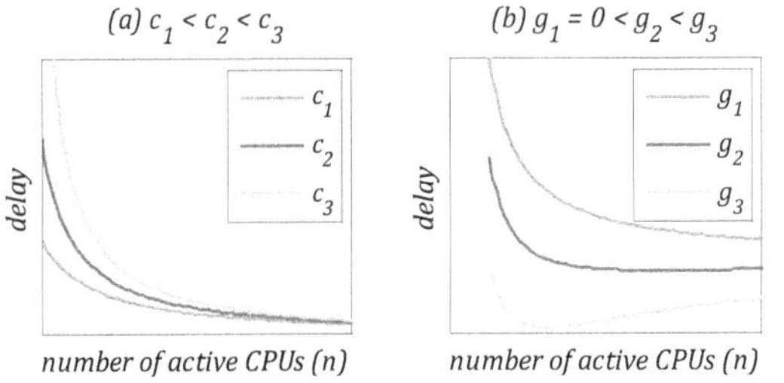

Figure 3 Delay vs. the number of active CPUs under the same overall workload (K tasks/s)

As seen from the above analysis, power saving and delay penalty effects of the consolidation depend on a few coefficients. Some of the coefficients are application-dependent, so there is a motivation to investigate the effectiveness of consolidation under different kinds of applications.

3. EXPERIMENTAL SETUP

3.1 Hardware Testbed

Hardware specification of our system under test is as follows. We have two Intel Xeon E5620 processors; each has four cores, all of which are running at the same clock frequency. Each core has its own dedicated L1/L2 caches but shares an L3 cache with the other cores. Each processor supports seven core frequency levels, from 1.6GHz to 2.4GHz. We cut the 12V CPU power lines and measure the amount of power supplied using a power analyzer.

3.2 XEN – *hypervisor-based virtualization product*

We chose XEN version 4.0.1 for constructing the virtualized system. XEN, which is an open source hypervisor-based virtualization product, provides APIs for managing virtual machines (VMs.) For this study, we run experiments under different configurations in terms of the number of vCPUs, clock frequencies, and the set of active CPUs.

3.3 Service model and Quality of Service

This paper targets a server/client service model where there are many clients, which send tasks to a server and the server responses when it completes the tasks. Turn-around time (from time when a task is sent from a client to the time when a service completion acknowledgement is received by the client) is considered as the delay of the task. For determining the *quality of service* (QoS), we use the 95th percentile delay (Figure 4 .) If the 95th percentile delay is less than the maximum allowed limit, we have met our QoS target. The limit itself is chosen as the 95th percentile delay of a fully-loaded base system (with no consolidation.) The term 'Fully loaded' means that the total CPU utilization is at 80% out of 100%. This is a reasonable value because servers are designed to produce high performance at around 80% utilization levels (performance level drops rapidly as the CPU utilization approaches 100 %.)

3.4 Benchmarks – *mcf, perl (SPEC2K), and SPECWeb*

As shown in Section II, the type of applications affects the effectiveness of consolidation. Hence, we do experiments for three common application types: CPU-bound, memory-bound, and I/O-bound. For the CPU-bound and memory-bound applications we use *perl* and *mcf* benchmarks, respectively. A *perl* shows high *Instruction per Cycle* (IPC) and low *Memory Access per Cycle* (MPC), but *mcf* has opposite characteristics [3]. We develop the WorkloadGen benchmark to control workload level of *perl/mcf* and gather system performance. For I/O-bound application *SPECWeb2009* benchmark is used.

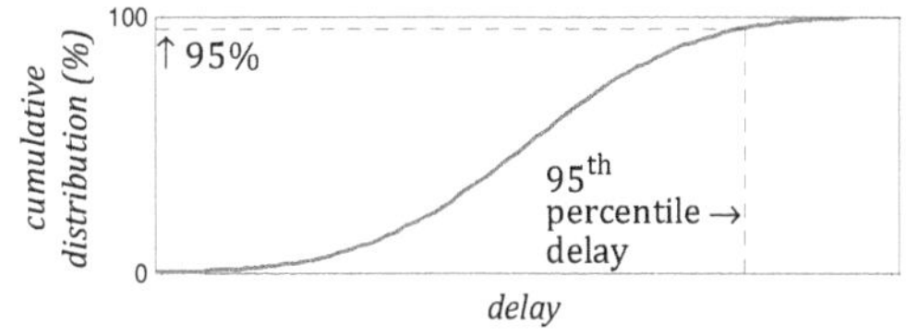

Figure 4 Delay cumulative distribution

3.5 WorkloadGen benchmark

We design and implement a benchmark program, *WorkloadGen*, to generate workload of desired characteristics and to measure performance of the system. It generates tasks and reports throughput and average response time (latency.) Type of tasks and workload intensity (defined as the number of tasks generated per second) are controllable.

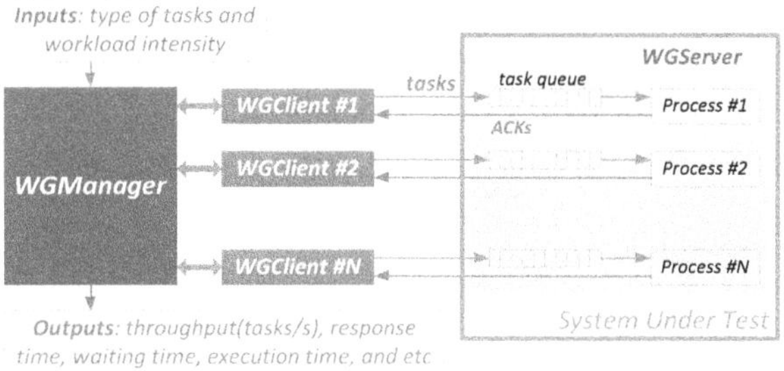

Figure 5 Block diagram of the WorkloadGen

The benchmark consists of three modules (*WGManager*, *WGClient*, and *WGServer*) as depicted in Figure 5 . *WGClients* request tasks to a *WGServer* and report performance statistics to a *WGManager*. *WGServer's* main role is to create workload by executing tasks requested by *WGClients*. When the *WGServer* completes a task, it sends an *ACK* packet to the *WGClient*. Based on the data in the *ACK*, *WGClients* can gather statistics of performance. There are a number of *WGClients*, so it is necessary to control them and gather statistics from them, and this is done by a *WGManager*. The *WorkloadGen* reports statistical performance data: average response time (turn-around time) per task, average waiting time in the queue, and average execution time per task, which are needed for analyzing the overall system performance.

4. RESULTS AND ALGORITHMS

In this section, we report experimental results of *perl*, *mcf*, and *SPECWeb*. Our purpose is to compute the energy saving of the consolidation technique, so all results correspond to an under-utilized server system, i.e., the CPU utilization is around 30% out of 100%. For *perl* and *mcf*, we investigate the delay and energy efficiency of different configurations (i.e., combinations of the number of vCPUs and active CPUs, and clock frequency)

We quantify the *energy efficiency* of a system as '*number of tasks served per unit of energy*.' For *SPECWeb*, we can specify the overall workload level, but instantaneous workload level changes dynamically. Hence, for *SPECWeb*, we verify the energy efficiency of the consolidation through dynamic management. We propose two very practical algorithms for dynamic management, and show up to 15% improvement in system's energy efficiency.

4.1 *perl* & *mcf* benchmark results

The number of vCPUs is an important parameter in a virtualized system: it determines how many CPUs can be utilized by a virtual domain at any time, so the performance of the domain is closely related to this parameter. At the same time, more vCPUs in a system cause higher power and performance overheads, so it will be detrimental if there are too many vCPUs. Figure 6 depicts the overhead caused by unnecessarily managing too many vCPUs. Utilization and delay of *perl* benchmark rapidly increase when the ratio of the number of vCPUs to the number of CPUs becomes greater or equal to three. The *mcf* benchmark shows the same trend. Note that workload level of all cases in Figure 6 is the same to each other. This means that the overhead of managing vCPUs causes this phenomenon. This result implies consolidation needs to be accompanied by dynamic vCPU count management. We maintain the ratio of vCPUs to CPUs to be around two for all test cases presented in this study through dynamic vCPU count control.

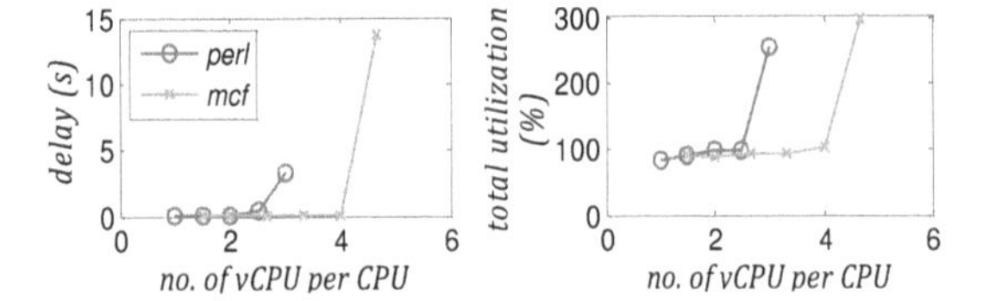

Figure 6 Delay and total utilization vs. the number of vCPUs per CPU (there are 4 active CPUs)

For consolidation, the choice about which set of CPUs is active can have an effect on performance and energy efficiency. Under multi-processor systems, several different CPU selection schemes are possible. In this study, we have only two processors, so we simply investigate two CPU selection schemes: 1) we select half of the required CPUs from one processor and other half from the other processor; and 2) we select all required CPUs from one processor, and only if more CPUs are needed than one processor can provide, we turn on the other processor and select the remaining CPUs from the other processor. We call these schemes as '*symmetric*' and '*asymmetric*' selection scheme.

Figure 7 shows the energy efficiency and delay of schemes. There is only small difference in energy efficiency and delay, but '*asymmetric*' selection scheme is a little bit better for the *perl* in terms of energy efficiency. On the other hand, there is no noticeable difference in energy efficiency and delay for the *mcf*. Hence, we choose '*asymmetric*' scheme for this study.

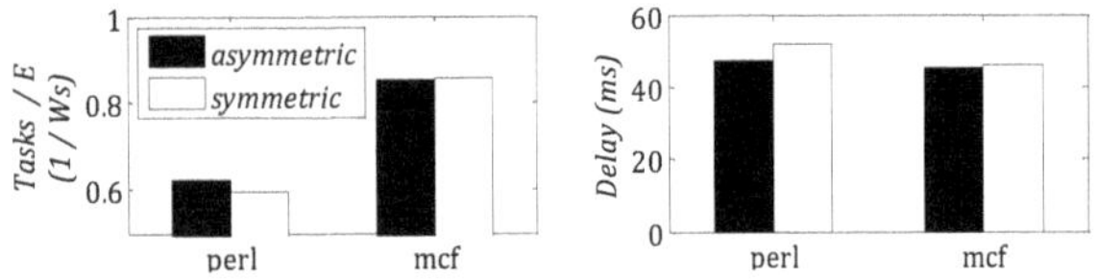

Figure 7 Energy efficiency & delay vs. CPU selection schemes

Figure 8 shows how much energy efficiency can be improved by consolidation without sacrificing QoS. Moreover, it shows that energy efficiency can increase even more if consolidation is accompanied by DVFS. The experimental setup is as follows: there are two guest domains. The '*perl*' case runs *perl* on both guest domains. '*mcf*' case runs *mcf* on both guest domains. '*mixed*' case runs both *perl* and *mcf*, i.e., one domain serves the *perl* while the other domain serves the *mcf*.

Energy efficiency of *perl* is dependent on both the number of active CPUs and clock frequency (Figure 8 a), i.e., fewer CPUs with slower frequency increases the energy efficiency. Delay of *perl* is more dependent on the clock frequency than the number of active CPUs (Figure 8 b), which means that consolidation can be done without performance degradation.

One observation is that the delay of the case with fewer active CPUs is sometimes even smaller than that of the case with larger number of CPUs. For example, 4CPU running at 2.2GHz shows smaller delay than 8CPU at the same frequency. In our model, coefficient 'g' in (4) represents this effect, i.e., a consolidated CPU may end up changing its power states less frequently because of higher utilization level, and this can reduce the delay. The energy efficiency of *mcf* seems to be independent of the number of active CPUs as well as frequency (Figure 8 c.) For different combination of clock frequency and the number of CPUs, there is only less than 3% difference in the energy efficiency. Delay of *mcf* is mainly affected by frequency, but again the difference in the delay is smaller than *perl* (Figure 8 d.) This result suggests DVFS and consolidation are not so effective for *mcf* because there is no noticeable improvement on energy efficiency. The *mixed* case exhibits similar trends to the *perl* case (Figure 8 e and f.)

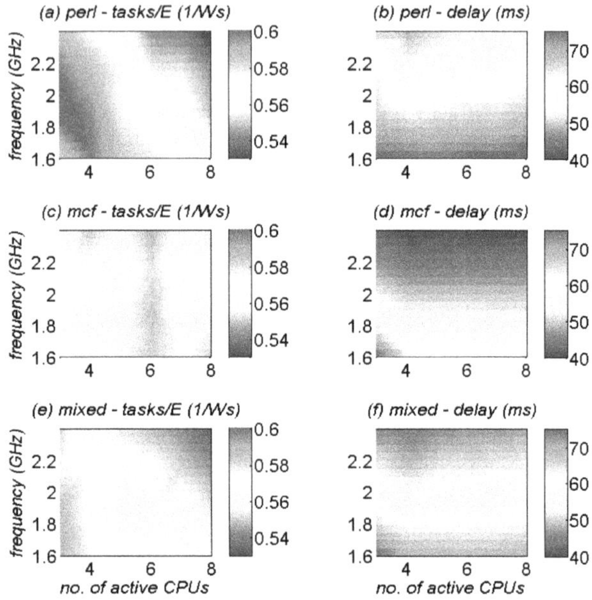

Figure 8 Energy efficiency and delay (Maximum allowed delay of *perl mcf*, and *mixed* is 73ms, 81ms, and 80ms respectively)

TABLE I. shows improvement on energy efficiency of *perl*, *mcf*, and *mixed* cases. Energy efficiency improvement is largest when both consolidation and DVFS are conducted. The *perl* shows biggest improvement and *mcf* shows smallest improvement. This result gives another insight; For VMM, if we consider types of application running on the VMs, we can enhance energy efficiency more. It suggests deploying heterogeneous VMs in a server machine. For example, we may have four domains (two domains serve CPU-bound tasks and others serve memory-bound tasks) and need to deploy them into two server machines. If homogeneous domains are mapped to the same server, energy improvement is 8.1% on average (13.8% from CPU-bound domains and 2.4% from memory-bound domains.) On the other hand, if heterogeneous domains are mapped to a server machine, we achieve 9.7% improvement for both server machines. It is not big difference though in this study, but it implies energy efficiency can be improved through sophisticated VM deployment. It is our future plan to find optimal VM deployments for energy efficiency.

TABLE II. shows coefficients of our model got from experimental results. Our model does not consider DVFS, so we fix frequency (2.4GHz) to find coefficients. R-square value represents how much the model is accurate. Power and utilization equation are quite accurate for *perl*. Delay model of *perl* is acceptable. However, power and delay model of *mcf* is not accurate. The only utilization model is accurate. It is because both power and delay does not change a lot for different number of active CPUs: difference in energy efficiency and delay is less than 3% (Figure 8 c and d.) The difference may be caused by some other factors such as uncertainty of measurement and noise, and the model does not consider those factors. Hence, our model does not fit to *mcf* result. '*c*' of both benchmarks is very small compared to '*d*', which means we can ignore '*c*'. Bigger '*c*' means larger consolidation overhead, but it is very small because we already reduce consolidation overhead by adjusting the number of vCPUs. '*g*' of *perl* is positive, and it implies delay can be reduced by consolidation (Figure 8 b.)

TABLE I. IMPROVEMENT ON ENERGY EFFICIENCY

	Improvement on energy efficiency (%)		
	perl	*mcf*	*mixed*
DVFS	7.4	0	6.2
Consolidation	10.7	1.9	8.8
DVFS +consolidation	13.8	2.4	9.7

TABLE II. COEFFICIENTS OF THE POWER AND DELAY MODELS

	coefficients	***perl***		***mcf***	
		value	R^2	**value**	R^2
power	*ac*	1.2E-05	0.977	-0.001	0.402
	ad	0.066		0.099	
	b	0.041		-0.022	
utilization	*c*	- 0.237	0.996	-0.046	0.999
	d	6.339		4.266	
delay	*e*	-143.938	0.913	-26.758	0.766
	f	45.796		35.419	
	g	0.071		-0.178	

4.2 SPECWeb2009 benchmark result

SPECWeb2009 is a well-developed benchmark suite for evaluating web servers which are I/O-bound, so its results can show how consolidation affects delay and energy efficiency of I/O-bound applications. *SPECWeb* requires Simultaneous User Sessions (SUS = 600 in this study) as input. We can specify level of workload by SUS count, but it is only overall workload. Instantaneous workload changes dynamically, so a dynamic management scheme is needed for consolidation. In this section, we start from understanding characteristics of *SPECWeb*. Next, we propose two algorithms for consolidation. Finally, we show experimental results for the proposed algorithms and compare them with the case without consolidation. The '*asymmetric*' scheme is used because it saves more energy than the other. Moreover, consolidation is accompanied by dynamic vCPU count management and DVFS.

4.2.1 Characteristics of the SPECWeb Suite

Web applications are typically not compute intensive [11]; hence, the performance (i.e., the response time per task) is less dependent on the clock frequency of CPUs (Figure 9 a.) One observation is that the number of active CPUs will not be an important factor in setting the web server performance if a sufficient number of CPUs is available. This is because the performance of web servers is highly sensitive to I/O, such as network and disk access. This result also implies that consolidation saves energy without noticeable performance loss for such applications.

Figure 9 b. depicts power dissipation, which is measured for different combinations of the number of active CPUs and clock frequency. One observation is that the power difference between *4CPU* case and *5CPU* case is biggest. This is because only one processor chip is active for the *4CPU* case (*asymmetric* scheme.)

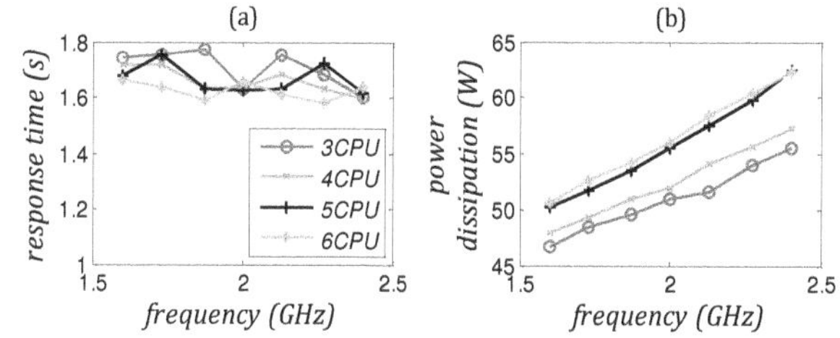

Figure 9 Response time and power dissipation

The relationship between frequency and utilization is needed for designing an effective consolidation controller. Our controller assumes that the workload level observed in the previous decision epoch persists into the current epoch (more sophisticated workload prediction schemes may be employed, but it falls outside the scope of present paper.) CPU utilization under the same workload, however, can be changed if the frequency changes. Hence, the controller must prevent an undesirable situation whereby the active CPUs are overloaded because the chosen frequency is too low. We assume the relationship between utilization and frequency is linear (5), and it is very accurate (Figure 10 .) β is relatively small and can be ignored, so we use a simpler approximated equation (6).

$$(u - \beta) \cdot f = \alpha \text{ , where } \alpha = 150.4, \beta = 29.9 \text{ and } 0 \leq u \leq 800 \text{ (8 pCPUs)} \quad (5)$$

$$f_i u_i = f_j u_j = \alpha \quad (6)$$

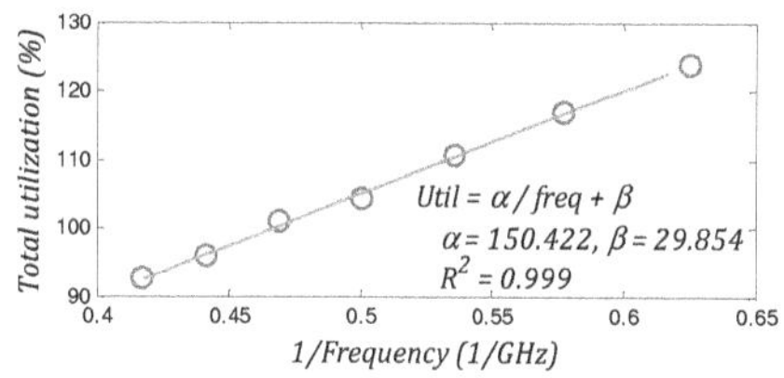

Figure 10 Frequency vs. total utilization

4.2.2 CPU Consolidation Controller

As shown in the previous section, clock frequency and the number of active CPUs are important factors. Hence, we have to carefully select these parameters. In this study, we propose two algorithms, which perform both DVFS and consolidation. They monitor CPU utilization every second, and change frequency and/or the number of active CPUs when desirable. Main idea of the algorithms is to utilize fewer CPUs at slower frequency as possible. It is reasonable for I/O bound applications because performance degradation is not significant unless CPU is almost fully utilized [9]. In addition, our main goal is not to maximize energy efficiency but to show potential of power saving by consolidation. Hence the simple algorithms are enough for this study. To reduce performance degradation, the algorithms change the system configuration conservatively: If a system is overloaded, they promptly increase frequency and/or the number of active CPUs. If, however, the system is under-utilized, they reduce frequency and/or turn off some CPUs when the situation persists for five seconds.

We present two algorithms and main idea of them is quite similar to each other: If the average utilization (u_n) of a CPU is greater than an upper threshold (T_h), they will assign more resource by increasing the clock frequency and/or the number of active CPUs. If the average utilization is less than a lower threshold (T_l), they will decrease frequency and/or the number of active CPUs.

Algorithm I favors using lower frequencies. If more CPU resource is required, it increases the number of active CPUs at the beginning (line 4 through 6.) Only when all CPUs are active and still more CPU resource is needed, it increases the frequency. In case of overbooking, it decreases the frequency first (line 12 through 14.) If the system is still overbooked at the lowest frequency, it will decrease the number of active CPUs (line 16 through 18.)

Algorithm II is similar to Algorithm I, which is not shown here, except that it favors using fewer active CPUs. If the average utilization of CPU is greater than T_h, the highest frequency will be selected. If still more CPUs are needed, it will increase the number of active CPUs. If the CPU is overbooked, it will keep the current frequency and decrease the number of active CPUs.

CPU Consolidation Algorithm I

Inputs: *N (total number of CPUs), f_n (frequency), c_n (the number of active CPUs), T_l/T_h (low/high threshold), and u_n (average utilization)*

Outputs: *f_{n+1} (frequency), c_{n+1} (the number of active CPUs)*

```
1:  if u_n > T_h then                          // more CPU resource is required
2:      persist = 0
3:      f_n+1 = f_n                            // keep the same frequency
4:      for c_n+1 in [c_n+1 , c_n+2 , ... , N] do  // increase # of active CPUs
5:          if u_n+1 < T_h then  // u_n+1 = u_n(f_n c_n) / (f_n+1 c_n+1) : new util.(6)
6:              break                          // stop here. enough CPU resource
7:      if u_n+1 > T_h then                    // still need more CPU resource (c_n+1=N)
8:          f_n+1 = f_max                      // increase frequency to the maximum
9:  else if u_n < T_l then                     // CPU resource is overbooked
10:     if ++persist >= 5 then                 // under-utilized for 5 decision period
11:         persist = 0, c_n+1 =c_n            // start from the same # of active CPUs
12:         for f_n+1 in all frequencies (ascending order) do
13:             if u_n+1 > T_l then
14:                 break                      // found the smallest frequency. stop
15:         if u_n+1 < T_l then                // still overbooked
16:             for c_n+1 in [c_n-1 , c_n-2 , ... , 1] do
17:                 if u_n+1 > T_l then
18:                     break                  // found minimum # of active CPUs. stop
```

Figure 11 consolidation algorithm I

TABLE III. shows the energy efficiency and QoS, which is defined as the percentage of packets that meet the performance specification. As defined before, we consider that there is no performance violation if QoS is greater than 95%. The first row shows results of the base system without consolidation and DVFS. The second test set uses Linux *ondemand* (DVFS) governor [12], which is a default governor of Linux, and it does not support consolidation in general. The last two rows show results from the proposed algorithms. Results are measured through measurements performed on our test bed hardware, so they also include the energy consumption and delay overheads of running the proposed consolidation algorithms. Algorithm I is the best one in terms of energy efficiency—it consumes around 15% less energy compared to the base case, which is better than 'Linux DVFS' case. QoS of Algorithm I is 96.9% which still meets performance requirement. Note that Algorithm I is better than 'Linux DVFS' in terms of performance as well as energy efficiency. Nevertheless additional overhead from turning on/off cores, Algorithm I shows better performance than 'Linux DVFS' because of its conservative nature: the algorithm infrequently changes frequency and active CPU count. Algorithm II improves energy efficiency by around 12% which is also greater than 'Linux DVFS', but it does not meet performance requirement. Algorithm I is outperforms II, and it shows that consolidation with DVFS improves energy efficiency of web servers.

5. CONCLUSION

DVFS has been a promising method for dynamic power management technique, but the energy saving leverage of DVFS decreases as the supply voltage level decreases with CMOS scaling. With HW support, such as *Core-level Power Gating*, consolidation becomes a promising power management technique. However, consolidation needs to be investigated by a more realistic scenario. In this study, we investigate effectiveness of consolidation for different configurations: types of applications, the number of active CPUs/vCPUs, and selection schemes. Through experiments, we presented suggestions for consolidation: 1) Control the number of vCPUs: we recommend that the ratio of the number of vCPUs to the number of CPUs to be less than 3. 2) Avoid aggressive consolidation for memory-bound applications: there is very small energy benefit. 3) Deploy heterogeneous VM domains in a single system: this saves more energy without violating performance. To summarize, this study shows energy saving of consolidation in virtualized systems. The proposed consolidation algorithm achieves about 15% of energy saving for I/O-bound application (*SPECWeb*.) Moreover, it increases the energy efficiency of CPU-bound application by 13.8%. Least improvement (2.4%) was achieved from Memory-bound applications.

TABLE III. SPECWEB BENCHMARK RESULT

	Tasks / E	*QoS (%)*	*Improvement*
Base	*1.098*	*98.0*	*n/a*
Linux DVFS	*1.212*	*93.4*	*10.4*
Algorithm I	*1.264*	*96.9*	*15.2*
Algorithm II	*1.228*	*90.3*	*11.9*

6. REFERENCES

[1] G. von Laszewski, *et al. Power-aware scheduling of virtual machines in DVFS-enabled clusters.* in *Cluster Computing and Workshops, 2009. CLUSTER '09. IEEE International Conference on.*

[2] P. Pillai and K.G. Shin. *Real-time dynamic voltage scaling for low-power embedded operating systems.* in *Proceedings of the eighteenth ACM symposium on Operating systems principles.*

[3] G. Dhiman, *et al. vGreen: a system for energy efficient computing in virtualized environments.* in *Proceedings of the 14th ACM/IEEE international symposium on Low power electronics and design.*

[4] R. Nathuji and K. Schwan. *VirtualPower: coordinated power management in virtualized enterprise systems.* in *Proceedings of twenty-first ACM SIGOPS symposium on Operating systems principles.*

[5] O. Bilgir, *et al.*, *Exploring the Potential of CMP Core Count Management on Data Center Energy Savings*, in *3rd Workshop on Energy Efficient Design.*

[6] N. Bobroff, *et al. Dynamic Placement of Virtual Machines for Managing SLA Violations.* in *Integrated Network Management, 2007. IM '07. 10th IFIP/IEEE International Symposium on.*

[7] H.N. Van, *et al. Autonomic virtual resource management for service hosting platforms.* in *Proceedings of the 2009 ICSE Workshop on Software Engineering Challenges of Cloud Computing.*

[8] C. Clark, *et al. Live migration of virtual machines.* in *Proceedings of the 2nd conference on Symposium on Networked Systems Design \& Implementation - Volume 2.*

[9] M. Pedram and I. Hwang. *Power and Performance Modeling in a Virtualized Server System.* in *Parallel Processing Workshops (ICPPW), 2010 39th International Conference on.*

[10] J. Leverich, *et al. Power Management of Datacenter Workloads Using Per-Core Power Gating.* in *Computer Architecture Letters.*

[11] D. Meisner, *et al. Power management of online data-intensive services.* in *Proceedings of the 38th annual international symposium on Computer architecture.*

[12] V. Pallipadi and A. Starikovskiy. *The Ondemand Governor.* in *Proceedings of the Linux Symposium.*

Energy-Efficient Scheduling on Heterogeneous Multi-Core Architectures

Jason Cong
Computer Science Department
University of California, Los Angeles
cong@cs.ucla.edu

Bo Yuan
Computer Science Department
University of California, Los Angeles
boyuan@cs.ucla.edu

ABSTRACT

The use of heterogeneous multi-core architectures has increased because of their potential energy efficiency compared to the homogeneous multi-core architectures. The shift from homogeneous multi-core to heterogeneous multi-core architectures creates many challenges for scheduling applications on the heterogeneous multi-core system. This paper studies the energy-efficient scheduling on Intel's QuickIA heterogeneous prototype platform [6]. A regression model is developed to estimate the energy consumption on the real heterogeneous multi-core platform. Our scheduling approach maps the program to the most appropriate core, based on program phases, through a combination of static analysis and runtime scheduling. We demonstrate the energy efficiency of our phase-based scheduling method by comparing it against the statical mapping approach proposed in [5] and the periodic sampling based approach proposed in [11], The experimental results show that our scheduling scheme can achieve an average 10.20% reduction in the energy delay product compared to [5] and an average 19.81% reduction in energy delay product compared to [11].

Categories and Subject Descriptors

C.1.3 [**Processor Architectures**]: Other Architecture Styles—*Heterogeneous (hybrid) systems*

Keywords

Energy Efficiency, Scheduling, Heterogeneous Multi-core

1. INTRODUCTION

Due to the increase in transistor budgets enabled by Moore's law, more and more cores are now integrated on chips. Accordingly, the on-chip power consumption becomes a critical issue. Conventional multi-core processors consist of identical cores. An alternative design approach for multi-core processors is to implement heterogeneous cores on a chip, which is a promising solution for power-efficient computing [8, 11].

Heterogeneous cores provide different power/performance tradeoffs. In order to benefit from heterogeneous multi-core architectures, the scheduler will need to consider the power/performance asymmetry of heterogeneous multi-core architectures when making a scheduling decision. Because the program has different performances and energy consumptions on different cores, scheduling the program to the most appropriate core is a challenging problem on heterogeneous multi-core architectures. To address these challenges, recent research proposes several scheduling schemes for heterogeneous multi-core architectures. Single-ISA heterogeneous multi-core architectures with simultaneous multi-threading processors are explored by Kumar et al. [10]. Dynamic core assignment policies are proposed to support the scheduling of multiprogrammed workloads. By adapting to inter-thread and intrathread diversities, heterogeneous multi-core architectures outperform homogeneous platforms in terms of performance. Lakshminarayana et al. [12] proposed an age-based scheduling policy, which schedules a task with a longer remaining execution time to a faster core. Shelepov et al. [16] collected the caching behaviors of applications via offline profiling, and predicted the performance of different threads on different cores based on a thread's caching behavior and a core's cache size and frequency. The OS scheduler assigned the threads to the cores based on the predicted performance. Srinivasan et al. [18] use the platform performance counters along with a performance prediction model to predict an application's execution time on different types of cores. With this information, the OS scheduler can schedule the applications to the suitable cores to improve system performance. Becchi and Crowley [4] assigned threads to cores based on the instructions-per-cycle (IPC) ratio to maximize the overall IPC on heterogeneous multiprocessor architectures.

The aforementioned work focuses on maximizing system performance such as the overall IPC and the overall performance gain. In this paper we focus on energy-efficiency scheduling, since the introduction of the heterogeneous platform is motivated by its potential energy efficiency. Kumar et al. [11] proposed single-ISA heterogeneous multi-core architectures to reduce power consumption. In order to optimize the energy-delay product, sampling-based dynamic switching heuristics are proposed to allow heterogeneous multi-core architectures to adapt to differences between applications or phases in the same application. Chen and John [5] project the core's configuration and the program's resource demands to a multi-dimensional space, and schedule programs to cores based on the weighted Euclidean distance between the core's configurations and the program's resource demands. The proposed approach in [5] statically maps programs to cores based on a program's resource demands for the entire execution without considering program phases.

However, less attention has been given to phase-based scheduling in previous work. Sondag and Rajan [17] identified the pro-

ISLPED'12, July 30–August 1, 2012, Redondo Beach, CA, USA.

gram phases through compiler analysis, and dynamically determined which type of core is most appropriate for each program phase by monitoring IPC at runtime. However, they identify program phases by clustering the basic blocks with similar properties, such as instruction types. Their approach considers the program phases at the granularity of basic block level, which might not capture program phases well without considering the program structure at the granularity of function calls and loops. Sawalha et al. [15] identified the program phases by tracking the program counters with a history table, which requires additional hardware support.

A previous temporal approach [11] sampled each type of core periodically and selected one type of core to schedule the program based on the measurements during sampling. The thread migration happens when one other core is sampled; sampling on both cores incurs certain overhead on both performance and energy. It is crucial to minimize the number of samplings to reduce the switching overhead and the energy consumed during samplings. Instead of periodic sampling, our approach explores the program phases based on program structures and proposes a phase-based sampling to guide scheduling to minimize the energy delay product. In contrast to prior research based on simulation, we evaluate our scheduling scheme using Intel's QuickIA heterogeneous prototype platform. The main contributions of our proposed energy-efficient scheduling method are as follows:

1. A regression model is developed to estimate the energy consumption on Intel's QuickIA heterogeneous prototype platform.

2. An energy-efficient scheduling approach is proposed to map the program to the most appropriate core based on program phases using a combination of static analysis and runtime scheduling.

The remainder of this paper is organized as follows. In Section 2 we characterize the Intel heterogeneous prototype platform in terms of microarchitectures and energy efficiency. Section 3 shows our regression model for energy consumption. Section 4 describes our energy-efficient scheduling approach. We show our evaluation methodology and experimental results in Section 5. Finally, we conclude this paper in Section 6.

2. HETEROGENEOUS MULTI-CORE ARCHITECTURES

2.1 Heterogeneous Prototype Platform

We conduct our experiments on the Intel QuickIA platform [6]. The heterogeneous platform consists of one Intel Atom Processor 330 and one Intel Xeon Processor E5450, as shown in Figure 1. This kind of experimental platform is considered to be a perfect heterogeneous system for evaluation [6]. The Intel Atom processor and Intel Xeon processor are representatives of two opposite types of microarchitecture. The Intel Xeon processor employs the high-performance server-class microarchitecture, while the Intel Atom processor employs the low-power microarchitecture targeted for mobile devices. We focused on these two types of microarchitectures because they capture most of the performance/power benefits from asymmetry [6].

2.2 Energy Efficiency

The heterogeneous cores lead to a variety of performance and energy consumptions. Here we propose the energy-efficient scheduling on heterogeneous multi-core architectures. The metric that we

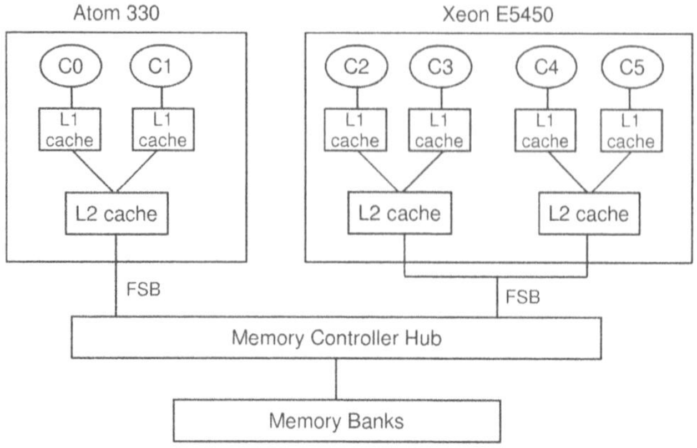

Figure 1: The configuration of the Intel QuickIA heterogeneous prototype platform

use to characterize energy efficiency is the energy delay product (EDP). Figure 2 shows the energy delay product over instruction intervals for the SPEC benchmark *473.astar*. Each instruction interval contains 50 million instructions.

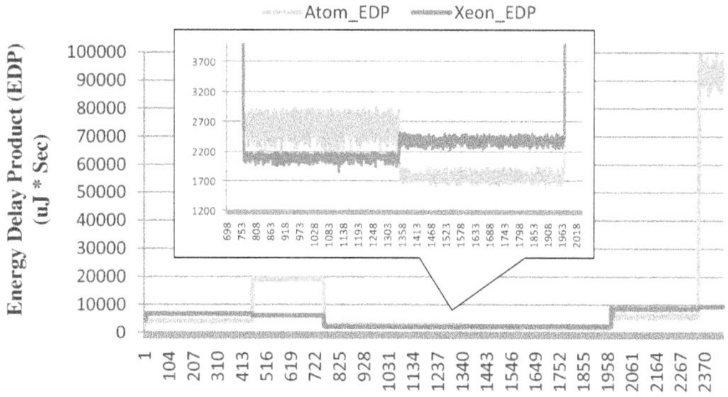

Figure 2: EDP on the Xeon processor and Atom processor vs. instruction intervals for the SPEC CPU2006 Benchmark *473.astar*

We can see that the program has different energy delay products on the Xeon core and Atom core in different program phases. The heterogeneous multi-core architectures have the potential to reduce the energy delay product by adaptively mapping the program to the most appropriate core. In order to make the correct scheduling decision, we need to know about the energy delay products on the Xeon processor and Atom processor. That means we need to obtain the execution time and energy consumption at runtime. We can obtain the execution time by using a system API such as *gettimeofday*. It is difficult to measure the energy consumption in real-time, although we can use thermal design power (TDP) to approximate the full-load energy consumption, CPU might not always work in the full-load situation during the entire execution. Therefore, we build a regression model to predict energy consumption at runtime.

3. REGRESSION MODELING FOR ENERGY

Regression modeling has been proposed to predict performance and power [13]. Here we use regression modeling to predict the energy consumption. To derive the regression model, we need to develop the training data to train the model. To generate the training data, we randomly select four benchmarks (*astar, bzip2, h264ref and hmmer*) from SPEC CPU2006 suite. For each benchmark, we collect fifteen pieces of hardware performance data (as shown in Table 1) on the Xeon core and Atom core for each instruction interval ranging from 50 million instructions to 25 billion instructions in 50-million instruction increments, 500 intervals in total. The hardware performance data is collected via hardware performance counters by using the Linux Perfctr library. Then, we feed this collected hardware performance data into McPAT [14] to calculate the energy consumption for each instruction interval. By using this approach, we can obtain 2000 training data samples.

Table 1: Collected hardware performance data

Category	Hardware performance data	
Performance	Cycles	
Instructions	Retired instructions	Floating point instructions
	Load instructions	Store instructions
Cache	L1 I-cache access	L1 I-cache miss
	L1 D-cache access	L1 D-cache miss
	L2 cache access	L2 cache miss
TLB	I-TLB miss	D-TLB miss
Branches	Branch instructions	Branch misprediction rate

With this training data, we build and evaluate our regression model by using statistical package R [2]. Given the fifteen hardware performance parameters shown in Table 1, we use variable clustering and correlation analysis [13] to identify the key hardware performance parameters that are most related to the energy consumption. The four selected key hardware performance parameters include cycles, the number of retired instructions, L1 D-cache access and L2 cache access. After identifying the four key hardware performance parameters, we specify a linear regression model as shown in equation (1) to predict the energy consumption with these four selected parameters. $\beta_1, \ldots, \beta_4$ denotes the corresponding regression coefficients, which can be interpreted as the expected change in energy per unit change in cycles, retired instructions, L1 D-cache access, or L2 cache access. The reason for using the linear regression model is that the static energy consumption is proportional to the execution time, and the dynamic energy consumption is proportional to the number of retired instructions, L1 D-cache access and L2 cache access.

$$\begin{aligned} Energy &= \beta_0 + \beta_1 \times Cycles + \beta_2 \times RetiredInsts \\ &\quad + \beta_3 \times L1DCacheAccess \\ &\quad + \beta_4 \times L2CacheAccess \qquad (1) \end{aligned}$$

Based on the training data, we use the ordinary least-squares method to determine the regression coefficients. We derive the regression model for the Xeon processor and Atom processor respectively. Our regression model with four parameters that include cycles (C), the number of retired instructions (I), L1 D-cache access (L1DCA), and L2 cache access (L2CA) achieves an average error of 7.6% across all training data samples. We also tried to derive the linear regression model with three parameters. For the Atom processor, eliminating L2 cache access from the regression model increased the average error to 200.2%, and eliminating L1 D-cache access from the regression model increased the average error to 68%. The comparisons are shown in Figure 3. So, we concluded that three variables are not sufficient for the regression model.

In order to evaluate our regression model, we use the derived regression model to predict the energy consumption of *lbm* (note that *lbm* is not used to generate the training data). We collect fifteen pieces of hardware performance data and calculate the energy consumption using McPAT to generate the observed energy consumption. We use our derived regression model with four parameters (C+I+L1DCA+L2CA) to generate the predicted energy consumption. Figure 4 shows the comparisons between the observed and predicted energy consumption of *lbm* over instruction intervals on both the Xeon processor and Atom processor. Each instruction interval contains 50 million instructions. The average error on the Xeon processor is 2.6%, while the average error on the Atom processor is 2.25%. These evaluations show that our regression model with four parameters (C+I+L1DCA+L2CA) can accurately predict energy consumption.

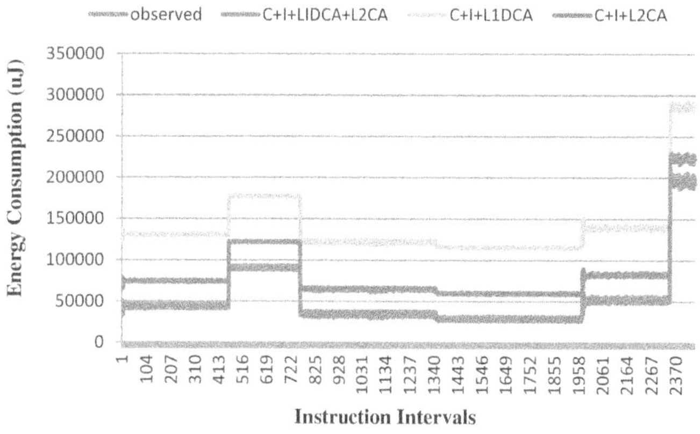

Figure 3: Comparisons between the observed and predicted energy consumption with training data on the Atom processor

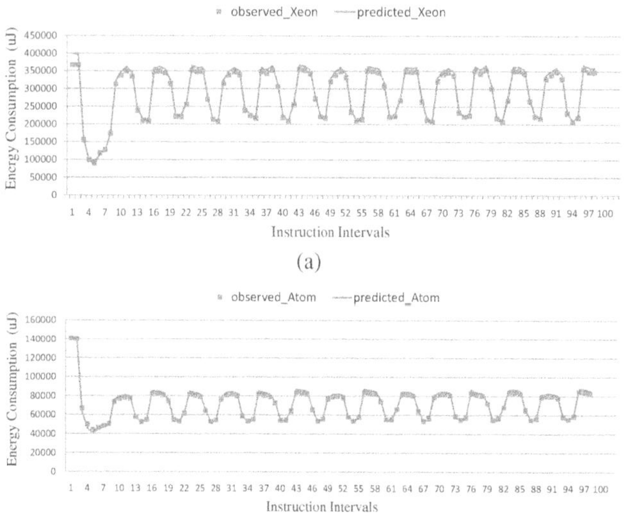

(a)

(b)

Figure 4: Comparisons between the observed and predicted energy consumption of *lbm* on both the Xeon processor (a) and Atom processor (b)

4. ENERGY-EFFICIENT SCHEDULING

Figure 5 shows *astar*'s EDP on the Atom processor and most executed function/loop ID over instruction intervals. Each instruction interval contains 50 million instructions. A unique function or loop

ID is assigned to each function and loop at the stage of static instrumentation (to be explained in Section 4.1). We can see that the program has a similar energy delay product across different invocations of the same functions and loops. Based on this observation, we identify the program phases based on the program structures

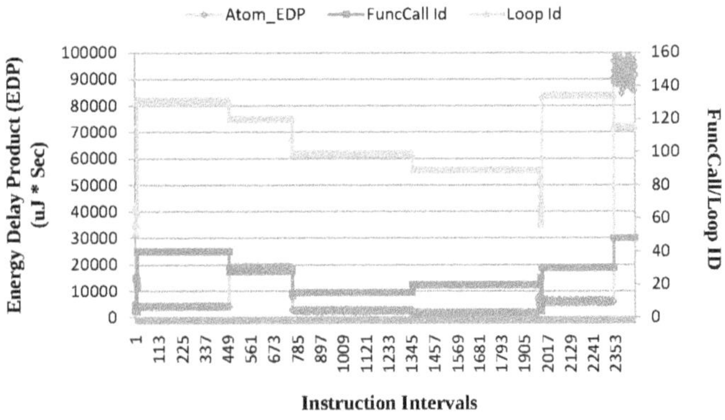

Figure 5: EDP and function/loop IDs vs. instruction intervals for SPEC CPU2006 Benchmark *473.astar*

Our scheduling method consists of four steps. Step one identifies the boundaries of the function calls and loops through static analysis. Step two profiles the program to construct the call graph that captures the caller-callee relationship among function calls and loops. Step three identifies the major program phases and instruments the selected function calls and loops. Step four estimates the energy consumption with the regression model built in Section 3 and predicts the energy delay product to make scheduling decisions.

4.1 Static Analysis and Instrumentation

We implement the static instrumentation in the LLVM compiler infrastructure [1]. LLVM takes the program source codes as inputs to generate LLVM intermediate representations. The LLVM intermediate representation (IR) is a static single-assignment (SSA)-based representation that essentially models a RISC processor with infinite registers. Starting from the LLVM IR, we identify the function calls and their returns through 'call' instruction and 'ret' instruction. We identify the loops and loop boundaries based on the LLVM loop analysis passes. During the static instrumentation stage, we insert calls to instrumentation functions at the boundaries of functions and loops. These instrumentation functions are used to calculate invocation times and the total number of instructions for each function call and loop, capture the caller and call sites, and track the caller-callee relationships among functions and loops. The reason for using static analysis and instrumentation is that it is difficult to identify the function boundaries and loop boundaries by using dynamic binary instrumentation tools such as Intel Pin tools.

4.2 Creating Call Graph

Our call graph is based on the calling context tree of Ammons et al. [3]. Each node of the call graph is a function call or loop. We label each node with the name of the function or loop. The path from the root of the tree to a node captures the callers and call sites. To differentiate the invocations of a function or loop in different paths, we extend the node name with the path ID. If a function is called from the inside of a loop, it will be considered a single node in the call graph, even if it might have different program behavior among different invocations. We profile the instrumented binary to count the number of invocations and the total number of executed instructions for each function call and loop. We annotate the node with the total number of instructions executed and annotate the edge with the number of invocations by callers.

Here we use the sample code shown in Figure 6 to illustrate how to create the call graph. Assume the number of instructions executed in *func2* is 1 million. The corresponding call graph of the sample code is shown in Figure 7. In the call graph, *main* has two children *func1*. Each *func1* has its own *Loop1*, *Loop2*, *Loop3*, and *func2*. Since *func2_1* is called 1000 times in path 1, *func1_1*, *Loop1_1*, *Loop2_1*, *Loop3_1*, and *func2_1* each have 1 billion instructions executed. In this way, we also can infer the average number of executed instructions in each invocation for other functions and loops, as shown in Figure 7.

```
void func1(float * a, int m, int n, int p)
{
Loop1:  for (i=0; i<m; i++)
Loop2:    for (j=0; j<n; j++)
Loop3:      for (k=0; k<p; k++)
              a[i] += func2(i+j+k);
}
int main()
{
  func1(A, 10, 10, 10);
  func1(B, 100, 100, 100);
}
```

Figure 6: Sample code

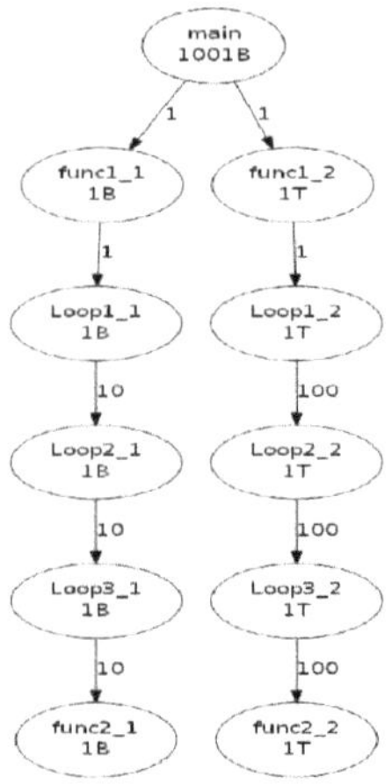

Figure 7: Call graph of the sample code

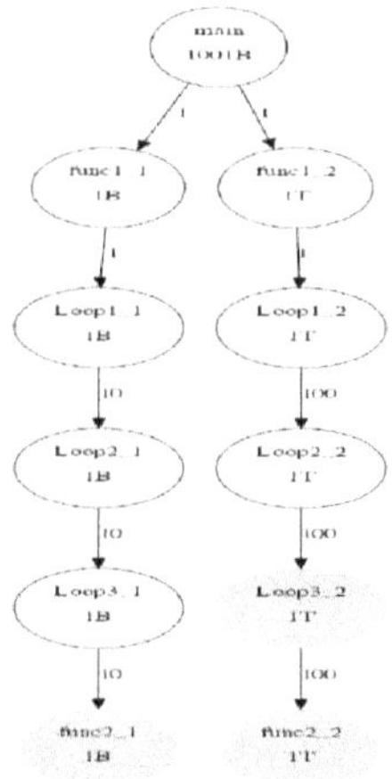

Figure 8: Major program phases on the call graph

4.3 Identifying Major Program Phases

The created call graph is used to identify the major program phases that run long enough. For function calls and loops to qualify as major program phases, the total number of instructions executed has to be larger than or equal to th_{inst}, and the number of invocations has to be larger than or equal to th_{invoke}. The reason for th_{inst} is that the thread migration will incur a certain overhead on performance and energy consumption. In order to obtain the benefits from scheduling, the function calls and loops need to have long execution time. The reason for th_{invoke} is that our sampling-based approach makes decisions based on the measurements during the first several invocations. In order to obtain the benefits from scheduling, the function calls and loops need to be invoked many times. The minor function calls and loops will be included as part of the closest major function calls and loops in the call graph. We insert calls to instrumentation functions at entry and exit points in the identified major function calls and loops by using the LLVM compiler infrastructure. These instrumentation functions are used to sample the hardware performance data, predict the energy delay products, and make the scheduling decisions.

Here we use the call graph shown in Figure 8 to illustrate how to identify the major program phases. Assume th_{inst} is equal to 1 billion and th_{invoke} is equal to 1000. We can obtain the number of invocations of a node by cumulative multiplication of all the edge values on the path from the root of the tree to the node. In Figure 8, the total number of instructions executed for every node is larger than or equal to th_{inst}. However, only node *func2_1*, node *Loop3_2*, and node *func2_2* are called more than or equal to th_{invoke}. So, *func2_1*, *Loop3_2*, and *func2_2* are identified as major function calls and loops. Since *func2_2* is called from the inside of *Loop3_2*, we can instrument either *func2_2* or *Loop3_2*.

4.4 Runtime Scheduling

The scheduling decision is made at runtime using the instrumented codes. In order to make the correct scheduling decision, we need to predict the energy delay products on the Xeon processor and Atom processor. That means we need to obtain the execution time and energy consumption at runtime. The regression model in Section 3 needs four key hardware performance parameters to accurately predict the energy consumption. But, there are only two hardware performance counters on the Xeon processor and on the Atom processor. During each invocation, we can only collect two pieces of hardware performance data by using the Linux Perfctr library. Therefore, we need two invocation to estimate the energy consumptions. The number of cycles is one parameter of the regression model; from this we can obtain the execution time. Then, we can multiply the estimated energy consumption by execution time to estimate the energy delay product.

Our heterogeneous prototype platform has two types of cores: the Xeon processor and Atom processor. In the first two invocations of each identified major function or loop, we run the invocations on the Xeon processor and estimate the energy delay product by using the above method. In the next two invocations, we run the invocations on the Atom processor and estimate the energy delay product. When an identified major function or loop finishes these runs, we select the core that has a lower energy delay product to execute the remaining invocations of this function or loop. The core switching is done using the Linux process affinity API.

4.5 Complexity and Scalability

The complexity of our scheduling scheme is $\mathcal{O}(P \times N)$, where P denotes the number of major program phases detected in the program, and N denotes the number of different types of processors on a chip. As the number of cores increases, scalability might become an issue for the sampling-based approach. In the future, the number of different types of processors on a chip might be much more than two. In that case, we need to do hierarchically sampling to reduce the sampling overhead. We can cluster the processor types into several categories based on the processors' characteristics, with each category having one representative processor type. Based on these representative processors, our scheduling scheme could determine which category the application or the program phase will be mapped into. Among the processors in that category, we can apply our scheduling method to determine onto which processor type the application or the program phases will be mapped.

5. EVALUATION

5.1 Evaluation Methodology

In the evaluation we use seven benchmarks (*astar, bzip2, h264ref, hmmer, lbm, libquantum and named*) from the SPEC CPU2006 suite and four benchmarks (*denoise, deblure, registration (reg) and segmentation (seg)*) from the medical imaging domain [7]. We evaluate our scheduling scheme on the Intel QuickIA heterogeneous platform [6]. To demonstrate the advantages of our scheduling scheme, we implement the following scheduling schemes for comparisons.

The statical mapping approach ***(SMap)***: This scheduling scheme proposed in [5] statically maps the programs to the cores, and does not switch the program among heterogeneous cores during program execution. We measure a set of program's inherent characteristics including instruction-level parallelism (ILP), branch predictability, and data locality by using the Pin tool, MICA [9]. Then, we use the projection functions [5] to identify the desired resource demands including issue width, branch predictor size, and data cache size. In the Intel QuickIA heterogeneous platform, two types of hardware resource are provided: one is the Intel Atom processor, the other is the Intel Xeon processor. We calculate the weighted Euclidean distances between the program's resource demands and the hardware resources of these two types of processors, respectively. The processor type with the smaller distance will be chosen as the preferred processor onto which the application will be mapped.

The periodic sampling approach ***(PS)***: This scheduling scheme proposed in [11] samples one or more cores for five intervals every 100 intervals. Each instruction interval contains 1 million instructions, and their scheduling scheme is implemented in the simulator. However, our evaluation is done on a heterogeneous prototype platform. Therefore, we implemented their method by using prior knowledge of the program execution. We first collect all hardware performance data (as shown in Table 1) using the Linux Perfctr library for each instruction interval. Then, we feed these hardware performance data into McPAT to calculate the energy consumption, and multiply the energy consumption by the execution time to obtain the energy delay product for each instruction interval. With this prior knowledge, we can use the scheduling method proposed in [11] to make the scheduling decision for each instruction interval. Given the scheduling decisions, we calculate the energy delay product by multiplying the total energy consumption with the total execution time.

The phase-based sampling approach ***(PhaseSamp)***: This is our scheduling scheme. Our approach first statically analyzes the program, detects the major program phases based on LLVM infrastructure, and inserts the instrumented codes at the boundaries of the program phases. We run the instrumented program on the Intel QuickIA prototype platform. Based on the instrumented codes, the program will dynamically select the most appropriate core in terms of energy delay product during the runtime.

5.2 Energy Delay Product Comparison

The comparisons of the energy delay product in our approach to the energy delay products in statical mapping and periodical sampling with different instruction intervals are shown in Figure 9. The results are normalized to our approach (PhaseSamp). In [11], one interval consists of 1 million instructions. In our evaluations, we also evaluated the impact of varying interval length on the scheduling scheme proposed in [11]. The interval lengths we evaluated contain 1 million instructions, 10 million instructions, and 50 million instructions.

Compared to the statical mapping approach, our approach can

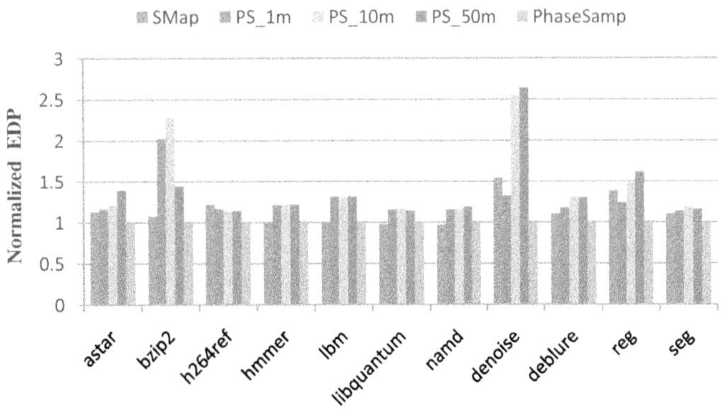

Figure 9: Comparison of the energy delay product in our approach to the energy delay products in statical mapping and periodical sampling with different instruction intervals

Table 2: Comparison of the energy delay product and the number of core switchings over benchmark sets

Benchmarks	PS_1m		PhaseSamp (Our Approach)	
	EDP (J * sec)	Total switches	EDP (J * sec)	Total switches
astar	27156.47	2452 (204X)	23346.32 (14.03%)	12
bzip2	489.13	468 (59X)	241.49 (50.63%)	8
h264ref	107048.22	9834 (307X)	92117.31 (13.95%)	32
hmmer	34060.29	5712 (238X)	28078.4 (17.56%)	24
lbm	65041.56	2062 (257X)	49550.21 (23.82%)	8
libquantum	90.43	246 (41X)	78.16 (13.57%)	6
namd	2988.97	1254 (313X)	2592.02 (13.28%)	4
denoise	1485.98	380 (47X)	1123.92 (24.37%)	8
deblure	1916.02	620 (78X)	1627.09 (15.08%)	8
reg	7716.94	1502 (57X)	6208.08 (19.55%)	26
seg	8642.92	1718 (286X)	7595.35 (12.12%)	6
average		171X	19.81%	

achieve an average 10.20% reduction up to 35.25% reduction in the energy delay product. We observe that the statical mapping approach works better than our approach for the benchmarks: *hmmer, lbm, libquantum, and namd.* We take a close look at these four benchmarks, and we find that they do not have obvious program phase changes in terms of energy-delay product. These four benchmarks all have a lower energy delay product on a specific processor (either the Xeon processor or the Atom processor) during the entire execution. The core switchings of our approach bring a certain overhead into the energy delay product. For other benchmarks, our approach outperforms the statical mapping approach. The other benchmarks have clear program phase changes in terms of energy-delay product. Some program phases have lower energy delay products on the Xeon processor, while other program phases have lower energy delay products on the Atom processor. And our approach can capture the program phases much better.

Compared to the periodical sampling approach, our approach can achieve an average 19.81% reduction in energy delay product when the interval length is 1 million instructions; it can achieve an average 26.25% reduction when the interval length is 10 million instructions; it can achieve an average 25.31% reduction when the interval length is 50 million instructions. The reduction in energy delay product comes from the improved capturing of program phases and the fewer number of switches needed (as shown in Table 2). The core switching will incur a certain overhead on performance and energy consumption. We measure the cost of core switching on the real platform, and the overhead of core switching on performance is around 20 μs per switching; the overhead of core switching on energy consumption is around 40 μJ per switching.

6. CONCLUSION

In this paper we propose an energy-efficient scheduling method for heterogeneous multi-core architectures. We develop a regression model to estimate the energy consumption. Our scheduling approach maps the program to the most appropriate core based on program phases using a combination of static analysis and runtime scheduling. We demonstrate the efficiency of our scheduling approach on the Intel QuickIA heterogeneous prototype platform [6]. Our approach achieves an average 10.20% reduction in energy delay product over the static mapping approach proposed in [5] and an average 19.81% reduction in energy delay product over the periodic-sampling approach proposed in [11].

7. ACKNOWLEDGMENTS

This research is partially supported by SRC contract 2009-TJ-1984, NSF grant CCF-0903541, and the NSF Expedition in Computing grant CCF-0926127.

We greatly appreciate research gifts and equipment donations from Intel, and helpful discussions with Ganapati Srinivasa, Ravishankar Iyer, and Mishali Naik from Intel concerning the Intel QuickIA system.

8. REFERENCES

[1] The llvm compiler infrastructure. `http://llvm.org/`.

[2] Statistical package R. `http://www.r-project.org/`.

[3] G. Ammons et al. Exploiting hardware performance counters with flow and context sensitive profiling. PLDI '97, pages 85–96.

[4] M. Becchi and P. Crowley. Dynamic thread assignment on heterogeneous multiprocessor architectures. CF '06, pages 29–40.

[5] J. Chen and L. K. John. Efficient program scheduling for heterogeneous multi-core processors. DAC '09, pages 927–930.

[6] N. Chitlur et al. QuickIA: Exploring heterogeneous architectures on real prototypes. HPCA '12, pages 1–8.

[7] J. Cong et al. Customizable domain-specific computing. *IEEE Design Test of Computers*, 28(2):6–15, 2011.

[8] M. Hill and M. Marty. Amdahl's law in the multicore era. *Computer*, 41(7):33 –38, 2008.

[9] K. Hoste and L. Eeckhout. Microarchitecture-independent workload characterization. *IEEE Micro*, 27(3):63 –72, 2007.

[10] R. Kumar et al. Single-ISA heterogeneous multi-core architectures for multithreaded workload performance. ISCA '04, pages 64–75.

[11] R. Kumar et al. Single-ISA heterogeneous multi-core architectures: the potential for processor power reduction. MICRO '03, pages 81 – 92.

[12] N. B. Lakshminarayana et al. Age based scheduling for asymmetric multiprocessors. SC '09, pages 25:1–25:12.

[13] B. C. Lee and D. M. Brooks. Accurate and efficient regression modeling for microarchitectural performance and power prediction. ASPLOS '06, pages 185–194.

[14] S. Li et al. McPAT: an integrated power, area, and timing modeling framework for multicore and manycore architectures. MICRO '09, pages 469–480.

[15] L. Sawalha et al. Phase-guided scheduling on single-ISA heterogeneous multicore processors. pages 736 –745, DSD '11.

[16] D. Shelepov et al. HASS: a scheduler for heterogeneous multicore systems. *SIGOPS Oper. Syst. Rev.*, 43:66–75, 2009.

[17] T. Sondag and H. Rajan. Phase-based tuning for better utilization of performance-asymmetric multicore processors. CGO '11, pages 11–20.

[18] S. Srinivasan et al. Heteroscouts: hardware assist for os scheduling in heterogeneous cmps. *SIGMETRICS Perform. Eval. Rev.*, 39:341–342, 2011.

MAC: Migration-Aware Compilation for STT-RAM based Hybrid Cache in Embedded Systems

Qingan Li
City University of Hong Kong,
Hong Kong
ww345ww@gmail.com

Jianhua Li
University of Science and Technology of China, Hefei, China
jianhual@mail.ustc.edu.cn

Liang Shi
University of Science and Technology of China, Hefei, China
shil0704@mail.ustc.edu.cn

Chun Jason Xue
City University of Hong Kong,
Hong Kong
jasonxue@cityu.edu.hk

Yanxiang He
Wuhan University, Wuhan,
China
yxhe@whu.edu.cn

ABSTRACT

Hybrid caches consisting of both STT-RAM and SRAM have been proposed recently for energy efficiency. To explore the advantages of hybrid cache, most work on hybrid caches employs migration based strategies to dynamically move write-intensive data from STT-RAM to SRAM. Migrations require additional read and write operations for data movement and may lead to significant overheads. To address this issue, this paper proposes a Migration-Aware Compilation (MAC) approach to improve the energy efficiency and performance of STT-RAM based hybrid cache. By re-arranging data layout, the data access pattern in memory blocks is changed such that the number of migrations is reduced without any hardware modification. The reduction of migration overheads in turn improves energy efficiency and performance. The experimental results show that with the proposed approach, on average, the number of write operations on STT-RAM is reduced by 13.4%, the number of migrations is reduced by 16.1%, the total dynamic energy is reduced by 8.5%, and the total latency is reduced by 12.1%.

Categories and Subject Descriptors

D.3.4 [**Program languages**]: Processors—*code generation, compilers*; B.3.2 [**Memory Structures**]: Design Styles—*cache memories*

Keywords

hybrid cache, migration, compiler

1. INTRODUCTION

As technology scaling continues, traditional SRAM-based caches are facing many challenges such as energy consumption and scalability. Recent advancements in technology present Spin-Torque Transfer RAM (STT-RAM) as a new candidate for building caches [6]. Compared to SRAM, STT-RAM has higher storage density and lower leakage power. However, write operations on STT-RAM have considerably longer latency and higher energy consumption than SRAM. To take advantage of both STT-RAM and SRAM, hybrid cache architectures have been studied and evaluated in recent works [14] [15] [8] [10]. These studies show that caches built with multiple memory technologies can outperform its counterpart of single technology. To explore the advantages of the hybrid cache, migration-based techniques are commonly used to dynamically move write-intensive or read-intensive data between STT-RAM and SRAM [14] [15] [8] [10]. Migrations require additional read and write operations for data movement and these additional access operations consume cycles as well as energy. These overheads, called migration overheads, can degrade the energy efficiency and performance of the hybrid cache. For embedded systems which often have more stringent power and performance constraints, this is a big problem. To solve the problem, this paper proposes a migration-aware compilation based approach to improve the energy efficiency and performance of embedded systems with STT-RAM based hybrid cache.

In this paper, a set of experiments is conducted and it is observed that migrations are frequently triggered in STT-RAM based hybrid cache for embedded systems. These migrations lead to significant overheads. It is also observed that migrations are sensitive to data access patterns, correlating closely with transition events in memory blocks, where a *transition event* is caused by a read operation following a write operation, or a write operation following a read operation. It is further observed that in embedded systems, most transition events in memory blocks come from stack area and global area, rather than heap area. With these observations, a Migration-Aware Compilation (MAC) approach is proposed in this paper to reduce the transition events in stack and global areas by re-arranging the data layout. The main concept is to place data with consecutively same access operations into the same memory block. The reduction of transition events leads to a reduction of migrations, which in turn improves the energy efficiency and performance of the hybrid cache. The experimental results show that with the proposed approach, on average, the number of write operations on STT-RAM is reduced by 13.4%, the number of migrations is reduced by 16.1%, the total latency is reduced by 12.1% and the total dynamic energy is reduced by 8.5%. The contributions of this paper include:

- A set of experiments is conducted to analyze migrations in STT-RAM based hybrid cache, and several key observations are presented.
- A compilation based approach is proposed to improve the efficiency of STT-RAM based hybrid cache by re-arranging data layout.
- Two Pin-based cache simulators are developed to evaluate the proposed approach. The experimental results show that the proposed approach can improve the energy efficiency and performance of STT-RAM based hybrid cache.

The rest of this paper is organized as follows. The analysis of migrations is illustrated in Section 2. A compilation based approach to improve the efficiency of hybrid cache is proposed in Section 3. In Section 4, a set of experiments is conducted to evaluate the propose approach for STT-RAM based hybrid cache. The related works are discussed in Section 5. Finally, Section 6 concludes this paper.

2. ANALYSIS OF MIGRATIONS

This section analyzes the migrations in STT-RAM based hybrid cache. This paper focuses on embedded systems where the configuration of one-level cache is widely applied, such as Freescale e300 family [1], Renesas VR-series [2], and Cortex-R based MCUs [3]. For the evaluations in this section, we consider a system with one-level on-chip hybrid data cache, where the cache size is 32K bytes, the associativity is 4-way (one way for SRAM, and three ways for STT-RAM), and the cache line size is 32 bytes. We defer the discussion of cache parameters to Section 4. The cache management strategy in [10] is implemented for this evaluation. There are two key observations:

1. Migrations are frequently triggered in STT-RAM based hybrid cache, which indicates significant migration overheads.
2. Migrations correlate closely with transition events in memory blocks.

Next we will discuss these two observations in detail.

2.1 overheads from migration

Migration based techniques are often proposed in recent works on hybrid caches [14] [15] [8] [10]. For STT-RAM based hybrid cache, the SRAM part is preferable for write operations and the STT-RAM part is preferable for read operations. Considering the high probability that a program writes data to a specific group of memory blocks repeatedly, data on STT-RAM should be migrated to SRAM if they are frequently written [14]. This kind of migration, triggered by write operations, is called *w-migrate* in this paper. Furthermore, it is observed that there is also high probability that programs read data from a specific group of memory blocks repeatedly. Li et al. [10] propose to migrate memory blocks from SRAM into STT-RAM if they are read frequently. In this paper, this kind of migration, triggered by read operations, is called *r-migrate*. Both *w-migrate* and *r-migrate* require additional read and write operations for data movement, and these extra overheads will degrade performance and energy efficiency of STT-RAM based hybrid cache.

Figure 1 shows the number of migrations, consisting of *w-migrate* and *r-migrate*, for the selected benchmarks. It is found that, on average, eight migrations are triggered per hundred memory accesses. In other words, assuming each migration involves a read and a write operation, for each hundred memory accesses, at lest sixteen additional operations are introduced by migrations. These migration overheads are detrimental to the system performance and energy consumption, and should be minimized.

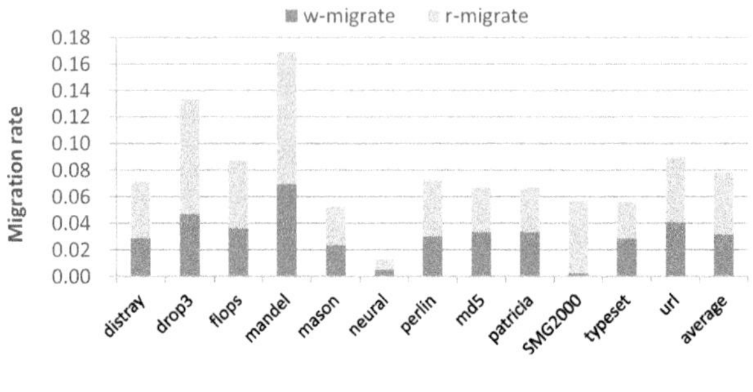

Figure 1: Migration rate. This rate is computed by dividing the number of migrations by the total number of memory accesses.

2.2 Relationship between migrations and transition events

It is observed that the migration overheads correlate closely with transition events in access sequences. A R/W *transition event* is caused by a read operation followed by a write, or a write operation followed by a read, in a memory block. Generally, more transition events lead to more migrations. This is because, a transition event followed by several write operations (indicating a trend of frequent write accesses) in an SRAM cache block, triggers a *r-migrate*, and a transition event followed by several read operations in an STT-RAM cache block, triggers a *w-migrate*. Therefore, if too many transition events occur in memory blocks, the migration mechanism will be triggered frequently. In this situation, benefits from migration will be offset by the migration overheads.

Figure 2 shows the correlation between the number of migrations and the number of transition events in memory blocks. It is found that transition-intensive memory blocks are often migration-intensive memory blocks. The number of migrations changes along with the number of transition events in memory blocks. For the selected benchmarks, the correlation coefficient between the number of migrations and the number of transition events is about 0.90. Therefore, the number of transition events, which is visible at compilation time, can be used as a good indicator to reduce the migration overheads.

Transition events from stack and global areas can be reduced by re-arranging the data layout, as illustrated in Figure 3. Assuming that there are four data objects, a, b, c, and d. For the access sequence in Figure 3(a), if using the default data layout as shown in Figure 3(b), both cache blocks will be written 20 times and read 20 times, so the total number of transition events will be 40. However, if we allocate a and

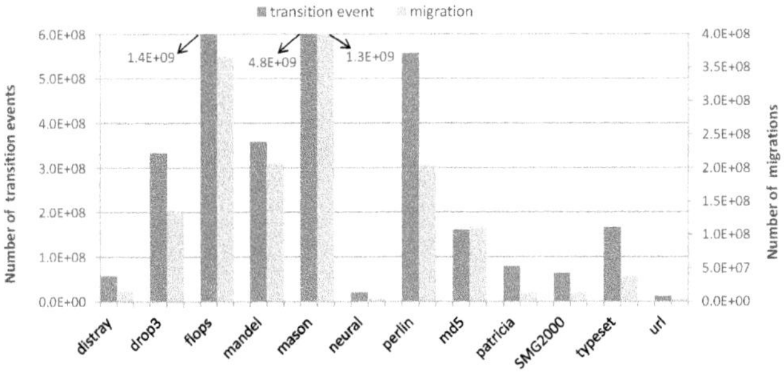

Figure 2: Relationship between transition events and migrations.

d to cache block cl_1, b and c to cache block cl_2, cl_1 will be read 40 times and cl_2 will be written 40 times. There will be no transition event in cache blocks as shown in Figure 3(c).

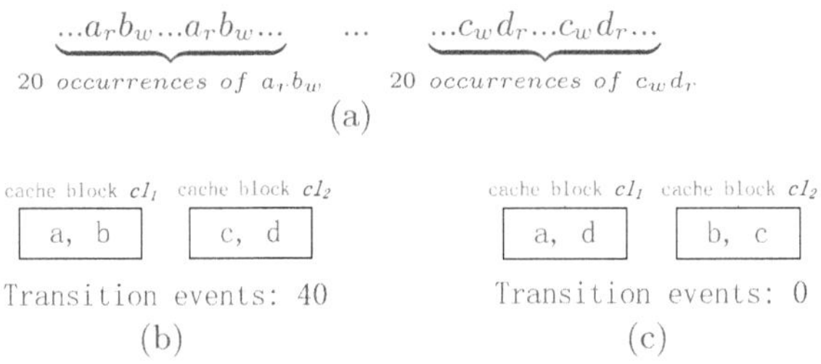

Figure 3: Data layout for reducing transition events. (a) Access sequence with read and write information. (b) Default layout results in transition events of 40. (c) Better layout results in 0 transition events.

These observations motivate the work in this paper. By re-arranging the data layout, it is possible to reduce the number of transition events in stack and global areas. A reduction of transition events will lead to a reduction of migration overheads, which can in turn improve both the energy efficiency and performance of STT-RAM hybrid caches. In this paper, a compilation based approach is proposed to improve the efficiency of the hybrid cache system by a judicious data layout of the stack and global areas.

3. MIGRATION-AWARE COMPILATION

This section introduces the proposed migration-aware compilation based approach to minimize the transition events in stack and global areas. The proposed approach consists of three steps to re-arrange data layout of stack area and global area separately:

Step 1. Obtain the data access sequence with data read-/write information.

Step 2. Assign the data into a set of memory blocks based on data read/write operations. The size of a memory block equals the cache line size.

Step 3. Finalize the data layout according to the previous data assignment.

In the first step, we obtain the estimated access sequence using a static profiling technique [16], and collect the information about how many times two objects appear adjacent to each other in the access sequence. This information is encoded into a Data Proximity Graph (DPG) and is used to guide the data assignment allocation. A DPG $G(V, E, w)$ is a weighted graph, where V is the set of vertices, E is the set of edges. Each vertex represents a data object. For an edge $e(a,b)$, the weight assigned to e, denoted as $w(e)$, is calculated by combining the occurrences of a and b appearing adjacent to each other as well as their corresponding read/write operation information. In the second step, we present a heuristic algorithm for the data assignment. In the weighted DPG, the weight of an edge indicates the benefits from assigning the related pair of objects into the same memory block. The assignment splits the DPG into its vertex-induced sub-graphs, where each sub-graph corresponds to a memory block, such that the total weight of all sub-graphs is maximized. In the last step, the data layout is finalized according to the previous data assignment. Note that the proposed algorithm does not handle objects with size greater than the cache line size. These objects are left to be placed using the default method. The proposed approach re-arranges data layout for global area and stack area in the similar fashion. Therefore, only data layout for stack area is discussed in detail in the rest of this paper.

3.1 Obtaining access sequence of stack objects

A static profiling technique [16] is employed in this paper to estimate the execution frequency of each statement and each basic block. Then, by visiting the control flow graph annotated with execution frequency, the frequency that each pair of data objects appear adjacent to each other can be obtained. At this time, all pairs of consecutive data accesses are taken into account. In the DPG, a larger weight indicates that there are more benefits if the related pair of data objects are assigned into the same memory block. Therefore, the weight assignment methods include:

- For each occurrence of two consecutively same operations, for example, a read followed by another read, a positive weight α is added. Assigning the related pair of data objects into the same memory block can improve the program locality and thus reduce cache misses. In other words, temporal locality between two data objects results in a positive weight.

- For each occurrence of two consecutive but different operations, for example, a read operation followed by a write, or a write operation followed by a read, a weight β is added. Assigning the related pair of objects in the same memory block has double-edged effects, improving the program locality while increasing the number of transition events in this block.

Here α and β are coefficients. To minimize the total number of transition events, β should be a negative number; to maximize the program locality, both α and β should be positive numbers. The choice of coefficients has impact on the effectiveness of the proposed approach. An example is presented in Figure 4 to illustrate the weight assignment process. Assume that there is a sequence of memory access as shown in Figure 4(a). The occurrences that two data objects appear adjacent to each other are shown in Figure 4(b). In Figure 4(b), the weight (x, y) of the edge (a, b) represents that the occurrences a and b appear adjacent to each other with the same operation is x, and the occurrences a and b

appear adjacent to each other with different operations is y. The following equation is used for computing the weight w: $w = \alpha x + \beta y$. With $(2, -3)$ as coefficients of (α, β), the weighted DPG is built as shown in Figure 4(c).

$$a_r b_w b_r a_r b_r c_w b_r c_r d_r c_w b_r c_r d_w$$

(a)

(b) (c)

Figure 4: An example for constructing the weighted DPG. (a) The memory access sequence. (b) The estimated data proximity. (c) The weighted DPG.

3.2 Data assignment

Data assignment is done by partitioning the DPG into its vertex-induced sub-graphs such that the total weight of all sub-graphs are maximized, with the constraint that the total size of each sub-graph is no more than the cache line size. A heuristic algorithm is proposed for this partitioning process, as illustrated in Algorithm 3.1. First, a list of empty memory blocks is built to be allocated. The size of each memory block equals the cache line size. Then, the allocation continues by iteratively retrieving the edge with the largest weight and allocating the related pair of data objects into the same memory block. After each allocation, the DPG is updated accordingly.

An example for the graph partitioning process is shown in Figure 5. In the proposed algorithm, the weighted graph is a complete graph, but edges with weight less-than or equal to zero are not shown for simplicity in this example. It is assumed that all data objects are of the same size, and each cache line (or memory block) can hold three objects. The graph to be partitioned is shown in Step 1 of Figure 5. Since there are nine objects, the memory blocks list is initialized with three (9/3) empty blocks. In Step 2, the edge connected by node a and node b has the largest weight, 100. So node a and node b are allocated into the first memory block and they are merged into node (a, b). The final result of allocation is shown in Step 7.

3.3 Data Layout Finalization

After the graph partition process, a list of memory blocks is obtained. The offset of each data object internal to the owner memory block is also obtained. If the stack base pointer for each function is aligned with the cache-line size, we can conduct the finalization of stack layout by mapping memory blocks into stack directly. In this work, two tasks are carried out to align the stack base pointer with the cache-line size. First, extra instructions are inserted at the entry of the main function to align the stack base pointer of the main function with the cache-line size. Second, the stack size of each user function is expanded to be a multiple of the cache-line size. Assume that the cache-line size is C, and the original stack size of a function is x. After expansion, the new stack size for this function is: $\lceil \frac{x}{C} \rceil \times C$. After

Algorithm 3.1 Graph Partitioning.

Input:

graph[V][V]: the DPG
blocks: a empty list of memory blocks
nStackSize: stack size for the function

Output:

blocks: a list of assigned memory blocks

```
// Step 1: initialize the list of memory blocks;
int nThreshold = nStackSize/CACHE_LINE_SIZE;
for i = 1 to nThreshold do
    build a new memory block b and add it into blocks;
end for
// Step 2: assign data object into the list of memory blocks;
while graph is not empty do
    bool bSucc ← false;
    retrieve an edge e(v1, v2) from graph with the largest
    weight (indicating benefits);
    // try to assign v1, v2 into the same block
    if both v1 and v2 are unassigned then
        if there is a block b can holding both v1 and v2 then
            bool bSucc ← true;
            assign v1 and v2 into the same block;
            merge vertex v1 and v2 in DPG;
        end if
    else if only one object, v2, is unassigned then
        denote the block holding v1 as b;
        if b can hold v2 then
            bool bSucc ← true;
            assign this object into the block holding the other
            object;
            merge vertex v1 and v2 in DPG;
        end if
    end if
    // if this edge can't guide a successful assignment, detete
    it;
    if ! bSucc then
        delete edge e(v1, v2) from DPG;
    end if
end while
// Step 3: assign the remaining data using the default
method;
assign the remaining data objects using the default method
return blocks;
```

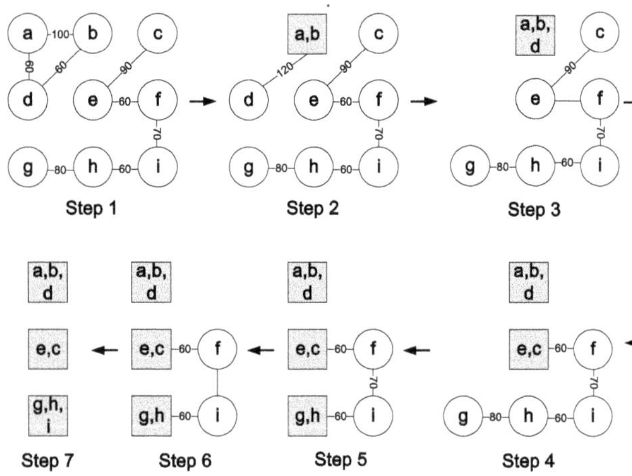

Figure 5: Graph partitioning.

these two tasks, it is guaranteed that the stack base pointer of each function instance is aligned with the cache-line size. Note that there are some special contents which belonging to the stack could not be rearranged freely. These contents include the space for saving registers and return address.

Table 1: Architecture parameters.

Parameter	Value
processor	single core
hybrid data cache	32KB, 32B line size, LRU, 4-way 1-way for sram (8KB) 3-way for stt-ram (24KB) write allocation, write back sram access latency: 6 cycles sram access dynamic energy: 0.388 nJ stt-ram read/write latency: 6/28 cycles stt-ram read/write dynamic energy: 0.4/2.3 nJ
main memory	latency: 300 cycles

Therefore, the mapping of memory blocks should be done after the space for all special contents is already reserved.

4. EXPERIMENTS

In this section, we first introduce the experimental setup, and then present the experimental results for evaluating STT-RAM based hybrid cache. A discussion of the experimental results is presented at the end of this section.

4.1 Experimental setup

The proposed compilation based approach is implemented based on LLVM compiler [9]. Two versions of executables are generated. The first version, called the original version, is compiled by LLVM compiler using the default method for data layout. The second version, called the optimized version, is compiled by LLVM compiler using the proposed approach for data layout. Both versions are compiled with "O3" optimization. The benchmarks are selected from the LLVM test suites, originated mainly from MiBench [7].

A PIN-based [4] simulator of STT-RAM based hybrid cache with the cache management strategy in [10] is developed. Note that the inter-core migration strategy in [10] is not included, since this work focuses on embedded systems with single-core processor. The target architecture is depicted in Table 1. The cache and memory parameters are obtained from a modified version of CACTI [11].

4.2 Evaluation of hybrid cache

Experimental results are normalized to the original method for data layout. The best configurations of (α, β) vary with benchmarks. The improvement on migration overheads and cache misses are shown in Figure 6(a) and Figure 6(c), respectively. On average, the number of migrations are reduced by 16.1%, and the cache misses are reduced by 12.7%. The proposed technique can also significantly reduce write operations on STT-RAM as shown in Figure 6(b). On average, write operations on STT-RAM are reduced by 13.4% compared with the original. As the proposed approach can reduce the number of migrations, cache misses, and write operations on STT-RAM, it can improve the energy efficiency and performance of STT-RAM based hybrid cache. We estimate the total latency using the cache and memory parameters as shown in Table 1, by counting the total number of read/write transactions for STT-RAM, SRAM and the main memory. To estimate the energy efficiency, we consider dynamic energy consumed by the hybrid cache, but not dynamic energy consumed by the main memory. Static energy is improved proportionally to the improvement in latency. As shown in Figure 6(d), on average, the total latency is reduced by 12.1% and the total dynamic energy is reduced by 8.5%.

4.3 Discussion of experimental results

As indicated in the experimental results, there are three observations of the proposed approach. First, for most benchmarks, it can improve the efficiency. Second, for several benchmarks, it works very well. Third, it does not work well for some benchmarks. In rare cases, such as *typeset*, it has a negative effect, although the negative effect is very limited.

These observations are mainly due to four reasons. First, the more concentrative the distribution of transition events (over memory blocks) is, the more benefits the proposed method of re-arranging data layout can achieve. This is the reason that the proposed approach works very well for *flops* and *md5*, which have highly concentrative distribution of transition events. Second, the static profiling technique works well for most benchmarks, but it can not guarantee a precise estimation of transition events for all benchmarks, which limits the effectiveness of the proposed data layout method, or even misleads the data layout method. Third, currently only pair-wise proximity relationships between data objects are considered, which also limits the effectiveness of data layout. Fourth, the effectiveness of stack layout is limited by the stack size. When the stack size is small relative to the cache line size, there is little opportunity to reduce transition events by re-arranging stack data layout.

5. RELATED WORK

There are several studies architecting energy efficient STT-RAM based hybrid cache using fully-hardware based techniques. Sun et al. [14] propose hybrid caches consisting of STT-RAM and SRAM, and employed migration based policy to mitigate the drawbacks of STT-RAM. Wu et al. [15] evaluate two types of hybrid cache architectures (HCAs), including inter cache level HCA and intra cache level HCA, both of which employ memory block migration policy. Their experiments show that these HCAs can improve both performance and energy efficiency. More migration based policies are explored in [8] [10] to further improve the efficiency of STT-RAM based hybrid cache.

The compilation based techniques for cache aware data placement have been studied for a long time. A general framework for cache conscious data placement is proposed in [5]. Petrank et al. [12] presents a good survey on cache aware data placement. Recently, compiler-assisted caching techniques are also proposed in the context of multi-threaded architectures and multiprocessors [13]. These studies focus on improving program locality for better performance, while this paper considers read/write asymmetry together with locality to improve efficiency of STT-RAM based cache.

6. CONCLUSIONS

Most work on STT-RAM based hybrid cache employs migration based strategies to improve the efficiency. It is found that migration based strategies can cause significant overheads. To address this issue, a compilation based approach is proposed in this paper to improve the energy efficiency and performance of STT-RAM based hybrid caches. The

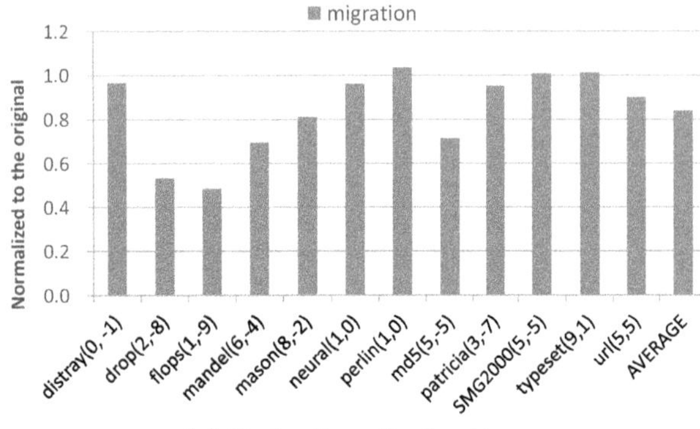

(a) Reduction of migrations.

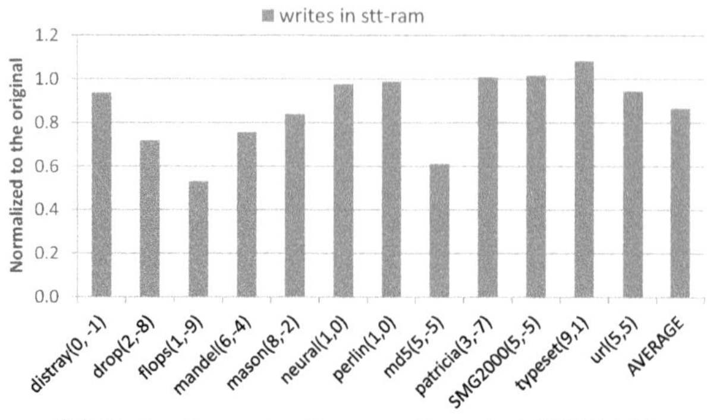

(b) Reduction of write operations to STT-RAM.

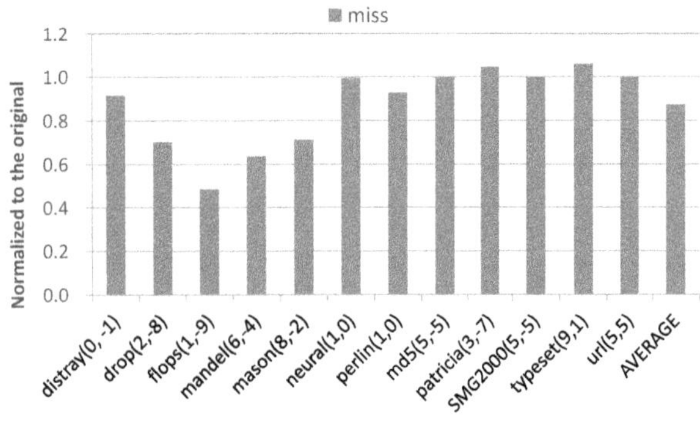

(c) Reduction of cache misses.

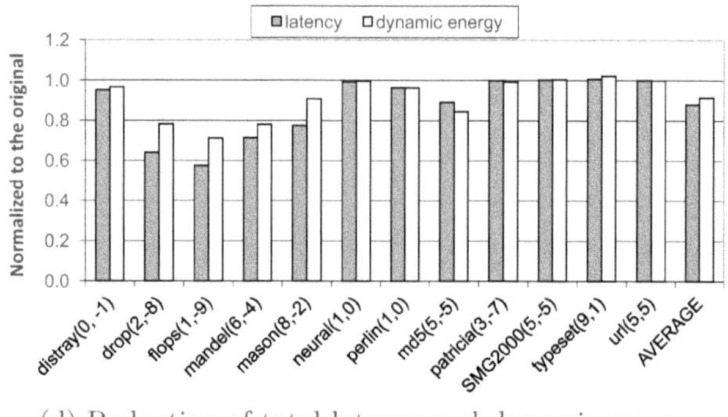

(d) Reduction of total latency and dynamic energy.

Figure 6: Experimental results for hybrid cache. The configuration of (α, β) for each benchmark are listed after benchmark names.

experimental results show that the proposed approach can significantly reduce the migration overhead and improve the energy efficiency and performance.

7. ACKNOWLEDGEMENTS

This work was partially supported by the Research Grants Council of the Hong Kong Special Administrative Region, China [Project No. CityU 123609, CityU 123210, CityU 123811], and the National Natural Science Found of China [Project No. 61170022, 61173014, 61133005, 91118003].

8. REFERENCES

[1] http://cache.freescale.com/files/32bit/doc/ref_manual/e300coreRM.pdf.

[2] http://www2.renesas.com/micro/en/product/vr/legacy.html.

[3] http://www.arm.com/products/processors/cortex-r/index.php.

[4] http://www.pintool.org.

[5] B. Calder, C. Krintz, S. John, and T. Austin. Cache-conscious data placement. In *Proceedings of the eighth international conference on Architectural support for programming languages and operating systems*, ASPLOS-VIII, pages 139–149, 1998.

[6] X. Dong, X. Wu, G. Sun, Y. Xie, H. Li, and Y. Chen. Circuit and microarchitecture evaluation of 3d stacking magnetic ram (mram) as a universal memory replacement. In *Proceedings of the 45th annual Design Automation Conference*, DAC '08, pages 554–559, 2008.

[7] M. Guthaus, J. Ringenberg, D. Ernst, T. Austin, T. Mudge, and R. Brown. Mibench: A free, commercially representative embedded benchmark suite. In *Workload Characterization, 2001. WWC-4. 2001 IEEE International Workshop on*, pages 3 – 14, dec. 2001.

[8] A. Jadidi, M. Arjomand, and H. Sarbazi-Azad. High-endurance and performance-efficient design of hybrid cache architectures through adaptive line replacement. In *Proceedings of the 17th IEEE/ACM international symposium on Low-power electronics and design*, ISLPED '11, pages 79–84, Piscataway, NJ, USA, 2011. IEEE Press.

[9] C. Lattner and V. Adve. Llvm: A compilation framework for lifelong program analysis & transformation. In *Proceedings of the international symposium on Code generation and optimization: feedback-directed and runtime optimization*, CGO '04, pages 75–, 2004.

[10] J. Li, C. J. Xue, and Y. Xu. Stt-ram based energy-efficiency hybrid cache for cmps. In *19th IFIP/IEEE International Conference on Very Large Scale Integration*, pages 31–36, 2011.

[11] N. Muralimanohar, R. Balasubramonian, and N. Jouppi. Optimizing nuca organizations and wiring alternatives for large caches with cacti 6.0. In *Proceedings of the 40th Annual IEEE/ACM International Symposium on Microarchitecture*, MICRO 40, pages 3–14, 2007.

[12] E. Petrank and D. Rawitz. The hardness of cache conscious data placement. *Nordic J. of Computing*, 12:275–307, June 2005.

[13] S. Sarkar and D. M. Tullsen. Compiler techniques for reducing data cache miss rate on a multithreaded architecture. In *Proceedings of the 3rd international conference on High performance embedded architectures and compilers*, HiPEAC'08, pages 353–368, 2008.

[14] G. Sun, X. Dong, Y. Xie, J. Li, and Y. Chen. A novel architecture of the 3d stacked mram l2 cache for cmps. In *High Performance Computer Architecture, 2009. HPCA 2009. IEEE 15th International Symposium on*, pages 239 –249, feb. 2009.

[15] X. Wu, J. Li, L. Zhang, E. Speight, R. Rajamony, and Y. Xie. Hybrid cache architecture with disparate memory technologies. In *Proceedings of the 36th annual international symposium on Computer architecture*, ISCA '09, pages 34–45, 2009.

[16] Y. Wu and J. R. Larus. Static branch frequency and program profile analysis. In *Proceedings of the 27th annual international symposium on Microarchitecture*, MICRO 27, pages 1–11, 1994.

Energy-Efficient Signal Processing in Wearable Embedded Systems: An Optimal Feature Selection Approach

Hassan Ghasemzadeh, Navid Amini, and Majid Sarrafzadeh
University of California, Los Angeles, California, United States
{hassan,amini,majid}@cs.ucla.edu

ABSTRACT

Many wearable embedded systems benefit from classification algorithms where statistical features extracted from physiological signals are mapped onto different user's states such as health status of a patient or type of activity performed by a subject. Conventionally selected features lead to rapid battery depletion in these battery-operated systems, mainly due to the absence of computing complexity criterion while selecting prominent features. In this paper, we introduce the notion of power-aware feature selection, which minimizes energy consumption of the signal processing for classification applications. Our approach takes into consideration the energy cost of individual features that are calculated in real-time. The problem is formulated using integer programming and a greedy approximation is presented to select the features in a power-efficient manner. Experimental results on thirty channels of activity data demonstrate that our approach can significantly reduce energy consumption of the computing module resulting in more than 30% energy savings while achieving 96.7% classification accuracy.

Categories and Subject Descriptors

C.3 [**Computer Systems Organization**]: Special Purpose and Application-Based Systems—*Real-time and embedded systems*; J.3 [**Computer Applications**]: Life and Medical Science—*Health*; H.1.2 [**Information Systems**]: Models and Principles—*User/Machine Systems Human information processing; Human factors.*

General Terms

Design, Algorithms, Experimentation.

Keywords

Wearable Monitoring, Embedded Signal Processing, Activity Recognition, Feature Selection, Power Optimization.

1. INTRODUCTION

Recent technology advances have led to the development of different sensing, computing and communication artifacts that are becoming essential part of our daily lives. These ubiquitous platforms have proved to be effective in a number of domains ranging from medical and well-being to military and smart vehicles. A special class of these platforms is wearable monitoring where computational elements are tightly coupled with the human body. As these embedded systems continue to become more pervasive, design and development of low-power architectures that enable their sustainable realization becomes much more crucial. Low power design is even more challenging in wearable systems that are battery-operated and are known as enabling technologies for many applications such as remote patient monitoring and personalized healthcare, gaming and sports, maintenance, production and process support [1, 2].

An important aspect of low power design is to optimize power consumption of the computing modules. Wearable monitoring systems often employ embedded signal processing and machine learning blocks that use sensor data (e.g. acceleration of body joints) to extract relevant information (e.g. human movements) about their subjects.

We take special interest in classification applications, where physiological signals from human body are used to classify different states of a subject. Examples of such applications include human action recognition and fall detection using accelerometer and gyroscope sensors, and arrhythmia detection from ECG signals. In the classification process, a set of representative features, such as 'signal amplitude' and 'root mean square' power, are typically extracted from the measured signals prior to executing the classification algorithm. Feature extraction is often time consuming and can deplete the battery if an exhaustive feature set is considered. In this paper, we propose the notion of power-aware feature selection which introduces a novel approach for optimizing the power efficiency of feature extraction mechanisms. Our approach combines three criteria, namely, feature relevance, feature redundancy, and computing complexity and builds a minimum cost feature set without sacrificing the classification performance of the system.

Unfortunately, none of the feature selection techniques studied in the past takes into consideration computing complexity of the selected features, an important measure in designing wearable monitoring systems. Similarly, none of the existing power-aware schemes in embedded system design has dealt with feature selection algorithms and how the energy constraints prevent some features from being accounted for classification. Our work embodies the innovation of the notion of power-aware feature selection; we model the problem of energy optimal feature set and prove that it constitutes a computational problem that is NP-hard and finally we provide an approximation to find the appropriate minimum cost feature set. Real human motion data sets are used in order to verify the efficacy of the proposed approach.

Our contributions can be summarized as follows: 1) we introduce the notion of power-aware feature selection by adding a new design dimension, computing complexity, to the feature selection problem; 2) we propose a graph model that embodies information regarding classical feature selection, relevance and redundancy; 3) we present an Integer Linear Programming (ILP) formulation of our optimization problem and provide a greedy approximation to solve it; 4) we use real data collected from a wearable sensor network designed for movement monitoring to evaluate the effectiveness of our power optimization approach.

2. RELATED WORK

Our study spans two broad research topics, feature selection and power-aware design, that are disjointedly explored in machine learning and embedded system design fields, respectively.

In general, feature selection [3, 4] aims to find an optimal set of features from an exhaustively extracted set of features. The optimality of the solution, however, is defined by two criteria, relevance and redundancy [5]. While relevance criterion focuses on eliminating features that are irrelevant to the classification task, redundancy criterion uses inter-feature correlation measures to eliminate features with high correlation.

The tight energy constraints of battery-operated wearable monitoring systems call for the development of energy-aware signal processing methods to preserve energy [6]. Depending on the application and signals involved, it may be more useful and energy-efficient to trade off radio communications energy for processing energy, or vice versa. Dynamic sensor selection [7] which trades off classification accuracy for battery lifetime is another power saving approach. The method proposed in [8] minimizes the number of nodes necessary to obtain a given classification ratio for activity recognition.

3. PRELIMINARIES

Several basic concepts are reviewed in this section with an emphasis on the architecture of wearable monitoring platforms, their signal processing, and information extraction from physiological signals.

3.1 Wearable Monitoring Systems

A wearable monitoring system, also called body sensor network, is composed of several body-worn sensor nodes and a gateway. Each sensor node is attached to the body to sample and process physiological signals, and transmit partial results to the gateway. A node usually has several sensors for capturing different user's states (e.g. body acceleration), an embedded processor to perform limited signal processing and information extraction, and a radio for data transmissions. The gateway is a more powerful unit such as a cell phone or a PDA that performs data fusion and makes conclusions about current state of the user (e.g. 'walking', 'running', and 'sitting'). The results are further transmitted, through the Internet, to a back-end server for storage, further processing, and clinical decision support.

Figure 1: Signal processing for classification applications

The focus of this study is primarily on power optimization of the wearable sensor nodes, where stringent constrained sensor units are used to process physiological signals in real-time. Other elements (e.g. gateway and back-end server) are usually powerful in terms of computing power.

Each sensor node processes sensor readings through a chain of embedded signal processing modules, each of which is intended to extract partial information from the signal and reduces the amount of sampled data. Figure 1 illustrates a typical signal processing flow for applications targeting classifying physiological signals into user's states. In physical movement monitoring applications, readings from motion sensors such as accelerometers, magnetometers, and gyroscopes undergo signal processing to classify human actions such as 'walking', 'sit to stand', and 'jumping'. The signals that are sampled by each sensor node are first passed through a filter to reduce high frequency noise. Segmentation is intended to identify 'start' and 'end' points of the actions being classified. Feature extraction module is responsible for calculating statistical and morphological characteristics of the signal segment. Finally, a classification algorithm is utilized to determine the current state of the user (e.g. type of actions being performed by the user). Signal processing and pattern recognition require a learning phase during which the system is trained based on a training data set. During this phase, parameters of the system used in different signal processing modules are adjusted. Such parameters define the training model and will be used throughout the operation of the system. Feature selection (as specified by dashed lines in Figure 1) is only part of the learning process. The selected features, however, determine complexity of the feature extraction block during execution of the system.

3.2 Feature Relevance and Redundancy

Feature selection is a learning process aimed to find significant features for the classification application. Initially, an exhaustive set of features is extracted due to lack of domain knowledge. Many features might be redundant due to their strong correlation with other features. Yet, non-redundant features can be irrelevant to a specific classification application. Feature selection is a mature field of study in machine learning with primary selection criteria being redundancy and relevance [5].

To better present the concept of relevance and redundancy, we give a simple example adapted from [5]. Assume that the feature set $\{f_1, f_2, \dots, f_5\}$ is given, where a classification decision is made based on a binary function g as $C = g(f_1,f_2)$. Furthermore, let $f_2=\bar{f}_3$ and $f_4=\bar{f}_5$. Clearly, f_1 is a required feature as it is necessary for classification and is not redundant with respect to other features. Also, features f_4 and f_5 are redundant because they are not used to make a classification decision. We note, however, that f_2 and f_3 can be interchanged as they are equally informative. Thus, an optimal feature set is either $\{f_1,f_2\}$ or $\{f_1,f_3\}$.

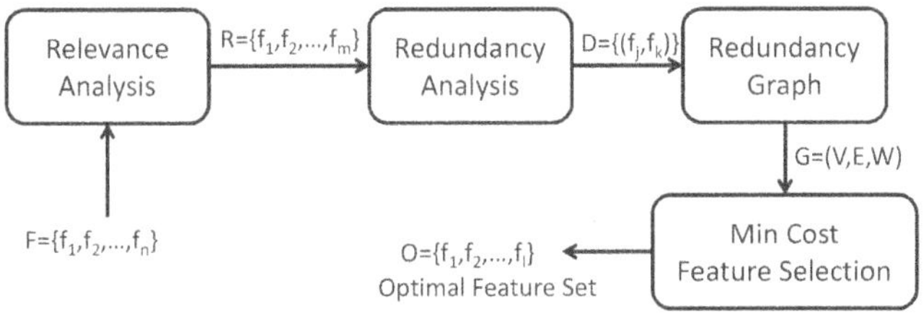

Figure 2: Proposed approach for optimal power-aware feature selection

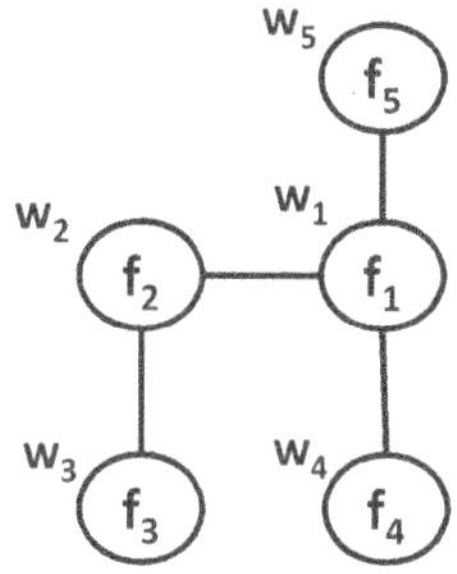

Figure 3: A motivational example for power-aware feature selection

4. PROPOSED APPROACH

Figure 2 illustrates our approach for power-aware feature selection. Given an initial feature set, F, irrelevant features are first eliminated from subsequent processing. A redundancy analysis is then performed to find features that are strongly correlated and can be substituted once power efficiency of the processing is taken into account. To this end, a graph model, called redundancy graph, is constructed based on the correlation information obtained during the redundancy analysis. Finally, the graph model is used to solve an optimization problem, called *Minimum Cost Feature Selection* (MCFS), meant for finding the optimal feature set.

4.1 Graph Model

Our relevance and redundancy analyses, which provide inputs to construct the graph model and formulate the problem, are based on the concept of symmetric uncertainty:

Definition 1 (SYMMETRIC UNCERTAINTY). *The symmetric uncertainty between two discrete random variables X and Y is given by:*

$$U(X,Y) = \frac{2I(X,Y)}{H(X)+H(Y)}, \tag{1}$$

where $H(X)$ and $H(Y)$ represent the entropy of random variables X and Y, respectively, and $I(X,Y)$ denotes the information gain between the two variables. $I(X,Y)$ is further defined by:

$$I(X,Y) = H(X) - H(X|Y) \tag{2}$$

The symmetric uncertainty is actually the normalized information gain and is always between 0 and 1, where U=1 means that knowing the value of either variable can completely predict the other variable, and U=0 indicates that the two variables are completely independent.

We note that the symmetric uncertainty is a measure of correlation between two random variables. Major advantage of this measure against other measures, such as correlation coefficient, is that the symmetric uncertainty can capture non-linear correlation between variables and therefore, is a safe measure for our feature analysis study.

Definition 2 (IRRELEVANT FEATURE). *Given an exhaustive set of n features F=$\{f_1, f_2, \ldots, f_n\}$ and a set of human actions $A = \{a_1, a_2, \ldots, a_h\}$ to be classified, a feature f_i is irrelevant to the classification task if*

$$min_j\,(U(f_i, a_j)) < \lambda_R, \tag{3}$$

where λ_R (relevance threshold) is a design parameter.

Relevance analysis will eliminate features that are irrelevant to the action recognition. The remaining m features (m<n) are subject to redundancy analysis whose main goal is to find strongly correlated features.

Definition 3 (STRONGLY CORRELATED FEATURES). *Two features f_i and f_k are considered to be strongly correlated if*

$$U(f_i, f_k) > \lambda_D, \tag{4}$$

where λ_D (redundancy threshold) is a design parameter.

The output of redundancy analysis is a set of feature pairs in the form of (f_i,f_k), which are strongly correlated and either of them can be eliminated according to the correlation analysis. However, these features are further examined for computing complexity using the graph model presented in the following section.

Definition 4 (REDUNDANCY GRAPH). *Given m relevant features introduced by the relevance analysis and a set of feature pairs $\{f_j,f_k\}$ generated according to the redundancy analysis, an undirected graph G=(V,E,W) is called redundancy graph, where V is a set of m vertices, $V = \{u_1, u_2, \ldots, u_m\}$ associated with the m relevant features, E=$\{e_1, e_2, \ldots, e_r\}$ is the set of r feature pairs that are strongly correlated, and W=$\{w_1, w_2, \ldots, w_m\}$ is the set of weights, assigned to the vertices, denoting the computing cost associated with each feature.*

4.2 Feature Selection

We now present a simple example to motivate our idea of finding the optimal feature set using MCFS. Assume that 10 features construct our exhaustive set of features, represented by $F = \{f_1, f_2, \ldots, f_{10}\}$. Further, assume that the relevance analysis decides to eliminate five features and hence, the redundancy graph will contain five features, i.e. $R = \{f_1, f_2, f_3, f_4, f_5\}$. The redundancy graph with each vertex representing one of the five features is shown in Figure 3. Note that the processing cost attributed to each feature is represented by the weight of each vertex, e.g., w_1 is the processing cost of f_1.

Let all the weights be equal to one unit, that is $W = \{w_1, w_2, w_3, w_4, w_5\} = \{1, 1, 1, 1, 1\}$. In this case, MCFS treats all features equally and thus, the optimal feature set consists of two vertices, specifically f_1 and f_3. However, if we modify the weight set to $W = \{10, 1, 1, 1, 1\}$, MCFS gives more consideration to vertices with lower weights and accordingly, features f_4 and f_5 will be favored over f_1. In the recent scenario the optimal feature set will contain three vertices, i.e. f_4, f_5, and f_3. As such, the computation energy cost will be decreased from 11 units to 3 units.

5. MINIMUM COST FEATURE SELECTION

In this section, we present an optimization problem that finds optimal feature set taking into account relevance, redundancy and complexity criteria discussed previously. The problem takes a redundancy graph as input and outputs a subset of relevant features that are optimal in terms of computing complexity and do not exhibit any redundancy.

Problem 1. *Given a redundancy graph G=(V,E,W), the minimum cost feature selection problem is to find a subset of vertices that are not dominated by any other vertex in the graph and their total cost is minimized.*

5.1 Problem Formulation and Complexity

Assume that a_{ij} is a given binary that encodes existence of edges in the redundancy graph:

$$a_{ij} = \begin{cases} 1, & \text{if } (u_i,u_j) \in \text{V} \\ 0, & \text{otherwise} \end{cases} \tag{5}$$

and x_i is a binary variable which determines whether or not a vertex u_i is chosen as a member of the final vertex set:

$$x_i = \begin{cases} 1, & \text{if vertex } u_i \text{ is chosen} \\ 0, & \text{otherwise} \end{cases} \tag{6}$$

The corresponding ILP formulation for the MCFS problem is as follows:

$$Minimize \sum_{i=1}^{m} w_i x_i, \tag{7}$$

subject to:

$$\sum_{j=1}^{m} x_i a_{ij} \geq 1 \quad \forall i \in \{1, 2, \ldots, m\} \tag{8}$$

$$x_i \in \{0, 1\} \tag{9}$$

The objective function in (7) is to minimize the total cost of the selected vertices (i.e. those with x_i=1). The constraint (8) guarantees that each selected vertex is adjacent to at least one more vertex and the constraint in (9) ensures that the variable x_i takes only binary values.

The MCFS problem is equivalent to the Minimum Cost Dominating Set (MCDS) problem. The MCDS problem is proved to be NP-hard by reduction from the Weighted Set Cover (WSC) problem.

Theorem 1. *The MCFS problem is NP-hard.*

PROOF. Proof is eliminated for brevity. □

5.2 Greedy Approach

Our greedy algorithm for solving MCFS problem is presented in Algorithm 1. For each vertex in the redundancy graph, the algorithm first finds all adjacent vertices (V_i). It then finds the best candidate vertex to include that vertex in the final vertex set (O). The best candidate is the one with maximum value of 'cardinality of V_i divided by vertex cost w_i'. The intuition behind selecting such vertex is that it has a large number of adjacent vertices and a small cost. Finally, the algorithm adds the candidate vertex (u_i) to O and eliminates u_i and all its neighbors from V_i as well as V. The algorithm iterates until there is no more vertex in V indicating that each vertex is either chosen as a final vertex or is dominated by a final vertex.

Algorithm 1 Greedy solution for MCFS problem

Require: Redundancy graph G=(V,E,W)
Ensure: Final vertex set O
  $O = \phi$
  for all $u_i \in V$ **do**
    V_i = {all vertices u_j adjacent to u_i}
  end for
  while $V \neq \phi$ **do**
    $V_i \leftarrow argmax_{V_i} \frac{|V_i|}{w_i}$ and $O \leftarrow O \cup u_i$
    Eliminate u_j from all V_i sets and V
  end while

Theorem 2. *Algorithm 1 achieves an* $\ln n$ *approximation to the MCFS problem.*

PROOF. Proof is eliminated for brevity. □

6. EXPERIMENTAL RESULTS

In this section, we demonstrate the performance of the proposed feature analysis techniques utilizing real data collected from three human subjects using wearable motion sensors. Motion sensors were used to measure acceleration and angular velocity of six different body segments (including upper and lower body limbs) while the subjects were instructed to perform 14 transitional movements listed in tblrefmvts. The collected data were partitioned into two disjoint data sets, one for solving the proposed optimization problem and the other for measuring classification accuracy of the system on the selected features. The energy consumption of the feature extraction module was estimated on TI MSP430 microcontroller.

Table 1: Experimental movements

No.	Movement
1	Stand to Sit
2	Sit to Stand
3	Sit to Lie
4	Lie to Sit
5	Bend to Grasp
6	Rising from Bending
7	Kneeling Right
8	Rising from Kneeling
9	Look Back
10	Return from look back
11	Turn Clockwise
12	Step Forward
13	Step Backward
14	Jumping

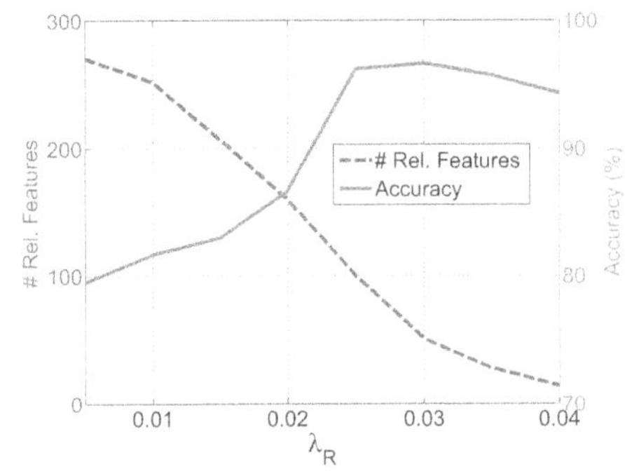

Figure 4: Number of relevant features and classification accuracy as a function of λ_R

6.1 Parameter Setting

One of the parameters used to solve our optimization problem is the energy consumed for processing each feature (w_i). Energy consumption variation for different features stems from different instruction types, circuit states, and memory address modes, as well as the overall complexity of each feature. We consider TI MSP430 processor in order to find each feature's processing energy. The MSP430 is widely used in wearable monitoring applications. These systems are obviously in need of low power consumption and MSP430 nicely meets such need (594 μW power consumption on average, which yields a performance of 37 μW/MIPS).

MSP430 is a 16-bit RISC CPU that uses the Von Neumann architecture. It has 48KB of Flash memory, 10KB of RAM, and uses an 8 MHz clock. The processor benefits from a 3-stage pipeline with 16 general purpose registers. Twenty-seven instructions comprise the instruction set with 7 memory addressing modes available.

As a multiplier is a peripheral and is not implemented in every member of the MSP430 family, we utilized a method based on the Horner's approach to implement multiplication only by means of shift and add instructions. It is worthy of attention that MSP430 can perform a register shift or add in one instruction cycle.

We calculated energy consumption of feature extraction block in Figure 1 for calculating each of the 9 features listed in Table 2 using the MSP430 microcontroller, which is available on the TelosB motes used in our experiments.

6.2 Relevant Features

Figure 4 shows the number of relevant feature and classification accuracy as a function of the relevance threshold (λ_R). Clearly, the number of relevant features decreases as the threshold in (3) increases. A very small value of λ_R results in all the original 270 features being evaluated as relevant. With the complete feature set, the classification accuracy is 79.5% at the beginning. As the threshold increases, the accuracy improves as a result of irrelevant features being eliminated from the classification process. The accuracy, however, starts decreasing at $\lambda_R = 0.03$ indicating that the threshold is exceeding its optimal value and some relevant features start being eliminated from the list. Thus, we consider $\lambda_R = 0.03$ as the optimal threshold value for our relevance analysis and perform the rest of our feature analysis with this value. Note that this threshold results in 51 relevant features.

An interesting observation is that irrelevant features can significantly reduce the performance of the action recognition system. We observe that by eliminating irrelevant features, the accuracy of the classification exhibits more than 21% improvement (i.e. from 79.5% with all 270 features to 96.7% with 51 features as obtained by $\lambda_R = 0.03$ in Figure 4).

Table 2: Energy consumption of feature extraction for individual features

Feature	Energy (nJ)	Description
Amp	16386	Signal amplitude (Max - Mean)
Med	405159	Median of signal segment
Mean	8126	Mean value of signal segment
Max	8103	Maximum amplitude of signal segment
P2P	16291	Peak to peak amplitude
Var	38846	Variance of signal segment
Std	40431	Standard deviation
RMS	29705	Root mean square power
S2E	83	Start to end value

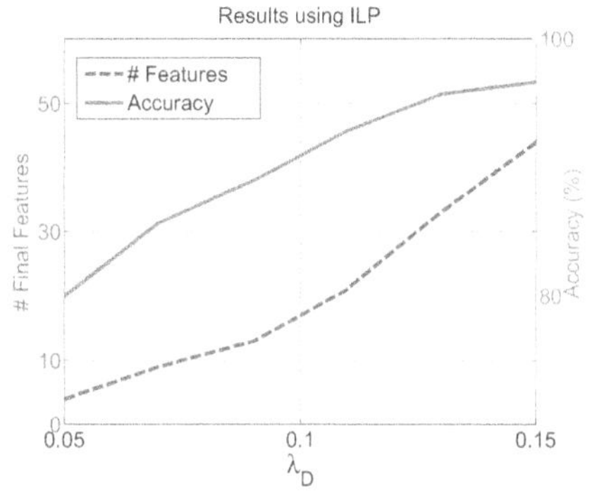

Figure 5: Optimal number of features and accuracy (using ILP solution)

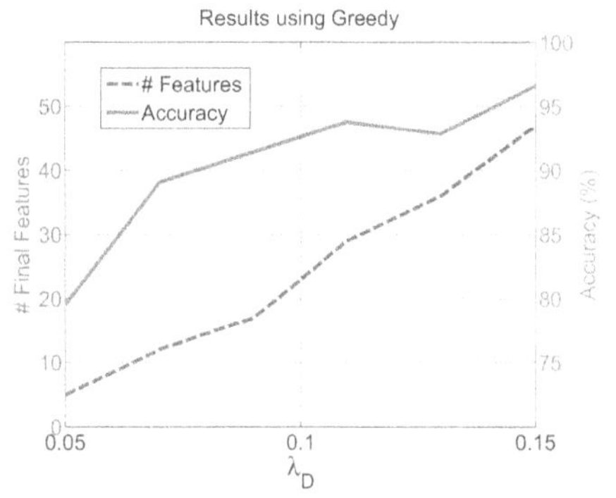

Figure 6: Number of features and accuracy for greedy algorithm

6.3 Optimal Feature Set

Figure 5 shows the number of optimal features and classification accuracy versus the redundancy threshold (λ_D) using the ILP approach. It should be noted that according to the redundancy criterion in (4), a larger value of λ_D results in less features being considered as strongly correlated and therefore, the redundancy graph will have smaller number

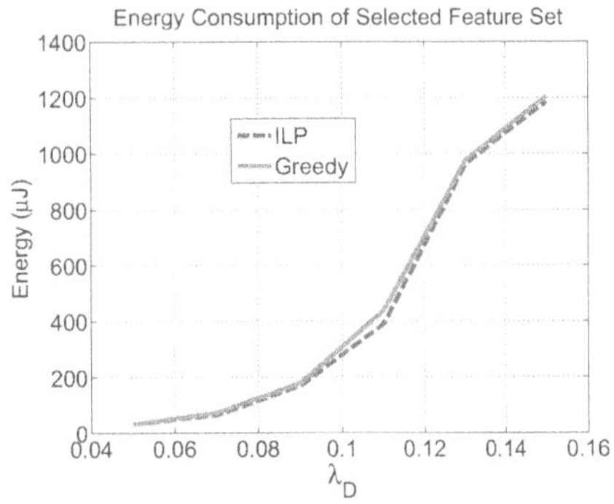

Figure 7: Energy consumption reported by optimal solution (ILP) as well as greedy approximation

of edges. With a small number of edges, more vertices need to be considered to cover all the vertices. This can result in a larger optimal set as Figure 5 shows. As the threshold increases from 0.05 to 0.15, the number of selected features grows from 4 to 44 with an average of 20.7 features. The classification accuracy ranges from 80% for λ=0.05 to 96.7% for λ=0.15. The classification accuracy is 90.1% on average.

The classification accuracy and number of final features reported by our greedy solution are illustrated in Figure 6. For the greedy approach, the number of selected features ranges from 5 to 47 depending on the design parameter λ_D. The optimal feature set has a length of 24.2 features on average. The accuracy ranges from 79.5% for the lowest threshold to 96.7% for λ=0.15, with an average accuracy of 90.0%. We note that, unlike the ILP solution, the greedy approach does not result in a monotonically increasing accuracy curve. The accuracy curve for the greedy approach has a local minimum at $\lambda_D = 0.13$. This is in fact due to the sub-optimality of the greedy approach, which does not necessarily select the optimal feature set at each step.

6.4 Energy Analysis

Figure 7 shows the total energy consumption of the selected features obtained by the ILP and greedy solutions. For optimal case, the energy values range from 33μJ for λ_D = 0.05 to 1184μJ for λ_D = 0.15, resulting in an energy consumption of 467μJ on average. The energy consumption results for the greedy approach range from 33μJ to 1204μJ with an average of 484μJ.

Table 3 lists energy savings obtained by applying our power-aware feature selection technique. For this specific table, only ILP results are reported. For greedy approach, however, similar results are achieved. The energy savings for the greedy solution range from 29.6% for the highest accuracy (96.7%) to 98.1% for the lowest performance (79.5%). This results in an average energy saving of 71.6% using the greedy solution.

Table 3: Energy savings using the proposed method

λ_D	Saving (%)	Accuracy (%)
0.05	98.1	80.8
0.07	96.2	85.6
0.09	90.3	89.1
0.11	77.2	92.9
0.13	43.6	95.7
0.15	30.7	96.7
Avg.	72.7	90.1

7. CONCLUSION AND FUTURE WORK

The accuracy and power trade-offs in wearable monitoring systems have been investigated, that is, guaranteed classification accuracy is required, while minimizing the system's power consumption. Our study in this paper accounts for energy consumption in the process of feature selection. To achieve that, we first formulated the problem as a weighted minimum set cover approximation, which is one of the oldest and most studied NP-hard problems. We then devised a greedy approach to select the features needed for the identification of activities being performed in a power-efficient manner. Experimental results on inertial data collected from real subjects demonstrated significant power savings. In the future, we will investigate dynamic feature selection and node activation based on the contextual information about the subject in real-time. That is, information such as subject's location and current activity can be used in real-time to further eliminate power-hungry features and deactivate sensor nodes not contributing to the classification.

8. REFERENCES

[1] P. Lukowicz, O. Amft, D. Roggen, and J. Cheng, "On-body sensing: From gesture-based input to activity-driven interaction," *Computer*, vol. 43, no. 10, pp. 92 –96, oct. 2010.

[2] U. Anliker and et. al., "Amon: a wearable multiparameter medical monitoring and alert system," *Information Technology in Biomedicine, IEEE Transactions on*, vol. 8, no. 4, pp. 415 –427, dec. 2004.

[3] M. Dash and H. Liu, "Feature selection for classification," *Intelligent data analysis*, vol. 1, no. 3, pp. 131–156, 1997.

[4] Q. Gu, Z. Li, and J. Han, "Generalized fisher score for feature selection," in *Proceedings of the Proceedings of the Twenty-Seventh Conference Annual Conference on Uncertainty in Artificial Intelligence (UAI-11)*. Corvallis, Oregon: AUAI Press, 2011, pp. 266–273.

[5] L. Yu and H. Liu, "Efficient feature selection via analysis of relevance and redundancy," *J. Mach. Learn. Res.*, vol. 5, pp. 1205–1224, December 2004.

[6] B. French, D. Siewiorek, A. Smailagic, and M. Deisher, "Selective sampling strategies to conserve power in context aware devices," in *Wearable Computers, 2007 11th IEEE International Symposium on*, oct. 2007, pp. 77 –80.

[7] H. Ghasemzadeh, V. Loseu, and R. Jafari, "Collaborative signal processing for action recognition in body sensor networks: a distributed classification algorithm using motion transcripts," in *Proceedings of the 9th ACM/IEEE International Conference on Information Processing in Sensor Networks*, ser. IPSN '10. New York, NY, USA: ACM, 2010, pp. 244–255.

[8] P. Zappi, C. Lombriser, T. Stiefmeier, E. Farella, D. Roggen, L. Benini, and G. Troster, "Activity recognition from on-body sensors: accuracy-power trade-off by dynamic sensor selection," *Lecture Notes in Computer Science*, vol. 4913, p. 17, 2008.

Advanced Power and Thermal Management for Low-Power, High-Performance Smartphones

Hwisung Jung
Mobile Platform Solutions
Broadcom Corporation
Irvine, CA, USA
hwisung.jung@broadcom.com

ABSTRACT
In the next ten years, data traffic from mobile devices is expected to grow at an exponential rate. Such growth, coupled with the ongoing advances in semiconductor technology and the demands of smartphone applications, makes managing the power and thermal conditions ever more challenging. In this paper, the various challenges of power and thermal management techniques/architecture for energy-efficient smartphones are reviewed, and design considerations for hardware and software power co-optimization are presented.

Categories and Subject Descriptors
C.1.3 [**Computer Systems Organization**]: Cellular Architecture

Keywords
Power and thermal management, Power optimization, Smartphone

1. INTRODUCTION
A close look at today's smartphone usage reveals that extending battery life has become a daunting, yet vital, task necessary for delivering a rich user experience. Furthermore, the introduction of GHz multicore microprocessors to handle the demand for high-performance applications has resulted in the emergence of hot spots and spatial temperature gradients that can cause accelerated circuit aging, timing error, or even physical damages to the chip [1]. Such conditions require the most innovative power and thermal management design techniques. In this paper, the challenges of power and thermal management techniques/architecture for mobile Systems-on-a-Chip (SoCs) are reviewed, and low-power design methodologies for hardware and software power co-optimization are discussed.

2. POWER MANAGEMENT
As multiple functionalities (e.g., 2G/3G/4G baseband, application processor, multimedia, GPS, Wi-Fi, Bluetooth, etc.) are integrated into a single die to provide cost-effective solutions for smartphones, the associated functional blocks are being built with various power-control mechanisms (e.g., dynamic voltage and frequency scaling (DVFS), adaptive voltage scaling, fine-grain power gating, clock gating, etc.) to reduce power consumption. Typically, an application processor that operates at GHz speed in a single, dual, or multicore configuration plays an important role in providing chip-level performance gains. This high-speed processing is driving a need for advanced power-management methods such as fine-grain pipeline shutdown, aggressive L2 power reduction, and fast-state save/restore capability [2], where power-management routines are controlled by software (e.g., Android/Linux power management framework) and/or hardware (e.g., microcontroller-based power manager).

ISLPED'12, July 30–August 1, 2012, Redondo Beach, California, USA.
ACM 978-1-4503-1249-3/12/07.

2.1 Variation-Aware Power Management
Conventional power-management methods have not been able to take full advantage of power-saving techniques due, in part, to the increasing levels of process, voltage, and temperature (PVT) variations (i.e., oxide thickness variation, channel length variation, dopant fluctuations, voltage drops, and temperature gradients). Uncertainties that arise from voltage and temperature variations give rise to additional implications to efforts aimed at optimizing energy consumption, especially for the application idle mode. Thus, PVT variations pose a major challenge to the design of low-power smartphone SoCs beyond the 40 nm technology node. When developing power-control strategies, it is prudent to account for various sources of variability even earlier in the design process. Improving the accuracy and robustness of decision making by modeling, and assessing the variability and uncertainty are important steps in guaranteeing the effectiveness of power-management techniques/architecture.

2.2 Learning-based Power Management
Although dynamic power management, a strategy to turn off the processing cores when they are idle and later wake them up, can be easily combined with DVFS, little attention has been paid to an improved decision-making strategy that minimizes the overhead of a power manager. Especially in the case of asymmetric application processors (e.g., big.Little architecture configuration [3]), the task of integrating multiple power-management policies is becoming one of the primary design challenges. One solution is to devise a learning-based power-management policy that can quickly analyze some of the easily available quantifiable features of the mobile device under consideration and accurately predict the system performance state, which can subsequently be used to choose and issue the optimal action, i.e., the DVFS value [4][5].

3. THERMAL MANAGEMENT
With IC process geometries shrinking below 40 nm and many smartphone applications requiring higher performance (e.g., gaming, 1080p video playback, etc.), thermal control is becoming the primary concern not only for IC designers but also for package, board, and system designers because of the increased power density as well as the increasing vulnerability of the system. If heat is not removed at a rate equal to or greater than its

rate of generation, the junction temperature will rise, which reduces the mean time to failure (MTTF) for the device. Typically, if package-on-package (PoP) is used for memory stacking configuration (e.g., LPDDR2/3) on top of a mobile SoC for example, the junction temperature of the mobile SoC is limited by the temperature of the stacked memory.

3.1 Thermal Control Techniques

Much of the past work has examined thermal control techniques that attempt to ensure thermal safety by employing run-time mechanisms to control power density and prohibit excessive local heating. Such mechanisms include proactive/reactive thermal management, stop-and-go, duty-cycle modulation, adaptive DVFS, and thermal chimney [6]. These techniques, however, may be ineffective in reducing the chip temperature of multicore processors inside smartphones if the uncertainty in measuring temperature with thermal sensors is not considered. Furthermore, these thermal-control techniques, which depend on the use of thermal sensors, can hardly observe the actual peak power dissipation and the resulting peak temperature due to the nonuniform power density across a chip. Thus, identifying and removing heat from these hot spots is a major task faced by design engineers concerned with effectively controlling system temperature.

3.2 Thermal Modeling Techniques

Thermal models based on equivalent circuit models cannot adequately model heat generation and diffusion in structures with complex shapes and boundary conditions. Indeed, it is extremely difficult to obtain the exact solution of the heat equations that arise from realistic die conditions [7]. Important aspects contributing to the advancement of thermal-managed smartphones is the ability to fully model and characterize the thermal behavior of the system. In particular, improving the accuracy of decision making in thermal management by modeling and assessing the uncertainty mentioned earlier is an important step to being able to guarantee the operational quality and longevity of smartphones.

4. HW/SW POWER CO-OPTIMIZATION

There is a strong motivation for hardware and software power co-optimization to meet architecture-level power objectives by exploiting synergism through their concurrent designs [8]. Power co-optimization problems with regard to smartphones vary depending on mobile application characteristics, low-power implementation technology, and design methodology. Figure 1 illustrates how power co-optimization works by identifying the dependency of power management from both a hardware and software perspective. This process is a first step toward optimizing the power consumption of smartphones. For example, the Android application framework and Linux kernel layers running on top of the mobile SoC have power-management functionalities, where user applications request processing resources through the Android application framework, followed by controlling DVFS through the Linux kernel. In addition, battery life is significantly affected by the use of the run-time PowerManager API inside the application framework. In this case, without capturing certain parameters inherited from the hardware (e.g., PVT variations) that are not transparent to software, it is hard to optimize total power/energy consumption of the devices, especially when PVT variations differentiate the dissipated total power consumption. Thus, it is important to assess the efficiency of power management from both hardware and software perspectives.

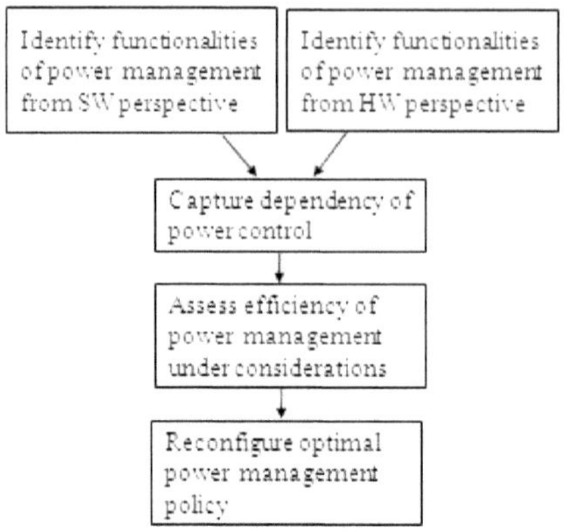

Figure 1. High-level flow of HW/SW power co-optimization.

5. SUMMARY

Various power and thermal management techniques along with related challenges are reviewed with regard to smartphone design. Being able to handle various sources of uncertainty would improve the accuracy and robustness of power and thermal management, thus ensuring low power consumption, thermal safety, and operational quality and reliability.

6. REFERENCES

[1] Jung, H., Rong, P., and Pedram, M. 2008. Stochastic Modeling of a Thermally-Managed Multi-Core System. In *Proceedings of the Design Automation Conference* (Anaheim, USA, June 08 - 13, 2008).

[2] Lanier, T. 2011. Exploring the Design of the Cortex-A15 Processor. http://www.arm.com/files/pdf/AT-Exploring_the_Design_of_the_Cortex-A15.pdf

[3] Greenhalgh P. 2011. Big.LITTLE Processing with ARM Cortex-A15 and Cortex-A7. White paper. http://www.arm.com/files/downloads/big_LITTLE_Final.pdf

[4] Jung, H. and Pedram, M. 2010. Supervised Learning based Power Management for Multicore Processors. *IEEE Trans. CAD of Integrated circuits and Systems*. 29, 9 (Sep. 2010).

[5] Rosing. T. and Dhiman, G. 2006. Dynamic Power Management Using Machine Learning. In *Proceedings of the Int'l Conference on Computer Aided Design* (San Jose, USA, Nov 5 - 9, 2006).

[6] Donald, J. and Martonosi, M. 2006. Techniques for Multicore Thermal Management: Classification and New Exploration. In *Proceedings of the Int'l Symposium on Computer Architecture* (Boston, USA, Jun 17 - 21, 2006).

[7] Janicki, M, and Napieralski, A. 2004. Inverse Heat Conduction Problems in Electronics with Special Considerations of Analytical Analysis Methods. In *Proceedings of the Int'l Semiconductor*. 2. 4. (Oct, 2004).

[8] Ayoub R., et al. 2011. OS-level Power Minimization under Tight Performance Constraints in General Purpose Systems. In *Proceedings of the Int'l Symposium on Low Power Electronics and Design* (Fukuoka, Japan, Aug 1 - 3, 2011).

Keynote

Ultra-low Power Challenges for the Next Generation ASIC

Speaker

Uming Ko
MediaTek USA, Inc.
San Jose, CA USA
uming.ko@mediatek.com

Abstract

While 40 years of semiconductor scaling and Moore's Law enabled system-on-a-chip (SoC) ASICs over the last decade, it is obvious that, fundamentally, both desktop and portable consumer electronics have hit the frequency and power (or thermal) wall. On the other hand, the insatiable demand of mobile computation and bandwidth, especially with the ubiquitous smartphones, has emerged and will continue to drive the technology. In this presentation, mobile computing performance explosion in host processors, multimedia co-processor engines and mobile bandwidth, as well as the opposing thermal and power limits will be addressed. As a result, even with the latest technology nodes, energy gap of a form-factor-sensitive smartphone between capability and demands continues to widen. The next wave of technology innovations from process, packaging, digital, and analog/RF circuits will be vital in solving ultra-low power challenges for the next-generation ASICs.

Categories and Subject Descriptors

C.1.0 [**Computer Systems Organization**]: Processor Architectures - *General*

Keywords

application-specific integrated circuits, energy gap, mobile computing processors, multimedia co-processors

Biography

Uming Ko (Fellow, IEEE) received BSEE from National Tsing Hua University, and MSEE and Ph.D. from University of Texas. He first worked for AMD and then joined Texas Instruments (TI) in 1986. He was elected TI Fellow in 2000 for innovating industry's most energy-efficient TMS320C55x DSP processor. In 2005, Dr. Ko was recognized as a TI Senior Fellow for his technical leadership in ultra low-power design, and innovations of the SmartReflex™ Power and Performance Management technology, which positioned TI as a leader in ultra-low-power wireless radios and OMAP products. As the director of Worldwide Chip Technology Center (CTC), he was responsible for the development of TI's 65nm, 45nm and 28nm technologies for Wireless products. Prior to his role at the CTC, he was responsible for the development of the first- and second-generation OMAP applications processors. Currently, Dr. Ko is the Chief Technical Advisor of MediaTek and is also responsible for the High Performance Technology Division leading MediaTek's 28nm and 20nm development. Dr. Ko has received 12 national-level and industrial awards, including 2006 Asian American Engineer of the Year Award. Uming holds 43 U.S. patents and has published 35 technical papers.

ISLPED'12, July 30– August 1, 2012, Redondo Beach, California, USA.
ACM 978-1-4503-1249-3/12/07.

The Core-C6 (CC6) Sleep State of the AMD Bobcat x86 Microprocessor

Aaron Rogers, David Kaplan, Eric Quinnell, Bill Kwan
AMD Austin
7171 Southwest Parkway
Austin, TX, USA 78735

ABSTRACT

This paper describes the architecture and physical design of the AMD Bobcat x86 microprocessor core-C6 (CC6) sleep state. Each core and dedicated 512KB L2 cache use control flows to save and restore the x86 architectural state in concert with a network of PFET sleep transistors that drive an internal voltage domain. Measured results show an approximate 92% reduction in leakage power at the cost of 1-3% die area with a 31μs restore latency. Each core and L2 cache pair use 98cm of total sleep transistor width in a TSMC 40nm bulk CMOS process.

Categories and Subject Descriptors

C.1.1 [Processor Architectures]: Single Data Stream Architectures – *CISC architectures*

General Terms

Algorithms, Measurement, Design

Keywords

AMD Bobcat, Sleep State, Sleep FET, C-State, leakage optimization, Power Gating

1. INTRODUCTION

The AMD Bobcat x86 microprocessor is a low-power, small area, fully-synthesized core with custom data arrays and a 512KB L2 cache, fabricated on the TSMC 40nm node [1][2][3][4]. The core design targets for small-form factor and low-power applications; therefore, a crucial feature of the core is an architectural and physical sleep-state to reduce sub-threshold leakage current when idle.

A well-known solution to the reduction of sub-threshold leakage is the use of "sleep" transistors, referenced here as SLPFETs. The solution uses an array of distributed transistors placed between logic and the power rail to create a virtual power supply. The virtual supply transistors are turned on and off via an always-on (AON) control network. Should a block be given a workload where no activity is required, the region may be put to "sleep" by turning off the SLPFET transistors, stopping nearly all sub-threshold leakage current in the virtual power domain.

In the Bobcat x86, two main design challenges are introduced when using integrated SLPFETs – state control of the architectural state and the physical design of a working virtual power plane. This paper describes the Bobcat microcontroller code and algorithms for preserving x86 architectural state during core-C6 state transitions, the physical design of 4 uniquely sized integrated PFET headers powering an internal virtual node, and the measured results seen when activating the CC6 feature in silicon.

ISLPED'12, July 30–August 1, 2012, Redondo Beach, California, USA.

2. BOBCAT C-STATE ARCHITECTURE

Three unique core power states (C-states) are supported by the Bobcat x86 processor, including: core-C0 (CC0), core-C1 (CC1), and core-C6 (CC6). CC0 is the normal operational state where instructions are being executed, while CC1/CC6 are sleep states. CC1 is a simple sleep state where the core is halted and the clock ramped down to a lower frequency. CC6 is a "deep sleep" state where internal core state is not retained. The CC6 state may be entered from the CC1 state.

Software may choose to enter a sleep core C-state by either executing the HLT (halt) instruction or by reading from specific I/O ports with the x86 IN instruction (handling C-states with I/O ports is a new feature in Bobcat called "I/O C-states"). The special I/O ports are specified in the model-specific-register (MSR) xC001_0073 (CStateAddr) [4] and include 8 locations – ranging from CStateAddr to CStateAddr+7. Each of these 8 locations may be independently programmed by software to cause the processor to take various actions, such as entering CC1 or CC6. For this purpose, the execution of the HLT instruction is equivalent to an I/O read of the special port CStateAddr and will follow the actions specified for port offset 0. Note that the hardware C-states described here are different than the ACPI defined C-states, which may be mapped to the hardware C-states at software's discretion.

I/O C-state handling is distinctly different compared to the HLT instruction. A core which has entered an I/O C-state will always be woken up from an external interrupt (INTR), but will conditionally transfer to the interrupt handler based on the EFLAGS.IF mask. The EFLAGS.IF flag controls whether the INTR will be taken, conditionally allowing software to perform clean-up tasks before executing the interrupt handler. For C-states entered via the HLT instruction, the EFLAGS.IF bit is used to conditionally mask external interrupts from waking up the core. If EFLAGS.IF is clear, no INTR will wake up the core, and if it is set, the interrupt handler will run immediately after the core resumes execution.

2.1.1 Sleep C-State Entry Decision

When software requests a sleep C-state transition, the hardware will always enter CC1, and may optionally enter CC6. Upon receiving the transition request, the northbridge (NB) will inspect the action field corresponding to the request. In Bobcat, this action field has a CC6 enable bit which specifies if the processor should attempt to enter the CC6 state. If this bit is clear, the core will stay in the CC1 state. If the bit is set, the NB checks various monitors to see if the CC6 transition should be allowed.

While the CC6 state saves more power than CC1, it also suffers from longer entry/exit latencies. The NB monitors check for conditions that indicate whether a CC6 transition would be harmful for performance or overall power consumption. This includes monitoring events like timer ticks, interrupt rates, and residency time. As an example, a high CC0 residency would indicate the system is in active use and entering CC6 might be harmful for performance.

If the hardware monitors indicate a CC6 transition is acceptable, the NB informs the core microcode to start saving off internal x86 state to DRAM. During this process, core state is written to a private DRAM area that is reserved specifically for CC6 data. All non-microcode access to this area is blocked for security. When the save routine completes, the physical power gating is activated.

2.1.2 Core C-State Entry

Although the NB ultimately decides which sleep C-state to transition to, the Bobcat core is responsible for implementing the software C-state request interface, as well as entering/exiting the state selected. When the core receives the software request for a sleep state, the microcode sets up the interrupt mask used for wakeup and then informs the NB of the C-state transition. As already described, the interrupt mask differs by the form of the C-state transition request – I/O C-states always unmask external interrupts (INTR), while the HLT instruction conditionally masks them.

After informing the NB of the C-state transition, the core enters the CC1 state by halting execution. Should the NB decide to keep the core in CC1, no further actions are taken until a wake-up event occurs. Should the NB decide to have the core enter CC6, it will send a special interrupt called "Stop-Clock" (StpClk) with information indicating a CC6 transition.

On the core side, when an interrupt arrives while in CC1, the microcode's first action is to check for StpClk. If StpClk is present, the core starts a CC6 transition. Otherwise, the core returns to x86 code, either returning to a HLT or finishing the I/O instruction (while restoring the interrupt flag). The core sleep C-state procedure is shown in Figure 1.

Upon receiving a CC6 transition request, microcode flushes both L1 and L2 caches in parallel (to reduce latency) and then saves core state to DRAM. The L2 cache on Bobcat has 16 ways, so each microcode cache flush loop flushes 1 full L2 way and 1/16th of the L1 cache. Each loop also checks for pending wake-up interrupts. Frequent interrupt checks are critical during the save routine to ensure the core can respond quickly.

When L1 and L2 cache flushing is complete, core microcode begins saving all internal architectural and micro-architectural state to the private DRAM space. This core state includes standard x86 registers, floating point registers, MSR values, and all other information necessary for resuming execution. As with the cache flushing, checks are continually made for pending wake-up interrupts. Should an interrupt happen, the state save process is aborted and the core begins processing the interrupt.

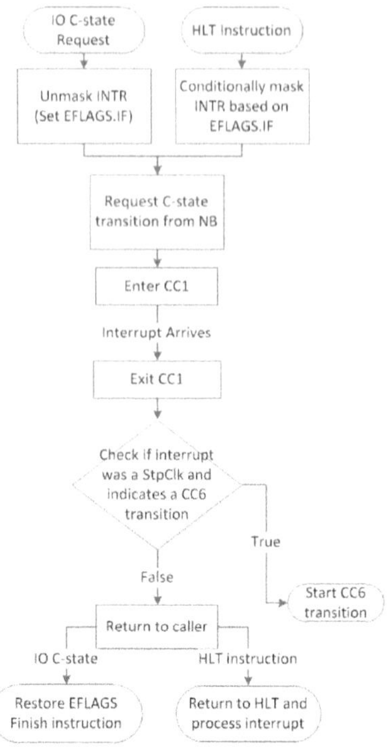

Figure 1. Core sleep C-state flowchart

Once all state is saved, core microcode informs the NB that the CC6 transition is complete. At this point the core is ready for power gating. Since the power gating request takes some time, a wake-up interrupt might still arrive during the gating request. Should the interrupt arrive before the core is physically power gated, it may be processed immediately like a normal interrupt. Once power gating has been applied, a wake-up interrupt causes a reset and full CC6 state restore. The flowchart of the CC6 entry routine is shown in Figure 2.

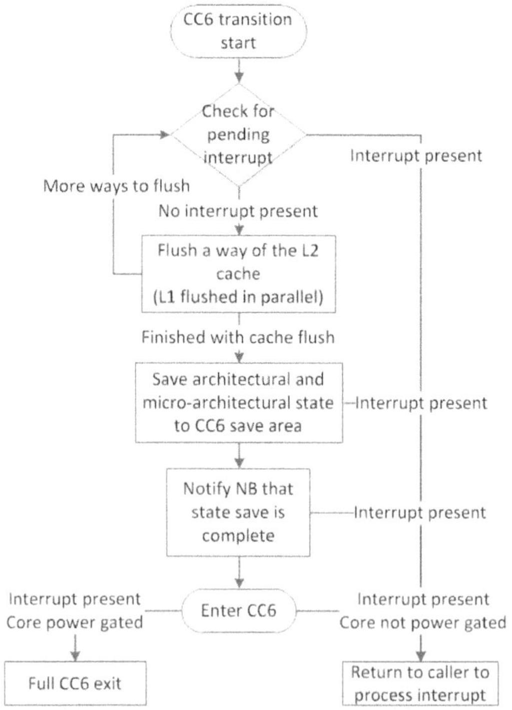

Figure 2. CC6 entry

2.1.3 CC6 Exit

Once power gating has been applied to the core and L2 pair, any wake-up interrupts require a reset. This reset behaves similar to a standard cold reset, with the exception that most built-in self-tests (BIST) are skipped to improve exit latency. When the core has been called to exit CC6, the microcode routine begins executing the standard reset routine while also checking to see if this reset is due to a CC6 exit. When this is true, microcode begins the process of restoring all the core state saved to DRAM during CC6 entry.

Chronologically, the first part of the core state that is restored includes the cache soft-repair information and data cache machine-check architecture (MCA) registers. Soft-repair information must be restored first to enable correct functionality of data arrays since BIST was not run during CC6 exit reset. MCA information also must be restored early, as MCA errors during the restore process must also be observed and combined with previous errors.

Once the soft-repair and MCA information is restored, the data cache may be used by the restore routine to speed up the restore process. Next, the routine restores the saved architectural and micro-architectural state, followed by invalidating the data cache to remove traces of the internal restore process. Finally, the NB is informed that the CC6 restore is complete.

After the CC6 restore, microcode deals with any MCA errors that may have been present. The first check is for MCA errors (such as a data cache parity error) that occurred during the CC6 restore routine itself. If such an error is present, the correctness of the restored state cannot be guaranteed, and the core state is considered corrupt. The control subsequently generates a "syncflood" event which will halt the core and shut down all external links to prevent data corruption from leaking to the system.

After this, microcode checks for less urgent MCA errors that occurred during the CC6 entry routine. For example, an MCA error could be a cache error that was found during the flush. Although such errors are critical, they do not imply corrupted internal core state, so no syncflood is generated. Instead, the MCA error is handled like a normal MCA error, running the appropriate x86 #MC handler.

If no errors are present after the restore, microcode jumps to the instruction pointer saved during CC6 entry and resumes execution. Additionally, any new present interrupts may be processed normally, running the software handlers as appropriate. A flowchart of the CC6 exit routine is shown in Figure 3.

2.1.4 SLPFET control

While the core handles the details of the C-state entry and exit for the core, the physical power gating itself is controlled by a state machine named the Power Gating (PG) Controller. Physically, the core SLPFETs are controlled from the PG by a 4-bit CC6 Sleep Enable bus (CC6SleepEn[3:0]). This bus is propagated throughout the core and L2 to directly control the effective width of sleep header PFETs.

Since the controlling SLPFETS are PFET headers, the control bus uses negative logic to turn the transistors ON and OFF. For example, when CC6SleepEn[3:0] has a value of 4'b1111, the virtual VDD node is in OFF; when CC6SleepEn[3:0] has a value of 4'b0000, the virtual VDD node is in ON state. Alternatively this may be viewed as positive logic that represents when the C-state is literally enabled.

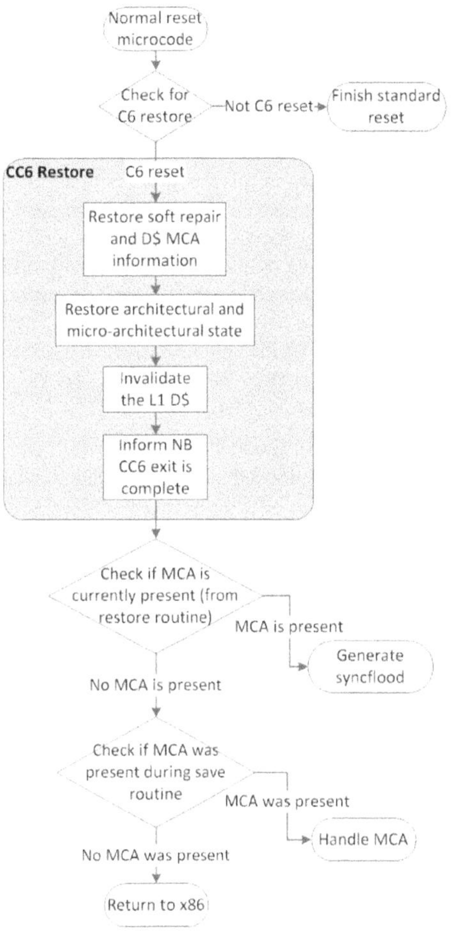

Figure 3. CC6 Exit

In addition to the 4-bit control bus, the PG also uses an electrical isolation signal (CC6Isolate) to latch and isolate the core outputs during physical sleep. When this signal is asserted, all the core outputs are forced to a known electrically safe state as to not corrupt downstream consumers of the signals. The drivers of the core output isolation cells are on a dedicated always-on voltage domain and not the internal virtual voltage domain.

When entering CC6, once the core microcode sequence finishes its routines and notifies the NB, the NB pairs up with the PG Controller to go through the sequence of events required to power gate the core. First, the core clock is ramped down and finally shut off. Next, the PG asserts the CC6Isolate signal to force core outputs to a known state. Finally, the PG sets the CC6SleepEn[3:0] to 4'b1111 turn OFF all integrated SLPFETs.

When exiting CC6, the NB and PG Controller coordinate to wake up the core through a sequence of different changes in the values of CC6SleepEn[3:0]. Each state of the PG exit has a residency requirement to guarantee a latency (programmable by fuses) between each transition of CC6SleepEn[3:0]. The programmable latency is to safeguard against any drastic voltage swing incurred from a di/dt event.

By default, the PG controller transitions CC6SleepEn[3:0] from OFF (4'b1111) to an initial wake state (4'b1110). The initial wake state turns on the smallest of the SLPFETs to slowly leak current into the virtual node for slow voltage ramp-up. After a programmable delay, the PG controller enters a second wake state (4'b0000) which activates all SLPFETs. After all SLPFETs are active, the CC6 hardware reset sequence is triggered, the core clock resumed, and the CC6Isolate signal de-asserted. Figure 4 shows the state diagram PG controller.

2.1.5 CC6 Latency

Minimizing CC6 entry/exit latency is critical to reducing the overall power consumption of the chip, while increasing the amount of time the core can be power gated. On CC6 entry, the latency is dominated by the L1/L2 cache flush loop. This flush time is variable depending on the amount of dirty (modified) lines in the cache, but in typical code with a 50% dirty cache, the flush takes ~75µs at 800MHz (minimum clock speed). The core state save time is constant, taking approximately 8.6µs at the same clock speed. Including control flow overhead and PG controller, the overall CC6 entry time is ~87µs.

CC6 exit is significantly faster than entry, taking approximately 31µs from the wake-up interrupt to the resumption of x86 code. This includes 10µs used for the hardware wake, reset and fuse propagation, with the microcode state restore taking the remaining 21µs.

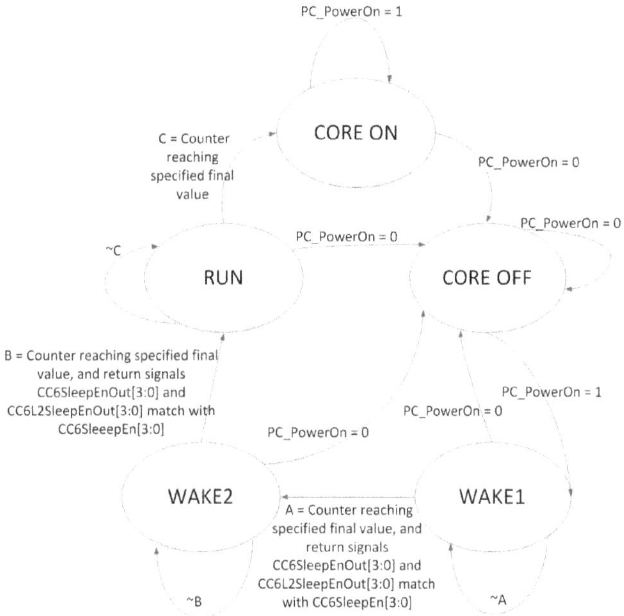

Figure 4. State machine for PG SLPFET control

3. BOBCAT SLPFET DESIGN

Bobcat is fabricated in a TSMC 40nm bulk CMOS process. The Bobcat SLPFETs use long-channel high-threshold voltage (HVT) PFET devices to drive a virtual VDD power plane while minimizing leakage current during an OFF state.

The Bobcat SLPFETs use 4 uniquely sized PFETs. The smallest of the 4 PFETs is designed primarily to slowly ramp-up the power supply while avoiding di/dt current in-rushes during the first wake sequence. Meanwhile, the sum width and distribution frequency of all 4 PFETs are designed to minimize the resistivity of the connection from the package VDD power plane to the internal VDD node during full RUN state.

The 4 uniquely sized PFETs serve an additional purpose outside of normal functional mode – programmable variable SLPFET widths during test and characterization. As described in Section II, the core uses the 4-bit bus CC6SleepEn[3:0] to control the 4 unique PFET headers. The 16 unique values of the bus allow for a programmable SLPFET effective width ranging from 1/15 * W to 15/15 * W with a 1/15 * W granularity, as well as an OFF state. A CC6SleepEn[3:0] value of 4'b1111 indicates virtual VDD state OFF, while value 4'b0000 indicates virtual VDD state full ON with SLPFET width 15/15 * W.

The 4 unique PFETs grouped together use a 6.44 µm x 2.52 µm area. Since the SLPFET body taps to VDD proper and not virtual VDD, additional space around each SLPFET group is needed for NWELL voltage isolation, pushing the group size to 8.96µm x 2.52µm. Figure 5 shows a concept schematic of the 4 Bobcat SLPFET headers. Figure 6 shows an electron microscope picture with visible implemented SLPFET columns. Table 1 shows the total sleep transistor width for a Bobcat core and L2 cache.

A common always-on (AON) network controls each core+L2 cache pair which is driven by the PG controller. AON logic is clustered in regional groups that share a local VDD power grid that bypasses the virtual VDD power grid and instead via-tunnels up to high-level VDD rails

Table 1. Bobcat SLPFET total transistor width

Block	Total SLPFET width
Bobcat Core	64.6 cm
512KB L2 Cache	33.5 cm
Core + L2 Total	98.2 cm

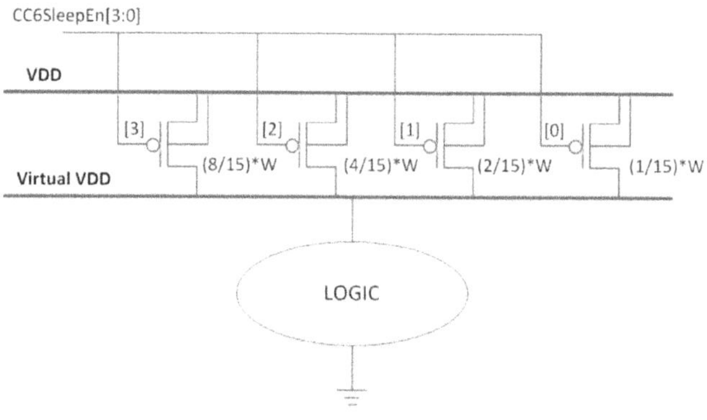

Figure 5. Bobcat SLPFET schematic

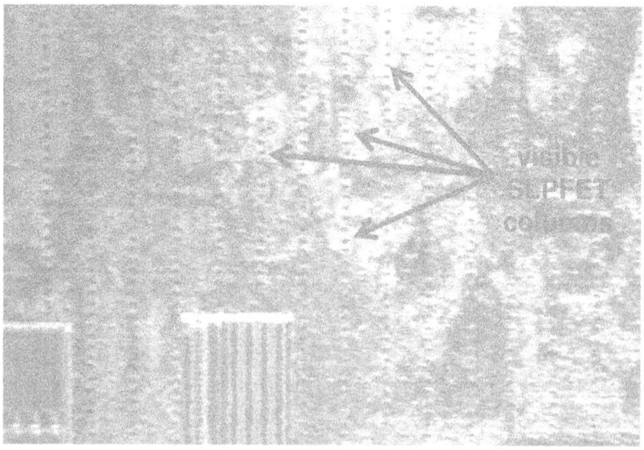

Figure 6. Bobcat SAPR region with distributed SLPFETs

The AON CC6SleepEn[3:0] input from the PG Controller drives a serial daisy-chain network of each SLPFET group that snakes up and down the core. Each nearby data array/ROM takes a branched input from the daisy chain control bus. At the end of the daisy-chain for the core, the control bus enters the adjacent L2 cache and again daisy-chains through the cache. When the control signal has been broadcast fully to the core and L2 cache, the bus returns to the PG as an output and is consumed by a logical control for the PG state machine.

The daisy-chain stitching of the CC6SleepEn[3:0] provides two advantages – first, during a CC6 restore, the sleeping power supply may turn on the smallest SLPFET (i.e. CC6SleepEn[0], or 1/15*W effective) in a time-delayed manner as described by Section 2, minimizing the number of transistors di/dt effects that are charging up the shared virtual VDD power rails; second, the locality of daisy-chained SLPFETs that turn on prevent the power rails from charging up on different parts of the silicon, preventing the threat of cross-over currents.

4. MEASURED RESULTS

The Bobcat x86 core was characterized using a variety of methods and tools over several silicon and metal stack revisions. Reported SLPFET and power characterizations use silicon-level (i.e. not system-level) measurements.

Measured full ON vs full OFF common VDD leakage power is shown in Table 2. The leakage power of the voltage plane during physical CC6 sleep reduces by roughly 92% -- in most cases by a factor of 10. The remaining leakage power is dominated by the NB core-interface modules (CIF) and clock controls that do not have integrated SLPFETs.

For SLPFET characterization, several batches of Bobcat silicon lots were measured across a series of voltages, temperatures, frequencies, and workloads while sweeping the effective total post-restore SLFPET width from 15/15 * W to 1/15 * W. The statistical averages and standard-deviations were gathered for dynamic current (DIDD), static current (SIDD), Vmin, Fmax, and dynamic core power during heavy workloads for selected workloads. Figs. 7 -11 show the collected parameterized results. Collectively, the data shows that no significant electrical degradation occurs during full functional mode until the (1/15 to 2/15) * W effective SLPFET width. CC6SleepEn[3:0] value 4'h15 is not shown, as that represents OFF.

The full AMD Ontario System-on-a-Chip (SoC) layout block diagram and die photo are shown in Figures 12 – 13. The Bobcat x86 cores with 512KB L2 Cache are on the far left of the SoC in a dual-core configuration. Each core+L2 cache pair is on a separate virtual VDD node and may enter/restore CC6 state independently. Figure 14 shows the Meridian photon-emission of an Ontario package with both Bobcat/L2 pairs and the GPU awake, either Core0 or Core 1 in CC6 state, and both Bobcat cores and the GPU in full sleep.

Table 2. VDD voltage plane measured leakage power

VDD corner	Leakage Power, ON	Leakage Power, OFF	CC6 Leakage Reduction
1.3V, 0C	429 mW	29 mW	93%
1.3V, 90C	1441 mW	96 mW	93%
0.9V, 0C	79 mW	6 mW	92%
0.9V, 90C	363 mW	29 mW	92%

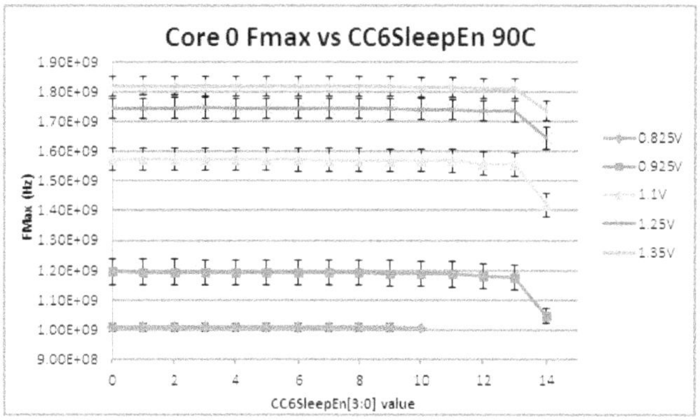

Figure 7. Single core Fmax sweep vs SLPFET width

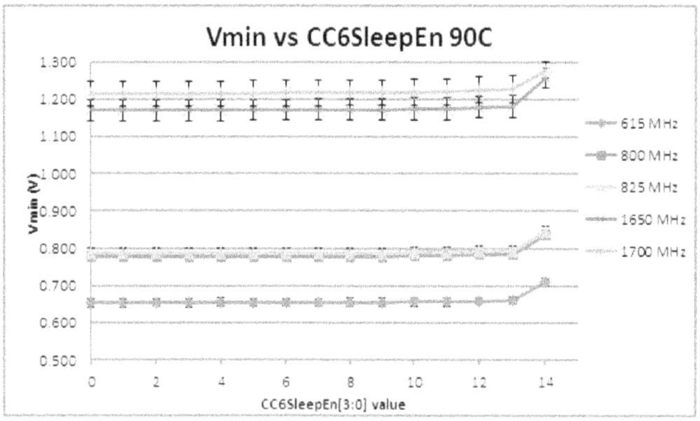

Figure 8. Dual core Vmin sweep vs SLPFET width

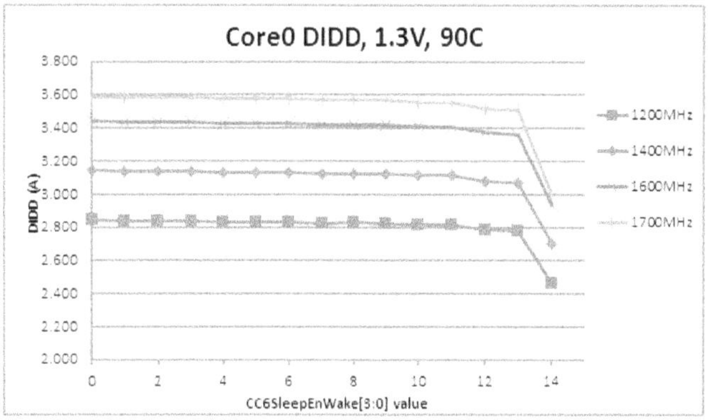

Figure 9. Single core DIDD sweep vs SLPFET width

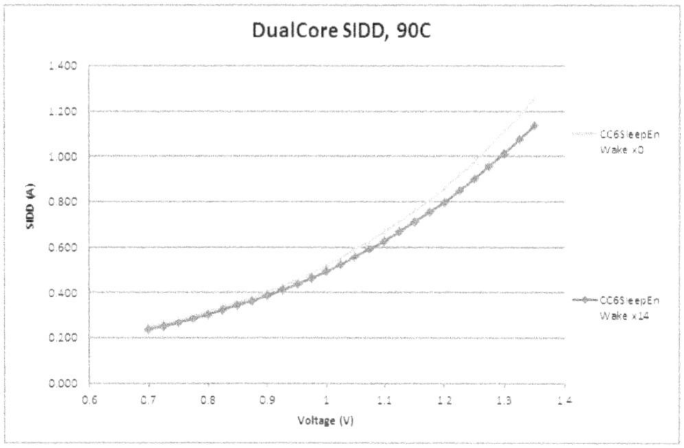

Figure 10. Dual core SIDD sweep vs SLPFET width

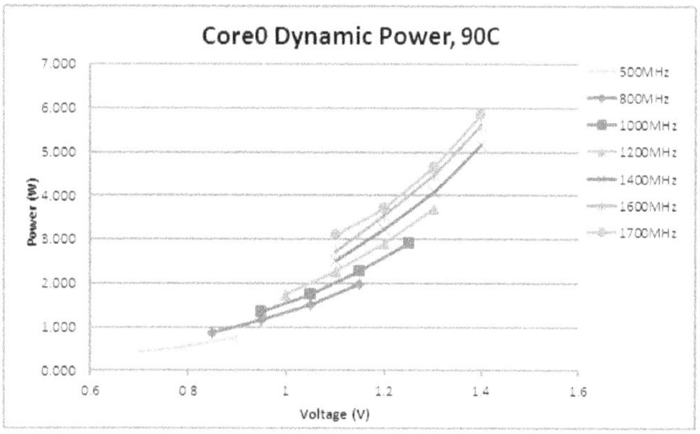

Figure 11. Single core dynamic power vs voltage

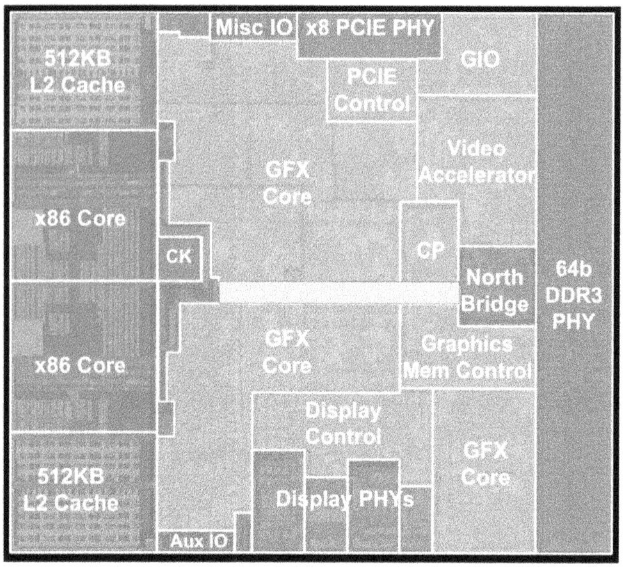

Figure 12. Block diagram of Ontario SoC (Bobcat cores + L2 cache seen on left) [2]

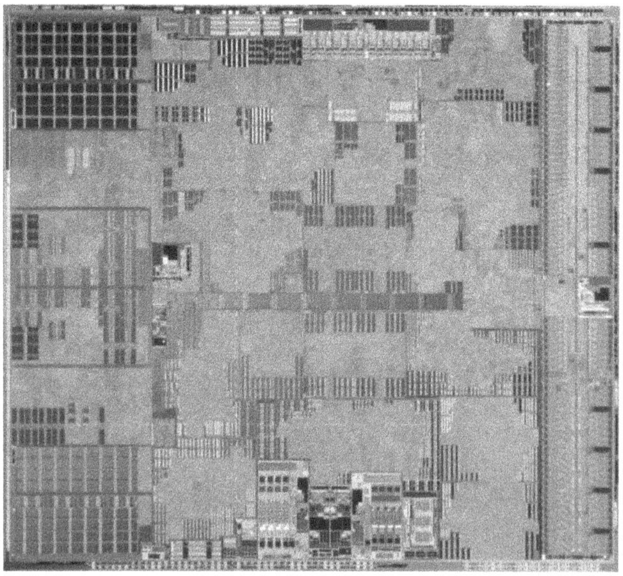

Figure 13. De-processed die photo of Ontario SoC

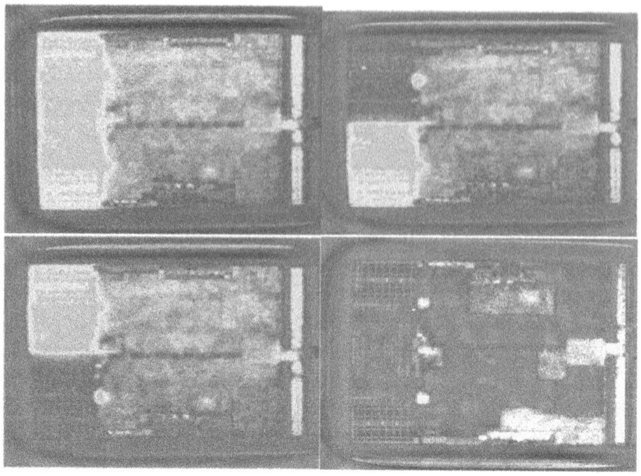

Figure 14. Photo emission of all core+GPU ON (upper left), single core power gated (upper right and low left), and all core+GPU OFF (low right) [2]

5. CONCLUSION

The Bobcat x86 microprocessor is designed in TSMC 40nm BULK CMOS process and uses a new core C6 (CC6) deep-sleep power state to reduce leakage during periods of inactivity. Each Bobcat core CC6 state uses a control sequence that saves or restores the x86 architectural state while handling interrupts, flushing caches, and detecting MCA errors. The core sleep transistors use 4 uniquely sized integrated PFET headers that are driven by a 4-bit CC6SleepEn PG bus. Measured results show a 92% reduction of VDD leakage at the cost of 1-3% die area and minimal disturbance to power supply and electrical characterizations of the system with a wake and state restore time of 31μs.

6. ACKNOWLEDGEMENTS

The authors would like to acknowledge the major contributions from Dan Bailey, Alex Branover, Andrew Balash, Joseph Piasecki and the entire CAT design team for their superb work.

7. REFERENCES

[1] Burgess, B.; Cohen, B.; Denman, M.; Dundas, J.; Kaplan, D.; Rupley, J.; "Bobcat: AMD's Low-Power x86 Processor," IEEE Micro, 2011, pp. 16-25.

[2] Foley, D.; Bansal, P.; Cherepacha, D.; Wasmuth, R.; Gunasekar, A.; Gutta, S.; Naini, A.; "A Low-Power Integrated x86–64 and Graphics Processor for Mobile Computing Devices," IEEE Journal of Solid-State Circuits, Vol. 47, Issue 1, 2012, pp. 220-231

[3] Gutta, S.R.; Foley, D.; Naini, A.; Wasmuth, R.; Cherepacha, D.; "A low-power integrated x86–64 and graphics processor for mobile computing devices," 2011 IEEE International Solid-State Circuits Conference (ISSCC), pp. 270-272

[4] AMD, BIOS and Kernel Developer Guide (BKDG) for AMD Family 14h Models 00h-0Fh Processors, 2011, http://www.amd.com

Evaluation of Voltage Stacking for Near-Threshold Multicore Computing

Sae Kyu Lee David Brooks Gu-Yeon Wei
Harvard University
33 Oxford St., Cambridge, MA
{saekyu, dbrooks, guyeon}@eecs.harvard.edu

ABSTRACT

This paper evaluates voltage stacking in the context of near-threshold multicore computing. Key attributes of voltage stacking are investigated using results from a test-chip prototype built in 150nm FDSOI CMOS. By "stacking" logic blocks on top of each other, voltage stacking reduces the chip current draw and simplifies off-chip power delivery but within-die voltage noise due to inter-layer current mismatch is an issue. Results show that unlike conventional power delivery schemes, supply rail impedance in voltage stacked systems depend on aggregate power consumption, leading to better noise immunity for high power (low impedance) operation for many-core processors.

Categories and Subject Descriptors

C.5.4 [**Computer System Implementation**]: VLSI Systems

General Terms

Design

Keywords

Near-threshold computing, power delivery, power management, voltage stacking

1. INTRODUCTION

Power consumption and delivery have emerged as one of the major challenges facing modern computing devices from mobile to server applications. In order to keep power consumption at manageable levels while continuing to offer performance enhancements, design trends have shifted towards multicore computing on a single chip. However, as advances in modern technology pack more and more transistors on a single chip, fixed power budgets due to thermal constraints effectively limit the amount of simultaneously active cores [4]. Near-threshold computing has been proposed as a way to improve energy efficiency for applications that have plenty of parallelism, the key idea being that lost performance due to reduction in voltage can be reclaimed through increase in aggregate throughput [3]. However, near-threshold operation leads to several challenges that must be overcome. Near-threshold computing suffers from higher susceptibility to V_{TH} and VDD fluctuations due to transistor current's exponential dependence on voltage. Also for a fixed power budget, decrease in supply voltage rapidly increases the amount of current driven through off-chip wire connections. Increase in chip current demand result in higher IR drop and I^2R power loss due to board and package resistance, exacerbated by the fact that off-chip power delivery impedance does not neccessarily scale. Increase in IR drop and susceptibility to voltage noise directly conflict with the more stringent voltage margins required in near-threshold operation to maintain performance, which aggravate the problem further.

Dynamic voltage and frequency scaling (DVFS) is another technique that can reduce power consumption and improve energy efficiency. However, as more cores are integrated on chip, traditional DVFS approaches are becoming more difficult to utilize due to the lack of temporal and spatial granularity of existing regulators. Off-chip DC-DC converters suffer from slow voltage transistion times, while the high board-level footprint and system cost limit the number of voltage domains that can be implemented. On-chip voltage regulators enable fast voltage transition times and facilitate multiple voltage domains [5], but such regulators suffer from low conversion efficiency, especially for high step-down ratios.

Voltage stacking is an alternative approach to power delivery that powers multiple low-voltage blocks off of a single higher voltage by "stacking" the logic blocks and recycling charge between the layers [8]. By charge recycling, voltage stacking significantly reduces the chip current demands, thereby alleviating the strain on the power delivery network. However, voltage stacking suffers from internal voltage noise issues unique to the voltage stacked power delivery scheme, caused by current mismatches between layers. To evaluate voltage stacking in the context of multicore computing, the underlying key attributes of voltage stacked power delivery system must be fully understood. This paper presents experimental results from a test-chip prototype built in 150nm fully-depleted silicon-on-insulator (FDSOI) process to evaluate voltage stacking for multicore computing. Experimental

ISLPED'12, July 30–August 1, 2012, Redondo Beach, CA, USA

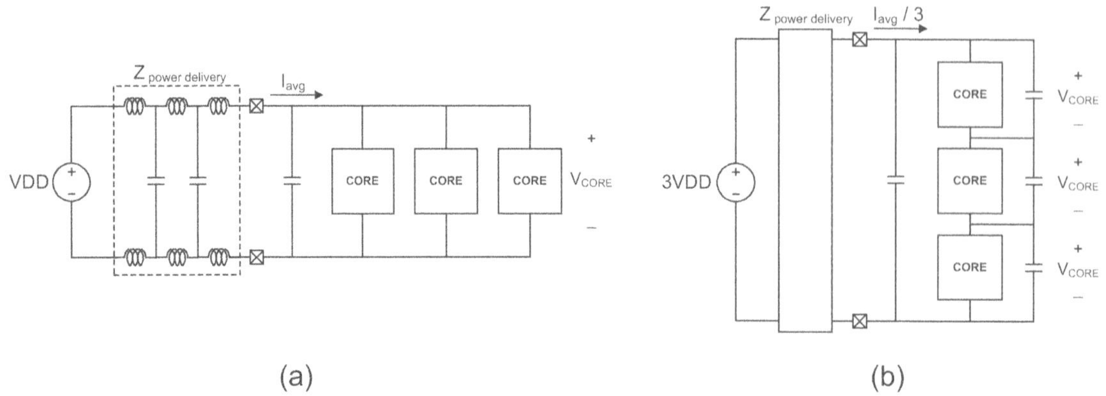

Figure 1: Block diagrams of (a) Conventional Power Delivery Scheme, (b) Voltage Stacking

results show that while susceptibility to internal voltage is a concern, voltage stacking has promising attributes for application in future many-core processors.

2. VOLTAGE STACKING

Voltage stacking is an energy-efficient approach for delivering power that provides high power-supply voltage to the chip and divides up the voltage consumed by the logic blocks via stacking with charge recycling between the layers. Figure 1 presents a conceptual block diagram of voltage stacking for a three core example. Instead of running the cores in parallel off of VDD, the cores are stacked on top of one another and a 3x higher voltage is applied across the stack. If each core consumes same amount of power, the stack voltage (3VDD) will distribute evenly between the cores, with each V_{CORE} settling at VDD and the average current flowing through the power-delivery system is reduced by a factor of 3. For a voltage-stacked design of n-stacks, the average current demand of the chip would decrease by a factor of n.

Voltage stackings addresses some of the key power delivery issues facing modern processors by delivering a higher voltage to the chip and thereby reducing the chip current draw.

- I^2R power loss due to board and package resistance reduces by a factor of n^2 compared to the conventional power delivery scheme of Figure 1(a), significantly reducing the power wasted through the power delivery subsystem.

- IR drop across the power delivery subsystem also reduces by a factor of n. However, in a conventional power delivery scheme, IR drop across the power delivery subsystem affects all cores equally, while in a voltage stacked design, the effect of the IR drop across the voltage stack is distributed across the n-stack layers. Therefore, IR drop across a core operating voltage (V_{CORE}) is ultimately reduced by a factor of n^2.

- Voltage regulators benefit from lower step-down ratio requirements of voltage stacking, leading to higher regulator efficiency and reduction in design complexity. This results in overall system-level power savings and potential reduction in board-level components and real estate.

While voltage stacking has the potential to alleviate the inefficiencies related to power delivery, voltage stacking introduces additional concerns that must be addressed. KCL dictates that identical current flows through each of the stack layers. However, current consumption of cores can fluctuate as a function of workload, various power-saving schemes (e.g., clock gating) and data, leading to inter-layer current imbalances. There is an implicit self-regulating loop built into the stack that distributes the internal voltage levels to compensate for any current imbalances between the stack layers. For example, if a stack layer consumes less current compared to other layers, its voltage increases to compensate at the expense of reducing the voltage across the other stack layers. Hence, any aggregate inter-layer core activity mismatch manifests itself as internal voltage noise. Therefore, balancing current consumption across the layers is a key challenge facing voltage stacking. Given the understanding of the potential problems of voltage stacking, a test-chip protoytpe, presented in the next section, was built to evaluate the key attributes and challenges of voltage stacked multicore systems.

3. TEST CHIP OVERVIEW

Block diagram of the test-chip prototype comprising a 3x3 array of power-consuming *cores* is presented in Figure 2. The test-chip is organized as three voltage-stack layers with three cores per layer and a total of four voltage rails for the array. Only VDDH and VSSC connect to an external voltage source ($V_{DD,EXT}$) and the internal rails VDDM and VDDL settle to levels dictated by the aggregate current consumption of the *cores*. *Peripheral support* circuitry, containing level shifters and voltage monitoring circuits, feeds signals into and reads signals out of the array, and operates off of separate rails VDDIO and VSSIO connected to a separate external source. The test-chip operates at VDDH=1.5V with each *core* ideally operating at 0.5V.

Each *core* contains two sets of power-consuming blocks shown in Figure 3. The *logic load* block contains three sets of four independent logic paths, all roughly 15 fanout-of-4 inverter (FO4) delays long. The logic paths operate at one

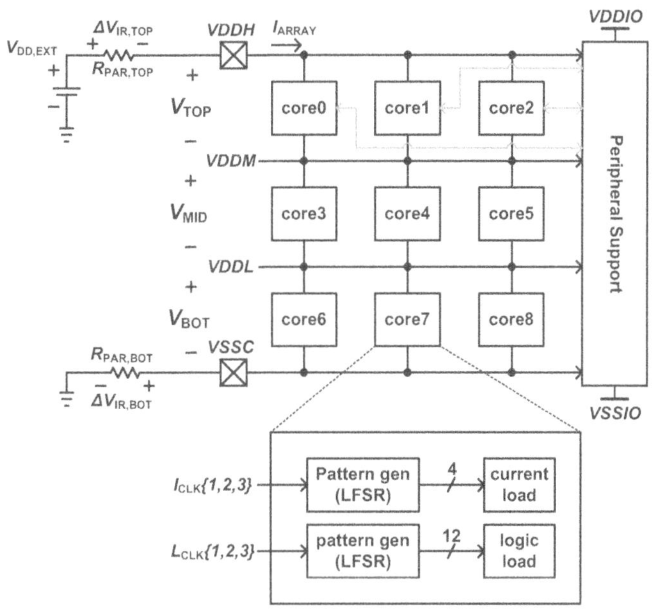

Figure 2: Block diagram of the test-chip prototype with a 3x3 voltage-stacked array of power-consuming blocks

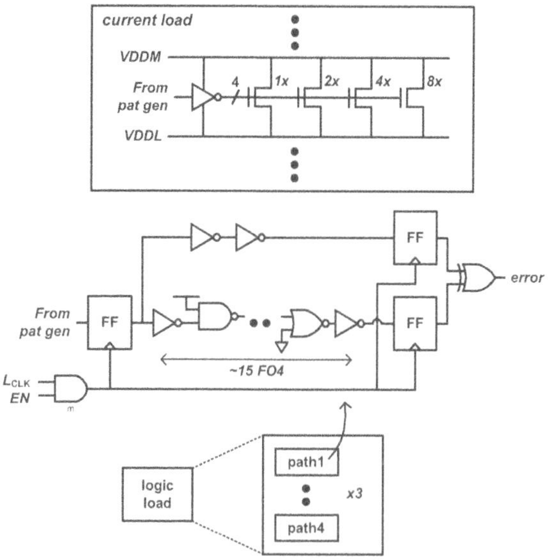

Figure 3: Detailed diagrams of power-consuming *current load* and *logic load* blocks within a core

of three clock frequencies ($I_{CLK}\{0,1,2\}$), with input data fed by pattern generators comprising a bank of linear feedback shift registers (LFSRs), programmable via scan chain. This enables configurations (i) to measure logic delay sensitivity to voltage and (ii) to inject fixed or pseudo-random patterns of switching noise into the array at different layers. The *current load* block contains a bank of nMOS switches operating as programmable 4-bit binary-weighted current sources, resulting in a total of 45 *current load* settings per stack layer. This block models average current consumption of processor logic and the switches (operating in saturation) approximate the impedance across the rails of digital CMOS circuits. Separate pattern generators allow the *cores* to mimic larger magnitudes of current fluctuations and coarser-grained switching activity (e.g, clock gating) at one of three clock frequencies ($L_{CLK}\{0,1,2\}$).

Voltage monitor circuit, shown in Figure 4, is used to capture the internal rail voltages in real time. The voltage monitor circuit consists of a source follower (SF) stage that senses the rail voltage followed by a transconductance (Gm) stage that converts the source follower output voltage into I_{OUT} [6]. The test-chip contains four separate voltage monitor circuits for each internal voltage rail (VDDH, VDDM, VDDL, VSSC).

The test-chip was fabricated in a 150nm FDSOI CMOS process from MIT Lincoln Laboratory. An annotated die photo of the chip is shown in Figure 5. Not only is voltage stacking easier with SOI process due to the absence of body effect, ultra-thin body SOI technologies are one of the candidates for future aggressively scaled process nodes beyond 32nm. Hence, this test-chip encompasses key attributes to evaluate the potential of using voltage stacking in future designs.

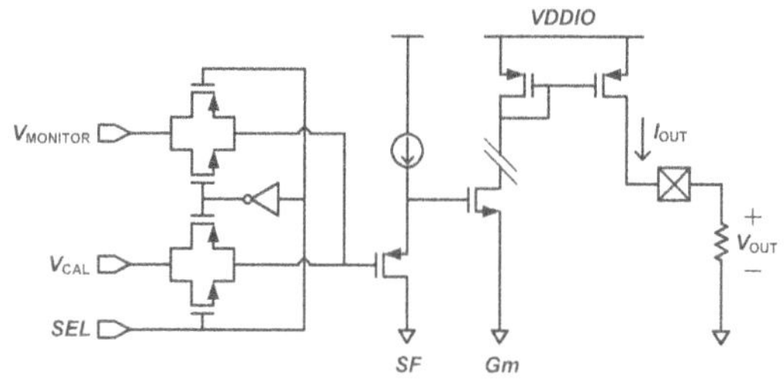

Figure 4: Block diagram of voltage monitor circuits for real-time voltage measurements

Figure 5: Annotated die photo

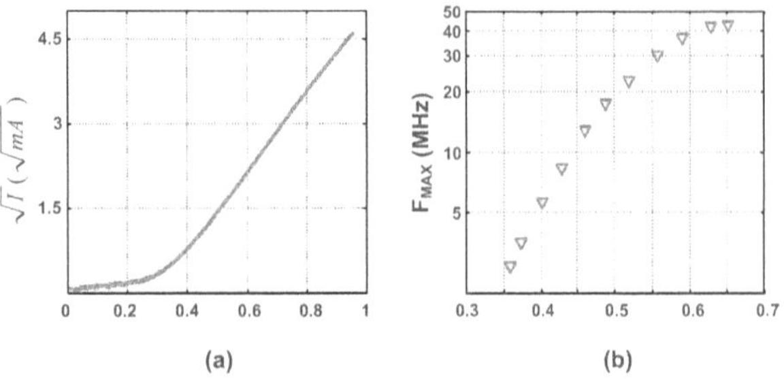

Figure 6: (a) *Current load* block's measured current and (b) *logic load* block's F_{MAX} vs voltage

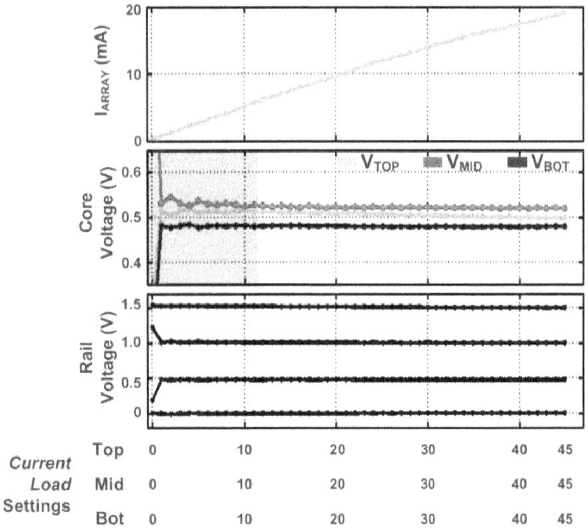

Figure 7: Measured static array current and voltage levels of voltage-stacked array for balanced *current load* setting

4. EXPERIMENTAL RESULTS

Measured current versus voltage relationship for the 8x nMOS switch of the *current load* is presented in Figure 6(a). The the maximum operation frequency (F_{MAX}) of a logic path in the *logic load* versus voltage across the stacks is shown in Figure 6(b). Exponential dependence of F_{MAX} on voltage (Figure 6(b)) clearly illustrate that the *cores* operate in the near-threshold regime at the intended *core* operating voltage of 0.5V.

Under static conditions, balanced inter-layer current consumption evenly distribute stack voltage across the stack layers as dictated by KCL. Figure 7 plots the measured static array current (I_{ARRAY}) and the voltage levels that the internal stack voltages settle to, all versus identical inter-layer static *current load* settings, as the *current load* setting is increased from 0 to 45. For identical current load block settings, the array evenly splits VDDH (at 1.5V) into ~0.5V levels. Inherent mismatches between the layers cause small disparities between core voltages. Furthermore, close examination of core voltages at low *current load* settings (shaded in gray) show that voltage deviations are larger at low *current load* setting; the largest difference seen when all

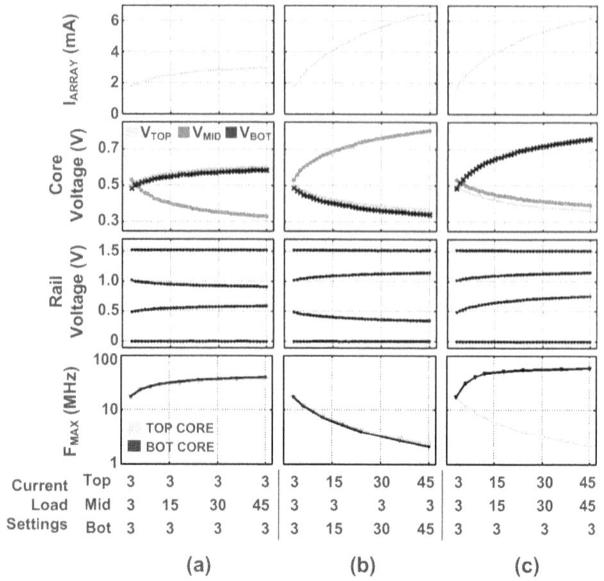

Figure 8: Measured static array current, voltage levels and F_{MAX} of voltage-stacked array for skewed *current load* settings with current increased only in (a) Middle Layer (b) Top and Bottom Layers (c) Top and Middle Layers

blocks only leak (settings of 0). This is due to internal rail impedance, which is further explained later on in the paper.

Figure 8 plots the measured static array current, the voltage levels and F_{MAX} of logic paths in the top and bottom layers, all versus various static skewed *current load* settings. A bug in the level-shifter circuit prevented measurement of F_{MAX} for logic paths in the middle layer. Figure 8 shows that skewed *current load* settings between layers leads to uneven distribution of voltages. For example, as *current load* settings in the middle layer increase relative to the top and the bottom, held at a setting of 3 (Figure 8(a)), voltage across the middle layer decreases. The results show that to maintain balanced voltage distribution across the stack layer as shown in Figure 7, inter-layer current difference must be absorbed using an active regulation scheme, such as push-pull regualtors [1], linear regulators [8] or switched-capacitor circuits [2]. However, using voltage regulators to absorb large inter-layer current difference can be costly for sustained current mismatches in terms of energy efficiency.

Conversely, Figure 8 also demonstrates how processor cores within a layer could operate at different voltage/frequency settings from cores at different layers by intentionally skeweing inter-layer current consumption. This could be achieved by enabling/disabling different number of cores per layer, essentially trading off core count for higher performance in a particular layer. Voltage stacking shows the potential to achieve workload heterogeneity while obviating voltage regulators.

To fully understand how internal voltages change with current mismatch, we investigate voltage deviation vs current mismatch relationship using a simple two-stack layer

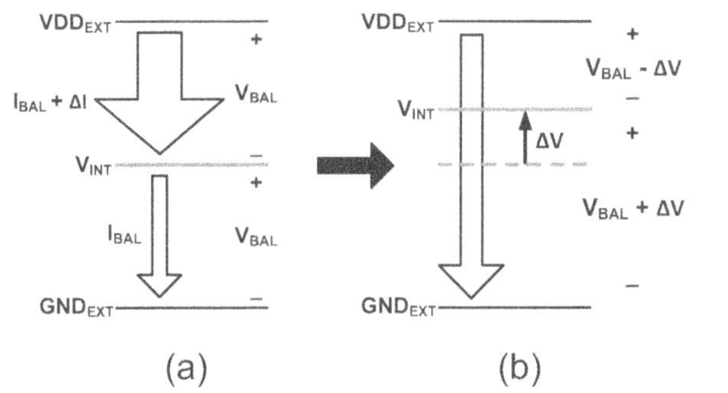

Figure 9: Diagram modeling voltage deviation for a simple two stack layer system

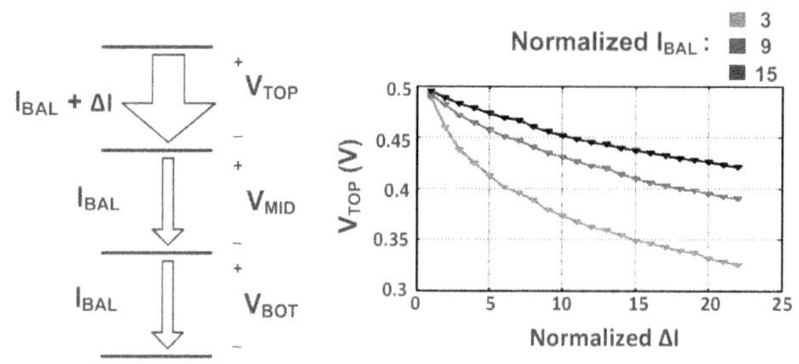

Figure 10: Measured V_{TOP} values for various initial power consumption conditions for inter-layer current skew (current increased only in top layer)

case illustrated in Figure 9, where external voltages VDD_{EXT} and GND_{EXT} are fixed and only V_{INT} change with current mismatch. Starting from a balanced condition ($V_{TOP} = V_{BOT} = V_{BAL}$ and $I_{TOP} = I_{BOT} = I_{BAL}$), if current in the top layer increase as shown in Figure 9(a), internal voltage V_{INT} increases by ΔV to balance the current flow through the stack. Assuming power consuming blocks in each layer are nMOS switches like in the test-chip prototype, the current vs voltage relationship can be modeled using generalized alpha-power law

$$I = K \cdot (V)^{\alpha} \tag{1}$$

where K represents aggregate width of the transistor and we have ignored V_{TH} for simplicity. Increase in current consumption in the top layer as shown in Figure 9(a) can be modeled as

$$I_{BAL} + \Delta I = (K + \Delta K) \cdot (V_{BAL})^{\alpha} \tag{2}$$

where ΔK represents the increase in aggregate nMOS device width in the top layer. ΔV due to current mismatch ΔI can then be calculated by solving the KCL equation

$$\begin{aligned}(K + \Delta K) \cdot (V_{BAL} - \Delta V)^{\alpha} \\ = (K) \cdot (V_{BAL} + \Delta V)^{\alpha}\end{aligned} \tag{3}$$

and substituting ΔI expression for V_{BAL}, which results in

$$\Delta V = \frac{\sqrt[\alpha]{K + \Delta K} - \sqrt[\alpha]{K}}{\sqrt[\alpha]{K + \Delta K} + \sqrt[\alpha]{K}} \cdot \sqrt[\alpha]{\frac{\Delta I}{\Delta K}} \tag{4}$$

Equations (1)–(4) can easily be expanded to other voltage stacked cases with more layers. Equation (4) illustrates how internal voltage deviations in voltage stacked systems relate to the mismatch in current. For nMOS switches of with quadratic current vs voltage relationship ($\alpha \approx 2$) (Figure 6(a)), the shape of VDDM and VDDL curves result in square root shape as seen in Figure 8. The equations also apply to synchronous CMOS logic blocks ($I \propto CVF$), where K in equation (1) corresponds to aggregate switched capacitance of the logic block, rather than transistor width. The value of α, which determines ΔV vs ΔI relationship, depends on clocking strategy for synchronous logic blocks. Constant frequency operation, corresponding to $\alpha \approx 1$ for equation (1), ignoring capacitance dependence on voltage, yields a close to linear ΔV vs ΔI relationship. On the other hand, current in self-clocked circuits[7], where operating frequency varies with voltage, varies more strongly with voltage, which would result in ΔV vs ΔI relationship more akin to the square root relationship of the test-chip prototype.

Unlike traditional power delivery schemes where circuitry has low-impedance connection to fixed supply rails, voltage stacking suffers from relatively high impedance paths for one or both rails, depending on the layer. Hence, the impedance of the internal supply rails of a voltage stack layer is highly dependent on the aggregate power consumption of the cores. Figure 10 plots the voltages that V_{TOP} settle to vs normalized inter-layer current difference when the *current load* settings are skewed by increasing only the top layer current, for three different values of balanced current I_{BAL}. Higher initial I_{BAL} values (higher chip power consumption), result in smaller voltage deviation for the same inter-layer current mismatch, due to lower impedance of the internal supply rails. This can also be seen from equation (4), where higher current consumption (for the same voltage conditions) translate to higher K value, leading to a smaller ΔV for the same change in current $\Delta I(\Delta K)$. This suggests that voltage stacked systems become more immune to inter-layer current mismatches as overall power consumption is increased.

This dependence is illustrated in Figure 11 by measuring the internal voltage variation in the test-chip when pseudo-random noise is injected into the array for different balanced current values. Noise is injected by toggling *current load* bits. Real-time internal voltage trace of $VDDM$ and $VDDL$, along with histogram plots of V_{TOP}, V_{MID} and V_{BOT} are shown. Pseudo-random noise of fixed magnitude was injected into all three stack layers for four different balanced *current load* settings. It can be clearly seen that as power consumption is increased, voltage stack becomes more robust to noise, demonstrating that the voltage variation depends on current mismatch to current consumption ratio, not the absolute value of current mismatch. This attribute suggests that for voltage stacked multicore systems, as the number of cores are increased, power fluctuation of a single core due to architectural events such as cache miss will have lesser effect on internal voltage. For many-core systems, the worst case scenario, when all cores in a stack layer experience power fluctuations at once can be easily avoided by workload scheduling via software. This suggests that for many-core voltage stacked systems, coordinated workload scheduling could provide efficient voltage regulation during

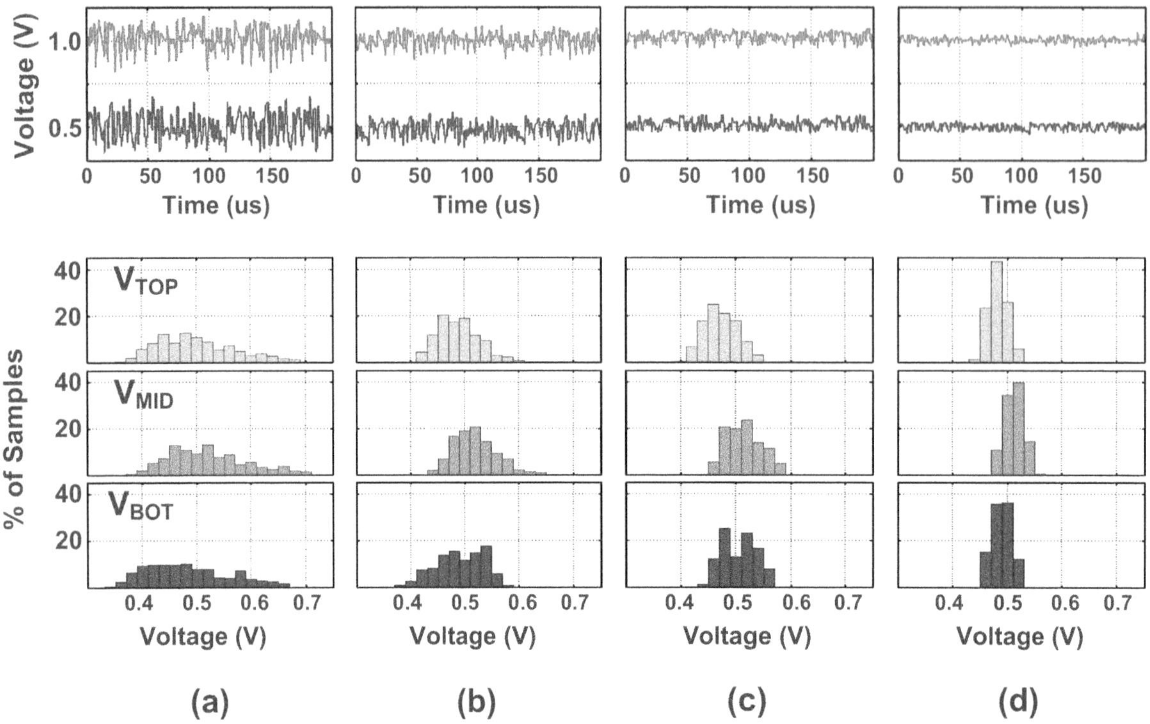

Figure 11: Measured VDDM and VDDL voltage traces (top) and histogram of V_{TOP}, V_{MID} and V_{BOT} when pseudo-random fixed-magnitude noise is injected in all three stack layers for balanced *current load* setting of (a) 3, (b) 8, (c) 15, and (d) 31

high-power states while on-die regulators may only be relied on during low-power states.

5. CONCLUSION

In this paper, an evaluation of voltage stacking is presented in the context of near-threshold multi-core computing. Key attributes of voltage stacking are shown using results from a test-chip prototype built in 150nm FDSOI CMOS. While voltage stacking simplifies off-chip power delivery, it adds extra complexity to combat within-die voltage noise caused by inter-layer current imbalance. Current mismatch between layers must be tightly regulated, especially for near-threshold systems more susceptible to voltage variation. The internal supply rail impedance in voltage stacked systems depends on aggregate power consumption of the stack layers unlike conventional power delivery schemes. Thus voltage noise depends on the current mismatch to chip current consumption ratio rather than the absolute magnitude of the mismatch, suggesting that voltage stacking will scale well with future many-core systems, where power fluctuations in a single core has less impact on voltage noise.

Acknowledgements

This work was partially supported under SRC contract with task ID 1973, as well as partially supported by National Science Foundation grants CCF-0903437, CSR-0720566 and CCF-0702344. Any opinions, findings, conclusions, or recommendations expressed in this material are those of the authors and do not necessarily reflect the views of the SRC or NSF. We also thank MIT Lincoln Labs for fabrication support.

6. REFERENCES

[1] E. Alon and M. Horowitz. Integrated regulation for energy-efficient digital circuits. *IEEE J. Solid-State Circuits*, 43(8):1795–1807, August 2008.

[2] L. Chang, R. K. Montoye, B. L. Ji, A. J. Weger, K. G. Stawiasz, and R. H. Dennard. A fully-integrated switched-capacitor 2:1 voltage converter with regulation capability and 90% efficiency at $2.3A/mm^2$. *IEEE Symp. VLSI Circuits*, pages 55–56, June 2010.

[3] R. G. Dreslinski, M. Wieckowski, D. Blaauw, D. Sylvester, and T. Mudge. Near-threshold computing: reclaiming Moore's law through energy efficient integrated circuits. *Proceedings of the IEEE*, 98(2):253–266, February 2010.

[4] H. Esmaeilzadeh, E. Blem, R. St.Amant, K. Sankaralingam, and D. Burger. Dark silicon and the end of multicore scaling. *Proc. 38th Int'l Symp. Computer Architecture, ACM Press*, pages 365–376, June 2011.

[5] W. Kim, D. Brooks, and G.-Y. Wei. A fully-integrated 3-level dc/dc converter for nanosecond-scale dvs with fast shunt regulation. *IEEE Int. Solid-State Circuits Conf. (ISSCC) Dig. Tech. Papers*, pages 268–270, February 2011.

[6] M. Nagata, T. Okumoto, and K. Taki. A built-in technique for probing power supply and ground noise distribution within large-scale digital integrated circuits. *IEEE J. Solid-State Circuits*, 40(4):813–819, April 2005.

[7] L. S. Nielsen, C. Niessen, J. Sparso, and K. van Berkel. Low-power operation using self-timed circuits and adaptive scaling of the supply voltage. *IEEE Transactions on VLSI Systems*, 2(4):391–397, 1994.

[8] S. Rajapandian, K. L. Shepard, P. Hazucha, and T. Karnik. High-voltage power delivery through charge recycling. *IEEE J. Solid-State Circuits*, 41(6):1400–1410, June 2006.

CHARM: A Composable Heterogeneous Accelerator-Rich Microprocessor

Jason Cong Mohammad Ali Ghodrat Michael Gill
Beayna Grigorian Glenn Reinman
UCLA CS Department and Center for Domain-Specific Computing
{cong, ghodrat, mgill, bgrigori, reinman}@cs.ucla.edu

ABSTRACT

This work discusses CHARM, a Composable Heterogeneous Accelerator-Rich Microprocessor design that provides scalability, flexibility, and design reuse in the space of accelerator-rich CMPs. CHARM features a hardware structure called the accelerator block composer (ABC), which can dynamically compose a set of accelerator building blocks (ABBs) into a loosely coupled accelerator (LCA) to provide orders of magnitude improvement in performance and power efficiency. Our software infrastructure provides a data flow graph to describe the composition, and our hardware components dynamically map available resources to the data flow graph to compose the accelerator from components that may be physically distributed across the CMP. Our ABC is also capable of providing load balancing among available compute resources to increase accelerator utilization. Running medical imaging benchmarks, our experimental results show an average speedup of 2.1X (best case 3.7X) compared to approaches that use LCAs together with a hardware resource manager. We also gain in terms of energy consumption (average 2.4X; best case 4.7X).

Categories and Subject Descriptors

C.1 [**PROCESSOR ARCHITECTURES**]: C.1.3—*Heterogeneous systems*

General Terms

Design

Keywords

Chip Multiprocessor, Hardware Accelerators, Accelerator Composition

1. INTRODUCTION

Domain-specific architectures [4] have emerged to satisfy demands for power-efficient and high-performance chip multiprocessors. Such architectures feature heterogeneous resources that are designed for a particular application domain. One important component in application-specific or domain-specific hardware is the *loosely coupled accelerator (LCA)*, which acts independently of individual cores and is more amenable to sharing. An example of this could be a dedicated encryption unit that runs a particular encryption algorithm on data it receives from different cores that share access to the accelerator [15].

LCAs have their own strengths and weaknesses. LCAs can help provide many orders of magnitude improvement in power-efficiency and performance by offloading computation from the less power-efficient cores entirely [13] [9]. However, it is difficult to integrate sufficient numbers of LCAs to satisfy all execution scenarios, particularly in application domains with aggressive performance requirements and a wide range of applications and implementations. Moreover, the degree of heterogeneity inherent in these *accelerator-rich* architectures becomes cumbersome to manage, especially considering the lack of block reusability and the lack of flexibility for future generations of software within the same domain.

We observe three key trends that motivate our proposed design. First, the tasks performed by LCAs tend to have a great deal of data parallelism, which can be exploited by the use of multiple LCAs of the same type. Second, there is often a variety of LCAs (with possible overlap), required by different applications, or regions of a single application, within a domain. This results in the utilization of any particular LCA being somewhat sporadic. These two trends illustrate that it would be expensive to have both sufficient diversity of LCAs to handle the various applications that comprise a domain and sufficient quantity of a particular LCA to handle the parallelism that exists within a particular accelerator task. Third, we observe that LCAs for a single domain have a large degree of similarity in terms of type of computation primitives being used, indicating that these LCAs can be built using a limited number of smaller, more general LCAs that we call *accelerator building blocks (ABBs)*.

There has been some preliminary work exploring hardware management of LCAs [13] [9], but they do not have a hardware mechanism to effectively leverage multiple accelerators for a single task (i.e. effectively load balance among cooperating accelerators), and therefore cannot leverage the parallelism we observe. Moreover, they do not exploit the sporadic nature of accelerator use, and would need to implement an LCA for each function they want to accelerate.

To address these concerns, we propose a multi-core design where loosely coupled ABBs are distributed in islands around the chip and may be dynamically composed together into an LCA. Thus, the ABBs are effectively the building blocks of our application-specific accelerators. Instead of having discrete LCAs for every desired accelerator function, the use of ABBs gives us tremendous flexibility in composing different accelerators at different times, thereby better matching application demand. In addition, we may better share resources at this fine granularity and improve ABB utilization. From a design perspective, the use of ABBs creates more regularity and static homogeneity in a design that still provides virtual heterogeneity through dynamic composition. Our paper makes the following contributions:

- We first analyze the LCA demands for the application domain of medical imaging. We then validate the system flexibility by targeting two independent application domains, namely navigation and computer vision. For all our benchmarks, we examine the potential for composability and accelerator sharing.

- We propose a hardware mechanism to control and compose ABBs that may be distributed anywhere in our system. This hardware provides virtualization of LCAs and allows software to interact with shared accelerators without the need to consider arbitration or contention.
- We further demonstrate how our approach provides load balancing to better distribute work among ABBs, thereby improving utilization for either a single application with multiple accelerator tasks or across multiple applications.

The rest of this paper is organized as follows: Section 2 provides analysis of our target domain and additional benchmarks. Section 3 details our approach to providing the flexible composition of ABBs into LCAs. Sections 4 and 5 outline our evaluation methodology and results. We compare our approach to prior work in Section 6 and conclude with Section 7.

2. THE CASE FOR COMPOSITION

In this section, we examine the characteristics of accelerator-rich architectures that will help motivate our design. Specifically, we show how LCAs from the medical imaging domain can be decomposed using ABBs. In Section 5.4 we will show how these ABBs are reusable in completely different domains.

2.1 Medical Imaging Application Domain

Medical imaging is an important tool for diagnosis and treatment, but is prohibitively time consuming for use in real-time clinical diagnostics. This need, combined with the fact that these applications feature a large amount of data parallelism and highly regular computation, make medical imaging an ideal candidate for accelerators. A more complete description of the applications in this domain, along with existing acceleration strategies, can be found in [5].

An interesting observation about all of these applications is that they can be decomposed into only 4 types of ABBs. These are listed in Table 1, along with the applications that utilize them.

Table 1: Accelerator Building Blocks (ABBs) used in medical imaging

ABBs	Denoise	Deblur	Registration	Segmentation
Float Reciprocal (FInv)	✓	✓		✓
Float Square-Root (FSqrt)	✓	✓	✓	✓
Float Polynomial-16 (Poly16)	✓	✓	✓	✓
Float Divide (FDiv)	✓	✓	✓	✓

2.2 Accelerator Demand

Rather than trying to provide sufficient LCAs for all possible execution scenarios, we propose providing a set of ABBs to use as a pool of building blocks to dynamically compose any desired LCAs. This idea is based on the following observations:

Overlapping Functionality - Table 1 emphasizes the considerable overlap in the functionality of ABBs and LCAs. We will leverage this commonality by using ABBs to construct LCAs on the fly.

LCA Utilization - Our experiments demonstrate that while LCAs have dramatic potential to improve performance and reduce power dissipation, their overall utilization is rather low. LCAs in the medical imaging domain, for instance, are utilized less than 10% of the time due to variations in accelerator resource demand. These relatively high idle times indicate an opportunity for accelerator sharing.

3. CHARM ARCHITECTURE

In this section, we address the design goals from Section 2 and propose a novel architecture that provides flexibility, scalability, and design reuse. Our Composable Heterogeneous Accelerator-Rich Microprocessor (CHARM) design includes both software and hardware components. This work is primarily focused on discussing the architectural contributions of CHARM, so discussion of the software component will be limited.

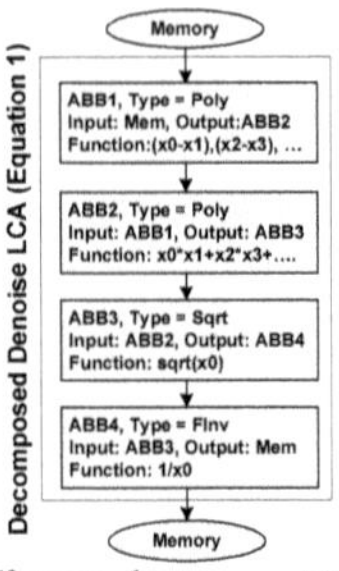

Figure 1: Composition

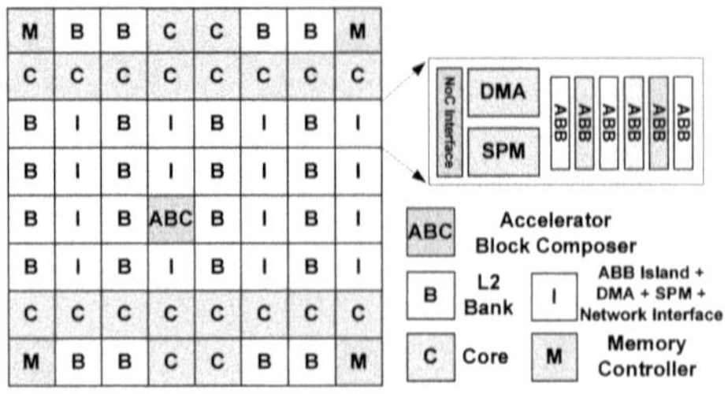

Figure 2: Architecture of CHARM

3.1 CHARM Software Infrastructure

CHARM's software component is responsible for: 1) LCA candidate selection - identifying program hotspots that would benefit from LCA implementation [13]; 2) ABB selection - generating a set of ABBs to cover a set of LCAs under physical design constraints (area, timing, power) [11] and 3) ABB flow graph creation - used to compose LCAs from ABBs. We have currently automated the second and third parts of the software component, and perform manual LCA candidate selection for now.

The structure of the ABB flow graph is the same as that of a task flow graph [14] (see Figure 1). Each node is a task that is represented by a desired ABB invocation, with edges representing memory transfers between ABBs. In memory, this graph consists of a list of ABBs that are part of the LCA, followed by an enumeration of memory transfers. Each ABB node consists of a type, an enumeration of starting addresses for locating argument streams in a virtually addressed private SPM region, and settings for any local configuration registers. Each memory transfer consists of an identifier for a source and destination device (either memory or an ABB node). It also includes a starting address and a series of size-stride pairs describing a polyhedral (regular, high-dimensional) space for both the source and destination. This graph is easy to parse, and consists mostly of values that are directly usable by various control registers on the ABBs and associated DMA. We further note that when a data flow graph is created, it is not tied to any physical instance of the ABBs. The flow graph simply connects virtual ABBs together as a template of an LCA. Figure 1 illustrates this for one of the LCAs used in Denoise, whose functionality is formulated by Equation 1:

$$1/\sqrt{\sum_{i=0}^{6}(x_c - x_i)^2} \quad (1)$$

The hardware will then map physical instances of ABBs to this LCA template on the fly to instantiate a virtual LCA.

3.2 Hardware Infrastructure

While the software is responsible for specifying candidates for acceleration and detailing how ABBs may be composed into particular LCAs, it is the hardware's responsibility to allocate ABB resources to particular threads to satisfy software demand. For this paper, we will restrict each ABB to be allocated to at most one LCA at a time. Our hardware will arbitrate use of the ABBs and LCAs among multiple competing threads/cores, and allocate resources in a way that maximizes the utilization of available resources (i.e. load balances requests from one or more cores among multiple LCAs). Note that additional complications exist due to variation in latency when streaming data to LCAs (i.e. caused by TLB and cache misses, congestion on the NoC, etc.) and varying contention for the use of any given ABB. A dynamic solution is preferable in order to adapt to nondeterminism in LCA memory latency and to the varying LCA demand across different cores.

Figure 2 shows an example of the CHARM architecture. It consists of cores, L2 cache banks, memory controllers, ABB islands and an *accelerator block composer* (ABC), which is the means of control for composing ABBs and essentially the mechanism by which we provide dynamic adaptation. We describe the ABC in more detail below.

Each ABB island has a small dedicated SPM, dedicated DMA engine and NoC interface. The SPM allows ABBs, when composed into an LCA, to have a fixed data access latency. By using memory streaming and task partitioning, and by overlapping communication with computation, the SPM size can be kept small. The allocation of SPM to each ABB is handled by the ABC.

The dedicated DMA engine in each ABB island is responsible for transferring data between the SPM and the L2 cache, and also between SPMs in different ABB islands (i.e. accelerator chaining or remote DMA [27]). In addition, each DMA has a small internal TLB, allowing LCAs to work with virtual addresses. In the event of a TLB miss, the DMA will forward its request to the ABC (see Section 3.2.1 for details on ABC TLB handling).

3.2.1 ABC Design

In our scheme, the ABC is contacted by cores that need access to an LCA. It then allocates ABBs to satisfy this request. An LCA can consist of any number of ABBs, provided that number is less than the number of ABBs that is available in the system. The ABC uses five components to manage its collection of ABBs: a Resource Table, a Composed LCA Table, a collection of Task Lists, a TLB, and a Data Flow Graph Interpreter.

Resource Table: The ABC has a Resource Table that it uses to track the allocation of different ABBs to LCAs. When a core requests the use of an ABB, the Resource Table is queried to determine which ABBs are available. If enough ABB resources are available, multiple instances of a particular type of LCA may be instantiated, assuming the computation to be done is large enough for these multiple instances to each perform non-trivial amounts of work. The ABC uses a two-tiered allocation policy to decide which ABBs to compose into a given LCA. First, the ABC will attempt to balance the concentration of memory-accessing ABBs across the entire system. The purpose of this is to limit contention in the DMA associated with each node. Second, the ABC will employ a simple greedy approach to select ABBs that are local to other ABBs they communicate with. This is done in order to minimize the cost of communication between ABBs. To further reduce latency, ABBs within the same island may use a common SPM for communication (rather than each using their own SPM in their respective islands) and eliminate the need to communicate through the NoC. When ABB resources are scarce, the above metrics degrade to greedily constructing LCAs out of any available ABBs, rather than waiting for more optimal choices to become available.

Composed LCA Table: To eliminate the need to repeatedly compose the same LCA out of the same ABBs when tasks are completed, a Composed LCA Table is introduced. This table tracks ABB allocation, and is used to remove the overhead of remapping patterns when an LCA is already composed.

Task Lists: When the ABC receives a request for an LCA, the requested computation is split into a number of fixed-size data chunks to enable efficient parallelism. Each of these is referred to as a task and the ABC maintains these in a Task List. Each entry in the task list consists of a marker identifying which LCA the task belongs to, which task of the whole computation the entry belongs to (for that specific LCA invocation), and a bit flag marking it as runnable or not runnable. As tasks are added to the Task List, the ABC iterates over the memory addressed by the task, and checks its local TLB. If all addresses in a task are resolvable by the internal TLB, the task is marked as runnable. Otherwise, it is marked as not runnable, and the ABC issues a TLB miss to the requesting core.

The ABC uses a round robin scheduling policy to iterate through all LCAs that have at least one task marked as runnable. So long as there are tasks that are marked as both runnable and for which there are enough ABBs to compose, the ABC continues attempting to compose more LCAs, and continues issuing tasks. We plan to implement more complex scheduling policies based on task priority and criticality in the future.

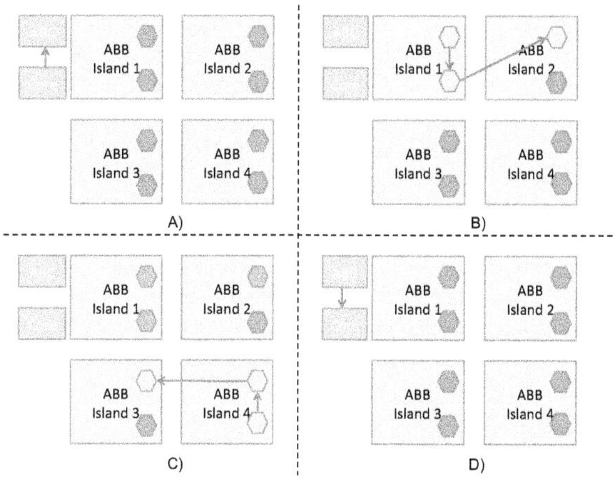

Figure 3: LCA composition example: A) A core sends a request for an LCA to the ABC; B) An LCA instance is allocated; C) An LCA instance is allocated with consideration for balancing DMA utilization; D) The ABC signals completion to the core.

TLB: The ABC maintains a shared TLB that caches address translations among all tasks in its task list. This allows the ABC to prescreen tasks for TLB misses prior to composition. If multiple ABBs under control of the ABC would have encountered the same TLB miss, the ABC can avoid sending duplicate requests to the corresponding core and simply satisfy these misses locally with its own TLB.

Data Flow Graph Interpreter: Our software framework provides composition instructions in the form of a data flow graph. These graphs are fed as resource instantiation templates from the cores to the ABC. Each node in the data flow graph needs to be allocated to a particular ABB, and each ABB is only assigned to a single graph node, and a single LCA, at a time.

When an ABB finishes with the work for a single task, it notifies the ABC that it is free for reassignment. If there are more tasks marked as runnable associated with the LCA to which the ABB was allocated, it is given another task from this set. If there are no runnable tasks associated with that LCA, the ABB becomes eligible for composition into a different LCA. We considered keeping LCAs composed for a longer duration to exploit potential locality of use of a particular LCA, but found that the overhead involved in mapping a set of ABBs to an LCA template is small enough such that releasing resources immediately is preferable due to the improved utilization of ABBs across multiple LCAs. This means that ABB utilization varies over the course of execution of a particular task, and it may be possible for there to be multiple constructed copies of a particular LCA at a given time, even if this is not possible when a given core initially requests an LCA. Therefore, as long as the ABC has runnable tasks in the Task Lists for a particular LCA, we allow it to attempt to compose additional copies of that LCA. In this way, the ABC can eventually make use of all available resources. In this paper, we do not allow ABB preemption (except in the event of error, such as an access violation in the requesting core), but we will explore this for future work.

3.2.2 Example of Composition

Figure 3 shows an example of LCA composition for an architecture with 4 ABB islands. This example architecture has eight ABBs (shown as hexagons), with two of them in each ABB island. We assume for the sake of simplicity that all ABBs in this example are of the same type. The ABC and a requesting core are shown in the upper left-hand side of the figure. In this example, the core requests the composition of an LCA consisting of three ABBs in sequence, with the first ABB reading from memory and the last ABB writing to memory. The core sends a data flow graph (DFG) of the desired LCA to the ABC (Figure 3A). The ABC then interprets the DFG, splits the request into tasks, and begins cycling over the addresses each computation will access. It puts each of these chunks in the task list. For this example, we assume there is more than one task associated with this LCA invocation, and that the ABC's local TLB has the required pages to make all tasks immediately runnable. The ABC then examines the availability of ABBs, discovering that they are all free, and begins allocating.

Since at least one task is made runnable, the ABC proceeds to execute the allocation algorithm described in Section 3.2.1. After finding a match, consisting of two ABBs in Island 1 and a single ABB in Island 2 (Figure 3B), the ABC makes an entry in its Composed LCA Table, marking these ABBs as belonging to this specific LCA. At this point, it chooses a runnable task from the task list belonging to this LCA type, and dispatches a task. The ABC then begins attempting to map another instance of the requested LCA to available ABBs, and finds two ABBs in Island 4 and one in either Island 2 or Island 3. The allocation algorithm chooses to use Island 3 for the last ABB instead of Island 2 to distribute load across more DMAs (Figure 3C). This process is then stopped since there are not enough ABBs to construct any additional LCAs. As ABBs finish their assigned work, they signal to the ABC that they are finished. Each time the first ABB in an LCA signals the ABC of completion, the ABC checks its task list for runnable tasks. If it finds a task, it begins sending a new task to each ABB in that composed LCA. If it does not find a task, it marks this LCA as retiring, and marks the associated ABB(s) as free. Each time an ABB that was part of a retiring LCA is marked as free, it is made available to be recomposed into a new LCA. When all ABBs of all clones of a retired LCA are freed in this manner, an interrupt is sent to the requesting core marking the completion of the requested computation (Figure 3D).

4. EVALUATION METHODOLOGY

To evaluate the CHARM architecture, we have modified Simics [24] and GEMS [25] to model accelerator-rich many-core architectures. Table 2 shows the parameters used in our simulations.

Table 2: Simulation parameters

Parameter	Value
Processor	Ultra-SPARC-III-i @ 2.0GHz
Operating system	Solaris 10
L1	32-KB, 4-way set-associative: 1-cycle
L2	8-MB, 8-way set-associative: 10-cycles
Coherence protocol	Shared banked L2-cache, L2:MOSI, L1:MSI
Memory	1000-cycles, Directory 6-cycles
Network topology	MESH, latencies: link 1-cycle, router 5-cycles

We have also implemented a series of supporting tools to automatically generate accelerators as well as application code that makes use of these accelerators. The complete sequence of steps involved in adding accelerators and generating programs that use accelerators in CHARM is shown in [12]. For calculating energy, we used the power result output from Synopsys for LCAs and ABBs, and used McPat [23] to generate power values for cores and caches.

Table 3 and Table 4 show the area and power overhead (using the Synopsys 32nm SAED library and CACTI 5.3 [1]) for the selected ABBs and LCAs corresponding to each benchmark. We have also included the synthesis results for the ABC that implements the ABB allocation algorithm mentioned in Section 3. To study the overhead of ABBs, we have synthesized the Poly16 ABB, the results of which are shown in Table 3. The internal structure of a Poly ABB is shown in Figure 4 (this Figure is actually showing a Poly8). It consists of Adder/Subtractor/Multiplier (ASM) modules, an SPM bank,

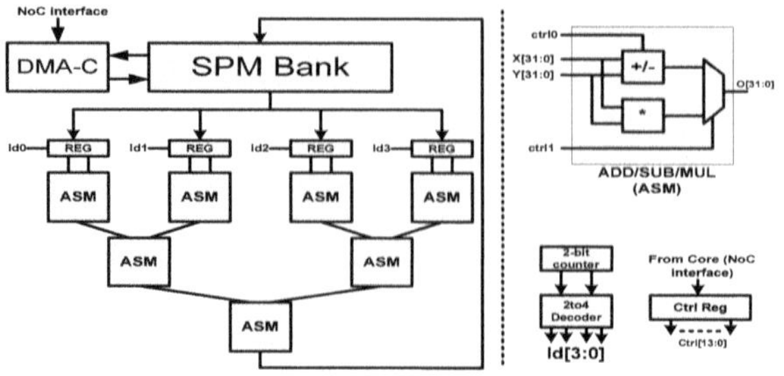

Figure 4: Poly ABB Details

and control logic which controls access to the SPM bank. The SPM bank has 3 sub-banks (for simultaneous read/compute/write) each one with 1 read/write port. One sub-bank is connected to the ASMs and two are ported to the DMA-C. For the experimental results, each ABB island in our design has 16 ABBs and 16 SPM banks to provide concurrent access to all the ABBs. We have used 128 ABBs in our design: 8 ABB islands, each having 3 FInv/FDiv, 1 FSqrt, and 12 Poly16 modules, along with 16 SPM banks. In [12] we have investigated sharing SPM between ABBs. Table 5 shows the area for the main components of the chip.

Table 3: Area/Power results - CHARM

Name	A(u^2)	P(mW)	Total#
FDiv	4949	0.264	12
Poly16	38276	1.608	96
FInv	3503	0.141	12
FSqrt	58683	1.83	8
SPM-4KB 1R/W	13591	17.6	288
SPM-768B 1R/W	2545	7	72
ABC	8383	0.066	1

Table 4: Area/Power results - LCAs

Name	A(u^2)	P(mW)	SPM Banks
Denoise	496908	16.5	6
Deblur	2013228	110.9	9
Segmentation	688298	27.3	6
Registeration	3853098	183.9	18
EKF-SLAM	1188252	42.0	24
LPCIP	239159	6.11	6
SPM-2KB 2R,1R/W	37043	17.5	-

Table 5: Total area(mm^2)- CHARM-HW is the total of ABB islands and ABC

Core	NoC	Cache & Dir	CHARM-HW	CHARM-Total	LCA HW	LCA Total
10.8 Ref[2](scaled to 32nm)	0.3 Ref[21]	39.8 Ref[1]	8.3(Table3) (14%)	59.2	8.5(Table4) (14.3%)	59.4

We modeled a system consisting of 1 to 8 processors, and a set of either physical LCAs or ABBs. When modeling a system consisting of physical LCAs, we included all the accelerators required to run a single instance of each benchmark, without contention. To illustrate the load balancing capacity of our ABC, we modeled instances in which we have some multiple of this number of accelerators. When modeling a system featuring ABBs, the number of ABBs corresponds to the total amount of area that would have otherwise been devoted to LCAs. As a baseline (i.e. 1x ABB area), the total area consumed by the ABBs equals the total area of all LCAs required to run a single instance of each benchmark (this can be verified by the number of ABBs in Table 3 and the LCA area numbers in Table 4). All ABB numbers are multiples of this base amount. We configured our system to have 8 ABB islands, and scaled the number of ABBs present on each island. We also scaled the amount of SPM space on each ABB island proportionally.

5. RESULTS

We compare the following architectures to evaluate CHARM:

Physical LCA sharing with Global Accelerator Manager (LCA+GAM): In this architecture, physical LCAs can be shared between multiple cores. Each benchmark in our domain is accelerated with special-purpose accelerators. A Global Accelerator Manager (GAM) [13] is implemented in hardware to dynamically allocate physical LCAs to cores. We examine cases where there are between 1 and 8 replicates of each required accelerator. This allows for the concurrent execution of multiple instances of any specific benchmark, one for each accelerator in the system. Also, LCAs are powered off when not in use. This approach is similar to the architecture in [13].

Physical LCA sharing with ABC (LCA+ABC): In this architecture, a core may share physical LCAs using a centralized hardware ABC. In addition, the ABC can load-balance the available physical LCAs. We examine cases where there are between 1 and 8 replicates of each required LCA. Since the ABC can split tasks among

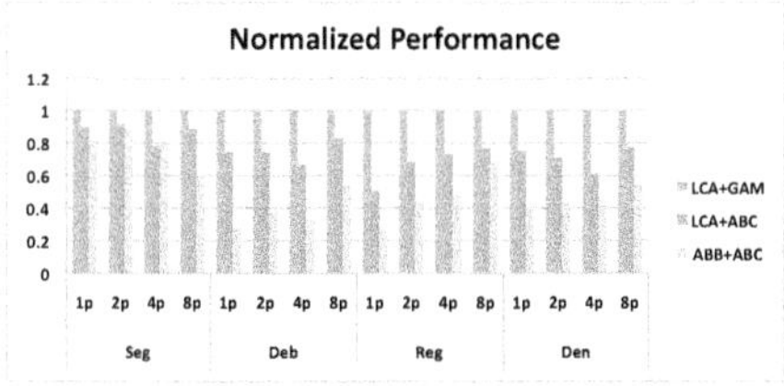

Figure 5: Performance improvement

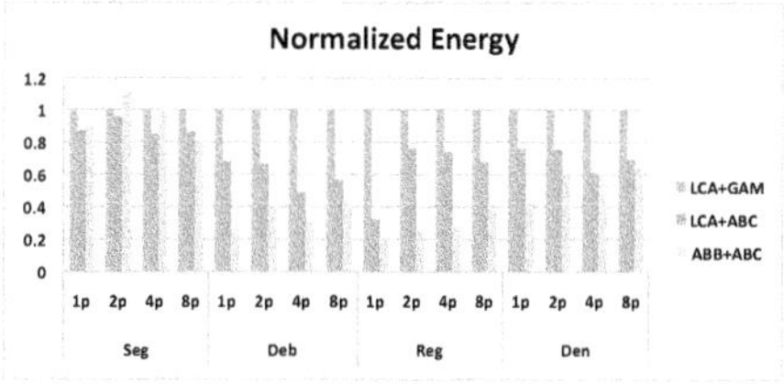

Figure 6: Energy improvement

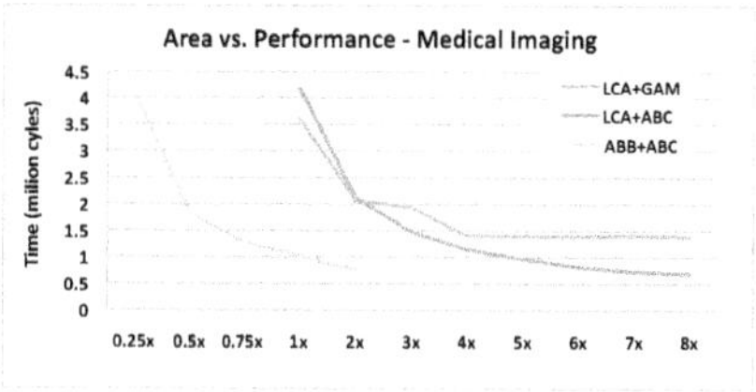

Figure 7: Effect of increasing accelerators

multiple LCAs, we are able to take advantage of all LCAs of a given type, even when only a single instance of that benchmark is executing. In the cases where there are more LCAs than can be allocated to available tasks, extra LCAs are left powered off.

ABB composition and sharing with ABC (ABB+ABC): In this architecture, a centralized hardware ABC is responsible for composing and managing available ABBs, load balancing the tasks, and managing TLB requests from ABBs. We examined multiple ABB quantities. For the purposes of making a comparison, we will refer to a quantity of ABBs with area equal to a single replicate of each LCA in the domain to be comparable to the case where we have one of each physical LCA in the domain. Typically these ABBs can be used to make multiple virtual LCAs, but this gives us a metric by which to make a fair comparison to the LCA+GAM and the LCA+ABC cases. In the cases where there are more ABBs than can be constructed into LCAs, the extra ABBs are left powered off.

For all cases, unless otherwise stated, we ran our 4 selected benchmarks from the medical imaging domain. The benchmarks were run with volumetric images of 32-pixel cubes in multiple iterations.

5.1 Improvement over LCA-based systems

Figure 5 and Figure 6 show the normalized performance and energy improvements we observed by using the ABB+ABC scheme compared to the LCA+ABC and LCA+GAM schemes. All numbers shown here are normalized to the corresponding LCA+GAM result. In each case, we have the same number of processors, threads and accelerators (e.g. the 4p case has 4 processors, 4 threads, and 4 accelerators). On average we observe more than 2.4X energy improvement over LCA+GAM (maximum 4.7X) and 1.6X energy improvement over LCA+ABC (maximum 3X). In general, as the number of independent tasks increases, ABB+ABC shows better performance because the ABC starts composing ABBs to create new LCAs (so long as ABBs are available in the system). This creates more parallel tasks, thereby achieving better performance and consuming less energy. We note that Segmentation-2p using ABB+ABC shows higher energy usage compared to other schemes. The reason for this is that the performance for segmentation improves only slightly. Therefore, the overhead of constructing LCAs and coordinating communication between ABBs consumes more energy than what is conserved by the slight reduction in execution time.

5.2 Effect of adding accelerators

Figure 7 shows the effect of adding more accelerator resources on the performance in all of our studied schemes. We observed a very similar result for energy improvement as well. For this experiment, we fix the number of processors and threads at 4. For LCA cases (LCA+GAM and LCA+ABC), the number of LCAs range from 1 to 8. For the ABC+ABB scheme, the quantity of ABBs range from 1/4 of the ABB number that area-wise matches 1 set of the LCAs in the domain, to 2 times that number.

There are several observations here. First, adding more accelerator resources in general improves speedup and energy. Second, as accelerator resources are increased, significant performance improvements are seen much earlier in the ABB+ABC case than in the LCA schemes (notice 1.5x-2x case in ABB+ABC vs. 6x-8x case in LCA+GAM and LCA+ABC). The reason behind this is that in the ABB+ABC scheme even 1x area allocation can reconstruct many copies of a virtual LCA to run concurrently. This is because a benchmark using physical LCAs only uses those of a specific type, and thus only a small number of the total LCAs. The ABB+ABC is free to replicate virtual LCAs out of the entire sum of accelerator resources, rather than leaving area unutilized. An implication of this is that the ABB+ABC case saturates much more quickly in the acceleration that it can offer, either exhausting potential parallelism or becoming memory bound. Third, after 4x, the LCA+ABC case still continues improving performance and energy, but LCA+GAM flattens. This is because ABC splits each individual LCA invocation into multiple tasks and load balances these tasks among accelerator resources, thereby benefitting from having more than one LCA per accelerator invocation. The GAM, which allocates accelerators directly to the calling thread, is not capable of doing this without the software actually having requested multiple accelerators.

5.3 Effect of changing task-grain

Task-grain is the maximum number of individual computations in each task assigned to a set of composed ABBs. The smaller the task-grain, the more parallelism in cases where computations can be performed independent of one another.

In order to measure the effect of task-grain on the ABB usage of each thread, we measured the number of ABBs allocated to LCA invocations by each thread for every moment of execution. For brevity, we are only showing results for registration, but all benchmarks we examined exhibited the same characteristics. We show this utilization for two cases: task-grain of 8 and task-grain of 128 as shown in Figure 8 and Figure 9, respectively. Each figure shows the ABB usage by each thread and the total number of ABBs used (the upper most curve). When the task-grain is 8, there is more parallelism and so more ABBs can be quickly allocated to a given thread (e.g. the initial spike seen in Figure 8). When the task-grain is 128, only one set of ABBs is used by each thread. The values shown in Figure 9 describe the ABBs allocated for a single LCA instance per thread. Also shown in Figure 8 is the impact of our round robin scheduling. This assures a measure of fairness when allocating ABBs. The jagged total use is the result of freeing ABBs prior to their reassignment.

5.4 Platform flexibility

An original argument we put forward as a justification for this approach was the reusability of this system in terms of block design as well as retargetability. To substantiate this argument, we examined two applications from two domains that are completely unrelated to medical imaging: computer vision and navigation. Computer vision and navigation require compute-intensive data processing, consisting heavily of linear algebra and floating point computation, to attain high levels of situational awareness. We examine Log-Polar Coordinate Image Patches (LPCIP) [20] from computer vision and Extended Kalman Filter-based Simultaneous Localization and Mapping (EKF-SLAM) [22, 3] from navigation. A more detailed description of these two applications and existing acceleration strategies can be found in [8]. Figure 10 shows results comparing the use of our medical imaging platform (unmodified and implementing virtual LCAs) against custom physical LCAs that specifically target these new domains. This illustrates that our platform is flexible, and is much more broadly targetable than a typical platform featuring custom LCAs.

6. RELATED WORK

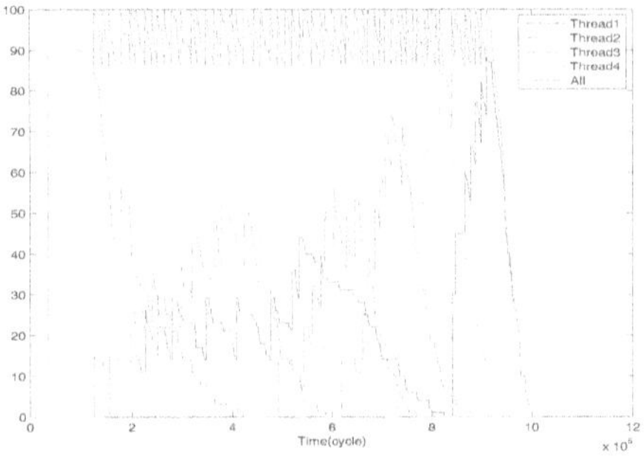

Figure 8: ABB utilization (Task-grain=8)

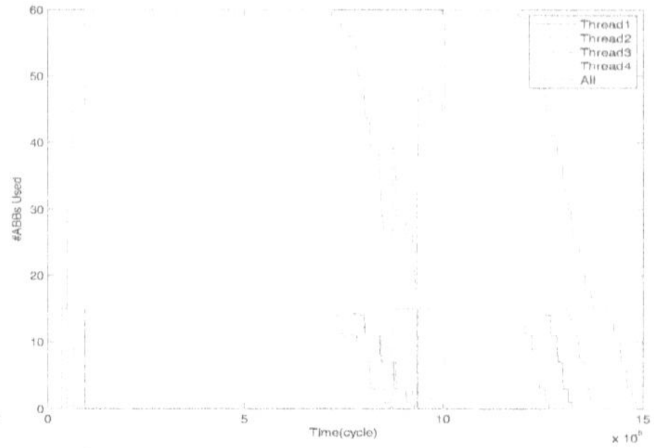

Figure 9: ABB utilization (Task-grain=128)

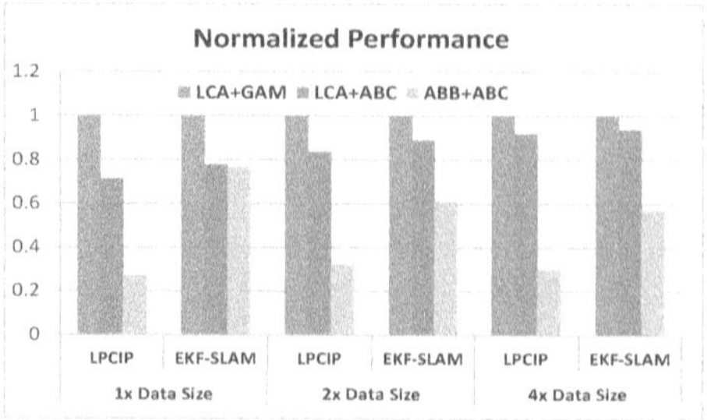

Figure 10: Performance improvement for computer vision and navigation

Some previous work has considered the on-chip integration of accelerators. Garp [17], UltraSPARC T2 [19], Intel's Larrabee [28] and IBM's WSP processor [15] are examples of this. Most of these platforms (except WSP) are tightly coupled with processor cores (or core-clusters). Our paper, on the other hand, focuses on loosely coupled accelerators that can be shared between multiple cores.

There have also been a number of recent designs of heterogeneous architectures, like EXOCHI [31], HiPPAI [29], SARC [27], and QsCores [30]. EXOCHI presents an architecture (exoskeleton sequencer) and a programming environment (C for heterogeneous integration) for a multi-core system. HiPPAI, like our work, eliminates system overhead involved in accessing accelerators through the use of a software layer. SARC also has a core and accelerator architecture similar to our work that relies on software for resource management. QsCores, although lacking dynamic accelerator management and load balancing, also relates to our work as it uses specialized cores to provide energy efficiency by exploiting similar code patterns within and across applications. We contrast with these software-based methodologies by advocating the use of hardware-based accelerator management.

Prior work has considered the composition of simple cores to form more complex general-purpose cores, as in core fusion [18], core spilling [7], and TRIPS [16]. However in those works, the composition is of coarser grain than ABBs, which allows for less flexibility in exploiting pipeline parallelism existing between ABBs. Also, in our approach there is no restriction on ABBs, which cannot be said for any of the above work. There is also some related work in accelerator virtualization, namely VEAL [6] and PPA [26]. VEAL [6] uses an architecture template for a loop accelerator and proposes a hybrid static-dynamic approach to map a given loop on that architecture. PPA [26] uses an array of PEs which can be reconfigured and programmed. It uses a technique called virtualized modulo scheduling which expands a given static schedule on available hardware resources. While for both of these works the target input is a schedule on a small nested loop, our approach is not limited in terms of either control flow or acceleratable problem size.

7. CONCLUSION

This work introduces CHARM - a design that provides scalability, flexibility, and design reuse in the space of accelerator-rich CMPs. CHARM features a hardware structure called the accelerator block composer (ABC), which dynamically composes accelerator building blocks (ABBs) into loosely coupled accelerators (LCAs), load balances tasks among the available ABBs, and includes a central TLB. Our experimental results show that on average we get a 2.1X speedup (best case 3.7X) running medical imaging benchmarks compared to approaches that use LCAs together with a hardware resource manager. We also gain in terms of energy consumption (on average 2.4X and best case 4.7X). These results are achieved with a platform that is flexible and easy to program, and which may be retargeted to a variety of domains. In [10], we have studied sharing SPMs in the NUCA for accelerator-rich CMPs.

8. ACKNOWLEDGEMENTS

This research is supported by the Center for Domain-Specific Computing (CDSC) funded by the NSF Expedition in Computing Award CCF-0926127, GSRC under contract 2009-TJ-1984 and NSF Graduate Research Fellowship Grant # DGE-0707424.

9. REFERENCES

[1] Cacti 5.3: http://quid.hpl.hp.com:9081/cacti/.

[2] http://en.wikipedia.org/wiki/ultrasparc_iii.

[3] J.L. Blanco. Derivation and implementation of a full 6d ekf-based solution to bearing-range slam. Technical report, University of Malaga, Spain, Mar 08.

[4] Alex Bui et al. Customizable domain-specific computing. *Design and Test of Computers, IEEE*, Mar/Apr 2011.

[5] Alex Bui et al. Platform characterization for domain-specific computing. In *ASPDAC*, 2012.

[6] Nathan Clark, , et al. Veal: Virtualized execution accelerator for loops. ISCA '08.

[7] J. Cong et al. Accelerating sequential applications on cmps using core spilling. *Parallel and Distributed Systems, IEEE Trans. on*, pages 1094 –1107, 2007.

[8] J. Cong et al. Accelerating vision and navigation applications on a customizable platform. In *ASAP*, 2011.

[9] Jason Cong et al. Architecture support for accelerator-rich cmps. In *Proceedings of the 49th Annual Design Automation Conference (DAC 2012)*.

[10] Jason Cong et al. Bin: A buffer-in-nuca scheme for accelerator-rich cmps. In *ISLPED 2012)*.

[11] Jason Cong et al. A generalized control-flow-aware pattern recognition algorithm for behavioral synthesis. DATE '10.

[12] Jason Cong et al. UCLA computer science department technical report #120008.

[13] Jason Cong et al. AXR-CMP: Architecture support in accelerator-rich cmps. In *2nd Workshop on SoC Architecture, Accelerators and Workloads*, Feb 2011.

[14] Kayvon Fatahalian et al. Sequoia: programming the memory hierarchy. In *Proceedings of the 2006 ACM/IEEE conference on Supercomputing*, 2006.

[15] H. Franke et al. Introduction to the wire-speed processor and architecture. *IBM Journal of Research and Development*, pages 3:1 –3:11, 2010.

[16] Mark Gebhart et al. An evaluation of the trips computer system. ASPLOS '09.

[17] J.R. Hauser and J. Wawrzynek. Garp: a MIPS processor with a reconfigurable coprocessor. FCCM'97, pages 12 –21, 1997.

[18] Engin Ipek et al. Core fusion: accommodating software diversity in chip multiprocessors. ISCA '07, pages 186–197.

[19] Tim Johnson and Umesh Nawathe. An 8-core, 64-thread, 64-bit power efficient sparc soc (niagara2). ISPD '07, pages 2–2, 2007.

[20] F. Jurie. A new log-polar mapping for space variant imaging : Application to face detection and tracking. *Pattern Recognition*, pages 865–875, 1999.

[21] A. B. Kahng et al. Orion 2.0: a fast and accurate noc power and area model for early-stage design space exploration. DATE '09, pages 423–428.

[22] R. E. Kalman. A new approach to linear filtering and prediction problems. *Journal of Basic Engineering*, pages 35–45, 1960.

[23] S. Li et al. Mcpat: An integrated power, area, and timing modeling framework for multicore and manycore architectures. MICRO '09.

[24] Peter S. Magnusson et al. Simics: A full system simulation platform. *Computer*, 35:50–58, 2002.

[25] M. Martin et al. Multifacet's general execution-driven multiprocessor simulator (gems) toolset. In *Computer Architecture New, Sep 2005*.

[26] Hyunchul Park et al. Polymorphic pipeline array:a flexible multicore accelerator with virtualized execution for mobile multimedia application. MICRO 42.

[27] A. Ramirez et al. The sarc architecture. *Micro, IEEE*, pages 16 –29, 2010.

[28] Larry Seiler et al. Larrabee: A many-core x86 architecture for visual computing. *IEEE Micro*, 29:10–21, 2009.

[29] P.M. Stillwell et al. HiPPAI: High performance portable accelerator interface for SoCs. HiPC 2009, pages 109–118, 2009.

[30] G. Venkatesh et al. QsCores: trading dark silicon for scalable energy efficiency with quasi-specific cores. MICRO-44 '11, pages 163–174.

[31] Perry H. Wang et al. EXOCHI: architecture and programming environment for a heterogeneous multi-core multithreaded system. PLDI '07, pages 156–166.

Something Old and Something New: P-states can Borrow Microarchitecture Techniques Too

Yasuko Eckert*† Srilatha Manne* Michael J. Schulte* David A. Wood†*
*AMD Research
Advanced Micro Devices, Inc.
{yasuko.watanabe, srilatha.manne, michael.schulte} @amd.com
†Department of Computer Sciences
University of Wisconsin – Madison
david@cs.wisc.edu

ABSTRACT

The limited utility of voltage scaling in nano-scale technologies has led high-performance processors to rely increasingly on frequency scaling for power management. However, frequency scaling provides only a linear dynamic power reduction.

In this paper, we make a case for dynamically disabling performance optimizations, leveraging previously proposed low-power techniques, for more efficient power-performance trade-offs. By carefully selecting which optimizations to turn off, our lowest P-state consumes less than half the power achieved by frequency scaling, on average, for comparable performance. For all workloads, our approach performs as well or better than DVFS, demonstrating the effectiveness of our approach.

Categories and Subject Descriptors

B.8.2 [**Performance Analysis and Design Aids**]

General Terms

Design. Performance.

Keywords

Dynamic voltage and frequency scaling, frequency scaling, P-states, power gating, dynamic microarchitectural power saving.

1. INTRODUCTION

Modern processors support performance states, or P-states, for dynamic power-performance management. The dominant P-state mechanism is dynamic voltage and frequency scaling (DVFS) because of its cubic relationship between (dynamic) power and performance: for a 3% reduction in power, DVFS reduces performance by about 1%. Unfortunately, the utility of DVFS has been decreasing due to transistor scaling limitations [6]. Although various circuit-level techniques have been proposed to address this problem, such as ultra-low voltage operation [5][6], many incur performance, leakage power, and/or reliability penalties.

Dynamic frequency scaling is an established alternative to DVFS, especially for high-performance commercial microprocessors [11]. Frequency scaling simply reduces the operating frequency and leaves the supply voltage unchanged. Although simple in design, frequency scaling provides only a linear reduction in dynamic power and has no direct impact on the increasingly important static power. Traditionally, frequency scaling was used to modestly extend the power-performance curve and *only after* DVFS had scaled the supply voltage to its minimum (Vmin), as the left side of Figure 1 depicts. However, future technology nodes increasingly will rely on frequency scaling as the gap between maximum (Vmax) and minimum supply voltages shrink (the right side of Figure 1). Replacing DVFS with frequency scaling will significantly flatten the achievable power-performance curve, especially at lower frequencies. Figure 2 demonstrates that a 36% power saving through frequency scaling comes at a cost of a 44% performance loss on average. Furthermore, memory-intensive workloads (e.g., *libquantum*) are power- and performance-insensitive to core frequency scaling because the majority of the baseline chip power comes from the static-power-dominated L2 and L3 caches and frequency scaling only has indirect impact on static power.

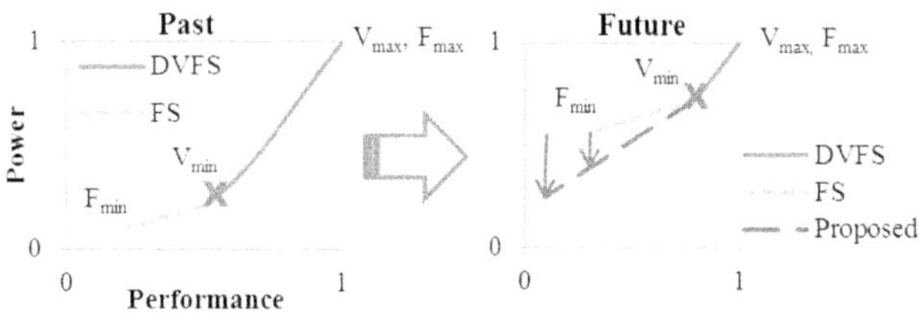

Figure 1. Conceptual example (lower-right is better).

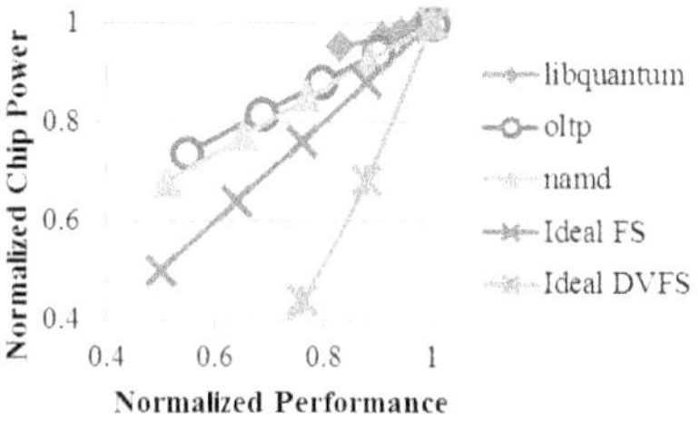

Figure 2. Core frequency scaling (FS).

This paper proposes new mechanisms to implement P-states that achieve a power-performance curve closer to DVFS than frequency scaling. Our mechanisms leverage previously proposed low-power techniques that result in a performance loss. Previously, these techniques might have been dismissed due to the effectiveness of DVFS; however, they become viable alternatives to frequency scaling. We focus on techniques that selectively disable or constrain microarchitectural performance optimizations, trading performance for power reductions. By coordinating the techniques rather than using them in the isolated fashion of previous implementations, we can gain much more power savings.

Our research draws insight from the performance optimization rule established by the Intel Pentium M: a 1% performance improvement should come with no more than a 3% power increase; otherwise, DVFS could do better [8]. If we use this rule *in reverse*, disabling optimizations that meet this 3:1 power-to-performance ratio, we can effectively extend the cubic DVFS power-performance curve. Although some optimizations may have much less than a 3:1 ratio, others may exceed the ratio for a given workload, allowing our techniques to perform better than DVFS.

For the initial evaluation of this concept, this paper selects, without in-depth analysis, two sets of previously introduced intuitive techniques to implement a pair of microarchitectural P-states: front-end and L2 power reduction. The former power-gates front-end structures designed for a worst-case scenario, including the entire checkpointing hardware and some of the fetch buffer and physical register file. In addition, we employ simplified speculation control [14] that selectively stalls the front-end with varying levels of aggressiveness.

The L2 P-states, on the other hand, exploit the fact that not all workloads require the optimized L2 cache access latency or the full cache capacity. We employ drowsy cache technology [7] and power-gate parts of the cache [17]. When combining these front-end and L2 P-states, we show that the lowest P-state consumes less than half the power of frequency scaling, on average, for comparable performance. The power savings possible with our P-states are bounded, on the low end, by the static power of the power-gated components. The upper bound, on the other hand, depends on the dynamic power of those components as well as the impact those components have on the rest of the system.

In this work, we assume these P-states are initiated as soon as we go below P0, which is the highest-performance state. However, our proposed P-states can also be applied in combination with DVFS or after reaching the Vmin. Compared to traditional DVFS-based P-states, our mechanisms have a longer transition delay (~87 μs [18]) due to power-gating overheads. We assume that we stay at a P-state long enough to compensate for the delay.

The contributions of this paper are:

- We make a case for using microarchitectural techniques to implement P-states. We selectively turn off power-dominant performance optimizations for more efficient scaling than frequency scaling. This approach can complement frequency scaling and be used beyond Vmin. To the best of our knowledge, this is the first paper to propose using microarchitectural mechanisms to implement P-states.
- We re-examine old (i.e., previously proposed) low-power techniques in a new context and apply them without complex policies or logic to obtain significant power savings. Our results demonstrate that these simplified techniques outperform frequency scaling for all workloads, and provide better power scaling than DVFS in most cases.

2. FRONT-END P-STATES

We first explore opportunities in the front-end for trading instruction-level parallelism (ILP) and/or memory-level parallelism (MLP) for power. The front-end is an attractive target: it is organized for quick retrieval of instructions after a pipeline flush, and it is generally underutilized once the scheduler is full. In addition, the rate of instruction flow to the back-end sets an upper bound on achievable ILP and MLP.

Hence, our aim is to reduce wasteful power from wrong-path instructions and lessen or turn off the capabilities of power-hungry front-end structures—even at the cost of performance. We propose a new synthesis of four old ideas (checkpoint removal, speculation control, and fetch buffer and register size reductions) to provide more efficient power-performance scaling than frequency scaling.

Checkpoint removal. Many high-performance out-of-order (OoO) processors checkpoint architectural state to recover from branch mispredictions. Although checkpoints facilitate fast misprediction recovery at detection [16], they result in modest performance improvement for highly-accurate branch predictors [19]. When allocating checkpoints to all branches, only 0.002-8.0% of all checkpoints are used to recover from mispredictions and the corresponding branches eventually commit (results not shown).

Moshovos proposed reducing checkpoints by predicting which branches are likely to require checkpoints and/or using OoO checkpoint release [16]. While both of these techniques still recover from mispredictions as soon as detected, we propose a novel technique that selects between recovery at detection and recovery at commit depending on P-states. During P0, we use the conventional checkpoint-based recovery at detection. To increase power efficiency during lower-performance P-states (i.e., P1 to Pn), we disable all checkpointing hardware and recover at commit by flushing any structures with corrupted states. Compared to the checkpoint-based recovery at detection, recovery at commit lengthens misprediction penalties by the cycles between detection and commit (unless it takes longer to re-fill the window), which can be significant. We mitigate the penalties by flushing wrong-path instructions in the OoO execution engine on a misprediction detection and re-fetching the correct-path instructions. We also stall the rename stage, which contains corrupted states, until the recovery action is taken.

Speculation control. To reduce wasteful power from wrong-path instructions during lower-performance P-states, we also apply a simplified version of speculation control that gates the rename stage when the amount of speculation in a window exceeds a certain threshold [14]. Prior proposals measure speculation by the number of unresolved low-confidence branches. We instead use the total number of unresolved branches regardless of their confidence as a proxy. This coarse approximation penalizes workloads with very low squashes per thousand instructions (SPKI) and phases with low SPKI; however, we obviate the need for a confidence estimator, thus reducing complexity and power consumption. The rest of the speculation-control mechanisms remain the same. The smaller the threshold, the more aggressively we suppress wrong-path instructions, with a greater risk of performance loss.

Fetch buffer resizing. A complementary method for regulating speculation is to lessen fetching aggressiveness by shrinking the fetch buffer [21]. The fetch buffer size is usually larger than the fetch width to mask I-cache miss latencies. However, a larger buffer size generally increases the probability of fetching wrong-path instructions. Even for workloads with small misprediction rates, the full fetching capability is not necessary most of the time

Table 1. Baseline configuration

Component	Configuration
Branch prediction	TAGE [19], 1K-entry tagged tables, 5 history bits, 4-way 256-entry BTB, 16-entry RAS
Pipeline width	4-wide fetch, rename, dispatch, issue, and commit
Front-end	7 stages, 16-entry fetch buffer, 16 renamer checkpoints, 128 physical registers
Execution	32-entry unified OoO IQ, 2IALU, 2 FPALU, and 2 AGEN, 128-entry window
Dis-ambiguation	NoSQ [35], 1024-entry predictor, 1024-entry double-buffered SSBF
L1-I cache	32KB, 4-way, 64B line, 2 cycles, per-core private, next-line prefetching
L1-D cache	32KB, 4-way, 64B line, 2 cycles, per-core private, write-invalidate, inclusive, 2 ports
L2 cache	1MB, 8-way, 2 banks, 64B line, 12 cycles, write back, inclusive, per-core private, high-performance device type
L3 cache	8MB, 16-way, 4 banks, 24 cycles, shared, low-standby-power device type
Main memory	2 QPI-like links (up to 64GB/s), 300 cycles
Coherence	MOESI-based directory protocol
Technology	3GHz, 0.9Vdd, 32-nm process

Table 2. Front-end and L2 P-state mechanisms

Front-end	A	B	C	D	L2	E	F	G	H
Base	*	16	16	128	Base	N	N	8	12
P1-FE	8	0	4	64	P1-L2	Y			13
P2-FE	4				P2-L2		Y		14
P3-FE	3				P3-L2			4	
P4-FE	2				P4-L2			2	
P5-FE	1				P5-L2			1	

A = Max in-flight unresolved branches. **B** = Checkpoint count. **C** = Fetch buffer size. **D** = Physical registers. **E** = Drowsy L2 data. **F** = Drowsy L2 tags. **G** = L2 associativity. **H** = L2 Access cycles. * = Unconstrained.

because the fetch is typically designed for infrequent window re-fills. Thus, in lower-performance P-states, we power-gate a portion of the fetch buffer, which in turn reduces accesses to the I-cache and instruction translation look-aside buffer.

Register file resizing. The preceding two techniques reduce in-flight instructions, creating opportunities to proportionally resize other structures as well. Although a more balanced approach is desirable, this initial work focuses on a major source of front-end power: physical registers. Dynamic physical register file resizing is not an easy task due to potentially scattered valid register mappings. Thus, when a lower P-state is requested, we stop allocating registers that will be power-gated (half of the registers in our experiments) and enter the P-state once those registers do not contain valid mappings. This operation occurs only once, and the power-gated portion remains fixed throughout the P-state.

3. L2 CACHE P-STATES

The second P-state implementation addresses static power in large caches. Static power imposes an upper limit on power savings achievable by frequency scaling, and large caches exacerbate the issue. As transistors become leakier in future technology nodes [7], it becomes more important to scale down static power along with dynamic power for large power savings.

This section examines existing circuit techniques for static power management and employs those techniques during lower P-states. Hence, we use them without any complex policies to fully enjoy the power-saving benefits of the techniques.

For gradual power-performance scaling, we enable five P-states (P1 through P5) in the L2 cache. P1 puts all the L2 data arrays into a drowsy mode to cut the L2 static power. Drowsy mode is an effective technique that supplies just enough voltage (drowsy voltage) to preserve the state of the memory cells and switches to a higher active voltage to safely read out the data [7]. By leveraging drowsy mode at the cache line granularity, significant static power reduction is possible. Accessing a drowsy line, however, takes an extra cycle to wake up, increasing the L2 access latency to 13 cycles in our implementation (Section 4). Given the goal of this work—providing a lower power-performance curve than frequency scaling—we are more interested in aggressive power reduction, and hence we simply apply drowsy mode uniformly to the cache.

P2 also puts the L2 tag arrays into drowsy mode, lengthening the access latency by another cycle (i.e., 14 cycles).

The remaining three P-states (P3 to P5) gradually trade off the L2 cache capacity for power by power-gating some of the associative ways, which is permanent during each level. Again, we apply this technique uniformly to all L2 sets for simplicity and maximum power savings, rather than optimizing for performance [17]. When entering each of the three P-states, any dirty lines in the ways that will be power-gated must be written back to the L3. Without attempts to minimize the associated power and latency cost [17], this work simply writes back dirty lines before gating. As expected, Section 5 shows that the performance impact of the reduced associativity is workload-dependent, depending on the workload's working set size, data locality, and re-use distance. However, compared to P1 and P2, power-gating a portion of the L2 completely eliminates the static power of the disabled portion *and* reduces the dynamic energy of L2 tag accesses.

4. METHODOLOGY

Our target machine is an 8-core CMP. Each node consists of a core, private L1 and L2 caches, and one bank of shared L3. We simulate the design with a full-system cycle-accurate simulator using Virtutech Simics [13], GEMS's Ruby [15], and in-house timing-first processor models. Because frequency scaling is often locally applied [11], we run single-threaded workloads from the SPEC CPU 2006 benchmark suite [9] and the Wisconsin commercial workloads [1] on a single core while power-gating the other cores for the initial investigation of our work. The workloads were compiled for the 64-bit SPARC ISA with the Sun Studio 11 compiler, and we simulate each workload for 100 million instructions on an unmodified SPARCv9 operating system after fast-forwarding the initialization phase and warming up architectural state for two million instructions. For architectural-level chip power approximations, we augmented our simulators with Wattch [3] and CACTI 5 [20]. Our power model assumes aggressive clock gating of logic structures not in use with no reactivation delay.

We use a 4-way OoO superscalar processor as our baseline core design. Table 1 details the configuration of the core and the

Table 3. Workload characterization

Workload	L3 MPKI	BPKI	SPKI	L2 Sensi-tivity	Workload	L3 MPKI	BPKI	SPKI	L2 Sensi-tivity	Workload	L3 MPKI	BPKI	SPKI	L2 Sensi-tivity
libquantum	36.1	251	.003	1.0	astar	2.9	129	23.8	2.1	h264ref	0.5	6.7	2.8	12.0
mcf	36.0	205	6.7	1.3	jbb	2.0	172	11.7	3.8	bzip2	0.4	166	2.4	7.9
bwaves	28.1	3	0.2	1.2	soplex	1.6	200	6.0	8.0	gobmk	0.4	140	19.0	40.0
milc	23.8	14	.009	1.0	xalancbmk	1.4	212	4.5	1.7	sjeng	0.3	137	18.4	14.9
lbm	22.9	12	0.1	1.5	oltp	1.2	155	20.1	8.1	tonto	0.3	74	8.2	13.4
Gems	15.1	6	0.1	2.9	gromacs	1.1	12	0.9	3.7	wrf	0.2	120	2.3	14.8
zeus	8.8	153	15.6	4.2	hmmer	0.9	6	0.2	1.4	namd	0.2	53	2.4	22.0
apache	8.1	154	20.7	4.0	omnetpp	0.9	231	7.9	12.4	povray	0.2	101	3.7	57.0
sphinx3	7.9	56	10.3	1.7	gcc	0.8	185	13.0	9.3	gamess	0.1	53	2.7	89.2
perlbench	5.8	176	5.2	2.3	leslie3d	0.7	116	13.2	19.6	HMean	1.7	17	0.1	13.8
cactus	4.2	1	0.1	2.2	calculix	0.6	45	1.4	5.5					
zeusmp	3.4	16	0.1	5.9	dealII	0.5	43	5.7	5.5					

L3 MPKI = L3 misses per thousand instructions. **BPKI** = Branches per thousand instructions. **SPKI** = Branch-misprediction-caused squashes per thousand instructions. **L2 sensitivity** = Increase in L2 miss rate from an 8-way L2 to a direct-mapped L2. Shaded cells indicate workloads presented in graphs.

memory hierarchy. Our frequency-scaling model is based on IBM's POWER7 [11], which implements per-core frequency scaling. We therefore assume that core frequency scaling affects only the associated L1 and L2; the frequencies of the L3 and the main memory remain the same. Furthermore, we idealize asynchronous domain crossing as having no overhead, resulting in an optimistic frequency-scaling model. Although POWER7 allows a fine-grained frequency step of 25 MHz, we simulate only the nominal and minimum frequencies and three intermediate frequencies while fixing the voltage, and use linear interpolation to estimate the remaining ones. Thus, the simulated normalized core frequency points are 1.0, 0.88, 0.76, 0.64, and 0.50.

We derive DVFS curves by assuming a 22% operating voltage range based on Intel Pentium M. We factor in temperature fluctuations when estimating the corresponding static power.

Table 2 summarizes our proposed microarchitectural P-states, using the notation of <P-state>-<mechanism type>. The left half of the table is for front-end P-states, which selectively stall renaming with varying degrees of aggressiveness. All the P-states, except P0, power-gate the entire checkpointing hardware, three quarters of the baseline fetch buffer, and half of the baseline physical register file and free list. The right half of Table 2 summarizes the L2 P-states that gradually trade the L2 cache performance for power. We model a drowsy cache based on the work by Flautner *et al.* [7].

5. EVALUATION

This section compares the power-performance of our microarchitectural P-states to both frequency scaling and DVFS. Our goal is to do better than frequency scaling's 1:1 power-to-performance ratio and approximate the 3:1 ratio of DVFS.

Table 3 lists the crucial characteristics of all workloads. As discussed earlier, frequency scaling is less effective in reducing power usage of memory-intensive workloads. Memory intensity, in this case, is measured by L3 misses per thousand instructions (MPKI). The front-end P-states exploit branches per thousand instructions (BPKI) and branch-misprediction-caused squashes per thousand instructions (SPKI) to gradually reduce the aggressiveness of the front-end. The performance sensitivity of the L2 P-states depends on the workload's L3 MPKI and L2 miss-rate sensitivity to smaller associativity (L2 Sensitivity). Due to space constraints, we show only the results of best, worst, and typical workloads for each P-state implementation, as well as the harmonic mean of all workloads.

Front-end P-states. Figure 3 presents power and performance normalized to the baseline for best (*gobmk*), typical (*povray*), and worst (*bwaves*) workloads as well as the harmonic mean of all SPEC and the commercial workloads. The data points labeled *Freq Scaling* are simulated results using the configurations in Section 4, and *P*-FE* represents *P1-FE* through *P5-FE* in Table 2. *Analytical DVFS* represents the derived DVFS curve for each workload. Although the degree and the scalability vary, all of our configurations yield lower power-performance curves than frequency scaling, indicating a more efficient power-performance trade-off. Furthermore, some of the *P*-FE* data points of the best and typical workloads even lay on top of the DVFS curves. We verified that the effectiveness of our configurations holds true for the rest of the workloads, and summarize the results using harmonic mean.

These favorable results demonstrate that the previously proposed techniques we have redeployed become viable alternatives to frequency scaling, even though the individual techniques by themselves may not have worked well under peak performance restrictions. Coordinating the techniques reduces the capability of structures designed for worst-case performance and decreases wasteful energy. In contrast, frequency scaling uniformly slows execution regardless of whether instructions are speculated or on the correct path.

The effectiveness of the front-end P-states, however, depends largely on the workloads' BPKI and SPKI due to the inherent dependency of branch characteristics. The higher the BPKI, the more unresolved branches exist in an instruction window, creating more opportunities for stalling the rename stage. Similarly, the higher the SPKI, the greater the reduction in wrong-path instructions, which in turn results in less wasted energy. Hence, high-BPKI and high-SPKI workloads (e.g., *gobmk*) exhibit the most power savings for the amount of performance loss and also scale well with different stalling aggressiveness. The converse—

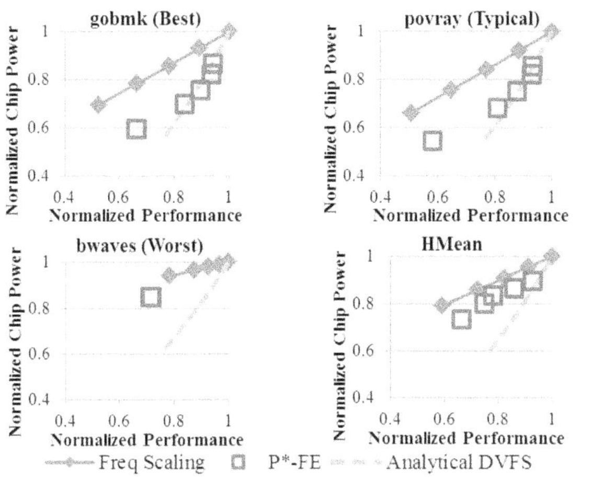

Figure 3. Power-performance of front-end P-states normalized to the baseline (lower-right is better).

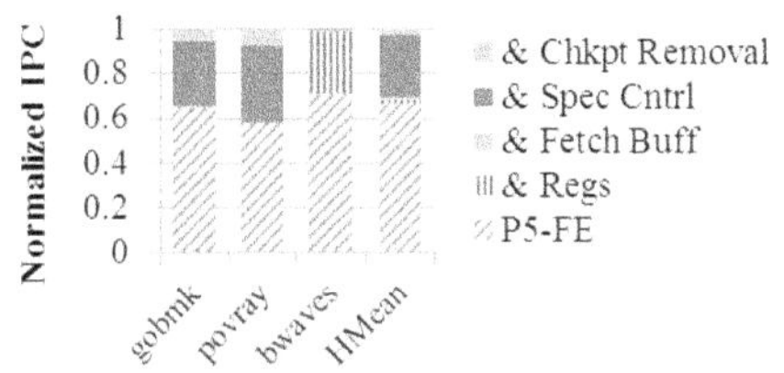

Figure 4. IPC impacts of the applied techniques with P5-FE.

that low BPKI and low SPKI exhibit less significant power savings—unfortunately also applies, a fact exemplified by *bwaves*. The low BPKI makes this workload category insensitive to front-end stalling; instead it suffers from fewer physical registers, resulting in one power-performance point for all P-states. However, only six out of 33 workloads fit in this category. Most workloads have enough BPKI for our technique to be effective, as represented by the typical workload *povray*.

To understand the performance impact of each front-end technique, we applied one technique at a time to the baseline and measured the resulting IPC degradations normalized to the baseline (Figure 4). In other words, each stack from top to bottom represents the following in cumulative form: IPCs when adding commit-time misprediction recovery (*& Chkpt Removal*); our most aggressive speculation control of allowing only one in-flight unresolved branch (*& Spec Cntrl*); fetch buffer resizing (*& Fetch Buff*); and register file resizing (*& Regs*). The height of the bottom *P5-1* stack represents the IPC of all techniques combined, normalized to the baseline.

As expected, disabling checkpoints has very small performance implications because of the low useful checkpoint rate in the baseline (Section 2). Even the high-SPKI workload, *gobmk*, has only 7% performance degradation from the longer commit-time misprediction penalties, and the fraction becomes negligible for low-SPKI workloads (*povray* and *bwaves*). Adding the most aggressive speculation control results in a more modest performance loss for high- to medium-BPKI workloads (*gobmk* and *povray*). In contrast, the low-BPKI *bwaves* allows few opportunities for the speculation control, thereby showing insensitivity to the technique.

Combining these two techniques with fetch buffer resizing has negligible performance impact for different reasons. On one hand,

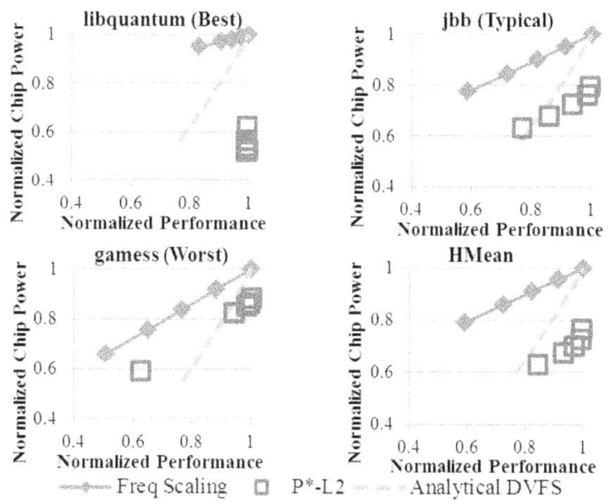

Figure 5. Power-performance of L2 P-states normalized to the baseline (lower-right is better).

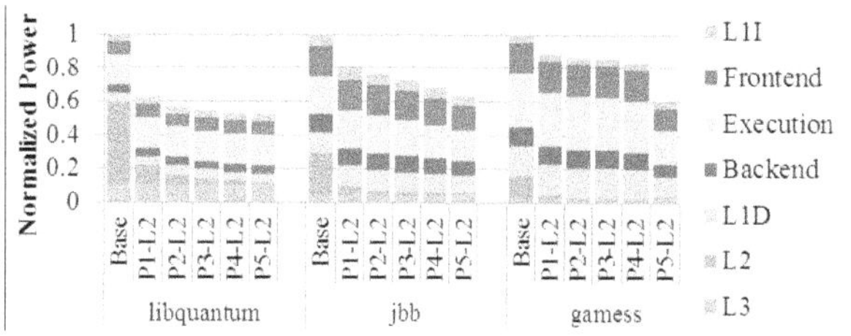

Figure 6. Power breakdown normalized to the baseline.

the speculation-control-sensitive workloads already have much fewer in-flight instructions than the baseline, and the reduced fetching capability does not significantly affect the overall performance. On the other hand, *bwaves* does not need a full-capability front-end to begin with because it has a mostly idle back-end due to the memory intensity as well as the infrequent misprediction-caused pipeline flushes. *bwaves* shows the most sensitivity (28%) to the reduced physical register count. The large IPC loss is also an indication that the baseline design is not skewed during P0.

Even though our approach targets only the core front-end, the rest of the pipeline and the caches also result in power reductions. *gobmk*, for example, lowers the power of the execution by 47%, back-end by 50%, and L1-D by 50% with *P5-FE* (results not shown) by scaling down the aggressiveness of the front-end.

L2 cache P-states. To evaluate the L2 P-states, we selected best (*libquantum*), typical (*jbb*), and worst (*gamess*) workloads. Figure 5 plots the power-performance scaling by our microarchitectural P-states (Table 2), frequency scaling, and the derived DVFS. With our mechanisms, all three workloads achieve much more power savings for the same performance than frequency scaling—and even more than DVFS in many cases—by addressing both dynamic and static power of the L2. We achieve these better power-performance trade-offs despite our simple, non-optimized use of drowsy mode and L2 way shrinking. The effectiveness depends on a workload's memory intensity and the L2 miss-rate sensitivity to smaller associativity. Memory-intensive workloads (e.g., *libquantum*) are the most significantly impacted because more than a third of the chip power comes from the static-power dominated L2 in the baseline. Hence, placing the L2 into drowsy mode reduces *libquantum*'s chip power by 43%. Although the rest of the workloads similarly benefit from drowsy mode, the much

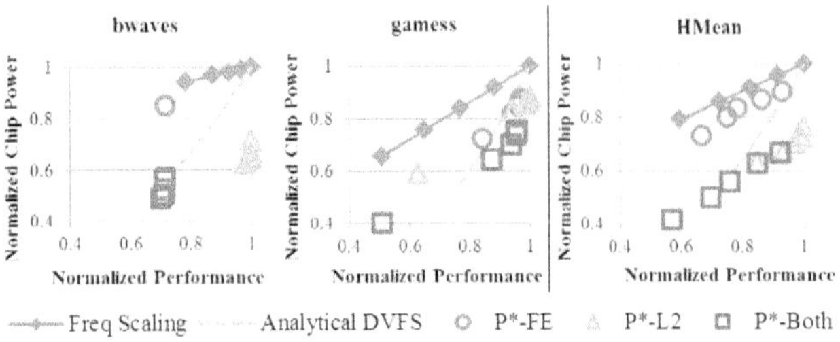

Figure 7. Power-performance of the combined P-states normalized to the baseline.

smaller L2 power fractions in the baseline cause the chip power reductions to be 24% and 14% for *jbb* and *gamess*, respectively.

P3-L2 through *P5-L2*, on the other hand, target both static and dynamic power of the L2 by power-gating some of the ways. Performance responds differently, depending on the L2 utilization and the miss-rate sensitivity to reduced associativity. Because the memory-intensive *libquantum* already has many misses to the memory, the workload has excellent power scaling with a 0.1% performance loss for up to 5% additional power savings. The compute-intensive *gamess* represents the opposite extreme. It has a small L2 power fraction, and its L2 miss rate remains largely insensitive as long as the L2 has at least two ways (*P4-L2*), resulting in the clustered four points in the upper-right region. However, *gamess* becomes the most sensitive workload once the associativity is reduced to one (*P5-L2*) and it experiences a significant decrease in performance. Yet our lowest P-state consumes 79% less power than frequency scaling for comparable performance. Most workloads (e.g., *jbb*) exhibit less L2 miss-rate sensitivity and yield more gradual power-performance trends, most of which outperform DVFS.

Combining front-end and L2 P-states. Because each of these two P-state implementations targets distinct parts of the processor, workloads that work well under one implementation do not always work as well under the other. Therefore, we applied the mechanisms together and plotted the resulting power-performance points in Figure 7. *P*-Both* represents the new combined P-state mechanisms, and the rest is the same as the data in Figure 3 and Figure 5. As expected, combining the mechanisms helps further optimize performance across all workloads. Our lowest P-state consumes less than half the power of frequency scaling, on average, for comparable performance. Even for *bwaves*, the worst-performing workload under the front-end P-state, implementing L2 P-states significantly decreases power with only a minor impact on performance, and implementing both front-end and L2 P-states reduces power even further, albeit with a larger impact on performance. For *gamess*, which is the worst-behaving benchmark when using only the L2 P-states, the combined lowest P-state uses only 61% as much power for comparable performance. In summary, all workloads perform as well or better than DVFS, making our P-state mechanisms a promising alternative to DVFS.

6. RELATED WORK

Previous microarchitectural designs also target low power [2] and/or a wider power range [10][12]. Most low-power proposals aim at improving power efficiency during peak performance, while we focus on efficient power and performance scale-down. Cebrian *et al.* use microarchitectural techniques to execute within a specific power budget [4], resulting in a complex control mechanism and ignoring performance impacts.

Our work incorporates many previously proposed techniques (Sections 2 and 3) but re-examines them in a different context: lowering the power scaling curve. Hence, we extract power savings those techniques offer for performance loss comparable to frequency scaling.

7. CONCLUSIONS

We proposed a new approach to implement P-states using previously proposed microarchitectural techniques. Our front-end and L2 P-states showed that we achieve more efficient power-performance scaling than frequency scaling across all workloads, and even exceed the power savings of DVFS in some cases.

8. ACKNOWLEDGMENTS

We thank John D. Davis, Mark D. Hill, and anonymous reviewers for their comments on the paper. This work was supported in part by the National Science Foundation (CNS-7020565, CNS-0916725, CCF-1017650, and CNS-1117280), Sandia/DOE (#MSN134324), and the University of Wisconsin. Wood has a significant financial interest in AMD.

9. REFERENCES

[1] Alameldeen et al. Variability in architectural simulation of multi-threaded workloads. In *Proc. of HPCA*, February 2003.
[2] Albonesi et al. Dynamically tuning processor resources with adaptive processing. *IEEE Computer*, 36(2):49–58, 2003.
[3] Brooks et al. Wattch: A framework for architecture-level power analysis and optimization. *Proc. of 27th ISCA*, 2000.
[4] Cebrian et al. Efficient microarchitecture policies for accurately adapting to power constraints. In *Proc. of the Intl. Symp. on Parallel & Distributed Processing*, 2009.
[5] Chandrakasan et al. Technologies for ultradynamic voltage scaling. In *Proc. of the IEEE*, 2010.
[6] Dreslinski et al. Near-threshold computing: reclaiming Moore's Law through energy efficient integrated circuits. *Proc. of the IEEE*, 2010.
[7] Flautner et al. Drowsy caches: simple techniques for reducing leakage power. In *Proc. of the 29th ISCA*, 2002.
[8] Gochman et al. The Intel Pentium M processor: microarchitecture and performance. *Intel Tech. J.*, 2003.
[9] Henning. SPEC CPU2006 Benchmark Descriptions. *Computer Architecture News*, 2006.
[10] Ipek et al. Core fusion: accommodating software diversity in chip multiprocessors. In *Proc. of the 34th ISCA*, 2007.
[11] Kalla et al. Power7: IBM's next-generation server processor. *IEEE Micro*, 2010.
[12] Kumar et al. Single-ISA heterogeneous multi-core architectures: the potential for processor power reduction. In *Proc. of the 36th Symp. on Microarchitecture*, 2003.
[13] Magnusson et al. Simics: A full system simulation platform. *IEEE Computer*, 2002.
[14] Manne et al. Pipeline gating: speculation control for energy reduction. In *Proc. of the 25th ISCA*, 1998.
[15] Martin et al. Multifacet's general execution-driven multiprocessor simulator (GEMS). *Comp. Arch. News*, 2005.
[16] Moshovos. Checkpointing alternatives for high performance, power-aware processors. In *Proc. of ISLPED* 2003.
[17] Naveh et al. Power and thermal management in the Intel Core Duo processor. *Intel Technology Journal*. 2006.
[18] Rogers et al. The Core-C6 (CC6) Sleep State of the AMD Bobcat x86 Microprocessor. In *Proc. of ISLPED*, 2012.
[19] Seznec et al. A case for (partially) Tagged Geometric history length branch prediction. *J. of Instr. Level Parallelism*, 2006.
[20] Shyamkumar et al. CACTI 5.1. Technical Report HPL-2008-20, Hewlett Packard Labs, 2008.
[21] Vandeputte et al. Offline phase analysis and optimization for multi-configuration processors. *Architectures, Modeling, and Simulation*, 2005.

Reducing L1 Caches Power By Exploiting Software Semantics *

Zhen Fang
NVIDIA Corp.
zfang@nvidia.com

Li Zhao
Intel Labs, Intel Corp.
li.zhao@intel.com

Xiaowei Jiang
Intel Labs, Intel Corp.
xiaowei.jiang@intel.com

Shih-lien Lu
Intel Labs, Intel Corp.
shih-lien.l.lu@intel.com

Ravi Iyer
Intel Labs, Intel Corp.
ravishankar.iyer@intel.com

Tong Li
Intel Arch. Group, Intel Corp.
tong.n.li@intel.com

Seung Eun Lee
Seoul National University of Science and Technology
seung.lee@seoultech.ac.kr

ABSTRACT

To access a set-associative L1 cache in a high-performance processor, all ways of the selected set are searched and fetched in parallel using physical address bits. Such a cache is oblivious of memory references' software semantics such as stack-heap bifurcation of the memory space, and user-kernel ring levels. This constitutes a waste of energy since e.g., a user-mode instruction fetch will never hit a cache block that contains kernel code. Similarly, a stack access will not hit a cacheline that contains heap data.

We propose to exploit software semantics in cache design to avoid unnecessary associative searches, thus reducing dynamic power consumption. Specifically, we utilize virtual memory region properties to optimize the data cache and ring level information to optimize the instruction cache. Our design does not impact performance, and incurs very small hardware cost. Simulations results using SPEC CPU and SPECjapps indicate that the proposed designs help to reduce cache block fetches from DL1 and IL1 by 27% and 57% respectively, resulting in average savings of 15% of DL1 power and more than 30% of IL1 power compared to an aggressively clock-gated baseline.

Categories and Subject Descriptors

B.3.2 [**Memory Structures**]: Design Styles

Keywords

First-level cache, virtual memory, ring level, simulation

1. BACKGROUND AND MOTIVATION

In a general-purpose processor, the first level cache is one of the most critical units: it has to both meet performance requirement and fit into the processor's power envelope. Most of the architecture-level power optimizations to caches have been designed for L2/L3 caches where slightly lengthened hit latencies and/or decreased hit rates are acceptable. However, these compromises are usually not acceptable for the L1 cache in a high-performance processor. Some of the power efficiency improvement techniques proposed for L1 caches [4, 6, 10] were adopted by earlier microprocessors. But latest commercial processors (Intel Nehalem/Sandy Bridge/Atom, IBM Power7[13], etc.), though more power-aware than before, have abandoned those designs in the L1 cache due to the hardware complexity they introduce and risks of lengthening the read hit cycle.

In this study, we take a novel approach to save the power consumption of the set-associative L1 cache without lengthening its access cycle or decreasing the hit rate. Specifically, we observe that a significant portion of the parallel accesses to the data/tag stores are unnecessary because many of the ways do not contain data whose software semantics are compatible with the memory request. By exploiting software semantics in the L1 cache design, we eliminate these unnecessary activities. The benefit is reduced dynamic power consumption in instruction fetches and data loads. The semantics that we will be using are 1) virtual memory, and 2) privilege ring levels.

Before we proceed to details of the proposed mechanism, we provide background on L1 caches in high-performance processors, as well as both software semantics.

Background on High-Performance L1 Caches

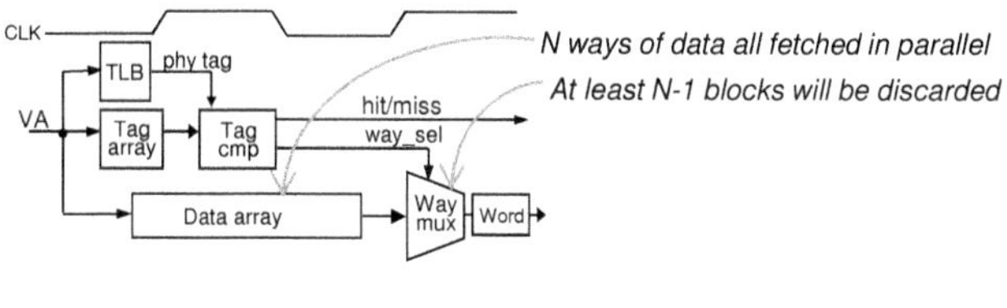

Figure 1: Read Pipeline in a High-Performance L1 Cache

L1 caches in high-performance processors have been aggressively optimized to achieve low latency for hits. Fig. 1 shows a representative timing diagram of a read hit in a physically-tagged set-associative L1 cache. Because the cache tags use physical address bits, tag comparison cannot start until the TLB has completed virtual-to-physical address translation. To meet the tight timing target, *in an N-way set-associative L1C, all N data blocks of the se-*

*This work was done when all authors were employed by Intel.

lected set are accessed in parallel, irrespective of the tag lookup result. At least N-1 blocks will be discarded later, resulting in low power efficiency. This is different than L2/L3 or low-power caches where tag check and data readout are serialized. L1 data and instruction caches of literally all modern general-purpose microprocessors are implemented in this manner.

Bifurcation of the Virtual Memory Space

Figure 2(a) shows a representative virtual memory map. The bifurcation design implies that in the virtual address space, the most significant bit (MSB) can differentiate stack and heap regions. However, this semantic is completely lost in the physical memory space, due to the way that the operating system manages page frames.

Figure 2(b) shows that on average, about 40% of all data memory references are accesses to the stack. The rest of the 60% consists of accesses to the heap, BSS and data segments, with heap being the dominant component. In the rest of the paper, we use the term 'heap' to refer to all three non-stack segments – heap, BSS and data. Although the stack accounts for 40% of all memory instructions, on average it only occupies 20% of the capacity on a 32KB, 8-way set-associative L1 data cache (DL1), as a result of the stack's relatively small footprint[7].

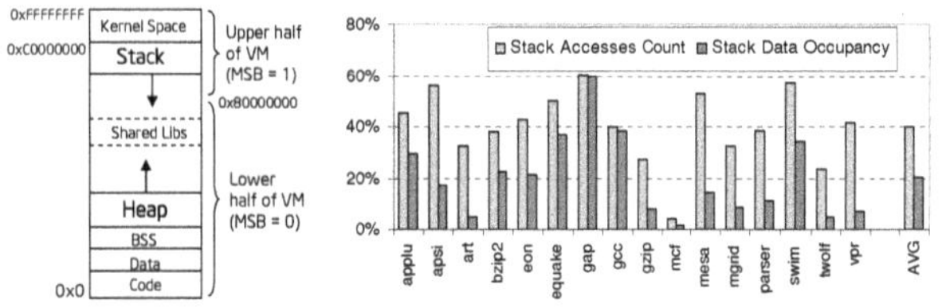

(a) Linux Addr Map (b) Stack Access Count and Avg Occupancy

Figure 2: Virtual Memory Map and Characteristics

Ring Levels and Instruction Cache Sharing

Ring level is a mechanism by which the operating system (OS) and the processor hardware cooperate to restrict what user-mode programs can do. All mainstream processors use similar mechanisms. For brevity, we base our discussion on the x86 architecture. When a user (kernel) program executes, the processor is set to the least (most) privileged level, ring 3 (ring 0). Rings 1 and 2 have been largely unused except by virtualization layers. The processor's current ring level is stored in the 2-bit CPL field of the Code Segment register.

User applications and the OS do not share code, but they share the instruction cache. There has been a number of work studying interaction of OS and microarchitecture[1, 9, 11]. They provide data on overall instruction count and cache miss rates of kernel and user codes, for example. One question remains unanswered – *at any given time, what percentage of the I-cache capacity is used by kernel code*?

One's intuition would suggest that in compute-intensive applications the I-cache be constantly filled with user application code, interfered only occasionally by kernel instructions brought in by events like system calls. We found that this intuition is not correct. Figure 3 shows histograms of dynamic instruction counts and occupancies of kernel-mode code in steady stages of two SPEC CPU benchmarks on a 32KB, 8-way L1 instruction cache (IL1). Each data point represents the average value in an interval of 25 μseconds. In each figure, we also give the algebraic mean values of both bands of data averaged over the 3400 intervals. As can be seen, although user code dominates instruction count, kernel instructions occupy a significant portion of the I-cache, clearly contradicting one's intuition. In *equake*, 76% of the cachelines actually contain kernel code.

The phenomenon can be explained by the different characteristics of kernel and user programs. User code usually spends the majority of the time in nested loops, while the OS is more likely to traverse non-iterative decision trees, occupying more I-cache entries. The tight loops in user code such as SPEC CPU benchmarks, on the other hand, have rather small static instruction memory sizes; user application execution tends to repeatedly fetch instruction from some hot ways of hot sets, which allows the (rarely used) kernel code to stay for a long time before being evicted.

I-caches do not take advantage of ring level information. Though a user-mode instruction fetch will never hit a cacheline that contains kernel-mode code, all ways in the selected cache set are still searched in parallel. For example, based on the profiling data of *equake* in Figure 3, statistically about $(1\text{-}2\%)\times 76\%+2\%\times(1\text{-}76\%)=75\%$ of all accesses to I-cache RAMs will result in tag mismatches and fetched instruction blocks being discarded. This observation motivates our optimization to the instruction cache.

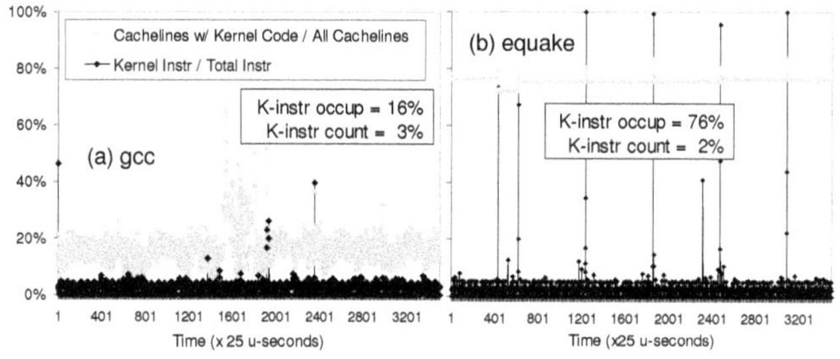

Figure 3: Kernel Code's Instr. Counts and Occupancies

2. TAGGING CACHELINES WITH SOFTWARE SEMANTICS

We propose to exploit software semantics in the cache design to avoid unnecessary associative searches, thus reducing dynamic power consumption. We present two software semantics-aware cache designs, for the data cache and instruction cache, respectively. In both designs, further search operations in the cache set are continued only if the *semantic tag* comparison gives a match.

Vtag (1 bit per cacheline): Tag each L1 data cacheline with virtual memory region information. Using the most significant bit of the virtual address can help to eliminate all the cross-checks between stack and heap data.

Rtag (2 bits per cacheline): Tag each L1 instruction cacheline with ring level information, represented by the processor's ring level status bits when the instruction is fetched and cached. For a user-level instruction fetch, this obviates the need to search the ways that contain kernel-level code, and vice versa.

A key advantage of the semantic tag over the physical tag is that the former can be checked concurrently with virtual-to-physical address translation, avoiding performance impact typically seen in a phased-access cache [6].

2.1 Augmenting DL1 with Vtags

Fig. 4 shows an implementation of a 32KB, 8-way L1 data cache augmented by Vtag. Thick lines denote new logic that we propose to add. The virtual page number is translated by the TLB to a physical page number, which will serve as the physical tag. When the N (N=8 in our example) tags are fetched out of the tag store, they are all compared with the incoming physical tag, and at most one comparison can be a hit. Data store accesses are performed in parallel with the N tag store accesses.

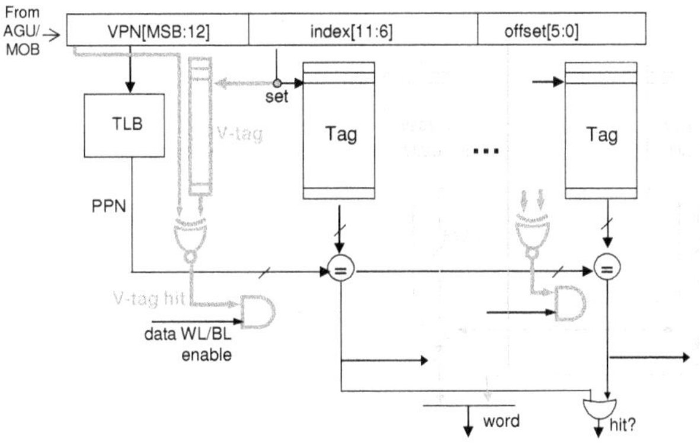

Figure 4: Eliminating Unnecessary DL1 Data Array Accesses using Vtags. Selective data readout from the SRAM arrays is realized by qualifying WL/BL operations with Vtag check results, eliminating unnecessary bitline discharges.

In a traditional cache, all N data subbanks have to be accessed. In a Vtag-optimized cache, we implement selective data readout from the SRAM arrays. To achieve this, we augment each cacheline with the most significant bit of its virtual address. When a data read arrives, Vtag check is performed in parallel with TLB access, in addition to the regular physical tag check. Data subbank i ($1 \leq i \leq N$) is accessed only if Vtag check = match for way i. The largest components of cache power dissipation, bitline discharge and sense amps ops, are eliminated for the Vtag mismatching ways. The Vtag design does not require changes to cache hit assertation and does not affect coherency protocols. Optimizing *tag* array accesses and DTLB is challenging because of the tight timing constraints. So regular physical tag check is performed regardless of whether Vtag is a hit or a miss. Fig. 5(a) shows the optimized DL1 pipeline for a read hit.

Using the example in Fig. 4, the Vtag logic consists of reading from a 64-bit store for each of the N subbanks followed by 1-bit XOR. We implemented the Vtag logic in 32nm CMOS technology. It can be completed in well under 0.08 nanoseconds. This gives enough time to propagate the way selection signals to the data store since it is in parallel with TLB look up. This extra delay will fit into the first phase of the clock even with frequency up to 4GHz.

Address Aliasing

Address aliasing refers to a special scenario in memory management where two virtual pages point to a same physical page. With respect to address aliasing, the Vtag mechanism does not change the cache's behavior except when a high virtual page and a low virtual page are backed by a same physical page. False match by Vtags will not occur since physical tag comparison is still performed. False miss, however, could occur and cause program execution errors. Because creation of an aliased mapping is performed by the OS memory manager, the OS can simply disable Vtag usage when such a mapping is created, and thus avoid false misses. Since intentional address aliasing is rare in practice (unintentional aliasing would indicate a bug in the OS), for the vast majority of applications, the Vtag feature will not need to be disabled.

2.2 Augmenting IL1 with Rtags

Conceptually, instruction cache's Rtag optimization is similar to data cache's Vtag optimization, except that Rtag uses the ring level bits of the *processor* while Vtag uses the highest order virtual address bit of the incoming *request*. One would derive an implementation of Rtag-augmented IL1 from Vtag in DL1. Fig. 6(a) shows how such a design can reduce accesses to the *data* store in a 4-way set-associative instruction cache. In such a design, similar to the Vtag case, we can not optimize IL1 *tag* array and ITLB accesses.

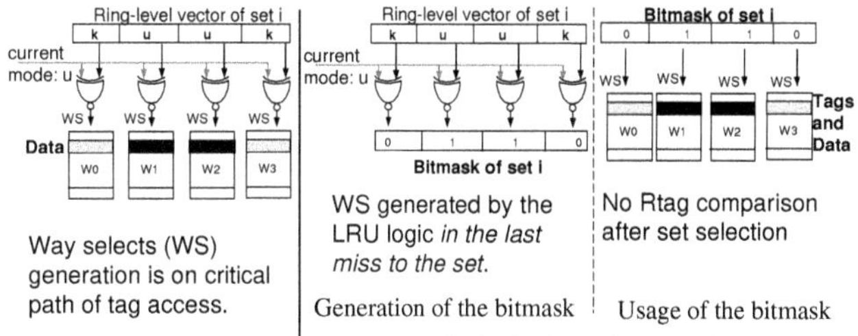

(a) Basic Implementation (b) Optimized Implementation

Figure 6: Using Rtags to Optimize ICache/ITLB Accesses

In Fig. 6(b), we present a better solution which removes ring-level check from the critical path of IL1 and ITLB tag array accesses. We make an observation that dynamically generating ring level match signals for every instruction fetch is not necessary, because these signals for a cache set do not change unless a cacheline replacement occurs in the set, or if the ring level of the current thread changes. We introduce a bitmask for each set, in addition to the ring level Rtag vector. The bitmask directly serves as way selection to enable/disable wordlines and bitlines. The left half of Fig. 6(b) illustrates how a bitmask is generated, while the right half shows how it is used. The bitmask generation and usage are decoupled. Compared with Fig. 6(a), the key difference is that Rtag comparison is performed when a cacheline is installed, not when it is looked up. When a cache set is selected by the indexing logic for an IL1 read, its bitmask is ready to use, *without the need for Rtag comparison*. In other words, the bitmask for each cache set is generated *before* any read has been made to the set. Compared with the basic implementation, in the decoupled mechanism, generation of the bitmask is no longer on the critical path of tag array reads; it is done in the cache miss handling process. When the processor's ring level changes, every bit in every bitmask is flipped.

With R ring levels in the processor, the total storage size for ring levels and bitmasks is $log(R) \times \frac{cache\_size}{cacheline\_size}$ bits and $log(R) \times Number\_of\_entries$ bits for the I-cache and I-TLB, respectively. Assuming $R = 4$, 32KB IL1, 64B/line, 128-entry ITLB, total storage overhead is 1024 bits + 256 bits = 160 bytes. The power consumption overhead of such a structure is minuscule compared with the number of subbank accesses that we will be saving.

Fig. 5(b) shows the pipeline of an IL1 hit with Rtag masking logic (the solid bars). The bitmask design allows us to place the Rtag check logic before IL1 physical tag and ITLB VPN processing, thus optimizing the power consumption of not only the data arrays, but also the tag arrays.

A Quick Summary on the Designs

Table 1 summarizes the optimizations that we have discussed. In DL1, Vtag uses a 1-bit XOR with the potential to eliminate a full cacheline data fetch. In IL1, for the vast majority of instruction fetches, Rtag uses a 1-bit way-selection mask with the potential to eliminate both the tag comparison and the data array access. Our Vtag and Rtag designs require no changes to cacheline replacement logic, does not change cache hit/miss rates, and has no impact on cache hit latency. The power consumption overhead of the new logic is minuscule compared with the number of subbank accesses that we will be saving. Being only a small extension to most L1

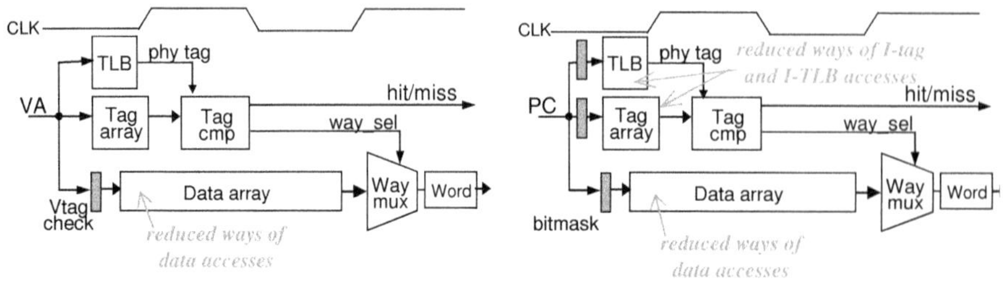

(a) Data Cache (b) Instruction Cache and Instruction TLB

Figure 5: Pipelines of Vtag- and Rtag-Optimized L1 Caches

cache implementations in general-purpose CPUs, we believe that the total cost of commercial processor adoption is low.

Table 1: Summary of cost and benefits. Storage overheads are for the entire DL1/IL1 based on an Intel Nehalem-like processor.

Extra Bits	Cache	Tag Array?	Data Array?	Overhead
Vtag: MSB of Virtual Addr	DL1	not optimized	optimized	64 bytes
Rtag: Ring Level, Bitmask	IL1	optimized	optimized	128 bytes

2.3 Qualitative Discussion on Power Savings

The effectiveness of the proposed optimizations primarily depends on the overall $selectivity$ of the semantic tags throughout the application execution. For a cache access, $selectivity$ is defined as the number of semantic tag matches divided by the cache associativity. For example, if the Vtag of an incoming data read matches the stored Vtags of 2 of the 8 ways in the set, the selectivity is 25% for this access. A lower selectivity value is desirable since more unnecessary data array accesses would be eliminated.

For an access to a Vtag-optimized D-cache,

$$P\_DL1 = P\_tag\_data\_lkg + P\_dec + P\_tag\_dyn + selectivity \times P\_data\_dyn + P\_vtag\_ovhd \quad (a)$$

For an access to a Rtag-optimized I-cache and I-TLB,

$$P\_IL1 = P\_tag\_data\_lkg + P\_dec + selectivity \times (P\_tag\_dyn + P\_data\_dyn) + P\_rtag\_ovhd \quad (b)$$

When there is no access to DL1, IL1 or I-TLB,

$$P\_idle = P\_tag\_data\_lkg + residual\_factor \times (P\_dec + P\_tag\_dyn + P\_data\_dyn) \quad (c)$$

Figure 7: Power Consumption Components. selectivity is the key parameter to be obtained through simulation.

Vtag and Rtag reduces power consumption of different structures, as is shown in Table 1. In an N-way set-associative array, accessing M of the N ways consumes approximately M/N of the baseline dynamic power by gating the clocks to the bitlines and segmented wordlines of the rest $N - M$ subarrays. When clock gated, the cache power consists of a constant leakage power and some residual dynamic power, given in Equation (c) in Fig. 7. Aggregating Equations (a), (b) and (c) for the whole application simulation gives us the overall average power of DL1 or IL1.

3. EVALUATION

3.1 Simulation Methodology

We use a trace-driven, cycle-accurate platform simulator [16] to model the performance aspect of the architecture. To estimate power dissipation, we integrated microarchitecture-level power estimation tool, Wattch[3], into ManySim. We plugged in 45nm CMOS technology parameters, obtained through MOSIS, into Wattch. Leakage is assumed to be 20% of the overall baseline power for L1 cache, and is not optimized by our design. We calculate energy consumption of the caches using the model in Section 2.3 on a cycle-by-cycle basis and integrate the numbers to derive the average power. Table 2 shows the key simulation parameters. In particular, the L1 cache parameters are representative of the latest high-performance processors like Intel Nehalem and Sandy Bridge.

Application traces of SPEC CPU2000 and SPECjappsServer04 were collected using full-system emulator SoftSDV[14] running Fedora 10. After skipping the warm-up phases, SoftSDV captured all fetched instructions and dumped them into traces. These traces faithfully contain the dynamic instruction sequence of all user and kernel instructions. In most experiments, we run each trace for 1 billion instructions in a steady stage of the applications.

Table 2: Simulation setup. (Bold fonts denote default values.)

Parameter	Value
Processor core	3.2GHz, out-of-order
L1 I-cache	32KB, 64B/line. **8-way** and 4-way, LRU
L1 D-cache	32KB, 64B/line. **8-way** and 4-way, LRU
TLB	Data: 96 entries, 8-way Instr: 64 entries, 8-way
L2 cache	64B/line, 8-way, 4MB, LRU
Latencies	L1=3 cycles; L2=12 cycles. DRAM=120ns

3.2 Experiment Results: Vtag on DL1

We present the selectivity profiles of SPEC CPU, shown in Figure 8. The data are collected on an 8-way set-associative DL1 setup. In the figure, for each application, the four segments represent the percentages of different selectivity ranges. For example, the '≤75%' segment of *applu* has a value of 28% (the yellow portion), indicating that 28% of all L1 data cache reads' Vtags hit either 5 or 6 of the Vtags in their respective selected sets. On average, about 19% of memory accesses have ≤25% Vtag matches in their respective cache sets, denoted by the bottom segment of the *Avg* bar. For these memory reads, their Vtags match 0, 1, or 2 of the Vtags in their respective selected sets. This means that for 19% of all DL1 accesses, only up to two data subbanks need to be activated on our 8-way cache. Because Vtag does not optimize write instructions, all write instructions count as 100% match in the statistics.

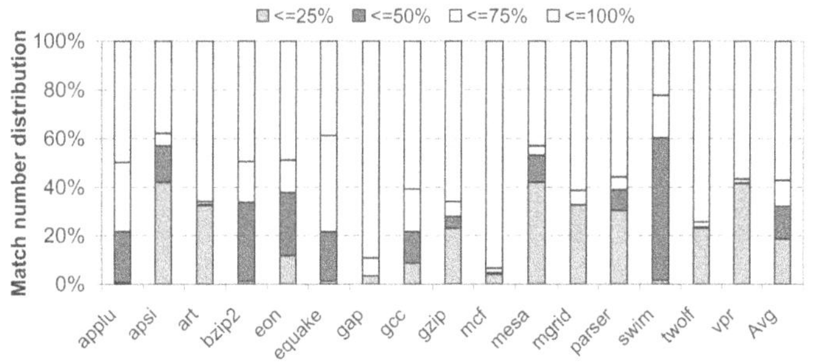

Figure 8: Distribution of Vtag Selectivity in DL1. Lower selectivity implies bigger power savings.

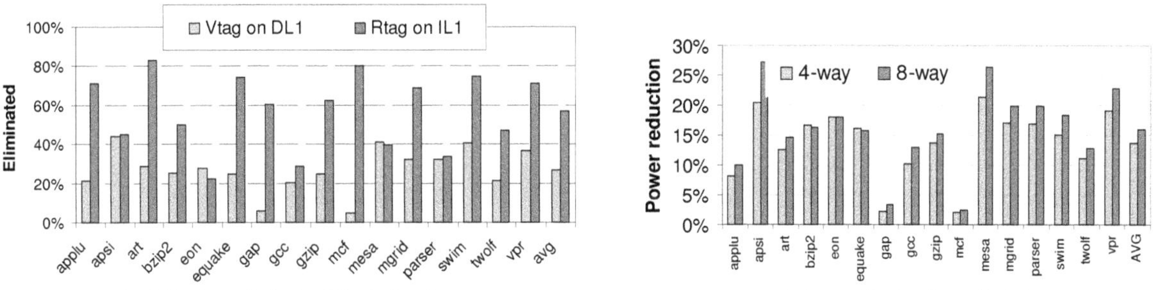

(a) Eliminated Cache Block Fetches

(b) Power Reduction of DL1

Figure 9: Eliminated Cache Block Fetches in DL1 and IL1, and Power Reduction in DL1

High match percentages are usually a result of unbalanced numbers of stack/heap accesses. For example, in *mcf*, stack reads only account for 4% of all memory references, and most cachelines contain heap data. Incoming heap reads will see their Vtag match most of the Vtags in the selected sets, while incoming stack reads will mostly have low selectivity. Since reads are now predominantly heap accesses, the result is dominance of the $\leq 100\%$ segment for *mcf*. By contrast, 56% of *apsi*'s DL1 accesses are user stack reads. User stack writes, user non-stack accesses and kernel data structures account for the rest, 44%. As a result, we observe a rather good distribution for *apsi*.

Using equations discussed in Section 2.3, we estimate total DL1 power consumption for the benchmarks. Overall, Vtag is able to eliminate 27% of all fetches from the DL1 data array for SPEC00, shown in Fig. 9(a). Fig. 9(b) shows estimated power reduction percentages. On average 13% and 17% of total cache power is reduced using Vtag on 4-way and 8-way L1 data caches for SPEC CPU.

3.3 Experiment Results: Rtag on IL1

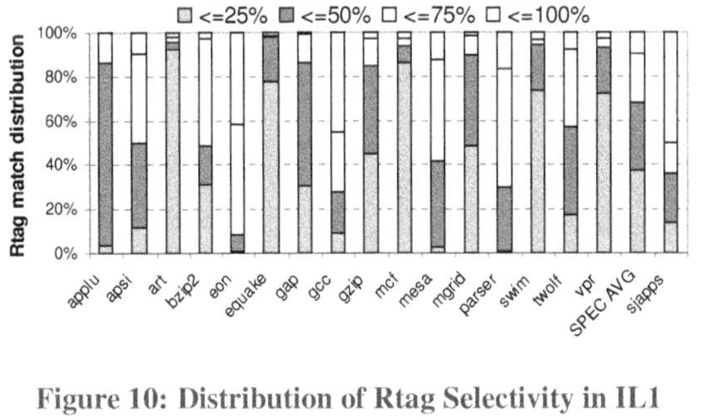

Figure 10: Distribution of Rtag Selectivity in IL1

Selectivities of Rtag are presented in Fig. 10. Apparently, Rtag is more effective on IL1 than Vtag is on DL1, indicated by the overall lower selectivity values. A direct comparison of eliminated cache structure accesses is shown in Fig. 9(a) where Rtag eliminates 57% of all fetches from the IL1 tag and data arrays for SPEC00. As a result, the relative power saving in the I-cache is even better than Vtag's effects on D-cache, shown in Fig. 11. For instance, SPEC CPU average power reduction is over 30% in IL1 compared to about 15% in DL1.

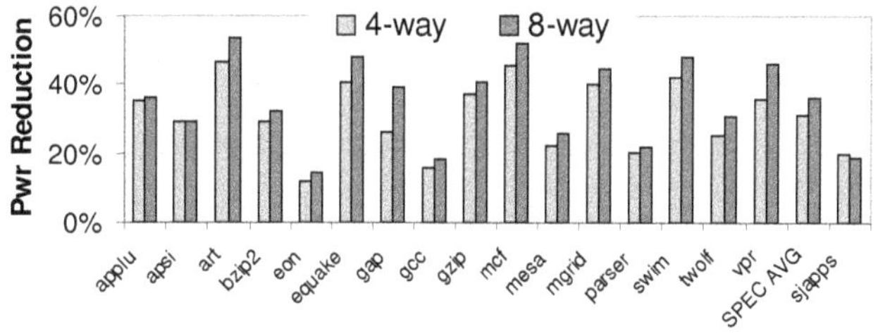

Figure 11: Power Reduction of IL1 using Rtag

4. RELATED WORK

The most closely related bodies of work on the data cache are Semantic-Aware Cache Partitioning and Way Prediction.

Semantic-Aware Partitioning (SAP) Lee [7] uses separate data caches for different virtual memory regions. Hard partitioning the L1 cache has two drawbacks. First, performance-wise the rigid partitioning will inevitably lead to low utilization of the precious L1 cache space. Second, virtual address aliasing would require expensive mechanisms to resolve.

Way Prediction (WP) Some commercial processors, including Pentium4 and UltraSPARC-III, use so-called microtag to achieve virtual address (VA) bits based way prediction [12], in order to reduce L1 cache power. The lower section of the tag ('loTag') is first checked and the hit way data is speculatively fetched. The remaining bits ('hiTag') are also checked to determine if the access was a

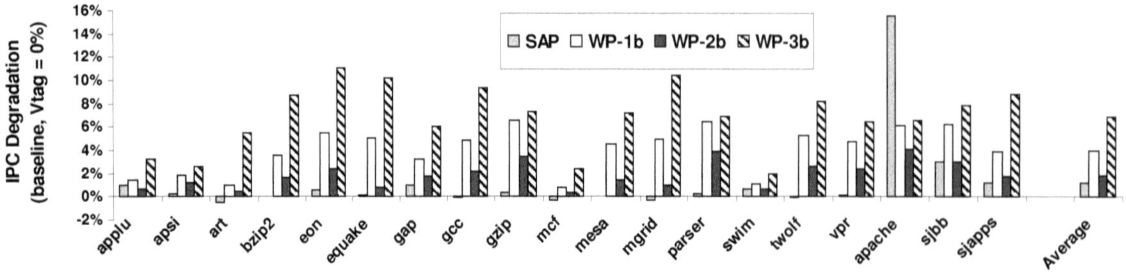

Figure 12: Performance Impact of Semantic Aware Partitioning and VA Based Way Prediction

true hit. To enforce uniqueness with loTags of all the ways in each set, if an incoming block has the same loTag bits as a block in the set, the latter has to be evicted. Premature evictions impact caching efficacy. One solution is to use more bits for loTag. But using more bits for loTag can add extra delay to L1 hits. In fact, our gate-level simulation in a 32nm, 3.0Ghz process revealed that 2-bit check can be finished early enough to control the wordline and bitline of the SRAM without adding an extra clock cycle, while loTag of 3 bits or more cannot.

Fig. 12 presents the performance of SAP and WP when compared to the baseline (Vtag and conventional cache). *WP-xb* denotes way prediction with a loTag of x bits. For SAP, the best performing partition(8KB for stack and 24KB for heap) causes performance loss as high as 15%, with an average of 1.2%. *WP-1b*, *WP-2b* and *WP-3b* use the lowest 1, 2, and 3 bits of the virtual page number as the way prediction source. Fig. 12 shows that conventional VA-based way predictions introduce perceivable performance loss. The IPC degradations of *WP-1b* and *WP-2b*, on average 4.2% and 1.9% respectively, have been caused by premature eviction of cachelines. The performance loss of 6.8% with *WP-3b* is primarily due to the extra clock cycle of read hits.

Besides SAP and WP, researchers[5, 8] designed specialized stack caches to exploit software semantics for performance or power. A phased-access cache[6] serializes tag check and data RAM aess, improving power efficiency at the cost of longer cache hit latency. Predictive way access is first used in the L2 cache in MIPS R10000 [15]. More complex mechanisms usually use a non-trivially sized prediction table to help to speculate on the way that is more likely to match an incoming request. Various prediction sources have been tried including the program counter (RSA [2]) and register numbers (PSA [4]). The fundamental difference than our work is that predictive way access is speculative in nature. Mispredictions result in longer hit latencies and also consume more of the valuable L1 cache read bandwidth due to additional lookups. In order to get satisfactory prediction accuracy, these methods have to introduce fairly high hardware complexity. For an 8KB, 32B/block cache, RSA would need about 2K bytes of storage for prediction [2], while PSA uses a 1024-entry table [4]. Extra storage for Vtag is only 256 bits for such a cache.

Compared with the large body of work dedicated to the data cache, the instruction cache in general-purpose CPUs is a relatively less explored area. On user/kernel instruction mix side, as we have discussed in Section 1, researchers presented detailed information that helps on design issues such as I-cache size and associativity [1, 9, 11]. However, these studies provide little insight on how user code and kernel code divide the I-cache capacity. To the best of our knowledge, this is the first published study that reveals the high *occupancies* of kernel-mode codes in the instruction cache, and also the first work that exploits user/kernel modes information to optimize the instruction cache.

5. SUMMARY AND FUTURE WORK

In this paper we present design and evaluation of software semantics-aware L1 caches. Our design is transparent to application software and is performance-neutral. A significant percentage of cache array accesses is eliminated, resulting in an average of over 15% and 30% power savings in L1 data and instruction caches. Overall, we believe that the cost to incorporate the optimizations into commercial products is low, with great power saving capabilities for literally all applications on a high-performance processor.

6. REFERENCES

[1] L. A. Barroso et al. Memory system characterization of commercial workloads. In *ISCA-25*, 1998.

[2] B. Batson and T. N. Vijaykumar. Reactive-associative caches. In *PACT*, 2001.

[3] D. Brooks, , V. Tiwari, and M. Martonosi. Wattch: A framework for architectural-level power analysis and optimizations. In *ISCA-27*, 2000.

[4] B. Calder, D. Grunwald, and J. Emer. Predictive sequential associative cache. In *HPCA*, 1996.

[5] M. Huang et al. L1d cache decomposition for energy efficiency. In *ISLPED*, 2001.

[6] R. Kessler et al. Inexpensive implementations of set-associativity. In *ISCA*, 1989.

[7] H. Lee and C. S. Ballapuram. Energy efficient D-TLB and D-cache using semantic-aware partitioning. In *ISLPED*, 2003.

[8] H. Lee et al. Stack value file: Custom microarchitecture for the stack. In *HPCA*, 2001.

[9] T. Li et al. Understanding and improving operating system effects in control flow prediction. In *ASPLOS*, 2002.

[10] M. D. Powell et al. Reducing cache energy via way-prediction and selective direct-mapping. In *MICRO*, 2001.

[11] J. A. Redstone et al. An analysis of operating system behavior on SMT architecture. In *ASPLOS*, 2000.

[12] G. W. Shen and C. Nelson. MicroTag for reducing power in a processor, US Patent 7117290, 2006.

[13] B. Sinharoy. POWER7 multi-core processor design. In *MICRO-42*, 2009.

[14] R. Uhlig et al. SoftSDV: A presilicon software development environment for the IA-64 architecture. *Intel Technology Journal*, 3(4), Nov. 1999.

[15] K. C. Yeager. The MIPS R10000 processor. *IEEE Micro*, 16(2), 1996.

[16] L. Zhao et al. Exploring large-scale CMP architecture using ManySim. *IEEE Micro*, 27(4):21–33, Apr. 2007.

DRAM Power-Aware Rank Scheduling

Sukki Kim
LG Electronics
Seoul, South Korea
sukki.kim@lge.com

Soontae Kim
Dept. of Computer Science, KAIST
Daejeon, South Korea
kims@kaist.ac.kr

Yebin Lee
Dept. of Computer Science, KAIST
Daejeon, South Korea
smartcode@kaist.ac.kr

ABSTRACT
Modern DRAMs provide multiple low-power states to save their energy consumption during idle times. The use of low-power states, however, can cause performance degradation because state transitions from low-power states to an active state incur time penalty. To effectively utilize the low-power states, we propose DRAM power-aware rank scheduling schemes applied to the last-level cache and the memory controller. Our scheme utilizing the last-level cache reduces write requests to DRAM and the state transitions by replacing cache blocks based on their dirty states and DRAM rank power states. Our scheme utilizing the memory controller decreases the state transitions with rank power state-aware batch writes. With the second scheme, the states transitions are reduced by 21.2%, on average. Consequently DRAM energy consumption is reduced by 11.2%, on average, with no performance loss.

Categories and Subject Descriptors
B.3.2 [**Memory Structures**]: Design Styles

Keywords
DRAM power and energy efficiency, low-power, rank scheduling.

1. INTRODUCTION
DRAMs are popularly used as the main memory in computing systems and consume up to 40% of total system energy [1]. Therefore, it is crucial to reduce DRAM energy consumption to reduce total system energy consumption. Modern DRAMs provide multiple low power states to reduce idle energy consumption. However, using the low power states without a proper strategy would incur large performance degradation and energy consumption.

There are many approaches proposed to utilize multiple low power states. Among them, actively reshaping memory traffic scheme [2] and the traffic skew scheme [3] utilize multiple ranks for power reduction. Their fundamental philosophy is the same; they divide ranks into hot (non-prioritized) ranks and cold (prioritized) ranks by force and then throttle memory traffic to cold (prioritized) ranks in order to combine short idle periods into longer ones. The actively reshaping memory traffic scheme is a software approach. It monitors memory traffic and then gathers frequently accessed pages into hot ranks by page migration. However, page migration overhead and memory request competition in hot ranks incur performance degradation. In contrast, the traffic skew scheme is a hardware approach. It chooses prioritized ranks in a round robin fashion, prohibits replacements of cache blocks going to the prioritized ranks, and lowers self-refresh threshold of the prioritized ranks. However, it suffers from severe performance degradation because prioritized ranks enter self-refresh state too early without considering read traffic going to them. In these schemes, performance degradation is not avoidable due to tradeoff between power consumption and performance. However, increased execution time requires more energy consumption not only in DRAM but also in all other parts of computer system, which may aggravate total system energy consumption.

To overcome the drawbacks of the above schemes, we propose DRAM power-aware rank scheduling schemes utilizing the last-level cache and the memory controller. Our scheme utilizing the last-level cache reduces write requests to DRAM and state transitions from a low-power state to an active state by selecting replaced cache blocks by the following priority rule (clean block > dirty block mapped to the active or idle rank > dirty block mapped to the power-down rank > dirty block mapped to the self-refresh rank) on cache misses. This priority rule is based on the following two facts. First, a clean cache block does not generate a write request to DRAM reducing write power consumption. Second, a dirty block mapped to an active rank does not incur a state transition from a low-power state to an active state. Our scheme utilizing the memory controller increases rank idleness and decreases state transitions by deferring write requests in the transaction queue and by batch-writing the delayed write requests considering rank power states. **Our fundamental philosophy is to use ranks naturally without dividing them into two groups by force as done by the traffic skew scheme.**

We evaluated our schemes and the traffic skew scheme. We used cycle-accurate CPU and DRAM simulators, and SPEC 2000 benchmark suite. With the traffic skew scheme, DRAM power consumption is reduced by 14.8% but execution time is increased by 25.5%. As a result, DRAM energy consumption is increased by 6%. In contrast, our scheme utilizing the last-level cache reduces write requests, DRAM power and energy consumptions by 6.8%, 7.1% and 4.7%, respectively, on average with small performance degradation. Our scheme utilizing the memory controller reduces state transitions, DRAM power and energy consumptions by 21.2%, 9.8% and 11.2%, respectively, on average, with no performance degradation.

This paper is organized as follows. Section 2 explains DRAM memory system organization and power states. Section 3 describes our DRAM power-aware rank scheduling. Section 4

ISLPED'12, July 30 - August 1, 2012, Redondo Beach, California, USA.

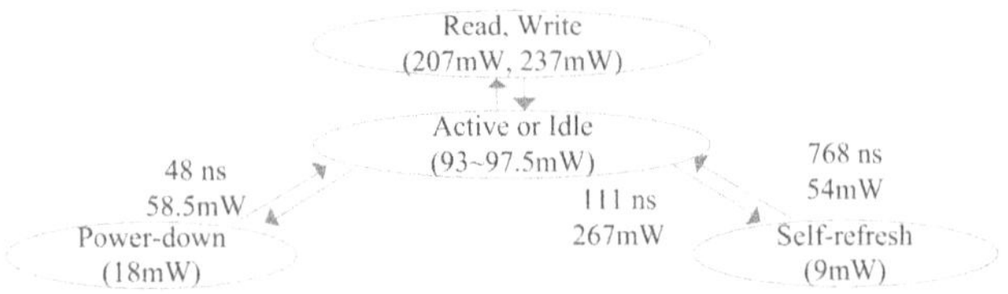

Figure 1. Simplified power states of 1Gb DDR3 SDRAM [13].

presents experimental results. Section 5 discusses conventional approaches and their limitations. Section 6 concludes the paper.

2. BACKGROUND

2.1 DRAM Memory System Organization

Modern DRAM memory system contains multiple ranks. A rank is the smallest power unit which can be controlled by the memory controller. Each rank consists of multiple DRAM devices. These multiple DRAM devices in the same rank are grouped together to form memory words. Each DRAM device consists of multiple banks, each of which has a row buffer for latching opened row data.

2.2 DRAM Power States

Modern DRAM devices provide multiple power states to reduce energy consumption during idle times. Figure 1 shows simplified power states of a DRAM device. The active or idle state is a standby state that can respond to read or write requests. The state of row buffers determines the active or idle state. When any row buffer is open, the power state of the rank is the active state. When all row buffers are closed, the power state of the DRAM rank is the idle state. In the active or idle state, each DRAM device consumes 93~97.5mW. Low-power states include the power-down state and the self-refresh state. During the power down state, DRAM deactivates all input and output buffers. The memory controller must be alive to deal with other commands such as refreshes. During the self-refresh state, DRAM deactivates all parts except the refresh logic which internally refreshes rows without external clocking and the memory controller is powered down. DRAM device can reduce idle power consumption when the device is in low-power states such as the power-down state and the self-refresh state. The power-down state consumes 18mW and the self-refresh state consumes 9mW. However, using the low-power states needs transition power and time. For example, transition from the power-down state to the active state needs power and time penalty of 58.5mW and 48ns, respectively. In addition to the power-down state, transition from the active state to the self-refresh state needs power and time penalty of 267mW and 111ns, respectively, because entering the self-refresh state requires at least one refresh operation [10, 11]. Transition from the self-refresh state to the active state needs power and time penalty of 54mW and 768ns, respectively. DRAM device cannot handle other requests and consumes power during the wake up from the low-power states to the active or idle state. Therefore, the use of the self-refresh state without care incurs severe performance degradation and excessive transition energy consumption.

3. DRAM POWER-AWARE RANK SCHEDULING

Requests to DRAM are generated from the last-level cache when cache misses occur. To reduce these requests, there are mainly two ways. The first way is to modify cache configuration. We can simply reduce cache miss rates by increasing cache size or the degree of associativity. We, however, cannot increase cache size infinitely because of cost and energy problems of larger caches. Similarly, we cannot increase the degree of associativity infinitely because of increasing hit time and power consumption. The second way is to change the cache replacement policy of the last-level cache. This is more flexible and cheaper than the first way. Hence, we redesigned the last-level cache replacement policy to be aware of DRAM rank power states. In addition to the last-level cache, we applied awareness of rank power states to the memory controller.

3.1 DRAM Power-Aware Rank Scheduling Utilizing the Last-Level Cache

The main idea of our DRAM power-aware rank scheduling utilizing the last-level cache is to select a victim cache block considering cache block state and rank power state on a cache miss in order to reduce write requests and transitions from a low-power state to an active or idle state. Figure 2 shows our proposed DRAM power-aware rank scheduling scheme utilizing the last-level cache in detail. We virtually divide n cache blocks of each set into two groups of non-LRU region and LRU region. The size of LRU region can be decided at design time as half, quarter, or eighth of n blocks. In Figure 2, the size of LRU region is half of 8 blocks. We restrict that cache blocks are evicted from the LRU region. Our scheme operates with two priority levels as follows. **Priority level 1** is **No Rank Scheduling**. **Priority level 2** is **Rank Power State-Aware Scheduling**.

Priority level 1. If any clean cache block exists in the LRU region, a clean cache block closest to the LRU position is evicted because the clean cache block does not incur a write request to DRAM.

Priority level 2. If no clean block exists in the LRU region, a dirty block is selected by the following rank state priority rule (active or idle > power-down > self-refresh) because a dirty block mapped to an active or idle state rank does not incur a transition from a low-power state to an active state. If there are dirty blocks which have the same priority level, the dirty block closest to the LRU position is evicted.

In addition to reducing write requests and state transitions, our

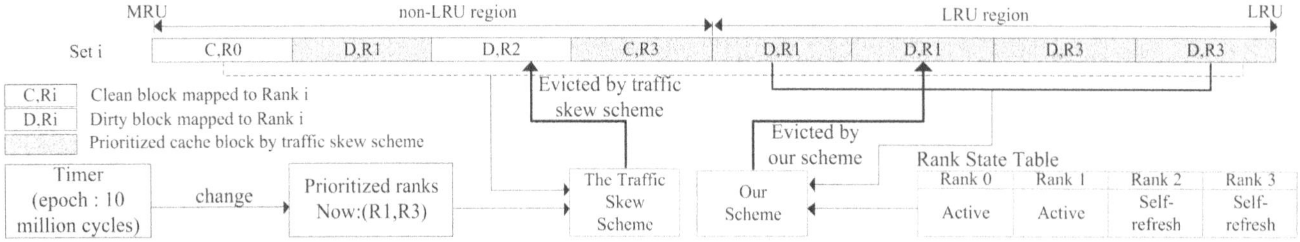

Figure 2. Traffic skew scheme and our power-aware rank scheduling utilizing the last-level cache.

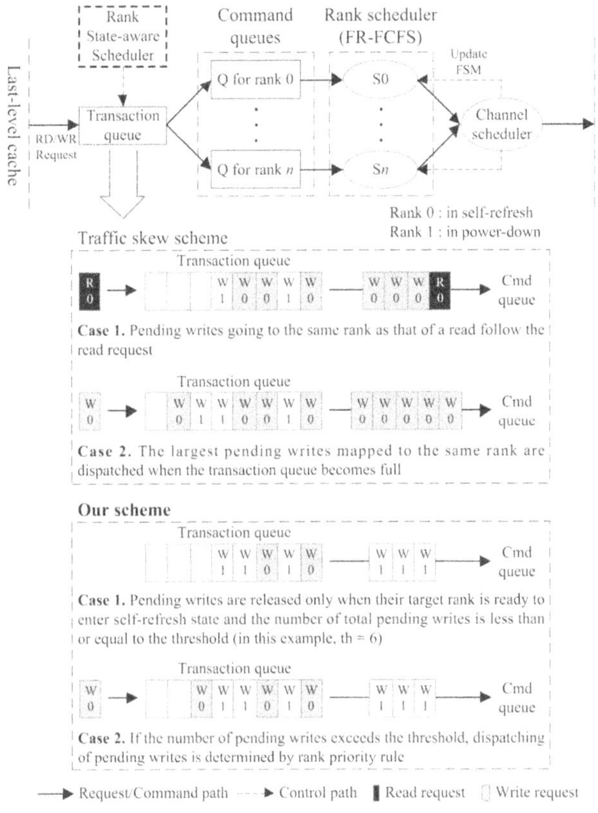

Figure 3. The memory controller with our DRAM power-aware rank scheduling.

scheme allows dirty blocks to stay longer in the last-level cache. If the dirty blocks are likely to be rewritten, the chance of write merging inside the last-level cache will be maximized. Performance of our scheme depends on the position of rank ID in physical address. For example, the rank ID can be positioned only in tag bits, across tag bits and set index bits, or only in set index bits. If rank ID is placed only in set index bits, cache blocks which are located in the same set are mapped to the same rank. In this case, our scheme does not reduce state transitions.

The most similar prior work to ours is the traffic skew scheme [3]. Figure 2 shows the traffic skew scheme in detail. The traffic skew scheme virtually divides cache blocks of each set into prioritized ones and non-prioritized ones by referencing prioritized ranks. Then, it selects a victim block among non-prioritized cache blocks using the LRU policy. If there is no non-prioritized cache block, the traffic skew scheme selects a victim block among prioritized cache blocks using the LRU policy. Timer expires every 10 million cycles. When the timer expires, prioritized ranks change using a simple counter. Then, the self-refresh threshold of new prioritized ranks is set to 500 memory cycles and the self-refresh threshold of non-prioritized ranks is set to 5000 memory cycles. Although this scheme prohibits write requests to prioritized ranks, it suffers from severe performance degradation because prioritized ranks are chosen in a round robin fashion and the prioritized ranks enter self-refresh state too early without considering read traffic going to the prioritized ranks. For example, rank 3 is selected as a prioritized rank. Hence, rank 3 enters the self-refresh state earlier. If there are read requests going to rank 3 in the self-refresh state, rank 3 is woken up to the active state to service the read requests, which degrades performance severely.

Our scheme and the traffic skew scheme are different as follows. First, our scheme effectively reduces write requests to DRAM while the traffic skew scheme does not reduce write requests. Second, our scheme slightly increases state transitions to the self-refresh state with increasing rank idleness while the traffic skew scheme increases state transitions to the self-refresh state tremendously. Third, our scheme is not disturbed by rank usage of programs while the traffic skew scheme is disturbed by read requests as explained above. Fourth, our scheme restricts that cache blocks are evicted from the LRU region while the traffic skew scheme can select a victim block from the non-LRU region as shown Figure 2.

3.2 DRAM Power-Aware Rank Scheduling Utilizing the Memory Controller

Before we explain our scheme, we briefly introduce operations of the conventional DRAM controller including transaction and command scheduling. As shown in Figure 3, read and write requests from the last-level cache are buffered by the transaction queue for arrival time assignment and address mapping. Then, those buffered requests are translated to appropriate DRAM commands and moved to the command queue. For example, if the first rank is idle, any request going to the first rank generates a row activation command prior to a read or write command. Rank scheduler (FR-FCFS [15] now releases generated commands in the command queues by the following priority: command targeting an already opened row first, then the oldest command. The channel scheduler arbitrates command issuing order among command queues in a round-robin manner. Once a command leaves the memory controller, the channel scheduler updates rank state information including bank state table, timer, etc. to preserve DRAM timing constraints.

The main idea of our DRAM power-aware rank scheduling scheme utilizing the memory controller is to defer write requests. Then, the memory controller batch-writes the delayed write requests considering rank power states. This scheduling strategy increases DRAM rank idleness and decreases transitions from low-power states to an active or idle state. Our scheme modifies conventional transaction queue operations as follows.

Case 1. If the number of pending transactions is less than or equal to the transaction queue threshold (the size of transaction queue – 2), our scheduler checks the current state and idle time of each rank. If the scheduler finds a power-down state rank which is ready to enter the self-refresh state (its idle time exceeds the self-refresh threshold), the scheduler batch-issues pending write transactions which are mapped to the power-down state rank. With the traffic skew scheme, write requests are released without buffering by the transaction queue if the target rank is in the active state. Otherwise, pending writes follow a read request that are targeting the same rank. Since the prioritized rank enters low-power state earlier than non-prioritized ranks, read requests to the prioritized rank incur rank state transitions frequently.

Case 2. If the number of pending transactions exceeds the transaction queue threshold, the scheduler checks the state, idle time, and the number of pending transactions of each rank. Then, our scheduler selects a rank by the following rank priority rule: active or idle > power-down (ready to enter self-refresh) > power-down (not yet ready to enter self-refresh) > self-refresh. If there are ranks which have the same priority, the scheduler selects a rank which has the most pending transactions. The scheduler batch-issues the pending transactions which are mapped to the

Table 1. Simulation parameters.

Parameter	Value
Processor	4GHz OOO, fetch / decode / issue 4 / 4 / 4, LSQ 32-entry, ROB 64-entry
L1 I- and D-cache	32KB 8-way set associative, LRU
Unified L2 cache	2MB 16-way set associative, 12 cycles
Memory controller	FR-FCFS [20] rank scheduler, transaction / command queue depth 64 / 32
Main memory	1Gb x8 DDR3-1333, channel / rank / bank 1 / 4 / 8
Power-down mode	Immediate power-down (0 cycle threshold), 32 memory cycles wake-up penalty
Self-refresh mode	5000 memory cycles threshold, 512 memory cycles wake-up penalty
Address mapping	Row[31:18], Rank[17:16], Bank[15:13], Column[12:3]

selected rank. Unlike our scheme, the traffic skew scheme dispatches the most pending writes mapped to the same rank regardless of its power state, which increases rank state transitions dramatically.

In order to support DRAM power-aware rank scheduling utilizing the memory controller, we need a counter to trade the number of pending requests for each rank. The rank state information is knowledgeable since the rank scheduler keeps track of rank/bank state information to preserve DRAM timing constraints. Pending request counter of each rank increases when a new request mapped to the rank arrives at the transaction queue. The counter decreases when a request mapped to the rank is issued to its corresponding command queue.

Our DRAM power-aware rank scheduling utilizing the memory controller differs from write buffering of the traffic skew scheme in the following aspects. First, our scheme defers all write requests except ones going to opened row buffers, while the traffic skew scheme defers only write requests going to a rank in a low-power state. Second, there is a difference in when write requests are issued. Our scheme issues delayed write requests when there is a rank in the power-down state which is ready to enter the self-refresh state. However, the traffic skew scheme issues delayed write requests following a read request going to the same rank. Third, our scheme considers penalties for waking up a rank in a low-power state when scheduling write requests while the traffic skew scheme does not consider the penalties.

4. EXPERIMENTAL RESULTS

4.1 Experimental Setup

We used MASE [6], DRAMSim2 [7] and SPEC2000 benchmark suite [8] for experiments to evaluate our schemes and the traffic skew scheme. MASE is an infrastructure that provides a flexible environment to model modern microarchitectures and supports interface for the memory system which has non-deterministic latency. DRAMSim is a detailed and highly-configurable memory system simulator. We integrated MASE with DRAMSim. We modified MASE and DRAMSim for our schemes. Since self-refresh mode and bus power are not modeled in DRAMSim, we implemented the self-refresh mode and bus power by referencing JEDEC standard of DDR3 [10] and XEEMU simulator [9] which models the self-refresh mode and bus power. In addition, state transition power is undisclosed in the datasheet of DDR3 [11] and JEDEC standard of DDR3 [10]. We modify the rank idle time reset mechanism. We reset the idle time only when a read request is issued. We exclude write requests in

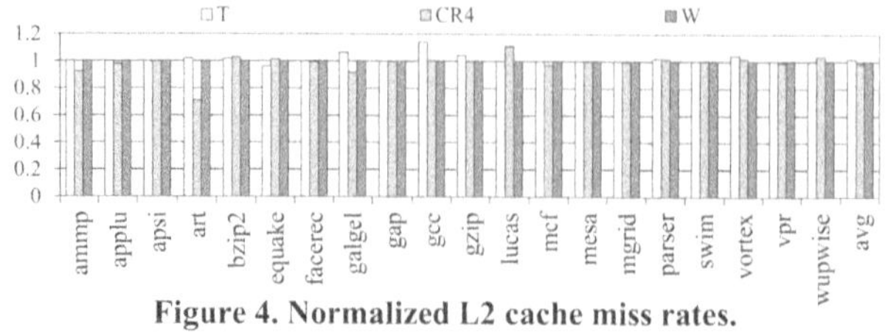

Figure 4. Normalized L2 cache miss rates.

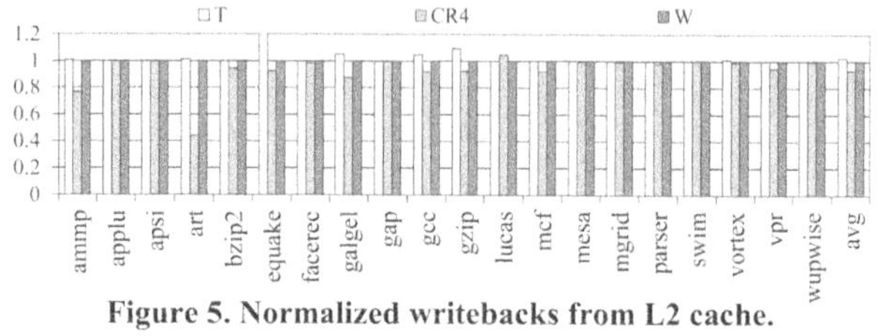

Figure 5. Normalized writebacks from L2 cache.

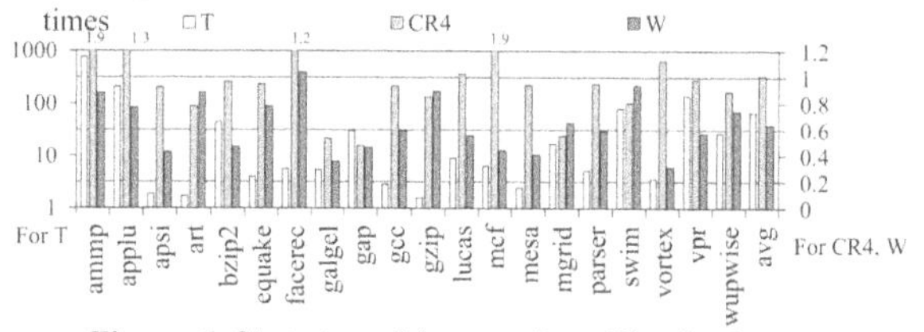

Figure 6. State transitions to the self-refresh state.

resetting the idle time because write requests can be controlled by our scheme. We reproduce the traffic skew scheme, which consists of cache replacement policy and write buffering scheme, to the last-level cache and the transaction queue of the memory controller, respectively, as explained in [3]. One epoch for altering prioritized ranks for the traffic skew scheme is 10 million cycles as mentioned in [3]. We experiment our schemes and the traffic skew scheme in the same environment. We assume 4GHz out-of-order processor and 4GB memory system. Detailed parameters of processor and DRAM are shown in Table 1. Baseline system uses immediately power-down policy and a self-refresh threshold of 5000 memory cycles because that threshold provides best results for our schemes and the traffic skew scheme. All benchmarks are fast-forwarded for one billion instructions to avoid initial start-up effects and then simulated for another one billion committed instructions.

4.2 Experimental Results

We use the following notations in explaining results. The notations **B**, **T**, **CR**, and **W** mean the baseline, the traffic skew scheme, our DRAM power-aware scheduling scheme utilizing the last-level cache, and our DRAM power-aware scheduling scheme utilizing the memory controller, respectively. **CR4** means that the size of LRU region is 4 blocks. All the results below are normalized to **B**.

We evaluated CR scheme with various sizes of LRU region. With larger LRU region size, execution time increases while DRAM power consumption decreases because larger LRU region incurs higher L2 cache miss rates but provides more opportunities for DRAM power reduction. DRAM energy and bus energy consumption are best with 4-block LRU region. Therefore, we plot results for CR4.

Figure 4 shows the results on the last-level cache miss rates. The cache miss rate of the traffic skew scheme compared to the baseline is increased by 1.6% on average. The cache miss rate of CR4 is decreased by 1.4% on average. The cache miss rate of W does not change. Therefore, our schemes and the traffic skew

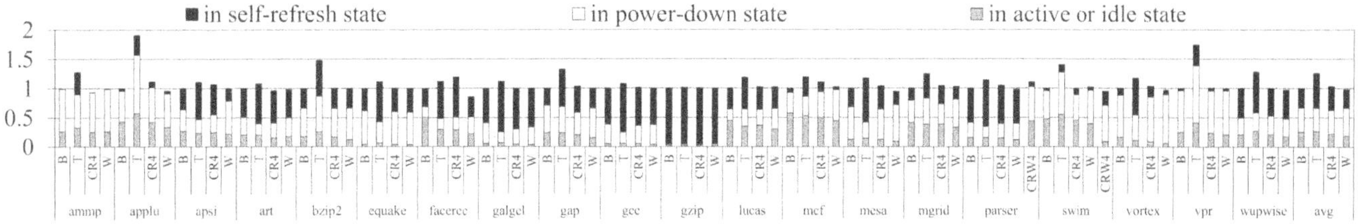

Figure 7. Residence times for each power state.

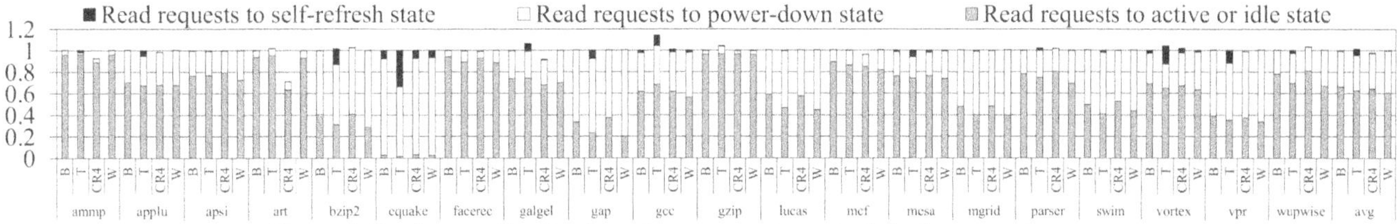

Figure 8. Rank state distribution when transaction scheduler issues read requests.

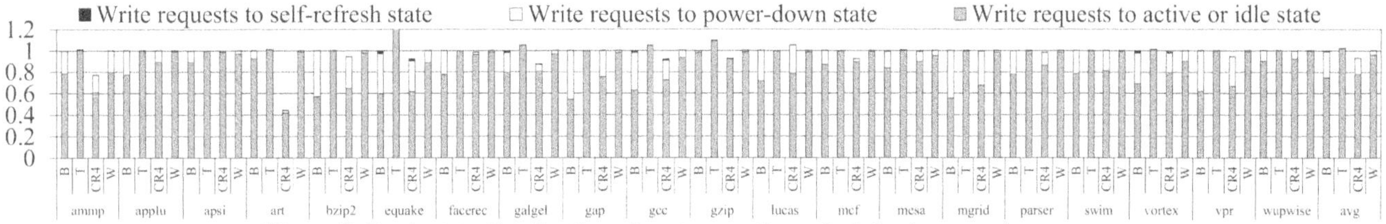

Figure 9. Rank state distribution when transaction scheduler issues write requests.

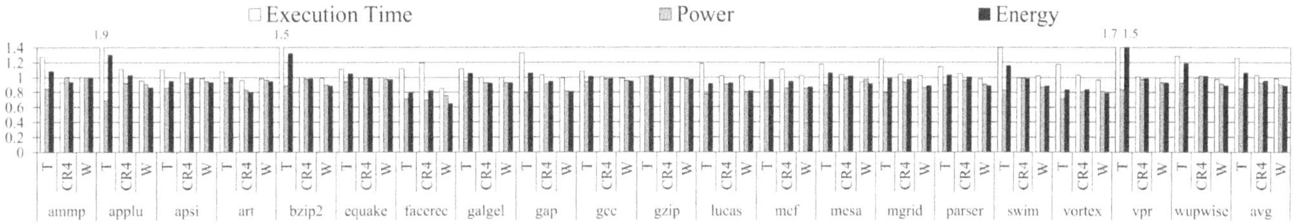

Figure 10. Execution time, DRAM power and energy consumptions.

scheme do not harm the last-level cache performance much. Figure 5 shows the results on write backs from the last-level cache. CR4 reduces write backs by 6.8% on average (up to 55.9%). However, W does not reduce write backs. The traffic skew scheme increases write backs by 2.4% on average because of increased L2 cache miss rate. Figure 6 shows the results on the number of state transitions to the self-refresh state. The traffic skew scheme, the number of state transitions to the self-refresh state is enormously increased by 68 times on average (up to 759 times) because this scheme lowers the self-refresh threshold by force. CR4 increases state transitions to the self-refresh state only by 0.8% on average. W reduces state transitions to the self-refresh state by 36.2%, on average, because pending write requests are issued to a power-down rank which is ready to go to the self-refresh state.

Figure 7 shows average residence time in each power state. The traffic skew scheme does not reduce the active or idle time. It means that the traffic skew scheme does not increase rank idleness. However, residence time in the self-refresh state is largely increased. This is because of decreased self-refresh threshold. CR4 reduces the active or idle time and power-down time by 3.3% and 0.9%, respectively, on average. Self-refresh time is increased by 7.1% on average because write requests are not issued to ranks in the self-refresh state. Reduced active time and increased self-refresh time mean that rank idleness is increased. W reduces the active or idle time by 6.5%. Power-down time and self-refresh time are increased by 5.8% and 1%, respectively, on average. These results mean that rank idleness is increased. Also we observe that once a rank enters the self-refresh state, it stays in the state as long as possible as shown in Figure 6 and Figure 7.

Figure 8 shows rank power states when the transaction scheduler issues read requests. The traffic skew scheme increases read requests to the self-refresh state by 5.6% on average because the traffic skew scheme lowers self-refresh threshold without considering read requests. This incurs severe performance degradation. In contrast, read requests going to the self-refresh state are reduced by our schemes because our schemes do not adjust self-refresh threshold by force and decrease state transitions to the self-refresh state. Figure 9 shows rank power states when the transaction scheduler issues write transactions. With baseline, write requests to active, power-down, and self-refresh states are 75%, 24.3%, 0.7%, respectively, on average. With the traffic skew scheme, these numbers are 102%, 0.1%, 0.3%, respectively, on average. In the result, there are more write requests to self-refresh state than write requests to power-down state because the traffic skew scheme does not discern different low-power states when they issue write requests. With CR4, these numbers are 78.2%, 14.7%, 0.3%, respectively, on average. With W, these numbers are 96.2%, 3.7%, 0.1%, respectively, on average. W reduces state transitions occurring by write requests by 21.2% on average. These results show that our schemes outperform the traffic skew scheme.

Figure 10 shows execution time, DRAM power and energy consumptions. With the traffic skew scheme, DRAM power consumption is reduced by 14.8% but execution time is increased by 25.5%. As a result, 6% more DRAM energy is consumed. With CR4, however, DRAM power and energy consumptions are reduced by 7.1% and 4.7%, respectively, on average but execution time is increased by 2.9% on average. Performance degradation comes from increased rank contention in active state ranks because write traffic to active ranks is increased. With W, execution time, DRAM power and energy consumptions are reduced by 1.7%, 9.8% and 11.2%, respectively, on average.

For the result of address and data bus energy consumption, with the traffic skew scheme, address and data bus energy consumptions are increased by 2.4% and 2.8%, respectively, on average because of increased L2 cache miss rate. CR4 reduces address and data bus energy consumptions by 1.3% and 3.4%, respectively, on average. These results are not directly proportional to reduced writes because bus energy consumption depends on previously transferred bus data. W reduces address and data bus energy consumptions by 6.6% and 1.9%, respectively, on average. W batch-writes write requests to the same rank, bank, row, and column so that address bus energy of bank, row, or column address is reduced.

From the above results, we observe that CR4 achieves DRAM power and energy gain from reduced write requests and state transitions. However, increased execution time (2.9%) of CR4 is much less than increased execution time (25.5%) of the traffic skew scheme. W achieves DRAM power and energy gain from reduced rank state transitions, which comes from the rank-state aware batch-writing.

5. RELATED WORK

There are many approaches proposed to effectively use row buffers. DRAM-aware last-level cache write-back is proposed in [4]. Its key idea is to aggressively send out last-level cache blocks that are expected to hit in DRAM row buffers. The Virtual Write Queue is proposed in [5]. Its key idea is to send out last-level cache blocks which can be written to the same DRAM row buffers as those of the write requests already in progress. These two schemes increase write requests to DRAM because dirty cache blocks, which are mapped to open row buffers, are written to DRAM by cleaning. Adaptive power and thermal management scheme is proposed in [14]. Reducing both row activation energy and DRAM peak temperature are the goals of these schemes by buffering write requests. Buffered write requests are executed when the corresponding row buffer is open. Prefetching-based memory traffic clustering to lengthen DRAM idle times for energy reduction is proposed by [16]. Adaptive history-based memory schedulers are proposed in [12]. They consider the history of recently scheduled operations when the memory controller selects next operation. Memory controller policy for DRAM power management is proposed in [13]. They demonstrate that an idle DRAM chip immediately switching to the low-power state is better than sophisticated policies.

6. CONCLUSIONS

We proposed DRAM power-aware rank scheduling schemes which can be implemented in the last-level cache and the memory controller. We demonstrated that our two schemes reduce DRAM power and energy consumptions and that their overheads are negligible. Our contributions are as follows. First, we reduce write requests to DRAM with negligible overhead. Our scheme utilizing the last-level cache reduces write requests, and DRAM power and energy consumptions by 6.8%, 7.1% and 4.7%, respectively, on average. Second, we increase rank idleness and decrease state transitions. Write requests are delayed in order to maximize rank idleness. Our scheme utilizing the memory controller reduces state transitions, DRAM power and energy consumptions by 21.2%, 9.8% and 11.2%, respectively, on average, with no performance degradation.

7. ACKNOWLEDGMENTS

This work was performed when Sukki Kim was with Dept. of Computer Science at KAIST. This research was supported by the National Research Foundation of Korea (NRF) grants funded by the Ministry of Education, Science and Technology (2011-0005378, 2012-0000980).

8. REFERENCES

[1] C. Lefurgy, K. Rajamani, F. Rawson, W. Felter, M. Kistler, and T. W. Keller, "Energy Management for Commercial Servers", *IEEE Computer*, vol. 36, pp. 39-48, December 2003.

[2] H. Huang, K. G. Shin, C. Lefurgy, and T. Keller, "Improving Energy Efficiency by Making DRAM Less Randomly Accessed", In *Proc. of the International Symposium on Low Power Electronics and Design*, 2005.

[3] A. M. Amin and Z. A. Chishti, "Rank-Aware Cache Replacement and Write Buffering to Improve DRAM Energy Efficiency", In *Proc. of the International Symposium on Low Power Electronics and Design*, 2010.

[4] C. Lee, V. Narasiman, E, Ebrahimi, O. Mutlu, and Y. N. Patt, "DRAM-Aware Last-Level Cache Writeback: Reducing Write-Caused Interference in Memory Systems", *HPS Technical Report*, TR-HPS-2010-002, April 2010.

[5] J. Stuecheli, D. Kaseridis, D. Daly, H. C. Hunter, and L. K. John, "The Virtual Write Queue: Coordinating DRAM and Last-Level Cache Policies", In *Proc. of the International Symposium on Computer Architecture*, 2010.

[6] E. Larson and S. Chatterjee, "MASE: A Novel Infrastructure for Detailed Microarchitectural Modeling", In *Proc. of the International Symposium on Performance Analysis of Systems and Software*, 2001.

[7] P. Rosenfeld, E. Cooper-Balis, and B. Jacob, "DRAMSim2: A Cycle Accurate Memory System Simulator", *Computer Architecture Letters*, vol. 10, no. 1, pp. 16-19, 2011.

[8] Standard Performance Evaluation Corporation, "SPEC CPU 2000 Benchmark Suite", http://www.specbench.org/osg/cpu2000/

[9] Z. Herczeg, D. Schmidt, A. Kiss, N. Wehn, and T. Gyimothy, "Energy Simulation of Embedded XScale Systems with XEEMU", *Jornal of Embedded Computing*, vol. 3, issue 3, 2009.

[10] JEDEC, "JEDEC Standard DDR3 SDRAM Specification", http://www.jedec.org/sites/default/files/docs/JESD79-3E.pdf

[11] Micron Inc. "1Gb: x4, x8, x16 DDR3 SDRAM", http://www.micron.com/products/dram/ddr3-sdram

[12] I. Hur and C. Lin, "Adaptive History-Based Memory Schedulers", In *Proc. of the International Symposium on Microarchitecture*, 2004.

[13] X. Fan, C. Ellis, and A. Lebeck, "Memory Controller Policies for DRAM Power Management", In *Proc. of the International Symposium on Low Power Electronics and Design*, 2001.

[14] S. Liu, S. O. Memik, Y. Zhang, and G. Memik, "A Power and Temperature Aware DRAM Architecture", In *Proc. of the Design Automation Conference*, 2008.

[15] S. Rixner, W. J. Dally, U. J. Kapasi, P. Mattson, and J. D. Owens, "Memory Access Scheduling", In *Proc. of the International Symposium on Computer Architecture*, 2000.

[16] Y. Lee and S. Kim, "DRAM Energy Reduction by Prefetching-Based Memory Traffic Clustering", In *Proc. of the Great Lakes Symposium on VLSI*, 2011.

Energy-efficient GPU Design with Reconfigurable In-package Graphics Memory

Jishen Zhao[1], Guangyu Sun[2], Gabriel H. Loh[3], and Yuan Xie[1,3]

[1]The Pennsylvania State University
Dept. of Computer Science and Engineering
{juz138, yuanxie}@cse.psu.edu

[2]Peking University
Center for Energy-efficient Computing and Applications
gsun@pku.edu.cn

[3]Advanced Micro Devices, Inc.
AMD Research
gabriel.loh@amd.com

ABSTRACT

We propose an energy-efficient reconfigurable in-package graphics memory design that integrates wide-interface graphics DRAMs with GPU on a silicon interposer. We reduce the memory power consumption by scaling down the supply voltage and frequency while maintaining the same or higher peak bandwidth. Furthermore, we design a reconfigurable memory interface and propose two reconfiguration mechanisms to optimize system energy efficiency and throughput. The proposed memory architecture can reduce memory power consumption up to 54%, without reconfiguration. The reconfigurable interface can improve system energy efficiency by 23% and throughput by 30% under a power budget of 240W.*

Categories and Subject Descriptors

B.3.1 [**Memory Structures**]: Semiconductor Memories

Keywords

Graphics Memory, 3D Packaging, Reconfigurable Interface

1. INTRODUCTION

Modern GPU systems have become an attractive solution for both graphics and general-purpose workloads that demand high computational performance. The graphics processing unit (GPU) exploit extreme multi-threading to target high-throughput [1, 15]. For example, the AMD Radeon™ HD 5870 employs 31,000 threads interleaved across 1600 stream processors [1]. To accommodate the high-throughput demands, the power consumption of GPU systems continues to increase. As existing and future integrated systems become power-limited, reducing system power consumption while maintaining high energy efficiency is a critical challenge for GPU system design.

To satisfy the demands of high-throughput computing, the GPUs require substantial amounts of off-chip memory (from hundreds of megabytes to gigabytes). Consequently, off-chip memory consumes a significant portion of power in a GPU system. Figure 1 evaluates the maximum power consumption of two GPU systems, the AMD Radeon™ HD 6990 [1] and NVIDIA Quadro® FX5800 [15] with the memory power model described in Section 5. We considered the

*This work is supported in part by SRC grants, and NSF 0903432, 0916887 and 1017277.

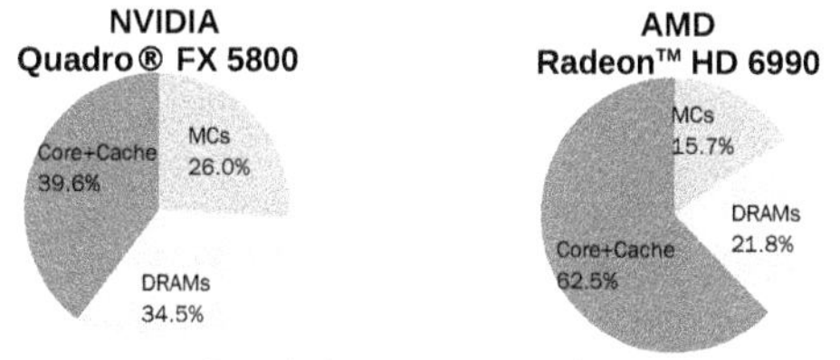

Figure 1: Power breakdown of NVIDIA and AMD GPUs.

memory bandwidth utilization (the fraction of all cycles when the data-bus is busy transferring data for reads or writes) from 10% to 50%. For both GPU systems, the off-chip memory consumes from 20% to over 30% of the total GPU system power. Note that for the workloads evaluated in this paper, the highest average bandwidth utilization observed was 35%. At this bandwidth level, the memory power consumption is 21.8% and 34.5% for the two GPU systems, respectively. These results match well with results provided by an industry partner, which indicated that the graphics memory consumes ~25% of the total GPU power. If we can reduce the memory power by half, 12.5% of system power can be saved; this may seem like a relatively small amount, but it is quite significant. For example, the maximum power consumption of the AMD Radeon™ HD 6990 is 375W; therefore a 12.5% power reduction saves 47W. Consequently, techniques that reduce the graphics memory power can be very effective at reducing the total system power.

In this paper, we propose to integrate GDDR-like graphics memory with the GPU processor and memory controllers in a single package with silicon interposer system-in-package (SiP) technology so the number of memory I/Os are not constrained by the package pins. By using the significantly larger number of data I/Os, we can provide a given amount of bandwidth while reducing the power consumption of memory by scaling down its supply voltage and frequency. Furthermore, we design a reconfigurable memory interface, and propose two reconfiguration mechanisms (EOpt and PerfOpt) to optimize (1) GPU system energy efficiency and (2) system throughput under a given power budget. With minor changes to the memory interface, our wide-bus in-package memory design is practical and easy to implement. With the flexibility of optimizing for either power or performance, our design can adapt to both high-performance computing which prefers high throughput, and power-constrained applications, which prefers better energy efficiency.

2. BACKGROUND AND RELATED WORK

Energy-efficient GPU Architectures: Various existing work explores how to address the GPU power challenge. Most of the existing studies explore either GPU shader cores and caches architecture [7, 17], or software optimization [13], and require both hardware and software modifications to current GPU processor design. In our work, we explore power reduction techniques by limiting the archi-

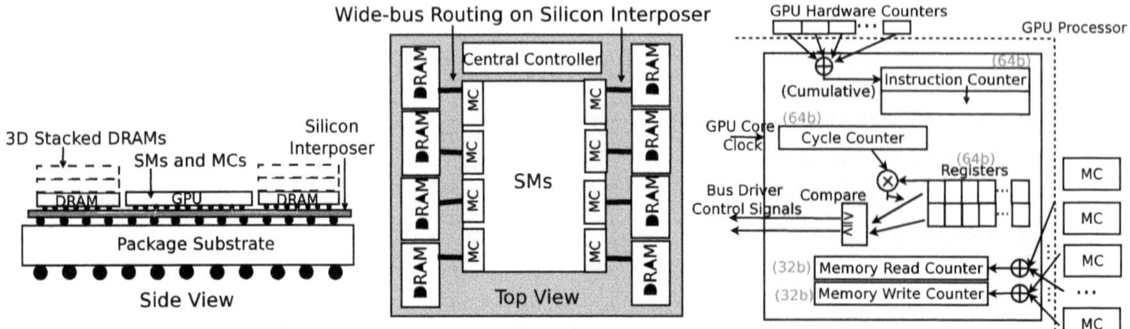

(a) Integrating GDDRs with GPU processor on a silicon interposer. (b) Central controller

Figure 2: Overview of the proposed in-package graphics memory design.

tectural modifications to the graphics memory interface with only minor changes to GPU compute-unit architecture.

Graphics memory: Graphics double data rate (GDDR) memories are specifically designed for graphics cards, game consoles, and high-performance computing systems. GDDR memories usually employ high frequencies to achieve very high bandwidths. Unfortunately, power consumption dramatically increases as well. A large power supply is required, and usually comes with high cost. Conventional GDDRs have employed VF scaling techniques to reduce power by adapting the memory interface to the memory bandwidth requirement of an application [6]. However, the power reduction comes at the expense of memory bandwidth degradation. For future high-performance GPGPU and advanced video processing (e.g., 3D HD multi-screen video games), existing power saving techniques may not suffice.

3D Package Using Silicon Interposer Technology: The new packaging approaches with silicon interposer attract much attention in high-performance multi-chip module (MCM) and SiP designs. A variety of work from academia and industry explores the development of MCM and SiP with silicon interposers [4, 5, 16]. A silicon interposer provides high-density interconnections, and is capable of providing a line width of less than $10\mu m$ and thousands of I/O pads per square centimeter. The coarse-pitch through-silicon vias (TSVs) provide the external connections between the package and individual dies/chips for the parallel and serial I/O, power/ground, clocking, configuration signals, etc. The length of the interconnection from the chip to the substrate is reduced. In our work, we leverage silicon interposer-based 3D packaging technology to explore the energy efficiency benefits for GPU systems.

3. IN-PACKAGE GRAPHICS MEMORY ARCHITECTURE

Many power-saving techniques for GPU memory come with the undesirable side effect of a degradation of the provided memory bandwidth [6]. In this section, we present the feasibility results and energy efficiency benefits of in-package wide-interface graphics memory, integrated side by side with a GPU processor. We show that VF scaling can be employed to reduce power consumption without affecting memory bandwidth by leveraging silicon interposer-based 3D packaging technology.

Figure 2 depicts an overview of our GPU system architecture with integrated graphics memory. The data I/Os of conventional graphics memory standards are 16 or 32 bits wide per channel, limited by the package pin count. Our in-package memory does not suffer from such limitations since the DRAMs are directly integrated with the GPU processor in the same package. In addition, the number of power/ground pins can be reduced at the much lower DRAM power consumption than conventional GDDRs. While we fix the DRAM capacity in this work, the capacity of graphics memory can be extended by stacking multiple DRAM dies together.

3.1 Feasible Design with 3D Packaging

DRAMs are integrated on the GPU processor package with silicon interposer technology. Our thermal analysis results showed that such integration does not incur significant temperature increase to the system. We studied a GPU system configuration based on NVIDIA Quadro® FX5800 [15]. We computed the maximum power consumption of GPU processors and memory controllers by subtracting the DRAM power from the reported maximum power consumption of Quadro® FX5800 [15], resulting in 124W. The power of 4GB DRAM is calculated as 60W, based on Hynix's GDDR5 memory [8]. The areas of different GPU components are obtained from the GPU die photo, which has a $470\,mm^2$ die area. We assume the ambient temperature to be $40\,^{\circ}C$. We used the HotSpot thermal simulation tool [14] to conduct our analysis. The maximum steady-state temperature of the GPU (without DRAMs) is $72.6\,^{\circ}C$. With 4GB interposer-mounted DRAMs placed beside the GPU processor, the maximum temperature is $78.4\,^{\circ}C$. Thus, it is feasible to employ interposer-based memory integration.

3.2 Energy Efficiency Benefits

We examine the peak-memory bandwidth (PBW) [2] and maximum total power consumption of 2GB graphics memories with different memory interface configurations, including per-channel bus widths (abbreviated as bus width in the following), memory frequency, and supply voltage (Figure 3). The memories considered are electrically similar to GDDR DRAMs except for the wider buses. The supply voltage is scaled appropriately to support the given memory clock frequency. DRAMs with bus widths of 16 and 32 bits per channel are evaluated as off-chip GDDR memory. DRAMs with wider buses are evaluated as in-package memory. The GDDR memory power model that we used will be described in detail in Section 5.

The bars in Figure 3(a)-(e) show the maximum power consumption for configurations with different bus widths and clock speeds, but maintaining the same PBW (i.e., bus-width × clock speed is kept constant). The opportunity for memory power reduction is exploited by scaling down the memory's supply voltage corresponding to the frequency reduction. It can be observed that the power consumption follows U-shaped bathtub curves. With wider memory buses, lower frequency allows us to scale down the supply voltage which directly results in a power reduction. However, the power consumed by I/O output drivers and on-die termination keeps increasing with the bus width. When the bus width is increased to 256 bits, the I/O power component starts to dominate the total memory power and finally overwhelms the power benefits of VF scaling. Figure 3(f) shows the potential benefit of achieving higher PBW with the in-package graphics memory at the same or lower memory power consump-

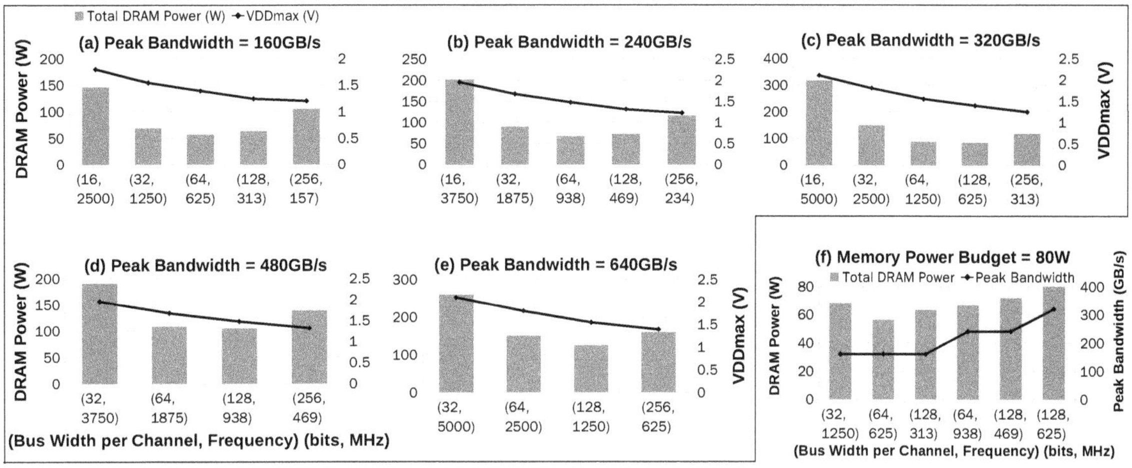

Figure 3: The PBW and maximum total memory power consumption of 16×1Gb GDDRs with various interface configurations of bus width and memory clock frequency. (a)-(e) show memory power at the same PBW. (f) shows different PBW that can be achieved at a fixed memory power consumption.

tion. With a fixed memory power consumption of 80W, the standard GDDR5 with 32-bit interfaces can only provide 160GB/s of PBW. However, the wide interface in-package memories can achieve up to 320GB/s bandwidth. In addition, the (64, 938) configuration can even provide higher bandwidth than the standard GDDR5 while consuming less power.

4. RECONFIGURABLE ARCHITECTURE

In this section, we present a reconfigurable memory interface that can dynamically adapt to the demands of various applications based on dynamically observed memory access and performance information. On top of the reconfigurable memory interface hardware, we further propose two reconfiguration mechanisms, EOpt and PerfOpt, that optimize system energy efficiency and performance (in terms of throughput), respectively.

4.1 Application Classification

We analyze the unique characteristics of GPGPU applications, and classify the applications into two types, memory-intensive and non-memory-intensive. Figure 5 illustrates the typical GPU system performance, power, and energy efficiency (applications *nw* and *sd2* are used as examples). Here we define energy efficiency as the performance per Watt. It is shown that higher PBW can directly lead to higher DRAM power consumption with both types of applications. The relationship between the system performance and memory interface configuration appears to be different with the two types of applications. The IPC of type-M (memory-intensive) is sensitive to the change of memory interface configuration, as shown in Figure 5(a). Decreasing the memory frequency (and consequently increasing the memory access latency) results in significant IPC degradation, even though we provide wider buses to keep the same PBW. Furthermore, the IPC of these applications typically stays low, due to the continuous memory demands that significantly slow down the instruction execution. As shown in Figure 5(b), the IPC of type-C (non-memory-intensive or compute-intensive) applications is much higher than that of type-M. The IPC curve remains stable with different memory interface configurations at the same PBW. Overall, varying the memory interface configuration will affect both IPC and DRAM power consumption in type-M applications. Trade-offs between the two must be considered to optimize the system energy efficiency. With type-C applications, we can improve the system energy efficiency by reducing the DRAM power without significant performance degradation. If we take a closer look at the details of each

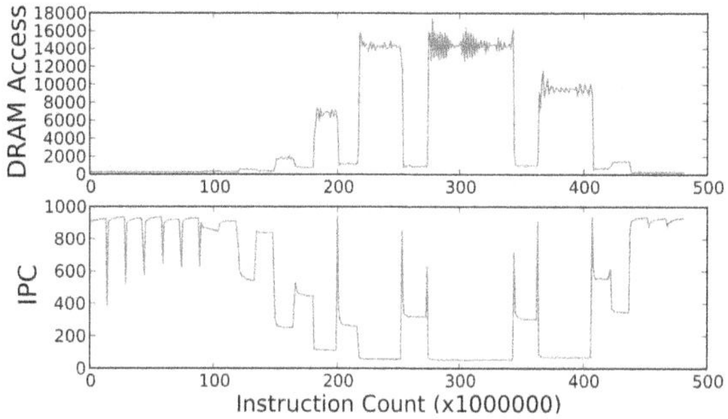

Figure 4: A single application can have different memory intensities during the execution.

application execution, both type-M and type-C periods can be observed in a single application (e.g., *bfs*) as illustrated in Figure 4. The rough analysis shows that this application is memory-intensive, due to the total amount of DRAM accesses. However, a non-trivial portion of the instruction execution is actually non-memory-intensive. For the best system energy efficiency or performance, different memory configurations should be adopted with the two levels of memory intensities.

4.2 Memory Interface Hardware Design

Our reconfigurable memory interface is aware of these different application behaviors, such that the bus width and memory frequency can be dynamically tuned according to the observed performance and memory access information of each workload.

Hardware implementation: At the interface between the GPU processor and the in-package graphics memories, we add a central controller, control signals to the bus drivers, and controls for dynamic VF scaling (Figure 2(b)). The central controller is used to collect global information of GPU performance and memory accesses. A vector of counters are maintained in the controller to collect performance and memory access information from either GPU hardware performance counters or memory controllers. A threshold register vector is used to store various thresholds and initial values described in reconfiguration mechanism. The calculator module calculates system energy efficiency based on the collected performance information and the estimated power consumption. The memory intensity is calculated as *the memory accesses per 1000 instructions*.

Overhead: The reconfigurable memory interface incurs both area and performance overheads. The total storage overhead is 128B, including various counters, registers, and arithmetic logic compo-

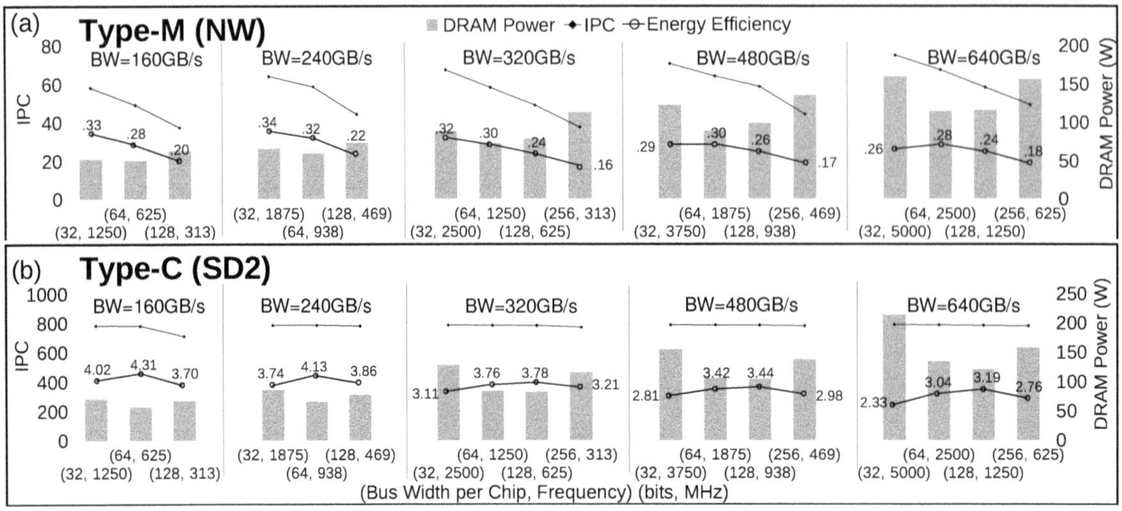

Figure 5: Performance, power, and energy efficiency of two types of applications, which are (a) memory-intensive (Type-M) and (b) non-memory-intensive (or compute-intensive) (Type-C).

nents shown in Figure 2(b). The bus transmission lines are routed on the silicon interposer, which has sufficient space for the buses. To estimate the performance overhead, we evaluated the total latency of memory access pattern change detection and reconfiguration from 100 to 1000 cycles, i.e., 300ns to $3\mu s$ at 325MHz GPU core clock frequency in our baseline. We measured the execution time of various applications, and the maximum and minimum values are 6,000,000 cycles (18ms) and 40,000 cycles ($120\mu s$), respectively. Therefore, the maximum performance overhead is estimated as $3\mu s$ / $120\mu s$, which is less than 2.5% of total execution time.

4.3 EOpt: Optimizing Energy Efficiency

A direct way of utilizing our reconfigurable memory interface is to optimize the system energy efficiency. Specifically, we strive to maintain performance that is competitive to a static memory interface approach, but dynamically choose different memory configurations to save power when possible. During type-M execution periods, both IPC and memory power consumption will be affected by the change of memory interface. Therefore, we choose configurations that maintain high memory clock frequencies to minimize the IPC degradation. Given the memory frequency constraint, the bus width is then configured to minimize the memory power consumption. During type-C execution periods, IPC is stable when we change the memory interface configuration. Consequently, we tend to adopt the memory frequency and bus width configuration that minimizes the memory power consumption.

Our reconfiguration mechanism for system energy efficiency optimization is composed of three steps: detection, comparison, and reconfiguration. During the execution of an application, we sample both IPC and the memory access count. If we detect a change of memory intensity, we compare the estimated system energy efficiency with different *(bus width, frequency)* configurations. The configuration that results in the highest system energy efficiency will be adopted. Sophisticated prediction schemes could be incorporated into the reconfiguration mechanism, although most GPU applications do not frequently change their memory access patterns. Therefore, we use a simple 1-bit prediction scheme that yields sufficient accuracy without much performance overhead.

4.4 PerfOpt: Optimizing System Performance Under a Given Power Budget

As long as the power consumption is affordable, the primary demand from the graphics and high performance computing markets is high performance, in terms of throughput. Therefore, we explore GPU system performance optimization under a given power budget. Our performance optimization target is the instruction throughput (the executed instructions per second).

Figure 6 shows the flow chart of our reconfiguration mechanism. Our reconfiguration mechanism addresses the type-M and -C execution periods of a workload with different strategies. During type-C periods, we always employ the memory interface configuration that minimizes the DRAM power consumption. Any saved power is transferred to scale up the GPU core clock frequency/supply voltage to improve system performance. During type-M periods, we consider two possible strategies. First, because the memory interface configuration directly affects system performance (during type-M phases), we choose the memory configuration that delivers the highest system performance while staying within the system power budget. Second, sometimes an application can be a relatively memory-intensive phase while still having significant compute needs as well. In these cases, reconfiguring the memory interface to free up more power for the GPU can still result in a net performance benefit despite the reduced raw memory performance. Based on the predicted benefit, our system will choose the better of these two strategies.

Note that short-duration violations can be tolerated if there is sufficient thermal headroom. Existing CPU systems (e.g., Intel's Turbo Boost technology) already "overclock" beyond their nominal power limits for short itervals. If violations last too long, on-chip thermal monitors can always trigger more aggressive VF scaling to prevent catastrophic damage to the chip [9]. Another possible issue with this reconfiguration mechanism is that the power budget may be violated within an instruction interval. In fact, it is a trade-off between the strictness of following the power budget constraint and the amount performance overhead. We evaluated various lengths of instruction intervals from 1000 to 1 billion, and find that 1 million instruction intervals can provide sufficient guarantee of power budget constraint with below 5% of reconfiguration overhead.

5. EXPERIMENTAL SETUP

Simulations are performed on GPGPU-Sim [2], a cycle accurate PTX-ISA simulator. The shader cores, caches, and interconnection network of the baseline GPU are configured based on NVIDIA Quadro® FX5800 [15]. We modify the simulator to implement our reconfiguration mechanisms. The instruction interval N is set to be 1 million. The baseline graphics memory is off-chip GDDR5 memory, reported as can support up to 3.5GHz frequency [8]. We

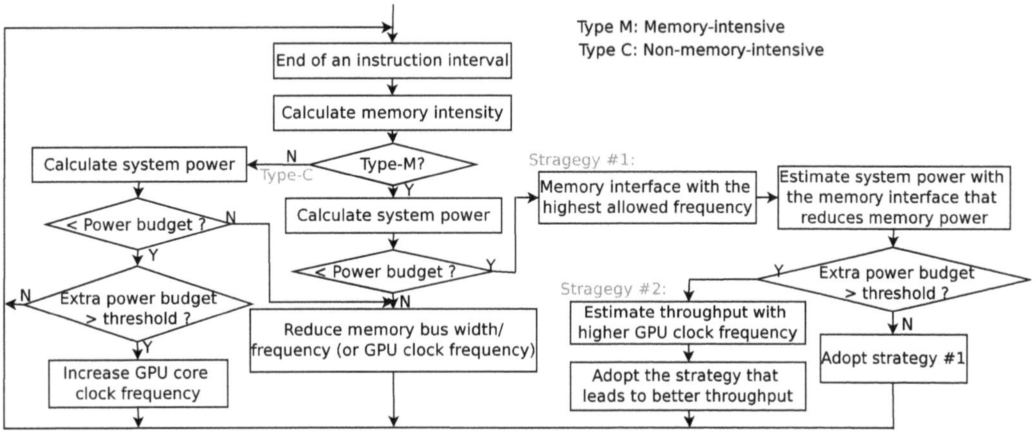

Figure 6: Flow chart for PerfOp.

Table 1: DRAM configurations.

Memory Clock	t_{RAS}	t_{CL}	t_{RP}	t_{RC}	t_{RCD}	t_{RRD}
Baseline: Off-chip GDDR5, 2GB, Bandwidth = 320GB/s						
2.5GHz	22ns	8ns	8ns	30ns	8ns	5ns
In-package Graphics Memory, 2GB, Bandwidth = 160GB/s, 240GB/s, 320GB/s, 480GB/s, 640GB/s						
3.75GHz	16ns	5ns	5ns	20ns	5ns	4ns
2.5GHz	18ns	7ns	7ns	24ns	7ns	4ns
1.875GHz	21ns	8ns	8ns	29ns	8ns	4ns
1.25GHz	23ns	10ns	10ns	32ns	10ns	5ns
938MHz	27ns	12ns	12ns	36ns	12ns	6ns
625MHz	31ns	14ns	14ns	38ns	14ns	7ns
469MHz	37ns	16ns	16ns	40ns	16ns	11ns
313MHz	47ns	20ns	20ns	50ns	20ns	14ns

scale the PBW to 320GB/s by increasing the memory frequency from 1.25GHz to 2.5GHz. We use curve fitting based on GDDR5 to obtain the low-level DRAM timing as shown in Table 1. The latency of signals passing through silicon interposer can be reduced to 1/5 of that with standard I/Os [5]. We conservatively assume 20% memory latency improvement compared to off-chip memories.

We evaluate various available GPU workloads from the NVIDIA CUDA SDK [12], Rodinia benchmarks [3], and applications distributed with GPGPU-sim [2]. Table 2 lists the characteristics of our 16 workloads. The memory intensity (MI) of some applications, such as *aes* and *sto*, is lower than 1.0. The three most memory-intensive benchmarks are *mum*, *bfs*, and *nw*.

We calculate the DRAM power based on the power model from Micron [10] and modify it to calculate the I/O power of GDDR memory with the pseudo-open drain (POD) signaling scheme. We calculate the power of GPU cores, caches and memory controllers based on the power model from McPAT [11], and modify the model to adapt to the configuration of GPU shader cores. The performance and memory-access information is fed into the power model to calculate the run-time system power consumption.

6. RESULTS

Static Interface: Figures 7 shows the performance and power of the in-package graphics memories with fixed bus widths of 64-bit (other bus width configurations also show the same trend in energy efficiency improvements). Our results of the integrated graphics memories are normalized to the baseline of off-chip GDDR5 memory that supports a peak bandwidth of 320GB/s. Some applications show IPC losses compared to the baseline. This is due to the fact that for a given fixed bandwidth, the larger buses provided by the in-package graphics memories are offset by lowering their clock speeds. At a system bandwidth allocation of 160GB/s, memory-intensive applications (e.g., *mum*, *bfs*, and *nw*), do incur some significant IPC

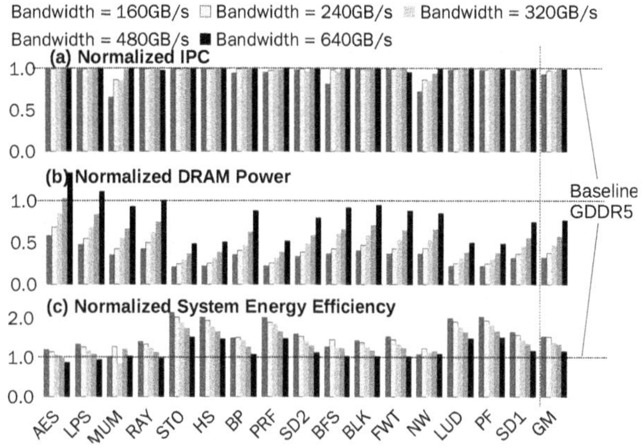

Figure 7: (a) IPC, (b) DRAM power, and (c) energy efficiency with in-package graphics memory (no reconfiguration).

degradations when using the in-package graphics memory. Not surprisingly, the non-memory-intensive applications (e.g., *aes* and *sto*), do not suffer from the reduction in memory clock speed. The in-package memory, however, provides a much more power-efficient implementation (middle plots), which in turn leads to better energy efficiency. Of course, fixing system bandwidth to be equal to the off-chip solution does not really take advantage of the wide interface provided by the in-package memory. By increasing the memory interface clock speed to provide bandwidths of 480GB/s and 640GB/s, performance on the memory-intensive applications can be brought back up. Even without reconfiguration, the in-package graphics memory solution significantly improves the overall energy efficiency of the GPU system by up to 54% on average.

EOpt: Figure 8 shows performance and power results of reconfigurable memory interface optimized for energy efficiency. Because we do not have any initial information about an application, the initial memory interface configuration is always set to a 128-bit bus width. All results are normalized to the case of in-package graphics memory using static 128-bit interfaces to demonstrate the additional benefit of dynamic reconfiguration on top of the benefits of integrated memories. As shown in Figure 8(a), the reconfiguration mechanism yields the greatest IPC improvement for the three most memory-intensive applications. The IPCs of non-memory-intensive applications are not affected by the change of memory interface, and execution pattern detection may cause performance overhead. Fortunately, most applications do not incur frequent execution pattern changes, and the IPC remains stable on the non-memory-intensive applications. Figure 8(b) shows that DRAM power with almost all applications is reduced, and by an average of 14%. Figure 8(c) il-

Table 2: **Characteristics of selected GPGPU benchmarks (the instruction count (IC) and the memory intensity (MI)).**

Abbrev.	Benchmarks	IC	MI	Abbrev.	Benchmarks	IC	Memory
HS	Hot Spot	80M	1.5	BP	Back Propagation	193M	4.3
PRF	Particle Filter	3.9G	4.4	BFS	Breadth First Search	484M	37.0
NW	Needleman Wunsch	218M	8.0	LUD	LU Decomposition	40M	2.5
PF	Path Finder	76M	1.9	FWT	Fast Walsh Transform	4.5G	5.2
AES	AES Encryption	30M	0.3	BLK	BlackScholes Option Pricing	196M	3.2
LPS	3D Laplace Solver	82M	3.8	MUM	MUMmerGPU	75M	43.7
RAY	Ray Tracing	65M	3.0	STO	StoreGPU	123M	0.6
SD1	Speckle Reducing Anisotropic Diffusion	8.4G	5.8	SD2	Speckle Reducing Anisotropic Diffusion	2.4G	3.8

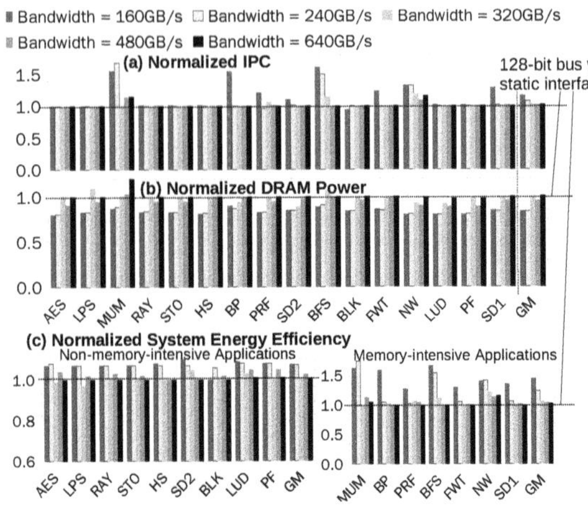

Figure 8: Results of using EOpt to all benchmarks. (a) IPC. (b) DRAM power. (c) System energy efficiency improvement.

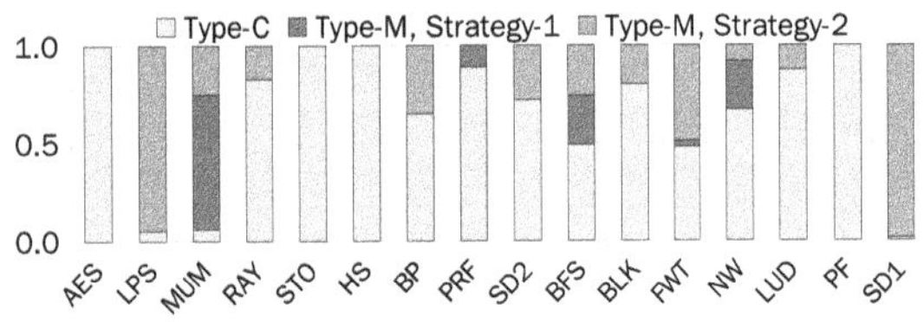

Figure 9: The fraction of instructions spent on different configuration modes, using PerfOpt.

lustrates that the overall system energy efficiency is improved for all the benchmarks. Across all low- and high-intensity applications, the energy efficiency improves by 23% on average.

PerfOpt: Figure 9 illustrates the fraction of instructions spent on each configuration mode under the power budget of 220W (similar trend can be observed with other power budgets). Even the most memory-intensive applications have a portion of less memory-intensive periods, which we can utilize to improve system performance. For example, *mum*, spends about 68% of its instructions with high memory clock speed to avoid system performance degradation (strategy-1). In 25% of the instructions, the memory clock is slowed down by strategy-2 to increase the GPU clock frequency. In the rest 7% of instructions, the memory power is minimized as type-C periods to further improve system performance. It is also shown that the configuration modes change from one to another for most applications. The exceptions are those non-memory-intensive applications, such as *aes*, *sto*,*hs*, and *pf*, which stay in type-C configuration all the time. Table 3 shows the results of the reconfiguration for improving overall GPU throughput given various system power budgets. All the performance results are normalized to the mean throughput of the static configuration that leads to the highest average performance with all the applications under the given power budget. The results demonstrate that the reconfiguration mechanism can adjust the memory power consumption to fit the application memory needs, and that

Table 3: Mean Throughput improvement of all tested benchmarks for each given GPU system power budget.

Budget(W)	200	210	220	230	240
Imprv.	10.0%	14.8%	21.8%	27.5%	30.3%

the saved power can be effectively redeployed to improve the GPU core performance.

7. CONCLUSION

We have presented a reconfigurable in-package wide interface graphics memory, integrated with a high-performance GPU on a silicon interposer. Our design is feasible and easy to implement. Almost all the hardware modification is limited to the memory interface and controllers and no modifications are required to the internal structures of the processors and the memory arrays. Reconfiguration is only applied to the memory interfaces, and internal bus widths are fixed. Therefore, the main extra manufacturing costs is the packaging cost. Another merit of our design is the flexibility of optimizing either power or performance. For high performance computing systems, performance is the primary design consideration. Yet, power constrained applications prefer high energy efficiency. Our reconfiguration mechanisms can be employed by both types of systems.

8. REFERENCES

[1] AMD. Radeon specifications. http://www.amd.com/us/products/desktop/graphics/.

[2] A. Bakhoda, G. Yuan, and W. Fung et al. Analyzing cuda workloads using a detailed gpu simulator. In *Proc. of Intl. Symp. on Performance Analysis of Systems and Software*, pages 163–174, 2009.

[3] S. Che, M. Boyer, and J. Meng et al. Rodinia: a benchmark suite for heterogeneous computing. In *Proc. of Intl. Symp. on Workload Characterization*, pages 44–54, 2009.

[4] X. Dong, Y. Xie, and N. Muralimanohar et al. Simple but effective heterogeneous main memory with on-chip memory controller support. In *Proc. of the International Conference for High Performance Computing*, pages 1–11, 2010.

[5] P. Dorsey. Xilinx stacked silicon interconnect technology delivers breakthrough FPGA capacity, bandwidth, and power efficiency. In *Xilinx White Papers*, 2010.

[6] Elpida. Introduction to GDDR5 SGRAM. http://www.elpida.com/pdfs/e1600e10.pdf. 2010.

[7] M. Gebhart, D. R. Johnson, and D. Tarjan et al. Energy-efficient mechanisms for managing thread context in throughput processors. In *Proc. of the Intl. Symp. on Computer architecture*, pages 235–246, 2011.

[8] Hynix. GDDR5 SGRAM datasheet. http://http://www.hynix.com/products/graphics/.

[9] Intel. Thermal protection and monitoring features: a software perspective. In *Intel Software Network*, pages 1–6.

[10] J. Janzen. The Micron power calculator. http://www.micron.com/products/dram/syscalc.html.

[11] S. Li, J. H. Ahn, and R. D. Strong et al. McPAT: an integrated power, area, and timing modeling framework for multicore and manycore architectures. In *Proc. of the Intl. Symp. on Microarchitecture*, 2009.

[12] NVIDIA. CUDA SDK. http://www.nvidia.com/object/cudasdks.html.

[13] D. Q. Ren and R. Suda. Modeling and optimizing the power performance of large matrices multiplication on multi-core and GPU platform with CUDA. In *Proc. of the International Conf. on Parallel Processing and Applied Mathematics*, pages 421–428, 2010.

[14] K. Skadron, M. R. Stan, and K. Sankaranarayanan et al. Temperature-aware microarchitecture: modeling and implementation. *ACM Trans. on Architecture and Code Optimization*, 1(1):94–125, 2004.

[15] N. specifications. http://www.nvidia.com/object/productquadrofx5800us.html.

[16] M. Sunohara, T. Tokunaga, T. Kurihara, and M. Higashi. Silicon interposer with TSVs (through silicon vias) and fine multilayer wiring. In *Proc. of the Electronic Components and Technology Conference*, pages 847–852, 2008.

[17] W.-K. S. Yu, R. Huang, and S. Q. Xu et al. SRAM-DRAM hybrid memory with applications to efficient register files in fine-grained multi-threading. In *Proc. of the Intl. Symp. on Computer Architecture*, pages 247–258, 2011.

Fan-Speed-Aware Scheduling of Data Intensive Jobs

Christine S. Chan† csc019@ucsd.edu
Yanqin Jin‡ y7jin@ucsd.edu
Yen-Kuan Wu† yew002@ucsd.edu
Kenny Gross§ kenny.gross@oracle.com
Kalyan Vaidyanathan§ kalyan.vaidyanathan@oracle.com
Tajana Šimunić Rosing†,‡ tajana@ucsd.edu

†Department of Electrical and Computer Engineering, UC San Diego, La Jolla, CA 92023
‡Department of Computer Science and Engineering, UC San Diego, La Jolla, CA 92023
§Oracle Physical Sciences Research Center, San Diego, CA 92121

ABSTRACT

As server processor power densities increase, the cost of air cooling also grows resulting from higher fan speeds. Our measurements show that vibrations induced by fans in high-end servers and its rack neighbors cause a dramatic drop in hard disk bandwidth, resulting in a corresponding decrease in application performance. In this paper we quantify the performance and energy cost effects of the fan vibrations and propose a disk performance aware thermal, energy and cooling technique. Results show that we can not only meet thermal constraints, but also improve performance by 1.35x as compared to the conventional methods.

Categories and Subject Descriptors

C.4 [**Performance of Systems**]: Reliability, availability, and serviceability; Modeling techniques

General Terms

Design, Performance, Reliability.

Keywords

Thermal, cooling, energy, vibration, disk I/O

1. INTRODUCTION

Datacenters have become a core component in modern computing. Companies like Google, Facebook, Amazon, etc., build large scale datacenters to perform data-intensive processing such as Map-Reduce and Hadoop, or provide virtual machine hosting such as Amazon EC2. For such services, it is crucial to provide a guarantee for *quality of service* (QoS) which is usually stipulated in the form of *service level agreement* (SLA). At the facility level, heating, ventilating and air conditioning (HVAC) moderates ambient temperature in the data center and inlet temperature of individual rack servers. A 2006 report [8] shows that servers and datacenters in the US consume 1.5% of the country's total electricity consumption, with power and cooling infrastructure consuming as much as 50% of total datacenter energy. Individual servers are equipped with a fan subsystem to reduce the temperature within the chassis. These fans consume power at a rate cubically related to the fan speed, making high fan speeds very expensive.

As transistor densities increase with Moore's Law, each generation of server fans grows more powerful to cope with increasingly pronounced on-chip thermal hotspots. Datacenters running mission-critical workloads (e.g. stock market servers) additionally over-provision fan subsystems to maintain a low set point. As recently as five years ago, fans in commercial servers ran at fixed, high speed. Currently, most enterprise servers have variable-speed fans, which enable server energy usage to be minimized when utilization or inlet air temperature is low, but allows fan speeds to rise as workload demand or ambient temperatures increase.

For commodity *hard disk drives* (HDDs), areal densities have been growing exponentially at a rate faster than Moore's law. Consequently, the read/write head on HDDs is now trying to hit a tiny track that is only 20 nanometers wide which is 1/1000th the width of a human hair while floating just 7 nanometers above the surface of a platter that is spinning at up to 15,000 RPM, so even very low levels of disturbance inside servers and storage arrays degrade IO throughput considerably. As fan speeds increase, internally generated vibrations rise as well. If a server is deployed in a metal rack with other servers that also have variable speed fans, then externally generated vibrations will grow. This can have a significant impact on overall throughput when servers are running data-intensive workloads, such as database lookups and updates. Ambient environment vibration-related impacts to storage IO throughput are shown in the widely viewed YouTube video [30], where hard disk latencies increase abruptly by over 100ms whenever a man shouts in the datacenter.

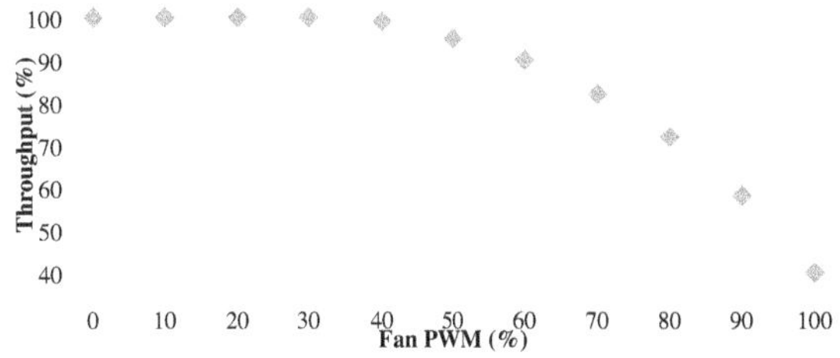

Figure 1 Disk throughput dependence on Fan PWM

Server fan specifications are usually listed by maximum power and maximum rate of airflow in units of *cubic feet per minute* (cfm). Fans are operated with *pulse width modulation* (PWM), where fan speed is adjusted by varying the duty cycle of a control signal. The duty cycle from 0-100% represents the ratio of time between when the fan motors are powered with a constant supply voltage, and when they are "off". This duty cycle is directly proportional to fan speed. Figure 1 visualizes the effect of fan speed on IO throughput measured on a server codenamed "E6" in this paper, with 2 sockets, 8 cores per socket, 64 GB memory, 2.3 TB disk and 6 fans with uniform voltage-RPM ratio. From Figure

1 we can see that disk throughput starts to decrease when fan PWM is at around 40%, and can fall to only 40% of the ideal bandwidth when fans are running at maximum speed.

Although *solid state disks* (SSD) performance is independent of vibrations, they are still too expensive to replace traditional hard drive as a sole solution to datacenters. Moreover, they lack the ability to provide sustained and high throughput [9]. Liquid cooling for 3D chips [31] is becoming a popular alternative cooling system. However, it is expensive. In most configurations the coolant pump is placed near the chip, thus inducing similar levels of vibrations that lead to disk performance degradation. Therefore fan speed control policies that address the vibration issue are still urgently needed.

Vibrations also have a secondary impact on server system energy, measured by the total integrated power needed to complete a fixed customer workload. If an application suffers from longer execution time due to disk performance degradation, the whole system will be kept running for a longer time, consuming more energy and cutting into the bottom line of commercial enterprises.

Today, most thermal and cooling management schemes optimize for energy efficiency in the CPU and memory subsystems separately, which leads to sub-optimal decisions. A recent work [20] combines thermal, energy and cooling management of CPU, memory and fan subsystem to improve *server power usage efficiency* (SPUE) with low performance overhead. However, these power management schemes largely focus on the direct effect of core temperature on computational performance and lack awareness for the indirect mechanical effects of cooling on disk IO throughput performance.

We are motivated by the desire to efficiently manage energy, thermal and cooling issues in servers running data intensive workloads, which are sensitive to disk performance. The contributions of this paper are outlined below.

- We model the effect of fan cooling on server performance. It is necessary to study this effect because there is a coupling between disk performance and fans as shown in Figure 1.
- We survey a range of low to high-end server configurations to study when vibration-caused disk throughput starts to affect application performance. Our experiments focus on servers with internal, local disks, though the analysis applies to externally attached storage arrays as well, since they have similar air-cooling systems.
- We design a disk-vibration-aware cooling controller which uses disk IO to adaptively adjust fan speed.
- Our joint thermal management framework can achieve up to 1.35x speedup comparing to the baseline.

2. RELATED WORK

Thermal and cooling management in server systems has been studied in recent years. However most of the previous research has focused on CPU and memory separately. Three reactive techniques have been used to resolve thermal issues: (1) clock or power gating, which eliminates switching activity by stopping the processor, (2) *dynamic voltage and frequency scaling* (DVFS), which allows processor to work under different voltage and frequency settings to reduce power consumption, thus resolving thermal emergencies, and (3) task migration, which balances power distribution across the die to remove thermal hotspots. All the benefits and drawbacks have been discussed in [10]. On the other hand, proactive thermal management techniques, presented in [13], [14], predict future temperature and take action when the processor is still running below the temperature threshold.

Besides energy and thermal management of CPU and memory subsystem, modern servers are equipped with a fan subsystem. Various fan speed control algorithms have been studied. Chiueh et al. [18] proposed a fan controller that integrates existing thermal management. Wang et al. [19] try to reduce blade server cooling to an optimization problem in which they try to minimize total energy consumption of fans while maintaining critical temperature below defined thresholds.

Nevertheless, all previous techniques perform energy, thermal and cooling management of CPU and memory subsystem separately, which may lead to sub-optimal results. Ayoub et al. [20] is the first to combine fan control, CPU and memory thermal management together. They propose a unified thermal and cooling model for CPU and memory subsystems, which is called JETC that reduces operational energy of memory by clustering memory accesses to a subset of memory modules while considering thermal and cooling metrics. It removes hotspots by performing socket level thread scheduling which involves migrating threads from hot CPUs to cool CPUs. Cooling costs are also reduced due to more balanced heat dissipation.

Relatively little work has focused on thermal management of disks. Kim et al. [22] study the effect of temperature on disk IO by presenting an integrated performance-thermal simulator. To the best of our knowledge, our work is the first to consider the effect of vibrations on disk performance. We not only present experimental results showing a relationship between fan speed and disk throughput drop to model the energy impacts of vibration, but also integrate performance feedback from disk into a novel thermal management system to address this issue.

Many new servers are designed with damping and isolation of vibration sources in mind. However, these enclosures are prohibitively expensive for typically cost-sensitive applications like web services. It would also be infeasible to retrofit new hardware enclosures for the tens of billions of dollars of servers already deployed in datacenters. Software innovations are a cost-sensitive alternative, which can be ported through firmware upgrades to legacy servers, benefiting both individual server performance and rack vibration levels. In addition, OS awareness of dynamic workload performance can adjust to vibrations at resonance in the racks, which is a physical attribute even damped enclosures have.

The rest of the paper is organized as follows. Section 3 presents a formal description of air cooling model and our joint thermal management framework considering disk throughput as a feedback. Section 4 discusses simulation results. Section 5 concludes this paper.

3. MODELING AND CONTROL

Modern servers have a shared fan subsystem to cool CPU and memory components. For example, in our 45nm Intel Quad Core dual socket Xeon E5440 server the fans generate air flow towards two CPU sockets, and then the warmed up air flows over memory modules which are positioned downstream of sockets, thus creating thermal dependency between the upstream and downstream components. The hard disks reside on the left side of the fans in our server. As the fans spin, their vibrations affect the reliability and throughput of the hard disk. In modern cooling systems with variable fan speeds, fan control decisions are made only according to the temperature of the hottest active core. Thus, the fan is over-provisioned most of the time due to the step-wise characteristic of fan control, where the speed is set at a maximum for some range of temperatures, even if the real current temperature is lower. This leads to higher than necessary vibrations that negatively impact disk performance.

A. Thermal model

We use electrical analogies to represent the thermal and cooling behavior of components [20]. Figure 2 shows such a thermal model of the CPU and memory, highlighting the thermal dependence between the two.

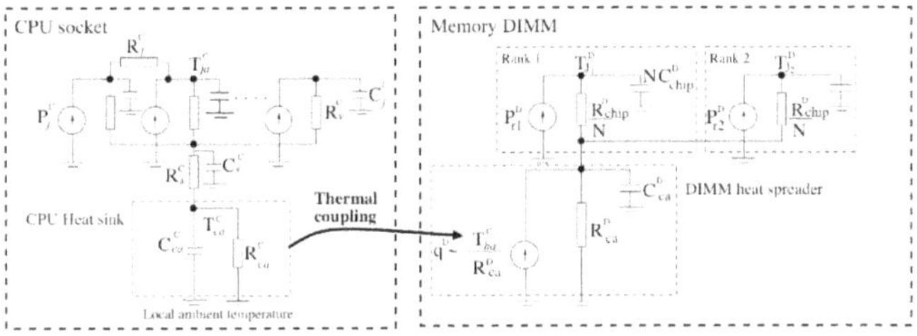

Figure 2 Thermal Coupling between CPU and Memory

The power dissipated at each component, P_j^c, acts as a heat source in the RC circuit. Heat exchange between components is modeled as resistances and capacitances. Each junction of the die components has lateral heat flow, modeled as R_j^c, vertical heat flow, modeled as R_v^c, and thermal capacitance C_j^c. The lateral heat flow among die components is negligible in our case, as the most exchange happens vertically between the components and the socket's heat spreader and heat sink, and between the CPUs and memory. Each junction's temperature relative to ambient is denoted T_{ja}^c while the case to ambient temperature is T_{ca}^c. Memory thermal model works in the same way as CPU; power dissipation due to activity is modeled as heat sources, and heat exchange is modeled as resistances and capacitances of each rank. There is no heat sink on memory, so the heat exchange between the interior and ambience are modeled as R_{ca}^D and R_{ca}^D , the thermal resistance and capacitance of its heat spreader. We represent the thermal coupling between the CPU and memory as an external power source q^D to the memory DIMMs, dependent on the temperature of the CPU and the case to ambient thermal resistance of the DIMMs:

$$q^D \propto \frac{T_{ha}^C}{R_{ca}^D} \tag{1}$$

The states of the memory heat spreader and the CPU heat sink are dominant factors for determining cooling control in the system. The case to ambient resistance measured from the heat sink to the ambient consists of the conductive resistance, a constant characteristic of the heat sink material, and the convective resistance, which is variable during runtime:

$$R_{ca}^C = R_{conv}^C + R_{hs}^C \tag{2}$$

A socket fan's speed influences its heat sink's ability to dissipate heat, so cooling is modeled through the convective resistance of the heat sink, calculated each time the fan speed updates as follows:

$$R_{conv} \propto \frac{1}{AV^\alpha} \tag{3}$$

where A is the effective area of heat sink and V is the air flow rate proportional to fan speed, and α is a coefficient ranging from 0.8 to 1.0. The air flow rate increases with fan speeds dictated by the duty cycle of the fan motors (PWM), and the driving fan power P_V which rises cubically with air flow rate:

$$V \propto PWM \qquad \frac{P_{V_2}}{P_{V_1}} = \left(\frac{V_2}{V_1}\right)^3 \tag{4}$$

The above relation is true for each individual fan, but in power calculations for scaled-up server configurations, air flow rate can increase linearly with power when varying the number of fans while keeping all fan speeds constant. Intuitively, air flow can be doubled by doubling the number of fans, which would also double the power drawn. This is more fan-energy efficient than doubling the speed of the original fans, which would cause an 8-fold increase in power consumption.

Vibrations are transmitted in typical enterprise servers from the fan motors by mechanical coupling to the housing for the disk drives. We have verified empirically that vibrational amplitudes go up linearly with fan motor PWM, so we model the vibrations as proportional to air flow rate in cubic-feet per minute. We model the effect on disk throughput as a function of fan speeds as shown in Figure 1. The aggregate effect on disk performance has been obtained empirically. In the future, we hope to optimize the model for the particular server design or storage configuration by also taking into account the rack geometry, mass loading, and resonance frequencies of components and chassis support structures.

B. Control framework

State of the art cooling controls are implemented independently from workload scheduling, so their algorithms are also off-limits to the OS. Engineering interfaces to the internal fan control algorithms are not exposed to the end user, and therefore cannot be directly modified for testing purposes. Since we cannot access every stage of thermal and cooling controls, we had to use simulations to investigate the combined effects of fine-grained fan control combined with workload scheduling. We characterized the effect of fan speeds on hard disk bandwidth through experiments on the E6 server (Figure 1), then separately characterized the effect of this hard disk bandwidth (Table 3) by modifying delays in the IO driver and monitoring the total execution time for a benchmark. With data from these experiments, we can estimate the dependence of application performance on fan speeds without an overhaul of the physical system design.

We developed an energy- and thermal-aware OS-level simulator to evaluate and demonstrate how different scheduling policies can mitigate this dependence. The scheduler makes decisions about power distribution to various system components based on the fan control policies mentioned at the end of this section, and passes them to another thermal simulator. For these experiments we used TPC-H[1] [6] as input workload to a large database server. The traces we obtained in this way are used as a part of our simulation system to quantify the application performance.

Figure 3 shows the components of our system: controller, actuators (workload schedulers, fan), and sensors (thermal, power, fan speed, disk throughput). The model is similar to the one proposed in [20], however we also add the effects of changing fan speed on disk throughput and application performance. The controller receives fan data from the fan subsystem and thermal data from the socket and memory module, and analyzes their thermal profile based on the current workload activity. It outputs a vector distributing power among the cores and memory, which dictate core and DIMM migrations or emergency throttling.

[1] TPC-H is only being used as an IO workload generator utility for the research reported in this paper to evaluate and compare relative IO rates versus vibration levels for our proposed thermal and energy policies.

Actuators translate the controller's decisions to real actions in the system, such as task migration among CPUs, DIMM activation, and fan speed settings.

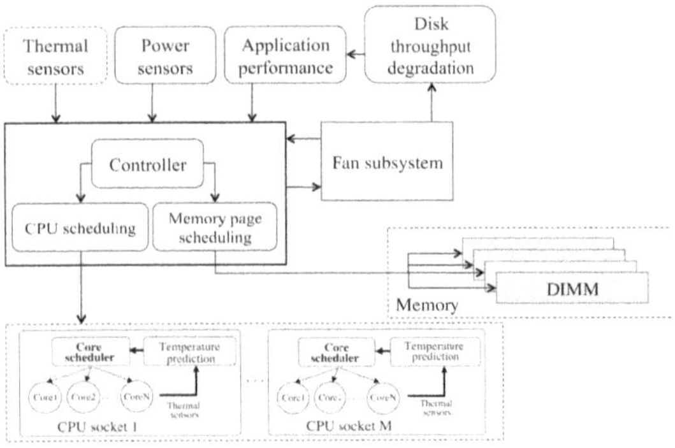

Figure 3 System Model

Mechanical vibrations from the fans slow down the hard disk, which translates into decreased disk throughput. To quantify the effects of fan vibrations on our system, the fan module monitors the fan PWM, and estimates the corresponding disk throughput degradation using physical measurements shown in Figure 1. The observed effect on application performance varies according to the frequency and duration of disk accesses. We empirically obtain it for the set of benchmarks that we run on the server.

$$\Delta R_{ca} = \frac{\Delta T}{P_{total}^{socket}} \tag{5}$$

At each scheduling tick, the controller observes the fan speed, socket power settings and core temperatures. The difference between the thermal set point and the current measured temperature, ΔT, determines the desired change in the case to ambient resistance ΔR_{ca}, which can be achieved with a proportional change in fan speed setting. This change may be bounded either by physical constraints (the maximum speed setting of the fan) or a management policy (e.g. limit on fan speed to ensure good disk throughput), and is enforced by the fan actuator. The CPU actuator migrates workloads across cores and sockets as a function of policies. The memory actuators perform DIMM-level thermal and performance management through activity migration or activating and deactivating individual modules. Below we describe three state of the art fan and workload control policies we compare with a new policy we developed which is aware of fan vibrations effect on disk throughput. All of the policies throttle activity via DVFS as the last resort if the cores reach the emergency temperature threshold.

Dynamic load balancing (DLB): This is a common policy in modern operating systems, so we use it as our default baseline. The operating system uses thread migration to minimize the length difference between task queues, but does not actually monitor temperatures. It uses a default proportional-integral (PI) fan controller and keeps all memory modules active.

Dynamic Thermal Management (DTM): Workloads are migrated proactively to alleviate the maximum core temperature [32], and memory activity is throttled when temperatures approach the emergency threshold. Again, the fan actuator has a PI controller and runs independently from workload scheduling.

Joint Energy, Thermal and Cooling (JETC): JETC is the combined solution for core scheduling, memory module activation and fan control proposed in [20]. A centralized controller oversees proactive core and memory page scheduling and adjusts the fan speed, DIMM power settings and workload assignment. This joint workload scheduling, memory resizing and cooling management is expected to yield higher stability and lower the frequency of socket migrations and fan speed adjustments. Although it is not explicitly performance-aware of disk intensive apps, the operating goal of lowering fan speeds is directly related to lowering costly disk vibrations. JETC tends to operate closer to peak thermal bound than the default, so disk-bound workloads suffer due to high fan speeds and vibrations, while CPU-bound workloads suffer if core temperatures hit the thermal threshold and emergency core throttling happens.

JETCXtension: Our extension adds awareness for disk performance to the thermal management framework. We track the disk intensity of workloads by monitoring IO utilization per second via the OS. The controller makes scheduling decisions at the granularity of 10ms while fan control and IO monitoring are done at 1 second intervals corresponding to server enclosure thermal time constants. At each fan control interval, there is a gap between heat dissipation of the cores and the thermal set point, which is usually closed by provisioning fan speed according to Equations (3)-(5). In this policy, the designer can specify an acceptable level of fan-induced IO degradation, which corresponds to a maximum fan speed shown in Figure 1. During periods of significant disk activity (over 1% utilization), we only allow fans to spin as fast as the specified limit, while progressively throttling the workloads with lower disk access rates to meet thermal constraints. During periods of low disk activity, the limit on fan speed is lifted, and control returns to the default JETC policy.

4. EVALUATION

The problems described arise in datacenters where workloads comprise both service jobs with high memory and disk access, and batch jobs with low average disk access. We model four different server system configurations, based on a Xeon and other high-end enterprise servers. Server Q (for "quad-core") is a 45nm Intel Quad-core dual socket Xeon E5440 [1] instrumented in our lab, providing thermal, cooling and power measurements for various workloads. Each socket has 4 cores and a thermal design power (TDP) of 80W, cooled with two fans that draw 60W at top speed. Server E (for "eight-core") is based on a higher end server targeted at reliability-sensitive customers. It has 45nm 8-core dual socket processors with TDP of 100W. Comparable processors on the market are the Intel Xeon X7560 in a SuperServer 8016B-6/T form factor with 7 fans, AMD Opteron 6140, and Oracle SPARC T4, all manufactured at 45nm, with 8 cores per socket and a similar TDP [2][3][4]. We have 3 variants – E2, 4 and 6. The effects of disk vibrations are quantified through physical measurements on Server E6, which has a full-sized cooling subsystem with 6 fans that total 300W at maximum speed. Alternate versions, E2 and E4, are simulated with scaled-down cooling subsystems of 2 and 4 fans respectively.

Table 1 Scaled server configurations

	Sockets	Cores/ socket	Heat sink width	TDP	Fan power	DRAM
Q	2	4	6cm	80W	60W	32GB
E2	2	8	10cm	100W	60W	64GB
E4	2	8	10cm	100W	200W	64GB
E6	2	8	10cm	100W	300W	64GB

We ran benchmarks on the Q server and collected the dynamic power consumption of the cores and memory DIMMs at runtime using NI DAQ [5]. We also changed the disk device driver by adding delays to each read/write request, emulating the total proportional decreases in IO throughput as observed in Figure 1.

By manually tuning these delays we emulated disk IO throughput degradation while running each disk-bound benchmark.

We tested the servers at 75% utilization, as this is a reasonable peak level a typical data center might run at. Each workload we run is provisioned as shown in **Table 2**. It contains a service job (TPC-H query) that must meet its response time requirements, coupled with batch jobs (for simplicity, we used SPEC benchmark *hmmer*). TPC-H is a well-known decision support benchmark set consisting of 22 business-oriented queries. Such decision support workloads are common in large-scale datacenters, especially in enterprise environments where companies need fast access to data for informed decision-making. These queries have intensive disk usage and relatively low CPU activity, but still represent a broad spectrum of operations present in databases. For example, TPC-H query 1 reports the amount of business that was billed, shipped and returned, queries 13 seeks relationships between customers and size of their orders, and query 19 reports the gross discounted revenue attributed to the sale of selected parts handled in a particular manner. We selected queries 1, 13 and 19 for experiments because they represent the range of disk bandwidth and CPU utilization across the benchmark set (Table 3).

Table 2 Workload Combinations of TPC-H and SPEC

Workload	Socket A	Socket B
Server Q		
W1	*2hmmer + tpchq1*	*3hmmer*
W2	*2hmmer + tpchq13*	*3hmmer*
W3	*2hmmer + tpchq19*	*3hmmer*
Server E2,4,6		
W4	*5hmmer + tpchq1*	*6hmmer*
W5	*5hmmer + tpchq13*	*6hmmer*
W6	*5hmmer + tpchq19*	*6hmmer*

We used an extended version of HotSpot Temperature Modeling Tool [25] to evaluate the policies described in the previous section. HotSpot is configured with a Xeon socket parameters which are scaled by the number of cores a particular server has. Critical temperature is set to 90°C for all our results. Given a power trace, HotSpot returns dynamic thermal changes to the controller, replacing the role of physical thermal sensors in Figure 3. The results we present are evaluated after warming up the server from ambient temperatures. Our workload scheduler has an interval of 10ms and the fan control response is every 1s as thermal time constants outside of the socket are on the order of seconds. Figure 1 shows the instantaneous disk throughput dependence on fan motor power obtained by stepping through the full fan operational range in 1% increments on server E6.

4.1 Results

We compare results for three thermal/cooling management policies, DTM, JETC and JETCX, against the default DLB policy. We also study the comparison across four server configurations shown in Table 2. For fair comparison across policies, the core thermal set point is 85°C and the emergency threshold 90°C, accounting for sensor error [33], so the small amount of reported amount of time spent at these temperatures do not violate actual device thermal constraints for reliability, which are much higher. Fans are triggered at 85°C, and if any core reaches 90°C, the frequency and corresponding power consumption are throttled progressively, with wake up time of 50μs.

We compare the benefits of policies by measuring the reduction of IO wait time for queries compared to DLB for server systems in which this reduction was measured. Note that system always has some IO delay, even when fans are not causing vibrations, due to various overheads in IO access. E4 fan causes some vibrations, but thermal management through core migration mitigates a significant portion, which is why DLB still performs well. A larger cooling system in E6 drives up IO delays, requiring policies to mitigate that effect. With JETCX, we do not tolerate fan-induced IO degradation, so the maximum PWM allowed in E6 is 40%. IO delay is reduced by 60%, to near minimum level possible in the system, as it detects disk access patterns and ensures that the fan speed stays low enough to minimize the effect on disk throughput.

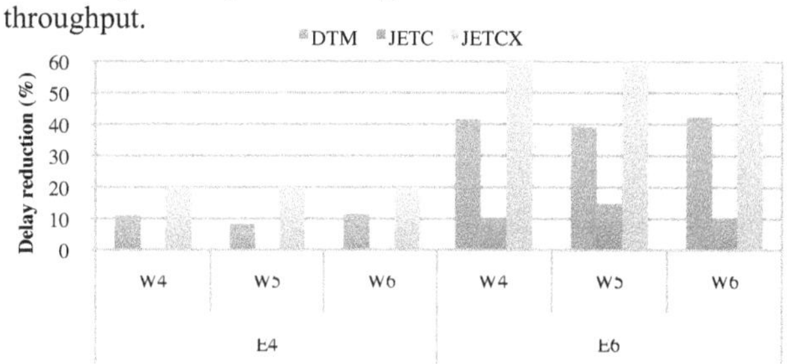

Figure 4 Reduction in hard disk IO wait times

Table 3 shows how TCP-H queries respond to disk BW degradation. When running queries on E6 system with DLB policy, the disk bandwidth degradation can reach as high as 60% (leaving 40% of the original bandwidth), resulting in large performance overhead. In contrast, our JETCX policy reduces this overhead, achieving an overall application speed up of 1.35x for query 19.

Table 3 Disk dependence of TPC-H queries

Query	1	13	19
Ave. disk BW utilization	59%	67%	99%
Performance hit with static disk bandwidth degradation:			
80% bandwidth available	24%	69%	20%
60% bandwidth available	81%	105%	57%
40% bandwidth available	108%	170%	100%

In Figure 5 we compare energy savings due to management policies. Most energy savings come from lower fan speeds and clustering of data in memory DIMMs, which JETC and JETCX take advantage of. CPU energy remains approximately constant as the workload does not fluctuate much, but DTM can get some energy savings by migrating cores to alleviate thermal hotspots instead of increasing fan speeds. JETCX and JETC optimizations for disk IO show little benefit in small systems, since there is little delay in the first place. Server E2 has a higher TDP, but the cooling system is under-provisioned, leading to higher core temperatures, bigger need for emergency throttling, and thus smaller energy saving opportunities. Although E4 and E6 cooling subsystems are more capable, core temperatures are still high because of the workload characteristics and non-linear increase of cooling capability. JETC's goal of reducing fan and overall energy cost by equalizing and lowering fans speeds also leads to lower fan vibrations, indirectly boosting disk throughput performance compared to policies that overprovision fan speeds such as DLB and DTM. However, JETC balances the fans at the expense of an elevated average core temperature. Energy savings in JETC are not as high because it hits emergency throttling too often – this extends the runtime of an application, keeping server components in on-state for longer and thus consuming more energy. In E6 system JETCX improves energy savings by 1.8x over JETC, and 4.5x as compared to DTM. JETCX gives the best IO throughput for data-intensive jobs at minimal cost to batch job performance. This cost is quantified by the amount of batch job slowdown shown in the second y-axis of Figure 5.

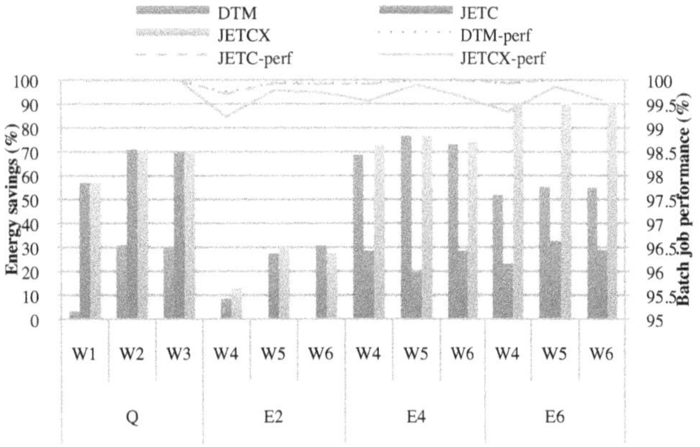

Figure 5 Energy savings in memory and cooling power, and normalized batch job performance *(higher is better)*

The highest level of throttling (0.77%) is seen on the E2 server due to a severely under-provisioned cooling system, where JETCX proactively throttles batch jobs to meet thermal constraints even with fans at maximum speed. In the realistic E6 server, the average performance hit due to throttling is 0.4%, with a maximum of 0.66% for workload 4. DTM and JETC only slow down the cores reactively at emergency temperatures, causing instability and lower efficiency of heat removal. JETC allows average temperatures to rise above the thermal set point in favor of lower fan power. While the performance cost is an unintended last resort for other policies, JETCX proactively reduces the performance of batch jobs before temperatures reach emergency, thus keeping fan speeds balanced and low and ultimately reducing disk vibrations.

5. SUMMARY

In this work we have shown for the first time that computer fans can have a significant impact on the IO throughput of hard disks, thus dramatically slowing down processing of data intensive workloads. Our work is the first to model such impact quantitatively and propose a policy capable of reducing the vibration-induced performance overhead. We conducted empirically-guided simulations to verify our policy. Our results show that compared to the dynamic load balancing (DLB) policy implemented in today's servers, JETCX gives up to 1.35x speed up in service jobs at only 0.41% performance hit to batch jobs, while saving as much as 89% in memory and cooling energy.

6. ACKNOWLEDGEMENTS

Our thanks to NSF Project GreenLight Grant 0821155, NSF Grants 0916127 and 1029783, MuSyC, Oracle, Google, Microsoft and UCSD CNS.

7. REFERENCES

[1] Quad-Core Intel Xeon Processor 5300 Series: Thermal/Mechanical Design Guidelines. 5, 22

[2] http://ark.intel.com/products/46499/

[3] http://products.amd.com/pages/OpteronCPUDetail.aspx?id=741

[4] http://www.oracle.com/us/products/servers-storage/servers/sparc-enterprise/t-series/sparc-t4-processor-ds-497205.pdf

[5] http://sine.ni.com/nips/cds/view/p/lang/en/nid/207099

[6] http://www.tpc.org/tpch/

[7] L. Barroso, U. Holzle, "The Datacenter as a Computer: an Introduction to the Design of Warehouse-scale Machines", 2009

[8] US Environment Protection Agency, "Report to Congress on Server and Data Center Energy Efficiency", 2007

[9] B. Debnath, S. Sengupta, J. Li, "FlashStore: High Throughput Persistent Key-Value Store", VLDB 2010

[10] J. Donald, M. Martonosi, "Techniques for Multicore Thermal Management: Classification and New Exploration", ISCA 2006

[11] J. Choi, C. Cher, H. Franke, H. H. A., Weger, and P. Bose, "Thermal-aware Task Scheduling at the System Software Level", ISLPED 2007, pp. 213-218

[12] S. Heo, K. Barr, K. Asanovic, "Reducing Power Density through Activity Migration", ISLPED 2003, pp 217-222

[13] A. Coskun, T. Rosing, K. Gross, "Proactive Temperature Management in MPSOCs", ISLPED 2008

[14] I. Yeo, C. Liu, E. Kim, "Predictive Dynamic Thermal Management for Multicore Systems", DAC 2008, pp 734-739

[15] X. Fan, C. Ellis, A. Lebeck, "Memory Controller Policies for DRAM Power Management", ISLPED 2001, pp 129-134

[16] H. Hai, S. Kang, L. Charles, K. Tom, "Improving Energy Efficiency by Making DRAM Less Randomly Accessed", ISLPED 2005, pp. 393-398

[17] J. Lin, H. Zheng, Z. Zhu, H. David, Z. Zhang, "Thermal Modeling And Management of DRAM Memory Systems", ISCA 2007, pp. 312-322

[18] H. Chiueh, L. Luh, J. Draper, J. Choma, "A Novel Fully Integrated Fan Controller for Advanced Computer Systems", SSMSD 2000, pp. 191-194

[19] Z. Wang, C. Bash, N. Tolia, M. Marwah, X. Zhu, P. Ranganathan, "Optimal Fan Speed Control For Thermal Management of Servers", IPAC 2009, pp. 1-10

[20] R. Ayoub, R. Nath, T. Rosing, "JETC: Joint Energy Thermal and Cooling Management for Memory and CPU Subsystems in Servers", HPCA 2012

[21] Y. Kim, J. Choi, S. Gurumurthi, A. Sivasubramaniam, "Managing Thermal Emergencies in Disk-based Storage Systems", Journal of Electronic Packaging, Transactions of the ASME, 130(4), 2008

[22] Y.Kim, S. Gurumurthi, A. Sivasubramaniam, "Understanding the performance-temperature interactions in disk I/O of server workloads", HPCA 2006

[23] M. Patterson, "The Effect of Data Center Temperature on Energy Efficiency", Proc. ITHERM 2008, pp. 1167-1174

[24] R. Ayoub, T. S. Rosing, "Cool And Save: Cooling Aware Dynamic Workload Scheduling in Multisocket CPU Systems" in ASP-DAC 2009, pp. 891-896

[25] K. Skadron, M. R. Stan, W. Huang, S. Velusamy, K. Sankaranarayanan, D. Tarjan, "Temperature-Aware Microarchitecture", ISCA 2003

[26] S. V. Garimela, V. Singhal, D. Liu, "On-Chip Thermal Management With Microchannel Heat Sinks and Integrated Micropumps" in *Proceedings of the IEEE*, vol.94, no.8, pp.1534-1548, Aug. 2006.

[27] D. B. Tuckerman, R. F. W. Pease, "High-Performance Heat Sinking for VLSI", *Electron Device Letter, IEEE*, vol.2, no.5, pp.126-129, May 1981.

[28] C. A. Soule, "Future Trend in Heat Sink Design", *Electronics Cooling Magazine, 2001.* http://www.electronics-cooling.com/2001/02/future-trends-in-heat-sink-design

[29] R. E. Simons, R. C. Chu, "Application of Thermoelectric Cooling to Electronic Equipment: A Review and Analysis", Semiconductor Thermal Measurement and Management Symposium, 2000. Sixteenth Annual IEEE, vol., no., pp.1-9, 2000.

[30] "Shouting in the Datacenter" [Video], 2008. Retrieved Feb 25, 2012 from http://www.youtube.com/watch?v=tDacjrSCeq4

[31] R. E. Simons, R. C. Chu, "Application of Thermoelectric Cooling to Electronic Equipment: A Review and Analysis", Semiconductor Thermal Measurement and Management Symposium, 2000. Sixteenth Annual IEEE, vol., no., pp.1-9, 2000.

[32] J. Choi, C. Cher, H. Franke, H. H. A., Weger, and P. Bose, "Thermal-aware task scheduling at the system software level," ISLPED 2007, pp. 213–218

[33] Y. Zhang and A. Srivastava. "Accurate temperature estimation using noisy thermal sensors", DAC 2009, pp. 472–477

Procedure Hopping: A Low Overhead Solution to Mitigate Variability in Shared-L1 Processor Clusters

Abbas Rahimi†, Luca Benini‡, Rajesh K. Gupta†

†Department of Computer Science and Engineering, University of California, San Diego, La Jolla, CA, USA

‡Dipartimento di Elettronica, Informatica e Sistemistica, Università di Bologna, Bologna, Italy

abbas@cs.ucsd.edu, luca.benini@unibo.it, gupta@cs.ucsd.edu

ABSTRACT

Variation in performance and power across manufactured parts and their operating conditions is a well-known issue in advanced CMOS processes. This paper proposes a resilient HW/SW architecture for shared-L1 processor clusters to combat both static and dynamic variations. We first introduce the notion of procedure-level vulnerability (*PLV*) to expose fast dynamic voltage variation and its effects to the software stack for use in *runtime* compensation. To assess *PLV*, we quantify the effect of full operating conditions on the dynamic voltage variation of a post-layout processor in 45nm TSMC technology. Based on our analysis, *PLV* shows a range of 18mV−63mV inter-corner variation among the maximum voltage droop of procedures. To exploit this variation we propose a low-cost procedure hopping technique within the processor clusters, utilizing *compile time* characterized metadata related to *PLV*. Our results show that procedure hopping avoids critical voltage droops during the execution of all procedures while incurring less than 1% latency penalty.

Categories and Subject Descriptors

C.4 [**PERFORMANCE OF SYSTEMS**]: Design studies; Reliability, availability, and serviceability

General Terms: Design, Reliability, Performance

Keywords: Variability, Dynamic IR-drop, Many-core

1. INTRODUCTION

Transistor miniaturization and increasing density continue to increase variability in transistor performance [1], and increasing gap between average and peak power [2]. These variations arise from different physical sources: (i) static inherent process parameter variations due to random dopant fluctuations and sub-wavelength lithography as well as device and interconnect degradation due to aging; (ii) dynamic environmental stresses such as temperature fluctuations and voltage droops that result from large current transients across the network power delivery system, commonly referred to as IR-drop. These issues are expected to worsen with technology scaling [3]. Designers commonly use conservative guard-bands for the operating frequency and voltage to handle these variations that significantly add to the cost and contribute to loss of energy efficiency.

Given the close relationship between power and temperature, and the increased importance of variability in the future, treatment of variability during pre-silicon and post-silicon design stages is crucially important. Indeed, several recent efforts have focused on measures to mitigate variability through innovations in design. On-die sensor circuits [4] have been widely used to detect process, voltage, and temperature (PVT) variations coupled with adaptive recovery methods leveraging CMOS knobs: voltage, frequency, and body bias. The process parameter variability information has been exposed to a dynamic voltage and frequency scaling (DVFS) controller that shifts workload from inefficient, leaky processor to efficient, less leaky ones [5]. IBM POWER7 prevents timing errors by integrating multiple critical path monitors (CPM) [4] per core to capture PVT variation, and employing two cooperating feedback controllers: (i) tight coupling between CPMs output and a phase-locked loop (PLL) enables DFS controller to react very quickly to voltage droops; (ii) the second controller dynamically adjusts the processor voltage (DVS) to achieve a desired performance level on a longer time scale [6]. Moreover, a dynamic fine-grain body biasing technique leveraging multiple CPMs is introduced in [7].

Going further up on the hardware-software stack, a notion of instruction-level vulnerability to dynamic voltage and temperature variations is defined to expose variation and its effects to the software stack [8]. Furthermore, it has been shown how collaborative design that encompasses microarchitecture and compiler can lead to a cost-effective solution for fast voltage droop avoidance in commodity processors [9]. A dynamic thread hopping scheme in conjunction with DVFS is proposed to mitigate the within-die variation across 80-core [10]. F. Paterna *et. al.* [11] propose a runtime variability-aware workload distribution technique for multimedia streaming applications running on parallel multiprocessor arrays. That platform has also been equipped against non-uniform aging by an adaptive idleness distribution technique [12].

This paper makes three contributions. First, we introduce the notion of procedure-level vulnerability (*PLV*) to capture the effects of dynamic IR-drop. Using characterized *PLV*, we enable a software preventive methods that build upon well-known hardware detection/correction techniques for process variability and aging. Second, we propose a low-cost *runtime* procedure hopping that facilitates migration of procedures within a processor cluster, utilizing *compile time* characterization (captured as metadata) of *PLV*. Third, an accurate gate-level analysis flow which leverages industrial design implementation tools and libraries to characterize IR-drop of individual procedures in the presence of variability is developed. We demonstrate our approach on a tightly-coupled shared-L1 multi-core cluster, representative of a large class of multi-core architectures (e.g. GP-GPUs, programmable multimedia accelerators). Full post place-and-route (P&R) results in 45nm TSMC technology confirm that the procedure hopping technique avoids the critical IR-drop during the execution of all procedures while incurring less than 1% latency penalty.

The rest of the paper is organized as follows. Section 2 surveys prior work in this topic area. Section 3 describes the architecture of the variation-tolerant shared-L1 processor clusters. The procedure hopping technique is presented in Section 4. In Section 5, we explain our methodology for the characterization of *PLV* to dynamic operating conditions. Section 6 details experimental results. Finally, Section 7 concludes the paper.

2. RELATED WORK

Core-level dynamic voltage scaling as well as discrete V_{DD}-hopping technique [13] has been widely used for multiprocessor systems [10][14], due to the quadratic relation between energy and voltage. Researchers have proposed the chip-wide voltage scaling to slow down aging [15] and dynamic voltage scaling for aging management [16]. Nevertheless, these techniques do not consider the dynamic temperature fluctuation and fast IR-drops.

Different runtime compensation methods that can adapt to dynamic supply voltage fluctuation have been proposed [9][17]-[21]. Although a direct link between voltage droop and processor activity has been provided [17][18], the temperature effects has not been considered. Thus, a thermal-aware adaptive frequency throttling method is proposed to combat voltage variation [19]. Since supply voltage can change rapidly and voltage compensation methods often require at least several clock cycles, a soft threshold is considered on voltage monitor to trigger compensation in advance [17][18], at the expense of frequent unnecessary compensations. Other approaches characterize voltage emergencies through processor activities [20], and problematic instruction sequences [21][9]. An early predictor of an impending voltage droop uses microarchitectural indicators, e.g. TLB/cache miss, and pipeline flush, as signatures of a voltage emergency [20].

Nevertheless, [9][17]-[21] use a very "generic" variability models on high-level architectural simulators which do not consider the detail of physical processor implementation on variability. For instance, the effect of fine-grain clock-gating on power as well as accurate switching activity, and details of power supply network (which simply modeled as a second-order linear system) have been largely overlooked [18]-[21]. On the other hand, real resilient silicon implementations either target a single-core [24], or loosely coupled processors [6][10], and hence their approach cannot fully exploit the possibility of low-cost procedure migrating from one core to another, for instance [10] has a migration overhead of transferring 1280 flits (entire instruction and data memory in one core) over a packet-switched router for threads that have a runtime of 35K cycles. Moreover, these circuit techniques suffer from power-hungry error recovery which is expensive for a many core fabric. In contrast, our approach is applicable to clusters of simple processors and exploits the opportunity given by tightly coupled architecture to dynamically shift work from one core to another with minimal overhead. Further progress in low-cost software-assist techniques requires a highly-accurate *design time* analysis on a physical implementation of cores with back-end details on a proven silicon technology with variability models and characterized operating conditions given by the semiconductor fabrication process. We believe a combination of *design time* and *runtime* is essential to achieve best results, considering only one of them leads to unacceptable overhead.

3. VARIATION-TOLERANT PROCESSOR CLUSTERS ARCHITECTURE

In this section, we describe the architectural detail of proposed variation-tolerant processing cluster. These clusters are the essential parallel components of many core fabrics, e.g., NVIDIA Fermi [22] features 512 CUDA processors organized into 16 groups of processing cluster. In our implementation, each cluster consists of sixteen 32-bit in-order RISC cores compliant with the SPARC V8 architecture, an intra-cluster shared level-one instruction cache (shared-L1 *I$*) [23], an on-chip tightly coupled data memory (TCDM), two fast logarithmic interconnections [25] for both

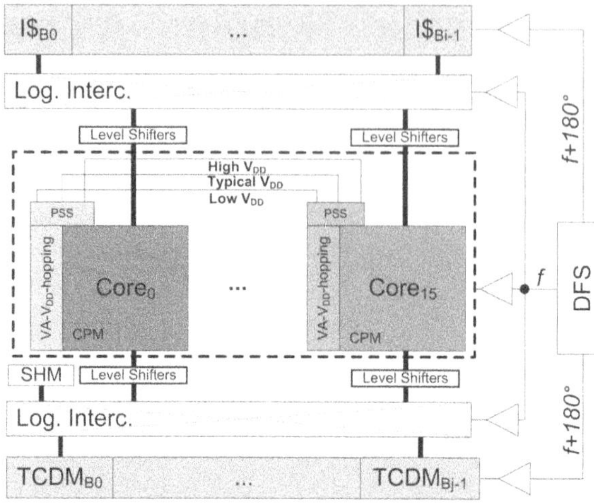

Figure 1. Variation-tolerant processor cluster.

instruction and data sides, and a hardware synchronization handler module (SHM). The shared-L1 *I$* for the MIMD cluster can achieve better performance, up to 60%, than the private *I$* per core [23]. On the data side, a multi-ported, multi-banked, level-one TCDM is directly connected to the interconnect. The number of memory ports is equal to the number of banks to have concurrent access to different memory locations. The logarithmic interconnection is composed of mesh-of-trees networks to support single cycle communication between processors and memories in L1-coupled processor clusters [25]. When a read/write request is brought to the memory interface, the data is available on the negative edge of the same clock cycle, leading to two clock cycles latency for a conflict-free TCDM access. The SHM acts as an extra slave device of the logarithmic interconnect to coordinate and synchronize cores for accessing shared data on TCDM [23].

All components of the cluster work with the same frequency (memories with a 180° phase shift) decided by DFS, while only the voltage of cores is isolated by the fast level shifters thus enabling core-level dynamic V_{DD}-hopping [13]. The V_{DD}-hopping uses three voltages provided by external DC-DC converters (no need of on-chip inductor and charge pump) to control the local voltage of the core based on the core's delay variation. To hop between three supply voltages, a device called power supply selector (PSS) is necessary. The V_{DD}-hopping utilizes an efficient voltage transition which allows changing the supply voltage following a controlled ramp, limiting wide current variations, avoiding any supply voltage under- or over-shoot and current flowing from one source to another [13]. Silicon results of a 65nm test-chip indicate that the core does not need to be stopped during V_{DD}-hopping thanks to smooth, and fast voltage transitions (less than 100ns), with no under-shoot or over-shoot [26]. The hopping unit and its power switches are fully integrated and are 20× smaller than the integrated buck-boost DC-DC converter [26]. As shown in Figure 1, the level shifter standard cells are utilized in the back-end with a fine-grain multi-V_{DD} design flow; each the high-to-low/low-to-high level shifter imposes only 12ps/42ps delay [27] (262nW/43nW average leakage power) for a load of fan-out-of-4, thus enabling single-cycle communications between cores and TCDM/shared-L1 *I$*.

3.1 Variation-Aware V_{DD}-Hopping

To observe the effect of process parameters variation on frequency of individual cores within a cluster, we have accurately analyzed how critical paths of each core are affected, considering the back-end details implementation of cores. Each core has been optimized during synthesis and P&R individually with a target frequency constraint of 830MHz, then a bottom-up synthesis approach is leveraged to form the physical implementation of the cluster. After parasitic extraction, in the sign-off stage, the process parameters are varied based on die-to-die and within-die characterized process parameters variations of 45nm TSMC models, derived from the first-level process created by principal component analysis. These standard industrial libraries and design process are supported by the state-of-the art commercial tools [28], thus the calculated cores' frequency accurately reflect the true results obtained in silicon. The maximum frequency variation of every core under different operating voltages is shown in Figure 2. Within a cluster, each core's maximum frequency varies significantly due to increasing within-die variations. For instance, at 0.81V, three cores (f_4, f_8, f_9) of out of 16-core cannot meet the *design time* target frequency of 830MHz.

To cope with this frequency variation problem there are three solutions: (i) limiting the frequency of cluster by the slowest core (f_8=820MHz); (ii) disabling the slowest cores and clocking the cluster with the next slowest core (f_4=826MHz); (iii) running each core at its maximum frequency independently. All these solutions impose non-negligible performance penalty; the first and second solutions directly diminish the throughput of cluster, and the last solution needs extra latency for synchronization of cores with different frequencies. Synchronization across multiple frequency islands increases the latency of interconnection which its performance impact can be as high as the cache miss.

On the other hand, we consider a core-level V_{DD}-hopping [13] for tuning the voltage of each core individually to compensate the impact of process variation. For instance, Figure 2 shows that all cores of the same cluster meet the target frequency of 830MHz when a higher V_{DD} (0.99V) is applied. Therefore, every core can have its own voltage domain, while all cores can work with the target frequency utilizing the fast level shifters. The critical paths delay of every core are measured in real-time by the less intrusive and low-overhead CPMs [4], hence the variation-aware V_{DD}-hopping (VA-V_{DD}-hopping) can accordingly tune the cores' voltage periodically at arbitrary post-silicon stages. It mitigates both process variation and even aging slows down. Consequently, the cores which are fabricated on a fast piece of silicon will work on a lower voltage than the boosted "high V_{DD}" voltage; this not only lowers their power but delays their aging. On the contrary, slow cores will supply at higher voltages to be able to meet the target frequency. As shown in Figure 2, the VA-V_{DD}-hopping elevates the voltage of slow cores (f_4, f_8, f_9) to 0.99V, while the rest of cores are supplying at 0.81V, therefore enabling the whole cluster to clock at the target frequency of 830MHz. Note that the VA-V_{DD}-hopping technique mitigates the within-cluster delay variations, but imposes voltage supply changes at the core-level that can affect core's aging. Therefore, to extend service life of the slow cores the ratio of stress to recovery time can be changed using core activity duty cycling techniques [12].

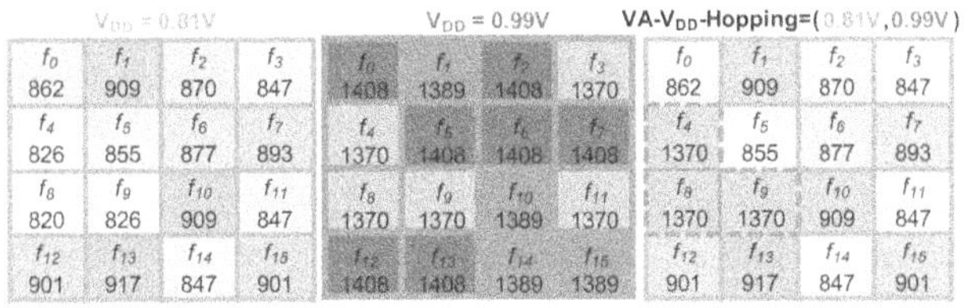

Figure 2. Frequency (MHz) variation of a 16-core cluster due to the process parameters variations under different voltages.

4. PROCEDURE HOPPING IN PRESENCE OF FAST DYNAMIC IR-DROP

In the previous section, we have shown that the variability-affected cluster can combat delay variation caused by the process parameter variations and aging, leveraging the real-time observers and voltage as the control knob. CPMs observe the available slack on paths, and VA-V_{DD}-hopping controls the voltage accordingly, this detection/correction control loop is a well-suited for those variations that: (i) have a slow time constant since compensation requires several clock cycles; (ii) contain low-frequency components to avoid the frequent cost of rollback and calibration. On the other hand, fast dynamic variations, like IR-drop, that contains high-frequency component cannot be countered by a reactive detection/correction loop. They need to be anticipated and prevented.

For this type of variations, we propose a technique consisting of two major phases: *design time* characterization of metadata related to *PLV*, and *runtime* preventive procedure hopping. During characterization, the probability of voltage droop/rise versus various voltage (V) and temperature (T) is characterized at the level of procedures, where the problematic sequences of instructions [9][21] exist. Therefore, the *PLV* is calculated for every procedure on different combinations of (V,T) of the core, then the metadata is generated as the result. The characterized metadata is attached to each procedure at the *compile time*, to be able to use for *runtime* decisions about finding the best location to run the procedure among the available (V,T)-islands within a cluster.

During *runtime*, the core can evaluate the *PLV* of every procedure just looking at the characterized metadata, and at the same time monitoring its current (V,T) using CPMs. If the calculated *PLV* is greater than a predefined threshold (PLV_threshold), this means that running procedure on the original core (caller) would likely cause critical IR-drops, thus the procedure hops to another core (callee) where its (V,T) is suitable for the procedure execution. As discussed in the next subsection, procedure hopping can be done remarkably fast and proactively enough thanks to the tightly coupled shared resources within a cluster.

4.1 Supporting Intra-Cluster Procedure Hopping

In this subsection, we describe the architectural HW/SW design to support the procedure hopping within a cluster. The goal is to facilitate fast and proactive migration of procedures from a caller core to the rest of cores, without special compiler support, minimal impact on the normal execution, and reasonable memory overhead. Figure 3 shows the HW/SW interactions, and steps of procedure hopping of the cluster. It is shown that accessing both data and instruction is facilitated by shared TCDM and L1 *I$*. The shared TCDM has four regions: (i) shared local: maintains variables explicitly defined to be shared at *compile time*; (ii) shared stack: maintains the parameters for passing among cores; (iii) stacks: region is defined to maintain the normal stack of all 16 cores; (iv) heap: is used for dynamically allocated structures.

For every procedure e.g. $Proc_X$, two variation-aware procedures, $Proc_X$@Caller and $Proc_X$@Callee, are considered to enable *runtime* accesses to the characterized metadata of $Proc_X$ in the caller and callee cores respectively. The only compiler transformation is to transform "call $Proc_X$" to "call $Proc_X$@Caller", as shown in the code of the caller core in Figure 3. Therefore, the $Proc_X$@Caller will first run on behalf of $Proc_X$ to decide whether current (V,T) of the caller core is suitable for running $Proc_X$ or not, utilizing the metadata and reading the operating condition monitors to calculate *PLV*. If *PLV* is less than/equal to the PLV_threshold, then "call $Proc_X$" will be executed; otherwise the procedure hopping will be applied to trigger migration of $Proc_X$ to a *favor* core. Once a procedure hops from the caller core to a callee core, its code is easily accessible via the shared-L1 *I$* (without paying the penalty of filling a private cache), but its parameters also needed to be visible for the callee core. Therefore, a shared stack layout is created on the stack region of TCDM which is accessible via a shared stack pointer (SSP). This 36-byte shared stack layout covers the eight *out* registers of SPARC for passing six 32-bit parameters (*%o0-%o5*), a pointer to extra parameters (*%o6*), a return address (*%o7*) as well as a pointer to the return data structure. The caller core needs to copy-out the *out* registers and extra parameters (if available) to TCDM before migration of procedure, and then copy-in the return value or structure form TCDM to the registers after finishing execution of the migrated procedure. In our implementation, we assume that procedures do not have any global variables, and all inter-procedure communications are done through parameters passing; otherwise the caller core needs to copy-out/in all context registers (32 current registers window) to/from TCDM.

To enable the callee core to access to the data and code of a migrated procedure, a procedure hopping information table (PHIT) is considered in the shared local area of TCDM. This table simply keeps the information of a migrated procedure, including its SSP, address, and status. Every core can have up to eight nested procedure calls (the window pointer is synthesized as a 3-bit register), and only one of them can migrate, since the in-order core is a single thread core, and needs to wait for returning the result of the migrated procedure. Therefore, the 192-byte PHIT has an entry for every core which keeps the following information for a migrated $Proc_X$: the shared stack pointer (SSP_X), the address of $Proc_X$@Callee ($ADDR_X$), status of $Proc_X$ (ST_X) ={*empty, waiting, running, done*}.

As shown in the code of the caller core in Figure 3, after filling the shared stack and PHIT, the core does a *broadcast_req* to inform the rest of cores about a waiting procedure for service. This broadcast triggers an interrupt for all cores except the caller core, as potential callee candidates, which can service the waiting procedure based on their programmable priorities– the core can be programmed to ignore this interrupt or trigger it only when the core is idle. In the corresponding interrupt service routine (ISR), the callee core resumes its normal execution, and then walks through PHIT circularly, starting from its neighbor core for minimizing contention, picks up a waiting procedure to assess it. For instance, if the callee core picks up the waiting $Proc_X$ for the service, it will jump to the $ADDR_X$, the address of $Proc_X$@Callee. The philosophy of $Proc_X$@Callee is like $Proc_X$@Caller, it essentially enables the callee core to assess *PLV* of the $Proc_X$ based on the current operating condition of the callee core. If *PLV* is less than/equal to the threshold, then the callee core will access to the code and data of $Proc_X$ for executing on behalf of the caller core; otherwise the callee will resume its normal execution. Particularly, the callee core changes the ST_X at PHIT from *waiting* to *running*, thus the rest of cores will not pick $Proc_X$ up for the assessment– SHM device coordinates multiple concurrent accesses to PHIT. The callee core then copies-in the procedure's parameters from the shared stack via SSP_X, and calls $Proc_X$ for its execution. After executing the procedure, the core copies-out the return value from register to the shared stack, sets the corresponding pointer to the return data structure (if any), sets the ST_X to *done*, and does a *broadcast_ack* to inform the caller about finishing execution of $Proc_X$.

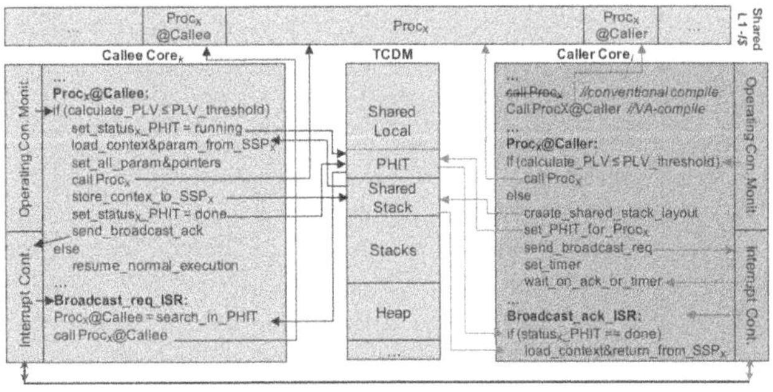

Figure 3. HW/SW collaborative architecture to support intra-cluster procedure hopping.

The caller core, in the corresponding interrupt service routine of *broadcast_ack*, checks the ST_X, if it is equal to *done*, it then copies-in the return value and structure (if any) from the shared stack to the caller core's registers. In the time between sending a *broadcast_req* until receiving a *broadcast_ack*, the caller core can service another waiting procedure available on PHIT, or can switch to an idle mode. If the caller core does not get any *ack* response after a programmable timer value (e.g. 100µs which is long enough to executing a procedure), this means that there is no better (V,T)-island (no *favor* core) within the cluster to prevent the voltage emergency during execution of the procedure. Therefore, the caller core sends a request to cluster's DFS controller to decrease the frequency of the whole cluster, thus lower the power density and temperature.

5. CHARACTERIZATION OF *PLV* TO DYNAMIC OPERATING CONDITIONS

In this section, we demonstrate an advanced CAD flow and methodology to address variation awareness for characterization of *PLV* to dynamic IR-drop (we separately consider both voltage droops on V_{DD} and voltage rises on V_{SS} power domains), under a full range of operating conditions. It consists of two stages as shown in Figure 4: (i) the *design time* stage which accurately analyzes the dynamic voltage droops/rises for individual procedures under full operating conditions; (ii) the *compile time* stage which generates *PLV* metadata and corresponding variation-aware procedures. Finally, the cluster benefits from the characterized *PLV* at the *runtime* stage.

Each core of the cluster is an open-source 32-bit in-order RISC LEON3 [29] processor which is synthesized with the normal V_{TH} cells of 45nm TSMC technology, the general purpose process. The back-end optimization is performed using Synopsys IC Compiler, and then the finalized net-list and parasitics are extracted for accurate power analysis. To generate the accurate gate-level switching activity factor for the vector-based power analysis, the procedure is simulated on top of the back-end extracted net-list with timing back-annotation using Mentor Graphics ModelSim. The instantaneous power of the procedure is then analyzed under four TSMC operating conditions [27] using Synopsys PrimeTime. Providing the signoff corner-based instantaneous power as well as the switching activity factor enables Synopsys PrimeRail for a fine-grain, time-based rail analysis of all resistive, capacitive and inductive components of the post-P&R processor. Consequently, the inter-corner dynamic voltage droop/rise of the power rails is analyzed as the output of the *design time* stage.

The quantification of the PLV_X (*PLV* of $Proc_X$) to dynamic IR-drops defined in (1), where N_X is the total number of clock cycles which takes to execute $Proc_X$, and $VolEmerg_i$ indicates whether there is at least a voltage emergency at the clock $cycle_i$ or not. The voltage fluctuations of greater than 4% are viewed as voltage emergencies [18][19][21] that can result in a malfunction within the processor, therefore the voltage droops/rises on V_{DD}/V_{SS} power rails are sampled k times during one clock cycle. The average signal activity is 70ps, so the k=15 for the target cycle time (1.2ns), while [18][19][21] sampled a second-order linear system as a model of power supply only once per cycle. The $VolEmerg_i$ is one if the maximum sampled voltage droop/rise is greater than 4% of V_{DD} during the clock $cycle_i$. In other words, PLV_X defines as the total number of cycles that have at least one voltage emergency over the total cycles for the $Proc_X$. Intuitively, if $Proc_X$ runs without any voltage emergency, PLV_X is zero; on the other hand, PLV_X is one if $Proc_X$ faces at least one voltage emergency in every cycle.

$$PLV_X = \frac{1}{N_X}\sum_{i=1}^{N_X} VolEmerg_i \qquad (1)$$

$$VolEmerg_i = \begin{cases} 1 & \text{If Max}\{drop(t), rise(t) \mid t=1,...,k\} \geq \frac{4\times V_{DD}}{100} \\ 0 & \text{otherwise} \end{cases}$$

PLV_X is characterized for the assigned voltages of VA-V_{DD}-hopping to various cores, {0.81V, 0.90V, 0.99V} representing {fast, typical, slow} cores on a variability-affected cluster. At *design time*, the slow cores and fast cores are distinguished based on their maximum frequency distribution as described in Section 3.1, then their voltage is tuned accordingly to meet the target cluster frequency. At *compile time*, the characterized *PLV* metadata of every $Proc_X$ is attached to the two variation-aware procedures, $Proc_X$@Caller and $Proc_X$@Callee, to be able to *runtime* access to the metadata on the caller and callee cores respectively. During *runtime*, the discretized (V,T) operating conditions are reported by CPMs thus enabling $Proc_X$@Caller/Callee to point to the corresponding characterized *PLV* metadata to assess the vulnerability of $Proc_X$ at the current (V,T).

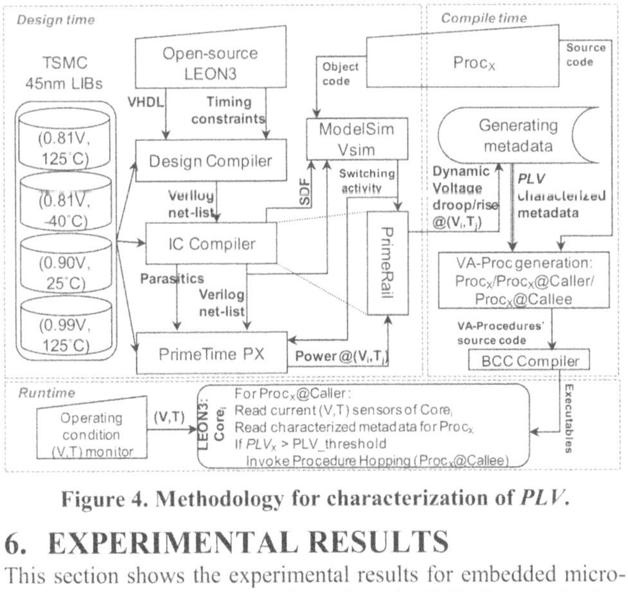

Figure 4. Methodology for characterization of *PLV*.

6. EXPERIMENTAL RESULTS

This section shows the experimental results for embedded microprocessor AutoBench suite of benchmarks [30] characterized at the *design time* flow of Figure 4. This section also evaluates the effectiveness of the procedure hopping technique to avoid voltage emergencies, and quantifies its latency overhead as well as the voltage droop/rise during the *runtime* stage. Every benchmark is a program consists of a "run" procedure for its major computation which is selected for characterization[1] and can be run on every core– the cluster is a multi-programmed environment. The inter-corner and intra-corner variations in the peak power of procedures are shown in Figure 5. The corner with higher (V,T) has higher power density which imposes higher peak power. It is shown that the maximum inter-corner peak power variation is 3.5× for *FIR*, while the maximum of 1.28× intra-corner peak power variation occurs between *IFFT* and *tblook* procedures at (0.81V,125°C). Furthermore, the maximum of 4.1× peak power variation is observed across corners and procedures, *a2time* at (0.81V,-40°C), and *IFFT* at (0.99V,125°C). We should point out that LEON3 is a simple in-order RISC processor, thus for fast and complex cores where the stress on the power grid is much higher [2], we expect to see even higher power variation.

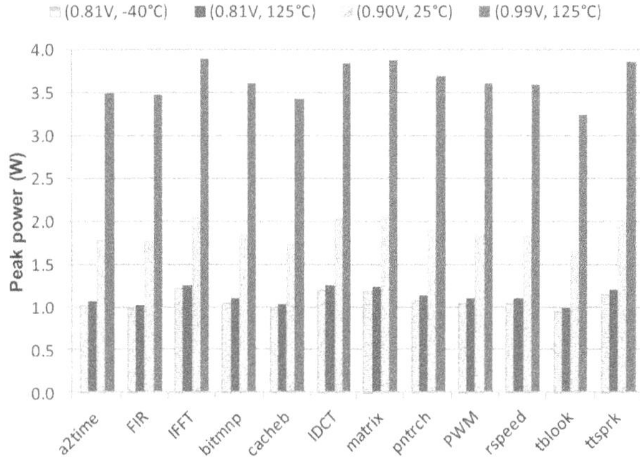

Figure 5. Intra-procedure peak power variation.

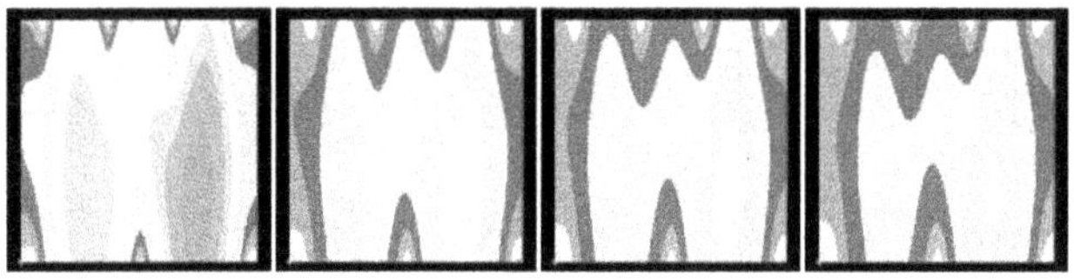

Figure 6. Voltage droop of *FIR* across corners: (0.99V,125°C), (0.90V,25°C), (0.81V,125°C), (0.81V,-40°C), left to right.

Increasing the (V,T) increases the power density as well as the peak power, consequently the power network of the core highly experiences the voltage emergencies in the high-power corner. The voltage droops of running *FIR* on the same core but various operating corners are shown in Figure 6. The core at the high-power corner (0.99V,125°C) faces the maximum voltage droop of 44mV and 41mV as the average of top-100 dynamic voltage droops, which are greater than 4% of V_{DD} (990mV), thus these voltage droops are considered as the voltage emergencies. As opposed to the high-power corner (0.99V,125°C), *FIR* does not face any voltage emergency at the corners with voltages of 0.90V/0.81V thanks to their lower power densities. The core has various power densities across the corners of Figure 6 (left to right): 0.66μW/μm^2, 0.21μW/μm^2, 0.18μW/μm^2, 0.16μW/μm^2.

Figure 7 illustrates the maximum voltage droop/rise that occurs during the execution of the procedures under the four characterized operating conditions. All procedures running at cores with 0.81V have the maximum voltage droop/rise less than 4% of V_{DD}. Increasing the power density by switching to (0.90V,25°C) causes only four procedures (*IFFT, IDCT, matrix, ttsprk*) to face the voltage emergencies. At the highest power corner, (0.99V,125°C), most of the procedures except *tblook* will face either voltage

[1] PLV_threshold is set at zero, since the procedures are not inherently resilient to any timing error and even a single IR-drop may cause a wrong result.

droop or voltage rise greater than 4% of V_{DD}. These results show that the procedure hopping technique can avoid the voltage emergency for all procedures by hopping them form a high-voltage (0.99V) core to a low-voltage (0.81V) core. Experimental results from the layout of variability-affected cluster, which are presented in Section 3.1, show that 13 low-power cores lie within a cluster of 16-core, thus providing enough callee cores to service the migrated procedures.

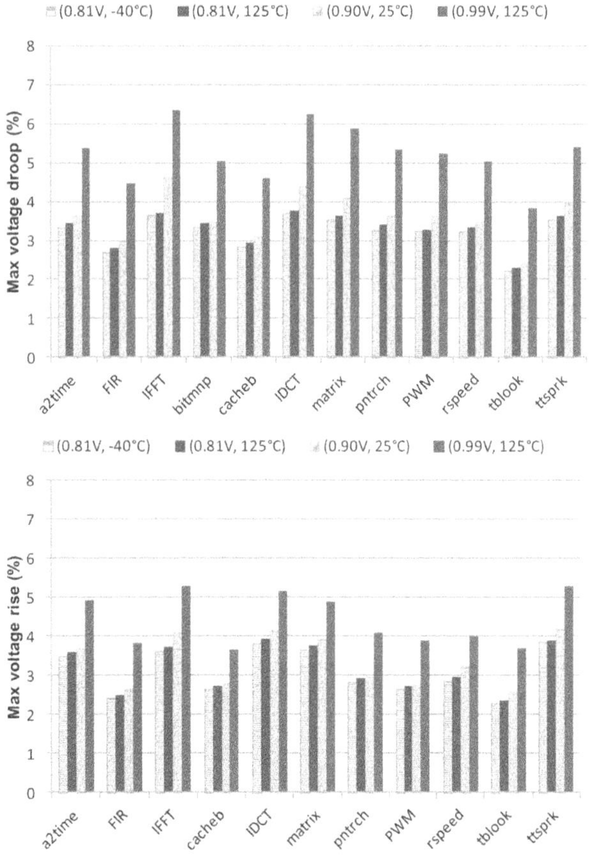

Figure 7. Percentage of the max voltage droop (top), and rise (bottom) across various corners and procedures.

6.1 Cost of Procedure Hopping

Table 1 lists the latency overhead of involving the procedure hopping both in the caller and the callee cores. The total roundtrip overhead of the hopping a procedure from the caller core and returning the results from the callee core is 793 cycles; this is less than 1% of the total cycles needed to execute any of the characterized procedures in [30], while [10] has at least a migration overhead of transferring 1280 flits only to transfer the instructions and data from one core to another. In particular, if a procedure has a runtime of 35K cycles, the amortized cost is only 2% and 0.2% latency penalty, in case of hopping procedure to another core, or keep running procedure on the same core respectively. This is accomplished through the advantage of shared-L1 *I$* and TCDM that eliminates the penalty of filling a private storage.

Moreover, during the procedure hopping no voltage emergency can occur even at (0.99V,125°C), neither in the caller nor the callee core, since the copy-in/out parameters from/to registers/TCDM does not cause any burst of activity. Consequently, the procedure hopping guarantees the voltage emergency-free migration of all procedures, fast and proactively enough.

Table 1. Latency overhead and IR-drops of procedure hopping

	Caller hopping	Caller not hopping	Callee service	Callee no service
Latency	218 cycles	88 cycles	575 cycles	342 cycles
Max droop	1.3%	0.6%	2.9%	1.8%

7. CONCLUSION

The notion of procedure-level vulnerability (*PLV*) to fast dynamic voltage variation is defined. Based on *PLV* metadata, a fully-software low-cost procedure hopping technique is proposed which facilitates fast and proactive migration of procedures within a shared-L1 processor cluster. Full post-P&R results in 45nm TSMC technology confirms that the procedure hopping avoids the voltage emergency across a variability-affected cluster, while imposing only an amortized cost of less than 1% latency for any of the characterized embedded procedures. Furthermore, the effectiveness of the variation-aware V_{DD}-hopping technique to combat intra-cluster static variation has been demonstrated.

8. ACKNOWLEDGMENTS

This research was supported by NSF Variability Expeditions Award CCF-1029783, and Virtical GA n. 288574.

9. REFERENCES

[1] S. Ghosh, et al., "Parameter Variation Tolerance and Error Resiliency: New Design Paradigm for the Nanoscale Era," Proc. *IEEE*, Vol.98, No.10, pp.1718-1751, Oct. 2010.

[2] C. Isci, et al., "An Analysis of Efficient Multi-Core Global Power Management Policies: Maximizing Performance for a Given Power Budget," Proc. *MICRO*, pp.347-358, 2006.

[3] ITRS [Online]. Available: http://public.itrs.net

[4] A. Drake, et al., "A Distributed Critical-Path Timing Monitor for a 65nm High-Performance Microprocessor," Proc. *ISSCC*, pp. 398-399, 2007.

[5] S. Herbert, et al., "Exploiting Process Variability in Voltage/Frequency Control," IEEE Trans. on Very Large Scale Integration (VLSI) Systems, 2011.

[6] C. R. Lefurgy, et al., "Active Management of Timing Guardband to Save Energy in POWER7," Proc. *MICRO*, 2011.

[7] R. Teodorescu, et al., "Mitigating Parameter Variation with Dynamic Fine-Grain Body Biasing," Proc. *MICRO*, pp. 27-42, 2007.

[8] A. Rahimi, et al., "Analysis of Instruction-level Vulnerability to Dynamic Voltage and Temperature Variations," Proc. *DATE*, pp.1102-1105, 2012.

[9] V.J. Reddi, et al., "Resilient Architectures via Collaborative Design: Maximizing Commodity Processor Performance in the Presence of Variations," IEEE Trans. on Computer-Aided Design of Integrated Circuits and Systems, Vol.30, No.10, pp.1429-1445, Oct. 2011.

[10] S. Dighe, et al., "Within-Die Variation-Aware Dynamic-Voltage-Frequency-Scaling With Optimal Core Allocation and Thread Hopping for the 80-Core TeraFLOPS Processor," IEEE J. of Solid-State Circuits, Vol.46, No.1, pp. 184-193, Jan. 2011.

[11] F. Paterna, et al., "Variability-Aware Task Allocation for Energy-Efficient Quality of Service Provisioning in Embedded Streaming Multimedia Applications," IEEE Trans. on Computers, 2011.

[12] F. Paterna, et al., "Adaptive Idleness Distribution for Non-Uniform Aging Tolerance in MultiProcessor Systems-on-Chip," Proc. *DATE*, pp. 906-909, 2009.

[13] S. Miermont, et al., "A power supply selector for energy- and area-efficient local dynamic voltage scaling," Proc. *PATMOS*, 2007.

[14] A. Rahimi, et al., "History-Based Dynamic Voltage Scaling with Few Number of Voltage Modes for GALS NoC," Proc. *FutureTech*, 2010.

[15] A. Tiwari, et al., "Facelift: Hiding and Slowing Down Aging in Multicores," Proc. *MICRO*, pp.129-140, 2008.

[16] U.R. Karpuzcu, et al., "The BubbleWrap many-core: Popping cores for sequential acceleration," Proc. *MICRO*, pp.447-458, 2009.

[17] E. Grochowski, et al., "Microarchitectural Simulation and Control of di/dt-induced Power Supply Voltage Variation," Proc. *HPCA*, pp. 7-16, 2002.

[18] R. Joseph, et al., "Control Techniques to Eliminate Voltage Emergencies in High-Performance Processors," Proc. *HPCA*, pp. 79-90, 2003.

[19] J. Zhao, et al., "Thermal-aware voltage droop compensation for multi-core architectures," Proc. *GLSVLSI*, 2010.

[20] V. Reddi, et al., "Voltage Emergency Prediction: A Signature-Based Approach To Reducing Voltage Emergencies," Proc. *HPCA*, pp. 18-27, 2009.

[21] K. Hazelwood, et al., "Eliminating Voltage Emergencies via Microarchitectural Voltag Control Feedback and Dynamic Optimization," Proc. *ISLPED*, pp. 326-331, 2004.

[22] NVIDIA's Next Generation CUDA Compute Architecture: Fermi, Whitepaper, V1.1, 2009.

[23] D. Bortolotti, et al., "Exploring instruction caching strategies for tightly-coupled shared-memory clusters," Proc. *Int. Sym. on SoC*, pp. 34-41, 2011.

[24] K. Bowman, et al. "A 45 nm Resilient Microprocessor Core for Dynamic Variation Tolerance," IEEE J. of Solid-State Circuits, Vol.46, No.1, pp.194-208, Jan. 2011.

[25] A. Rahimi, et al., "A Fully-Synthesizable Single-Cycle Interconnection Network for Shared-L1 Processor Clusters," Proc. *DATE*, pp.1-6, 2011.

[26] E. Beigne, et al., "An Asynchronous Power Aware and Adaptive NoC Based Circuit," IEEE J. of Solid-State Circuits, Vol.44, No.4, pp.1167-1177, April 2009.

[27] TSMC 45nm standard cell library release note, TCBN45GSBWP, version 120A, Nov. 2009.

[28] Synopsys PrimeTime® VX User Guide, June 2011.

[29] LEON3 [Online]. Available: http://www.gaisler.com/cms/

[30] EEMBC benchmark Consortium [Online]. Available: http://www.eembc.org

Author Index

A Low-power "Near-threshold" Epileptic Seizure Detection Processor with Multiple Algorithm Programmability

Himanshu Markandeya[1], Shriram Raghunathan[2], Pedro Irazoqui[3,1], Kaushik Roy[1,3]

[1]School of ECE
Purdue University
West Lafayette, IN 47907, USA
{hmarkand, kaushik}@purdue.edu

[2]Cyberonics Inc.
100 Cyberonics Blvd.
Houston, TX 77058, USA
shriram.raghunathan@cyberonics.com

[3]Center for Implantable Devices
Purdue University
West Lafayette, IN 47907, USA
pip@purdue.edu

ABSTRACT
In this paper, we developed and implemented a low-power multiple-algorithm processor for detection of epileptic seizures with high efficacy. Four different algorithms -- The Coastline, Hjorth parameter, Energy and Non-linear energy metric based algorithms -- are used in our analysis. The Boolean logical combinations of these algorithms allow programmability according to the patient specific needs and improve efficacy by almost 10%. The system implemented in 65-nm TSMC technology consumes about 13.1 μW (measured) at scaled supply voltage of 400mV. Programmability using Boolean logic helps in improving the efficacy resulting in almost an error-free detection.

Categories and Subject Descriptors
B.7.1 [**Integrated Circuits**]: Types and Design Styles – *algorithms implemented in hardware, VLSI (very large scale integration), Input/output circuit.*

General Terms
Algorithms, Measurement, Design.

Keywords
Biomedical, ASIC, Epilepsy, Low-power Implants.

1. INTRODUCTION
Epilepsy is the third most common neurological disorder in United States after Alzheimer's disease and stroke. It is characterized by recurrent involuntary physical convulsions viz. epileptic seizures [1] [2]. A third of the 50 million people affected worldwide do not respond well to surgical or pharmacological therapy. Neuromodulation by means of electrical, chemical or even optical stimulation is rapidly emerging as a promising alternative therapy for these patients [3] [4]. At present, the Cyberonics Vagus Nerve Stimulator (VNS) is the only commercially available anti-epileptic prosthesis [5]. VNS is based on continuous neurostimulation which has been argued to be detrimental and ineffective in the long run [6]. Responsive neurostimulation impacts the efficacy of this therapy by critically increasing its temporal and spatial specificity. Neuropace device (Medtronics Inc.) is developed based on this principle and awaits Food and Drug Administration (FDA) approval [4] [7]. However, responsive neurostimulation implies that the epileptic seizure be detected with high efficacy. Implantable epilepsy prosthesis that can deliver neurostimulation at the epileptogenic focus is rapidly emerging as an effective solution for treatment of epilepsy, especially for the population that is medically refractory.

ISLPED'12, July 30–August 1, 2012, Redondo Beach, California, USA.

In recent decades, scaling of integrated circuit technology has allowed the development of implantable biomedical applications in order to provide alternative solutions for biological ailments. Scaling has the advantage of implementing complex functionalities on the silicon chip and operating them at lower supply voltages. However, the exponential increase in the leakage implies higher power consumption and poses a significant design challenge. This is undesirable in an implantable application such as epilepsy prosthesis. This is because the implanted battery is a limited source for energy. A wasteful usage of energy by the implant will severely deplete the battery. This may necessitate an expensive and often risk-prone surgery to replace the battery. Apart from this, it is also mandatory to keep the power dissipation of the implant to a minimal to avoid any tissue damage by heat. Thus, the design of any implantable biomedical application is subjected to stringent design constraints. Efficient and low-power seizure detection algorithms are the key to realizing a fully integrated implantable prosthesis which can responsively deliver the therapy to the epileptogenic focus. An efficacious detection will also enable the clinicians and researchers to better understand the disorder.

In this paper, we have implemented a seizure detection processor with multiple algorithms using mathematical metric based algorithm in 65-nm bulk-Si technology. Due to inter-patient variability of neural recordings, it is difficult to maintain high efficacy over a range of patients. We have addressed this by providing programmability in the system to operate according to patient-specific needs. A built-in self-test (BIST) block has been implemented to check the functional nature of each algorithm as well as the peripheral blocks which aide in improved functionality. It will be shown in the results that there is a distinct improvement in efficacy of detection due to provision of low cost programmable logic. Due to low frequency of operation, it is possible to operate the system at a scaled power supply in the vicinity of threshold voltage for the transistor. This greatly reduces the power consumption of the system without affecting the efficacy.

2. BACKGROUND
The literature is abundant with numerous algorithms for detection of the epileptic seizure [8-11]. The characteristic of the seizure is a gradual surge in amplitude of the signal in specific frequency bands. This signature is exploited by number of existing algorithms for seizure detection. Signal processing tools such as

Fast Fourier Transform (FFT), Short-Time Fourier Transform (STFT) and Discrete Wavelet Transform (DWT) have been efficiently utilized to this effect [8] [9]. Some of the FFT and DWT based algorithms utilize the artificial neural networks in order to learn the seizure pattern for detection [8] [9]. This results in a power hungry architecture that is not suitable for implantable application. Furthermore, majority of these algorithms are based on neural recordings from the scalp electrodes i.e. Electroencephalogram (EEG) or Electrocorticogram (ECoG). This implies that the seizure has already spread through the brain before being detected at the surface. In this paper, we use the Local Field Potential (LFP) recordings from a small volume of brain tissue near the epileptogenic focus. This helps in reducing the latency of detection. The event- based algorithm utilizing LFP provides a low power solution for implantable epilepsy prosthesis [10]. However, its efficacy may be debatable over a range of patients [11]. Utilizing the DWT with a quasi-averaging technique helps in providing the programmability and maintain high efficacy as well [11]. However, it has a large number of redundant computations and hence is wasteful in terms of energy consumption. Statistical mathematical parameters have also been devised in order to quantify the amplitude and frequency variations. Some of these parameters are energy, line-length or coastline, hjorth parameter, non-linear autocorrelation etc. [12]. These parameters can also be used to detect the seizures with a fair degree of accuracy [12]. They can be used to demarcate the recorded data into baseline and seizure states. The parameters may be linear or non-linear in nature and may be operated in the time or the frequency domains. Although these mathematical metrics detect seizure with reasonable accuracy, their usage is limited by the fact that, they may not be efficacious over a range of patients due to inter-patient variability of neural recordings. However, by logically combining the detections based upon these parameters, the efficacy can be improved over different patients. In this paper, we have implemented the processor in which the outputs of the implemented algorithms can be combined logically by using simple Boolean AND's and OR's. Hence, the best possible combination can be selected on patient-specific basis to maintain high efficacy of detection.

The paper is organized as follows. Section 3 explains the parameter based algorithms implemented in the processor. It is followed by Section 4 for the explanation about the implemented hardware. Section 5 includes the results from efficacy and power efficiency perspective. Section 6 concludes the paper.

3. SYSTEM OVERVIEW

The implemented system has an algorithm bank consisting of four algorithms. Each of these algorithms is based on a mathematical parameter extracting the seizure feature. These algorithms are selected based on the ability to extract onset feature and proven functionality for seizure detection. These seizure features are used to demarcate the seizure from the baseline. The algorithms operate on real-time input data and produce processed signals. The processed signals are then compared with prefixed threshold values to cause a detection of the onset of seizures. The detection from the four algorithms can be logically combined according to needs of the patient. This maintains high efficacy operation of the processor on a wider range of patients. The thresholds for each of the algorithms and the Boolean logical combination to be used are decided in the training phase. The algorithms, logical combination technique and the training phase are explained below.

3.1 Algorithms

The system consists of Energy, Coastline, Non-Linear Energy and Hjorth parameter based algorithms. These algorithms are chosen on the basis of their onset feature isolation properties. While Energy and Coastline are simple arithmetic parameters which measure the energy content and the line-length of the LFP, Hjorth variance and Non-Linear Energy are more statistical in nature. They capture the non-linear features in the recorded LFP. All these algorithms operate in the time domain on real-time data stream. The recorded LFP data is digitized into 10-bit 2's complement format and streamed into each of the four algorithms. The definition of the four parameters is explained below.

3.1.1 Energy

The "energy" parameter calculates the energy content of the signal in a prefixed window. It is a well known fact that as the onset of seizure approaches, there is gradual surge in the amplitude of the signal in specific frequency bands. This results in an increase in energy content of the signal. This parameter operates on adjacent windows of the streaming data. Hence, a gradual increase in the energy value implies that the signal is fast approaching a seizure onset. The size of the window is taken as 1024 samples in this case. By appropriately thresholding the energy value in the window, a detection of the onset can be made. The mathematical description of the energy metric is given in equation 1. Here E is the energy parameter, x is the input data, N is the size of the window and n is the window index.

$$E_{AVG}[n] = \frac{1}{N}\sum_{i=1}^{N} E(i+(n-1)*N) \qquad (1)$$

where $E(i) = x^2(i)$

3.1.2 Coastline

The coastline parameter is also known as line-length algorithm. Its definition has been adapted from [13]. This feature measures the distance between adjacent data points x in the signal and accumulates them over a window of width N. In the baseline section of the recording, this parameter remains more or less constant. However as we approach the onset of seizure, due to increase in amplitude and frequency of spikes in the signal, the distance between the neighboring data points will increase. The threshold for this parameter can be adjusted accordingly in the training phase. The Neuropace responsive stimulation device by Medtronics Inc. utilizes the coastline or "line-length" as one of the parameters for seizure detection [4] [12]. The mathematical expression for computing the coastline (CL) for k^{th} window is given in equation 2.

$$CL(k) = \sum \left| x[i+(k-1)*N] - x[i-1+(k-1)*N] \right| \qquad (2)$$

3.1.3 Non-Linear Energy

The non-linear energy parameter is based on Teager's algorithm [14]. In the conventional energy parameter, all the frequency components in the signal were weighed equally. In comparison, the non-linear energy operator weighs the components at different frequencies non-uniformly. Using square-law weighting, it emphasizes higher frequencies over lower frequencies. Since there is a surge in amplitude at higher frequencies in the seizure phase of the LFP recordings, it expends greater energy during the seizure or ictal phase. This non-uniformity is compensated in the

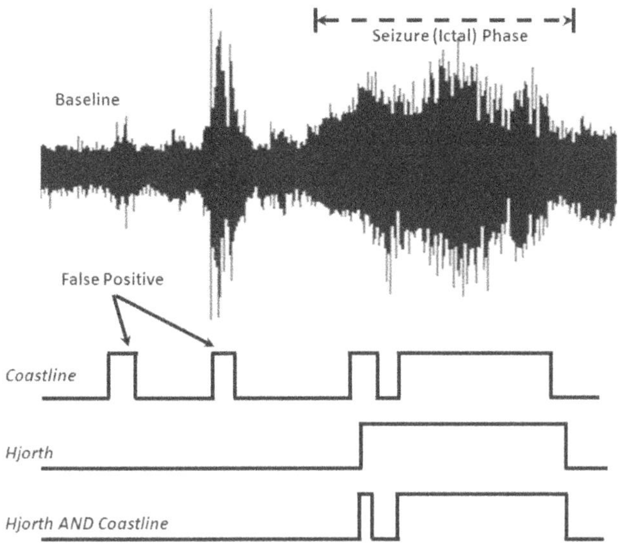

Figure 1 Seizure Detection with independent algorithm and Boolean combination

Teager's algorithm. It has been shown that Teager's algorithm to calculate energy attributes to higher contrast between the energy content in the baseline and the ictal phase [15]. This results in better seizure detection. The non-linear energy Ψ for k^{th} window of size N for data x is given by equation 3.

$$\psi(k) = \frac{1}{N}\sum_{i=1}^{N} x^2[i+(k-1)*N] - x[i+(k-1)*N-1]*x[i+(k-1)*N+1] \quad (3)$$

3.1.4 Hjorth Variance Parameter

Hjorth parameter has been widely used for statistical analysis of EEG [16] [17]. The variance of the signal amplitude also termed "activity" is the first hjorth parameter. It calculates the variance of a window of N samples and averages it to obtain mean variance. The mean of the kth window, μ_k, is also calculated to obtain the first hjorth parameter. The mathematical expression for computing the hjorth parameter is given in equation 4.

$$Var(k) = \frac{1}{N}\sum_{i=1}^{N} x^2[i+(k-1)*N] - \mu \quad (4)$$

$$\text{where } \mu = \frac{1}{N}\sum_{i=1}^{N} x[i+(k-1)*N]$$

3.2 Boolean Logical Combination

The four algorithms described above are used to process the input data and independently detect the onset of the seizure. The system can be programmed to provide a Boolean combination of the detections of any two of these algorithms. By using simple Boolean AND or OR operation, it is shown that the efficacy of detection of seizure is improved with negligible increment in hardware cost. For instance, it can be seen in Figure 1 that the false positives detected in the coastline algorithm are eliminated by using the Boolean ANDing of the coastline and the Hjorth parameter detections. It can also be seen that this might lead to increased delay of detection. Depending upon the nature of training data and neural recording for each patient the best possible combination for the patient can be identified. The detection could be independent using individual algorithm or by the Boolean combinations of multiple algorithms. This ability of

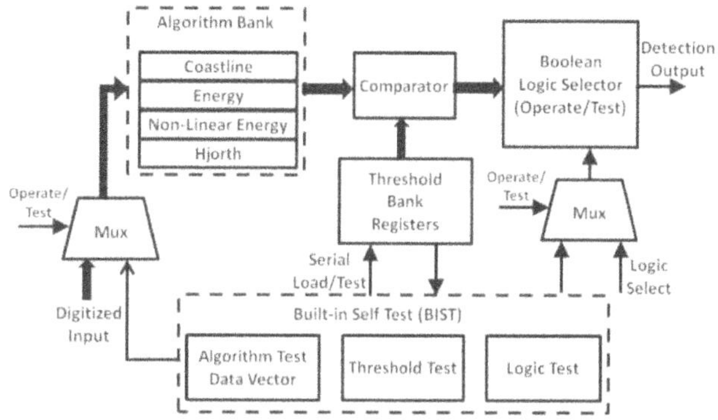

Figure 2 Block Diagram of the Seizure Detection Processor

programming enables the use of the processor over wide range of patients, thereby increasing the scope of its application.

3.3 Training Phase

The algorithms generate processed values from the input data according to equations 1-4. These values have to be compared with thresholds in order to classify the data as a baseline or seizure. The thresholds are prefixed in the training phase. The training data is randomly selected from the LFP recordings consisting of an equal duration of baseline and seizure phase. This data is then processed using each of the four described algorithm and the threshold is adjusted in order to have the highest efficacy of detection. The definition and method for calculation of efficacy is discussed in Section 5. The training phase is also used to decide the Boolean logic to be used for the combination of multiple algorithms. This is because a patient may not respond positively to individual algorithm but may show a significant improvement if a combination of algorithms is used.

4. HARDWARE IMPLEMENTATION

The multi-algorithm programmable seizure detection processor is implemented as shown in the block diagram in Figure 2. The system basically consists of the Algorithm Bank, Threshold Bank, Boolean logic selector and the Built-in self-test (BIST) blocks. The system has two modes of functions viz. operate mode and the test mode. The operate mode puts the system in the seizure detection mode. The test mode activates the BIST block and performs a self test on the system to verify its integrity. The individual implementation and function of each of these blocks is described in this section.

4.1 Algorithm Bank

The algorithm bank consists of four parameter based algorithms as described in the section 3 viz. Energy, Coastline, Non-linear energy and Hjorth parameter. The hardware implementation is based on the equations 1-4 respectively. The block diagram for each of these algorithms is shown in Figure 3.

The Energy parameter uses the multiplier to square the 10-bit input data and an adder to accumulate it over a window. The window width is 1024 and is controlled by the counter. The counter generates a signal at the end of the window which resets the registers and loads the computed average energy to the output register. Since the window width chosen is a power of two, the average energy is calculated by simple dropping the last 10 bits of the accumulated sum. This avoids a power hungry divider for computing the average. The accumulated sum is passed on to the threshold register bank for comparison and detection.

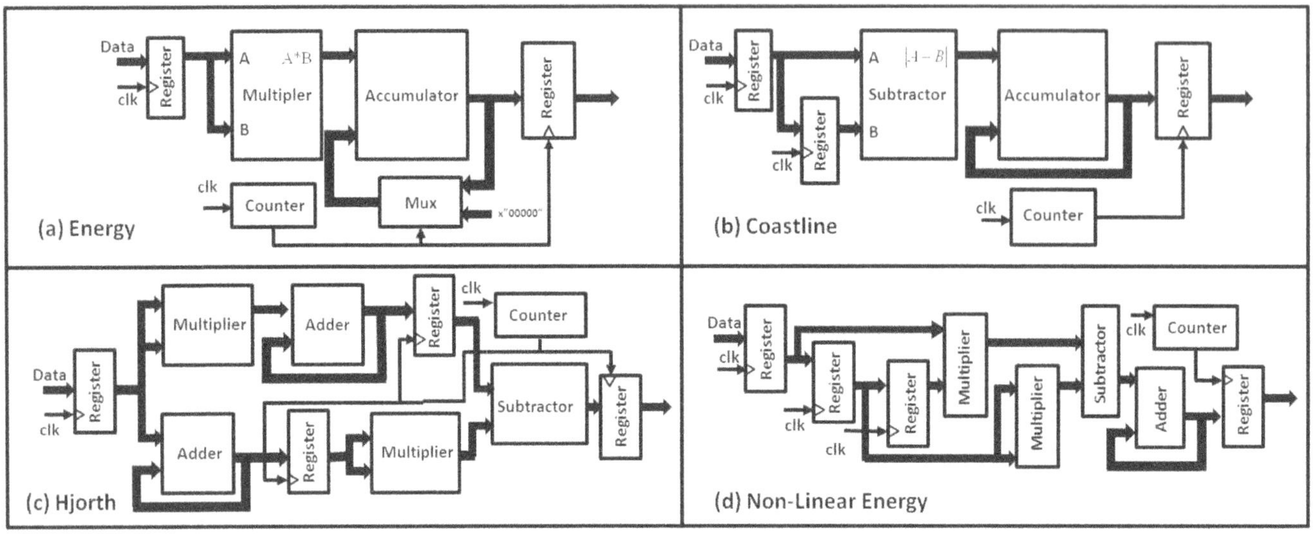

Figure 3 Block Diagram of the Algorithms in the Algorithm Bank

The implementation for coastline parameter is shown in Figure 3(b). The registers at the input store the two neighboring values of the signal. The difference is then accumulated by the adder. The width of the window is controlled using a counter which generates a signal for passing the accumulated difference to the threshold bank.

The Non-linear energy parameter is computed using the values preceding and succeeding each data point as seen in equation 3. This is implemented in hardware by introducing a delay of one clock cycle. The first value is computed only after 3 samples of data are streamed in the system. The counter is used to control the window size and pass the output to the threshold bank.

The Hjorth parameter is calculated by finding the difference between the accumulated squared data and the squared mean in a window of 1024 samples. The counter controls the window size. The Hjorth first parameter is compared to threshold in the threshold bank.

In the operate mode, all the algorithms are simultaneously processing on the 10-bit input data for detecting the seizure. However in the test mode, the algorithms are provided with previously stored data vectors from the BIST block. The intermediate processed values indicate the correctness of computation in each of the algorithms. It should be noted that, the algorithm bank can be generalized to encompass more algorithms. This paper limits the number of algorithm to four for testing the concept.

4.2 Threshold Bank

The processed signals from the algorithm bank are compared to thresholds stored in the threshold bank. The threshold bank consists of four 20-bit registers. These registers hold the threshold values for each of the algorithms. The values in the threshold registers are decided in the training phase as described in the previous section. In the operate mode, the threshold bank is loaded with these threshold values by serially streaming in the bits using a dedicated clock. The processed signals are compared with the threshold values to detect the onset of the seizure. In the test mode, the clock can be used to serially stream out the stored threshold value to verify its integrity. The system has a dedicated clock to operate with threshold registers. This is independent of the system clock and can be turned off after the loading of the threshold registers is complete.

4.3 Boolean Logic Selector

The Boolean logic selector is the programmable block which can be used to select the type of output needed for seizure detection. Depending on the select lines of this block, the output can be either the detection by individual algorithms or their Boolean AND/OR combination. Since the system is designed to be efficacious over a range of patients, the type of logic used for seizure detection is decided in the training phase. In the test mode, by appropriately applying the select values the Boolean logic selector can be tested for its functionality of multiplexing the correct data to the output.

4.4 Built-in Self-Test (BIST)

The seizure detection system implemented on the ASIC includes a built in self testing block. The BIST block is provided to have an on chip testing circuitry to verify the functionality of the system. The BIST block is activated by pulling the test/operate signal low. It provides set of prefixed data vectors to the algorithm bank to test the functionality of the algorithms. The intermediate outputs of each of the algorithms are compared to stored values to verify their correctness. The comparisons generate signals which are given to the Boolean logic selector. By appropriate select signals the output of the system can be used to check the functionality of each of the algorithms. The BIST tests the threshold bank registers by streaming out the stored threshold values through a serial output port by a dedicated threshold clock. The Boolean logic selector block is tested by providing logical inputs to the block and testing the selectivity of the block. The BIST covers all the significant sections of the system to maintain sustained and correct functionality.

5. RESULTS

The seizure detection processor was implemented in 65 nm bulk-Si TSMC technology. Figure 4(a) shows the die photograph of the system. Figure 4(b) shows the test setup. The functionality of the system was tested using the internal BIST block. The efficacy obtained was also verified by comparing it with a MATLAB simulation result. The LFP neural data was recorded in vivo from

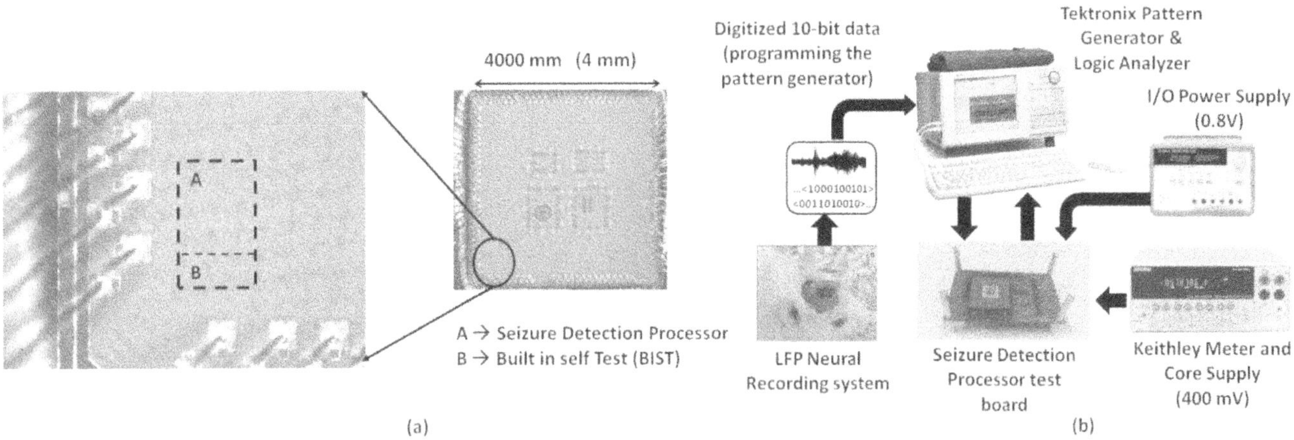

Figure 4 (a) Die Photo of Seizure Detection Processor (b) System Test Setup

rats at a sampling rate of 1.526 KHz along with video. The seizures were induced in the animals using periodic kainic acid injections. The data was classified into the baseline and seizure sections by electrographic and visual evidence. A section of the data was randomly selected as training data set for each animal and used to fix the threshold values. The recorded data was streamed in a digitized form into the 10-bit system input. The system was put in the "operate mode" and the output was observed for detection of the onset of seizure. The Boolean logic selector is programmed according to the training phase to select the best configuration for each set of data.

The efficacy was calculated based on the Specificity, Sensitivity and Average Detection Rate (ADR) as defined by equations 5-7.

$$\text{Sensitivity}=\frac{TP}{TP+FN}\times 100\% \quad (5)$$

$$\text{Specificity}=\frac{TN}{TN+FP}\times 100\% \quad (6)$$

$$\text{ADR}=\frac{\text{Sensitivity}+\text{Specificity}}{2}\times 100\% \quad (7)$$

The Sensitivity is calculated using true positives (TP) and false negatives (FN). It represents the number of false detections caused in the algorithm due to seizure-like events or due to motion artifacts. A high sensitivity indicates that the algorithm would flag any seizure like event and may cause undesirable triggering of neurostimulation. Specificity represents the ability of the algorithm to reject a FN and is calculated using true negative (TN) and false positive (FP). A higher specificity indicates that the algorithm may reject a non-seizure event with higher confidence and may result in missing a seizure. An increase in the sensitivity indicates a higher percentage of false positive and hence lower specificity and vice versa. Hence, the ADR is used as a fair measure of the efficacy of the algorithm and is the average of the sensitivity and specificity. Neural recordings from five animals were used to compute the efficacy of detection. In Figure 5, the efficacy results are summarized for the four implemented algorithms. It can be seen that the efficacy of the algorithms vary over different animals. On an average, the Hjorth parameter algorithm performs better than the others. However, if we use a Boolean combination of two of these algorithms, the efficacy can be increased for a nominal increase in hardware. This is shown in Figure 6. In this figure, we have plotted the average efficacy of the four algorithms with respect to normalized hardware cost. The hardware cost is calculated by normalizing total power consumption of the algorithm and its area as obtained from simulation using Synopsys and Cadence tools. As seen in Figure 6 we combine the coastline and hjorth parameter using a single OR gate. This configuration increases the average efficacy. By configuring the system for each patient, the best possible efficacy can be achieved.

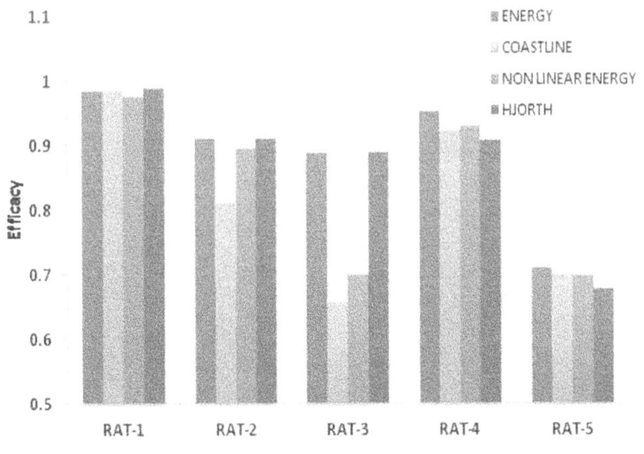

Figure 5 Efficacy of the Algorithm for Large Animal Study

The seizure detection processor was operated with a clock of 10 KHz. This allowed for voltage scaling in order to reduce the energy consumption of the system. The supply voltage was scaled to 400 mV which is very close to the transistor threshold voltage of 310 mV. This is the most energy efficient region of operation. The power consumption is significantly reduces due to quadratic

Table 1 Measured Results for Power

Technology	65 nm bulk-Si (TSMC)
Dimension	200 μm X 340 μm
I/O supply	0.8 V
Core supply	400 mV
f_{clk}	10 KHz
$P_{leakage}$	12.8 μW
$P_{dynamic}$	360 nW

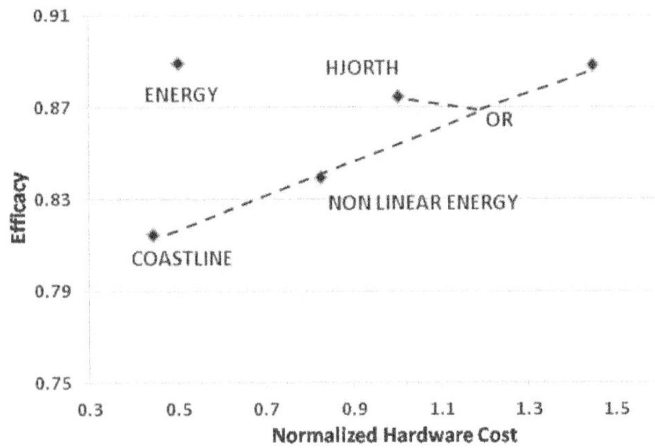

Figure 6 Average Efficacy vs. Normalized Hardware Cost

dependence on the supply voltage. The processor functionality was verified. The power consumption of the system is listed in Table 1. The I/O voltage was scaled to 0.8V. All four algorithms are operational at all times and consume about 12.8μW as leakage power. The dynamic power is 360 nW. This is almost 80% lower than the wavelet based algorithm [11]. Although the parameter based algorithm do not show higher efficacy than the wavelet based algorithm, the programmability can be utilized to improve the efficacy with negligible change in power consumption. The power consumption reported in this paper is without utilization of any leakage control techniques. Programmable logic and near-threshold operation make this seizure detection processor efficient as well as practical for implantable prosthesis.

6. CONCLUSION

In this paper, we have implemented a seizure detection "near threshold" processor which is low in power consumption and has a high efficacy. Furthermore, it is programmable to be adaptable to wider range of patients. The seizure detection processor successfully addresses two major concerns in development of implantable epilepsy prosthesis viz. low power consumption and functionality over inter-patient variation. The processor operates at near threshold voltage supply. This reduces the power consumption significantly and hence increases the longevity of the battery. In future, leakage control techniques such as powering off the unused algorithms can be applied to further reduce the power. Such low-power efficient implementations are the basis for development of practical implantable epilepsy prosthesis

7. ACKNOWLEDGEMENTS

This research was funded in part by Cyberonics Inc. and by National Science Foundation.

8. REFERENCES

[1] http://www.eilepsyfoundation.org

[2] Begley C. E., Famulari M., Annegers J. F., Lairson D. R., Reynolds T. F., Coan S, et al., 2000, The cost of epilepsy in the United States: an estimate from population-based clinical and survey data. *Epilepsia*, 41(3), 342–351.

[3] Ghai R. S., Bikson M., Durand D. M., 2000, Effects of applied electric fields on low-calcium epileptiform activity in the CA1 region of rat hippocampal slices. *J. of Neurophysiology*, 84,274–280.

[4] Sun F. T., Morrell M. J., Wharen R. E., 2008, Responsive cortical stimulation for the treatment of epilepsy. *Neurotherapeutics*, 5, 68–74.

[5] Labar D., Murphy J. and Tecoma E., 1999, Vagus nerve stimulation for medication-resistant generalized epilepsy, *Neurology*, 52, 1510–1512.

[6] Politsky J. M., Estellar R., Murro A. M., Smith J. R., Ray P., Park Y. D., et al., 2005, Effects of electrical stimulation paradigm on seizure frequency in medically intractable partial seizure patients with a cranially implanted responsive cortical neurostimulator. In *Proc. of the Annual Meeting of American Epilepsy Society (AES)*.

[7] D'Allesandro M., Esteller R., Vachtsevanos G., Hinson A., Echauz J., Litt B., 2003, Epileptic seizure prediction using hybrid feature selection over multiple intracranial EEG electrode contacts: a report of four patients. *IEEE Transactions on Biomedical Engineering*; 50, 603–613.

[8] Zarjam P.; Mesbah M. Discrete wavelet transform based seizure detection in newborns EEG signals, 2003, In: *Proc. 7th International Symposium on Signal Processing and Its Applications*, 2, 459- 462.

[9] Webber W. R. S., Ronald P., Lesser, Richardson R. T., Wilson K., An approach to seizure detection using an artificial neural network (ANN), 1996, *Electroencephalography and Clinical Neurophysiology*, 98(4), 250-272.

[10] Raghunathan S., Gupta S. K., Ward M. P., Worth R. M., Roy K., Irazoqui P. P., 2009, The design and hardware implementation of low-power real-time seizure detection algorithm, *J. Neural engineering*, 6(5).

[11] Markandeya H., Karakonstantis G., Raghunathan S., Irazoqui P. P., and Roy K., 2010, Low-power DWT-based quasi-averaging algorithm and architecture for epileptic seizure detection, *Proc. of the 16th ACM/IEEE international symposium on Low power electronics and design (ISLPED '10)*,301-306.

[12] Raghunathan S., Gupta S. K., Markandeya H. S., Roy K., Irazoqui P. P., 2010, A hardware-algorithm co-design approach to optimize seizure detection algorithms for implantable applications, *J. of Neuroscience Methods*, 193(1), 106-117.

[13] White A.M., Williams P.A., Ferraro D.J., Clark S., Kadam S.D., Dudek F.E. et al., 2006, Efficient unsupervised algorithms for the detection of seizures in continuous EEG recordings from rats after brain injury, *J. of Neuroscience Methods*, 152, 255–266.

[14] Kaiser J. F., 1990 "On a simple algorithm to calculate the 'energy' of a signal," *International Conference on Acoustics, Speech, and Signal Processing (ICASSP), 1*, 381-384.

[15] Zaveri H. P., W. J. Williams W. J., and Sackellares J. C., Energy Based Detection of Seizures, 1993, *15th Annual International Conference on Engineering and Medicine in Biology*, 363-364.

[16] Hjorth B., An on-line transformation of EEG scalp potentials into orthogonal source derivations, 1975, *Electroencephalography and Clinical Neurophysiology*, 39,526–530.

[17] Pfurtscheller G., Lopes da Silva F.H., 1999, Event-related EEG/MEG synchronization and desynchronization: basic principles, *Clinical Neurophysiology*, 110, 1842–1857.

www.ingramcontent.com/pod-product-compliance
Ingram Content Group UK Ltd.
Pitfield, Milton Keynes, MK11 3LW, UK
UKHW051412290926
15471UKWH00009B/1564

9 781450 317382